Handbook of Practical X-Ray Fluorescence Analysis

B. Beckhoff B. Kanngießer
N. Langhoff R. Wedell H. Wolff
(Eds.)

# Handbook of Practical X-Ray Fluorescence Analysis

With 385 Figures and 53 Tables

Springer

Dr. Burkhard Beckhoff
Physikalisch-Technische Bundesanstalt, X-ray Spectrometry
Abbestr. 2–12, 10587 Berlin, Germany
E-mail: burkhard.beckhoff@ptb.de

Dr. habil. Birgit Kanngießer
Technische Universität Berlin, Fakultät 2, Institut für Atomare Physik und Fachdidaktik
Hardenbergstr. 36, 10623 Berlin, Germany
E-mail: bk@atom.physik.tu-berlin.de

Professor Dr. Norbert Langhoff
IfG-Institute for Scientific
Instruments GmbH
Rudower Chaussee 29/31, 12489
Berlin, Germany
E-mail: langhoff@ifg-adlershof.de

Dr. sc. nat. Reiner Wedell
Dr. sc. nat. Helmut Wolff
Institut für Angewandte Photonik e.V.
Rudower Chaussee 29/31, 12489 Berlin, Germany
E-mail: wedell-iap@ifg-adlershof.de, HFW.Wolff@t-online.de

Library of Congress Control Number: 2005934096

ISBN-10 3-540-28603-9 Springer Berlin Heidelberg New York
ISBN-13 978-3-540-28603-5 Springer Berlin Heidelberg New York



Springer is a part of Springer Science+Business Media.

springer.com


Printed in Germany



Typesetting by the Authors and SPI Publisher Services using a Springer LaTeX macro package
Cover design: *design & production* GmbH, Heidelberg

Printed on acid-free paper SPIN 10755097 57/3141/SPI 5 4 3 2 1 0

# Preface

X-Ray fluorescence analysis (XRF) has developed into a well-established multi-elemental analysis technique with a very wide field of practical applications, especially those requiring nondestructive analytical methods. Over a long period of time, steady progress of XRF was made, both methodological and instrumental. Within the last decade, however, advancements in technology, software development, and methodologies for quantification have provided an impetus to XRF research and application, leading to striking new improvements. The recent technological advances, including table-top instruments that take advantage of novel low-power micro-focus tubes, novel X-ray optics and detectors, as well as simplified access to synchrotron radiation, have made it possible to extend XRF to low Z elements and to obtain two- and three-dimensional information from a sample on a micrometer-scale. The development of portable and hand-held devices has enabled a more flexible use of XRF in a variety of new situations, such as archaeometry and process control. Furthermore, synchrotron radiation provides high excitation flux and even speciation capabilities due to energetically tunable radiation.

Because of these recent advancements, the editors decided to compile a practical handbook of XRF as a resource for scientists and industrial users that provides enough information to conceive and set up modern XRF experiments for use in a wide range of practical applications. Additionally, selected sections consist of a concise summary of background information for readers who wish to gain a more in-depth understanding of the topics without conducting a lengthy search of the literature. The present handbook is not intended to be a textbook with interdependent chapters, rather a reference in which the information in each section is largely self-contained. In this way, the reader is not required to read the handbook from cover to cover, but can refer to any section without a lot of additional background.

The handbook is organized as follows. The first chapter provides a historical account of XRF and an introduction reflecting the extension of XRF to modern fields of methodology and applications. Chapters 2, 3, and 4 follow the path of a photon in an XRF instrument, originating at a source (2), being modified by an optic (3), and registered by a detector (4). Chapter 5 deals with the various aspects of quantifying the spectra obtained from a sample by this instrument. Expert information on how to prepare the sample is the theme of Chapter 6. Chapter 7 is devoted to a variety of applications: micro-, trace, and layer analysis; environmental, geological, archaeometric, forensic, and biomedical applications; and process control. The handbook concludes with a discussion on safety regulations and useful links to physical data (Chapter 8).

We would like to take this opportunity to express our gratitude to all of the authors, especially those who completed their contributions at an early stage in the preparation of this book and patiently awaited its completion. Our thanks also go to Katherine Roegner of the Technical University, Berlin for her support in matters of language.

We hope you enjoy this practical handbook and that it contributes to the continued development of XRF. We also hope that it encourages and inspires newcomers to the field in exploring the multifaceted aspects of XRF.

Berlin,
November 2005

*Burkhard Beckhoff,*
*Birgit Kanngießer,*
*Norbert Langhoff,*
*Reiner Wedell*
*and Helmut Wolff*

# Contents

# List of Contributors

**Dr. Peter Ambrosi**
Physikalisch-Technische Bundesanstalt (National Metrology Institute) Department of Radiation Protection Dosimetry
Bundesallee 100
38116 Braunschweig, Germany

**Dr. Alexander Antonov**
Optigraph GmbH
Rudower Chaussee 29/31
12489 Berlin, Germany

**Dr. Tomoya Arai**
Rigaku Corporation
X-Ray Research Lab
Akaoji 14-8, Takatsuki
Osaka 569-1146, Japan

**Dr. Vladimir Arkadiev**
Institut für angewandte Photonik e.V.
Rudower Chaussee 29/31
12489 Berlin, Germany

**Dr. Ray Barrett**
European Synchrotron Radiation Facility (ESRF)
BP 220
38043 Grenoble, France

**Dr. Burkhard Beckhoff**
Physikalisch-Technische Bundesanstalt (National Metrology Institute) X-Ray Spectrometry group
Abbestraße 2-12
10587 Berlin, Germany

**Dr. Maria Betti**
European Commission
JRC, Institute for Transuranium Elements
P.O. Box 2340
76125 Karlsruhe, Germany

**Dr. Aniouar Bjeoumikhov**
IfG – Institute for Scientific Instruments GmbH
Rudower Chaussee 29/31
12489 Berlin, Germany

**Prof. Dr. Avram Blank**
Institute for Single Crystals of the National Academy of Sciences of Ukraine
Ave. Lenin 60
Kharkov, 6001, Ukraine

**Ir. Peter N. Brouwer**
PANalytical B.V.
Lelyweg 1, 7602 EA Almelo
The Netherlands

**Dr. Victor Buhrke**
The Buhrke Company
10 Sandstone, Portola Valley
California 94028, USA

**Dr. Tomáš Čechák**
Czech Technical University
Faculty of Nuclear Sciences and Physical Engineering
V Holešovičkách 2
18000 Prague 8, Czech Republic

**Dr. Joël Chavanne**
European Synchrotron Radiation Facility (ESRF)
Insertion Group
BP 220
38043 Grenoble, France

**Dr. Pierre Chevallier**
LPS, CEA/CNRS, CE Saclay
91191 Gif-sur-Yvette
France, and LURE
Centre Universitaire Paris-Sud
BP34, 91898 Orsay, France

**Dr. Fabio Comin**
European Synchrotron Radiation Facility (ESRF)
BP 220
38043 Grenoble, France

**Dr. Jörg Engelhardt**
Landesanstalt für Personendosimetrie und Strahlenschutzausbildung
im Innovationspark Wuhlheide
Haus 41, Köpenicker Straße 325
12555 Berlin, Germany

**Dr. Lyudmyla Eksperiandova**
Institute for Single Crystals of the National Academy of Sciences of Ukraine
Ave. Lenin 60
Kharkov, 6001, Ukraine

**Prof. Dr. Alexei Erko**
Senior Scientist
Berliner Elektronenspeicherring-Gesellschaft für Synchrotronstrahlung mbH (BESSY)
Albert-Einstein-Straße 15
12489 Berlin, Germany

**Dr. Laszlo Fabry**
Siltronic AG
Central Research & Development
84479 Burghausen, Germany

**Dipl.-Chem. Joachim Feuerborn**
Heußweg 102
20255 Hamburg, Germany

**Prof. Dr. Carlo Fiorini**
Politecnico di Milano
Dipartimento di Elettronica e Informazione
Via Ponzio 34/5
20133 Milano, Italy

**Prof. Dr. René Van Grieken**
University of Antwerpen
Department of Chemistry
2610 Wilrijk-Antwerpen, Belgium

**Dr. Inna Grigorieva**
Optigraph GmbH
Rudower Chaussee 29/31
12489 Berlin, Germany

**Dr. Oliver Hahn**
Bundesanstalt für Materialprüfung (BAM)
Labor IV.22 Unter den Eichen 44-46
12203 Berlin, Germany

**Dr. Robert Hartmann**
PNSensor GmbH
Römerstraße 28
80803 München and MPI Halbleiterlabor
Otto-Hahn-Ring 6
81739 München, Germany

**Dr. Michael Haschke**
IfG – Institute for Scientific Instruments GmbH
Rudower Chaussee 29/31
12489 Berlin, Germany

**Dr. Peter Hoffmann**
Technical University Darmstadt
Department of Materials
and Earth Sciences
Chemical Analytics
Petersenstraße 23
64287 Darmstadt, Germany

**Dr. Jasna Injuk**
University of Antwerpen
Department of Chemistry
2610 Wilrijk-Antwerpen, Belgium

**Dr. habil. Birgit Kanngießer**
Technische Universität Berlin
Institut für Atomare Physik und Fachdidaktik
Analytical X-Ray Spectroscopy group
Sekr. PN 3-2, Hardenbergstraße 36
10623 Berlin, Germany

**MSc. Naoki Kawahara**
Rigaku Corporation
X-Ray Research Laboratory
Akaoji 14-8, Takatsuki
Osaka 569-1146, Japan

**Prof. Dr. Wolfgang Knüpfer**
Siemens AG
Medizinische Technik
Henkestraße 127
91052 Erlangen, Germany

**Dr. Sandor Kurunczi**
KFKI Atomic Energy Research Institute
P.O. Box 49
H-1525 Budapest, Hungary

**Gerald Lachance**
St-Felix, 1100
K0A 2A0, Hammond
Ontario, Canada

**Prof. Dr. Norbert Langhoff**
IfG – Institute for Scientific Instruments GmbH
Rudower Chaussee 29/31
12489 Berlin, Germany

**Prof. Dr. Jürgen Leonhardt**
IUT Institut für Umwelttechnologien GmbH
Volmerstraße 9b
12489 Berlin, Germany

**Prof. Dr. Antonio Longoni**
Politecnico di Milano
Dipartimento di Elettronica e Informazione
Via Ponzio 34/5
20133 Milano, Italy

**Dr. Katharina Lüning**
Stanford Synchrotron Radiation Laboratory (SSRL)
2575 Sand Hill Road, Menlo Park
CA 94025, USA

**Dr. Wolfgang Malzer**
Technische Universität Berlin,
Institut für Atomare Physik und Fachdidaktik
Analytical X-Ray Spectroscopy group
Sekr. PN 3-2
Hardenbergstraße 36
10623 Berlin, Germany

**Prof. Dr. Michael Mantler**
Vienna University of Technology
Wiedner Hauptstraße 8-10/138
A-1040 Vienna, Austria

**Dr. Karl-Eugen Mauser**
Bruker AXS GmbH,
Östliche Rheinbrückenstraße 49
76187 Karlsruhe, Germany

**Dr. Norbert Meidinger**
Max-Planck-Institut für
extraterrestrische Physik
85741 Garching and MPI
Halbleiterlabor
Otto-Hahn-Ring 6
81739 München, Germany

**Dr. Andreas Möller**
Institut für Geowissenschaften
Universität Potsdam
Postfach 601553
14415 Potsdam, Germany

**Dr. Bernd Nensel**
Helmut Fischer GmbH & Co KG
R&D, Industriestraße 21
71069 Sindelfingen, Germany

**Dr. János Osán**
KFKI Atomic Energy Research
Institute
P.O. Box 49
H-1525 Budapest, Hungary

**Dipl.-Ing. Siegfried Pahlke**
Siltronic AG
Central Research & Development
84479 Burghausen, Germany

**Prof. Dr. Pietro Pianetta**
Stanford Synchrotron Radiation
Laboratory (SSRL)
2575 Sand Hill Road, Menlo Park
CA 94025, USA

**Dr. Dieter Rammlmair**
Bundesanstalt für Geowissenschaften
und Rohstoffe
Stilleweg 2
30655 Hannover, Germany

**Dr. Ina Reiche**
Centre de recherche et de restauration des musées de France (CRRMF)
rue de Pyramides 6
F-75041 Paris CEDEX 01, France

**Dr. Karen Rickers**
GeoForschungsZentrum Potsdam
P.B. 4.1, Telegrafenberg
14473 Potsdam, Germany

**Dr. Volker Rößiger**
Helmut Fischer GmbH & Co KG
R&D, Industriestraße 21
71069 Sindelfingen, Germany

**Richard M. Rousseau, Ph. D.**
Geological Survey of Canada
601 Booth St., Ottawa
Ontario, K1A 0E8, Canada

**Dr. Frank Scholze**
Physikalisch-Technische
Bundesanstalt (National Metrology
Institute), EUV Radiometry group
Abbestraße 2-12
10587 Berlin, Germany

**Prof. Dr. Robert A. Schwarzer**
Institut für Physik und Physikalische
Technologien
TU Clausthal, Leibnizstraße 4
38678 Clausthal-Zellerfeld, Germany

**BSc. Takashi Shoji**
Rigaku Corporation
X-Ray Research Lab
Akaoji 14-8, Takatsuki
Osaka 569-1146, Japan

**Dr. Alexandre Simionovici**
LST, École Normale Supérieure de
Lyon
46 Allée d'Italie
69007 Lyon, France

**Dr. Susanne Staub**
Zolltechnische Prüfungs- und
Lehranstalt Hamburg
Baumacker 3
22532 Hamburg, Germany

**Dr. habil. Heike Stege**
Doerner-Institut
Bayrische Staatsgemäldesammlungen
Barer Straße 29
80799 München, Germany

**Prof. Dr. Christina Streli**
Atominstitut der Österreichischen Universitäten
Stadionallee 2
A-1020 Wien, Austria

**Prof. Dr. Lothar Strüder**
Max-Planck-Institut für extraterrestrische Physik
85741 Garching and MPI Halbleiterlabor
Otto-Hahn-Ring 6
81739 München, Germany

**Dr. habil. Szabina Török**
KFKI Atomic Energy Research Institute
P.O. Box 49
H-1525 Budapest, Hungary

**Dr. Bernard Vidal**
Directeur de Recherche au CNRS
L2MP, UMR 6137
Faculté des Sciences de St. Jérôme
13397 Marseille CEDEX 20, France

**Dr. Bruno A. R. Vrebos**
PANalytical B.V.
Lelyweg 1
7602 EA Almelo
The Netherlands

**Dr. sc. nat. Reiner Wedell**
Institut für angewandte Photonik e.V.
Rudower Chaussee 29/31, 12489 Berlin, Germany

**Dr. Gundolf Weseloh**
Max Planck Institute of Colloids and Interfaces
14424 Potsdam, Germany

**Dr. Max Wilke**
Institut für Geowissenschaften
Mineralogie – Petrologie
Universität Potsdam
Karl-Liebknecht-Str. 24
14476 Golm, Germany

**Prof. Dr. James Willis**
Department of Geological Sciences
University of Cape Town
Private Bag
7701, Rondebosch, South Africa

**Dr. Antje Wittenberg**
Bundesanstalt für Geowissenschaften und Rohstoffe
Stilleweg 2
30655 Hannover, Germany

**Prof. Dr. Peter Wobrauschek**
Atominstitut der Österreichischen Universitäten
Stadionallee 2
A-1020 Wien, Austria

**Dr. Janina Ziêba-Palus**
Institute of Forensic Research
Westerplatte Street 9
31-033 Krakow, Poland

# List of selected abbreviations used in the handbook

| | |
|---|---|
| ADC | analog to digital converter |
| AD | Alzheimer's disease |
| ALS | amyothrophic lateral sclerosis |
| APDC | ammonium pyrrolidine-dithiocarbamate |
| APP | analog pulse processing |
| APS | active pixel sensor |
| APS | Argonne Photon Source |
| BFL | Bragg-Fresnel lens |
| BLR | baseline restorer |
| BM | bending magnet (synchrotron ID) |
| CAMEX | pn-CCD camera on XMM and ABRIXAS space missions |
| CL | cathodoluminescence |
| CMC | carboxymethyl cellulose |
| CNS | central nervous sytem |
| CRL | compound refractive lens |
| CRM | certified reference material |
| CVD | chemical vapour deposition |
| DAC | digital to analog converter |
| DAC | diamond anvil cell |
| DefMA | definition of measurement and application conditions |
| DL | detection limit |
| DPP | digital pulse processing |
| DSP | digital signal processing |
| DDTC | sodium diethyldithio-carbamate |
| DU | depleted uranium |
| EDS | energy dispersive system (spectroscopy) |
| EDXRS | energy-dispersive X-ray spectroscopy |
| EDXRF | energy-dispersive X-ray fluorescence |
| EDXRD | energy-dispersive X-ray diffraction |
| EFEM | equipment front end module |
| EIC | empirical influence coefficient |
| EMMA (*see* micro-XRF) | energy dispersive miniprobe multielement analyzer |

| | |
|---|---|
| ENC | equivalent noise charge |
| EPMA | electron probe micro analysis |
| EMPA | electron microprobe analyser |
| EXAFS | extended X-ray absorption fine structure |
| FAAS | flameless atomic absorption spectrometry |
| FEL | free electron laser |
| FET | field effect transistor |
| FOUP | front opening unified pods (plastic box used for wafers) |
| F-PC | flow proportional counter |
| FP | fundamental parameter |
| FWHM | full width at half maximum |
| GEXRF | grazing emission X-ray fluorescence |
| GPS | global positioning system |
| GPSC | gas proportional scintillation counter |
| HCA | hierarchical cluster analysis |
| HDAC | hydrothermal diamond anvil cell |
| HPGe | high purity germanium |
| HOPG | highly oriented pyrolytic graphite |
| ICP-AES | inductively coupled plasma - Auger electron spectroscopy |
| ICP-MS | inductively coupled plasma - mass spectroscopy |
| IEF | isoelectric focusing |
| ID | insertion device (synchrotron, e.g. wiggler, bending magnet) |
| IDX | 4'-iodo-4'-deoxydoxorubicin (anticancer drug) |
| LA-ICP-MS | Laser Ablation – Inductively Coupled Plasma – Mass Spectrometry |
| LOD | limit of detection |
| LLD | lower level of detection |
| MDL | minimum detection limit |
| MIBK | methylisobutylketone |
| micro-PIXE | *see* PIXE |
| micro-XRF (μ-XRF) | micro-X-ray fluorescence (analysis) |
| micro-SRXRF (also SXRF) | micro-synchrotron X-ray fluorescence |
| ML grating | multilayer grating |
| NDXRF | non-dispersive X-ray fluorescence |
| NaDDTC | sodium diethyldithiocarbamate |
| NDXRF | nondispersive X-ray fluorescence |
| NEXAFS | near edge extended X-ray absorption fine structure |
| PHA | pulse height analyzer |
| poly-CCC | polycapillary conical collimator |
| PCA | principle component analysis |
| PCs | principal components |
| PD | Parkinson's disease |
| PIN-diode | **p**ositive /**i**ntrinsic/**n**egative detector |
| PIXE | proton induced X-ray emission |
| pn-CCD | pn-charge coupled device |

| | |
|---|---|
| PSD | position sensitive detector |
| PSPC | position sensitive proportional counter |
| PTFE | polytetrafluorethylene |
| P-Z | pole-zero compensator |
| QXAS | quantitative X-ray analysis system (spectroscopy) |
| RDA | regularized discriminant analysis |
| REE | rare earth element |
| RI | refraction (refractive) index |
| RM | reference material |
| ROI | region of interest |
| ROSITA, XEUS, XMM, ABRIXAS | space missions of ESA |
| RTM | rhenium-tungsten-molybdenum composite material |
| SAXS | small angle X-ray scattering |
| SDD | silicon drift detector |
| SDS-PAGE | sodium dodecyl sulphate polyacrylamide gel electrophoresis |
| SEM | scanning electron microscopy |
| SEM/WDX | scanning electron microscopy/wavelength-dispersive X-ray analysis |
| SEM/EDX | scanning electron microscopy/energy-dispersive X-ray analysis |
| SIMS | secondary ion mass spectrometry |
| SHA | shaping amplifier |
| SMIF | standard mechanical interface (plastic box used for wafers) |
| SML | synthetic multilayer |
| S-PC | sealed proportional counter |
| SPC | statistical process control |
| SPE | solid phase extraction |
| SR | synchrotron radiation |
| SRXRF (SR-XRF) | synchrotron radiation X-ray fluorescence |
| SRXRFA | synchrotron radiation X-ray fluorescence analysis |
| SR-TXRF, (SRTXRF) | synchrotron radiation total reflection X-ray fluorescence |
| SRW | software package for synchrotrons 'SRW' developed by ESRF |
| STJ | superconducting tunnel junction |
| TIC | theoretical influence coefficient |
| TXRF | total reflection X-ray fluorescence |
| TXRFA | total reflection X-ray fluorescence analysis |
| TZM-anode | Mo + W anode with admixtures of Ti and Zr |
| VLS grating | variable line spacing grating |
| VPD | vapour phase decomposition |
| VPD-SR-TXRF | vapour phase decomposition - synchrotron radiation - total reflection X-ray fluorescence |
| WDXRF | wavelength dispersive X-ray fluorescence |
| WDS | wavelength dispersive system |
| WDX | wavelength dispersive X-ray analysis |
| XAFS | X-ray absorption edge fine structure |

| | |
|---|---|
| XANES | X-ray absorption near-edge structure |
| XAS | X-ray analysis system (spectroscopy) |
| XPS | X-ray photo electron spectroscopy |
| XRD | X-ray diffraction |
| XRF | X-ray fluorescence |
| XRFA | X-ray fluorescence analysis |
| XRGS | X-ray geo scanner (geoscanner) |
| XSI | X-ray scanning instrument |
| *ADP, EDDT, KAP, PET, RAP, TlAP* | Special crystals used in X-ray diffraction and WDS: (*see* Eugene P. Bertin, Principles and Practice of X-Ray Spectrometric Analysis, Plenum Press New York 1975 (ISBN: 0-306-30809-6)) |

# 1

# Introduction

T. Arai

## 1.1 The Discovery of X-Rays and Origin of X-Ray Fluorescence Analysis

The development of the modern theory of atomic structures was initiated based on the discovery of X-rays (1895). It was further triggered by the awareness of the existence of electrons in the atom, which was clarified by the line splitting observed when applying an external magnetic field (1896) and by the scattering of alpha particles at the atomic nucleus (1910). During the past few decades, X-ray physics has not only inspired and supported various research and development in the natural sciences, but has also had a beneficial impact on medical applications [1]. In today's civilized world, X-ray technology continues to play an important role in the advancement of material science, inspections in production processes, and diagnostics for medical treatment.

Cited below are two evolutional events in the history of X-ray science.

Watson and Crick proposed the DNA structure based on biological and structural chemistry including X-ray crystal structure analysis. Wilkins precisely studied the crystal structure using a rotating crystal method. The consolidation of their works led to the determination of the double helical structure of DNA, which has a three-dimensional structure of a screwed ladder and a regular arrangement of the four bases: adenine, thymine, guanine, and cytosine in the space between the two ladder poles [2].

The combination of the high X-ray transparency of the human body and its use for medical treatment brought about a notable advance in the use of X-rays for medical applications. Oldendorf planned to develop a relevant instrument in 1960. Then, Cormack presented his idea that included a mathematical treatment for three-dimensional imaging in 1963 and 1964. As Hounsfield used a radioactive source, a long time was required for taking a picture. Finally, he developed a computer-assisted tomogram using the consolidated technology of X-ray tube radiation, X-ray detectors, and computer calculations for the preparation of three-dimensional pictures of X-ray intensity and indications for easy and precise diagnosis. Clinical data were presented in 1972 and 1973. Many instrumental improvements led to high-grade medical treatment that was founded on the present X-ray machine [3].

On November 8, 1895, Wilhelm Conrad Röntgen discovered X-rays in his laboratory at the physics institute of Julius-Maximilians University of Würzburg in Bavaria. He had studied cathode rays using an air-filled Hittorf-Crooks tube, which was shaded with a black paper. The tube wall was hit by electrons and emitted light. In his darkened room, he noticed a weak luminescence which radiated from a fluorescent screen located near the tube. He recognized "eine neue Art von Strahlen" (a new type of rays), which originated from the tube. After changing the experimental and surrounding conditions, he was able to observe the emission of weak rays of light on the fluorescent screen. He announced the new experimental results. It was immediately recognized that this discovery might be used to look into the structure of a living human body and the interior of constructed materials [4, 5].

After the announcement by Röntgen, two further important discoveries were made: radioactivity from uranium by Becquerel (1896) as well as radium and polonium by Marie and Pierre Curie (1898).

Using an aluminum filter method for the separation of X-rays and an ionization chamber for X-rays detection, Barkla studied the nature of X-rays relative to the atomic structure. Observing the secondary X-rays which were radiated from a target sample, he discovered the polarization of X-rays (1906), the gaps in atomic absorption (1909), and the distinction between continuous and characteristic X-rays, which consisted of several series of X-rays, named the K, L, M ... series (1911). The intensity and distribution of continuous X-rays were dependent on the number of electrons in an atom, and the characteristic X-rays were related to the electron energy configuration in the atom [6]. In succession to Barkla's works, the wave properties of X-rays were investigated by von Laue, who exhibited X-rays diffraction from a single crystal, which was composed of a three-dimensional structure with a regularly repeating pattern (1912). The experimental results showed the comparability of the wavelength of X-rays with the atomic distances and confirmed the wave properties of X-rays.

W. H. Bragg, who derived the famous Bragg's formula, was interested in von Laue's experiments. Using a Bragg spectrometer, the X-ray reflection patterns from single crystals of NaCl and KCl were observed to be the regular patterns of an isometric system showing differences in the X-ray intensity when comparing sodium and potassium. This was the starting point of crystal structure analysis with X-rays [7].

For the expansion of radiographic technology, the need for a heavy-duty X-ray tube emerged. After the tungsten filament (1908) and the tungsten incandescent lamp (1911) were invented, Coolidge developed a new type of tube setting, successfully solving the problem of low power and instabilities of a gas-filled discharge tube. In this new tube, thermal electrons emitted from a hot filament hit the target, which was an emission source of X-rays (1913) [8].

Following the investigation of the properties of X-rays by Barkla, Moseley studied characteristic X-rays in an exchange of communications with W.L. and W.H. Bragg. He put target samples into a gas-filled discharge tube, which were then irradiated with electrons for the generation of characteristic x-rays. The narrow collimated characteristic X-rays hit the cleaved surface of a $K_4Fe(CN)_6{\cdot}H_2O$ crystal and the third-order lines of Bragg reflection X-rays were obtained, which were shown in the famous photograph taken in 1913 [9]. Moseley elucidated the relationship between the characteristic X-rays and the measured elements, and communicated his experimental results to Bohr [10].

Siegbahn produced an X-ray spectrometer for a wider range of characteristic X-rays. He measured the wavelengths of characteristic X-rays precisely and classified them into $\alpha$, $\beta$, $\gamma$ ... according to the X-ray intensities in the respective series. X-ray spectroscopy was established with these works (1913–1923).

In the next advances, Hadding tried to analyze rare earth elements using the X-ray method. His work was supported by Siegbahn.

Due to the establishment of the structure of atoms, it became possible to predict the existence of elements that had yet to be discovered. This was based on the assumption that undiscovered elements belonging to the same family of elements in the periodic table have the same chemical features. In this respect, hafnium was isolated by von Hevesy and Coster (1923) [11] and rhenium by Noddack and Tacke with the support of Berg (1925).

During the initial stage of the use of X-ray spectroscopy for chemical analysis, the samples being analyzed were modified (or even destroyed) when electron excitation was applied, leading to changes in the X-ray intensities. Hadding, Glocker, and Frohnmayer pointed out the analytical problems of inter-element effects in quantitative analysis. When electron excitation was used, Coster and Nishina noticed sample evaporation because of the induced heat in the sample (1925), and Glocker and Schreiber found concentration changes in the constituent elements (1928). For the emission of characteristic X-rays in X-ray spectrochemical analysis, the X-ray excitation method was adopted as a non-destructive analysis method. Although the relative distance between the X-ray source and sample was reduced to increase the primary X-ray intensity, the resultant X-ray intensity was still insufficient to be measured [12]. On the positive side background X-rays became lower and, as a result for quantitative analysis, low intensity peaks could be measured easily.

For the measurement of X-ray intensities an ionization chamber or a photographic plate had been used. Perrin invented the ionization chamber in 1896, which was used in Barkla's works and adopted in Bragg spectrometers. In 1928, Geiger and Müller produced a new useful counter for the detection of $\gamma$-rays and X-rays, called the Geiger-Müller counter. Although analytical principles and procedures had been studied in the academic field, the development of X-ray analytical instruments for general use had to wait until the end of the Second World War.

During the Second World War, the precise measurement of the cutting angle of quartz was required in mass production of oscillation plates. For this purpose, an X-ray apparatus was built by Parrish and Gordon based on a modified Bragg ionization chamber spectrometer (1945) [13]. Based on production experiences of the X-ray apparatus, Friedman introduced detectors for $\gamma$-rays and X-rays, as well as sensing head systems for various X-ray applications (1947) [14]. Then he started to develop a prototype X-ray spectrometer for the measurement of diffracted X-ray intensities and Bragg angles. Adopting a new Geiger-Müller counter and an electronic pulse counting unit (1947), a quartz plate which was located at the rotation center of a goniometer was replaced with a solidified powder sample (1945) [15].

When the iron-containing samples were measured using a Cu target X-ray tube, an abnormal increase in X-ray intensity was found, because the iron fluorescent X-rays radiating from a diffraction sample had strayed into the detector. As a result of the realization that fluorescent X-rays could be detected easily by this measuring system, a new X-ray fluorescence spectrometer was built. Analytical problems of measuring weak fluorescent X-ray intensities were encountered in the 1920s and the 1930s, which changed the requirements for the improvement of instrumentation. Friedman and Birks adopted a high power X-ray tube with a large window which gave rise to an increase in the effective solid angle, contributing to a short distance between the x-ray source in the x-ray tube and the sample to be analyzed. On the goniometer that is used in X-ray diffraction measurements, a bundle of narrow nickel pipes was used for collimation. Based on the need for a large reflecting surface, high reflecting power and relatively small lattice spacing, NaCl and fluorite crystals were chosen (1948) [16]. Figure 1.1 shows the X-ray fluorescence spectrometer developed by Friedman and Birks.

Expanding upon Friedman and Birks' work, Abbott was successful in building the first commercial X-ray fluorescence spectrometer in 1948 [17]. These works can be regarded as the beginning of modern X-ray spectrometry.

## 1.2 Historical Progress of Laboratory X-ray Fluorescence Spectrometers

In this section, the historical progress and important developments of the wavelength dispersive method in laboratory X-ray instruments are briefly reviewed.

In 1964, Birks, one of the pioneers of the X-ray fluorescence spectrometer, visited Japan and delivered a lecture "X-ray fluorescence: Present limitations and future trends." In his lecture, the analytical limits achievable in those days were reviewed covering elements from sodium to uranium, the detectability of

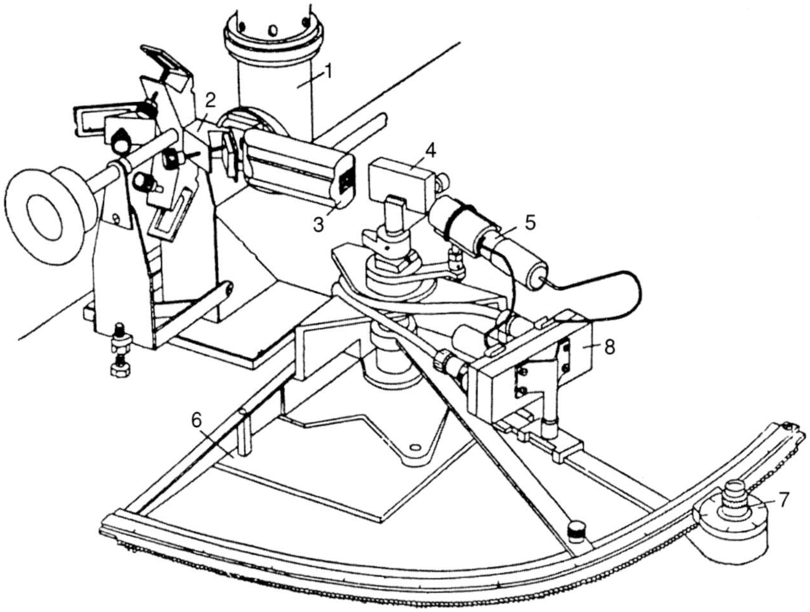


**Fig. 1.1.** The X-ray fluorescence spectrometer by Friedman and Birks [16]. 1, X-ray tube; 2, specimen holder; 3, Soller collimator; 4, crystal; 5, Geiger counter; 6, base plate; 7, vernier for setting and reading the angular position of the crystal; 8, pre-amplifier

ppm order, the analytical precision of about 1%, and the possible analytical error caused by the matrix effects [18]. Concerning the anticipated further progress of the analytical method, he directed attention to the measurement of light and ultralight elements, sample preparation, improvement of excitation and detection of X-rays, utilization of computers for spectrometer control and analytical calculation as well as to the energy dispersive method appearing just then.

In 1976, Birks reviewed again the principles of X-ray fluorescence analysis and the progress of analyzing techniques including the instrumentation and the evaluation of the new methods. In this review, he emphasized the progress in the matrix correction method and the fundamental parameter method, which were led by the evaluation of the X-ray tube spectrum. In addition, some applications and future expectations were discussed [19].

In 1990, Gilfrich made a survey of X-ray fluorescence analysis. He directed attention to the new X-ray source, namely, the synchrotron radiation and to the introduction of synthetic mutilayers as analyzing crystals, to advanced X-ray technologies such as TXRF and EXAFS, semiconductor detectors for energy dispersive techniques, and to the significant progress of data handling with small computers [20].

Furthermore, in 1997, Gilfrich [21] gave a retrospect on the historical development of X-ray analysis during the past 100 years in commemoration of the discovery of X-ray by W.C. Roentgen.

Against the backdrop of such constructive remarks and the popularization of the X-ray fluorescence spectrometer, there have been many kinds of X-ray instruments developed for various measuring purposes, so that the instruments available today have gone ahead of the prediction by Birks in terms of type and number.

An X-ray analysis system configured with a X-ray diffractometer and an X-ray fluorescence spectrometer was introduced by Parrish [22]. As both equipments were provided with a high voltage power supply for an X-ray tube and shared a pulse counting system , it was widely utilized in laboratory applications. In addition, this spectrometer was equipped with a helium attachment for measuring soft X-rays.

Spielberg, Parrish, and Lowitzsch described the functional elements of nonfocusing optics and the geometrical condition for their harmonizing combinations [23]. They used a closer coupling of an X-ray tube for the sample and a large solid angle of primary X-rays for higher fluorescent intensity. Consequently, the inhomogeneity of fluorescent intensity distribution arising from the change of irradiating density of primary X-rays on the sample was brought forth. Their equipment was based on the Bragg spectrometer and had a parallel beam optics composed of a flat analyzing crystal and a double Soller collimator. Furthermore, the X-ray tube, the composed X-ray optics, and a sample container for primary X-ray irradiation were assembled in such a way as to embody convenience of use.

Arai pointed out that the total reflection from a metal sheet of a Soller collimator broadened a peak profile in its tails [24]. In addition, he studied the aberration of peak profiles caused by the vertical (the direction parallel to the goniometer rotation axis) divergence, reflection profiles from imperfect single crystals, and spectral overlapping [25].

Campbell and Spielberg, and Parrish and Lowitzsch studied a double Soller collimator on the basis of a flat crystal X-ray optics [23, 26].

Arai proposed as a practical solution that the horizontal divergence of a sub-Soller collimator should be two or three times larger than that of the main collimator, which was dependent on the grade of mosaic structure of analyzing single crystals.

For the analysis of the light elements, a helium and vacuum path spectrometer was offered by Miller and Zingaro for laboratory-use instruments [27].

An X-ray spectrometer for industrial applications, equipped with parallel beam optics and named Autrometer, was developed by Miller and Kiley in 1958. It was equipped with a step scanning goniometer, a tandem detector connecting a scintillation counter and a gas flow proportional counter, and adapted with a helium path for light element analysis. The spectrometer further incorporated the intensity ratio method designed to maintain the X-ray intensity stability upon referring to the standard sample intensity for quantitative determination [28, 29].

Equipments other than the scanning (and parallel-beam-optics) spectrometers, pursuing the basic features of rapid and high precision analysis for industrial applications, are the spectrometers equipped with multi-channel fixed goniometers.

Kemp developed the first combination model of scanning and fixed channel multi-element X-ray spectrometer based on the development and production of an optical emission spectrometer [30].

Furthermore, Jones, Paschen, Swain, and Andermann proceeded with the development of this advanced X-ray equipment, which adopted the focusing circle optical system with curved crystals, the detectors with the gas discrimination, and the direct capacitor accumulation of electric signals of the detector [31]. In order to obtain a higher intensity of measuring X-rays, focusing optics were adopted using Johann or Johansson curved crystals. In the case of the scanning goniometer, a curved single crystal moved in a straight line away from the entrance slit on the focusing circle, and then the detector slit on the focusing circle crawled along the four-leaf rose locus. The distance between the entrance slit and the curved crystal center was proportional to the wavelength of the measuring X-rays. The gas discrimination in the detector had a favorable effect on the intensity reduction of backgrounds and overlapping X-rays. The capacitor accumulation method was effective to measure a high counting rate of analyzing X-rays.

For light element analysis of cement samples, a helium path was adopted by Andermann, Jones, and Davidson [32], and then Andermann and Allen intensified the X-ray analysis of various materials related to cement industry. Additionally, a vacuum spectrometer was developed for light element analysis of cement and steel production applications by Dryer, Davidson, and Andermann [33]. In order to procure high intensity stability of the measuring X-rays and compensate the matrix effect, an X-ray monitoring method to detect scattered X-rays from the sample was introduced into the intensity measuring system by Andermann and Kemp [34]. However, the aforementioned capacitor accumulation and the monitoring method were changed to the absolute intensity measurement using pulse-counting circuits with a clock timer and the pulse selection method later.

Anzelmo and Buman presented a combined instrument which contained a scanning goniometer and several fixed goniometers in one spectrometer, in 1983 at the Pittsburgh Conference. This was a new concept of adaptable use in an analytical laboratory [35].

In 1995, Kansai, Toda, Kohno, Arai, and Wilson developed a fixed channel-multi-element spectrometer provided with 40 fixed goniometers by adopting logarithmic-spiral curved crystal monochromators. For high speed analysis, high counting rate X-ray intensity measurement of 10 to 50 million counts per second was carried out with a pure material by means of a combination of an X-ray beam attenuator and high speed electronic circuits with fast counting rate response. In the meanwhile, for the impurity analysis of various ores or high purity materials, two receiving slits located beside each other, one for a fluorescent peak and the other for the background, were equipped in a goniometer for background correction calculation [36].

The core technology of an X-ray spectrometer consists of the excitation of fluorescent X-rays, the X-ray optics, and matrix correction calculations based on the fundamental parameter method. Described in this section are the remarkable progress and development in X-ray optics. Other features will be touched upon in the following section.

## 1.3 Measurement of Soft and Ultrasoft X-Rays

The purpose of conducting soft and ultrasoft X-ray measurements is to study the emission spectra influenced by chemical bonding or to make a quantitative determination of low atomic number elements. For the study of emission spectra, a high resolution spectrometer, and for quantitative determination, a high intensity one are required, respectively. The analytical problems in a quantitative determination of low atomic number elements originate from the inherent performance caused by the low excitation efficiency of soft X-rays and low reflectivity of spectroscopic device.

### 1.3.1 X-Ray Tubes for Soft and Ultrasoft X-Rays

In earlier days, most of the X-ray tube manufacturers supplied a side window tube with a thick beryllium window (about 1 mm thick) for spectrometer use. Inasmuch as these X-ray tubes are almost inefficient for X-ray measurement of light elements owing to the low excitation efficiency, new X-ray tubes with chromium and scandium target were developed on the basis of the side window structure by Kikkert and Hendry [37]. Characteristic K-radiation from this new tube passing through a relatively thin beryllium window can effectively excite the fluorescent X-rays of light elements.

Caldwell used a General Electric XRD 700 spectrometer equipped with a dual target (W, Cr) tube [38]. For heavy element measurement, the tungsten target, and for light element measurement like titanium and silicon in high alloy steel, the chromium target, were used, respectively. The analytical errors for titanium and silicon could be reduced. It demonstrated an improvement in the analytical accuracy of light elements by increasing the soft X-ray excitation efficiency.

Mahn of Machlett Laboratories Inc. developed an end window X-ray tube with a thin beryllium window and a rhodium target [39]. In order to minimize the secondary electron bombardment effect on the thin beryllium window, the target surface was charged with positive potential and the cathode filament was earth-grounded. The L series X-rays from the rhodium target are effective for excitation in soft and ultrasoft X-ray regions while the K series X-rays from the target are effective for heavy element analysis.

Gurvich compared various X-ray tubes and emphasized the advantage of the end window X-ray tube for light element analysis [40].

In scientific works by Henke, a specially designed demountable X-ray tube for soft and ultrasoft X-ray excitation was developed [41]. Using a very thin window or working without a window, an aluminum target for emission of Al-K lines or a copper target for Cu-L lines was adopted. The target was charged with positive potential and the filament was earth-grounded to protect the tube from window damage.

Indispensable features of dispersive devices are a large d-spacing according to the Bragg equation, high reflectivity due to the crystal structure, and low absorption. Single crystals having relatively large d-spacing as in EDDT, ADP, and KAP have been used at the beginning, and later on, PET and TlAP came into use for soft X-ray measurement.

In ultrasoft X-ray measurement, the role of dispersive devices has been classified into two categories: to the first category belong high resolution optics for profile studies based on single crystals, soap multilayered pseudo crystals, and grating dispersive analyzers, while the second category consists of high reflectivity devices for measurement of elemental concentration using total reflection mirrors, and later on, synthetic multilayer analyzers. X-ray analyzers and dispersing principles are shown in Fig. 1.2.

### 1.3.2 Scientific Research Work on Soft and Ultrasoft X-Rays

Holliday studied the fine structure of emission lines of O-K$\alpha$, C-K$\alpha$ and B-K$\alpha$ and of the L-band of titanium, vanadium, and manganese, upon which chemical bonding has direct effects, using electron bombardment excitation of X-ray emission lines, curved grating optics, and a gas flow proportional counter for X-ray detection [42]. C-K$\alpha$ X-ray profiles of the F-C system were studied, to which belong graphite, $Fe_3C$, and mixtures of martensite and austenite.

Fischer and Baun investigated X-ray emission lines influenced by chemical bonding using electron bombardment excitation [43]. Studies of K-series of beryllium, boron, carbon, and nitrogen, of the L-band of sulfur, chlorine, and potassium, K-series of magnesium, aluminum, and silicon, and self-absorption effects were carried out by use of flat crystals of EDDT, ADP, and a soap pseudo crystal, and a gas flow proportional counter.

Henke observed narrow profiles with low background intensity owing to the lack of short wavelength X-ray components in the primary beam, using the previously mentioned X-ray tube, flat soap multilayer pseudo crystals which were developed by means of the Langmuir–Blodgett dipping method, and a gas flow proportional counter with an ultrathin window [44].

### 1.3.3 Synthetic Multilayer Analyzers

As reviewed by Barbee, the attempts to develop synthetic multilayer (SML) analyzers have a long history [45]. Today's SML analyzers had been introduced into practical use in 1975–1985. Spiller tried to produce SML analyzers using

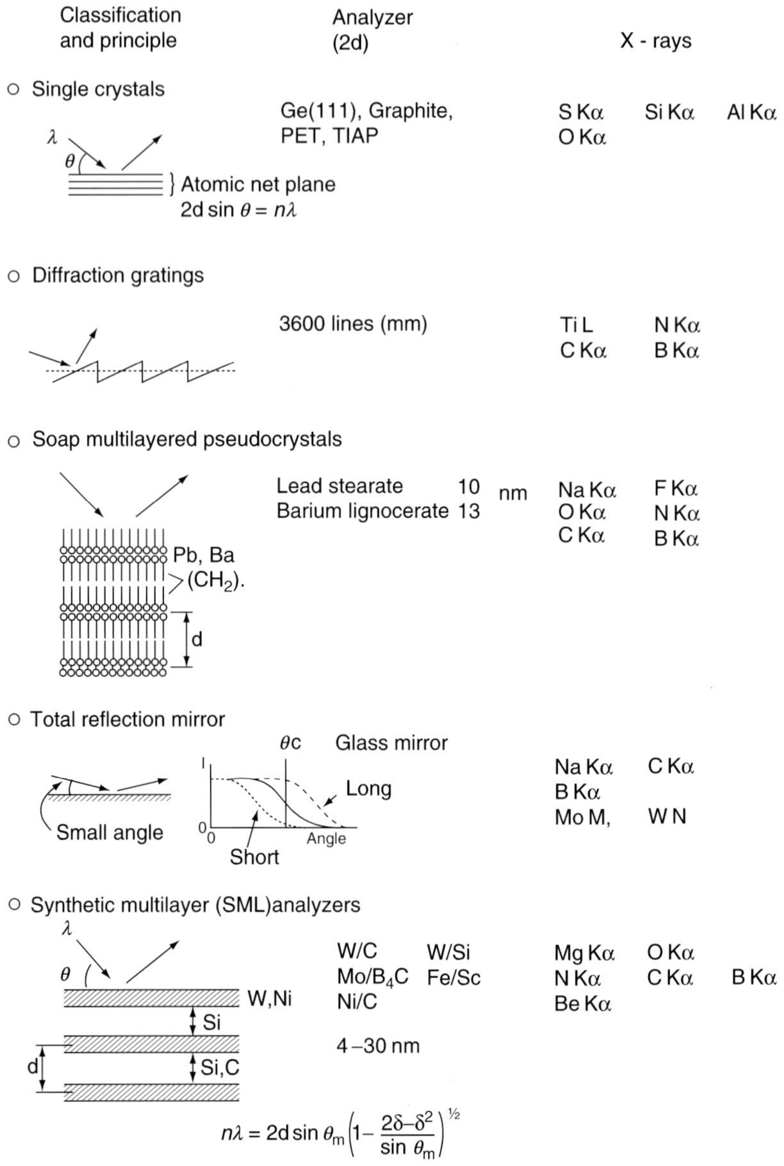


**Fig. 1.2.** Various X-ray analyzers for soft and ultrasoft X-rays and their principles

a thermal evaporation method controlled by a quartz crystal oscillator [46]. Barbee and Underwood exhibited the theory of X-ray reflection and observed reflection profiles of various X-rays including C-K$\alpha$ X-rays using an SML analyzer [47].

Using the sputtering method, thin layers of reflecting and spacing films were arranged alternately on a substrate with smooth surface. Reflecting material selected from high reflectivity substances and spacing material having low absorption within the considered X-ray region were chosen. The total layer thickness equal to the sum of the thickness of reflecting and spacing layers is 2–20 nm and the total number of accumulated multilayers is 30–100 layers.

For soft X-ray measurement, Gilfrich, Nagel, Loter, and Barbee studied the feasibility of the development of SML analyzers by comparing a RAP crystal with a W/C SML upon observation of the peak profiles and reflecting intensity of the Al-K$\alpha$ line [48].

Arai, Shoji, and Ryon carried out an ultrasoft X-ray measurement, using a standard laboratory spectrometer equipped with a rhodium target end window tube and a W/C analyzer [49]. The first detection of Be-K$\alpha$ X-rays from beryllium–copper alloys was performed by Toda, Kohno, Araki, Arai, and Hamill [50]. The analytical precision at 1.75 wt% beryllium was 0.034 wt% and the accuracy in the concentration range of 0.2–2.0 wt% was 0.01 wt%.

Boron oxide in glasses in the concentration range of 2.5–19 wt% was analyzed with a precision of 0.4 wt% at the concentration of 10 wt% and an accuracy of 0.9 wt% under the condition that B-K$\alpha$ lines were superimposed by the third order lines of O-K$\alpha$ X-rays.

Boron in boron-contained stainless steel used for a radioactive material container was analyzed with a precision of 0.01wt% at the concentration of 0.76 wt% and an accuracy of 0.03 wt% in the range of 0.1–1.6 wt%.

Carbon concentrations in steel were measured with a precision of 0.015 wt% at 1 wt% and the accuracy was two or three times higher than the precision. It was necessary to correct for the overlapping interference of W-N, Mo-M and Fe-L lines and for the influence of absorption of silicon in matrix material. In the case of X-ray measurement of oxygen and nitrogen, the strong absorption by carbon in matrix constituents was discovered.

Anzelmo and Boyer investigated the analytical performance of SML analyzers using an end window tube with a rhodium target and showed the potential for ultrasoft X-ray measurements [51].

Huang, Fung, and White evaluated various SML analyzers for the measurement of B-K$\alpha$, C-K$\alpha$, N-K$\alpha$ and O-K$\alpha$ X-rays. They showed an increase in the reflective intensity corresponding to the increase of d-spacing [52].

The increase of the observed Bragg angle of SML analyzers due to refraction effects of long wavelength X-rays was pointed out by Martins and Urch [53].

Based on the work of Huang and his co-workers on thin film characterization using the fundamental parameter method [54], Arai accomplished the simultaneous determination of thickness and constituent elements in thin layered materials using soft and ultrasoft X-rays [55]. The analysis of PSG and BPSG layers, which are used as covering materials for semiconductor memory devices, was also carried out. The achieved analytical precision and

**Table 1.1.** Analytical precision and accuracy of PSG, BPSG, and $SiO_2$ films [55]

| | | Thickness (nm) | $P_2O_5$ (wt%) |
|---|---|---|---|
| PSG | Precision | 3.6–3.8 | 0.07–0.09 |
| | (range) | (700) | (14–22) |
| | Accuracy | 6.4–15.5 | 0.15–0.4 |
| | (range) | (1100–1500) | (2.2–25) |

| | | Thickness (nm) | $P_2O_5$ (wt%) | $B_2O_3$ (wt%) |
|---|---|---|---|---|
| BPSG | Precision | 1.6–2.8 | 0.02–0.06 | 0.06–0.14 |
| | (range) | (530–1040) | (12–14) | (2–7) |
| | Accuracy | 6.2–9.2 | 0.1–0.5 | 0.12–0.35 |
| | (range) | (420–1080) | (0–27) | (0.02–13) |

| | | Thickness (nm) |
|---|---|---|
| $SiO_2$ | Precision | 0.07 |
| | (range) | (7.6) |
| | Accuracy | 0.2 |
| | (range) | (1.0–15) |

accuracy of film thickness and concentration of constituent materials are given in Table 1.1. O-K$\alpha$ X-rays were used for the sake of film thickness measurement of a thin-layered $SiO_2$ on a silicon wafer. It was demonstrated that very thin layers of $SiO_2$ can be measured.

White and Huang carried out thickness measurements of a carbon film in a double-layered structure of carbon and Co Cr alloy on a silicon wafer. The analytical precision of 2% for a 25 nm thick layer could be obtained using a W/C SML analyzer with 4 min of counting time [56].

Kobayashi and his co-workers studied analytical problems regarding the use of an SML analyzer and made the following observations [57].

Pure metals of aluminum or silicon emit the fluorescent X-rays of the Al-L or Si-L series. However, when oxide materials are irradiated, the fluorescent X-rays of the Al-L or Si-L series disappear because an outer electron of silicon or aluminum metal moves to oxygen. Covering a metallic surface of aluminum or silicon with a thin layered film of their oxide, their L lines emitted from the metallic substance penetrate the oxide layer and lead to a long tail of radiation at the short wavelength side. The latter interferes with the measuring X-rays increasing the background radiation.

If short wavelength X-rays emitted from an analyzed sample hit an SML analyzer, fluorescent X-rays generated from constituent elements of the SML stray into the X-ray detector and eventually contribute to the increase of the SML background radiation. Furthermore, some part of the fluorescent radiation interferes with reflecting net planes yielding Bragg reflection radiation.

Long wavelength X-rays emitted from an analyzed sample are reflected totally from the surface of an SML analyzer and as a result, an increase in background radiation can be seen occasionally.

When higher order reflections of short wavelength X-rays emitted from a sample interfere with the analyzing radiation, an SML analyzer modified by an appropriate choice of the thickness ratio of reflecting to spacing layers is used to suppress the higher order X-ray reflections.

### 1.3.4 Total Reflection Mirrors

In order to obtain a higher intensity of measuring X-rays, electron bombardment excitation was adopted. For soft X-ray measurement, Franks and Braybrook used X-ray optics consisting of the combination of a collimator, a total reflection mirror, and a windowless photomultiplier, as shown in Fig. 1.3 [58].

Herglotz improved the measuring system for soft X-ray measurement, shown in Fig. 1.3 [59]. Using electron bombardment excitation, the measuring system was made up of a curved surface paraffin mirror, receiving slit, and a windowless photomultiplier. Later on, an X-ray excitation method was developed for routine applications [60].

Arai sought for practical solutions to measure ultrasoft X-rays and demonstrated that the combination of a rhodium target end window tube with a thin beryllium window and a total reflection mirror along with a selected filter could be applied for the detection of B-K$\alpha$, C-K$\alpha$, N-K$\alpha$ and O-K$\alpha$ lines using a standard X-ray fluorescence spectrometer. The quantitative determination of boron oxide in boron glasses, carbon in steel, nitrogen in various chemical compound substances, and oxygen in coal were carried out [60]. Boron oxide in glasses in the range of 1 to 20 wt% could be analyzed with a precision of 0.2 wt% at lower concentration and 0.42 wt% at 19 wt%; the analytical accuracy was 0.65 wt% corrected with $K_2O$ and PbO. Carbon in carbon steel, cast iron, and stainless steel were measured. The analytical error composed of X-ray intensity precision and grinding effect error on the sample surface was about 0.01% in carbon and stainless steel and 0.01 wt% at the concentration of 3.82 wt% in cast iron. The accuracy was 0.01–0.02 wt% in carbon and stainless steel and 0.05 wt% in the range of 2–4 wt% in cast iron. Overlapping influences of molybdenum, niobium, tungsten, and tantalum and an absorption influence of matrix effect of silicon were found [60]. Oxygen measurement in coal or iron ores and nitrogen measurement in various materials were carried out and matrix effects were studied for the improvement of accuracy.

For industrial application, Sugimoto, Akiyoshi, and Kondou studied the determination of carbon in pig iron and obtained analytical results for carbon in the range of 3.5–4.7 wt% with a reproducibility of 0.05 wt% and an accuracy of 0.083 wt% using a Rigaku multi-fixed channel spectrometer provided with a total reflection mirror for the detection of C-K$\alpha$ X-rays [61].

Comparing an SML analyzer with a total reflection mirror along with an optimized filter in the X-ray separating system, it can be said that the SML analyzer offers a wider applicable method because of high resolution profiles and low background X-ray intensity.

Franks and Braybrook (1959) [58]

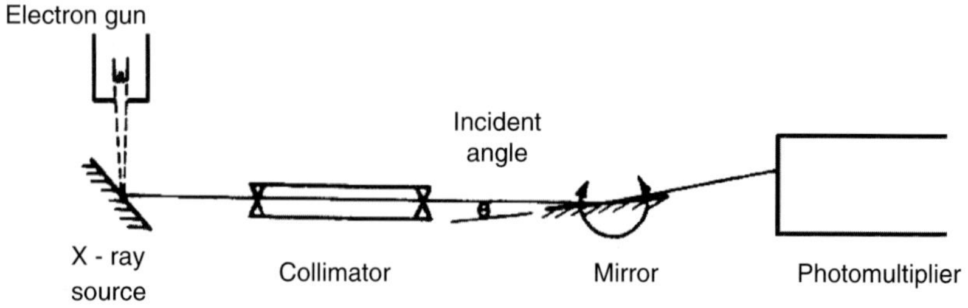


Herglotz (1967) [59]

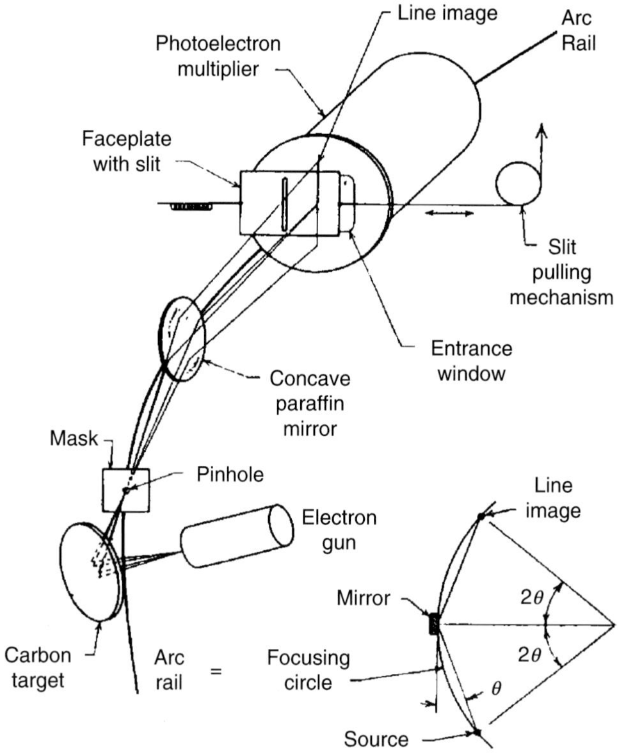


**Fig. 1.3.** Historical succession of the development of X-ray fluorescence spectrometers for light element analysis

Additionally, to conclude this section, it can be stated that at present the quantitative potentiality of soft and ultrasoft X-ray fluorescence analysis depends decisively on the efficiency of the end window X-ray tube with a thin beryllium window and a rhodium target as well as on the high reflectivity of the SML analyzer.

Henke (1964) [41][44]

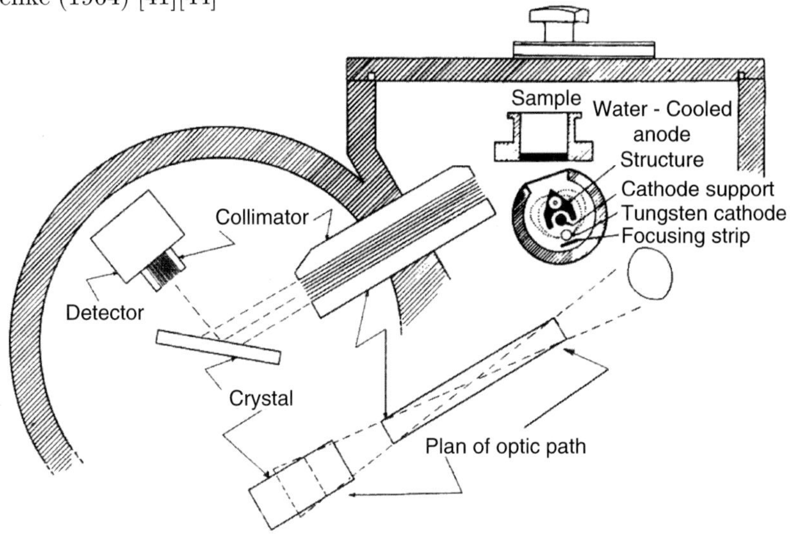


Rigaku (1982) [60]

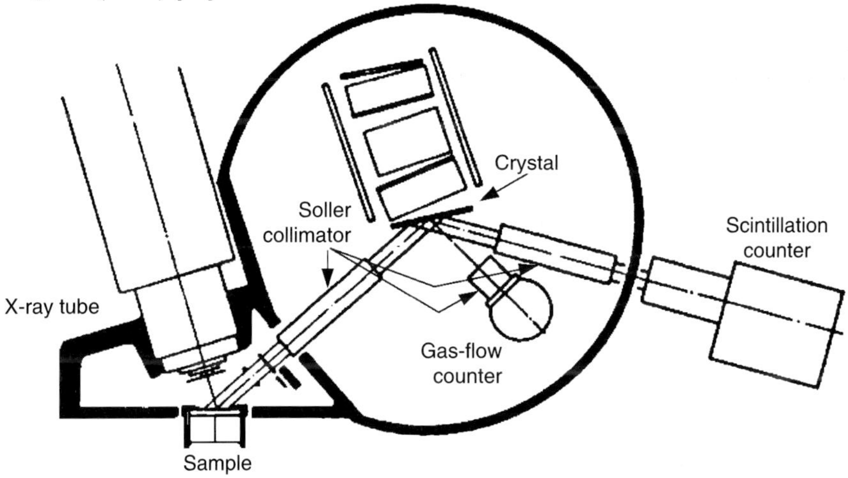


**Fig. 1.3.** *Continued*

## 1.4 Analytical Precision and Accuracy in X-Ray Fluorescence Analysis

When an analytical sample is irradiated with X-rays emitted from an X-ray tube or radioactive source, fluorescent X-rays are generated in the sample and can be measured for quantitative analysis of its constituent elements. X-ray fluorescence analysis is rapid, precise, and nondestructive.

From the standpoint of X-ray intensity measurement, Zemany made a summary of precision and accuracy [62]. In this section, the X-ray matrix effect, which is the most basic and the largest component in analytical accuracy, is concisely discussed by comparing X-ray accuracy with X-ray precision (see subchapter 5.9 for further details).

X-ray intensity, or the accumulated count of measured X-ray photons per unit time, is always accompanied by a statistical fluctuation which conforms to the Gaussian distribution with a standard deviation equal to the square root of the total counts. The precision of an X-ray measurement can, therefore, be predicted by the measured intensity. For example, an accumulated intensity of 1,000,000 counted X-ray photons has a standard deviation of 0.1%, and for 100,000,000 counts the standard deviation is 0.01%.

When an X-ray beam propagates through a sample, its intensity is modified by matrix element effects, concurring with the generation of characteristic X-rays, absorption of the emitted X-rays along their paths, and the enhancement effect due to secondary excitation. Studies of these modification processes and related X-ray physical phenomena lead to the derivation of mathematical correction formulae. The development of these X-ray correction methods dominates the analytical performance of the X-ray fluorescence method.

### 1.4.1 Correction of Matrix Element Effects

The advances in X-ray fluorescence instruments and applications have led to the need for the development of practical and effective mathematical correction formulae. A number of correction methods have been developed (see for example, Lachance and Traill [63], Rassberry and Heinrich [64], etc.). Beattie and Brissey derived a basic correction formula for the relationship between the intensity of characteristic X-rays and the weight fraction of constituent elements, which was the product of a term containing the intensity of measured analytical X-rays and a correction factor containing the concentrations of the constituent elements [65].

A classification of the correction equations published in the literature is carried out from the standpoint of mathematical simplicity and shown in the following:

1. The correction term attributed to constituent elements consists of a constant plus the sum of the products of an X-ray intensity and a correction coefficient, or the sum of the products of a weight fraction and a correction coefficient.
2. The correction coefficients may or may not include the term with the analytic element.
3. The correction coefficients are mostly treated as constants, and this assumption is efficient in the case of small concentration changes of constituent elements.

4. In order to develop wider applicable correction equations and improve the elimination of analytical errors, terms with variable correction coefficients are used in the correction formulae, which are affected with the third or the fourth constituent elements.
5. Least-squares methods have been used for the determination of correction coefficients and correction equations by using experimental data from a large number of standard samples. However, after the development of the fundamental parameter method, calculated intensities have been used for the derivation of correction coefficients and equations as well as for the verification of experimentally determined coefficients and equations. Since there exist many correction methods for quantitative analysis, it is necessary for practical applications to know about the characteristics of matrix correction equations to select the proper fitting algorithm for the analyzed sample.

Rousseau reviewed the concept of the influence of coefficients in matrix correction method from the standpoint of theoretical and experimental approaches and he admonished the essence of a fundamental parameter method [66].

The development of the fundamental parameter method has been carried out by a number of X-ray scientists. At first, Sherman [67] studied the generation of characteristic X-rays theoretically. Shiraiwa and Fujino [68] proceeded with this method even more accurately and verified it experimentally. For the spectral distribution of a primary X-ray source, they combined Kulenkampff's formula [69] of continuous X-rays with their own measured intensity ratios of continuous X-rays and tungsten L series X-rays from a side window X-ray tube. Criss and Birks [70] developed the method further by measuring the primary X-ray intensity distributions from side window X-ray tubes and using mini-computer systems to control X-ray fluorescence spectrometers [71].

To improve the performance of an X-ray spectrometer, a high-power end window X-ray tube with a thin beryllium window was developed by Machlett Laboratories, Inc. [39]. A remarkable improvement in the analytical performances for light elements was achieved by a close coupling of the X-ray source with the sample and a high transmittance window. In order to accomplish a reliable fundamental parameter method, the primary X-ray distributions from end window X-ray tubes were measured by Arai, Shoji, and Omote. It was found that the output of the X-ray spectral distribution in the long wavelength region was increased [72].

Figure 1.4 shows the comparison between measured and calculated intensities of various steels and alloy metals. At low concentrations background intensity corrections should be applied and at the higher intensity ranges the measured intensity requires a counting loss correction. Samples used in Fig. 1.4 are shown in Table 1.2.

Using calculated X-ray intensities, matrix correction coefficients and correction equations have been inspected, and methods using variable correction

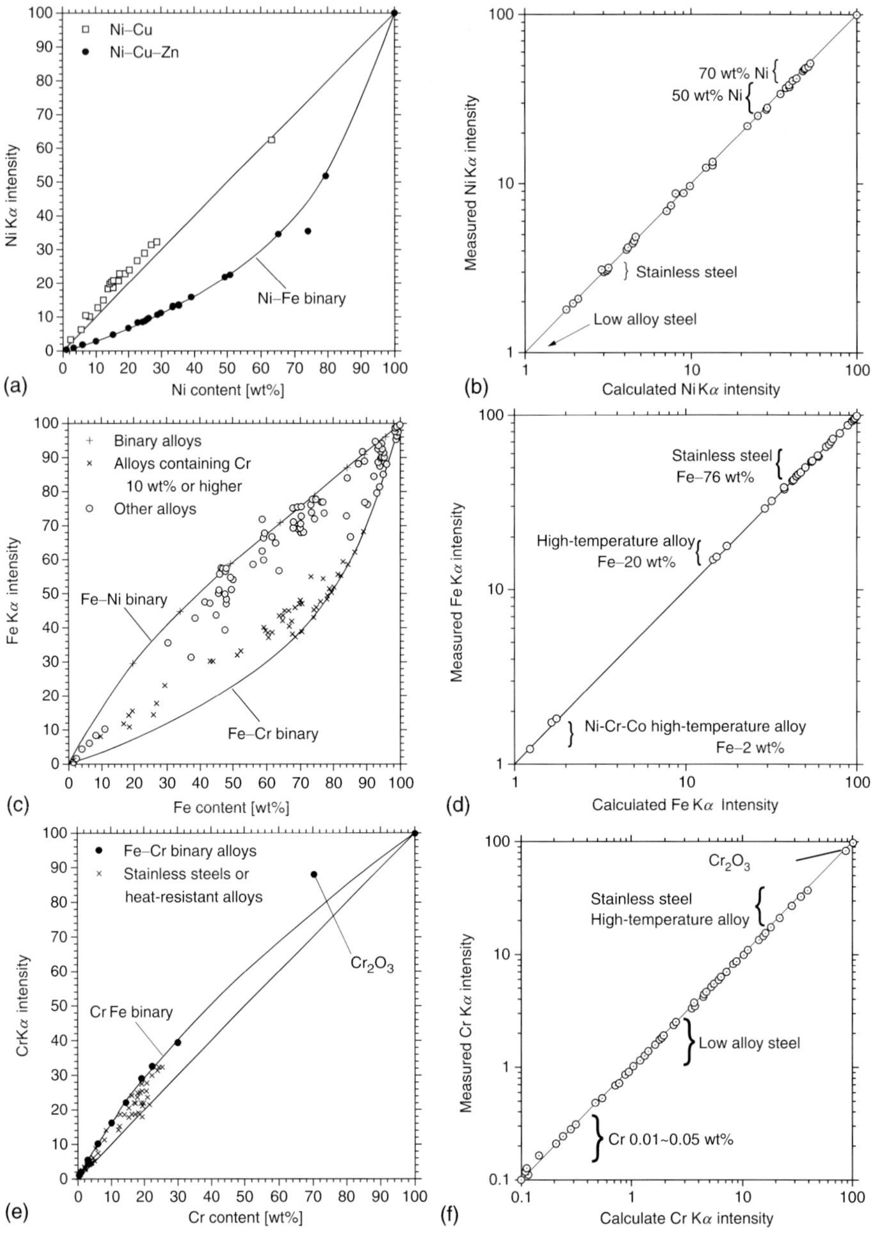


**Fig. 1.4.** (**a**) Relationship between nickel concentration and Ni-K$\alpha$ intensities, (**b**) Relationship between iron concentration and Fe-K$\alpha$ intensities, (**c**) Relationship between chromium concentration and Cr-K$\alpha$ intensities, (**d**) Comparison between calculated and measured Fe-K$\alpha$ intensities, (**e**) Comparison between calculated and measured Fe-K$\alpha$ intensities, (**f**) Comparison between calculated and measured Cr-K$\alpha$ intensities

**Table 1.2.** Measured samples in Fig. 1.4

| | | | |
|---|---|---|---|
| Low alloy steel | Heat-resistant steel | Tool steel | Magnetic alloy |
| Stainless steel | High-speed steel | Binary alloy | German silver |
| High-temperature alloy | Magnetic steel | | |

coefficients have been derived for high accuracy analysis and wider applicable correction equations. Rousseau [73] showed comparative figures of measured and calculated X-ray intensities and applied them to the correction equations developed by Claisse and Quintin [74] and Criss and Birks [70].

Furthermore, Rigaku Industrial Corporation.. tried to compare the measured intensity with calculated intensities based on its own developed fundamental parameter method. Using the primary X-ray distribution from the end window X-ray tube, precisely matching calibration curves were obtained. Using these curves, direct quantitative analysis was then carried out by iterative computer algorithms without the need of matrix correction equations. The analytical results for high alloy analysis are shown in the following section.

### 1.4.2 Quantitative Analysis of Heat-Resistant and High-Temperature Alloys

Gould [75] summarized metal analysis with X-ray spectrochemical analysis. In this section, matrix correction and some segregation influencing analytical accuracy are discussed in detail.

Abbott who was the first developer of a commercial X-ray fluorescence spectrometer, presented a strip chart record of high alloy steel (16-25-6) shown in Fig. 1.5 [17]. Compared to modern equipment, his spectrometer gave much weaker intensities and poorer spectral resolutions due to adoption of a Geiger counter and a NaCl analyzing crystal. Figure 1.6 is a spectrum of NBS 1155 high alloy steel measured with a modern instrument, which equips a scintillation counter and a LiF analyzing crystal. Since the fluorescent intensities and spectral resolution are sufficiently high for practical applications, the difference between these two pictures exhibits the historical progress of 50-years development.

As pointed out by Abbott, the X-ray method is well suited for analyzing heat-resistant and high temperature alloys which consist of nickel, cobalt, iron, and chromium as major constituents, and low concentrations of various other elements. Because the concentrations of the constituent elements influence the metallurgical properties of high temperature and heat-resistant alloys, high precision is required in quantitative determination. In order to review the analytical accuracy of numerous reports, a comparison parameter is introduced, defined as the root mean square between chemical analysis and non-corrected or corrected X-ray values (abbreviated as RMS-difference). Studies of RMS-differences were performed and are shown in Table 1.3.

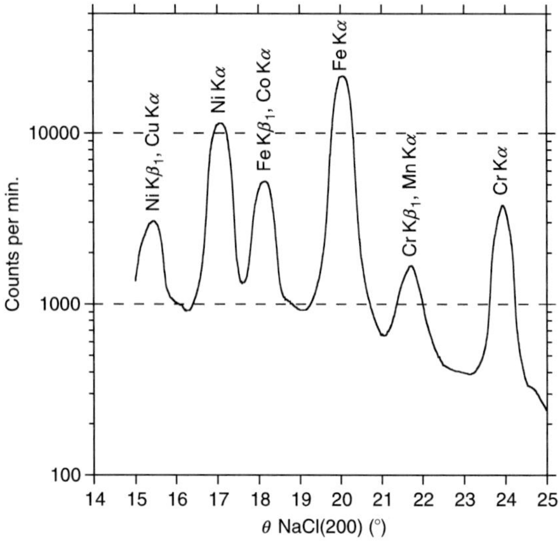


**Fig. 1.5.** Spectrum of 16-25-6 alloy taken by Abbott [17]

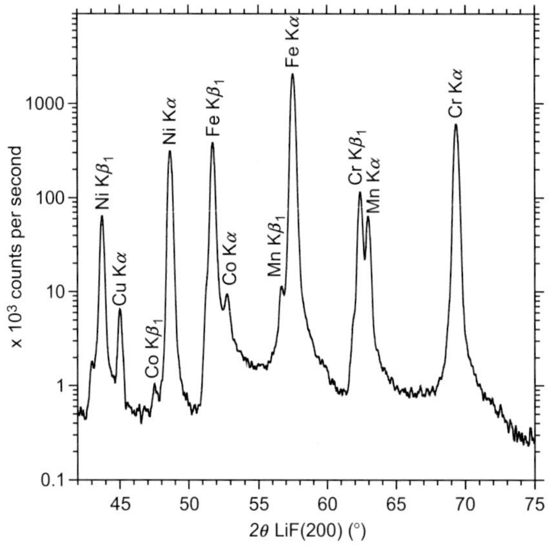


**Fig. 1.6.** Sepctrum of NBS 1155 taken by using Rigaku ZSX 100e

Rickenbach [76] showed the precision and accuracy of nickel and chromium analysis. The measured precision of nickel and chromium in an A286 metal was shown as the composite error within a day and extending to several days. The mean value of the nickel error was 0.03 wt% at the concentration

**Table 1.3.** RMS-difference in heat resistance steel and high temperature alloy analysis. The upper value is the RMS-difference and the lower is the concentration range of the element in each cell. "Sum difference" means the RMS-differences between wt% sums of chemical and X-ray analysis of the analyzing elements. All values are in wt%

| # | Measurements | Analyzed elements | | | | | | | |
|---|---|---|---|---|---|---|---|---|---|
| | | Al | Si | P | S | Ti | V | Cr | Mn |
| 1 | Rickenbach (1963) | | | | | | | 0.11<br>14 ~15 | |
| 2 | Lucas-Tooth and Pyne (1964) | | | | | | | 0.07<br>7 ~ 26 | 0.032<br>4 ~ 10 |
| 3 | Gillieson, Reed, Milliken and Young (1964) | | | | | 0.034<br>0.0 ~ 3.6 | | 0.36<br>0.3 ~33 | 0.029<br>0.0 ~ 1.6 |
| 4 | Lachance and Traill (1966) | | | | | 0.10<br>0.0 ~ 3.6 | | 0.97<br>4 ~ 33 | 0.11<br>0.0 ~ 2 |
| 5a | Caldwell (1976) | | 0.014<br>0.1 ~ 1.1 | | | 0.016<br>0.1 ~ 0.5 | | 0.19<br>0.05 ~ 25 | 0.045<br>0.2 ~ 1.9 |
| 5b | *ditto* | | 0.013<br>0.1 ~ 1.1 | | | 0.016<br>0.1 ~ 0.4 | | 0.098<br>0.05 ~ 25 | 0.023<br>0.2 ~ 1.9 |
| 6 | Ito, Sato and Narita (1983) | 0.056<br>0.9 ~ 10 | 0.024<br>0.2 ~ 2.1 | 0.002<br>0.~0.022 | | 0.052<br>1.0 ~ 11 | 0.021<br>0.2 ~ 2.2 | 0.14<br>5 ~ 28 | 0.023<br>0.5 ~ 3.5 |
| 7a | Griffiths and Webster (1986) | 0.026<br>0.4 ~ 5.6 | 0.039<br>0.0 ~ 0.6 | | | 0.026<br>0.0 ~ 3.8 | | 0.28<br>8 ~ 25 | 0.013<br>0.0 ~ 0.6 |
| 7b | *ditto* | 0.025<br>0.0 ~ 5.6 | 0.039<br>0.0 ~ 0.6 | | | 0.027<br>0.0 ~ 3.8 | | 0.078<br>8.8 ~ 25 | 0.012<br>0.0 ~ 0.6 |
| 8 | Itoh, Sato, Ide and Okochi(1986) | 0.047<br>0.8 ~ 5.8 | | | | 0.13<br>0.2 ~ 4 | | 0.29<br>8 ~ 22 | 0.006<br>0.0~0.07 |
| 9 | Rigaku (2000) | 0.024<br>0.0 ~ 1.4 | 0.023<br>0.07~1.0 | 0.0009<br>0.0~0.02 | 0.0009<br>0.0~0.02 | 0.026<br>0.0 ~ 3.1 | | 0.10<br>15 ~ 22 | 0.010<br>0.08~1.4 |

**Table 1.3.** *Continued*

| # | Analyzed elements | | | | | | | | | Sum |
|---|---|---|---|---|---|---|---|---|---|---|
| | Fe | Co | Ni | Cu | Zr | Nb | Mo | Ta | W | difference |
| 1 | | | 0.19 | | | | | | | |
| | | | 25 ∼ 27 | | | | | | | |
| 2 | | | | | | | | | | |
| 3 | 0.38 | 0.28 | 0.45 | | 0.014 | 0.046 | 0.05 | 0.014 | 0.09 | 0.51 |
| | 0.1 ∼ 96 | 0.0 ∼ 64 | 0.0 ∼ 78 | | 0.0 ∼ 0.5 | 0.0 ∼ 2.0 | 0.0 ∼ 17 | 0.0 ∼ 0.5 | 0.0 ∼ 18 | |
| 4 | 1.1 | 3.7 | 1.9 | | | 0.22 | 0.29 | | 0.55 | 3.1 |
| | 0.1 ∼ 28 | 0.0 ∼ 64 | 0.7 ∼ 78 | | | 0.0 ∼ 4.7 | 0.0 ∼ 25 | | 0.0 ∼ 18 | |
| 5a | 0.47 | | 0.40 | 0.031 | | 0.014 | 0.10 | | | 0.20 |
| | 14 ∼ 90 | | 0.3 ∼ 81 | 0.1 ∼ 0.6 | | 0.2 ∼ 0.7 | 0.0 ∼ 4.4 | | | |
| 5b | 0.34 | | 0.06 | 0.031 | | 0.014 | 0.013 | | | 0.22 |
| | 14 ∼ 90 | | 0.3 ∼ 81 | 0.0 ∼ 0.6 | | 0.2 ∼ 0.7 | 0.01∼4.4 | | | |
| 6 | 0.16 | 0.073 | 0.23 | 0.032 | 0.009 | 0.066 | 0.088 | 0.047 | 0.069 | |
| | 4.6 ∼ 56 | 2.0 ∼ 30 | 31 ∼ 96 | 0.2 ∼ 6.4 | 0.1 ∼ 0.8 | 0.8 ∼ 9.4 | 1 ∼ 28 | 0.2 ∼ 10 | 1 ∼ 12 | |
| 7a | 0.039 | 0.11 | | 0.055 | 0.057 | 0.015 | 0.009 | 0.015 | 0.058 | 0.38 |
| | 0.1 ∼ 3.5 | 0.5 ∼ 20 | (48 ∼75) | 0.0 ∼ 0.3 | 0.0 ∼ 0.3 | 0.0 ∼ 1.0 | 0.0 ∼ 11 | 1.3 ∼ 2.6 | 0.0 ∼ 10 | |
| 7b | 0.044 | 0.12 | 0.17 | 0.015 | 0.010 | 0.009 | 0.009 | 0.016 | 0.031 | 0.39 |
| | 0.0 ∼ 3.5 | 0.5 ∼ 20 | 47 ∼ 76 | 0.0 ∼ 0.3 | 0.0 ∼ 0.3 | 0.0 ∼ 1.0 | 0.0 ∼ 11 | 0.0 ∼ 2.6 | 0.0 ∼ 10 | |
| 8 | 0.074 | 0.38 | | | | 0.025 | 0.053 | | 0.030 | |
| | 0.0 ∼ 7.2 | 0.2 ∼ 13 | | | | 0.1 ∼ 2.0 | 0.1 ∼ 9. | | 0.03 ∼ 10 | |
| 9 | 0.19 | 0.035 | 0.29 | 0.005 | | 0.020 | 0.014 | 0.007 | 0.015 | 0.48 |
| | 1.4 ∼ 67 | 0.0 ∼ 21 | 9.4 ∼ 74 | 0.0 ∼ 0.1 | | 0.0 ∼ 5.4 | 0.0 ∼ 5.4 | 0.0 ∼ 0.4 | 0.0 ∼ 2.4 | |

**Table 1.3.** *Continued*

| # | X-ray instrument | Tube target | Matrix correction | Number of Samples | Reference |
|---|---|---|---|---|---|
| 1 | GE XRD-5 (Sequential) | - | Simple calibration method. No matrix correction | 22 | [76] |
| 2 | Solartron automatic vacuum spectrometer XZ1030 | - | Lucas-Tooth and Pyne method. Intensity correction | 60 (Cr) 3 (Mn) | [77] |
| 3 | Norelco 100kV | W | Lachance - Traill method Fixed a correction | 25 | [78] |
| 4 | Philips 100kV | W | | 80 | [81] |
| 5a | GE XRD-700 (Sequential) | Cr (for Si and Ti) W (heavy elements) | Lachance - Traill method Fixed a correction | 80 | [80] |
| 5b | *ditto* | *ditto* | Lachance - Traill method Variable a correction | *ditto* | *ibid.* |
| 6 | Rigaku Simultix 4B (Fixed channels) | Rh | JIS correction | 80 | [82] |
| 7a | Philips 1410 (Sequential) | Cr | Lachance - Traill method with small modification (Standard) | 14 | [83] |
| 7b | *ditto* | *ditto* | Lachance - Traill method with small modification ("60:20:20" correction) | *ditto* | *ibid.* |
| 8 | Philips 1400 (Sequential) | Rh | de Jongh's theoretical correction coefficients | 7 | [85] |
| 9 | RigakuZSX100e (Sequential) | Rh | Confrontation method between measured and calculated intensity based on FP | 12 | - |

of 26.2 wt% and for chromium it was 0.023 wt% at 14.5 wt%. They are one-fifth of the RMS-differences in Table 1.3. It was noted that no matrix corrections were required for specimens with only small concentration variations.

Lucas-Tooth and Pyne discussed a formula where the correction factor was a constant plus the sum of products of the individual X-ray intensities of the constituent elements with correction coefficients. RMS-differences of 0.07 wt% in chromium and 0.032 wt% in manganese were reported [77].

A third report sponsored by the ASTM committee in 1964 was presented by Gillieson, Reed, Milliken, and Young [78]. Simultaneously, a report about spectrochemical analysis of high temperature alloys by spark excitation was presented. Referring to the matrix correction methods by Lucas-Tooth and Price [79] and Lucas-Tooth and Pyne, they applied a correction on the basis of X-ray intensities of the constituent elements. The measured intensities of aluminum and silicon constituents should be added in order to improve the matrix correction, thereby increasing the accuracy.

Lachance and Traill studied simple matrix correction equations that were one plus the sum of products of the weight fraction of constituent elements and correction coefficients [63]. Based on the analysis of high nickel alloys that were selected from the application report, RMS-differences were calculated and are shown in Table 1.3.

On the basis of the Lachance–Traill equations, Caldwell derived two kinds of correction equations [38]. The first one was a fixed correction coefficient equation and the second one was a variable correction coefficient equation for wider concentration applications, on which the third or fourth constituent elements exerted reform. RMS-differences of major constituents in variable correction coefficient calculations improved those of the fixed correction coefficient method.

Ito, Sato, and Narita [82] studied the JIS correction equations that consisted of the product of a factor containing a quadratic polynomial of the intensity of the measured X-rays, and a matrix correction factor which was authorized by the JIS Committee. The coefficients of the intensity part were determined by least-squares algorithms from binary alloys with known chemical composition or from mathematical models, and the second factor was one plus the sum of products of the weight fractions of the constituent elements and correction coefficients, in which the terms containing the base component and the analyzing elements were excluded. In practical applications for nickel base alloys the correction coefficients of light and heavy elements from iron-based alloys were used; for the analysis of the major constituent elements, chromium, iron, and cobalt in nickel base alloys, the correction coefficients were determined experimentally.

Griffiths and Webster discussed the derivation of matrix correction equations in detail. They adopted the modified Lachance–Traill equation [83]. The calibration constants of the X-ray intensity terms were determined by regression analysis and the correction factors were calculated with the ALPHAS program which was theoretically derived by de Jongh [84]. The two kinds of RMS-differences of authenticated sample analysis are shown in Table 1.3. The values in the upper line are the calculated results based on normal matrix correction in the ALPHAS program, and the second values in the lower line are derived with the use of correction coefficients calculated under the condition of a fixed 60:20:20 constituent sample.

Itoh, Sato, Ide, and Okochi studied the analysis of high alloys using the theoretically calculated correction coefficients of the ALPHAS program. They also compared the ALPHAS program and the JIS correction method and clarified that there were no differences between them. The results of analytical accuracy were two or more times higher than those of X-ray analytical precision [85].

Rigaku Industrial Corporation analyzed high nickel alloys of NBS and specially prepared samples. For the matrix correction, they adopted their own fundamental parameter method and prepared calibration curves, which were used for the determination of the constituent elements. The RMS-differences in these studies were fairly small.

### 1.4.3 Segregation Influencing Analytical Accuracy

The influence of inhomogeneity phenomena on analytical accuracy is one of the most important factors. The internal soundness of an ingot which was studied by Marburg [86] conferred that the inhomogeneity induced in the cooling process from molten metals intimated strong effects to analytical problems to be solved.

Stoops and McKee studied the reduction of analytical accuracy for titanium concentrations owing to segregation of nickel base alloys of M252 (19 wt% chromium, 10 wt% cobalt, 10 wt% molybdenum, 2.5 wt% titanium, 3 wt% iron, 1 wt% aluminum, 0.15 wt % carbon, and 0.35 wt% silicon). As a major portion of the titanium could be found in the grain boundaries of carbide or carbon-nitride particles, the differences between regular chemical and X-ray analysis values indicated a wider distribution, which was 5–10% of the amount of titanium present. When chemical analysis was carried out using samples scooped up from the X-ray analytical surface, X-ray values were nearly equal to that of chemical analysis, i.e., 0.5–1% of the amount of titanium present [87].

It is well known that in rapidly cooled steel, a low concentration of manganese and impurity sulfur are dispersed and that a high analytical accuracy in manganese and sulfur determination can be found. In sufficiently annealed steel, small particles of manganese sulfide are precipitated in the grain boundaries of steel grains and large deviations of Mn-K$\alpha$ and S-K$\alpha$ X-rays are found [88].

Free cutting metals that are among the most widely used industrial materials are typical examples for exhibiting segregation phenomena in metals. Small metal particles like lead in steels and copper alloys influence machine processing of high-speed cutting.

Iwasaki and Hiyoshi studied lead segregation in bronze alloys. Microscopic observations of the surface pictures of a cast bronze showed a mixture of ground metal of copper and lead precipitated particles, and the recast pictures showed the surface with scattering of small lead particles. Comparing surfaces of recast and cast samples, Cu-K$\alpha$ X-rays from cast sample surface exhibited higher intensities and Pb-L$\alpha$ X-rays showed lower intensities. In recast sample surfaces, Cu-K$\alpha$ X-rays showed lower intensities and Pb-L$\alpha$ X-rays higher intensities because of absorption of Cu-K$\alpha$ X-rays and the exciting of bare surfaces of many small particles of lead on analyzing surface [89].

In the analysis of lead free cutting steels (Pb content: 0.1–0.3 wt%), small particles of lead metal (1–15 μm) are scattered in steel. Arai reported that simple repeatability of Pb-L$\alpha$ X-rays was 0.003 wt% and RMS-difference of lead was 0.018 wt% [90].

The characteristics of segregation or inhomogeneity have been recognized as one of natural phenomena or as discoveries through experimental works. If some studies or works have attained success after long or hard work, possibilities of more success or discovery will be increased on the basis of research work and their process.

## 1.5 Concluding Remarks

Elemental analysis of materials is absolute or relative. Gravimetric analysis is a typical example of the former, while X-ray fluorescence and optical emission methods represent the latter. In the case of a relative analysis, standard samples are required, which are attached to reliable or authenticated analytical values supported by absolute analysis. The values guaranteed by absolute analysis are the mean values of volume analysis, while X-ray analytical values represent surface analysis with an information depth of 5–50 μm. In order to reduce the analytical uncertainty, which orginates from the differences between volume and surface analysis, homogeneous samples samples should be used. For the sake of reducing the effect of inhomogeneity on the sample surface, large analyzed surfaces of 3–10 $cm^2$ are recomended.

Since the analytical accuracy is defined by a combination of errors of chemical analysis, uncertainty in the measured X-ray intensity, and uncorrected matrix effects by the constituent elements,the observed accuracy can be reduced by effective matrix corrections adapted for X-ray analysis, and elimination of other systematic errors may be activated.

## References

1. Segrè E, From X-rays to quarks: Modern physicists and their discoveries. W.H. Freeman & Co., New York (1980)
2. The Nobel Prize in Physiology or Medicine 1962 to Crick FHC, Watson JD, Wilkins MHF, for their discoveries concerning the molecular structure of nucleic acids and its significance for information transfer in living material

3. The Nobel Prize in Physiology or Medicine 1979 to Cormack AM and Hounsfield GN, for the development of computer assisted tomography. Cormack AM, Representation of a function by its line integrals, with some radiological applications. J Appl Phys **34**, 2722–2727 (1963)
4. Glasser O, Wilhelm Conrad Röntgen and the early history of the Röntgen rays. Norman Publishing, San Francisco (1993)
5. Mould RF, A century of X-rays and radioactivity in medicine: with emphasis on photographic records of the early years. Institute of Physics Publishing, Bristol and Philadelphia (1993)
6. Barkla CG, Polarisation in secondary Röntgen radiation. Proc R Soc Sect **A** **77**, 247–255 (1906)
   Barkla CG, Sadler CA, The absorption of Röntgen rays. Philos Mag **17**, 739–760 (1909)
   Barkla CG, The spectra of the fluorescent Röntgen radiations. Philos Mag **22**, 396–412 (1911)
7. Bragg WL, The crystalline state volume I: A general survey. Bell & Heyman, London (1949)
8. Coolidge WD, A powerful Röntgen ray tube with a pure electron discharge. Phys Rev **2**, 409–430 (1913)
9. Jaffe B, Moseley and the numbering of the elements. Doubleday & Co., Garden city, New York (1971)
10. Rozental S, Niels Bohr: His life and work as seen by his friends and colleagues. Elsevier, New York (1985)
    Moore RE, Niels Bohr: The man, his science, and the world they changed. Knopf/Random House, New York (1966)
11. von Hevesy G, Chemical analysis by X-rays and its applications. McGraw-Hill, New York (1932)
12. Glocker R, Materialprüfung mit Röntgenstrahlen unter besonderer Berücksichtigung der Röntgenmetallkunde. Springer Verlag, Berlin (1936)
13. Parrish W, Gordon SG, Precise angular control of quartz-cutting with X-rays. Am Mineral **30**, 326–346 (1945)
14. Friedman H, Applications of electronic methods to the measurement of X-ray and gamma ray intensities. Ind Radiography **Summer**, 9–20 (1947)
15. Friedman H, Geiger counter spectrometer for industrial research. Electronics, 132–137 (1945)
16. Friedman H, Birks LS, A Geiger counter spectrometer for X-ray fluorescence analysis. Rev Scientific Instrum **19**, 323–330 (1948)
17. Abbott JL, X-ray fluorescence analysis. The Iron Age, 58–62 (1948); Abbott JL, X-ray fluorescence analysis. The Iron Age, 121–124 (1948)
18. Birks LS, X-ray fluorescence: Present limitations and future trends. X-ray Anal, Japan, **2**, 5–13 (1965) [in Japanese]
19. Birks LS, Current capabilities and future goals in X-ray spectroscopy. Pure Appl Chem **48**, 45–52 (1976)
20. Gilfrich JV, New horizons in X-ray fluorescence analysis. X-ray Spectrom **19**, 45–51 (1990)
21. Gilfrich JV, 100 years of progress in X-ray analysis. **39**, 29–39 (1997)
22. Parrish W, X-ray spectrochemical analysis. Philips Technical Rev **17**, 269–286 (1956)
23. Spielberg N, Parrish W, Lowitzsch K, Geometry of the non-focusing X-ray fluorescence spectrograph. Spectrochim Acta **13**, 564–583 (1959)

24. Arai T, The profile of the parallel optical system having a Soller slit in X-ray fluorescence spectrograph. X-ray Anal, Japan **2**, 139–153 (1965) [in Japanese]
25. Arai T, Studies of $2\theta$ values and peak profiles in non-focusing X-ray fluorescent spectrometer. Adv X-ray Chem Anal, Japan **19**, 293–305 (1988) [in Japanese]
26. Campbell WJ, Leon M, Thatcher J, Flat crystal X-ray optics. Adv X-ray Anal **1**, 193–206 (1958)
27. Miller DC, Zingaro PW, The universal vacuum spectrograph and comparative data on the intensities observed in an air, helium, and vacuum path. Adv X-ray Anal **3**, 49–56 (1960)
28. Miller DC, Results obtained with the modified Norelco Autrometer. Adv X-ray Anal **1**, 283–295 (1958)
29. Kiley WR, A universal detector for the X-ray spectrograph. Adv X-ray Anal **2**, 293–301 (1959)
30. Kemp JW, Instrumentation for rapid spectrochemical analysis–Optical and X-ray emission monochromators and polychromators. Anal Chem **28**, 1838–1843 (1956)
31. Jones JL, Paschen KW, Swain HH, Andermann G, Components for X-ray polychromators. Adv X-ray Anal **1**, 471–482 (1958)
32. Andermann G, Jones JL, Davidson E, The evaluation of the PXQ for the analysis of cements and related materials. Adv X-ray Anal **2**, 215–227 (1959)
    Andermann G, Allen JD, The evaluation and improvement of X-ray emission analysis of raw-mix and finished cements. Adv X-ray Anal **4**, 414–432 (1961)
33. Dryer HT, Davidson E, Andermann G, Vacuum X-ray instrumentation and its application to mill-products control. Adv X-ray Anal **5**, 477–485 (1962)
34. Andermann G, Kemp JW, Scattered X-rays as internal standards in X-ray emission spectroscopy. Anal Chem **30**, 1306–1309 (1958)
35. Anzelmo JA, Buman AI, Evaluation criteria for wavelength dispersive XRF instrumentation. Am Lab, 65–73 (1988)
36. Kansai K, Toda K, Kohno H, Arai T, Wilson R, Accurate measurement of trace elements using an innovative fixed goniometer for a simultaneous spectrometer. Adv X-ray Anal **38**, 691–698 (1995)
37. Kikkert J, Hendry G, Comparison of experimental and theoretical intensities for a new X-ray tube for light element analysis. Adv X-ray Anal **27**, 423–426 (1984)
38. Caldwell VE, A practical method for the accurate analysis of high alloy steels by X-ray emission. X-ray Spectrom **5**, 31–35 (1976)
39. Mahn GR, The OEG-75G—A new high power tube for X-ray fluorescent analysis. Cathode Press **22**, 41–46 (1965)
40. Gurvich YM, Comparison of various X-ray tube types for XRF analysis. Adv X-ray Anal **28**, 113–118 (1985)
    Gurvich YM, An advanced end-window. XRF spectrometer for analysis of light elements. Am Lab, 103–111 (1984)
41. Henke BL, X-ray fluorescence analysis for sodium, fluorine, oxygen, nitrogen, carbon, and boron. Adv X-ray Anal **7**, 460–488 (1964)
    Henke BL, Tester MA, Techniques of low energy X-ray spectroscopy (0.1 to 2 keV region). Adv X-ray Anal **18**, 76–106 (1975)
    Henke BL, Techniques of low energy X-ray and electron physics. 50 to 1000 ev region. Norelco Reporter **14**, 75–83, 98 (1967)
42. Holliday JE, The use of soft X-ray fine structure in bonding determination and light element analysis. Norelco Reporter **16**, 84–91, 116 (1967)

43. Fischer DW, Baun WL, The influences of chemical combination and sample self absorption on some long wavelength X-ray emission spectra. Norelco Reporter **14**, 92–98 (1967)
44. Henke BL, Measurement in the 10 to 100 angstrom X-ray region. Adv X-ray Anal **4**, 244–279 (1961)
Henke BL, Microanalysis with ultrasoft x-radiations. Adv X-ray Anal **5**, 285–305 (1962)
Henke BL, Application of multilayer analyzers to 15–150 angstrom fluorescence spectroscopy for chemical and valence band analysis. Adv X-ray Anal **9**, 430–440 (1966)
Henke BL, Lent RE, Some recent work in low energy X-ray and electron analysis. Adv X-ray Anal **12**, 480–495 (1969)
45. Barbee TW, Sputtered layered synthetic microstructure (LSM) dispersion elements. AIP Conf Proc **75**, 131–145 (1981)
46. Spiller E, Evaporated multilayer dispersion elements for soft X-rays. AIP Conf Proc **75**, 124–130 (1981)
47. Underwood JH, Barbee TW, Synthetic multilayers as Bragg diffractors for X-rays and extreme ultraviolet: Calculations of performance. AIP Conf Proc **75**, 170–178 (1981)
48. Gilfrich JV, Nagel DJ, Loter NG, Barbee TW, X-ray characteristics and applications of layered synthetic microstructures. Adv X-ray Anal **25**, 355–364 (1982)
Gilfrich JV, Nagel DJ, Barbee TW, Layered synthetic microstructures as dispersing devices in X-ray spectrometers. Appl Spectrosc **36**, 58–61 (1982)
Gilfrich JV, Multilayered structures as dispersing devices in X-ray spectrometry. Analytica Chim Acta **188**, 51–57 (1986)
49. Arai T, Shoji T, Ryon RW, Wavelength dispersing devices for soft and ultrasoft X-ray spectrometers. Adv X-ray Anal **28**, 137–144 (1985)
50. Toda K, Kohno H, Arai T, Araki Y, Hamill G, Sensitivity improvement and stabilization for ultra light element analysis by X-ray spectrometry. Adv X-ray Anal **37**, 629–637 (1994)
51. Anzelmo JA, Boyer BW, The analysis of carbon and other light elements using layered synthetic microstructures. Adv X-ray Anal **30**, 193–200 (1987)
52. Huang TC, Fung A, White RL, Recent measurements of long wavelength X-rays using synthetic multilayers. X-ray Spectrom **18**, 53–56 (1989)
53. Martins E, Urch DS, Problems in use of multilayers for soft X-ray spectroscopy and analysis: A comparison of theoretically and experimentally determined refraction effects. Adv X-ray Anal **35**, 1069–1078 (1992)
54. Huang TC, Thin film characterization by X-ray fluorescence. X-ray Spectrom **20**, 29–33 (1991)
55. Arai T, An X-ray fluorescent spectrometer for the measurement of thin layered materials on silicon wafers. Adv X-ray Anal **30**, 315–323 (1987)
56. White RL, Huang TC, Determination of ultra-thin carbon coating thickness by X-ray fluorescence technique. Adv X-ray Anal **32**, 331–339 (1989)
57. Kobayashi H, Toda K, Kohno H, Arai T, Wilson R, The study of some peculiar phenomena in ultra-soft X-ray measurements using synthetic multilayer crystals. Adv X-ray Anal **38**, 307–312 (1995)

Kobayashi H, Toda K, Kohno H, The study of some peculiar phenomena in ultra-soft X-ray measurements using synthetic multilayer crystals. Adv X-ray Chem Anal, Japan **26**, 45–58 (1995) [in Japanese]

58. Franks A, Braybrook RF, Analysis of the lighter elements by total reflexion of their characteristic X-ray wavelengths. Br J Appl Phys **10**, 190–191 (1959)
59. Herglotz HK, Wavelength identification of ultrasoft X-rays by the critical angle of total reflection. J Appl Phys **38**, 4565–4568 (1967);
Davies RD, Herglotz HK, A total reflection X-ray spectrograph for fluorescence analysis of light elements. Adv X-ray Anal **12**, 496–505 (1969)
60. Arai T, Soft and ultrasoft X-ray fluorescent spectrometer using total reflection monochromator. Japanese J Appl Phys **21**, 1347–1358 (1982)
Arai T, Sohmura T, Tamenori H, Determination of boron oxide in glass by X-ray fluorescence analysis. Adv X-ray Anal **26**, 423–430 (1983)
Arai T, Ohara S, Determination of oxygen and nitrogen in various materials by X-ray fluorescence spectrometry. Adv X-ray Anal **27**, 547–556 (1984)
Arai T, Measurement of soft and ultrasoft X-rays with total reflection monochromator. Adv X-ray Anal **30**, 213–223 (1987)
61. Sugimoto K, Akiyoshi T, Kondou T, XRF determination of carbon in pig iron. Bunseki Kagaku **37**, 589–594 (1988) [in Japanese]
62. Zemany PD, X-ray spectrometry Chapter 4. Precision and Accuracy. Marcel Dekker, New York, 69–110 (1978)
63. Lachance GR, Traill RJ, A practical solution to the matrix problem in X-ray analysis. Part 1: Method. Can Spectrosc **11**, 43–48 (1966)
64. Rassberry SD, Heinrich KFJ, Calibration for interelement effects in X-ray fluorescence analysis. Anal Chem **46**, 81–89 (1974)
65. Beattie HJ, Brissey RM, Calibration method for X-ray fluorescence spectrometry. Anal Chem **26**, 980–983 (1954)
66. Rousseau RM, Why the fundamental algorithm is so fundamental. Adv X-ray Anal **37**, 639–646 (1994)
67. Sherman J, The theoretical derivation of fluorescent X-ray intensities from mixtures. Spectrochim Acta **7**, 283–306 (1955)
68. Shiraiwa T, Fujino N, Theoretical calculation of fluorescent X-ray intensities in fluorescent X-ray spectrochemical analysis. Japanese J Appl Phys **5**, 54–67 (1966)
69. Kulenkampff H, Annalen der Physik **69**, 548 (1922)
70. Criss JW, Birks LS, Calculation methods for fluorescent X-ray spectrometry empirical coefficients vs. fundamental parameters. Anal Chem **40**, 1080–1086 (1968)
Criss JW, Birks LS, Gilfrich JV, Versatile X-ray analysis program combining fundamental parameters and empirical coefficients. Anal Chem **50**, 33–37 (1978)
71. Gilfrich JV, Birks LS, Spectral distribution of X-ray tubes for quantitative X-ray fluorescence analysis. Anal Chem **40**, 1077–1080 (1968)
Gilfrich JV, Burkhalter PG, Whitlock RR, Warden ES, Birks LS, Spectral distribution of a thin window rhodium target X-ray spectrographic tube. Anal Chem **43**, 934–936 (1971)
Brown DB, Gilfrich JV, Measurement and Calculation of Absolute X-Ray Intensities. J Appl Phys **71**, 4044–4046 (1971)
Brown DB, Gilfrich JV, Peckerar MC, Measurement and calculation of absolute intensities of X-ray spectra. J Appl Phys **46**, 4537–4540 (1975)

72. Arai T, T. Shoji, K. Omote, Measurement of the spectral distribution emitted from X-ray spectrographic tubes. Adv X-ray Anal **29**, 413–422 (1985)
73. Rousseau RM, Fundamental algorithm between concentration and intensity in XRF analysis. 1-Theory. X-ray Spectrom **13**, 115–120 (1984)
Rousseau RM, Fundamental algorithm between concentration and intensity in XRF analysis. 2-Practical application. X-ray Spectrom **13**, 121–125 (1984)
74. Claisse F, Quintin M, Generalization of the Lachance-Traill method for the correction of the matrix effect in X-ray fluorescence analysis. Can Spectrosc **12**, 129–146 (1967)
75. Gould RW, X-ray spectrometry Chapter 11. Metals and alloys. Marcel Dekker, New York, 277–295 (1978)
76. Rickenbach JR, Some aspects of nondestructive X-ray spectrochemical analysis of alloys. Adv X-ray Anal **6**, 352–360 (1962)
77. Lucas-Tooth J, Pyne C, The accurate determination of major constituents by X-ray fluorescent analysis in the presence of large interelement effects. Adv X-ray Anal **7**, 523–541 (1963)
78. Gillieson, AH, Reed DJ, Milliken KS, Young MJ, X-ray spectrochemical analysis of high-temperature alloys, ASTM Special Technical Publication No.376, 3–22 (1964)
79. Lucas-Tooth HJ, Price BJ, A mathematical method for the investigation of inter-element effects in X-ray fluorescence analysis. Metallurgia **64**, 149–152 (1961)
80. Lachance GR, Traill RJ, A practical solution to the matrix problem in X-ray analysis. Part 2: Application to a multielement alloy system. Can Spectrosc **11**, 63–71 (1966)
81. Caldwell VE, A practical method for the accurate analysis of high alloy steels by X-ray emission. X-ray Spectrom **5**, 31–35 (1976)
82. Ito M, Sato S, Narita M, X-ray fluorescence analysis of nickel base alloys. Tetsu-to-Hagan **69**, 169–176 (1983) [in Japanese]
Ito M, Sato S, Narita M, Comparison of Japanese industrial standard and a-correction method for X-ray fluorescence analysis of steel. X-ray Spectrom **10**, 103–105 (1981)
83. Griffiths JM, Webster HWM, X-ray analysis of nickel-base alloys with theoretically derived inter-element correction coefficients. X-ray Spectrom **15**, 61–72 (1986)
84. de Jongh WK, X-ray fluorescence analysis applying theoretical matrix corrections. Stainless steel. X-ray Spectrom **2**, 151–158 (1973)
85. Itoh S, Sato K, Ide K, Okochi H, X-ray fluorescence analysis of nickel-based heat-resisting alloys by matrix correction using theoretical alpha coefficients. Bunseki Kagaku **35**, 33–37 (1986) [in Japanese]
86. Marburg E, Accelerated solidification in ingots: Its influence on ingot soundness. J Metals, 157–172 (1953)
87. Stoops RF, McKee KH, Sampling errors in the X-ray fluorescent determination of titanium in a high temperature alloy. Anal Chem **33**, 589–592 (1961)
88. Bain EC, Functions of the alloying elements in steel, U S Steel Corporations, Pittsburgh PA, USA (1945)
89. Iwasaki K, Hiyoshi K, XRF analysis of bronze castings after recasting. Bunseki Kagaku **37**, 152–156 (1988) [in Japanese]
90. Arai T, Quantitative determination of lead in lead-free cutting steels. The Rigaku Denki J **3**, 207–208 (1961) [in Japanese]

# 2

# X-Ray Sources

## 2.1 Introduction

*N. Langhoff and A. Simionovici*

X-rays, first detected by Wilhelm Conrad Röntgen in 1895 [1], are electromagnetic waves with a spectrum spanning from about 80 nm wavelength (about 15 eV) adjacent to the vacuum ultraviolet down to about 0.001 nm (about 1.2 MeV) overlapping to some extent the region of $\gamma$-rays (Fig. 2.1). Electromagnetic radiation above 1 MeV generated by nuclear processes is usually called $\gamma$-radiation while the radiation below 80 nm wavelength, generated by electrons slowed down in the outer field of an atomic nucleus or by changes of bound states of electrons in the electronic shell of an atom, is called x-radiation. The division into different regions of X-rays and $\gamma$-rays is to some extent artificial and sometimes misleading.

The x-radiation generated by the slowing down processes of electrons is called "Bremsstrahlung" and has a continuous spectrum with a sharp termination $\lambda_t$ at the short wavelength side corresponding with the maximum kinetic energy $E_{\text{kin}} = e \cdot U$ of the electron

$$\lambda_t[\text{nm}] = \frac{c}{\nu_{\text{t}}} = \frac{ch}{E_{\text{kin}}} = \frac{1.24}{U[\text{kV}]}.$$

Electron transitions between inner electron shells of the atom may generate the so-called characteristic x-radiation, that is a line emission which can be attributed unambiguously to the elements. According to the electronic shell structure of the elements such characteristic radiation is to be expected beginning with sodium (e.g., Na-$L_\alpha$ with an excitation energy of 55 eV).

In X-radiation sources, for practical purposes, mainly two different principles are used: first, the spontaneous radioactive decay of certain isotopes, and second, the interaction of artificially accelerated electrons with matter in X-ray tubes or with magnetic fields in accelerators or storage rings.

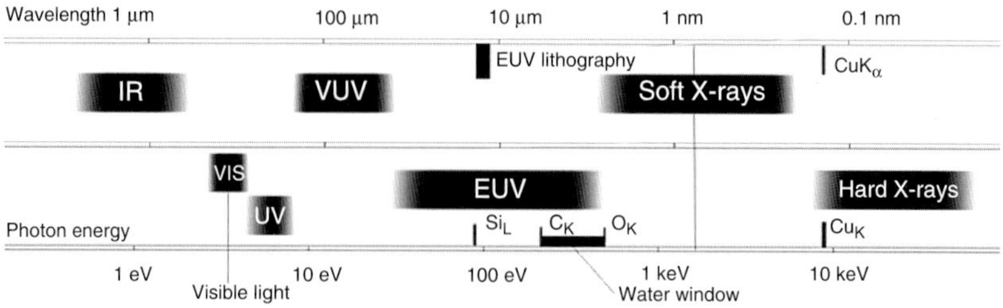


**Fig. 2.1.** Spectrum of electromagnetic radiation from infrared radiation to hard X-rays (IR, infrared; VIS, visible light; UV, ultraviolet; VUV, vacuum ultraviolet; EUV, extreme ultraviolet)

Historically, the first X-ray generator used by W.C. Röntgen [1] was an ion tube (Hittorf–Crooks-tube, air filled at reduced pressure). Electrons were produced by ion bombardment of a cold cathode in a high voltage gas discharge. The electrons are accelerated by the tube voltage onto the anode generating X-rays. These ion tubes are almost out of use since they have a short lifetime, are instable and difficult to control. W.D. Coolidge [2] introduced the hot filament electron emitter in a high vacuum X-ray tube. This schema turned out to be very effective and is also widely used at present. A very extensive and specific development of X-ray tubes of all kinds for every particular purpose started in the past and is still progressing. New technologies and new materials are forming the basis for new highly specialized X-ray tubes.

Field electron emitters instead of thermo-electron emitters were tried repeatedly in DC X-ray tubes, because it appeared to be attractive to get rid of the clumsy hot cathode with its heating circuit, which in many cases has to run at high potential. Recently, promising results with carbon nano tubes as field electron emitters have been obtained and their technical application seems within reach [3].

A wide variety of X-ray tubes is now available, from low power of a few Watt up to very high power to tens of kW for DC-, intermittent or short pulse operation specialized for application in all fields. An overview on X-ray tubes with emphasis on the requirement of X-ray analysis is given in Part 2.2.

A particular motivation for new developments of X-ray sources was provided by the recent progress in X-ray optics such as glass capillary optics, Fresnel and Fresnel–Bragg optics as well as HOPG crystals. The efficiency of these optics depends critically on the brilliance of the X-ray source and the compact design of the source. The anode spot should be well focussed and the optics should be mounted as close as possible to the anode spot with the aim to catch as much of the diverging radiation as possible. This allows to reduce the tube power well below 50 W and to achieve short measuring times at improved local and energy resolution. In consequence, different manufacturers are now offering first samples of miniaturized low power microfocus X-ray tubes. At present the state of the art is not yet quite satisfying and the improvement of various parameters is highly desirable, e.g., the focus size should be reduced at higher current density, end-window design is preferred and the use of field emission electron emitters would be very attractive.

Recent developments in X-ray microscopy and extreme ultraviolet lithography demand bright table-top sources of extreme soft X-rays, in particular in the region of the water window. Most desirable are sources with a high degree of coherency and ultra short pulse duration. At present laser-based plasma X-ray sources meet these demands best but they are still bulky, expensive and in an experimental state [4–7].

Natural or artificial radioactive isotopes are in principle convenient X-ray sources. They are compact, low cost, continuously radiating at high constancy, independent on surrounding conditions and do not need any power supply. These advantages originated many engineering applications in XRF devices, in particular in portable instruments. The major problem is their radiation hazard potential leading to very stringent safety conditions making these sources quite unpopular now. But they are still irreplaceable in many industrial applications.

In radioactive radiation sources the decay products are used for direct excitation of the studied samples. For direct excitation in XRF applications mainly $\gamma$-radiating isotopes are used for decaying by K-electron capture (EC) processes which are accompanied by emission of K- and L-lines of X-rays. The latter allows an effective excitation of K-lines and L-lines of numerous elements. The $^{55}$Fe electron capture source which, as an exception, does not emit $\gamma$-radiation is widely used as a standard source for calibration of X-ray detectors.

Other sources use the generation of mono-energetic X-rays in a special target material where secondary fluorescence is excited by the emitted $\gamma$-radiation. However, the photon output of such indirect excitation sources is rather low.

A third group of sources used for XRF are the $\beta$-excited X-ray sources having a continuous spectrum in a wide range of energies. A detailed overview on commonly used radioisotope sources is given in Part 2.3.

A further mechanism which is capable of generating X-rays among other wavelengths is the electromagnetic emission of fast charged particles moving in strong magnetic fields. This kind of emission is produced by any charged particle undergoing centripetal acceleration and has already been predicted theoretically at the end of the nineteenth century by Liénard [8]. Mostly known now is the synchrotron radiation (SR) emitted by electrons or positrons in storage rings or similar circular high energy particle accelerator facilities.

SR is a relatively novel electromagnetic radiation, with a continuum spectrum spanning from the infrared to gamma rays. It is a highly polarized, intense radiation emitted in very short pulses of a few ps to ns duration, with a repetition rate of about $10^6$ pulses per second. The radiation is concentrated in a cone tangent to the curved trajectory of the charged particle initiating it. Postulated at the end of the nineteenth century by Liénard [8] (1898) and further developed by Wiechert [9] (1900), Schott [10] (1912), Ivanenko and Pomeranchuk [11] (1944) the theory of radiation emitted by relativistic charged particles in fixed curvature orbits was definitively solved by Schwinger [12] in 1949. The first experimental observation of SR was made in 1947 by Elder [13], on a 70 MeV synchrotron at the General Electric Laboratories in Schenectady, New York.

Actually, one of the motivations to develop the theory underlying the SR was the minimization of radiative energy loss of charged particles on curved trajectories in accelerators. Later, after the advent of light sources based on SR, the motivation shifted to generating optimized SR of specific wavelength for use in spectroscopy or photon scattering studies and in applied research in chemistry, physics or biology.

The basics of SR and the state of the art of SR generation are considered in Part 2.4. The modern sources of SR are based on circular accelerators (storage rings) where electron/positron beams are circulated at constant relativistic energies and which progressively lose their energy by emitting light.

After the proof-of-principle experiments of the 1940s, SR has gradually moved to rings of increasing energies, going from the hundreds of MeV infrared/visible light machines, to the current 2–8 GeV ones, adapted for X-ray generation. With over 40 dedicated synchrotrons operating worldwide nowadays, and at least five more upcoming (Diamond, UK; Soleil, France; CLS, Canada; Australia's Synchrotron; LLS, Spain;), SR has become a straightforward and "democratic" technology, allowing access to several tens of experiments simultaneously, on the dedicated beamlines of 2nd and 3rd generation machines.

Currently, light sources labeled as 4th generation are under study, following the free electron laser (FEL) concept, the energy recovering linear accelerators (ERL) or the ultimate storage rings.

## 2.2 X-Ray Tubes

*V. Arkadiev, W. Knüpfer and N. Langhoff*

### 2.2.1 Basic Physical Principles

In an X-ray tube electrons are emitted from a cathode and accelerated towards an anode in an electric field created by a positive potential of the anode relative to the cathode. They strike the anode target, interact with its atoms and lose their energy through a number of processes. First, incident electrons can undergo elastic or inelastic scatter on target nuclei and can be even backscattered in the opposite direction. Elastic scattering dominates inelastic scattering, not being connected with energy losses. Only a relatively small part of incident electrons takes part in inelastic scattering decelerating continuously in a strong Coulomb field of nuclei and gradually losing their energy. This process gives rise to the *continuous spectrum*, which is also known as Bremsstrahlung (a German term for "braking radiation"). Second, incident electrons interact with electrons of the target transferring them their energy. The dominant process consists of subsequent collisions with outermost electrons with small energy losses at every collision. However, sometimes an inner electron is removed from its orbit as a result of a collision so that an atom is ionized. Then another orbital electron fills the vacancy radiating an X-ray quantum. This process gives rise to the *characteristic lines* in the X-ray emission spectrum.

Figure 2.2 shows a typical spectra of X-ray tubes with a tungsten anode for different anode voltages U [14]. Every spectrum is a superposition of a continuous part and characteristic lines. The form, intensity and the maximum energy of a continuous spectrum depend on the value of the high voltage U, while the position of discrete lines is characteristic for an anode material. For every line there is an excitation threshold equal to a corresponding absorption edge, so that a line is not present in a spectrum unless an accelerating voltage does not exceed this threshold. The intensity of characteristic lines increases with the accelerating voltage. The falloff of the spectral distribution at low energies is usually due to X-ray absorption in the exit window and subsequently in the air. Some X-ray tubes (e.g., for medical applications) have additional filtering at low energies due to built-in filters. The real intensity emitted from an anode can be measured in vacuum for an open type X-ray tube [15]. The spectral distribution extends to very low energies and can contain strong characteristic lines in this region (Fig. 2.3).

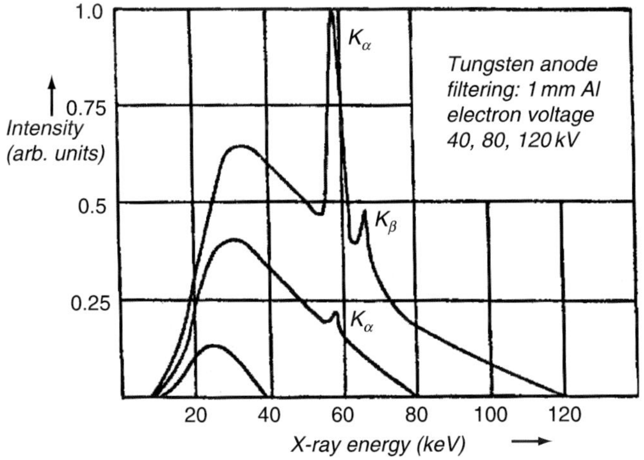


**Fig. 2.2.** Typical spectra of an X-ray tube with a tungsten anode

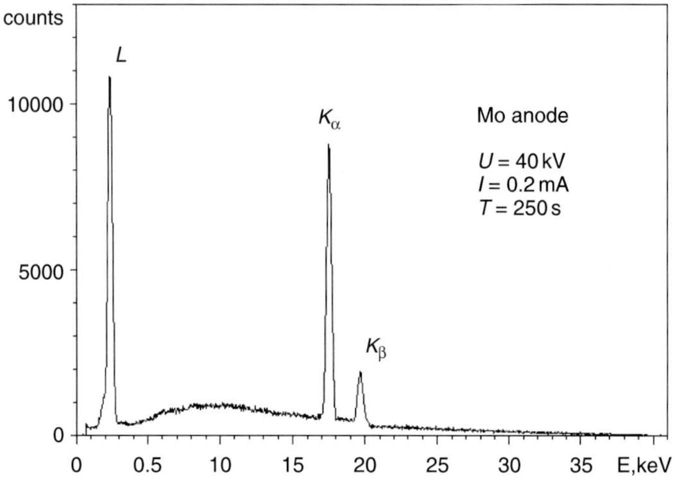


**Fig. 2.3.** Measured spectrum of a windowless X-ray tube with a Mo anode

The efficiency of the X-ray excitation is very low. The total irradiated X-ray power P[kW] can be estimated according to the following expression [16, 17]:

$$P = CZIU^2,$$

where $Z$ is the order number of the anode material, $I$ is the anode current (A); $U$ is the high voltage (kV); $C$ is a constant equal approximately to $10^{-6}$ (kV$^{-1}$). It means that the efficiency coefficient $\eta$ is approx.

$$\eta \equiv P/IU \approx 10^{-6} ZU \text{ [kV]}.$$

In practice, the value of the efficiency is of the order of 0.1–1%, so that a high thermal energy dissipation occurs in anodes of the X-ray tubes. The maximum electric power of the X-ray tubes can reach several kW for continuous operations and up to 100 kW in a pulsed mode (e.g., in medical diagnostics) thus making anode cooling a very serious technical problem.

### 2.2.2 Technology of the Components

#### General Design

Modern X-ray tubes contain a cathode and an anode mounted in a high vacuum ($<10^{-6}$ Torr) chamber. The tubes may be either sealed or continuously pumped. A cathode is a heated filament (usually tungsten), which provides control of the emission current, and an anode is normally an elemental metal (chromium, copper, molybdenum, tungsten, etc.).

Most X-ray tubes are similar in their design to diodes as illustrated in Fig. 2.4 for a stationary tube in dental applications. A directly heated electron emitter is arranged in a so-called Wehnelt electrode (introduced by the German scientist A.Wehnelt in 1908), which enables focusing of electrons on the anode target. The target is a thin tungsten disk or a plating imbedded into a massive copper block, which conducts dissipated heat away from the target. Emitted electrons are accelerated in an electric field between the cathode and the anode. In practice, this field is usually created by applying a negative potential to the cathode while keeping the anode at the ground, which makes anode cooling easier (e.g., with water circulation). The glass housing of the tube serves for vacuum as well as for high voltage isolation. X-rays are emitted from the anode target in all directions; however, they can emerge only through a special exit window. The arrangement of the window when the X-ray beam goes out perpendicular to the tube axis is called a side-window geometry. For medical applications the exit window can be a part of a glass envelope as in Fig. 2.4. However, in spectroscopic X-ray tubes the exit window is usually made from a thin beryllium film to minimize X-ray absorption at low energies. To increase the intensity of the exit beam, it is also reasonable to reduce the anode–window distance. Yet, this is possible only to some extent since scattered anode electrons can reach the window and cause its heating. One cannot ignore this effect when both the anode and the window are grounded, which is the usual case for side-window tubes. The angle between the target surface and the normal to the window (anode angle) can amount to 15–20°, so that the total aperture of the exit X-ray beam can reach 30–40°.

Dual-anode geometry can be realized in side-window tubes. For that, a thin layer of a light element covers a target made of a heavy element (e.g., Rh

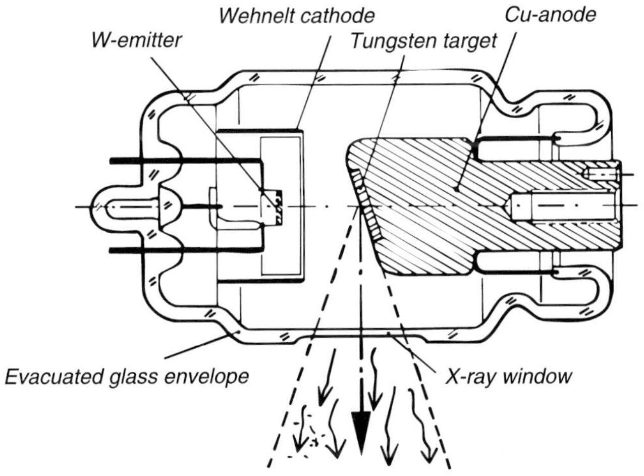


**Fig. 2.4.** Design of a side-window X-ray tube for dental applications

on W). Depending on the value of the high voltage, two types of spectra can be obtained with the same X-ray tube. When the accelerating voltage is low, the penetration depth of electrons is also small and x-radiation is excited mainly within the upper layer consisting of the light element. At large values of the accelerating voltage electrons penetrate deep into the second target, exciting the spectrum of the heavy element, which dominates in this case.

Another possible arrangement of an exit window is shown in Fig. 2.5. Here, the cathode has a ring shape, the takeoff angle amounts to 90° and the exit window is normal to the tube axis. Generally, such end-window geometry allows realizing smaller anode-window (respectively anode-sample) distances. Besides, the cathode and the window are usually grounded while the anode has a high positive potential. As a result, the bombardment of the window by the scattered on the anode electrons is strongly suppressed. However, the anode–window distance remains restricted, mainly because of the danger of a disruptive discharge between the anode and the window. Self-absorption of the excited X-radiation in the target is minimized due to the large takeoff angle. This fact makes end-window tubes especially attractive for low-energy applications. Besides, the tubes are considered to give more isotropic emission spectra. On the other hand, small anode-cathode distances typical for the design of the end-window tubes limit the value of the high voltage, which usually does not exceed 60 kV.

In transmission-anode X-ray tubes the target is a thin layer deposited directly on an internal side of an exit window made of a beryllium film. The electron beam strikes the target at right angle and the excited photons go through the target and the beryllium window further in the same direction. Usually, a continuous spectrum of transmission-anode X-ray tubes is suppressed to a certain extent due to the absorption of bremsstrahlung in the target foil, especially at low energies and above the absorption edge of the excited line, so that one can regard these tubes as possessing an intrinsic filtration. Low contribution of the continuous spectrum leads to the reduction of the back-

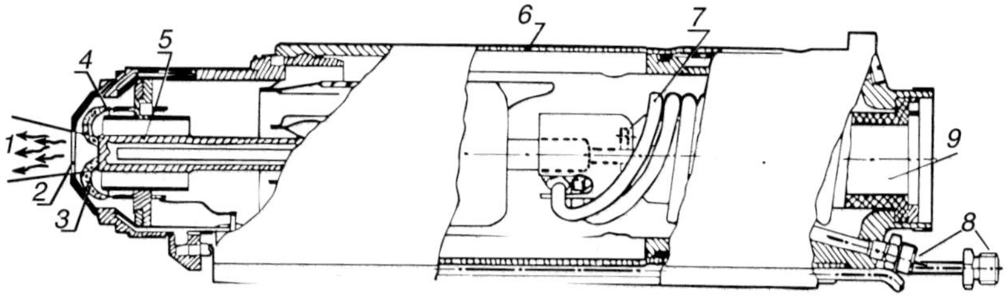


**Fig. 2.5.** Design of an end-window X-ray tube for fluorescence analysis

ground and consequently to the improvement of detection limits in the X-ray fluorescence analysis. Transmission-target X-ray tubes belong to low-power sources (<100 W) because thin anode foils cannot withstand large heat loads.

Immense heat dissipation on the anode is the main problem in achieving small focal spots while preserving the large power of the X-ray tube. At first glance, obtaining small focal spots is possible only at the cost of power reduction. There are two standard methods to partially overcome this difficulty.

The first method is based on the fact that the optical focal spot may differ from the electronic focal spot on the anode (Fig. 2.6). In most applications (e.g., X-ray imaging or while combining with suitable optics) only the optical focal spot is of primary importance. When the anode angle $\alpha$ amounts to 6°, the length of the optical spot is approximately ten times smaller than the real length of the electronic spot on the anode. For example, fine focus tubes for diffractometry have the optical spot of ca. 0.4 mm × 0.8 mm, while the real spot size on the anode amounts to about 0.4 mm × 8 mm. The power of fine focus tubes can reach approx. 3 kW. The total aperture of the exit X-ray beam does not exceed 10–12°.

In order to operate at still higher power, it is necessary to use a design with a rotating anode (Fig. 2.7). A disk-like anode is fixed on a rotor with a bearing system supporting rotation. A stator that drives the rotor electrically via a rotating field is located outside the vacuum housing. The target rotates under a space-fixed electron beam. The trajectory of the electron beam on the anode during its rotation is a circle, so that the heat load is distributed over the

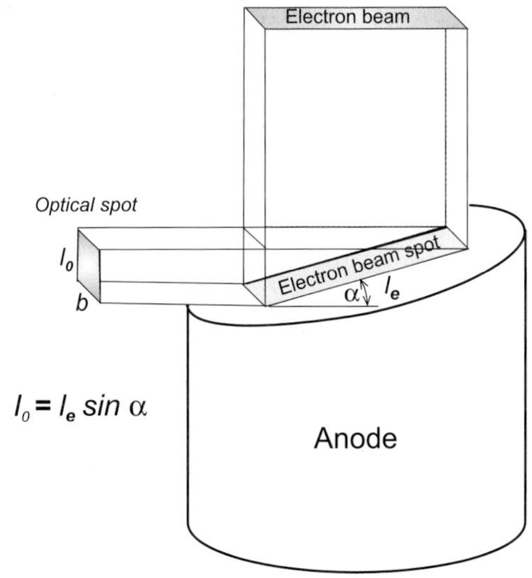


**Fig. 2.6.** Electronic and optical focal spots for a specific anode angle $\alpha$

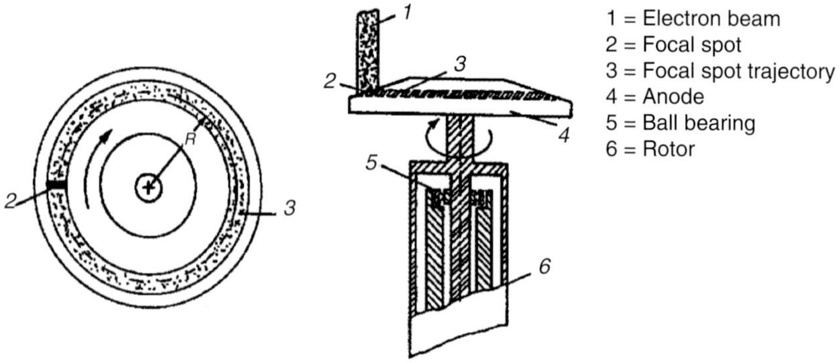


**Fig. 2.7.** Principle of a rotating anode tube

area, which is significantly larger than an instant focal spot. The dissipation of thermal energy away from the anode is achieved mostly by radiation and is accompanied optionally by air-cooling [18, 19]. Typical power densities for line-type focal spots of approx. 1 mm × 10 mm are 5–10 kW/mm$^2$, for focal spots with areas smaller than 1 mm$^2$ are about 25 kW/mm$^2$ at rotational frequencies of 17000 min$^{-1}$. Water-cooled rotating anodes exit also in special systems for material investigations thus enabling continuous operations at very high powers.

## Technology and Design of the Cathode

Generally, emission currents in X-ray tubes are controlled by the temperature of the cathode, which depends in its turn on the value of the filament current. Here, it is important that the size of the focal spot does not depend on the anode current over a range of more than 500 : 1 and on the applied accelerating voltage over the range of 3 : 1.

The control of the emission current through the cathode temperature requires emitters with very low heat capacity for time constants in the ms region. This can be realized in a simple manner by using thin tungsten wires and foils. Depending on the required shape of the focal spot emitters are used in the form of a coil, helix, hair pin, ring or a flat meandering foil. (Fig. 2.8).

As seen from Fig. 2.9, operating emitters is possible only within certain limits. At small filament currents (below some threshold value) there is no emission. At large filament currents a saturation of the anode current occurs. The saturation value depends on the high voltage applied. The temperature of the emitter under extreme conditions can achieve 2300°C. At this temperature the effect of metal evaporation can take place, so that the useful operation time is limited to 30–50 h. This is the reason that cathodes are switched on only for a few s at such high temperatures to realize short exposure times. The anode current depends also on the value of the high voltage due to the effect of partial screening by a space charge.

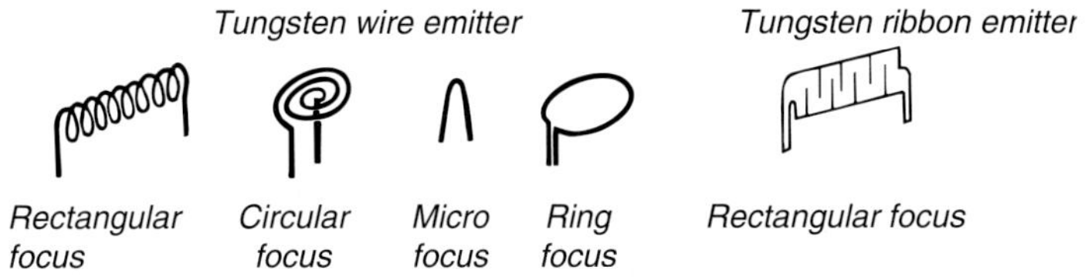


**Fig. 2.8.** Different types of emitters for X-ray tube cathodes

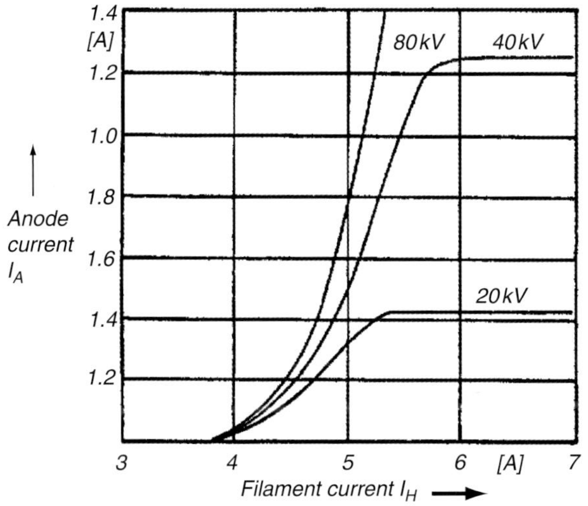


**Fig. 2.9.** Anode current as a function of filament current with anode voltage as parameter

Flat and thin emitters cut in meandering structures are used for tubes at low anode voltages and small focal spots (<0.3 mm). The reason consists in better focusing properties of electrons emitted from flat surfaces. Besides, it is possible to realize relatively high emission currents already at low cathode temperatures.

The Wehnelt electrode is used to control the electron pattern in the vicinity of the filament and to shape properly the electron beam. Now the form of the Wehnelt electrode is determined generally through suitable computer calculations (see Fig. 2.10 for the case of a rectangular focal spot).

## Technology and Design of the Anode

As mentioned above, a standard anode consists of a thin target embedded into a massive copper body (see Fig. 2.4). The most common target materials are copper, molybdenum and tungsten. Other metals such as magnesium, aluminium, chromium, iron, nickel, rhodium and silver are also used.

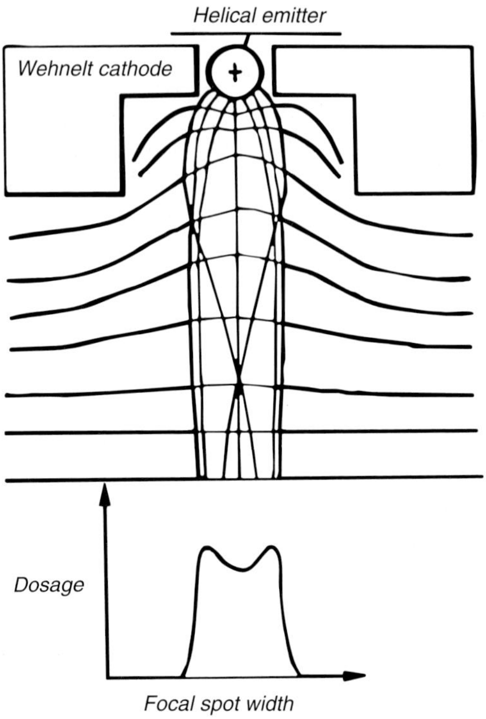


**Fig. 2.10.** Schematic design of the cathode for generating a rectangular focal spot. The dosage distribution versus width of the focal spot is shown

The technology of the rotating anodes is more complicated. Rotating tungsten anodes are produced on the basis of standard powder metallurgy. A disadvantage of this technology is a rapid abrasion in the surface region due to the high thermo-mechanical stress. This effect leads to a fast falloff of the X-ray intensity and causes the so-called heel effect. Here, a part of the radiation in the deeper lying layers of the anode is absorbed before it leaves the anode. In addition mechanical deformations occur, which can destroy the anode. Since the specific heat of tungsten is relatively low, the heat storage capacity is also small for an acceptable weight of the anode. Thus high temperatures of the anode are achieved at relatively low loads.

The enhanced technology is based on the Rhenium–Tungsten–Molybdenum composite material (RTM). A schematic cross-section of an RTM-anode is shown in Fig. 2.11. Here an admixture of up to 10% Re in the 1–2 mm thick W-cover is applied. This improves effectively the elastic properties of the anode and considerably reduces the abrasion behaviour. Admixing a few percents of Ti and Zr (TZM-anode) in the Mo + W-body nearly doubles the heat storage capacity of the anode with a given weight and inertia moment [20].

An optimal solution for the management of heat storage capacity and heat radiation can be achieved in an anode bounded with a graphite body

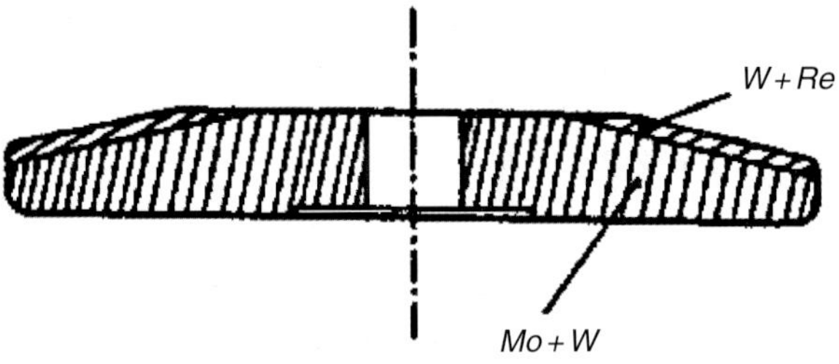


**Fig. 2.11.** Design of an RTM anode

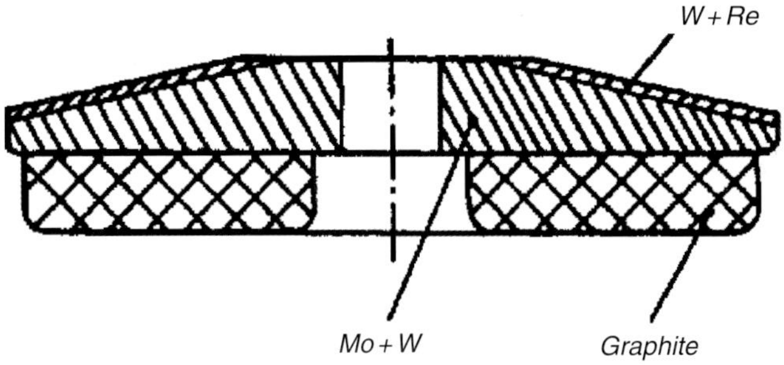


**Fig. 2.12.** Design of an RTM-graphite anode

(Fig. 2.12). With Calorex anodes, where an RTM-body is soldered to a graphite bulk, tubes are produced with capacities larger than $3 \times 10^6$ J operating at an average thermal power of 4 kW.

### Bearing System for a Rotating Anode

To match the requirements of an operating ball bearing system in vacuum at temperatures not higher than 400°C, high temperature compatible steels have to be used. As a rule, dry-lubricated ball bearings are now used, whereby different manufacturers of X-ray tubes carefully guard the secrets of their designs and, above all, of their lubricating methods.

Thus, there are spring-loaded and nonspring-loaded bearings in use, both with and without bearing cages, lubricated with molybdenum bisulphate, silver or lead. The methods of applying lubricants also differ. Some designs make use of sputtering, others of physical coating and still others use a galvanic process or a running process under powder.

Properties affected by different designs and processes are the rotational frequency of the anode attainable, useful lifetime (depending on the power requirements up to 200 working h) and the operating noise level.

### 2.2.3 Vacuum Envelope of X-Ray Tubes

The above-mentioned components of an X-ray tube are positioned within a tube envelope, which serves two main purposes: vacuum isolation and electric insulation. The common materials used for manufacturing tube envelopes are glass, metal and ceramic.

#### Glass Tubes

The material mostly used for the tube envelope is glass. The main demands on a suitable type of glass are high specific resistance, good dielectric strength and capability to withstand temperature changes. These requirements are met to a large extent by borosilicate glasses. Additionally, these glasses can be amalgamated with Ni–CO–Fe alloys (Vacon or Kovar) in order to ensure the necessary electrical contacts. Since the glass envelope must be a high voltage insulator between the anode and cathode potentials, its smooth and easy to clean surface is also an important advantage.

#### Metal-glass Tubes

An anode and a cathode tend to evaporate their material under high loads. A thin metal layer deposited on the glass envelope influences negatively its dielectric strength. To avoid this effect and to extend the lifetime of tubes, the middle part of the glass envelope is made of metal to establish a definite electrical potential. In addition, a fraction of secondary electrons from the anode can be removed via the middle part of the envelope, which is on the middle potential between the anode and the cathode. In this way, a part of the so-called extra focal radiation can be reduced.

#### Metal-ceramic Tubes

Metal-ceramic X-ray tubes have been in use since the 1960s. Their distinguishing feature consists in the use of ceramic materials instead of glass for the high voltage insulator and the vacuum envelope. Metal-ceramic tubes have a number of advantages over standard glass tubes. In particular, ceramics allow easier mechanical treatment: cutting and drilling are possible. Ceramic envelopes can be manufactured with a high precision of their form. As a result, there is more freedom in shaping the vacuum envelope. Tube metal parts can be joint vacuum tight to the ceramic insulator. Surface conductivity of ceramic is low, enabling short distances to insulate high voltages. Ceramic insulator improves the position of the focal spot. Compact and robust design, reduced weight, extended lifetime are the benefits of ceramic X-ray tubes [21].

### 2.2.4 Tube Housing Assembly

X-ray tubes are enclosed in a protective casing to ensure their safe operation. Two kinds of design are distinguished: a simple assembly, which houses an X-ray tube and a few control components, and a single-tank assembly where an X-ray tube is integrated with a high voltage generator. The second kind of X-ray assembly was originally used almost exclusively for stationary anode tubes, but in the meantime was also used for rotating anode tubes because of the application of high frequency technology in generating high voltages.

The tube casing performs the following main functions: high voltage insulation, cooling, protection against implosion, protection against radiation. The radiation protection must be checked for each individual X-ray tube assembly by means of a suitable $4\pi$-measurement.

### 2.2.5 Modern X-Ray Tubes

Commercially available X-ray tubes can be divided roughly into three large categories according to their power:

- Low-power X-ray tubes (<1 kW);
- High-power X-ray tubes (1–5 kW);
- High power X-ray tubes with rotating anode (>5 kW).

The current tendency of the X-ray tube progress consists, on the one hand, in development of compact low-power sources with a small focal spot and high brilliancy and, on the other hand, in development of huge power (>30 kW) X-ray tubes with rotating anode (e.g., for medical applications or for defectoscopy).

#### Low-Power X-Ray Tubes

Until recently the main application of microfocus X-ray tubes was concentrated on X-ray imaging with a very high spatial resolution. This objective remains as important as before but has considerably expanded owing to the necessity of other applications. Now, the importance of microfocus X-ray tubes has grown significantly due to the latest developments in X-ray optics, primarily capillary optics [22–24]. Polycapillary concentrators (so-called "X-ray lenses") allow focusing of X-ray beams onto small spots (approx. 10–100 µm) on a sample. However, the efficiency of capillary optics depends significantly on a source size. Generally, the source size should have approximately the same dimension as the desired focal spot on a sample. In this case, a combination of a low-power high-brilliant X-ray tube with a polycapillary concentrator enables achieving such intensities within a small spot on a sample that are otherwise obtained only with high-power tubes without optics. This fact makes microfocus tubes equipped with a suitable focusing optics indispensable for performing X-ray fluorescence microanalysis.

The typical power of a microfocus X-ray tube (in Watt) can be estimated as

$$P[\mathrm{W}] \approx D[\mu\mathrm{m}],$$

where $D$ is the diameter of the anode focal spot in microns [25, 26].

Among the different microfocus X-ray tubes available one may mention, first of all, the Hamamatsu products [27]. For example, an X-ray tube L9181 has three spot modes: 5 µm at 4 W, 20 µm at 16 W and 40 µm at 39 W. It is a sealed end-window X-ray tube with a tungsten anode, 500 µm Be exit window and air-cooling, which can be operated with high voltage up to 130 kV. The minimum distance between an anode and a sample (optics) is 13 mm. The latest development of Hamamatsu is an open type tube L8321, which has a focal spot of about 1 µm and can be operated up to 160 kV. The minimum focus-object distance can be as small as 0.5 mm.

The company "Kevex X-Ray" produces a number of high-resolution X-ray sources for X-ray imaging [28]. The end-window X-ray tube PXS5-926EA has a minimum focal spot of 5 µm, maximum power 8 W and maximum high voltage 90 kV.

The company "Oxford Instruments" offers also an end-window microfocus X-ray tube XTG UltraBtight [29]. The target material can be Cr, Cu, Mo, Rh and W. The target voltage is changeable in the range 20–60 kV (10–90 kV for W). The focal spot size is 12–40 µm and the maximum output power is 10–80 W. The tube is air-cooled. The minimum anode-object distance of 4 mm allows coupling with suitable X-ray optics.

The company "rtw Röntgen-Technik Dr. Warrikhoff KG" [30] produces MCBM series of metal-ceramic X-ray tubes for high resolution radiography. They are side-window X-ray tubes with minimum focal spot of approx. 50 µm (optical), maximum power 30 W, high voltage up to 50 kV and air-cooling. Different anode materials can be used: Cr, Fe, Co, Cu, Mo, W. The minimum anode-object distance of 14.5 mm allows coupling with polycapillary optics. A modular system, which combines such an X-ray tube with integrated exchangeable capillary optics (focusing or parallelizing), was developed by IfG - Institute for Scientific Instruments GmbH and is available commercially [31, 32]. The following types of capillary optics are offered: cylindrical and elliptical monocapillaries, polycapillary lenses, polycapillary zoom optics with variable focal distance. Focal spot sizes of 20–100 µm with polycapillary lenses and of 5–10 µm with elliptical monocapillaries are achievable on a sample. Low energy consumption, small weight and compact dimensions allow realizing a portable XRF spectrometer [33]. Besides rtw, also PANalytical [34], Siemens [35], YXLON [36], etc. produce metal-ceramics tubes.

### High-Power X-Ray Tubes

High-power X-ray tubes (1–5 kW) are produced by a number of manufacturers: PANalytical [34], Pantak [37], Siemens [35], rtw Röntgen-Technik [30], Varian [38] etc. The X-ray tubes are usually water-cooled. The maximum value of high voltage amounts to 60–120 kV. The nominal focal spot can be normal (1 mm × 10 mm), fine (0.4 mm × 8 mm), long fine (0.4 mm × 12 mm) or broad (2 mm × 12 mm). Under the anode angle of 6° the optical focal spot is effectively reduced in one direction by a factor of 10. Depending on the viewpoint, it leads either to a point focus or line focus. For example, the optical size of the fine focus can be 0.4 mm × 0.8 mm (point option) or 0.04 mm × 8 mm (line option). The tubes find their standard application in X-ray fluorescence analysis and X-ray diffractometry.

### High-power X-Ray Tubes with Rotating Anode

Rotating anode tubes allow obtaining huge power values. The progress of the last few years has been dominated mostly by the development of a powerful X-ray source for medical applications. It is worth mentioning the OPTILIX series of X-ray tubes manufactured by Siemens [35]. Depending on the modification, these tubes can operate at high voltage up to 150 kV and achieve the maximum power up to 80 kW. The anode disk is made of rhenium-alloyed tungsten with molybdenum and graphite. The focal spots are in the range between 0.2 mm and 2 mm. The field of applications covers the whole of general diagnostics. Tubes with a small focus (0.2–0.3 mm) are suitable for the display of fine details. Tubes with a large focus and maximum power are used for short-duration exposures involving high loads.

For X-ray diffractometry the company Rigaku [39] has developed an ultra high power X-ray generator SupreX with the output of up to 60 kV and 1500 mA. The rotating anode assembly includes Cu target (option: Cr, Fe, Co, Ni, Mo, Ag, Au). The available focus sizes are 1 mm × 10 mm, 0.3 mm × 3 mm and 0.1 mm × 1 mm.

### Field Emission X-Ray Tubes

As mentioned above, two main trends prevail now in the development of modern X-ray tubes. On the one hand, high-power tubes with rotating anodes are designed and persistently improved (e.g., for medical applications). On the other hand, compact low-power X-ray sources coupled with suitable X-ray optics are created for applications in fluorescence microanalysis and microdiffractometry. The development of the X-ray tubes with field emission cathodes follows logically the second tendency.

The action principle of a tube with a cold cathode is based on the phenomenon of field emission and not on thermo-emission as in conventional tubes. The first samples of field emission tubes used a cathode in the form of a metal needle. A high potential between a cathode and anode produces a strong electric field at the tip of the needle sufficient to pull electrons out of the cold cathode. Then the electrons are accelerated towards an anode as in standard X-ray tubes. Generally, a vacuum of $10^{-8}$ Torr or better is required because bombardment of the cathode by positive ions of the rest gas deteriorates its sharp tip and reduces drastically the electron emission. The latest developments use carbon-based cathodes in the form of a single fibre, a bundle of fibres (multifibre), a sharp edge (e.g., a small hole in a thin graphite film), graphite nanotubes or so-called "bucky paper" [40–44]. The main advantage of the carbon cathodes results from the fact that their bombardment by positive ions does not "smooth" sharp tips and edges. On the contrary, such a bombardment creates new sharp tips and edges due to the graphite structure. Therefore, the lifetime of carbon-based cathodes is significantly larger. Besides, one can work with lower vacuum (approx. $10^{-6}$ Torr).

Field emission tubes are similar in their design to triodes and contain a cathode, an anode and a gate electrode (modulator). The potential between the cathode and the modulator serves for pulling electrons out of the cathode. Depending on the cathode design, this potential can lie between 2 kV and ca. 150 V. Changing this potential regulates the tube current. Emitted electrons are accelerated in the electrical field created by a high voltage between the modulator and the anode. Normally, the tube has an end window in the form of a thin beryllium film with a deposited anode layer (transmission-anode geometry).

One possible design of the cathode component is presented in Fig. 2.13 (E. Sheshin, private communication). The emitter is a bundle of carbon fibres and has the diameter of approx. 150–200 μm (a single fibre has a diameter of approx. 1 μm). The bundle is coated with a glass mantle, which is fixed inside the modulator case by means of metal-ceramic technology. The modulator has an exit aperture of approx. 0.8 mm. The whole component has a diameter of approx. 4 mm and a length of approx. 8 mm. The values of the necessary gate voltage depend mainly on the distance between the cathode tip and the modulator aperture (typically 0.5 mm) and lie usually in the interval of 1500–2000 eV. The cathode component gives stable emission currents of approx. 100 μA.

Another possible scheme with a bucky paper as an emitter (Fig. 2.14) allows realization of very small emitter-gate distances and, consequently, low gate voltages of ca. 150–200 eV [43, 44]. The emission current can achieve the value of 100 μA and remain stable over large time periods. The scheme is very promising due to its compact realization and low gate voltages.

Since a field emission cathode does not need heating, a filament power supply is not necessary, which makes the tube design more simple and compact in comparison with conventional X-ray tubes. Cooling system is also

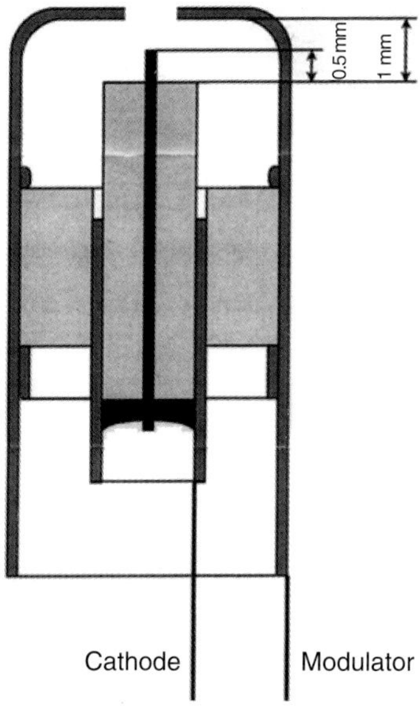


**Fig. 2.13.** Scheme of the cathode component with a bundle of carbon fibres as an emitter

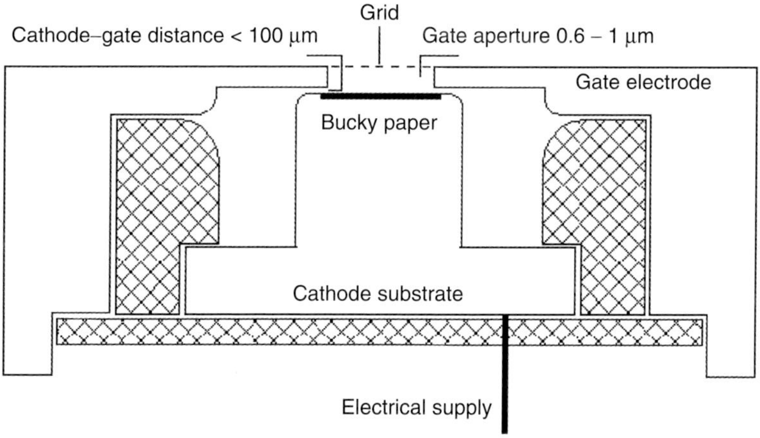


**Fig. 2.14.** Scheme of the cathode component with bucky paper as an emitter

not required because the power does not exceed several watts. The direction of further development consists in systematic improvement of the most important parameters: stability and lifetime. It is also supposed to combine field emission X-ray tubes with capillary optics within a compact module for applications in material microanalysis.

The first portable EDXRF instrument on the basis of field emission X-ray tubes was offered by the company Oxford Instruments [45]. The tube has a rhodium anode and can be operated up to 30 kV and 100 µA.

### 2.2.6 Some Applications

X-ray tubes for particular applications may have certain particular features of design. The most common applications today are material analysis (e.g., XRF, XRD, XPS) and medical investigations.

For X-ray fluorescence analysis X-ray sources are needed, which emit constant characteristic radiation over a long time. Until recently tubes with several kW power were needed to shorten measurement time and to enhance precision. Also desirable is that tube design enables short anode-sample distances. A typical end-window tube for fluorescence analysis is shown in Fig. 2.5. The anode is located in the central position and has the high potential. The anode is cooled with deionized water through a system of plastic tubes. The Wehnelt electrode in the form of a thread ring surrounds the anode concentrically to shape a ring focal spot. The exit window from beryllium is at cathode potential, which reduces the thermal load due to secondary electrons. In general, about 30% of the electrons impinging on the focal spot are backscattered and under the influence of the electric field fall again onto the anode, however, outside the focal spot, producing the so-called undesired extra-focal radiation. In many tube designs there is a special copper cup near the anode for capturing backscattered electrons as shown in Fig. 2.15. The tube operates with voltages up to 60 kV and currents up to 100 mA at a maximum power of 3 kW.

In contrast to XRF, where standard tubes may have broad focal spots, diffractometry requires sharply defined small focal spots. An example of a

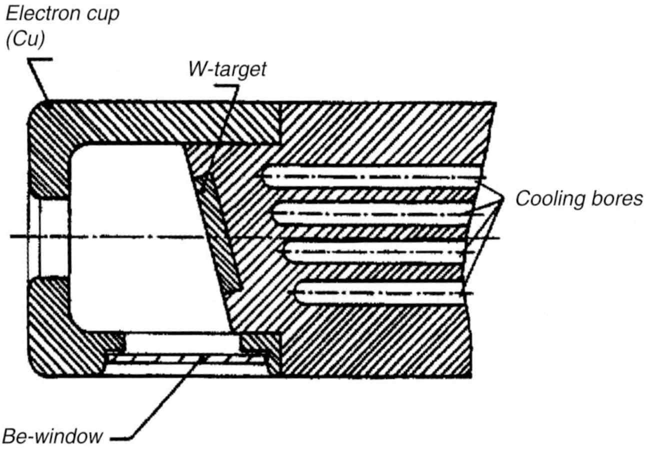


**Fig. 2.15.** Schema of a fluid-cooled stationary anode for continuous high load

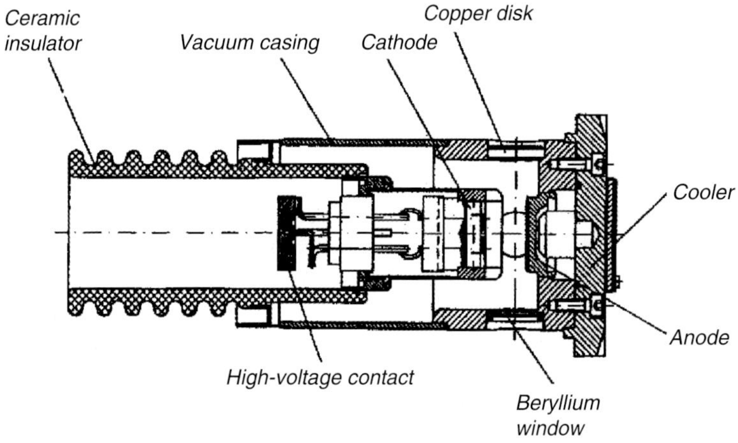


**Fig. 2.16.** Siemens X-ray tube for XRD

Siemens X-ray tube for XRD is shown in Fig. 2.16. The cathode generates a line-shaped focus in order to spread the power dissipation over a large area. The anode is grounded and cooled via injection cooling. A power of more than 1 kW is achievable in continuous operations in spite of the small size of the focal spot. Partial monochromatization of the exit radiation is achieved through the use of absorption filters (e.g., Cu-filter for Cu-anode). The filter thickness may be between 30 µm and 60 µm depending on the penetration depth of the photons.

Figure 2.17 shows the design of the anode of the forward emission X-ray tube, which generates x-radiation on the back of the anode in the direction of the impinging electron beam. Here, the anodes are only 10–30 µm thick. In

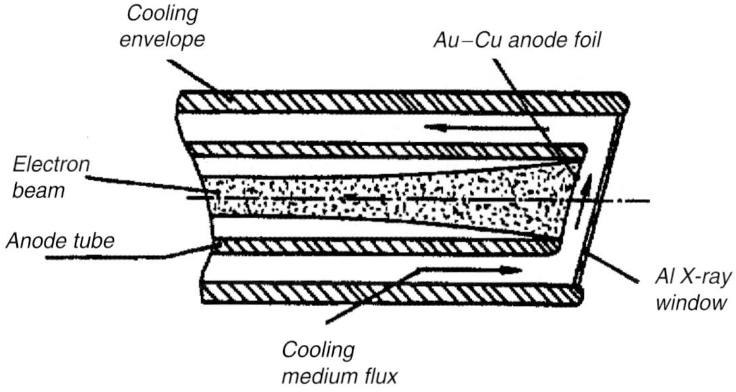


**Fig. 2.17.** Water-cooling of an anode of a forward emission X-ray tube

practice, a diffusion barrier layer of 3–5 μm Ni on the copper base is deposited. The front side of the anode is cooled with flowing water through a 1-mm thick slit.

X-ray tubes for coarse structure analysis should withstand voltages up to 400 kV (cathode: −200 kV, anode: +200 kV) in order to manage X-ray transmission through thick welds. Generally, the problem of motion blurring is not significant, so that extremely short pulses are not required. The power level for continuous mode operations is of the order of several kW. This power is sufficient to expose the film material properly and in a reasonable time.

Generally, X-ray sources for photoelectron spectroscopy do not need a point focal spot, but forward emission and the possibility of changing the anode target and hence the energy of the characteristic line are required.

Some tubes for microradiography use hairpin tungsten filament and a focusing electromagnetic lens to give an electron spot with a 10-μm diameter. At 100 kV of high voltage and 100 μA of beam currents this corresponds to the power of 10 W. A tungsten target is used to give a high-brilliance microfocus source. The power loading at the anode reaches 130 kW $mm^{-2}$. The source lifetime amounts to about 2 h despite the presence of a rotating anode and intensive water-cooling.

The main limitation to the X-ray intensity in the high-power X-ray tubes is the amount of power that can be delivered to the anode without causing any damage. In rotating anode devices, such as Rigaku FRC, this difficulty is overcome by rapidly rotating a cooled anode. The main technical problem here is achieving a vacuum seal on a water-cooled rotating target. Normally, this is done by using oil seals. The characteristics of the Rigaku FRC device (anode diameter 250 mm, rotation speed 9000 $min^{-1}$, focal spot size 0.1 mm × 0.1 mm, maximum power 3.5 kW, maximum power loading 35 kW $mm^{-2}$) are suitable for performing X-ray lithography. Hereby, a variety of target materials can be used by coating the rotating cooper anode, e.g., with 20 μm aluminium.

## 2.3 Radioisotope Sources

*T. Čechák* and *J. Leonhardt*

Radioactive radiation sources are devices in which radioactive material is sealed by means of one or more inactive capsules. The tightness and strength of the sealing material are such that no radioactive compound can be freed at use or during storage. The type of sealing material used depends not only on the radiation that is emitted and the chemical properties of the radionuclide used, but also on the conditions of application. The use as excitation source in the XRF is mainly realized by means of gamma radiation emitters. Sometimes beta emitters are combined with secondary targets, giving Bremsstrahlung a wide range of energy. In the past there were some tests to use alpha and beta particles directly but these did not succeed.

The high constancy of radioactive decay and its nondependence on external variables like temperature, pressure, chemical infected surrounding, and the nonrequirement of power supply have originated many industrial XRF applications over the years. Other advantages are those of small source dimensions and relatively low costs. But safety regulations of radiation did limit the user's possibilities at that time.

Nevertheless, the excitation of characteristic X-rays by means of radioisotope sources has led to a large number of interesting applications, especially for handheld and portable devices, for example, for the determination of the sulphur content in gasoline and coal or of the metal content in scrap.

### 2.3.1 Basic Physical Principles

Radioisotope sources may be characterized by

- Type of radioactive decay
- Activity of the sources
- Half-life of the sources
- Energy of emitted radiation
- Properties connected with the production of the sources.

#### Types of Radioactive Decay

A survey of radioactive decay types can be found, e.g., in [46, 47]. For the XRF $\alpha$-decay, $\beta^-$-decay, $\beta^+$-decay, and K capture are most important. A-decay effect occurs when the nucleus emits an $\alpha$ particle consisting of two protons and two neutrons. $\beta^-$ decay is characterized by the emission of an electron and antineutrino. The energy spectrum of electrons, i.e., $\beta$ particles, is continuous because the antineutrino takes part of the energy. $\beta^+$ decay is characterized by the emission of $e^+$ particles, positrons, and as with $\beta^-$decay, $\beta^+$ decay produces a continuous spectrum. K capture is the capture of an electron from the inner shell (K shell). The nucleus changes its proton number; in the nucleus one proton transforms into the neutron.

The nuclear decay can be followed by the emission of gamma radiation, emission of X-rays or internal conversion. Gamma radiation is emitted during transitions of excited nuclei from their nuclear levels to lower lying nuclear levels. The excited nucleus emits one or several photons. Emission of the cascade of photons depends on the existence of intermediate states in the nucleus. The internal conversion process is an alternative process to gamma emission. The excitation energy is transferred to one of the orbital electrons. This electron receives the energy $E_e$ given by:

$$E_e = E_{ex} - E_b, \tag{2.1}$$

where $E_{ex}$ is the excitation energy, and $E_b$ is the binding energy of the electron. Emission of X-rays can be followed by filling the holes in the atomic shells with an electron from higher shells. The energy of emitted photons is given as a difference of the energies of these shells. The energy of emitted photons is a function of the proton number $Z$.

### Activity

Activity $A$ of a radioisotope source is defined as its rate of decay—the number of disintegration per time unit

$$A = \frac{\mathrm{d}N}{\mathrm{d}t} \quad [\mathrm{Bq}]. \tag{2.2}$$

Bq (Becquerel) replaced the older unit of activity Ci (Curie). 1 Ci=$3.7\times10^{10}$ Bq.

The number of disintegrations per unit of time is equal

$$\frac{\mathrm{d}N}{\mathrm{d}t} = -\lambda N, \tag{2.3}$$

where $\lambda$ is the decay constant of the radionuclide.

Integrating the equation, we obtain the radioactive decay law:

$$N = N_0\,\mathrm{e}^{-\lambda t}, \tag{2.4}$$

where $N$ is the number of nuclei of a given radioisotope after a time $t$, and $N_0$ is the number of nuclei for the time $t = 0$.

### Half-Life of Sources

Half-life $T_{1/2}$ is defined as the time after which the initial number of radioactive nuclei is reduced by one half, i.e.,

$$T_{1/2} = \frac{\ln 2}{\lambda} \tag{2.5}$$

because

$$\frac{1}{2}N_0 = N_0\,\mathrm{e}^{-\lambda t} \quad \text{for } t = T_{1/2}. \tag{2.6}$$

For practical purposes the application of radionuclides with a short $T_{1/2}$ has a disadvantage. Their activity decreases rapidly, and therefore the number of emitted photons also decreases.

### Energy of Emitted Radiation

The energy of gamma radiation or X-rays emitted by the radioactive sources used by XRF can be linear or continuous. Linear spectra are typical of sources if the nuclear decay is followed by emission of gamma rays or X-rays. Continuous spectra are typical of sources emitted by the bremsstrahlung radiation produced when fast electrons interact in matter. Part of their energy can be converted into electromagnetic radiation called bremsstrahlung. This process is used to produce bremssstrahlung X-rays from X-rays tubes (see Sect. 2.1). However, a combination of the β source with the target material also creates a source of bremsstrahlung radiation. These sources with a continuous spectrum of bremsstrahlung can increase the set of available sources for excitation of characteristic radiation. The shape of the energy spectrum depends on the spectrum of β particles emitted by a radioactive source. The maximum energy in the spectrum of bremsstrahlung radiation is determined by the maximum energy of β particles.

### Properties Connected with the Production of the Source

The properties of the sources my be differentiated according to the type of manufacture, dimensions, shapes, type of coating, etc.The type of coating influences the shape of the spectrum. Thicker coating or coating of a material with a higher Z can absorb low energy radiation from the source.

## 2.3.2 Radioisotope Sources

Source selection for different elements is based on the energy of source radiation. The atoms can be excited by photons having energies higher than the binding energy of the electron on the determined shells (K, L etc.). To excite $K\alpha$ line of, e.g., Sn, the energy of an excited radiation must be greater than the binding energy on the Sn K shell, i.e., 29.19 keV. On the other hand, the exciting energy should also not be too high. The cross section of the photoeffect decreases if the energy rises, and, proportionally the probability of exciting the characteristic radiation also decreases. Every source has a group of elements for which the source is best suited [51].

The photons from the source can interact in the sample by Coherent scattering and Compton scattering too. This scattered radiation can be absorbed in the detector, and it can raise the background in a part of the spectrum and complicate the measurement of low intensity line in this part of the spectrum. A similar problem can cause escape peaks in the spectrum [49]. All these circumstances have to be taken into account when choosing the source.

### Radioisotope Gamma and X-Ray Sources

The widely employed gamma and X-ray sources are given in Table 2.1, showing also the preferred element ranges. $^{55}$Fe is a very useful source for elements with a low proton number. Its energy spectrum is shown in Fig. 2.18. $^{238}$Pu, $^{244}$Cm, and $^{109}$Cd sources are used for exciting elements with $Z = 20-42$. The energy spectrum of $^{238}$Pu and $^{109}$Cd are shown in Figs. 2.19 and 2.20, respectively. $^{109}$Cd is a source which emits X-rays due to K capture. Figure 2.21 shows the decay scheme of $^{109}$Cd. Higher energies of photons are emitted by $^{241}$Am and $^{57}$Co. The energy spectrum of $^{241}$Am and its decay scheme are shown in Figs. 2.22 and 2.23, respectively.

**Table 2.1.** Radioisotope gamma and X-ray sources [48, 52, 56]

| Source | Half-life | Type of decay | Particle energy | Gamma rays Energy (keV) | Gamma rays Yield (%) | Characteristic X-rays Assignment | Characteristic X-rays Energy (keV) | Characteristic X-rays Yield (%) | Typical elemental range K lines | Typical elemental range L lines |
|---|---|---|---|---|---|---|---|---|---|---|
| $^{55}$Fe | 2.73 y | EC | – | – | – | Mn $K_{\alpha 2}$ | 5.89 | 8.5 | Si–V | Zr–Ce |
| | | | | | | Mn $K_{\alpha 1}$ | 5.90 | 16.9 | | |
| | | | | | | Mn $K_{\beta 1}$ | 6.49 | 2.0 | | |
| | | | | | | Mn $K_{\beta 3}$ | 6.49 | 1.01 | | |
| $^{57}$Co | 271.8 d | EC | – | 14.41 | 9.16 | Fe $K_{\alpha 2}$ | 6.39 | 16.4 | Yb–U | |
| | | | | 122.06 | 85.60 | Fe $K_{\alpha 1}$ | 6.40 | 32.6 | | |
| | | | | 136.47 | 10.68 | Fe $K_{\beta 3}$ | 7.06 | 2.0 | | |
| | | | | | | Fe $K_{\beta 1}$ | 7.06 | 3.9 | | |
| $^{109}$Cd | 462.6 d | EC | – | 88.04 | 3.61 | Ag $K_{\alpha 2}$ | 21.99 | 29.5 | Cr–Mo | Tb–U |
| | | | | | | Ag $K_{\alpha 1}$ | 22.16 | 55.7 | | |
| | | | | | | Ag $K_{\beta 3}$ | 24.91 | 4.8 | | |
| | | | | | | Ag $K_{\beta 1}$ | 24.94 | 9.2 | | |
| | | | | | | Ag $K_{\beta 2}$ | 25.46 | 2.3 | | |
| $^{125}$I | 59.41 d | EC | – | 35.49 | 6.68 | Te $K_{\alpha 2}$ | 27.20 | 40.6 | As–Cd | Tb–U |
| | | | | | | Te $K_{\alpha 1}$ | 27.47 | 75.7 | | |
| | | | | | | Te $K_{\beta 3}$ | 30.94 | 6.8 | | |
| | | | | | | Te $K_{\beta 1}$ | 31.00 | 13.2 | | |
| | | | | | | Te $K_{\beta 2}$ | 31.70 | 3.8 | | |
| $^{145}$Sm | 340 d | EC | – | 61.25 | 12.00 | Pm $K_{\alpha 2}$ | 38.17 | 39.9 | Ga–Tb | |
| | | | | | | Pm $K_{\alpha 1}$ | 38.73 | 72.4 | | |
| | | | | | | Pm $K_{\beta 3}$ | 43.71 | 7.0 | | |
| | | | | | | Pm $K_{\beta 1}$ | 43.83 | 13.6 | | |
| | | | | | | Pm $K_{\beta 2}$ | 44.94 | 4.5 | | |
| $^{155}$Eu | 4.76 y | β– | 134.1 keV | 45.30 | 1.33 | Gd $L_{\alpha 1}$ | 6.06 | 3.0 | Pd–Ra | |
| | | | 146.8 keV | 60.01 | 1.13 | Gd $L_{\beta 1}$ | 6.71 | 2.1 | | |
| | | | 165.6 keV | 86.55 | 30.70 | Gd $K_{\alpha 2}$ | 42.31 | 6.6 | | |
| | | | 192.1 keV | 105.31 | 21.20 | Gd $K_{\alpha 1}$ | 43.00 | 11.9 | | |
| | | | 252.1 keV | | | Gd $K_{\beta 3}$ | 48.55 | 1.2 | | |
| | | | | | | Gd $K_{\beta 1}$ | 48.70 | 2.3 | | |
| $^{153}$Gd | 240.4 d | EC | – | 69.67 | 2.42 | Eu $L_{\alpha 1}$ | 5.85 | 8.8 | Ba–Fr | |
| | | | | 97.43 | 29.00 | Eu $L_{\beta 1}$ | 6.46 | 5.6 | | |
| | | | | 103.18 | 21.11 | Eu $L_{\beta 2}$ | 6.84 | 1.9 | | |
| | | | | | | Eu $K_{\alpha 2}$ | 40.90 | 35.2 | | |
| | | | | | | Eu $K_{\alpha 1}$ | 41.54 | 63.5 | | |
| | | | | | | Eu $K_{\beta 3}$ | 46.91 | 6.3 | | |
| | | | | | | Eu $K_{\beta 1}$ | 47.04 | 12.1 | | |
| | | | | | | Eu $K_{\beta 2}$ | 48.25 | 4.0 | | |
| $^{170}$Tm | 128.6 d | β– | 883.3 keV | 84.25 | 2.50 | Yb $L_{\alpha 1}$ | 7.42 | 1.1 | Pd–Hg | |
| | | | 968.0 keV | | | Yb $L_{\beta 1}$ | 8.40 | 1.0 | | |
| | | | | | | Yb $K_{\alpha 2}$ | 51.35 | 0.94 | | |
| | | | | | | Yb $K_{\alpha 1}$ | 52.39 | 1.7 | | |
| | | | | | | Yb $K_{\beta 3}$ | 59.16 | 0.18 | | |
| | | | | | | Yb $K_{\beta 1}$ | 59.38 | 0.34 | | |
| | | | | | | Yb $K_{\beta 2}$ | 60.96 | 0.12 | | |

**Table 2.1.** *Continued*

| Source | Half-life | Type of decay | Particle energy | Gamma rays: Energy (keV) | Gamma rays: Yield (%) | Characteristic X-rays: Assignment | Characteristic X-rays: Energy (keV) | Characteristic X-rays: Yield (%) | Typical elemental range: K lines | Typical elemental range: L lines |
|---|---|---|---|---|---|---|---|---|---|---|
| $^{238}Pu$ | 87.7 y | α | 5.456 MeV | — | — | U $L_l$ | 11.62 | 0.26 | Ca–Sr | Sn–At |
| | | | 5.499 MeV | | | U $L_{\alpha2}$ | 13.44 | 0.42 | | |
| | | | | | | U $L_{\alpha1}$ | 13.62 | 3.8 | | |
| | | | | | | U $L_{\beta2}$ | 16.41 | 1.00 | | |
| | | | | | | U $L_{\beta5}$ | 17.07 | 0.21 | | |
| | | | | | | U $L_{\beta1}$ | 17.22 | 3.9 | | |
| | | | | | | U $L_{\gamma1}$ | 20.17 | 0.94 | | |
| | | | | | | U $L_{\gamma6}$ | 20.84 | 0.20 | | |
| $^{241}Am$ | 432.2 y | α | 5.485 MeV | 26.34 | 2.40 | Np $L_{\alpha2}$ | 13.76 | 1.1 | Zn–Nd | W–U |
| | | | 5.422 MeV | 33.20 | 0.13 | Np $L_{\alpha1}$ | 13.95 | 9.6 | | |
| | | | 5.388 MeV | 59.54 | 35.90 | Np $L_{\beta2}$ | 16.82 | 2.5 | | |
| | | | | | | Np $L_{\beta1}$ | 17.75 | 5.7 | | |
| | | | | | | Np $L_{\beta3}$ | 17.99 | 1.4 | | |
| | | | | | | Np $L_{\gamma1}$ | 20.78 | 1.4 | | |
| $^{244}Cm$ | 18.10 y | α | 5.762 MeV | — | — | Pu $L_{\alpha1}$ | 14.28 | 3.1 | Ti–Se | Ba–Bi |
| | | | 5.804 MeV | | | Pu $L_{\eta}$ | 16.33 | 0.08 | | |
| | | | | | | Pu $L_{\beta6}$ | 16.50 | 0.06 | | |
| | | | | | | Pu $L_{\beta2}$ | 17.24 | 0.82 | | |
| | | | | | | Pu $L_{\beta5}$ | 17.95 | 0.18 | | |
| | | | | | | Pu $L_{\beta1}$ | 18.30 | 3.0 | | |
| | | | | | | Pu $L_{\gamma1}$ | 21.42 | 0.73 | | |
| | | | | | | Pu $L_{\gamma6}$ | 22.15 | 0.15 | | |

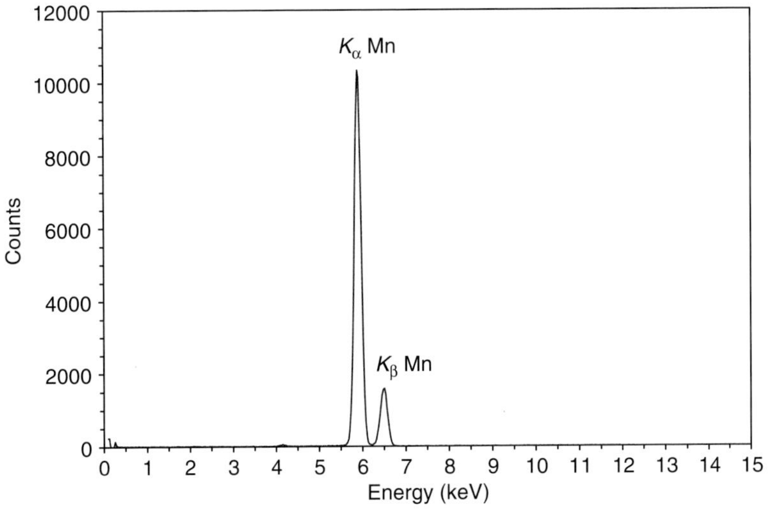


**Fig. 2.18.** X-ray spectrum from $^{55}Fe$

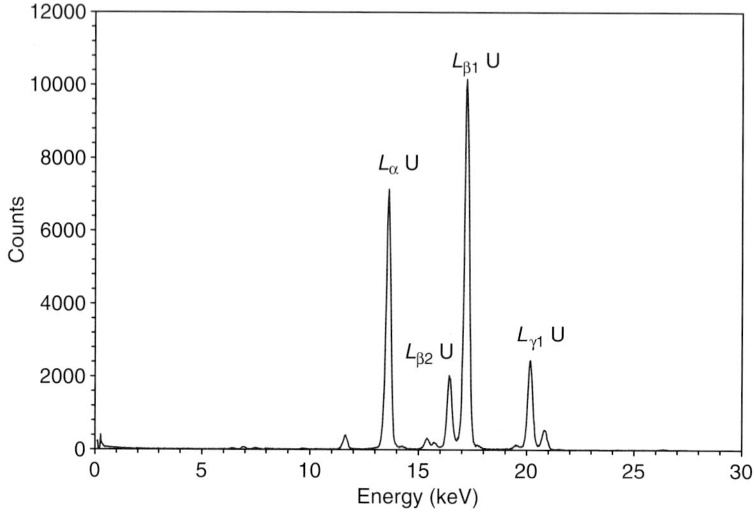


**Fig. 2.19.** X-ray spectrum from $^{238}$Pu

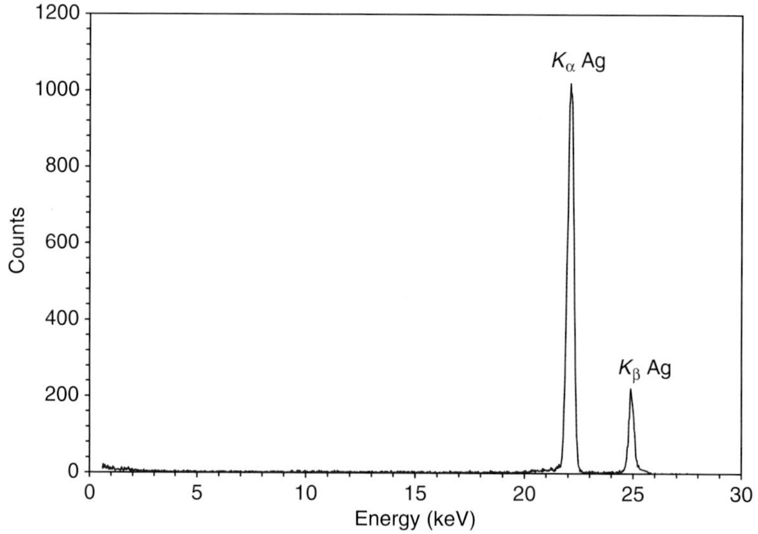


**Fig. 2.20.** X-ray spectrum from $^{109}$Cd

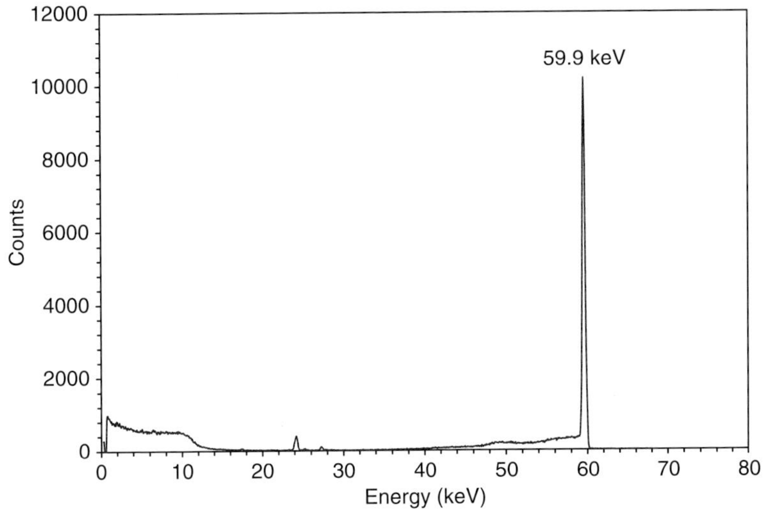


**Fig. 2.21.** X-ray and gamma ray spectrum from $^{241}$Am

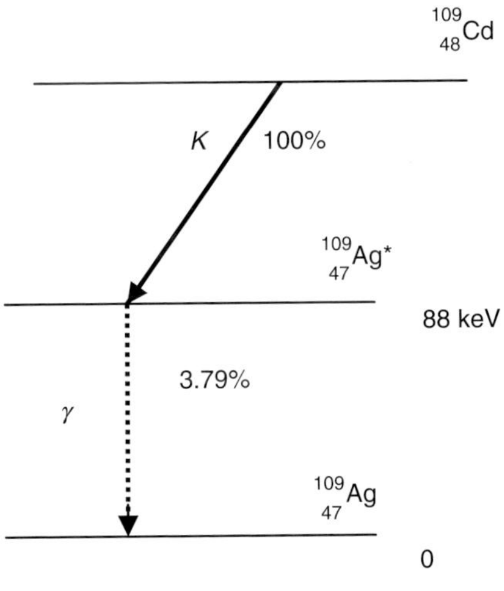


**Fig. 2.22.** Decay schema of the $^{109}$Cd

## Radioisotope Beta-excited X-Ray Sources

Beta excited X-ray sources complement the set of the sources available for XRA. The most important beta-excited X-ray sources are given in Table 2.2. They have a continuous spectrum in a wide range of energies. The bremsstrahlung radiation from the source is added to the characteristic radiation of the target material excited by beta particles or by the bremsstrahlung directly.

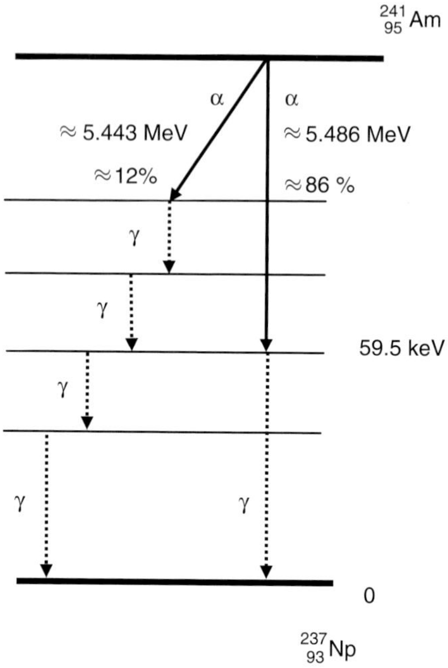


**Fig. 2.23.** Decay schema of the $^{214}$Am

The relative high background of the scattered radiation in the measured spectrum is the disadvantage of beta-excited sources.The energy spectrum from $^{3}$H/Zr is shown in Fig. 2.24.

## Photon-Excited X-Ray Sources

The principle of photon-excited X-ray sources is based on the excitation of the target material by gamma radiation from the isotopic source. The advantage of these sources is that it is possible to choose the energy of the emitted radiation by selecting the target material. However, as compared with the

**Table 2.2.** Radioisotope beta-excited X-ray sources [48, 52]

| Source | Half-life | Type of decay | Particle energy [keV] | Usable energetic range [keV] | Typical elemental range: K line | Typical elemental range: L lines |
|---|---|---|---|---|---|---|
| $^{3}$H/Ti | 12.33 y | β− | 18.59 | 4–8 | Si-Cr | Ag-Sm |
| $^{3}$H/Zr | 12.33 y | β− | 18.50 | 5–9 | Si-Zn | Ag-Tb |
| $^{147}$Pm/Al | 2.62 y | β− | 224.1 | 10–45 | Mn-Nd | Tb-U |

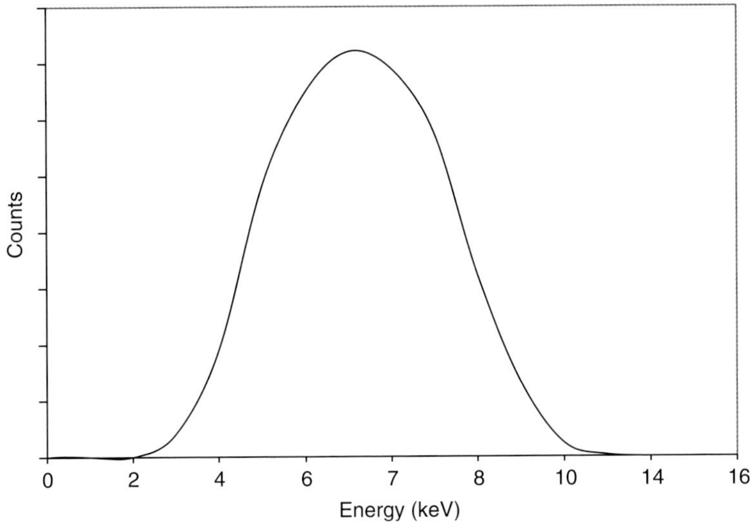


**Fig. 2.24.** Bremsstrahlung spectrum from $^{3}H/Zr$

primary photons, the main disadvantage is the low output of photons emitted from the target material. Using such sources makes it possible to replace the missing gamma and X-ray sources [53].

At present, photon excited X-ray sources are used for calibration purposes. The source can contain one radioactive source, e.g., $^{241}Am$ and several target materials. The source can emit characteristic radiation of elements in a wide range of energies. Figure 2.25 shows the spectra of Cu, Rb, Mo, Ag, Ba, and Tb emitted by such calibration variable energy source, presented in one diagram.

### 2.3.3 Production of Radioactive Sources

Radioisotope X-ray sources are produced by a number of manufacturers, e.g., [50, 54, 55, 58, 59]. For $^{55}Fe$ the radioactive material is electrodeposited as iron metal on a metal ring and sealed in a welded stainless steel capsule with a beryllium window. $^{238}Pu$, $^{244}Cm$, and $^{241}Am$ are incorporated in a ceramic enamel sealed in a stainless steal capsule with or without a Be window, etc. Figures 2.26 and 2.27 show a typical disk source and annular source used by XRF. The use of annular sources is typical of devices with detectors having a small window (e.g., semiconductor detectors). Mostly ring sources are used because of higher photon fluxes. A photo of an Amersham [57] ring source is given in Fig. 2.28.

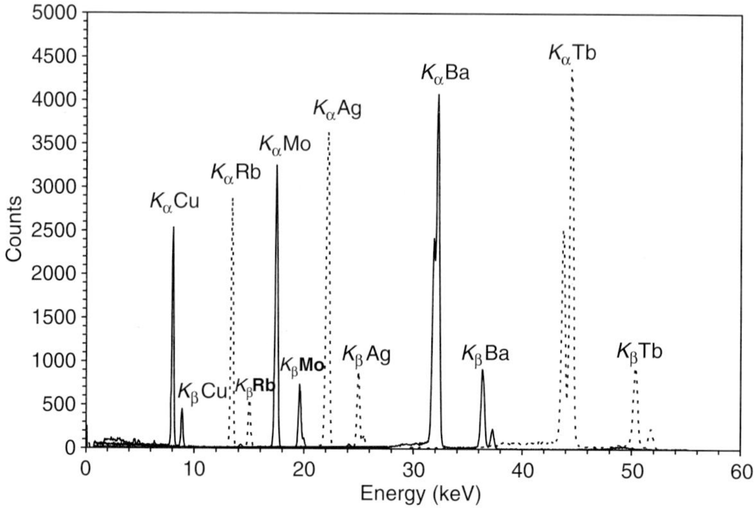


**Fig. 2.25.** X-ray spectra from calibration variable energy source 3837LA Amersham

### 2.3.4 Radiation Protection Regulations

The use of radioactive radiation sources in XRF equipment is approved and controlled by the appropriate National Radiation Safety Agency. The basic principle is that any person working with this type of equipment should not receive a dose of more than 1 millisievert yearly. The radioactive sources are generally required to be inspected every two years by independent experts in order to control the tightness of the seals.

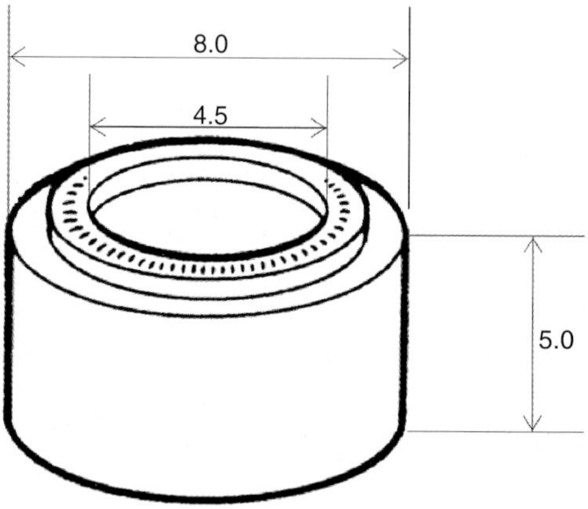


**Fig. 2.26.** Example of a typical disc source

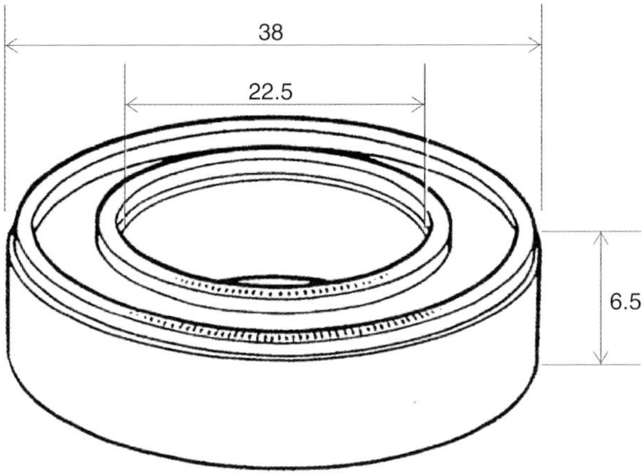


**Fig. 2.27.** Example of a typical annular source

**Fig. 2.28.** The photo of a $^{241}$Am-ringshaped XRF-source [12]

## 2.4 Synchrotron Radiation Sources

*A. Simionovici* and *J. Chavanne*

### 2.4.1 SR Basics

*Radiation from a Charged Particle*

Following the pioneering work of Ivanenko and Pomeranchuk, Schwinger tackled in great detail the generation of synchrotron radiation (SR). More recently, Jackson [60] gave an exhaustive account of SR in his seminal textbook "Classical Electrodynamics." Here, we follow his lead, and start with the total power radiated by a relativistic monoenergetic electron on a circular orbit (CGS units):

$$P = \iint I(\lambda, \Psi) \mathrm{d}\lambda\, \mathrm{d}\Psi = \frac{2}{3}\frac{e^2 c}{R^2}\left(\frac{E}{m_0 c^2}\right)^4 , \tag{2.7}$$

where $e$ is the electron charge, $c$ the speed of light, $E$ the electron energy, $m_0c^2$ its rest mass energy, $R$ the orbit radius, $\lambda$ the radiated photon wavelength (cm), $\Psi$ the azimuthal angle (vertical – away from the orbital plane) and $I$ is:

$$I(\lambda, \Psi) = \frac{27}{32\pi^3}\frac{e^2 c}{R^3}\left(\frac{\lambda_\mathrm{c}}{\lambda}\right)^4 \gamma^8 \left[1 + (\Psi\gamma)^2\right]^2 \left[K_{2/3}^2(\xi) + \frac{(\Psi\gamma)^2}{1 + (\Psi\gamma)^2} K_{1/3}^2(\xi)\right] . \tag{2.8}$$

Here $\gamma$ is the relativistic factor $E/m_0c^2$ and $K_{1/2}$ and $K_{2/3}$ are the modified Bessel function of the second kind. $\lambda_c$ is the "cut-off" wavelength, given by:

$$\lambda_\mathrm{c} = \frac{4}{3}\pi R \gamma^{-3}; \quad \text{and } \xi = \frac{\lambda_\mathrm{c}}{2\lambda}\left[1 + (\Psi\gamma)^2\right]^{3/2} . \tag{2.9}$$

Note the $E^4$ (2.7) dependence of the total radiated power, which explains why storage rings optimized for high electron energies must compensate this by a rather large ring radius $R$ in order to keep the power $P$ in the required range. From (2.7) it can be seen that owing to the rest mass energy denominator, electrons radiate about $(2000)^4$ as much power as protons. For practical vacuum properties of the storage rings, positrons are sometimes used instead of electrons as they are less sensitive to recombination with trapped ions, which contributes to the longer lifetime of a stored beam.

Other practical expressions routinely used in SR are:

$$E_\mathrm{c}(\mathrm{keV}) = 0.665\, E^2(\mathrm{GeV}) \times B(T), \tag{2.10}$$

which defines the critical energy, and which in most cases represents the median energy in the spectrum for the power distribution scheme as a function of the electron energy $E$ and of the peak magnetic field $B$ in the orbit. Also, the total radiated power is:

$$P_{\mathrm{tot}}(\mathrm{kW}) = 26.6\, E^3(\mathrm{GeV}) \times B(T) \times I(A), \tag{2.11}$$

as a function of $E$, $B$, and the storage ring current $I$.

The emission angle describing the cone into which all the photons are generated by a relativistic electron is equal to $1/\gamma$ (rad) around the critical energy $E_c$. For relativistic energies this is in the microradian range and represents one of the main advantages of SR. The great directionality and wide spectral range of the SR combine to yield a high spectral brilliance over a large range, which is the parameter of interest in the calculation of the heatload absorbed by the various beam components, such as monochromators.

One of the most interesting aspects of SR is its linear polarization, which is exactly 100% in the orbit plane, with the electric vector parallel to the plane. Above and below the orbit plane, the radiation is elliptically polarized to a degree depending on the angle of observation $\Psi$. For "real" beams of nonzero emittance, the linear polarization in the orbit plane is slightly decreased.

Another particular advantage of SR is its fast time structure. In a storage ring, electrons travel in bunches, alternatively losing their energy by emitting SR radiation, then replenishing it in the radio-frequency cavities. The length of bunches is in the *ps* to *ns* and is highly reproducible as a function of the bunch structure. Different operation modes, detailed in Sect. 2.4.2, take advantage of this "lighthouse" effect for experiments requiring a time structure in the SR excitation, such as Nuclear Resonance Spectroscopy or Time Resolved Spectroscopy. For the ESRF electron energy of 6 GeV, the time per revolution is about 3 μs.

A striking characteristic of third generation SR sources is their coherence properties which initiate the emergence of a totally new field of "coherent imaging" which is of considerable interest for Life or Materials Sciences, as it allows the observation using the "phase contrast" of properties undetectable so far.

### 2.4.2 Storage Ring Description

Generation of SR is done in a storage ring, by passing the electron (positron) beam through various magnetic field structures, in a curved trajectory. The orbit is closed by alternating bending magnets which make the $e^-$-beam turn and insertion devices which are straight beam sections. Where some experiments require X-ray beams of low divergence (virtually parallel), others need a small initial source size to demagnify it by focusing on a submicron beam. As is the case with all charged particles beams, Liouville's theorem applies, which describes the conservation of the total beam emittance for a constant velocity. The emittance is $\varepsilon = \sigma \times \sigma'$ where $\sigma$ and $\sigma'$ represent the $e^-$-beam source size and divergence (the derivative of the source size with respect to the longitudinal coordinate $s$), in the horizontal ($x$) or vertical ($z$) directions. The beamsize and divergence of the photon beams generated by the electron beam will obviously depend on the above parameters, as well as the radiation cone $1/\gamma$. Practically, the orbit is modulated by the bending magnets which impose the closing of the orbit and by alternating quadrupole magnets which act as magnetic lenses, producing focusing and defocusing in the two coordinates $x$ and $z$. Insertion devices act as perturbing devices in a storage ring so the whole magnetic structure (lattice) is designed so as to minimize transverse oscillations, also called "betatron" oscillations. Consequently, beamlines of high and low $\beta$ values (the amplitude of the betatron stable oscillations) are distributed across the ring, separated by the dipole/quadrupole steering magnetic devices. As a guide of the values of these parameters, the ESRF lattice RMS values are:

| | Bending magnet | High $\beta$ (undulators) | Low $\beta$ (wigglers) |
|---|---|---|---|
| $\beta_x$(m) | 1.4 | 35.2 | 0.5 |
| $\beta_z$(m) | 35 | 2.5 | 2.7 |
| $\sigma_x$(µm) | 100 | 400 | 60 |
| $\sigma_z$(µm) | 30 | 8 | 8 |
| $\sigma'_x$(µrad) | 115 | 11 | 90 |
| $\sigma'_z$(µrad) | 0.8 | 3.2 | 3 |

The high $\beta$ beamlines are optimized for undulators whereas the low $\beta$ ones cover both undulators and wigglers. The emittance values do not depend on the longitudinal coordinate $s$ and they are linked to the $\beta$ oscillations and the size/divergence $\sigma/\sigma'$ by the following equations:

$$\sigma_x(s) = \sqrt{\varepsilon_x \beta_x(s)} \quad \text{and} \quad \sigma_z(s) = \sqrt{\varepsilon_z \beta_z(s)}. \tag{2.12}$$

Finally, in order to combine the advantages of high flux with those of the time structure of the beam, operation of the storage rings is made in different modes. At ESRF they are: the so-called *uniform* mode for the even filling of the ring with electron packets, yielding maximum flux; *single bunch*, used for time-resolved experiments which make use of a single electron bunch with a repetition rate of about 300 kHz but with low-intensity; *16 bunch* where the ring is injected with 16 equally spaced electron bunches in order to produce a moderate intensity while allowing certain time-resolved experiments, and *hybrid*, which combines a one-third filling of the storage ring with normal electron bunches in opposition to a single bunch. Thus, intensity is maximized while allowing time-resolved experiments requiring a time structure of about 1 µs.

### 2.4.3 Generation of SR

*Bending Magnets*

Bending magnets (BM) are not only mandatory devices for closing the $e^-$-beam orbit but are also reduced heatload devices, capable of producing a smooth continuum spectrum extending over several tens of KeV. For practical purposes the ESRF uses permanent magnets made of SmCo for BMs, with a remanent magnetic field of approximately 0.85 T which emit light in a cone of about $\Psi = 1.3/\gamma$ (FWHM) or about 0.11 mrad vertical opening. The magnetic field yields a critical energy $E_c$ of about 19.2 keV for these devices. The radiation emitted by a BM has a uniform distribution as a function of the photon energy in the horizontal plane. The flux per unit of time per emitted angle in the horizontal and vertical planes, in a 0.1% energy bandwidth, the so-called spectral brightness varies strongly with the vertical viewing angle $\Psi$ as:

$$N\,(\text{photons/s/mrad}_\theta\ \text{mrad}_\Psi) = 1.32 \times 10^{13} I[\text{A}] E^2[\text{GeV}] F(E_{\text{ph}}/E_c, \Psi), \tag{2.13}$$

where $E_{\text{ph}}$ is the emitted photon energy, $E_c$ the critical energy and $F$ is defined as the sum of the contributions of the horizontal/vertical polarizations:

$$F(y, \Psi) = F_\sigma + F_\pi = y^2(1 + \gamma^2\Psi^2)K^2_{2/3}(y/2) + y^2\gamma^2\Psi^2(1 + \gamma^2\Psi^2)K^2_{1/3}(y/2), \tag{2.14}$$

and $K_{1(2)/3}$ are the modified Bessel functions. Beamlines making use of BM at ESRF employ X-rays in the range 4–100 keV. In a BM the X-rays are collected from a zone where the electrons are in the magnetic field which is approximately 15 cm at ESRF.

*Insertion Devices*

Insertion devices (IDs) are the standard top-of-the-line radiation sources at third generation storage rings. Their detailed description represents a subject far out of the scope of the present work. Our purpose here is to give a summary description of their main properties and parameters of interest for the physicist or engineer dealing with these devices for their routine spectroscopy work. For an in-depth account, see the book by Onuki and Elleaume [61] and references therein. IDs are assemblies of alternating magnetic elements (see Fig. 2.29) inserted in the straight sections of the ring.

Their advantage is the generation of many undulations in the $e^-$-beam which produce photons at each bend, thereby considerably increasing the total emitted flux. Secondly, contrary to BM the magnetic field values of which are fixed, IDs can use variable field values, thus covering a very large spectral range. They are described by their remanent field $B_r$ in Teslas which for SmCo pure permanent magnets (PPM) reaches 0.85 T while for NdFeB ones 1.17 T. The period of these IDs called $\lambda_0$ is the separation between identical magnetic structure units in the $s$ direction. The main IDs are: the wavelength shifter, the multipole wiggler, and the undulator. We will only present here the wiggler and the undulator, which are the most frequently used IDs.

*Wigglers*

Wigglers are assemblies of alternative magnetic field regions which induce fairly large deviations of the electrons from the straight trajectory. These

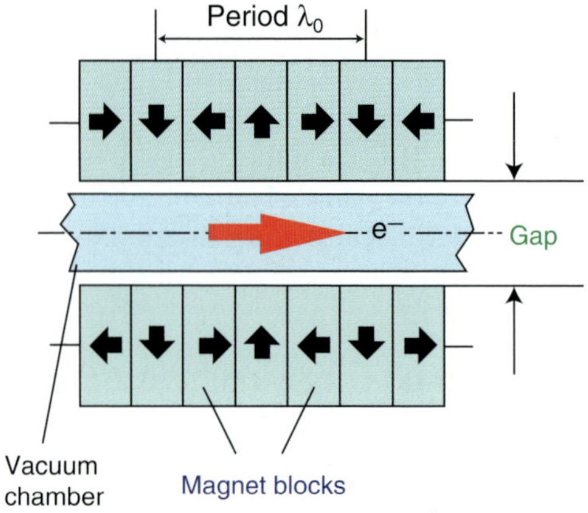


**Fig. 2.29.** Scheme of the magnetic structure in an insertion device segment of period $\lambda_0$

multipole devices act on the $e^-$-beam as a succession of BM which generate photons incoherently. The total intensity generated by $N$ poles spaced by $\lambda_0$ is equal to that of $2N$ bending magnets (a period comprises two bends in the $e^-$-beam trajectory which generate X-rays). The vertical magnetic field inside a wiggler (end-effects neglected) is, to a good approximation, a sine wave of period $\lambda_0$:

$$B_Z = B_0 \times \sin(2\pi s/\lambda_0) \quad \text{and} \quad B_0 = 1.43 \times B_\mathrm{r} \times \mathrm{e}^{-(\pi g/\lambda_0)} \tag{2.15}$$

with $B_\mathrm{r}$ the remanent magnetic field in T and $g$ the gap in cm, while the transverse velocity in the horizontal plane is

$$v_X/c = \frac{0.3}{E(\mathrm{GeV})} \int B_Z(s)\mathrm{d}s = \frac{K}{\gamma}\cos(2\pi s/\lambda_0), \tag{2.16}$$

where distances are in $m$ and $B$ in T. $K$ is the dimensionless deflection parameter which governs the regime of the ID:

$$K = 93.4 B_0(T)\lambda_0(m). \tag{2.17}$$

For $K < 1$ values, photons are emitted in a cone smaller than $1/\gamma$ and they interfere, yielding the "undulator" case. For $K \gg 1$, virtually no interference occurs and the emitted flux can be obtained summing up the contributions of the trajectory points tangent to the emitted cones.

*Undulators*

Undulators (Fig. 2.30) are thus "multipole" wigglers of small $K$ values and a large number of poles. Due to the small (but frequent) trajectory deviations, the flux can be collected from the whole trajectory and due to interference

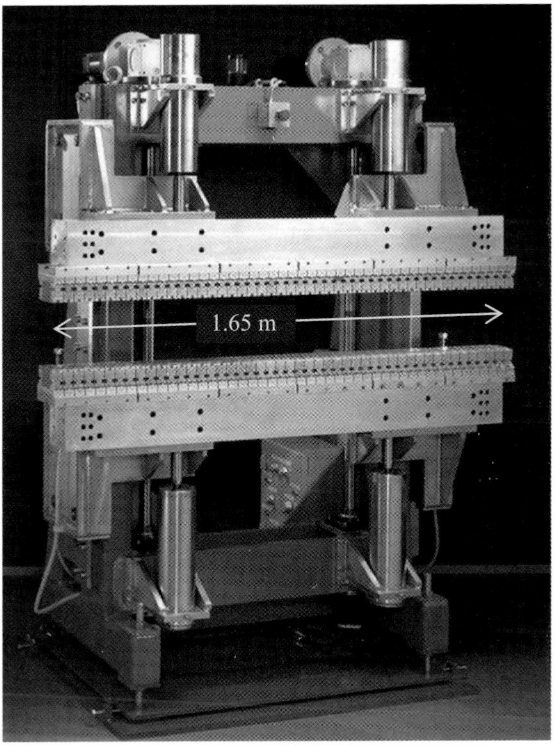


**Fig. 2.30.** ESRF undulator carriage

effects it is proportional to $N^2$ ($N$ = number of poles). The wavelength of the spectrum emitted by an undulator is the Fourier transform of its time structure and thus exhibits a line spectrum corresponding to the periodicity of the emission in the time domain and is given by:

$$\lambda = \frac{\lambda_0}{2i\gamma^2}(1 + \frac{K^2}{2} + \gamma^2\theta^2), \quad i = 1, 2, 3, \ldots, \tag{2.18}$$

where $i$ is the harmonic number and $\theta$ the observation angle in the horizontal plane. The fundamental wavelength is then:

$$E_{\text{fund}} = 0.949\frac{E^2[\text{GeV}]}{\lambda_0[\text{cm}]}(1 + \frac{K^2}{2} + \gamma^2\theta^2). \tag{2.19}$$

The odd harmonics are the ones of maximum yield and their flux is given by:

$$I(\text{phot/s/0.1\%bw}) = 1.43 \times 10^{14} NI[A]Q_i, \tag{2.20}$$

where $Q_i$ is a function of the harmonic number $i$:

$$Q_i = \frac{iK^2}{(1 + \frac{K^2}{2})}\left(J_{(i-1)/2}\left(\frac{iK^2}{4 + 2K^2}\right) - J_{(i+1)/2}\left(\frac{iK^2}{4 + 2K^2}\right)\right)^2 \tag{2.21}$$

and $J$ the standard Bessel function.

### 2.4.4 SRW Package

When addressing the SR fundamentals, it is more straightforward to treat the matter from the point of view of a user not particularly familiar with X-ray sources but with a good understanding of X-ray spectroscopy and its relevant parameters. Consequently, we will continue our exploration of SR by making use of a calculation package optimized for designing and comparing synchrotron radiation sources, entitled SRW (*Synchrotron Radiation Workshop*).

SRW is a software package developed at the ESRF by O. Chubar and P. Elleaume from the Insertion Device Group. The package is freely downloadable at http://www.esrf.fr/Accelerators/Groups/InsertionDevices/ Software/SRW/SRW download and runs under the multiplatform graphical program Igor Pro (http://www. wavemetrics.com/) for Macintosh or PC. The current version of SRW (3.7) enables a large variety of calculations of SR generated by relativistic electron beams in magnetic fields of arbitrary configurations as well as standard bending magnets, insertion devices, or even Free Electron Laser(FEL) devices. In the rest of this chapter, we will be dealing with SR generated from the three basic elements – bending magnets, wigglers, and undulators – and we will estimate its spectral content as well as the power density or spectral brilliance produced by such devices. We will adopt a practical point of view, dealing only with the mandatory parameters to fully describe the SR, and we will do so from the experimentalist's point of view. Interested readers are directed to use the SRW package in-depth and go through the full set of examples therein.

Following the detailed procedures described in the introduction of the SRW package, let us use the default SR source which is the ESRF storage ring. All further calculations are performed after installing the SRW package and starting the Igor Pro program. All the following SRW calculations are accessible from the SRWE menu of Igor Pro. One activates the SRW "Initialize" item from the submenu "Utilities," then defines the electron beam parameters from the submenu "Electron Beam." Interesting parameters here, which users might change to reproduce conditions prevailing on different machines, are respectively:

1. Electron beam energy in GeV (6 GeV for the ESRF)
2. Electron current in A (0.2 is the average for the ESRF normal model)
3. RMS energy spread (0.001 or 1‰ of the energy beam is the ESRF standard)
4. Emittance in the horizontal/vertical planes in nm (3.9 and 0.039 are usual at ESRF)
5. Horizontal/vertical amplitude of orbit beta oscillations in m (35.6 and 2.5 are usual at ESRF).

Then, "Radiation sampling" is activated from the corresponding submenu with the following parameters:

1. Slit position at which the flux is estimated, counted from the source, in m (30).
2. Horizontal/vertical size in mm (1 and 1) and center of the slit, in mm (0 and 0) with respect to the beam position. Here we will calculate the spatial distribution of the flux around the beam center by using a slit size of 20 by 20 $mm^2$.
3. Horizontal/vertical number of points used to calculate the distribution of flux through the slit, in mm (1 and 1). Let us take 100 points for both the horizontal/vertical direction, in order to obtain a good spatial distribution of the spectral flux around the beam center.
4. Initial/final energy, in keV of the spectrum to be generated, as well as the number of points (0.1, 20, and 1000 by default). We will change the maximum energy value to 100 keV for the sake of clarity.

*Bending Magnet*

Next comes the choice of sources. For the first example, let us choose a bending magnet, from the "Bending Magnet" menu. The rest of this paragraph will be dealing with options of this menu. This part of SRW computes spectral flux per unit surface of radiation emitted by an electron beam with nonzero emittance. The following general assumptions are made:

1. Radiation from different electrons is incoherent: i.e., the flux is proportional to the number of electrons.
2. Only transverse SR polarization components are considered.
3. Spectral-angular distribution of the bending magnet radiation is computed in the far-field approximation.

Let us then define a standard magnetic structure made of permanent magnets from NdFeB or SmCo. To do so, one chooses from the "Constant magnetic field" submenu the default values for the intensity of the vertical/horizontal magnetic field, in Tesla (0.85 and 0). We can now calculate the Stokes components of radiation produced by the electron beam by selecting "Compute Stokes" from the corresponding submenu.

*Spectrum Through a Slit*

Now we can directly plot the photon flux per unit surface by selecting the item "Visualize" and leaving the default values. One obtains thus a plot of the spectral flux in units of photons/s/1% energy bandwidth/$mm^2$ (which is a traditional synchrotron unit) as a function of the energy of the emitted photons, in the previously selected energy range in keV (Fig. 2.31).

*Power Density*

Alternatively, one can calculate the total power density/surface emitted by the bending magnet in the given energy range (0.1–100 keV) through the

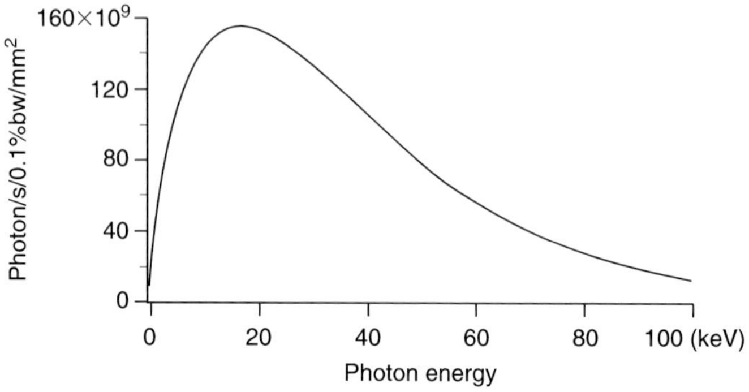


**Fig. 2.31.** Spectral flux/surface at 30 m from a standard ESRF bending magnet in a $20 \times 20\,\text{mm}^2$ slit, centered on the photon beam

predefined $20 \times 20\,\text{mm}^2$ slit. This is a very useful calculation when designing SR beamlines and dealing with the heatload that devices exposed to the beam have to withstand, or when performing shielding estimates for radiation safety.

In general, one alternates between the spatial distribution and the integral of the power density as a function of energy, by selecting various slit sizes and/or number of sampling points. Thus, if the "Radiation Sampling" is changed and only one sampling point is taken for both the vertical and horizontal directions, one obtains the integral of the total power density through the slit, which in our case gives $1.32692\,\text{W/mm}^2$ (Fig. 2.32)

*Brilliance*

Brilliance is one of the important parameters for rating the performance of a source of radiation in general. It synthesizes the source's properties of delivering a flux of photons through a unit surface in a unit solid an-

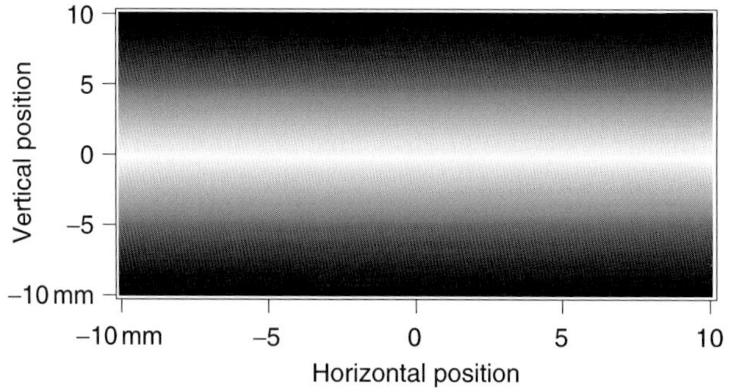


**Fig. 2.32.** Power density at 30 m from a standard ESRF bending magnet as a function of the vertical and horizontal offsets from the beam center

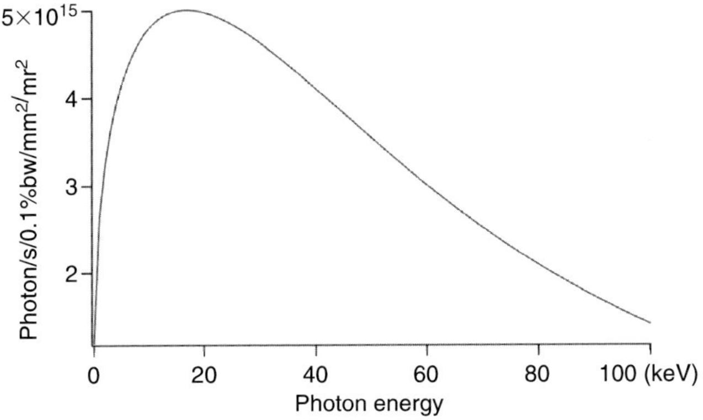


**Fig. 2.33.** Brilliance at 30 m of a standard ESRF bending magnet

gle. This has the merit of describing both the source emittance properties and the particular size and distance from the source of the surface where radiation is measured. The radiation has the customary SR bandwidth of 1‰. To perform a brilliance calculation one chooses then the submenu item "Estimate brilliance" and obtains a plot of the brilliance in units of photons/seconds/0.1% bandwidth/mm$^2$/mr$^2$ (Fig. 2.33).

*Insertion Devices*

SRW handles standard insertion devices (ID) such as wigglers and undulators, but also hybrid cases known as "wundulators" and even exotic FEL devices. These are all accessible either directly through the submenus WIGGLER/UNDULATOR/FEL or through the ARBITRARY MAGNETIC FIELD, and then customizing a particular device. In the following section, we will use standard IDs.

*Wiggler*

Now we reproduce the previous calculations for the case of a planar wiggler. This part of SRW computes the spectral flux per unit surface of radiation emitted by an electron beam with nonzero emittance. The computation is performed in the near-field approximation which is quite accurate. The particular assumptions applicable to this case are as follows.

1. The radiation is generally emitted from distinct separate parts of the electron trajectory, and phase shifts of the radiation between these trajectory parts is much larger than $\pi$.
2. The effect of the wiggler terminations is taken into account.

To implement the ESRF wiggler case, one uses the settings of a "low $\beta$" beamline, that is, in the "Electron Beam" submenu the following parameters:

1. Emittance in the horizontal/vertical planes 4 and 0.03 nm
2. Horizontal/vertical amplitudes of orbit beta oscillations 0.5 and 2.73 m

Then, in the "Radiation sampling" submenu, one changes the following parameters:

1. Horizontal range and number of points: 1 and 1.
2. Vertical range and number of points: 20 and 250.
3. Energy final/initial/number of points: 1, 100 keV and 1000 points.

*Spectrum Through a Slit*

One selects the menu "Wiggler," then submenus "Magnetic Field" then "Create and Modify" and sets the length of the field to 1 m. Then from the item "Sinusoidal" the following values are changed from the default ones for the "MagBZfld" field component:

1. Zero the field before = Yes.
2. Period = 150 mm for a standard ESRF wiggler.
3. Number of periods = 3.5 for a three-period wiggler.
4. Peak field = 1.8 T.

From the "Visualize" submenu one chooses energy as the abscissa of the graph and plot the wiggler spectrum. The same calculation is then performed using the Bending Magnet case, which approximates fairly well the wiggler case if multiplied by $2N$ where $N$ is the number of poles (here 3). One performs the calculation selecting "Bending Magnet," then "Constant Magnetic Field" and entering 1.8 T as the field value. The wave values are then multiplied by $2N = 6$ using the "Analysis, Compose expression" item. Finally, the resulting wave is added to the previous plot, using the "Graph, Append Traces to Graph" menus and one obtains a plot as in Fig. 2.34. Here the wiggler case is too close to the undulator approximation ($K$ too small) so the bending magnet case is preferred which yields a realistic smooth spectrum.

*Power Density*

One then selects "Compute power density" from the "Wiggler" menu and obtains the power density as a function of the vertical position as in Fig. 2.35. If one selects the "Horizontal Position" mode for the "Visualize" menu options, the integral power density of the wiggler is calculated $-14.2568\,\mathrm{W/mm^2}$ in our case.

*Brilliance*

The brilliance is then obtained by selecting the item "Estimate brilliance" from the "Wiggler" submenu. One just accepts the parameters in the first dialog box, then changes the value for the length of the device to 1 m instead of the default 3.2 m in the second dialog box, then accepts the parameters in the final dialog box. The graph of the brilliance will then look as in Fig. 2.36.

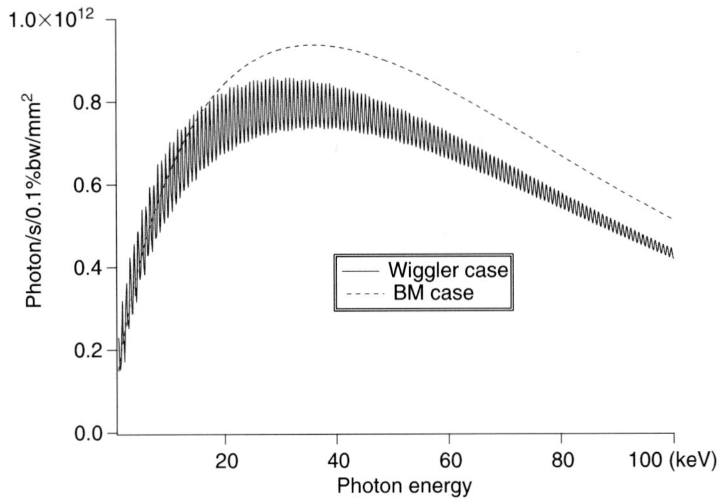


**Fig. 2.34.** Spectral flux/surface at 30 m from an ESRF wiggler (see text for specifications)

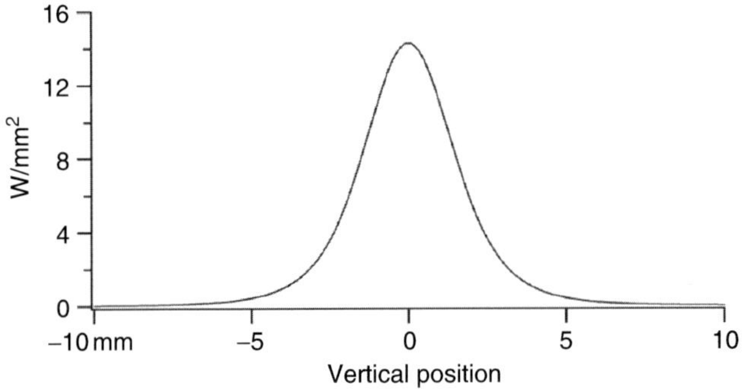


**Fig. 2.35.** Power density/surface at 30 m from an ESRF wiggler as a function of the vertical position offset from the beam center

*Undulator*

This part of SRW covers undulator radiation emitted by an electron beam with nonzero emittance and energy spread in a periodic magnetic field and observed through a rectangular aperture. The typical computation is a spectrum integrated through a slit or an image of the radiation pattern observed at a single energy at some distance from the source. The computation is performed in the far-field approximation. This type of computation is probably less accurate than the near-field computation. Its main advantage is speed.

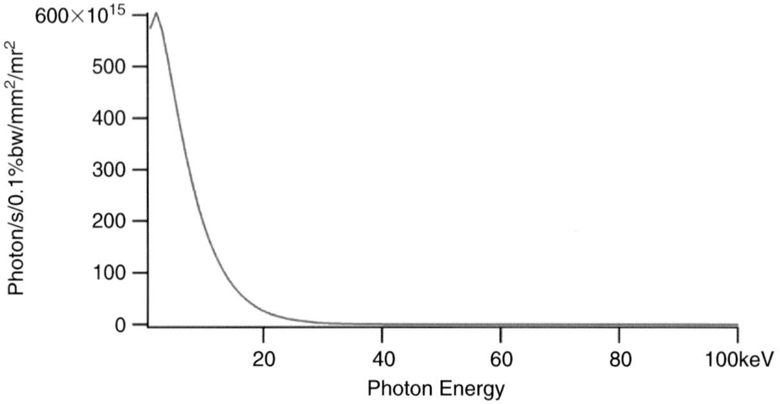


**Fig. 2.36.** Brilliance at 30 m from an ESRF wiggler as a function of the photon energy

*Spectrum Through a Slit*

This method of computation has a slightly different meaning compared to the other methods implemented in SRW, since it produces spectral flux integrated within a slit, while most of the other methods in SRW produce spectral flux per unit surface. The specific assumptions in effect for the undulator calculations are:

1. The distance from undulator to observation plane is considerably larger than the length of the undulator.
2. The number of periods in the undulator is significantly larger than 1.
3. The effect of terminating poles of the undulator is neglected.
4. The angle between the electron velocity and the direction of observation is comparable to the angular size of the central cone of the undulator radiation.
5. The SR emission is not affected by the conductive walls of the ID vacuum chamber.
6. The diffraction effects due to the vacuum chamber are neglected.

One activates the "Electron Beam" menu and accepts the default values for an ESRF "high $\beta$" beamline. Then from the "Radiation sampling" menu, one changes the values of the horizontal/vertical slit sizes to $0.5 \times 0.5$ mm, and the energy initial/final to 1 and 35 keV. From the "Utilities" item, one chooses "Undulator:PPM" in order to calculate the $K$ deflection parameter corresponding to our undulator magnetic field. In the popup dialog box one changes only the values of the gap and period for a standard ESRF undulator, the period is 35 mm, and the minimal gap is 16 mm. This yields a $K$ value of 1.44794. In the "Undulator" item we then select "Periodic Magnetic Field,"

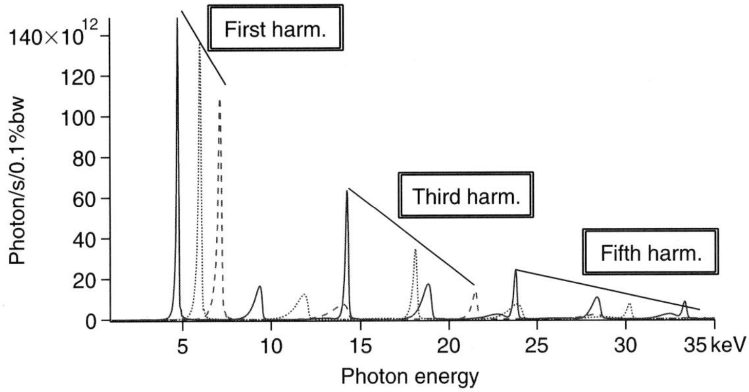


**Fig. 2.37.** Spectral flux integrated on a $0.5 \times 0.5\,\mathrm{mm}^2$ slit at 30 m from an ESRF undulator (see text for specifications)

then "Create and Modify" and we enter this $K$ value as well as the value of 1.65 m for the length of a standard ESRF undulator, and 35 mm for the undulator period. In the "Compute Stokes" menu item, we only change the value of the "Final harmonic of spectrum" to 6. Then we visualize the results as previously. If we repeat the procedure for gaps of 19 ($K = 1.106$) and 22 ($K = 0.845$) mm, we obtain the plot as in Fig. 2.37, which describes the variation of the intensities of the first, third and fifth harmonics when the gap varies from 16 to 19 (dotted line) and then 22 mm (dashed line). Notice the drop in intensity which goes along with the energy increase.

This plot is in units of photons/s/0.1% bandwidth as it is integrated over the slit surface. Notice the narrower and higher odd $(1, 3, 5)$ harmonics which are the ones preferentially used.

*Power Density*

Let us restart IGOR fresh and go through "Initialize" and "Electron beam" then from "Radiation sampling" use a slit size of 20 by 20 mm, 100 points for both the horizontal/vertical directions, and the energy range 1 to 30 keV. From the item "Undulator" we again select "Periodic Magnetic Field," then "Create and Modify" and we enter the previous $K$ value (1.4479) as well as the value of 1.65 m for the length of a standard ESRF undulator, and 35 mm for the undulator period. We then use "Compute power density" from the "Undulator" menu and accept the default values, and finally obtain an image of the power density profile as a function of the horizontal/vertical offsets from the beam center. If one uses only one point for both horizontal/vertical "Radiation sampling," after the appropriate steps one obtains the total power density integrated over the $20 \times 20\,\mathrm{mm}^2$ slit surface as $63.1051\,\mathrm{W/mm}^2$ (Fig. 2.38).

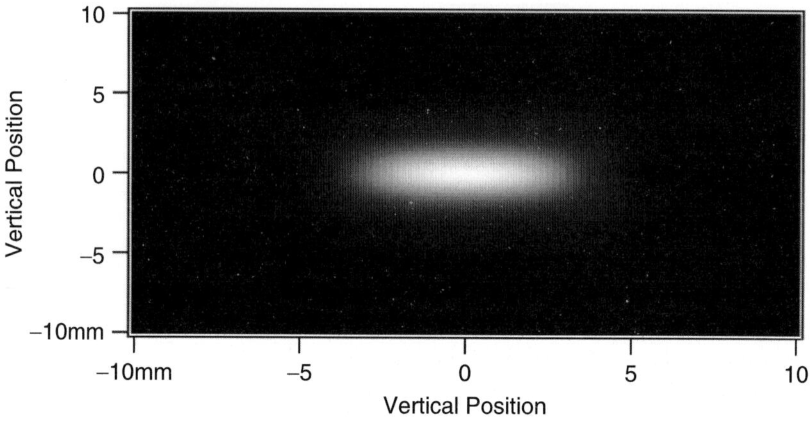


**Fig. 2.38.** Power density/surface at 30 m from an ESRF undulator as a function of the vertical and horizontal offsets from the beam center

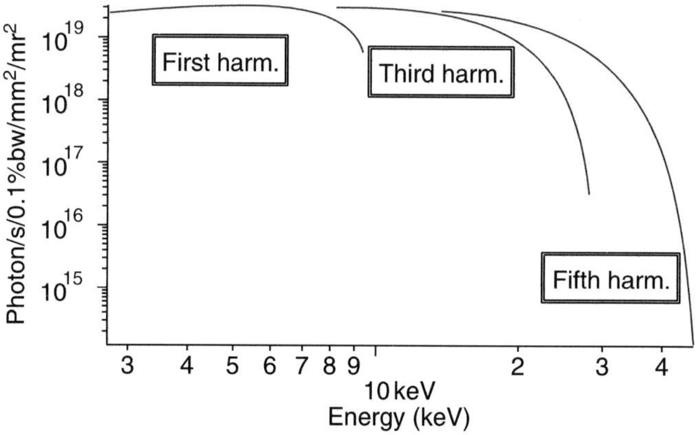


**Fig. 2.39.** Brilliance at 30 m from an ESRF undulator as a function of the photon energy for the first three odd harmonics

*Brilliance*

For the brilliance calculation one follows the appropriate steps for an undulator of 11 mm gap ($K = 2.268$) except for the "Final harmonic" in the "Estimate brilliance" submenu, which is set to 6. Then the plot of the brilliance will be displayed for only the first three odd harmonics (Fig. 2.39).

## References


1. Röntgen WC, Annalen der Physik und Chemie, Neue Folge **64**, 1 (1898)
2. Coolidge WD, A powerful Röntgen ray tube with a pure electron discharge. Phys Rev **2**, 409–430 (1913)
3. Institut für Gerätebau GmbH, Berlin: Miniaturisierte Röntgenquelle mit nanostrukturierter Feldemissions-Elektronenquelle, Berlin, TIB Hannover (2003) www.tib.uni-hannover.de
4. Mourou GA, Barty CPJ, Perry MD, Ultrahigh – Intensity Lasers: Physics of the Extreme on a Tabletop. Phys Today, **January 1998**, 22
5. Umstadter D, Phys Plas **8**, 1774 (2001)
6. Kruer WL, Phys Plas **10**, 2087 (2003)
7. Sokolowski-Tinten K, von der Linde D, Physik J **3**, 49 (2004)
8. Liénard A, L'éclairage électrique **16**, 5 (1898)
9. Wiechert JE, Arch Néerlandaises **549** (1900)
10. Schott GA, Electromagnetic radiation and the mechanical reactions arising from it, Cambridge (1912)
11. Ivanenko D, Pomeranchuk I, Phys Rev **65**, 343 (1944)
12. Schwinger J, Phys Rev **75**, 1912 (1949)
13. Elder FR, Gurewitsch AM, Langmuir RV, Pollock HC, Phys Rev **71**, 829 (1947)
14. Agarwal BK, *X-ray Spectroscopy*, Springer, Berlin Heidelberg New York (1991) [Springer Series in Optical Sciences, Vol. **15**]
15. Arkadiev V, Bräuninger H, Burkert W, Bzhaumikhov A, Gorny HE, Langhoff N, Oppitz A, Rabe J, Monochromatic X-ray source for calibrating X-ray telescopes. Nucl Instrum Methods Phys Res A **455**, 589–595 (2000)
16. Bertin EP, *Principles and Practice of X-ray Spectrometric Analysis*, New York: Plenum Press, 1079 (1975)
17. *Handbook of X-ray Spectrometry*, René E. Van Grieken and Andrzej Markowicz A (eds), New York: Marcel Dekker Inc, 983 (2002) [Practical Spectrometry Series, Vol **29**]
18. Krestel E, *Imaging System for Medical Diagnosis*, Wiley, New York (1990)
19. Baumann W, Dietz H, Geldner E, *Temperature Distribution in X-ray Rotating Anodes*, Mathematical computation for compound anodes, Siemens Forschung und Entwicklung, Bericht **7**, 111 (1978)
20. Elsas A, Zimmer Th, *Höhere Belastungen von Drehanodenröhren durch Verwendung von legierten Anoden*, Fortschritte auf dem Gebiet der Röntgenstrahlen und der Nuklearmedizin, Vol. **97**, 511 (1962)
21. Windeck C, Warrikhoff H, A New Series of Metal-Ceramic X-ray Tubes for Low Power Applications, Proceedings of the 7th European Conference on Non-Destructive Testing, Vol. **3**, Broendby, Denmark, 2790–2795 (1998)
22. Arkadiev VA, Kolomiitsev AI, Kumakhov MA, Ponomarev IYu, Khodeev IA, Chertov YuP, Usp Fiz Nauk **157**, 529 (1989) [in Russian] [Sov Phys Usp **32** (1989) 271]
23. Arkadiev V, Beloglazov V, Bjeoumikhov A, Gorny H-E, Langhoff N, Wedell R, Application of capillary optics in modern scientific instrumentation. Surface **1**, 48–54 (2000)
24. Bjeoumikhov A, Langhoff N, Wedell R, Beloglazov V, Lebedev N, Skibina N, New generation of polycapillary lenses: manufacture and applications. X-Ray Spectrom **32**, 172–178 (2003)

25. Grider DE, Wrigth A, Ausburn PK, J Phys D: Appl Phys **9**, 2281–2292 (1986)
26. Brodersen P, Liebram A, Niemann W, Nielsen J, *The e-journal of Nondestructive Testing*, **7**(No. 10) (2002) [www.ndt.net]
27. Website of Hamamatsu, http://www.hamamatsu.com
28. Website of Kevex, http://www.kevex-X-ray.com
29. Website of Oxford Instruments, http://www.oxfordxtg.com
30. Website of Röntgen-Technik Dr. Warrikhoff KG, http://www.rtwxray.de
31. Website of IfG, http://www.ifg-adlershof.de
32. Bjeoumikhov A, Langhoff N, Rabe J, Wedell R, A modular system for XRF and XRD applications consisting of a microfocus X-ray source and different capillary optics. X-Ray Spectrom **33**, 312–316, (2004)
33. Bronk H, Röhrs S., Bjeoumikhov A, Langhoff N, Schmalz J, Wedell R, Gorny HE, Herold A, Waldschläger U, ArtTAX–a new mobile spectrometer for energy-dispersive micro X-ray fluorescence spectrometry on archaelogical objects, Fresenius J Anal Chemistry, **371**, 307–316 (2001)
34. Website of PANalytical, http://www.panalytical.com
35. Website of Siemens, http://www.med.siemens.com
36. Website of YXLON, http://www.yxlon.com
37. Website of Pantak, http://ndt.agfa.com
38. Website of Varian medical systems, http://www.varian.com
39. Website of Rigaku, http://www.rigaku.com
40. Sheshin EP, Surface Structure and Field-emission Properties of Carbon Materials, Moscow (2001)
41. Knapp W, Schleußner D, Baturin AS, Yeskin IN, Sheshin EP, CRT lighting element with carbon field emitters. Vacuum **69**, 339–344 (2003)
42. Knapp W, Bischoff L, Teichert J, Formation of a nano-emitter for electron field emission on a liquid metal ion source tip after solidification of the alloy. Vacuum **69**, 345–349 (2003)
43. Knapp W, Schleußner D, Carbon buckypaper field emission investigations. Vacuum **69**, 333–338 (2003)
44. Knapp W, Schleußner D, Field emission characteristics of carbon buckypaper. J Vac Sci Technol, (B) **21**(1), 557–561 (2003)
45. Price B, The Application of Portable EDXRF Instrumentation Using Cold Cathode X-ray Tubes, Proceedings of PRORA 2003, Berlin, 3–4 April (2003)
46. Lilley J, *Nuclear Physics—Principles and Applications*, Wiley, England (2001)
47. Knoll GF, *Radiation Detection and Measurement*, Wiley, England, 3–28 (2000)
48. Lederer MC, Shirley VS, *Table of Isotopes*, Wiley, England (1978)
49. Knoll GF, *Radiation Detection and Measurement*, Wiley, England, 430 (2000)
50. http://www.atom.nw.ru/isotope
51. Thomsen V, Schatzlein D, Advances in Field-Portable XRF. Spectroscopy **17**, 14–21, (2002)
52. http://nucleardata.nuclear.lu.se/nucleardata/toi/nucSearch.asp
53. Dziunikowski B, *Energy Dispersive X-ray Fluorescence Analysis*, Polish Scientific Publisher, Warsaw (1989)
54. http://isotopeproduct.com
55. http://www.qsa.aeat.com/prod_sources.html
56. Zschornack G, Atomdaten für die Röntgenspektralanalyse (Atomic Data for X-Ray Spectral Analysis in German). Deutscher Verlag für die Grundstoffindustrie, Leipzig (1989)

57. Amersham Catalogue of Calibration standards and instruments for measuring radioactivity, 1st edn.
58. J. Leonhardt BI (ed), *Taschenlexikon Radioaktivität*, Bibliographisches Institut, Leipzig (1982)
59. Rogozev BI, Ritverz Catalogue of Radiation Sources, especially for XRF (2001)
60. Jackson JD, *Classical Electrodynamics*, $2^{nd}$ edn, Wiley, New York, 627 (1975)
61. Undulators, wigglers and their applications, Onuki H, Elleaume P (eds), Taylor & Francis: London (2003)

# 3

# X-Ray Optics

## 3.1 Introduction

*A. Erko*

There is a growing need in many fields of human activity for the determination and characterization of elements at trace concentrations which can be well below one part per million by weight (ppm). For qualitative as well as quantitative investigations X-ray microanalysis is often used as an established method of acquiring elemental composition. However, this first step of characterization is nowadays becoming gradually insufficient and a distribution map of each element is more and more desirable. The latter aim can be achieved only with the *large flux, optimal excitation energy*, and *high lateral resolution*. For that appropriate optical elements should be developed and placed in between a source and a sample to fulfill the demands for powerful, monochromatic and highly concentrated X-ray beams. As a result *X-ray optics* has grown rapidly in the last years as a particular branch of physics and technology.

Under the title of "X-ray optics" one can gather a wide range of X-ray optical elements exploiting diffraction and refraction phenomena on sub-µm and sub-nm artificial structures and natural crystals as well as the total external reflection phenomenon. These elements are intended to focus, monochromatize or reflect X-rays. Natural crystals could be regarded as prototypes of many artificial structures.

A compromise between the flux and the necessary energy resolution can be achieved by the choice of a suitable crystal or multilayer monochromator. Low-resolution monochromators can be also built from transmission and reflection zone plates. New possibilities are offered by broadband monochromators on the basis of highly oriented pyrolytic graphite (HOPG) crystals and high-resolution monochromators using laterally graded SiGe crystals.

HOPG is an artificial crystal with a mosaic structure. Typical values of the mosaic spread amount to approximately 0.4°. HOPG crystals can be used in the wide energy range from 2 up to 100 keV and more, revealing a peak reflectivity of 40–50% at 10–30 keV. One distinguishing feature of HOPG crystals is the largest integral reflectivity among all known crystals, which is the direct outcome of the large mosaic spread. Another important property is the principal possibility to produce HOPG crystals as thin flexible foils, which can be stacked on a substrate of arbitrary form. In this way different focusing geometries can be realized. Presently HOPG optics belong to standard elements of focusing monochromators with high intensity output and moderate energy resolution.

The use of materials with spatial variation of optical parameters (absorption, refraction index, etc.) is well known. A large number of X-ray optical devices on the basis of multilayers and crystals have been designed in recent years employing this phenomenon [1]. *Laterally graded crystals* are to be used as neutron-, $\gamma$-ray and X-ray monochromators when spectral intensity of a divergent beam has to be enhanced relative to spectral intensities from perfect periodic crystals [2]. They also gain importance with the development of new high-brilliance X-ray synchrotron radiation sources and are optimized for a very small source size of about ten to twenty micrometers and a vertical beam divergence less than $5 \times 10^{-4}$ rad.

Laterally graded aperiodic crystals on the basis of a SiGe mixed crystal represent a novel type of optical elements for synchrotron radiation applications in the X-ray region [3]. The lattice parameter of such gradient crystals containing up to several atomic percent (at. %) of Ge in a Si matrix changes almost linearly along the plane of diffraction. Thus the variation of the Bragg angle of a divergent incident beam on the crystal can be compensated. This opens up the possibility to operate a crystal monochromator with nearly crystal-limited resolution ($\lambda/\Delta\lambda \sim 5 \times 10^3$–$15 \times 10^3$) in the whole energy range above 2 keV making the most of the full vertical synchrotron source divergence working and avoiding a collimating premirror. Simultaneously, the reflected spectral intensity can be increased considerably as compared with a conventional Si-crystal monochromator [4].

The development of multilayer interference mirrors for the nanometer range providing efficient reflection at angles close to normal incidence [5] was a great step forward. These mirrors are able to support ultra-high power X-ray beams and can be also used for broadband high-flux monochromators providing an energy resolution in the order of $\lambda/\Delta\lambda \sim 25$ in the energy range 100 eV–15 keV.

Further development of the microstructure methods, in particular the fabrication of planar submicron structures with sizes in the order of the X-ray wavelengths, as well as the deposition and growth of thin films of different materials, enabled manufacturing of diffraction optical elements of the nanometer range.

With the development of nanotechnology, mostly for microelectronics applications, the way was found to create nanostructures and multilayer films for effective control of X-rays and achieving submicron spatial resolution.

These are mainly Fresnel optical elements (zone plates) with the lateral resolution up to 15 nm [6] and diffraction gratings in combination with natural crystals or artificial multilayer structures. The way to produce metal structures is to generate a zone plate pattern by electron-beam lithography. This is followed by dry or wet etching to transfer the pattern onto a highly cross-linked copolymer, which is filled by electroplating with gold or nickel [7]. Free-standing structures are obtained by removing a supporting layer or a substrate. Sliced zone plates are produced by the successive deposition of materials with different properties onto a rotating wire [8].

Gold zone plates are extensively used in hard X-ray microprobes with a submicron resolution at ESRF [9, 10], ALS [11], APS [12], and ELETTRA [13]. Using a gold structure and a synchrotron radiation beam it becomes possible to achieve spatial resolution down to 150 nm at the photon flux of about $10^9$ photons per second. Kinoform (blazed) structure was also realized. With a three-step profile efficiency as high as 45% at 8 keV photon energy was experimentally obtained [14]. The photon flux density gain of 15,000 was achieved. Zone plates made of Ni [15], Ta [16], and Si [17, 18] to improve diffraction efficiency in a particular energy range have also been successfully tested. Another direction of zone plate technology, "sliced" zone plates (exploiting Cu–Al multilayer structure), also found extensive applications at SPring-8 [19]. The NiCr–$SiO_2$ sliced zone plates were also tested in the energy range of 4–6 keV. Experimental efficiency of 15% at 25 keV and 0.3 μm resolution was obtained. Zone plates as focusing elements and X-ray waveguides as sources of nanometre size are recognized as the main optical elements in the nanoworld. On the other hand, during the last decade, conventional zone plate technology has reached the theoretical limit of spatial resolution. Volume diffraction effects in the outer zones with sizes comparable to X-ray wavelengths were found to be the fundamental limitation of zone plate resolution. Reducing wavelengths towards hard X-rays increases the zone plate thickness and leads to the same volume of diffraction effects and resolution limitation as for low-energy zone plates.

However, the use of transparent zone plates comes up against a number of fundamental limitations. Primarily, it is the above-mentioned limitation on resolution. In addition, the insufficient mechanical and radiation stability of zone plates makes it impossible to apply them to very powerful sources. These problems can be solved using the principles of three-dimensional diffraction in combination with reflection, which leads to the so-called reflection zone plates. The present state of developing new optics for X-ray beams calls for creating effective focusing elements with the structure of two- and three-dimensional Fresnel zones: combined microstructure X-ray optics [20] or Bragg–Fresnel optics [21, 22], graded multilayer structures and multilayer gratings. Recently, the new development of graded crystals was considered [23] which allowed simultaneously to focus the beam and to enhance spectral flux at the sample by several orders of magnitude. All these optical elements are united by the basic principle of Bragg, Bragg–Laue, or Bragg–Fresnel diffraction on artificially made volume structure and differ from the other elements by the unique combination of useful properties.

The first successful demonstrations of the Bragg–Fresnel principle were reported in 1985–1986 using multilayer [20, 22] and crystalline [21] substrates. The general principles of Bragg–Fresnel diffraction were first formulated in the work of Aristov [24]. The use of Fresnel focusing in combination with total external reflection was demonstrated in 1994 [25]. Since that time Bragg–Fresnel lenses have been used at several synchrotron radiation facilities to construct microprobes [26, 27] imaging beam monitors [28] and time-resolved systems [29]. The theory of Bragg–Fresnel optics has already been published in several works [7, 30]. Unfortunately, the main advantage of the Bragg–Fresnel optics, namely the combination of a monochromator and a focusing element in the same device, restricts the field of its potential applications. The necessity to design an optical element for a particular geometry and a fixed energy conflicts with the desired flexibility of the experimental arrangement and limits the number of possible experimental methods mainly to microfluorescence analysis and microdiffraction. The situation is additionally aggravated by the complexity of the Bragg–Fresnel optics technology and the very high cost of its production.

Modern technology for Bragg–Fresnel and Fresnel reflection optics is based on evaporating or sputtering metals onto surfaces of crystals and multilayers [31]. The fabrication and successful tests of a synthesized X-ray hologram made with Ni phase-shift layer on a surface of Si [111] crystal was reported [32]. A linear Bragg–Fresnel lens placed onto the second crystal of a double-crystal monochromator was tested and will serve as a basic sagittal focusing element for the small angle scattering facility at the BESSY microfocus beamline project [33].

*Capillary X-ray optics* has been successfully applied with conventional X-ray sources. Straight glass monocapillaries can effectively transport X-radiation from the source to the sample thus increasing radiation intensity on the sample [34]. Special polycapillary arrays with curved channels can be used for transforming divergent radiation from a point source into a quasiparallel beam or for focusing a divergent beam onto a small spot [35, 36]. Straight polycapillary arrays have been used for X-ray imaging and for beam splitting and filtering [37]. Capillary optics in synchrotron beamlines is mainly applied for focusing radiation onto a spot of μm and sub-μm size by means of tapered monocapillaries. The first experimental results on the beam focusing and filtering properties of mono- and polycapillaries using synchrotron radiation were obtained at BESSY (Berlin) [38] and LURE (Orsay).

Recently, elliptically bent *Kirkpatrick-Baez mirrors* have been used to produce sub-micrometre size X-ray beams [39, 40]. This optics is achromatic and has, in comparison with capillaries, relatively long focal distance. The last property can be very important for microfluorescence applications in special environments, when a sample is placed in a chamber filled with gases having a low or high temperature.

*Refractive X-ray optics* in a form of compound refractive lens appeared in 1996 and was successfully used with the third-generation synchrotron radiation sources [41]. This kind of optical element is ideal for high-energy X-ray undulator radiation characterized by low divergence in both vertical and horizontal directions.

The great majority of X-ray microscopes or microprobes use high-energy photons as projectiles. By using diffraction X-ray optics, photon microprobes have become available not only as laboratory instruments, but also at synchrotron radiation facilities with a resolution down to 100 nm [42].

## 3.2 Mirror Optics

*V. Arkadiev and A. Bjeoumikhov*

### 3.2.1 Total External Reflection Mirrors

X-ray optical systems based on a total external reflection of radiation from a metal surface are now most widely used. The main principle of these optics is based on the physical nature of optical properties of materials in the X-ray energy range: the refractive index of all materials is less than the in-vacuum value. Consequence from this fact is the mirror reflection of X-rays from the super-polished surfaces at angles less than the so-called "critical angle". The reflection coefficient is very high and reaches 90%, but due to the very small value of the refractive index "delta" the critical angle is also small, in the order of 1–10 mrad at typical X-ray energies. Reflection at grazing angle is used in different optical devices, such as capillary optics, focusing systems. To provide reasonable angular acceptance of these optics, mirrors must be very long and therefore they are very expensive. An example of an X-ray mirror is installed at the BESSY beamline BM 9.2. This mirror is used to focus X-rays on the sample to provide a high flux density. The mirror length is 1.4 m, the surface is polished to a quality of 0.4 nm RMS and coated with Rhenium of 60 nm thickness. The effect of total external reflection of X-rays is used in different types of X-ray optical elements and will be discussed below in detail.

### 3.2.2 Capillary Optical Systems

Very often large intensities of the primary beam are necessary for performing X-ray fluorescence analysis. This is usually the case when elements with a very low concentration in a sample are to be detected. Another typical example is local fluorescence analysis with a high spatial resolution (e.g., below 100 µm). In this case primary radiation can be collimated with a system of pinholes for obtaining tiny illuminated spots on a sample. As a result, only a very small part of the primary beam intensity is used for exciting fluorescence radiation.

In both cases, intensity sufficient for the analysis of the exciting radiation is normally achieved by using very powerful X-ray tubes, typically with a rotating anode.

Alternative methods of increasing the beam intensity on a sample employ focusing systems, which can capture radiation emitted from a source within a relatively large solid angle and concentrate the radiation captured on a sample. In this way a more efficient use of the emitted radiation is realized allowing to obtain such intensities within a small focal spot which cannot be achieved with collimated direct beams from powerful X-ray tubes.

The main difficulty of the second approach is connected with the fact that X-ray beams are difficult to control because of a negligibly small reflectivity of all substances in the X-ray range at large angles of incidence. Only grazing incidence mirrors (see e.g., [42, 43]) have relatively large reflection coefficients but the aperture of these mirrors is small due to extremely small values of the angle of the total reflection. Development of refractive optics [41] analogous to the optics of visible light faces also a lot of problems, first, due to a small X-ray refraction ability of all substances and, second, due to a strong absorption of X-rays in matter. Refractive lenses are used for focusing synchrotron radiation but their application in laboratory set-ups is hardly comprehensible at present. On the other hand, diffraction and interference elements such as crystals, multilayers, gratings, and zone plates [44, 45] are spectral selective and therefore are not suitable for concentrating X-ray radiation within a wide energy range. The mentioned difficulty was partially overcome with the help of capillary optics, which has been developed intensively during the last 10–15 years [46].

## Physical Foundations and Principal Possibilities of X-ray Capillary Optics

It is well known that interaction of X-ray radiation with matter can be described by means of a refraction index with a real part smaller than 1:

$$n = 1 - (f_1 + i f_2) N r \lambda^2/(2\pi) \equiv 1 - \alpha/2 + i\beta/2 \quad (0 < \alpha, \beta \ll 1), \tag{3.1}$$

where $N$ is the concentration of atoms in medium, $r$ is the classical electron radius, $\lambda$ is the radiation wave length and $f_1 + i f_2$ is the forward scattering amplitude of a photon on an atom. A small deviation of the refractive index from the value 1 explains low reflection and refraction abilities of matter in the X-ray range. On the other hand, the fact that the real part of a refraction index is smaller than 1 means that the optical density of any substance in the X-ray range is smaller than that of vacuum and the phenomenon of total external reflection can take place at grazing angles of incidence [47, 48]. Parameter $\alpha$ in the expression (3.1) determines the value of the critical angle of total external reflection $\theta_{\mathrm{cr}} = \alpha^{1/2}$, while the imaginary part of the refraction index

$\beta$ describes absorption of radiation in substance: $\beta = \lambda\mu/2\pi$, where $\mu$ is a linear absorption coefficient. The values of parameters $\alpha$ and $\beta$ can be calculated for different energies using the tables of scattering amplitudes [49]. The typical values of these parameters lie in the range of $10^{-4}$–$10^{-8}$ for $\alpha$ and $10^{-4}$–$10^{-10}$ for $\beta$, so that the critical angle of total external reflection $\theta_{cr}$ is of the order of several mrad. The reflection coefficient for an X-ray beam falling at an angle $\theta$ onto a solid body surface can be described by the Fresnel formula [50]:

$$R(\theta) = \frac{(\theta - \theta_1)^2 + \theta_2^2}{(\theta + \theta_1)^2 + \theta_2^2},$$

where

$$\theta_1 = 2^{-1/2}[((\theta^2 - \alpha)^2 + \beta^2)^{1/2} + \theta^2 - \alpha]^{1/2},$$

$$\theta_2 = 2^{-1/2}[((\theta^2 - \alpha)^2 + \beta^2)^{1/2} - \theta^2 + \alpha]^{1/2}.$$

Figure 3.1 shows a typical angle dependence of the reflection coefficient $R$ on a smooth glass surface at different energy values. Reflection coefficient $R$ is close to unity at $\theta < \theta_{cr}$ and decreases rapidly to zero at $\theta > \theta_{cr}$. The range of incidence angles below the critical value $\theta_{cr}$ is exactly the sphere of validity for the optics of total external reflection and, in particular, for capillary optics. The critical angle $\theta_{cr}$ depends on the composition of the reflecting surface and is larger for heavy elements. Its value decreases as $E^{-1}$ with the increase of the photon energy $E$. Figure 3.2 shows the energy dependence of the critical angle for glass (which contains the relatively light elements oxygen and silicon in its composition) and for platinum (heavy element).

A single reflection of an X-ray beam falling at grazing incidence onto a smooth surface changes its propagation direction. This change is very small because it does not exceed $2\theta_{cr}$. The main idea of the capillary optics is to use

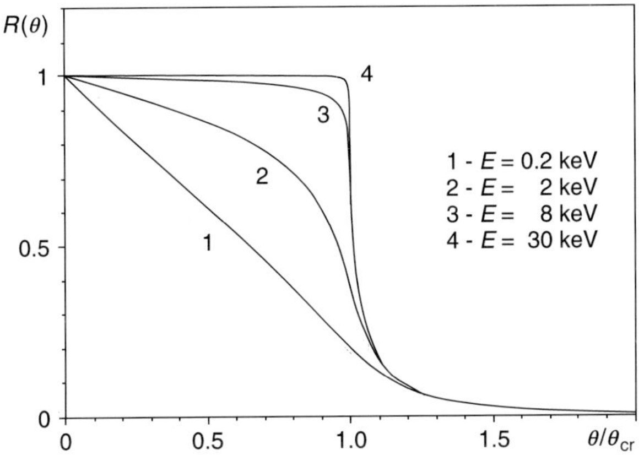


**Fig. 3.1.** Angle dependence of the reflection coefficient on a glass surface

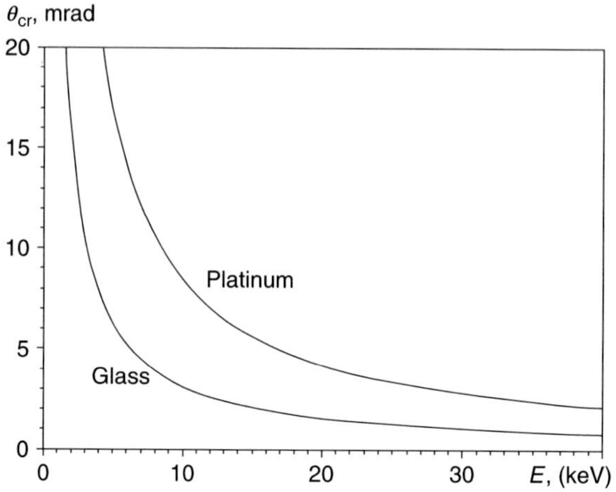


**Fig. 3.2.** Energy dependence of the critical angle of total external reflection

multiple reflections of X-ray beams on specially curved surfaces [35, 51, 52]. As a result of $N$ subsequent reflections, a beam can be reflected to a relatively large angle of the order of $2N\theta_{\rm cr}$. Naturally there is some radiation absorption at every reflection. Although a single reflection coefficient is close to unity, a large number of subsequent reflections can in principle lead to significant intensity reduction. Therefore, it is reasonable to estimate the intensity losses which take place while reflecting beams to large angles in order to understand the principal possibilities of the optics of multiple reflections at grazing incidence [53, 54]. In the region of total external reflection $\theta < \theta_{\rm cr}$ the value of the reflection coefficient can be approximated by the expression

$$R(\theta) \approx \exp\{-2\beta\theta(\alpha - \theta^2)^{-3/2}\},$$

which is especially suitable at small incidence angles $\theta \ll \theta_{\rm cr}$. To reflect a beam to a large angle $\Phi$ one needs $N \approx \Phi/2\theta$ reflections, so that the resulting coefficient of multiple reflections equals to

$$R_\Phi \approx R(\theta)^N \approx \exp\{-\Phi\beta(\alpha - \theta^2)^{-3/2}\}.$$

In particular, for $\theta \ll \theta_{\rm cr}$ one gets a very simple formula:

$$R_\Phi \approx \exp\{-\Phi\beta\alpha^{-3/2}\}. \tag{3.2}$$

The last expression does not depend on the incidence angle and on the number of reflections and contains only material constants and the angle of reflection. Therefore it can be used as a characteristic, which describes the intensity losses while reflecting X-ray beams to large angles. The values of $R_\Phi$ calculated for different reflectors are presented in Fig. 3.3 as functions of the photon energy.

In the region of relatively small energies reflectivity $R_\Phi$ depends strongly on the mirror material and on the photon energy due to the presence of

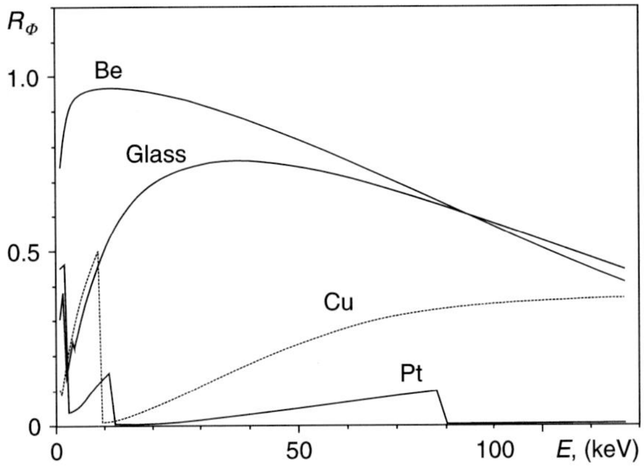


**Fig. 3.3.** X-ray reflectivity for turning radiation to the angle $\Phi = 0.25$ rad

absorption lines. The best reflectors are light elements because their K-lines of absorption lie at still smaller energies. Above K-lines of absorption reflectivity of all substances increases monotonically, reaches its maximum and then begins to decrease. Therefore, every substance has its natural energy range where its reflectivity is maximal. The observed regularity can be explained by the following simple considerations. From the relation (3.1) and the optical theorem follows: $\alpha \sim E^{-2}$, $\beta \sim E^{-1}\,\sigma(E)$, so that

$$\beta\alpha^{-3/2} \sim E^2\sigma(E), \tag{3.3}$$

where $\sigma(E)$ is the total cross-section of photon–atom interaction. Just above the K-edge of absorption the main contribute to the cross-section $\sigma(E)$ comes from the photo-effect. The value of the photo-effect cross-section $\sigma_F(E)$ decreases rapidly with the increase of the photon energy: $\sigma_F(E) \sim E^{-7/2}$, so that the value of $\beta\alpha^{-3/2}$ decreases as $E^{-3/2}$ and the reflectivity $R_\Phi$ becomes larger according to the expression (3.2). At still larger photon energies the cross-section $\sigma(E)$ practically coincides with the cross-section of the Compton scattering and the value of $\beta\alpha^{-3/2}$ begins to decrease due to the presence of the factor $E^2$ (see (3.3)) which leads in its turn to the reduction of the reflectivity $R_\Phi$ (3.2).

It is worth mentioning that the position of the reflectivity maximum moves towards higher energies with the increase of the order number of the element, whereby its peak value decreases. Therefore, relatively light elements reveal themselves as the best reflectors in the wide energy range. In particular, different sorts of glass (main components: silicon and oxygen) can be used as suitable substances for the optics of multiple reflections. Heavy elements may have some advantages in the high-energy range due to a larger angle of the total external reflection especially when the number of reflections remains small.

These simple considerations demonstrate the principal possibility of turning X-ray beams to relatively large angles by using multiple reflections on specially curved surfaces at grazing angle incidence and form the theoretical basis of the capillary optics.

## Main Elements of the Capillary Optics

Preparation of a surface for total external reflection reveals itself very often as a very difficult problem. On the one hand, the surface must be very smooth. On the other hand, the form of the surface must guarantee that the incidence angle remains smaller than the critical angle of the total external reflection. The demands on grazing incidence mirrors are so high that these mirrors are usually very expensive. It is no wonder that preparation of a surface form to multiple subsequent reflections could be a much harder problem.

The main idea of X-ray capillary optics is to use hollow channels in glass for transporting X-ray radiation [55–57]. Such an approach enables to solve several difficult problems at once. First, the surface of glass is smooth enough for reflections with a high reflection coefficient. Second, the form of the channels usually guarantees that the second and all subsequent reflections take place automatically at incidence angles smaller than the critical angle $\theta_{\rm cr}$ if the incidence angle at the first reflection does not exceed $\theta_{\rm cr}$. Third, there is an established technology of manufacturing glass capillaries.

The simplest element of the X-ray capillary optics is a straight cylindrical monocapillary. Due to the axial symmetry, after the first reflection at an incidence angle $\theta < \theta_{\rm cr}$ all the subsequent reflections occur also at the same angle $\theta < \theta_{\rm cr}$ so that the radiation captured at the capillary entrance is transported further along its length with minimum losses. Usually a direct beam from an X-ray source is divergent and its intensity falls down as the inverse square of the source–sample distance. In contrast to it, radiation captured by a capillary is confined inside this capillary and can be transported to large distances without spreading. This can result in considerable radiation density increase on a sample. The efficiency of applying a capillary is usually described by the notion "intensity gain" relative to a pinhole collimator of the same diameter placed instead of a capillary at its end. Intensity gain originates from a larger angular aperture of a capillary as compared with a pinhole and can be easily estimated in the following way. The angular aperture of a capillary is $2\theta_{\rm cr}$, while the aperture of a pinhole collimator is $d/R$, where $d$ is a collimator (capillary) diameter and $R$ is a source–collimator distance. Neglecting intensity losses at reflections, one can obtain the following simple expression for the intensity gain:

$$\text{gain} \approx \left(\frac{2\theta_{cr}R}{d}\right)^2 .$$

The average number of reflections $N_{\rm cap}$ inside a capillary is proportional to $\theta_{\rm cr}R/d$, so that the intensity gain increases rapidly with the number of reflections:

$$\text{gain} \sim N_{\rm cap}^2 .$$

The value of the spot size on a sample can be estimated according to the following simple formula:

$$s_2 \approx d + 2\theta_{\rm cr} F_2, \tag{3.4}$$

where $F_2$ is the capillary–sample distance. Exit beam divergence leads to a certain spread of the beam after the capillary end. The relative contribution of the second term can become dominant for small capillary diameters. In this case it is necessary to work with small capillary–sample distances.

In many applications a single-pinhole collimator (or a double-pinhole collimator) can be easily substituted for the monocapillary of the same diameter $d$. In this case, the monocapillary performs the following main functions:

1. It collimates the beam spatially (exit beam size equals to a capillary diameter $d$);
2. It collimates the beam angularly (exit beam divergence is of the order of the critical angle of total external reflection $\theta_{\rm cr}$);
3. It can produce significant intensity gain on the sample relative to conventional pinhole collimators.

One must remember that intensity gain is achieved at the expense of a certain increase of the beam divergence up to $\sim \theta_{\rm cr}$ but in many cases considerations of large total intensity on a sample prevail over this factor. Monocapillaries can be successfully used when small irradiated spots on the sample are needed while the distance between the source and the sample is large and cannot be reduced. This is a typical situation in diffractometry where the positions of the source and the sample are fixed by the diameter of the goniometer [58]. In some investigations (e.g., texture investigations) one does not need very high angular collimation of the beam and angular divergence $\sim 4$ mrad for Cu K$\alpha$-line ($E = 8.0$ keV) and $\sim 2$ mrad for Mo K$\alpha$-line ($E = 17.4$ keV) obtained with a glass capillary is quite suitable. In the last several years, cylindrical monocapillaries have found their wide application in diffractometers produced by GE Inspection Technologies Ahrensburg GmbH Co. KG (Ahrensburg, Germany), PANalytical B.V. (Almelo, Netherlands) and Bruker AXS (Karlsruhe, Germany) as a convenient substitute for conventional pinhole collimators. Typical intensity gains are 3–10 for Cu-anode fine-focus tube (effective anode spot 0.4 mm $\times$ 0.8 mm) with the capillary diameters 0.1–1.0 mm.

Monocapillaries with small diameters can be also applied as a substitute for pinhole collimators in local fluorescence analysis for obtaining small excitation spots on the sample [59]. They can give significant intensity gain, especially when used in combination with a microfocus tube (diameter of the anode spot is of the order of capillary diameter). Of course, focusing capillary optics (tapered capillaries and X-ray lenses) could give still larger intensity gain and monocapillaries can be regarded only as a first step in this direction, especially taking into account their simplicity and relative cheapness.

However, obtaining microspot beams with monocapillaries can meet certain difficulties. First of all, the number of reflections inside a capillary becomes large and absorption losses increase significantly. In this case a special combination of a monocapillary with a pinhole may be useful [60]. One can use a monocapillary with a large diameter $\sim 0.3$–$1.0$ mm and place a small pinhole $\leq 0.1$ mm at its end. A monocapillary of a large diameter captures radiation near the source and transports it to a required distance without spread while a small pinhole at its end makes the final collimation of the beam. Usually such a combination works more efficiently than direct application of a monocapillary with a tiny diameter.

In contrast to straight capillaries, bent capillaries not only transport radiation without spreading but also turn X-ray beams to the angles defined by the capillary bend. The angles of turning can reach relatively large values as was demonstrated in "Physical foundations and principal possibilities of X-ray capillary optics". Two main factors determine transmission of X-ray radiation through a bent capillary: radiation capture at a capillary entrance and intensity loss due to multiple reflections inside a capillary. The role of the second factor has already been discussed above where it was shown that the intensity losses are determined by the angle of turning. The first factor is specific for bent capillaries because the condition of the total external reflection is not automatically fulfilled for all incoming rays at the first reflection (Fig. 3.4a).

From the simple geometrical considerations one can easily obtain that all the rays of a quasiparallel beam entering a capillary fall onto its inner wall at incidence angles smaller than the critical angle $\theta_{\rm cr}$ only if the following condition is fulfilled [54]:

$$\gamma \equiv \frac{r\theta_{\rm cr}^2}{2d} \geq 1,$$

where $r$ is the curvature radius and $d$ is the capillary diameter. Otherwise (at $\gamma < 1$) a certain part of the beam strikes the channel walls at glancing angles larger than $\theta_{\rm cr}$ and is therefore lost at the very first reflection. Consequently at $\gamma < 1$ radiation is not totally captured by a bent capillary and the capillary cross-section is only partly filled with radiation (Fig. 3.4b). The distribution of radiation over the capillary cross-section in the form of "semimoons" was observed experimentally [54] at the capillary exit for various values of the parameter $\gamma < 1$. Parameter $\gamma \equiv r\theta_{\rm cr}^2/2d$ decreases fast with the increase of the radiation energy because $\theta_{\rm cr}^2 \sim E^{-2}$. Therefore radiation transmission through a bent capillary can become negligible in the high-energy region

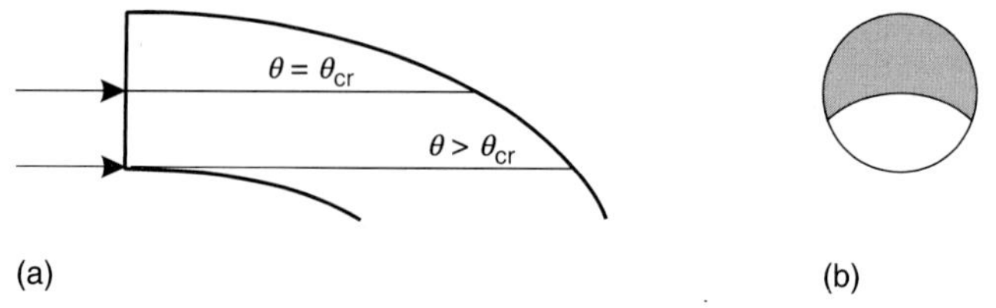


**Fig. 3.4.** Radiation capture for a bent capillary

due to very large losses at a capillary entrance. To preserve the value of the parameter $\gamma$ at the level $\gamma > 1$ within a certain energy range, one must choose capillaries with sufficiently small channel diameter $d$. The necessity of applying capillaries with small channel diameters for deflecting X-ray radiation limits possibilities of capillary optics in the high energy range. At present monocapillaries and polycapillaries with channel diameters down to ca. 1 μm are available.

Although reflecting X-ray beams to large angles may seem to be very attractive, bent capillaries have hardly found their application as independent optical elements up to date. Instead they serve as the most essential part of complicated optical elements—X-ray concentrators, which are also often called as "X-ray lenses" [61]. A typical X-ray lens contains a large number of specially arranged individual channels (capillaries). Usually the initial parts of all channels are oriented towards an X-ray source while the ends of the channels can be either mutually parallel (Fig. 3.5a) or directed towards one and the same spot on a sample (Fig. 3.5b). In the first case a divergent beam from a source is transformed into a quasiparallel one, while in the second case divergent radiation becomes convergent and is concentrated on a sample within a small spot. The first type of this polycapillary optics is often called "semilens", while the second type is usually known as "full lens" or simply "lens", although both elements have hardly anything to do with real optical focusing based on a point-to-point correspondence. Both types of X-ray lenses have a large angular aperture, which can many times exceed the value of the critical angle of the total external reflection. X-ray semilenses form quasiparallel beams with a divergence of the order of the critical angle $\theta_{\mathrm{cr}}$ and with a high angular radiation density. They find their main application in diffractometry [62, 63]. X-ray full lenses are in fact X-ray concentrators, which enable obtaining small irradiated spots on a sample with a very high spatial radiation density. Their main application lies in local X-ray fluorescence analysis [64]. One of the most important characteristics of X-ray full lenses is again the so-called "intensity gain", which is defined as the intensity increase in the focal spot of the lens relative to the intensity of the direct beam through a corresponding

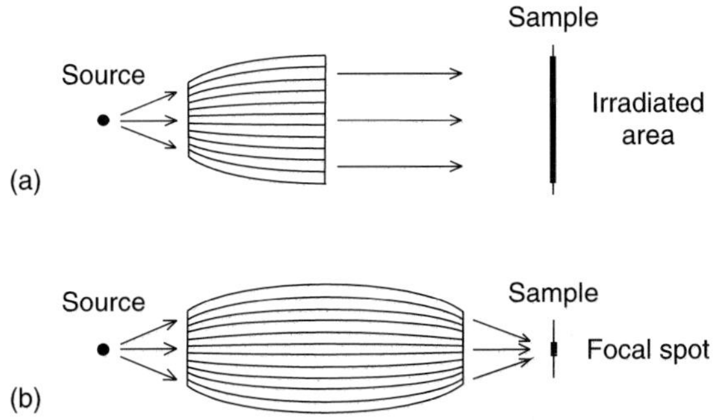


**Fig. 3.5.** Different types of X-ray lenses. (a) "semilens"; (b) "full lens"

pinhole. At present intensity gains obtained with some types of X-ray lenses can achieve the value of several thousand times in the energy range of about 5–20 keV. These values exceed significantly typical intensity gains obtainable with straight and tapered monocapillaries due to much larger aperture values of polycapillary optics.

The notion of intensity gain is very expressive but can sometimes be misleading because it does not take into account the distance to a sample where the gain is achieved. The situation can be better described with the help of the "effective distance" which is defined as the source–sample distance at which the direct beam gives the same intensity as in the focal spot of a lens. The effective distance of a lens can be calculated according to the formula:

$$D_{\mathrm{eff}} = \frac{D}{\sqrt{\mathrm{gain}}}, \tag{3.5}$$

where $D$ is a source–sample distance with a lens. The size of the spot on a sample can be estimated according to the expression (3.4), where $d$ is now the diameter of an individual channel. The contribution of the second term is usually dominant for polycapillary lenses with channel diameters below ca. 10 µm, so that the size of the focal spot is determined by the divergence of the beam from individual channels. Typical spot sizes obtained in the energy range 8–17 keV are of the order of ca. 20–100 µm for the working distances $F_2$ of ca. 10–40 mm. Further reduction of the spot size can be achieved by manufacturing lenses with smaller working distances $F_2$ [65, 66].

As mentioned above, a polycapillary X-ray lens has a relatively wide passband with a maximum at some fixed energy. This optimal energy depends on the lens parameters, first of all, on channel diameters. One can achieve more homogeneous transmission in a wide energy range, combining channels with different diameters within the same lens. Such lenses with a complicated polycapillary structure containing different channel diameters are being manufactured now [67].

A spatially resolved X-ray fluorescence analysis can also be done by the use of a capillary optics on the detector side. For this a polycapillary conic collimator (Poly-CCC), proposed in [68], can be applied. In this approach focusing is not necessary for obtaining space resolution and a primary beam irradiates the whole of a sample, while a Poly-CCC is placed between a sample and a detector. Polycapillary "sees" only a small spot on a sample and collects fluorescence signal from it. Spatial resolutions between 150 and 20 µm were experimentally obtained [68, 69]. On the other hand, the combination of a focusing optics and a Poly-CCC made it possible to realize 3-D micro-XRF with spatial resolution not only laterally on the sample surface, but also into the sample depth [70] (see also section "Micro-XRF").

While polycapillary optics concentrates X-ray radiation by means of multiple reflections and does not possess true focusing properties, certain types of tapered monocapillaries can really focus X-ray radiation by means of single reflections. Two main types of focusing capillary optics are elliptic and parabolic monocapillaries [71–73].

An elliptic capillary has two foci and must be aligned relative to a source so that the position of a source coincides with the first focus. The radiation captured by the capillary is reflected towards the second focus, which lies on a sample. The angular aperture of elliptic capillaries is determined by the critical angle of the total external reflection as it is typical for optics with single reflection. Therefore, total intensity on a sample is usually smaller than that obtained with polycapillary optics due to a smaller solid angle of radiation capture. On the other hand, the size of a spot can also be much smaller due to pure focusing. Asymmetric cutting of an elliptic capillary so that a source–capillary distance is much larger than a capillary–sample distance can additionally reduce the size of a focal spot on a sample, because in this case an elliptic capillary images a source onto a sample with some reduction factor in accordance with optics laws. In this way it is possible to obtain focal spots on a sample smaller than the size of a source. Elliptic capillaries are especially suitable for focusing soft X-ray radiation with energies below ca. 3 keV due to the large value of the critical angle $\theta_{cr}$, which determines the value of the aperture. Polycapillary optics has in the soft X-ray range some limitations connected with the increasing absorption, which becomes especially significant at multiple reflections. Besides, the size of the spot for an X-ray lens is dominated by the large exit divergence (the second term in (3.4)) and can be as large as several hundreds of μm in the soft energy range.

Parabolic capillaries have one focus and are suitable for focusing quasi-parallel beams, especially on synchrotron beam lines. Spots below 1 μm are reported to have been realized with an intensity gain of several tens. The required working distances $F_2$ are well below 1 mm.

The main advantage of elliptic and parabolic capillaries for focusing radiation is the existence of a focus, which lies at some distance from the capillary exit so that reasonable capillary–sample working distances can be realized. Other types of tapered capillaries (e.g., conical capillaries) are less suitable for obtaining microspots because they have no focal point at all, so that radiation, confined in a capillary, spreads very quickly after leaving it.

### Designing Capillary Optics

Designing capillary optics involves a number of steps. The choice of the optimal type of an optical element is dictated first of all by the concrete application under consideration: energy range, source size, source–sample distance, desired size of a focal spot on a sample, etc. Large intensity gains are achievable when the capture angle of an optical element is large. As a rule of thumb, in this respect polycapillary optics has potential advantages over monocapillary optics because its angular aperture can considerably exceed the value of the critical angle of the total external reflection, which determines the angular aperture of a monocapillary.

The angular aperture of a polycapillary lens can be increased either by increasing its entrance diameter or by reducing its focal distance to a source. The first way meets certain technological problems connected with manufacturing polycapillary optics with a large diameter. At present the second way of reducing the focal distance is preferred. However, small focal distances put several restrictions on a source size. First of all, each channel of an X-ray lens can capture radiation only from a spot with the dimensions of ca. $d + 2\theta_{\mathrm{cr}}F_1$ on a source. For the efficient use of the total source its size $s$ must be smaller than this spot, i.e., fulfil the condition $s < s_1$, where

$$s_1 = d + 2\theta_{\mathrm{cr}}F_1. \tag{3.6}$$

Usually microfocus X-ray tubes are needed for polycapillary optics with a small focal distance $F_1$. When, on the other hand, the optics–sample distance $F_2$ is also small ($F_2 \approx F_1$), one can concentrate X-ray radiation on a spot with the dimensions of the order of a source size $s$. As a result, such a combination of compact polycapillary concentrators with low-powered microfocus X-ray tubes can produce microspots on a sample with intensities that are usually achievable only with high-powered X-ray tubes.

These simple considerations show the necessity of optimizing not only parameters of capillary optics relative to a source but also parameters of a source relative to capillary optics. The best results are achieved when all the components of a system (source, optics, detectors) are mutually optimized.

Radiation transmission through mono- and polycapillaries can be estimated by computer simulation. The calculation algorithm is based on the following main assumptions [74].

*Physical assumptions.* Radiation transmission through capillaries can be completely described within geometrical optics. Wave corrections are negligible since $\lambda/d \ll 1$, where $\lambda$ is radiation wave length and $d$ is channel diameter. Reflection of an X-ray beam from capillary walls obeys the Fresnel laws. Specular reflection coefficient from the ideal surface can be calculated according to the Fresnel formula. Radiation absorption enters this formula through the nonzero imaginary part of the glass refraction index.

*Mathematical assumptions.* The underlying model of ray tracing is basically a three-dimensional one. One starts with a random point ($x_0$, $y_0$, $z_0$) on a source and a random point ($x_{\mathrm{ent}}$, $y_{\mathrm{ent}}$, $z_{\mathrm{ent}}$) on a channel entrance. These two points define an X-ray initial direction, which can be described by a unit vector ($\boldsymbol{p}_{0x}$, $\boldsymbol{p}_{0y}$, $\boldsymbol{p}_{0z}$). At the first step one finds the incidence point ($x_1$, $y_1$, $z_1$) of the X-ray on the channel wall and the new direction of propagation ($\boldsymbol{p}_{1x}$, $\boldsymbol{p}_{1y}$, $\boldsymbol{p}_{1z}$) after the reflection. This procedure is multiple repeated for all subsequent points of reflection ($x_n$, $y_n$, $z_n$) along the channel, till the channel end is achieved.

The problem of finding a point of reflection is therefore reduced to a mathematical problem of finding an intersection of a straight line with a 2-dimensional surface. For some special forms of capillaries (e.g., straight and uniformly bent cylindrical capillaries, conical, elliptical and parabolic capillaries, etc.) explicit analytical solution is possible. In general case of capillaries with arbitrary form and cross-section iterative algorithms were elaborated which proved to be efficient.

Simulation algorithm should also contain corrections for possible imperfection of real capillaries. Usually two main effects are considered: roughness and waviness.

Roughness of the surface reduces the value of the specular reflection coefficient. The simplest way of taking roughness into account is to use the following corrected version for the reflection coefficient [45]:

$$R_{\mathrm{rough}}(\theta) = R(\theta)\,\exp\left(-4q^2\sigma^2 \sin^2\theta\right),$$

where $R(\theta)$ is the Fresnel reflection coefficient for the ideal surface, $\sigma$ – roughness root-mean-square height, $q = 2\pi/\lambda$ – radiation wave vector. More elaborate formulas take into account also the roughness correlation length [45, 75].

The effect of waviness can be explained in the following way [74, 76]. During its manufacture, a capillary is usually subjected to some uncontrollable perturbations (e.g., temperature fluctuations, deviations of pulling velocities from constant values, etc.). As a result, the radius of a manufactured capillary can experience smooth fluctuations along the capillary length. One can usually neglect the absolute value of the surface displacement from its averaged value but one cannot neglect the tilt $\omega$ of the surface because the value of the slope may become comparable to the critical angle of the total external reflection. Waviness is easily incorporated into the simulation algorithm by making correction to the direction of the reflected beam: the reflection angle $\theta'$ relative to the unperturbed surface is no longer equal to the incidence angle $\theta$ but acquires a random quantity $\delta\theta$:

$$\theta' = \theta + \delta\theta, \qquad \delta\theta \equiv 2\omega,$$

where $|\omega| < \omega_{\mathrm{max}}$. The effect of waviness is described by a single parameter $\omega_{\mathrm{max}}$, which is the maximal angle of the possible local tilt of the surface. The role of waviness increases towards higher energies and becomes significant when $\theta_{\mathrm{cr}}(E) \sim \omega_{\mathrm{max}}$.

Glass composition enters the calculations through the parameters $\alpha$ and $\beta$ of the glass refraction index. Usually light glass (e.g., borosilicate glass) is used for manufacturing capillary optics. But in some cases heavy glass (lead glass) may have advantages because of the larger value of the critical angle $\theta_{\mathrm{cr}}(E)$, although the absorption becomes also larger. Computer simulation can help to choose the best type of glass for an application under consideration. Heavy glass is better suitable for capillary optics with small number of reflections in the high-energy range.

### Performance of Capillary Optics

The first X-ray concentrator [35, 77] was assembled mechanically from 2,000 glass monocapillaries with an outside diameter of 0.4 mm and a channel diameter of 0.36 mm. The system was 98 cm long with equal focal distances of 5 cm and had an angular aperture of 23°. The required curvature of capillaries was ensured by a special supporting system, which consisted of seven plates with 2,000 holes each placed at certain distances one from another. The form of each capillary was fixed by passing every capillary through a corresponding hole in all the seven plates. As was expected, a focal spot of 1 mm was observed in the experiments at the distance of 108 cm from the source. The measured intensity gain in this spot was ca. 3,000 for the energy 4 keV and decreased rapidly towards higher energies (down to ca. 500 at 8 keV). Large values of the measured gains are not to be overestimated because they were obtained at very large distances from the source. According to the formula (3.5) the effective distance of the lens was ca. 2 cm for 4 keV and ca. 5 cm for 8 keV. In fact the average transmission coefficient through the capillaries was ca. 2.5% for 4 keV and only ca. 0.4% for 8 keV while the optimal energy was estimated to be at ca. 1.7 keV. Nevertheless, the first system demonstrated in practice the possibility of concentrating X-ray radiation over a broad spectral range by the methods of capillary optics. Later a number of X-ray concentrators were manufactured according to this technology, first of all for applications in the soft energy region (e.g., for photoelectron spectroscopy which uses Al $K\alpha$-line as the exciting radiation).

Another type of X-ray concentrators is presented by monolithic optics. Monolithic X-ray lenses are specially tapered polycapillaries, which contain thousands or millions of bent channels arranged according to the schemes in Fig. 3.5. The form and dimensions of monolithic optics are application dependent and can be diverse. Figure 3.6 shows concrete examples of assembled and monolithic X-ray lenses.

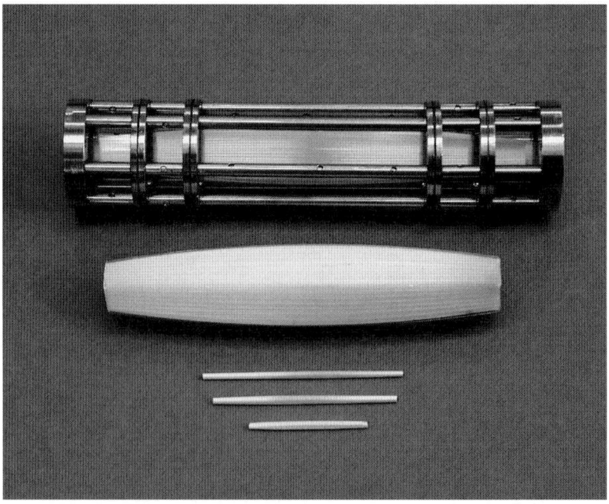

**Fig. 3.6.** Examples of X-ray lenses

Two large lenses were optimized and manufactured for the soft energy range, had the source–sample distance of 212 and 225 mm and the intensity gain of 76 and 57, respectively, for Al K$\alpha$-line at 1.5 keV [78]. The other three lenses were intended for applications in the higher energy range.

In contrast to mechanically assembled X-ray lenses, monolithic lenses can be very compact. The main geometrical parameters of a lens are as follows: $F_1$ is source-optics distance, i.e., the entrance focal distance, $L$ the length of the lens, $F_2$ the optics–sample distance, i.e., the exit focal distance. The optimum size of the source $s_1$ and the expected size of the focal spot $s_2$ are estimated according to (3.4) and (3.6), respectively. Since the size of the capillary channels is usually much smaller than the contribution of the beam divergence term in these formulas, the ratio between these lens parameters can be presented as

$$\frac{s_2}{s_1} \approx \frac{F_2}{F_1} . \tag{3.7}$$

Both the optimum source size and the focal spot size depend on the photon energy due to the energy dependence of the critical angle of the total reflection. However, their ratio (3.7) remains approximately constant.

For an optimum lens design the entrance focal distance $F_1$ should be chosen in such a way that the parameter $s_1$ corresponds to the anode spot size of the X-ray tube. The exit focal distance $F_2$ should ensure the required focal spot $s_2$ on the sample. Since the optics is to work efficiently in some energy interval, the entrance focal distance $F_1$ is optimized for the high-energy part of the spectrum. On the other hand, the exit focal distance $F_2$ has to be selected for the more important part of the spectrum.

There are numerous experimental results of the test measurements of lenses (see e.g., [79]). As an example, we present here the characteristics of two different monolithic lenses. The main geometrical parameters of both lenses are collected in Table 3.1, while the meaning of these parameters is displayed explicitly in Fig. 3.7. Two values are specified for the lens diameters: the external diameter and the diameter of the internal polycapillary core. The terms "minilens" and "microlens" are stipulated to denote the difference in

**Table 3.1.** Main geometrical parameters of two lenses

| | "minilens" | "microlens" |
|---|---|---|
| $L$ mm | 101.5 | 36.8 |
| $F_1$ mm | 47.5 | 39.8 |
| $F_2$ mm | 15.3 | 16.8 |
| $D_{\text{in}}$ (structure/shell) mm | 4.6/5.3 | 1.7/3.2 |
| $D_{\text{max}}$ (structure/shell) mm | 6.7/7.5 | 2.3/4.2 |
| $D_{\text{out}}$ (structure/shell) mm | 2.5/2.7 | 1.5/2.8 |
| $\Phi$ rad | 6.7/7.5 | 0.043 |
| $R = F_1 + L + F_2$ mm | 164.3 | 93.4 |

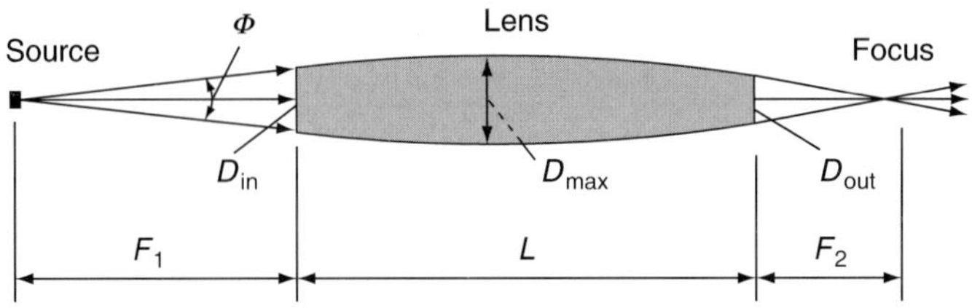


**Fig. 3.7.** Main geometrical parameters of an X-ray lens

the length and focal distances which leads in its turn to a different spot size on a sample. "Microlenses" are more compact and have smaller focal distances (less than 10 mm) which allows to obtain spot sizes below 20 µm. Typical spot sizes for "minilenses" lie in the range of 20–100 µm. Both lenses under consideration are asymmetric, the entrance focal distance being three times larger than the exit one. Therefore, a focal spot size on a sample is expected to be smaller than the source size. Table 3.2 contains the measured physical parameters of the "minilens": the focal spot size and the gain factor.

A microfocus X-ray tube with a molybdenum anode (spot size 50 µm) was used as source and a SDD X-Flash detector (manufacturer Bruker AXS Microanalysis GmbH, Germany) was used for detection of X-rays. The intensity distribution in the focal spot and the spot size were measured by scanning with a 5-µm pinhole at different energies. The intensity gain factor was calculated as the ratio between the intensities on a sample with and without the lens.

The values of the intensity gain are very large and correspond to the effective distance $D_{\mathrm{eff}}$ of less than 1.8 mm for the energy range 15–20 keV. Such intensities are hardly achievable without optics, because it is practically impossible to realize the anode–sample distance of 1.8 mm.

Figure 3.8 shows the exit spectrum of the "minilens" measured with a scattering target made of PMMA (plexiglas). The spectrum of the direct beam through a 50-µm pinhole at the same distance from the X-ray source is presented for comparison. Both spectra give a good idea of the intensity gain achieved. Besides, they demonstrate clearly how the lens modifies the primary spectrum. In this case a Si(Li) detector with an active area of 10 mm$^2$ was used for the measurement.

The measured characteristics of the "microlens" are shown in Table 3.3. Due to the smaller focal distance $F_2$, the spot size on a sample is also smaller. The effective distance $D_{\mathrm{eff}}$ is ca. 2.7 mm for the energy range 16.7–23.2 keV.

**Table 3.2.** Measured characteristics of the "minilens"

| E/keV | 3–5 | 5–7.5 | 7.5–10 | 10–15 | 15–20 | 20–25 | 25–30 |
|---|---|---|---|---|---|---|---|
| Spot size/µm | 38 | 40 | 41 | 41 | 35 | 31 | 33 |
| Intensity gain | 2,892 | 7,078 | 7,929 | 8,070 | 8,502 | 5,631 | 1,550 |

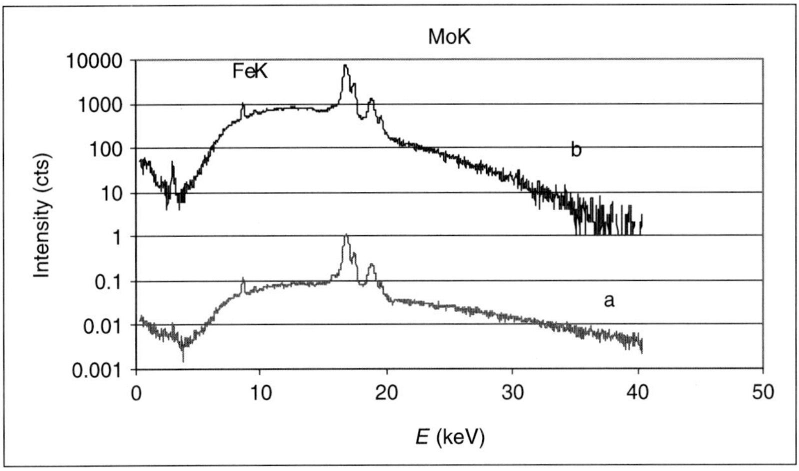


**Fig. 3.8.** Spectra obtained by scattering on PMMA. (a) direct beam; (b) beam at the exit of the optics. Measurement condition: 50 kV, 50 $\mu$A, Si(Li) detector

**Table 3.3.** Measured characteristics of the "microlens"

| E keV | 2.5–6.5 | 6.5–10.2 | 10.2–16.7 | 16.7–23.2 | 23.2–30.0 |
|---|---|---|---|---|---|
| Spot size μm | 25 | 25 | 22 | 20 | 20 |
| Intensity gain | 1,380 | 1,460 | 1,590 | 1,160 | 345 |

Figure 3.9 shows the energy dependence of the focal spot size measured and calculated according to (3.4), respectively. As discussed above, the focal spot size is expected to increase when radiation energy decreases. However, the measured values depend weakly on the energy and they decrease even below 4 keV. Only in the region of 25 keV does the measured value approach that of the calculated one. The result can be explained by the fact that X-rays with relatively large incidence angles become quickly extinct while propagating through the channels due to the stronger absorption. This effect leads to a certain "self-collimation" of the beam and is especially significant in the low-energy region, where absorption becomes stronger. In spite of large values of the critical angle at low energies, only photons with small incidence angles can transmit through the lens channels.

Some increase of the focal spot size in the energy interval 25–30 keV as compared to 20–25 keV in Fig. 3.9 is obviously connected with the edge effects at the pinhole. The effective size of the pinhole increases due to smaller absorption of high-energy photons in the pinhole material. To test this assumption, a spot scan with a 10-μm wire containing 94% Cu and 6% Sn was carried out (see Fig. 3.10).

This experiment was important to show if the so-called "halo effect" may be responsible for the increase of the focal spot at high energies. In many

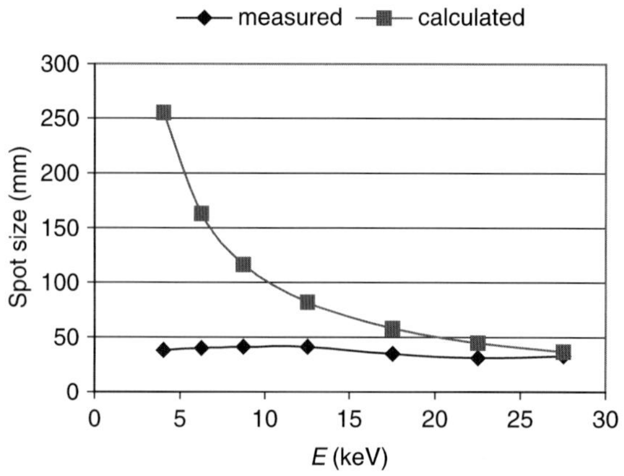


**Fig. 3.9.** Calculated and measured energy dependences of the focal spot size for the "minilens"

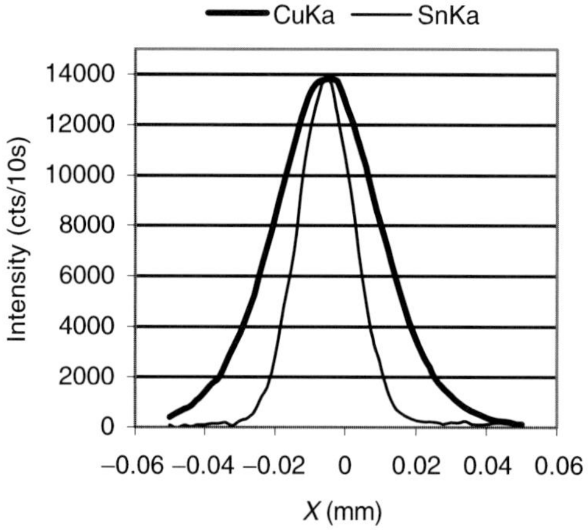


**Fig. 3.10.** Spatial distributions of Cu K$\alpha$ and Sn K$\alpha$-yields while scanning the lens focal spot with a 10- μm wire (Ua = 50 kV)

publications the existence of such a transmission halo around the main sharp focus of polycapillary lenses is mentioned. This effect occurs in the high-energy region and deteriorates the spatial resolution of the analysis of heavy elements [80]. The measured Sn K$\alpha$-distribution in Fig. 3.10 does not give any evidence of the "halo effect" for the lens under consideration: the focal spot does not become blurred for radiation energies above the Sn K-edge of absorption ($>$ 29.2 keV). This result confirms our assumption that in our pinhole scan the edge effect influences mainly the measured value of the focal spot size.

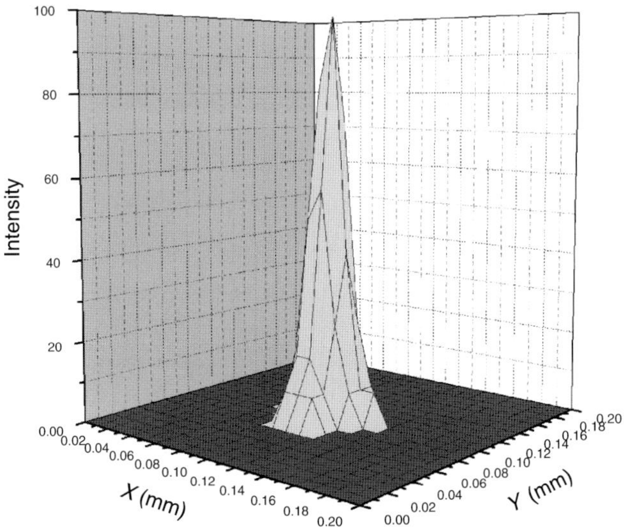


**Fig. 3.11.** Two-dimensional intensity distribution in the focal plane of the "minilens"

Therefore, the actual size of the focal spot in the energy region above 29 keV is not larger than 19 μm and the spatial resolution is not deteriorated.

Figure 3.11 shows the intensity distribution in the focal spot of the "minilens" under consideration in the energy interval 15–20 keV. The measurement was performed by a two-dimensional scan with a 5-μm pinhole. The focal spot is relative symmetric without any distortions and artefacts.

## Manufacturing Capillary Optics

There are many different methods of manufacturing glass capillary optics. The standard technology uses large pulling machines with a furnace for heating glass and two drives (upper and lower) for pulling (see Fig. 3.12).

If the velocity of the lower drive is much larger than the velocity of the upper drive, capillaries with small channel diameters can be obtained from glass tubes with large diameters. This is also a standard method of manufacturing polycapillaries from a bundle of monocapillaries. Applying the method once more to a bundle of polycapillaries, it is possible to reduce further the channel diameters and to produce in this way polycapillaries with very small channels. Figure 3.13 presents a typical polycapillary structure with channel diameters of ca. 2 μm. The image was obtained with a scanning electronic microscope. At present channel diameters of ca. 600 nm are achieved.

The pulling method also allows producing tapered monocapillaries and polycapillaries (including X-ray lenses), if the velocities of the drives do not

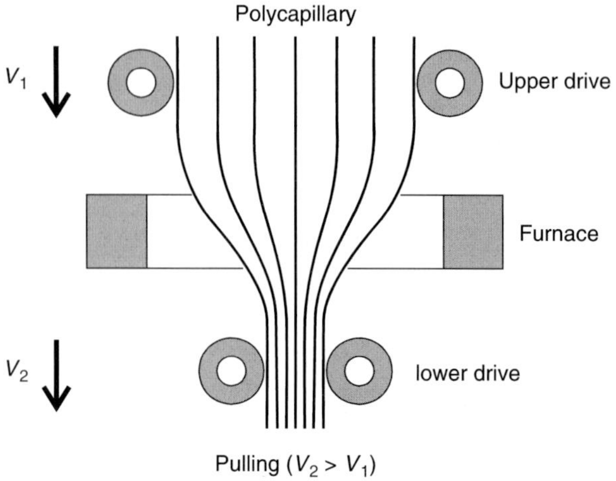


**Fig. 3.12.** Scheme of manufacturing capillary optics by pulling

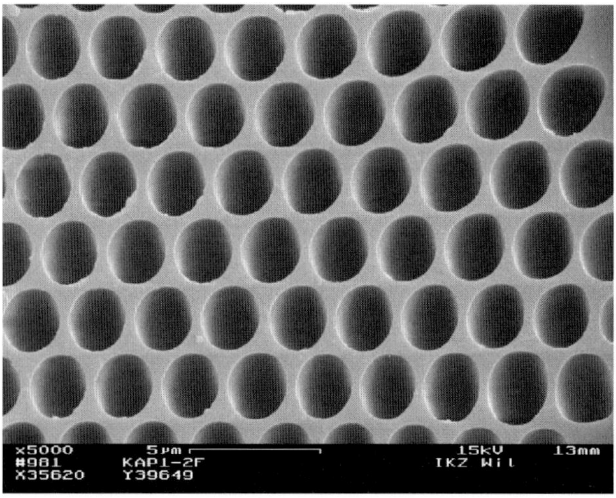


**Fig. 3.13.** Cross-section of a polycapillary structure

remain constant but vary according to a given law. Usually the law of velocity variation has to be found empirically for obtaining the required shape of the glass array. Two main problems here are the shape accuracy and the reproducibility of the shape in a series of subsequent pullings. These problems can be avoided by another method [81], which is presented schematically in Fig. 3.14.

A cylindrical monocapillary or polycapillary with closed (fused) ends is placed in a specially prepared form. While heating in an oven, the glass becomes soft and the air pressure inside the channels increases. As a result,

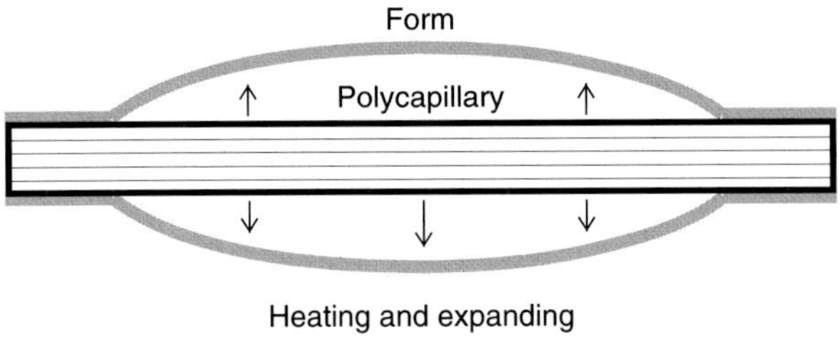


**Fig. 3.14.** Scheme of manufacturing capillary optics by expanding

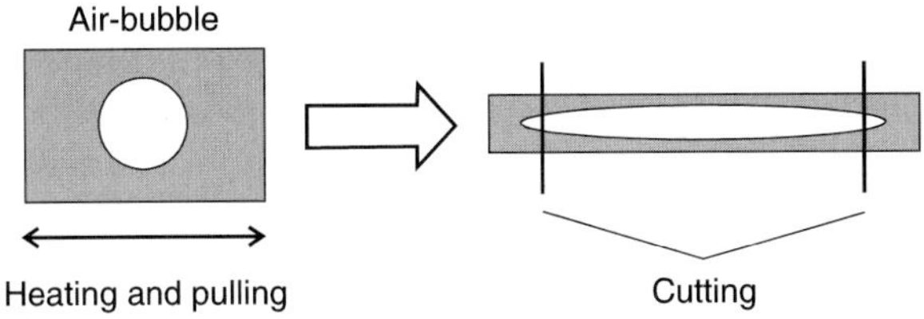


**Fig. 3.15.** Scheme of manufacturing elliptical capillaries

the glass array expands and accepts the fixed external form. The method is suitable for manufacturing tapered capillary arrays (X-ray lenses) or tapered monocapillaries with high accuracy and reproducibility of the shape.

One more technology seems to be very promising for one special case, namely, manufacturing elliptical monocapillaries [82]. One starts with a spherical air-bubble inside a massive glass tube (Fig. 3.15).

After heating and pulling the tube in two opposite directions the air-bubble accepts the elliptical form. What remains is to cut the tube at both its ends at appropriate points. The existence of such a relatively simple technology increases significantly the role of elliptical capillaries among other tapered monocapillaries. In particular, elliptical capillaries can serve in many cases as a suitable substitute for parabolic capillaries. Indeed a properly designed and asymmetrically cut elliptical capillary can form a weakly convergent X-ray beam when the second focal point lies far away from its end. And such a beam can be regarded as a good approximation to a quasiparallel beam. Experiments show also that, on the contrary, a quasiparallel beam can be successfully focussed with an elliptic capillary. The complexity of manufacturing parabolic capillaries and the inevitable deviations from the exact shape required are the main arguments in favour of applying elliptical capillaries instead of parabolic ones.

### Combinations of Capillary Optics with Other Optical Elements

Combining different optical elements may lead to interesting results [83]. One of the most promising applications of capillary optics for monochromatizing

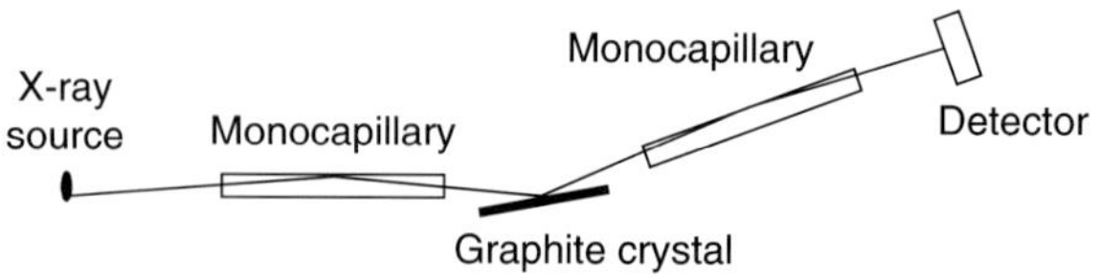


**Fig. 3.16.** Scheme of a graphite-capillary filter

X-ray radiation is the combination of monocapillaries with a graphite crystal (Fig. 3.16).

Radiation is captured near the source by the first monocapillary and transported to the graphite crystal. After reflection from the graphite crystal radiation is captured by the second monocapillary and transported to the sample (detector in our experiments). The graphite crystal used is a mosaic crystal with $\sim$ 4 mrad halfwidth of the rocking curve. The critical angle of total external reflection from glass is also $\sim$ 4 mrad for $E = 8\,\mathrm{keV}$ (Cu K$\alpha$-line radiation). The most important point in the presented schema is the match of two values: the critical angle of total external reflection and the half width of the rocking curve of the crystal. This avoids intensity losses when X-rays are reflected from the graphite crystals and then captured again by the second monocapillary. Naturally, monochromatization is achieved by means of a graphite crystal. Monocapillaries fulfil in this schema their usual role: they transport the radiation and prevent the beam from spreading. In this way one can get larger intensity on a sample relative to a standard monochromator without capillaries.

Both the direct beam intensity and the intensity of the beam obtained with graphite-capillary filter are presented in Fig. 3.17. The resulting spectrum after the filter (curve 2) contains practically only Cu K$\alpha$-line and an additional peak at the energy 16 keV which corresponds to the second order reflection. The intensity of Cu K$\alpha$-line radiation is larger than in the direct beam (curve 1) due to the application of monocapillaries.

A similar filter can be also used for obtaining monochromatic MoK$\alpha$-line radiation but the parameters of the graphite crystal and the geometry of the filter must be carefully optimized in that case. In particular, one has to use heavy glass in order to match the values of the critical angle of total external reflection for $E = 17.4\,\mathrm{keV}$ and the half width of the rocking curve of the graphite crystal.

A considerable intensity gain is obtained with an axially symmetric system, which consists of a cylindrically formed HOPG crystal and two sets of capillaries [84]. The diverging X-ray radiation from the source is collected and transported by the first set of capillaries to the cylindrically shaped HOPG film, where the radiation is monochromatized due to the Bragg effect. The second set of capillaries collects the Bragg reflected radiation and concentrates it on a focal spot. An intensity gain of ca. 25 in comparison with the direct beam has been observed within a focal spot of 150 µm for the Mo K$\alpha$-line.

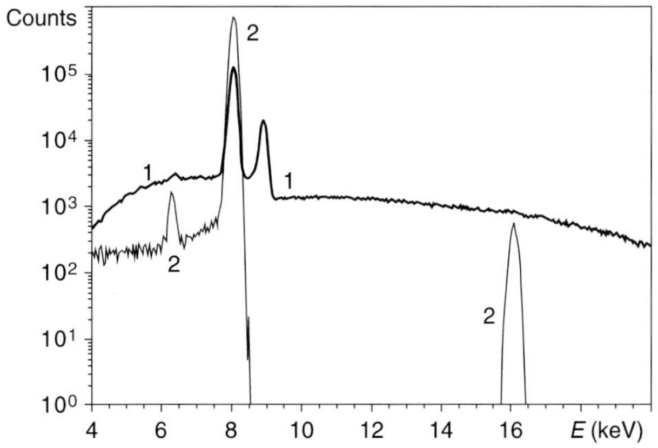


**Fig. 3.17.** Intensity of the direct (curve 1) and monochromatized (curve 2) beams

## 3.3 Diffraction Optics - Elements of Diffraction Theory

*A. Erko*

### 3.3.1 Electromagnetic Wave Propagation

In the case of electromagnetic waves, wave equation can be written for electrical or magnetic vector in the form [85]:

$$\frac{\partial^2 \boldsymbol{E}}{\partial x^2} + \frac{\partial^2 \boldsymbol{E}}{\partial y^2} + \frac{\partial^2 \boldsymbol{E}}{\partial z^2} \equiv \nabla^2 \boldsymbol{E} = \varepsilon\mu \frac{\partial^2 \boldsymbol{E}}{\partial t^2}, \tag{3.8}$$

where $\varepsilon$ and $\mu$ are the dielectric and magnetic constants of the material, $\boldsymbol{E}$ is the vector of electric field. Normally, the vector property of the field is not very important, especially when non-polarized radiation is used. One can simplify the (3.8) supposing scalar amplitude $A(\boldsymbol{r})$, the solution of the equation:

$$\nabla^2 A(\boldsymbol{r}) = \varepsilon\mu \frac{\partial^2 A(\boldsymbol{r})}{\partial t^2}. \tag{3.9}$$

The solution of such a differential equation can be found in the form of the exponential term:

$$A(\boldsymbol{r}) = A_0(\boldsymbol{r}) \exp(\mathrm{i}\omega t). \tag{3.10}$$

Combining (3.9) and (3.10) one can write the following wave equation in the form:

$$\nabla^2 A(\boldsymbol{r}) + k^2 A(\boldsymbol{r}) = 0, \tag{3.11}$$

where $k = \omega\sqrt{\varepsilon\mu} = 2\pi/\lambda$ called wave number. The fundamental solution of this equation for the point source is the spherical wave in the form of:

$$A(\boldsymbol{r}) = A_0(\boldsymbol{r})\frac{\exp\{\mathrm{i}(\omega t - kr)\}}{r}. \tag{3.12}$$

Huygens's principle describes the diffraction pattern in an image plane as a superposition of spherical waves integrated over the whole surface of the object plane. This principle is mathematically represented in the form of the Fresnel–Kirchhoff diffraction integral [50]. Referring to Fig. 3.18, this gives the complex amplitude $A(x', y', z')$ as

$$A(x', y', R_0) = -\frac{\mathrm{i}}{2\lambda}\iint A(x, y, 0)\frac{\exp[\mathrm{i}k(\boldsymbol{r})]}{(\boldsymbol{r})}\,\mathrm{d}x\,\mathrm{d}y\,\mathrm{d}z, \tag{3.13}$$

where $A(x, y, z)$ and $A(x', y', z')$ are the amplitude in the object and image planes, $k$ is the wave number ($k = 2\pi/\lambda$), $\bar{r}$ and $\bar{s}$ are the propagation vectors, $\mathrm{d}x\mathrm{d}y\mathrm{d}z$ is an element of area of the aperture, and the integral is carried out over the whole object aperture.

In practice, in order to get the intensity distribution behind a zone plate one must take the double integral in (3.13) for each point $A(x', y', z')$ on the screen and that would be just the distribution for a point source at $A_0(x, y, z)$. In turn, if the source is not just a point, this procedure must be repeated for each point of the source. This method can give good results but the time needed for such calculations is too long.

Here we are supposing that the wavelength $\lambda$ is much smaller than object and image sizes and therefore one can use a so-called "small angle approximation" that corresponds to $\cos(\boldsymbol{Zr}) \approx 1$ (see Fig. 3.18).

Equation (3.13) can not be solved analytically in the general form. To simplify the task one can use two approximations of the diffraction equation: in the Fresnel (near-field) and Fraunhofer (far-field) forms described below.

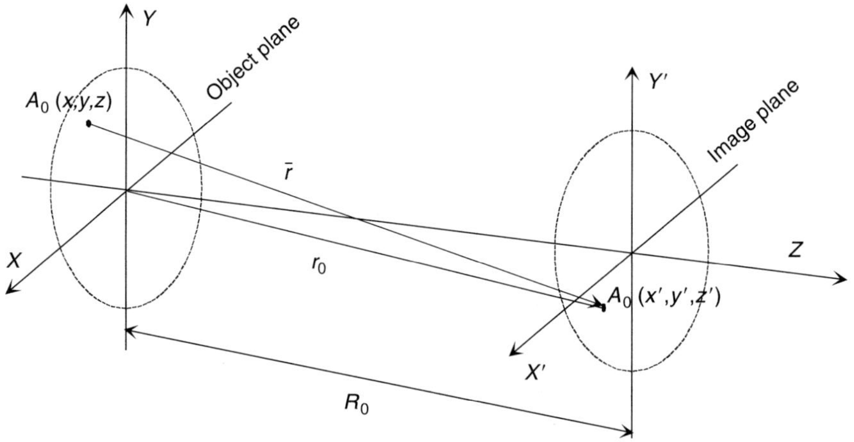


**Fig. 3.18.** The coordinate system used in the calculation of the diffraction pattern of a circular aperture

A simple equation, which corresponds to typical diffraction conditions, can be derivated on the basis of Kirchhoff formula.

Let us take the value of the radius in Cartesian coordinates as:

$$r^2 = R_0^2 + (x' - x)^2 + (y' - y)^2. \tag{3.14}$$

The different approximations of the $r$ value are used to define different types of diffraction.

### 3.3.2 Fraunhofer Approximation

In an approximation of the Kirchhoff formula which defines conditions of Fraunhofer diffraction, the overall dimensions of the object are much smaller than the distances to the source or to the point of observation. Supposing the distances in (3.14) having the following values:

$$R_0 \gg x', y' \gg x, y, \tag{3.15}$$

which corresponds to the far-field approximation. In this case the value of the radius-vector will be:

$$r \approx r_0 - \frac{x'}{r_0}x - \frac{y'}{r_0}y \tag{3.16}$$

and defining the diffraction angles as

$$\sin\varphi = \frac{x'}{r_0} \quad \text{and} \quad \sin\psi = \frac{y'}{r_0},$$

one gets the amplitude in the image plane as a function of angles for the plane parallel incident wave:

$$A(\varphi, \psi) = \frac{\mathrm{i}\exp(-ikr_0)}{r_0\lambda} \iint A\,(x, y, 0)\exp[ik\,(x\,\sin\varphi + y\,\sin\psi)]\,\mathrm{d}x\,\mathrm{d}y. \tag{3.17}$$

This equation shows that the amplitude in the image plane corresponds to the Fourier transform of the amplitude in the object plane in angular coordinates. From this equation one can easily derive well-known diffraction properties of gratings, slits, and apertures in Fraunhofer approximation.

### 3.3.3 Fresnel Approximation

Fresnel diffraction normally corresponds to the diffraction phenomena observed close to two-dimensional objects illuminated by plane parallel incident wave. As a criterion of this type of diffraction one can use the following relation:

$$R_0 \gg x', y' \approx x, y, \tag{3.18}$$

which corresponds to the near-field approximation. In this case the value of the radius-vector in (3.14) can be approximated:

$$r \approx R_0 - \frac{(x' - x)^2 - (y' - y)^2}{2R_0}. \tag{3.19}$$

If an object plane is perpendicular to the plane parallel incident wave with the amplitude equal to one in the point $Z = 0$, than the amplitude in an image plane can be written as:

$$A(x', y') = \frac{\mathrm{i}\exp(-\mathrm{i}\boldsymbol{kr})}{R_0\lambda} \iint A(x, y, 0) \exp\left[\mathrm{i}k\,\frac{(x' - x)^2 - (y' - y)^2}{2R_0}\right] dx\,dy. \tag{3.20}$$

Equation (3.20) describes the most important properties of the diffraction in the near-field, Fresnel region. An analytical solution of (3.20) is possible only for a few functions $A(x, y, 0)$, but they include very important classes of periodical and quasiperiodical functions. Periodical objects can be used for the transmission of the so-called "Fourier images", pure diffraction lensless high-resolution image formation in X-rays. One of the first successful experiments on the production of two-dimensional Fourier images in soft X-rays with spatial resolution down to 100 nm was published in 1985 [86].

The argument in the exponential function under integral in (3.20) can be used for the calculation of the lateral positions on the object plane, the so-called *Fresnel zone* in which the function $A(x, y, 0)$ changes the sign periodically. These positions are indicated by the argument in the exponent equal to $n\pi$.

$$k\frac{(x' - x)^2 - (y' - y)^2}{2R_0} = n\pi, \tag{3.21}$$

where $n$ is the number of the Fresnel zone. According to (3.21), if the point $A(x', y')$ in the image plane is fixed, the Fresnel zones on the object plane are circles, with a centre in this point. One can define the centre point position on an axis. In this case the central point will have the coordinates $A(0, 0)$ and Fresnel zone positions are indicated by circles with the centre on the optical axis and the radii:

$$r_n^2 = x^2 + y^2 = nR_0\lambda. \tag{3.22}$$

The last equation defines a two-dimensional structure of a zone plate. The transmission function of this optical element can be written in the form:

$$A(r) = \begin{cases} 1 & \text{if} \quad r \subset r_{2n} - r_{2n-1} \\ 0 & \text{if} \quad r \subset r_{2n-1} - r_{2n-2} \end{cases}. \tag{3.23}$$

Such a zone plate structure concentrates the beam intensity on the optical axis at the distance of $R_0$ from the object plane. The same structure can be used also for image transfer as conventional lens.

### 3.3.4 Bragg Diffraction

For a three-dimensional diffraction structure the conditions for diffraction can be written in terms of the concept of atomic lattice planes in crystals or artificial multilayers. The periodicity of the crystal ensures that sets of parallel planes may be drawn to pass through atom centres at regular intervals. These sets of planes are denoted by the Miller indexes *hkl*. If one plane is drawn through an atom at the unit cell origin the intercepts of the next plane of the set on the axes are $a/h$, $a/k$, $a/l$. Here $a$ is the lattice parameter. The distances between planes of the *hkl* set is $d_{hkl}$ and:

$$d_{hkl} = \frac{a}{\sqrt{h^2 + k^2 + l^2}}. \tag{3.24}$$

Taking into account (3.17), one can write the so-called Bragg law:

$$2d_{hkl} \sin(\theta_{hkl}) = \lambda. \tag{3.25}$$

For X-rays, the full kinematic theory of diffraction by a crystal was developed between 1912 and 1920 by Laue, Ewald and Bragg [87]. The possibility of optical coatings was discovered after 1930, and matrix theory for optical multilayers, which is equivalent to the dynamical theory of diffraction on crystals, was published by Abeles [88] and later by P. Berning [89].

## 3.4 Optics for Monochromators

### 3.4.1 Diffraction Gratings

*A. Erko*

Diffraction gratings are the dispersive elements used to produce low-energy monochromatic light from broadband or "white" beam. We will discuss the properties of low-energy total external reflection and transmission gratings, mostly used in combination with a focusing mirror and precise slit [90]. Other types of gratings are under development and have been tested in various laboratories, such as variable line spacing (VLS) gratings [91], and gratings for high-energy beams: multilayer (ML) gratings [92] and Bragg–Fresnel gratings [93].

#### Reflection Diffraction Grating

Grating definitions are given in Fig. 3.19. Diffraction on the plane grating, the so-called "fixed grating" geometry is described by the "grating equation" as follows:

$$m\lambda = d(\sin\alpha + \sin\beta), \tag{3.26}$$

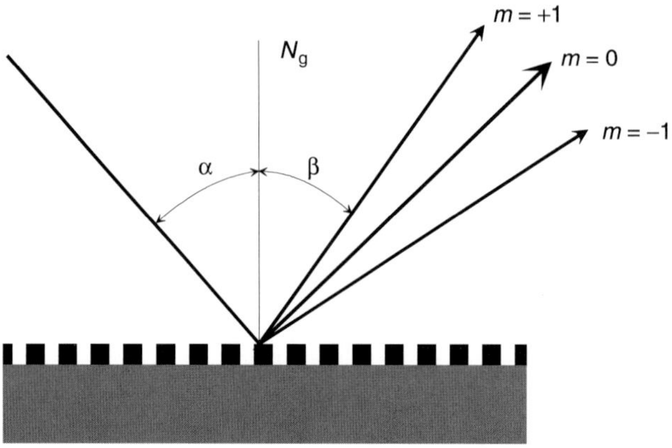


**Fig. 3.19.** Grating orders definition. The sign for the angle $\beta$ is positive, when $\beta$ is on the same side of the grating normal $N_g$ as $\alpha$

where $m$ is diffraction order, $d$ is the grating period, $\alpha$ and $\beta$ corresponding angles. For some applications it is convenient to use the so-called "fixed exit" geometry, when the sum of $\alpha + \beta$ is fixed and the energy of interest can be found by grating rotation. In this case grating equation can be written as:

$$m\lambda = 2d \sin\theta \sin\phi, \tag{3.27}$$

where $\theta = 0.5\ (\alpha - \beta)$ and $\phi = 0.5\ (\alpha + \beta)$. All equations above are valid for the infinite grating with non-limited large number of grooves. For the real case one has to take into account grating lateral dimensions $L_x L_y$. The diffraction pattern of finite aperture in one-dimension has a form of the function $\sin(x)/x$:

$$E(\beta) = \frac{L_x L_y E_{in} \sin^2 \left[\frac{\pi(\Delta\beta) L_x \cos(\beta)}{\lambda}\right]}{\left[\frac{\pi(\Delta\beta) L_x \cos(\beta)}{\lambda}\right]^2}, \tag{3.28}$$

where $L_x$ and $L_y$ are the irradiated grating dimensions along and perpendicular to the beam direction, $E_{in}$ is the flux density ph mm$^{-2}$. Intensity reaches zero values for the $\Delta\beta$ values:

$$(\Delta\beta) = \frac{\lambda}{L_x \cos(\beta)}. \tag{3.29}$$

With the combination of (3.28) and (3.29) one can calculate diffraction limited energy resolution:

$$(\Delta\lambda) = \frac{\lambda d}{2mL_x}. \tag{3.30}$$

### Angular Dispersion by Reflection Grating

Practically, it is also important to calculate the value of the angular dispersion. For the plane grating, using (3.26) one gets:

$$\left(\frac{\Delta\lambda}{\Delta\alpha}\right)_{\beta\text{-const.}} = \frac{d\cos(\alpha)}{m} \tag{3.31}$$

$$\left(\frac{\Delta\lambda}{\Delta\beta}\right)_{\alpha\text{-const.}} = \frac{d\cos(\beta)}{m}$$

For a "fixed grating" case the incidence angle is constant and the angular dispersion is

$$\frac{(\Delta\beta)}{(\Delta\lambda)} = \frac{\sin\alpha + \sin\beta}{\lambda\cos\beta}. \tag{3.32}$$

For a "fixed exit" case following the (3.27), angular dispersion is:

$$\frac{\Delta\alpha}{\Delta\lambda} = \frac{m}{2d\left[\cos\theta + \cos(\phi - \alpha)\right]}. \tag{3.33}$$

### Reflection Grating Diffraction Efficiency

Diffraction efficiency for a periodic grating can be estimated, depending on a groove profile and grating amplitude or phase origin. *Laminar Grating* is the simplest form of an X-ray grating shown in Fig. 3.20a. The path difference $\Delta L_x$ for a laminar grating is equal to:

$$(\Delta L_x) = t(\cos\alpha + \cos\beta), \tag{3.34}$$

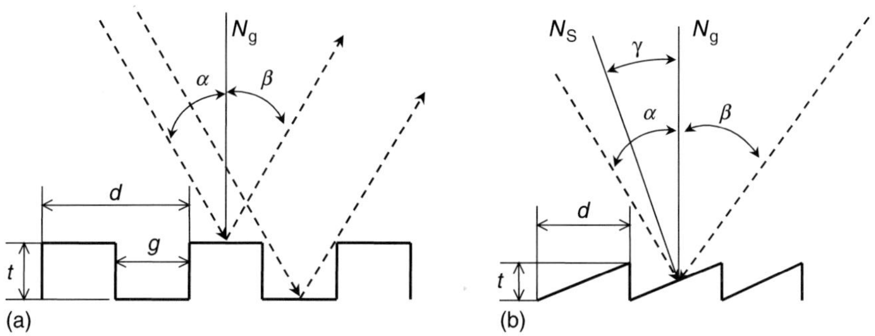


**Fig. 3.20.** **(a)** Rectangular profile of a laminar grating, **(b)** Blazed grating

where $t$ is a grating bars height. The diffraction efficiency in the $m^{th}$ order $E_m$ is proportional to the Fourier coefficients of the diffraction integral in Fraunhofer approximation:

$$E_m = \frac{4}{(m\pi)^2} \sin^2\left(\frac{m\pi g}{d}\right) \sin^2\left[\frac{\pi t}{\lambda}(\cos\alpha + \cos\beta_m)\right], \tag{3.35}$$

where $g$ is the grove width and $\beta_m$ is the angle of the $m$th diffraction order. For any given incidence angle $\alpha$ the groove depth can be optimized, making the 0th diffraction order as small as possible and maximizing the higher diffraction orders. In this case the groove depth should be

$$t = \frac{\lambda}{2\left\{\cos\alpha + \cos\left[a \sin\left(\sin\alpha - \frac{m\lambda}{d}\right)\right]\right\}}. \tag{3.36}$$

Using (3.29) it is possible to estimate the maximum diffraction-tolerable roughness $\sigma_{\text{diff}}$ of a grating applying the $\lambda/4$ criterion:

$$\sigma_{\text{diff}} = \frac{\lambda_{\min}}{8\cos\alpha_{\mathrm{c}}}, \tag{3.37}$$

where $\lambda_{\min}$ is the minimum wavelength of used radiation and $\alpha_{\mathrm{c}}$ is the critical angle. The mean of reflectivity compared to that from a smooth surface is given by the so-called Debye–Waller factor. The maximum value of a surface roughness parameter can be calculated by the equation:

$$\sigma < \frac{\lambda\sqrt{\ln(R_{\mathrm{f}}/R_{\mathrm{r}})}}{4\pi\cos\alpha}, \tag{3.38}$$

where $R_{\mathrm{r}}$ is the demanded reflectivity of a "rough" surface, $R_{\mathrm{f}}$ is the flat surface reflectivity. In a real situation the $\sigma_{\text{diff}}$ criterion is not strong enough in order to reach a good reflectivity and $\sigma$ should be much less than the diffraction-limited value of (3.37).

Blazed grating can be realized by the use of a wedge-shaped profile, shown in Fig. 3.20b. For a grating operated so that the angle $\theta = 0.5\,(\alpha - \beta)$ is fixed, the grating equation may be written as

$$m\lambda = 2d\sin\gamma\cos\theta. \tag{3.39}$$

The blazed condition is optimal for only a small wavelength range for a given grating and diffraction order. The efficiency of a blazed grating in an ideal case may be estimated for some integer order $m_k$ when the maximum of the diffraction intensity does not correspond to the zero order. In particular, the diffraction efficiency can be written as:

$$E_m = R_{\mathrm{r}}\frac{\sin^2[\pi^2(m_k - m)^2]}{\pi^2(m_k - m)^2}, \tag{3.40}$$

where $R_{\mathrm{r}}$ is the surface reflectivity at the used wavelength and $k$ is the optimized order of diffraction.

### Transmission Diffraction Grating Efficiency

The transmission gratings at normal incidence are used for low-energy X-ray beam monochromatization. They are produced by evaporation of a phase-shifting material on a thin film or $Si_3N_4$ membrane. There is also the possibility to produce a free-standing structure to reduce absorption in the grating.

To produce an optimum phase-shift of $\Delta\Phi$ between adjacent zones (i.e., to ensure the maximum diffraction efficiency) for a transmission grating at normal incidence a structure with thickness $t_{\mathrm{opt}}$ is required

$$t_{\mathrm{opt}} = \frac{\Phi_{\mathrm{opt}}\lambda}{2\pi\delta}, \tag{3.41}$$

where $\lambda$ is the wavelength and $\delta$ is the refractive index decrement of the zone plate material. The complex refractive index is

$$\tilde{n} = 1 - \delta - \mathrm{i}\beta \tag{3.42}$$

$$\chi = \frac{\beta}{\delta}, \tag{3.43}$$

with $\beta$ the absorption index and $\chi$ the optical characteristic of the material. The amplitude of the $m$th Fourier component of the wave in the plane of a diffraction grating $c_m$ can be written in a simple form, following from the Fourier analysis of a plane diffraction grating:

$$c_m = \frac{1}{m\pi}. \tag{3.44}$$

Therefore, the amplitude coefficient of the wave, coming through the "free", "positive" zones at the output surface of a diffraction grating can be written as:

$$S_{m1} = \frac{1}{m\pi}\exp(-\mathrm{i}kt). \tag{3.45}$$

The amplitude coefficient of the wave coming though a phase shifting material in negative zones can be written as:

$$S_{m2} = \frac{1}{m\pi}\exp(-\mathrm{i}kt\tilde{n}). \tag{3.46}$$

Taking into account (3.45) and (3.46), one can write an equation for the amplitude coefficient on the output surface of a $m$-order $S_m$ as:

$$S_m(r) = \begin{cases} \frac{1}{m\pi}\exp(-\mathrm{i}kt) & \text{if} \quad r \subset r_{2n} - r_{2n-1} \\ \frac{1}{m\pi}\exp(-\mathrm{i}kt + i\Delta\Phi - \chi\Delta\Phi) & \text{if} \quad r \subset r_{2n-1} - r_{2n-2} \end{cases}. \tag{3.47}$$

For r and $\mathrm{r_n}$ see equations (3.19) to (3.23).

where $\chi = \beta/\delta$. Combining the amplitudes of the waves, transmitted through the transparent and covered zone, one can calculate the intensity $E_m$ diffraction efficiency in the $m$th order [94]:

$$E_m = \frac{1}{(m\pi)^2}[1 + \exp(-2\chi\Delta\Phi) - 2\cos(\Delta\Phi)\exp(-\chi\Delta\Phi)] \qquad (3.48)$$

for the odd orders $m = \pm1; \pm3; \pm5; \ldots$ Even orders do not exist.

The maximum of the function $E_m$ can be found taking derivative of the (3.48):

$$\sin(\Delta\Phi_{\mathrm{opt}}) + \chi[\cos(\Delta\Phi_{\mathrm{opt}}) - \exp(-\chi\Delta\Phi_{\mathrm{opt}})] = 0, \qquad (3.49)$$

where the value of $\Delta\Phi_{\mathrm{opt}}$ is the optimum phase shift in the diffraction grating material. This equation can be solved numerically and an optimum phase shift can be approximated by the sum of two exponential curves:

$$\Delta\Phi_{\mathrm{opt}} \approx 1.69 + 0.71\exp\left(-\frac{\chi}{0.55}\right) + 0.74\exp\left(-\frac{\chi}{2.56}\right). \qquad (3.50)$$

The optimum thickness of the zone plate can be calculated using (3.41). Analogous to (3.48), the efficiencies of the zero order diffraction $E_0$ and the part absorbed in the material of a diffraction grating $E_{\mathrm{abs}}$ can be calculated using expressions:

$$E_0 = 0.25[1 + \exp(-2\chi\Delta\Phi) + 2\cos(\Delta\Phi)\exp(-\chi\Delta\Phi)] \qquad (3.51)$$

and

$$E_{\mathrm{abs}} = 0.5[1 - \exp(-2\chi\Delta\Phi)]. \qquad (3.52)$$

Depending on the value of parameter $\chi$ all materials can be called as a "phase" material or an "amplitude" one. Usually effective phase-shifting materials could be characterized by the value of $\chi < 0.1$.

### Bragg–Fresnel Grating

The properties of a diffraction grating fabricated on a total external reflection surface are almost isotropic in respect to the angle between ingoing beam direction and groove orientation. The depth of profile for such a grating is of the order of several nanometers and shadowing effect is important only at very small grazing angles. For such a case the rigorous theory is developed, which takes into account the grating material and grove depth [95].

In comparison with a conventional grating, the diffraction grating evaporated on a surface of a Bragg reflector, i.e. a Bragg–Fresnel grating, has a very remarkable anisotropy to the ingoing beam direction. One can define two main cases with different diffraction properties: sagittal and meridional directions.

The investigations of Bragg–Fresnel gratings show the difference in diffraction properties of sagittal and meridional grating structures. Dispersion angles of a sagittal grating are similar to those of a transmission grating and can be calculated using the simple diffraction equation. The main difference of a Bragg–Fresnel grating in comparison with a transmission grating is the value of optimal mask thickness. For a Bragg–Fresnel grating at a small grazing incidence angle the thickness can be an order of magnitude less at the same high diffraction efficiency. For example the grating with a gold thickness of 20 nm at 0.37° angle has an efficiency of 16%. The same efficient grating at a normal incidence should have the thickness of 1350 nm. The grating material thickness (aspect ratio) is the most important limitation in production of gratings with nanometer period by modern technology methods. Reduction of an optimal thickness to the nanometer scale opens the possibility of production of nanometer period gratings with an efficiency of up to 30% for hard X-rays.

The investigation of a meridional Bragg–Fresnel grating shows completely different diffraction properties in a comparison with conventional reflection grating. First, for a monochromatic beam, an angular spectrum of the input beam is limited by the value of crystal angular acceptance (rocking curve). All the other angular spectral components are rejected. The output angular spectrum is not limited and all higher diffraction orders are present. The inclined geometry at small grazing incidence angles makes it possible to enlarge the angle of dispersion of a meridional grating. To check the value of the angular dispersion in the meridional direction the performance of a meridional Bragg–Fresnel grating has been measured. A 2-µm period, 20 nm thick nickel grating was evaporated on the surface of an asymmetrically cut Si [111] crystal with an asymmetry angle of 13°. A diffraction efficiency of 1% was obtained at the energy of 8.2 keV and grazing angle of 0.9°. This value corresponds to an effective period for a transmission grating at normal incidence of 25.5 nm. This property of a meridional Bragg–Fresnel grating opens up the possibility of producing a Bragg–Fresnel diffraction grating with nanometer spatial resolution.

The definition of a sagittal and meridional grating is given in Fig. 3.21 and Fig. 3.24. In the sagittal geometry, the grating grooves are aligned parallel to the optical plane, as defined by the ingoing and outgoing X-ray beams.

In this case diffraction takes place perpendicular to the optical plane. In meridional geometry, the grating grooves are aligned perpendicular to the optical plane and diffraction takes place in the optical plane.

### Sagittal Bragg–Fresnel Gratings on a Symmetrical Crystal

Angular diffraction spectra of a sagittal grating can be described with the same formulas as for transmission gratings. The main difference is only in the optimal thickness of a phase-shifting profile.

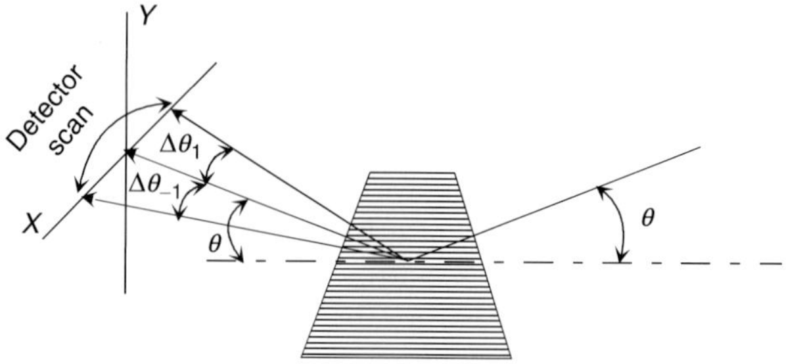


**Fig. 3.21.** Experimental geometry of sagittal grating measurements

The Bragg reflector, in the case of sagittal grating, employs e.g., a perfect Si [111] crystal. One can take gold as the material of a grating structure. The integral diffraction efficiency can be estimated using formulas for transmission gratings as in (3.48). The optical constants for a gold layer at the energy of 8500 eV are: refraction constant $\delta = 4.2 \times 10^{-5}$, absorption constant $\beta = 4 \times 10^{-6}$. In our particular case for the grating with a groove width equal to the free space, the relative integral diffraction efficiency of a transmission gold grating should be of the order of 31% with an optimal groove thickness of 1650 nm. Taking into account the Bragg angle of a Si [111] reflection at 8500 eV ($\lambda = 0.146$ nm, $\theta_B = 13.43°$) one can define the optimal groove thickness for the sagittal Bragg–Fresnel grating equal to 192 nm.

The angular dispersion of the diffraction orders relative to zero reflection can be written in the small angle approximation:

$$\Delta\theta_m = \arcsin\left(\frac{\lambda m}{d}\right), \tag{3.53}$$

where $\Delta\theta_m$ is the diffraction angle for the $m$th diffraction order. The results of the experimental measurements are shown in Fig. 3.22. The gratings S1 and S2 have the same value for the thickness of phase-shifting material (gold). The grating S1 has a higher integral efficiency than S2 because of the equal value of groove and space width in the period. A decrease in efficiency can be roughly estimated by the ratio between the open and covered part of the period. In the case of S2 this ratio is 0.72. Therefore the theoretical value of maximum diffraction efficiency is reduced by the factor of $\sin^2(m\pi g/d) = \sin^2(0.72m\pi)$ (see Fig. 3.22 and (3.35)). Furthermore, for a grating with non-unit groove-to-space ratio, odd and even diffraction orders exist.

The results of S1 and S2 measurements demonstrate the validity of our theoretical estimate of the optimal thickness of phase-shifting material of the Bragg–Fresnel grating in the sagittal direction. With an inclined grazing incident beam it is possible to achieve the theoretical maximum of diffraction efficiency for a grating using a considerably reduced mask thickness. For a reflection sagittal grating the optimum thickness of material is defined by the equation

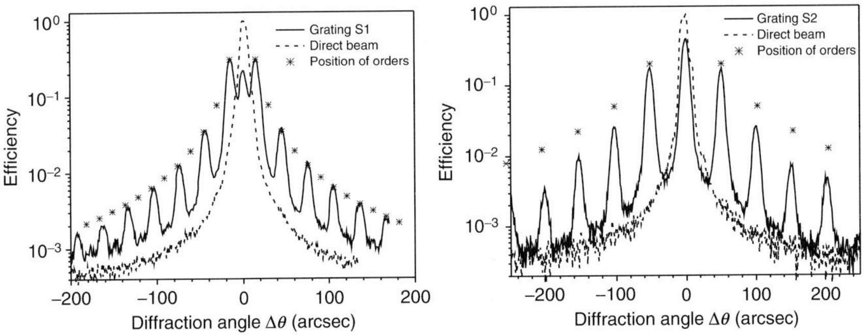


**Fig. 3.22.** The diffraction spectra of the sagittal gratings S1 (left) and S2 (right). Theoretical position and efficiencies of $m$th diffraction orders (*)

$$t_{\text{opt}} = \frac{\Phi_{\text{opt}} \lambda \sin\theta_B}{4\pi\delta} . \tag{3.54}$$

The X-ray beam is transmitted twice through the thickness $t_{\text{opt}}$ at the angle $\theta_{\text{B}}$. In comparison with a transmission grating at normal incidence, $t_{\text{opt}}$ is reduced by a factor of $0.5 \sin\theta_{\text{B}}$.

The difference in optimal thicknesses is most remarkable for high energies. A comparison of the theoretical value of the optimal thickness for a transmission diffraction grating and two types of Bragg–Fresnel lenses is shown in Fig. 3.23.

For this plot we have chosen a Si [111] crystal with the lattice period of 0.31 nm and a multilayer mirror with a period of 3 nm as the substrate for a

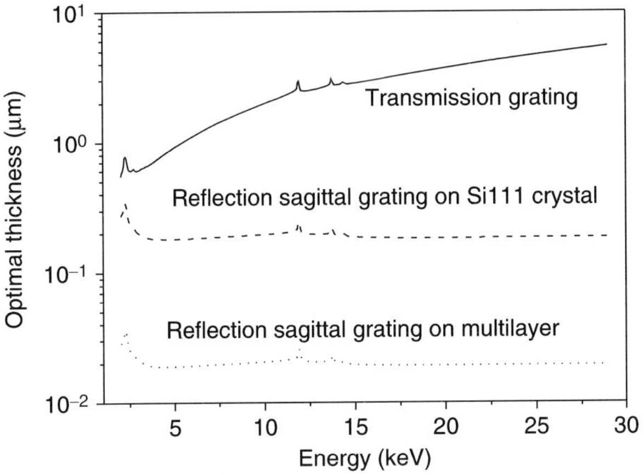


**Fig. 3.23.** Optimal thickness of a gold layer vs. photon energy for a transmission grating and two sagittal Bragg–Fresnel gratings on a Si [111] and a multilayer substrate with 3-nm period

Bragg–Fresnel lens. According to Fig. 3.23 even with the technological limit for an aspect ratio of about 1:1 (layer thickness equal to a groove width) an outer zone width of the Bragg–Fresnel grating at 30 keV can reach 0.2 µm. It can be easily fabricated for photon energies as high as 100–200 keV, which is a great problem for conventional gratings due to the very high material thickness required.

## Meridional Bragg–Fresnel Gratings on a Symmetrical Crystal

The diffraction properties of a meridional grating on a mirror surface can be described in first approximation using the conventional grating equation. The basic grating equation may be written as:

$$\lambda m = d[\cos\theta - \cos(\theta + \triangle\,\theta_m)] \tag{3.55}$$

with the grazing incidence angle $\theta$ and $\Delta\theta_m$ the angular dispersion. The experimental geometry and angular definitions are shown in Fig. 3.24. A detector scan was done in the optical plane, which corresponds to the meridional diffraction plane.

In the case of a meridional grating the angular dispersion in the $m$th order can be defined using the formula

$$\triangle\,\theta_m = \arcsin\left(\cos\,\theta + \frac{\lambda m}{d}\right) - 90^\circ + \theta, \tag{3.56}$$

where $\theta$ is the grazing angle of incidence. The positions of the positive and negative orders are defined in Fig. 3.24. Equation (3.56) leads us to the two important consecutive results

- dispersion angles for the $+\,m$ and $-\,m$ orders have different absolute values. This difference is increased considerable for the small grazing angles.
- The angle of dispersion increases with a decrease of grazing angle.

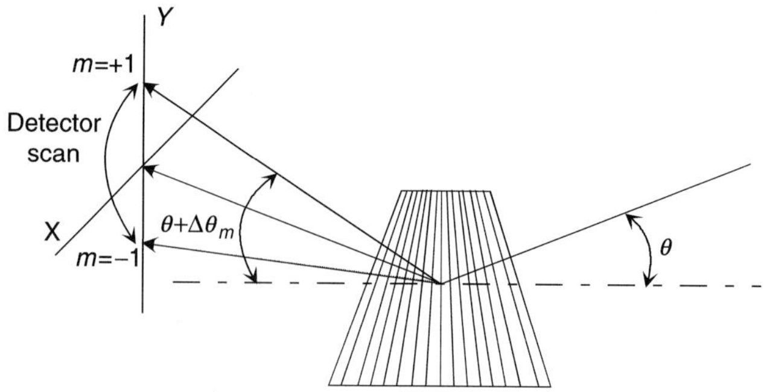


**Fig. 3.24.** Experimental geometry for meridional grating

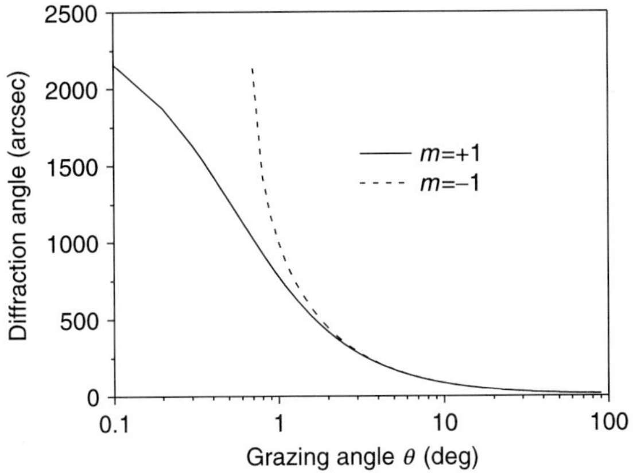


**Fig. 3.25.** Diffraction angles for the first order vs. grazing incidence angle $\theta$. Grating period of 2 μm, energy of 8,500 eV

Figure 3.25 shows an example of diffraction angle values for the first (minus first) orders at different grazing incidence angles. The value of a grating period used for the calculations is 2 μm, the wavelength is 0.146 nm (8,500 eV).

The difference between the values of positive and negative orders is important for small grazing angles $\alpha$. The value of the angle of negative diffraction orders is limited by the grazing angle of incidence of the incoming beam. The efficiency of diffraction orders for a meridional grating with a crystal substrate cannot be calculated using equations for a sagittal grating. These formulas are valid for meridional gratings on a mirror substrate, for example, a total external reflection mirror. A crystalline or multilayer substrate has an angular selectivity according to Bragg law. Such a mirror can reflect a monochromatic beam only in the limited angular interval inside the so-called *rocking curve*. The diffraction on a meridional Bragg–Fresnel grating can be divided into three steps (Fig. 3.26).

First, the incoming beam is diffracted on a surface transmission grating and penetrates into the crystal. If an angular dispersion of a grating $\Delta\theta_m$ is larger than the intrinsic rocking curve of the crystal, only the zero order, corresponding to the Bragg condition, is reflected by the crystal. The intensities of the diffraction orders outside of the Bragg peak will be rejected and absorbed in the substrate. The reflected zero order beam is diffracted the second time on a surface diffraction grating. Finally, the crystal substrate does not limit the angular spectrum of the outgoing beam but limits the angular acceptance of the Bragg–Fresnel grating.

The first diffraction process on a surface grating leads to a loss of output intensity. The accepted angular spectrum of the input beam is limited by a crystal rocking curve. After the second diffraction on the surface the grating produces higher diffraction orders, observed in the experiment.

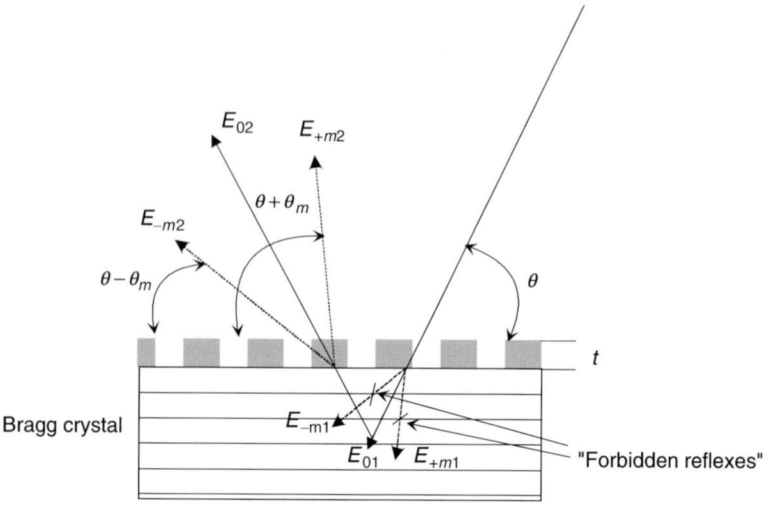


**Fig. 3.26.** X-ray diffraction on a meridional BraggFresnel grating with the thickness $t$. Bragg diffraction inside of the crystal rejects all diffraction orders that do not fulfill the Bragg conditions

If the grazing incidence angle corresponds to the Bragg conditions of the crystal, then only zero order of the primary diffracted beam will be diffracted a second time on a surface grating after reflection on a crystal. The total efficiency of the diffraction orders on a Bragg–Fresnel grating can be written using (3.48) and (3.51) as

$$E_m = \frac{1}{4m^2\pi^2}\left\{(1+\exp(-2\chi\Delta\Phi_{\mathrm{opt}}))^2 - 4\cos^2(\Delta\Phi_{\mathrm{opt}})\exp(-2\chi\Delta\Phi_{\mathrm{opt}})\right\}. \tag{3.57}$$

The maximum diffraction efficiency of a meridional grating is limited according to (3.57) to the value of 0.1 in comparison with 0.4 for a sagittal Bragg–Fresnel grating. Consequently, the diffraction efficiency of an amplitude (absorbing) meridional grating is limited by the value of 0.025 in comparison with 0.1 for a sagittal Bragg–Fresnel grating. The value of the optimal phase-shift related to a meridional grating is less than in the case of a sagittal grating (e.g. Fig. 3.27).

The optimal phase-shift for a gold mask at 8,500 eV is equal to 2.89 rad for a sagittal grating and 1.48 rad for a meridional one. The optimal mask thickness has in both cases almost the same value because of the factor of 2 in the equation for the optimal thickness of meridional grating

$$t_{\mathrm{opt}} = \frac{\Phi_{\mathrm{opt}}\lambda\sin\theta_{\mathrm{B}}}{2\pi\delta}. \tag{3.58}$$

Figure 3.27 shows the relative efficiency of sagittal and meridional gratings on a surface of Si [111] crystal. The gratings material is gold. The optimal value of a phase-shift in a gold mask of 1.48 rad yields the theoretical maximum

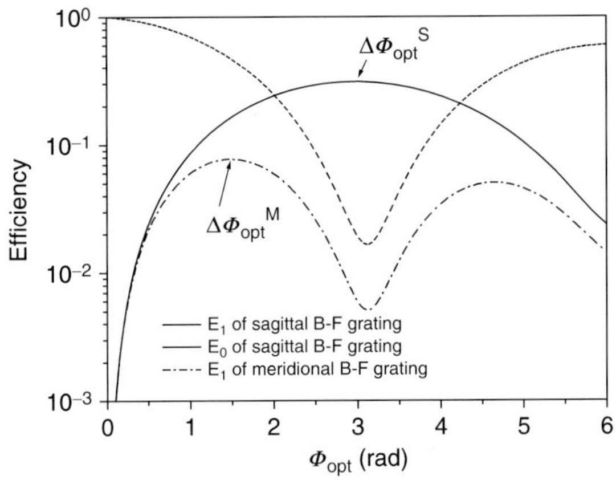


**Fig. 3.27.** Diffraction efficiency of sagittal and meridional Bragg–Fresnel gratings with a gold structure ($\chi = 0.095$) on a Si [111] crystal substrate at 8,500 eV

grating efficiency of 7.7%. Taking into account the corresponding Bragg angle of 13.43° for Si [111] reflection and the optical parameters of gold one obtains the value of optimal thickness: 190 nm.

As in the previous case of sagittal grating a gold mask for the meridional grating on a surface of Si [111] crystal was fabricated by electron-beam lithography and lift-off technology. Three gratings were investigated. The grating periods were equal to 5 μm (2.5 μm ×2.5 μm) for the grating M1, 0.6 μm (0.24 μm ×0.36 μm) for the grating M2 and 2 μm (1 μm ×1 μm) for the grating M3. A symmetric Si [111] crystal was used as a grating substrate. According to the experimental scheme shown in Fig. 3.21 and Fig. 3.24, a detector scan by the crystal analyzer was done in the vertical (optical) plane, which corresponds to the meridional diffraction plane.

In spite of the same value of the thickness of a phase-shifting material (gold) for the gratings M1, M2 and M3, the gratings M2 and M3 have lower integral efficiency because of the shadow effect at the grazing incidence angle. As in the case of a sagittal grating, for the meridional grating with groove-to-space ratio not equal to one, odd and even orders of diffraction exist (Fig. 3.28, right).

The results of M1 and M2 measurements approved the validity of our theoretical estimate of the optimal thickness of phase-shifting material of the Bragg–Fresnel grating in the meridional direction (3.58). Using inclined grazing incident beam it is possible to achieve a dispersion angle much larger than for a transmission angle.

In the case of sample M2, an effective period of the grating at a normal incidence is equivalent to 139 nm. An experimental verification of the concept of *the three-step diffraction process* on a meridional Bragg–Fresnel grating is represented in Fig. 3.29.

A two-dimensional $\theta - 2\theta$ mapping for the grating with period of 2 μm has been done to record the diffraction spectra in reciprocal space. The diffraction

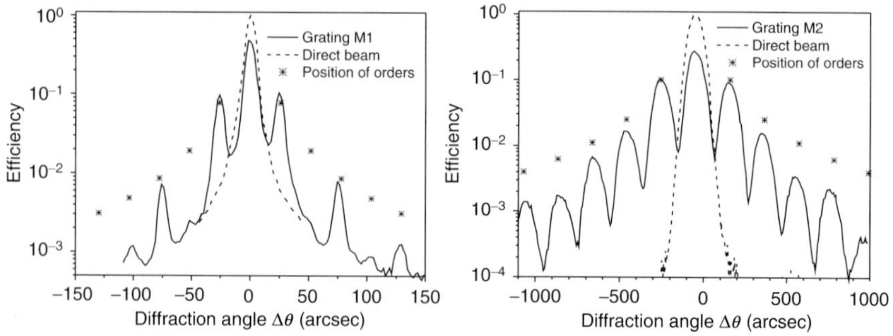


**Fig. 3.28.** The diffraction spectra of the meridional gratings M1 and M2. (*) are the theoretical positions and efficiencies of the $m$th diffraction orders

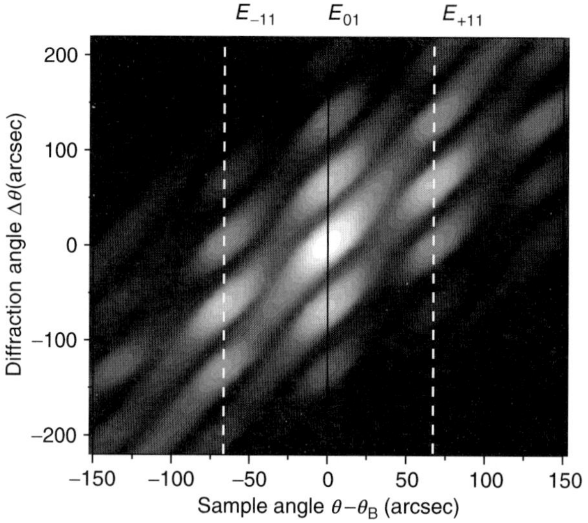


**Fig. 3.29.** Intensity distribution in a reciprocal space for the meridional Bragg–Fresnel grating. Grating period of 2 µm

angles were measured by a detector scan ($2\theta$ scan) for each value of grazing incidence angle around the Bragg maxima of the crystal. In accordance with Fig. 3.26 crystals do not reflect the X-ray beam at the angles which do not correspond to Bragg conditions. The secondary resonances that appear when +1st and -1st diffraction orders fulfill Bragg conditions are shown in Fig. 3.29 as $E_{-11}$ and $E_{+11}$. At these angles in the detector plane one can measure the same angular spectrum as at exact Bragg conditions for the primary beam analogous to the one in Fig. 3.28.

## Meridional Bragg–Fresnel Gratings on the Asymmetrical Crystal

According to (3.55) and (3.56) to increase the angular dispersion of a diffraction grating and to decrease the optimal grating thickness one has to use

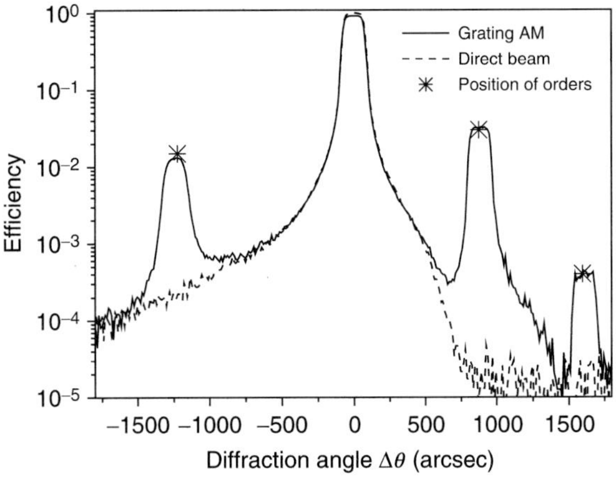


**Fig. 3.30.** The diffraction spectra of the meridional gratings (AM) on a surface of an asymmetrical crystal. (*) are the theoretical positions and efficiencies of the $m$th diffraction orders

a small grazing angle for the primary beam. Conventional crystals such as Si or Ge provide Bragg angles on the order of 10–15° in the energy range around 10 keV. Smaller reflection grazing angles can be realized using asymmetrical crystals with a high asymmetry parameter or multilayer mirrors with nanometer-scale periods. In our investigation, we measured samples of the grating placed on an asymmetrically cut Si [111] crystal with 13.06° asymmetry angle. Both sagittal and meridional gratings were investigated.

The meridional grating, AM, and sagittal grating, AS, on an asymmetrical crystal Si [111] were irradiated at the grazing angle of 27 and 26.5° at the energy of 8,200 and 8,500 eV, respectively. In both cases, the mask thickness was of 20 nm of evaporated nickel.

The spectra, shown in Fig. 3.30, support the results of theoretical calculations of the optimal mask thickness. A small grazing angle on a meridional grating provides a dispersion angle as high as 1, 224 arcsec (Fig. 3.30, left).

This value corresponds to the grating period of 25.5 nm at normal incidence. The spectra for the gratings M2, M3 and AM were measured using 50-μm slit placed at a distance of 1,000 mm from the sample. For this the crystal analyser was removed.

### 3.4.2 Multilayers for X-Ray Optics

*B. Vidal*

#### X-Ray Refraction

The real part of the refractive index of all materials in the X-ray domain is very close to unity. Even for heavy metals such as tungsten or gold a refractive

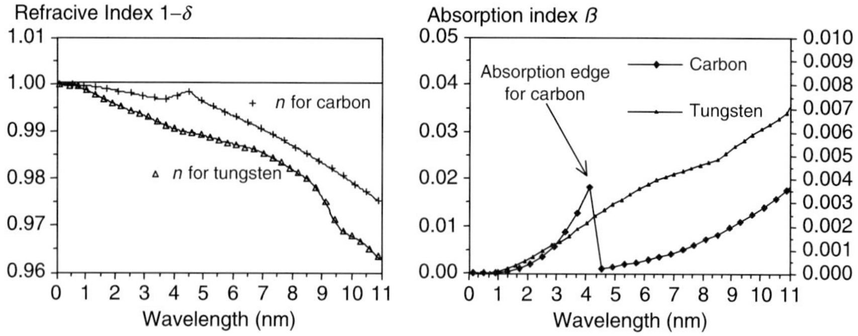


**Fig. 3.31.** 1-$\delta$ and $\beta$ in the X-UV wavelength range

index $\delta$ has a value on the order of $10^{-4}$ at the energy of 10 keV. Figure 3.31 gives $\delta$ and $\beta$ values for the wavelength range from 0.1 to 11 nm, tabulated in [49]. The limitations of the total external reflectance phenomena at grazing incidence angles were discussed in detail in Sect. 3.2.

Here is shown one simple equation which relates a complex refractive index $\tilde{n}$ and reflectance at grazing incidence angles $R$. The reflectance can be deduced from the well-known Fresnel formula

$$R = \left| \frac{\tilde{n} - 1}{\tilde{n} + 1} \right| . \tag{3.59}$$

The reflectivity $R$ decreases proportional to $\lambda^{-4}$ for the shorter wavelengths. This reflectivity is too low for practical use at a large incidence angle (see (3.1) and (3.2)) for X-ray optics. A reflection takes place only at the angles below the critical angle $\theta_c \approx \sqrt{2\delta}$.

Enhancement of a reflectance at large incidence angles can be obtained by adding coherently and in phases all the reflections from several thin layers: multilayer structure. A combination of layers with different optical properties can lead to a structure similar to that of crystal. Effective Bragg diffraction can be obtained even at normal incidence. These multilayers can be considered as an extension of natural crystals to larger lattice spacing. Such "artificial" crystals became, in the last decade, the most important optical devices for a soft and middle photon energy range. Having a high reflectance which reaches 70–80% at normal incidence angles multilayer structures can provide also a beam monochromatization on the order of $\lambda/\Delta\lambda \sim 25$. The diffraction properties of a multilayer mirror can be described by the theory of Bragg diffraction on a crystal. The basic elements of this theory are given above in Chaps. 3.3, 3.4.3 and 3.4.4. Here one can find a description of the special properties of multilayer optics connected with the choice of materials and technological aspects.

The microscopic model used in the kinematic theory of X-ray diffraction on a crystal describes a medium atom by atom. This theory can be applied only to the ultrathin layers with a few atoms stacked in a layer. An application of the kinematic theory for the calculations of X-ray diffraction on a "thick" multilayer mirror cannot be done without considerable approximations due to the complexity of this model. In comparison with crystals the X-ray diffraction on a multilayer mirror needs "averaged" characteristics of the media, such as refractive and absorption indexes. The optical (also called dynamical) theory operates with average optical properties of the complete layer. In a real case one has to consider properly which theory should be applied to a particular structure. For example one can apply the Fresnel Equation (3.59) even to mono-atomic films for calculation of reflectance.

Reflection on an interface can be achieved using the difference in refraction indexes between two materials as it was first suggested in the 1960s to produce multilayer optical coatings for visible light. An effective reflection takes place due to the phase-matched reflection on the interfaces of two materials. These mirrors have a maximum reflectance theoretically up to 100%. In reality absorption considerably reduces the reflectance.

### X-Ray Absorption

The reflection on an interface can also be achieved using the difference in absorption indexes between two materials. An effective reflection can be obtained if the layers of a high-absorbing material are located in the nodes of standing waves. This technique was developed in the 1970s to reduce the intensity of electric fields at layer boundaries and therefore increases the damage threshold of high-power laser optics [96]. In X-ray diffraction on crystals the reduction of absorption losses at Bragg angles (Bormann effect) was already known since 1920. Spiller has demonstrated the possibility of the fabrication of specially tailored reflection curves using nonperiodic multilayers [97]. As already mentioned in Chap. 3.4.1 optical properties of materials are characterized by the complex refractive index $\tilde{n}$, (3.42). The values of $\tilde{n}$ components are energy dependent. This dependence is nonmonotonic and characterized by many fast changes, jumps in a spectrum, which are called absorption edges of materials. They are known as K, L, M, etc. edges, which are fundamental properties of materials. They must be taken into account in multilayer design to provide demanded reflection properties: enough turning energy range or special properties of polarization transmission (reflection).

The linear absorption coefficient, $\mu_1 = 4\pi\beta(\lambda)/\lambda$ describes the decrease in intensity of a propagating wave through a material thickness $t$ that can be estimated by Snell's law

$$\frac{I}{I_0} = \exp(-\mu_1 t). \tag{3.60}$$

### Multilayer Design

The task of a multilayer design can be divided into two problems: direct and inverse, summarized in Fig. 3.32.

The first one is what we call the "analysis" problem. This problem corresponds to computing of the optical properties of a given stack with a finite number of layers $N$. The refractive indices $\tilde{n}$ of each material and each layer thickness $d_{i,j} = 1, \ldots, N$ are the computation parameters. Using these parameters and personal computer it is possible to determine the spectral reflectivity (or angular reflectivity) of a multilayer [98].

The second problem called the "synthesis" problem is more complicated than the first one. This problem corresponds to computing of the stack parameters to obtain the required spectral properties of the optical element. In order to obtain a specific spectral profile, we have to determine the materials which must be used (two or more), the number of layers to be deposited and

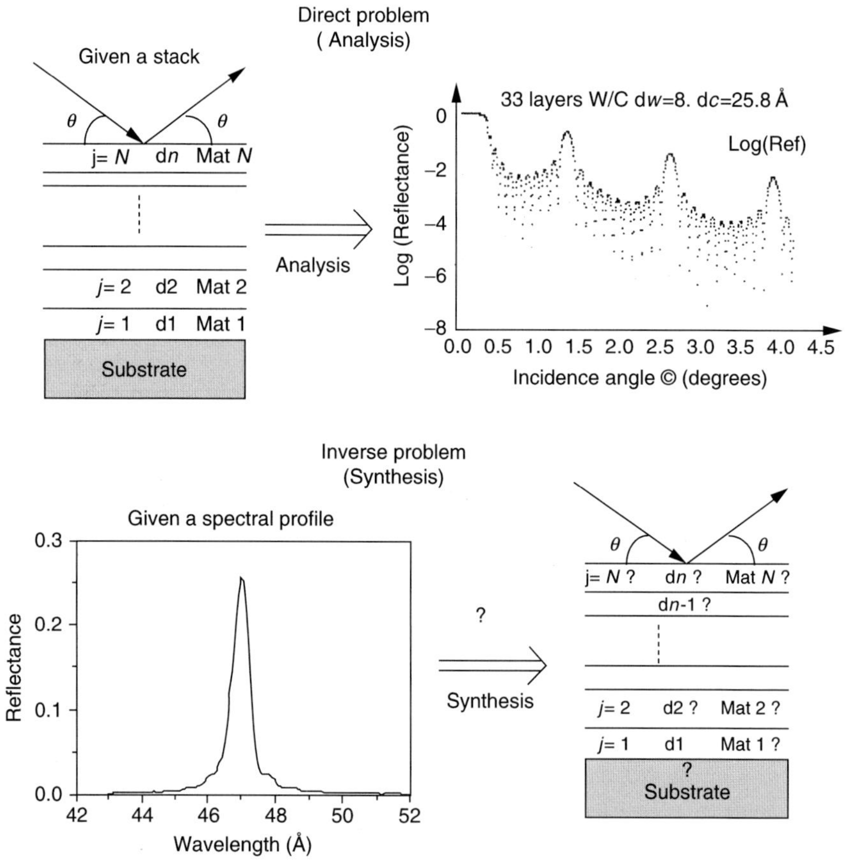


**Fig. 3.32.** Principle of the direct and inverse problems in multiplayer computation

their appropriate thickness. There are probably more than one single mathematical solution and one needs several years of computation to check all these different combinations of parameters. However, some rules can help to find a reasonable solution.

## Direct Problem of Multilayer Design

Two well-known methods can be applied to solve the problem of analysis calculation. One is called the impedance method and the other is the matrix method. Computations are recursive and must start from the first layer deposited on a substrate assuming no reflected wave is coming back from the substrate. The first method introduced by P. Berning [89] does not use complex values. The matrix method is more precise and needs powerful computers. This method allows one to apply direct formalism that is at present generally used.

Matrix method. Let's consider a simple interface between two media as shown in Fig. 3.33. Above this interface, the electric field amplitude can be written as

$$E(x, y) = [A \exp(-\mathrm{i}\zeta_{2y}) + B \exp(\mathrm{i}\zeta_{2y})] \exp(\mathrm{i}k_y x). \tag{3.61}$$

In the same manner for $y < y_0$

$$E(x, y) = [C \exp(-\mathrm{i}\zeta_{1y}) + D \exp(\mathrm{i}\zeta_{1y})] \exp(\mathrm{i}k_y x), \tag{3.62}$$

with the wave vector $k_0 = 2\pi/\lambda$, $k_y = k_0\ cos\ \theta$:

$$\zeta_1 = \sqrt{(k_0\tilde{n}_1)^2 - k_y^2} \quad \text{and} \quad \zeta_2 = \sqrt{(k_0\tilde{n}_2)^2 - k_y^2}. \tag{3.63}$$

The continuity of the electric field and its first derivates across the interface gives for "s-polarization" (electric field perpendicular to the incident plane)

$$\begin{bmatrix} A \\ B \end{bmatrix} = 0.5 \begin{bmatrix} p_{11}\ p_{12} \\ p_{21}\ p_{22} \end{bmatrix} \times \begin{bmatrix} C \\ D \end{bmatrix}, \tag{3.64}$$

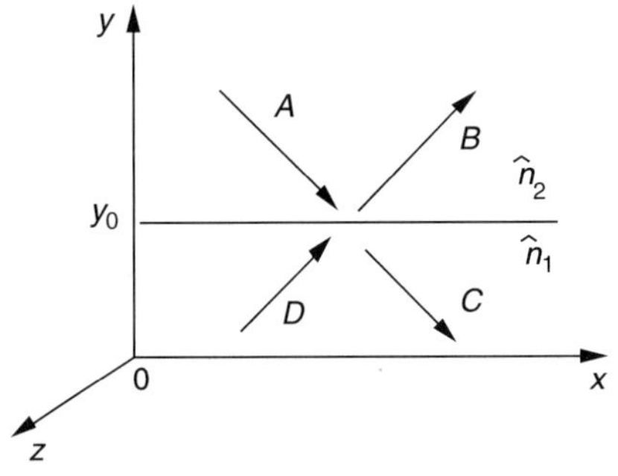


**Fig. 3.33.** Definition of the wave directions on an interface between two media with the refractive indexes $\tilde{n}_1$ and $\tilde{n}_2$

where

$$
\begin{aligned}
p_{11} &= a_1 \exp\{-\mathrm{i}(\zeta_1 - \zeta_2)y\} \quad \text{(a)} \\
p_{12} &= a_2 \exp\{+\mathrm{i}(\zeta_1 + \zeta_2)y\} \quad \text{(b)} \\
p_{21} &= a_2 \exp\{-\mathrm{i}(\zeta_1 + \zeta_2)y\} \quad \text{(c)} \\
p_{22} &= a_1 \exp\{+\mathrm{i}(\zeta_1 - \zeta_2)y\} \quad \text{(d)}
\end{aligned}
\tag{3.65}
$$

and

$$
a_1 = 1 + \frac{\zeta_1}{\zeta_2}; \; a_2 = 1 - \frac{\zeta_1}{\zeta_2}. \tag{3.66}
$$

The reflectance coefficient is

$$
R = \left| \frac{p_{21}}{p_{11}} \right|^2 . \tag{3.67}
$$

For "p-polarization", the continuity of the magnetic field at the interface is obtained by changing $\zeta_1/\zeta_2$ from (3.66) into

$$
\frac{\zeta_1}{\zeta_2} \times \frac{\varepsilon_1}{\varepsilon_2} \text{ where } \varepsilon_{1,2} = \tilde{n}_{1,2}^2 . \tag{3.68}
$$

For a single boundary, situated at $Y = 0$, (3.65) (a, b, c, and d) are much more simple and leads us to the Snell–Descartes law. They can be used to calculate the angle $Q_i$ between the interface plane and the direction of propagation in the $i$th medium in a stratified structure like a multilayer. For a stratified medium of $N$ layers

$$
n_i \cos \theta_i = \cos \theta \text{ for } i = 1, 2, \ldots, N. \tag{3.69}
$$

Consequence of this relation for a single boundary is the existence of a critical angle $\theta_\mathrm{c}$ below which the reflectivity is 1 (total reflection) for a nonabsorbing medium. For a metallic interface absorption of the wave in the surface (evanescent wave) gives less reflectivity around several percents.

For very thin transparent layers, if the imaginary part is neglected we get

$$
\sin \theta_c = \sqrt{2\delta}, \tag{3.70}
$$

which is true for each interface but is also an approximate relation for a single layer.

The reflection coefficient $S_{12}$ from a single boundary between two materials for the propagation angle $\theta_1$ and $\theta_2$ is deduced from (3.65), the so-called Fresnel equations. For the boundary of two materials with real part of refractive indexes $n_1$ and $n_2$ (See Fig. 3.33)

$$\text{for s-polarization}: \quad S_{12}^{s} = \frac{n_1 \cos\theta_1 - n_2 \cos\theta_2}{n_1 \cos\theta_1 + n_2 \cos\theta_2} \quad \text{and}$$

$$\text{for p-polarization}: \quad S_{12}^{s} = \frac{n_1 \cos\theta_1 - n_2 \cos\theta_2}{n_1 \cos\theta_1 + n_2 \cos\theta_2}. \tag{3.71}$$

In this case one has to take into account interferences which occur in the layers due to the second interface. However, the reflectance measurements allow to determine the main of characteristics of the layer as thickness, refractive indices ($1 - \delta$, $\beta$) and the roughness $\sigma$ at the interfaces [99]. The interference fringes are normally well visible in a logarithmic scale.

### Inverse Problem of Multilayer Design

The design of a multilayer mirror structure to produce a maximum reflectivity at a particular wavelength or on a large spectral range can be defined as inverse problem of multilayer design. As it was mentioned above, the general synthesis problem is complicated and a final solution needs too much computing time to check every possible value of each parameter. As parameters one can define the type of material (two or more), the thickness of each layer and the number of layers. This problem can be solved using a minimization method analogous to that which has been used in the case of dielectric mirrors [100].

The method is based on the property of the refractive index of materials in the X-ray range. For mirror efficiency optimization, the minimization method needs to define an appropriate "defect function" which characterizes the difference between the desired optical property (reflectance or energy band) and those obtained at each step of the computation.

The minimization method needs also the knowledge of the so-called "initial values" of the parameters which must be defined at the beginning of the computation. The initial values must be found as near as possible of the optimal one to reduce the computation time.

There are two well-known basic geometries which utilize interference effects to produce high reflectance mirrors: the quarter wave stack, and the ideal Bragg crystal (Fig. 3.34a and b). The first structure mainly used for nonabsorbing multilayers consists of alternating layers of high (h) and low (l) refractive index, each of the same optical thickness $nd_{\mathrm{l}} = nd_{\mathrm{h}} = \lambda/4$ (for normal incidence to the surface) such that all boundaries add with equal phase to the reflected layer.

If both the materials are completely absorption free, the quarter wave stack gives the highest reflectivity approaching $R = 100\%$. However, according to the scheme of Fig. 3.34a and b, the electric field has a maximum at the interfaces. Practically it means, in case of rough interfaces significant light scattering takes place, which leads to the damage of the mirror and considerably reduces multilayer reflectance.

In a quarter wave stack the thickness of materials are of the order of half a period. In an ideal Bragg crystal, the atomic planes thickness is usually

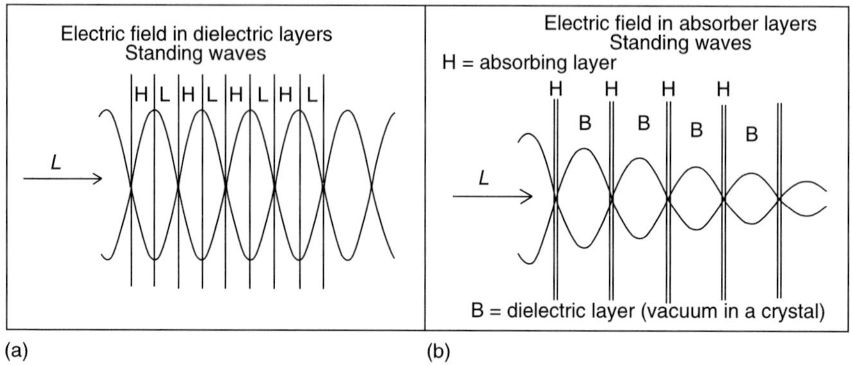


**Fig. 3.34.** Two basic designs for a high reflectivity mirror. (**a**) The quarter-wave stack. (**b**) Crystal lattice structure

much thinner than a period value. The different planes are spaced $\lambda/2$ apart (for normal incidence) and contribute in phase to the reflected wave. The absorption losses can be minimized due to the fact that the atomic planes are located at nodes of standing waves generated by the superposition of incidence and reflected waves. Reducing the absorption layer thickness the absorption losses can be made arbitrarily small compared to the reflectivity. In the limit a very thin film of absorbing material is equivalent to an absorption-free film of low reflectivity. Therefore, the number of layers can be increased considerably in comparison to the number of layers of the dielectric quarter wave stack. This fact leads not only to a reflection efficiency comparable with a dielectric multilayer mirror, but also to higher spectral selectivity (energy resolution) of the mirror.

However, the crystal-like multilayer design has important experimental limitations. First, it is impossible to fabricate infinitely thin layers and the second, the absorption coefficient $\beta$ of the spacer layer, where the electric field is high, is not negligible.

Practically, almost more than 200 layers are necessary to produce high-efficiency reflection with hard and middle X-rays, about 100 layers with soft X-rays and only around 10–20 layers with far UV at about 30 nm.

## Material Selection Rules

The following rules can be used to select two materials of a bilayer stacking in order to have the highest possible reflectivity [101]

- Select a first material with the lowest possible absorption $\beta$ as a spacer material.
- Find a second material with the largest possible reflection coefficient at the boundary with the first material.
- If several materials give similar reflection coefficients, choose the one with the smaller absorption coefficient.

In addition, the choice of materials depends on interface quality, roughness and small diffusion rate, as well as on the exclusion of hazardous materials. The absorption constant $\beta$ of the spacer material determines the maximum number of $N_{\max}$ bilayers. An easy equation can be used for the estimation of the maximum number of layers involved in diffraction process

$$N_{\max} = \frac{\sin^2\theta}{2\pi\tilde{\beta}}, \tag{3.72}$$

where $\theta$ is the angle of incidence. The value $\tilde{\beta}$ is the average absorption index of a multilayer structure (see below).

This limitation in the number of periods due to the absorption of the spacer layer and the thinnest possible absorbing layer defines also an upper limit of the *maximum reflectivity and the highest spectral resolution* $\lambda/\Delta\lambda$. These optical properties of multilayers are very important in practical applications of multilayers in X-ray monochromators.

The Fresnel reflection coefficient $S$ at the boundary of two coating materials defines the effective number $N_{\mathrm{eff}}$ of periods required to obtain a substantial reflectivity. Considering the quarter wave stack, and neglecting absorption losses and multiple reflections, one can estimate

$$N_{\mathrm{eff}} = \frac{1}{2\,|S|}. \tag{3.73}$$

At a normal incidence using the fact that all refractive indices are close to 1, we can express

$$N_{\mathrm{eff}} = \frac{1}{\sqrt{\Delta\,\delta^2 + \Delta\,\beta^2}}, \tag{3.74}$$

where $\Delta\delta$ and $\Delta\beta$ are the difference in the optical constants of coatings materials

On the basis of the discussion above one can choose the proper design for a multilayer for particular application. To reach the highest possible energy resolution one should use the design similar to a Bragg crystal. To have the largest possible integrated reflectivity one should use the design similar to the quarter wave stack.

### Periodic Designs

In the case of the periodic multilayer structure one has to define only three parameters: the multilayer period $d = d_{\mathrm{h}} + d_{\mathrm{l}}$ defined by Bragg relation (3.25), the distribution $\gamma = d_{\mathrm{h}}/d_{\mathrm{l}}$ of the material's thickness within one period and the total number of periods. In this formula $d_{\mathrm{h}}$ and $d_{\mathrm{l}}$ correspond to the thickness value of high and low index material and $d$ to the period of the stack. The total number of layers which have to be deposited onto a substrate is not critical if it is possible to realize a large number of layers with a high degree of stability in the deposition process. Maximum reflectivity is obtained within the first Bragg peak and the mirror performances in hard X-rays are quite insensitive to changes in the thickness ratio.

To find the best $\gamma$ ratio the analytical formulas for optimum design and performance of periodic multilayers with a large number of periods are developed by Vinogradov and Zeldovich [102]

$$\tan(\pi\gamma_{\mathrm{opt}}) = \pi\left[\gamma_{\mathrm{opt}} + \frac{(1-\delta_{\mathrm{h}})\beta_{\mathrm{h}}}{(1-\delta_{\mathrm{l}})\beta_{\mathrm{l}} - (1-\delta_{\mathrm{h}})\beta_{\mathrm{h}}}\right]. \tag{3.75}$$

For $\beta_{\mathrm{l}} \ll \beta_{\mathrm{h}}$ an approximate solution for small values of $\gamma_{\mathrm{opt}}$ is

$$\gamma_{\mathrm{opt}} = \frac{1}{\pi}\sqrt[3]{\frac{3\pi(1-\delta_{\mathrm{h}})\beta_{\mathrm{h}}}{(1-\delta_{\mathrm{l}})\beta_{\mathrm{l}}}}. \tag{3.76}$$

For a practical consideration of a periodic multilayer design it is important to estimate energy resolution and the effective number of layer pairs $N_{\mathrm{eff}}$ involved in the diffraction process. This can be done using simple formulas connecting major multilayer parameters. The spectral resolving power $\lambda/\Delta\lambda$ of a multilayer mirror can be estimated as [44]

$$\frac{\lambda}{\Delta\lambda} \approx mN_{\mathrm{eff}}, \tag{3.77}$$

where $m$ is the diffraction order.

The effective full width of a half maximum (FWHM) of the multilayer mirror rocking curve is approximated by

$$\Delta\theta \approx 2\sqrt{\frac{\Delta\lambda}{\lambda}}. \tag{3.78}$$

For a real multilayer mirror, the angle of peak reflectivity is greater than that predicted by the conventional Bragg formula. The formula which takes into account the average refraction index $\tilde{\delta}$ of the multilayer materials is given by equation

$$\sin\theta_m = \frac{m\lambda}{2d} + \frac{2d\tilde{\delta}}{m\lambda}, \tag{3.79}$$

where $\tilde{\delta} = (1-\gamma_{\mathrm{opt}})\delta_{\mathrm{l}} + \gamma_{\mathrm{opt}}\delta_{\mathrm{h}}$.

## Quasiperiodic Designs

To make a more precise optimization of a multilayer reflectivity one has to take into account the fact that all materials in the X-ray range have considerable absorption. The optimum design which maximizes the peak reflectivity with the minimum number of layers is quasiperiodic. The thickness of a heavy material should be decreased from the bottom to the top layer. Practically for the X-ray multilayers with a large number of periods the difference between optimum quasiperiodic design and simple periodic design is very small. The quasiperiodic design affects only the bottom layers whose contribution in the final efficiency is relatively small. This was shown for example using minimization programs in [103].

### Aperiodic Designs

In the case of transparent materials direct methods exploring Fourier transform are useful to find quickly the best solution [104]. Computer algorithms for the inverse problem solution, which have been developed for visible light, in some cases can be applied to design an X-ray mirror. However, the design freedom is much more limited due to the considerable absorption in all materials in the X-ray range. The most important application of these methods is the design of aperiodic structures to increase the energy bandwidth of reflection. The method supposes to divide a multilayer stack on several period regions in depth. Each region has a period which corresponds to a defined reflected energy at Bragg angle. A superposition of several regions with slightly different periods will provide an enlarged bandwidth of reflected light in comparison with a perfect periodical structure. In this way minimization programs can enlarge the bandwidth keeping the reflectance value as high as possible. In addition, the total external reflection phenomena can be used to reflect low-energy photons.

The case of aperiodic design is called "supermirror" as opposed to the mirrors corresponding to periodic stack and which can give a high reflectivity at a total reflectance angle for low energy photons. Such a supermirror has been fabricated for the BESSY II beamline. The results of the mirror tests are shown in Fig. 3.35. A multilayer mirror with 50 W/Si bilayers of different

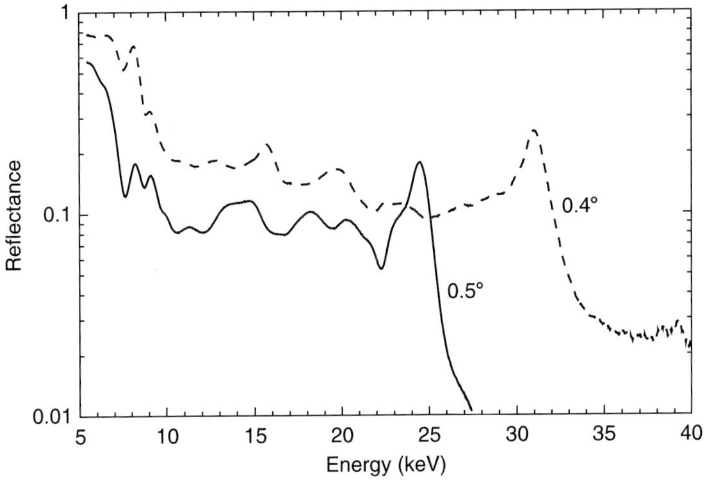


**Fig. 3.35.** Spectral characteristics of the BESSY "supermirror" measured at two different grazing incidence angles

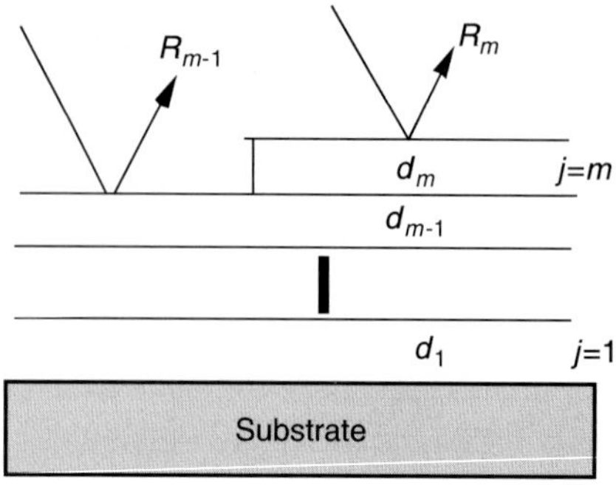


**Fig. 3.36.** Optical scheme of a multilayer mirror

thickness on a Si substrate has a smooth reflectivity of 12–30% in the whole energy range from 5 to 25 keV at a grazing incidence angle of 0.5° [105].

### Reflectance Computations in Multilayer Design

An interesting work on material pairs selection for high reflectance X-ray multilayers as well as layer-by-layer optimization was done in [106]. In order to select the best materials one can deduce the relation from the Berning's formula according to the scheme in Fig 3.36

$$R_j = \frac{S_j(1 - S_j R_{j-1}) + (R_{j-1} - S_j)\exp(-i\varphi_j)}{1 - S_j R_{j-1} + S_j(R_{j-1} - S_j)\exp(-i\varphi_j)} \tag{3.80}$$

with $\varphi_j = \frac{4\pi n_j d_j \cos\theta_j}{\lambda}$, $S_j$ are the Fresnel coefficients of an interface between the vacuum and the $j$th layer defined by (3.71) according to the polarization of the light.

Starting from $j = 1$ with $R_0 = S_0$ to $j = m$ one can rewrite (3.80) into

$$\frac{R_j - S_j}{1 - S_j R_j} = \frac{R_{j-1} - S_j}{1 - S_j R_{j-1}} \exp(-i\varphi_j). \tag{3.81}$$

The denominators of this equation represent the contribution of multiple reflections in the multilayer, which can be ignored in the following considerations, because $R$, $S \ll 1$ holds in the X-ray region. Thus we get

$$R_j - S_j = R_{j-1} - S_j \exp(-i\varphi_j). \tag{3.82}$$

This simplified formula can be used as an optical criterion for selection of material pairs. The formulas given above are valid in both cases of periodic and aperiodic design.

For an estimation of a peak reflectance $R_0$ in the case of periodic design one can use the following simplified equation

$$R_0 = \frac{1 - u}{1 + u}, \tag{3.83}$$

where

$$u = \sqrt{\frac{1 - \cos^2 \pi\gamma_{\text{opt}}}{1 + \left(\frac{\delta_\text{h} - \delta_\text{l}}{\beta_\text{h} - \beta_\text{l}}\right)^2 \cos^2 \pi\gamma_{\text{opt}}^2}} . \tag{3.84}$$

## Imperfection Effects

Three main types of imperfection errors must be considered in the production of a multilayer mirror.

1. Thickness errors or variations in lattice parameters. This is mainly the case if all parameters of the deposition process are not perfectly maintained and controlled during all the time of the multilayer fabrication.
2. Material interdiffusion in interface region or interfaces roughness which can reduce reflectance dramatically.
3. Inhomogeneities in the mirror layers such as porosity, impurities and local fluctuation of composition. This type of error is the most complicated because it can be caused by many technological reasons.

*Thickness Errors*

For an angle of incidence larger than the critical angle, the thickness errors lead to Bragg peak widening while the roughness imperfections lead to a decrease of the efficiency or peak reflectivity of a multilayer. To calculate analytically the influences of thickness errors one can use an expression introduced by Piecuch from a kinematical approach. This equation which gives for reflectivity at a Bragg peak the value [107]

$$\langle R \rangle = R_{0L} + 2R_1 \sum_{l=1}^{L-1} (L-1) \exp\left\{(-4l \left\langle \Delta\, \Phi^2 \right\rangle) \cos(2l\Phi_\text{p})\right\} , \tag{3.85}$$

where $\left\langle \Delta\Phi^2 \right\rangle$ corresponds to the variance in phase $\Phi$ due to the thickness errors and $\Phi_\text{p}$ is the mean phase change through the bilayer

$$R_0 = 2S_{21}^2 \left[1 - \cos(2\Phi_\text{p}) \exp(-2 \left\langle \Delta\, \Phi_\text{p}^2 \right\rangle)\right] \text{ and}$$

$$R_1 = 2S_{21}^2 \left[1 - \cos(2\Phi_\text{p}) \exp(2 \left\langle \Delta\, \Phi_\text{p}^2 \right\rangle)\right] , \tag{3.86}$$

where $S_{21}$ corresponds to the Fresnel coefficient between the two media of the bilayer.

*Interface Roughness*

Computations of the influence of an interfacial roughness on multilayer reflectance were published in the classical paper of Beckmann and Spizzichino [108].

In these computations for the visible light photon energy range several assumptions have been made which cannot be easily transformed into X-rays range. They are

1. The slope of the surface irregularity is small enough to neglect the influence of polarization of the light due to local variations in the angle of incidence.
2. The radius of curvature of any surface irregularity is large compared to the wavelength.
3. Scattering of the bulk media bounding the interface is negligible.

The main mathematical parameters which are involved in the theory are: the variance $s$ of the interface profile height distribution, which is assumed to be a Gaussian function and the coherence length $l_{\mathrm{coh}}$ of the variation of the interface positions. If $L_{\mathrm{tot}}$ is the total length of the sample in the direction along the x-axis parallel to the surface, one can write the self-correlation roughness function in the classical form

$$C(x) = \frac{1}{L_{\mathrm{tot}}} \int_0^{L_{\mathrm{tot}}} z(s)(x+s)\,\mathrm{d}s, \tag{3.87}$$

where $C(x)$ is assumed to have an exponential behaviour: $C(x) = \exp(-x/l_{\mathrm{coh}})$.

In case of specular reflectivity, the theory of Beckmann and Spizzichino leads to a change of the Fresnel coefficients $S_{12}$ to

$$S'_{12} = S_{12} \exp\left(-\frac{k_{y1}^2 \sigma^2}{2}\right), \tag{3.88}$$

where $k_{y1}$ is the $z$ component of the wave vector $k_0$ in the first medium.

Using the Wronskian method, one can find also a variation in the electric field transfer matrix of (3.64). For the s-polarization corresponding to an electric field parallel to the surface, one can write a new transfer matrix

$$\begin{bmatrix} A \\ B \end{bmatrix} = 0.5 \begin{bmatrix} p_{11} \exp\left[-(\zeta_1-\zeta_2)^2 \frac{\sigma^2}{2}\right] & p_{12} \exp\left[-(\zeta_1+\zeta_2)^2 \frac{\sigma^2}{2}\right] \\ p_{21} \exp\left[-(\zeta_1+\zeta_2)^2 \frac{\sigma^2}{2}\right] & p_{22} \exp\left[-(\zeta_1-\zeta_2)^2 \frac{\sigma^2}{2}\right] \end{bmatrix} \times \begin{bmatrix} C \\ D \end{bmatrix}, \tag{3.89}$$

where $\zeta_1$ and $\zeta_2$ have been previously expressed on page 133, equation (3.63). The theory developed here ignored roughness-induced plasmon waves which is not quite correct according to the fact that hypothesis $\beta$ is certainly not verified.

Approximations of Beckmann's matrix method [108] lead to the same well-known expression suggested in crystallography, the so-called Debye–Waller factor

$$R = R_0 \exp\left(-\frac{8\pi^2\sigma^2 \sin^2\theta}{\lambda^2}\right),$$

where $R_0$ is the reflectivity of a perfect multilayer without roughness at the interfaces and $\sigma$ is the root mean square value of effective roughness.

At this step it should be pointed out that all theories developed by the majority of workers in the X-ray area used several approximations which are not quite correct because they consider only the specular reflectivity and do not take into account the self-correlation function of the interface position. All these approximations are equivalent to the insertion of a transition layer to which the roughness is attributed.

In practice such models are useful to evaluate the mean roughness of the layer and not a particular roughness at one layer of the stack. In some cases, the computations using the transition layer model can lead to an increase of the reflectance or transmittance value. It is completely in disagreement with the physical reality. In practice, both transmittance and reflectance values must be lower than those computed for a perfect multilayer because the light is scattered in all directions.

### 3.4.3 HOPG-based Optics

*A. Antonov, I. Grigorieva, B. Kanngießer, V. Arkadiev, B. Beckhoff*

#### The Concept of Monochromators

Monochromators nowadays constitute an essential part of many modern setups for X-ray diffractometry and X-ray fluorescence analysis (XRF). They determine not only the energy and the bandwidth of the beam, but also to a large extent its spatial and angular characteristics. Monochromators can be used to modify both the primary and the secondary beams in XRF. In the first case, monochromators contribute to form a primary beam with the spectral parameters, which are optimal for a particular problem of material analysis or medical investigation. They eliminate undesirable parts of a continuous spectrum and disturbing characteristic lines, thus suppressing the background and increasing the sensitivity of the analytical method. In the second case, monochromators are applied for collecting and analysing the secondary (fluorescence) radiation detected. They are mostly used as dispersive elements for identifying characteristic lines of the elements, which may be present in a sample.

Crystal monochromators are based on the phenomenon of X-ray diffraction. Spectral selective reflection decreases significantly the total intensity of the beam. First, large parts of a spectrum are drastically reduced. Usually these parts are considered to be useless or even disturbing for an application under consideration and constitute an undesirable background. Second, the intensity of the line or narrow energetic band to be monochromatized decreases because the reflection coefficient is always smaller than 100%. To compensate these losses of the useful radiation, one can utilize a focusing geometry, which enables obtaining large intensities of monochrochromatized beams within small focal spots.

Decisive demands for effective focusing are a high reflectivity and a large angular acceptance of the optics employed. Perfect crystals usually have large values of reflection coefficients but they reflect X-rays in a very narrow spectral region and have small acceptance angles. High spectral resolution $\Delta E/E \sim 10^{-4}$ to $10^{-5}$ of these crystals leads to a relatively poor effectiveness for X-ray focusing. Besides, realizing a focusing geometry with strongly bent perfect crystals is a nontrivial task. For applications with moderate demands on monochromatization and high requirements on the intensity of the reflected radiation mosaic crystals reveal themselves as the most suitable candidate. Among the known mosaic crystals, highly oriented pyrolytic graphite (HOPG) crystals, both bent and flat ones, belong to the category most widely used.

## Monochromators in XRF

X-ray fluorescence spectroscopy is based on the measurements of fluorescent radiation emitted by the element, which is excited by X-rays having a photon energy higher than the corresponding absorption edge of the respective element. Although qualitative interpretation of the spectra obtained is rather simple, the reliable quantitative analysis of the elemental concentrations is often a nonlinear optimization problem due to various mutual absorption and secondary enhancement effects modifying the fluorescence intensities. In addition, XRF detection limits (DL) are affected by the shape of spectral background partially induced by scattered excitation radiation and its intensity depending on the major matrix constituents of the sample.

Monochromatic excitation can drastically decrease the spectral background below fluorescence lines associated with photon energies considerably lower than that of the excitation radiation. Furthermore, a fundamental parameter based XRF quantification is, in general, more reliable in the case of a monochromatic than polychromatic excitation radiation. This is due to the fact that the shape of the spectral distribution of the excitation radiation is often not known well enough when X-ray tubes, in particular in conjunction with nonmonochromatizing optics, are employed. For a given count rate capability of the energy-dispersive detection system, the sensitivity of XRF can be improved in the case of a monochromatic excitation if enough primary intensity is available to achieve the highest count rate allowed by the given detector.

In energy-dispersive X-ray fluorescence analysis (EDXRF), the conventional techniques aimed at either a quasimonochromatic excitation or a reduction of scattered excitation radiation involve the use of secondary targets, selective filtration of the source primary radiation or the production of polarized primary radiation by means of Barkla scattering. However, all these approaches substantially decrease the intensity of primary X-ray tube radiation by several orders of magnitude, thus considerably reducing the sensitivity of EDXRF analysis.

An alternate way to modify the spectrum of the primary radiation is offered by diffraction-based X-ray optics that selectively reflects a narrow band of the beam spectrum. Especially important for XRF is the fact that

Bragg-reflection monochromators are meanwhile available as doubly curved crystals with an application-adapted geometry and permit to compensate the reflection-caused losses of the useful radiation by collecting the primary radiation within a large solid angle of acceptance. Besides, such monochromators can focus radiation onto a small spot on a sample, thus increasing both the spatial resolution and enhancing the absolute sensitivity of the EDXRF analysis by modifying the characteristics of the excitation channel.

In the detection channel, singly and doubly curved monochromators can enhance the solid angle of detection and modify the spectral distribution of the fluorescence radiation detected in favour of the lines of interest. This selection of a particular energetic region of interest reduces the burden on the detection system and permits to enhance the count rate of fluorescence line detected by increasing the excitation intensity. The focusing ability of the crystal allows for the use of a small-area X-ray detector possibly associated with both an improved peak-to-background ratio and an improved energy resolution.

### HOPG Crystals

HOPG is an artificial crystalline structure with unique X-ray reflection properties. HOPG crystals reveal a good peak reflectivity of 40% to 50% in the energy region of 10 keV to 30 keV. However, their most distinguishing feature is a large mosaicity with a typical value of about 0.4°, which leads to the largest integral reflectivity among all known crystals. For example, the integrated reflectivity of the main 002 reflection of a HOPG crystal is almost an order of magnitude higher than that of other crystals and can achieve approximately $1.3 \times 10^{-2}$ rad. Even the second order 004 reflection has an integrated reflectivity comparable with that of the first order reflection of many crystals, providing practically the same high intensities ($R_{\text{int}} = 2 \times 10^{-3}$ rad) as lithium fluorite and silicon [109, 110].

HOPG is formed by thermocracking of hydrocarbon gas on a heated substrate and subsequent high temperature annealing. HOPG has a very high thermoconductivity along the CC plane (about 1600 to 1800 $\mathrm{Wm^{-1}K^{-1}}$), as well as excellent thermo- and radioactive stability. HOPG consists of carbon having a purity of about 99.99%.

The deposited material named pyrocarbon (PC) has a turbostrate structure with the interlayer distance of 3.44 Å. Annealing of the deposit under pressure at a temperature of over 2,800° C results in formation of HOPG possessing a 3-dimensional highly anisotropic laminar structure with the interlayer distance of 3.354 Å. Although the material remains polycrystalline in its base plane, it becomes almost monocrystalline in the perpendicular direction. The angular distribution of the $c$-axes of crystallites called mosaic spread decreases during annealing under pressure from 30° for pyrocarbon to less then 1° for HOPG [111].

Early studies reported HOPG to be an ideal mosaic crystal, but later it was discovered, that HOPG has a more complex, two-level mosaic structure [112–114]. Relatively large blocks are disoriented relative to each other

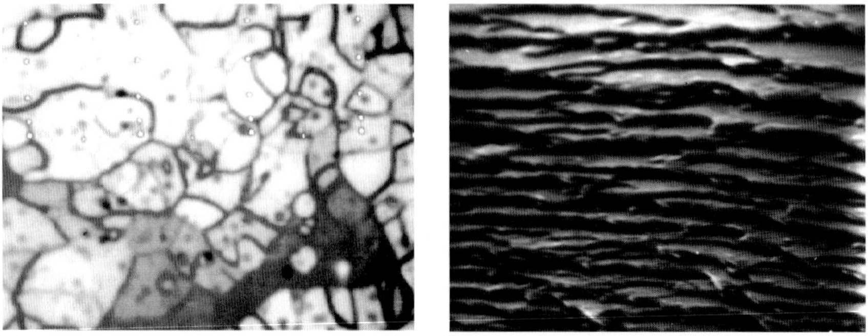

(a) (b)

**Fig. 3.37.** HOPG block structure: (**a**) in the plane of CC net, size 1 mm ×1 mm, 125 µm under the surface; acoustic microscopy; (**b**) in perpendicular direction; electron microscopy after laser etching

according to a certain distribution, which determines the HOPG mosaic spread. These blocks contain a second, small-scale mosaic structure with a smaller mosaic spread. Mosaic spread of crystallites within individual blocks can be as low as 0.1°. The form of the HOPG blocks is very anisotropic: their length along the CC plane is about two orders larger than in the perpendicular direction. Block structure can be visualized by acoustical microscopy [115] and by laser etching (Fig. 3.37a and b). It causes irregularities of the intensity, shape and symmetry of the diffraction patterns when a HOPG plate is scanned with a collimated X-ray beam.

### Bent HOPG Crystals

The traditional technology of manufacturing bent HOPG crystals includes the process of annealing under pressure on concave or convex pistons [111]. However, the anisotropy of thermal expansion and the laminar structure are impediments to obtaining the required profiles with a proper accuracy, especially when the shape is significantly more complicated than a simple cylinder, and the curvature radii are smaller than 100 mm. Here, process difficulties can lead to a drastic increase of the production costs. As a result, the first experiments with bent crystals used HOPG cylinders [116] or toroids assembled from a set of flat or singly curved HOPG crystals [117, 118]. To a certain extent, this approach involved both deviations of the shape obtained from the shape required and decreased reflectivities at the edges of separate crystals due to some damage during cutting. Some of these principle difficulties that arose during assembly restricted the range of applications of bent HOPG in analytical instrumentation.

The difficulties of this traditional approach to produce bent crystal initiated dedicated investigations aimed at a way to modify flat HOPG crystals into a bent form under chemical and thermomechanical treatment. The main difficulty was that such a treatment (e.g., intercalation by strong oxidizing agents [119]) led to a destruction of the material structure and a dramatic deterioration of the crystal mosaicity.

Dedicated investigations of the HOPG structure [115] resulted in the modification procedure [120] offering HOPG with both an excellent flexibility and an improved mosaic spread. The flexible HOPG can be easily stacked and mounted on a substrate of any form at room temperature [120]. In many cases, the crystals can be fixed on a substrate due to adhesion even without glue, if the substrate surface is smooth enough (e.g., polished). The method allows the creation of a wide range of crystal geometries ranging from conical and cylindrical [119], toroidal [121], paraboloidal [122] to spherical ones [123]. The crystal curvature radius can be as small as 5 mm. The mosaic spread depends on the film thickness. It varies from 0.1° for 10 μm thick films up to 0.4° ± 0.1° for the 200 to 400 μm thick HOPG layers mostly used in focusing optics [124].

Another advantage of the new technology refers to the relatively low production costs of these HOPG crystals. Usually, the costs of a flexible HOPG in an X-ray device does not exceed those of a similar standard flat HOPG crystal of a high quality and is notably smaller than the cost of an analogous standard bent HOPG crystal. Figure 3.38 shows a set of cylindrically bent HOPG crystals with different curvature radii from 5 to 20 mm [120].

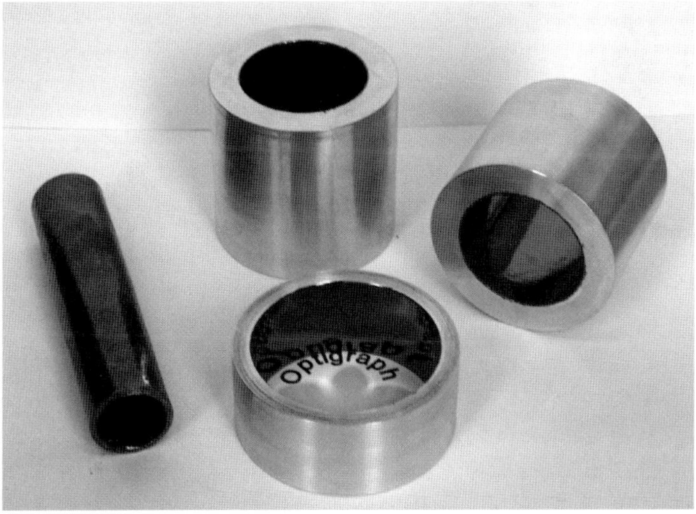


**Fig. 3.38.** Cylindrically bent HOPG crystals (radii from 5 to 20 mm) on aluminum substrates

## Main Physical and Geometrical Relations

The diffraction condition for an X-ray beam impinging on a crystal under the incidence angle $\theta$ is described by the well-known Bragg equation

$$2d \sin\theta = n\lambda,$$

where $d$ is a crystal interplanar spacing for a given lattice plane, $\lambda$ is the radiation wavelength and $n$ is the order of the reflection. The maximum value of the wavelength, at which this relation can still remain valid, is determined by the condition $\lambda_{\max} = 2d$.

HOPG crystals have a hexagonal structure with the lattice constant $d = 3.354\,\text{Å}$ for the 002 lattice planes. Therefore, the maximum wavelength, which can meet the Bragg equation, amounts to 6.708 Å and the diffraction is possible only for photon energies above approximately 1.85 keV.

HOPG crystals can be operated in a broad energy range (Fig. 3.39). Although the peak reflectivity reaches its maximum at about 17 to 20 keV, it preserves relatively large values up to very high energies where other optical elements (e.g., multilayers, capillaries, etc.) lose their efficiency drastically. Therefore, HOPG crystals can be also used for hard x-radiation in applications such as medicine and X-ray astronomy.

Focusing with bent crystals should meet normally two conflicting requirements: the Bragg condition and the condition of geometrical focusing which imply different meridional curvature radii of the lattice planes involved and of

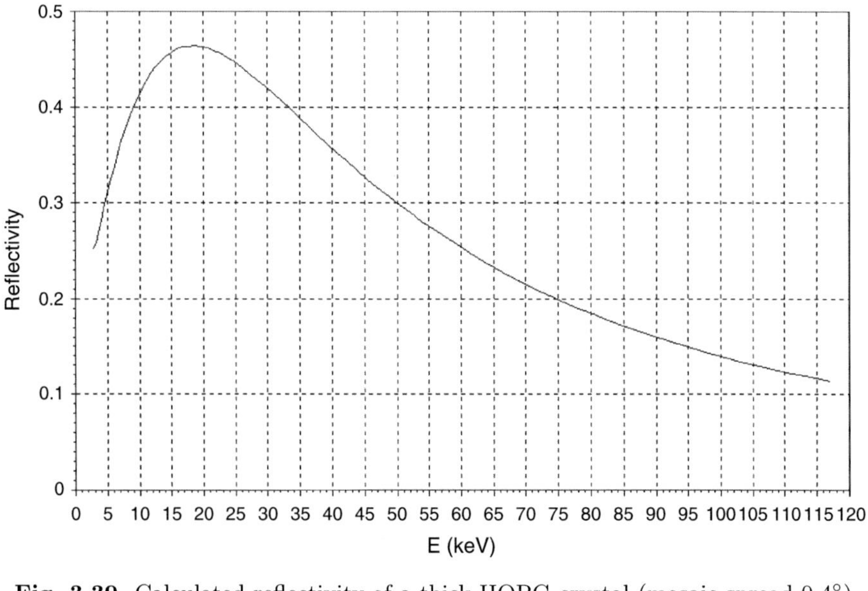


**Fig. 3.39.** Calculated reflectivity of a thick HOPG crystal (mosaic spread 0.4°)

the crystal surface. It is well known that the curvature radius $R_F$ for optimum focusing should be twice the Bragg radius $R_B$. Application of the Johansson scheme, which is the best solution for single crystals, is not directly possible for the mosaic case. However, one can use the crystal with the curvature radius $R_F$ for optimum focusing. Small angular deviations from the exact Bragg condition are mostly compensated by the mosaic spread of individual blocks.

The best results can be obtained with doubly bent crystals, when focusing takes place in both meridional and sagittal planes. The relationship for the meridional $R_1$ and sagittal $R_2$ curvature radii can be established geometrically (see Fig. 3.40) as

$$R_1 = D/2\sin\theta_B,\ R_2 = D\sin\theta_B,$$

where $D$ is the distance between the point source and the crystal centre (≡ distance between the focus and the crystal centre); $\theta_B$ is the Bragg angle.

The condition $R_1 = R_2$ (spherical shape) is valid only for $\theta_B = 45°$, which corresponds to $E = 2.61\,\text{keV}$. In a general situation $R_1 \neq R_2$, and the ellipsoidal shape could be offered as the most suitable one.

Being a mosaic crystal, HOPG has some peculiarities that should be taken into account while designing X-ray optics. First, there is a spatial beam smearing induced by the deep penetration of the reflected beam into the bulk of the mosaic crystal [125] aggravated by the fact that HOPG consists of light carbon. Two main effects determine the penetration depth of the X-ray radiation into a crystal: true absorption and interference extinction [125, 126]. In ideal crystals extinction due to interference is very strong because of a perfect arrangement of reflecting crystal lattices. As a result, one can assume that Bragg reflection takes place practically on a surface of an ideal crystal. This is not the case for mosaic crystals. Although extinction in HOPG dominates over true absorption, it is not so strong as in ideal crystals due to the spread of mosaic blocks [126]. Typical values of the total penetration depth lie in the interval 20–100 μm for the energies 6–30 keV and increase further

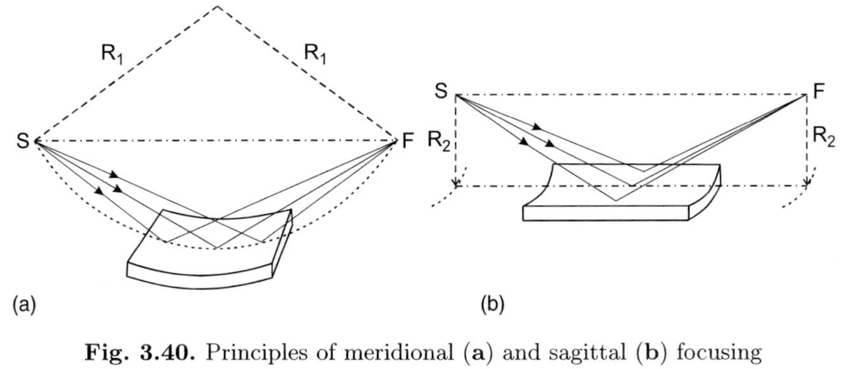


**Fig. 3.40.** Principles of meridional (**a**) and sagittal (**b**) focusing

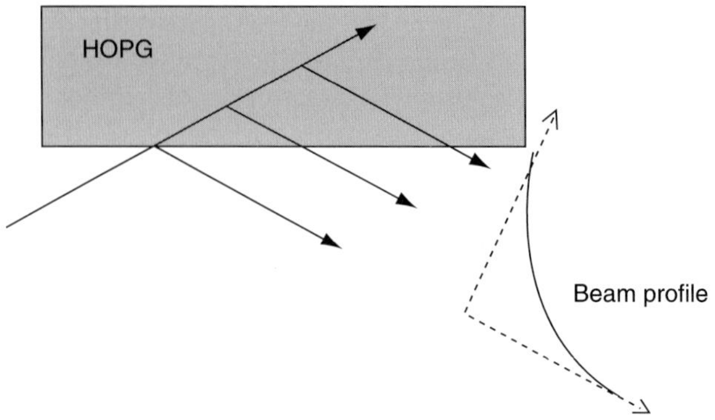


**Fig. 3.41.** Spatial broadening of the beam after reflection

with the raise of the energy. This fact has two simple consequences. First, a well-collimated beam becomes wider after the Bragg reflection (Fig. 3.41).

This effect limits the possibility of exact focusing with HOPG crystals and should be always taken into account when estimating the spot size on a sample. Second, the thickness of standard HOPG crystals (usually 1–2 mm) is superfluous for forming a diffracted beam of the highest intensity. In practice the thickness of 2–3 times larger than the penetration depth for a given energy is already sufficient for a strong diffraction. Therefore flexible HOPG foils with a thickness optimized for a high effective reflectivity or for a spatially narrow reflected beam can be successfully employed offering the respective advantages in comparison to standard thick crystals.

The second effect, which is characteristic for HOPG crystals and can be named as mosaic focusing, is illustrated in Fig. 3.42. A beam can be reflected even when its incidence angle has a certain deviation from the exact value determined by the Bragg condition. Such a beam penetrates into a crystal and somewhere in the depth meets a mosaic block with the necessary orientation. As a result, a flat mosaic crystal focuses incident radiation with the same energy into a spot. The angular aperture of mosaic focusing amounts approximately to the value of the mosaic spread of the crystal.

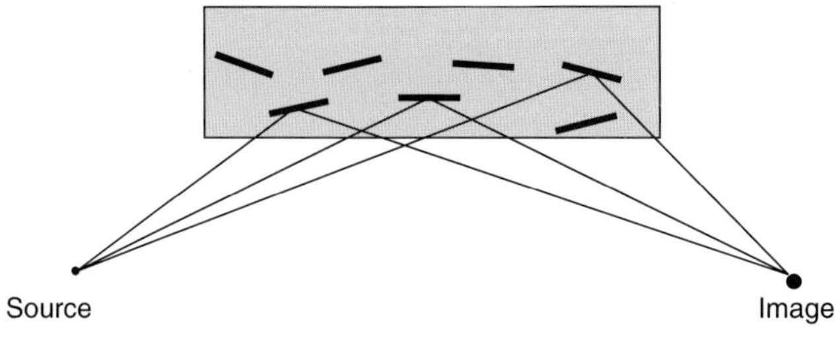


**Fig. 3.42.** The effect of mosaic focusing with HOPG

Designing HOPG optics consists in optimizing both the crystal form and local thickness for a given application involving specific constraints such as the source size, angular emission characteristics, desirable distance to the optics as well as the conflicting requirements (size, effective intensity gain, and energetic resolving power) with respect to the image focal spot. This optimization procedure can be performed best by a ray-tracing computer simulation [109, 126–128] involving all constraints and focal spot requirements.

## Applications in XRF: Excitation Mode

The use of HOPG monochromators for the primary beam in XRF, i.e., in the excitation channel, allows solving several problems at once. First, the background in the most interesting parts of a fluorescence spectrum can be significantly reduced due to the monochromatization, mostly due to eliminating bremsstrahlung contributions originating from the source and preserving only a strong characteristic line of the primary beam. Besides, employing a focusing device in the excitation channel can considerably increase the intensity of the excitation line due to a larger angular acceptance. In addition, focusing primary radiation onto a small spot enables performing local analysis with a high spatial resolution or to improve absolute detection limits, which is of particular interest in case of inhomogeneous contaminants.

Thus the use of focusing HOPG optics can also contribute to an improved analysability of small-sized specimens by EDXRF. Of decisive importance is a very high angular acceptance of focusing HOPG crystals for the incident radiation in order to ensure the highest possible intensity of the reflected beam. This requirement is due to the fact that as long as the absolute count rate capacity of the detection system is not saturated, i.e., mainly in the case of thin specimens, an intensity as high as possible in the fluorescence lines is of first priority for the minimum detection limits according to [129] and [130]. Only in the second place one can profit by a reduced background count rate.

Both requirements can be met by applying monochromatic exciting radiation with a high intensity. Reducing the spatial size of a specimen spot on any kind of backing, combined with a corresponding reduction of beam diameter, will also contribute to decrease the background, because the scattering at the backing is decreased.

A systematic selection among 55 single crystals undertaken in [131] revealed HOPG crystals to be superior candidates for the characteristic K$\alpha$ line of a Mo tube yielding the highest intensities. Another important result of this work consists in experimental proof that even singly curved HOPG crystals (either sagitally or meridionally bent) give a remarkable intensity gain of the reflected beam in comparison with flat crystals. The intensity gains of 2–10 times for HOPG cylinders compared with flat HOPG was obtained also in [120]. Using doubly bent toroidally shaped HOPG crystals [109, 121], it was possible to concentrate the characteristic radiation of a Mo tube onto a small focal spot of about 1 mm, achieving an intensity gain of 6.2 relative to the corresponding intensity of the Mo K$\alpha$ line of the radiation passing directly.

**Table 3.4.** EDXRF analysis of an aerosol loaded filter and detection limits (DL) for different excitation modes

| Fluorescence line | Line energy (keV) | Relative concentration (µg cm$^{-2}$) | DL for Sn secondary target excit. (ng cm$^{-2}$) | DL for direct tube excitation (ng cm$^{-2}$) | DL for HOPG focused tube radiation (ng cm$^{-2}$) |
|---|---|---|---|---|---|
| Ti Kα | 4.51 | $3.38 \pm 0.20$ | $301.0 \pm 19.3$ | $36.4 \pm 2.2$ | $26.9 \pm 1.6$ |
| Mn Kα | 5.89 | $36.73 \pm 1.36$ | $163.6 \pm 6.4$ | $39.7 \pm 1.6$ | $22.6 \pm 1.0$ |
| Fe Kα | 6.40 | $78.93 \pm 2.68$ | $134.7 \pm 4.9$ | $39.6 \pm 1.5$ | $20.1 \pm 0.8$ |
| Cu Kα | 8.04 | $1.06 \pm 0.06$ | $63.1 \pm 4.2$ | $27.5 \pm 1.7$ | $15.9 \pm 1.0$ |
| Zn Kα | 8.63 | $22.03 \pm 0.79$ | $52.8 \pm 2.2$ | $39.3 \pm 1.4$ | $12.7 \pm 0.7$ |
| Pb Kα | 10.55 | $2.65 \pm 0.12$ | $58.9 \pm 3.0$ | $72.5 \pm 3.4$ | $19.1 \pm 1.1$ |
| Br Kα | 11.91 | $3.30 \pm 0.14$ | $25.3 \pm 1.2$ | $35.2 \pm 1.8$ | $13.8 \pm 1.0$ |
| Rb Kα | 13.38 | $0.58 \pm 0.03$ | $18.7 \pm 1.1$ | $30.9 \pm 1.7$ | $8.0 \pm 0.6$ |
| Zr Kα | 15.75 | $0.33 \pm 0.02$ | $13.7 \pm 0.9$ | | |
| Nb Kα | 15.75 | $0.07 \pm 0.01$ | $13.1 \pm 1.2$ | | |
| Mo Kα | 17.44 | $0.04 \pm 0.01$ | $12.7 \pm 1.3$ | | |

A comparison of various excitation methods for aerosol-loaded filter specimens [109] revealed an enhancement of the detection sensitivity by a factor of 2 to 11 when implying a focusing HOPG toroid in comparison with the direct beam excitation or with the secondary fluorescence (Sn Kα) target excitation (see Table 3.4.) For the Bragg reflected Mo Kα radiation focused by the HOPG toroid the X-ray tube could be operated at a current being 25 times lower (i.e., 2 mA instead of 50 mA) than in a routine set-up with the Sn target, still using the full count-rate capacity of the detection system.

Optimization of the crystal thickness together with the crystal geometry can lead to the reduction of a focal spot size. A spot size of 150 µm × 250 µm was reported for thin HOPG toroids with fine focus tubes at Cr Kα and Rh $L_\alpha$ lines [132].

The results obtained demonstrate the possibility to simultaneously monochromatize and focus primary radiation onto small spots by means of doubly bent HOPG crystals, thus enabling local XRF analysis with a sub-mm resolution and improving absolute detection limits.

### Applications in XRF: Detection Mode

In the detection mode the HOPG optics can be used as a broadband filter or as a radiation collector between a specimen and a detector. The cylindrical geometry is the most commonly used geometry for dispersion filters. Varying the detector-to-sample distance and the size of the beam-stop, one can tune the position and the width of the energy window. The advantages of the use of a cylindrically bent HOPG in the detection channel of an EDXRF set-up consist mainly in the possibility of modifying the spectral distribution to enhance the fluorescence lines of interest. Furthermore, the high burden on the capacity of the detection system by scattered radiation can be reduced correspondingly.

Application of such types of dispersion filters is of special interest for samples with a heavy or radioactive matrix. The analysis of the group of light platinoid metals (Ru, Ro, Pd) and Ag in the presence of massive amount of Cu and Ni with a HOPG cylinder as a dispersion filter indicated the possibility to detect a 20 times smaller content of trace elements than in the conventional scheme ($2\times10^{-9}\,\mathrm{g\,g^{-1}}$ for SR source) [133]. Similar improvement of sensitivity was obtained by Chevalier at SR source [134] and by Beckhoff with Mo K$\alpha$ radiation of an X-ray tube (60 kV, 50 μA) [109].

By varying both the local curvature radii and the crystal thickness employing a Monte Carlo ray-tracing method [109] a HOPG focusing optics [135] was designed to ensure a quasi-rectangular bandpass shape in the energy range from 9 to 16 keV. This device leads in an EDXRF arrangement with Mo K$\alpha$ excitation to a drastic modification of the detected spectral distribution. Thus, both Rayleigh and Compton scattered excitation radiation and the fluorescence radiation of matrix elements having photon energy below 9 keV are drastically reduced while the fluorescence lines between 9 and 16 keV are enhanced correspondingly. The XRF detection sensitivities of the latter elements of interest are improved considerably.

In the second approach, a HOPG monochromator is adjusted to enhance a particular fluorescence line of interest collecting as much its intensity as possible. One of the best forms of the crystal for this task is a logarithmic spiral of revolution. As it is known, a logarithmic spiral has the remarkable property that all the rays coming from a certain point meet its surface under the same angle. Therefore, such a spiral can collect very effectively a particular fluorescence line from a small spot on a sample within a large capture angle and direct it towards a detector. In this case a detector without high spectral resolution can be used (e.g., an ion chamber, a photo diode or a PIN diode). However, its window and entrance apertures should be large enough to accept all the collected radiation because a log spiral does not provide point-to-point focusing. Realization of such an idea would be completely impossible without flexible HOPG crystals, because other known crystals could hardly be shaped in the form of a logarithmic spiral of revolution.

Pease et al. [136] have reported the first successful use of a log spiral HOPG for investigating the fluorescence X-ray absorption edge fine structure (XAFS). With the HOPG form optimized for the Cr K$\alpha$ line, the Cr X-ray absorption fine structure was investigated in an alloy of 1% Cr in a V matrix. The results showed a significant improvement of data quality due to the drastic suppression of the scattering background and the fluorescence radiation from the matrix.

A log spiral HOPG optimized for Au L$\alpha$1 line was used recently for enhancing the fluorescence microprobe sensitivity for precious metals [137].

The results reveal an enhancement of the Au signal, a rejection of close interfaces (As) and both relatively easy alignment and exploitation.

### Applications in XRF: Combined Mode

The above-mentioned merits of the primary and secondary monochromatization can be combined by the simultaneous use of two monochromators in the excitation and detection channels correspondingly which may lead to the development of a new type of the X-ray analytical instrumentation [138].

One of the first applications of HOPG crystals within this approach was presented in [109] where detection limits of the scheme using a HOPG toroid in the excitation line and a cylindrical HOPG crystal for collecting the secondary radiation were estimated. Improvement of the detection limits by the factor of about 5 compared to the data in Table 3.4 was reported. It was pointed out that further optimization and improvement of the detection limits could enhance the sensitivity of the EDXRF analysis up to the level of PIXE analysis.

The scheme employing two HOPG toroids in excitation and detection channels, respectively (Fig. 3.43) was realized for the geochemical analysis of Rb, Sr, I, Zr and Nb in separate grains of clinopyroxene [139]. An X-ray tube with an Ag anode was operated at 50 kV high voltage and 30 mA current. It was reported that the primary toroidal HOPG monochromator gave

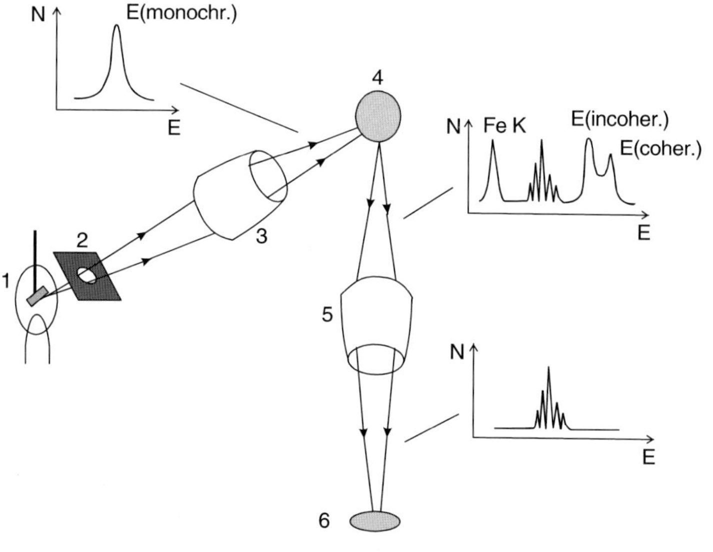


**Fig. 3.43.** X-ray optic scheme of the EDXRF spectrometer equipped with a focusing toroidal monochromator and a dispersion HOPG Johann filter [139] (1 – x-ray tube; 2 – Pb diaphragm; 3 – primary monochromator; 4 – sample; 5 – secondary monochromator; 6 – semiconductor Si(Li) detector)

**Table 3.5.** DL for five elements in a BCR-1 basalt sample [139]

| Optical scheme | Parameter | Rb | Sr | Y | Zr | Nb | Total intensity, counts |
|---|---|---|---|---|---|---|---|
| Flat | $I_p$, counts | 5,477 | 49,950 | 5,604 | 38,076 | 2,109 | |
| HOPG in | $I_b$, counts | 1,622 | 1,550 | 1,684 | 2,237 | 2,386 | 270,9447 |
| excitation mode | DL | 1.04 | 0.77 | 0.85 | 0.71 | 0.97 | |
| HOPG toroids | $I_p$, counts | 65,227 | 902,038 | 12,339 | 589,056 | 22,721 | |
| in excitation | $I_b$, counts | 8,627 | 12, 339 | 14,153 | 14,633 | 12,548 | 2,766,383 |
| and detection mode | DL | 0.20 | 0.12 | 0.11 | 0.12 | 0.21 | |

an intensity gain of 20 relative to a primary flat HOPG monochromator. Under these conditions the count rate of the secondary radiation reached $5 \times 10^4$ count $s^{-1}$, which exceeded the maximum count rate of the detector used. Therefore the secondary toroidal HOPG monochromator was installed as a broadband filter offering a passband of about 4 keV. In this way the detector load was reduced to an acceptable level due to suppressing scattered radiation and some disturbing lines (e.g., Fe). The passband realized allowed the simultaneous registration of at least five elements. Detection limits of the order of 0.1 ppm are reported (see Table 3.5, where $I_p$ is the intensity of the peak; $I_b$ is the background intensity), which is almost an order of magnitude better than in the case of a flat HOPG monochromator in the excitation channel.

It is worth mentioning that usually for mineral samples with volumetric weight in the range of 0.1–10 mg only a few methods meet the analytical requirements: ionic microprobe methods, inductively coupled plasma mass spectrometry (ICP-MS) and EDXRS with synchrotron radiation. The spectrometer presented in [139] is a simple realization of EDXRF with conventional X-ray tubes and HOPG monochromators.

### Other Applications

The possibility of realizing Bragg optics with arbitrary geometry offered by flexible HOPG crystals revives a new interest in old classical schemes, based on strongly bent crystals. A typical example is a von Hamos spectrometer [140–143] with a cylindrically bent crystal. Up to now nearly the only suitable candidate for realizing small curvature radii has been mica. With flexible HOPG manufacturing cylindrical reflectors with any dimension and any curvature (above 5 mm radii) is no longer a problem. Due to the effect of mosaic focusing one can expect significantly larger intensities with HOPG crystals relative to the ones achievable with mica. Two von Hamos spectrometers with mica and HOPG correspondingly (20 mm radius, see Fig. 3.44) were built up and tested at a wavelength of 2 Å using a $^{55}$Fe source [144].

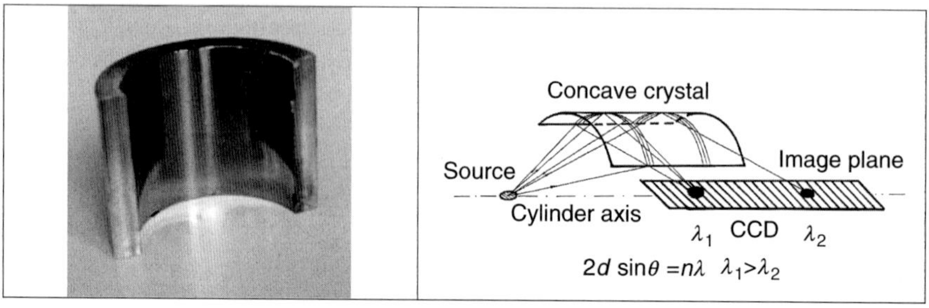


**Fig. 3.44.** HOPG cylindrical crystal (20 mm radius) and the scheme of von Hamos spectrometer [144]

As expected, the HOPG-based spectrometer revealed 70 times larger intensity than the mica-based spectrometer. Used for determination of sulfur in oil (7.5 W tube, PIN diode detector), the HOPG spectrometer demonstrated detection limits of 2 ppm in comparison with 14 ppm for mica [145]. A very important conclusion consists also in the fact that the energy resolution of a HOPG spectrometer is much higher than one could expect from the typical mosaic spread of HOPG (0.3–0.4°). In [146] the energy resolution better than 15 eV is reported, which enables to resolve clearly Cu Kα1 and Kα2 lines.

Recent investigations showed the applicability of HOPG optics for ultrafast time resolved X-ray measurements [147]. An ellipsoidally bent HOPG crystal was used for focusing monochromatic X-rays having a photon energy of 4.5 keV (Ti Kα-line) with an efficiency 30 times larger than that of the previously used bent crystals (Fig. 3.45) The main aim of the investigation was to maximize the available photon flux while keeping the focal spot size well below 1 mm. The spot size measured proved to be below 500 μm. The crystal was designed for a plasma X-ray source induced by a femtosecond laser. The layer thickness was only 50 μm to minimize a temporal spread of an ultrashort impulse. The temporal resolution was theoretically estimated as 300 fs, which (together with the high intensity) renders HOPG optics attractive for ultrafast time resolved

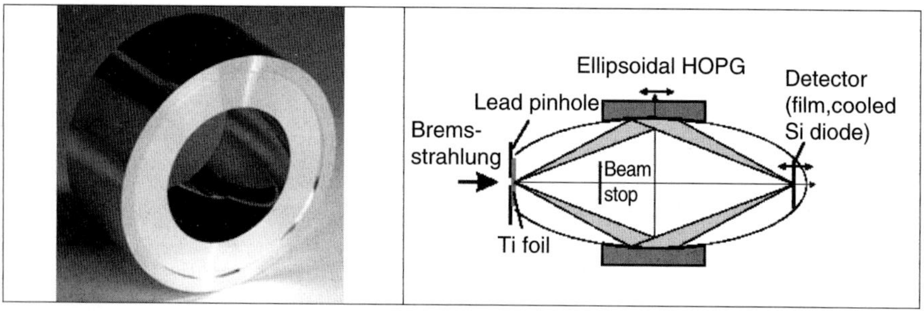


**Fig. 3.45.** Ellipsoidal HOPG crystal and the optical test scheme using Ti $K_\alpha$ radiation [147]

experiments. Although in this particular case the main application in mind was femtosecond X-ray diffraction recording many simultaneous reflections [148], the same scheme of focusing with a HOPG ellipsoid in the primary beam seems to be promising for performing μ-XRF. Depending on both the source diameter employed and the optics design, focal spot sizes of 150–300 μm are expected to be quite realistic.

#### Conclusions

HOPG optics can provide monochromatic or quasimonochromatic beams with the required spectral properties, divergence and dimensions. Suppressing continuous background contributions leads to a significant enhancement of the sensitivity of various analytical methods.

Focusing HOPG optics can find its main application in material analysis. In combination with low-power X-ray tubes it enables obtaining on a sample photon fluxes normally achieved only with high-power sources. As a result compact and sensitive spectrometers can be designed.

The development of HOPG-based focusing optics follows the modern trend towards miniaturization of X-ray analytical devices, which dominates nowadays in scientific instrumentation. Especially promising is the realization of different focusing geometries on the basis of strongly bent crystals. Such schemes of focusing were completely impossible before. One should expect that the application of these new schemes will contribute to the further development and improvement of different analytical methods.

### 3.4.4 Laterally Graded SiGe Crystals

*A. Erko*

In this chapter, the results of our investigations of laterally graded crystals on the basis of the $Si_{1-x}Ge_x$ alloy are presented. This work began in 1995 with the goal to develop a new monochromator for the BESSY II facility [149]. The gradient was produced during the crystal growth due to the different concentration of the Ge component in a Si matrix. The specially optimized laterally graded crystals with a *d*-spacing gradient *along* the dispersion direction have been used in an X-ray monochromator in a divergent beam for an improved spectral reflectance. The $Si_{1-x}Ge_x$ graded crystals were also successfully used at HASYLAB (Hamburg) for the energy range higher than 100 keV to enhance the integral flux through a monochromator [150].

#### Crystal Monochromator Energy Resolution

As is well known from the basic principles of crystal optics, the energy resolution of a double crystal monochromator with symmetrical Bragg reflection depends on the Bragg angle and divergence of the incident radiation

$$\triangle E/E \approx \triangle\theta \frac{1}{\tan\theta_B} \tag{3.90}$$

with $E$, photon energy, $\theta_B$, Bragg angle, $\triangle\theta$, incoming beam divergence. For a divergent incident beam, the approximated expression for $\triangle\theta$ is as follows

$$\triangle\theta = \sqrt{(\triangle\theta_n)^2 + \psi^2} \tag{3.91}$$

with $\psi$, the beam divergence in the plane of diffraction, vertical divergence, $\triangle\theta_n$, the Darwin width of a perfect crystal, which depends on the material, energy and crystal orientation.

For realistic situations the influence of the horizontal beam divergence can normally be neglected. This is however not the case in the vertical direction.

Figure 3.46 represents a comparison of the effective angular acceptance of the natural opening angle for BESSY II dipole radiation (electron energy: 1.7 GeV, critical energy: 2.5 keV) and the Darwin widths of perfect Si crystals in (111) and (220) directions.

According to (3.90) and (3.91) crystal limited resolution can only be achieved for parallel incident light. Thus the energy resolution of a symmetrical crystal monochromator strongly depends on the incident beam divergence. Except for the lowest photon energies the resolution is dominated by the divergence, if no collimating pre-mirror is used.

## Diffraction by a Laterally Graded Crystal

Whenever the crystal has a variation of the lattice parameter along its surface in dispersion direction $d(z)$, the influence of the incident beam divergence on the energy resolution can be avoided. The condition to be fulfilled according

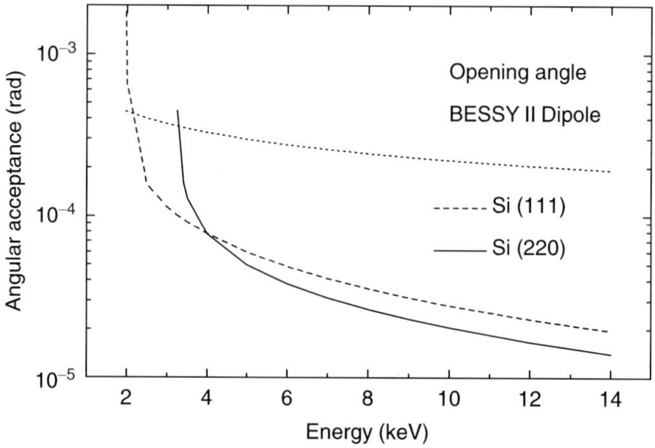


**Fig. 3.46.** Comparison between the natural opening angle of BESSY II dipole radiation and the Darwin width for Si (111) and Si (220) reflections

to the Bragg-equation is

$$\triangle\theta = d(z)\sin(\theta_{\mathrm{B}}). \tag{3.92}$$

In other words, the change of the Bragg angle along the crystal must be compensated for by a corresponding change in its d-spacing. This simple idea can now be realized on the basis of $\mathrm{Si}_{1-x}\mathrm{Ge}_x$ single crystals [151].

As is well known, silicon and germanium form a continuous series of solid solutions with gradually varying properties. The variation of the lattice parameter going from pure Ge to pure Si is shown according to references [152]. Evidently, a significant variation can be obtained with a very low concentration of Ge in Si. In the first approximation this curve can be approximated linearly by the following equation, Vegard's law

$$\frac{\triangle d}{d_{si}} = JC, \tag{3.93}$$

where $C$ is concentration of Ge in atomic % and the coefficient $J$ is derived from data given in Fig. 3.47.

To adapt a real crystal to the beamline monochromator, one needs to measure precisely the dependence between the lattice parameter and Ge concentration in a defined concentration region [153]. From the basic Bragg formula with optimum lattice parameter

$$\lambda = 2d\,(1+\triangle d/d)\sin(\theta_{\mathrm{B}} - \triangle) \tag{3.94}$$

one can easily calculate the optimum lattice parameter variation along the crystal and the energy resolution

$$\triangle E/E = (1+\triangle d/d)\sin(\theta_{\mathrm{B}})\,/\sin(\theta_{\mathrm{B}} - \triangle) - 1. \tag{3.95}$$

According to (3.95) the energy resolution can be optimized for a given energy $E_0$ with Bragg angle $\theta_{\mathrm{B}_0}$, assuming a linear Ge concentration gradient along

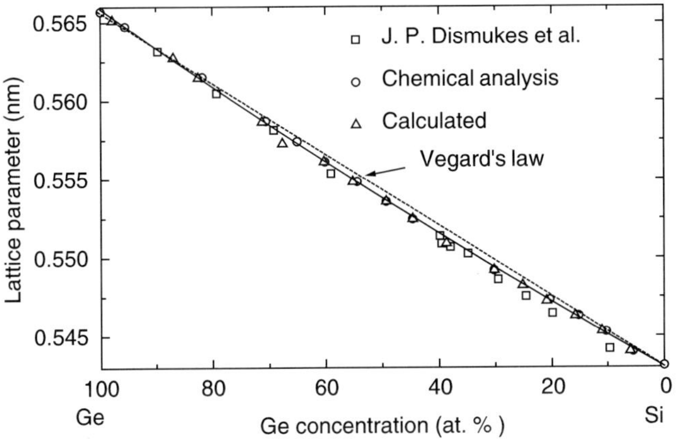


**Fig. 3.47.** Variation of lattice parameter with composition in the Si–Ge alloy system

the surface of the crystal z with total length $L$ according to

$$d(l) = d_{\mathrm{Si}}(1 + lz). \tag{3.96}$$

Here $l$ is a linear gradient of lattice parameter and using (3.93)

$$\ell = J\frac{(C_1 - C_2)}{L}, \tag{3.97}$$

where $C_1$ and $C_2$ are minimum and maximum Ge concentrations on the distance $L$ along the crystal. Then the energy resolution becomes

$$\Delta E/E = \frac{L\left(\ell R - \cos\left(\theta_{\mathrm{B}_0}\right)\right)}{R + L\cos\left(\theta_{\mathrm{B}_0}\right)} + (\Delta E/E)_n \tag{3.98}$$

within $(\Delta E/E)_n$ being the natural energy resolution corresponding to the Darwin width. This term will be minimized, if

$$\ell = \frac{\cos\left(\theta_{\mathrm{B}_0}\right)}{R} \tag{3.99}$$

$R$ is the source to crystal distance. So, the optimum concentration gradient corresponds to

$$\frac{C_1 - C_2}{L} = \frac{\cos\theta_{\mathrm{B}}}{R}. \tag{3.100}$$

The optimum concentration gradient is plotted as function of photon energy (3.99) in Fig. 3.48 for two different reflections: (111), (220) assuming a distance of 20 m between source and crystal.

Using these data one can effectively minimize the bandpass for one given energy $E_0$. For all energies the resolution is given by the equation

$$\Delta E/E = \frac{\Delta\left[\cos\left(\theta_{\mathrm{B}_0}\right) - \cos\left(\theta_{\mathrm{B}}\right)\right]}{\sin\left(\theta_{\mathrm{B}}\right)} + (\Delta E/E)_n\,. \tag{3.101}$$

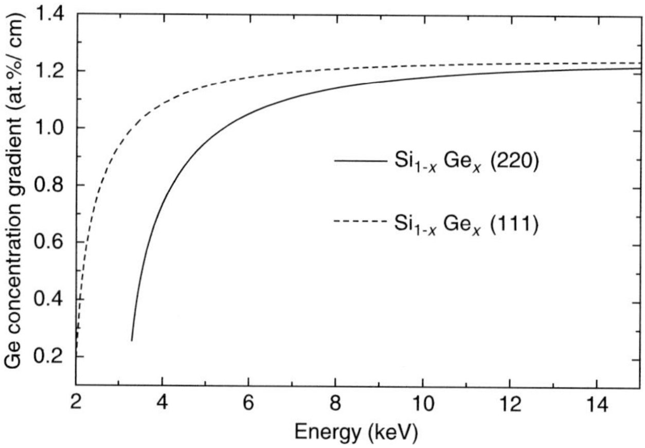


**Fig. 3.48.** The optimum concentration gradient of Ge in Si as function of photon energy

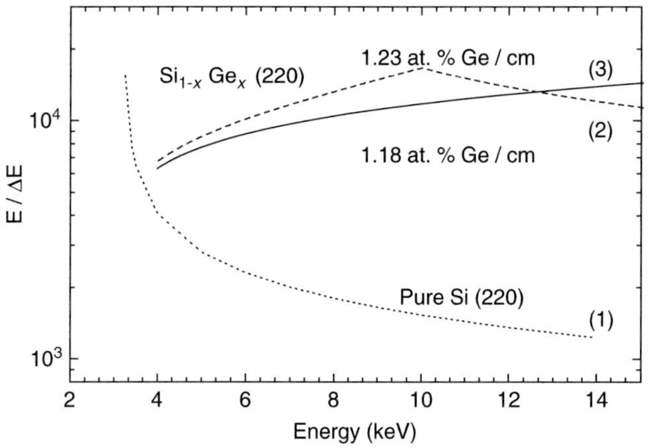


**Fig. 3.49.** Single crystal energy resolution as function of photon energy for: 1 pure Si crystal in a divergent beam; 2 gradient crystal optimized for 10 keV; 3 gradient crystal optimized for 20 keV. All calculations are for BESSY II dipole radiation

It is clearly seen that for all energies away from the optimized energy the resolution is also considerably improved. In Fig. 3.49 the calculated energy resolution for two different gradient crystals optimized for two different energies is shown in comparison with a pure Si crystal. The calculations were done for a single crystal at a distance of 20 m accepting the divergent BESSY II dipole radiation.

In addition to the improvement in energy resolution a gradient crystal gives much higher spectral reflectance. The integral reflectance does not change, but the flux is concentrated in a smaller energy range. The gain in the spectral flux is approximately given by the ratio of the vertical divergence of the radiation accepted by the crystal and the Darwin width of the crystal

$$G \approx \frac{\psi}{\Delta\theta_n} \tag{3.102}$$

### The Real Parameters of SiGe Crystals

The Czochralski (Cz) technique was used for the growth of $Si_{1-x}Ge_x$ gradient single crystals. There are two possibilities to change the crystal composition during the growth: passive – using the natural segregation of one of the components and active – using the melt replenishment with one of the component. The last method was used to grow the Ge-rich $Ge_{1-x}Si_x$ gradient single crystals by continuous feeding a melt with a number of Si rods [154, 155]. The application of such feeding technique for the growth of Si-rich crystals is more complicated, therefore the conventional Czochralski technique was chosen.

The measurements of the crystals were done in several laboratories [156–158]. According to Vegard's law ((3.93) and Fig. 3.47) the lattice parameter increases linearly with Ge concentration. At a rough estimate, this is a suitable approach. However, measurements on the lattice parameter revealed deviations from Vegard's law. For the measurements of the absolute lattice parameter variation was used, the Bond method in combination with two-dimensional scanning X-ray diffractometry [159]. Measurements of the lattice parameter done on the test samples with the known Ge concentration are shown in Fig. 3.50a. A similar result was published in [160]. One can see very clearly a difference between the linear dependence (Vegard's law) and experimental value. The dependence of the coefficient $J$ (3.93) from the Ge concentration can be derived by the empirical formula

$$J \approx 3.67 \cdot 10^{-4} + 1 \cdot 10^{-6} C \tag{3.103}$$

and is shown in Fig. 3.50b. Here $C$ is the Ge concentration in at.%.

A scanning double-crystal X-ray diffractometer was used at BESSY for two-dimensional local lattice parameter measurements. The beam size in diffraction plane was about 100 µm. The sample holder can be translated automatically parallel and perpendicular to the incident beam in the range of ±50 mm in horizontal and ±10 mm in vertical-direction with a linear reproducibility of 10 µm. Using this diffractometer it was possible to measure the Bragg angle variation at different points on the sample. The Ge-concentration was derived from the experimental diffraction data for the $Si_{1-x}Ge_x$-samples with (111) and (220) reflecting planes orientations. Furthermore, the width of

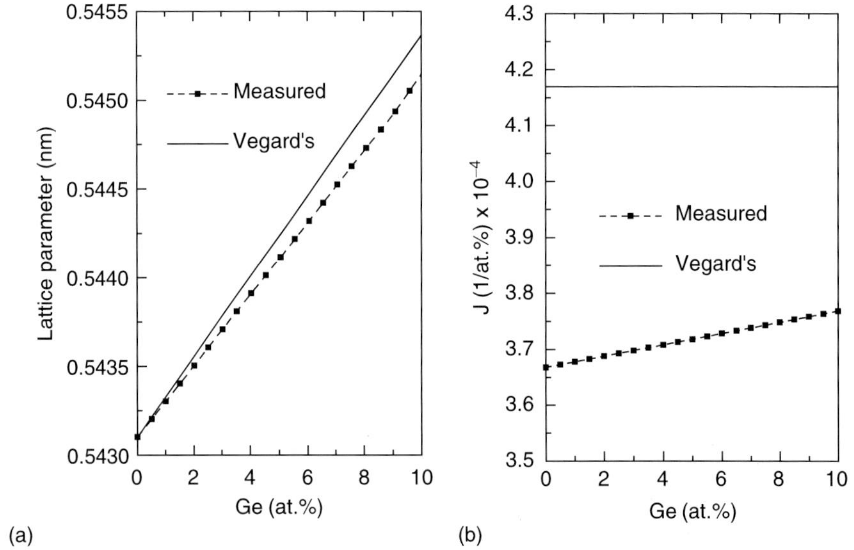


**Fig. 3.50.** The measured SiGe lattice parameter variation vs. Ge concentration (**a**) and measured value of the $J$ coefficient in comparison with Vegard's linear dependence (**b**)

the peak permits some conclusions about the quality of the sample-crystal in comparison to Si-reference crystals. In addition, the absolute reflectance of the specimen can be measured, providing further information about the quality and the homogeneity of the crystal. In Fig. 3.51a a two-dimension scan of the $Si_{1-x}Ge_x$ sample is shown. The grey scale variation in this figure corresponds to a different value of the lattice parameter in the sample.

From only one experiment we obtained information simultaneously about the Ge distribution and the sample curvature. The same procedure was used for the calculations of Si (111) and Si (333) reflections. Figure 3.51b shows a comparison of an infrared spectroscopy method [161] and diffraction measurements of a sample with <111> surface orientation for the horizontal dash line shown in Fig. 3.51a.

All results obtained with graded samples were compared with reference silicon single-crystal diffraction curves at the same experimental conditions. For the rocking curve measurements $Si_{1-x}Ge_x$ samples with uniform Ge surface concentrations of 1.5–7 at.% were examined. The surface orientation of those crystals was <110>. The rocking curve width of a $Si_{1-x}Ge_x$ crystal with a different Ge concentration is identical to a pure Si crystal. Both crystals show full width of half maximum (FWHM) of 4.5 arcsec, which is only slightly above the theoretical predicted width of 4.2 arcsec for this configuration for a perfect crystal. The intensity of the beam reflected by the sample was normalized to the flux from the crystal monochromator.

The measurements were performed at a fixed Bragg angle and with variable beam energies [162]. Energy scans were made by the double-crystal monochromator with two perfect Si (111) crystals. The energy spectra with a small beam divergence ($\sim$ 1 arcsec in both directions) and large divergence ($>$ 50 arcsec in horizontal and 1 arcsec vertical directions) were compared. We also compared the energy resolution from a reference Si (220) crystal.

The experimental results are shown in Fig. 3.52a,b. One can see the similar energy resolution of the $Si_{1-x}Ge_x$ and Si crystals for a beam with the divergence less than intrinsic rocking curve width. In case of high-beam divergence the energy resolution is very different. For the graded crystal angular

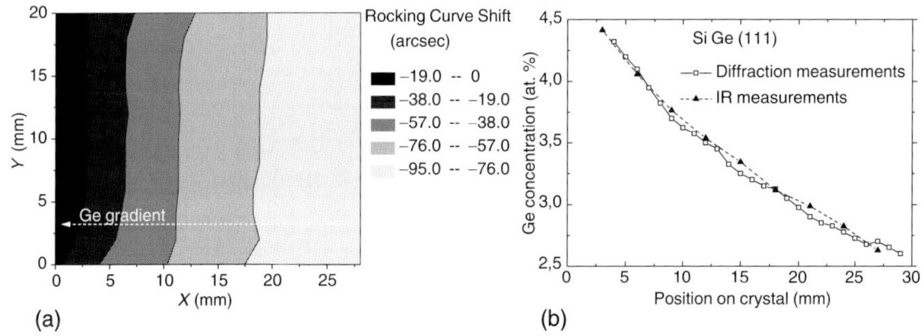


**Fig. 3.51.** The two-dimensional scanning diffractogram of the SiGe crystal (**a**) and the Ge concentration along the $X$-axis (**b**) measured with two different methods

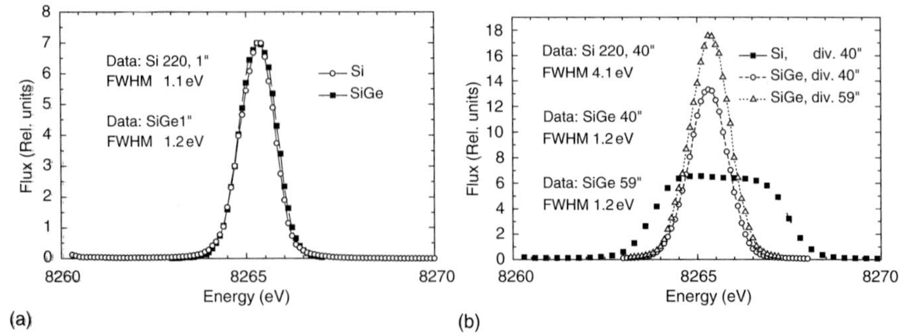


**Fig. 3.52.** Energy spectrum from the SiGe and Si (220) crystals, reflected in horizontal plane. (**a**) Beam divergence 1 arcsec, (**b**) beam divergence 40–59 arcsec

divergence is compensated by the lattice parameter variation along the crystal surface. Energy resolution of the graded crystal is the same, as for Si reference crystal in a parallel beam.

### Beam Collimation and Focusing by a Graded Crystal

It was found experimentally that the Ge concentration gradient is always followed by the lattice plane bending. Depending on the cut orientation, the average Ge concentration and the value of the Ge gradient one can choose optimal crystal parameters: gradient and crystal plane curvature for a particular beamline geometry.

One can find predictions of this effect in the literature [163]. The measured linear gradient of Ge concentration of 0.8 at.%/cm followed a curvature of the lattice plane with a radius of 102 m. Such crystal parameters fulfil the conditions of an effective compensation of a natural synchrotron beam divergence at a distance of 22 m from the source. For the crystals chosen we found perfect beam collimation at the energy of about 8 keV using simultaneously a linear lattice parameter gradient and a bent lattice structure. Double-crystal diffractometer measurements of the divergence of the beam diffracted from the graded crystal at 8 keV photon energy are shown in Fig. 3.53.

Here is plotted the angular variation of the diffracted beam measured with the beam directed along the concentration gradient (1) and anti-parallel to the gradient (2). The monochromatic parallel beam with a divergence of less than 1 arcsec and cross-section of $0.1 \times 0.1\,\mathrm{mm}^2$ was used as the input beam in this experiment. Diffraction angle variations were measured by scans over the entire crystal surface with 1 mm steps. One can see no variation of the Bragg angle for the beam coming anti-parallel to the gradient direction along the crystal surface.

For energies lower than 8 keV, due to the larger Bragg angle, the diffracted beam becomes slightly divergent. For energies higher than 8 keV, the diffracted beam is convergent and focused at the corresponding distance. The effect of this convergence/divergence gives us an additional, very important

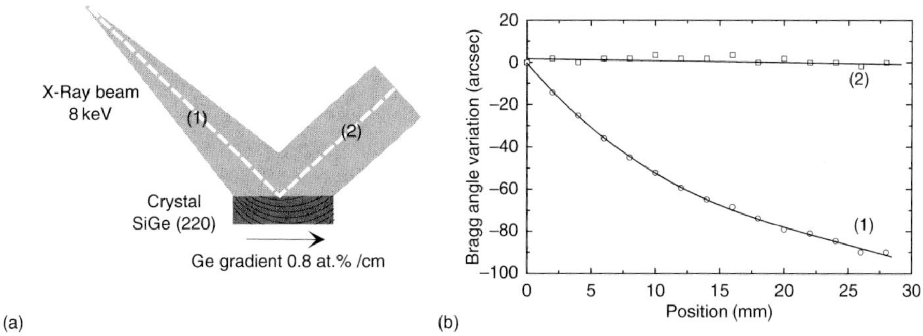


**Fig. 3.53.** X-ray beam collimation by a graded SiGe crystal: Optical arrangement (**a**) and Bragg angle variation in parallel (1) and anti-parallel (2) incident beam directions (**b**)

advantage. For a crystal with a linear parallel lattice and a parameter gradient, the theory predicts a very sharp energy of maximum energy resolution, especially for the higher indices of reflection (3.101). In the case of a bent lattice this range becomes relatively broad, as is shown experimentally in this paper. The broadening of the "resonant" curve is produced by a combination of the lattice parameter gradient and bent lattice planes.

## Double-Crystal Monochromator Performance

The graded crystals were mounted in a double-crystal monochromator at the BESSY KMC-2 beamline. Carefully polished crystals with (220) orientation were placed in positions according to the calculations as it is shown in Fig. 3.54.

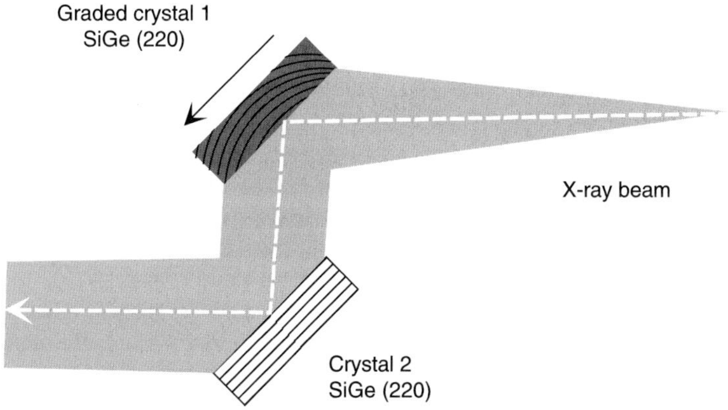


**Fig. 3.54.** Optical scheme of the KMC-2 monochromator with a graded SiGe crystal as the first element. The second crystal is SiGe without gradient

The first graded crystal has curved lattice planes in addition to the lattice parameter gradient. The gradient direction is shown in Fig. 3.54 by an arrow. Thus diffraction focusing takes place in the meridional direction and changes the beam divergence. A monochromatic beam, slightly divergent or convergent depending on the energy is reflected by the second crystal. To keep the same direction of the output beam, a SiGe crystal with uniform Ge concentration distribution was used as the second crystal in a monochromator. Beam divergence, photon flux and energy resolution were first calculated and optimized using the ray-tracing program RAY [164].

The measurements of the energy resolution of the double-crystal graded monochromator were done in the energy range from 5,930 up to 15,000 eV. For lower energies (5,931 and 7,908 eV) a normal incidence PIN diode produced on a Si (111) single crystal substrate was used. The diode area (20 mm × 10 mm) allows to accept the full beam aperture and to measure its spectral band. Reflection of a part of the incoming beam by the substrate lattice under Bragg conditions is the cause of the minimum in the photocurrent of the PIN diode. In comparison with expected monochromator energy band the energy range of this reflection is very narrow in the order of 50 mV. We have a change in current of the order of 5–10%. Similar measurements were made at energies corresponding to the Si (333), Si (220), Si (555) and Si (880) reflections at normal incidence. A calibrated GaAs photodiode was used to measure the photon flux density as for Si (220) and SiGe (220) graded monochromator. The comparison of both measurements is shown in Fig. 3.55. In spite of the fact that a lower photon flux density in the case of a graded crystal (220) monochromator the spectral flux density in the case of graded monochromator is higher due to better resolution and lower reflection coefficient. Simultaneously, we obtained an energy resolution, $E/\Delta E$, on the order of 10,000 over the entire energy range from 5 to 15 keV which is comparable with an intrinsic Darwin width of a perfect Si crystal. For most applications the increase of the spectral flux density is of great importance. The graded crystal monochromator allows one to obtain experimentally up to six times higher spectral flux density than with a normal crystal.

The maximum of the spectral intensity enhancement curve is at 12,000 eV and corresponds to the value of the Ge concentration gradient (0.8 at.%) and the lattice radius of curvature of 102 m. Maximum spectral flux performance can be optimized by variation of the gradient value. It is important that the low average concentration of Ge of about 3 at.% is not the cause of the decrease of crystal reflectance around the Ge absorption K-edge. We did not observe any decrease of monochromator intensity in the region of this energy.

Physical principles of the lateral changes of the crystal lattice parameter were discussed. The application of laterally graded crystals was a subject of several works [165]. A high-resolution monochromator system for synchrotron radiation with a lateral gradient of the d-spacing produced by a thermal gradient has been suggested [166]. In several papers it was suggested to apply crystals with a lattice space gradient as focusing elements in X-ray and neutron optics [167]. Successful attempts to grow KCl–RbCl laterally graded

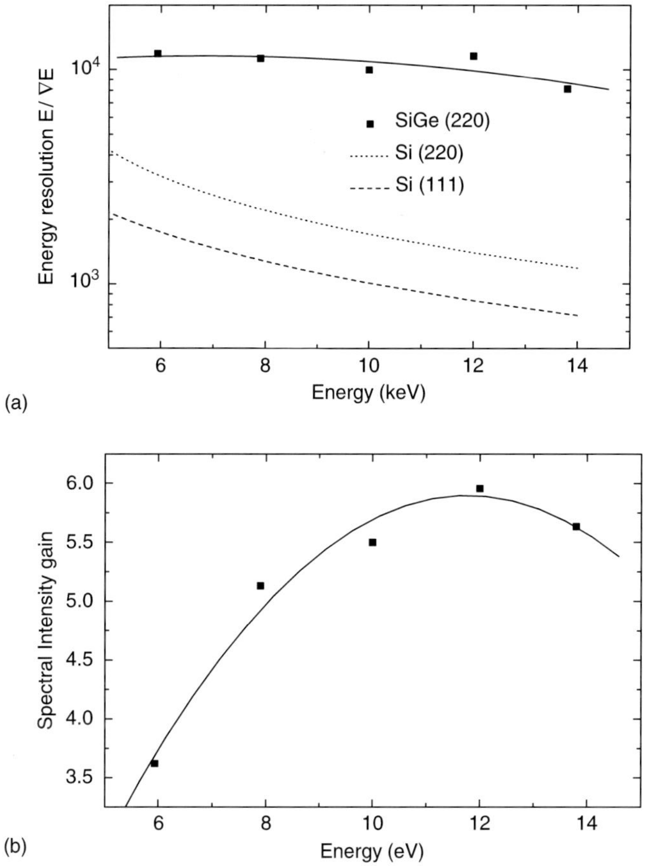


**Fig. 3.55.** Energy resolution (**a**) and spectral flux enhancement (**b**) of the graded SiGe monochromator in comparison with double Si (220) monochromator

crystals were also reported [168, 169]. The idea of a focusing device on the basis of a laterally graded asymmetric crystal was published [167].

# 3.5 Focusing Diffraction Optics

*A. Erko*

## 3.5.1 Zone Plates

### Zone Plate as an Imaging System

The principles of 3-D diffraction focusing optics can be described using the simple scheme of a transmission zone plate shown in Fig. 3.56.

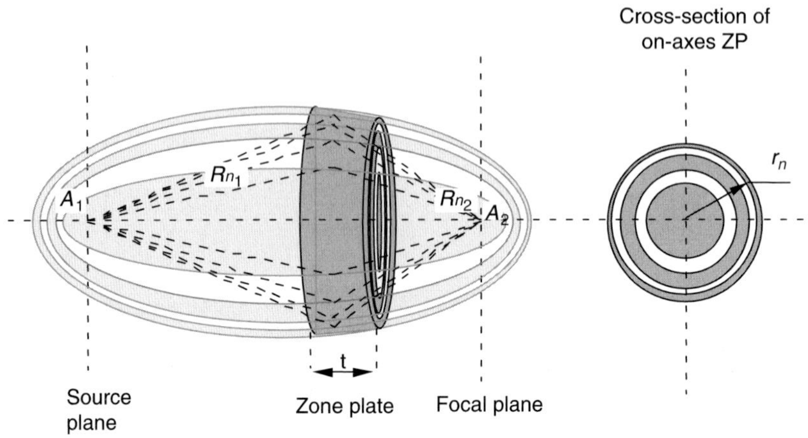


**Fig. 3.56.** Three-dimensional zone-structure of an "on-axis" transmission zone plate. The volume cross-section of the ellipsoids of revolution (left) is represented perpendicular to optical axis projection (right)

Shown here is an interference pattern produced upon the interaction of two spherical waves irradiated in points $A_1$ and $A_2$. Each rotational ellipsoid in Fig. 3.56 corresponds to a surface of equal phase, according to the fundamental property of an ellipsoid: the equal sum of the paths $R_1 + R_2$ for each point of the ellipsoidal surface. In Fig. 3.56 the path difference for the different surfaces corresponds to phase shift, $\pi$, of $n\lambda/2$. A cross-section of this system of ellipsoids perpendicular to the optical axis represents a zone plate. A real zone plate has a finite thickness, which increases with X-ray energy. Therefore, according to Fig. 3.56 a zone plate is a 3-D object with an ellipsoidal shape of the outer zones. Each zone is a volume section of equal phase surfaces, which lie on rotational ellipsoids.

If point $A_1$ is at infinity, the rotational ellipsoid degenerates into rotational paraboloid. Any scattering point C on a given ellipsoid surface forms the same wave phase at observation in point $A_2$ if point $A_1$ is a radiation source. The phase difference between the ellipsoidal surfaces is chosen to be $\pi$, so the optical path difference:

$$(R_{n1} + R_{n2}) - (R_{01} + R_{02}) = \frac{n\lambda}{2}, \tag{3.104}$$

where $(R_{01} + R_{02})$ is the distance between $A_1$ and $A_2$ along the axis.

If rays penetrate the screen at appropriate points one can obtain an on-axis Fresnel zone plate. In this case only those regions of the screen are opened that contribute to the image of point $A_2$ with the same phase sign. This is a transparent Fresnel zone plate, characterized by the Fresnel zone radii $r_n$ and thickness $t$. The Fresnel $n$-zone radius is determined by the equation:

$$r_n = \sqrt{\frac{nR_{01}R_{02}\lambda}{R_{01} + R_{02}}}, \tag{3.105}$$

where $n$ is the zone number. Here $R_1$ and $R_2$ are the corresponding distances from the screen to radiation source and the image. Using the thin lens formula $(1/R_{01}) + (1/R_{02}) = 1/F$, (3.105) gs written as:

$$r_n = \sqrt{Fn\lambda} \tag{3.106}$$

with $F$ denoting the focal distance.

## Zone Plate Aberrations

The aberrations of optical elements are conventionally analysed in terms of $\Delta L$: the optical path difference:

$$\Delta L = \sqrt{R_1^2 + r_n^2} + \sqrt{R_2^2 + r_n^2} - \sqrt{R_1^2 + R_2^2}. \tag{3.107}$$

These aberrations become noticeable when the term $\Delta L$ is comparable with $\lambda/4$.

*(a) Spherical aberrations:*
For the regular zone plate constructed in accordance with (3.105), the term responsible for the spherical aberration is:

$$\frac{1}{8}n^2\lambda^2 \left[1 - \frac{R_1^3 + R_2^3}{R_1 + R_2}\right] < \frac{\lambda}{4}. \tag{3.108}$$

One finds that the spherical aberration is a maximum for $R_1 = 2F$, i.e., for the unit magnification. The maximum zone number $N$ is limited by the value of:

$$N < \sqrt{\frac{8F}{3\lambda}}. \tag{3.109}$$

*(b) Chromatic aberrations*
Since the focal length of a zone plate is a function of wavelength, the chromatic aberrations are very strong. In some schemes of X-ray microscopes [170] holographically made zone plates in combination with pinhole in focal plane are used as the monochromators. Chromatic aberration is significant for:

$$\Delta L = \frac{n\lambda}{2} \pm \frac{\lambda}{4}. \tag{3.110}$$

The maximum number of zones for which chromatic aberration can still be neglected is given by:

$$N < \frac{\lambda}{\Delta\lambda}. \tag{3.111}$$

This equation is also valid for the calculation of an energy resolution when the pinhole of the diffraction-limited size $\beta$ is used in a focal plane.

*(c) Off-axis aberrations*
These aberrations arise due to off-axis source position or ellipticity in zone-plate structure. If the primary beam is parallel and $\alpha$ is a beam incidence

angle (angle to the optical axes), then the optical pass difference will have aberration terms, as:

$$\Delta L = \frac{r_n^3 \alpha}{4F^2} - \frac{3r_n^2 a^2}{4F} + \frac{n\lambda}{2}. \tag{3.112}$$

Using (3.112), one can estimate tolerance on zone plate orientation for the definite resolution in microprobe application. The first term in (3.112) is responsible for the coma aberration and second for the astigmatism and field curvature. Applying the $\lambda/4$ criteria for the above mentioned equation, one can write alignment requirements. In case of a zone plate with a number of zones $N > 100$, coma is the main aberration:

$$\text{-coma:} \quad \alpha < N^{-1.5}\sqrt{\frac{F}{\lambda}}. \tag{3.113}$$

In case of a zone plate with a number of zones $N < 100$, astigmatism and field curvature are the main aberrations:

$$\text{-astigmatism and field curvature:} \quad \alpha \leq \frac{1}{\sqrt{3N}}. \tag{3.114}$$

In comparison with a conventional lens a zone plate forms a distortion-free image.

Another technological requirement to the optics designed for high-resolution image transmission is that it should be free of wave-front phase distortions caused by zone plate *surface roughness* and refraction index fluctuations within the medium. The approximate maximum value of the surface roughness could be roughly estimated by equation:

$$\varsigma < 0.1 t_{\text{opt}}. \tag{3.115}$$

For example, the surface distortions for transparent silicon zone plates can practically reach $\varsigma = 0.1$–$0.3\,\mu\text{m}$ without affecting the image quality substantially.

### Intensity Gain

Intensity gain is defined as the ratio of the intensity in the focal spot of an optical element and the intensity behind a pinhole of size equals the spot size of the lens. Using the thin lens approximation, one can define the intensity gain $G_{\text{zp}}$ as:

$$G_{\text{zp}} = \varepsilon(2r(R_1 - F)/b_s F)^2, \tag{3.116}$$

where $b_s$ is the source diameter, $2r$ is the lens aperture and $\varepsilon$ is the lens efficiency.

### Zone Plate Aperture and Depth of Focus

The zone plate aperture can be easily calculated from (3.106). Thus, the opening aperture is equal to:

$$2r = \frac{F\lambda}{(\delta r)}, \tag{3.117}$$

where $(\delta r)$ is the minimum zone width. The focus depth of a zone plate also has a strong wavelength dependence:

$$\Delta F = \pm 2\frac{(\delta r)^2}{\lambda}. \tag{3.118}$$

It can be introduced for another definition of focus depth using a numerical aperture (NA):

$$\Delta F = \pm\frac{\lambda}{2(\mathrm{NA})^2}, \text{ where } \mathrm{NA} = \frac{\lambda}{2(\delta r)}. \tag{3.119}$$

### Integral Diffraction Efficiencies

Consider a zone plate in which odd zones are transparent and even zones are covered by a material with a rectangular form of grooves (phase-amplitude zone plate). The phase shift and the amplitude attenuation of the input light in the even zones analogous to transmission diffraction gratings is given by the (3.41–3.52) [7].

### Diffraction Limited Resolution

Although the above analysis indicates the positions and diffraction efficiency of the foci of a zone plate, it does not give the form of diffraction maxima, which can be obtained using the Fresnel–Kirchhoff diffraction integral.

The solution of the two-dimensional Fresnel–Kirchhoff diffraction integral can be found in a form of the Bessel or $\sin(x)/(x)$ function of first order with an argument:

$$\nu_m - r_N k\frac{r'}{F_m}, \tag{3.120}$$

where $r' = \sqrt{x'^2 + y'^2}$, the radial distance between the optical axis and an arbitrary point in the image plane $2r_N$ is the aperture of a zone plate.

The radial intensity distribution at the focus of a circular zone plate is well described by an Airy pattern analogous to a perfect thin lens:

$$I'_m(\nu_m) = \left[\frac{2J(\nu_m)}{\nu_m}\right]^2. \tag{3.121}$$

The solution of the one-dimensional diffraction integral can be found in a form of the sinus function of first order with an argument:

$$\nu_m = x_N k \frac{x'}{F_m}, \tag{3.122}$$

where $x'$ or $y'$ are the linear distance between the optical axis and an arbitrary point in the image plane.

The linear intensity distribution at the focus of a linear zone plate is well described by a pattern analogous to a perfect thin lens:

$$I'_m(\nu_m) = \left[\frac{2\,\sin(\nu_m)}{\nu_m}\right]^2 . \tag{3.123}$$

There is, however, a significant difference in the intensity distribution at the focus of a zone plate compared to the focus of a perfect refractive lens, which is not shown up by (3.121) and (3.123). For a zone plate there is always a low-intensity background caused by the zero order and high-diffraction orders.

## Thickness-limited Resolution of a Real 3-D Zone Plate

High-resolution transmission zone plates widely used as focusing X-ray optical elements have reached the theoretical limit of spatial resolution of about 15 nm [6]. This limit is defined not only by technological possibilities to produce small outer zone widths or continuous thin layers, but by the effects of volume diffraction/refraction in the structure of the optical element. The spatial resolution $\Delta$ of a standard zone plate is estimated in scalar diffraction theory by the Rayleigh criterion [50] which leads to:

$$\Delta = \frac{1.22(\delta r)}{m}, \tag{3.124}$$

where $(\delta r)$ is the width of the outermost zone and $m$ is the diffraction order. The resolution can be further improved by using curved zone profiles [171], but in practice even if zone plates with very small outer zone widths (say $\sim 10$ nm) could be fabricated, it will be not possible to reach the resolution expected from (3.124) even by using a high-diffraction order. In a real zone plate the thickness of material limits the validity of the (3.124) through, as in the case of waveguides, a volume diffraction phenomenon. To produce an optimum phase-shift of $\Delta\Phi_{\mathrm{opt}}$ between adjacent zones (to ensure the maximum diffraction efficiency) requires a structure with thickness $t_{\mathrm{opt}}$, (3.41).

Unfortunately, for X-rays with wavelengths less than $\sim 0.2$ nm the value of optimal thickness $t_{\mathrm{opt}}$ is of the order of several μm for all materials. Thus, zone plate technology in this energy range is complicated even at a sub-μm spatial resolution, due to the high-aspect ratios (line height/width) required.

An expression for the minimum possible zone width, introduced as the validity criterion of scalar diffraction theory is [7]:

$$(\delta r)_{\mathrm{min}} = \sqrt{m\ \lambda t_{\mathrm{opt}}}. \tag{3.125}$$

This approximation is in good agreement with the rigorous electromagnetic theory and with the theory of volume holograms [172]. For zones with spacings less than $(\delta r)_{\mathrm{min}}$ the scalar diffraction theory is not valid due to multiple diffraction of radiation at the zone plate structure. For high-energy X-rays greater than 1000 eV, the value of $t_{\mathrm{opt}}$ is of the order of several microns. A small focal spot size down to 0.1 µm × 0.1 µm can be achieved at a focal distance of 10–100 mm using a so-called modified zone plate [173] exploiting the first and higher orders of diffraction simultaneously.

## Ray-Tracing Model of a Zone Plate

Ray tracing is an indispensable tool for the design of optical systems for synchrotron radiation sources, and various programs have been developed during the last decades [174, 175]. By using a general-purpose ray-trace program, it is possible to obtain detailed information about the overall performance of the beamline optical system. Usually, the optical elements that are treated by a ray-trace program are slits and screens, mirrors and gratings, Bragg crystals and multilayers. Modifications of the wave front of light produced by these optical elements are described in the frame of geometrical optics and analytical equations rather well. However, the weak point of the ray optics is microfocusing with diffraction limited imaging.

In the paper [176], a ray-tracing code for zone plates incorporated into the program RAY [164] which is extensively used for beamline performance calculations at BESSY is described. The mathematical model allows one to follow a chromatic blurring of the focal spot as well as the smearing of the focus due to unevenness of the incident wave front (described by rays). Another advantage of the model is that it gives an intensity distribution, including auxiliary maxima and background radiation in the focal position.

The intensity distribution in the focal plane of a zone plate for a point source can be calculated analytically. Supposing a non-coherent irradiation, that is valid for all existing X-ray sources, one can reconstruct the image as a superposition of the images of the point sources, distributed in the object plane. Each point source will be transferred through a zone plate with a resolution defined by the zone plate aperture, the so-called diffraction limited resolution. At this step one can replace a wave front presentation with a ray-presentation. In the case of ray transmittance, each point of the source (or preceding element on an optical arrangement) produces a ray with defined parameters: spatial coordinates and energy. After interaction with a zone plate the ray angular coordinates are changed according to the defined probability. The probability distribution can be calculated using analytical formulas, represented in the following parts of the paper.

### Ray Propagation Probabilities

Tracing the zone plate, the program first solves the standard ray-tracing task of the ray surviving probability. The ray, which falls into the aperture of the zone plate, is considered to be partially absorbed by zone plate material. Together with the rays diffracted to negative ($m < 0$) and high ($m > 5$) orders, this ray is considered as "lost" because its intensity at the first order focus is infinitesimally weak. So, the ray must be thrown away with the probability:

$$E_{\mathrm{lost}} = E_{\mathrm{abs}} + \sum_{m=-\infty}^{m=-1} E_m + \sum_{m=5}^{m=\infty} E_m \tag{3.126}$$

for all $m < 0$ and $m > 5$.

If the ray is still considered a survived one, then its destiny has also two ways:

1. A ray is not diffracted (zero order) and its angle $\xi$ to the optical axis remain unchanged with the probability of $E_0$.
2. The remaining probabilities for the ray to be diffracted into first, third and fifth positive orders according to (3.126) are: $E_1$, $E_3$, $E_5$.

### Diffraction Limited Resolution

For the diffracted ray the probability to be deflected by the diffraction angles $\delta\varphi$, $\delta\psi$ and $\delta\xi$ to the $X$, $Y$ and $Z$ axis respectively, is defined by $I'_m(\nu_m)$ and calculated by the (3.120–3.123). The definition of the RAY coordinate system is shown in Fig. 3.57.

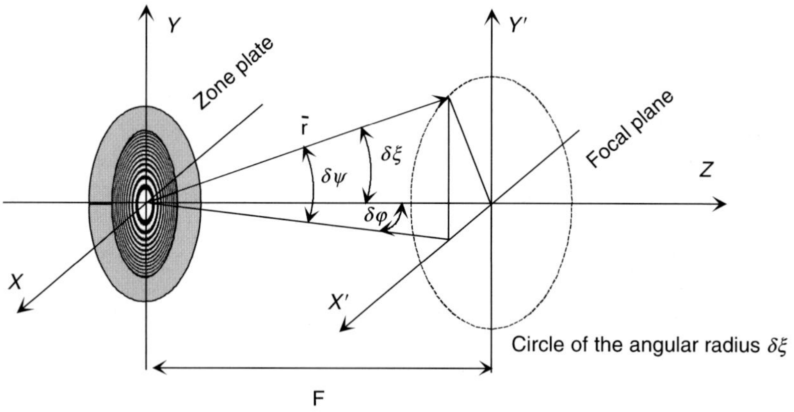


**Fig. 3.57.** The reference frame of the program, the angles of diffraction of a ray and the circle of the angular radius $\delta\xi$ at the position of the first order maximum of a zone plate

The ray is deflected randomly by the angle of $0 \leq \delta\xi \leq \delta\xi_{\text{max}}$ . According to Fig. 3.57 the values of the diffraction angle $\delta\varphi$ and $\delta\psi$ are defined in small angle approximation by the expression:

$$(\delta\xi)^2 = (\delta\xi)^2 + (\delta\varphi)^2. \tag{3.127}$$

For each ray the values $\delta\psi$ and $\delta\varphi$ are randomly selected within the angular range of $0 \leq \delta\psi,\ \delta\varphi \leq 0.707\,\delta\xi_{\text{max}}$. Than values of $\delta\xi$ calculated according to (3.127) pass through the probability generator in accordance to (3.121) and (3.123). Only those values of angles which passed the filter and have a Bessel function like probability distribution are used in the further procedure. With respect to the real intersection point of the ray with the zone plate, the real angles of its deflection are calculated as follows:

$$\Delta\psi = \delta\psi - \frac{x_{\text{zp}}}{\text{F}_m} \quad \text{and} \quad \Delta\varphi = \Delta\varphi - \frac{y_{\text{zp}}}{F_m}, \tag{3.128}$$

where $y_{\text{zp}}$ and $x_{\text{zp}}$ are the coordinates, where the ray hits the zone plate.

### Examples of Ray-tracing Calculations

*Diffraction limited resolution* has been checked using a point source and a divergent beam. An image aperture was placed in a distance, corresponding to the image plane calculated by the thin-lens formula. The results, obtained with the RAY program are shown in Fig. 3.58 for a linear (left) and circular (right) zone plate, respectively. The same zone plate was calculated by means of the diffraction program used for the calculation of X-ray holograms. The results were found to be identical.

The calculations were performed for a monochromatic point source with the energy of 100 eV, located on the optical axes of the zone plate. The zone

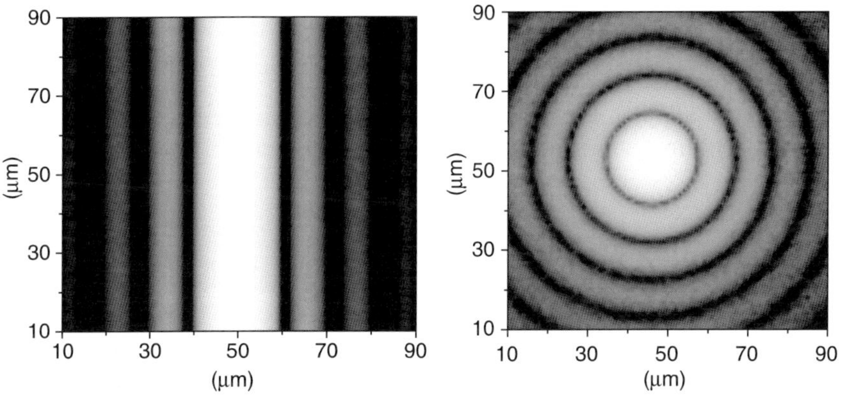


**Fig. 3.58.** Diffraction limited resolution for a linear (left) and a circular (right) zone plate

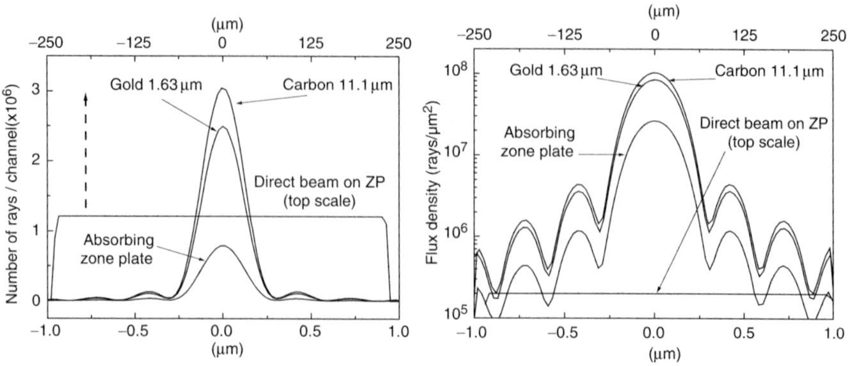


**Fig. 3.59.** The number of rays per channel (**a**) and the flux density (**b**) in the focal plane of a zone plate made of different materials (bottom scale). Here is also shown the direct beam flux (**a**) and flux density (**b**) on a zone plate (top scale). The total number of rays is $10^8$.

plate material was carbon with an optimal thickness of 171 nm. The two-dimensional intensity profile in the focal plane for the zone plates, shown in Fig. 3.58 , was obtained by ray-tracing of $10^8$ rays. *The integral efficiency* of zone plates made of different materials is shown in Fig. 3.59.

In this case, to demonstrate the RAY performance in different energy ranges, a parallel beam with photon energy of 8,500 eV has been chosen. The linear zone plate focuses in horizontal (X) direction. The outer zone width is equal to 0.3 μm. At the focal distance of 1,000 mm the corresponding lens aperture is 500 μm. Three materials were chosen: gold with an optimum thickness of 1629.5 nm, carbon with optimum thickness of 11.1 μm and finally a non-transparent material for the simulation of an amplitude zone plate. Parameters of zone plates are shown in Table 3.6.

RAY presents the results in the form of a 100 × 100 pixel array of a multi-channel analyser. The size of each channel depends on the aperture, covered with scattered rays. This aperture can be defined by a slit placed in the image plane, as was done for calculations shown in Fig. 3.59. Figure 3.59a represents the number of rays, collected in channels in $X$ direction, integrated vertically, for three-zone plate materials (bottom scale). The slit is placed in

**Table 3.6.** Parameters of zone plates used for the integral efficiency calculations. Zone plate type: linear; first order focal length $F_1$=1,000 mm; aperture: 500 μm × 500 μm

| Zone plate material/thickness | Flux density gain, rays/μm$^2$ | First order integral efficiency (%) |
|---|---|---|
| Carbon/11.11 μm | 507 | 40 |
| Gold/1.63 μm | 416 | 31 |
| Absorbing material | 131 | 10.1 |

the focal plane, the corresponding width of the channel is 0.03 µm. Shown also is the direct beam, measured in the plane of the optical element (upper scale). Zone plate aperture is 500 µm and the corresponding width of each channel in horizontal ($X$) direction is 6.05 µm. The area covered by each curve in Fig. 3.59a corresponds to the integral diffraction efficiency. In Table 3.6 these areas are normalized to the incident number of rays of $10^8$. In Fig. 3.59b the corresponding flux density in rays/µm$^2$ is shown. A flux density gain in rays/µm$^2$ can be obtained using the data in Fig. 3.59b.

*Image transfer* has been examined using an object consisting of four square frames 1 µm × 1 µm in size (Fig. 3.60a). The object is located at a distance of 2,000 mm from the zone plate. The focal length of the zone plate is 1,000 mm and the minimum zone width is 0.3 µm.

## Angular Spectral Analysis of a Zone Plate

Conventional zone plate as well as reflection or Bragg–Fresnel zone plate can be described as a superposition of diffraction gratings with different periods (spatial frequencies). This model is very common in mathematical analysis of holograms and diffraction images [7]. An illustration of this model is shown in Fig. 3.61.

A zone plate irradiated with a parallel beam produces diffracted limited images of different orders $F_1$ $F_3$ $F_5$, etc. at the distances of $F_1$, $F_2$, $F_3$ from the optical element. Each grating, located on the radius $r_i$ from the optical axis has a local period of $d$ and directed the diffraction orders to point $F_n$. In Fig. 3.61 are shown three periods $d_1, d_2$ and $d_3$, but in reality we have a continuous change of period within the zone plate radius.

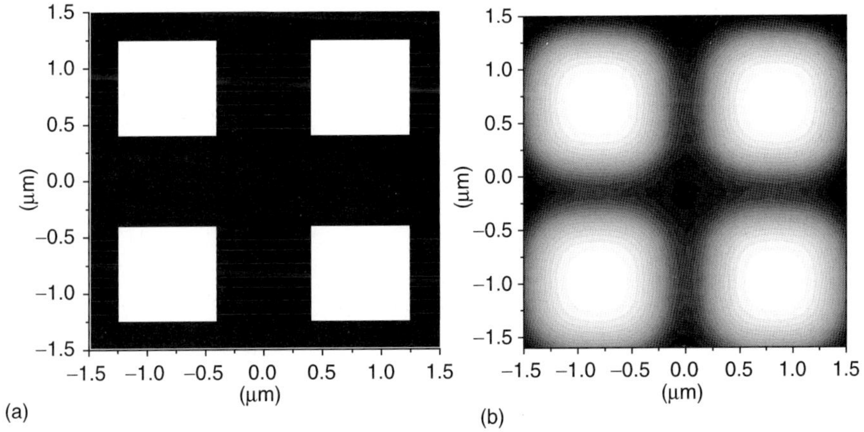


**Fig. 3.60.** An object consisting of four square frames with the size of 1 µm × 1 µm (**a**) and its ray-tracing image obtained with a gold zone plate at a distance of 2,000 mm (**b**)

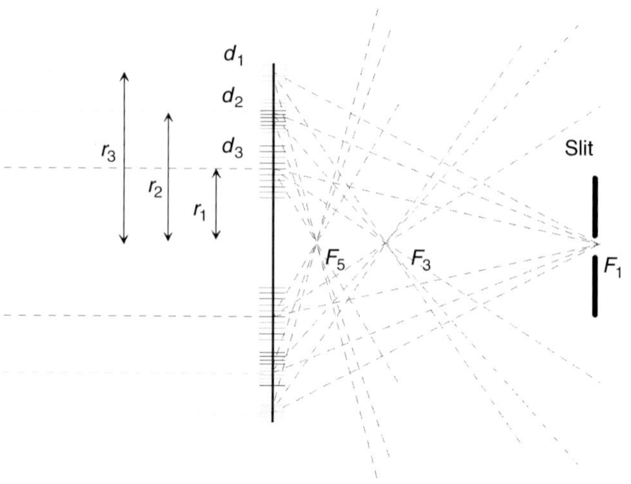


**Fig. 3.61.** Presentation of a zone plate as a superposition of diffraction gratings with variable period

Using a ray-tracing presentation, one can define each direction of diffraction as a ray, deflected by the optical element toward the focal points of the diffraction orders. Therefore the angular spectra of these rays will correspond to the angular spectra of diffracted beams in diffraction theory. According to general principles of diffraction optics analysis this spectra can be used for the calculation of focal spot size by Fourier transform of the angular distribution. In Fig. 3.62 is shown an angular spectrum calculated by the RAY program for the zone plate parameters listed in Table 3.6.

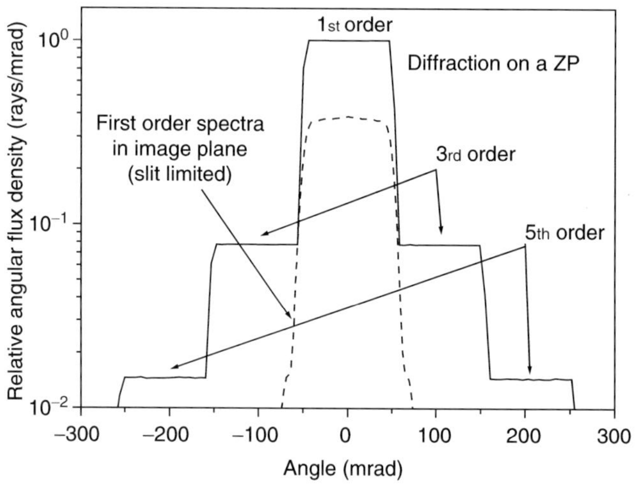


**Fig. 3.62.** The angular spectrum, calculated just behind a ZP (solid line) and in the focal plane limited by a slit of 10 μm (dashed line)

The ray-trace program RAY for a zone plate was used for the development of a practical method for measurements of zone plate parameters: efficiency and resolution. In Fig. 3.62 is shown an angular spectrum of the rays, diffracted on a zone plate and calculated just behind an optical element. One can evidently see different angular regions, responsible for focusing in different diffraction orders: +1; +3; +5. These spectra can be used for direct measurements of a zone plate resolution and efficiency. With the help of the Fast Fourier Transform program this spatial frequency spectrum can be converted into an intensity distribution in a focal plane of a zone plate. Our calculations show identical results obtained by Fourier transformation of the spectra and ray-tracing. The parameters measured by this method are free from experimental errors, arising from the source size and quality of the optics, placed in front of a zone plate. A parallel beam, spatially filtered through a pinhole can be used as a source in this experiment.

### 3.5.2 Reflection Zone Plate and Bragg–Fresnel Optics

#### Reflection Zone Plate Principle

Here the possibility of using a reflection zone plate is discussed. This consists of a reflector with an elliptical phase-shifting Fresnel structure on the surface (Fig. 3.63). A Bragg reflector on a crystal or multilayer structure or a total

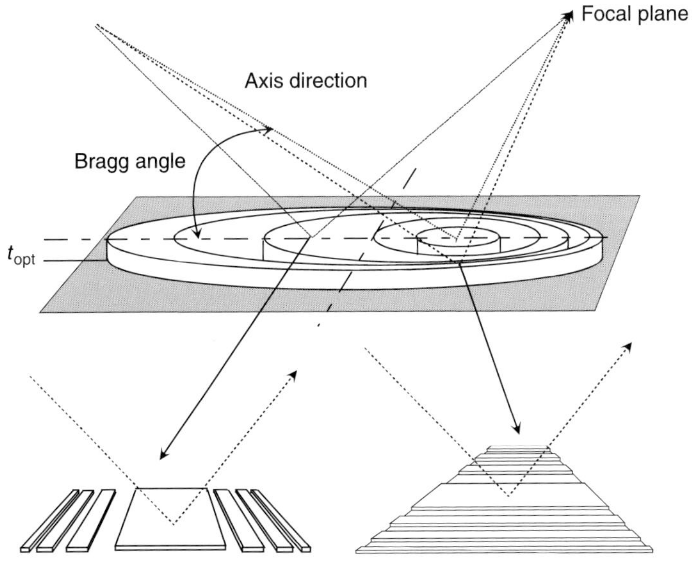


**Fig. 3.63.** Diffraction on the meridional (**a**) and sagittal (**b**) directions for reflection zone plates with optimum thickness $t_{\mathrm{opt}}$

external reflector on a super-polished surface can be used as reflective substrate. The optical path, $l$, of the beam in the phase-shifting material must be the same for both reflection zone plates and transmission zone plates. Consequently, the zone thickness required to give the optimum phase shift for reflection zone plates is smaller than for transmission zone plates since the radiation passes twice through the zone material at an angle.

Unlike for the previously reported Bragg–Fresnel lenses [177] the Fresnel structure is evaporated or sputtered onto the surface of a reflector. The structure is approximated by the formulae:

$$\left[x_n + \frac{n\lambda \cos\theta_B}{2\nu \sin\theta_B}\right]^2 + \frac{y_n^2}{\sin^2\theta_B} = \frac{Fn\lambda}{\sin^2\theta_B} \tag{3.129}$$

with

$$\nu = \frac{R_1/R_2 + 1}{R_1/R_2 - 1}, \tag{3.130}$$

where $x_n$ and $y_n$ are the coordinates of the $n$th elliptical zone on the surface, $\theta_B$ is the Bragg or grazing incidence angle, $F$ is the focal distance and $R_1$ and $R_2$ are the source to lens and source to image distances respectively.

For a reflection zone plate the optimum thickness of material is defined by the equation

$$t_{opt} = \frac{\Phi_{opt}\lambda \sin\theta_B}{4\pi\delta} \tag{3.131}$$

and the beam is transmitted twice through the thickness $t_{opt}$ at the angle $\theta_B$. In comparison with a transmission zone plate at normal incidence $t_{opt}$ is reduced by a factor of $0.5 \sin\theta_B$.

According to (3.131) a reflection zone plate the resolution limit is

$$\Delta_{mer} = 1.22\lambda\sqrt{\frac{\Phi_{opt} \sin\theta_B}{4\pi\delta}}. \tag{3.132}$$

The improvement in resolution is best for high energies and/or small angles of incidence. A comparison of the theoretical limits of spatial resolution for three types of gold zone plates is shown in Fig. 3.64. For the reflection zone plate substrates a Si (111) crystal with lattice period 0.313 nm and a multilayer mirror with a period of 3 nm have been assumed.

The values of $\Phi_{opt}$ were calculated from the non-trivial solutions of (3.49) [178] using tabulated values of the refractive index decrement and absorption index [49]. Due to the decrease of the Bragg angle at higher energies, the optimum thickness for the reflection zone plate remains constant for all the energies higher than about 2 keV, which also demonstrates the dramatically lower optimum thickness, and hence aspect ratios, for reflection zone plates. The resolution limit in the first diffraction order for energies higher than 15 keV can reach nanometer values, especially for small grazing angles on a

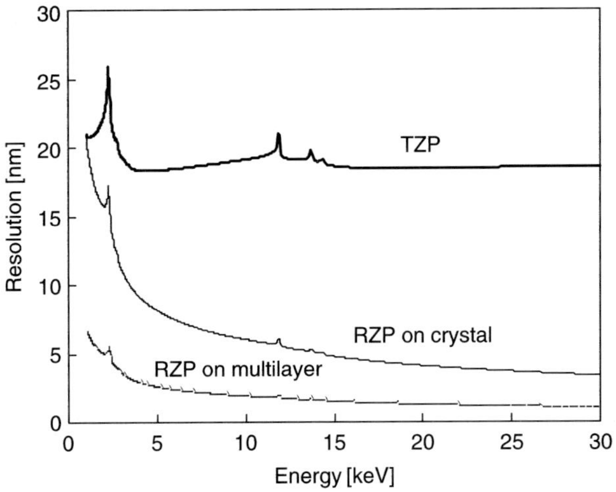


**Fig. 3.64.** Theoretical resolution limits for a transmission zone plate and reflection zone plates (RZP) on Si (111) crystal and 3 nm period multilayer substrates with optimum thickness gold zones

multilayer substrate and for larger period multilayers than in the example here [179].

### Reflection Zone Plate Aberrations

To analyse the off-axis aberrations, one can use an optical path difference assuming a rectangular groove profile. Using the optical path representation it is possible to find the analytical expression for main off-axis aberrations: coma, astigmatism and field curvature. According to the Rayleigh criterion, angle of view, limited by coma, astigmatism and field curvature are equal to:

$$\alpha_{\mathrm{coma}} = \frac{(\delta r)^2}{\lambda F}\left\{\frac{\lambda}{4(\delta r)\left[1+\frac{\lambda}{2(\delta r)}\cot(\theta_{\mathrm{B}})\right]^3} + \frac{\cot(\theta_{\mathrm{B}})}{1+\frac{\lambda}{2(\delta r)}\cot(\theta_{\mathrm{B}})}\right\}^{-1} \tag{3.133}$$

$$\alpha_{\mathrm{ast.},\,f.cur} = \sqrt{\frac{(\delta r)}{F}\left\{\frac{3\lambda}{4(\delta r)\left[1+\frac{\lambda}{2(\delta r)}\cot(\theta_{\mathrm{B}})\right]^3} + \frac{\cot(\theta_{\mathrm{B}})}{1+\frac{\lambda}{2(\delta r)}\cot(\theta_{\mathrm{B}})}\right\}^{-1}}\,, \tag{3.134}$$

where $(\delta r)$ is the minimum zone size; $F$ is the focal length; $\theta_{\mathrm{B}}$ is the Bragg angle.

As an example, the field of view of the off-axes lens at the grazing angle of 1° is about 12 μrad in the plane of reflection. Those parameters were chosen to demonstrate the importance of coma aberration at small angles and field curvature domination close to normal incidence. All other properties of a reflection zone plate are similar to those of on-axes zone plates.

### Bragg–Fresnel Lens Structure

Natural crystals or multilayer X-ray mirrors comprising alternating layers with different atomic numbers are extensively used as X-ray optics elements. X-ray radiation incident on the crystal or multilayer structure under a certain angle diffracts reflecting in a narrow spectral and angle range according to the Bragg law.

A reflection zone plate on a reflecting crystalline or multilayer substrate according to (3.129–3.130) enables simultaneous monochromatization and focusing of incident radiation. Optical elements of this type are known as Bragg–Fresnel lens [180]. In a general case the form of zone plates on a multilayer mirror surface was found to be rather complicated also in depth of the structure. The BFL structure fully replicates the interference pattern profile of reflection volume holograms. The layer structure intersects volume Fresnel zones and forms the BFL volume structure. Fabrication of such elements requires development of a spatial zone profile. Calculations require algorithms written in ellipsoidal coordinates [7].

Moreover, in the case of a crystal substrate the properties of a Bragg–Fresnel lens cannot be described in kinematic approximation and should be calculated with regard to the effects of multiple reflections involving several thousands of reflection layers. Due to this theoretical and technological complexity, crystal Bragg–Fresnel lens were realized only in a linear or backscattered geometry, where a simple circular structure is used [27]. In the case of a multilayer substrate the task is much easier. In the medium X-ray energy range the optimal number of layers of a Bragg–Fresnel multilayer lens structure does not exceed hundred layers. Therefore, the kinematic approximation may be used for qualitative description of the diffraction processes.

### Bragg–Fresnel Lens Efficiency

Focusing properties of a Bragg–Fresnel lens are the same as for off-axes reflection zone plates, excluding chromatic aberration, which is very limited due to the energy selection of the crystal or multilayer substrate. Due to the volume structure of a Bragg–Fresnel lens a 2-D focal spot with submicron size can be obtained only in one position in space at one fixed energy [181].

There are two possible methods of the Bragg–Fresnel lens fabrication: etching of the diffraction structure in crystal or multilayer material or masking of a part of the substrate surface by a phase-shifting material, e.g., gold, to produce a proper zone structure. In the first case, a Bragg–Fresnel zone plate achieves high efficiency, but needs a very large depth of the profile, for a crystal on the order of several μm. In the second case, the structure thickness can be small and a much higher spatial resolution is obtained.

*Etched Bragg–Fresnel lenses* on a multilayer substrate can be used successfully at small Bragg angles ($\sim 1°$), because the depth of the profile for such lenses is of the order of 0.1 μm and the shadowing effect is not very strong. The etched Bragg–Fresnel lenses on a crystal substrate can be used only at the normal incidence geometry, because the needed depth of profile is of the order of several microns.

*Evaporated Bragg–Fresnel lenses* have a much smaller thickness but a different efficiency in different diffraction directions. The efficiency can be calculated using the equations for the sagittal and meridional Bragg–Fresnel gratings.

### Reflection Zone Plate Resolution Enhancement

According to (3.132) the Bragg–Fresnel lens has higher spatial resolution in comparison with a transmission zone plate with the same outer zone width. This fact was demonstrated by measurements of the angular spectra of a meridional Bragg–Fresnel lens [182]. A monochromatic beam with $\lambda/\Delta\lambda \sim 5,000$ was collimated by a toroidal mirror and filtered by a pinhole of 100 μm in diameter. An angular spectrum of a zone plate was measured at the distance of $\sim 1,000$ mm with one-crystal Si [111] Bragg spectrometer in non-dispersive geometry. An input beam divergence after the pinhole, measured with a crystal spectrometer, was of the order of 4.5 arcsec.

Due to the small beam size the sample was scanned along the beam by a translation stage. The spectra from the different parts of the zone plate have been measured and composed on the same graph to build a complete angular spectrum of the Bragg–Fresnel lens. The reflection at the crystal surface free from the lens has also been measured to produce a reference spectrum of the direct beam.

The diffraction spectrum of the lens is shown in Fig. 3.65. All the measured spectra were superimposed in one graph and normalized to the reflection from the free crystal surface. On the same figure is shown the calculated spectrum, obtained by ray-tracing the zone plate with an input beam divergence of 4.5 arcsec. A minimum zone width for the best fit was found at 0.34 μm and a gold thickness of 1,350 nm. These values are in good agreement with the effective outer zone width and thickness of the BFL.

To prove the possible resolution in the meridional direction the performance of a meridional Bragg–Fresnel grating has been measured. A 2-μm period, 20-nm thick nickel grating was evaporated on the surface of an asymmetrically cut Si (111) crystal with an asymmetry angle of 13°. A diffraction efficiency of 1% was obtained at the energy of 8.45 keV at a grazing angle of 0.48°, corresponding to an effective period for a transmission grating at normal incidence of 16.8 nm or, for a zone plate, an outer zone width of 8.4 nm.

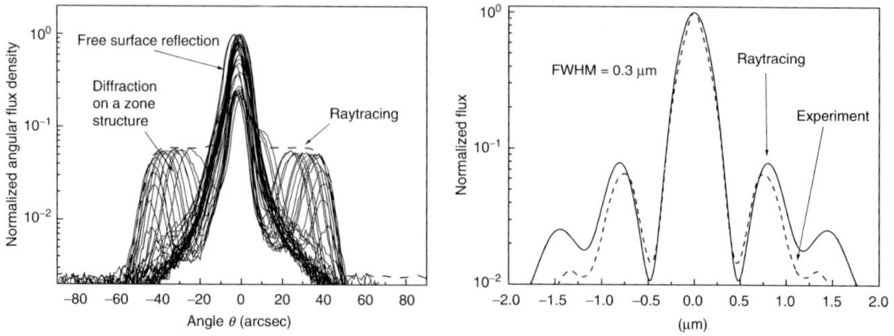


**Fig. 3.65.** The measured angular diffraction spectra in comparison with ray-tracing calculations (left) and a reconstructed flux distribution in a focal plane (right)

In Fig. 3.65 (right) is shown a reconstructed flux distribution in a focal plane obtained by Fourier transformation of the calculated ray-tracing spectra and the shape of the experimental spectra. Both curves have the same Full Width of Half Maximum (FWHM) of 0.3 μm.

Analysis of an angular distribution of diffracted rays leads to the possibility to fit experimental spectra and find zone plate characteristics, such as outer zone width and thickness of a phase-shift layer. These characteristics can be obtained almost independently from the experimental conditions, taking into account only beam divergence. Using this method considerable resolution improvement has been demonstrated for the meridional Bragg–Fresnel lens in comparison with a normal-incidence zone plate with the same outer zone width. The resolution enhancement of 4.3 times was measured experimentally. The measured efficiency of the Bragg–Fresnel zone plate with a thickness of 190 nm corresponds to the efficiency of a transmission zone plate with a thickness of 1,350 nm.

## Bragg–Fresnel Optics Applications

At the new-designed BESSY Microfocus Beamline, the horizontal beam focusing is achieved with the help of a sagittal Bragg–Fresnel lens for micro small angle X-ray scattering (μSAXS) experiments (Fig. 3.66).

Several Bragg–Fresnel lenses are placed on the second monochromator crystal, in our case Ge (111), which provides about 30% higher flux in comparison with a Si (111) crystal.

A linear Bragg–Fresnel lens on a crystal substrate was suggested and tested several years ago [183]. The first tests were done on the structures etched in a Si crystal to produce a π phase shift, necessary for an effective diffraction focusing. The main problem in using such a structure for microfocusing was a very large depth of profile necessary to achieve the optimal phase shift. For Ge (111) at 10 keV the depth reaches 0.6 μm. One needs to use an etching

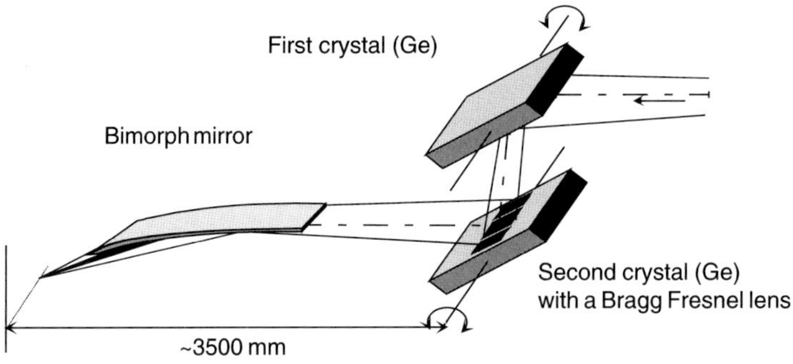


**Fig. 3.66.** Small Angle X-Ray Scattering focusing system employing a Bragg–Fresnel lens and bimorph mirror at BESSY

technology with a very high aspect ratio to obtain a reasonable lens aperture, which is defined by the minimum zone width. Several attempts were made to produce a special "incline" profile to improve zone plate acceptance and resolution. Recently this problem has been solved using metal coating technology.

A Fresnel structure was fabricated using a metal coating on a perfect Si (111) crystal in the shape of a Fresnel zone plate. In the case of such a coated zone plate the beam is transmitted twice through the thickness of a metal coating at the Bragg grazing angle $\theta_B$. In comparison with transmission zone plates the value of the optimal thickness is reduced by the factor of 0.5 $\sin(\theta_B)$.

The difference in optimal thickness is increased for high energies. A comparison of the theoretical value of the optimal thickness for the transmission zone plate and two types of Bragg–Fresnel lens is shown in Fig. 3.23. As the Bragg–Fresnel lens substrate we chose a Si (111) crystal with the lattice period of 0.31 nm and a multilayer mirror with a period of 3 nm. Even with the technology with an aspect ratio of about 1:1 (layer thickness equal to a groove width) an outer zone width of the Bragg–Fresnel lens at 30 keV can reach 0.2 μm. It can be easily fabricated for photon energies as high as 100–200 keV, which is a great problem for conventional zone plates due to the very high material thickness required.

The combination of the linear Bragg–Fresnel lens and a bimorph mirror in meridional direction allows a small-range energy scan without loss in spatial resolution. Several lenses fabricated on the same substrate will cover the entire operational energy range of the beamline. Special attention must be paid to beam position stabilization for the microfocus experiments.

### 3.5.3 Bragg–Fresnel Holographic Optics

In 1948, the Nobel Prize Laureate Denis Gabor proposed his famous "method of wave-front reconstruction" which began the era of holography [184]. In the

1960s with the development of optical lasers, E.N. Leith and J. Upatnieks, demonstrated the first wave-front reconstruction by a hologram with a reference beam [185]. Since this first successful hologram demonstration the method of holographic reconstruction has found applications in many different fields of art, science and technology. The possibility of creating holograms in the X-ray range has also been discussed in a number of works. However, such holograms have neither been fabricated nor tested until now except in the case of simple holograms (focusing zone plates). Due to the absence of sensitive and high-resolution materials for X-ray holography, it seems to be more effective to use X-rays only in the reconstruction stage to produce images with micron resolution. A so-called synthetic hologram [186] can be generated by computer and transferred into material using modern methods of microelectronics technology [187]. The holographic optics is becoming even more important with the construction of X-ray lasers in Germany and USA. The first successful reconstruction of a synthetic X-ray hologram was reported in 2001 [32].

A hologram structure can be calculated using specially developed computer software and fabricated on the surface of Si (111) monocrystal using e-beam lithography and a metal coating technique. A "white" broad-band synchrotron radiation beam from the BESSY bending magnet source was used for the hologram reconstruction.

The computer hologram generation comprises iterative calculations of the hologram phase distribution under the condition of constant value of intensity of the complex amplitude $A\ (x,\ y)$ in the hologram plane using direct Fourier transformations for definition of the image amplitude to be "as close as possible" to the demanded image.

For the hologram fabrication we employ the principle proposed for Bragg–Fresnel optical elements: a combination of Bragg diffraction on a single crystal and Fresnel focusing on a surface profile. The synthetic hologram has been designed as a phase shift mask placed on a polished surface of a Si (111) crystal (Fig. 3.67).

An initial hologram structure on resist was fabricated by electron-beam lithography. The synthetic hologram is fabricated assuming an ideal coherent source. The only limitation on spatial resolution is the minimum structure period of the e-beam writing process. The resolution of a synthetic hologram

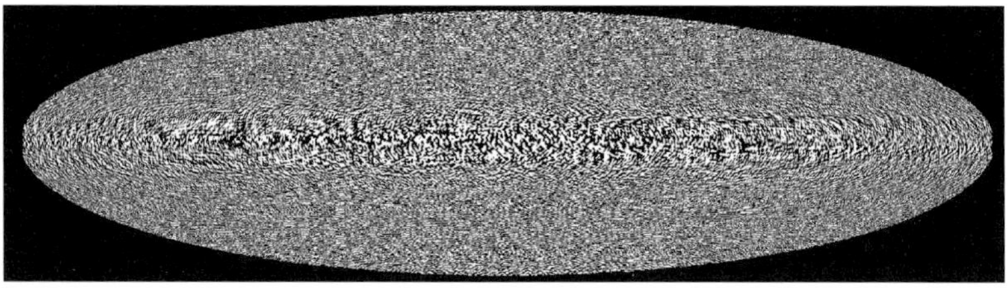

**Fig. 3.67.** Synthetic hologram structure of Ni layer on a Si 111 substrate. The minimum diffraction feature on the hologram is equal to 0.3 μm.

**Fig. 3.68.** Reconstructed holographic image with an aperture of 239 µm. The image resolution is of the order of 0.5 µm vertically and 2.4 µm horizontally

reconstruction can be estimated to be similar to the resolution of an X-ray Fresnel zone plate in partially coherent illumination. Each point of the source reconstructs the whole hologram image with a resolution defined by the size of the emitting source area.

The first experimental results of a new method of high-resolution imaging in the X-ray energy range, namely synthetic X-ray holography are shown in Fig. 3.68. The transverse and longitudinal spatial resolution of $\sim 2.5\,\mu m$ was experimentally measured. The resolution limitations appear due to the source size and angular reflection range of the mono-crystalline substrate of the BFL.

Extrapolations for X-ray laser sources give us a possible resolution limit of a few tenths of micrometers. This method should find applications in high-energy lithography (LIGA technology), for projection printing of micron size structures in very deep resists layers as well as in microobjects imaging, material diagnostics (e.g., in X-ray differential phase contrast microscopy) of solid-state and biological samples building X-ray optical systems with very special intensity and phase distribution in a focal spot of special form.

## 3.6 Refraction X-Ray Optics

*A. Erko*

### 3.6.1 Compound Refractive Lens

All the techniques of X-ray high-resolution optics mentioned above are based on total reflection or (and) diffraction. Focusing by refractive multiple lenses are based on refraction of X-ray beam on spherical or elliptical-shaped holes made inside of solids. One can obtain a convergent beam after propagation

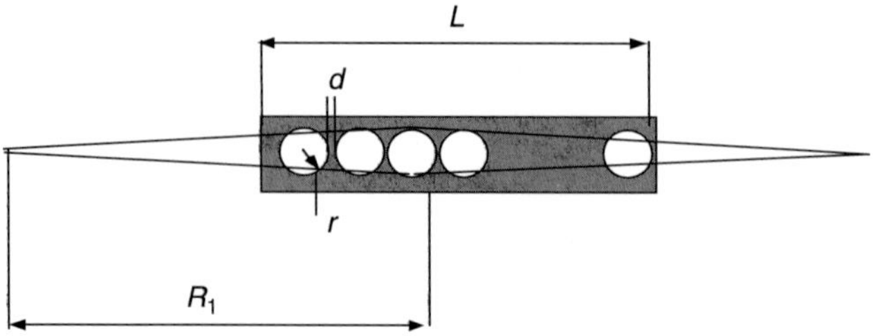


**Fig. 3.69.** Compound refractive lens with linear array of holes

through such a structure due to the fact that the real part (1-$\delta$) of the optical index is smaller than 1. Since the value of $\delta$ is of the order of $10^{-6}$, one has to use a series of holes to obtain a high enough optical phase shift, as shown in Fig. 3.69.

The focal length of a compound refractive lens (CRL) with $N$ individual lenses is:

$$F = \frac{r}{2\delta N_{\text{holes}}}, \tag{3.135}$$

where $r$ is the radius of cylindrical holes or the radius of curvature at the tip of a parabola for a rotational paraboloid. Parabolic lenses are more difficult in fabrication, but show no spherical aberration. The total length of the compound lens with the space $d$ between the holes is:

$$\begin{aligned} &\text{for cylindrical holes } L_{\text{cyl}} = N_{\text{holes}}(2r + d) \\ &\text{for parabolic holes } L_{\text{par}} = N_{\text{holes}}\left(\tfrac{r_0^2}{2r} + d\right) \end{aligned} \tag{3.136}$$

where $2r_0$ is the parabolic lens aperture or slit size in front of cylindrical lens.

### Compound Refractive Lens Efficiency

According to the model, developed in [188] the overall transmission $\varepsilon_{\text{cyl}}$ and $\varepsilon_{\text{par}}$ of a linear array of cylindrical and parabolic lenses respectively is:

$$\begin{aligned} \varepsilon_{\text{cyl}} &= 0.5\sqrt{\tfrac{\pi}{\mu N_{\text{holes}} r}}\left(\tfrac{r}{r_0}\right)\exp(-\mu N_{\text{holes}} d) \\ \varepsilon_{\text{par}} &= \left(\tfrac{r}{\mu N_{\text{holes}} r_0^2}\right)\left[1 - \exp\left(-\tfrac{\mu N_{\text{holes}} r_0^2}{r}\right)\right]\exp(-\mu N_{\text{holes}} d), \end{aligned} \tag{3.137}$$

where $\mu$ (µm$^{-1}$) is the linear absorption coefficient.

The gain of intensity in the focus of cylindrical $G_{\text{cyl}}$ and parabolic $G_{\text{par}}$ linear lens arrays can be evaluated using (3.137) as follows:

$$G_{\text{cyl,par}} = \varepsilon_{\text{cyl,par}}\frac{2r_0(R_1 - F)}{b_\nu F}. \tag{3.138}$$

For focusing in two dimensions (2-D) crossed cylindrical lenses can be used. In this case transmission and gain can be calculated as follows:

$$\varepsilon_{\mathrm{cyl,par}}^{2D} = (\varepsilon_{\mathrm{cyl,par}})^2 \tag{3.139}$$

$$G_{\mathrm{cyl,par}}^{2D} = (\varepsilon_{\mathrm{cyl,par}})^2 \left( \frac{2r_0(R_1 - F)}{F} \right)^2 \frac{1}{b_\nu b_h}, \tag{3.140}$$

where the indexes <cyl> and <par> correspond to cylindrical and parabolic lenses and $b_v$, $b_h$ are vertical and horizontal source sizes.

Absorption is the most important limiting factor for the refractive lens. One can define an effective aperture of the parabolic refractive lens, similarly to the one defined for a zone plate, using the following equation:

$$2r_{\mathrm{eff}} = 2\sqrt{\frac{r}{\mu N_{\mathrm{holes}} r_0^2} \left[ 1 - \exp\left( -\frac{\mu N_{\mathrm{holes}} r_0^2}{r} \right) \right]}. \tag{3.141}$$

### Compound Refractive Lens Aberrations

As a conventional refractive lens, an X-ray compound lens has strong aberration, which can be minimized by a special form of the lens profile. The simplest cylindrical lens has a very strong spherical aberration, which strongly limits the lens aperture. All these values can be calculated using optical lens theory. As it was shown using a parabolic or an even more complicated profile the lens aperture could be considerably increased. The compound refractive lens has one considerable advantage, in comparison with the Bragg–Fresnel lens, namely the possibility of a chromatic aberration correction using two lenses of the same material and different focal lengths $F_1$ and $F_2$ separated by a distance $d$ [189]. Chromatic aberration compensation could be achieved by letting:

$$d = 0.5[F_1(\lambda) + F_2(\lambda)]. \tag{3.142}$$

Off-axes misalignment of parabolic lenses does not alter considerably the focusing properties of the compound lens.

### Compound Refractive Lens Technology

For the CRL fabrication a number of different materials can be used. Depending on the photon energy range Be, Al, Si, etc. lenses can be fabricated. For example in the energy range of 10–60 keV the Al lens was found to be a "genuine imaging device" like a glass lens in visible light. A proposed modular system of a CRL allows one to assemble lenses to meet the specific requirements of a given experiment. Microbeams from several 10 µm down to 0.5 µm can be routinely produced [190]. In this case the lenses were made of polycrystalline aluminum.

# References


1. Bradaczek H, Hildebrandt G, Uebach W, Abstracts XII ECM, Moscow, **1**, 304 (1989)
2. Bradaczek H, Hildebrandt G, Uebach W, Acta Cryst **A49**, Suppl 374 (1993)
3. Erko A, Abrosimov NV, Alex V, Laterally-graded SiGe crystals for high resolution synchrotron optics, Cryst Res Technol, **37**(7), 685–704 (2002)
4. Erko A, Packe I, Gudat W, Abrosimov N, Firsov A, The crystal monochromator based on graded SiGe crystals, Nucl Instrum Methods Phys Res **A467–468**, 358–361 (2001)
5. Dhez D, Soft X-Ray optics and technology, SPIE Proc **733**, 308 (1986)
6. Chao W, Harteneck BD, Liddle JA, Anderson EH, Attwood DT, Nature **435**, 1210–1213, (2005)
7. Erko AI, Aristov VV, Vidal B, Diffraction X-Ray Optics, IOP Publishing, Bristol 150 (1996)
8. Duevel A, Rudolph D, Schmahl G, XRM-99, AIP Conf Proc **507**, 607–614 (1999)
9. Susuni J, Barret R, Kaulich B, Oestreich S, Salome M, XRM-99, AIP Conf Proc **507**, 19–26 (1999)
10. Drakopoulos M, Snigireva I, Snigirev A, Castelnau O, Chauveau T, Bacroix B, Schroer C, Ungar T, XRM-99, AIP Conf Proc **507**, 263–268 (1999)
11. Yun W, Howels MR, Feng J, Celestre R, Chang C-H, MacDowell AA, Padmore HA, Spence J, XRM-99, AIP Conf Proc **507**, 529–534 (1999)
12. Kemner KM, Lai B, Maser J, Schneegurt A, Cai Z, Ilinski PP, Kulpa CF, Legnini DG, Nealson KH, Pratt ST, Rodriges W, Lee Tischler M, Yun W, XRM-99, AIP Conf Proc **507**, 319–322 (1999)
13. Di Fabrizio E, Gentili M, Romanato F, XRM-99, AIP Conf Proc **507**, 635–640 (1999)
14. Yun W, Lay B, Cai Z, Maser J, Legnini D, Gluskin E, Chen Z, Krasnoperova AA, Vladimirsky Y, Cerrina F, Di Fabrizio E, Gentili M, Rev Sci Instr **70**, 2238 (1999)
15. Peuker M, XRM-99, AIP Conf Proc **507**, 682–687 (1999)
16. Kagoshima Y, Takai K, Ibuki T, Yokoyama K, Takeda S, Urakawa M, Tsusaka Y, Matsui J, XRM-99, AIP Conf Proc **507**, 668–671 (1999)
17. Davydov AV, Panchenko LA, Erko AI, Pis'ma Zh Tech Fiz **13**, 1017–1020 (1987)
18. Rehbein S, Doak RB, Grisenti RE, Schmahl G, Toennies JP, Woell Ch, XRM-99, AIP Conf Proc
19. Suzuki Y, Awaji M, Kohmura Y, Takeuchi A, Kamijo N, Tamura S, Handa K, XRM-99, AIP Conf Proc **507**, 535–538 (1999)
20. Barbee Jr, T W, SPIE Proc **911**, 169 (1988)
21. Aristov VV, Snigirev AA, Basov YuA, Nikulin AYu, AIP Conf Proc **147**, 253 (1986)
22. Aristov VV, Gaponov SV, Genkin VM, Gorbatov YuA, Erko AI, Martynov VV, Matveeva LA, Salaschenko NN, JETP Lett **44**(4), 265–267 (1986)
23. Erko A, Schäfers F, Gudat W, Abrosimov NV, Rossolenko SN, Alex V, Groth S, Schröder W, Nucl Instrum Methods Phys Res **A347**, 408–412 (1990)

24. Aristov VV, in X-Ray Microscopy II, Sayre D (ed.), Howells M, Kirz J, and Rarback H, Springer Series in Optical Sciences, Springer Berlin HeidelBerg, Newyork, 108–117 (1988)
25. Basov Yu A, Roshchupkin DV, Yuakshin AE, Optics Commun **109**, 324–327, (1994)
26. Erko A, Agafonov Yu, Panchenko LA, Yuakshin A, Chevallier P, Dhez P, Legrand F, Optics Commun **106**, 146–150 (1994)
27. Chevallier P, Dhez P, Legrand F, Idir M, Soullie G, Mirone A, Erko A, Snigirev A, Snigireva I, Suvorov A, Freund A, Engstroem P, Als Nielsen J, Gruebel G, Nucl Instrum Methods Phys Res **A354**, 584–587 (1995)
28. Holldack K, Erko AI, Noll T, Peatman WB, Nucl Instrum Methods Phys Res **A365**, 40–45 (1995)
29. Tucoulou R, Roshchupkin DV, Schelokov IA, Brunel M, Ortega L, Zigler E, Lingham M, Mouget C, Douillet S, Nucl Instrum Methods Phys Res **B132**, 207–213 (1997)
30. Kohn VG, The theory of x-ray Bragg-Fresnel Focusing by flat and elastically bent lens, Russian research Centre Kurchatov Institute, Preprint IAE-5878/9. M. (1995)
31. Firsov A, Svintsov A, Erko A, Gudat W, Kuznetsov S, Grigoriev M, Asryan A, Ferstl M, Shapoval S, Aristov V, Nucl Instrum Methods Phys Res **A467–A468**, 366–369 (2001)
32. Firsov A, Svintsov A, Zaitsev SI, Erko A, Aristov V, The first synthetic x-ray hologram: results, Optics Commun **202**, 55–59 (2002)
33. Erko A, Schäfers F, Firsov A, Peatman WB, Eberhardt W, Signorado R, The BESSY X-Ray microfocus beamline project, Spectrochim Acta A **Part B 59**, 1543–1548 (2004)
34. Rindby A, Nucl Instrum Methods Phys Res **A249**, 536–540 (1986)
35. Arkadiev VA, Kolomiitsev AI, Kumakhov MA, Ponomarev IYu, Khodeev IA, Chertov YuP, Shakhparonov IM, Sov Phys Usp **32**, 271–276 (1989)
36. Arkadiev VA, Bzhaumikhov AA, Gornyv H-E, Ibraimov NS, Proc SPIE **2515**, 103–113 (1995)
37. Erko A, Langhoff N, Bjeoumikhov AA, Beloglasov VI, High-order harmonic suppression by a glass capillary array, Nucl Instrum Methods Phys Res **A 467–468**, 832–835 (2001)
38. Erko A, Firsov A, Kanngießer B, Malzer W, Bartoli J, Röhrs S, Hahn O, Bjeoumikhov A, Langhoff N, A Capillary Focusing System for Micro-XANES/EXAFS Measurements, 17 Congress on X-Ray Optics and Microanalysis, Chamonix, September 22–27 (2003) Abstracts, 29
39. Marcus MA, MacDowell AA, Celestre R, Manceau A, Miller T, Padmore HA, Sublett RE, Beamline 10.3.2 at ALS: a hard x-ray microprobe for environmental and materials sciences, J Synchrotron Rad **11**, 239–247 (2004)
40. Ortega R, Bohic S, Tucoulou R, Somogyi A, Deves G, Microchemical element imaging of yeast and human cells using synchrotron x-ray microprobe with Kirkpatrick-Baez optics, Anal Chem 15; **76**(2), 309–314 (2004)
41. Snigirev A, Kohn V, Snigireva I, Lengeler B, Nature **384**, 49–51 (1996)
42. Hignette O, Cloetens P, Rostaing G, Bernard P, Morawe C, Review of Scientific Instruments **76**, 063709, (2005)
43. Wolter H, Spiegelsysteme streifenden Einfalls als abbildende Optiken für Röntgenstrahlen, Ann der Phys **10**(1), 94–114 (1952)

44. Michette AG, Backley CJ (eds), X-Ray Science and Technology, Plenum/IOP Publishing, Bristol (1993)
45. Vinogradov AV, Zorev NN, Kozhevnikov IV, et al. Reflective X-Ray Optics, Leningrad, Mashinovedenie, (1989) [in Russian]
46. Arkadiev VA, Beloglazov VI, Bjeoumikhov AA, Gorny H-E, Langhoff N, Wedell R, Application of capillary optics in modern scientific instrumentation, Poverkhnost (Physics, Chemistry and Mechanics of Surface), No. **1**, 48–54 (2000)
47. Compton AH, The total reflection of x-rays, Phil Mag **45**(270), 1121–1131 (1923)
48. Parrat LG, Surface studies of solids by total reflection of x-rays, Phys Rev **95**(4), 359–369 (1954)
49. Henke BL, Gullikson EM, Davis JC, x-ray interactions: photoabsorption, scattering, transmission, and reflection at E = 50–30000 eV, Z = 1–92, At Data Nucl Data Tables, **54**(2), 181–342 (1993). Data available on-line at http://www-cxro.lbl.gov/optical_constants
50. Born M, Wolf E, Principles of Optics, $5^{th}$ ed., Pergamon Press, Elmsford, NY (1975)
51. Kumakhov MA, Radiation by channeled particles in crystals, Energoatomizdad, Moscow (1986) [in Russian]
52. Kumakhov MA, Komarov FF, Phys Rep **191**(5), 289 (1990)
53. Vinogradov AV, Kovalev VF, Kozhevnikov IV, Pustovalov VB, V.B, Zh Tekh Fiz **55**(2), 244–250 (1985) [Sov Tech Phys **30**, 145 (1985)]
54. Arkadiev VA, Kumakhov MA, Fayazov RF, Pisma Zh Tech Phys **14**(3), 226–230 (1988) [Sov Tech Phys Lett **14**, 101 (1988)]
55. Jentzsch F, Phys Zeitschrift **30**, 268–273 (1929)
56. Jentzsch F, Nähring E, Die Fortleitung von Licht-und Röntgen-Strahlen durch Röhren, Zeitschrift für Technische Physik, **12**(3), 185–189 (1931)
57. Arkadiev VA, Kumakhov MA, Multiple reflection of x-rays on a curved surface, Poverkhnost, No. **10**, 25–32 (1986) [Phys Chem Mech Surf (1986)]
58. Müller JJ, Gorny H-E, Schmalz J, Heinemann UJ, Glass-capillary collimator for distance compensation and partial monochromatization at rotating anode x-ray generators, Appl Cryst **28**, 853–855 (1995)
59. Carpenter DA, Lawson RL, Taylor MA, et al., Microbeam Analysis, Ed. Newbury DE, San Francisco Press (1988)
60. Arkadiev VA, Bzhaumikhov AA, X-ray focusing in capillary systems, Proc SPIE, **2859**, 131–139 (1996)
61. Arkadiev VA, Bzhaumikhov AA, Gorny H-E, Ibraimov NS, Experimental investigation of Kumakhov lenses, Proc SPIE, **2515**, 103–113 (1995)
62. Ullrich JB, Ponomarev IYu, Gubarev MV, Gao N, Xiao Q-F, Gibson WM, Development of monolithic capillary optics for x-ray diffraction applications, Proc SPIE, **2278**, 148–155 (1994)
63. Owens SM, Ullrich JB, Ponomarev IYu, Carter DC, Sisk RC, Ho JX, Gibson WM, Polycapillary x-ray optics for macromolecular crystallography, Proc SPIE, **2859**, 200–209 (1996)
64. Gao N, Ponomarev I, Xiao QF, Gibson WM, Carpenter DA, Application of monolithic polycapillary focusing optics in MXRF, Proc SPIE, **2859**, 140–147 (1996)
65. Gao N, Ponomarev I, Xiao QF, Gibson WM, Carpenter DA, Appl Phys Lett **71**, 3441 (1998)

66. Gao, N, Janssens K, 3.3. Polycapillary In: Tsuj K, Injuk J, Van Grieken R (eds), X-Ray Optics in X-Ray Spectrometry, Wiley, New York, 89–110 (2004)
67. Bjeoumikhov AA, Langhoff N, Wedell R, Beloglazov V, Lebedev NF, Skibina N, X-Ray Spectrom **32**, 172–178 (2003)
68. Bzhaumikhov AA, Langhoff N, Schmalz J, Wedell R, Beloglasov VI, Lebedev NF, Polycapillary conic collimator for micro-XRF, Proc SPIE, **3444**, 430–435 (1998)
69. Fiorini C, Longoni A, Bjeoumikhov A, IEEE Trans Nucl Sci **48**(3), 268–271 (2001)
70. Kanngießer B, Malzer W, Reiche I, A new 3D micro x-ray fluorescence analysis set-up – First archeometric applications, Nucl Instrum Meth Phys Res **B 211**, 259–264 (2003)
71. Bilderback DH, Thiel DJ, Microbeam generation with capillary optics, Rev Sci Instrum **66**(2), 2059–2063 (1995)
72. Hosokawa Y, Ozawa S, Nakazawa H, Nakayama Y, An x-ray guide tube and a desk-top scanning x-ray analytical microscope, x-ray Spectrom, **26**, 380–387 (1997)
73. Pahl R, Bilderback D, Development of capillary optics for microbeam applications with synchrotron radiation, Proc SPIE, **2805**, 202–211 (1996)
74. Arkadiev VA, Bzhaumikhov AA, Method of investigating polycapillary performance, Proc SPIE, **3115**, 132–142 (1997)
75. Kimball JC, Bittel D, Surface roughness and the scattering of glancing-angle x-rays: Application to x-ray lenses, J Appl Phys **74**(2), 877–883 (1993)
76. Wang L, Rath BK, Gibson WM, Kimball JC, MacDonald CA, Analysis of capillary optics performance for hard x-rays, Proc SPIE, **2859**, 170–181 (1996)
77. Arkadiev VA, Kolomiitsev AI, Kumakhov MA, Labusov VV, Ponomarev IYu, Khodeev IA, Chertov YuP, Shakhparonov IM, Pisma Zh Tekh Fiz **14**, 97 (1988) [Sov Tech Phys Lett **14**, 48 (1988)]
78. Arkadiev VA, Beckhoff B, Bzhaumikhov AA, Gorny H-E, Kanngießer B, Experimental investigation of capillaries and focusing systems in soft energy range, Proc SPIE, **2859**, 123–130 (1996)
79. Bjeoumikhov A, Bjeoumikhova S, Langhoff N, Wedell R, Capillary optics for micro x-ray fluorescence analysis (to be published)
80. Rindby A, Janssens K, Microbeam XRF, in Handbook of x-ray Spectrometry, Marcel Dekker, Inc New York, 631–717 (2002)
81. Bjeoumikhov AA, Verfahren und Vorrichtung zur Herstellung optischer Elemente für die Kapillaroptik, Patentschrift DE 195 27 794 C2 (filed in 1995)
82. Knöchel A, Gaul G, Lechtenberg F, Verfahren zur Herstellung von Glaskapillaren und deren Verwendung, Patentschrift DE 44 44 102 C1 (filed in 1994)
83. Arkadiev VA, Bzhaumikhov AA, Capillary-based x-ray filters, Proc SPIE, **2859**, 220–226 (1996)
84. Bzhaumikhov AA, Monochromatizing x-ray concentrator (MONO-XC), Proc SPIE, **3444**, 512–516 (1998)
85. Cowley JM, Diffraction physics, North – Holland, Amsterdam, 290 (1975)
86. Aristov VV, Aoki S, Erko AI, Kikuta S, Martynov VV, Observation of Fourier Images in Soft X-Ray Radiation, Optics Commun **56**(4), 223–255 (1985)
87. James RW, The Optical Principles of the diffraction of X-rays, London, G. Bell and Sons Ltd., (1962)

88. Abeles F, Recherches sur la propagation des ondes electromagnétiques sinusoidales dans les milieux stratifiés. Application aux couches minces Ann Physique 12th series **5**, 596–640, 706–784 (1950)
89. Berning PH, Theory and calculation of optical thin Films Physics of thin films, Hass G (ed.), Vol. **1**, 69–75 (1963)
90. Peatman WB, Gratings, Mirrors and Slits, Gordon and Breach Science Publishers (1997)
91. Harada T, Itou M, Kita T, SPIE Proc **503**, 114–118 (1984)
92. Barbee Jr. TW, Rev Sci Instrum **60**, 1588–1595 (1989)
93. Erko A, Vidal B, Vincent P, Agafonov Yu, Martynov VV, Roshchupkin DV, Bunel M, Multilayer gratings efficiency: numerical and physical experiment, Nucl Instrum Methods Phys Res **A333**, 559–606 (1993)
94. Kirz J, J Opt Soc Am **64**, 301–309 (1974)
95. Attwood D, Soft x-rays and Extreme Ultraviolet Radiation. Principles and Applications, Cambridge (1999)
96. Apfel JH, Electric fields in multilayers at oblique incidence Appl Opt **15**, 2339 (1976)
97. Spiller E, Evaporated multilayer dispersion elements for soft x-rays. In: Attwood DT and Henke BL (eds) Low Energy X-Rays Diagnostics 1981AIP Conf proc no. 75, 124
98. Vidal B, Vincent P, Metallic multilayers for x-Rays using classical thin film theory, Appl Optics, **23**(No. 11) 1794–1801 (1984)
99. Croce P, Nevot L, Pardo B, Contribution l'Etude des couches minces par reflexion speculaire de rayons X, Nouv Rev d'Optique applique, **3**(1), 37 (1972)
100. Pelletier E, et al., Nouvelle revue d'Optique Applique, **2**(5), 247–254 (1972)
101. Yamamoto M, Namioka T, Best Materials and design of high Reflectance Multilayers for Soft X-Ray Microscope in X-Ray Microscopy in biology and medicine Shinohara K (ed.), et al.; Japan Sci Soc Press, Tokyo/Springer, Berlin, 91–97 (1990)
102. Vinogradov AV, Zeldovich BA, Appl Optics, **16**, 89 (1977)
103. Vidal B, Vincent P, Dhez P, Neviere M, X-Ray Optics, **563**, 78–85 (1985)
104. Schaefers F, Mertins HCh, Schmolla F, Packe I, Salashenko N, Shamov E, Appl Optics, **37**(No. 4), 719–728 (1978)
105. Erko A, Schäfers F, Vidal B, Yakshin A, Pietsch U, Mahler W, x-ray Supermirrors for BESSY II, Rev Sci Instr **66**(No. 10), 4845–4846 (1995)
106. Yamamoto M, Namioka T, Layer-by-layer design method for soft X-Ray multilayers, Appl Optics, **31**, 1622–1630 (1992)
107. Piechuch M, Nevot, Metallics Multilayers, Chamberod A (ed.), and Hillaret J, Mat Sci Forum, (**59** and **60**) (1990)
108. Beckmann P, Spizzichino A, The scattering of electromagnetic waves from rough surfaces, Pergamon New York, (1963)
109. Beckhoff B, X-ray focusing with Strongly Curved HOPG Crystals in the Energy-Dispersive x-ray Fluorescence Analysis, Ph.D. thesis, University of Bremen (1995)
110. Kanngießer B, Beckhoff B, Scheer J, Swoboda W, The comparison of three excitation modes in the energy dispersive x-ray fluorescence analysis, Adv X-Ray Anal **35**, 1001–1007 (1992)
111. Moore AW, Highly oriented pyrolytic graphite, in Chemistry and Physics of Carbon, vol. 11, Marcel Dekker Inc., New York (1973)

112. Ohler M, Baruchel J, Moore AW, Galez Ph, Freund AK, Direct observation of mosaic blocks in highly oriented pyrolytic graphite, Nucl Instr Meth Phys Res **B129**, 257–260 (1997)
113. Kan XB, Misenheimer ME, Forster K, Moss SC, x-ray study of planar defects in highly oriented pyrolytic graphite (HOPG), Acta Crystallogr Sect A: Found Crystallogr **43**, 418–425 (1987)
114. Tuffanelli A, Sanchez Del Rio M, Pareschi G, Gambaccini M, Taibi A, Fantini A, Ohler M, Comparative characterization of highly oriented pyrolytic graphite by means of diffraction topography, Proc SPIE **3773**, 192–198 (1999)
115. Levin VM, Anotonov A, Grigorieva IG, et al., Investigation of acoustical properties of carbon materials and composites, Institute of Chemical Physics of RAS, Report No. 30–89, Moscow (1990)
116. Ice GE, Sparks CJ, Jr, Mosaic crystal x-ray spectrometer to resolve inelastic background from anomalous scattering experiments, Nucl Instr Meth Phys Res **A291**, 110–116 (1990)
117. Furnas, Jr. TC, Kunts GS, Furnas RE, Toroidal monochromators in hybrid XRF system improve effectiveness tenfold, Adv X-Ray Anal **25**, 59–62 (1981)
118. Furnas, Jr. TC, Lambert MC, Furnas RE, Use of toroidal monochromators in hybrid XRF system to obtain increased effectiveness ratios, Nucl Instr Meth Phys Res, special issue, 245–249 (1982)
119. Dowell MB, Howard BA, Tensile and compressive properties of flexible graphite foils, Carbon **24**, 311–323 (1986)
120. Antonov AA, Baryshev VB, Grigorieva IG, Kulipanov GN, Shipkov NN, Focusing shaped pyrographite monochromators in synchrotron radiation experiments, Nucl Instr Meth Phys Res **A308**, 442–446 (1991)
121. Beckhoff B, Kanngiesser B, Scheer J, Swoboda W, Laursen J, Toroidally shaped HOPG crystals as strongly focusing Bragg reflectors of characteristic x-ray tube radiation for EDXRF analysis, Adv X-Ray Anal **37**, 523–533 (1994)
122. Dolbnya IP, Zolotarev K, Sheromov MA, Antonov AA, Grigorieva IG, Focusing parabolic pyrolytic graphite x-ray monochromator, Nucl Instr Meth Phys Res **A359**, 141–145; (1995) *ibid.* **A370** 646–647 (1996)
123. Heckel J, Ryon RW, Polarized Beam x-ray Fluorescence Analysis, Handbook of X-Ray Spectrometry, 2nd edn, Marcel Dekker Inc, New York Basel, 603–630 (2002)
124. Grigorieva IG, Antonov AA, HOPG as powerful x-ray optics, X-Ray Spectrom **32**, 64–68 (2003)
125. Zachariasen WH, Theory of X-Ray Diffraction in Crystals, Dover Publications, New York (1967)
126. Freund AK, Munkholm A, Brennan S, X-Ray Diffraction Properties of Highly Oriented Pyrolytic Graphite, SPIE Proc **2856**, 68–79 (1996)
127. Sánchez del Rio M, Gambaccini M, Pareschi G, Taibi A, Tuffanelli A, Freund A, Focusing properties of mosaic crystal, Proc SPIE **3448**, 246–255 (1998)
128. Sanchez del Rio M, Boscherini F, Savoia A, Ray tracing results for a doubly focusing independent crystal x-ray monochromator, Nucl Instr Meth Phys Res **A291**, 209–212 (1990)
129. Ryon RW, Adv x-ray Anal **20**, 575 (1977)
130. Kanngiesser B, Beckhoff B, Scheer J, Swoboda W, Comparison of highly oriented pyrolytic and of ordinary graphite as polarizer of Mo-K$\alpha$ in EDXRF, X-Ray Spectrom **20**, 331–336 (1991)

131. Beckhoff B, Laursen J, Bent and Flat Highly oriented Pyrolytic Graphite Crystals as Small Bragg Angle Monochromators in Thin-Specimen Energy-Dispersive XRF Analysis, X-Ray Spectrom **23**, 7–18 (1994)
132. Heckel J, Report on the European Conference on EDXRF, Berlin, Germany, 16–22 June (2002)
133. Antonov AA, Baryshev VB, Grigorieva IG, Kulipanov GN, Terekhov YaV, Shipkov NN, First results on application of short-focus monochromators from formed pyrolytic graphite for x-ray fluorescent analysis using synchrotron radiation, Rev Sci Instr, **60**, 2462–2463 (1989)
134. Chevallier P, Brissand I, Wang JX, Quantitative analysis by synchrotron radiation induced x-ray fluorescence at LURE, Nucl Instr Meth Phys Res **B49**, 551–554 (1990)
135. Beckhoff B, Kanngiesser B, Malzer W, New broad-band filtering device with rectangular efficiency shape based on x-ray focusing by strongly curved HOPG crystals, SPIE **2859**, 190–199 (1996)
136. Pease DM, Daniel M, Budnick JI, Rhodes T, Hammes M, Potrepka DM, Sills K, Nelson C, Heald SM, Brewe DI, Frenkel A, Grigorieva I, Antonov A, Log spiral of revolution highly oriented pyrolytic graphite monochromator for fluorescence x-ray absorption edge fine structure, Rev Scien Instr, **7**, 3267–3273 (2000)
137. Etschmann B, Ryan C, Vogt S, Maser J, Harland C, Brugger J, Selective x-ray Bragg spectrometry: optimizing fluorescence microprobe sensitivity for precious metals, Regolith 2004, 81–85 (2004)
138. Beumer B, Radley I, New x-ray fluorescence methods enable real-time detection of low sulfur concentration in hydrocarbons, Hydrocarbon Processing, 49–53 (2004)
139. Kolmogorov Y, Trounova V, Analytical potential of EDXRF using toroidal focusing systems of highly oriented pyrolytic graphite (HOPG), X-Ray Spectrom **31**, 432–436 (2002)
140. von Hamos L, Naturwiss **20**, 705–706 (1932)
141. von Hamos L, Röntgenspektroskopie und Abbildung mittels gekrümmter Kristallreflektoren, Ann Phys **17**, 716–724 (1933)
142. von Hamos L, X-Ray image method of chemical analysis, Am Miner **23**, 215–226 (1938)
143. von Hamos L, X-Ray Micro-Analyzer, J Sci Instrum **15**, 87–94 (1938)
144. Shevelko A, Antonov A, Grigorieva I, et. al., x-ray focusing crystal von Hamos spectrometer with a CCD linear array as a detector, Proc SPIE, **4144**, 148–154 (2000)
145. Shevelko AP, Kasyanov YuS, Yakushev OF, Knight LV, Compact focusing von Hamos spectrometer for quantitative x-ray spectroscopy, Rev Sci Instr **73**, 3458–3463 (2002)
146. Legall H, Stiel H, Nickles PV, Bjeoumikhov AA, Langhoff N, Haschke M, Arkadiev VA, Wedell R: Applications of highly oriented pyrolytic graphite (HOPG) for x-ray diagnostics and spectroscopy, Proc SPIE **5918**, 11–21 (2005)
147. Uschmann I, Nothelle U, Förster E, et al., High efficiency, high quality x-ray optics based on ellipsoidally bent HOPG crystals for ultrafast x-ray diffraction experiments, Appl Optics **44**, 5069–5075 (2005)
148. Boschetto D, Rischel Ch, Uschmann I, Perez J, Fourmaux S, Hulin D, Förster E and Rousse A, Large-angle convergent-beam setup for femtosecond x-ray crystallography, J Appl Cryst **36**, 348–349 (2003)

149. Erko A, Veldkamp M, Gudat W, Abrosimov NV, Rossolenko SN, Shekhtman V, Khasanov S, Alex V, Groth S, Schröder W, Vidal B, Yakshin A, J Synchr Rad **5**, 239 (1998)
150. Keitel S, Retsch CC, Niemöller T, Schneider JR, Abrosimov NV, Rossolenko SN, Riemann H, Nucl Instr Meth Phys Res **A 414**, 427 (1998)
151. Erko A, Schäfers F, Gudat W, Abrosimov NV, Rossolenko SN, Alex W, Groth S, Schröder W, Nucl Instr Meth Phys Res **A 374**, 408 (1996)
152. Dismukes JP, Ekstorm L, Paff RJ, J Phys Chem **68**, 3021 (1964)
153. Veldkamp M, Erko A, Gudat W, Abrosimov NV, Alex V, Khasanov S, Shekhtman V, Neissendorfer F, Pietsch U, Proc 2$^{nd}$ Intern Conf on Synchrotron Radiation in Material Science, Jpn J Appl Phys Suppl **38-1**, 612 (1999)
154. Abrosimov NV, Rossolenko SN, Thieme W, Gerhard A, Schröder W, J Cryst Growth **174**, (1997)
155. Abrosimov NV, Rossolenko SN, Alex V, Gerhard A, Schröder W, J Cryst Growth, **166**, 657 (1996)
156. Shulakov E, Khasanov S, Smirnova I, Materials Science Forum Vols **308-311**, 597 (1999)
157. Veldkamp M, Erko A, Gudat W, Abrosimov NV, Alex V, Proc 7$^{th}$ Intern Conf on Defect Recognition and Image Processing in Semiconductors (DRIP VII), Institute of Physics Conference Series No. 160 (1997) 195
158. Souvorov A, Drakopoulos M, Freund A, Snigireva I, Snigirev A, Erko A, Gudat W, Abrosimov N, Rossolenko , Schröder W, Nucl Instr Meth Phys Res **A 413**, 427 (1998)
159. Veldkamp M, Erko A, Schäfers F, Gudat W, Abrosimov N, Alex V, Rossolenko S, Shekhtman V, Shulakov E, Khasanov S, Smirnova I, Materials Science Forum Vols. **308-311**, 597 (1999)
160. Herzog HJ, in: Properties of strained and relaxed Silicon Germanium, EMIS Data reviews series No. 12, Eds Kasper E, INSPEC, London 1993
161. Donecker J, Gerhardt A, Wollweber J, Mater Sci Eng **B 28**, 18 (1994)
162. Erko A, Schäfers F, Gudat W, Sawhney KJ, Abrosimov NV, Rossolenko SN, Alex V, Groth S, Schröder W, Proc SPIE, **2856**, 110 (1996)
163. Petrashen P, Erko A, Nucl Instr Meth Phys Res **A 467–468**, 358 (2001)
164. Schäfers F, RAY - the BESSY raytrace program to calculate synchrotron radiation beamlines, Technischer Bericht, BESSY TB 202, 1–37 (1996)
165. Smither RK, Rev Sci Instr **63**, 131 (1982)
166. Knapp GS, Smither RK, Nucl Instr Meth Phys Res **A 246**, 365 (1986)
167. Bradaczek H, Hildebrandt G, J Optoelectr Adv Mater **1**, 3 (1999)
168. Uebach W, Acta Cryst **A46.2**, 439 (1990)
169. Uebach W, Bradaczek H, Hildebrandt G, Acta Crystallor **A49**, Supp 392 (1993)
170. Rudolph D, Niemann B, Schmahl G, Phys Scr **17**, 201–203 (1987)
171. Schneider G, Appl Phys Lett **71**, 2242–2244 (1997)
172. Collier RJ, Burckhardt ChB, Lin LH, Optical Holography, Academic Press, New York & London (1971)
173. Simpson MJ, Michette AG, Opt Acta **31**, 403–413 (1984)
174. Welnak C, Chen GJ, Cerrina F, Nucl Inst Meth Phys Res **A347**, 344–347 (1994)
175. Yamada T, Kawada N, Doi M, Shoji T, Tsuruoka N, Iwasaki H, J Synchrotron Rad **8**, 1047–1050 (2001)

176. Erko A, Firsov A, Investigation of the properties of Bragg-Fresnel gratings. Design and Microfabrication of Novel X-Ray Optics II; Anatoly Snigirev A, Derrick C. Mancini; Eds. Proc SPIE **5539**, 148–159 (2004)
177. Aristov VV, Gaponov SV, Genkin VM, Gorbatov YuA, Erko AI, Martynov VV, Matveeva LA, Salaschenko NN, Focusing properties of profiled multilayer mirrors, JETP Lett **44**, 265–267 (1986)
178. Miyamoto K, The phase Fresnel lens, J Opt Soc Am **51**, 17–21 (1961)
179. Michette AG, Pfauntsch SJ, Erko A, Firsov A, Svintsov A, Nanometer focusing of x-rays with modified reflection zone plates, Optics Commun **245**, 249–253 (2005)
180. Aristov VV, in X-Ray Microscopy II, Sayre D (ed.), Howells M, Kirz J, Rarback H, Springer Series in Optical Sciences, Springer Berlin Heidelberg, New York, 108–117 (1988)
181. Erko A, Agafonov Yu, Panchenko LA, Yuakshin A, Chevallier P, Dhez P, Legrand F, Optics Commun **106**, 146–150 (1994)
182. Erko A, Schaefers F, Artemiev N, A ray-tracing code for zone plates, Advances in Computational Methods for X-Ray and Neutron Optics; Manuel Sanchez del Rio (Ed.), Proc SPIE **5536**, 61–70 (2004)
183. Aristov VV, Snigirev AA, Basov YuA, Nikulin AYu, AIP Conf Proc **147**, 253 (1986)
184. Gabor D, Nature **161**, 777 (1948)
185. Leith EN, Upatnieks J, J Opt Soc Am **54**, 11 (1964) 1295
186. Lohmann AW, Paris DP, Appl Opt **6**, 1739 (1967)
187. Jacobsen C, Howells M, J Vac Sci Technol **B10**, 3177 (1992)
188. Lengeler B, Tuemmler J, Snigirev A, Snigireva I, Raven C, J Appl Phy **84**, 5855–5861 (1998)
189. Pantel RH, Feinstein J, Beguristan HR, Piestrup MA, Gary CK, Cremer JT, Characteristics of the thick, compound refractive lens, Appl Opt 1; **42**(4), 719–723 (2003)
190. Schroer CG, Lengeler B, Benner B, Tümmel J, Günzel F, Drakopoulos M, Simionovici AS, Snigirev A, Snigireva I, XRM-99, AIP Conf Proc, **507** 694–699 (1999)

# 4

# X-Ray Detectors and XRF Detection Channels

## 4.1 Introduction

*F. Scholze*

The discovery of X-rays by Wilhelm Conrad Röntgen was based on the observation of fluorescence in materials well away from his tube. Soon it was recognized that photographic plates and gas-filled ionization chambers (electrometers) are also sensitive to X-rays. With these principles the basis for X-ray detection was laid almost 100 years ago. After the initial observation of X-ray and $\gamma$-ray with photographic plates the development of gas filled counters [1] enabled the instantaneous detection of radiation. Proportional counters provided the capability to measure the photon energy of X-rays.

Although proportional counters have been in use for almost 100 years, they are still valuable detectors, which offer strong advantages. First, they are cheap and easy to operate because the internal gas-amplification yields a high charge output, which can be quickly and easily processed by the following electronics. These detectors are specifically insensitive to electronic interference and thus suited for harsh environments. Gas detectors are also a good choice if large areas are needed for detection, and they are capable of spatial resolution if, e.g. a resistive anode is used. More sophisticated, though still easy to produce, structures like multiwire or multistrip anode designs allow for simultaneous spatial and energy resolution over large areas [2]. The main disadvantage of gas detectors is their low density that results in a low efficiency for X-ray detection unless the detectors are made rather large.

An alternative with much higher detection efficiency for high energy photons is the scintillation detector. While early observations of ionizing radiation were made by looking at fluorescence screens in darkened laboratories, state-of-the-art scintillation detectors use a scintillating material coupled to photo multipliers, photodiodes or CCDs for the detection of fluorescence light. The benefit of these detectors is that they can be made very small and effective for hard X-rays by using scintillating crystals of heavy elements. Depending on the scintillator, the speed of these detectors can also be increased. When large areas of scintillating material are coupled to a large number of light detectors, a high-efficiency imaging detector is obtained which can be used for, e.g. medical applications. The disadvantage of scintillation detectors is their poor energy resolution.

Therefore, there has constantly been a search for a way to realize the principle of the ionization chamber with a high-density material. Consequently, shortly after high-purity germanium and silicon became available, the first Ge(Li) [3] and Si(Li) detectors were produced in the early 1960s. These detectors offered a much higher resolution than the gas detectors and scintillators. The reason is that the mean energy needed to create one primary elementary charge is about 20 to 30 eV for gases while for semiconductors it is only a few eVs. Thus, a ten times higher number of primary charge carriers is produced reducing the statistical fluctuations by about a factor of 3. Furthermore, the higher charge generated allows for the direct measurement by low-noise electronics. For the proportional counters, the gas amplification process adds additional noise.

The planar Si(Li) and Ge(Li) detectors are the semiconductor analogue to the ionization chamber. In principle a solid-state proportional counter, the avalanche diode also exists. Due to the high noise of the avalanche process, however, it is not used for high-resolution applications. The search for a solid-state analogue to the multiwire chambers, resulted in a versatile device – the silicon drift detector (SDD). Originally introduced for spatially resolving detectors for high-energy particle detection, the device became one of the most promising new X-ray detectors.

One of the main disadvantages of the planar Si(Li) and Ge(Li) detectors is the need for liquid nitrogen cooling. The vacuum-isolated cryostats make these detection systems rather large and difficult to handle. The reason for the cooling is the high leakage current at room temperature combined with a rather high read-out capacitance, because these detectors form a plate condenser with the capacitance proportional to the area. The high capacitance results in a high contribution of the serial noise component that requires longer shaping time. For long shaping time, however, the shot-noise contribution from the leakage current increases. In the SDD, the charge is drifted from a large area into a small read-out node with low capacitance. Thus, the serial noise decreases and shorter shaping time can be used. This offers two advantages: first faster counting is enabled and at the same time, higher leakage current can be accepted, drastically reducing the need for cooling. Presently, SDD at room temperature achieves an energy resolution of the order of 180 eV [4], comparable to the early liquid nitrogen cooled systems, with shaping times of 250 ns. A detailed discussion of these detectors is given in the first section of the chapter. In the second section, the perspective of a combination of energy and spatial resolution using silicon detectors is discussed.

A consequent next step on the path to higher energy resolution was the search for "materials" with even lower "ionization" energy. The electrons of superconductors bound in Cooper pairs are such a material. The binding energy of the Cooper pairs is of the order of a few meVs. Thus about 1,000 times more charge is generated per absorbed energy. This charge can be detected with a superconducting tunnel junction (STJ), a Josephson contact where the super conducting Josephson current is suppressed by a magnetic field. Potentially, the resolution limit should be about 30 times lower as compared to semiconductor detectors. An energy resolution in the range of 12 eV for manganese X-rays (6 keV photon energy) has, indeed, been demonstrated [5]. Even more impressive, the energy dispersive spectra of optical photons recorded with these detectors which look very much like the spectra of few-keVs X-ray photons measured with a Si(Li) detector [6]. The price to be paid is to go to an even lower temperature. Si(Li) detectors are operated at about 100 K and STJs below a few 100 mK. If detectors are cooled to about 100 mK or below, the heat capacitance becomes low enough so that single photons create a measurable increase of the detector temperature. The best energy resolution demonstrated so far, 4.5 eV for Mn K$\alpha$, was performed with the so-called micro-bolometer [7]. A micro-bolometer consists of an absorber at low temperature coupled to a heat sink and a thermometer to measure the temperature change. A large variety of possibilities [8] exists for the realization of either the absorber or the thermometer, and further subgroups of detectors can be defined in terms of whether the signal detected corresponds to thermal equilibrium (classical bolometer) or nonequilibrium excitations, e.g. hot electrons or phonons which are detected. Based on such indirect detection schemes, spatially resolving devices have also been demonstrated for low-temperature detectors [9]. Although mK-operation sounds difficult, as liquid nitrogen already poses a disadvantage for many applications, these systems are no longer exotic. The first low-temperature detectors of the STJ and bolometer type are already being operated at scanning electron microscopes and there is an increasing number of commercial suppliers. A comprehensive overview of the recent achievements of low-temperature detectors for X-ray spectroscopy was presented in a paper by M. Frank et al. [10].

This is a short summary of the means to ever better energy resolution. There are, however, other criteria that are even more important for many applications. A main issue for all detectors is the purity of the spectra recorded. X-ray fluorescence spectroscopy is based on the excitation with higher energy X-rays and the detection of the lower energy fluorescence X-rays. If monochromatic X-rays are used for excitation, low detection limits are principally possible because there is no background radiation at the fluorescence line energies. On the other hand, for some of the detected X-rays, energy loss processes in the detector can occur, leading to a shift in the measured energy toward a lower energy. These events create an artificial low energy background in the energy region of the fluorescence lines. Especially for total reflection X-ray fluorescence, this low-energy background from higher-energy photons incident at the detector deteriorates the detection limits of trace elements. Therefore, for many applications the so-called peak-to-background ratio, which defines the relative height of the full-energy peak in a spectrum relative to the low energy background, is very important. An unavoidable effect, which causes an energy loss, is that caused by the escape of photoelectrons or Auger electrons from the active detector volume. X-rays are indirectly ionizing, i.e. after absorption of an X-ray the energy is either fully or partially transferred to a Compton electron or photoelectron, respectively. These electrons create the charge carriers that are subsequently detected by scattering processes during their slowdown. During this process, the electrons travel a certain distance depending on their initial energy. If such an electron is created close to the boundary of the active detector volume it might escape from the detector and a corresponding part of the initial energy would be lost. There is also the possibility for the opposite process, that is, energy from photons absorbed close to the active detector volume is partially transferred into the detector. It is obvious that the probability of these processes scales with the ratio of the active detector volume to the surface. This effect strongly compromises its practical use, especially for the small low-energy detectors. This is especially true for the pure STJ detectors, which are made of thin layers of superconductors serving also as the absorber. For photon energies above about 1 keV, the majority of pulses detected with such a device are background pulses most often originating from photons absorbed in the substrate close to the active region [11]. Therefore, advanced designs use an absorber made of a different material from where the energy is transferred to the STJ junction where it is detected [5]. If several junctions are used for reading the signal, spatial resolution is also achieved [9].

Besides these intrinsic unavoidable effects, there are several others which cause partial detection of the charge generated. For semiconductor detectors, field distortions might force the charge to drift to inactive regions. This was a problem for the early Ge(Li) and Si(Li) detectors caused by a charging of the side-walls of the crystal. Improved passivation techniques solved this problem, and Si(Li) detectors with a Ni-front contact achieve the best peak-to-background ratios. Basically, SDDs could also achieve approximately the same performance. The background in the present SDDs is not caused by processes at the front side but by charges that have partially drifted to the outer region and an inactive region around the integrated read-out transistor. New designs with the transistor shifted to the outer region of the detector significantly reduced the background [12]. Using small collimated photon beams in the area well away from the edge of the active region, a background comparable to Si(Li) detectors is demonstrated for SDD.

For some applications, however, it might also be advantageous to filter the spectrum before detection. So, the high energy photons of a high rate (e.g., scattered excitation radiation or intense fluorescence of the main matrix elements) which do not need to be analysed, can be removed from the spectrum to reduce the total count rate, thus enabling higher count rates for the analysed elements. Such pre-dispersion can be done using mirrors or broadband synthetic crystals or multilayer reflectors.

A consequent application of such pre-dispersion approaches leads back to the so-called wavelength-dispersive analysis (WDX), covered in the last section of this chapter. The classical set-up of WDX uses a crystal that disperses the X-rays and a detector, which is placed at the position where the respective energy is diffracted, to count the photons. In most cases, proportional counters or scintillators are used for this purpose. Two obvious ways of obtaining more sophisticated solutions are the following. First, the dispersing crystal could be replaced by a broader bandpass filter; for example, the multilayer mirrors or mosaic crystals such as highly orientated pyrolytic graphite (HOPG) can be used. Both of these solutions also offer the opportunity to use focusing optics, increasing the collection angle for the fluorescence radiation originating from the sample. Second, the detector could be replaced by e.g. a spatially resolving device for performing multichannel WDX, or a detector with higher energy resolution could be used to detect diffusely scattered background radiation and higher diffraction orders.

This chapter provides an overview of the vast variety of X-ray detectors and wavelength-dispersive arrangements which could be used as building blocks for intelligent solutions adapted to the specific needs of an application.

## 4.2 X-Ray Detectors and Signal Processing

*A. Longoni and C. Fiorini*

### 4.2.1 Introduction

The aim of this Section 4.2 is to introduce briefly the main detector types used in X-ray fluorescence analysis, present their principles of operation and performances, and discuss how to read out and process their signals in order to measure "at the best" the energy of the X-ray photons. Emphasis is given to semiconductor detectors and, in particular, to the silicon drift detector (SDD), successfully introduced in an increasing number of X-ray fluorescence applications. The theoretical bases of the signal processing of detectors as well as practical considerations about the electronic acquisition and processing systems are given.

### 4.2.2 Basic Properties of X-Ray Detectors

An X-ray detector is used to convert the energy released by an X-ray photon in the detector material into an electric signal. The readout and processing of this electric signal by means of a suitable electronics chain is used to measure the energy released by the photon and the arrival time of the event.

## Radiation Interaction, Signal Generation

The interaction of X-rays in the detector material is mainly determined by *Photoelectric effect*, *Compton scattering*, and *Rayleigh scattering*. For a detailed description of these effects, the reader could refer to chapter 5 of this book. For most of the detector materials, the *Photoelectric effect* is dominant for photon energies up to 100 keV. The photoelectric absorption coefficient can be roughly approximated as $const \times Z^n/E^{3.5}$ ($n \sim$4–5). This relationship explains therefore the high efficiency of high-$Z$ detectors. *Compton scattering* is significant in radiation absorption at high energy (> 100 keV) and increases linearly with $Z$. *Rayleigh scattering* does not usually exceed 10–20% of the total absorption probability and it is higher than *Compton* at low energies (up to 20–50 keV, depending on the atomic number). In Fig. 4.1, the relative contribution of the three absorption effects versus the photon energy for three typical detector materials like Ar, Si, and Ge is reported [13]. The absorption factors are expressed in terms of mass absorption coefficients, defined in Section "Detection Efficiency".

The energy released in the detector material during the radiation interaction is converted into an electrical signal by means of a direct or an indirect process and with a conversion factor which depends on the specific type of detector considered. In semiconductor and gas-filled detectors, the energy is directly converted in a certain amount of charge which is then suitably collected at an output electrode of the device. In *scintillation* detectors, the interaction occurs in a scintillation crystal where the radiation energy is converted into a shower of optical photons. These photons are then collected and converted in charges by means of a photodetector (a photomultiplier tube or silicon photodetector). According to this "two-steps" conversion mechanism,

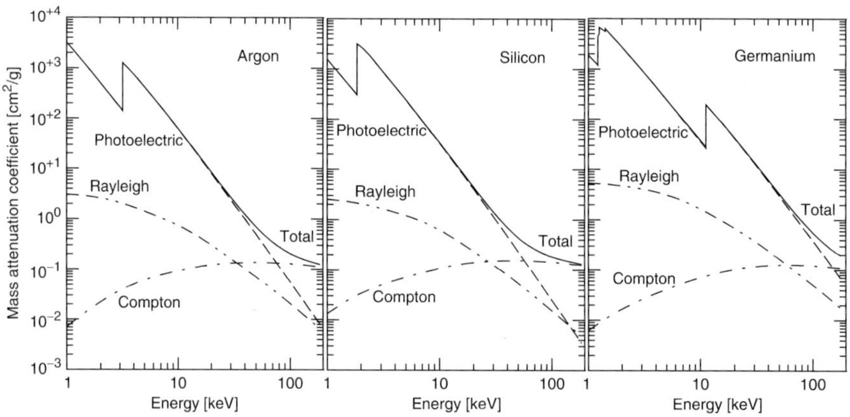


**Fig. 4.1.** Mass attenuation coefficients for Argon, Silicon, and Germanium. The photoelectric, Compton, and Rayleigh components of the total attenuation are also indicated in the graphs

scintillator-based detectors are called *indirect* detectors. In *cryogenic* detectors, the energy of the radiation is converted into (a) heat (phonons) and measured by coupling a suitable thermometer (thermistor, superconducting strip) to the absorbing material, (b) quasiparticles in a superconductor material and measured with a superconducting tunnel junction (STJ).

In most X-ray detectors, the charge collected at the output electrode of the device represents the basic electrical signal available for further processing. The amount of charge $Q$ is proportional to the photon energy $E$ by:

$$Q = E/\varepsilon. \tag{4.1}$$

The conversion factor $\varepsilon$ differs considerably among different detectors, according to the specific physics mechanisms involved in the charge generation. Typical values for $\varepsilon$ are 26 eV per each ion–electron pair generated in argon-filled detectors, 3.6 eV per electron–hole pair in silicon, of the order of 25 eV per collected electron in a CsI(Tl) scintillator coupled to a silicon photodetector and ~1 meV per quasiparticle produced in a superconducting detector. The statistical deviations from the average value expressed by (4.1) depend also on the specific interaction process, as discussed later in this chapter.

The *generation time* of the carriers ranges from a few ps in semiconductors up to a few μs for the photon emission of some inorganic scintillators. In most detectors, like gas-filled or semiconductor detectors, suitable electric fields are applied in order to separate electrons from positive charges (ions or holes) and to collect the carriers at the output of the device. The *collection time* depends on the mobility of the carriers in the detector material, on the travelling distances and on the applied fields and could range from a few nanoseconds in conventional silicon detectors up to several microseconds in gas-filled detectors. The signal shape at the output of the detector could be therefore represented as a current pulse, whose integral is equal to charge $Q$ and width equal to the collection time, supposing negligible, as in most cases, generation time. The pulse width is smaller than the charge collection time in those detectors, like the SDDs (described later in this Section 4.2), where the signal is induced at the output electrode only when the charge, in its path, is in the proximity of this electrode.

Statistical fluctuations on the shape of the signal and the collection time contribute to limiting the timing resolution of the detection system. However, the analysis of the timing properties of X-ray detectors is beyond the purpose of the present Section 4.2 and will not be considered here.

### Detection Efficiency

The *detection efficiency* is defined as the fraction of the total number of photons emitted by the source, which actually interacts in the detector volume and is completely absorbed. A first limitation to the detection efficiency arises from the limited active area of the detector and from the distance with respect to the source (*geometrical efficiency*). Moreover, the specific absorption coefficient of the detector material at a given photon energy and its thickness limit the number of photons which are actually absorbed (*intrinsic efficiency*). Finally, events that deposit only part of their energy in the detector (by Compton scattering, escape events), do not contribute to the full energy peak in the spectrum (*photopeak efficiency*).

The *geometrical efficiency* is the fraction of the photons emitted by the source which enter the detector volume. It can be calculated, in the case of a point source, as:

$$\eta_G = \Omega/4\pi, \tag{4.2}$$

where $\Omega$ is the solid angle under which the detector is seen from the source. In the case of a thin cylindrical detector, with the axis passing through the point source, $\Omega = S/h^2$ where $S$ is the detector cross-section and $h$ the source–detector distance.

The *intrinsic efficiency* is the fraction of the number of photons entering the detector which interact in the detector material. The intrinsic efficiency depends on the absorption properties of the detector material and on the thickness according to the well-known formula:

$$\eta_i = 1 - \exp(-\mu\rho x), \tag{4.3}$$

where $\mu$ is the *mass attenuation coefficient* ($\mathrm{cm}^2\ \mathrm{g}^{-1}$) at the given photon energy, $\rho$ is the *density* ($\mathrm{g\,cm}^{-3}$) of the material, and $x$ is the detector *thickness* (cm). In Fig. 4.2, the *linear attenuation coefficient* ($\mathrm{cm}^{-1}$), defined as $\mu\rho$, for some common detector materials is plotted in the energy range 1–120 keV. The graphs show clearly that high-$Z$ materials like Ge or CsI have higher absorption capabilities with respect to Si and especially, to Ar. For a given detector material, the intrinsic efficiency depends also on the detector thickness, according to (4.3). As an example, in Fig. 4.3, the intrinsic efficiency for Ar, Si, and Ge detectors of different thicknesses are compared.

The *photopeak efficiency* is the fraction of the photons interacting in the detector, which deposit their full energy in the material. For X-ray energies up to 100 keV, photoelectric process is dominant with respect to Compton effect and the probability of re-absorption of the Compton scattered photon in the detector itself is also rather high. When distinguished from other background components (e.g. tails of fluorescence peaks) in the X-ray fluorescence spectrum, a Compton continuum extends from 0 up to $2E\alpha/(1 + 2\alpha)$. This last value corresponds to the maximum energy released to the scattered electron ($\alpha = E/m_0c^2$, where $m_0c^2$ is the rest-mass energy of the electron (511 keV) and $E$ is the primary X-ray energy).

A second limitation to the *photopeak efficiency* arises from the *escape* process. When an external electron fills the vacancy left in an atom as a consequence of the absorption of an X-ray photon, another X-ray photon of an energy equal to the energy difference between the two levels involved could be

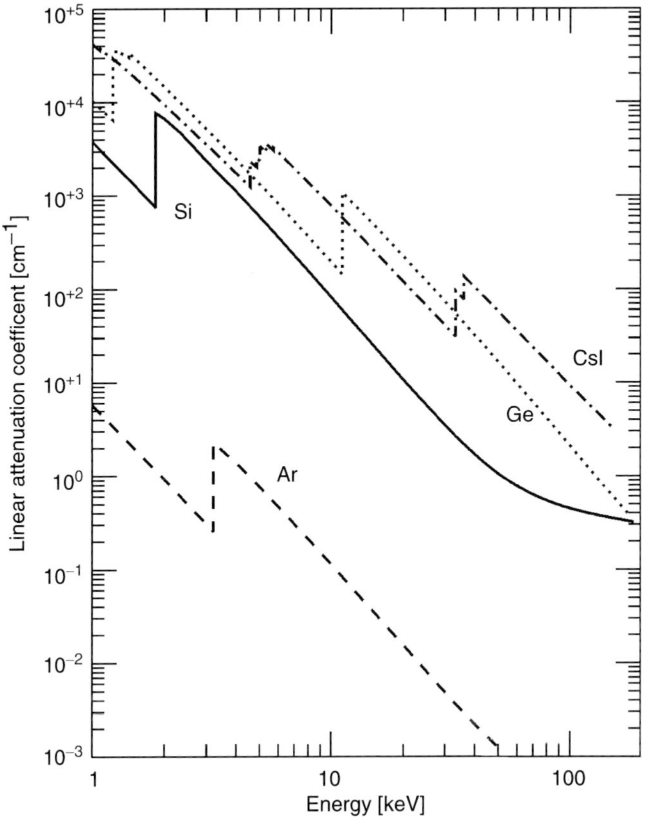


**Fig. 4.2.** Linear absorption coefficient for Ar (gas, 1 atm), Si, Ge, and CsI detector materials

emitted. This photon has a high probability to be re-absorbed in the material itself leading to a full-energy absorption of the primary photon. However, when the incoming X-ray photon is absorbed close to the detector surface (as in the case of low-energy X-rays), there is a not negligible probability that the re-emitted photon *escapes* from the detector without being absorbed. The consequence of this effect is the presence of an *escape peak* in the spectrum whose energy is equal to the difference between the energy of the primary photon $E$ and the energy of the escaping photon $E_{\mathrm{f}}$. Typical values for $E_{\mathrm{f}}$ are 3 keV in Ar, 1.74 keV in Si, 9.9 keV in Ge, 23 keV (Cd K$\alpha$) and 27.5 keV (Te K$\alpha$) in CdTe, 28.6 keV (I K$\alpha$) and 31.0 keV (Cs K$\alpha$) in CsI. The escape effect reduces the amount of primary photons counted in the full-energy peak more or less significantly, according to the specific detector material and the primary photon energy. In Fig. 4.4, the ratio between escape peak and full-energy peak calculated for different detectors and for energies up to 30 keV is reported. The experimental data measured with a silicon drift detector 300 μm thick are also reported [14].

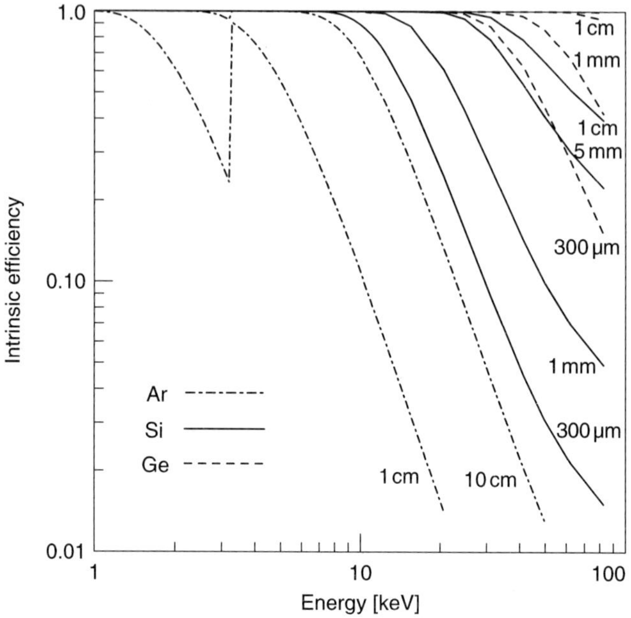


**Fig. 4.3.** Intrinsic efficiency for Ar, Si, and Ge detectors of different thickness

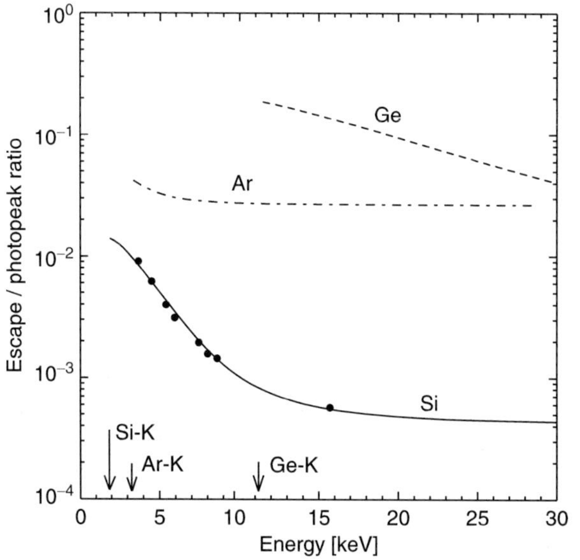


**Fig. 4.4.** Escape/photopeak ratio calculated for three detectors: Ar (2.54 cm thick), Si (300 µm thick), Ge (5 mm thick). The dots represent the experimental values measured by means of an SDD 300 µm thick

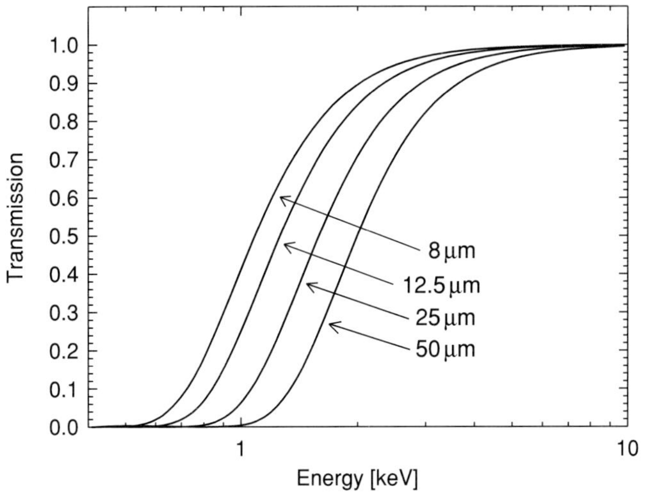


**Fig. 4.5.** X-ray transmission of Be windows of different thickness in the 0.4–10 keV energy range

In the low X-ray energy range (100 eV–1 keV), the detection efficiency is generally limited by: (a) the absorption in the window material, when used in front of the detector, (b) the absorption in a dead layer of the detector material itself.

The probability of absorption in the detection module entrance window can be expressed as exp($-\mu\rho s$) where $\mu$, $\rho$, and $s$ are, respectively, the mass attenuation coefficient, the density, and the thickness of the window material. Beryllium is a very common material used for windows. In Fig. 4.5, the transmission probability versus the energy for different Be window thickness is plotted, while in Fig. 4.6 the high transmission capability of an ultra-thin polymer window (from MOXTEK) is reported [15].

For "window-less" detection modules, the detection efficiency in the low energy range is limited by the unavoidable presence of a dead layer in the radiation entrance region of the detector itself. This limitation involves especially semiconductor detectors where the presence of an electrode deposited or implanted on the detector surface determines a thin region which could absorb the photon with a partial or incomplete release of signal charge. A model for the entrance window for silicon X-ray detectors is presented in [16]. An example of quantum efficiency measured in the low energy range for a silicon detector is shown in Fig. 4.20. on p. 227.

## Energy Resolution

In most X-ray fluorescence applications an accurate energy distribution of the incident photons has to be measured. The *energy resolution* $\Delta E$ of a spectroscopy system (detector + readout electronics) represents the capability of

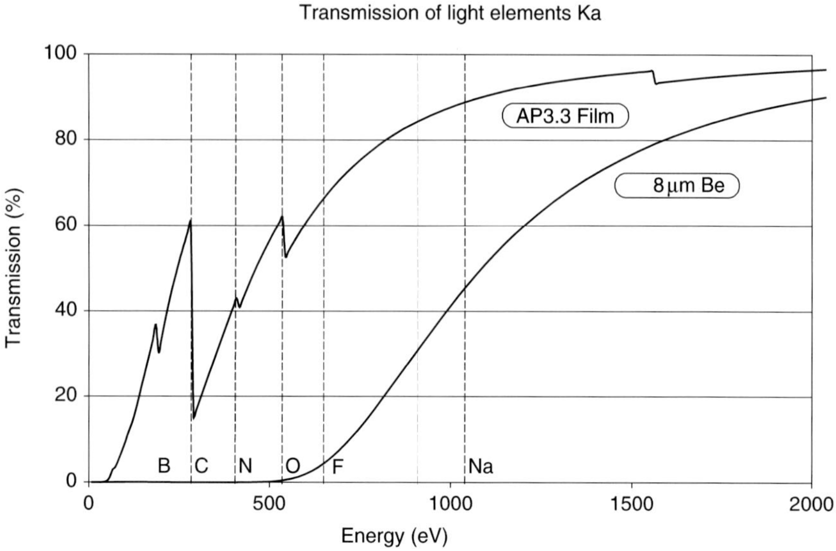


**Fig. 4.6.** X-ray transmission of an ultrathin polymer window (graph courtesy of MOXTEK, Inc.). For comparison, the transmission of an 8 μm Be window is also reported

the system to distinguish photons closely separated in energy. As shown in Fig. 4.7, the pulse height distribution in a typical measurement of photons of energy $E_0$ is broadened with respect to an ideal Dirac-$\delta$-like distribution due to several sources of statistical fluctuation affecting the measurement. The energy resolution is commonly expressed as *full-width-at-half-maximum* (FWHM) of the measured distribution. The larger the FWHM, the more difficult will be the identification of peaks corresponding to photons of close energies. Alternatively, the energy resolution can be expressed also as percentage $R$, defined as the ratio between the FWHM and the centroid value of the distribution:

$$R = \Delta E_{\mathrm{FWHM}} / E_0 . \tag{4.4}$$

Very often, the measured distribution can be described by means of a Gaussian function, whose expression is given by:

$$G(E) = \frac{N_0}{\sigma\sqrt{2\pi}} \exp\left(-\frac{(E-E_0)^2}{2\sigma^2}\right) , \tag{4.5}$$

where $\sigma$ is the standard deviation, $N_0$ is peak area, and $E_0$ is the peak centroid. For the Gaussian distribution the FWHM results related to the $\sigma$ as FWHM $= 2.35\,\sigma$.

Different sources of statistical fluctuation limit the energy resolution $\Delta E_{\mathrm{meas}}$ of a detection system, as indicated in the following expression:

$$\Delta E^2_{\mathrm{meas}} = \Delta E^2_{\mathrm{statistical}} + \Delta E^2_{\mathrm{el.noise}} + \Delta E^2_{\mathrm{multiplication}} + \Delta E^2_{\mathrm{collection}} . \tag{4.6}$$

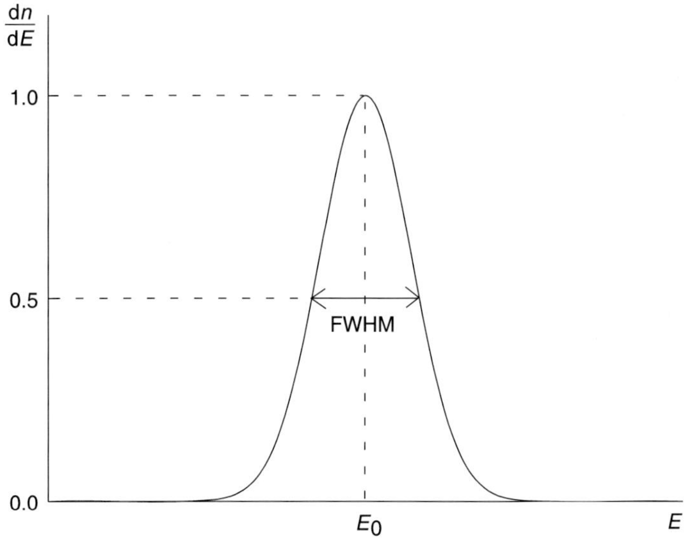


**Fig. 4.7.** Statistical broadening of the energy peak centred at $E_0$ measured by a detection system. The FWHM of the peak is commonly used as indication of the energy resolution of the system

The *statistical resolution* $\Delta E_{\text{statistical}}$ is related to the statistical fluctuations in the conversion process of the photon energy into charge carriers. This contribution is unavoidable and cannot be reduced with any refinement of the detection system, for a given detector material. It is generally characterized by a variance ($\sigma_E^2$) proportional to the number of carriers generated in the detector. For detectors like scintillation-based detectors, Poisson statistics well applies, leading to a variance $\sigma_n^2$ in the number of generated carriers equal to the average number of these carriers $\bar{n}$ (standard deviation $\sigma_n = \sqrt{\bar{n}}$). For a given conversion factor $\varepsilon$ from energy to number of carriers, the standard deviation of carriers is $\sigma_n = \sqrt{(E/\varepsilon)}$, the resolution $\Delta E_{\text{statistical}}(\text{FWHM}) = 2.35\sigma_E = 2.35\varepsilon\sigma_n = 2.35\sqrt{(E\varepsilon)}$, and $R = 2.35 \times 1/\sqrt{(E/\varepsilon)}$.

For detectors like semiconductor detectors and gas proportional counters the variance of the generated charge is smaller with respect to that predicted by pure Poisson statistics, because of the correlation in the processes of generation of the individual carriers. The deviation from Poisson statistics is taken into account by introducing the *Fano factor* (from Fano who first proposed a theory in 1947 [17]) as follows:

$$\sigma = F\bar{n}. \tag{4.7}$$

The corresponding expressions for the energy resolution are $\Delta E_{\text{statistical}}$ (FWHM) $= 2.35\sqrt{(FE\varepsilon)}$ and $R = 2.35 \times \sqrt{F}/\sqrt{(E/\varepsilon)}$. Measurements of the Fano factors have been carried out for different materials by several authors. Values of 0.11 for Si and 0.08 for Ge can be considered as good reference values.

The second term in (4.6) $\Delta E_{\text{el.noise}}$ is the contribution to the energy resolution of the *electronics noise* of the detector–amplifier system. The electronics noise of a detection system (described later in Section 4.2.6 (pp. 235–249)) is expressed in most of the cases by equivalent noise charge (ENC), defined as the charge delivered by the detector which produces at the end of the electronic chain a pulse amplitude equal to the noise rms value (signal-to-noise ratio $S/N = 1$). Once the value of ENC for a given detector–amplifier system is known, the $\Delta E_{\text{el.noise}}$ can be determined as:

$$\Delta E_{\text{el.noise}} = 2.35\varepsilon\,\text{ENC}/q, \tag{4.8}$$

where $q$ is the electron charge.

In the case of detectors with internal multiplication gain $M$, like gas proportional counters, photomultiplier tubes or silicon avalanche photodiodes, the effect of the electronics noise on the energy resolution is reduced by a factor $M$, $\text{ENC}/M$ being the number of noise electrons to be compared with the number of primary carriers generated in the detector before multiplication. However, for detectors with internal multiplication, a broadening contribution $\Delta E_{\text{multiplication}}$ due to the statistics of the multiplication process has to be included, according to the specific multiplication process involved. Finally, among other sources of statistical fluctuation, $\Delta E_{\text{collection}}$ is related to the partial collection of the signal charge, and its contribution could be relevant especially for semiconductor detectors.

In X-ray fluorescence spectroscopy, the energy resolution offered by the detection system should allow a satisfactory identification of characteristic fluorescence lines in the spectrum. This is required, for instance, in the qualitative/quantitative determination of the elements contained in a sample by means of the X-ray fluorescence (XRF) analysis technique. In Fig. 4.8, a main limitation for the identification of characteristic X-ray fluorescence lines is shown. The fluorescence peak of interest (centred at $E_{\text{f1}}$) could be partially overlapped by a contiguous peak (centred at $E_{\text{f2}}$). The overlapping between the two peaks introduces both a systematic and a statistical error in the determination of the peak area or could even prevent the identification of the peak in the spectrum.

Let us consider the case of two partially overlapped fluorescence peaks at the energies $E_{\text{f1}}$ and $E_{\text{f2}}$ and with areas $N_{\text{f1}}$ and $N_{\text{f2}}$, respectively. If we consider, for simplicity, the same energy resolution $\Delta E_{\text{meas}}$ (FWHM) for both peaks (we suppose that the resolution does not change significantly for contiguous peaks), the ratio between the number of counts $n_{f1}$ of the first peak within an energy window centred at $E_{\text{f1}}$ and large $\Delta E_{\text{meas}}$ (see Fig. 4.8) and the number of counts $n_{\text{f2}}$ of the second peak within the same window is equal to:

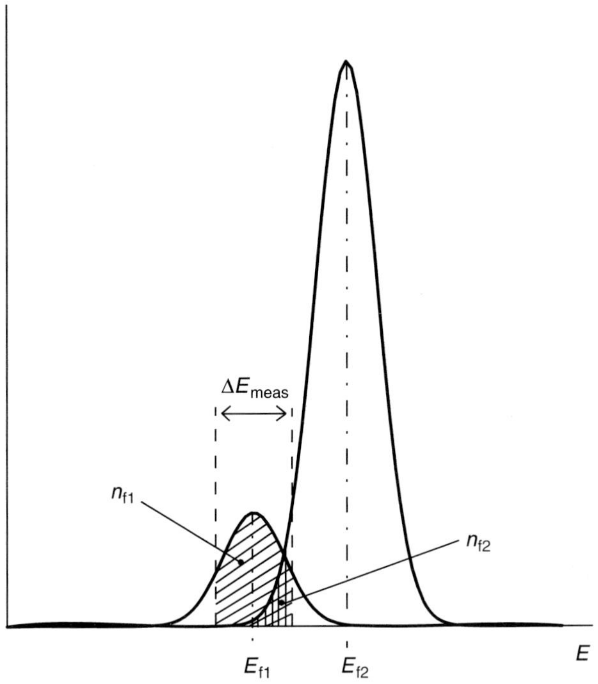


**Fig. 4.8.** Partial overlapping between two contiguous lines in the X-ray spectrum. $\Delta E_{meas}$ represents an energy window centred at the first peak ($E_{f1}$) and having the same width of the energy resolution (FWHM) of the detection system. The quantities $n_{f1}$ and $n_{f2}$ are the number of counts in the spectrum, respectively of the peak $E_{f1}$ and the peak $E_{f2}$, within the interval $\Delta E_{meas}$

$$\frac{n_{f1}}{n_{f2}} = \frac{N_{f1}}{N_{f2}} \frac{\mathrm{erf}\left(\sqrt{\ln 2}\right)}{\mathrm{erf}\left(\frac{E_{f2}-E_{f1}}{\Delta E_{meas}} 2\sqrt{\ln 2} + \sqrt{\ln 2}\right) - \mathrm{erf}\left(\frac{E_{f2}-E_{f1}}{\Delta E_{meas}} 2\sqrt{\ln 2} - \sqrt{\ln 2}\right)} . \tag{4.9}$$

If the energies $E_{f2}$ and $E_{f1}$ and the ratio between the peak areas $N_{f1}/N_{f2}$ (which are related to the concentrations of the corresponding elements) are known, the necessary energy resolution $\Delta E_{meas}$ of the detection system could be estimated by means of (4.9) for a required ratio $n_{f1}/n_{f2}$. This ratio has to be chosen according to the requirements of the specific application. In Fig. 4.9, the energy resolution $\Delta E_{meas}$ is plotted versus the energy difference $E_{f2} - E_{f1}$ for different values of $N_{f2}/N_{f1}$, supposing $n_{f1} = 3\,n_{f2}$. Just as an example, in the same plot, the energy difference between the K$\alpha$ lines of a few couples of contiguous elements (C–B, Cl–S, Mn–Fe and Ag–Pd) are reported. For a given couple of peaks to be distinguished, the required energy resolution can be determined from the graph, for different intensity ratios $N_{f2}/N_{f1}$.

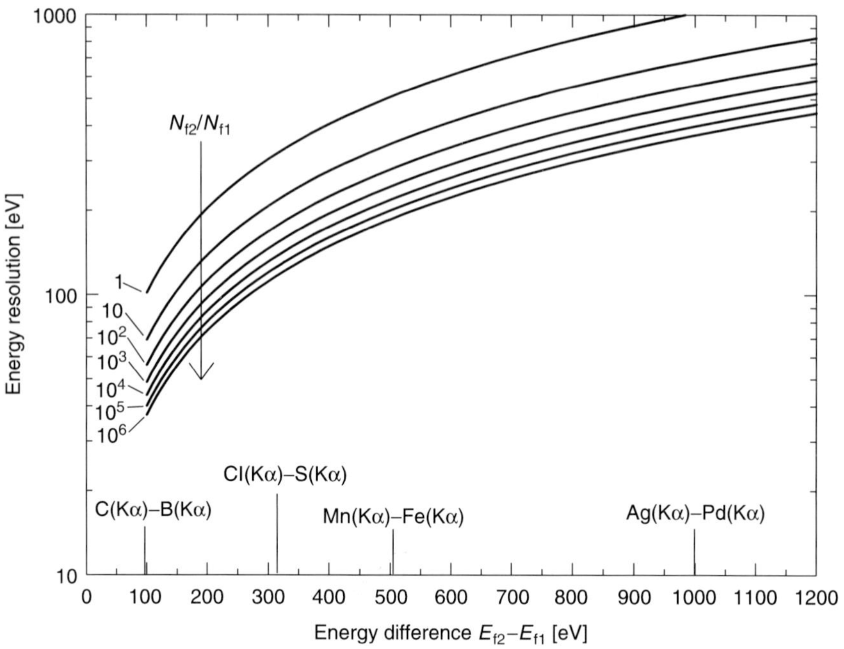


**Fig. 4.9.** Energy resolution required to suitably distinguish two fluorescence lines centred at $E_{f1}$ and $E_{f2}$, calculated with respect to the energy difference $E_{f2} - E_{f1}$ and for different ratios $N_{f2}/N_{f1}$ of the total areas of the two peaks. In this example, as a criterion for the separation of the two peaks, the number of counts $n_{f1}$ of the first peak within an energy window $\Delta E_{meas}$ centred at $E_{f1}$ has been required to be three times the number of counts $n_{f2}$ of the second peak within the same window

## Other Factors Affecting the Energy Spectrum

An additional limitation for the identification of characteristic X-ray fluorescence lines is shown in Fig. 4.10. The tail of an intense peak at higher energy ($E_{f2}$) in the spectrum could significantly cover the peak of interest ($E_{f1}$). A limited peak-to-valley ratio of the detector could therefore prevent the identification of weak fluorescence lines in a spectrum containing one or several intense lines at higher energy. This problem is crucial, for instance, in the quantitative analysis of elements at very low concentrations (e.g. tens of ppm) or in X-ray Absorption Fine Structure (XAFS) measurements in fluorescence mode on diluted sample where the tail of the intense scattering peak from the primary beam could overcome the weak fluorescence line of interest.

Similar to the evaluation of the required energy resolution, the effect on a XRF measurement due to a peak-tail background in the spectrum can be estimated (see Fig. 4.10). Although models for peak tailing for semiconductor detectors can be found in literature [18], we consider, as a first approximation,

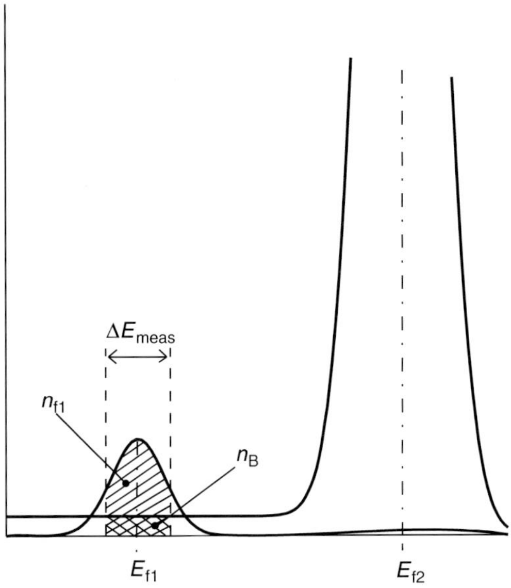


**Fig. 4.10.** Effect of the superposition of the tail of a peak at energy $E_{f2}$ on a peak at lower energy $E_{f1}$. The quantities $n_{f1}$ and $n_B$ are the number of counts in the spectrum, respectively of the peak $E_{f1}$ and of the tail of the peak $E_{f2}$, within the interval $\Delta E_{meas}$

that the total number of counts in the tail $N_{B2}$ is proportional to the peak area $N_{f2}$ by a factor $\rho(E_{f2})$, dependent on the peak energy $E_{f2}$ (physical phenomena leading to tailing of the peak, e.g. incomplete collection of the signal charge generated close to the entrance window in a semiconductor detector, are energy-dependent):

$$N_{B2} = \rho(E_{f2})N_{f2}. \tag{4.10}$$

Correspondingly, the background level (counts/eV) in the energy spectrum is given by:

$$N_{f2}\rho(E_{f2})/E_{f2} \tag{4.11}$$

and the number of counts in the window centred at $E_{f1}$ and with width $\Delta E_{meas}$ is:

$$n_B = N_{f2}\rho(E_{f2})/E_{f2} \times \Delta E_{meas}. \tag{4.12}$$

It has to be observed here that in this simple model the number of background counts $n_B$ within the window of interest centred at $E_{f1}$ depends on both the *peak-tailing* limitation of the detector, through $\rho(E_{f2})$, and on the *energy resolution* of the detector, through $\Delta E_{meas}$. To achieve a required ratio $n_{f1}/n_B$ for a satisfactory identification of a peak at $E_{f1}$ of a given intensity $N_{f1}$, both peak-tailing and energy resolution of the detector have therefore to be properly chosen.

The previously described effect of the escaping process in the detector could also be the reason of misunderstanding in the interpretation of an XRF spectrum. In fact, escape peaks which could have a not-negligible intensity especially in the soft X-ray region of the spectrum could appear at energies where possible fluorescence lines of interest are located.

When Compton interactions take place in the detector material, a continuum background could also be present in the spectrum. However, in most detector materials, this effect takes place with a much lower probability than the one of the photoelectric effect for photon energies below 100 keV. Furthermore, it can be reduced by a suitable choice of the detector size and shape in order to increase the photoelectric absorption, also with respect to the Compton scattered photons.

Another source of misunderstanding in the analysis of an XRF spectrum could be represented by spurious fluorescence lines due to the excitation of materials contained in the detector head or in the collimator for the radiation.

### 4.2.3 Classification of the Most Commonly Used X-Ray Detectors

In X-ray fluorescence techniques, different types of detectors can be employed like gas proportional counters, scintillation detectors and nitrogen-cooled or Peltier-cooled semiconductor detectors. We will limit ourselves here to a short description of the general features of the main detector classes, only in sufficient detail, to understand their different performances and operating characteristics. In Sect. 4.2.4, we will describe in some more detail the most commonly used semiconductor detector and, finally, we will dedicate a complete section to a particular kind of silicon detector, the SDD, which has been recently introduced successfully in an increasing number of X-ray fluorescence applications.

Many X-ray detectors are also used to detect electrons and heavy charged particles, but we will limit our discussion to their use as photon detectors. For a more extensive description of detector characteristics we would recommend the book by Knoll [19].

The *gas proportional counters* are X-ray detectors characterized by moderate energy resolution, room temperature operation, large detection areas, and high counting rates capability. A gas proportional counter consists of a volume of gas with an electric field applied across it. Typically, the container of the gas is cylindrical with one electrode being the outside surface (cathode) and the other a wire along the axis of the cylinder (anode), positively biased with respect to the cathode. The electrons produced in the gas under irradiation with X-rays, mainly by photoelectric effect, are accelerated on their way to the anode by a suitable voltage difference applied between the two electrodes (proportional regime of a gas detector), producing ionization by their collision with atoms of the gas. The amplitude of the resulting signal is therefore proportional to the energy of the incoming photon (and to the applied voltage) by means of the number of electrons generated by the ionization and a multiplication factor $M$ for these electrons. The most commonly used gases are high purity argon, xenon, neon, and krypton. Typical values for $\varepsilon$ (see (4.1)) range from 25 to 35 eV/ion pair while $M$ could range from 10 to $10^4$.

A typical proportional counter diameter ranges between 2 and 5 cm. The low density of the gas and the relatively small thickness of these detectors lead to a rather small efficiency for X-rays of medium and high energies, if compared with other detectors (see Fig. 4.3). For instance, in a counter of 5 cm thickness, a transmission of 50% of the incoming photons is reached at about 10 keV, using Ar, and 40 keV, using Xe, both at atmospheric temperature. The energy resolution of these detectors is intermediate between the high resolution offered by the semiconductor detectors and the poor resolution offered by the scintillation detectors. The energy resolution of a gas proportional is of the order of 12% at 5.9 keV.

The *gas proportional scintillation counters* (GPSC) represent also an attractive solution for energy-dispersive X-ray fluorescence analysis [20]. In a GPSC, the applied electric field allows the primary electrons to excite but not ionize the gas atoms (as occurs in a conventional gas proportional counter), producing a light pulse as a result of the gas atoms de-excitation. These secondary scintillation photons are then collected by a suitable photodetector, like a photomultiplier tube or an avalanche photodiode. The signal amplification is therefore achieved without spatial charge accumulation effects due to the generation of electron–ion pairs, as occurs in gas proportional counters. GPSCs have shown to reach remarkable performances for large detection areas in a 0.1–100 keV X-ray energy range. The energy resolution amounts 7.8% at 6 keV. For X-ray energies below 2 keV the energy resolution of GPSCs can even approach values which are obtained with nitrogen-cooled semiconductor detectors, such as Si(Li) and HPGe detectors (see Section 4.2.4 pp. 218–222). An example of spectra recorded in the soft X-ray region is shown in Fig. 4.11.

In *scintillation detectors* a considerable number of visible light photons is generated during interaction of X-rays with crystal scintillators, such as NaI(Tl) or CsI(Tl), and the following de-excitation process. These photons

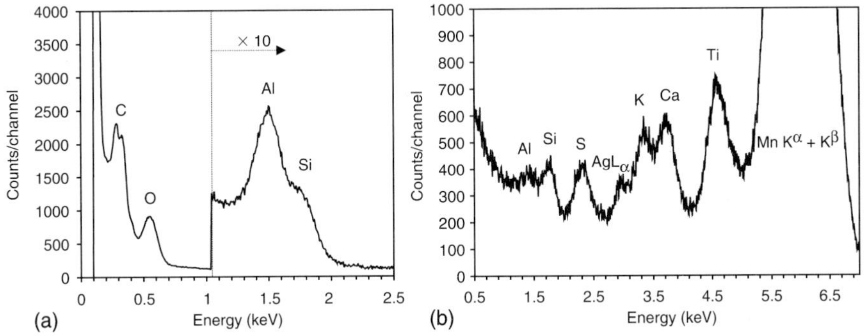


**Fig. 4.11.** Typical X-ray fluorescence spectra of andalusite **(a)** and anthracite **(b)** measured with a Gas Proportional Scintillation Counter (figure from Ref. [20])

are converted into electrons by means of a photodetector, typically a photomultiplier tube. In comparison to the gas detectors, the scintillation detectors are based on a higher $Z$ material with a higher density and often they have a larger thickness. This means that they have a high probability of detecting photons in a wider X-ray energy range. However, the energy resolution offered by this detector is the poorest among X-ray detectors, especially in the low energy range. When separated from the noise, a 5.9 keV peak can be measured with a resolution of the order of 30%. They are mostly used for low- resolution applications in an energy range from a few tens of keV up to hundreds of keV or a few MeVs.

In *semiconductor detectors*, electron–hole pairs are generated by direct interaction of the photon inside the detector material, similar to the primary electron–ion pairs generation in a gas proportional counter. In contrast, the output signal is generated by the collection of this primary charge without any multiplication process (except in the silicon avalanche photodiode). With respect to gas detectors, semiconductor detectors have higher density and $Z$. Moreover, in a semiconductor material the average energy required to produce a charge–carrier pair is of few electron-volts (3.62 eV for Si, 2.96 eV for Ge) while for gases this quantity is about 30 eV and for a scintillator-photodetector system could be, at best, of the order of 25 eV. According to these energy/charge conversion factors, the number of charge carriers generated for a given energy is higher for semiconductor detectors than for gases, leading to a much smaller statistical broadening of the peaks produced in the X-ray spectrum. For this reason, semiconductor detectors are nowadays the most preferred detectors in X-ray spectroscopy, especially when energy resolution is of primary concern.

Energy resolutions much better than the ones achievable even by the best nitrogen-cooled semiconductor detectors can be provided by *cryogenic detectors*.

### 4.2.4 Semiconductor Detectors

In this section, we will briefly discuss the main characteristics and performances of semiconductor detectors, with special focus on Si and Ge detectors. Within the scope of this presentation, we cannot discuss in depth the fundamental solid-state physics of these detectors. Interested readers could refer to specialized books, such as those of Knoll [19] and Lutz [21].

In a single crystal of semiconductor material such as silicon or germanium, the sharply defined atomic electron states are broadened into bands of energy states. The outer electrons are kept in the valence band while the next higher states lie in the conduction band, separated from the valence band by an energy gap (Fig. 4.12). The band gaps are 1.12 eV and 0.74 eV, respectively, in silicon and germanium. In pure semiconductors without impurities the gap contains forbidden states. An electron can be promoted from the valence band to the conduction band if it receives an energy at least equal to that of the

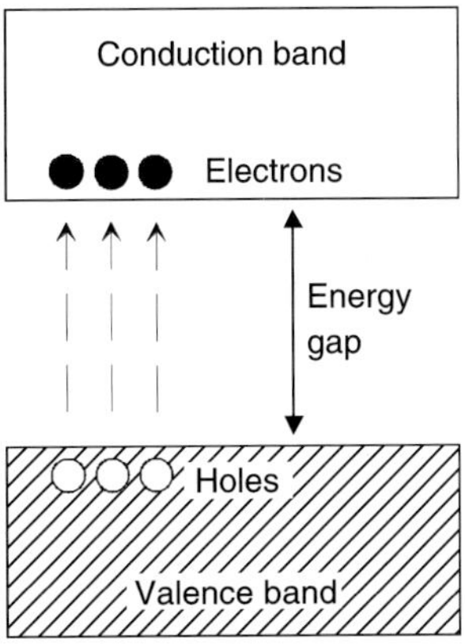


**Fig. 4.12.** Band structure of a semiconductor material

band gap. Once promoted in the conduction band, an electron can move under the influence of an externally applied electric field and can be collected at an electrode. The vacancy or hole left in the valence band by the promoted electron can also move by the applied field in the opposite direction of the electron. The velocities of electrons and holes are different according to their different mobilities in the crystal (e.g. $1450\,\mathrm{cm^2\,V^{-1}\,s^{-1}}$ for electrons and $505\,\mathrm{cm^2\,V^{-1}\,s^{-1}}$ for holes in silicon at room temperature [22]).

When an X-ray photon interacts in the crystal, the primary electron created by the ionization process excites bound electrons to the conduction band. These secondary electrons, if sufficiently energetic, can further create additional electrons by a cascading process which finally leads to a large number of electron–hole pairs that can be collected at the electrodes of the device.

Impurity-free Si and Ge materials are called *intrinsic* semiconductors. When *dopants* are added to an intrinsic semiconductor, the conduction properties of the material can be modified. Adding pentavalent dopants like P or As into a tetravalent Si (or Ge) semiconductor will increase the conductivity of the material because only four electrons of the pentavalent atom are involved in binding it with the Si (or Ge) atoms and therefore the fifth electron can easily be promoted to the conduction band with just a small amount of energy. Dopants of this kind are called *donors* because they introduce free electrons, and the doped semiconductor is referred to as n-type material. On the contrary, trivalent dopands like B, called *acceptors*, provide free holes, increasing the conductivity of the material. Semiconductors doped with acceptors are called p-type materials.

The basic principle of a simple *Si diode detector* based on a junction between two p and n semiconductor materials is shown in Fig. 4.13. When the pn diode is reverse biased, the bulk material (n-type in our example) is depleted from the free charge and the carriers produced by the interaction of photon in the bulk can be suitably collected at the electrodes by means

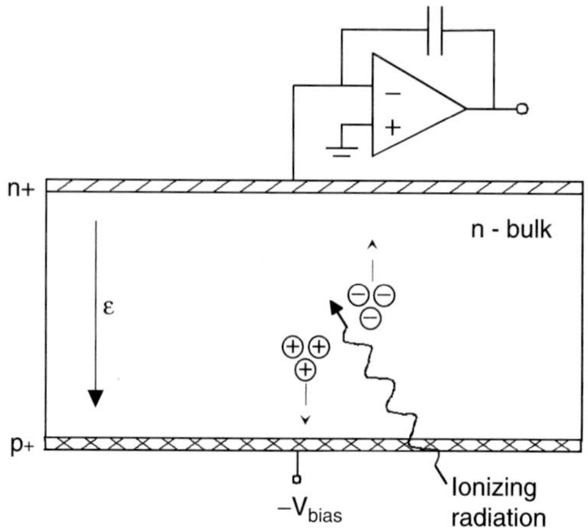


**Fig. 4.13.** Working principle of a pn diode detector. The same voltage applied to the device provides the depletion of the semiconductor bulk and the drift field $\varepsilon$ responsible for the collection of the charge carriers, created by the ionizing radiation, by the electrodes

of the applied field. A continuous flow of charge thermally generated within the bulk material (dark current) is also collected at the detector electrodes, contributing to a statistical spread in the measurement of the signal charge. However, the dark current value can be highly reduced by suitably cooling the detector. The contribution of the various parameters of the detector to the overall electronics noise of the detector–amplifier system will be discussed in detail in Section 4.2.6 (pp. 235–249)

The depletion depth that can be achieved by reverse biasing a conventional silicon pn detector is usually limited to 0.3–1 mm. Thicker depletion depths (5–10 mm) can be reached by means of the *lithium drifting* process. In a *lithium-drifted silicon detector* or Si(Li), lithium ions, which act as donors, are driven through a large volume of a high purity silicon crystal, which tends to be p-type, in order to obtain an "intrinsic"-like bulk material by means of the *compensation* of the donors and acceptors impurities concentrations. The excess lithium on the surface of deposition on the crystal results in a highly doped n+ layer which acts as an electrical contact, while the uncompensated p region on the opposite is contacted by either a metallic contact or a thin p+ layer. In a Si(Li) detector, the lithium continues to drift significantly at room temperature. Therefore, in order to prevent an undesired redistribution of the lithium dopants, the detector must always be kept cold (usually at liquid nitrogen temperature), even when not operated. The Si(Li) detector is currently the most popular X-ray detector in the energy range from a few hundreds eV up to about 40 keV. The typical energy resolution of commercial detectors is of the order of 135 eV at 6 keV. An energy resolution of 128 eV has been measured by using a Pentafet front-end transistor [23].

Nitrogen-cooled high purity germanium HPGe detectors can be also used for high-resolution X-ray detection. Unlike Si(Li) detectors, Ge detectors need not be cooled permanently, but only when they are operated. With respect to Si, Ge is characterized by a better detection efficiency for higher energy photons, because of both higher Z (32 for Ge, 14 for Si) and higher density ($2.33\,\mathrm{g\,cm^{-3}}$ for Ge, $5.46\,\mathrm{g\,cm^{-3}}$ for Si). For a given thickness, the better efficiency of a Ge detector with respect to a Si detector is shown in Fig. 4.3.

HPGe detectors exhibit superior statistical energy resolution ($\Delta E_{\mathrm{statistical}}$ in (4.6)) compared with Si(Li) detectors of the same geometry. The two factors which contribute to the superior statistical contribution to the energy resolution are the lower values respectively of mean energy required for an electron–hole pair generation (2.96 eV with respect to 3.62 eV for Si) and of Fano factor (0.08 with respect to 0.11 for Si). The statistical term in the energy resolution, typically dominating at high energies ($> 10\,\mathrm{keV}$) with respect to the electronics noise contribution, is then about 25% better for a Ge detector than for a Si detector. Moreover, the possibility of fabricating thicker HPGe detectors, compared with Si(Li), translates directly into a smaller output capacitance for the former ones with a correspondingly smaller electronics noise and higher energy resolution.

However, other factors limit the spectroscopic performances of HPGe detectors with respct to Si(Li) especially in X-ray detection lower than about 20 keV. In the low energy range ($< 3\,\mathrm{keV}$), the X-ray spectrum measured with a Ge detector can be affected by distortions and higher background resulting from the thicker entrance window with respect to Si(Li) detectors. In the energy region above the Ge K-absorption edge at 11.1 keV, a complex X-ray spectrum can be significantly disturbed by the presence of escape peaks which are much more intense in Ge detectors, compared to Si detectors. The higher escape/photopeak ratio for Ge than for Si (shown in Fig. 4.4) is due to the smaller average penetration distance of the incident radiation in the detector, the higher fluorescence yield, and the higher escape probability of the more energetic Ge-K fluorescence lines (9.9 and 11 keV). Escape peaks in spectra measured with Si(Li) detectors are observed at lower energies ($> 1.84\,\mathrm{keV}$, Si K-absorption edge) with respect to Ge. However, they are much less intense and therefore overlap only very weak photopeaks. The presence of the escape peaks on a $^{241}$Am spectrum measured with HPGe detector is shown in Fig. 4.14, where, for comparison, the spectrum of the same source measured with a Si(Li) detector is also reported. A more detailed comparison between Ge and Si(Li) detectors in the 2–20 keV energy range is presented by Rossington et al. [24].

In some applications, like portable X-ray spectrometers, the difficulties arising from the use of a liquid nitrogen cryostat (large size of the cryostat, nitrogen refill) can be overcome by the use of a Peltier-cooled silicon detector. Si-PIN diodes from AMPTEK ($25\,\mathrm{mm^2}$ active area, $500\,\mu\mathrm{m}$ thickness) [25],

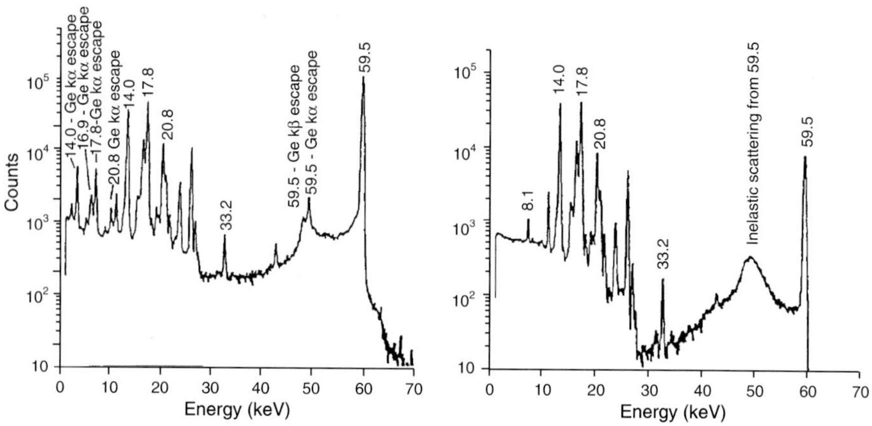


**Fig. 4.14.** $^{241}$Am spectrum measured with an HPGe detector (left) and a Si(Li) detector (right) (figure from [24])

cooled at about −30°, offer an energy resolution of the order of 250 eV at 6 keV at 20 μs shaping time. Smaller devices (5 mm$^2$, 500 μm thickness) have recently reached energy resolutions as good as 158 eV FWHM at 6 keV at 20 μs shaping time. Also Peltier-cooled SDDs, described in Section 4.2.5, have reached resolutions not far from the best ones of Si(Li) and Ge detectors, with a shaping time of the order of 1 μs.

Room temperature or Peltier-cooled high-$Z$ semiconductor detectors like $HgI_2$ [26], CdTe [27], CdZnTe (CZT) [28], GaAs [29] can also be used for X-ray detection without the need for liquid nitrogen cryostat. Due to the high mean atomic number, a thickness of a few mm is sufficient for these detectors to provide a good efficiency in a wide X-ray energy range, up to a few hundred keV.

### 4.2.5 Silicon Drift Detectors

#### The Working Principle

The Silicon Drift Detector (SDD), introduced by E. Gatti and P. Rehak in 1983, is a detector of ionizing radiation characterized by a very low capacitance of the electrode collecting the signal charge. This capacitance is, moreover, independent of the active area of the device. This feature allows to achieve a low electronics noise, obtained, moreover, by using short shaping times, as it will be shown in Section "SDDs for X-Ray Spectroscopy" (p. 225). Since its invention, the SDD has been developed in a large variety of topologies for applications in the field of high-energy physics and in the field of X-ray spectroscopy.

The working principle of this device can be understood starting from the *sideward depletion* concept shown in Fig. 4.15 [30]. With respect to a conventional p-n diode detector (Fig. 4.15a), where the ohmic n+ contact extends

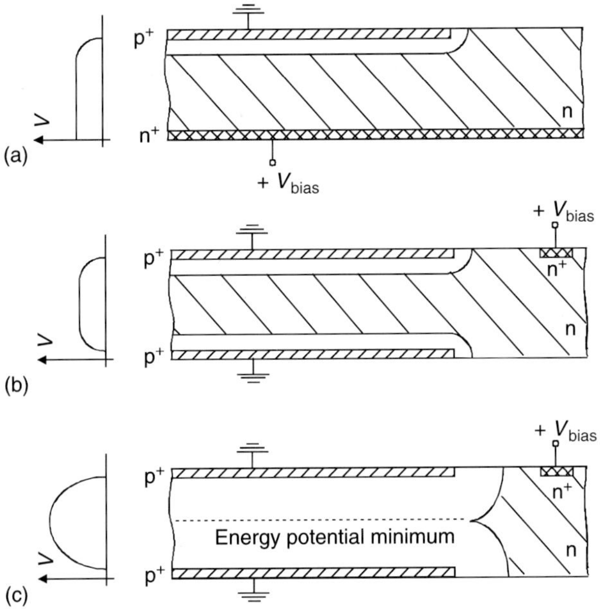


**Fig. 4.15.** Sideward depletion concept

over the full area on one wafer side, in the structure shown in Fig. 4.15b the depletion of the bulk can be also achieved by positively biasing a small n+ electrode with respect to p+ electrodes covering both sides of the wafer. When the n+ voltage is high enough, the two space-charge regions separated by the undepleted bulk shown in Fig. 4.2.15b touch each other (Fig. 4.15c), leading to a small undepleted bulk region only close to the n+ electrode. The depletion of the bulk is achieved in this way with an applied voltage which is four times lower with respect to the voltage needed for a conventional diode of the same thickness. According to the biasing mechanism described, the diagram of the electron potential energy perpendicular to the wafer surface has a parabolic shape, with a minimum located in the middle of the wafer.

In the SDD, based on the principle of the sideward depletion described above, an additional electric field parallel to the surface of the wafer is added in order to force the electrons in the energy potential minimum to drift towards the n+ anode. This is achieved by implanting two arrays of p+ electrodes on both sides of the wafer (Fig. 4.16), instead of the single p+ implants shown in Fig. 4.15b. These electrodes are suitably biased with a voltage gradient in order to provide an electric field parallel to the surface. A drawing of the potential energy in the drifting region is shown in Fig. 4.17a. Once generated by the ionizing radiation, the electrons are focused in the bottom of the potential channel and driven towards the anode region of the detector while the holes, driven by the depletion field, are quickly collected by the nearest p+ electrodes. In the region close to the collecting anode (Fig. 4.17b), the bottom

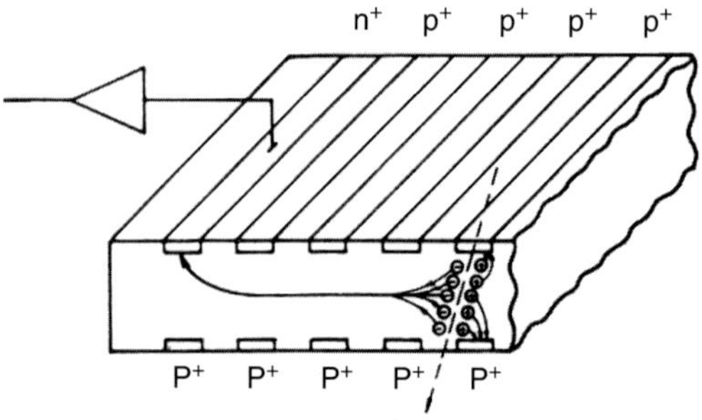


**Fig. 4.16.** The SDD working principle

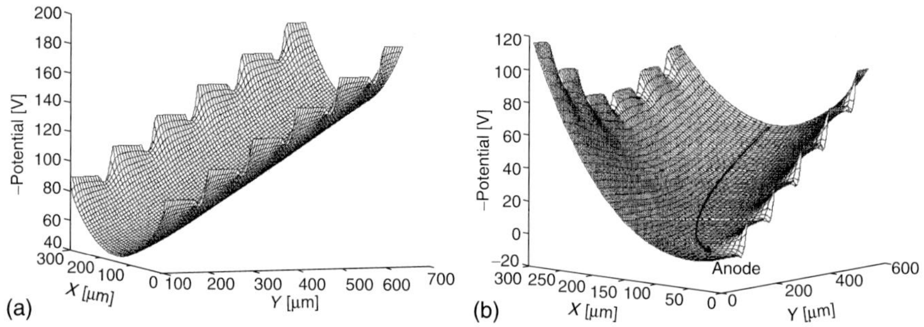


**Fig. 4.17.** Electron energy potential diagrams in the drifting region of the SDD (**a**), and in the region close to the anode where the potential valley is directed towards the surface (**b**)

of the potential channel is shifted towards the surface where the anode is placed, by suitably biasing the electrodes on the opposite side.

The cloud of electrons induces to the anode an output pulse only when the electrons arrive close to it because of the electrostatic shield of the p+ electrodes. The drift time of the electrons may be used to measure one of the interaction coordinates while the collected charge allows to measure the energy released by the incident ionizing event [31].

The main advantage of an SDD with respect to a conventional p-n diode of equivalent active area and thickness is the low value of the capacitance of the collecting anode, which is of the order of 100 fF. This value is moreover independent of the active area of the device. As will be shown later in the section describing the electronics noise of a detector-preamplifier system, this feature allows to reduce both the electronics noise and the value of the shaping time to be used for the processing of the signal.

Originally, the SDD was designed to be used as a position sensitive detector for high-energy physics experiments, by using the measurement of the signal electrons' drift time to reconstruct one coordinate of the particle's interaction point. The second coordinate is given by a suitable segmentation of the collecting anode. For instance, two-dimensional ($x$,$y$) position detectors with an active area of $6.3 \times 6.3\,\mathrm{cm}^2$ and a row of 240 anodes with a pitch of 250 µm have been produced [32] as well as a cylindrical ($r$, $\varphi$) position detector with a diameter of 10 cm and an angular resolution of 1°, realized by a radial drift field and 360 anodes placed along the edge of a 4-in. wafer [33].

## SDDs for X-Ray Spectroscopy

The low electronics noise achievable with an SDD, thanks to the low value of output capacitance, fully exploited by the integration of the front-end transistor on the detector chip (see Section "The On-chip Electronics", p. 227), has made the SDD also an ideal detector for high-resolution X-ray spectroscopy measurements. For this purpose, in order to enhance the quantum efficiency in the soft X-ray region ($E < 5\,\mathrm{keV}$), particular care has been used in the design of the radiation entrance window, limiting as much as possible the insensitive area and optimizing the doping profile to minimize the charge loss.

In a detector of the type shown in Fig. 4.16, the surface areas between p+ strips are covered by thermally grown $SiO_2$. The fixed positive charges, always present within the oxide, bend the potential distribution downward at the detector surface and create local potential minima for electrons which could collect the signal electrons generated close to the surface [34]. This is an important limitation for the detection of soft X-rays, which are totally absorbed within a few µm from the surface.

For soft X-ray detection a suitable topology has been designed with the radiation entrance window of the detector consisting of a continuous p+ implant without oxide gaps [35]. The schematic view of an SDD for X-ray spectroscopy based on this design is shown in Fig. 4.18 [36]. By using an equipotential electrode on the p-side, only the potential on the opposite side of the detector is varied to provide the drifting field, as shown in Fig. 4.19. As can be easily understood from the figure, also for this device, as in the case of the SDD shown previously, wherever the electrons are generated inside the detector volume, they are driven to the small collecting anode. An integrated voltage divider can be used to bias the p+ rings by just contacting externally the first ring next to the anode and the last one at the edge of the detector [37].

In order to reach a good response in the low energy range (a few hundred eVs) a very shallow implantation of the p+ back contact that acts as radiation entrance window has been obtained. A careful tailoring of this implantation in the fabrication technology of the Max Planck Institut (MPI) Halbleiterlabor has allowed to obtain devices with a pn-junction at the backside located at a depth of approximately 40 nm [38]. The curve of the quantum efficiency of a 300-µm thick SDD is reported in Fig. 4.20 [39], showing a quantum efficiency higher than 60% above 200 eV. In Fig. 4.21 [35], the low energy spectrum of a Macrolon filter containing several light elements, irradiated with an X-ray beam and measured with an SDD, is shown. The energy resolution in correspondence of the oxygen peak (524.9 eV) is 92 eV FWHM.

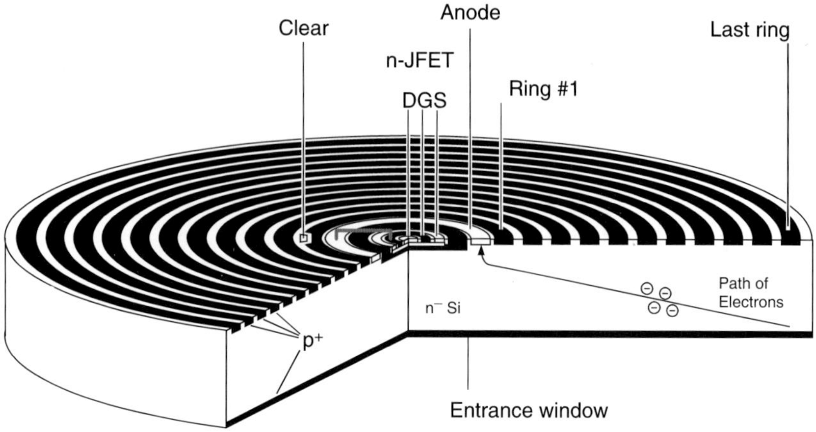


**Fig. 4.18.** Schematic diagram of the SDD for X-ray spectroscopy with integrated n-channel JFET. The transistor gate is connected to the anode ring by a metal strip

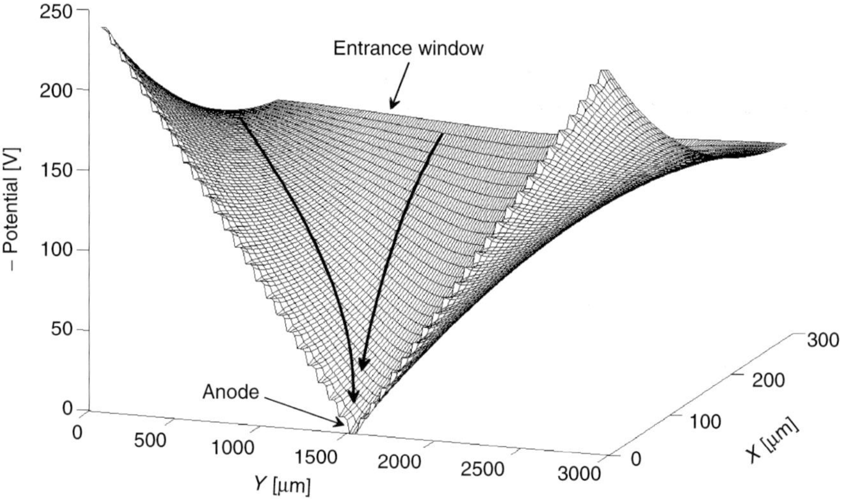


**Fig. 4.19.** Energy potential for electrons inside an SDD with homogeneous entrance window. Possible electrons paths are shown in figure

At high energy, the X-ray detection efficiency of the SDDs, limited by the total thickness of the wafer, typically 300 μm, is about 90% at 10 keV and 50% at 15 keV.

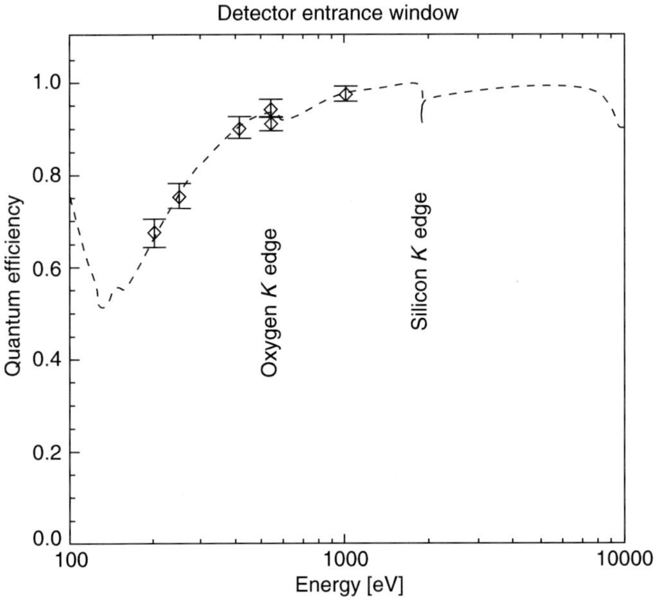


**Fig. 4.20.** Quantum efficiency of a 300-µm thick SDD. The curve has been calculated taking into account the absorption coefficients and the dead layer introduced by the backside p+ implantation. The points represent measured data (figure from [39])

## The On-chip Electronics

To fully exploit the benefit in terms of energy resolution and short shaping time arising from the low output capacitance typical of the SDD, both the preamplifier's input capacitance and stray capacitances of the connection between detector and preamplifier have to be kept as small as possible. This goal can be achieved by means of the integration of the front-end transistor of the amplifying electronics directly on the detector wafer [40–43]. This solution allows to minimize the stray capacitance of the connections because the bond wire connecting the detector and an external amplifier is substituted by a short metal strip on the chip. By a proper design of the input transistor, also the capacitative matching condition, $C_{\text{detector}} = C_{\text{FET}}$, can be achieved. Moreover, the detector–preamplifier sensitivity to microphonic noise (mechanical vibrations) and electrical pickups is highly reduced. To fully exploit this solution, in addition to the first transistor of the preamplifying electronics,

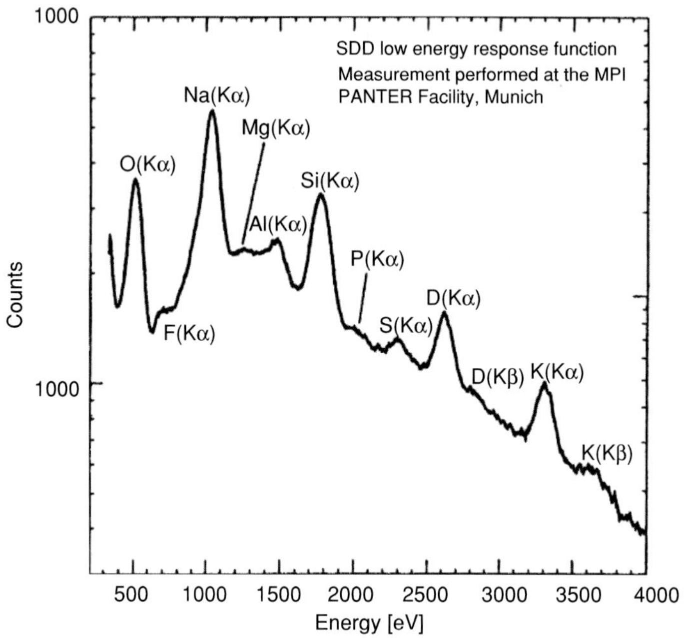


**Fig. 4.21.** Low energy spectrum of a Macrolon target acquired by means of a SDD. The energy resolution measured on the oxygen K$\alpha$ line (524.9 eV) is 92 eV (figure from [39])

all devices required for the discharge of the signal charge and leakage current have to be integrated on the detector chip [44].

An example of integration of the input transistor in an SDD is shown in Fig. 4.18. The transistor is a nonconventional n-channel JFET, designed to be operated on completely depleted high resistivity silicon, placed inside the ring shaped anode. A narrow metal strip connects the JFET gate to the anode. A circular deep p implantation, biased through a guard ring, divides the transistor region from the collecting region of the detector. The transistor works in a source–follower configuration with an external current supply connected to the source. The discharge of the leakage current from the detector and the reset of the signal charge accumulated on the anode are done continuously by means of the gate-to-drain current of the FET, which is provided by a weak avalanche mechanism occurring in the high-field region of the transistor channel [44]. The mechanism is self-adapting to new values of leakage current avoiding the need of external circuitry.

SDDs with on-chip JFET have allowed to reach state-of-the art energy resolutions in X-ray spectroscopy at room temperature or with moderate cooling by a single-stage Peltier cooler. In Fig. 4.22, two $^{55}$Fe spectra measured by using a selected SDD of 5 mm$^2$ of active area produced at MPI are shown [39].

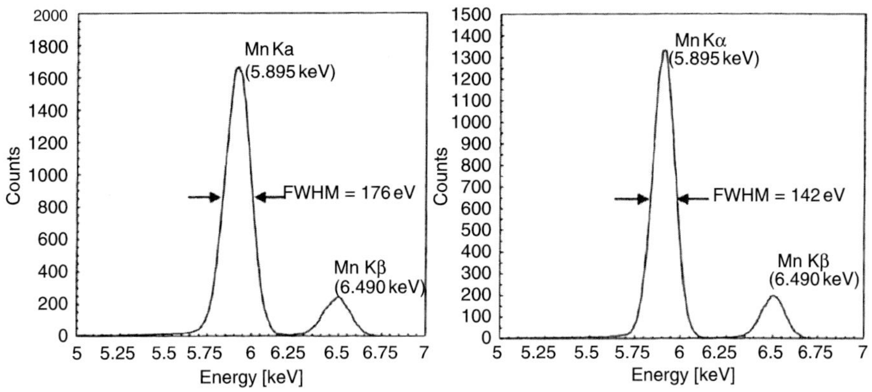


**Fig. 4.22.** $^{55}$Fe spectra measured with an SDD (5 mm$^2$), respectively, at 25°C with 0.25 μs shaping time (left) and at −10°C with 0.5 μs shaping time (right) (figure from [39])

The energy resolution at the Mn K$\alpha$ line (5.898 keV) is 176 eV FWHM at room temperature and 142 eV FWHM at −10°C, measured with 0.5 μs shaping time.

The short value of shaping time used to reach the best energy resolution makes the SDD with on-chip JFET the fastest X-ray spectroscopy detector when compared with conventional systems (Si(Li), Ge, PIN diode). This feature, related to the low value of output capacitance (detector + JFET total capacitance below 250 fF), makes this detector very attractive in those applications where a good energy resolution is required at high counting rates. In Fig. 4.23, the plot of the energy resolution of an SDD (5 mm$^2$ of active area) cooled at about −10° is shown as a function of the X-ray ($^{55}$Fe source) count rate, at two shaping times, 100 ns and 250 ns [45]. It can be noted that up to about 100.000 cps no relevant degradation of the energy resolution occurs.

Recently, a new layout of SDDs allows to optimize the peak-to-background ratio and to further improve the energy resolution [46]. In this new SDD, named SDD Droplet or SDD$^3$, the anode and the integrated JFET have been placed at the margin of the active area where they can be shielded from direct irradiation by the use of a collimator (Fig. 4.24). This feature allows to eliminate the contribution to a limited peak-to-background ratio due to the charge created under the JFET region and collected by the transistor electrodes instead of the anode. The peak-to-background value is enhanced to values higher than 6,000. This new layout allows, moreover, to design a readout anode with a smaller area and therefore to reduce the total capacitance to a value of about 120 fF to be compared with the 230 fF of a conventional SDD. The result is an improvement in energy resolution to 128 eV at −15°C.

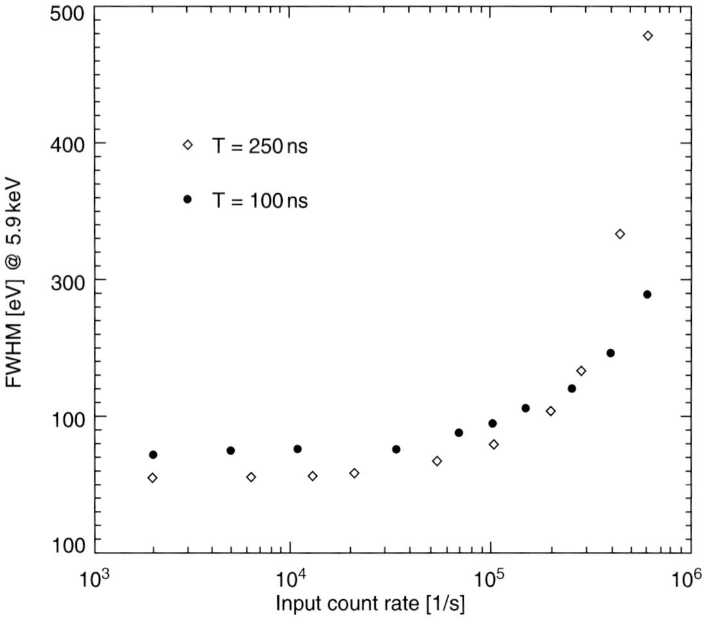


**Fig. 4.23.** Energy resolution of an SDD (5 mm$^2$) measured at room temperature as a function of the X-ray ($^{55}$Fe source) count rate. The measurements have been carried out at 100 ns (black points) and 250 ns (white points) shaping time (figure from [45])

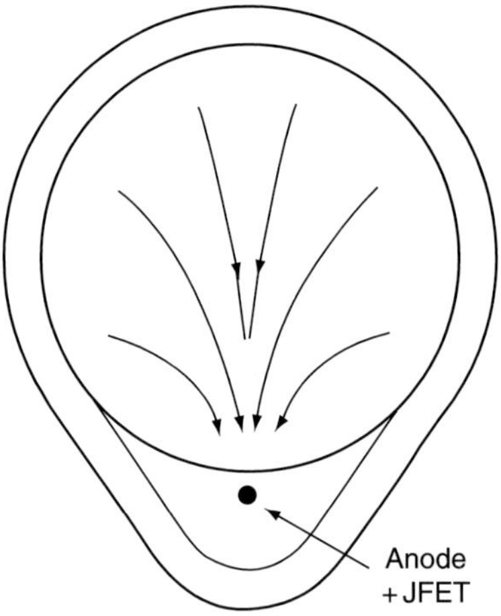


**Fig. 4.24.** Layout of the SDD Droplet (SDD$^3$) detector. The paths of the signal electrons towards the anode placed at the border of the detector are shown. The inner circle in the figure represents the region where the irradiation of the detector can be shielded to improve the peak-to-background ratio

## Multichannel SDDs

Single SDDs of large area, ranging from a few tens of $mm^2$ up to few $cm^2$, have also been produced. When cooled, these detectors offer an energy resolution similar to the one achievable with SDD of smaller area, thanks to the reduction of the leakage current by cooling and to the independence of the detector capacitance with respect to the active area. However, when a large active area is required in measurements near room temperature, the contribution of the leakage current to the electronics noise can be still relevant. In this case, the energy resolution of the detector can still maintain good by means of a segmentation of the detector area in single smaller units, each one with its own amplifying chain. This solution offers also two additional advantages. If a single unit (i.e. $5\,mm^2$) can operate without significant degradation of the energy resolution up to 100.000 cps, a detection system based on arrays of several units could operate with a very high counting rate capability ($2 \times 10^6$ cps $cm^{-2}$ in our example). As a second advantage, several single units can be suitably arranged in almost any kind of shape in one array in order to match the special geometrical configurations required by specific applications. In Fig. 4.25, some examples of multicell SDDs are shown [48]. They consist of several SDDs, each with individual on-chip JFET, monolithically arranged in a single chip. All units have common voltage supplies, entrance window and guard ring structure. An array could be used to fill a given area with a 100% efficiency (a), or to provide a specific two-dimensional shape like a linear chain (b) or a ring (c). These detectors can be suitably mounted in a compact module with Peltier cooler for near room temperature applications.

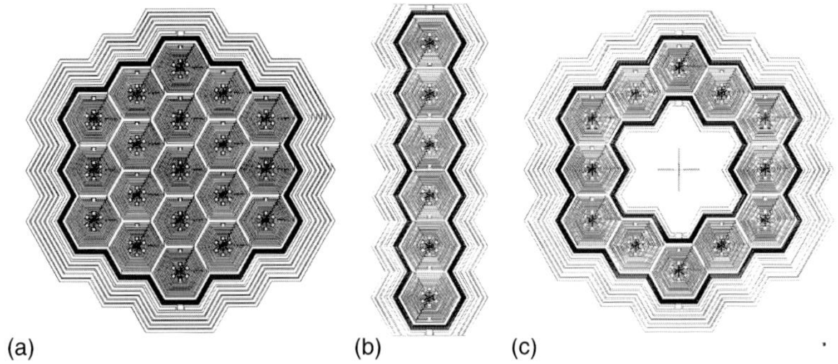


**Fig. 4.25.** Examples of Multicell SDD detectors: (**a**) an array of 19 cells for a total area of $95\,mm^2$, (**b**) a linear array of 6 cells for a total area of $30\,mm^2$, (**c**) a ring-shaped array of 12 cells for a total area of $60\,mm^2$. All plots show the layout of the detectors front side with the field strip system and the readout transistor in each cell's centre (figure from [48])

**Fig. 4.26.** Photograph of a 7-channel SDD module shown without cover. The detector chip is glued on and bonded to a ceramic carrier, which is mounted on the cold side of a Peltier cooler and bonded to the electrical feed-through pins of the housing (figure from [48])

In Fig. 4.26, a photograph of a seven-cell SDD module ($35\,\mathrm{mm}^2$ active area) manufactured by KETEK GmbH (Germany) is shown [48].

## Applications of SDDs

SDDs with on-chip JFET cooled by a single-stage Peltier element and mounted in a standard TO8 package are used in industrial scale for fast elemental mapping in scanning electron microscopes. Spectrometers based on the SDD module can measure X-ray fluorescence spectra at about ten times higher count rates than conventional energy dispersive spectrometers. Spatially resolved elemental mappings, like the one shown for example in Fig. 4.27, can be recorded with a high dynamic range (i.e. several hundred grey levels) and with short measurement times [45].

The high resolution achievable at high counting rates make the SDD an ideal detector also for X-ray spectroscopy experiments with synchrotron light, like EXAFS (Extended X-ray Absorption Fine Structure) and X-ray Holography, where the measurement times can be also highly reduced with respect to conventional detectors. Small SDD arrays have been already tested for EXAFS [49] and for X-ray Holography. A complete sphere composed of more than 1,000 SDD cells grouped in arrays of 61 units each has been conceived to detect all fluorescence photons emitted by a sample placed in its centre. The principle scheme of the system is shown in Fig. 4.28 [48].

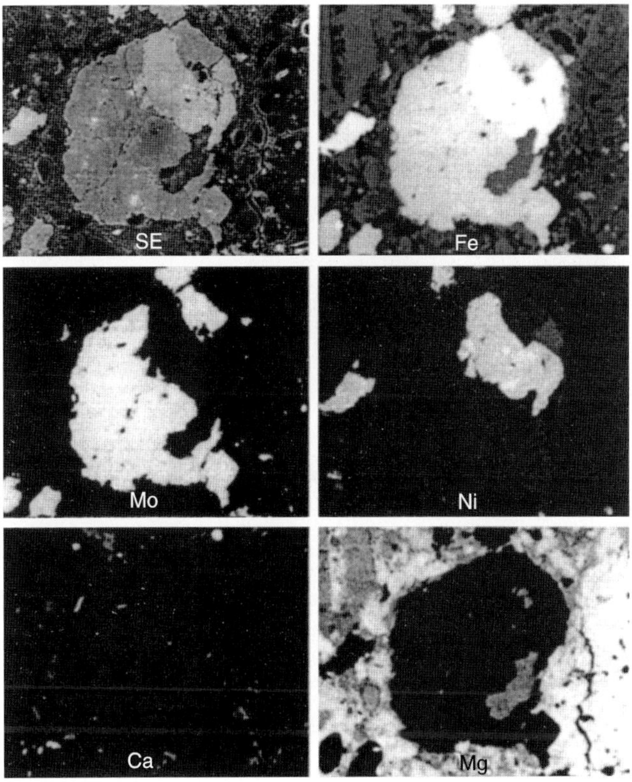


**Fig. 4.27.** Element imaging performed with an SDD in a ROENTEC Xflash$^{TM}$. The upper left picture shows the topological image acquired with the scanning electron microscope. All other pictures are element images, showing the special distribution of Fe, Mo, Ni, Ca, and Mg. The six pictures were acquired simultaneously, with an overall measuring time of 10 min and an average output count rate of 250.000 cps (figure from [45])

The 12-SDDs ring detector shown in Fig. 4.25c is the basic component of a new XRF spectrometer for material analysis, whose principle scheme is shown in Fig. 4.29 [50]. The sample is excited by X-rays emitted by a source which is intensified and focussed by a capillary fibre passing through the central hole of the detector chip. With this geometrical arrangement, the SDD units of the array can detect the X-ray fluorescence emitted by the sample covering a large solid angle. The size of the irradiation spot after the capillary optics could be as small as a few tens of μm. This system can therefore be employed also as a table-top instrument for fast elemental mappings. Figure 4.30 shows an example of elemental mapping obtained with this spectrometer. A $2 \times 2$ mm area of a nickel wire gauze (the one of the blade of an electric shaver) put over an iron foil (99.99%) has been scanned with the spectrometer and the fluorescence has been sampled in $20 \times 20$ points. The measurement at each point lasted 5 s.

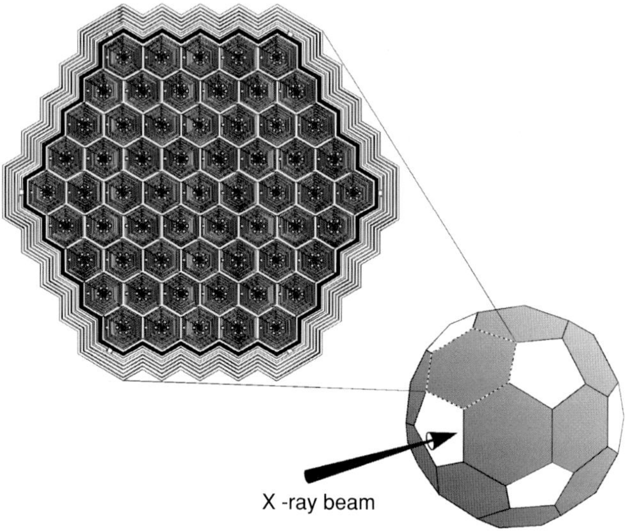


**Fig. 4.28.** Scheme of a spectrometer conceived for X-ray Holography experiments. The irradiated sample is surrounded by an almost complete sphere of detectors. Each hexagon of the football like configuration is a 61-channel SDD. The whole detector system consists of more than 1,000 channels (figure from [48])

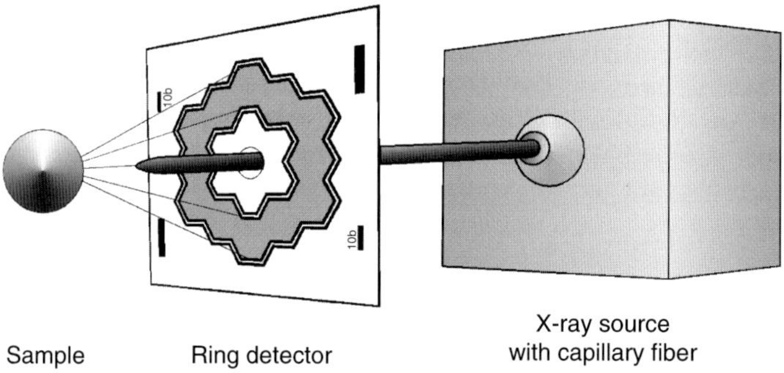


**Fig. 4.29.** Principle of an X-ray spectrometer based on the 12-SDD ring array. The use of a capillary optics to irradiate the sample through a hole located in the centre of the array allows to reach an intense photon flux on the sample and to achieve a high geometrical efficiency in the detection of the X-ray fluorescence (figure from [50])

A portable spectrometer based on a Peltier-cooled SDD has been developed for nondestructive analyses on the cultural heritage [51]. With respect to other portable XRF spectrometers based either on nitrogen-cooled detectors (Si(Li), Ge) or Peltier-cooled PIN diodes, this instrument ensures a high

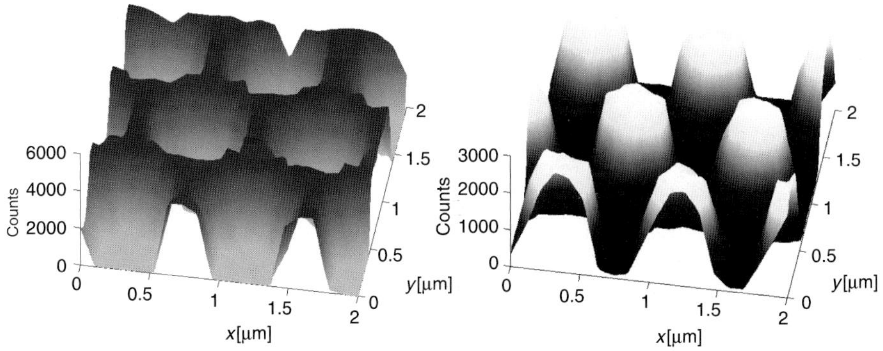


**Fig. 4.30.** Mapping of nickel (left) and iron (right) distributions in the sample, measured with the 12-SDDs ring spectrometer

energy resolution achievable with a SDD. Hence a liquid nitrogen cryostat is no longer required. The spectrometer can be employed for an in situ identification of chemical elements in pigments, stones, metal alloys and other materials of artistic and archaeological interest.

Single SDDs and monolithic arrays of SDDs have also been used as photodetectors for the readout of scintillators in hard X-ray and $\gamma$-ray detection. With respect to the conventional photomultiplier tubes, the SDDs are more compact and are characterized by a higher quantum efficiency for the scintillation light typical of silicon photodetectors. With respect to conventional PIN diodes, the SDDs offer a lower electronics noise. When coupled to CsI(Tl) scintillators, SDDs have shown to achieve both good energy resolution and low energy threshold in hard X-ray and $\gamma$-ray spectroscopy [52] as well as a sub-mm position resolution in $\gamma$-ray imaging [53].

### 4.2.6 Basics of Signal Electronics

The purpose of the electronic chain is to measure "at the best" the value of the charge $Q$ delivered by the detector when an ionizing particle interacts in its active volume. "At the best" means with the maximum accuracy and precision compatible with many physical and practical constraints [54–57].

The charge $Q$ is supposed to be proportional to the energy released by the ionizing particle in the detector. We do not consider in this section the statistical fluctuations in the amount of charge $Q$ delivered by the detector for a fixed energy $E$ released by the particle in its active volume.

The presence of several unavoidable noise sources in the detector and in the electronic circuits affects the precision of the measurement of the charge $Q$. The precision of the measurement is usually defined in terms of the $S/N$ (signal-to-noise) ratio. An alternative concept, very often used in the field of radiation detection, is the equivalent noise charge (ENC). The ENC is, simply, the charge delivered by the detector which makes the $S/N$ equal to one. We will base our presentation on the ENC concept.

## Basic Electronics Blocks of an XRF Detection System

An XRF detection system is basically composed of a detector, of a signal processor and of a unit for storing and analysing the acquired spectra. The detector can be usually modelled as capacitance in parallel with a current generator, which delivers a current pulse carrying a charge $Q$ proportional to the energy $E$ released by the ionizing particle in the detector. The charge $Q$ is affected by statistical fluctuations considered above in this chapter.

In order to introduce in the simplest way the basic concepts of signal processing, we will here consider "time-invariant analog signal processors". By definition, a time-invariant analog signal processor is a circuit which produces at its output, when a signal pulse $s_{\mathrm{in}}(t - t_0)$ is applied at its input, a signal $s_{\mathrm{out}}(t - t_0)$ whose shape and amplitude is independent of the value of the time $t_0$.

A time-invariant analog signal processor for radiation detection is usually composed of the following basic units: a charge preamplifier, a "pulse shaping amplifier", a peak sampler, an analog-to-digital converter and a digital unit for the storing and processing of the acquired data.

The *charge preamplifier* usually provides the first level of amplification of the detector signal. The detector is connected to the virtual ground of the amplifier and the charge $Q$ is integrated on its feedback capacitance $C_{\mathrm{F}}$. The preamplifier output signal is a step-like voltage waveform whose amplitude is proportional to the charge $Q$ delivered by the detector. Alternatively, a *voltage preamplifier* can be used. In this case, the charge $Q$ is integrated on a capacitance $C_{\mathrm{IN}}$ which is the sum of the capacitance $C_{\mathrm{D}}$ of the detector and the input capacitance $C_{\mathrm{A}}$ of the voltage amplifier. The configuration based on the charge preamplifier has the main advantage of making the gain of the input stage insensitive to the variations of the detector capacitance.

The charge preamplifier is followed by the *pulse-shaping amplifier*, (also called "filtering amplifier"), whose main duties are a further amplification of the signal and the improvement of the $S/N$ (and equivalently, of the ENC). It provides an output voltage pulse whose peak amplitude is proportional to the amplitude of the step-like waveform provided by the charge preamplifier (or by the voltage preamplifier).

The electronic stages which follow the pulse-shaping amplifier are related to the digitalization of its output peak value. A sampler detects the peak value of the output pulse and provides a stable signal suitable for the operation of the analog-to-digital converter (ADC). We will here suppose that the sampling and the digitalization processes do not affect the $S/N$ at the output of the pulse-shaping amplifier.

## Signal Processing Theory

*Preliminary Considerations*

Figure 4.31 shows schematically the detection system. The detector and the input field effect transistor (FET) of the electronic chain are of particular importance because the resolution of the system is strongly affected by the noise performances of these two components.

The detector is modelled as a current source, in parallel to the detector capacitance $C_{\mathrm{D}}$, which delivers an indefinitely short current pulse carrying a charge $Q$. This signal is mathematically represented by a delta-like pulse of area $Q$.

$$i_{\mathrm{D}}(t) = Q\delta(t). \tag{4.13}$$

The $S/N$ of the measurement is defined as the ratio between the peak amplitude of the signal and the root mean square value of the noise, both measured at the output of the pulse shaping amplifier.

$$\frac{S}{N} = \frac{v_{\mathrm{so\ peak}}}{\sqrt{\langle v_{\mathrm{no}}^2 \rangle}} = \frac{Q\,\mathrm{Max}[v_{\mathrm{so}\ u\delta}(t)]}{\sqrt{\langle v_{\mathrm{no}}^2 \rangle}}, \tag{4.14}$$

where $v_{\mathrm{so}\ u\delta}(t)$ is the pulse at the output of the shaping amplifier when the detector delivers a unitary charge.

The ENC (which is, by definition, the charge delivered by the detector which makes the $S/N$ equal to one) is therefore

$$\mathrm{ENC} = \frac{\sqrt{\langle v_{\mathrm{no}}^2 \rangle}}{\mathrm{Max}[v_{\mathrm{so}\ u\delta}(t)]}. \tag{4.15}$$

For what concerns the evaluation of the $S/N$ and of the ENC, it can be shown that the charge-sensitive configuration at the input of the electronic chain is completely equivalent to the voltage-sensitive one. We will consider in the following the voltage-sensitive configuration, which leads to a more easy understanding of the phenomena involved.

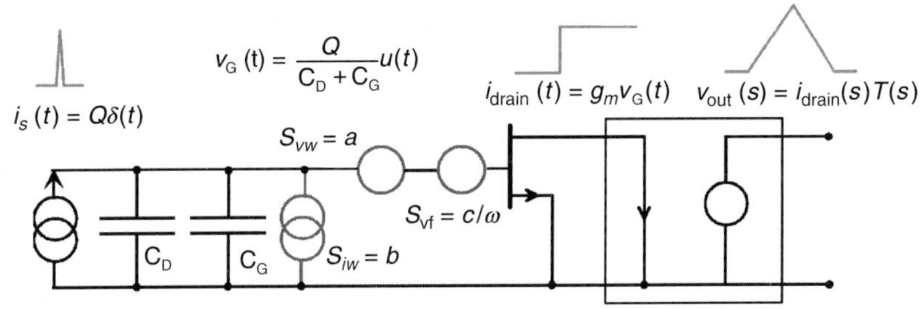


**Fig. 4.31.** Basic elements of the detection system

**The Noise Sources**

Only the noise sources localized at the input of the electronic chain are usually relevant in the determination of the ENC. In fact, due to the usually large gain of the input stage, the noise contribution of the sources localized in the latter stages is negligible. We will therefore here consider only the noise sources present at the input of the detection chain (see Fig. 4.31).

The noise sources considered are [58, 59]:

- The thermal noise of the FET channel, represented by a voltage generator placed in series to the gate terminal of the FET, with a white spectral power density[a]

$$S_{\mathrm{vw}} = \alpha \frac{2kT}{g_m} = \alpha \frac{2kT}{C_{\mathrm{G}}} \frac{1}{\omega_{\mathrm{T}}}, \tag{4.16}$$

  where $\omega_T$ is its cut-off angular frequency defined as $\omega_{\mathrm{T}} = g_m / C_{\mathrm{G}}$ and $\alpha$ is a constant equal to about 2/3 for Silicon JFET.
- The shot noise associated with the leakage current $I_{\mathrm{D}}$ of the detector, represented by a current generator placed in parallel to the detector, with a white spectral power density

$$S_{\mathrm{iD}} = qI_{\mathrm{D}}. \tag{4.17}$$

- The shot noise associated with the gate leakage current $I_{\mathrm{G}}$ of the input FET, represented by a current generator placed between the gate and the source terminals of the FET, with a white spectral power density

$$S_{\mathrm{iG}} = qI_{\mathrm{G}}. \tag{4.18}$$

- The thermal noise of any resistor connected to the gate of the FET (for biasing or feedback purposes) represented by a current generator placed between the gate and ground

$$S_{\mathrm{iR}} = qI_{\mathrm{R\ eq}} = \frac{2kT}{R}. \tag{4.19}$$

- The $1/f$ noise of the FET, mainly related to trapping phenomena in the channel region, represented by a voltage generator placed in series to the gate terminal of the FET, with a $\omega^{-1}$ spectral power density

$$S_{\mathrm{vf}}(\omega) = \frac{1}{2} \frac{A_{\mathrm{f}}}{|f|} = \frac{\pi A_{\mathrm{f}}}{|\omega|} = \alpha \frac{2kT}{C_{\mathrm{G}}} \frac{\omega_1}{\omega_{\mathrm{T}}} \frac{1}{|\omega|}, \tag{4.20}$$

  where $A_{\mathrm{f}}$ is the usual 1/f noise coefficient and $\omega_1$ is defined as the angular frequency at which the 1/f series noise spectrum equals the white series noise spectrum.

---

[a] The mathematical bilateral representation of the noise power spectra is here considered, with the angular frequency $\omega$ ranging from $-\infty$ to $+\infty$.

The three equivalent noise generators of Fig. 4.32 take into account the considered noise sources, being

$$\begin{aligned} S_{\mathrm{vw}} &= a = \alpha \frac{2kT}{C_{\mathrm{G}}} \frac{1}{\omega_{\mathrm{T}}} \\ S_{\mathrm{iw}} &= b = S_{\mathrm{iD}} + S_{\mathrm{iG}} + S_{\mathrm{iR}} = q(I_{\mathrm{D}} + I_{\mathrm{G}} + I_{\mathrm{R\,eq}}) = qI_{\mathrm{L}} \\ S_{\mathrm{vf}}(\omega) &= c/\omega = \alpha \frac{2kT}{C_{\mathrm{G}}} \frac{\omega_1}{\omega_{\mathrm{T}}} \frac{1}{|\omega|} . \end{aligned} \tag{4.21}$$

*The ENC of the Detection System*

In the following section the way to calculate the ENC will be outlined. The main results are here presented and discussed. See Appendices 1 and 2 (pp. 256–259) for more details.

It turns out that the squared value of the ENC can be expressed as the sum of three independent contributions:

$$\mathrm{ENC}^2 = (C_{\mathrm{D}} + C_{\mathrm{G}})^2 a \frac{1}{\tau} A_1 + (C_{\mathrm{D}} + C_{\mathrm{G}})^2 c A_2 + b\tau A_3 . \tag{4.22}$$

The first contribution is due to the channel thermal noise of the input FET. The second one is due to the 1/f noise associated with its drain current. The third contribution is due to the shot noise of the leakage current of the detector and of the FET and to the thermal noise of any resistor connected to the gate of the input FET. The noise power spectra $a$, $c$, and $b$ (introduced in Section "The Noise Sources" see (4.21)) correspond, respectively, to the three equivalent noise generators at the input of the signal processor: the white series noise generator, the 1/f series noise generator and the white parallel noise generator. For this reason, the three contributions to the ENC are usually called respectively white series noise contribution, 1/f series noise contribution and parallel noise contribution.

The capacitances $C_{\mathrm{D}}$ and $C_{\mathrm{G}}$ are the detector and the gate to source capacitance of the transistor, respectivly. It must be noted that any parasitic capacitance in parallel to the detector must be included in $C_{\mathrm{D}}$ (for instance the parasitic capacitance due to the connection between the detector and gate of the input transistor). In the case of the charge-preamplifier configuration, the feedback capacitance has also to be included in $C_{\mathrm{D}}$.

The characteristic time $\tau$ represents the width of the output pulse (for instance $\tau$ can be the peaking time, or the time width at half height, or a time constant of the filter). The characteristic time $\tau$ is also called "shaping time" of the filter.

The coefficients $A_1$, $A_2$, and $A_3$ are constants which depend only on the "shape" of the output pulse of the filter and not on its "width". If a different choice of the characteristic time $\tau$ is performed (for instance if the peaking time instead of the time width at half height is chosen) the value of the coefficients $A_1$, $A_2$, and $A_3$ changes. In section 4.2.10 (Appendix 2) it will be shown that, if $\tau'' = k\tau'$, the new set of coefficients is given by

$$\begin{aligned} A_1(\tau'') &= kA_1(\tau') \\ A_2(\tau'') &= A_2(\tau') \\ A_3(\tau'') &= \frac{1}{k}A_3(\tau') \end{aligned} \quad . \tag{4.23}$$

By using the explicit expressions for the noise power spectra given in (4.21), (4.22) can be written in the following useful way

$$\begin{aligned} \mathrm{ENC}^2 = {} & A_1 C_\mathrm{D}\left(\sqrt{\frac{C_\mathrm{D}}{C_\mathrm{G}}}+\sqrt{\frac{C_\mathrm{G}}{C_\mathrm{D}}}\right)^2 \alpha\frac{2kT}{\omega_\mathrm{T}}\frac{1}{\tau} \\ & + A_2 C_\mathrm{D}\left(\sqrt{\frac{C_\mathrm{D}}{C_\mathrm{G}}}+\sqrt{\frac{C_\mathrm{G}}{C_\mathrm{D}}}\right)^2 \alpha\frac{2kT}{\omega_\mathrm{T}}\omega_1 + A_3 q I_L \tau . \end{aligned} \tag{4.24}$$

Figure 4.32 shows the $\mathrm{ENC}^2$ versus the shaping time $\tau$ and the contributions of the series, parallel and 1/f noise sources.

*Optimization of the ENC*

The question now is: how to obtain the best ENC in a given measurement condition? [55, 56]

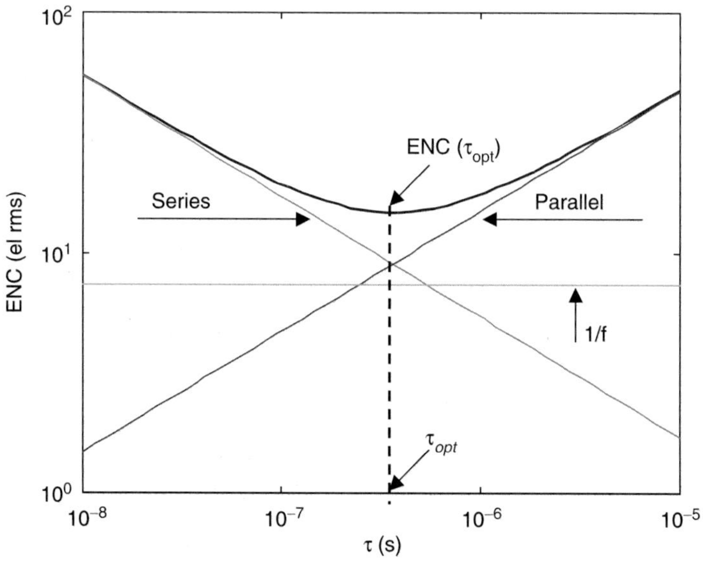


**Fig. 4.32.** Contributions to the ENC (the ENC is expressed in terms of electrons rms and it has been evaluated for a Gaussian filter and for $C_\mathrm{D} = 150\,\mathrm{fF}$, $C_\mathrm{G} = 150\,\mathrm{fF}$, $g_\mathrm{m} = 0.5\,\mathrm{mS}$, $A_\mathrm{f} = 5\times10^{-12}\,\mathrm{V}^2$, $I_\mathrm{D} = 10\,\mathrm{pA}$, $I_\mathrm{G} = 10\,\mathrm{pA}$)

*Choice of the Optimum "shaping time"*

Let us first suppose that we have chosen a given "shape" of the output pulse of the filter (or, more practically, let us suppose that only one kind of filtering amplifier is available for our measurement). In this case, the coefficients $A_1$, $A_2$, and $A_3$ of the filter are determined. In order to optimize the ENC we have the freedom to choose the most suitable "shaping time" $\tau$.

By considering (4.24) we can observe that the contribution of the series noise is inversely proportional to the shaping time $\tau$, while the contribution of the parallel noise is directly proportional to $\tau$. The contribution of the $1/f$ noise is independent of $\tau$. The optimum choice for the shaping time $\tau$ is the one for which the series and parallel noise contributions to the $\mathrm{ENC}^2$ are equal. Let us call $\tau_{\mathrm{opt}}$ this particular value of the shaping time. By using (4.24) its value turns out to be

$$\tau_{opt} = (C_D + C_G)\sqrt{\frac{a}{b}}\sqrt{\frac{A_1}{A_3}} = \sqrt{C_D}\left(\sqrt{\frac{C_D}{C_G}} + \sqrt{\frac{C_G}{C_D}}\right)\sqrt{\alpha\frac{2kT}{\omega_T}\frac{1}{qI_L}}\sqrt{\frac{A_1}{A_3}}. \tag{4.25}$$

The ENC obtainable at the optimum shaping time is the best one with the considered shaping amplifier. The squared value of the ENC is given by

$$\begin{aligned}\mathrm{ENC}^2(\tau_{opt}) &= 2(C_D + C_G)\sqrt{ab}\sqrt{A_1A_3} + (C_D + C_G)^2 cA_2 \\ &= 2\sqrt{C_D}\left(\sqrt{\frac{C_D}{C_G}} + \sqrt{\frac{C_G}{C_D}}\right)\sqrt{\alpha\frac{2kT}{\omega_T}\frac{1}{qI_L}}\sqrt{A_1A_3} \\ &+ C_D\left(\sqrt{\frac{C_D}{C_G}} + \sqrt{\frac{C_G}{C_D}}\right)^2 \alpha\frac{2kT}{\omega_T}\omega_1 A_2. \end{aligned} \tag{4.26}$$

The first term represents the sum of the contributions of the white series and white parallel noise, the second term represents the contribution of the $1/f$ noise. If the $1/f$ noise contribution is negligible, the ratio between the ENC that can be obtained by choosing any $\tau$ and the value that can be obtained by choosing $\tau_{\mathrm{opt}}$, is a universal function independent of the choice of the pulse shape (we use the symbol $\mathrm{ENC}_w$ for representing the ideal case in which the $1/f$ noise is absent or negligible):

$$\frac{\mathrm{ENC}_w(\tau)}{\mathrm{ENC}_w(\tau_{opt})} = \frac{1}{\sqrt{2}}\sqrt{\frac{\tau_{opt}}{\tau} + \frac{\tau}{\tau_{opt}}}. \tag{4.27}$$

Figure 4.33 shows a plot of this ratio. It is evident that the $\mathrm{ENC}_w(\tau)$ worsens only about 10% with respect to $\mathrm{ENC}_w(\tau_{\mathrm{opt}})$ in the interval $0.5\,\tau_{\mathrm{opt}} < \tau < 2\,\tau_{\mathrm{opt}}$.

In nearly all commercial shaping amplifiers there is a reasonably wide possibility of choosing a shaping time $\tau$ close to the optimum one for the

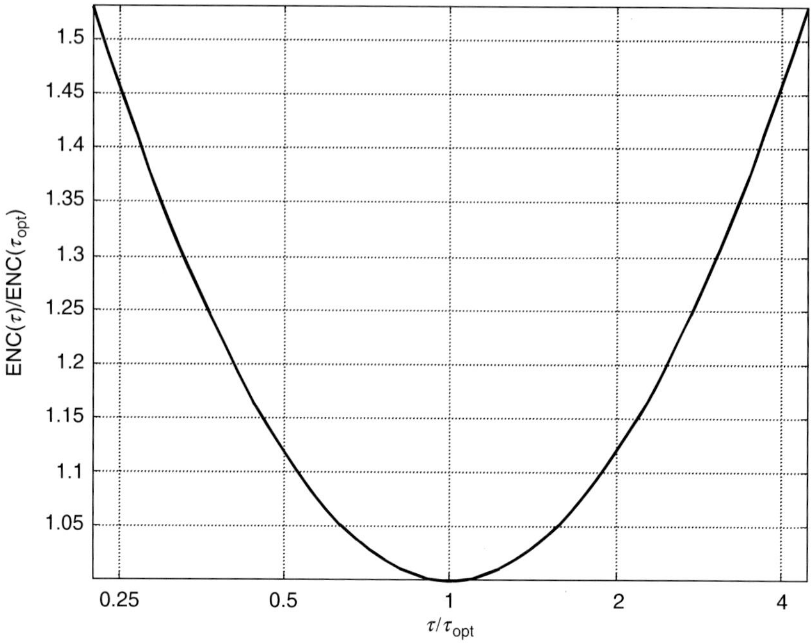


**Fig. 4.33.** Ratio of the ENC obtained with any filter type at shaping time $\tau$ and that obtained with the same filter type at the optimum shaping time $\tau_{\rm opt}$

considered application. The available shaping times are usually set in such a way that it will be possible to select a value of $\tau$ which lies in the interval $0.5\,\tau_{\rm opt} < \tau < 2\,\tau_{\rm opt}$.

*Detector Capacitance and Detector–Transistor Capacitance Matching*

Coming back to the (4.24), we can observe that the white series noise contribution to the ENC and the $1/f$ series noise contribution depend on the values of the detector and transistor capacitances $C_{\rm D}$ and $C_{\rm G}$. The white parallel noise contribution is, on the contrary, not explicitly dependent on the value of these parameters.

In order to minimize the white and $1/f$ series noise contributions the detector capacitance $C_{\rm D}$ should be as low as possible. In practice, the detector capacitance is determined by the physical structure of the detector adopted. In the semiconductor PIN-diode detectors, for instance, the detector sensitive area is proportional to the detector capacitance. Therefore, the sensitive area should be kept as small as possible, compatible with the requirement of the considered application. In the SDD, on the contrary, the output capacitance is independent of the sensitive area and is moreover very small.

Once the detector capacitance and the cut-off frequency of the transistor have been defined, the gate capacitance of the input FET of the signal processor should be matched with the detector capacitance in order to minimize the white and the $1/f$ series noise contributions. In fact the factor $M = \left(\sqrt{C_{\mathrm{D}}/C_{\mathrm{G}}} + \sqrt{C_{\mathrm{G}}/C_{\mathrm{D}}}\right)$, called "matching factor", is minimized for $C_{\mathrm{G}} = C_{\mathrm{D}}$.

In Fig. 4.34 the matching factor is plotted as a function of the ratio $C_{\mathrm{G}}/C_{\mathrm{D}}$. This plot shows the influence of the mismatching between detector and transistor on the white and 1/f series contributions finally leading to a worsening of the ENC (comp. (4.24)).

From (4.25) and (4.26) it is evident how the matching of the detector and transistor capacitances reduces the value of $\tau_{\mathrm{opt}}$ (this is generally an advantage because it improves the maximum counting rate of the system) and improves $\mathrm{ENC}(\tau_{\mathrm{opt}})$.

Figure 4.35 shows a comparison of the ENC obtainable with detectors of different capacitance. The simulation has been carried out by using realistic parameters for detectors and transistors and by supposing that the matching condition is satisfied ($C_G = C_D$). A PIN diode detector with an active area of $10\,\mathrm{mm}^2$ and a thickness of $300\,\mu\mathrm{m}$ having a capacitance of about 3 pF and an SDD with the same active area and thickness having a capacitance of 150 fF are considered. A detector leakage current of about 2.5 pA at $-10°$C has been considered for both detectors. In the case of the PIN diode the parameters of the matched FET are extrapolated from the parameters of typical discrete FETs (a transconductance $g_m = 6\,mS$ and an $A_{\mathrm{f}} = 4 \times 10^{-15}\,\mathrm{V}^2$ have been considered), while for the case of the SDD the matched FET is supposed to be integrated on the detector wafer (a $g_m = 0.3\,\mathrm{mS}$ and an $A_{\mathrm{f}} = 5 \times 10^{-12}\,\mathrm{V}^2$ have been considered). In both cases, the gate current of the FET is supposed to be equal to that of the detector and the noise of biasing or feedback resistors has been neglected.

From Fig. 4.35 it is evident that the reduction of the detector capacitance (in matched conditions) reduces both the $\tau_{\mathrm{opt}}$ and the $\mathrm{ENC}(\tau_{\mathrm{opt}})$. Note that the two considered transistors have the same $\omega_{\mathrm{T}}$ and different $A_{\mathrm{f}}$.

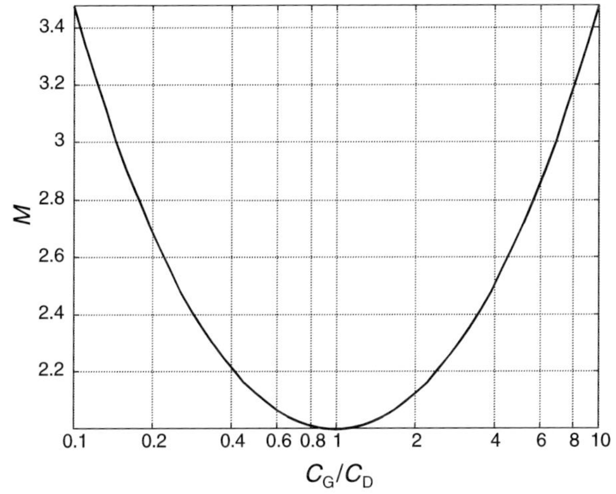


**Fig. 4.34.** The matching factor $M$ as a function of the ratio $C_{\mathrm{G}}/C_{\mathrm{D}}$

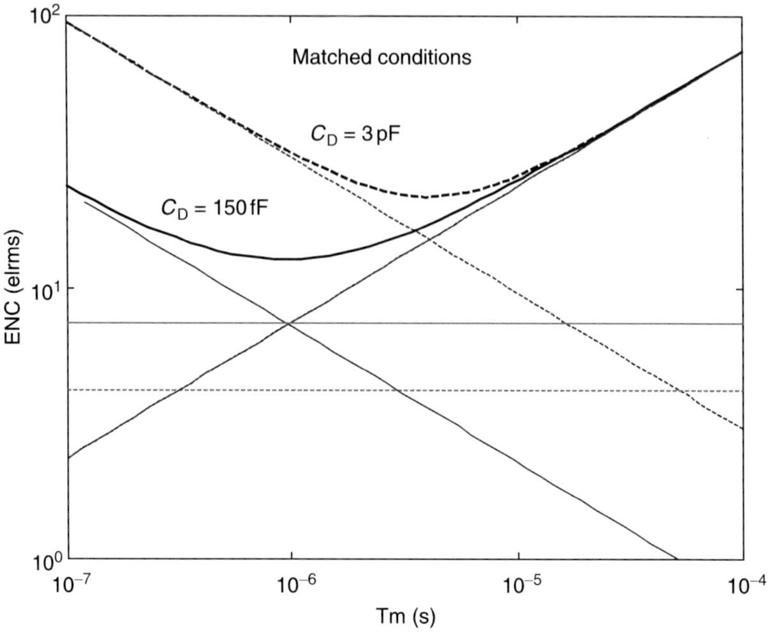


**Fig. 4.35.** Comparison of the ENC obtainable with detectors of different capacitance (see text) in matched conditions (solid lines $C_D = 150\,\text{fF}$, dashed lines $C_D = 3\,\text{pF}$). The serial, parallel and $1/f$ noise contributions are also shown

*Transistor Cut-off Frequency*

Once the FET capacitance is supposed to be matched to the detector capacitance, other transistor parameters can be considered in order to improve both $\tau_{\text{opt}}$ and $\text{ENC}(\tau_{\text{opt}})$. In particular, the cut-off frequency $\omega_T$ of the transistor appears explicitly in the expression of the ENC given by (4.24) both in the white series noise contribution and in the $1/f$ series noise contribution.

For what concerns the white series noise contribution, it is usually an advantage to have a FET with a high cut-off frequency, because, for a given $C_G$, a higher $\omega_T$ means a higher transconductance $g_m$ and therefore a lower series noise. If transistors fabricated with the same technology are considered, $\omega_T$ grows by reducing the gate length.

For what concerns the $1/f$ series noise contribution the critical parameter is the ratio $\omega_1/\omega_T$ (see (4.24)). The angular frequency $\omega_1$ is not independent of $\omega_T$. This ratio can be easily written as

$$\frac{\omega_1}{\omega_T} = \frac{\pi A_f C_G}{\alpha 2kT}. \tag{4.28}$$

The parameter $H_f = A_f C_G$ is nearly constant for a given fabrication technology of the transistor (in particular it is independent of the gate width) and

**Table 4.1.** Value of quality factor $H_f$

| Device | $H_f$(J) |
|---|---|
| JFET, n-channel, discrete | $2 \times 10^{-26}$ |
| JFET, n-channel, in CMOS process | $10^{-25}$ |
| MOSFET, p channel, in CMOS process | $6 \times 10^{-25}$ |
| MOSFET, n channel, in CMOS process | $2.5 \times 10^{-23}$ |
| MESFET, GaAs, discrete | $10^{-23}$ |

it is considered as a quality parameter of the corresponding transistor class. It often happens that a transistor with a better $\omega_T$ has a worse ratio $\omega_1/\omega_T$. In Table 4.1 the value of the quality factor $H_f$ for different classes of FETs is reported [60]. From this reference we recall here, for sake of exemplification, a few values of $H_f$. If transistors fabricated with the same technology are considered, the ratio $\omega_1/\omega_T$ is nearly independent of the gate length.

*Optimum Signal Processing*

A basic question is: which shape of the output pulse is better to choose in order to improve the ENC? Or, in other words, is there a "shape" of the output pulse of the filtering amplifier which guarantees the best ENC? This is the problem of the "optimum signal processing theory". We will synthetically report here only the main results of this theory.

It has been shown [55, 56] that, if only the white noise sources (series and parallel) are present at the input of the detection circuit, the best ENC can be obtained by using an ideal filtering amplifier which gives at its output, when it is fed by a unitary amplitude step-like pulse, an "infinite cusp" -shaped pulse:

$$v_{so}(t) = \exp\left(-\frac{|t|}{\tau}\right), \tag{4.29}$$

with a shaping time $\tau$ set equal to the "noise corner" time constant $\tau_c$. The "noise corner" time constant $\tau_c$ is, by definition, the reciprocal of the angular frequency $\omega_c$ at which the contribution of the white series noise and of the white parallel noise at the preamplifier input are equal

$$\tau_c = (C_D + C_G)\sqrt{\frac{a}{b}}. \tag{4.30}$$

Figure 4.36 shows the infinite-cusp shaped pulse at the output of the filtering amplifier. This shaper is clearly nonrealistic. The peak value of the output pulse happens after an infinite time delay from the delta pulse delivered by the detector and the duration of the pulse is infinite. The ENC that would be obtained with this ideal shaper is the best obtainable, if only white noise source is present.

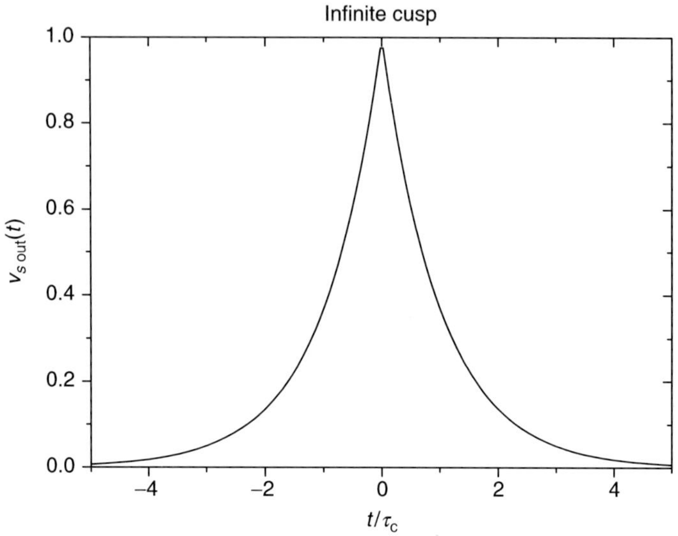


**Fig. 4.36.** Optimum shape of the output pulse of the filter with white series and white parallel noise

The "shape factors" for this filter are

$$A_1 = 1 \quad A_2 = \frac{2}{\pi} \simeq 0.64 \quad A_3 = 1. \tag{4.31}$$

From (4.25) it comes out that $\tau_\mathrm{c} = \tau_\mathrm{opt}$ for this filter.

The ENC that can be obtained in this situation (remember: without $1/f$ noise) is the "optimum" one for the considered noise sources. From (4.26) we have for the "infinite cusp" filter for $\tau = \tau_\mathrm{opt}$

$$\mathrm{ENC}^2_{w\ \mathrm{opt}} = \mathrm{ENC}^2_w(\tau_\mathrm{opt}) = 2\,(C_\mathrm{D} + C_\mathrm{G})\,\sqrt{ab}. \tag{4.32}$$

If also the $1/f$ noise is present, this filter is no more the "optimum" one [61]. Figure 4.37 shows the shape of the "optimum" filter output pulse when the $1/f$ noise is present.

The parameter $K$ of the different curves is given by the following formula with $a_\mathrm{f} = A_\mathrm{f}/2$:

$$K = \frac{\mathrm{ENC}^2_{1/\mathrm{f}}(\tau_\mathrm{opt})}{\mathrm{ENC}^2_w(\tau_\mathrm{opt})} = 2\,(C_\mathrm{D} + C_\mathrm{G})\,\frac{a_\mathrm{f}}{\sqrt{ab}}, \tag{4.33}$$

where $\mathrm{ENC}^2_{1/\mathrm{f}}(\tau_\mathrm{opt})$ is the $\mathrm{ENC}^2$ obtainable with the ideal infinite-cusp filter when *only* the $1/f$ noise is present and the shaping time is set according to the values $a$ and $b$ of the white power spectra considered.

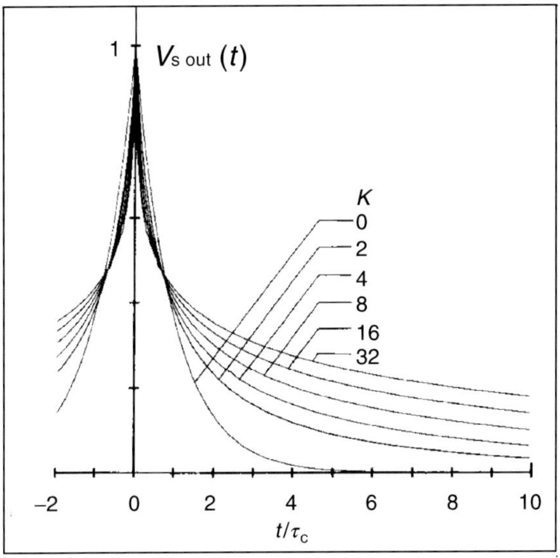


**Fig. 4.37.** Optimum shape of the output pulse of the filter with white series, white parallel and $1/f$ series noise. The parameter $K$ (defined in the text) refers to different values of the $1/f$ noise source [61]

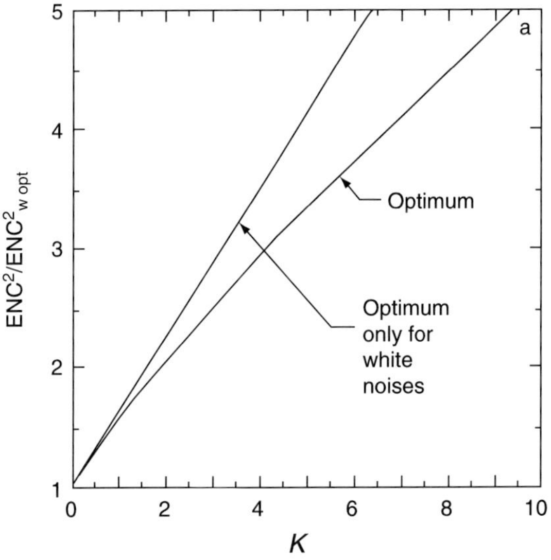


**Fig. 4.38.** Plots of $ENC^2/ENC^2_{w\,opt}$ versus $K$ for different pulse processing (see text) [61]

Figure 4.38 shows the $\mathrm{ENC}^2$ obtainable in the presence of $1/f$ noise if an optimum filter for each value of the parameter K according to Fig. 4.37, is used. Figure 4.38 shows also the $\mathrm{ENC}^2$ obtainable in the presence of $1/f$ noise when the filter optimum for white noise only (the infinite-cusp) is used. The $\mathrm{ENC}^2$ is normalized to the value of $\mathrm{ENC}^2_{w\ \mathrm{opt}}$.

We must observe also that this kind of shaping is nonrealistic, due to the bilateral infinite width. Nevertheless, the results of the "optimum filtering" theory are very useful because they represent the limit to be approached by the practical filters. A discussion about practical filter is given in Section "Practical Signal Processing".

## Practical Signal Processing

A practical signal processor can only "approach" the optimum ENC obtained with the ideal optimum filters described in Section "Optimum Signal Processing" (p. 245). for any filter, characterized by the shape factors $A_1$, $A_2$, and $A_3$, it can be seen from (4.25) and (4.30) that its optimum shaping time $\tau_{\mathrm{opt}}$ is related to the noise corner time constant $\tau_c$ (which is the optimum shaping time for the ideal cusp-shaped filter) by

$$\frac{\tau_{\mathrm{opt}}}{\tau_{\mathrm{c}}} = \sqrt{\frac{A_1}{A_3}}. \tag{4.34}$$

By comparing (4.26) and (4.32) it is possible to see how worse is the ENC obtainable with the considered filter with respect to the one obtainable with an ideal filter. In particular, if only white noise is present, it turns out that

$$\frac{\mathrm{ENC}^2_w(\tau_{\mathrm{opt}})}{\mathrm{ENC}^2_{w\ \mathrm{opt}}} = \sqrt{A_1 A_3}. \tag{4.35}$$

If also $1/f$ noise is present, the worsening of the ENC with respect to the ideal case is less direct to see. It can be shown that the ENC obtainable with the considered filter satisfies the following inequality

$$\frac{\mathrm{ENC}^2(\tau_{\mathrm{opt}})}{\mathrm{ENC}^2_{w\ \mathrm{opt}}} \leq (1+K)\sqrt{A_1 A_3}, \tag{4.36}$$

where the factor $K$, which is related to the amount of the $1/f$ noise present, has been defined in (4.33).

In the following the shape factors of some practical filters are given. Nowadays, the most used filtering amplifier is the "semi-Gaussian" one based on a constellation of complex poles. Semi-Gaussian filters based on real poles and CR–RC filters can still be found in many laboratories. An important filter is the trapezoidal one, which can nowadays be also implemented as a time-variant parameter circuit. The triangular filter is important for didactical reasons (it is easy to make mathematics and determine the shape factors with this filter as well as with CR–RC filter).

It should be noted that nowadays digital signal processing (DSP) techniques allow practically the implementation of any weighting function, with any kind of shape constraint (e.g. finite width of the impulse response, flat top, etc.). See 4.2.11 Appendix 3 (pp. 259–262) for a short introduction to the digital filtering techniques in X-ray spectroscopy.

### 4.2.7 Shape Factors of some Filtering Amplifiers

The ENC can be easily evaluated in the case of some practical signal processors, once the "shape factors" are known. We consider a few typical cases and the values of the shape factors for each filter are given in Table 4.2.

*CR–RC Shaping*

The CR–RC filter is the simplest among the shaping amplifiers. Nowadays, it is very seldom used because other higher order filters giving a better ENC can be easily implemented. The output pulse of this shaper, fed by a unitary amplitude step-like pulse, is

$$v_{\mathrm{uso}}(t) = \frac{t}{\tau_{\mathrm{c}}} \exp\left(-\frac{t}{\tau_{\mathrm{c}}}\right), \tag{4.37}$$

where $\tau_{\mathrm{c}} = RC$. The peak value of the output signal is obtained at $\tau_{\mathrm{p}} = \tau_{\mathrm{c}}$.

*Triangular Shaping*

It is interesting from a conceptual point of view. The obtainable resolution approaches that of the ideal cusp filter, with the advantage of having an output

**Table 4.2.** Shape factors of filters

| Shape Factors | | $A_1$ | $A_2$ | $A_3$ | $\sqrt{A_1 A_3}$ | $\sqrt{A_1/A_3}$ |
|---|---|---|---|---|---|---|
| Infinite cusp | $\tau$ | 1.00 | 0.64 | 1.00 | 1.00 | 1.00 |
| Triangular | $T_{\mathrm{base}}/2$ | 2.00 | 0.88 | 0.67 | 1.16 | 1.73 |
| Gaussian | $\sigma$ | 0.89 | 1.00 | 1.77 | 1.26 | 0.71 |
| CR–RC | RC | 1.85 | 1.18 | 1.85 | 1.85 | 1.00 |
| CR–RC$^6$ | RC | 0.40 | 1.03 | 4.37 | 1.35 | 0.38 |
| CR–RC$^6$ | $\tau_{\mathrm{peak}}$ | 2.38 | 1.03 | 0.73 | 1.35 | 2.26 |
| Semi-Gaussian 7 poles | $\sigma$ | 0.92 | 1.03 | 1.83 | 1.30 | 0.71 |
| Semi-Gaussian 7 poles | $\tau_{\mathrm{peak}}$ | 2.70 | 1.03 | 0.62 | 1.30 | 2.08 |
| Trapezoidal ($T_{\mathrm{f}} = 0.5 \times T_{\mathrm{r}}$) | $T_{\mathrm{r}}$ | 2.00 | 1.18 | 1.16 | 1.52 | 1.31 |
| Trapezoidal ($T_{\mathrm{f}} = T_{\mathrm{r}}$) | $T_{\mathrm{r}}$ | 2.00 | 1.38 | 1.67 | 1.83 | 1.09 |
| Trapezoidal ($T_{\mathrm{f}} = 2 \times T_{\mathrm{r}}$) | $T_{\mathrm{r}}$ | 2.00 | 1.64 | 2.67 | 2.31 | 0.87 |

Notes:
- For CR–RC$^6$ filter, $\tau_{\mathrm{peak}} = 6$ RC. For semi-Gaussian 7 poles filter, $\tau_{\mathrm{peak}} = 2.93\,\sigma$.
- For trapezoidal filter, $T_r$ is the duration of the leading and of the trailing part of the waveform, while $T_{\mathrm{f}}$ is the duration of the flat top.

pulse with a finite duration. Nevertheless, a practical realization of the triangular shaping is very difficult, because it requires delay lines in a time-invariant implementation. On the other hand, a time-variant implementation is usually affected by synchronization problems between the random occurrence of the detector pulse and the processing sequence of the filter. Moreover, the ideal triangular shape of the pulse (completely lacking of a flat top or, at least, of a top with a zero time-derivative) generates difficulties in the acquisition of its peak value. The output pulse of this filter, fed by a unitary amplitude step-like pulse, is an isosceles triangle whose peaking time is $\tau_p$ and whose base width is equal to $2\tau_p$.

*Semi-Gaussian Shaper CR–RC$^n$*

A semi-Gaussian shaper of the $n$th order is constituted by an approximated differentiator of time-constant $\tau_c$ followed by $n$ approximated integrators of the same time-constant $\tau_c$. An $n$th order filter presents $n+1$ real negative poles and a single zero at the origin of the complex plane. This kind of filter was extremely diffused, due to its simplicity of construction and use, but is nowadays generally substituted by the semi-Gaussian filter based on active filters characterized by complex poles (see later).

For illustration the output pulse of a sixth order semi-Gaussian shaper fed by a unitary amplitude step-like pulse is presented by the following formula

$$v_{uso} = \frac{1}{720}\left(\frac{t}{\tau_c}\right)^6 \exp\left(-\frac{t}{\tau_c}\right). \tag{4.38}$$

The peak value of its output signal is obtained at $\tau_p = 6\tau_c$.

*Semi-Gaussian Shaper with Complex Poles*

Nowadays most of the commercial shaping amplifiers are characterized by an overall transfer function which has a constellation of poles evenly spaced along a curve in the complex frequency plane. This solution allows the output pulse to reach a good approximation of the Gaussian shape. Moreover, this constellation of poles allows to reduce the problem of undershoots of the output pulse due to nonidealities of the components used. In Fig. 4.39, the constellation of the seven poles used in a semi-Gaussian shaper is shown.

In Fig. 4.40 the output pulse of such a seven-poles semi-Gaussian shaper is shown and compared with the output pulse of a CR–RC shaper which has the same amplitude and the same area.

The characteristic time-constant $\tau$ of the semi-Gaussian filter is usually defined as equal to the time constant $\tau_c = RC$ of such a CR–RC filter. It can be shown that

$$\tau_c = 0.9221\sigma, \tag{4.39}$$

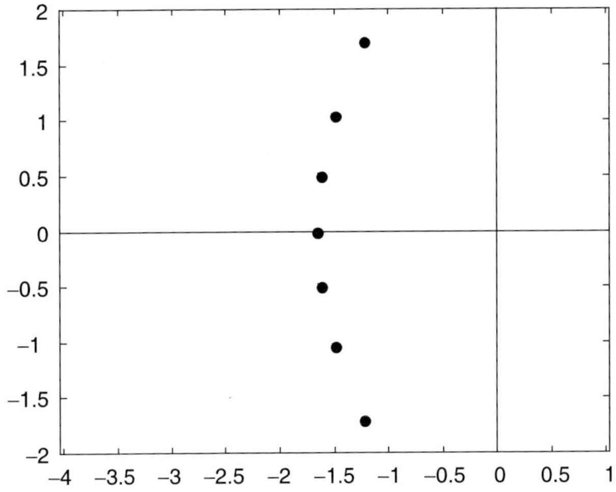


**Fig. 4.39.** Constellation of poles of the semi-Gaussian filter (the axes are normalized to the reciprocal of the $\sigma$ of the Gaussian which is approximated

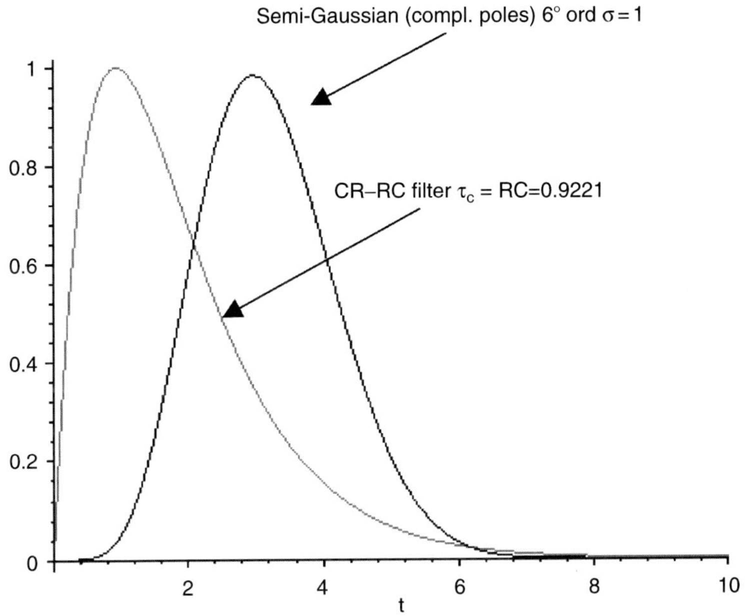


**Fig. 4.40.** Output pulse of a 7-poles semi-Gaussian filter having a constellation of complex poles (see Fig. 4.39) compared with the output pulse of a CR–RC filter with the same peak amplitude and the same area

while the peaking time $\tau_p$ is

$$\tau_p = 2.93\sigma. \tag{4.40}$$

*Trapezoidal Filter*

As for the case of the triangular filter, it can be implemented in a time-invariant approach with delay lines or it can be implemented in a time-variant approach. The flat top allows easy acquisition and digitalization of the output peak value. Recently, a VLSI implementation of a time-variant trapezoidal filter has been proposed which overcomes synchronization problems [62].

*Shape Factors*

In Table 4.2, the shape factors $A_1$, $A_2$, and $A_3$ of some filters relevant from the theoretical or practical point of view are given. In some case, for an easier comparison of the filters, the shape factors are given for different choices of the characteristic time $\tau_c$ of the filter. The factor $\sqrt{A_1 A_3}$ gives the ratio between the best ENC that can be obtained with the considered filter at its optimum shaping time and optimum ENC that can be obtained with the ideal cusp filter (only white noise sources are supposed to be present). The factor $\sqrt{A_1/A_3}$ gives the ratio between the optimum shaping time of the considered filter and the time constant of the ideal cusp filter.

### 4.2.8 Auxiliary Functions

Filtering amplifiers usually provide auxiliary functions that are very often of great importance for high resolution spectroscopy. We deal here with the pole-zero compensation, with the baseline restoration and with the pile-up rejection.

*Pole-zero Compensation*

The considered shape of the output pulses of the filtering amplifiers is obtained for an ideal step-like signal at their input. In practice, the signal at the output of the charge preamplifiers (which is fed to the input of the shaping amplifier) is usually affected by a long exponential decay

$$v_{pa}(t) = u(t)\exp(-t/\tau_{pa}), \tag{4.41}$$

where $u(t)$ is the unitary step function. Generally, this decay is due to the discharge of the feedback capacitor of the preamplifier through the resistor enclosed in the feedback network in order to avoid the saturation of the preamplifier. The time constant of this decay is usually much larger than the characteristic times of the shaping amplifiers (usually $\tau_{pa}$ exceeds 50 µs). The value of the discharge resistor is in fact kept as high as possible in order to reduce its thermal noise contribution.

A differentiator stage is always included in the shaping network of a filtering amplifier (at least if it is a constant-parameter analog filter) in order to avoid its saturation due to the pile-up of the long-lasting preamplifier pulses. Due to the differentiation, the exponentially decaying preamplifier pulse produces undershoots at the output of the shaping amplifier (see Fig. 4.41). If a subsequent pulse arrives during the period in which the undershoot is sensible (this period lasts a few $\tau_{pa}$) the measurement of its peak value is affected by an error due to the superposition of the tail of the previous pulse. Due to the random distribution in time of the detected pulses, this phenomenon worsens the resolution of the detection system, mainly in the case of high detection rate.

In order to overcome the problem, the differentiator stage is modified with respect to the simple CR configuration. A differentiator network is used which introduces a real negative zero and a real negative pole, called pole-zero compensator (P-Z). By means of an external control (usually a potentiometer) the operator can set the value of the time-constant of the zero to be equal to the decay time-constant of the preamplifier. In this way the pole of the preamplifier waveform is cancelled and the signal at the input of the filtering network can be considered ideally step-like. The pole introduced by the differentiator turns out to have necessarily a time-constant shorter than that of zero. It can be set to the value required by the pole configuration of the filter. For instance, this pole can be used as the real pole of the constellation of poles of the semi-Gaussian filter. Care must be spent in compensating exactly

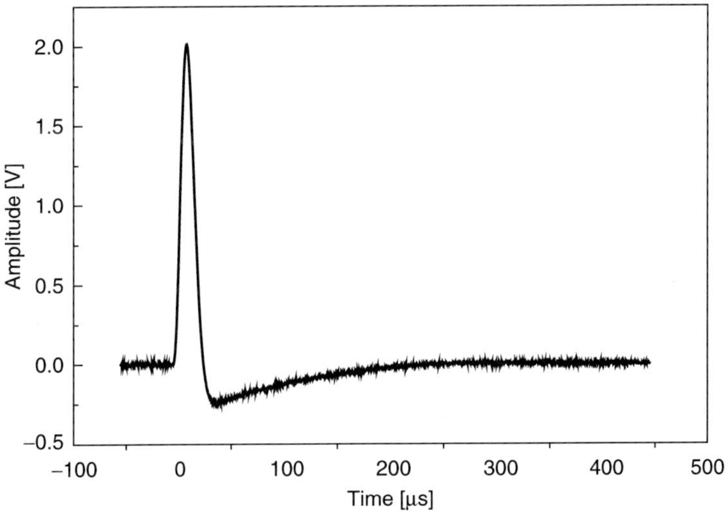


**Fig. 4.41.** Output pulse of a shaping amplifier affected by undershoot

the undershoot. Overcompensation causes a long tail of the same polarity of the main pulse, which worsens the resolution for the same reasons seen for the undershoot.

*Baseline Restoration*

Due to the AC coupling in the filtering network, the average value of the waveform at the output of the filtering amplifier must be ideally zero. Therefore, if the output pulses are unipolar, it must be assumed that the baseline is affected by a shift of the opposite polarity in order to assure a zero average value of the whole waveform. Because of the random occurrence of the signal pulses this shift has a variable amplitude. The measurement of the peak amplitude of the pulses is therefore affected by an error whose amplitude depends on the detection rate. Moreover, the baseline of the output waveform can also be affected by slow drifts due to low frequency noise in the final stages of the amplifier. All these phenomena worsen the peak position stability and the resolution of the acquired spectrum. In order to overcome these problems suitable time-variant differentiator circuits have been developed (called baseline restorers, BLR).

The general concept is that the baseline of the output waveform is differentiated with a relatively short time-constant, while the differentiation time-constant is switched to a very high value (ideally infinite) as soon as a signal pulse overcomes a suitably set threshold. The baseline is therefore kept to nearly zero at the output of the BLR, while the signal passes undisturbed. Figure 4.42 shows the basic principle of the BLR circuits. The drawback of this circuit is an increase of the output noise of the filtering amplifier, but in several experimental situations this disadvantage is compensated by a better stability of the peak positions.

*Pile-up Rejection*

The pile-up of the output pulses of the filtering amplifier is due to the finite width of the pulses and to their random distribution in time. This effect is particularly important at high count rates and it exists also with a perfect pole/zero compensation. Two coincident events produce a peak in the spectrum at an energy which is the sum of that of the two individual events (see Fig. 4.43 a and b). This ghost peak is not distinguishable from a possible peak at the same energy. Its intensity increases with the detection rate. Two events which are not strictly coincident can produce output pulses which partially superpose each other. In this case, the peak amplitude of each pulse is randomly increased by the presence of the leading or trailing edge of the other pulse. The shape of the spectrum is therefore distorted (see Fig. 4.43 c and d).

Techniques have been developed capable of rejecting the measurement of the pulse height if the distance between two contiguous pulses is less than a

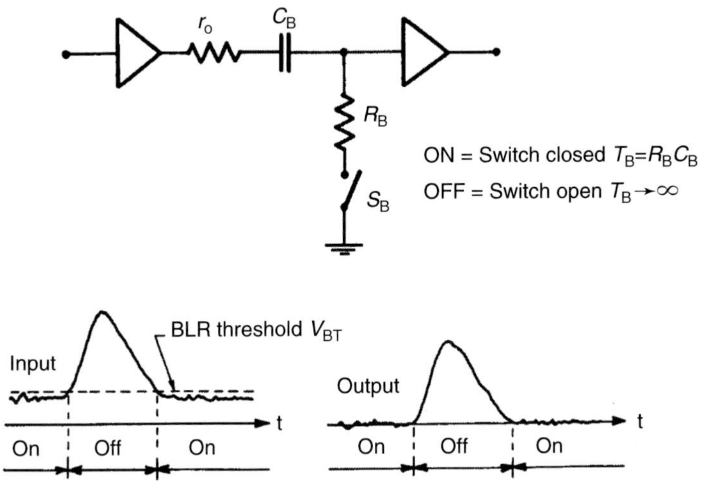


**Fig. 4.42.** The basic principle of the BLR circuits

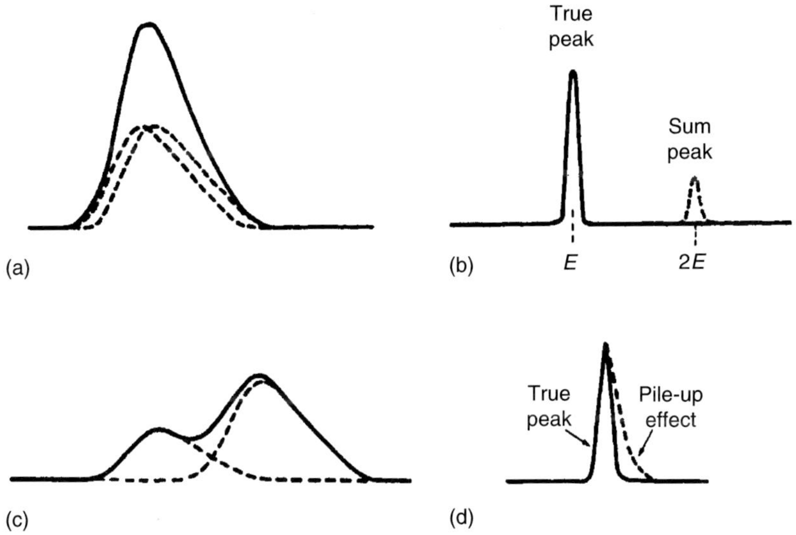


**Fig. 4.43.** Effects of the pile-up of pulses (a and c) on the recorded spectrum (b and d)

prefixed value. Pile-up rejection circuits are usually included in the filtering amplifier unit. A very fast auxiliary shaping of the preamplifier signal is usually adopted in order to detect the time of arrival of the pulses. Logic circuits control the time interval between successive pulses. Strictly coincident events cannot be rejected.

### 4.2.9 Appendix 1 – The Laplace Transform

The Laplace transform $F(s)$ of a signal $f(t)$ is

$$F(s) = \int_{-\infty}^{+\infty} f(t)\mathrm{e}^{-st}\,\mathrm{d}t \tag{4.42}$$

and the signal $f(t)$ can be expressed as

$$f(t) = \frac{1}{2\pi j} \int_{b-\mathrm{j}\infty}^{b+\mathrm{j}\infty} F(s)\mathrm{e}^{st}\mathrm{d}s. \tag{4.43}$$

The decomposition of the signals on the basis of the generalized sinusoids

$$\mathrm{e}^{st} = \mathrm{e}^{\alpha t} \cdot \mathrm{e}^{\mathrm{j}\omega t} \tag{4.44}$$

is of a wider applicability, in handling pulsed signals, than the more common decomposition on the basis of constant amplitude sinusoids $\mathrm{e}^{\mathrm{j}\omega t}$ that is adopted when using the Fourier transform.

The ratio between the Laplace transforms of the output and input signals of a linear network defines the transfer function $T(s)$ of the systems in the domain of the complex frequency $s$. For instance, if the input signal is a current and the output signal is a voltage

$$\frac{V_{\mathrm{out}}(s)}{I_{\mathrm{in}}(s)} = T(s). \tag{4.45}$$

If the structure of the linear network is known, the expression of the transfer function $T(s)$ can be easily determined by associating to each reactive elements its complex impedance (e.g. $1/sC$ for the capacitor and $sL$ for the inductor) and performing standard circuit analysis.

When analysing the noise, the transfer function of the network is expressed in the domain of the physical frequency $\omega$ (simply by substituting $\mathrm{j}\omega$ to the complex frequency $s$).

The ratio between noise power spectrum at the output of the network (supposed to be ideally noiseless) and the noise power spectrum at the input is equal to the squared modulus of the transfer function. For instance, if the input noise signal is a current and the output noise signal is a voltage

$$\frac{S_{v\,\mathrm{out}}(\omega)}{S_{i\,\mathrm{in}}(\omega)} = |T(\mathrm{j}\omega)|^2 . \tag{4.46}$$

If several noise sources are present at the input of the network, the total power spectrum at the output is the sum of the power spectra due to each noise source, if each one is supposed to be uncorrelated with the others.

### 4.2.10 Appendix 2 – Calculation of the ENC

In order to calculate the ENC we will represent the signal in the domain of the complex frequencies $s = \alpha + j\omega$ and we will therefore use the mathematical instrument of the Laplace transform.[b] The evaluation of the root-mean-square value of the noise will be performed in the real frequencies domain and we will therefore use the Fourier transform.

#### The Signal

The detector is modelled as a current source, in parallel to the detector capacitance $C_D$, which delivers an indefinitely short current pulse carrying a charge $Q$. This signal is mathematically represented by a delta-like pulse of area $Q$.

$$i_{\mathrm{D}}(t) = Q\delta(t) \quad I_{\mathrm{D}}(s) = Q. \tag{4.47}$$

The charge $Q$ is stored in the parallel of the two capacitances $C_{\mathrm{D}}$ and $C_{\mathrm{G}}$, representing, respectively, the detector capacitance and the gate to source capacitance of the input FET.

The step voltage $V_{\mathrm{G}}$ across the input terminals of the FET modulates its drain current

$$i_{\mathrm{DRAIN}}(t) = \frac{Q}{C_{\mathrm{D}} + C_{\mathrm{G}}} g_m u(t), \quad I_{\mathrm{DRAIN}}(s) = \frac{1}{s} \frac{Q}{C_{\mathrm{D}} + C_{\mathrm{G}}} g_m, \tag{4.48}$$

where $g_m$ is the FET transconductance and $u(t)$ is the unitary step function.

The drain current is fed to a linear amplifier (the pulse shaping amplifier) which provides an output pulse whose amplitude is proportional to the amplitude of the current step of the drain current of the FET and, therefore, to the charge $Q$ delivered by the detector.

Let $T(s)$ be the transfer function of the shaper that is the Laplace transform of the output signal of the shaper to a unitary area delta current pulse fed at its input. The output voltage pulse is:

$$v_{\mathrm{so}}(t) = \frac{Q}{C_{\mathrm{D}} + C_{\mathrm{G}}} g_m L^{-1}\left[\frac{1}{s} T(s)\right] \quad V_{\mathrm{so}}(s) = \frac{Q}{C_{\mathrm{D}} + C_{\mathrm{G}}} g_m \frac{1}{s} T(s), \tag{4.49}$$

where $L^{-1}$ represents the inverse Laplace transform operator and $L^{-1}[\frac{1}{s}T(s)]$ represents the output voltage of the pulse shaping amplifier when a unitary amplitude step-like current waveform is fed at its input. The amplitude of the maximum of the output pulse is assumed as the measurement of the charge delivered by the detector.

---

[b] The signals in the time domain are here represented by small letters (e.g. $i_{\mathrm{D}}(t)$ is the current waveform at the output of the detector), while the Laplace transforms of the signals are represented by capital letters (e.g. $I_{\mathrm{D}}(s)$ is the Laplace transform of the current waveform at the output of the detector).

*The Output Noise*

The output root mean square noise can be evaluated in the frequency domain. By taking into account the noise sources considered in Section "The Noise Sources" on pp. 283–239, the output noise power spectrum is

$$S_{\mathrm{no}}(\omega) = \left( a + \frac{2\pi a_{\mathrm{f}}}{|\omega|} + \frac{b}{\omega^2 \left(C_{\mathrm{D}} + C_{\mathrm{G}}\right)^2} \right) g_m^2 \left| T\left(\mathrm{j}\omega\right) \right|^2 \tag{4.50}$$

and the output root mean square noise is

$$\left\langle v_{\mathrm{no}}^2 \right\rangle = \frac{1}{2\pi} \int_{-\infty}^{+\infty} S_{\mathrm{no}}(\omega)\, \mathrm{d}\omega. \tag{4.51}$$

*The ENC*

The ENC is therefore

$$\mathrm{ENC}^2 = \frac{\left\langle v_{\mathrm{no}}^2 \right\rangle}{\mathrm{Max}^2 \left(v_{\mathrm{so}\ u\delta}\right)} = \frac{\frac{1}{2\pi} \int_{-\infty}^{+\infty} \left( a + \frac{2\pi a_{\mathrm{f}}}{|\omega|} + \frac{b}{\omega^2 (C_{\mathrm{D}} + C_{\mathrm{G}})^2} \right) g_m^2 \left| T(\mathrm{j}\omega) \right|^2 \mathrm{d}\omega}{\frac{g_m^2}{(C_{\mathrm{D}} + C_{\mathrm{G}})^2} \mathrm{Max}^2 \left( L^{-1} \left[ \frac{1}{s} T(s) \right] \right)}. \tag{4.52}$$

The ENC does not depend on the gain of the shaper (it affects both the signal and the noise in the same way). It is therefore convenient to assume, in the previous equation, a shaper gain which makes equal to one the term $\mathrm{Max}\left\{L^{-1}[T(s)/s]\right\}$. Moreover, the angular frequency $\omega$ can be conveniently normalised to a characteristic frequency $\omega_{\mathrm{c}} = 1/\tau$, where $\tau$ is a characteristic time which represents the width of the output pulse (for instance the peaking time, or the time width at half height, or a characteristic time constant of the filter). The characteristic time $\tau$ is also called "shaping time" of the filter. In this case the Fourier transform of the delta response of the shaper becomes a function of the a-dimensional variable $x = \omega\tau$. The $\mathrm{ENC}^2$ can be, in this way, expressed as

$$\mathrm{ENC}^2 = (C_{\mathrm{D}} + C_{\mathrm{G}})^2 a \frac{1}{\tau} \frac{1}{2\pi} \int_{-\infty}^{+\infty} |T(x)|^2 \, \mathrm{d}x + (C_{\mathrm{D}} + C_{\mathrm{G}})^2 2\pi a_{\mathrm{f}} \frac{1}{2\pi}$$
$$\times \int_{-\infty}^{+\infty} \frac{1}{|x|} |T(x)|^2 \, \mathrm{d}x + b\tau \frac{1}{2\pi} \int_{-\infty}^{+\infty} \frac{1}{x^2} |T(x)|^2 \, \mathrm{d}x. \tag{4.53}$$

The values of the three integrals appearing in this equation depend only on the "shape" of the output pulse of the filter and not on its "width". The "shape factors" $A_1$, $A_2$, and $A_3$ of the filter are defined as

$$\begin{aligned} A_1 &= \frac{1}{2\pi} \int_{-\infty}^{+\infty} |T(x)|^2 \,\mathrm{d}x \\ A_2 &= \frac{1}{2\pi} \int_{-\infty}^{+\infty} \frac{1}{|x|} |T(x)|^2 \,\mathrm{d}x\,. \\ A_3 &= \frac{1}{2\pi} \int_{-\infty}^{+\infty} \frac{1}{x^2} |T(x)|^2 \,\mathrm{d}x \end{aligned} \quad (4.54)$$

For a given filter and for a given width of its output pulse, the values of the three factors $A_1$, $A_2$, and $A_3$ depend on the particular choice of the characteristic time $\tau$ which has been chosen in order to normalize the angular frequency $\omega$. Clearly, the value of the ENC does not depend on the particular choice of the characteristic time $\tau$ adopted in order to define the pulse width. If $A_1(\tau')$, $A_2(\tau')$, and $A_3(\tau')$ are the values of the shape factors related to a given choice of the characteristic time $\tau'$, it turns out that the values $A_1(\tau)$, $A_2(\tau'')$ and $A_3(\tau'')$ related to a different choice $\tau''$, with $\tau'' = k\tau'$, are given by

$$\begin{aligned} A_1(\tau'') &= kA_1(\tau\prime) \\ A_2(\tau'') &= A_2(\tau\prime) \quad . \\ A_3(\tau'') &= \frac{1}{k}A_3(\tau\prime) \end{aligned} \quad (4.55)$$

The $\mathrm{ENC}^2$ can be finally written as

$$\mathrm{ENC}^2 = (C_\mathrm{D} + C_\mathrm{G})^2 a \frac{1}{\tau} A_1 + (C_\mathrm{D} + C_\mathrm{G})^2 2\pi a_\mathrm{f} A_2 + b\tau A_3. \quad (4.56)$$

This is the fundamental equation of the detector signals processing theory. In this equation, the first term represents the contribution to the $\mathrm{ENC}^2$ of the white series noise, the second one that of the $1/f$ series noise and the third that of the parallel noise. It is worth noting that the contribution of the series noise is inversely proportional to the shaping time $\tau$, while the contribution of the parallel noise is directly proportional to $\tau$. The contribution of the $1/f$ noise is independent of $\tau$.

### 4.2.11 Appendix 3 – Digital Pulse Processing

It goes beyond the purview of Section 4.2 to deal with the digital filtering of the detector signals. Nevertheless, the growing diffusion of digital pulse processing (DPP) in X-ray spectroscopy requires at least a few remarks on this technique. For a comprehensive understanding of this subject the reader should first address one of the textbooks on digital signal processing (DSP) to learn the basics of this discipline [63], DPP being one of the many applications of DSP (DSP applies for instance to speech and music filtering, radar pulses analysis, image processing).

The basic concepts of analog pulse processing (APP) introduced in Section 4.2 (impulse response, optimum shaping, ENC, etc.) fully apply also to DPP. What is different is the starting point. While in APP the processed signal is a continuous waveform, in DPP the processed signal is a sequence of samples of the waveform of interest. The value of each sample is (ideally) identical to the value of the waveform at the corresponding sampling time. The time interval between samples is usually (but not necessarily) constant. The sequence of samples allows an unambiguous reconstruction of the waveform if the conditions set by the "sampling theorem" (also called Nyquist or Shannon theorem) are satisfied, that is, if the sampling frequency is at least two times higher than the highest frequency component contained in the spectrum of the analog waveform. For this reason the waveform to be sampled is "pre-filtered" in order to cut-off its frequency components higher than one half of the foreseen sampling frequency (this filter is also called "antialiasing filter"). The sampled values are then digitized by a suitable analog to digital converter (ADC) and the sequence of digital data is fed to the DSP. For simplicity we do not consider in these introductory notes the quantization noise introduced by the ADC (we can, at least in principle, suppose that this added noise is negligible with respect to the other noise components present in the input waveform if the number of quantization bits of the chosen ADC is sufficiently large).

In a typical APP chain for X-ray spectroscopy *only the peak value* of the pulse at the output of the shaping amplifier (SHA), whose amplitude is proportional to the charge delivered by the ionizing event, is sampled and digitized. The obtained digital value is the measure of the charge delivered by the ionizing event. In a typical DPP chain for X-ray spectroscopy *the full waveform* at the output of the anti-aliasing filter is sampled at a frequency satisfying the Nyquist theorem; each sample is digitized and the sequence of digital values is sent to the DSP unit. The DSP unit provides an output pulse whose peak value is proportional to the charge delivered by the ionizing event. Note that the output of the DSP unit is a sequence of digital values that can be converted (or not, according to the user needs) in an analog pulse by means of a digital to analog converter (DAC). In the case of APP the filtering action is performed analogically by the SHA, while in the case of DPP the filtering is performed digitally by the DSP unit.

In APP as well as in DPP the noise properties of the detection system depend on the shape and on the characteristic time of the *weighting function* of the filter (the SHA in the former case and the DSP unit in the latter one). The weighting function of a filter is its impulse response reversed in time. As is well known, the output waveform $x_o(t)$ of an analog filter is given by the convolution of the input waveform $x_i(t)$ with the impulse response $h(t)$ of the filter

$$x_{\mathrm{o}}(t) = x_{\mathrm{i}}(t)h(t) = \int_{-\infty}^{t} x_i(\tau)h(t-\tau)\mathrm{d}\tau .$$

In the case of a digital filter, the input and the output signals are sequences of digitized data $x_i(i)$ and $x_o(i)$. The impulse response of the DSP unit is a (usually finite) sequence of $M$ values $h(0),\ldots,h(M-1)$. The output signal is given by the convolution of the input sequence of data with the sequence describing the filter response

$$x_{\mathrm{o}}(i) = x_{\mathrm{i}}(i)h(i) = \sum_{j=0}^{M-1} x_i(j)h(i-j)$$

that is, each sample of the output signal is obtained by weighting the samples of the input signal and by adding them together. If the input signal is a pulse of finite duration and the impulse response of the filter is given by a finite sequence of values, then the output signal has a finite duration. By a proper design of the sequence of values describing the impulse response of the DSP unit, performed by taking into account the shape of the waveform at the output of the analog prefilter, it is possible to synthesize any weighting function for the whole processing chain [64]. The ENC achievable is the same as that achievable with a fully analog processing chain with the same overall weighting function. The great advantage of DPP is that "any" weighting function can be easily synthesized, in particular the weighting function of finite duration and with flat-tops (e.g. trapezoidal w.f.) with advantages in the reduction of pile-up and ballistic deficit effects, mainly at high detection rates.

The described DSP filters are time-domain filters of the "convolution type". A sometime more efficient way to synthesize the same weighting function, based on a different algorithm, is the one adopted by the "recursion type" filters. These filters use also previously calculated values of the output to calculate each sample of the output signal. A comparison of these procedures can be found in textbooks of DSP. We do not consider here DSP filters working in the "frequency domain", based on Fourier analysis of the sampled values.

A different approach has led to the "optimum" digital processing of detector signals. This method has been developed [65] for signals known a priori in shape (the signals at the output of the prefilter), their amplitude and time of occurrence being unknown and to be determined. In this case, because the shape of the signal is known, the waveform can be to some extent undersampled with respect to the Nyqist criterion without losing the unambiguous reconstruction capability. The processing is based on the maximum-likelihood method. The method states that the best estimate of the amplitude $A$ and of the time of occurrence $T$ of a signal in presence of noise is the one which maximizes the probability density of the deviations between the set of the experimental samples and the shape of the signal, known as a function of the estimates $A$ and $T$. This probability is a functional of the values $A$ and $T$ that can be maximized once the auto-correlation function of the noise at the output of the prefilter is known.

A general method for synthesizing optimum digital filters with arbitrary constraints in shape, time and frequency and with arbitrary noise and interferences has also been developed [66]. The optimum weighting function is determined on the basis of the direct measurement of the noise and of any interference possibly affecting the signal (for instance power supply ripples or digital interferences). The determination of the optimum weighting function of the digital filter is quite fast. Therefore, it is possible to track online slowly varying noises and disturbances and automatically adapt the weighting function of the digital filter.

## 4.3 High Resolution Imaging X-Ray CCD Spectrometers

*L. Strüder, N. Meidinger and R. Hartmann*

### 4.3.1 Introduction

X-ray astronomy has been pushing for several years the instrumentation for broadband imaging nondispersive X-ray spectrometers. Since the launch of the European XMM-Newton satellite in December 1999, reliably operating X-ray CCDs have been delivering extraordinary images, recorded in a single photon counting mode, imaged through the largest X-ray telescope ever built. Related applications in other fields of basic and applied science will equally be mentioned: X-ray microscopy, quantum optics, X-ray plasma diagnostics and others. State of the art X-ray pn-CCD detectors with energy, position and time resolution at high quantum efficiency from the near infrared up to 20 keV are described in detail. They have been developed for astrophysics experiments in space, for material analysis, and for experiments at synchrotron radiation facilities. The functional principles of the silicon devices are derived from basic solid state device physics. The spatial resolution, the spectroscopic performance of the systems, the long-term stability, and the limitations of the detectors are described in detail. The various fields of application show the unique utility of silicon radiation detectors.

Imaging of photons is best known in the visible domain, ranging from a wavelength of 3,500 Å up to 6,000 Å. Optics and detectors are equally well developed for those applications. But all these imaging systems do not count the incoming photons individually to measure their position, energy, and arrival time. The photon information is either integrated in the grains of a photographic film that is developed chemically afterwards or the photons are collected in individual *picture cells* (pixels) and sequentially read out after a given time. The photonic or electronic content of each grain or pixel is then "counted" to measure the intensity of the incident photon flux. Traditionally, the energy of the photons is determined by an arrangement of various filters, transparent only for a narrow, well-defined bandwidth of the incoming photons. In this sense, the image is a static, integrated reconstruction of a local photon intensity distribution.

Single "optical" photons cannot be counted up to now in a practical manner, i.e., with large arrays. The energy of the photons is too small to detect them individually with non-cryogenic detectors: it is a fraction of an electron volt in the near infrared and up to 4 eV for the violet part of the visible spectrum.[1] In gas detectors, more than 20 eV are needed for the ionization of a detector gas atom, and room temperature silicon detectors need around 1 eV for the generation of an electron–hole pair in the optical range and 3.7 eV on average for ionizing particles with sufficiently large energy. For a proper electronic extraction of the very weak signal of one optical photon, readout electronics should operate below $0.1e^-$ equivalent noise charge (ENC). This is by far not reached today in the state-of-the-art silicon sensor systems. From approximately 11,000 Å to 3,000 Å only one electron–hole pair per photon is generated due to the ionization process and its statistics in silicon. In this sense, direct spectroscopic information in the optical region is physically not available from silicon detectors.

The X-ray imaging detector systems that are described below record simultaneously the energy, position, and arrival time of each individual X-ray photon without using selective absorbers. The physical reasons for being able to make truly energy-dispersive X-ray detectors are the low average electron–hole pair creation energy of about 3.7 eV in silicon at room temperature and the very thin radiation entrance windows of only a few tens of partially insensitive atomic layers of silicon and native $SiO_2$, which can be penetrated by (even soft) X-rays. For a good quantum efficiency at higher X-ray energies, only the depleted thickness of silicon (charge collection depth) is of relevance. At 500 μm sensitive detector thickness, e.g., a fraction of 25% of 25 keV X-rays is converted in electron–hole pairs and can be collected and detected. For two-dimensional silicon detectors with high position and energy resolution, the fabrication by a planar process – comparable to the fabrication in state-of-the-art microelectronics – is obligatory. Depletion thicknesses of 1,000 μm are a practical limit for detector fabrication. State-of-the-art imaging silicon detector systems are an ideal instrument for direct detection in the energy band between 0.1 keV and 30 keV with high quantum efficiency, position and energy resolution.

---

[1] Cryogenic detectors are able to perform single photon counting in the near infrared, visible and soft X-ray domain [67]. The band gap for this kind of detectors is in the millielectron volt range as compared to 1.1 eV for Si. But cooling down to the order of 100 mK is required.

The astrophysical requirements have driven the development of the high resolution X-ray detectors from 0.1 keV to 10 keV in the last 10 years. The X-ray multi mirror (XMM) mission of the European Space Agency (ESA) was successfully launched in a highly eccentric orbit with three large X-ray telescopes and reflecting grating spectrometers all having specially designed X-ray charge coupled devices (CCDs) in their focal planes [68, 69]. The energy dispersive gratings are read out with the more conventional backside illuminated 30 μm deep depleted MOS CCDs for energies up to 4 keV [70].

The pn-CCD detectors, formed only by rectifying pn-junctions, deliver excellent position (in the order of tens of microns), energy (about 140 eV FWHM at 6 keV), and time resolution (below 100 μs in dedicated operating modes) with high quantum efficiency (above 90%) at soft and medium energy X-rays from 0.5 keV to 10 keV.

All experimental results shown here are from devices which have been designed, fabricated and tested at the MPI Halbleiterlabor.

### 4.3.2 Fully Depleted Backside Illuminated pn-CCDs

Conceptually, the pn-CCD is a derivative of the silicon drift detector [30]. The development of the pn-CCDs started in 1985. In the following years, the basic concept was simulated, modified, and designed in detail [71]. N-channel JFET electronics was integrated in 1992 [40, 72] and the first reasonably fine working devices were produced in 1993. Up to then, all presented devices were "small" devices, i.e., 3 $cm^2$ in sensitive area [68].

The flight-type large area detectors were produced from 1995 to 1997, with a sufficiently high yield to equip the X-ray satellite missions ABRIXAS and XMM [73–75] with defect free focal plane pn-CCDs.

XMM was launched from Kourou in French-Guiana on December 10, 1999. Commissioning of the scientific payload was completed in the middle of March 2000. In this overview, the basic instrument features as measured on ground and in orbit will be shown.

#### The Concept of Fully Depleted, Back Illuminated, Radiation Hard pn-CCDs

For ESA's XMM mission, we have developed a $6{\times}6\,cm^2$ large monolithic X-ray CCD [76] with high detection efficiency up to 15 keV, low noise level (ENC $\approx 5e^-$(rms) at an operating temperature of $-90°C$) and an ultrafast readout time of 4.6 ms per $3{\times}1\,cm^2$ large subunit (see Figs. 4.45 and 4.50). A schematic cross-section, already showing some of the advantages of the concept, is displayed in Fig. 4.44.

The pn-CCD concept and the fabrication technology allow for an optimum adaption of the pixel size to the X-ray optics, varying from 30 μm up to 300 μm pixel size. Up to now systems with 50–200 μm have been produced. The XMM telescope performance of 13 arcsec half energy width (HEW) translates

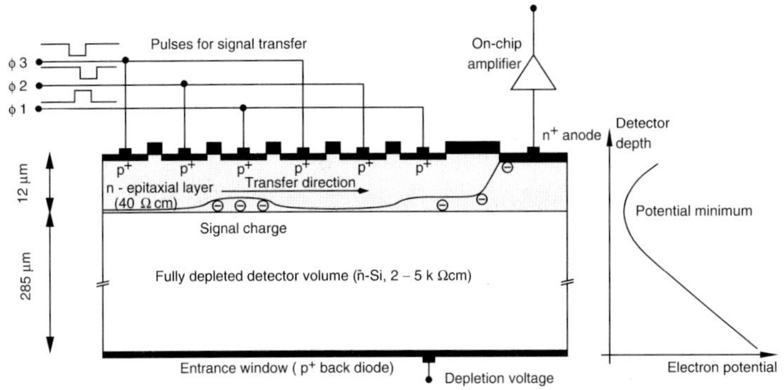


**Fig. 4.44.** A schematic cross-section through the pn-CCD along a transfer channel. The device is back illuminated and fully depleted over 300µm thickness. The electron potential perpendicular to the wafer surface is shown on the right-hand side

to 470 µm position resolution in the focal plane. The FWHM of the point spread function (PSF) is about 7 arcsec. A pixel size of 150 µm × 150 µm was chosen, giving a position resolution of $\leq$120 µm, resulting in an equivalent spatial resolving capability of $\leq$3.3 arcsec. This is sufficient to fully conserve the positional information of the X-rays from the mirrors. The energy response is higher than 90% at 10 keV because of the sensitive thickness of 300 µm according to the wafer thickness. The low energy response is given by the very shallow implant of the $p^+$back contact; the effective "dead" layer is smaller than 200 Å [77]. The excellent time resolution is achieved by the parallel readout of 64 channels per subunit, 768 channels for the entire camera. A high radiation hardness is built in by avoiding active MOS structures and by the fast transfer of the charge in a depth of more than 10 µm. The spatially uniform detector quality over the entire field of view is realized by the monolithic fabrication of the pn-CCD on a single wafer. For redundancy reasons, 12 individually operated $3 \times 1\,\text{cm}^2$ large pn-CCDs subunits were defined. Inhomogeneities were not observed over the entire sensitive area in the calibration energy band from 0.5 keV up to 8 keV, within the precision of the measurements limited by Poisson statistics. The insensitive gap in the vertical separation of the pn-CCDs is about 40 µm, neighboring CCDs in horizontal direction have insensitive border regions of 190 µm (see Fig. 4.50).

The basic concept of the pn-CCD is shown in Fig. 4.44 and is closely related to the functional principle of silicon drift detectors (SDDs). A double-sided polished high resistivity n-type silicon wafer has both surfaces covered with a rectifying $p^+$-boron implant. On the edge of the schematic device structure (see Fig. 4.44) an $n^+$-phosphorus implant (readout anode) still keeps an ohmic connection to the nondepleted bulk of the silicon. A reverse bias is now applied to both $p^+$ junctions, i.e., a negative voltage is applied with respect to the $n^+$ anode. For simplicity, let us assume that the silicon bulk is homogeneously doped with phosphorus with a concentration of $1 \times 10^{12}$ per cm$^3$.

Depletion zones in the high ohmic substrate, with a resistivity of about 4 k$\Omega$ cm, develop from both surfaces, until they touch in the middle of the wafer. The potential minimum for electrons is now located in the middle of the wafer. An additional negative voltage on the $p^+$ back diode shifts the potential minimum for electrons out from the center toward the surface containing the pixel structure. Typical depletion voltages on the backside are −150 V. To make a CCD-type detector, the upper $p^+$ implant must be divided in $p^+$ strips as shown in Figs. 4.44 and 4.49. Adequate voltages should now be applied to the three shift registers such that they form local potential minima for $e^-$ in a distance of approximately 10 µm from the surface. Three $p^+$ strips (shift registers) with the potentials $\phi_1$, $\phi_2$, and $\phi_3$ comprise one pixel. Charges are collected under $\phi_3$, the potential minimum for electrons. A reasonable change with time of the applied voltages transfers the charges to the new local $e^-$ potential minimum in a discrete way toward the $n^+$ readout node. In reality the side having the $p^+$ shift registers has an additional phosphorus-doped epitaxial layer, 12 µm thick, with a concentration of approximately $10^{14}$ donors per $cm^3$. The interface of the epi-layer and the high resistivity bulk silicon fixes the electron potential minimum to a distance of about 10 µm below the surface. As can be seen in Fig. 4.45, one pn-CCD subunit consists of 64 individual transfer channels each terminated by an on-chip JFET amplifier. Figures 4.46, 4.47 and 4.48 show the charge transfer mechanism in a depth

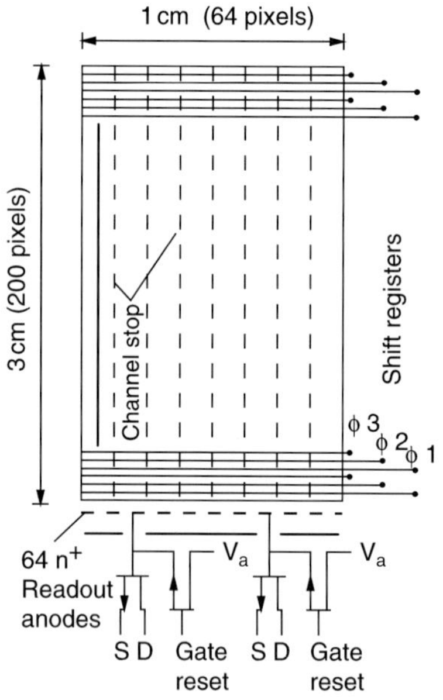


**Fig. 4.45.** One pn-CCD subunit with 64 on-chip amplifiers and a size of $3\times1\,cm^2$. Each of the 64 columns is terminated by an on-chip JFET amplifier

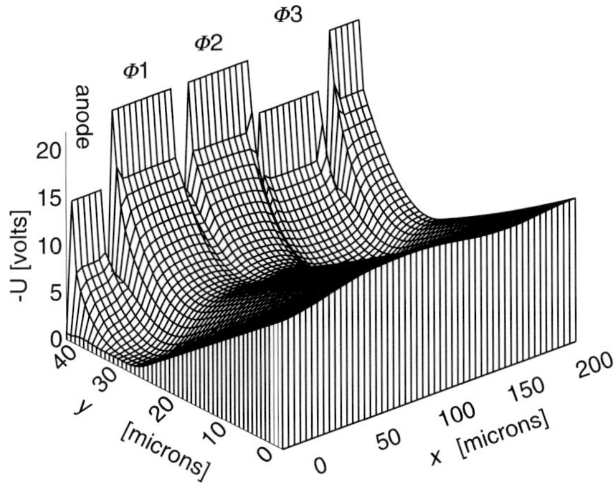


**Fig. 4.46.** Negative potential distribution inside the pn-CCD pixel. In this operating condition the signal charges are stored under the register $\phi_3$ only. The $p^+$backside potential is only shown up to the depth of 40 μm for clarity. It expands to –150 V at 300 μm distance from the pixel surface

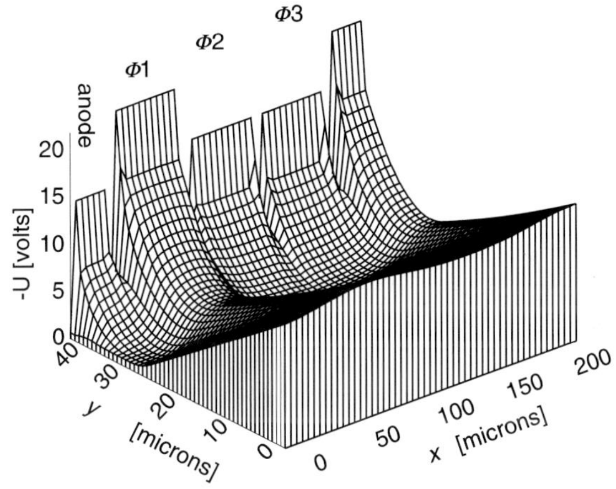


**Fig. 4.47.** Negative potential distribution inside the pn-CCD pixel according to charge transfer sequence. In this operating condition the signal charges are stored under the registers $\phi_2$ and $\phi_3$. The electrons now share a larger volume for a short time. Note that the electrons are still nicely confined in the potential well

of approximately 10 μm below the shift registers. The $p^+$ back contact is not shown: it expands quite uniformly an additional 260 μm toward a negative potential of −150 V. The sequence of changing potentials shows nicely the controlled transfer from register $\Phi_3$ to register $\Phi_2$, one-third of a pixel.

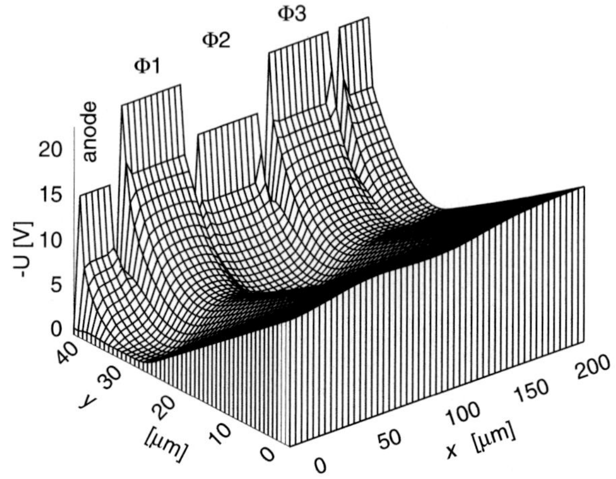


**Fig. 4.48.** Negative potential distribution inside the pn-CCD pixel at next transfer step. In this operating condition the signal charges are stored under the register $\phi_2$ only. The charge was transferred by one-third of the pixel length in approximately 150 ns

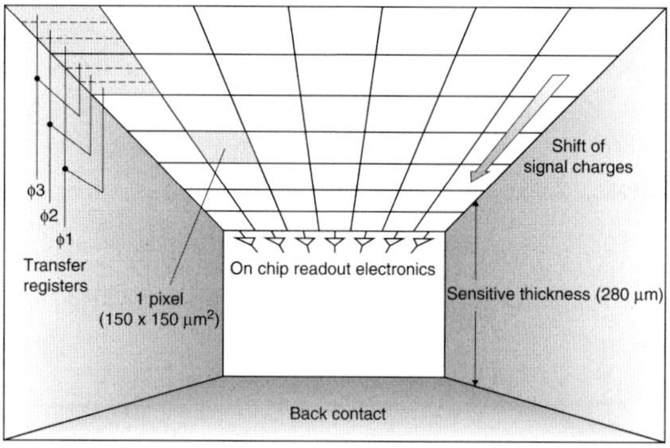


**Fig. 4.49.** Inside the pn-CCD. The X-rays hit the device from the back (bottom). The charges are collected in the pixel well close to the surface having the pixel structure. After integration, they are transferred to the on-chip amplifier

This concept is seen from a different point of view in Fig. 4.49, seen from the inside of a pn-CCD: X-rays hit the detector from the back (indicated as back contact in Fig. 4.49). The positively charged holes move to the negatively biased back contact, electrons to their local potential minimum in the transfer channel, located about 10 μm below the surface having the pixel structure. The electrons are fully collected in the pixels after 5 ns at most, the collection of holes is completed in no more than 15 ns because of their reduced mobility. As can be seen in Fig. 4.49, each CCD column is terminated by a readout

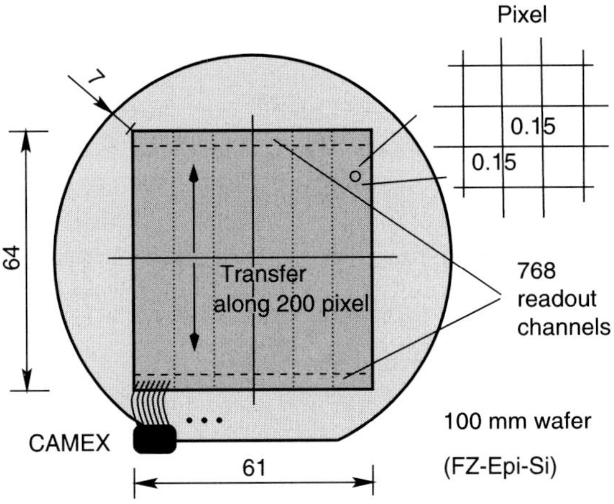


**Fig. 4.50.** The focal plane of the pn-CCD camera on XMM and ABRIXAS consists of 12 independent, monolithically integrated pn-CCDs with a total image area of approximately $60 \times 60\,\mathrm{mm}^2$. In total 768 on-chip amplifiers process the signals and transfer them to a VLSI JFET-CMOS amplifier array. The 12 output nodes of the CAMEX arrays are fed into four ADCs, i.e., one ADC is dedicated to a quadrant

amplifier. The on-chip single-sided JFET has already been described in the chapter written by Longoni and Fiorini in this book as the first amplifying element in the SDD.

The focal plane layout of XMM is depicted in Fig. 4.50. Four individual quadrants each having three pn-CCD subunits are operated in parallel. The camera housing and its mechanical, thermal and electrical properties are described in reference [78].

## Limitations of the CCD Performance

The performance of the CCDs is subject to several limitations: physical and technical. We will treat some of the limitations.

*Quantum efficiency.* The quantum efficiency at the lowest energies from several tens of electron volts to slightly above the Si K edge at 2 keV is determined by the transmission of insensitive or partially insensitive layers. This directly leads to incomplete charge collection resulting in an asymmetric signal peak with a "shoulder" on the low energetic side of the spectrum and a shift of the peak position. The remedy for this effect is to avoid lattice perturbations which could lead to charge trapping and recombination. Additionally, electrons from the ionization process in the insensitive layers may reach the partially or totally sensitive regions and leave their remaining energy there. Both effects reduce the proper detection of the X-ray energy and worsen the peak-to-valley ratio.

The high energy response is simply given by the depleted and thus sensitive thickness $d$.

$$d = \sqrt{\frac{2\epsilon_o\epsilon_r(N_\mathrm{A}+N_\mathrm{D})}{qN_\mathrm{D}N_\mathrm{A}}V_\mathrm{bias}} \approx \sqrt{\frac{2\epsilon_o\epsilon_r}{q}\frac{V_\mathrm{bias}}{N_\mathrm{D}}} \quad \text{for} \quad N_\mathrm{A} \gg N_\mathrm{D}\,. \tag{4.57}$$

The depleted device thickness $d$ can be increased by using silicon with higher resistivity, i.e., with small donor concentration $N_\mathrm{D}$ in the case of n-type material. Or the reverse bias $V_\mathrm{bias}$ must be increased. But for long-term stable operation $V_\mathrm{bias}$ should be kept below, e.g., 500 V with given resistivity that limits the achievable sensitive thickness. With 4.5 k$\Omega$cm n-type silicon ($N_\mathrm{D} = 10^{12}$ donors per cm$^3$), a depletion width of 800 μm can be achieved at a reverse bias of 500 V. In the above equation, $\epsilon_o\epsilon_r$ are the dielectric constants of silicon, $q$ the charge of one electron and $N_\mathrm{A}$ the high acceptor concentration from the boron implant.

*Ionization statistics.* Once above 50 eV of photon energy, UV and X-rays need approximately $w = 3.7$ eV (average) of energy to generate an electron–hole pair, despite the fact that the band gap of silicon is only 1.1 eV. Most of the incident energy is converted into phonons, only about 30% goes into the creation of electron–hole pairs. In addition, the ionization cascade is not an uncorrelated process, therefore straightforward Poisson statistics does not apply. The energy response of incident monochromatic X-rays is broadened by the competing relaxation processes in silicon. The resultant ENC contribution $\mathrm{ENC}_\mathrm{fano}$ is given by

$$\mathrm{ENC}^2_\mathrm{fano} = \frac{FE}{w}, \tag{4.58}$$

where $E$ is the photon energy, $F$ the Fano factor and $w$ the electron–hole pair creation energy. Assuming a material dependent Fano factor $F$ of 0.12 for silicon and a pair creation energy of 3.7 eV, the intrinsic line width of a photon of 5.9 keV cannot be better than 121 eV (FWHM). According to Fig. 4.51, the Fano noise is dominant for energies above 1 keV if the electronic ENC contribution amounts to five electrons. For 1 electron noise ($\mathrm{ENC}_\mathrm{el}$) this threshold is lowered to 50 eV only.

*Charge transfer noise.* In chargeeff coupled devices, where the signal electrons are transferred over many pixels, the charge shifting mechanism from pixel to pixel must be excellent. Leaving charges behind during the transfer means a reduction in signal amplitude. This loss can be corrected, but adds noise to the signal amplitude measurement.

$$\mathrm{ENC}^2_\mathrm{trans} \approx \frac{E}{w}(1-\mathrm{CTE})N_\mathrm{trans}. \tag{4.59}$$

In a simple model, the lost charges can be parametrized according to (4.59), where the left behinds can be considered as a backward flow loss of electrons.

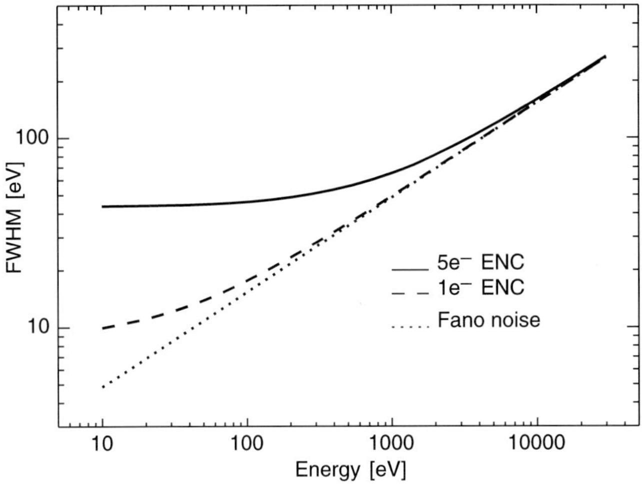


**Fig. 4.51.** Energy resolution as a function of the photon energy. The Fano noise is taken into account as well as a $5e^-$ and $1e^-$ electronic equivalent noise charge ($ENC_{el}$), respectively

As the loss process is of statistical nature, it is treated similar to a signal leakage current. $N_{trans}$ denotes the number of pixel transfers in the CCD and CTE is the charge transfer efficiency – a number close to one, so that (1−CTE) is in the order of $10^{-5}$, depending on CCD type, radiation damage, temperature, etc.

*Electronic noise.* The origin of electronic noise and the processing of signals have already been treated in detail in the Sect. 4.2 "X-ray Detectors and Signal Processing" of Longoni and Fiorini. As the on-chip JFET is almost identical to the one described in a previous chapter, we simply summarize the result:

$$\begin{aligned} \mathrm{ENC}_{\mathrm{el}}^2 &= (\alpha \frac{2kT}{g_m} C_{\mathrm{tot}}^2 A_1)\frac{1}{\tau} + && \text{series noise} \\ &+ [(2\pi a_{\mathrm{f}} C_{\mathrm{tot}}^2 + \frac{b_f}{2\pi})A_2] + && \text{low frequency noise} \\ &+ A_3(qI_{\mathrm{l}} + \frac{2kT}{R_{\mathrm{f}}})\tau && \text{parallel noise.} \end{aligned} \qquad (4.60)$$

$C_{tot}$ is the total input capacitance, $kT$ Boltzmann's constant and absolute temperature, $\tau$ is the shaping time constant, $I_l$ the leakage current and $R_f$ the feedback resistor of the charge-sensitive amplifier, $a_f$ and $b_f$ parametrize the low frequency $1/f$ noise. $A_1$, $A_2$ and $A_3$ are constants depending on the shaper's filter function.

For the pn-CCD (as any other silicon drift detector type system) the electronic noise can be reduced by reducing the readnode capacitance, by lowering the leakage current and the $1/f$ noise constants and by optimizing the shaping time constant $\tau$. The total read noise, if not correlated, can be added quadratically and delivers the total ENC:

$$\mathrm{ENC}^2_{\mathrm{tot}} = \mathrm{ENC}^2_{\mathrm{el}} + \mathrm{ENC}^2_{\mathrm{fano}} + \mathrm{ENC}^2_{\mathrm{trans}} + \cdots . \tag{4.61}$$

State-of-the-art systems of today exhibit electronic noise figures around three electrons and a readout speed of more than one pixel per µs and operating temperatures higher than $-90°$C .

In pixellated detectors, the signal charges of one single X-ray photon will not always be collected in a single pixel, so the electronic content of several pixels must be added. This increases for the so-called split events, the electronic noise floor by $\sqrt{N}$, with $N$ as the number of pixels involved.

### Detector Performance (on Ground and in Orbit)

The best value for the readout noise of the on-chip electronics is $2e^-$ rms at $-90°$C for the most recent devices, whereas typical values scatter around $5e^-$ rms, e.g., for the XMM system. This includes all noise contributions described in (4.60). The charge transfer properties of the pn-CCDs on XMM are reasonably good, in the order of a several % signal loss from the last to the first pixel over a distance of 3cm charge transfer. As the charge transfer losses describe the position dependent energy resolution, it is one of the key parameters for the spectroscopic performance, especially after radiation damage may have occurred. Figure 4.55 shows a $^{55}$Fe spectrum of a pn-CCD in a flat field measurement resulting in a typical energy resolution of 130 eV at an operating temperature of $-120°$C [68]. The XMM flight camera was operated at $-90°$C during calibration on ground with a resolution of about 145 eV (FWHM) over the entire area of 36 cm$^2$. The main effect on the degradation of energy resolution was the reduction of the charge transfer efficiency at warmer temperatures. Leakage currents and on-chip JFET properties only played a minor role. The impact of the material properties of silicon and related impurities and their consequences for the operation of scientific grade X-ray pn-CCDs including the effects of radiation damage, are treated in detail in the references [79, 80]. The radiation damage accumulated over the expected life time of XMM is equivalent to a 10 MeV proton fluence of $4 - 5 \times 10^8$ p/cm$^2$. Figures 4.52 and 4.53 show the results of the irradiation tests with 10 MeV protons: the expected decrease of energy resolution over the 10-year dose is from 146 eV to 164 eV at an operating temperature of $-100°$C. At the actual operating temperature of $-90°$C , the expected effect of trapping and detrapping at A-centers, generated by the radiation, is even more reduced.

In a single photon counting mode the quantum efficiency was measured with respect to a calibrated solid state detector. Figure 4.54 shows measurements from the synchrotron radiation facilities in Berlin and Orsay. At 525 eV, a 5% dip can be seen due to the absorption at the oxygen edge in the $SiO_2$

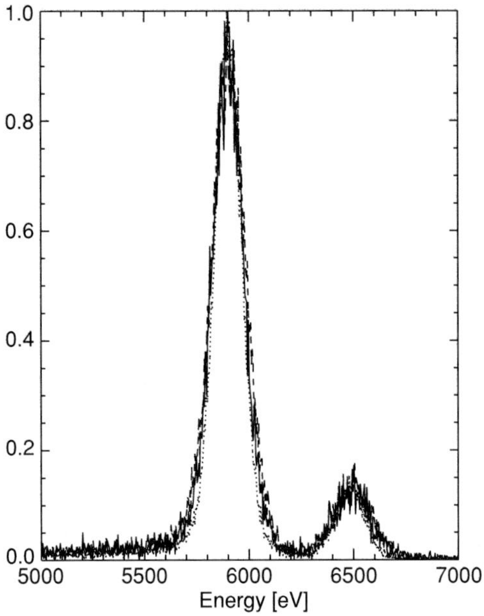


**Fig. 4.52.** Energy spectrum from an $Fe^{55}$ source after different 10 MeV proton fluences of 0 p/cm$^2$ (dotted line), $4.1 \times 10^8$ p/cm$^2$ (solid line), $6.1 \times 10^8$ p/cm$^2$ (dashed line), measured at the low (and after irradiation unfavorable) temperature of 142 K. The expected dose for the pn-CCD aboard XMM over a life time of 10 years is $5 \times 10^8$ 10 MeV p/cm$^2$

layers. The same happens at the Si K edge at 1840 eV showing the fine structure of a typical XAFS spectrum (see insert of Fig. 4.54). For all energies the quantum efficiency is nicely represented by a model using the photo absorption coefficients from the atomic data tables. The quantum efficiency on the low energy side can be further improved with respect to the measurements shown in Fig. 4.54, by increasing the drift field at the $p^+$- junction entrance window [77] and by using < 100 > silicon instead of < 111 > silicon. The useful dynamic range of the pn-CCD camera on XMM was adjusted to the energy band from 100 eV to 15 keV (see Fig. 4.56).

Split events, i.e., events with electrons in more than one pixel, originating from one single photon, were reconstructed and summed to one photon event. In total, about 70% of all events are single pixel events, 28% are two pixel events and 2% are events with three and four pixels involved. In the case of the XMM pn-CCDs, one single X-ray photon spreads the generated signal charge never over more than four pixels.

The readout electronics of the pn-CCD system is described in references [76, 81]. A charge sensing amplifier followed by a multicorrelated sampling stage, multiplexer and output amplifier (CAMEX64B JFET/CMOS ASIC chip) guide the pn-CCD pixel content as a voltage signal to a 10 MHz 12-bit

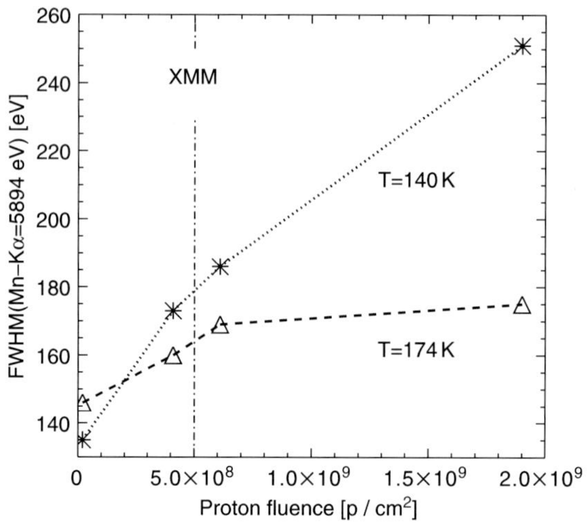


**Fig. 4.53.** FWHM of the Mn-$K_\alpha$-spectrum (5894 eV) in dependence on proton fluence and temperature. Before proton exposure the lower operating temperature of 140 K gains better results. After a 10 MeV proton fluence of more than $2 \times 10^8$ cm$^{-2}$, the higher temperature of 174 K results in a better energy resolution. The FWHM is degraded from 135 eV (140 K) to 160 eV and 175 eV (174 K) after $4.1 \times 10^8$ p/cm$^2$ and $1.9 \times 10^9$ p/cm$^2$, respectively. A FWHM of 164 eV is expected after the 10-year XMM mission

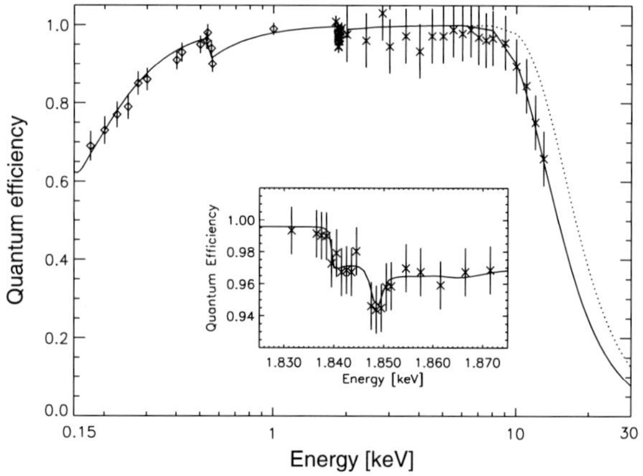


**Fig. 4.54.** Quantum efficiency of the pn-CCD as a function of the incident photon energy. The energy scale ranges from 0.15 keV to 30 keV. The solid line represents a 300μm thick sensitive volume, the dotted line 500 μm

flash ADC system. The whole system, i.e., CCD and CAMEX64B amplifier array dissipates a power of 0.7 W for the entire camera (768 readout channels), a value which is acceptable in terms of thermal budget on XMM realized through passive cooling. A further increase of the readout speed can be made

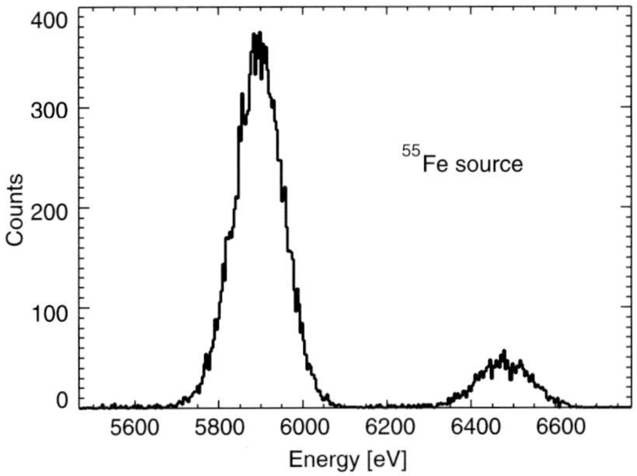


**Fig. 4.55.** Mn $K_\alpha$ and Mn $K_\beta$ spectrum of an $^{55}$Fe source. The measured FWHM is 130 eV at $-120°$C

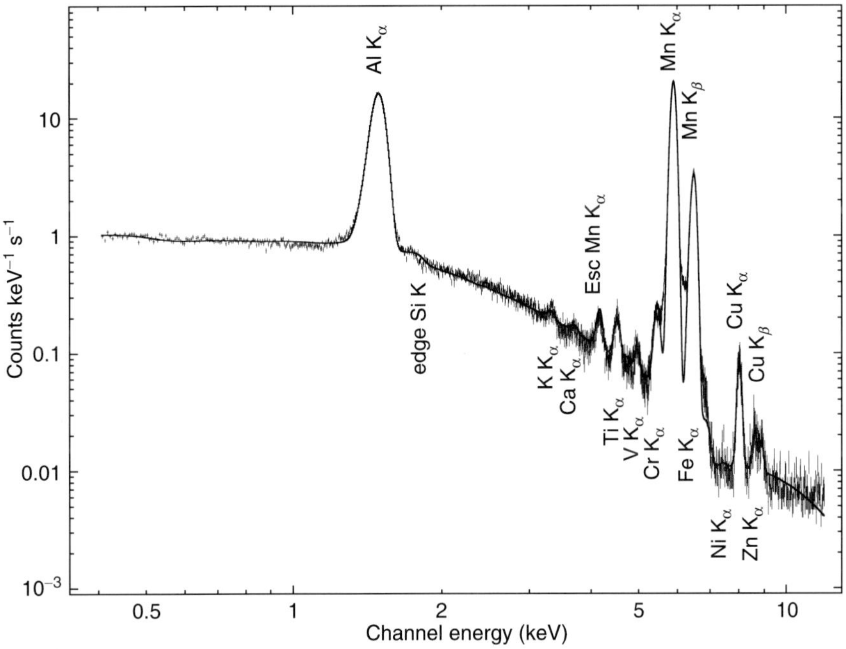


**Fig. 4.56.** Calibration spectrum measured with the pn-CCD on XMM in orbit. The spectrum is composed of the X-rays from the internal radioactive source and the instrument background as the filter wheel is moved to closed position. The continuous background below the Mn lines arises mainly from photoelectrons stimulated from the $^{55}$Fe source in the Al target. The iron $K_\alpha$ line between Mn $K_\alpha$ and Mn $K_\beta$ is not resolved. The additional lines are due to X-ray fluorescent background from the camera structure

only at the expense of further increase of power, or a degradation of the noise performance.

The charge handling capacity of the individual pixels was tested with the 5.5 MeV alpha particles from a radioactive $^{241}$Am-source. Around $10^6$ electrons can be properly transferred in every pixel. The spatial resolution was intensively tested in the PANTER facility with the flight mirror module in front of the focal plane. The first light image of the Large Magellanic Cloud in Fig. 4.57, as well as the quantitative analysis of the point spread function have shown a perfect alignment of the telescope system as on ground; the spatial resolution of the entire telescope system measured on ground corresponds exactly to the performance in orbit.

The operating temperature of the pn-CCD on XMM in orbit is −90°C. This temperature optimizes on one side the requirement of "warm" operating

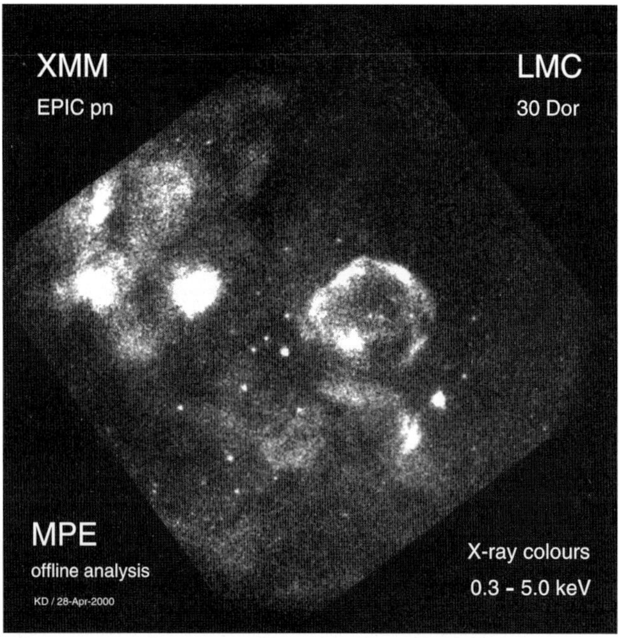


**Fig. 4.57.** The Large Magellanic Cloud in X-rays. It shows the first light image of the pn-CCD camera on XMM. This data set comprises the time dependent, spatially resolved spectroscopic X-ray emission within a field of view of 30 arcmin. This corresponds approximately to our perception of the size of the moon. The image shows the area of 30 Doradus, a supernova remnant as an extended source of X-rays. The 'north-east' of 30 Dor shows an emission of X-rays up to 5 keV, while the 'south-west' rims appears much softer in X-rays. The supernova 1987A is the bright source 'south-east' of 30 Dor. About 40 new X-ray objects have been found in this exposure. The exposure time was about 10 h.

conditions to avoid contamination and to relax stress to the mechanical structures. On the other side, it matches the need of "cold" temperatures because of leakage current reduction and efficient charge transfer.

Up to now, more than 6 years after launch, no instrumentational surprise has occurred, the energy resolution is almost equal to the ground measurements as is the case for the charge transfer efficiency [82]. To date, the electrical stability of the instrument is perfect. The first light image in Fig. 4.57 qualitatively summarizes the above enthusiastic statements.

### 4.3.3 Frame Store pn-CCDs for ROSITA, and XEUS

Future missions and other applications require pn-CCDs with smaller pixels and even faster readout. Potential applications are the German/ European ROSITA mission and ESA's XEUS mission (Table 4.3). The first two missions shall be launched at the end of this decade, and the XEUS satellite around 2015.

As in conventional CCDs, pn-CCDs equally can be designed in a frame store format. This optimizes the ratio of exposure to transfer time, but requires more space on a chip because the store area does not serve as active area but as an analog storage region (see Figs. 4.58 and 4.59).

The area to be processed in a quasi defect-free manner increases by the size of the store area. A $7.5 \times 7.5\,\mathrm{cm}^2$ large image area can be realized monolithically on a 6-in. wafer (see Fig. 4.58). By that technique the whole field of view of XEUS could be covered with a minimum of insensitive gaps between adjacent devices. The central part, the inner diameter of 7 cm, would be homogeneously sensitive.

**Table 4.3.**

| Property | XMM | ROSITA | XEUS |
|---|---|---|---|
| Status | Operating | Prototyping | Research |
| Type | Full frame | Frame store | Frame store or APS |
| Format | 400 × 384 | 256 × 256 | 1024 × 1024 |
| Pixel size | $150 \times 150\,\mu\mathrm{m}^2$ | $75 \times 75\,\mu\mathrm{m}^2$ | $50 \times 50\,\mu\mathrm{m}^2$ or $75 \times 75\,\mu\mathrm{m}^2$ |
| Readout noise | 5 electrons | 3 electrons | 1 electron |
| Sensitive thickness | 300 μm | 450 μm | 450 μm |
| Frame rate | 14 frames/s | 20 frames/s | 200–1000 frames/s |
| Readout speed | 360 ns/pix | 100 ns/pix | 50 ns/pix |
| Output nodes per CCD | 12 | 2 | 32 |
| Energy resolution[a] at Mn $K_\alpha$ | 140 eV | 130 eV | 125 eV |
| Energy resolution[a] at C $K_\alpha$ | 130 eV | 80 eV | 45 eV |
| Energy range | 0.15–15 keV | 0.1–20 keV | 0.05–20 keV |

[a] The energy resolution (FWHM) refers to incident X-rays of the Mn $K_\alpha$ line at 5.9 keV and C $K_\alpha$ at 277 eV measured at temperatures around −100°C

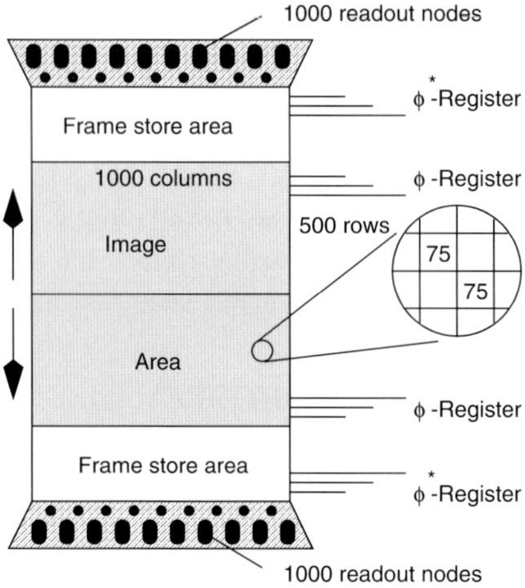


**Fig. 4.58.** Example for a pn-CCD operated in a frame store mode. The image area may have a pixel size of $50 \times 50\,\mu\text{m}^2$ or $75 \times 75\,\mu\text{m}^2$ and the store area pixels with smaller dimensions. The image area comprises $1000 \times 1000$ pixels

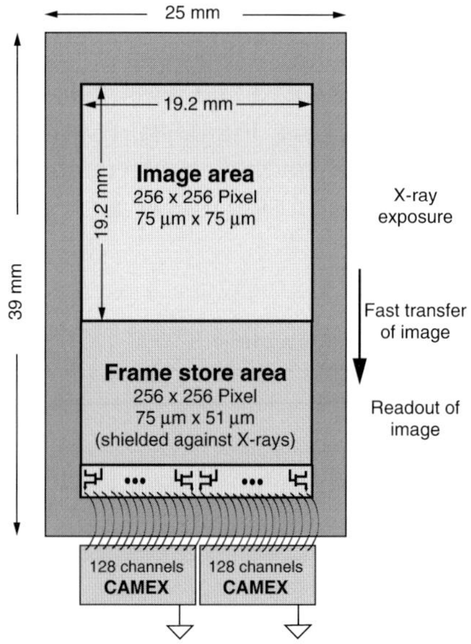


**Fig. 4.59.** Frame store pn-CCD for the ROSITA mission, having a format of $256 \times 256$ pixels in the imaging area. The pixel size there is $75 \times 75\,\mu\text{m}^2$, while it is $75 \times 51\,\mu\text{m}^2$ in the store area

As will be shown later, we expect to get a frame rate of the whole camera of at least 200 images per s. That will lead to a count rate capability of more than 200 counts per s and half energy width. An alternative approach are active pixel sensors (APS) where every pixel has its own amplifier [83].

Conceptually, they do not need an analog store area, i.e., no charge transfer, leading potentially to an even higher readout speed.

As the pixel size shrinks, the number of read nodes and transfers increases for the pn-CCD. At the same time, the system will be requiring more readout time and being more sensitive to radiation damage due to the higher number of transfers. If spreading of signal charges over more than one pixel is needed for the improvement of position resolution (see Figs. 4.60 and 4.61), the effective read noise per event will be higher by a factor $\sqrt{N}$ ($N$ is the number of pixels involved). The readout noise of every pixel involved must be quadratically added to get the total noise for one photon event.

For the pn-CCD aboard the XMM satellite the signals of one row (64 pixels) are processed in parallel in 23 µs. The extension to 128 channels on the CAMEX amplifiers, to match the smaller pixel pitch, was already realized for first prototype devices. In addition, the signal processing time must be shortened by a factor of 2 to obtain the same readout time per row. The increased readout speed will certainly have an impact on the power consumption which is actually below 1 W for the 36 cm$^2$ array.

If 128 channels are read out with 12.8 MHz, 10 µs would be required for the parallel readout of one pixel line. For the parallel transfer from the image to the storage area 100 ns are needed for one line transfer. A device of

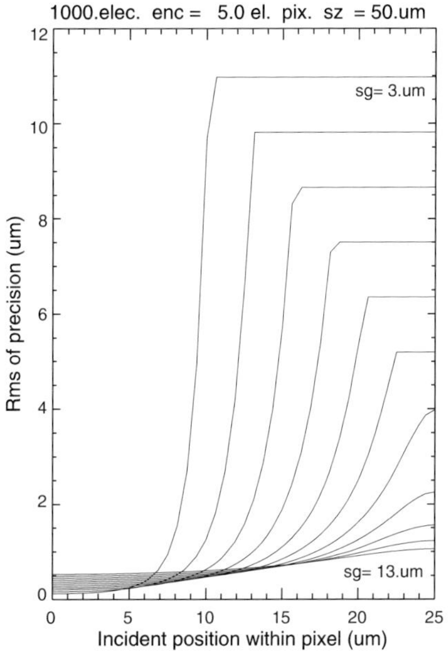


**Fig. 4.60.** Improvement of the position precision as a function of the gaussian spreading of the electron charge cloud. The typical sigma of the gaussian ("sg") electron distribution is 7 µm. The assumed pixel size is 50 µm. One-thousand electrons have been generated and processed with a noise level of 5e$^-$ (rms)

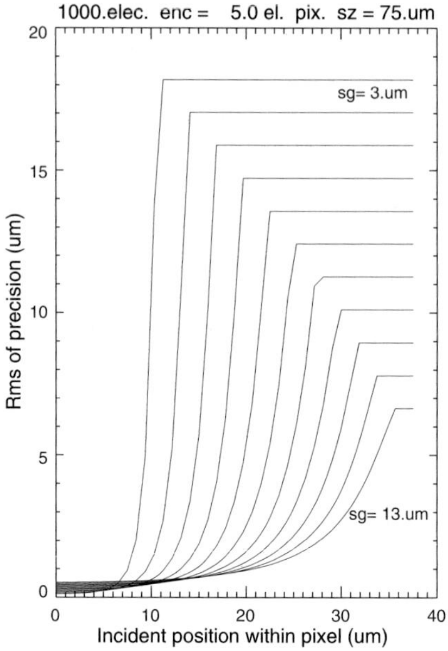


**Fig. 4.61.** The same calculation is made for a pixel size of 75 μm. The pixel edge is located at the $x$-coordinate 0, while the center of the pixel is located at 25 μm in Fig. 4.60, respectively at 37.5 μm in this figure

1000 × 1000 pixels would be divided (as in the XMM-EPIC case) in two identical halves of the image area, i.e., 500 × 1000 pixels each (see Fig. 4.58). For the parallel 500 shifts thus 50 μs would be needed for the entire transfer from the image to the shielded storage area. The readout time for the storage area while integrating X-rays in the image part, would then be $500 \times 10\,\mu s = 5\,ms$. That means, within 5 ms the whole focal plane would be read out. The out-of-time probability for the X-ray events will then be 1:100. In this operation mode 200 image frames can be taken in one second with a full frame time resolution of 5 ms.

According to the progress of the development for both detector systems – pn-CCDs and APS – a decision about the final choice has to be taken at a later stage.

## New Devices

The first prototypes of the pn-CCDs for the ROSITA mission (see Fig. 4.64) have been tested with excellent results [84]. The charge transfer efficiency was improved at least by a factor of 10 at the critical lower energies (see Fig. 4.62). In addition, the low energy response was significantly improved (see Fig. 4.63). The trigger threshold was as low as 60 eV, the peak-to-background ratio was

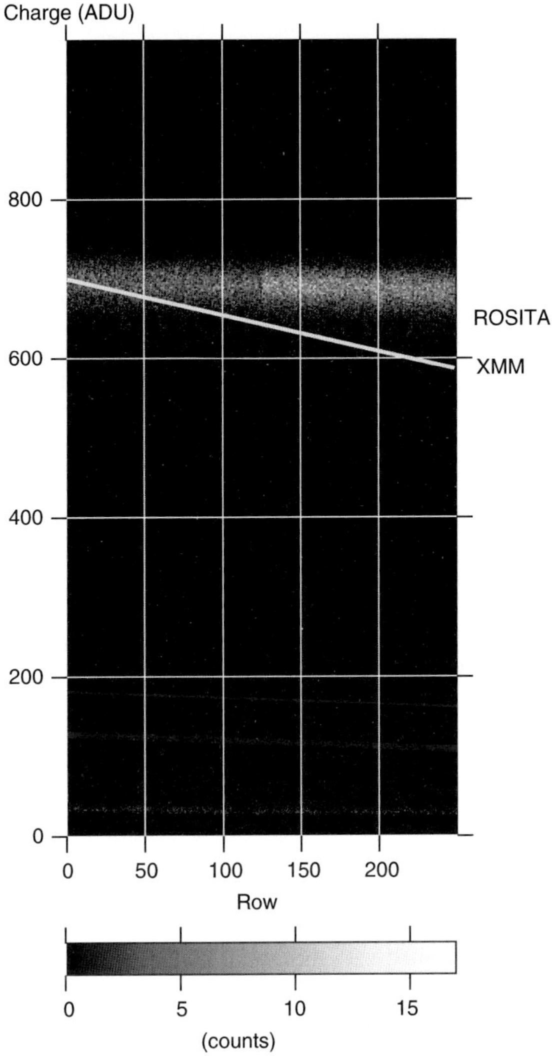


**Fig. 4.62.** Improvement of charge transfer efficiency from the XMM devices to the ROSITA devices. For Al $K_\alpha$ X-rays (1.49 keV) the XMM signal charge losses over 256 pixel transfers accumulate to about 17%, while in the new devices the loss is about 1.3%

50:1 and the FWHM for C K (277 eV) was around 80 eV. This width is still larger as the theoretical limit (around 40 eV) and reveals some additional improvements to be done in the near future. At Al $K_\alpha$ the FWHM is also 80 eV. These improvements enable us to get closer to the limits given by silicon as a detector material.

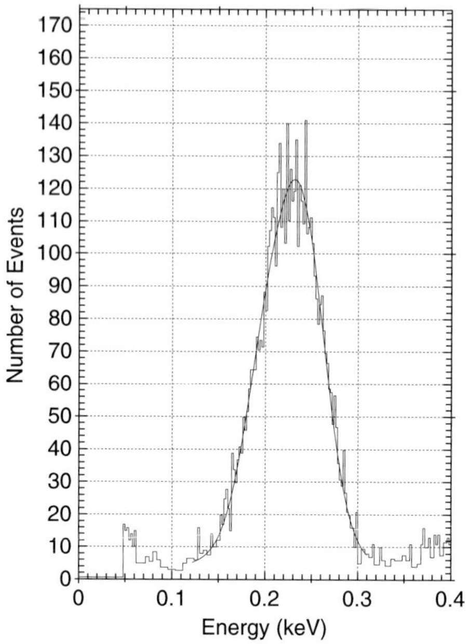


**Fig. 4.63.** Carbon spectrum recorded with a frame store pn-CCD. The C-line energy is at 277 eV corresponding to 75 electrons generated by the incoming low energy X-ray. The measured FWHM is around 80 eV. Due to partial absorption of signal carriers, the peak is shifted by 30 eV toward lower energies, if compared to the peak position of the Mn $K_\alpha$ line at 5.9 keV

## Other Applications

In recent years, many other applications with pn-CCDs have been realized. The fields of application are very different as well as the appreciated advantages of the pn-CCD. Some examples:

1. ***X-ray microscopy.*** At BESSY II a new generation of X-ray microscope has been installed, equipped with a pn-CCD system. The high efficiency for X-rays below 1 keV, the ultrafast readout and the high radiation hardness were the key properties to switch from MOS-type CCDs to fully depleted back-illuminated pn-CCDs [85].
2. ***Plasma diagnostics.*** X-ray spectroscopy is frequently used for plasma diagnostics in fusion reactors. The temperature can be determined quite precisely (the black body radiation of the plasma exhibits X-ray energies up to 10 keV with high flux) and the contamination of the plasma can be analyzed [86]. It is foreseen to install a pn-CCD system at the new fusion reactor in Greifswald, Germany. The preferred properties were the high quantum efficiency at the high energies and the fast readout.

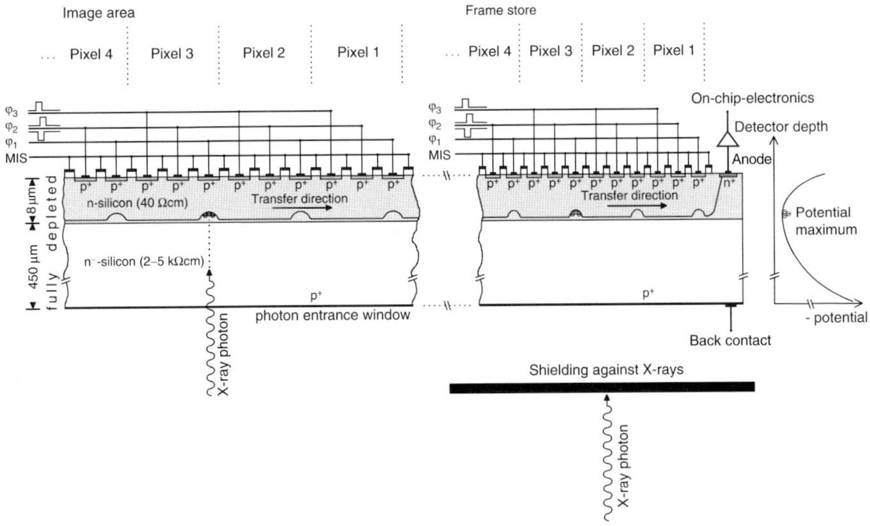


**Fig. 4.64.** The frame store pn-CCD as designed for the German ROSITA mission will have a format of 256 × 256 pixel with a size of $75 \times 75\,\mu m^2$ in the image area and 256 × 256 pixel with a size of $75 \times 51\,\mu m^2$ in the frame store

3. ***Quantum optics.*** Multi-ionization processes have been studied with a pn-CCD system at the MPI for quantum optics. They could improve the sensitivity of the experiment by the high efficiency of the pn-CCD in the VUV range at energies around 25 eV. The radiation, stimulated with a 6 ms laser was recorded in an integration mode, i.e., not in a single photon counting mode. [87]
4. ***Electron emission channeling spectroscopy.*** The emission channeling spectroscopy technique allows the direct determination of the lattice sites of radioactive impurity atoms that are incorporated into single crystalline solids [88]. Electrons from a few keV to several hundreds of keV were recorded with high precision in energy and position.
5. ***CAST.*** The CAST experiment is dedicated to the search for solar axions [89]. Its main components are a large superconducting magnet of 10 T, an X-ray telescope mounted behind the magnet and a pn-CCD camera system behind the telescope. The pn-CCD is used for the detection of X-ray radiation between 1 keV and 8 keV caused by the conversion of axions into X-rays through the Primakov effect.
6. ***Transition radiation.*** A novel K-edge imaging method has been developed aiming at the efficient use of transition radiation generated by a high energy electron beam. The method is applied in material science, biology and medicine [90].

### 4.3.4 Conclusion

Since the invention of the silicon drift detector by Gatti and Rehak in 1983, a large variety of new detector structures based on the principle of sideward depletion have been developed. Those detectors have left their initial fields of applications in high energy physics, astrophysics and synchrotron radiation research. They are now a mature technology and open many new industrial applications. Experiments in basic research have driven the performance parameters toward the optimum for the specific applications: high quantum efficiency, excellent energy resolution, high radiation tolerance, good position resolution, high speed, large and almost cosmetic defect free devices, homogeneous response of the full bandwidth of radiation and high background rejection efficiency. It will be the aim of future developments to approach the physical limits in radiation detection and to breathe in additional intelligence into the local detector systems to face the steadily increasing amount of data and power dissipation.

**Acknowledgments**

The XMM focal plane configuration has profited from many discussions with scientists of the Max–Planck-Institut für extraterrestrische Physik and of the MPI-Halbleiterlabor. In particular, we are indebted to Heike Soltau, Gerhard Lutz, Rainer Richter, and Heinrich Bräuninger. The frequent discussions with our friends and colleagues from the Politecnico di Milano, the BNL, New York, and the University of Tübingen, were always stimulating. Special thanks to Antonio Longoni, Carlo Fiorini, Marco Sampietro, Emilio Gatti, Pavel Rehak, and Eckhard Kendziorra.

## 4.4 Wavelength Dispersive XRF and a Comparison with EDS

*N. Kawahara and T. Shoji*

Wavelength dispersive XRF (WDXRF) is the method that utilizes Bragg diffraction of single crystal or synthetic multilayer to disperse fluorescent X-rays.

WDXRF has the following advantages compared to energy dispersive XRF (EDXRF):

- Higher spectral resolution especially in longer wavelength, or low energy region
- Lower background
- Wider dynamic range
- Measuring ability of ultralight elements down to beryllium

On the other hand, the disadvantages are:

- More complex system
- Lower detection efficiency and requiring X-ray source of higher power

In this section, the components, the varieties of equipments and some applications of WDXRF are briefly described.

### 4.4.1 Dispersion Materials for WDXRF

The WDXRF spectrometer utilizes Bragg diffraction with crystals to disperse X-rays. Bragg's equation is written as (4.62).

$$n\lambda = 2d \sin\theta. \tag{4.62}$$

Here, $n$ is the reflection order, $\lambda$ is the wavelength of incident X-rays, $d$ is the lattice spacing of the crystal and $\theta$ is the incident angle. Equation (4.62) shows $\lambda$ must be smaller than $2d$. Practically, the scanning range of the goniometer (usually $2\theta < 150°$) limits $\lambda < 2d \sin 75°$. Angular dispersion of a spectrometer is obtained by differentiating (4.62).

$$\frac{\mathrm{d}\lambda}{d\theta} = \frac{2d \cos\theta}{n}. \tag{4.63}$$

Equation (4.63) illustrates that higher reflection angle $\theta$ and/or higher diffraction order $n$ gives better wavelength resolution.

The relationship between angular resolution and relative wavelength resolution is derived from (4.62) and (4.63) as (4.64).

$$\frac{\mathrm{d}\lambda}{\lambda} = \frac{\mathrm{d}\theta}{\tan\theta}. \tag{4.64}$$

Equation (4.64) illustrates that higher reflection angle $\theta$, in other words, using smaller $2d$ value crystal gives higher wavelength resolution.

#### Crystals

There are several crystals commonly used in WDXRF spectrometers (Table 4.4). These are chosen by the $2d$ value (spectral range) and resolution or reflectivity. For the finite thermal expansion coefficient, spectrometers are usually equipped with temperature-stabilizing systems to fix the $2d$ value of the crystal.

#### Synthetic Multilayers

For analyses of lighter elements, i.e., detecting longer wavelength X-rays, we need a dispersion material with a larger $2d$ value as mentioned above. For example, $N$-$K\alpha$'s wavelength is 3.16 nm. This cannot be diffracted even with

**Table 4.4.** Common crystals for WDXRF

| Crystal | Chemical formula | Miller indices | $2d$ (nm) | Lightest measurable element | Linear thermal expansion coefficient $\beta(10^{-6}\,\mathrm{deg}^{-1}.)$ | Typical energy resolution in a spectrometer | Remarks |
|---|---|---|---|---|---|---|---|
| Lithium Fluoride | LiF | (220) | 0.2810 | $_{24}$Cr | 34 [107] | 14 eV @Mn-K$\alpha$ | High resolution |
| Lithium Fluoride | LiF | (200) | 0.4028 | $_{19}$K | 34 [107] | 25 eV @Mn-K$\alpha$ | General use |
| Sodium Nitride | NaCl | (200) | 0.5640 | $_{16}$S | 40.4 [107] | 5.5 eV @S-K$\alpha$ | For sulfur<br>High resolution<br>Deliquescent |
| Germanium | Ge | (111) | 0.65327 | $_{15}$P | 6.1 [108] | 5 eV @P-K$\alpha$ | Eliminates second order reflection |
| Graphite | C | (0002) | 0.6705 | $_{15}$P | 28.2 [109] | 8.5 eV @P-K$\alpha$ | Higher reflectivity and poorer resolution than Germanium |
| Indium Antimonide | InSb | (111) | 0.74806 | $_{14}$Si | 4.7 [108] | 4.5 eV @Si-K$\alpha$ | For silicon |
| PET (Penthaerythritol) | $C(CH_2OH)_4$ | (002) | 0.876 | $_{13}$Al | 131 [109] | 4 eV @Al-K$\alpha$ | Higher reflectivity but larger $\beta$ than EDDT<br>Deteriorative |
| EDDT (Ethylene diamine dextrotartrate) | $C_6H_{14}N_2O_6$ | (020) | 0.8808 | $_{13}$Al | 20 [110] | 4 eV @Al-K$\alpha$ | Deliquescent |
| ADP (Ammonium dihydrogen phosphate) | $NH_4H_2PO_4$ | (101) | 1.0648 | $_{12}$Mg | 17 [107] | 4 eV @Mg-K$\alpha$ | For magnesium<br>Deliquescent |
| TlAP (Thallium acid phthalate) | $CO_2HC_6H_4CO_2Tl$ | (001) | 2.57626 | $_{8}$O | 32.7 [110] | 15 eV @Na-K$\alpha$ | Deliquescent |

TlAP (2$d$:2.576 nm), though Langmuir–Blodgett films such as lead stearate (2$d$:10.04 nm) have been utilized for ultralight elements, with very low reflectivity.

Synthetic multilayers (SML) have been utilized since the 1980s as dispersion devices for soft X-rays or EUV [91]. Typically, SML is formed as a multiple periodic stack of refraction layers, which consist of a heavy element in most cases, and transmission, or spacer layers which consist of a light element, as illustrated in Fig. 4.65.

There are factors which influence the characteristics of an SML, such as the refraction coefficients of layers, the thickness ratio of the refraction layer and transmission layer (called $\Gamma$ value, see Fig. 4.65 for the definition), total layer number, sharpness of the layer boundaries, etc.

The optimal substance pair should have low absorption and large Fresnel coefficient, i.e., a large difference of refraction indices between the spacer and refraction layer. The complex refraction index $\mathbf{n}$ is described using atomic scattering factor $f_1 + \mathrm{i}f_2$ as (4.65)

$$\mathbf{n} = 1 - \delta + \mathrm{i}\beta = 1 - \frac{r_0\lambda^2}{2\pi}\frac{N_\mathrm{A}\rho}{A}(f_1 - \mathrm{i}f_2). \tag{4.65}$$

Here, $\delta$ and $\beta$ are the real and imaginary part of refractive index, $r_0$ is the classical electron radius, $\lambda$ is the wavelength of incident X-rays, $N_\mathrm{A}$ is Avogadro number, $\rho$ is the density, and $A$ is the atomic weight.

Yamamoto and Namioka [92] have presented a way to find such a proper pair using a plot of complex refraction indices. Figure 4.66 is an example of the plot for C-K$\alpha$. The scattering factors used are that of Henke et al. [93]. Yamamoto and Namioka also proposed designing methods of each layer thickness and total layer number using complex amplitude reflectance.

It should be mentioned that realizing a sharp interface between layers is also an important property for good SML and it limits the choice of layer materials. SMLs have a number of advantages over natural crystals; much higher reflectivity, physical stability, small thermal expansion coefficient, and suppression of higher order reflection. Regarding higher order reflection, drastic suppression is achieved by adjusting $\Gamma$ value. For example, a second (third) order reflection is virtually eliminated with an SML of which $\Gamma$ is 1/2 (1/3).

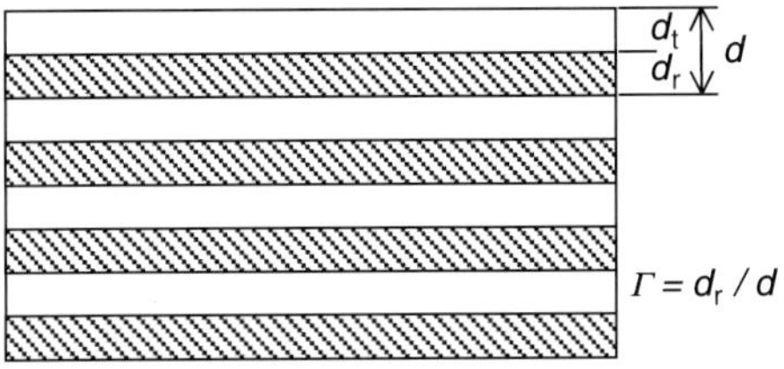


**Fig. 4.65.** Schematics of an synthetic multilayer (SML)

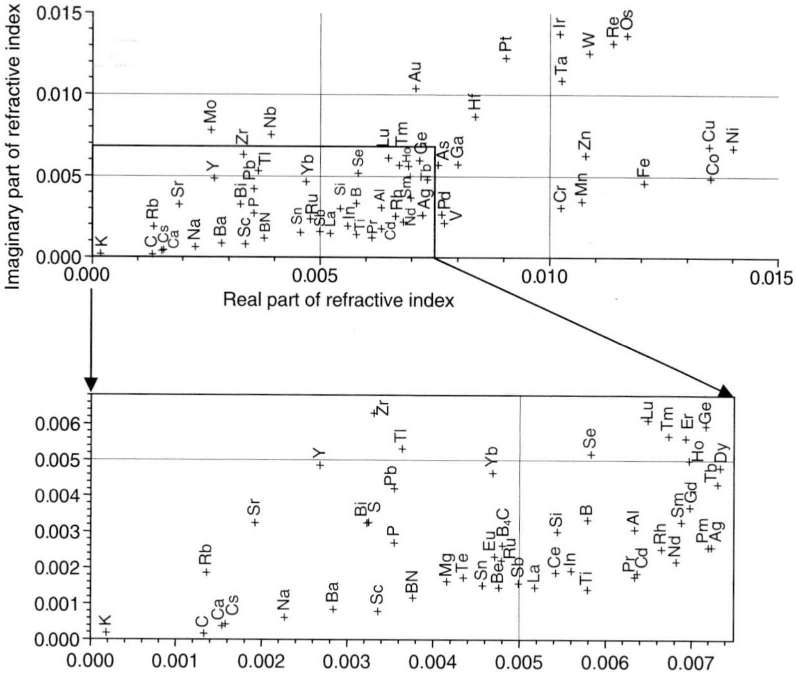


**Fig. 4.66.** Material map in complex refractive index plane at 277 eV (C-K$\alpha$). Bulk densities at room temperature are used

A noteworthy matter on the use of SML is that the dispersion may not obey Bragg's equation. Considering the refraction index (4.62) has to be modified as

$$n\lambda = 2d \sin\theta \left(1 - \frac{\delta - \mathrm{i}\beta}{\sin^2\theta}\right) . \tag{4.66}$$

As refraction index depends on the wavelength of incident X-rays, the apparent $2d$ value calculated by using "simple" Bragg's equation varies depending on wavelength.

The other issue is the background from SML. The best choice of the spacer layer material is often the element to be analyzed. In this case, the fluorescence from this spacer layer diffracted in the SML can shape a weak peak.

Some commercial SMLs are listed in Table 4.5. A useful database for performances of various SMLs is available at a web site of The Center for X-Ray Optics, Lawrence Berkeley National Laboratory [94].

### 4.4.2 Detectors and Electronics

#### Scintillation Counter

A scintillation counter (SC) is one of the most common X-ray photon counters because of ease of handling, low noise characteristic and high counting rate.

**Table 4.5.** List of some commercial SMLs [95, 96]

| Name | $2d$ (nm)* | Suitable elements |
|---|---|---|
| OV-045A | 4.5 | F ∼ Mg |
| OV-050A | 5.0 | F ∼ Mg |
| OV-055A | 5.5 | F ∼ Mg |
| OV-120N | 12 | B ∼ O |
| OV-160N | 16 | B ∼ O |
| OV-160H | 16 | B |
| OV-200H | 20 | B |
| PX1 | 5 | O ∼ Mg |
| PX2 | 12 | B, C |
| PX3 | 20 | B |
| PX4 | 12 | C |
| PX5 | | N |
| PX6 | | Be |
| RX25 | 3 | O ∼ Mg |
| RX35 | 5.5 | O ∼ Mg |
| RX40 | 8 | O, N |
| RX45 | 11 | N |
| RX61 | 16 | C, B |
| RX75 | 16 | B |
| RX80 | 20 | Be |

*$2d$ values may have ranges.

The SC consists of a scintillator and a photomultiplier (Fig 4.67). The most widely used scintillator is sodium iodide crystal doped with thallium 0.1 mol% (NaI(Tl)). An X-ray photon penetrating into a NaI(Tl) scintillator produces a light pulse of around 410 nm. The intensity of the light pulse is proportional to the incident X-ray photon energy. The light pulse is converted to electric pulse, the height of which is proportional to the X-ray photon energy.

SC is utilized for X-rays of 4 keV or higher energy due to poor energy resolution in the low energy region.

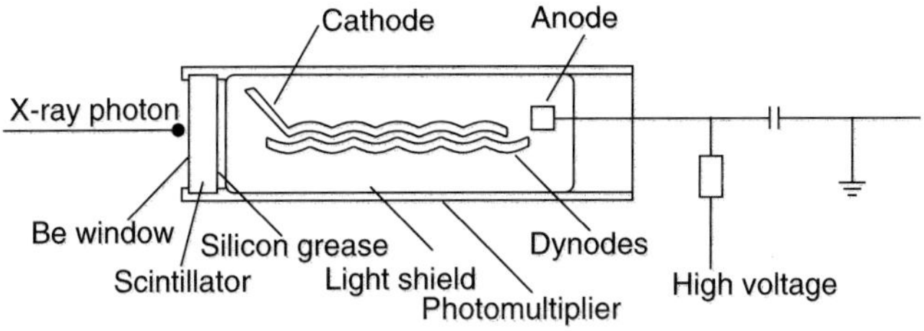


**Fig. 4.67.** Structure of scintillation counter

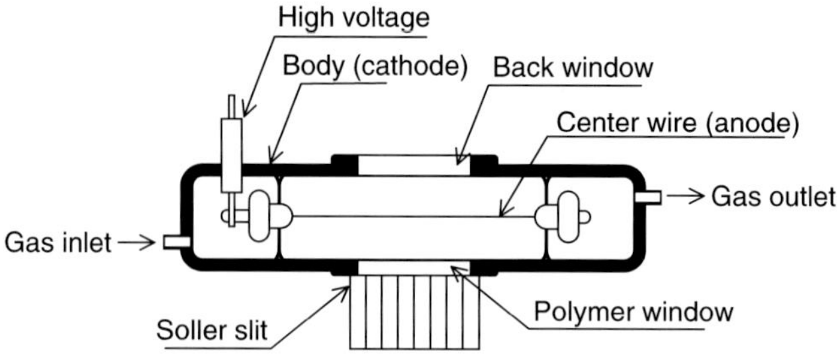


**Fig. 4.68.** Structure of gas-flow proportional counter

## Gas Proportional Counter

The gas proportional counter consists of a gas chamber filled with noble gas and a center wire applied with a high positive voltage (Fig 4.68). An incident X-ray photon ionizes a gas atom and the cation and electron formed move toward the body (cathode) and the center wire, respectively. The electron gives rise to an avalanche effect. Finally, a current pulse which height is proportional to the energy of the incident photon is formed.

Gas proportional counters are used for the energy range of 100 eV to 20 keV. When the transmission rate of the window is critical to detecting efficiency in low energy region, thin (less than 1 μm) polymer film is used as the incident window. Unavoidable gas leakage from such a thin film requires constant gas flow into the chamber (flow proportional counter—F-PC ). The most common gas for F-PC is Ar + 10% $CH_4$ (called P-10 gas).

For the energy region higher than 1 keV, a sealed proportional counter (S-PC) can be used. S-PC has a window of beryllium foil, which has sufficient transmission rate and practically no gas leakage. The detecting efficiency in this region depends on the absorption of the gas and the window transmission rate. Optimization for a certain energy of X-rays can be done by adjusting the gas pressure and choice of gas. As proportional counters have a simple structure and design flexibility, one with a wide detecting area can easily be realized.

## Selection of Detector for WDXRF

The detectors commonly used for the WDXRF system are listed in Table 4.6 for selection criterion. In this table, some characteristics of the detectors are compared and the element range that is suitable for measurement by each detector is shown.

## One-Dimensional Position Sensitive Detector

In WDXRF, one-dimensional position sensitive proportional counter (PSPC) may be used. This type of detector has a subdivided cathode that is combined

**Table 4.6.** Comparison for various detectors in the WDXRF system

| | | SC | S-PC | F-PC |
|---|---|---|---|---|
| Applicable | K-line | Ti ~ Sn | Mg ~ Mo | B ~ Ti |
| Elements | L-line | Sn ~ U | Sn ~ U | — |
| Quantum Efficiency | | High for short wavelength | High for middle wavelength (depends on the filled gas and window thickness) | High for long wavelength |
| Counting Rate ($c/s$) | | Up to $1\times10^6$ | Up to $4\times10^6$ | Up to $4\times10^6$ |
| Energy resolution of detector itself (at Mn-K$\alpha$) | | 40% | 18% | 18% |
| Elimination of higher order reflection | | Using PHA is not so effective | Using PHA and the gas discrimination are effective. | Using PHA and the gas discrimination are effective. |
| Note | | | Measures to peak shift in PHA at high counting rate are required. | Gas density should be stable |

with the position analyzing circuit so as to acquire information of the incident position of each X-ray photon. The typical type of the detector has the spatial resolution of about 0.1 mm.

Figure 4.69 schematically shows an example in which this detector is used for a WDXRF measurement. The simultaneous measurement of wavelengths is possible without moving the crystal and the detector by means of combining this PSPC with the divergent fluorescent X-rays from a point-like source and with a flat monochromator crystal [97].

### Detector Electronics

Figure 4.70 is a block diagram of detector electronics. The most different point from the EDXRF system is the pulse height analyzer (PHA) unit, while the multichannel analyzer takes the place in the EDXRF system. In addition, the time constant of the system is tuned corresponding to the characteristic pulse shape of the detector. Examples of output pulses from S-PCs are shown in Fig. 4.71 This pulse is integrated by the preamplifier and then formed to a pulse of which the width is 0.3 µs to 1 µs, by the pulse shaper.

The PHA unit is utilized to eliminate higher order reflection by energy difference from the first order reflection. Figure 4.72 shows the effect of the

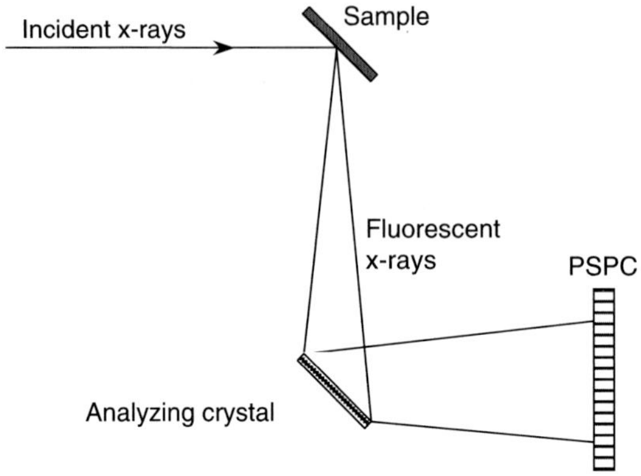


**Fig. 4.69.** Schematics of a PSPC optics

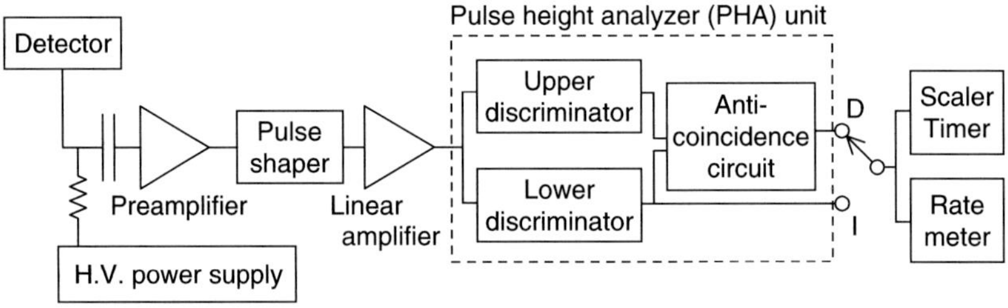


**Fig. 4.70.** Block diagram of detector electronics

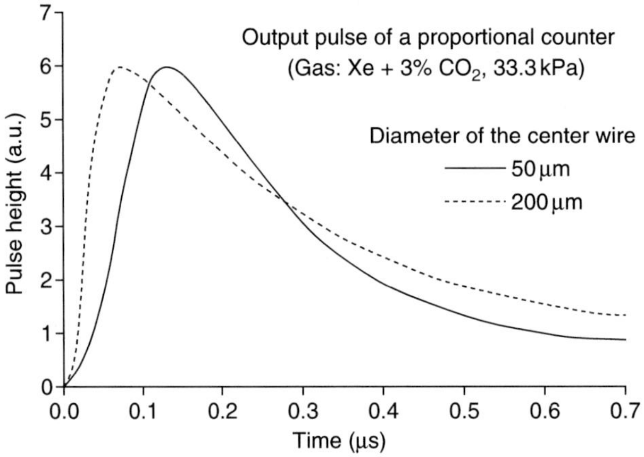


**Fig. 4.71.** Output pulse of S-PCs

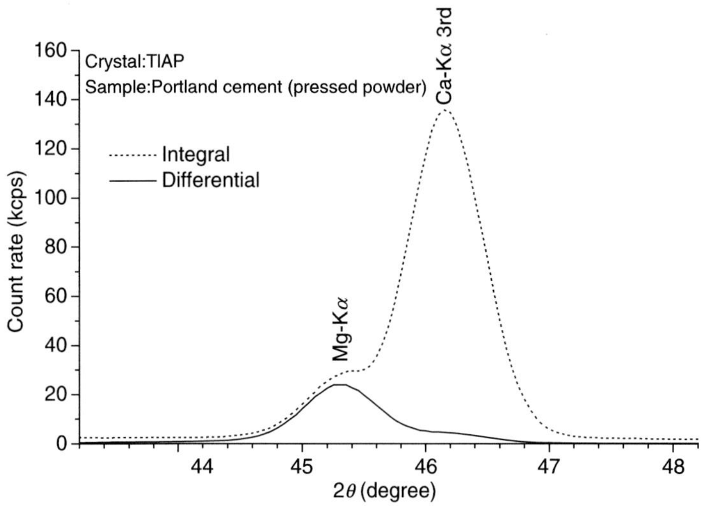


**Fig. 4.72.** Mg-K$\alpha$ spectrum from a Portland cement sample in integral mode (dotted) and differential mode (solid)

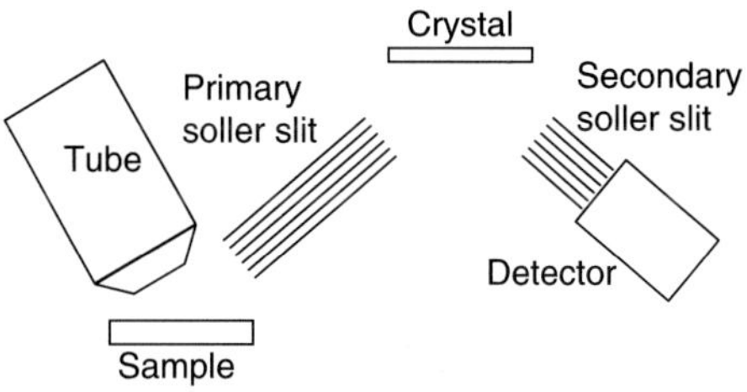


**Fig. 4.73.** Schematics of the parallel beam optics

PHA unit. Interfering third order reflection of Ca-K$\alpha$ in the integral mode measurement (I in Fig. 4.70) is eliminated in the differential mode (D in Fig. 4.70).

### 4.4.3 Optics Used for the WD Spectrometer and its Components

#### Parallel Beam (Bragg–Brentano) Optics

Parallel beam optics consists of a flat crystal or a synthetic multilayer and usually two Soller slits. Figure 4.73 illustrates the schematics. The Soller slit [98] is utilized to define the divergence of fluorescent X-rays. It is formed by thin metal foils (typical thickness: 0.05 mm) stacked in equal pitch. The divergence of the X-ray beam passed through a Soller slit $\alpha$ is written as

$$\alpha = 2S/l. \tag{4.67}$$

Here $S$ is the gap of the foils and $l$ is the length of the foils. In reality, $\alpha$ gets tailing due to the total reflection on the foil surface. Typical values for $\alpha$ range from 0.15 to 3°. Most commercial spectrometers come with several Soller slits of various divergences due to a Soller slit with smaller $\alpha$ (i.e., finer $S$ and/or longer $l$) gives higher angle resolution and lower flux.

The situation can be visualized and evaluated using a DuMond diagram [99], that is a dispersion diagram in the angle-wavelength plane shown in Fig. 4.74. The horizontal axis represents the $2\theta$ angle and the vertical axis the wavelength. This diagram shows the line of an expanded part of the function given by Bragg's equation (4.62) with the width $\omega$, which is the intrinsic diffraction width of the analyzing crystal in the reflectivity curve (so-called "rocking curve" (b)). $\alpha$ in Fig. 4.74(a) stands for the divergence in (4.67).

In addition to the Soller slit parameter $\alpha$, the wavelength resolution is affected by $\omega$ and $\theta$ as well. If the X-ray beam incident on the crystal at angle $\theta_0$ has a divergence $\alpha$, reflection of X-rays occurs only in the shaded portion in Fig. 4.74(a). The wavelength range $\Delta\lambda$ of the reflected X-rays is easily determined from the figure.

## Convergent Optics

Convergent optics has the advantage of a larger capture angle when compared to parallel beam optics. A geometry in which the incident angle to the crystalline plane is constant from a point improves the convergent optics. One solution is to use a crystal bent in a logarithmic spiral shape [100], which is defined by (4.68) in polar coordinates ($r$, $\phi$)

$$r = A \exp(\phi \cot \theta). \tag{4.68}$$

Here $A$ is the scale factor and $\theta$ is the incident angle. The angle between the tangent line to the crystal surface and the line from the origin $O$ has the constant angle $\theta$ at any point on the curvature. That means there is no

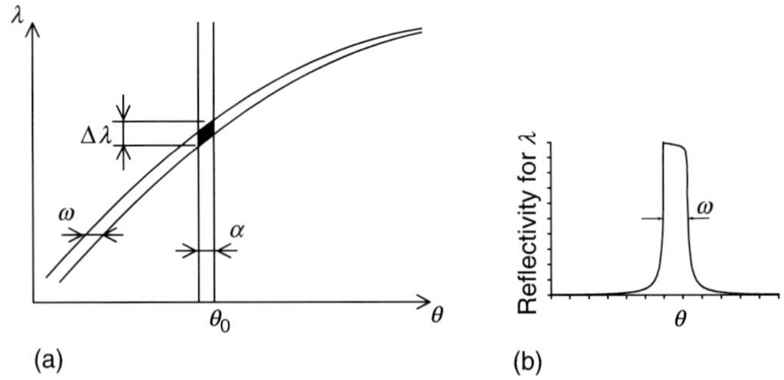


**Fig. 4.74.** (**a**) DuMond diagram of a parallel beam optics (**b**) Rocking curve of a crystal

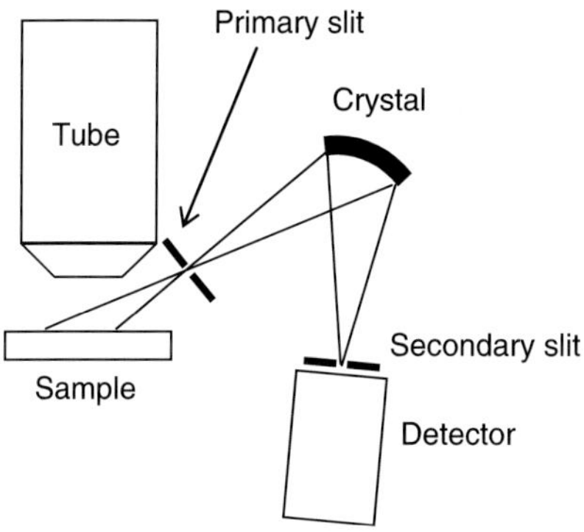


**Fig. 4.75.** Schematics of the convergent optics

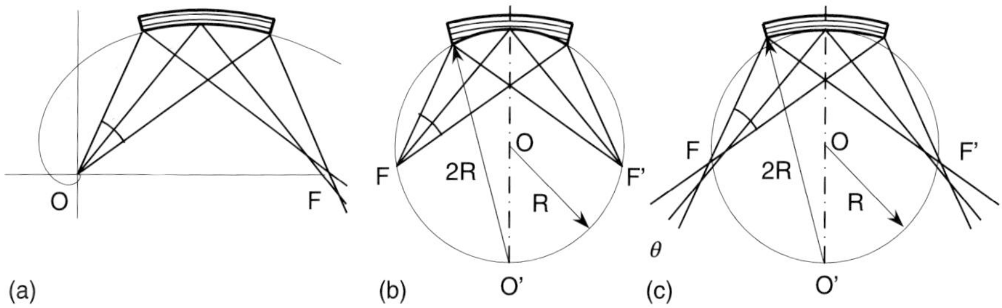


**Fig. 4.76.** Focusing of beam by (**a**) logarithmic spiral, (**b**) Johansson and (**c**) Johann shaped crystal

aberration at the origin $O$, while it has aberration at the other focal point F (Fig. 4.76a). Therefore, by putting the logarithmic curved crystal as F to be at the primary slit and $O$ at the secondary slit in Fig. 4.75, the wavelength resolution can be tuned effectively with the gap of the secondary slit.

A focusing crystal with no aberration was developed by Johansson [101]. This Johansson type crystal has a crystal surface shaped as Rowland circle of radius $R$ and the crystalline plane curved with radius of $2R$ (Fig. 4.76b). Thus, all the X-rays from the focal point $F$ to the crystal have the identical Bragg angle and, then, converge at the other focal point $F'$. Although the Johansson type realizes an ideal convergent optics with no aberration, it is not so widely used due to the difficulty to give crystals the necessary shape.

Another crystal shape commonly used for convergent optics is the Johann type (Fig.4.76c), which is curved with radius $2R$ and put so as to be tangent to the Rowland circle of radius $R$ [102]. Unlike the Johansson crystal, in the case of Johann geometry the crystal surface is on the circle of radius $2R$. Therefore, the Johann crystal produces aberration at both focal points. This is a major practical difference from the case of the logarithmic spiral crystal, which has the aberration only on one side. For both geometries, the Johann and the logarithmic spiral crystal, the aberration is smaller at higher Bragg angles and it is almost negligible in $2\theta$ range larger than 90°.

### 4.4.4 Types of WDXRF Spectrometer

#### Sequential Spectrometer

The sequential WDXRF spectrometer has single parallel beam optics, as described in Sect. 4.4.2. This disperses X-rays sequentially by rotating the crystal and the detector, keeping these rotating angles as $\theta$ and $2\theta$, respectively. The sequential spectrometer is capable of detecting arbitrary wavelength of X-rays in the range covered by the crystals with which the spectrometer is equipped. There are two types of layouts, tube-above (as shown in Fig. 4.73) and tube-below (turn Fig. 4.73 upside down). Tube-above layout is robust to contamination from samples while tube-below layout suits liquid samples.

#### Simultaneous Spectrometer

The simultaneous WDXRF spectrometer has a multiple number of convergent optics, one for each X-ray line to be analyzed. Each optic can detect a certain X-ray wavelength. The main fields of use of simultaneous spectrometers are for quality control in material industries such as steel, cement, semiconductors, etc. In these applications, accurate and rapid analyses of limited elements are required. There are also tube-above and tube-below layout simultaneous spectrometers, which are adapted to the particular application as sequential spectrometers.

#### Double Crystal Spectrometer

When a single crystal is insufficient for a required wavelength resolution, an arrangement using two (or more) crystals in sequence can be used. This technique is explained using the DuMond diagrams below. First, we consider the arrangement of two crystals, A and B, as given in Fig. 4.77a. Rotation of crystal A to give an increase in X-ray incidence angle will decrease the incidence angle on B.

Giving the DuMond curves of A and B on a single graph, the directions are opposite, as shown in Fig. 4.77a. The distance between the origins of the two curves is given as the difference in the set crystal angles $\omega$. Conditions for reflection are given at the intersection of the curves in Fig. 4.78a. Wavelength spectrum resolution is much better than for the single crystal case.

The configuration of Fig. 4.77a is called as a $(++)$ configuration, while Fig. 4.77b is known as a $(+-)$ configuration. When using three or more crystals, individual DuMond diagrams may be layered together. An actual example of the optics of a double crystal spectrometer with a $(++)$ arrangement is given in Fig. 4.79 [103]. It is designed to measure chemical shifts.

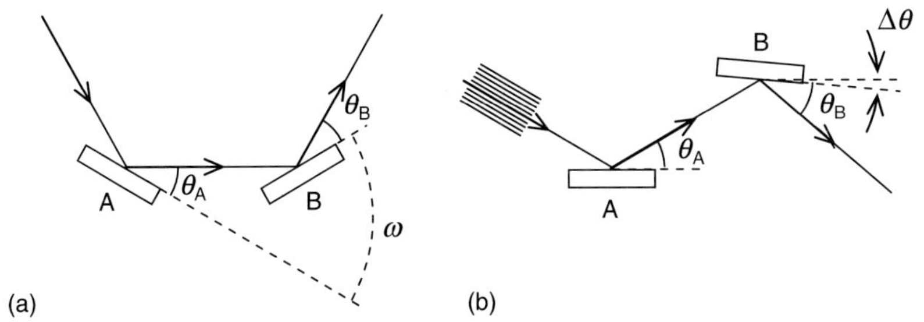


**Fig. 4.77.** Schematics of double crystal spectrometer. **(a)**: (+ +) configuration, **(b)**: (+ −) configuration

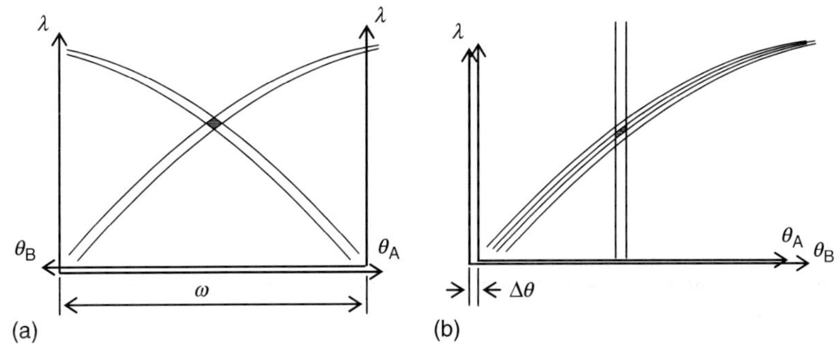


**Fig. 4.78.** DuMond diagrams for **(a)**: (+ +) configuration, **(b)**: (+ −) configuration

A planetary gear $A1$ meshing with a stationary gear $B1$ carries an arm $C$. The arm holds a second crystal $D$. A coaxial shaft $E1$ of the gear $A1$ is directly connected to a first crystal $F$ at an equal distance from the second crystal $D$. The center of the first crystal $F$ is positioned at the center of a main wheel $G$, which is driven through a worm after 2:1 reduction by a stepping motor $M$. The stepping motor $M$ is positioned outside the vacuum spectrometer chamber. A detector $P$ is kept in track with an independent gearing system. Another coaxial shaft $E2$ drives a gear $A2$ meshing with a stationary gear $B2$. The shaft is directly connected to a gear $A3$ and the wheel $G$. The gear $A2$ has an arm, which carries a shaft $J$ that follows the same track as the second crystal. A planetary gear $K$, which meshes with the gear $A3$, generates the $4\theta$ motion needed to receive the X-rays diffracted by the second crystal. A Soller slit $S$ is inserted between the sample and the first crystal to avoid vertical divergence of the X-rays.

Figure 4.80 is an example of the comparison of the energy resolution of the double crystal spectrometer to the single crystal spectrometer. The high wavelength resolution of the double crystal spectrometer is useful for chemical state analyses. There are studies using double crystal spectrometers, such as

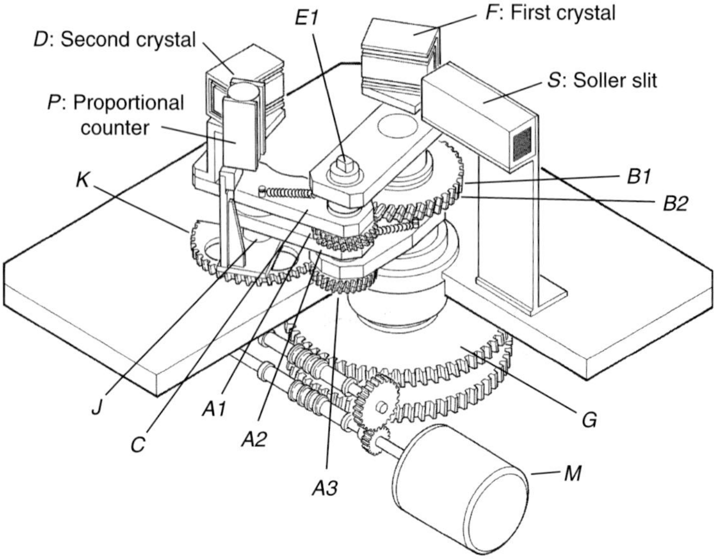


**Fig. 4.79.** A double crystal spectrometer of (+ +) configuration [103]

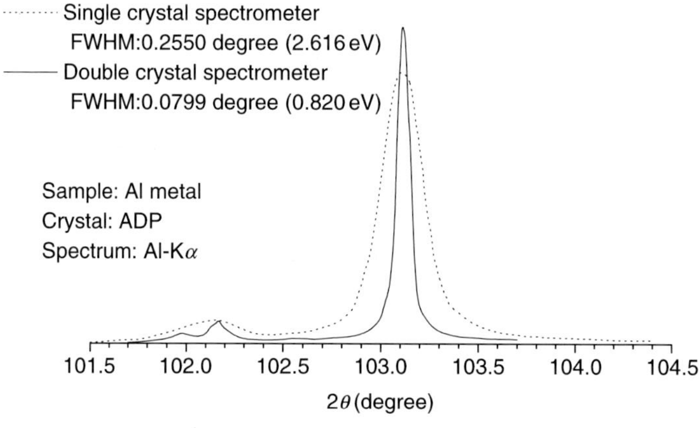


**Fig. 4.80.** Al-K$\alpha$ spectrum taken with a single crystal spectrometer (dotted line) and with a double crystal spectrometer (solid line)

valence analysis of metal compounds [104] and, as a recent study, electronic state analysis of cobalt in lithium ion battery [105].

Double crystal spectrometers have the following advantages compared to X-ray photoelectron spectrometry (XPS or ESCA):

- Ultra high vacuum is not required.
- Insulators are analyzable as well as conductors.
- It is not a surface analysis but a bulk measurement.

### 4.4.5 Selected Applications Suitable for WDXRF

#### Boron in BPSG (Boro-Phospho-Silicate Glass)

BPSG has been a widely used material as interlayer insulation films in the semiconductor industry. As controlling the composition of the material is critical for the property, the analysis of the content of boron and phosphorus in BPSG layers takes a key roll in the process. Equipping a gas-flow proportional counter with an ultrathin (sub-$\mu$m) polymer window for detecting and using a synthetic multilayer for dispersing, WDXRF spectrometer realizes sufficient analyzing efficiency [106].

#### Phosphorus in Silicon

The measurement of dopant content such as phosphorus in silicon is another important application in the semiconductor field. However, to detect P-K$\alpha$ (2.013 keV) in the huge silicon peaks (Si-K$\alpha$: 1.74 keV, Si-K$\beta$: 1.836 keV) is not practical with an EDXRF spectrometer. On the other hand, when using a WDXRF spectrometer equipped with germanium (111) ($2d = 0.65327$ nm) or graphite (0002) ($2d = 0.6705$ nm), there is no interference of the silicon fluorescence with the phosphorus fluorescence. This fact is explained by the Bragg's equation (4.62). The equation shows that when the X-ray wavelength is longer than the $2d$ value of the crystal, no Bragg diffraction occurs. This is the case for Si-K$\alpha$ (0.7126 nm) or Si-K$\beta$ (0.6753 nm) with the crystals given above but not for P-K$\alpha$ (0.6159 nm).

#### Vanadium in Titanium Alloy

Owing to high wavelength resolution, V-K$\alpha$ can be separated from Ti-K$\beta$ with a WDXRF spectrometer.

Figure 4.81 illustrates spectra of Ti alloy SRMs taken with a WDXRF and an EDXRF spectrometer. Besides this example, interference of Cr with Mn or Fe with Co can be resolved with WDXRF spectrometers. Generally, in the lower energy region, higher relative energy resolution is achieved with a WDXRF spectrometer.

#### Mol Ratio Analysis of $BaTiO_3$

Barium titanate ($BaTiO_3$) is one of the most widely used dielectric materials for capacitors. The stoichiometry is a key parameter for the dielectric property of $BaTiO_3$ and Ba/Ti mol ratio analysis is required to satisfy the precision of 0.05 C.V.% or better. This means that even the lower one of Ba and Ti fluorescence counts is required to be larger than $4 \times 10^6$ by statistics, practically both are to be of the order of $10^7$. Furthermore, Ti-K$\alpha$ (0.2750 nm, 4.509 keV) and Ba-L$\alpha$ (0.2776 nm, 4.466 keV) must be well separated from each other.

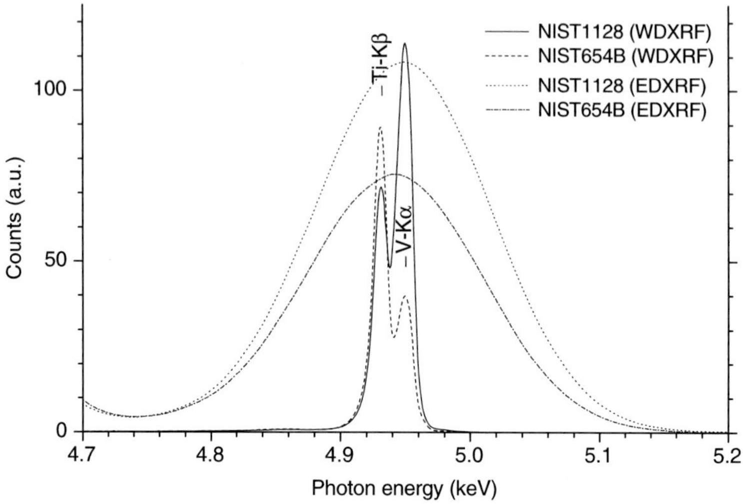


**Fig. 4.81.** Spectra of two Ti alloys taken by a WDXRF spectrometer and a EDXRF spectrometer

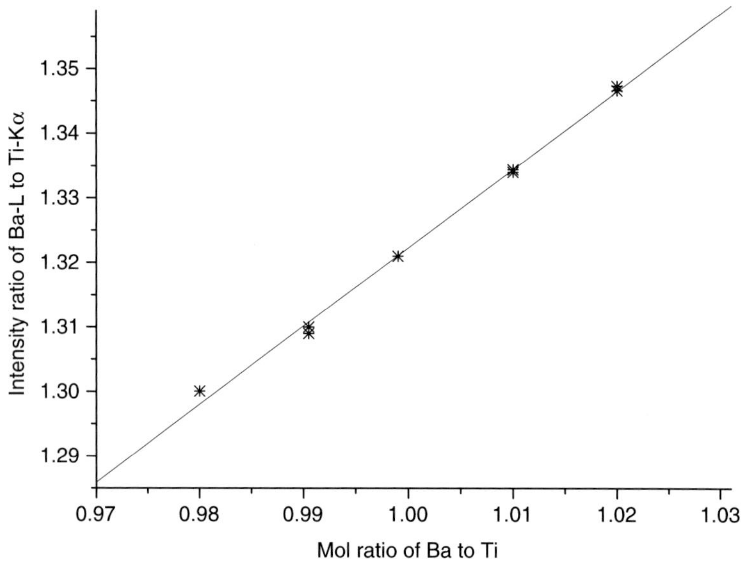


**Fig. 4.82.** Calibration line of Ba/Ti mol ratio for $BaTiO_3$ using a WDXRF spectrometer. Samples are prepared as fusion beads

Figure 4.82 shows a calibration curve of Ba/Ti ratio of barium titanate measured by using a simultaneous spectrometer equipped with high-resolution optics and a 4 kW tube. The measuring time is 200 s and each measured point of Ti-K$\alpha$ and Ba-L$\alpha$ amounts to more than $10^8$ counts (i.e., count rates are more than $5 \times 10^5$ cps). The achieved repeatability is 0.02 C.V.%. In addition, the samples are prepared as fusion beads (flux: $Li_2B_4O_7$, flux/sample ratio = 10/1). It should be mentioned that a reproducible sample preparation technique is another key issue of a precise analysis.

### 4.4.6 Comparison of WDXRF and EDXRF

As the summary of this section, the characteristics of WDXRF (**W**) and EDXRF (**E**) are summarized as follows.

#### Energy Resolution and Peak to Background(PB) Ratio

**W**: WDXRF is excellent for detecting light and middle elements. It can perform separation of proximity lines and with a high energy resolution delivers a high PB ratio. Therefore, it is suitable for analysis of a multi-ingredient sample.

**E**: As compared with WDXRF, since energy resolution is poor especially in middle and light element range, it tends to be influenced by peak overlapping. In case of the analysis of a light element that appears near the main ingredients, the tail of the main-ingredients peak serves as a strong background, and PB ratio may become poor.

#### Dynamic Range

**W**: Since the dynamic range of counting rate is wide compared to EDXRF, the measurement is possible under the same excitation conditions of X-ray source from a very-small-quantity ingredient to the main ingredient. In order to realize the high precision analysis of the main ingredients, it is necessary to suppress statistical error in X-ray counting. The upper limit of counting rate is 1 or 2 million cps or more in WDXRF system. Since this value is large compared to EDXRF, the measurement that suppresses the error becomes easy.

**E**: The saturation of counting rate of the ED detector at high-count rate restricts the performance in a certain kind of application. It is not suitable when high intensity is needed in order to take high precision measurement of the main ingredients etc.

#### Measurable Element Range

**W**: Measurable element range of WDXRF is down to beryllium owing to the ultrathin (sub-μm) polymer detector window and a synthetic multilayer

**E**: Practical element range is sodium and heavier for usual EDXRF, except for EDXRF with special specification.

### Equipment Composition and Dimension

**W**: Generally equipment size becomes large and is expensive because of the using of an optical system for wavelength dispersion and high power unit for X-ray tube.

**E**: Construction of EDXRF equipment is simple and can be miniaturized. Portable types and even handheld types have been already realized.

### Detecting Efficiency, Smallest Limit of Sample Size

**W**: Since the X-ray acceptance solid angle from a sample is small because of the using of a collimator system including a monochromator crystal that has low reflectivity of X-rays, the detection efficiency of fluorescent X-rays from a sample is very small. Therefore, generally it is unsuitable for measurement of feeble fluorescent X-rays, i.e., analysis of a small quantity sample or a small area. Since detection efficiency is so small, a high power source of X-rays is required.

**E**: It is possible to make the detector position closer to a sample and the acceptance solid angle of the fluorescent X-rays from the sample can be enlarged. Therefore high sensitivity measurement for a small area or a small amount of sample quantity can be performed.

## References

1. Rutherford E, Geiger F, Proc R Soc **81**, 141 (1908)
2. Fiorini C, Longoni A, Perotti F, Labanti C, Rossi E, Lechner P, Soltau H, Strüder L, A monolithic array of silicon drift detectors coupled to a single scintillator for $\gamma$-ray imaging with sub-millimeter position resolution, Nucl Instrum Meth A **512**, 265–271 (2003)
3. Pell EM, J Appl Phys **31**, 1675 (1960)
4. Strüder L, Meidinger N, Stotter D, Kemmer J, Lechner P, Leutenegger P, Soltau H, Eggert F, Rhode M, Schülein T, High-resolution X-ray spectroscopy close to room temperature, Microsc Microanal **4**, 622–631 (1999)
5. Angloher G, Hettl P, Huber M, Jochum J, v. Feilitzsch F, Mößbauer RL, Energy resolution of 12 eV at 5.9 keV from Al-superconducting tunnel junction detectors, J Appl Phys **89**, 1425–1429 (1999)
6. Peacock A, Verhoeve P, Rando N, van Dordrecht A, Taylor BG, Erd C, Perryman MAC, Venn R, Howlett J, Goldie DJ, Lumley J, Wallis M, Single photon detection with a superconducting tunnel junction, Nature **381**, 135–137 (1996); Peacock A, On the development of superconducting tunnel junctions as photon counting spectrometers for applications in astrophysics, Physica B **263–264**, 595–603 (1999)
7. Irwin KD, Hilton GC, Martinis JM, Deiker S, Bergren N, Nam SW, Rudman DA, Wollman DA, A Mo–Cu superconducting transition-edge microcalorimeter with 4.5 eV energy resolution at 6 keV, Nucl Instrum Meth A **444**, 184–187 (2000)

8. Angloher G, Low-temperature detectors, Nucl Instrum Meth A **512**, 401–407 (2003)
9. den Hartog R, Verhove P, Martin D, Rando N, Peacock A, Krumrey M, Goldie DJ, An X-ray photon-counting imaging spectrometer based on a Ta absorber with four superconducting tunnel junctions, Nucl Instrum Meth A **444**, 278–282 (2000)
10. Frank M, Friedrich S, Höhne J, Jochum J, Cryogenic detectors and their application to X-ray fluorescence analysis, J X-ray Sci Technol **11**, 83–112 (2003)
11. Bechstein S, Beckhoff B, Fliegauf R, Weser J, Ulm G, Characterization of an $Nb/Al/AlO_x/Al/Nb$ superconducting tunnel junction detector with a very high spatial resolution in the soft X-ray range, Spectrochim Acta Part B **59**, 215–221 (2004)
12. Eggert T, Goldstrass P, Kemmer J, Pahlke A, Analysis of background events in Silicon Drift Detectors. Nucl Instrum Meth A **512**, 257–264 (2003)
13. Berger MJ, Hubbel JH, XCOM: photon cross sections on a personal computer, US Dept of Commerce, NBSIR 87–3597
14. Measurements from the authors
15. MOXTEK, Inc, Orem, Utah, USA
16. Lechner P, Strüder L, Ionization statistics in silicon X-ray detectors – new experimental results, Nucl Instrum Meth A **354**, 464–474 (1995)
17. Fano U, Phys Rev **27**, 26 (1947)
18. Debertin D, Helmer RG, Gamma- and X-ray Spectrometry with Semiconductor Detectors, North-Holland, Amsterdam (1988)
19. Knoll G, Radiation Detection and Measurement, 3rd edn, Wiley, New York (1999)
20. dos Santos JMF, et al., The development of portable gas proportional scintillation counters for X-ray spectrometry, X-Ray Spectrom **30**, 373–381 (2001)
21. Lutz G, Semiconductor Radiation Detectors, Springer, Berlin (1999)
22. Landolt-Börnstein, Semiconductors, Physics of Group IV Elements and III–V Compounds, vol. **III/17a**, Springer, Berlin (1982)
23. Nashashibi T, White G, IEEE Trans Nucl Sci **NS-37**(No. 2), 452 (1990)
24. Rossington CS, Giauque RD, Jaklevic JM, A direct comparison of Ge Na Si(Li) detectors in the 2–20 keV range, IEEE Trans Nucl Sci **39**(No. 4), 570–576 (1992)
25. Amptek, Inc., Bedford, MA, USA
26. Wang YJ, Iwanczyk JS, Graham WR, Evaluation of $HgI_2$ detectors for lead detection in paint, IEEE Trans Nucl Sci **40**, 846 (1993)
27. Squillante MR, Entine G, New applications of CdTe nuclear detectors, Nucl Instrum Meth **A 322**, 569 (1992)
28. McConnell ML, et al., Three-dimensional imaging and detection efficiency performance of orthogonal coplanar CZT strip detectors, Proc SPIE **4141**, 157 (2000)
29. Bertuccio G, et al., Pixel X-ray detectors in epitaxial gallium arsenide with high-energy resolution capabilities (Fano factor experimental determination), IEEE Trans Nucl Sci **44**, 1–5 (1997)
30. Gatti E, Rehak P, Semiconductor drift chamber – An application of a novel charge transport scheme, Nucl Instrum Meth **A 225**, 608–614 (1984)
31. Rehak P, et al., Semiconductor drift chambers for position and energy measurements, Nucl Instrum Meth A **235**, 224–234 (1985)

32. Takahshi J, et al., Silicon drift detectors for the STAR/SVT experiment at RHIC, Nucl Instrum Meth A **439**, 497–506 (2000)
33. Holl P, et al., A 55 $cm^2$ cylindrical silicon drift detector, Nucl Instrum Meth A **377**, 367–374 (1996)
34. Bertuccio G, Castoldi A, Longoni A, Sampietro M, Gauthier C, New electrode geometry and potential distribution for X-ray drift detectors, Nucl Instrum Meth A **312**, 613 (1992)
35. Kemmer J, Lutz G, Belau E, Prechtel U, Welser W, Low capacity drift diode, Nucl Instrum Meth A **253**, 378–381 (1987)
36. Lechner P, et al., Silicon drift detectors for high resolution room temperature X-ray spectroscopy, Nucl Instrum Meth A **377**, 346–351 (1996)
37. Lechner P, et al., New DEPMOS applications, Nucl Instrum Meth A **326**, 284–289 (1993)
38. Hartmann R, et al., Ultrathin entrance windows for silicon drift detectors, Nucl Instrum Meth A **387**, 250–254 (1997)
39. Leutenegger P, et al., Silicon drift detectors as radiation monitor for x-, gamma rays and particle, Proc SPIE **4012**, 579 (2000)
40. Radeka V, et al., Implanted silicon JFET on completely depleted high-resistivity devices, IEEE Electron Device Lett **10**(No. 2) 91 (1989)
41. Kemmer J, et al., Experimental confirmation of a new semiconductor detector principle, Nucl Instrum Meth A **288**, 92 (1990)
42. Pinotti E, et al., The pn-CCD on-chip electronics, Nucl Instrum Meth A **326**, 85–91 (1993)
43. Sampietro M, Fasoli L, Rehak P, Strüder L, Novel pJFET embedded in silicon radiation detectors that avoids preamplifier feedback resistor, IEEE Electron Device Lett **16**(No. 5), 142 (1995)
44. Fiorini C, Lechner P, Continuous charge restoration in semiconductor detectors by means of the gate-to-drain current of the integrated front-end JFET, IEEE Trans Nucl Sci **46**(No. 3), 761 (1999)
45. Strüder L, et al., Room temperature x- and gamma-ray spectroscopy with silicon drift detectors, Proc SPIE **4141**, 29 (2000)
46. Lechner P, Pahlke A, Soltau H, Novel high-resolution silicon drift detectors, X-Ray Spectrom **33**(No. 4), 256–261 (2004)
47. Gauthier C, et al., A high-resolution silicon drift chamber for X-ray spectroscopy, Nucl Instrum Meth A **349**, 258 (1994)
48. Lechner P, et al., Multichannel silicon drift detectors for X-ray spectroscopy, Proc SPIE **4012**, 592 (2000)
49. Gauthier Ch, et al., A high resolution, 6 channels, silicon drift detector array with integrated JFET's designed for XAFS spectroscopy: first X-ray fluorescence excitation spectra recorded at the ESRF, Nucl Instrum Meth A **382**, 524–532 (1996)
50. Longoni A, Fiorini C, Guazzoni C, Gianoncelli A, Strüder L, Soltau H, Lechner P, Bjeoumikhov A, Schmalz J, Langhoff N, Wedell R, A New XRF Spectrometer Based on a Ring-Shaped Multi-Element Silicon Drift Detector and on X-Ray Capillary Optics IEEE Trans on Nucl Sci, **49**, 1001–1005 (2002)
51. Fiorini C, Longoni A, Application of a new non-cryogenic X-ray detector in portable instruments for archaeometric analyses, Rev Sci Instrum **69**(No. 3) 1523 (1998)

52. Fiorini C, Longoni A, Perotti F, Labanti C, Lechner P, Strüder L, Gamma ray spectroscopy with CsI(Tl) scintillator coupled to silicon drift chamber, IEEE Trans Nucl Sci **44**(No. 6), 2553, (1997)
53. Fiorini C, Longoni A, Perotti F, Labanti C, Rossi E, Lechner P, Strüder L, Detectors for high-resolution gamma-ray imaging based on a single CsI(Tl) scintillator coupled to an array of silicon drift detector array, IEEE Trans Nucl Sci **48**, No. 3, 346–350 (2001)
54. Nicholson PW, Nuclear Electronics, New York, NY (1974)
55. Gatti E, Manfredi PF, Processing the signals from solid-state detectors in elementary-particle physics, Riv Nuovo Cimento **9**(3) (1986)
56. Radeka V, Low-noise techniques in detectors, Annu Rev Nucl Part Sci **38**, 217 (1988)
57. Radeka V, Karlovac N, Nucl Instrum Meth **52**, 86 (1967)
58. Gatti E, Manfredi PF, Nucl Instrum Meth **226**, 142 (1984)
59. De Geronimo G, Longoni A, IEEE Trans Nucl Sci **45**(No. 3), 1656, (1998)
60. Bertuccio G, Pullia A, De Geronimo G, Nucl Instrum Meth **380**, 301 (1996)
61. Gatti E, Sampietro M, Manfredi F, Nucl Instrum Meth **287**, 513 (1990)
62. Fiorini C, Pullia A, Gatti E, Longoni A, Buttler W, IEEE Trans Nucl Sci **47**(No. 3), 823–828 (2000)
63. Smith SW, The Scientist and Engineer's Guide to Digital Signal Processing, California Technical Publishing, San Diego, California (1999)
64. Jordanov VT, Knoll GF, Digital synthesis of pulse shapes in real time for high resolution radiation spectroscopy, Nucl Instrum Meth A **345**, 337–345 (1994)
65. Bertuccio G, Gatti E, Sampietro M, Rehak P, Rescia S, Sampling and optimum processing of detector signals, Nucl Instrum Meth A **322**, 271–279 (1992)
66. Gatti E, Geraci A, Ripamonti G, Automatic synthesis of optimum filters with arbitrary constraints and noises: a new method, Nucl Instrum Meth A **381**, 117–127 (1996)
67. Rando N, Peacock A, Favata F, Perryman M, S-Cam: an imaging spectrophotometer based on superconducting tunnel junctions, Exp Astron **10**(4), 499–517 (2000)
68. Soltau H, Holl P, Krisch S, von Zanthier C, Hauff D, Richter R, Bräuninger H, Hartmann R, Hartner G, Krause N, Meidinger N, Pfeffermann E, Reppin C, Schwaab G, Strüder L, Trümper J, Kendziorra E, Krämer J, Performance of the pn-CCD X-ray detector system designed for the XMM satellite mission, Nucl Instrum Meth A **377**, 340–345 (1996)
69. Holland A, et al., X-ray spectroscopy using MOS CCDs, Nucl Instrum Meth A **377**, 334–339 (1996)
70. van den Berg ML, et al., Back illuminated CCDs made by gas immersion laser doping, Nucl Instrum Meth A **377**, 312–320 (1996)
71. Strüder L, Holl P, Lutz G, Kemmer J, Development of fully depletable CCDs for high energy Physics Applications, Nucl Instrum Meth A **257**, 594–602 (1987)
72. Pinotti E, Bräuninger H, Findeis N, Gorke H, Hauff D, Holl P, Kemmer J, Lechner P, Lutz G, Kink W, Meidinger N, Metzner G, Predehl P, Reppin C, Strüder L, Trümper J, von Zanthier C, Kendziorra E, Staubert R, Radeka V, Rehak P, Bertuccio G, Gatti E, Longoni A, Pullia A, Sampietro M, The pn-CCD on-chip electronics, Nucl Instrum Meth A **326**(1,2), 85–92 (1993)

73. Meidinger N, Bräuninger H, Briel U, Hartmann R, Hartner G, Holl P, Kemmer J, Kendziorra E, Krause N, Lutz G, Pfeffermann E, Popp M, Reppin C, Richter R, Soltau H, Stötter D, Strüder L, Trümper J, von Zanthier C, The PN-CCD detector for XMM and ABRIXAS, SPIE **3765** 192–203 (1999)
74. Soltau H, Kemmer J, Meidinger N, Stötter D, Strüder L, Trümper J, von Zanthier C, Bräuninger H, Briel U, Carathanassis D, Dennerl K, Haberl F, Hartmann R, Hartner G, Hauff D, Hippmann H, Holl P, Kendziorra E, Krause N, Lechner P, Pfeffermann E, Popp M, Reppin C, Seitz H, Solc P, Stadlbauer T, Weber U, Weichert U, Fabrication, test and performance of very large X-ray CCDs designed for astrophysical applications, Nucl Instrum Meth A **439**(2,3), 547–559 (2000)
75. Strüder L, Bräuninger H, Briel U, Hartmann R, Hartner G, Hauff D, Krause N, Maier B, Meidinger N, Pfeffermann E, Popp M, Reppin C, Richter R, Stötter D, Trümper J, Weber U, Holl P, Kemmer J, Soltau H, Viehl A, von Zanthier C, A 36 $cm^2$ large Monolythic pn-CCD X-ray Detector for the European XMM Satellite Mission, Rev Sci Instrum **68**(11), 4271–4274 (1997)
76. Strüder L, Bräuninger H, Meier M, Predehl P, Reppin C, Sterzik M, Trümper J, Cattaneo P, Hauff D, Lutz G, Schuster KF, Schwarz A, Kendziorra E, Staubert A, Gatti E, Longoni A, Sampietro M, Radeka V, Rehak P, Rescia S, Manfredi PF, Buttler W, Holl P, Kemmer J, Prechtel U, Ziemann T, The MPI/AIT X-ray imager — high speed pn-CCDs for X-ray detection, Nucl Instrum Meth A **288**, 227–235 (1990)
77. Hartmann R, Strüder L, Kemmer J, Lechner P, Fries O, Lorenz E, Mirzoyan R, Ultrathin entrance windows for silicon drift detectors, Nucl Instrum Meth A **387**(1,2), 250–254 (1997)
78. Pfeffermann E, Meidinger N, Bräuninger H, Hartmann R, Hartner G, Krause N, Metzner G, Popp M, Reppin C, Riedl J, Stötter D, Strüder L, Trümper J, Weber U, Hauff D, Holl P, Kemmer J, Krisch S, Soltau H, von Zanthier C, Bihler E, Böttcher H, Kendziorra E, Krämer J, Pflüger B, Staubert R, PN-CCD detector for XMM and ABRIXAS, SPIE **3765**, 184–191 (1999)
79. Krause N, Soltau H, Hauff D, Kemmer J, Stötter D, Strüder L, Weber J, Metal contamination analysis of the epitaxial starting material for scientific CCDs, Nucl Instrum Meth A **439**, 228–238 (2000)
80. Meidinger N, Schmalhofer B, Strüder L, Particle and X-ray damage in pn-CCDs, Nucl Instrum Meth A **439**, 319–336 (2000)
81. Buttler W, Lutz G, Cesura G, Manfredi PF, Speziali V, Tomasini A, Short channel, CMOS-compatible JFET in low noise applications, Nucl Instrum Meth A, **326**(1,2), 63–70 (1993)
82. Strüder L, Englhauser J, Hartmann R, Holl P, Meidinger N, Soltau H, Briel U, Dennerl K, Freyberg M, Haberl F, Hartner G, Pfeffermann E, Stadlbauer T, Kendziorra E, pnCCDs on XMM-Newton—42 months in orbit, Nucl Instrum Meth A **512**, 386–400 (2003)
83. Strüder L, High resolution imaging X-ray spectrometers, Nucl Instrum Meth A **454**, 73–113 (2000)
84. Meidinger N, Bonerz S, Bräuninger H, Eckhardt R, Englhauser J, Hartmann R, Hasinger G, Holl P, Krause N, Lutz G, Pfeffermann E, Richter R, Soltau H, Strüder L, Trümper J, Frame store pn-CCD detector for the ROSITA mission, SPIE **4851**, 1040–1047 (2002)

85. Wiesemann U, Thieme J, Guttmann P, Niemann B, Rudolph D, Schmahl G, New scanning transmission X-ray microscope at BESSY II, Proc Sixth Int Conf on X-ray microscopy, AIP (2000)
86. Bertschinger G, Biel W, Herzog O, Weinhammer J, Kunze H-J, Bitter M, X-ray spectroscopy at the TEXTOR–94 Tokamak, Phys Script **T83**, 132–141 (1999)
87. Lindner F, Stremme W, Schätzel MG, Grasbon F, Paulus GG, Walther H, Hartmann R, Strüder L, High-order harmonic generation at a repetition rate of 100 kHz, Phys Rev A **68**, 013814 (2003)
88. Ronning C, Vetter U, Uhrmacher M, Hofsäss H, Bharut-Ram K, Hartmann R, Strüder L, Electron emission channeling spectroscopy using X-ray CCD detectors, Nucl Instrum Meth A **512**(1,2), 378–385 (2003)
89. Lutz G, Bräuninger H, Englhauser J, Hartmann R, Kang D, Kotthaus R, Kuster M, Serber W, Strüder L, An application of space technology to the terrestrial search for axions: the X-ray mirror telescope at CAST, Nucl Instrum Meth A **518**(1,2), 201–206 (2004)
90. Hagenbuck F, Backe H, Clawitter N, Euteneuer H, Görgen F, Holl P, Johann K, Kaiser K-H, Kemmer J, Kerschner T, Kettig O, Koch H, Kube G, Lauth W, Mattäy H, Schütrumpf M, Stötter R, Strüder L, Walcher T, Wilms A, von Zanthier C, Zemter M, Novel digital K-edge imaging system with transition radiation from an 855 MeV electron beam, IEEE–Trans Nucl Sci **48**(3), 843–848 (2001)
91. Spiller E, Soft X-ray Optics, SPIE press, Bellingham, USA (1994)
92. Yamamoto M, Namioka T, Layer-by-layer design method for soft-X-ray multilayers Appl Opt **31**, 1622–1630 (1992)
93. Henke BL, Gullikson EM, Davis J, X-ray interactions: photoabsorption, scattering, transmission, and reflection at E = 50-30000 eV, Z = 1-92. Atomic Data and Nucl Data Tables **54**, 181–342 (1993)
94. http://www.cxro.lbl.gov/multilayer/survey.html
95. http://www.osmic.com/
96. Kato M, X-sen Bunko Bunseki (X-ray spectrochemical analysis), Uchida Rokakuho Publishing, Tokyo, 152 (1998)
97. Ohashi K, Iida A, Gohshi Y, Kishimoto S, Takahashi M, Synchrotron radiation X-ray fluorescence analysis with a crystal spectrometer Adv X-ray Anal **35**, 1027–1033 (1992)
98. Soller W, A new precision X-ray spectrometer Phys Rev **24**, 158–167 (1924)
99. DuMond JWM, Theory of the use of more than two successive X-ray crystal reflections to obtain increased resolving power Phys Rev **52**, 872–883 (1937)
100. Barraud J, Monochromateur-focalisateur donnant un faisceau d'ouverture notable. C R Acad Sci Paris **214**, 795–797 (1942); de Wolff PM, An adjustable curved crystal monochromator for X-ray diffraction analysis Appl Sci Res **B1**, 119–126 (1950)
101. Johansson T, Selektive Fokussierung der Röentgenstrahlen Naturwiss **20**, 758–759 (1932)
102. Johann HH, Die Erzeugung lichtstarker Röentgenspektren mit Hilfe von Konkavkristallen Z Phys **69**, 185–206 (1931)
103. Gohshi Y, Kamada H, Kohra K, Utaka T, Arai T, Wide range two-crystal vacuum X-ray spectrometer for chemical state analysis Appl Spectroscopy **36**, 171–174 (1982)

104. Gohshi Y, Hirao O, Suzuki I, Chemical state analysis of sulfur, chromium and tin by high resolution X-ray spectrometry Adv X-ray Anal **18**, 406–414 (1975)
105. Yamashita M, Motoyama M, Fukushima S, Vlaicu AM, Nisawa A, Ozaki S, Ito Y, Shigeoka N, Ohashi K, X-ray emission spectrum analysis of CO in Li-ion battery using a high-resolution XRF Adv X-ray Chem Anal Japan **35**, (2004) [to be published]
106. Kohno H, Arai T, Araki Y, Wilson R, High accuracy analysis of BPSG thin films on silicon wafers by X-ray wafer analyzer Adv X-ray Anal **37**, 229–234 (1993)
107. Lee FS, Campbell WJ, Variation of X-ray spectral line position with ambient-temperature change: a source of error in X-ray spectrography Adv X-ray Anal **8**, 431–442 (1964)
108. CRC Handbook of chemistry and physics, 75th edition, CRC Press, Boca Raton, USA, 12-87–12-96 (1995)
109. International tables for X-ray crystallography, volume III, Kynoch press, Birmingham, 125–129 (1962)
110. Measured data taken at Rigaku Industrial Corp, Unpublished

# 5

# Quantitative Analysis

## 5.1 Overview

*M. Mantler*

The observed photon rate from an analyte element in a specimen is a function of many factors. Most notably it depends on the weight fraction (in bulk materials) or mass (in thin layers) of that element, the matrix (i.e., the accompanying elements), the specimen type (bulk, powder, liquid, thin film structure, etc.), irradiated size, shape, thickness, method of manufacturing and preparation, geometrical setup of the spectrometer, flux and spectral distribution of the exciting radiation, and the detection system. In most cases the analytical goal of quantitative XRF is the determination of concentrations of elements (or stoichiometric compounds), but in many applications the thickness of single or multiple layers, or individual thicknesses in layered structures are of interest as well, possibly in addition to the concentrations in each layer and/or the substrate. Theoretical as well as empirical approaches are available and used in quantitative XRF (Table 5.1).

The theoretical methods are based on mathematical models for the excitation of atoms and subsequent relaxation processes, the absorption (attenuation) of radiation within the specimen by the analyte atoms and the matrix, and possibly indirect excitation effects by certain matrix atoms. Simplifying assumptions are often made, for example that the specimen is perfectly flat and homogeneous, and that each, incident and fluorescent beam, is parallel. Because of the involved basic physical parameters, such as attenuation coefficients, transition probabilities, and fluorescent yields, this is called the fundamental parameter (FP) method. An introduction to FP theories based on the mathematical foundations laid by Sherman [51, 52] and by Shiraiwa and Fujino [53] is given in Sect. 5.2. This is then extended to the analysis of multiple layer structures (Sect. 5.5), and a discussion of light elements analysis with their specific indirect excitation effects (Sect. 5.6).

**Table 5.1.** Main applications of the various analytical methods

| Method | Bulk | Single layers | Multiple layers | Inhomogeneous materials |
|---|---|---|---|---|
| EIC | Yes | Very limited | No | No |
| TIC | Yes | No | No | No |
| FP | Yes | Yes | Yes | No |
| MC | Possible | Possible | Possible | Yes |

(EIC: empirical influence coefficients, TIC: theoretical influence coefficients, FP: classical fundamental parameters, MC: Monte Carlo)

An alternative is to use empirical methods, which employ relatively simple mathematical descriptions of the relationship between photon counts and concentration (working curves or calibration curves). They may be limited in their applicability to a certain range of concentrations and work best with carefully selected and prepared sets of standard specimens of similar composition as the unknown specimen. Methods with empirical calibration curves can also be used for the determination of the thickness of (single) thin layers.

The concept of *influence coefficients* is used to describe the relationship between count rates and concentrations in a way that directly reveals the effect of the matrix on the count rates from the analyte element (Sect. 5.3). The general principle is that the "ideal" calibration curve is assumed to be in principle linear, however affected by matrix effects and thereby distorted (i.e., made nonlinear). By applying corresponding counteracting "corrections," the experimental curve is again linearized. The coefficients, which account for the extent of these (mathematical) corrections for each matrix element, are called *empirical* or *theoretical influence coefficients*, respectively, depending on the method of their determination.

The accuracy of empirical methods (where the influence coefficients are determined by measuring standards) can be very high and is only limited by experimental issues (mainly the repeatability and reproducibility of the measurements and the specimen preparation procedure) and the quality of standards (reliability of the provided data and a close similarity of their composition and preparatory method to that of the unknown material). In practice, it is often difficult (and expensive) to maintain the required quality and sometimes impossible to obtain or manufacture the standard specimens.

Theoretical influence coefficient methods are based on the classical set of FPs and their accuracy is comparable to that of any other FP method. Since the main time-consuming step is the computation of the influence coefficients, routine analysis with precomputed sets of coefficients is very fast. Compared to the empirical parameter method, the accuracy can be similar if the same set of standards is used.

The high long-term stability of modern instruments allows precalibration for all elements in the factory with simple drift corrections on site. Analysis can thereby be carried out without actual standards (Sect. 5.7).

Monte Carlo methods are sometimes employed for analyses of inhomogeneous materials and for studying complex indirect excitation effects, such as the excitation of light elements by photo-electrons and Auger-electrons. They are by their nature an FP method, but based on a statistical description of interactions rather than on a deterministic equation system. A brief introduction to the principles of the method is given in Sect. 5.8.

An important aspect of any analytical task is to estimate the resulting error in view of the chosen methodology (achievable trueness and precision) and the probable statistical error made in an individual measurement (Sect. 5.9).

Following the demands of modern industrial networks for precise criteria, rules, guidelines, and definitions, internationally accepted standards have been set up not only for materials and manufacturing processes but also for analytical procedures. An introduction to standardized methods in XRF, as they are used for example in the petrochemical industry and in the cement industries, is given in Sect. 5.10.

## 5.2 Basic Fundamental Parameter Equations

*M. Mantler*

Many of the roots of the FP method can been found in the early works by Barkla and Sadler [3], Beatty [5], Kramers [30], Moseley [40], Siegbahn [54], von Hevesy [59], and Glocker and Schreiber [21]. The immediate ancestors of modern methods are, however, the investigations by Gillam and Heal [20], Beattie and Brissey [4], and finally Sherman [51, 52] as well as Shiraiwa and Fujino [53], who laid the mathematical foundations to the FP method as it is used today. Gilfrich and Birks [19] measured and published spectral distributions of radiation from X-ray tubes, which for the first time allowed the computation of realistic numerical count-rate values from FP equations, and Criss and Birks [12] took advantage of the development of fast and affordable computers and developed the first widely available FP software package for quantitative analysis. A survey of historical landmarks in X-ray analysis has been given by Gilfrich [18].

This section introduces the basic method and the involved FPs. It covers in some mathematical detail a description of the basic interactions, direct and indirect excitation of fluorescent radiation in bulk materials, and direct excitation in thin layers. The equations are extended in following sections dealing with multiple thin film layers [13, 36, 37], complex interactions in light element analysis [28, 29, 38], and the theoretical influence coefficients methods. In view of the complexity of the subject, the theory presented in this chapter remains nevertheless incomplete. It is limited to those components which are usually included in commercial software packages and is complemented by brief discussions of scattering, light elements, and Monte Carlo techniques. A new database of FPs for XRF has been recently compiled by Elam et al. [16], and data from various sources (mainly of attenuation coefficients) compared by Ebel et al. [15].

### 5.2.1 Fundamental Parameter Equations for Bulk Materials

#### Basic Interactions Between Photons and Atoms

*Note 1: Interaction and absorption.* In analytically oriented discussions, *interaction* is often not clearly distinguished from *absorption.* There is, however, an important difference: *interaction* encompasses all events in which a photon can participate, while *absorption* relates only to those, where a transfer of energy from a photon to an atom occurs. Photo-effect and Compton effect (incoherent scattering) transfer energy, but coherent scattering does not. Absorbed energy is responsible not only for the creation of fluorescent radiation (by photo-absorption), but also for changes in the chemical state of absorbing atoms, which are bound into molecules, due to the primary photon absorption as well as to exposure to secondary photons and electrons originating from the primary interaction. The amount of energy transferred to the material is fundamentally associated with the concept of radiation doses, which is used in health sciences to describe the effect of radiation on (human) biological systems. Gray (Gy) is the base SI unit of absorbed dose and equivalent to $1\,\mathrm{J\ kg^{-1}}$. The more generic term *attenuation coefficient* is used for unspecified interactions (integrating photo-absorption, Compton- and coherent scattering in the analytical range of X-rays), while *absorption coefficient* applies specifically to photo-absorption.

*Note 2: Photon counts and intensities.* This and the following sections deal with the excitation of fluorescent photons and present models to compute their numbers as a function of specimen properties. These numbers correspond to, or are proportional to, experimentally observed photon counts and are in practice often called *intensities.* This is in contradiction to the exact physical definition of an intensity, which refers to a quantity measured in units of ($\mathrm{W\ m^{-2}}$) or equivalent. In order to provide a consistent notation, the terms counts, count rate, and (photon-) flux should be preferred and used according to their definitions, and the term *intensity* avoided except where physically appropriate.

*Atomic cross-sections.* Within the energy range of XRF photons interact with atoms by coherent scattering, incoherent scattering, and photo-absorption. The probability $p$ that a single photon participates in any of these possible interactions with a single atom is quantified by the total atomic cross-section, $\mu_{\mathrm{Atom}}$, and depends upon the properties of the atom as well as on the photon energy. In a simplistic visual representation,[1] the atom covers a virtual area

[1]The term cross section was originally used in particle physics to describe the probability of a neutron to interact with an uranium nucleus (which is "big as a barn" compared to the tiny neutron). The apparent areal size of an uranium nucleus is around $10^{-24}$ cm$^2$

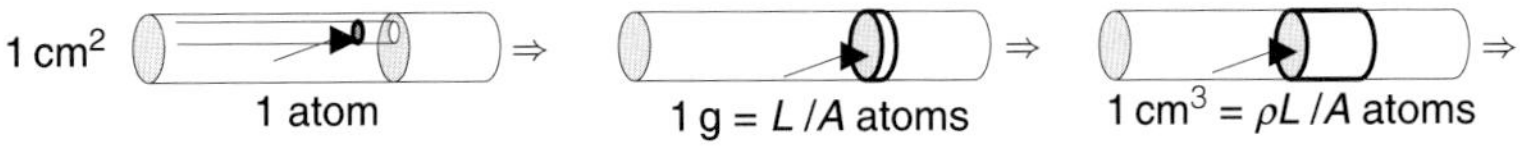


**Fig. 5.1.** Atomic cross-section, mass attenuation, linear attenuation

of $\mu_{\mathrm{Atom}}$ (cm$^2$) within a beam of monochromatic photons with a cross-section of 1 cm$^2$. The practical unit is ($10^{-24}$ cm$^2$) = (barns/atom) (Fig. 5.1).

The *total atomic cross-section* comprises all three possible ways of interaction and is thereby an *attenuation coefficient*:

$$\mu_{\mathrm{Atom,\,total}} = \tau_{\mathrm{Atom}} + \sigma_{\mathrm{Atom,\,coh}} + \sigma_{\mathrm{Atom,\,incoh}} \tag{5.1}$$

$\tau$ is the usual denotation for photo-absorption and $\sigma$ for scattering, where coherent and incoherent scattering are distinguished.

If $p_{\mathrm{Atom}}$ is the probability for a photon to interact with a single atom located anywhere within the area of 1 cm$^2$, then the presence of two atoms will evidently double the probability, and $n$ atoms increase it to $p_n = np_{\mathrm{Atom}}$. This is, however, only true for small values of $n$, and a meaningless probability $np_{\mathrm{Atom}} > 1$ would result from large values of $n$. The correct model is to virtually arrange the atoms in a row, one after the other, and to compute the probability that the first interaction of the photon occurs at or with the $n$th atom. The successful event terminates the sequence and no second or further interaction is permitted for a photon (which is absorbed and thereby annihilated, or scattered out of the beam into another direction). It is helpful to ask first for the complementary probability of *no interaction* while passing $n$ atoms, $1 - p_n = (1 - p_{\mathrm{Atom}})^n$, from which follows $p_n = 1 - (1 - p_{\mathrm{Atom}})^n$.

For very large numbers of $n$ and small values of $p_{\mathrm{Atom}}$ it is valid to apply the approximation $(1 - p_{\mathrm{Atom}})^n \approx \exp(-np_{\mathrm{Atom}})$. This term is equal to the ratio of the number of photons arriving behind an absorber, $N$, and those entering, $N_o$:

$$1 - p_n = \frac{N}{N_o} = e^{-np_{\mathrm{Atom}}}.$$

Both $n$ and $p_{\mathrm{Atom}}$ are dimensionless. If the attenuation coefficient $\mu_{\mathrm{Atom}}$ with a dimension of (cm$^2$) is substituted for $p_{\mathrm{Atom}}$, $n$ must be replaced by a matching variable of dimension (cm$^{-2}$), $\hat{n}$, which specifies the number of atoms per cm$^2$ and ensures a dimensionless exponent:

$$1 - p_n = \frac{N}{N_o} = e^{-\hat{n}\mu_{\mathrm{Atom}}}.$$

*Mass attenuation and linear attenuation.* Since there are $L/A$ atoms in 1 g of atoms ($L$ is Avogadro's number and $A$ the atomic weight or relative atomic mass), a *mass attenuation coefficient* can be defined by $\mu_{\mathrm{Mass}} = \mu_{\mathrm{Atom}} L/A$ (cm$^2$ g$^{-1}$). By introducing $\hat{m} = m/A$ for the mass thickness (mass $m$ per area

$\Lambda$; the unusual symbol $\Lambda$ is used while $A$ stands for the atomic weight), one obtains $\hat{n} = \hat{m}L/A$, and:

$$\hat{n}\mu_{\text{Atom}} = \hat{m}\mu_{\text{Atom}}\frac{L}{A} = \hat{m}\mu_{\text{Mass}}$$

$$\frac{N}{N_0} = e^{-\hat{m}\mu_{\text{Mass}}}.$$

Along a path of 1 cm, there are $\rho \cdot L/A$ atoms per cm$^2$ ($\rho$ is the density of the material in g cm$^{-3}$) interacting with the photons. We define

$$\mu_{\text{linear}} = \mu_{\text{Mass}}\rho = \mu_{\text{Atom}}\rho\frac{L}{A}\ (\text{cm}^{-1})$$

$\mu_{\text{linear}}$ is called *linear attenuation coefficient.* It is used to compute the attenuation along a path length $T$ (cm) by $\hat{n} = T\rho L/A$ atoms/cm$^2$:

$$\hat{n}\mu_{\text{Atom}} = T\mu_{\text{Atom}}\rho\frac{L}{A} = T\mu_{\text{linear}}$$

$$\frac{N}{N_0} = e^{-T\mu_{\text{linear}}}.$$

Symbols for $\mu_{\text{Atom}}$, $\mu_{\text{Mass}}$, and $\mu_{\text{linear}}$ are not used in a standardized manner in the literature. We write $\bar{\mu} = \mu_{\text{linear}}$ for the linear coefficient and $\mu = \bar{\mu}/\rho$ for the mass attenuation coefficient (as well as $\bar{\tau}$, $\bar{\sigma}_{\text{coh}}$, $\bar{\sigma}_{\text{incoh}}$, and $\tau$, $\sigma_{\text{coh}}$, $\sigma_{\text{incoh}}$, respectively). Literature data are generally given as mass attenuation coefficients (cm$^2$ g$^{-1}$) or occasionally, in more theoretically inclined publications, as atomic cross-sections (barns/atom).

*Attenuation (absorption) laws.* The photon counts $N$ after an attenuator (absorber) in the equations given earlier are obviously a function of penetrated thickness, $T$, or mass thickness, $\hat{m}$:

$$N(T) = N_0 e^{-\bar{\mu}T} \quad \text{and} \quad N(\hat{m}) = N_0 e^{-\mu\hat{m}}. \tag{5.2a}$$

These equations describe the (*integral*) *attenuation* (*absorption*) *law.*

The differentially small amount of photons, d$N$, which participate in an interaction while penetrating a differentially small path length, d$T$ (or mass thickness, d$\hat{m}$) is obtained by differentiation of $N(T)$ and $N(\hat{m})$, respectively:

$$\mathrm{d}N = -N\bar{\mu}\mathrm{d}T \quad \text{and} \quad \mathrm{d}N = -N\mu\mathrm{d}\hat{m}. \tag{5.2b}$$

This is the *differential attenuation* (*absorption*) *law.* We distinguish the differentially small number d$N$ from a small real number (as it is used in numerical calculations) by using the symbol $\Delta$ instead of the differential operator, d, and omitting the negative sign:

$$\begin{aligned} \Delta N &= N\bar{\mu}\Delta T = N\mu\Delta\hat{m} = N\mu_{\text{Atom}}\Delta\hat{n} \\ p_n &= \Delta N/N = \bar{\mu}\Delta T = \mu\Delta\hat{m} = \mu_{\text{Atom}}\Delta\hat{n} \end{aligned}. \tag{5.2c}$$

Note that because $\Delta T$, $\Delta\hat{m}$, and $\Delta\hat{n}$ are inherently associated with small numbers of atoms, $p_n$ is again the *probability* for an interaction with the atoms along the path through $\Delta T$, $\Delta\hat{m}$, or $\Delta\hat{n}$. This follows also from the integral attenuation law by applying the approximation $\exp(-\alpha) \approx 1 - \alpha$ for $\alpha \ll 1$:

$$\begin{aligned}1 - p_n &= \exp(-\hat{n}\mu_{\text{Atom}}) \approx 1 - \hat{n}\mu_{\text{Atom}} \quad \text{for} \quad \hat{n}\mu_{\text{Atom}} \ll 1\\ &= \exp(-T\bar{\mu}) \approx 1 - T\bar{\mu} \quad \text{for} \quad T\bar{\mu} \ll 1\\ &= \exp(-\hat{m}\mu) \approx 1 - \hat{m}\mu \quad \text{for} \quad \hat{m}\mu \ll 1.\end{aligned}$$

*Attenuation by materials composed of several kinds of atoms.* Real materials, such as chemical compounds, alloys, or (homogenized) mixtures, often consist of several elements. If the fractional numbers of atoms, $C_{j,\text{Atom}}$, (where $C_{j,\text{Atom}} \cdot N$ is the number of $j$-atoms and $\Sigma C_{j,\text{Atom}} = 1$) are known for each element, $j$, then the total (average) atomic attenuation coefficient of the material is the weighted sum of the individual atomic attenuation coefficients:

$$\mu_{\text{Atom}} = \sum_j C_{j,\text{Atom}}\mu_{j,\text{Atom}}. \tag{5.3a}$$

If instead of atomic fractions the mass fractions (concentrations), $C_j$, of the components are known ($\Sigma C_j = 1$), then the mass attenuation coefficient of the composed material is the weighted sum of the individual mass attenuation coefficients, $\mu_j$, with weight fractions as weight factors:

$$\mu = \sum_j C_j\mu_j. \tag{5.3b}$$

The same applies to $\tau$ and $\sigma$ (and $\tau_{\text{Atom}}$ and $\sigma_{\text{Atom}}$). It should be noted that for simple mixtures, the density of the composed material is given by:

$$\frac{1}{\rho} = \sum_j C_j \frac{1}{\rho_j}.$$

However, the density of chemical compounds or crystalline materials cannot usually be accurately computed from the densities of the atomic constituents.

*Photo-absorption in a specific shell or subshell.* The photo-absorption coefficient $\tau$ is composed of the absorption coefficients of the individual shells and in general given by: $\tau = \tau_{\text{K}} + \tau_{\text{L1}} + \tau_{\text{L2}} + \tau_{\text{L3}} + \tau_{\text{M1}} + \cdots$ These coefficients for the individual shells (for example, $\tau_{\text{K}}$) are usually not readily known, but the ratios $\tau_{\text{K}}/\tau$, $\tau_{\text{L1}}/\tau$, $\tau_{\text{L2}}/\tau$, etc. can be computed from tabulated values of *absorption edge jump ratios*, $S_{\text{K}}, S_{\text{L1}}, S_{\text{L2}}$, etc. They are defined by the ratio of the photo-absorption coefficients at the higher and lower energy sides of an edge, for example for the K-edge:

$$S_{\text{K}} = \frac{\tau}{\tau - \tau_{\text{K}}} \quad \text{or} \quad \tau_{\text{K}} = \frac{S_{\text{K}} - 1}{S_{\text{K}}}\tau \tag{5.4a}$$

or for the $\text{L}_2$-edge:

$$S_{\text{L2}} = \frac{\tau - \tau_{\text{K}} - \tau_{\text{L1}}}{\tau - \tau_{\text{K}} - \tau_{\text{L1}} - \tau_{\text{L2}}} \quad \text{or} \quad \tau_{\text{L2}} = \frac{S_{\text{L2}} - 1}{S_{\text{K}}S_{\text{L1}}S_{\text{L2}}}\tau \tag{5.4b}$$

and correspondingly for all other subshells. These equations are valid for the case that $E_{\text{K-edge}} < E_{\text{Photon}}$. For cases where $E_{\text{L1-edge}} < E_{\text{Photon}} < E_{\text{K-edge}}$, set $\tau_K = 0$ and $S_K = 1$; for $E_{\text{L2-edge}} < E_{\text{Photon}} < E_{\text{L1-edge}}$, set $\tau_K = \tau_{L1} = 0$ and $S_K = S_{L1} = 1$, etc.

*Excitation and relaxation of an atom.* Photo-absorption is characterized by a complete energy transfer of the photon's energy to the atom, whereby the photon is annihilated. In turn, the atom emits a (photo-) electron, which carries a kinetic energy $E_{\text{kin}} = E_{\text{Photon}} - E_{\text{Binding}}$, as well as a set of characteristic photons and (Auger-) electrons with altogether a sum energy of $E_{\text{Binding}}$. Given a particular X-ray photon energy and kind of atom, this effect favors interactions with electronic shells for which $E_{\text{kin}}$ is small but $E_{\text{Photon}} > E_{\text{Binding}}$ (i.e., photon energies just above an absorption edge are most effectively absorbed by that shell).

This differs from optical excitation where no initial ionization occurs ($E_{\text{Photon}} < E_{\text{Binding}}$) and the energy of the absorbed photon must be *exactly* the energy difference between two electronic states of the atom, often the ground state and an excited state. Relaxation leads then to the emission of a photon with the same energy as the originally absorbed photon, however into a random direction. Such processes are not entirely impossible for X-rays, but highly improbable.

The excited (ionic) atomic state relaxes by refilling the vacancy from an outer shell. This is often associated with the emission of a characteristic photon and the creation of a new vacancy, which is in turn filled from a further outer shell under emission of another photon. This cascade continues until all allowed transitions are exhausted. Finally, a free electron fills the remaining vacancy in an outer shell.

The transitions are governed by quantum-mechanical selection rules. Within the regime of X-rays, fluorescent radiation is observed only from those transitions with reasonable intensity (probability) where

$$\Delta j = -1 \quad \text{Or} \quad \Delta j = +1 \quad \text{Or} \quad \Delta j = 0 \quad \text{and} \quad \Delta l = -1 \quad \text{Or} \quad \Delta l = +1 \tag{5.5}$$

($l$ and $j$ are the quantum numbers with the usual meaning). For example, if the K-shell was ionized, the most probable electron transition into the K-hole is $K \leftarrow L_3$. Also possible, however with lower probability, are $K \leftarrow L_2$, and $K \leftarrow M_3$, etc. The vacancy in $L_3$ after a $K \leftarrow L_3$ transition can be filled by $L_3 \leftarrow M_5$ or $L_3 \leftarrow M_4$ and so forth. Note that $K \leftarrow L_1$ and $L_1 \leftarrow M_1$ are forbidden transitions.

Any of these transitions allows the emission of a photon or of an Auger-electron, as discussed later. The designation of emission lines follows the *transition level designation* (TLD) similar to the scheme mentioned earlier (but without arrows, e.g., $KL_2$ or $L_3M_4$), or the *Siegbahn notation*, where $KL_3$ corresponds to $K\alpha_1$, $KL_2$ to $K\alpha_2$, etc. Not all allowed transitions have a Siegbahn designation.

*Auger-effect and Coster–Kronig transitions.* The relaxation of an excited state by an electron transition is not necessarily associated with the emission of a photon. In a complex quantum-mechanical process, the energy can be transferred to another electron of the atom, which is then emitted from the atom and carries a kinetic energy of $E_{\text{kin}} = E_{\text{ExcitedState}} - E_{\text{GroundState}} - E_{\text{Binding}}$. Altogether there are three electrons involved: the first is the photo-electron, which is emitted when the photon is absorbed. In the subsequent relaxation process, an outer electron fills the vacancy and appears to release the energy $E_{\text{ExcitedState}} - E_{\text{GroundState}}$ by transferring it to a third electron, which is bound to the atom by $E_{\text{Binding}}$. However, quantum-theoretical models deny the existence of separable intermediate states during relaxation and treat the relaxation and emission of the Auger-electron as a single process. The atom is left in a doubly ionized state.

The Auger-electrons have characteristic energies, which can be measured and related to the elemental composition of surface-near layers (Auger-electron spectroscopy). Auger-probability, $p_{\text{Auger}}$, and probability of photon emission (fluorescent yield) [1, 2], $\omega$, complement each other: $p_{\text{Auger}} + \omega = 1$. Both are a function of atomic number and the (sub-) shell; $\omega_K$ is around 50% for $Z = 32$ and higher for heavier atoms. The Auger-probability dominates significantly over the emission of fluorescent radiation at light elements. For example, $\omega_{\text{K}} \approx 1.4 \times 10^{-3}$ for carbon and $\omega_{\text{K}} \approx 7 \times 10^{-4}$ for beryllium [16]. The poor photon count rates resulting from such low fluorescent yields are among the main reasons for the difficulties arising in light element analysis by XRF.

Note that for the L-shell (and higher shells), the subshell should be specified, such as $\omega_{\text{L1}}$, $\omega_{\text{L2}}$, $\omega_{\text{L3}}$. Each of the subshell-yields is much lower than $\omega_{\text{K}}$.

Another effect can take place within the L−, M−, and higher subshells, where electrons can move from outer subshells to unoccupied states of inner subshells, e.g., from $L_3$ to $L_2$ after $L_2$-ionization or $KL_2$ relaxation. This is called a Coster–Kronig transition; it differs from the Auger-effect only by its confinement to a set of subshells. The (small) excessive energy is transferred to an outer electron or to the environment (lattice). The fact that Coster–Kronig transitions effectively move vacancies into outer subshells has a considerable effect on the intensities of the corresponding emission lines, which are proportional to the number of vacancies.

### 5.2.2 Direct Excitation

*Direct (primary) excitation* is a process by which *primary fluorescent radiation* from atoms in a specimen is excited by *primary photons* from an external source, such as an X-ray tube, radioactive source, or synchrotron beam (the theory of excitation by electrons or other charged particles is not covered in this chapter) (Fig. 5.2). The alternative is indirect excitation, where the observed fluorescent radiation is excited as a secondary process by photons or particles (electrons) originating from direct excitation or other secondary processes within the specimen.

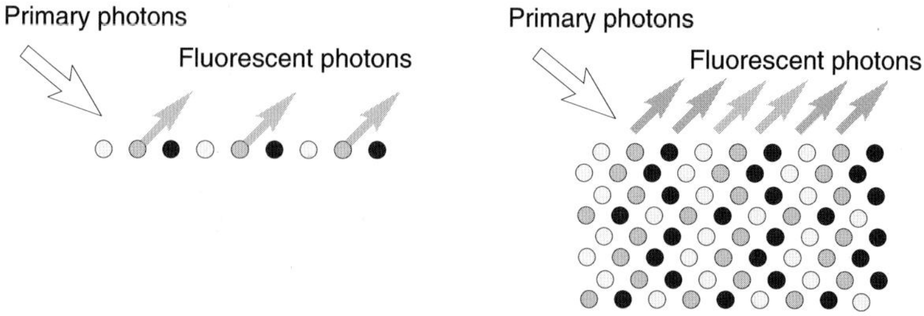


**Fig. 5.2.** Left: Direct fluorescent excitation of a single layer of atoms. The number of the observed photons is proportional to the number of emitting atoms. Right: A stack of atoms, where element and matrix dependent absorption effects must be accounted for

*Fluorescent radiation from a small number of identical atoms.* From (5.2c) it follows that $\Delta N = N_{\mathrm{prim}}\mu_{\mathrm{Atom}}\Delta\hat{n}$ interactions occur, when $N_{\mathrm{prim}}$ monochromatic primary photons hit a small number $\Delta\hat{n}$ of atoms/cm$^2$. Thereof the probability for photo-absorption (and not scattering) is $\tau_{\mathrm{Atom}}/\mu_{\mathrm{Atom}}$, whereby any shell can be ionized. For the electron transition associated for example with K$\alpha_1$-lines this must be the K-shell and the corresponding probability is $\tau_{\mathrm{K,Atom}}/\tau_{\mathrm{Atom}}$. After ionization, relaxation by the proper transition, e.g., KL$_3$, takes place with probability $p_{\mathrm{K}\alpha 1}$. The probability for the subsequent emission of a photon rather than of an Auger-electron is $\omega_{\mathrm{K}}$ (fluorescent yield). Combining these factors gives the number of emitted K$\alpha$1-photons: $\Delta N_{\mathrm{K}\alpha 1} = N_{\mathrm{prim}}\tau_{\mathrm{K,\,Atom}}\Delta\hat{n}p_{\mathrm{K}\alpha 1}\omega_{\mathrm{K}}$ and

$$\begin{aligned}\Delta N_{\mathrm{K}\alpha 1} &= N_{\mathrm{prim}}\frac{S_{\mathrm{K}}-1}{S_{\mathrm{K}}}\Delta\hat{n}\tau_{\mathrm{Atom}}p_{\mathrm{K}\alpha 1}\omega_{\mathrm{K}}\\ &= N_{\mathrm{prim}}\frac{S_{\mathrm{K}}-1}{S_{\mathrm{K}}}\Delta\hat{m}\tau p_{\mathrm{K}\alpha 1}\omega_{\mathrm{K}}\\ &= N_{\mathrm{prim}}\frac{S_{\mathrm{K}}-1}{S_{\mathrm{K}}}\Delta T\bar{\tau}p_{\mathrm{K}\alpha 1}\omega_{\mathrm{K}}.\end{aligned}\tag{5.6}$$

*Fluorescent radiation from a small number of mixed atoms.* If only a certain fraction of the interacting atoms is of element $i$ (which emits the analyte photons, $\Delta N_i$), the relative amount of $i$-atoms must be known. The number of photons absorbed by element $i$ is $N_{\mathrm{prim}}C_i\Delta\hat{m}\tau_i = N_{\mathrm{prim}}C_{i,\,\mathrm{Atom}}\Delta\hat{n}\tau_{i,\,\mathrm{Atom}}$, respectively, and by using $\Delta\hat{m} = \Delta m/A = \rho\Delta T$, the emitted number of photons follows as:

$$\begin{aligned}\Delta N_i &= N_{\mathrm{prim}}\frac{S_i-1}{S_i}C_{i,\,\mathrm{Atom}}\Delta\hat{n}\tau_{i,\,\mathrm{Atom}}p_i\omega_i\\ &= N_{\mathrm{prim}}\frac{S_i-1}{S_i}C_i\Delta\hat{m}\tau_i p_i\omega_i\\ &= N_{\mathrm{prim}}\frac{S_i-1}{S_i}C_i\Delta T\frac{\rho}{\rho_i}\bar{\tau}_i p_i\omega_i.\end{aligned}\tag{5.7}$$

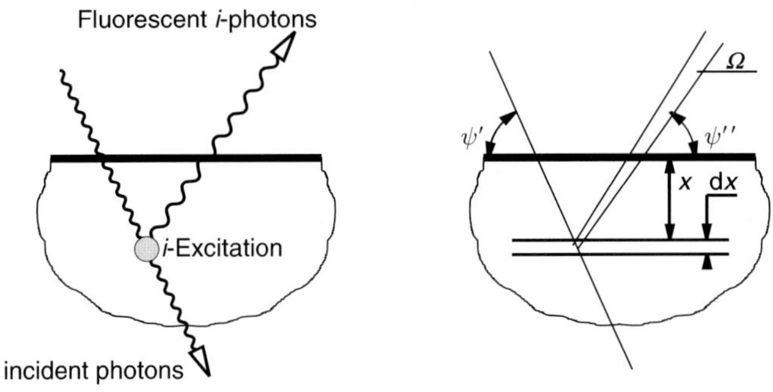


**Fig. 5.3.** Direct excitation: beam geometry within sample

*Fluorescent radiation from a large number of mixed atoms.* The derivation of the equation for a thick stack of layers extends these equations in two steps: assuming that the layer from which the fluorescent photons originate is shielded by a stack of atoms with thickness $x$ (Fig. 5.3), the first step is to account for the attenuation of the primary and fluorescent photons on their path from and to the surface. The second step is to add up all layers which emit fluorescent radiation.

The occurrence of path lengths suggests employing the third line of (5.7) to determine the number of photons emitted from a thin layer of thickness $\Delta x$ at depth $x$, and those observed at the surface (note that this does not account for the detection efficiency of the actual detector and/or spectrometer):

$$\begin{aligned}
\Delta N_{i,\,\text{emitted at}\,\Delta x} &= N_{\text{prim}} \exp\left(-\frac{x}{\sin\psi'}\bar{\mu}_{S,E}\right) \frac{S_i - 1}{S_i} C_i \frac{\Delta x}{\sin\psi'} \frac{\rho}{\rho_i} \bar{\tau}_{i,E} p_i \omega_i \\
\Delta N_{i,\,\text{observed}} &= \Delta N_{i,\,\text{emitted at}\,\Delta x} \exp\left(-\frac{x}{\sin\psi''}\bar{\mu}_{S,i}\right) \frac{\Omega}{4\pi} \\
\Delta N_{i,\,\text{observed}} &= N_{\text{prim}} G_i C_i \frac{\rho}{\rho_i} \bar{\tau}_{i,E} \exp(-x \cdot \bar{\mu}^*) \Delta x
\end{aligned} \tag{5.8}$$

$$\begin{aligned}
g_i &= \frac{S_i - 1}{S_i} p_i \omega_i \\
G_i &= \frac{S_i - 1}{S_i} p_i \omega_i \frac{\Omega}{4\pi \sin\psi'} = g_i \frac{\Omega}{4\pi \sin\psi'} \\
\bar{\mu}^* &= \frac{\bar{\mu}_{S,E}}{\sin\psi'} + \frac{\bar{\mu}_{S,i}}{\sin\psi''} = \bar{\mu}'_S + \bar{\mu}''_S \quad \text{where} \quad \bar{\mu}'_S = \frac{\bar{\mu}_{S,E}}{\sin\psi'} \\
\text{and} \quad \bar{\mu}''_S &= \frac{\bar{\mu}_{S,i}}{\sin\psi''} \quad \text{and} \quad \mu^* = \frac{\mu_{S,E}}{\sin\psi'} + \frac{\mu_{S,i}}{\sin\psi''} = \mu'_S + \mu''_S
\end{aligned} \tag{5.9}$$

$\mu_{S,E}$ accounts for the attenuation of the primary photons in the (in general, multi-element) specimen $S$, $\mu_{S,i}$ for the attenuation of the fluorescent photons from element $i$, $\Omega$ is the solid angle under which the photons are measured at the surface. The index $S$ is often omitted. See Fig. 5.3 for the definitions of $x$, $\Delta x$, $\psi'$, and $\psi''$.

The second step requires adding up all layers $\Delta x$. This is solved by using again the differential notations, $\mathrm{d}N$ and $\mathrm{d}x$ (including the negative signs), and integrating over $x$. $T$ is the thickness of the specimen and, alternatively, $\hat{m}$ its mass per area:

$$\begin{aligned} \mathrm{d}N_{i,\,\text{observed}} &= -N_{\text{prim}} G_i C_i \frac{\rho}{\rho_i} \bar{\tau}_{i,E} \exp\left(-\bar{\mu}^* x\right) \mathrm{d}x \\ N_{i,\,\text{observed}} &= \int_{x=0}^{x=T} \mathrm{d}N_{i,\,\text{observed}} = N_{\text{prim}} G_i C_i \frac{\rho}{\rho_i} \bar{\tau}_{i,E} \frac{1-\exp\left(-\bar{\mu}^* T\right)}{\bar{\mu}^*} \\ &= N_{\text{prim}} G_i C_i \tau_{i,E} \frac{1-\exp\left(-\mu^* \cdot \hat{m}\right)}{\mu^*} . \end{aligned} \tag{5.10}$$

Note that in the last line only mass attenuation coefficients are used instead of linear coefficients. For bulk materials ($T \to \infty$ or $\hat{m} \to \infty$) follows:

$$N_{i,\,\text{observed}} = N_{\text{prim}} G_i C_i \frac{\rho}{\rho_i} \bar{\tau}_i \frac{1}{\bar{\mu}^*} = N_{\text{prim}} G_i C_i \tau_i \frac{1}{\mu^*} . \tag{5.11}$$

*Fluorescent radiation excited by polychromatic sources: bulk material.* For polychromatic sources, the sum of photons excited by all primary photons of energies within $E_{\text{edge},i}$ and $E_{\max}$ must be computed by integration of (5.11). $E_{\max}$ is the maximum photon energy in the spectrum ($E_{\max} = eU$ for X-ray tubes with the operating voltage $U$) and $E_{\text{edge},i}$ is the absorption edge energy of the analyte line of element, $i$.

For a polychromatic primary energy spectrum, the number of monochromatic primary photons $N_{\text{prim}}$ of energy $E$ must be replaced by the infinitesimally small number $\mathrm{d}N_{\text{prim}}(E)$ of (pseudo-) monochromatic primary photons of energies within the infinitesimally narrow interval $(E, E+\mathrm{d}E]$. The energy dependence of $\mathrm{d}N_{\text{prim}}(E)$ is given by a probability density (*spectral distribution function*), $N_0(E)$, such that $\mathrm{d}N_{\text{prim}}(E) = N_0(E)\,\mathrm{d}E$ and $N(\lambda)\,\mathrm{d}\lambda \leftrightarrow N(E)\,\mathrm{d}E$. The highest energy in the spectrum, $E_{\max}$, corresponds to $\lambda_{\min} = hc/E_{\max}$; building derivatives gives $\mathrm{d}\lambda/\lambda = -\mathrm{d}E/E$ and $\mathrm{d}E = -\mathrm{d}\lambda\, hc/\lambda^2$.

$$\begin{aligned} \mathrm{d}N_{i,\,\text{observed}} &= \mathrm{d}N_{\text{prim}}(E) G_i C_i \tau_i \frac{1}{\mu^*} \\ N_{i,\,\text{observed}} &= G_i C_i \int_{E=E_{\text{Edge},i}}^{E=E_{\max}} \frac{\tau_i N_0(E)\,\mathrm{d}E}{\mu^*} \\ &= G_i C_i \int_{\lambda=\lambda_{\min}}^{\lambda=\lambda_{\text{Edge},i}} \frac{\tau_i N_0(\lambda)\,\mathrm{d}\lambda}{\mu^*} . \end{aligned} \tag{5.12}$$

*Notes regarding the spectral distribution of photons from X-ray tubes.* The integral over $E$ (or $\lambda$) is practically solved by approximating it by a sum:

$$
\begin{aligned}
N_{i,\,\text{observed}} &\approx G_i C_i \sum_{E=E_{\text{Edge},i}}^{E=E_{\max}} \frac{\tau_i N_0(E)\Delta E}{\mu^*} \\
&\approx G_i C_i \sum_{\lambda=\lambda_{\min}}^{\lambda=\lambda_{\text{Edge},i}} \frac{\tau_i N_0(\lambda)\Delta\lambda}{\mu^*}. \qquad (5.13)
\end{aligned}
$$

Attention must be paid to the fact that tube spectra usually contain continuous radiation as well as characteristic tube lines (following general practice, the wavelength notation is used here): Both are part of $N_0(\lambda)\Delta\lambda$. However, photon counts for the tube lines, $N_{0,\text{lines}}(\lambda)$, are often provided in the literature as integral photon counts (because the natural line width is so small that a distribution function makes no practical sense), while data for the continuous component, $N_{0,\text{continuum}}(\lambda)$, originating from measurements or from theoretical computations, are generally given in photons per wavelength interval $\Delta\lambda$. They are therefore possibly of different physical dimension and care must be taken to add them appropriately: $N_0(\lambda)$ $\Delta\lambda = [N_{0,\text{continuum}}(\lambda)\ \Delta\lambda + N_{0,\text{lines}}(\lambda)]$ for the lines in that interval given in absolute numbers of photons, and $N_0(\lambda)\Delta\lambda = [N_{0,\text{continuum}}(\lambda)+N_{0,\text{lines}}(\lambda)]\Delta\lambda$ for lines given in photons per spectral interval. The same applies to $N_0(E)$.

*Fluorescent radiation excited by polychromatic sources: thin films.* For a single, freestanding thin layer follows (multilayer structures and films on substrates are discussed in Sect. 5.5):

$$
\begin{aligned}
N_{i,\text{observed}}^{\text{Layer}} &= G_i C_i \int_{E=E_{\text{Edge},i}}^{E=E_{\max}} \frac{\tau_{i,E}\,[1-\exp(-\mu^*\hat{m})]\,N_0(E)\,\mathrm{d}E}{\mu^*} \\
&= G_i C_i \int_{E=E_{\text{Edge},i}}^{E=E_{\max}} \frac{\tau_{i,E}\,[1-\exp\left(-\mu^*\rho T\right)]\,N_0(E)\,\mathrm{d}E}{\mu^*}. \qquad (5.14)
\end{aligned}
$$

In cases of very thin layers, the approximation $1-\exp\left(-\mu^*\hat{m}\right)\approx\mu^*\hat{m}$ and correspondingly $1-\exp\left(-\mu^*\rho T\right)\approx\mu^*\rho T)$ can be applied:

$$
\begin{aligned}
N_{i,\text{observed}}^{\text{Layer}} &\approx G_i C_i \int_{E=E_{\text{Edge},i}}^{E=E_{\max}} \tau_{i,E}\hat{m}N_0(E)\,\mathrm{d}E \\
&\approx G_i C_i \int_{E=E_{\text{Edge},i}}^{E=E_{\max}} \tau_{i,E}\rho T N_0(E)\,\mathrm{d}E. \qquad (5.15)
\end{aligned}
$$

### 5.2.3 Indirect Excitation

The term *indirect excitation* comprises a variety of excitation mechanisms within a specimen, where fluorescent photons, scattered photons, or particles (photo-electrons, Auger-electrons), which originate from within the specimen by interaction of primary photons with atoms, excite additional fluorescent radiation.

As mentioned earlier, excitation of (primary) fluorescent photons by primary (external) photons is called direct or primary excitation. When primary fluorescence photons have sufficient energy to excite fluorescent radiation of other atoms in the specimen, the interaction is called *secondary excitation* and the excited photons are secondary fluorescent photons. A second level of interaction can be observed, when secondary fluorescent photons excite tertiary fluorescent photons. Higher levels of interactions are possible but have no practical importance.

In the range of transition elements, secondary excitation is particularly strong for elements differing by 2 in atomic numbers, and stainless steel (Cr–Fe–Ni) is the usual example for demonstrating this (Fe → Cr, Ni → Cr, and Ni → Fe) as well as tertiary excitation (Ni → Fe → Cr). The actual contributions amount up to 30% from secondary excitation and up to 2.5% from tertiary excitation, relative to the photons excited by direct excitation.

If a primary photon of sufficient energy is (coherently or incoherently) scattered, these photons can still excite fluorescent photons. Scattered fluorescent radiation can enhance the measured count rate as well, if the interaction changes their direction toward the detector (or analyzer). Scattering effects are usually neglected. However, the approximation $\mu \approx \tau$, which is sometimes used to simplify the FP equations, has the opportune side effect of providing a partial correction.

Excitation effects in the range of light elements (cascade effects, excitation by photo- and Auger-electrons) appear to be of considerable importance, when direct excitation is inefficient (which is usually the case when X-ray tubes are used), but their mathematical treatment is complex. They are discussed in Sect. 5.6 in the context of light element analysis.

All excitation processes, which create fluorescence in addition to primary fluorescent photons, are referred to as *indirect excitation.* While the term *secondary excitation* is widely used synonymously with *indirect excitation* (including all secondary processes), preference should be given to reserve it for the first level of indirect excitation, by photons.

#### Secondary Excitation

Mathematical models for secondary excitation are normally included in commercial general purpose FP computer algorithms. The following treatment assumes that $i$ denotes the analyte line and $j$ the primary fluorescent photons, which excite secondary fluorescent photons of $i$. See Fig. 5.4 for details

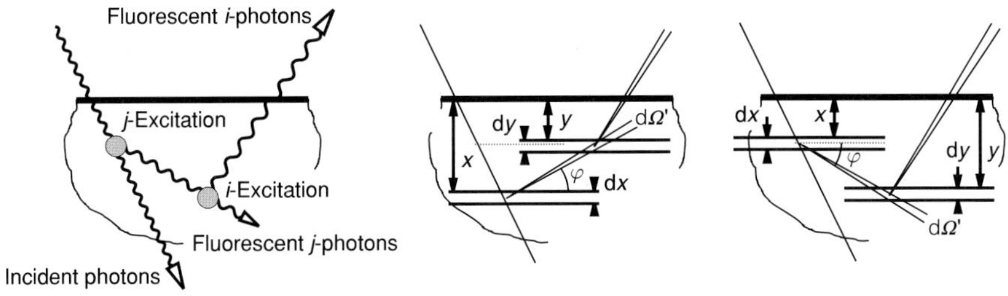


**Fig. 5.4.** Indirect (secondary) excitation: beam geometry within the specimen

of the geometrical notations. It is assumed that $E_j > E_{\text{edge},i}$.

$$\begin{aligned}
\mathrm{d}N_{j,\,\text{emitted at }\mathrm{d}x} &= -N_{\text{prim}} \exp\left(-\frac{x}{\sin\psi'}\bar{\mu}_{S,E}\right) \\
&\quad \times \frac{S_j - 1}{S_j} C_j \frac{\rho}{\rho_j} \bar{\tau}_{j,E} \frac{\mathrm{d}x}{\sin\psi'} p_j \omega_j \\
\mathrm{d}^2 N_{j,\,\text{arriving at }\mathrm{d}y} &= \mathrm{d}N_{j,\,\text{emitted at }\mathrm{d}x} \exp\left(-\frac{|x-y|}{\sin\varphi}\bar{\mu}_{S,j}\right)\frac{\mathrm{d}\Omega'}{4\pi} \\
\mathrm{d}^3 N_{i,\,\text{observed}} &= -\mathrm{d}^2 N_{j,\,\text{arriving at }\mathrm{d}y} \frac{S_i - 1}{S_i} C_i \frac{\rho}{\rho_i} \bar{\tau}_{i,j} \frac{\mathrm{d}y}{\sin\varphi} p_i \omega_i \\
&\quad \times \exp\left(-\frac{y}{\sin\psi''}\bar{\mu}_{S,i}\right)\frac{\Omega}{4\pi}.
\end{aligned}$$

For integration over $x$ and $y$, the boundaries are set to $0 < x < \infty$ and $0 < y < \infty$ for bulk materials. For the differential solid angle the substitution $\mathrm{d}\Omega' = 2\pi \cos\varphi \,\mathrm{d}\varphi$ is used with the boundaries $-\pi/2 < \varphi < \pi/2$:

$$\begin{aligned}
N_{i,\,\text{observed}} &= N_{\text{prim}} \frac{S_j - 1}{S_j} C_j \frac{\rho}{\rho_j} \bar{\tau}_{j,E} \frac{S_i - 1}{S_i} C_i \frac{\rho}{\rho_i} \bar{\tau}_{i,j} \\
&\quad \times p_j \omega_j p_i \omega_i \frac{\Omega}{4\pi \sin\psi'} I \\
&= N_{\text{prim}} G_i g_j C_i C_j \frac{\rho}{\rho_j} \bar{\tau}_{j,E} \frac{\rho}{\rho_i} \bar{\tau}_{i,j} I, \qquad (5.16)
\end{aligned}$$

where

$$\begin{aligned}
I = &\int_{\varphi=0}^{\varphi=\pi/2} \frac{\mathrm{d}\varphi}{2} \cot\varphi \int_{x=0}^{x=\infty} \exp\left[-x\left(\frac{\bar{\mu}_{S,E}}{\sin\psi'} + \frac{\bar{\mu}_{S,j}}{\sin\varphi}\right)\right] \mathrm{d}x \\
&\times \int_{y=0}^{y=x} \exp\left[-y\left(\frac{\bar{\mu}_{S,i}}{\sin\psi''} - \frac{\bar{\mu}_{S,j}}{\sin\varphi}\right)\right] \mathrm{d}y \\
&+ \int_{\varphi=0}^{\varphi=\pi/2} \frac{\mathrm{d}\varphi}{2} \cot\varphi \int_{x=0}^{x=\infty} \exp\left[-x\left(\frac{\bar{\mu}_{S,E}}{\sin\psi'} - \frac{\bar{\mu}_{S,j}}{\sin\varphi}\right)\right] \mathrm{d}x \\
&\times \int_{y=x}^{y=\infty} \exp\left[-y\left(\frac{\bar{\mu}_{S,i}}{\sin\psi''} + \frac{\bar{\mu}_{S,j}}{\sin\varphi}\right)\right] \mathrm{d}y.
\end{aligned}$$

See (5.9) for the definitions of $G_i$ and $g_j$. Solving the integrals and accounting for the fact that the total contribution of secondary excitation is the sum of the contribution of all lines, $J$, from all elements, $j$, with $E_J > E_{\mathrm{Edge},i}$, one obtains for a polychromatic primary spectrum:

$$N_{i,\,\mathrm{sec,\,observed}} = G_i C_i \sum_j \sum_J C_j g_J \tau_{i,J} \int_{E=E_{\mathrm{Edge}\,J}}^{E=E_{\max}} \frac{\tau_{j,E} N_0(E)}{\mu^*} L_{iJ}\,\mathrm{d}E$$

$$L_{iJ} = \frac{1}{2}\left[\frac{1}{\mu'_{S,E}} \ln\left(1 + \frac{\mu'_{S,E}}{\mu_{S,J}}\right) + \frac{1}{\mu''_{S,i}} \ln\left(1 + \frac{\mu''_{S,i}}{\mu_{S,J}}\right)\right]. \quad (5.17)$$

The sum over all elements, $j$, may include the analyte element $i$ (self-enhancement) if for example $i$-L$\alpha$ is chosen as analyte line and $i$-K lines are present as well. Such contributions are often neglected, when direct excitation is strong. Note that no line can excite another line of the same subshell of a given atom, but that in some cases secondary excitation by lines of another subshell of the same main quantum number, $n$, is possible.

### Tertiary Excitation

While secondary excitation is appropriately accounted for in most professional environments in XRF, the possibility of computing the contribution of tertiary excitation is a feature of only the more advanced implementations. The reason is that such contributions are often assumed to be negligible except for a few (in general well-known) cases, and that computing them is disproportionately time consuming due to nested integrations, which require numerical evaluation.

The mathematical derivation of the equation for tertiary excitation follows the same basic concept as for secondary excitation and can be carried out without any principal difficulty [53]. There remains, however, an additional integral (nested into that over the energy), which requires numerical solution. The resulting equation assumes that lines $K$ of elements $k$ are excited by primary radiation (direct excitation). They excite lines $J$ of elements $j$ (secondary excitation step), which in turn excite the analyte line of element $i$ (tertiary excitation). The dimensionless variable $t$ is temporarily used for the integrations.

$$N_{i,\,\mathrm{sec,\,observed}} = \frac{1}{4} G_i C_i \sum_k \sum_K C_k g_K$$

$$\times \int_{E=E_{\mathrm{Edge}\,K}}^{E=E_{\max}} \sum_j \sum_J C_j g_J \tau_{i,K_J} \tau_{j,K} \frac{\tau_k N_o(E)}{\mu^*} L_{iJK}\,\mathrm{d}E$$

$$
\begin{aligned}
L_{iJK} = & \frac{1}{(\mu'_{S,E})^2} \ln\left(1+\frac{\mu'_{S,E}}{\mu_{S,K}}\right) \ln\left(1+\frac{\mu'_{S,E}}{\mu_{S,J}}\right) \\
& + \frac{1}{\mu'_{S,E}\cdot\mu''_{S,i}} \ln\left(1+\frac{\mu'_{S,E}}{\mu_{S,K}}\right) \ln\left(1+\frac{\mu''_{S,i}}{\mu_{S,J}}\right) \\
& + \frac{1}{(\mu''_{S,i})^2} \ln\left(1+\frac{\mu''_{S,i}}{\mu_{S,K}}\right) \ln\left(1+\frac{\mu''_{S,i}}{\mu_{S,J}}\right) \\
& + \left(\frac{1}{\mu'_{S,E}}+\frac{1}{\mu''_{S,i}}\right)\left(\frac{1}{\mu_{S,J}}\ln\left(1+\frac{\mu_{S,J}}{\mu_{S,K}}\right)+\frac{1}{\mu_{S,K}}\ln\left(1+\frac{\mu_{S,K}}{\mu_{S,J}}\right)\right) \\
& - \frac{1}{\mu'_{S,E}}\int_0^{\mu_K/\mu_J}\frac{1}{\mu'_{S,E}t+\mu_{S,K}}\log\left(\frac{1+t}{t}\right)dt \\
& - \frac{1}{\mu''_{S,i}}\int_0^{\mu_J/\mu_K}\frac{1}{\mu''_{S,i}t+\mu_{S,J}}\log\left(\frac{1+t}{t}\right)dt. \qquad (5.18)
\end{aligned}
$$

### 5.2.4 Use of Standards [24]

Standards are made from reference materials (see Sect. 5.9 for types and definitions) with accurately known composition, by following the same preparatory steps as for the unknown specimen. The effective size as well as the surface quality of standard and unknown specimens should be the same. In light element analysis, the microstructure of a specimen as well as the chemical state of the analyzed atoms may affect the measured count rates and must therefore be identical for standard(s) and unknown.

Thin films can be used as standards without principal difficulties, but in practice the error frames associated with their thickness and density are limiting factors; they are, however, very useful for the analysis of other thin layers made by the same technique.

The purpose of using standards is twofold: By building count-rate ratios, unknown factors in the FP equations cancel and are thereby eliminated (see later). The second is more subtle. The FP model is not a perfect description of reality. Besides the fact that the parameters themselves are subject to inaccuracies, most models are incomplete by neglecting the real geometry (e.g., beam divergences), several types of interactions (such as scattering), and other influence factors (microstructure, chemical state, varying spectral distribution of the primary beam across the specimen, etc.). If the assumption holds that the measured line intensities from unknown and standards are affected in the same manner and by the same factor by these errors, this factor may cancel by building appropriate ratios. This is, however, not a matter of course and a careful analysis of the problem is strongly advised, because many errors depend in a complex way on concentrations and the qualitative composition.

*Count-rate ratios.* In almost all practical applications, count-rate ratios are used rather than absolute counts. Such ratios are built with the count rates from the same element in another specimen, which can be a standard of any composition, or a pure element. The main advantage is that thereby a number of unknown or less accurately known factors cancel, namely all factors included in $G_i$ of (5.11), as well as any scaling constant in the absolute photon flux of the primary radiation (which is rarely ever known in practice), and the detection efficiency, which is omitted in (5.11). Note that the factors $g_J$ and $g_K$ in the secondary and tertiary excitation terms in (5.17) and (5.18) do not cancel. In this chapter, all count-rate ratios are relative to pure elements unless otherwise indicated:

$$R_i = \frac{N_i}{N_{(i)}}. \tag{5.19}$$

*(Non-) Necessity of measuring pure element counts.* It is not necessary to actually measure pure elements in order to obtain count-rate ratios relative to pure elements. This is an important point because several elements cannot be produced or analyzed by reasonable means, in pure form. Assume that the counts from a specimen, $N_{i,S}$, and a standard, $N_{i,St}$, have been measured. Then the count-rate ratio of the standard, $R_{i,St}$, relative to a pure element can be computed by FP methods, and the count-rate ratio of the specimen, $R_{i,S}$, relative to a pure element obtained from (see also Sect. 5.3.1 and (5.31)):

$$R_{i,S} = \frac{N_{i,S}^{\text{measured}}}{N_{(i)}^{\text{extrapolated}}} = \frac{N_{i,S}^{\text{measured}}}{N_{i,St}^{\text{measured}}} \cdot R_{i,St}^{\text{computed}}. \tag{5.20}$$

*Requirement and selection of additional standards.* While one standard is required in order to build relative intensities, additional standards can improve the accuracy of the analysis. Standards with a similar composition as the unknown are called *local standards.* A single local standard pins the calibration curve to the point defined by this standard and its measured count rates. When several standards are used, an average calibration factor can be computed. This is achieved by computing pure element count rates for each analyte line and standard, and averaging these values. The observed standard deviation of these data should match the expected uncertainty of the certified chemical composition, eventual preparatory inconsistencies, and the error introduced by counting statistics. Sets of standards that give larger errors should not be used.

Instead of local standards, standards with widely varying composition may be employed with advantage. These are *global standards.* The same as given earlier applies with respect to averaging and consistency. A weighting factor may be introduced for each standard, which matches the differing statistical reliability of low and high element counts. Note that reanalyzing sets of standards will not necessarily deliver the certified concentrations.

*Back-calculation schemes.* The concentrations in FP equations for polychromatic excitation cannot be directly computed and iterative methods must be applied. Practical algorithms are subject to a compromise in speed, reliability, and independence of initial values for the iteration. The usual procedure takes advantage of fact that the count rate from element $i$, $N_i$, (and the count-rate ratio, $R_i$) is roughly proportional to its concentration, $C_i$, $R_i = C_i f(C_1, \ldots, C_n)$ with a locally constant value for $f$. An initial assumption is required in the iteration for the concentrations, usually $C_i = R_i/\Sigma_n R_n$ for each element $i$. This is used to compute all count-rate ratios $R_n^{\text{comp}}$ for the unknown, which will in general differ from the measured values by factors $R_n^{\text{meas}}/R_n^{\text{comp}}$. The iteration is continued by updating the concentrations using $C_n^{\text{new}}/C_n^{\text{old}} = R_n^{\text{meas}}/R_n^{\text{comp}}$, and terminated when no significant further improvement is achieved or some other criteria are met. Occasionally the result delivers a set of concentrations with a sum differing essentially from one. In influence coefficients methods this indicates an error which should be investigated and – in case of $\Sigma c_i < 1$ – possibly (not necessarily) a missing element. Note, however, that count rates from FP equations for bulk materials are independent of a normalization of concentrations to any sum.

Iterations are in general a slow process and only small steps are allowed, for which $f(C_1, \ldots, C_n)$ remains practically constant. It can be considerably improved by using intelligent nonlinear approximations for the function $f(C_1, \ldots, C_n)$. These are for example some of the theoretical influence coefficients equations, which are generally seen as basis of independent methods of their own kind, but can also serve exactly this purpose.

## 5.3 Matrix Correction Methods and Influence Coefficients

### 5.3.1 The Nature of Influence Coefficients

*M. Mantler*

Experimental evidence as well as theoretical considerations show that the photon count rate $N_i$ from an analyte element $i$ increases in first approximation proportional to $C_i$, but it is also a distinct function of the matrix, i.e., the elements accompanying the analyte element in the sample.

This is illustrated in Fig. 5.5d, where (computed) relative count rates for Fe K$\alpha$ photons, $R_{Fe}$, in a FeCrNi alloy are drawn versus the iron concentration, for varying concentrations of chromium and nickel. The data points scatter considerably within enveloping curves of 0%Cr (system Fe–Ni) and 0%Ni (system Fe–Cr). This is, however, not a random scatter but an overlay of data points from various curves of iron counts from iron in different Ni–Cr matrices, as shown in Fig. 5.5c. Note that the Ni/Cr ratio is constant for each of these curves (as along the lines ending at the point "100%Fe" in Fig. 5.5a).

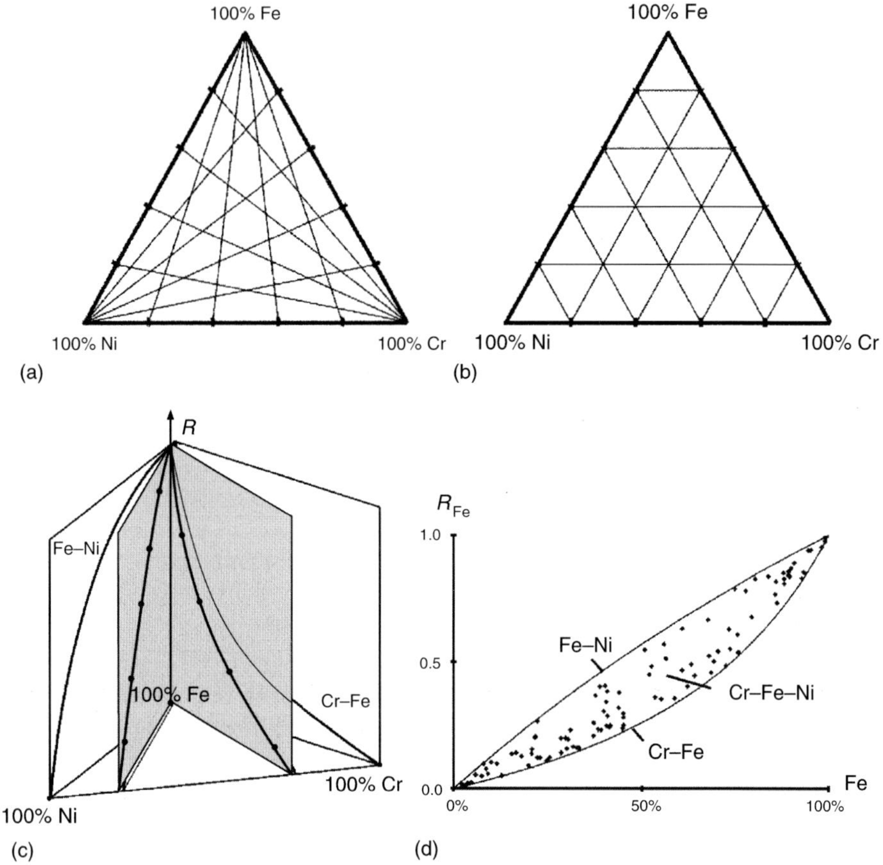


**Fig. 5.5.** (**a**) Base concentration coordinates for a ternary system, with intermediate lines indicating constant concentration ratios (fixed matrix), and (**b**) constant values of a concentration. (**c**) Curves of counts versus concentration of the analyte element in matrices of fixed composition. (**d**) Counts versus concentration from various matrix compositions, i.e., superposed datapoints from various curves as in (**c**)

The influence coefficients methods assume a linear relationship between an element count rate and its concentration, and attribute all deviations from linearity to the influence of the matrix. Corresponding counteracting "corrections" must therefore be applied to the measured data in order to transform them into the desired linear relationship. Thereby each element of the matrix is assigned a certain power (mathematically, an *influence coefficient*) of attenuating or enhancing the measured count rate from the analyte element. The mathematical starting point is the formal relationship $C_i = R_i M(C_1, C_2, \ldots, C_n)$. This is interpreted as a dominating linearity $C_i = R_i$, but modified by a function $M(C_1, C_2, \ldots, C_n)$ that takes the corrective role of moving the scattered data-points to a straight line.

Various functions $M(C_1, C_2, \ldots, C_n)$ have been proposed in the literature, with differing complexity and methods of defining their parameters (i.e., the influence coefficients). The following paragraphs show relationships between influence coefficients and FPs.

*Direct excitation.* The relationship between directly excited photon count rates, $N_i$ (or count-rate ratios $R_i$), and the concentration of the analyte element in a bulk specimen has been derived in Sect. 5.2. The indication $S$ for the specimen and $\lambda$ for the incident radiation is now omitted (see Appendix):

$$N_i = G_i C_i \int_{\lambda=\lambda_{\min}}^{\lambda=\lambda_{\mathrm{Edge},\, i}} \frac{\tau_i N_0(\lambda)\,\mathrm{d}\lambda}{\mu^*}$$

$$R_i = \frac{N_i}{N_{(i)}} = C_i \frac{\displaystyle\int_{\lambda=\lambda_{\min}}^{\lambda=\lambda_{\mathrm{Edge},\, i}} \frac{\tau_i N_0(\lambda)\,\mathrm{d}\lambda}{\mu^*}}{\displaystyle\int_{\lambda=\lambda_{\min}}^{\lambda=\lambda_{\mathrm{Edge},\, i}} \frac{\tau_i N_0(\lambda)\,\mathrm{d}\lambda}{\mu_i^*}} .$$

Introducing a term $M$, which is interpreted as the influence of the matrix on the analyte count rate, $N_i$ (or count-rate ratio, $R_i$), leads to

$$M = M(C_1, C_2, \ldots, C_n) = \frac{\displaystyle\int_{\lambda=\lambda_{\min}}^{\lambda=\lambda_{\mathrm{Edge},\, i}} \frac{\tau_i N_0(\lambda)\,\mathrm{d}\lambda}{\mu_i^*}}{\displaystyle\int_{\lambda=\lambda_{\min}}^{\lambda=\lambda_{\mathrm{Edge},\, i}} \frac{\tau_i N_0(\lambda)\,\mathrm{d}\lambda}{\mu^*}} \tag{5.21}$$

$$R_i = \frac{N_i}{N_{(i)}} = \frac{C_i}{M}$$

$$C_i = M R_i .$$

In many publications the term "intensity" ($I_i$ and $I_{(i)}$) is used synonymously for the count rates $N_i$ and $N_{(i)}$, respectively, from element $i$, and the third equation given here can then be rewritten accordingly as *"KIM" equation* [34]:

$$C_i = K\, I_i\, M \tag{5.22}$$

$$\left( K = \frac{1}{I_{(i)}} \right) .$$

In the count-rate equation, $\mu^*$ can be substituted by

$$\begin{aligned}\mu^* &= \sum_{j=1}^{n} C_j(\mu'_j + \mu''_j) = \sum_{j=1}^{n} C_j \left(\frac{\mu_{j,\lambda}}{\sin\psi'} + \frac{\mu_{j,i}}{\sin\psi''}\right) \\ &= C_i\mu_i^* + \sum_{j=1,\, j\neq i}^{n} C_j\mu_j^* = \left(1 - \sum_{j=1,\, j\neq i}^{n} C_j\right)\mu_i^* + \sum_{j=1,\, j\neq i}^{n} C_j\mu_j^* \\ &= \mu_i^*\left(1 + \sum_{j=1,\, j\neq i}^{n} C_j \frac{\mu_j^* - \mu_i^*}{\mu_i^*}\right) \\ \mu^* &= \mu_i^*\left(1 + \sum_{j=1,\, j\neq i}^{n} C_j\alpha_{ij}\right) \end{aligned} \tag{5.23}$$

with

$$\alpha_{ij} = \frac{\mu_j^* - \mu_i^*}{\mu_i^*} = \frac{\left(\frac{\mu_{j,\lambda}}{\sin\psi'} + \frac{\mu_{j,i}}{\sin\psi''}\right) - \left(\frac{\mu_{i,\lambda}}{\sin\psi'} + \frac{\mu_{i,i}}{\sin\psi''}\right)}{\left(\frac{\mu_{i,\lambda}}{\sin\psi'} + \frac{\mu_{i,i}}{\sin\psi''}\right)}. \tag{5.24}$$

Note that by this definition all $\alpha_{ii}$ are 0. The specification of $i \neq j$ in the sum is therefore not necessary, but kept for clarity. The matrix-influence factor, $M$, can be expressed by

$$M = \frac{\displaystyle\int_{\lambda=\lambda_{\min}}^{\lambda=\lambda_{\mathrm{Edge},\,i}} \frac{\tau_i N_0(\lambda)\,\mathrm{d}\lambda}{\mu_i^*}}{\displaystyle\int_{\lambda=\lambda_{\min}}^{\lambda=\lambda_{\mathrm{Edge},\,i}} \frac{\tau_i N_0(\lambda)\,\mathrm{d}\lambda}{\mu_i^*\left(1 + \sum_{j=1,\, j\neq i}^{n} C_j\alpha_{ij}\right)}}. \tag{5.25}$$

The $\alpha_{ij}$-terms are of course a function of the sample composition (concentrations) and of wavelength in the integral. The expression for $M$ simplifies greatly by assuming monochromatic incident radiation:

$$M_{\mathrm{mono}} = 1 + \sum_{j=1,\, j\neq i}^{n} C_j\alpha_{ij}. \tag{5.26}$$

In the resulting equation for the concentration of the analyte element

$$C_i = R_i\left(1 + \sum_{j\neq i} \alpha_{ij}C_j\right) = C_{i,\,\mathrm{apparent}}(1 + \text{correction terms}) \tag{5.27}$$

the $\alpha_{ij}$-terms are constants and can be interpreted as influence coefficients of the matrix element $j$ upon the analyte element, $i$. The *uncorrected concentration*, $C_i = R_i$, is often referred to as *apparent concentration*. Involved assumptions are that the incident radiation is monochromatic and that no contributions by indirect excitation occur.

For polychromatic incident radiation it is possible to apply the mean value theorem for integrals, which states that there exists a value $\lambda_{\text{equiv}}$ within the integration interval $\lambda_{\min}$, $\lambda_{\text{Edge},i}$ such that

$$\int_{\lambda=\lambda_{\min}}^{\lambda=\lambda_{\text{Edge},\,i}} f(\lambda)\, g(\lambda)\, \mathrm{d}\lambda = f(\lambda_{\text{equiv}}) \int_{\lambda=\lambda_{\min}}^{\lambda=\lambda_{\text{Edge},\,i}} g(\lambda)\, \mathrm{d}\lambda$$

hence[2]

$$M = \frac{\displaystyle\int_{\lambda=\lambda_{\min}}^{\lambda=\lambda_{\text{Edge},\,i}} \frac{\tau_i N_0(\lambda)\, \mathrm{d}\lambda}{\mu_i^*}}{\displaystyle\int_{\lambda=\lambda_{\min}}^{\lambda=\lambda_{\text{Edge},i}} \frac{\tau_i N_0(\lambda)\, \mathrm{d}\lambda}{\mu_i^* \left(1 + \sum\limits_{j=1,\, j\neq i}^{n} C_j \alpha_{ij}\right)}} = \left(1 + \sum_{j=1,\, j\neq i}^{n} C_j \alpha_{ij}(\lambda_{\text{equiv}})\right). \tag{5.28}$$

In many practical cases, the equivalent wavelength, $\lambda_{\text{equiv}}$ varies only slowly with composition and can be treated as constant. The concept of equivalent wavelengths (often called *effective wavelengths*) was an important step in the development of the influence coefficient methods [56]. Note that the above equation for $M$ contains no approximations (but the restriction to direct excitation). Despite its simple appearance, an accurate (theoretical) determination of the equivalent wavelength and of $\alpha_{ij}(\lambda_{\text{equiv}})$ is not a simple matter and requires solving the integrals.

*Indirect excitation.* In analogy to direct excitation, the starting point is (5.17) for indirect (secondary) excitation. $k$ encompasses all elements in the specimen and $K$ all lines of element $k$ which are capable of exciting the analyte line of element $i$:

$$\begin{aligned}
N_{i,\,\text{sec}} &= G_i C_i \sum_k \sum_K C_k g_K \tau_{i,K} \int_{\lambda=\lambda_{\min}}^{\lambda=\lambda_{\text{Edge}\,K}} \frac{\tau_k N_0(\lambda)}{\mu^*} \\
&\quad \times \frac{1}{2}\left[\frac{1}{\mu_E'} \ln\left(1 + \frac{\mu_E'}{\mu_K}\right) + \frac{1}{\mu_i''} \ln\left(1 + \frac{\mu_i''}{\mu_K}\right)\right] \mathrm{d}\lambda \\
&= G_i C_i \sum_k \sum_K C_k g_K \tau_{i,K} \\
&\quad \times \int_{\lambda=\lambda_{\min}}^{\lambda=\lambda_{\text{Edge}\,K}} \frac{\tau_k N_0(\lambda)}{\mu_i^* \left(1 + \sum\limits_{j=1,\, j\neq i}^{n} C_j \alpha_{ij}\right)} L_{iK}\, \mathrm{d}\lambda .
\end{aligned}$$

---

[2] A mathematical requirement for general validity is that the integrand is a continuous function. Since the integrand is (practically) discontinuous at the $\lambda$-values of absorption edges and/or tube lines, a mean value ($\lambda_{\text{equiv}}$) may not necessarily exist.

For monochromatic radiation, this simplifies similar to direct excitation, and by using the previously derived expression for $\alpha$-coefficients one obtains:

$$\frac{N_{i,\,\text{sec, mono}}}{N_{(i)}} = \frac{C_i \sum\limits_k \sum\limits_K C_k \beta_{iK}}{1 + \sum\limits_{j=1,\,j\neq i}^{n} C_j \alpha_{ij}}$$

$$\beta_{iK} = g_K \tau_{i,K} \frac{\tau_k/\tau_i}{\left(1 + \sum\limits_{j=1,\,j\neq i}^{n} C_j \alpha_{ij}\right)} L_{iK}$$

$$R_i = \frac{N_{i,\,\text{prim, mono}} + N_{i,\,\text{sec, mono}}}{N_{(i)}} = C_i \frac{1 + \sum\limits_k \sum\limits_K C_k \beta_{iK}}{1 + \sum\limits_j C_j \alpha_{ij}}$$

$$C_i = R_i \frac{1 + \sum\limits_j C_j \alpha_{ij}}{1 + \sum\limits_k \sum\limits_K C_k \beta_{iK}}. \tag{5.29}$$

Therein $\beta_{iK}$ is an influence coefficient accounting for secondary excitation (which can be 0). The computation of the coefficients $\alpha_{ij}$ and $\beta_{iK}$ is straightforward for monochromatic radiation, but the extension to polychromatic excitation requires a number of additional algebraic transformations; this is discussed in Sect. 5.3.8. A disadvantage of the equation for $C_i$ as written earlier is that it leads to a system of quadratic equations for the concentrations.

Under the assumption that the correction terms $C_k\beta_{iK}$ and $C_j\alpha_{ij}$ are small in comparison to unity, the following approximation is valid:

$$\begin{aligned} C_i &= R_i \frac{1 + \sum\limits_j C_j \alpha_{ij}}{1 + \sum\limits_k C_k \beta_{ik}} \\ &\approx R_i \left(1 + \sum_j C_j \alpha_{ij}\right)\left(1 - \sum_k C_k \beta_{ik}\right) \quad \text{if} \quad \sum_k C_k \beta_{ik} \ll 1 \qquad (5.30) \\ &\approx R_i \left(1 + \sum_j C_j (\alpha_{ij} - \beta_{ij})\right). \end{aligned}$$

The summations are now over all matrix elements. The coefficients $\beta_{ik}$ comprise all secondary excitation contributions from an element $k$ (accounting for all of its lines with sufficient energy) and may be 0. Similar and additional approaches are described in more detail and in context with historical and recent developments in Sect. 5.3.

*Normalization of concentrations.* In the derivation of (5.24) it was assumed and used that the concentrations in a specimen add up to 1. Instead, any other normalization factor NF can be chosen, e.g., NF = 100 (%). This leads to

$$M^{\mathrm{NF}} = \mathrm{NF} + \sum_{j \neq i} C_j^{\mathrm{NF}} \alpha_{ij}$$

$$C_{\mathrm{i}}^{\mathrm{NF}} = R_i M^{\mathrm{NF}}$$

e.g.,

$$M^{\Sigma=100\%} = 100\% + \sum_{j \neq i} C_j^{\Sigma=100\%} \alpha_{ij}$$

$$C_i^{\Sigma=100\%} = R_i \left( 100\% + \sum_{j \neq i} C_j^{\Sigma=100\%} \alpha_{ij} \right) .$$

Therein the $\alpha$-coefficients remain unaffected by the choice of normalization of the concentrations. This is not true for equations accounting for indirect (secondary) excitation:

$$C_i^{\Sigma=100\%} = R_i \frac{100\% + \sum_j C_j^{\Sigma=100\%} \alpha_{ij}}{1 + \sum_k C_k^{\Sigma=100\%} \beta'_{ik}}$$

$$(\beta'_{ik} = \beta_{ik}/100) .$$

*Types of influence coefficients.* Influence coefficients depend on:

- The analyte element and the analyte line;
- The matrix elements and, for polychromatic radiation, their concentrations;
- The wavelength (or energy) of the monochromatic incident radiation or, in case of polychromatic excitation, the spectral distribution of the incident radiation (including the excitation voltage in the case of X-ray tubes);
- On the geometry ($\psi'$ and $\psi''$).

Influence coefficients can be categorized accordingly in a number of ways (some authors follow a corresponding naming scheme, but not in a standardized manner):

- By the incident radiation: monochromatic or polychromatic. This is only meaningful for theoretically determined coefficients. The assumption of monochromatic incident radiation makes the derivation of the equations from FP theories much simpler. This has been historically helpful when powerful computers were not available, and it still is for tutorial purposes. However, monochromatic X-rays are rarely ever used in mainstream practical applications and influence coefficients for polychromatic radiation should be appropriately preferred in real environments;

- By the way of determination: theoretical or empirical.
  - Theoretical influence coefficients are computed from FP equations. One possible way is in a first step to compute relative intensities by conventional FP methods and in a second step to use them in the influence coefficients equations (any model) for the actual determination of the coefficients. This second step is simple and consists merely in solving a linear system of equations. The other method is to employ dedicated FP equations, which directly relate influence coefficients to FPs, as for example shown in (5.24) or, more elaborate, in Sect. 5.3.8;
  - Empirical influence coefficients are exclusively determined from experimental data. A sufficient number of standards and measured count rates must be available; the usual recommendation is that at least twice the number of the required minimum should be used. The coefficients are then determined by least squares methods. Note that this makes the empirical coefficients equation a fit-function through experimental data. This may produce a linear working curve even in the case of inhomogeneous samples, pressed or loose powders, specimens with uneven surfaces, etc. where other methods fail. This may, however, turn out to be a dangerous feature and deliver wrong results unless these factors are fully controlled and have identical effects in all standards and the analyte material.
- By the analytical context of their determination: binary or multielement. This applies to theoretical as well as empirical coefficients. The basic concept of influence coefficients aims to describe the influence of an element $j$ on (the count rate of) an analyte element, $i$; and is strictly "between two." This would allow the determination of influence coefficients $\alpha_{ij}$ (or $\beta_{ij}, \ldots$) from any sample containing the pair $i$ and $j$ of elements (including binary compounds) and to apply the derived coefficients to any other sample containing these two elements. As shown earlier, the mathematical verification of this method is possible by FP models, however only for monochromatic excitation. In environments of polychromatic excitation, the influence coefficients are a function of the whole matrix;
- By the effect of the correction: absorption by the matrix (direct excitation), enhancement (indirect excitation), or unspecified. Some theoretical models allow the mathematical quantification of the individual influence factors. Empirically determined influence coefficients are by their nature fit-coefficients and evaluation in other terms should not be exaggerated.

*Intensity ratios without pure element standards.* For many chemical elements the preparation of pure element standards is impossible or impractical so that employment of compound standards is necessary or preferable. For theoretical influence coefficient methods, a good estimate for the pure element count rates can be obtained by computing the count-rate ratio for analyte element $i$ in a standard specimen, St, and measuring its count rate. From that, the pure element count rate can be extrapolated and substituted for the (un-) measured $N_{(i)}$. The count-rate ratio, $R_{i,S}$, for a sample $S$ computed by using a single compound standard $St$ is:

$$R_{i,S} = \frac{N_{i,S}}{N_{(i)}}$$

$$R_{i,St}^{\text{measured}} = \frac{N_{i,St}^{\text{measured}}}{N_{(i)}^{\text{measured}}} \quad \text{and} \quad R_{i,St}^{\text{computed}} = \frac{N_{i,St}^{\text{computed}}}{N_{(i)}^{\text{computed}}}$$

$$\text{assume: } R_{i,St}^{\text{measured}} \approx R_{i,St}^{\text{computed}} \tag{5.31}$$

$$N_{(i)}^{\text{measured}} = \frac{N_{i,St}^{\text{measured}}}{R_{i,St}^{\text{measured}}} \approx \frac{N_{i,St}^{\text{measured}}}{R_{i,St}^{\text{computed}}} = N_{(i)}^{\text{extrapolated}}$$

$$R_{i,S} \approx \frac{N_{i,S}}{N_{(i)}^{\text{extrapolated}}} = \frac{N_{i,S}}{N_{i,St}^{\text{measured}}} R_{i,St}^{\text{computed}} .$$

Several extrapolated values can be obtained from different standards, *St*, and simply averaged. For empirical coefficients the unmeasured pure element counts can be treated as an additional unknown in the course of the determination of the influence coefficients, for example in the following way where an additional parameter $\alpha'_0$ is added to the unknown $\alpha'$-coefficients:

$$C_i = N_i M'$$

$$M' = \frac{1}{N_{(i)}} + \sum_j \frac{\alpha_{ij}}{N_{(i)}} C_j = \alpha'_0 + \sum_j \alpha'_{ij} C_j .$$

### 5.3.2 The Lachance–Traill Algorithm

*J.P. Willis and G.R. Lachance*

The concept of deriving algorithms to compensate for matrix effects in quantitative XRF analysis and defining influence coefficients to be used in mathematical expressions was contemplated very early in the development of analytical methods. The influence coefficients were visualized as *values* that, when multiplied by the concentration of the matrix element, quantify the matrix (interelement) effect of that element. The following is an abridged examination of algorithms and definitions that preceded or were contemporary with the algorithm proposed by Lachance and Traill (L–T) [33].

In 1954 Beattie and Brissey [4] had proposed the following expression:

$$(1 - R_a) W_a + \sum A_{ab} W_b = 0, \tag{5.32}$$

where

$$R_a = \frac{I_{aa}}{I_{as}} \quad A_{ab} = \frac{W_a}{W_b}(R_a - 1)$$

$I_{aa}$ is the net intensity measured for the pure element, $I_{as}$ is the net intensity for a specimen. $W_a$ and $W_b$ are weight fractions and $A_{ab}$ are the influence coefficients. In retrospect, the definition of a relative intensity as $I_{i,\mathrm{pure}}/I_{i,\mathrm{specimen}}$ and the fact that the first expression involves an "over determined" system of equations made this approach difficult to apply in practice and it is not currently used.

Also in 1954, Sherman [51] proposed a suite of equations of the type:

$$(a_{11} - t_1)c_1 + a_{21}c_2 + a_{31}c_3 = 0, \tag{5.33}$$

where $t$ is the time to accumulate $N$ counts; $a$ are coefficients, and $c$ are concentrations. Once again analysts were faced with a suite of equations that are equal to 0 and could only be solved given the premise that the sum of the components is equal to unity.

In 1955 Sherman [52], following an in-depth study of the correlation between intensity and concentration, proposed the following expressions/definitions:

$$Q = \frac{I_{f3}(0,0,1)}{I_{f3}(c_1,c_2,c_3)} = \frac{T_{f3}(c_1,c_2,c_3)}{T_{f3}(0,0,1)} \tag{5.34}$$

and

$$T_{f3}(c_1,c_2,c_3) = u_0 + \frac{c_1}{c_3}u_1 + \frac{c_2}{c_3}u_2,$$

where $T$ is the time to accumulate $N$ counts; $u_0$, $u_1$, and $u_2$ are coefficients, and $c$ are concentrations. The drawback in the case of the latter expression is that the correction terms involve multiplication by concentration *ratios* involving $c_3$ (the analyte) in the denominator of each of these ratios.

In 1968 Criss and Birks [12] proposed an FP approach, which is still currently used in one form or another. They also proposed an empirical influence coefficient approach involving a suite of equations equal to 0. However, they adopted the definition for relative intensity as $I_i/I_{(i)}$ (intensity of the specimen divided by the intensity of the pure element $i$), which is commonly used today, given that a relative intensity is in fact an apparent (i.e., uncorrected for matrix effects) concentration. Thus,

$$(R_\mathrm{A}\alpha_\mathrm{AA} - 1)\,C_\mathrm{A} + R_\mathrm{A}\alpha_\mathrm{AB}C_\mathrm{B} + R_\mathrm{A}\alpha_\mathrm{AC}\cdot C_\mathrm{C} = 0, \tag{5.35}$$

where $R_\mathrm{A} = I_i/I_{(i)}$; $\alpha$ are influence coefficients and $C$ are concentrations.

In 1973 Tertian [56] proposed equations of the following type relating *concentration* to *relative intensity*

$$c_\mathrm{A} = R_\mathrm{A}\left[c_\mathrm{A} + \alpha_\mathrm{A}^\mathrm{B}c_\mathrm{B} + \alpha_\mathrm{A}^\mathrm{C}c_\mathrm{C}\right]$$

$$R_A = \frac{I_\mathrm{A}}{I_{\mathrm{A}_1}} \quad \text{and} \quad K = \frac{1 - R_\mathrm{A}}{R_\mathrm{A}}\frac{c_\mathrm{A}}{1 - c_\mathrm{A}} \tag{5.36}$$

for binary contexts. $A_1$ is the pure element, $K$ is an influence coefficient, and $c_\mathrm{A}$ is the concentration of the analyte in the binary system.

In 1966 Lachance and Traill [33] proposed the following expression wherein the concentration of the analyte does not appear in the correction expression term quantifying the concentration of the analyte, i.e.:

$$R_{\mathrm{A}} = \frac{C_{\mathrm{A}}}{1 + C_{\mathrm{B}}\alpha_{\mathrm{AB}} + C_{\mathrm{C}}\alpha_{\mathrm{AC}} + \cdots + C_n\alpha_{\mathrm{A}n}}, \tag{5.37a}$$

which can be transformed for analysis as:

$$C_{\mathrm{A}} = R_{\mathrm{A}}\left[1 + \sum_{\mathrm{B}} \alpha_{\mathrm{AB}} C_{\mathrm{B}}\right], \tag{5.37b}$$

where the influence coefficient $\alpha_{\mathrm{AB}}$ is defined as

$$\alpha_{\mathrm{AB}\lambda} = \frac{\mu_{\mathrm{B}}^* - \mu_{\mathrm{A}}^*}{\mu_{\mathrm{A}}^*};$$

the subscript $\lambda$ is used to indicate that the definition is for a monochromatic incident source, and $\mu^*$ is the total effective mass absorption coefficient taking into account the spectrometer geometry.

The proposed expression can be transformed in order to define and compute the intensity of the pure analyte, namely,

$$I_{(\mathrm{A})} = \frac{I_{\mathrm{A}}[1 + C_{\mathrm{B}}\alpha_{\mathrm{AB}} + C_{\mathrm{C}}\alpha_{\mathrm{AC}} + \cdots]}{C_{\mathrm{A}}}. \tag{5.38}$$

By taking advantage of the fact that Criss and Birks' FPs approach made it possible to generate $R_i$ values, and therefore possible to calculate influence coefficients for binary contexts, Lachance [31] defined a *theoretical* influence coefficient for the context {*polychromatic incident source, matrix effect* (*absorption and enhancement*), *binary system*}, namely,

$$\alpha_{ij,\,\mathrm{bin}} = \frac{C_i - R_{i,\,\mathrm{bin}}}{C_j R_{i,\,\mathrm{bin}}}. \tag{5.39}$$

Given that $R_{i,\mathrm{bin}}$ could be computed from fundamental theory for both absorption and enhancement contexts confirmed the currently held general observation that influence coefficients are not constants when the incident excitation source is *polychromatic.*

In 1995 another step was taken by Lachance and Claisse [34] in expanding the generalization of influence coefficients in order to address the analytical context mainly used in practice, namely, {*polychromatic incident sources, matrix effects, multielement, explicitly from theory*}, by taking advantage of the contribution of Broll and Tertian [10] who proposed a valid expression for the long sought "weighting factor," which was none other than the monochromatic component of Criss and Birks [12] equation numbered "7." This led to defining influence coefficients as:

$$m_{ij} = \frac{A_{ij} - E_{ij}}{I_{i\mathrm{A}} + I_{i\mathrm{E}}}, \tag{5.40}$$

where $I_{i\mathrm{A}}$ is the theoretical intensity excited by absorption of the primary radiation and $I_{i\mathrm{E}}$ is the theoretical intensity resulting from enhancement by matrix elements, and the terms $A_{ij}$ and $E_{ij}$ are based on the summation of related monochromatic components. The definitions are given here.

The process of defining influence coefficients for the comprehensive context, i.e.

- Polychromatic incident excitation source;
- Matrix effect(s), absorption and enhancement;
- Multielement systems;
- Defined explicitly as a function of FPs;

can be visualized as consisting of four steps:

1. Define monochromatic absorption and enhancement influence coefficients

$$\alpha_{ij\lambda} = \frac{\mu_j^* - \mu_i^*}{\mu_i^*}$$

$$e_{ij\lambda} = e_j \left(e_j' + e_j''\right) \tag{5.41}$$

in which

$$e_j' \text{ relates to incident } \lambda,\ \psi',\ \mu_s',\ \text{and } \mu_{s\lambda_j},$$
$$e_j'' \text{ relates to emitted } \lambda_i,\ \psi'',\ \mu_s'',\ \text{and } \mu_{s\lambda_j},$$

where the superscripts $'$ and $''$ refer to incident and emergent radiation, respectively, and where

$$e_j = \frac{0.5(p_{\lambda_j}\mu_{i\lambda_j}\mu_{j\lambda})}{\mu_{i\lambda}}$$

$p_{\lambda j}$ is the probability that a $\lambda_j$ photon will be emitted, which in turn is the product of three probabilities; $\mu_{i\lambda_j}$ is the mass absorption coefficient of the analyte element for the wavelength of the enhancing matrix element line; $\mu_{j\lambda}$ is the mass absorption coefficient of the enhancing element for the incident monochromatic wavelength; $\mu_{i\lambda}$ is the mass absorption coefficient of the analyte element for the incident monochromatic wavelength.

For example, in the case of the K$\alpha$ line,

$$p_{\lambda_j} = \frac{r_\mathrm{K} - 1}{r_\mathrm{K}} \omega_\mathrm{K} p_{f_{\mathrm{K}\alpha}},$$

where $(r_\mathrm{K} - 1)/r_\mathrm{K}$ is the probability that the absorbed incident photon ejects a K shell electron rather than an L or M shell electron, and $r_\mathrm{K}$ is the

$K_{edge}$ absorption jump ratio; $\omega_K$ (the fluorescent yield) is the probability that the electronic transition leads to the emission of a characteristic photon rather than ejecting an Auger-electron; $p_{f_{K\alpha}}$ is the probability that a $K\alpha$ photon is emitted rather than a $K\beta$, and

$$e'_j = \frac{1}{\mu'_s} \ln\left(1 + \frac{\mu'_s}{\mu_{s\lambda_j}}\right)$$

$$e''_j = \frac{1}{\mu''_s} \ln\left(1 + \frac{\mu''_s}{\mu_{s\lambda_j}}\right).$$

2. Define $A_{ij\lambda}$ and $E_{ij\lambda}$

$$A_{ij\lambda} = I_{i\mathrm{A}\lambda} a_{ij\lambda} \quad E_{ij\lambda} = I_{i\mathrm{A}\lambda} e_{ij\lambda}. \tag{5.42}$$

3. Define $A_{ij}$ and $E_{ij}$

$$A_{ij} = \sum_{\lambda_{\min}}^{\lambda_{\mathrm{Edge}}} A_{ij\lambda}\Delta\lambda \quad E_{ij} = \sum_{\lambda_{\min}}^{\lambda_{\mathrm{Edge}}} E_{ij\lambda}\Delta\lambda. \tag{5.43}$$

4. Define $a_{ij}$ and $e_{ij}$

$$a_{ij} = \frac{A_{ij}}{I_{i\mathrm{A}} + I_{i\mathrm{E}}} \quad e_{ij} = \frac{E_{ij}}{I_{i\mathrm{A}} + I_{i\mathrm{E}}}. \tag{5.44}$$

In summary, therefore,

$$\begin{matrix} a_{ij\lambda} \Rightarrow A_{ij\lambda} \Rightarrow A_{ij,\mathrm{poly}} \Rightarrow a_{ij,\mathrm{poly}} & \text{(absorption)} \\ e_{ij\lambda} \Rightarrow E_{ij\lambda} \Rightarrow E_{ij,\mathrm{poly}} \Rightarrow e_{ij,\mathrm{poly}} & \text{(enhancement)} \end{matrix}.$$

Thus, in the L–C algorithm, the comprehensive influence coefficient $m_{ij}$ is simply equal to the difference $a_{ij} - e_{ij}$ and the coefficient $m_{ij}$ is inversely proportional to the total emitted intensity, which is none other than equation numbered "9" in Criss and Birks [12].

$$C_i = C_{i,\,\mathrm{app}}\left[1 + \sum\{m_{ij}\}C_j\right] = C_{i,\,\mathrm{app}}\left[1 + \sum\{a_{ij} - e_{ij}\}C_j\right]. \tag{5.45}$$

While the above formulation retains the original formulation first proposed by Lachance and Traill as far as the algorithm is concerned, the definition of the influence coefficient $\{\ldots\}_{ij}$ is an entirely different expression due to the much more comprehensive expression defining the coefficient, namely from $\{\ldots\}_{ij} = \alpha_{ij\lambda}$ to $\{\ldots\}_{ij} = m_{ij,\,\mathrm{poly}}$.

The L–T model readily lends itself to be reformulated for an empirical approach to generate influence coefficients from experimental data, something that is commonly done in practice.

For a three-element system, these equations are:

$$\sum_n C_j^2 \alpha_{ij} + \sum_n C_j C_k \alpha_{ik} = \sum_n C_j \left( \frac{C_i}{C_{i,\,\mathrm{app}}} - 1 \right)$$

$$\sum_n C_k^2 \alpha_{ik} + \sum_n C_j C_k \alpha_{ij} = \sum_n C_k \left( \frac{C_i}{C_{i,\,\mathrm{app}}} - 1 \right). \quad (5.46)$$

These equations can be extended for a $4, 5, \ldots$ element system in which case the number of standards must be increased so as to have enough "degrees of freedom."

### 5.3.3 The Claisse–Quintin Algorithm

*J.P. Willis and G.R. Lachance*

In 1967 Claisse and Quintin [11] addressed the definition of absorption influence coefficients for the context of a polychromatic incident source. They concluded from both their theoretical and experimental treatment that the influence coefficients were no longer constants, but vary in a well-defined (somewhat narrow) concentration range as a function of specimen composition. This led to an algorithm in which this deviation/curvature of $\alpha_\mathrm{B}$ is approximated by a linear expression

$$C_\mathrm{A} = R_\mathrm{A} \left( 1 + \alpha_\mathrm{B} C_\mathrm{B} + \alpha_\mathrm{C} C_\mathrm{C} + \cdots + \alpha_\mathrm{BB} C_\mathrm{B}^2 + \cdots + \alpha_\mathrm{BC} C_\mathrm{B} C_\mathrm{C} \right). \quad (5.47)$$

The fact that the total matrix effect cannot be expressed as a sum of binary influence coefficients led to the introduction of a "third element effect" influence coefficient times the product $C_j C_k$.

$$C_\mathrm{A} = R_\mathrm{A} \left( 1 + \sum_\mathrm{B} \left\{ \alpha_\mathrm{B} + \alpha_\mathrm{BB} C_\mathrm{B} + \sum_\mathrm{C} \alpha_\mathrm{BC} C_\mathrm{C} \right\} C_\mathrm{B} \right). \quad (5.48)$$

This was a major contribution in that it extended the range for which the L–T model was applicable. They also proposed the concept of a "weighting factor" for the calculation of the $\alpha$ influence coefficient, which eventually proved not to be applicable in practice.

In 1974 Rousseau and Claisse [48], based on observation of theoretical data, proposed that, in practice, the curve could be approximated on the basis of two binary compositions $C_\mathrm{A} = 0.8$ and 0.2, the balance being any other element as shown in Table 5.2 and Fig. 5.6.

In 1976 Tertian [57] noted that it was much more advantageous to calculate the values of the coefficient $\alpha_\mathrm{AB}$ *at the* $C_\mathrm{A}$ *concentration*. It was shown that in this case the coefficient $\alpha_\mathrm{BC}$ essentially became $\sim$0 when enhancement was absent, and that the values of $\alpha_\mathrm{BC}$ were lower by a factor $\sim$3 if enhancement is present. This translated into the following expression if the C–Q model is retained, i.e., model C–Q+:

**Table 5.2.** Influence coefficients for binary mixtures

| A = Fe (26) | | | | | | |
|---|---|---|---|---|---|---|
| $C_A$ | B = O (8) | B = Si (14) | B = Cr (24) | B = Ni (28) | B = Zr (40) | B = Sn (50) |
| 0.80 | −0.83 | −0.20 | 2.19 | −0.20 | 0.69 | 2.20 |
| 0.20 | −0.84 | −0.22 | 2.10 | −0.36 | 0.63 | 2.11 |

| A = Mo (42) | | | | | | |
|---|---|---|---|---|---|---|
| $C_A$ | B = O (8) | B = Si (14) | B = Cr (24) | B = Ni (28) | B = Sr (38) | B = Sn (50) |
| 0.80 | −0.96 | −0.81 | −0.19 | 0.21 | 1.65 | −0.24 |
| 0.20 | −0.96 | −0.81 | −0.20 | 0.20 | 1.62 | −0.24 |

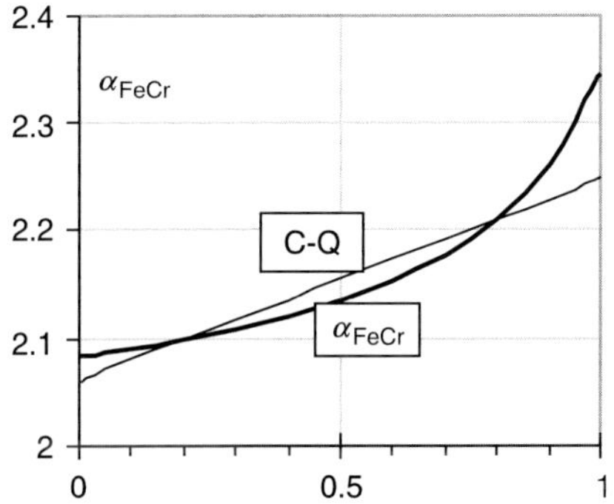


**Fig. 5.6.** $\alpha_{\mathrm{FeCr}}$ plotted against $C_{\mathrm{Fe}}$ indicating a hyperbolic fit, and a linear approximation through $C_{\mathrm{Fe}} = 0.2$ and 0.8

$$C_A = R_A\left(1 + \sum_B \left\{\alpha_B + \alpha_{BB}C_M + \sum_C \alpha_{BC}C_C\right\} C_B\right), \qquad (5.49)$$

where $C_M = C_j + C_k + \cdots + C_n$, i.e., equal to the sum of the matrix components. The following is a numerical example of the calculation of the influence coefficients from theory principles. The analytical context is that of Rasberry and Heinrich [44], namely: W target, 45 kV, incident angle 63°, emergent angle 33°. The analyte is Fe in the system Cr–Fe–Ni.

Detailed example of the calculation of $\alpha_B$ and $\alpha_{BB}$ using data from Table 5.3; A = Fe, B = Cr:

(1) $C_{\mathrm{Fe}} = 0.2 \quad C_{\mathrm{Cr}} = 0.8 \quad \alpha_{\mathrm{Cr}} = 2.0988$
(2) $C_{\mathrm{Fe}} = 0.8 \quad C_{\mathrm{Cr}} = 0.2 \quad \alpha_{\mathrm{Cr}} = 2.2112$
From (1) and (2) one obtains:
(3) $2.0988 = \alpha_{\mathrm{Cr}} + 0.8\,\alpha_{\mathrm{CrCr}}$
(4) $2.2112 = \alpha_{\mathrm{Cr}} + 0.2\,\alpha_{\mathrm{CrCr}}$

**Table 5.3.** Computed influence coefficients

| | A = Fe (26) | |
|---|---|---|
| $C_A$ | B = Cr (24) | C = Ni (28) |
| 0.8 | 2.2112 | −0.2368 |
| 0.2 | 2.0988 | −0.4010 |
| $\alpha_B$ | 2.2487 | −0.1821 |
| $\alpha_{BB}$ | −0.1873 | −0.2737 |

Subtracting (4) from (3) yields

$$-0.1124 = 0.6\,\alpha_{CrCr}.$$

Solving for $\alpha_{CrCr}$ yields: $\alpha_{CrCr} = -0.1124/0.6 = -0.1873$, from which: $\alpha_{Cr} = 2.2487$.

Thus, the approximation results in a linear variation from a maximum value equal to 2.2487 to a minimum value of 2.2487 – 0.1873, equal to 2.0614. The equivalent theoretical values are 2.3456 and 2.0837. Note the large difference in the maximum value.

The value for $\alpha_{CrNi}$ is calculated as follows, component $C$ is Ni:
Specimen: $C_{Fe} = 0.30 \quad C_{Cr} = 0.35 \quad C_{Ni} = 0.35$

(1) Generate the value for the correction term from FPs approach: 1.5722.
(2) Generate the value for the correction term using the C–Q+ model: 1.6094.

Subtract (2) from (1) and divide by the product $C_j C_k$:

$$(1.5722 - 1.6094)/(0.35 \times 0.35) = -0.0372/0.1225 = -0.304 = \alpha_{CrNi}.$$

Thus, the correction term for matrix effects using the C–Q+ model for analyte Fe ($R_{Fe} = 0.1908$) would take the form:

$$\begin{aligned} C_{Fe} &= 0.1908\,[1 + \{2.2487 + (-0.1873)\,0.70\}0.35 \\ &\quad + \{-0.1821 - 0.2734\,(0.70)\}0.35 + (-0.304)\,0.1225] \\ &= 0.1908\,[1.5732] \\ &= 0.3002. \end{aligned}$$

## Empirical Influence Coefficients Approach

The aim of the Claisse–Quintin [11] approach was to show that the influence coefficients are *not* constants. Therefore it would be somewhat counterproductive to recommend an empirical approach using this algorithm. The FP approach proposed by Criss and Birks [12] provided a theoretical method for calculating the C–Q influence coefficients and this remains by far the better approach.

### 5.3.4 The COLA Algorithm

*J.P. Willis and G.R. Lachance*

The concept proposed by Claisse and Quintin [11] of deriving a mathematical expression for the L–T algorithm in which the values of the influence coefficient varied as a function of specimen composition, in the context of a polychromatic incident excitation source, was a major advancement. It extended the concentration range of application and provided an approach to computed influence coefficients from theory. However, their approach involved approximating a curve by linear fitting. They also introduced a "third element term" to compensate for the fact that the algorithm cannot be simply "a sum of binary correction terms."

The COLA model sought to replace the linear curve fitting with a hyperbolic curve fitting, involving a correction algorithm of the type:

$$\left[1 + \sum \{\alpha_{ij,\,\mathrm{hyp}} + \alpha_{ijk} C_k\} C_j\right]$$

where the subscript "hyp" is used to denote that the binary influence coefficients are approximated by an hyperbola (compare Figs. 5.6 and 5.7).

In 1981 Lachance [32] proposed the following expression for quantifying the correction term, namely:

$$C_i = R_i \left[1 + \sum_j \left\{\alpha_1 + \frac{\alpha_2 C_\mathrm{M}}{1 + \alpha_3(1 - C_\mathrm{M})} + \alpha_{ijk} C_k\right\} C_j\right] \tag{5.50}$$

and the better quality of the hyperbolic fit to the binary influence coefficient values is shown in Fig. 5.7.

The COLA model also takes into consideration the observation by Tertian that the influence coefficient $\alpha_{ij}$ should be calculated at the $C_i$ concentration level. For the Claisse–Quintin model, this translates into the introduction of $C_\mathrm{M}$, where $C_\mathrm{M}$ is equal to the sum of the matrix elements, i.e., $C_\mathrm{M} = C_j + C_k + \cdots + C_n$.

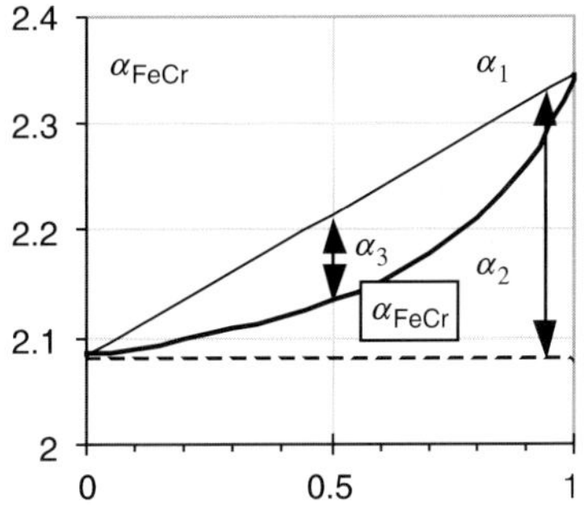


**Fig. 5.7.** Explanation of the three influence coefficients in the COLA algorithm

This approach meant that, in the absorption context, $\{\ldots\}_{ijk}$ values were for all practical purposes equal to 0, and in the cases where enhancement was present, the values as computed by C–Q were reduced by as much as a factor of 2 to 3. If the C–Q model is retained, this is equivalent to using the formulation

$$[1+\{\alpha_{ij}+\alpha_{ijj}C_{\mathrm{M}}\}C_j] \text{ instead of}$$

$$[1+\{\alpha_{ij}+\alpha_{ijj}C_j\}C_j] \text{ in the case of a linear approximation.}$$

Tao et al. [55] published "NBSGSC – A FORTRAN ..." program for performing quantitative analysis of bulk specimens by X-ray fluorescence spectrometry. This program corrects for absorption/enhancement phenomena using the comprehensive alpha coefficient algorithm proposed by Lachance (COLA). NBSGSC is a revision of the program ALPHA and CARECAL originally developed by R.M. Rousseau of the Geological Survey of Canada.

The process can be described in the following steps:

- The value of the correction term, i.e., [ ... ], is calculated using the classical Criss and Birks approach;
- The value of [ ... ] is calculated using the six binary influence coefficients $\alpha_1$, $\alpha_2$, and $\alpha_3$ in Table 5.4, where
  - $\alpha_1$ are the values when $C_i$ approaches the limit 1.0 and are calculated for $C_i = 0.999$: 2.3456 and −0.1954 for $\alpha_{\mathrm{FeCr}}$ and $\alpha_{\mathrm{FeNi}}$, respectively;
  - $\alpha_2$ is the difference between those values calculated for $C_i = 0.001$ and the above values, i.e., $2.0837 - 2.3456 = -0.2619$ and $-0.4861 - (-0.1954) = -0.2906$, respectively;
  - $\alpha_3$ is the value that defines the "degree of curvature" of the hyperbola (deviation from the straight line at $C_i = 0.5$) and causes $\alpha_{ij,hyp}$ to match the theoretical value at the mid point. $\alpha_3$ is defined as

$$\alpha_3 = \frac{\alpha_2}{m_{ij,\mathrm{bin},C_i=0.5} - \alpha_1} - 2.$$

For example, $\alpha_3$ for analyte Fe and matrix element Cr:

$$\alpha_3 = \frac{-0.2619}{2.1348 - 2.3456} - 2 = -0.7578.$$

**Table 5.4.** Influence coefficients used to generate {...} for the COLA model

| | $i$ = Fe (26) | |
|---|---|---|
| $C_i$ | $j$ = Cr (24) | $k$ = Ni (28) |
| 0.999 | 2.3456 | −0.1954 |
| 0.5 | 2.1348 | −0.3092 |
| 0.001 | 2.0837 | −0.4861 |
| $\alpha_1$ | 2.3456 | −0.1954 |
| $\alpha_2$ | −0.2619 | −0.2906 |
| $\alpha_3$ | −0.7578 | 0.5542 |

This means that a hyperbolic fit will be concordant with theory at the two-end limits and at the concentration of 0.5 for $C_i$.

The software program NBSGSC, which used the COLA algorithm, was developed in the public domain but not commercially as a "ready to use" product. Martin [39] remedied this and the product was described as "... the NBSGSC documentation that ... developed a package which is in scope far beyond that of the original mainframe NBSGSC." This software was/is available from Rainier Software. The software offered the following options: alloy systems, oxide systems (pressed powders), and fused disk systems. The software also included a utility program "that corrects intensities to their 'infinite thickness or critical thickness ... of the sample'."

### 5.3.5 The de Jongh Algorithm

*B.A.R. Vrebos*

The influence coefficient algorithm proposed in 1973 by de Jongh was probably the first influence coefficient algorithm [14] explicitly based on FP calculations. The publication focused more on the calculation of the influence coefficients, rather than justifying its format. The general formulation of his equation can be written, in original notation, as

$$W_i = E_i R_i \left( 1 + \sum_{\substack{j=1 \\ j \neq e}}^{n} \alpha_{ij} W_j \right), \tag{5.51}$$

where $E_i$ is a proportionality constant (and is usually determined during the calibration), $R_i$ is the count rate and $W_i$ are the element concentrations (weight fractions). At first sight, the equation looks similar to the classic Lachance–Traill equation: The summation covers $n-1$ elements (as is the case with the algorithm of Lachance and Traill), but the eliminated element, $e$, is the same for *all* equations. The following set of equations is obtained if for a ternary specimen (with compounds A, B, and C) element C is eliminated:

$$\begin{aligned} W_\mathrm{A} &= E_\mathrm{A} R_\mathrm{A} \left(1 + \alpha_\mathrm{AA} W_\mathrm{A} + \alpha_\mathrm{AB} W_\mathrm{B}\right) \\ W_\mathrm{B} &= E_\mathrm{B} R_\mathrm{B} \left(1 + \alpha_\mathrm{BA} W_\mathrm{A} + \alpha_\mathrm{BB} W_\mathrm{B}\right) \\ W_\mathrm{C} &= E_\mathrm{C} R_\mathrm{C} \left(1 + \alpha_\mathrm{CA} W_\mathrm{A} + \alpha_\mathrm{CB} W_\mathrm{B}\right). \end{aligned} \tag{5.52}$$

From this set of equations, one of the major differences with the equation of Lachance–Traill can be clearly seen: In order to obtain the concentration of elements A and B, the concentration of C, $W_\mathrm{C}$, is not required.

#### Calculation of the Coefficients

The basis for the calculation of the coefficients is an approximation of $W_i/R_i$ by a Taylor series with the expansion limited to the first-order terms:

$$\frac{W_i}{R_i} = E_i + \delta_{i1}\Delta W_1 + \delta_{i2}\Delta W_2 + \cdots + \delta_{in}\Delta W_n, \tag{5.53}$$

where $E_i$ is a constant given by

$$E_i = \left(\frac{W_i}{R_i}\right)_{\text{Average}} \tag{5.54}$$

and

$$\Delta W_i = W_i - W_{i,\,\text{average}} \tag{5.55}$$

$\delta_{ij}$ are the partial derivatives of $W_i/R_i$ with respect to a concentration:

$$\delta_{ij} = \frac{\partial \frac{W_i}{R_i}}{\partial W_j}. \tag{5.56}$$

In 1973, the matrix effect was not yet given the status of a separate symbol. The symbol $W_i/R_i$would now be replaced by $M_i$. For practical reasons the partial derivatives are calculated as finite differences, and not by using the equations for the first derivative as derived by Shiraiwa and Fujino [53].

The series expansion is written around a given (or calculated) average composition $W_{1,\,\text{average}}$, $W_{2,\,\text{average}}, \ldots,$ $W_{n,\,\text{average}}$ (represented by the expression $[W_i/R_i]_{\text{average}}$). The concentration of each element in turn is increased by 0.1%, and $W_i/R_i$ is now calculated again; the expression $[W_i/R_i]_{Wj+0.001}$ is used. The partial derivative is thus calculated as follows:

$$\delta_{ij} = \frac{\partial \frac{W_i}{R_i}}{\partial W_j} = \frac{\left[\frac{W_i}{R_i}\right]_{W_j+0.001} - \left[\frac{W_i}{R_i}\right]_{\text{average}}}{0.001}.$$

The calculation of all the $\delta$s for a given analyte $i$ requires thus the calculation of $[W_i/R_i]_{Wj+0.001}$ for each of the constituent compounds. Since the summation of all deviations $\Delta W$ from the average must equal to 0, i.e.,

$$\sum_{j=1}^{n} \Delta W_j = 0$$

one compound can be eliminated. Let the eliminated compound be denoted by the subscript $e$:

$$\Delta W_e = -\Delta W_1 - \Delta W_2 - \cdots - \Delta W_n.$$

Substitution yields:

$$\frac{W_i}{R_i} = E_i + \beta_{i1}\Delta W_1 + \beta_2\Delta W_2 + \cdots + \beta_{in}\Delta W_n, \tag{5.57}$$

which has only $(n-1)$ terms, since the term in $\Delta W_e$ has been eliminated. The values for the $\beta$ coefficients can be calculated from

$$\beta_{ij} = \delta_{ij} - \delta_{ie}.$$

Obviously, $\beta_{ij}$ equals 0 when in (5.57) $j = e$. Equation (5.57) has (n-1) terms, but they are still in $\Delta W$, rather than $W$, so the average composition is still required. Conversion of $\Delta W$ to $W$ is done by substituting (5.55) in (5.57):

$$\frac{W_i}{R_i} = E_i - \sum_{j \neq e}^{n} \beta_{ij} W_{j,\,\text{average}} + \sum_{j \neq e}^{n} \beta_{ij} W_j. \tag{5.58}$$

Defining the $\alpha$-coefficients as

$$\alpha_{ij} = \frac{\beta_{ij}}{E_i - \sum\limits_{j \neq e}^{n} \beta_{ij} W_{j,\,\text{average}}} \tag{5.59}$$

yields finally

$$W_i = M_{i,\,\text{Average}} R_i \left( 1 + \sum_{\substack{j=1 \\ j \neq e}}^{n} \alpha_{ij} W_j \right). \tag{5.60}$$

This indicates that the weight fraction of the analyte is calculated from its relative intensity, a matrix correction term, and the matrix correction term for the average composition (which for a given composition is a constant). The value for $M_{i,\,\text{Average}}$ can be calculated using the influence coefficients calculated and taking the composition ($W_i$, $W_j$) equal to the average composition. This equation is very similar at first sight to the Lachance–Traill equation except for the term $M_{i,\,\text{Average}}$.

### 5.3.6 The Broll–Tertian Algorithm

*K.-E. Mauser*

In 1983 Broll and Tertian [10] outlined the general principles of "Quantitative X-ray fluorescence analysis by use of fundamental influence coefficients." More detailed descriptions and slightly different names for the same basic algorithm can be found in Broll [8] ("fundamental coefficient method") and Broll et al. [9] ("effective coefficient method"). The algorithm has been designed to combine the theoretical exactness of the FP approach with the flexibility of the usual Lachance–Traill formulation.

#### Fundamental Coefficients Versus Fundamental Parameters

Both FP and fundamental coefficient methods rely on the calculation of theoretical fluorescence intensities after Sherman's equations. Tertian established a basic relationship between these approaches and postulated the existence of *effective* influence coefficients, which, for any given composition $C_i, C_j, C_k, \ldots$, satisfy exactly the Lachance–Traill equations:

$$C_i = R_i\,(1 + \bar{\alpha}_{ij}C_j + \bar{\alpha}_{ik}C_k + \cdots). \tag{5.61}$$

These effective influence coefficients $\bar{\alpha}_{ij}$ can be calculated using Tertian's identities

$$\bar{\alpha}_{ij} = \alpha'_{ij} - h_{ij}\frac{C_i}{R_i}, \tag{5.62}$$

where $\alpha'_{ij}$ accounts for primary fluorescence and $h_{ij}$ is a factor due to secondary fluorescence (enhancement). For any given composition, the effective coefficients, as defined by Tertian's identities, can be derived from the calculation of the theoretical fluorescence intensities, as used in the FP approach. The effective coefficients are calculated for a given composition, and of practical value at or near this composition, but they are more or less composition dependent. In a typical case of a multicomponent system, it is necessary to work with the effective coefficients corresponding to the composition of each sample.

### The Algorithm of Standard Comparison

Broll and Tertian [10] proposed an algorithm to compare the unknown sample with a standard sample. This procedure compensates for uncertainties in FPs and instrumental factors and avoids the use of relative intensities, $R_i$. This comparison algorithm is applied as follows:

The effective coefficients $\alpha^*_{ij}$, corresponding to the standard sample, are introduced into the Lachance–Traill equation for the standard

$$C^*_i = R^*_i\left(1 + \sum_{j\neq i}\alpha^*_{ij}C^*_j\right). \tag{5.63a}$$

The same coefficients $\alpha^*_{ij}$ are used for the unknown sample in the following equation

$$C_i = R_i\left(1 + \sum_{j\neq i}\alpha^*_{ij}C_j\right). \tag{5.63b}$$

The algorithm for standard comparison is obtained by combining the two equations and thus eliminating the fluorescence intensity of the pure elements $I_{i,1}$:

$$C^n_i = C^*_i\frac{I_i}{I^*_i}\frac{1 + \sum_{j\neq i}\alpha^*_{ij}C_j}{1 + \sum_{j\neq i}\alpha^*_{ij}C^*_j}. \tag{5.64a}$$

In this equation, the influence coefficients calculated for the composition of the standard are used for the unknown sample, too. If, as usual in the practical application of X-ray fluorescence analysis, the composition of the standard $C_i^*$ and that of the unknown sample $C_i$ are different, the composition obtained by (5.63) can be used as an approximate composition $C_i^n$ to calculate new effective coefficients $\alpha_{ij}^n$, which are then reintroduced in the numerator of the algorithm for standard comparison allowing the calculation of a composition with a higher precision:

$$C_i^n = C_i^* \frac{I_i}{I_i^*} \frac{1 + \sum\limits_{j \neq i} \alpha_{ij}^n C_j}{1 + \sum\limits_{j \neq i} \alpha_{ij}^* C_j^*} . \tag{5.64b}$$

Broll et al. [9] expanded this method to be employed with several standards and modified the algorithm for better precision in trace element analysis. The algorithm of effective fundamental coefficients has been further refined and is applied in the FP calculations of a current commercial software package. There it serves exactly the purpose described at the end of Sect. 5.2.4 about back-calculation schemes: To improve speed and reliability of the iteration process by giving intelligent, nonlinear approximations for the function $f(C_1, \ldots, C_n)$. On the other hand, in this implementation the Broll–Tertian algorithm is no longer visible as an independent method of its own kind.

### 5.3.7 The Japanese Industrial Standard Method

*N. Kawahara*

The Japanese industrial standard (JIS) method was developed for high-accuracy analyses of stainless steel by Ito et al. [25]. Revised several times by the Technical Committee on XRF Analysis in the Iron and Steel Institute of Japan after its first edition in 1973, this method was standardized as the latest revision of JIS G1256-1997 [26].

The JIS method is an empirical method to account for matrix effects and line overlaps. Calibration curves of binary systems consisting of iron (or the principal element) and the analyte elements are prepared. Each apparent calibration curve is approximated by a parabolic function to the fluorescent intensity of the analyte:

$$C_{i,\,\mathrm{app}} = b_0 + b_1 R_i + b_2 R_i^2 , \tag{5.65}$$

where $b_0$, $b_1$, and $b_2$ are constants. Such calibration curves of binaries are obtained experimentally using standard materials that have concentrations covering the range of interest.

The concentration of the analyte $i$ in a ternary system, which consists of Fe and elements $i$ and $j$, is expressed as:

$$C_i = C_{i,\,\mathrm{app}}(1 + d_{ij} C_j) , \tag{5.66}$$

where $d_{ij}$ is the matrix coefficient of element $j$ to $i$. The general form of the JISmodel for multicomponent materials is written as

$$C_i = C_{i,\,\mathrm{app}} \left(1 + \sum_{j \neq i,\,\mathrm{Fe}} d_{ij} C_j\right) \tag{5.67}$$

or

$$C_i = (b_0 + b_1 R_i + b_2 R_i^2) \left(1 + \sum_{j \neq i,\,\mathrm{Fe}} d_{ij} C_j\right) . \tag{5.68}$$

The matrix coefficients $d_{ij}$ are derived by multiple regression from experimental intensities using multicomponent standard samples or ternary standard samples, by applying (5.66) or (5.67). Many values for $d_{ij}$ have been obtained and are summarized in JIS G1256-1997, and they are widely used in iron and steel industries in Japan and other countries.

The analyst can either prepare the reference materials over the composition-range of interest for obtaining the apparent calibration curves, or derive them from the measurements of multielement standard materials and employ $d_{ij}$ values published by JIS or other resources.

Use of parabolic functions has the practical advantage that experimental factors such as the counting linearity of the detector and effects of higher order matrix coefficients can be included in them, although this may not be applicable to the entire composition range (0–100%).

For comparison of the JIS method to de Jongh's equation, (5.51) can be rewritten as

$$\begin{aligned} W_i &= E_i R_i \left(1 + \sum_{j \neq \mathrm{Fe}} \alpha_{ij} W_j\right) \\ W_i - E_i R_i \alpha_{ii} W_i &= E_i R_i \left(1 + \sum_{j \neq i,\,\mathrm{Fe}} \alpha_{ij} W_j\right) \\ W_i &= \frac{E_i R_i}{(1 - \alpha_{ii} E_i R_i)} \left(1 + \sum_{j \neq i,\,\mathrm{Fe}} \alpha_{ij} W_j\right) . \end{aligned} \tag{5.69}$$

Note that the parameter $e$ of (5.51) is here denoted as Fe. The final form of (5.69) shows that the JIS method and de Jongh's equation are mathematically equivalent as far as a hyperbolic curve can be well approximated to a parabolic function in the composition range of interest.

### 5.3.8 The Fundamental Algorithm

*R.M. Rousseau*

In 1984, Rousseau proposed the *Fundamental Algorithm* [46, 47] to correct for all matrix effects that modify the measured net intensity emitted by an *analyte i* in a given specimen. It is deduced from the Sherman equation [51, 52] by algebraic manipulation without any approximation and is thus the same equation expressed in another form, namely that usually used for influence coefficient algorithms. It considers direct and indirect (secondary) excitations by the polychromatic incident radiation, calculates influence coefficients for each sample to be analyzed, and separately takes into account both matrix effects, *absorption* and *enhancement.* It can thus be considered as an extension of the Sherman equation.

## Deduction of the Fundamental Algorithm

The equation describing the observed photon count rates originating from direct and secondary excitations as discussed in Sect. 5.2 is the starting point of this section. Rousseau's original notation is kept in this section (Table 5.5):

$$
\begin{aligned}
N_{i,\,\text{observed}} &= N_{i,\,\text{prim}} + N_{i,\,\text{sec}} \qquad (5.70)\\
&= g_i C_i \int_{\lambda=\lambda_0}^{\lambda=\lambda_{\text{Edge},i}} \frac{\tau_i(\lambda) I_0(\lambda)}{\mu'_S(\lambda) + \mu''_S(\lambda_i)}\,\mathrm{d}\lambda\\
&\quad + \sum\nolimits_{\text{Lines}\,j} \frac{1}{2} g_i C_i C_j \frac{r_j - 1}{r_j} p_j \omega_j \tau_i(\lambda_j)\\
&\quad \times \int_{\lambda=\lambda_0}^{\lambda=\lambda_{\text{Edge},j}} \frac{\mu_j(\lambda) I_0(\lambda)}{\mu'_S(\lambda) + \mu''_S(\lambda_j)} \times \left\{ \left[ \frac{1}{\mu'_S(\lambda)} \ln\left(1 + \frac{\mu'_S(\lambda)}{\mu_S(\lambda_j)}\right) \right] \right.\\
&\quad \left. + \left[ \frac{1}{\mu''_S(\lambda_i)} \ln\left(1 + \frac{\mu''_S(\lambda_i)}{\mu_S(\lambda_j)}\right) \right] \right\} \mathrm{d}\lambda .
\end{aligned}
$$

The tertiary fluorescent X-ray photon count rate is not taken into account because its contribution is usually very small and its mathematical treatment is complex. Note that in this equation, the terms for primary and secondary excitations contain integrals with the same upper boundary (the maximum photon energy in the tube spectrum) but the lower boundaries are element

**Table 5.5.** Notation used by Rousseau

| Rousseau | Sect. 5.2 |
|---|---|
| $I_0$ | $N_0$ |
| $\mu_i(\lambda)$, $\mu_i(\lambda_j)$ | $\mu_i$, $\mu_{ij}$ |
| $\mu_S$ | $\mu$ or $\mu_S$ |
| $\tau \approx \mu$ | |
| $g_i$ | $G_i$ |
| $r$ | S |

dependent (absorption edge energies). By introducing two variable coefficients, $D_i$ and $D_{i,j}$,

$$D_i(E) = \begin{cases} 0 & \text{for} \quad E < E_{\text{Edge},i} \\ 1 & \text{for} \quad E \geq E_{\text{Edge},i} \end{cases}$$

$$D_{i,j} = \begin{cases} 0 & \text{for} \quad E_{\text{Edge},j} < E_{\text{Edge},i} \\ 1 & \text{for} \quad E_{\text{Edge},j} \geq E_{\text{Edge},i} \end{cases}$$

the equation can be rewritten as a single integral expression. $D_i$ sets the range of those integrals to 0 when the running energy, $E$, is below the absorption edge of the current element, and $D_{ij}$ sets the integrals of the secondary excitation term to 0 when the matrix elements $j$ cannot cause secondary excitation of the analyte line $i$:

$$\begin{aligned} N_{i,\,\text{observed}} &= N_{i,\,\text{prim}} + N_{i,\,\text{sec}} \\ &= g_i C_i \int_{\lambda=\lambda_0}^{\lambda=\lambda_{\text{Edge},i}} \frac{\tau_i(\lambda) I_0(\lambda)}{\mu'_S(\lambda) + \mu''_S(\lambda_i)} \left(1 + \sum_{\text{Elements}\, j} C_j \delta_{ij}(\lambda)\right) \mathrm{d}\lambda , \end{aligned}$$

where

$$\begin{aligned} \delta_{ij}(\lambda) &= \frac{1}{2} D_{ij}(\lambda)\, D_i(\lambda_j) \frac{r_j - 1}{r_j} p_j \omega_j \mu_j(\lambda) \frac{\tau_i(\lambda_j)}{\tau_i(\lambda)} \\ &\times \left\{ \left[ \frac{1}{\mu'_S(\lambda)} \ln\left(1 + \frac{\mu'_S(\lambda)}{\mu_S(\lambda_j)}\right) \right] + \left[ \frac{1}{\mu''_S(\lambda_i)} \ln\left(1 + \frac{\mu''_S(\lambda_i)}{\mu_S(\lambda_j)}\right) \right] \right\} . \end{aligned} \tag{5.71}$$

As opposed to (5.70), the sum in the above equation, $\Sigma_j$, is taken over all matrix elements, not lines. The lines emitted by each element $j$ are separately accounted for by the sum in the expression for $\delta_{i,j}$.

*Note:* When a K$\alpha$ line may enhance the measured L$\alpha_1$ or the L$\beta_1$ line of the **same** element $j$, this contribution (at the most 2–3% of total intensity) is neglected because of the complexity of the calculation. For example, the Ba K$\alpha$ line enhances the measured Ba L$\alpha_1$ line when the tube voltage is $\geq$40 kV. This omission is more or less compensated by the calibration procedure [49] owing to the use of a "comparison standard" approach. The enhancement effect of a K$\alpha$ line on the L$\alpha_1$ line of the same element can also be eliminated by reducing the tube voltage to a value lower than the binding energy of the K shell or simply by using the K$\alpha$ line rather than the L$\alpha_1$ line as the analytical line for the element to be determined.

The first fundamental information needed to calculate the intensity from element $i$ using (5.71) is the spectral distribution of the incident radiation. However, there is no analytical expression available to generate an incident spectrum in an essentially exact way, for any experimental conditions. Also, in order to be able to use the equations of the X-ray tube output proposed by Pella et al. [42], the integral over $\lambda$ is replaced here by a finite sum of $k$ values of wavelength intervals $\Delta\lambda_k$ with corresponding incident intensities $I_0(\lambda_k)$:

$$\int_{\lambda=\lambda_{\min}}^{\lambda_{\mathrm{Edge},i}} I_0(\lambda)\,\mathrm{d}\lambda \rightarrow \sum_{k=\lambda_{\min}}^{\lambda_{\mathrm{Edge},i}} I_0(\lambda_k)\,\Delta\lambda_k . \tag{5.72}$$

To simplify (5.71), two new variables ($\beta_{ij}$ and $W_i'$) are introduced based upon *effective mass absorption coefficients*. The word "effective" designates the quantification of the total absorption of specific wavelengths by an element or a specimen in a given analytical context. It therefore takes into consideration not only the fact that it is the sum of absorption of the incident and emergent radiations but also their path lengths in the specimen. Thus, the following definitions are used (compare 5.9):

$$\begin{aligned} \mu_i' &= \mu_i(\lambda_k)\csc\phi' & \mu_j' &= \mu_j(\lambda_k)\csc\phi' \\ \mu_i'' &= \mu_i(\lambda_i)\csc\phi'' & \mu_j'' &= \mu_j(\lambda_i)\csc\phi'' \end{aligned} \tag{5.73}$$

$$\begin{aligned} \mu_S' &= \sum_{n=i}^{N} C_n\mu_n' \\ \mu_S'' &= \sum_{n=i}^{N} C_n\mu_n'' \\ \mu_i^* &= \mu_i' + \mu_i'' \quad \mu_j^* = \mu_j' + \mu_j'' \quad \mu_S^* = \mu_S' + \mu_S'' . \end{aligned}$$

The equation of the effective mass absorption coefficient can be transformed as follows:

$$\begin{aligned} \mu_s^* &= \mu_S'(\lambda_k) + \mu_S''(\lambda_i) \\ &= \sum_n C_n\mu_n(\lambda_k)\ \csc\,\phi' + \sum_n C_n\mu_n(\lambda_i)\ \csc\,\phi'' \\ &= \sum_n C_n[\mu_n(\lambda_k)\ \csc\,\phi' + \mu_n(\lambda_i)\ \csc\,\phi''] \\ &= \sum_n C_n[\mu_n^*] = C_i\mu_i^* + C_j\mu_j^* + C_k\mu_k^* + \cdots + C_N\mu_N^* \\ &= \mu_i^*\left(C_i + C_j\frac{\mu_j^*}{\mu_i^*} + C_k\frac{\mu_k^*}{\mu_i^*} + \cdots + C_N\frac{\mu_N^*}{\mu_i^*}\right) . \end{aligned}$$

Now, bearing in mind a given specimen of $N$ elements:

$$\sum_{n=i}^{N} C_n = 1$$

we obtain

$$C_i = 1 - C_j - C_k - \cdots - C_N$$

and

$$\mu_s^* = \mu_i^* \left[1 + C_j \left(\frac{\mu_j^*}{\mu_i^*} - 1\right) + C_k \left(\frac{\mu_k^*}{\mu_i^*} - 1\right) + \cdots + C_N \left(\frac{\mu_N^*}{\mu_i^*} - 1\right)\right].$$

Defining the variable

$$\beta_{ij}(\lambda_k) = \frac{\mu_j^*}{\mu_i^*} - 1, \tag{5.74}$$

where

$$\mu_j^* = \mu_j(\lambda_k) \csc \phi' + \mu_j(\lambda_i) \csc \phi''$$
$$\mu_i^* = \mu_j(\lambda_k) \csc \phi' + \mu_i(\lambda_i) \csc \phi''$$

produces

$$\mu_s^* = \mu_i^* \left[1 + \sum_j C_j \beta_{ij}(\lambda_k)\right].$$

Note that the variable $\beta_{ij}(\lambda_k)$, as we will see later, is defined as the *absorption influence coefficient* in the case of a monochromatic incident source of wavelength $\lambda_k$. This coefficient corrects for the absorption effects of the matrix element $j$ on the analyte $i$ and can be positive or negative. If (5.74) is rewritten in the following form:

$$\beta_{ij}(\lambda_k) = \frac{\mu_j^* - \mu_i^*}{\mu_i^*}, \tag{5.75a}$$

it is easier to determine when the influence coefficient $\beta_{ij}(\lambda_k)$ is positive or negative. The values for $\beta_{ij}(\lambda_k)$ depend on the matrix composition. For example, if Fe is determined in the presence of Mg (a lighter matrix element), then $\mu_i^* > \mu_j^*$ and $\beta_{ij}(\lambda_k)$ is negative. If Fe is determined in the presence of Ni (a heavier matrix element), then $\mu_i^* < \mu_j^*$ and $\beta_{ij}(\lambda_k)$ is positive.

Furthermore, the coefficient $\beta_{ij}(\lambda_k)$ is the ratio of the difference between the mass absorption coefficients of elements $j$ and $i$ relative to the mass absorption coefficient of element $i$. In other words, the coefficient $\beta_{ij}(\lambda_k)$ shows, in a relative way, how much greater or smaller the absorption of element $j$ is compared to that of element $i$. The relative absorption of element $i$ compared to itself is therefore equal to 0. Indeed,

$$\beta_{ii}(\lambda_k) = \frac{\mu_i^* - \mu_i^*}{\mu_i^*} = 0. \tag{5.75b}$$

In the following paragraphs, we continue to modify Sherman's equation to make it easier to understand and manipulate. Defining a second variable:

$$W_i'(\lambda_k) = \frac{\mu_i(\lambda_k)}{\mu_i^*} I_o(\lambda_k)\,\Delta\lambda_k$$

and combining with the above equations leads to

$$I_i(\lambda_i) = g_i C_i \sum_{\lambda_k=\lambda_{\min}}^{\lambda=\lambda_{\text{Edge } i}} W_i'(\lambda_k) \frac{1+\sum\limits_j C_j \delta_{ij}(\lambda_k)}{1+\sum\limits_j C_j \beta_{ij}(\lambda_k)}. \tag{5.76}$$

This preliminary *modified Sherman equation* is already simpler and more revealing. Indeed, the intensity is still proportional to the concentration $C_i$ and also to a ratio on the right-hand side. The numerator contains all the enhancement coefficients $\delta_{ij}$ of each element $j$ of the matrix, and the denominator contains all the absorption coefficients $\beta_{ij}$ of each element $j$. Thus, $I_i$ will *increase* with the enhancement effects and *decrease* with the absorption effects (if $\beta_{ij}$ is positive). Furthermore, all these matrix effects are weighted by the factor $W_i'$, which takes into account the polychromaticity of the incident spectrum. We will return to this subject later.

Because of the difficulty in determining the experimental constant $g_i$ and for making the measured intensities independent of the instrument, the second important step is to replace the absolute intensity $I_i$ of element $i$ by the *relative X-ray intensity*, $R_i$, which is defined as follows:

$$R_i = \frac{I_i(\lambda_i)}{I_{(i)}(\lambda_i)}, \tag{5.77}$$

where $I_{(i)}(\lambda_i)$ is the intensity emitted by the pure element $i$. For a specimen composed only of the pure analyte $i$, $C_i = 1$, all $C_j = 0$, and the modified Sherman equation (5.76) becomes

$$I_{(i)} = g_i \sum_k W_i'(\lambda_k). \tag{5.78}$$

The combination of (5.76), (5.77), and (5.78), after some algebraic manipulations, leads to

$$R_i = C_i \frac{1+\sum\limits_j \varepsilon_{ij} C_j}{1+\sum\limits_j \alpha_{ij} C_j}, \tag{5.79a}$$

where

$$\varepsilon_{ij} = \frac{\sum\limits_k W_i(\lambda_k)\,\delta_{ij}(\lambda_k)}{\sum\limits_k W_i(\lambda_k)} \tag{5.79b}$$

$$\alpha_{ij} = \frac{\sum\limits_k W_i(\lambda_k)\,\beta_{ij}(\lambda_k)}{\sum\limits_k W_i(\lambda_k)} \tag{5.79c}$$

and where the new *weighting factor*, $W_i(\lambda_k)$, present in the two previous equations is defined by

$$W_i(\lambda_k) = \frac{\mu_i(\lambda_k) I_0(\lambda_k)\, \Delta\lambda_k}{\mu_i^* \left[1 + \sum_j C_j \beta_{ij}(\lambda_k)\right]}. \tag{5.79d}$$

Equation (5.79a) is still the same Sherman equation (5.70), except that it calculates a count-rate ratio, $R$. Indeed, the intensity $R_i$ is still proportional to the concentration $C_i$ but also to a ratio on the right-hand side. The numerator contains all the enhancement coefficients $\delta_{ij}$ (or $\varepsilon_{ij}$) of each element $j$ of the matrix, and the denominator contains all the absorption coefficients $\beta_{ij}$ (or $\alpha_{ij}$) of each element $j$. Thus, here again the count-rate ratio $R_i$ will increase with the enhancement effects and decrease with the absorption effects (if $\beta_{ij}$ is positive). Furthermore, all these matrix effects are weighted by the factor $W_i$, which takes into account the polychromaticity of the incident spectrum and the matrix composition of the sample.

Regarding the count-rate ratio, $R$, the same applies as to the conventional FP methods of Sect. 5.2. When the pure specimen is not available, the calibration procedure [49] enables to determine the intensity of the pure analyte from multielement standards (5.20). This intensity is simply equal to the slope of the calibration line.

As analysts are interested to calculate concentrations rather than intensities, which are measured, (5.79a) must be reversed:

$$C_i = R_i \frac{1 + \sum_j \alpha_{ij} C_j}{1 + \sum_j \varepsilon_{ij} C_j}. \tag{5.80}$$

If we accept the fundamental nature of the Sherman equation, and since (5.79a) is the only equivalent equation that respects the Sherman equation in every respect (algebraically, mathematically, and physically), the above reversed expression (5.80) of (5.79a) can be called the *fundamental algorithm.* Consequently, the $\alpha_{ij}$ and $\varepsilon_{ij}$ coefficients defined by the explicit (5.79c) and (5.79b) are the *fundamental influence coefficients* correcting for absorption and enhancement effects, respectively. We will see the reason in the next section.

**Physical Interpretation**

At first approximation, (5.80) reveals that the concentration of the analyte $i$ is proportional to its measured relative intensity, i.e., $C_i \propto R_i$, which is multiplied by a ratio correcting for all matrix effects. In fact, the coefficient $\alpha_{ij}$, calculated from the $\beta_{ij}$ and $W_{ij}$ coefficients, includes all mass absorption coefficients $\mu_S^*$ (when there is no enhancement) of the Sherman equation.

Therefore, we can affirm that the $\alpha_{ij}$ coefficient corrects for all absorption effects caused by element $j$ on analyte $i$ and the numerator of the ratio corrects thus for all absorption effects of the matrix, each element $j$ bringing its contribution to the total correction in a proportion $C_j$. If the numerator is greater than unity (it could be lower if the matrix is less absorbent than the analyte), the intensity $R_i$ will increase by a quantity equivalent to that absorbed by the matrix. Briefly then, $\alpha_{ij}$ is the *fundamental influence coefficient* that corrects for absorption effects, and its theoretical definition is given by (5.79c).

Now, if some elements of the matrix are able to enhance the analyte $i$, the corresponding coefficient $\varepsilon_{ij}$, deducted from $\delta_{ij}$, will be different from 0 and always positive. Therefore, we can affirm that the $\varepsilon_{ij}$ coefficient corrects for all enhancement effects caused by element $j$ on analyte $i$, and the denominator of the ratio corrects thus for all enhancement effects of the matrix, each element $j$ contributing to the total correction in a proportion $C_j$. In this case, the denominator will be greater than unity and the intensity $R_i$ will be reduced by a quantity equivalent to that caused by the enhancement. Thus, $\varepsilon_{ij}$ is the *fundamental influence coefficient* that corrects for enhancement effects, and its theoretical definition is given by (5.79b).

Note that the concentration $C_i$ of the analyte $i$ does not explicitly appear in the ratio of the Fundamental Algorithm except by the intermediary of the coefficients, which are concentration dependent. Indeed, at the numerator, the $\alpha_{ii} \cdot C_i$ term is equal to 0 because the coefficient $\beta_{ij}$ expresses a difference between the mass absorption coefficients and therefore $\beta_{ii} = 0$ as shown in (5.75b). In the denominator, the $\varepsilon_{ii}C_i$ term is also equal to 0 because an element cannot enhance itself, except in the case (rare in most analytical situations) where a K$\alpha$ line can enhance significantly the L$\alpha_1$ analytical line of the same element. However, as explained previously, such a situation is ignored in this algorithm.

The mathematical definitions ((5.79c) and (5.79b)) of the $\alpha_{ij}$ and $\varepsilon_{ij}$ coefficients have the form of a weighted mean. Thus, these coefficients can be considered as the *weighted means* of all absorption and enhancement effects, respectively, caused by element $j$ on analyte $i$, where to each incident wavelength $\lambda_k$ is given a weight $W_i(\lambda_k)$, which can be seen as the product of two terms:

$$W_i(\lambda_k) = \left[\frac{\mu_i(\lambda_k)}{\mu_i^*} I_0(\lambda_k)\Delta\lambda_k\right] \left[\frac{1}{1 + \sum_j C_j\beta_{ij}(\lambda_k)}\right]. \tag{5.81}$$

The first gives a weight to each value of the incident photon flux, while the second weights for each binary combination of elements $i$ and $j$ by the total absorption effect of the matrix. These considerations concerning $W_i$ lead to the two following important conclusions:

- The $\alpha_{ij}$ and $\varepsilon_{ij}$ coefficients are not constants for a binary system $i - j$ and vary not only with the concentrations of element $j$ but also with the nature and concentration of all other elements in the matrix.
- The total matrix effect on the analyte $i$ is not equal to the sum of the binary effects of every element $j$ of the matrix, each effect being calculated independent of each other. In other words, each element $j$ contributes to the total matrix effect on the analyte $i$, but the individual contribution of each element $j$ cannot be isolated and cannot be calculated independently of all other matrix elements.

  For the practical application of the Fundamental Algorithm (5.80), see [47]. For its experimental verification, see [50].

## 5.4 Compensation Methods

*B.A.R. Vrebos*

### 5.4.1 Internal Standards

The internal standard method is based on the addition of an element (or a compound) in a known, constant proportion to a (series of) specimen(s), in order to minimize the effect of variations in matrix effects. The same proportion has to be added to the standards as well as to the unknowns. The compound added is called an *added internal standard.* For the added internal standard, pure elements, pure compounds, mixtures, and solutions can be used. Care must be taken to avoid adding a compound with a significant amount of one of the analytes of interest; this will lead to erroneous results. When using solutions or mixtures, it is vital that their composition be constant.

There is basically only one selection criterion for the compound added: the added internal standard should have at least one characteristic line of sufficient intensity, which is subject to similar matrix effects as the analytical line(s) of interest. This means that both the analytical line and the line of the added internal standard are affected in the same way by the absorption and enhancement phenomena in the matrix. This requirement is generally fulfilled when there are no absorption edges (leading to a difference in absorption) or characteristic lines including scattered tube lines (leading to a difference due to enhancement) between the two wavelengths considered. For the concentration $C_i$ of an analyte $i$ in a sample, (5.22) can be used:

$$C_i = K_i\, I_i\, M_i \tag{5.82a}$$

and similarly for a standard, $s$:

$$C_s = K_s\, I_s\, M_s . \tag{5.82b}$$

The calibration constants $K_i$ and $K_s$ are not the same, and neither are the intensities $I_i$ and $I_s$ nor the matrix effects $M_i$ and $M_s$. By dividing (5.82a) by (5.82b), the ratio

$$\frac{I_i}{I_s} = K_{is} C_i \tag{5.83}$$

is obtained with

$$K_{is} = \frac{K_s M_s}{K_i M_i C_s}. \tag{5.84}$$

In the above equation, $C_s$ is a constant (the same amount is added to each standard and unknown specimen) and can be included in the constant $K_{is}$. The matrix effect term $M_i$, however, is not a constant for the case on hand. If $M_i$ were a constant over the range of concentrations concerned, then the simple linear regression (5.82a) could be used, and there would be no reason to use an internal standard. The trick of the method is now to select the internal standard element in such a way that the matrix effect on the analyte and on the characteristic line of the internal standard varies in the same way. When enhancement is dominant on the analyte's emission line and increases its intensity, then the intensity of the internal standard should also be affected in the same way. When the intensity of the analyte is reduced by absorption effects, the intensity of the internal standard should be reduced as well. By observing these guidelines, both $M_i$ and $M_s$ will vary in the same direction. The ratio $M_i/M_s$ is thus less sensitive to variation in the matrix effect, and, in many cases, can be considered a constant. In practice, the internal standard is added to all specimens (standards and unknowns) and the intensity of both the analyte and the internal standard is measured. Using linear regression of the ratio of the intensities against the concentration of the analyte in the standard specimens, the value of the calibration constant $K_{is}$ in (5.83) is calculated.

The difference between the internal standard method and the method using scattered radiation in their effectiveness to correct for matrix effects is due to the different origin of the "ratio" intensity (i.e., Compton scattered tube radiation or characteristic fluorescence from an element in the specimen). This is the reason why the internal standard method is capable of correcting for absorption as well as for enhancement. Variations in density of pressed specimens are also partially corrected for. To ensure proper compensation of the matrix effects, it is imperative that there are no absorption edges and no characteristic lines from other elements located between the two lines of interest. Also, in the ideal case, none of the characteristic lines of the analyte should enhance the internal standard or vice versa. This is difficult to realize. In practice, the requirement is reduced to avoid using an internal standard that contains elements, which change the matrix effects on the analyte significantly. For analytes with atomic number $Z$ higher than 23, the elements with atomic numbers $Z - 1$ and $Z + 1$ are very well suited to be used as internal standards. In those cases, there are no absorption edges and no K emission lines of other elements between the two elements considered. The element with atomic number $Z$ is not enhanced by the K$\alpha$ radiation from element $Z + 1$, but only by the (much) weaker K$\beta$ radiation. Some degree of enhancement between the internal standard element and the analyte is allowed as the concentration of the internal standard is constant and the method is based on a ratio of intensities. Please note that this guideline does not take into account the presence (or absence) of other elements with L absorption edges and L emission lines.

The method will break down, or produce results of less than optimal quality, when (1) a major emission line of a major matrix element with a variable concentration is present between the absorption edges of the analyte and the internal standard element. In this case, only one of the two elements (analyte or internal standard) is enhanced in a variable way. Or (2) when a major absorption edge of a major element of variable concentration is situated between the energies of the measured characteristic lines of the analyte and the internal standard element. In this case, one of the two lines considered is absorbed in a variable way. In both cases, varying concentrations of the matrix element lead to different effects on the intensities of the analyte and of the internal standard element, but, the ratio of (5.83) will not compensate for this variability.

The method of the added internal standard has also several drawbacks.

- The internal standard must be added to each of the standard specimens and all of the unknowns. Specimen preparation is thus made more complicated, and is more susceptible to errors;
- The addition of reagents and the fact that the specimens must be homogeneous limit the practical application of the method to the analysis of liquids and fused specimens; although it sometimes finds application in the analysis of pressed powders. When dealing with liquids, care must be taken to ensure that the additive is in a stable solution over time, without precipitating; this is less of a concern when fused beads are used. Ideally, in this case, the added internal standard should be part of the flux;
- The rule $(Z+1)$ or $(Z-1)$ can serve as a rule of thumb for simple cases. In samples where many elements are to be quantified, or in samples where only few elements need to be quantified but in a matrix whose composition varies from specimen to specimen it is quite clear that a suitable internal standard cannot be found for every analyte element. Sometimes, more elements are used in one internal standard solution to provide a suitable internal standard for more analytes.

The internal standard method is most easily applied to liquids. However, even in this favorable case some problems persist. Elements with a higher atomic number (for example Mo) are more difficult to analyze using this method, because liquid specimens are generally not of infinite thickness for the K wavelengths of such elements. In these cases, the L line can be used with an appropriate internal standard. The method will also provide some compensation for non-infinite thickness, especially in cases where the wavelength of the internal standard selected is very similar to that of the analyte line.

In principle, L lines of a given element can be used as internal standards for K lines of other elements and vice versa. The same requirements are to be satisfied as when dealing with K lines.

The method allows the analysis of one or two elements in a specimen without requiring analysis or knowledge of the composition of the matrix. In favorable situations, the method can provide accurate results over a concentration range of up to 10%, and even 20% in some cases; it is most effective at lower concentrations and still finds widespread applications. For example, ISO prescribes the method for the determination of Ni and V in petroleum products [24]. In this case, Mn K$\alpha$ is used as an internal standard. Note that only one internal standard element is used for the two analyte elements. Accurate determination of the analytes is obtained over a much wider variety of petroleum products than a simple linear calibration would do.

### 5.4.2 Standard Addition Methods

The standard addition method is based on the addition of small quantities of the analyte element to the specimens. The method is of particular interest when the analyte's concentration is low, there are no standards similar to the unknowns available, and the analyst is only interested in one (or few) constituent elements in the specimens. The method is based on the following: by adding a known quantity $\Delta C_i$ of the analyte element $i$ to the unknown specimen, the intensity $I_i$ of the analyte's radiation will be increased by an amount $\Delta I_i$. Working under the assumption of a linear calibration relationship, the following equations apply:

$$C_i = K_i I_i \tag{5.85a}$$

for the original specimen and

$$C_i + \Delta C_i = K_i(I_i + \Delta I_i) \tag{5.85b}$$

for the specimen with the addition. In these equations all quantities are known, except the concentration $C_i$ of the analyte in the specimen. This can be obtained by solving the above equations for $C_i$. Please note that the use of net (background corrected) intensities is required to minimize bias.

The method assumes that increasing the concentration of the analyte by a fraction $\Delta C_i$ causes the intensity to increase by an amount $K_i \Delta I_I$; it thus relies explicitly on the linear calibration. To check if this assumption is valid, several different additions can be made. If the measured intensities are then plotted against the concentrations added to the original specimen, a straight line should be obtained. The concentration in the unaltered specimen (the unknown) can also be read from such a graph: The intercept of the "calibration" line with the concentration axis equals $-C_i$.

The concentration of the analyte is effectively found by extrapolating the "calibration" line to 0. In order to minimize the effect of the extrapolation (the uncertainty increases with increasing degree of extrapolation), it is vital to add amounts of $\Delta C_i$ that are in proportion to the amount $C_i$ itself. If more than one addition is made, the top point should have an addition $\Delta C_i$ at least equal to $C_i$ and preferably two to three times higher. On the other hand, it is important to avoid nonlinearity caused by adding substantial amounts, which cause the matrix effect to change considerably. This method is thus ideally suited for the accurate determination of trace levels of one (or a few) analyte(s) at low concentration in a specimen of an unknown matrix.

In terms of the additions themselves, both solutions and compounds can be used. If the analyte in the original sample is in a different phase than in the additions, the correct stoichiometric proportions must be taken into account in the calculation of the concentration. This is equally true if the analyte is present under elemental or ionic form in the additions and in a compound phase in the original specimen (or vice versa).

If the analyte element is not available or considered too expensive, the concentration of the analyte can also be altered by diluting the sample. By diluting the sample one to one, the concentration of the analyte is halved, etc. A calibration line can be established using different dilutions. This approach is to be preferred when the concentration of more than one analyte is to be determined. Diluting the specimens affects all analytes in the same way. The disadvantage is that the intensity and thus the counting statistical precision will suffer, and that the matrix must be known.

### 5.4.3 Dilution Methods

The method of the added internal standard and the standard addition method essentially *compensate* for matrix effects. They do not *calculate* the matrix effect, nor do they *alter* the matrix effect in any significant way. The dilution method is based on actually altering the matrix effect to such an extent that it becomes constant for the concentration range considered. The basis of the dilution method is (5.11) expressing intensity (photon counts), $N_{is}$, in sample $s$, as a function of the mass attenuation coefficients. $m_{i,s}$ denotes the mass of element $i$ in the specimen and $m_s$ the total mass of the specimen:

$$N_{i,s} = G_i C_i \frac{\tau_i}{\mu_s^*} = G_i \frac{m_{i,s}}{m_s} \frac{\tau_i}{\mu_s^*}. \tag{5.86}$$

Adding a diluent, $d$, with mass $m_d$ to a sample $s$ with original mass $m_s$ leads to

$$\begin{aligned} N_{i,s+d} &= G_i m_{i,s} \frac{\tau_i}{m_s \mu_s^* + m_d \mu_d^*} \\ &= G_i C_{[i \text{ in mixture}]} \frac{\tau_i}{\dfrac{1}{1+D}\mu_s^* + \dfrac{D}{1+D}\mu_d^*} \\ &= G_i C_{[i \text{ in } s]} \frac{\tau_i}{\mu_s^* + \dfrac{C_{[d \text{ in mixture}]}}{C_{[s \text{ in mixture}]}}\mu_d^*} \end{aligned} \tag{5.87}$$

with $D = m_d/m_s$. Deviations from linearity in the calibration lines are expected due to variations in the denominator. Enhancement is ignored at this stage.

The basis of the dilution method is to add a fixed amount $D$ of a diluent $d$ to each sample. The factors $1/(1 + D)$ and $D/(1 + D)$ account for the changes in specimen composition. The first term describes the absorption by the compounds in the original specimen, while the second term is for the diluent. This second term is constant for each specimen, as the same diluent is added in the same proportion to each specimen. By making the second term $D/(1+D)\mu_d^*$ significantly larger than the first one, the denominator can be made nearly constant: the only source of variation is due to changes in composition of the specimens. There are essentially two distinct methods to achieve this:

- By making $D$ very large;
- By using a diluent with large absorption coefficients, i.e., a large $\mu_d^*$.

The latter case is also known as the technique of the heavy absorber.

In both cases, reagents must be added to the specimens, standards, and unknowns. Obviously, this is much easier when dealing with liquids. The method of the heavy absorber has found wide acceptance in the past with fusions: a lithium-based salt (lithium tetraborate or lithium metaborate) is then mixed with $La_2O_3$ (90:10 or 80:20) and is then used to fuse the specimen. Lanthanum is a material with a large attenuation coefficient over a wide range of wavelengths and it is not commonly measured; it is thus an excellent diluent in many cases.

This method aims at *reducing* the effect caused by the variation in specimen composition. In theory, it could also *eliminate* the variation of matrix effect completely by adding very large quantities of diluent, but at the same time, the sensitivity is reduced significantly. A compromise must thus be sought between reduction of matrix variability and loss of sensitivity.

### 5.4.4 Scattered Radiation – Compton Scatter

Compton scatter or incoherent scatter involves interaction of a photon with a single free (i.e., weakly bound) electron, in which part of the photon energy is transferred to the electron. The wavelength of the scattered photon is thus longer (has less energy) than that of the incident photon. It can be shown that the Compton shift $\Delta\lambda$ is given by

$$\Delta\lambda = \lambda_s - \lambda_0 = \frac{h}{m_e c}(1 - \cos\psi) \cong 0.00243\,(1 - \cos\psi), \tag{5.88}$$

where $h$ is Planck's constant, $m_e$ is the rest mass of the electron, and $c$ is the velocity of light in vacuum. The angle $\psi$ is the angle between the direction of the photon after scattering and its original direction. When $\Delta\lambda$ is calculated in nm, the constants evaluate to 0.00243 nm.

The intensity of the Compton scattered radiation from a tube line can be used to obtain an estimate of the absorption coefficient of the specimen at the wavelength $\lambda_s$ of the scattered photons. Furthermore, it can be shown that the intensity of the scattered radiation is inversely proportional to the mass attenuation coefficient $\mu_s$ of the specimen:

$$I_s(\lambda_s) \sim \frac{1}{\mu_s(\lambda_s)}, \tag{5.89}$$

where $\lambda_s$ is the wavelength of the scattered radiation and $\mu_s(\lambda_s)$ is the mass attenuation coefficient of the specimen for wavelength $\lambda_s$. This relationship is illustrated in Fig. 5.8 for a number of standard specimens. The agreement is striking. This property can be used to the analysts' advantage when dealing with analyses of elements whose characteristic radiation is mainly subject to absorption effects. In such cases, the variation in the magnitude of the matrix effect between specimens is mainly due to changes in the absorption properties. If enhancement can be neglected and the intensity of the fluorescent radiation, $I_i$, is inversely proportional to the mass attenuation coefficient (monochromatic excitation is assumed):

$$I_i \propto \frac{C_i}{\mu_s^*}. \tag{5.90}$$

Knowledge of the value of $\mu_s^*$ for each specimen thus enables quantitative analysis. The intensity of the X-rays scattered by the specimen can be used to determine the value of the mass attenuation coefficient at one wavelength. Mass attenuation coefficients at two different wavelengths are virtually proportional, independent of matrix composition, provided there are no significant absorption edges between these two wavelengths [23].

In Fig. 5.9, the mass attenuation coefficient as a function of wavelength is given for three pure compounds: $Fe_2O_3$, $TiO_2$, and $SiO_2$. Note the rapid

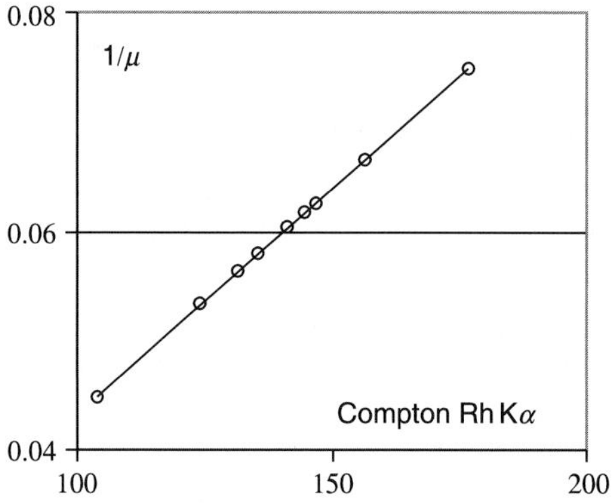


**Fig. 5.8.** Inverse of the mass attenuation coefficient at the wavelength of Rb Kα, versus the intensity of Compton scattered Rh Kα tube line, for a variety of oxide matrices. The linear relationship of (5.89) can clearly be seen

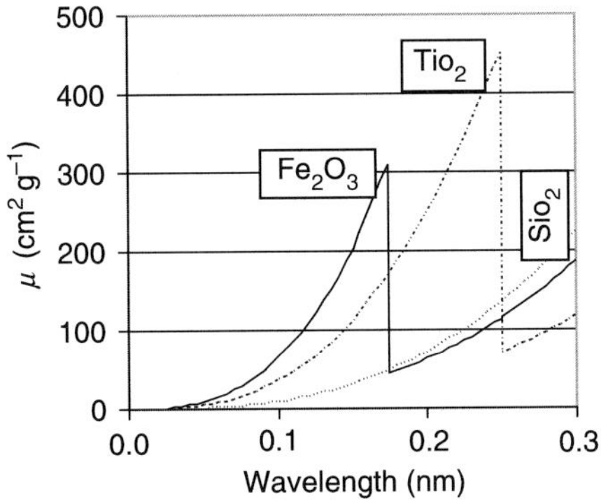


**Fig. 5.9.** Mass attenuation coefficient as a function of wavelength for $Fe_2O_3$, $TiO_2$ and $SiO_2$. Note the large differences between the curves for the different materials and the absorption edges

increase with wavelength (approximately proportional to $\lambda^3$), as well as the K-absorption edges of Fe and Ti. The K-absorption edge for Si is at a wavelength not covered by the scale of Fig. 5.9. At any given wavelength, there is a considerable difference between the value of the mass attenuation coefficients for $Fe_2O_3$, $TiO_2$, and $SiO_2$ (except for $Fe_2O_3$ and $SiO_2$ on the long wavelength side of the Fe K-edge). However, when the mass attenuation coefficient at any wavelength is divided by the value at a given wavelength, this ratio is similar, irrespective of the compound. This is illustrated in Fig. 5.10 where the ratio of the mass attenuation coefficients is plotted for the same compounds. From Fig. 5.10 it appears that the ratio of the value of the mass attenuation coefficient at any given wavelength and its value at 0.07 nm is very similar for all three compounds considered. At about 0.17 nm (the wavelength of the Fe K-edge) the plot for $Fe_2O_3$ diverges rapidly, while the values for $TiO_2$ and $SiO_2$ are still very similar, until the Ti K-edge (at 0.25 nm) is crossed.

The method is thus limited to those cases where only trace elements have absorption edges in that wavelength range. Under these conditions, the ratio $I_i/I_s$ is proportional to the concentration of the analyte. Combination of (5.89) and (5.90) yields

$$I_i \propto \frac{C_i}{\mu_s^*} \sim C_i I_s(\lambda_s). \qquad (5.91)$$

In Fig. 5.11, the intensity of Rb K$\alpha$ is plotted against the chemical concentration (circles); the crosses represent the ratio of the intensity of Rb K$\alpha$ and the scattered Rh K$\alpha$. The spread around the line is much reduced in the latter case: the root mean square in the first case is about 14 ppm, while – with Compton corrected intensities – it is reduced to about 3 ppm.

In practice, both coherently and incoherently scattered primary radiation, such as tube lines, as well as the scattered continuous radiation can be used. The contribution of Compton scatter to the total scatter at a given wavelength

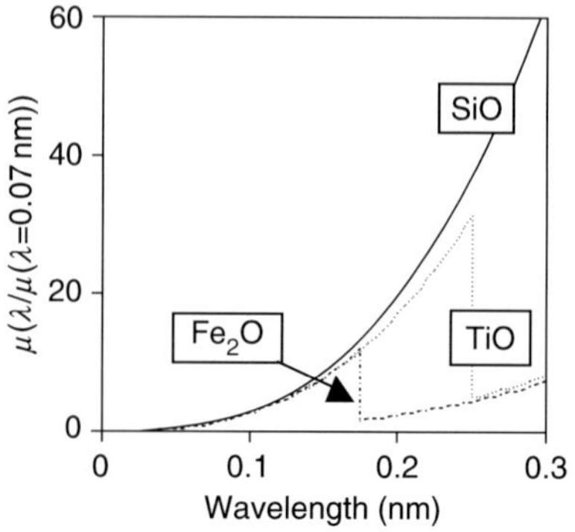


**Fig. 5.10.** Ratio of the mass attenuation coefficient for $Fe_2O_3$, $TiO_2$, and $SiO_2$ divided by the mass attenuation coefficient at 0.07 nm. Note the agreement between the curves in the region up to the absorption edge of Fe (0.174 nm)

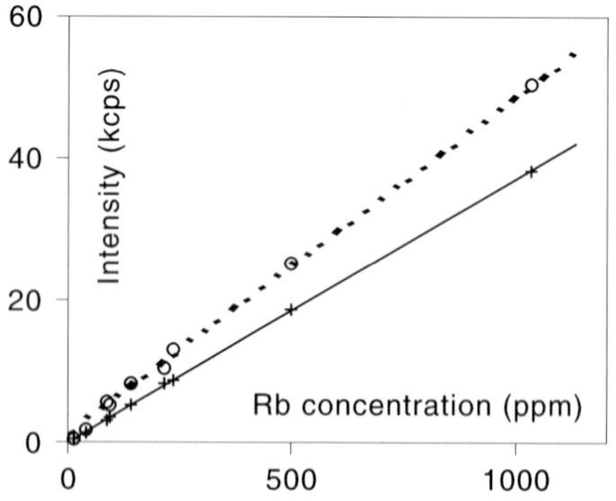


**Fig. 5.11.** Intensity of Rb K$\alpha$ versus chemical concentration (∘) and the ratio of the Rb K$\alpha$ intensity with the Compton scattered Rh K$\alpha$ (+). The Compton corrected data show reduced spread around the regression lines. *Note:* The Compton corrected data have been scaled by a factor of 100 for display reasons

increases with decreasing wavelength. For example, the scattered intensity at 0.05 nm consists for more than 80% of Compton scatter.

It has been shown that the method corrects also to some degree for surface finish, grain size effects, and variations in tube voltage and current [6, 27]. Unfortunately, the method of scattered radiation does not correct for enhancement, its use is thus limited to analytes that are mainly influenced by absorption only.

Another limitation of the method is that there should be no absorption edges of major elements between the analytical wavelength and the wavelength of the scattered radiation. This is to ensure that the ratio between the mass attenuation coefficients for those wavelengths remains constant between specimens. Several authors (e.g., Reynolds [45], Walker [60], and Nesbitt et al. [41]) have presented work-arounds to extend the method beyond major absorption edges.

These methods are based on the assumption that the intensity of the major element is proportional to its concentration, which in turn is proportional to its contribution to the absorption. It is most easily implemented by determining the relationship between the *ratio* of the mass attenuation coefficients at two different wavelengths (one at either side of the absorption edge) and the intensity of the characteristic radiation of the element related to the edge. Figure 5.12 shows for example the relationship between the ratio of the mass attenuation coefficients at 0.07 nm and 0.23 nm and the intensity of Fe K$\beta$. The edge is Fe K (at 0.174 nm):

$$\frac{\mu_{0.07}}{\mu_{0.23}} \cong I_{\mathrm{Fe\,K}\beta} = f(I_{\mathrm{Fe\,K}\beta}). \tag{5.92}$$

By using this ratio and the mass attenuation coefficient at 0.07 nm (from the Compton scattered intensity), the mass attenuation coefficient at 0.23 nm, $\mu_{0.23}$, can be obtained and then be used in (5.90):

$$\mu_{0.23} = \frac{\mu_{0.23}}{\mu_{0.07}} \mu_{0.07}. \tag{5.93}$$

In combination with (5.89) and (5.92), this can be rewritten in terms of intensities.

$$\mu_{0.23} \cong \frac{1}{f(I_{\mathrm{Fe\,K}\beta})} \frac{1}{I_{\mathrm{S}}(\lambda_{\mathrm{S}})}. \tag{5.94}$$

In Table 5.6, some data regarding cobalt in geological matrices are collated to show the benefits from using jump edge corrections. In its second part, the concentrations determined by linear regression (no matrix correction), Compton correction (5.91), and a combination of Compton correction and jump edge correction (5.89 and 5.93) are compared.

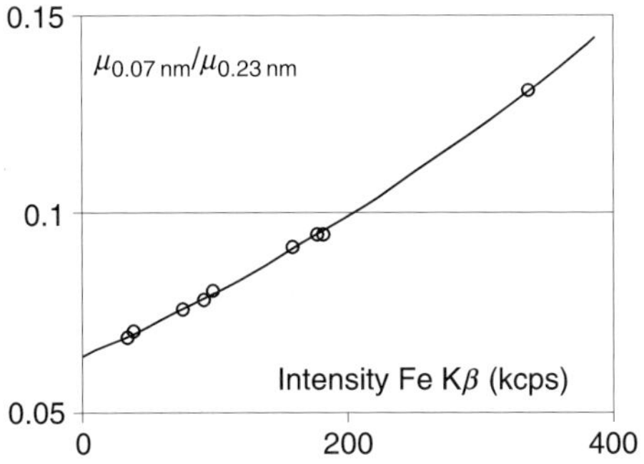


**Fig. 5.12.** Ratio of the mass attenuation coefficients at 0.07 nm and 0.23 nm versus the intensity of Fe K$\beta$. The data can be approximated very well by a second degree polynomial as indicated by the curve

**Table 5.6.** Determination of cobalt concentrations

| Standard | C(Chem) ppm | Co K$\alpha$ kcps | $\mu_{0.07}$ cm$^2$/g | Rh$_{\text{Compton}}$ kcps | $\mu_{0.23}$ cm$^2$/g | Fe K$\beta$ kcps |
|---|---|---|---|---|---|---|
| SY-2 | 8.6 | 0.121 | 18.75 | 124.29 | 239.3 | 92.836 |
| SO-3 W | 5.5 | 0.074 | 16.57 | 141.34 | 240.9 | 33.490 |
| 73326 | 12.7 | 0.210 | 16.17 | 144.92 | 213.2 | 75.326 |
| 73325 | 97 | 1.943 | 22.27 | 103.86 | 169.8 | 335.818 |
| 73324 | 7.6 | 0.099 | 15.96 | 146.92 | 173.9 | 158.535 |
| 73321 | 5.5 | 0.090 | 13.36 | 176.75 | 189.4 | 38.186 |
| 73319 | 14.2 | 0.256 | 15.02 | 156.51 | 186.9 | 98.150 |
| TR-006 | 431.0 | 8.267 | 17.21 | 134.62 | 181.0 | 180.785 |
| TR-013 | 889.0 | 16.396 | 17.77 | 131.32 | 186.9 | 177.684 |

Chemical compositions (reference values), intensities, and mass attenuation coefficients

| Standard | C(Chem) ppm | Concentration determined ppm | | |
|---|---|---|---|---|
| | | LINREG | COMPT | JEC |
| SY-2 | 8.6 | 6.5 | 6.9 | 8.4 |
| SO-3 W | 5.5 | 4.0 | 3.7 | 5.3 |
| 73326 | 12.7 | 11.3 | 10.3 | 13.1 |
| 73325 | 97 | 104.6 | 132.7 | 100.1 |
| 73324 | 7.6 | 5.3 | 4.8 | 5.0 |
| 73321 | 5.5 | 4.8 | 3.6 | 5.1 |
| 73319 | 14.2 | 13.8 | 11.6 | 14.0 |
| TR-006 | 431.0 | 444.9 | 435.8 | 433.7 |
| TR-013 | 889.0 | 882.4 | 886.0 | 887.5 |
| RMS (ppm) | | 6.2 | 12.6 | 1.8 |

Values of the concentrations found using simple linear regression (LINREG), Compton correction (COMPT) and jump edge correction (JEC).

The analysis of cobalt in the samples concerned is hampered by the fact that the Fe K-absorption edge is located between the wavelength of the Compton scattered tube line and the Co K$\alpha$. This is the reason why the Compton correction does not work. It actually makes the results worse: the RMS increases from 6.2 ppm (no correction) to 12.6 ppm. This is not surprising because the Compton correction would account for a change in absorption, based on the assumption that there are no significant absorption edges between the wavelength of the Compton scattered intensity ($\lambda_s$) and the analyte wavelength ($\lambda_i$). In the case at hand, however, the presence of a large amount of iron is responsible for a large absorption jump between $\lambda_s$ and $\lambda_i$. By using both, the Compton scattered intensity and the jump edge correction based on the Fe K$\beta$ intensity, the RMS is reduced to 1.8 ppm.

This method for jump edge correction requires, however, additional measurements and the calculation of mass attenuation coefficients for reference samples. Also, even with jump edge correction, the method still does not correct for enhancement.

## 5.5 Thin and Layered Samples

*P.N. Brouwer*

XRF is an attractive technique for the analysis of thin and layered samples, because both the composition and thickness of layers can be determined in a fast and nondestructive way. The application area is growing; examples are optical coatings, protective coatings, magnetic recording materials, optical recording materials, superconductive films, as well as wafers for IC technology.

The photon rates originating from thin and layered specimens depend on the composition of each of the layers as well as on the thickness of the layers. It is therefore possible to determine simultaneously the composition as well as the thickness(es). Just as in the case of bulk samples, quantification can be accomplished using calibration standards with composition close to that of the sample. Since the layer thicknesses are additional unknown quantities, additional analytical lines may be required to analyze a sample. Producing accurate thin film or layered standards is very difficult, thus more than in the case of bulk samples, FP-based analysis is demanded for thin and layered samples, where reference samples are only required to determine the spectrometer sensitivity for an analytical X-ray line. The reference sample can be any type of sample of known composition: bulk, thin film, or layered. Using reference samples close to the sample composition in some cases eliminates systematic errors, which can improve the accuracy.

### 5.5.1 Direct Excitation by Polychromatic Sources

In Sect. 5.2 the fluorescence of a single thin layer is briefly discussed. For the primary fluorescence, a layered sample is treated as a stack of thin films, taking into account the absorption of layers that are above the layer of interest (Fig. 5.13).

Before the incoming radiation reaches a layer $n$, it is partly absorbed by the layers above it and the same holds for the outgoing characteristic radiation of the analyte. The absorption by all the layers above layer $n$ for radiation with wavelength $\lambda$ under an angle $\psi'$ is given by

$$A_{\lambda,\psi'}^{1,n-1} = \prod_{k=1}^{n-1} \exp\left(-\frac{\rho_k T_k \mu_\lambda^k}{\sin\psi'}\right) = \exp\left(-\sum_{k=1}^{n-1} \frac{\rho_k T_k \mu_\lambda^k}{\sin\psi'}\right). \tag{5.95}$$

The primary fluorescence for element $i$ coming from layer $n$ due an incoming photon with wavelength $\lambda$ is given by

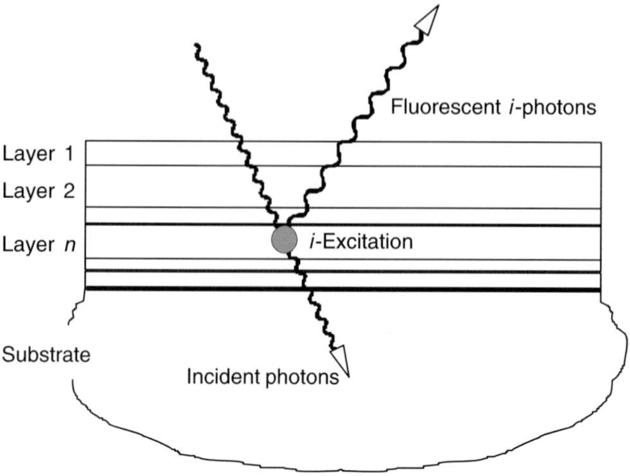


**Fig. 5.13.** Multilayer structure on a substrate

$$N_{iP\lambda}^{n} = \frac{\Omega}{4\pi \sin \psi'} \varepsilon_{i\lambda} A_{\lambda,\psi'}^{1,n-1} \frac{C_i^n \tau_{i\lambda}}{\mu_\lambda'^n + \mu_i''^n} A_{\lambda,\psi''}^{1,n-1} \left(1 - \exp\left(-\rho_n T_n \left(\mu_\lambda'^n + \mu_i''^n\right)\right)\right) \tag{5.96}$$

with

$$\varepsilon_{i\lambda} = \frac{S_i - 1}{S_i} p_i \omega_i .$$

Note that $S_i$ is wavelength dependent for lines of the L- and higher series; otherwise $\varepsilon_{i\lambda}$ corresponds to $g_i$ in (5.9). For films with a very small thickness $T_k \to 0$ the exponential term can be approximated by

$$\left(1 - \exp\left(-\rho_n T_n \left(\mu_\lambda'^n + \mu_i''^n\right)\right)\right) \approx \rho_n T_n \left(\mu_\lambda'^n + \mu_i''^n\right) .$$

In that case, the primary fluorescence is given by

$$N_{iP\lambda}^{n} = \frac{\Omega}{4\pi \sin \psi'} \varepsilon_{i\lambda} A_{\lambda,\psi'}^{1,n-1} C_i^n \tau_{i\lambda} \rho_n T_n A_{\lambda_i,\psi''}^{1,n-1} . \tag{5.97}$$

This means that the primary fluorescence of a very thin layer is linear proportional to the thickness of the layer.

For very thick layers with $T_k \to \infty$ follows

$$\left(1 - \exp\left(-\rho_n T_n \left(\mu_\lambda'^n + \mu_i''^n\right)\right)\right) \approx 1$$

and in that case, the primary fluorescence is given by

$$N_{iP\lambda}^{n} = \frac{\Omega}{4\pi \sin \psi'} \varepsilon_{i\lambda} A_{\lambda,\psi'}^{1,n-1} \frac{C_i^n \tau_{i,\lambda}}{\mu_\lambda'^n + \mu_i''^n} A_{\lambda_i,\psi''}^{1,n-1} . \tag{5.98}$$

The primary fluorescence of layer $n$ leaving the sample due to the continuum and characteristic lines of a tube, is obtained by integration over the continuum and summing over all tube lines (see also (5.13)):

$$N_{i\mathrm{P}}^{n} = \int_{\lambda_0}^{\lambda_{\mathrm{abs},\, i}} N_{i\mathrm{P}\lambda}^{n} N_0(\lambda)\, \mathrm{d}\lambda + \sum_{t=1}^{\#\mathrm{tubelines}} N_{i\mathrm{P}\lambda_t}^{n} N_0(\lambda_t). \qquad (5.99)$$

In order to obtain the total primary fluorescence of analyte $i$, this has to be summed over all layers, $n$:

$$N_{i\mathrm{P}} = \sum\nolimits_n N_{i\mathrm{P}}^{n}. \qquad (5.100)$$

### 5.5.2 Indirect Excitation by Polychromatic Sources

Secondary fluorescence can be produced within one and the same layer, and also across different layers. In the former case a characteristic photon is absorbed in the same layer where it is created, leading to secondary fluorescence in that layer. In the latter case, the characteristic photon crosses at least one interlayer boundary and is absorbed in another layer, leading to secondary fluorescence in that layer. There is no means of indicating which of these effects is more dominant without taking into account the sample model (Fig. 5.14).

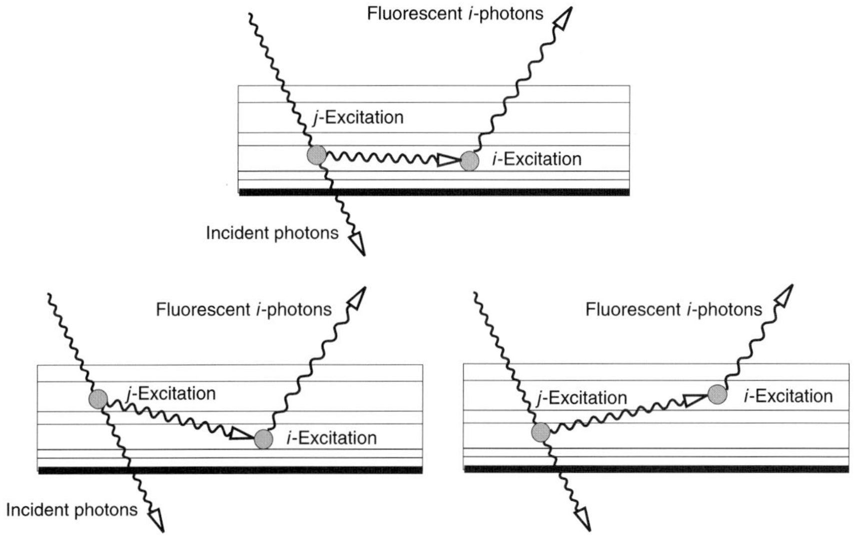


**Fig. 5.14.** Multilayer structure: intralayer and interlayer secondary fluorescence

**Intralayer Secondary Fluorescence**

The secondary fluorescence from element $i$ due to excitation by element $j$ in the same layer $n$ due an incoming photon with wavelength $\lambda$ is given by

$$N_{ij\lambda}^{n} = \frac{1}{2} A_{i,\psi''}^{1,n-1} A_{\lambda,\psi'}^{1,n-1} \varepsilon_{j\lambda} \varepsilon_{ij} \times \frac{\Omega}{4\pi \sin\psi'} \frac{C_i^n C_j^n}{\mu_\lambda'^n + \mu_i''^n} \frac{\tau_{j\lambda}\tau_{ij}}{\mu_j^n} L \tag{5.101}$$

$L$ is a function of the mass absorption coefficients of the layer $n$ for the incoming radiation, the radiation of analyte $i$ and of element $j$, the thickness of layer $n$ and the angles of the incoming and outgoing radiation, $\psi'$ and $\psi''$, respectively. The factor $L$ can be written as the sum of a downward (element $j$ below element $i$) and upward (element $j$ above element $i$) contributions:

$$L\left(\mu_i^n,\, \mu_j^n,\, \mu_\lambda^n,\, T_n,\, \psi',\, \psi''\right) = L_0\left(\frac{\mu_i^n}{\sin\psi''},\, \frac{\mu_\lambda^n}{\sin\psi'},\, \mu_j^n,\, T_n\right) + L_0\left(\frac{\mu_\lambda^n}{\sin\psi'},\, \frac{\mu_i^n}{\sin\psi''},\, \mu_j^n,\, T_n\right), \tag{5.102}$$

where $L_0$ is defined as

$$\begin{aligned} L_0(\mu_1,\, \mu_2,\, \mu_n,\, T) = \frac{\mu_n}{\mu_2}\Bigg( & -\frac{\mu_1+\mu_2}{\mu_1} \exp\left(-\mu_1 T\right) E_1(\mu_n T) \\ & + \exp\left(-(\mu_1+\mu_2)T\right) E_1((\mu_n-\mu_2)T) \\ & + \exp\left(-(\mu_1+\mu_2)T\right) \ln\left(\left|1-\frac{\mu_2}{\mu_2}\right|\right) \\ & + \frac{\mu_2}{\mu_1} \ln\left(\left|1+\frac{\mu_1}{\mu_n}\right|\right) + \frac{\mu_2}{\mu_1} E_1((\mu_1+\mu_n)T)\Bigg) \end{aligned}$$

$E_1$ is the *exponential integral*:

$$E_1(x) = \int_x^\infty \frac{e^{-t}}{t}\, \mathrm{d}t.$$

The secondary fluorescence of element $i$ due to element $j$ in the same layer $n$ due to the continuum and characteristic lines of a tube is again obtained by integration over the continuum and summing over the lines as in (5.99). In order to obtain the total intralayer secondary fluorescence of element $i$ originating from a sample, the secondary fluorescence has to be summed over all layers and all elements $j$ in each layer:

$$N_{i,\,\text{intra-layer}} = \sum_n \sum_j \left( \int_{\lambda_0}^{\lambda_{\text{abs},\,i}} N_{ij\lambda}^n N_0(\lambda)\, \mathrm{d}\lambda + \sum_{t=1}^{\#\text{tubelines}} N_{ij\lambda_t}^n N_0(\lambda_t) \right). \tag{5.103}$$

For infinitely thick layers $L$ and $L_0$ reduce to

$$L^\infty\left(\mu_i^n,\,\mu_j^n,\,\mu_\lambda^n,\,\psi',\,\psi''\right) = \frac{\mu_j^n\,\sin\psi'}{\mu_\lambda^n}\,\ln\left(1+\frac{\mu_\lambda^n}{\mu_j^n\,\sin\psi'}\right) + \frac{\mu_j^n\,\sin\psi''}{\mu_i^n}\,\ln\left(1+\frac{\mu_i^n}{\mu_j^n\,\sin\psi''}\right) \tag{5.104a}$$

$$L_0(\mu_1,\,\mu_2,\,\mu_n,\,T\to\infty) = \frac{\mu_n}{\mu_1}\,\ln\left(\left|1+\frac{\mu_1}{\mu_n}\right|\right). \tag{5.104b}$$

For a bulk sample $N_{ij}$ reduces to

$$N_{ij} = \frac{1}{2}\varepsilon_{ij}\varepsilon_{j\lambda}\frac{\Omega}{4\pi\sin\psi'}L^\infty(\mu_i,\,\mu_j,\,\mu_\lambda,\,\psi',\,\psi'')\,\frac{C_iC_j}{\mu'_\lambda+\mu''_i}\,\frac{\tau_{j\lambda}\tau_{ij}}{\mu_{sj}}. \tag{5.105}$$

### Interlayer Secondary Fluorescence

For interlayer secondary fluorescence, two cases are distinguished: enhancement by an element in a layer below the layer of the analyte or by an element in a layer above the layer of the analyte. The first is denoted by $N^\uparrow$ and the latter by $N^\downarrow$:

$$\begin{aligned}
N_{ij}^{\uparrow kn} &= \frac{1}{2}\varepsilon_{j\lambda}\varepsilon_{ij}\frac{\Omega}{4\pi\sin\psi'}A_{\lambda_i,\psi''}^{1,n-1}C_i^nC_j^k\tau_{j\lambda}\tau_{ij}A_{\lambda,\psi'}^{1,k-1} \\
&\quad X\left(\frac{\mu_\lambda^k}{\sin\psi'},\,\frac{\mu_i^n}{\sin\psi''},\,\mu_j^n,\,d_n,\,\mu_j^k,\,T_k,\,\sum_{b=n+1}^{k-1}\mu_j^bT_b\right) \\
N_{ij}^{\downarrow kn} &= \tfrac{1}{2}\varepsilon_{j\lambda}\varepsilon_{ij}\tfrac{\Omega}{4\pi\sin\psi'}A_{\lambda_i,\psi''}^{1,n-1}C_i^nC_j^k\tau_{j\lambda}\tau_{ij}A_{\lambda,\psi'}^{1,k-1} \\
&\quad X\left(\frac{\mu_i^n}{\sin\psi''},\,\frac{\mu_\lambda^k}{\sin\psi'},\,\mu_j^k,\,T_k,\,\mu_j^n,\,T_n,\,\sum_{b=k+1}^{n-1}\mu_j^bT_b\right)
\end{aligned} \tag{5.106a}$$

with

$$\begin{aligned}
&X(p,\,q,\,\mu_1,\,T_1,\,\mu_2,\,T_2,\,M) = \\
&\quad\frac{1}{q}\left[Y\left(-p,\,\mu_2,\,\frac{\mu_1T_1+M}{\mu_2},\,T_2\right) - \exp(-qT_1)\,Y\left(-p,\,\mu_2,\,\frac{M}{\mu_2},\,T_2\right)\right. \\
&\quad+\exp\left(-qT_1-q\frac{M}{\mu_1}\right)Y\left(-q\frac{\mu_2}{\mu_1}-p,\,\mu_2\left(1-\frac{q}{\mu_1}\right),\,\frac{M}{\mu_2},\,T_2\right) \\
&\quad\left.-\exp\left(-qT_1-q\frac{M}{\mu_1}\right)Y\left(-q\frac{\mu_2}{\mu_1}-p,\,\mu_2\left(1-\frac{q}{\mu_1}\right),\,\frac{\mu_1T_1+M}{\mu_2},\,T_2\right)\right]
\end{aligned} \tag{5.106b}$$

$Y$ is defined as

$$Y(a,\,b,\,c,\,d) = \int_0^d \exp(ax)\,E_1(b\,(x+c))\,\mathrm{d}x. \tag{5.106c}$$

The integration can be carried out analytically and leads to

$$\begin{aligned} Y(a,\, b,\, c,\, d) = \frac{1}{a}\ &[\exp{(ad)} E_1(b\,(c+d)) \\ &- \exp{(-ac)} E_1\,((-a+b)\,(c+d)) \\ &- E_1(bc) + \exp{(-ac)} E_1\,((-a+b)c)] \end{aligned} \quad . \tag{5.107a}$$

This general expression cannot be used if $c = 0$ and/or $d = \infty$. In that case

$$Y(a,\, b,\, 0,\, d) = \frac{1}{a}\ \left[\exp{(ad)} E_1(bd) - E_1((-a+b)d) - \ln\left|1 - \frac{a}{b}\right|\right] \tag{5.107b}$$

and

$$\begin{aligned} Y(a,\, b,\, c,\, \infty) &= \tfrac{1}{a}\ [-E_1(bc) + \exp{(-ac)} E_1((-a+b)c)] \\ Y(a,\, b,\, 0,\, \infty) &= -\tfrac{1}{a}\ \ln\left|1 - \tfrac{a}{b}\right| \end{aligned} \quad . \tag{5.107c}$$

The total interlayer secondary fluorescence of element $i$ in layer $n$ due to excitation by all elements $j$ in all layers $k \neq n$ is obtained by summation over all layers and all elements $j$ in each layer, and for excitation by polychromatic radiation from X-ray tubes, by integration over the continuum and summing over the tube-lines:

$$\begin{aligned} N_{i,\,\text{inter layer}} = \sum_{n}\sum_{j}\sum_{k\uparrow} \Bigg( &\int_{\lambda_0}^{\lambda_{\text{abs},\,i}} N_{ij\lambda}^{nk\uparrow} N_0(\lambda)\,\mathrm{d}\lambda + \sum_{t=1}^{\#\text{tubelines}} N_{ij\lambda_t}^{nk\uparrow} N_0(\lambda_t) \\ &+ \sum_{k\downarrow} \int_{\lambda_0}^{\lambda_{\text{abs},\,i}} N_{ij\lambda}^{nk\downarrow} N_0(\lambda)\,\mathrm{d}\lambda + \sum_{t=1}^{\#\text{tubelines}} N_{ij\lambda_t}^{nk\downarrow} N_0(\lambda_t) \Bigg) \end{aligned} \tag{5.108}$$

## Calibration Procedure

The equations for primary and secondary fluorescence describe the photon rate coming from a layered specimen in arbitrary units. The detected count rate differs from this value because of the response of the detection system like crystals, collimators, and detectors. The total response is the product of the response of the individual components.

$$N_i^c = S_{\text{xtal},\,i} S_{\text{collimator},\,i} \cdots S_{\text{detector},\,i} N_i^t = S_{\text{detection},\,i} N_i^t. \tag{5.109}$$

For some components like detectors it is possible to calculate the response, but for others like crystals it is much more difficult. The total response can be split into a known and an unknown part.

$$N_i^c = S_{\text{known},\,i} S_{\text{unknown},\,i} N_i^t.$$

The unknown scaling factor can be determined empirically by one or more reference samples. Sample(s) of known composition is measured and the scaling factor is defined as the ratio of the measured intensity over the theoretical intensity. When more than one sample is used, a least squares fit is used to determine the relation between the measured and theoretical count rate. So, if one reference sample is used, $S_{\mathrm{unknown}}$ is calculated as

$$S_{\mathrm{unknown},\,i} = \frac{N_i^m}{S_{\mathrm{known},\,i} N_i^t}. \tag{5.110}$$

It should be noted that the determination of the scaling factors could be done on any type of samples of known composition. The scaling factor is an instrument parameter and is independent of the used samples. The reference samples can be pure elements, bulk, and layered samples as long as the composition and the layer thicknesses are known. The response of the unknown part can be energy dependent. In that case, it has to be determined for each used analytical line. In case it is not energy dependent, it is the same for all lines, and only one reference sample has to be measured to calibrate all lines.

### 5.5.3 Back-Calculation Schemes

With the formulas discussed earlier, it is possible to calculate the photon rates originating from thin and layered samples of known composition. In general, the back-calculation schemes for bulk samples cannot be used for layered samples.

The back-calculation schemes for bulk samples are based on a one-to-one relation between the concentration of an analyte and its measured count rate. Thereby it is assumed that the count rate of the analyte is proportional to the concentration of the analyte. If the same element occurs in more than one layer, then there exists no one-to-one relationship between the concentrations in the different layers and the measured count rate. The measured count rate is the sum of the rates produced in the different layers. Besides the concentrations, the thickness of the layers is also a parameter that influences the count rate.

For layered samples, an iterative method is used to calculate the composition and the thickness of the layers. This is done with a least squares method like the Levenberg–Marquardt algorithm. This method minimizes the difference between the measured and the calculated count rates of the analytes. The iteration process starts with an initial guess of the composition and thicknesses of the layers. The composition and thicknesses are varied until the weighted difference between the calculated and the observed intensities is minimal. For the weighting, the counting statistical error of the measured rate, $\sigma$, can be used. The objective is:

$$\text{Minimize} \sum_i \left( \frac{N_i^m - N_i^c(C,T)}{\sigma_i} \right)^2 , \tag{5.111}$$

where $C$ and $T$ are all unknown concentrations and thicknesses.

In each layer, the total fraction equals 1, which has to be taken into account in the minimization. This can be done by adding constraints to the least squares algorithm. A layer cannot have *one* unknown concentration. In that case, the value is set 1 minus the sum of all the other fractions in the layer. The constraint that for each layer $n$ the total fraction equals 1 is expressed as

$$\forall n : \left[\sum_j C_j^n - 1 = 0\right] . \tag{5.112}$$

The constraints can be added to the minimization function with their own weight $w_l$. After adding the constraints, the object function is still a sum of squares, which can be solved by standard techniques. The concentrations $C$ can be element concentrations, but also compound concentrations. The final optimization function is given by:

$$\min \left( \sum_i \left( \frac{N_i^m - N_i^c(C,T)}{\sigma_i} \right)^2 + \sum_l \left( w_l \left( \sum_j C_j^l - 1 \right) \right)^2 \right) . \tag{5.113}$$

A general disadvantage of the method is that it requires an initial guess of the composition and thicknesses. On the other hand, analysis of thin and layered samples is mainly done in production control environments, where the average composition is known and can be used as initial guess. Another disadvantage is that it can iterate to a local minimum, which gives the wrong composition and/or thicknesses. The method can also be used for bulk samples, with the advantage that more analytical lines can be used and that weights can be used. The value of the reached minimum can be used as indication of the accuracy. High values indicate that the analysis is probably not correct. A disadvantage of the method used for bulk samples is the required calculation time, which is much longer than the methods dedicated for bulk samples.

### 5.5.4 Solvability

The minimization can be carried out if the measured analytical lines contain sufficient information. The minimum number of analytical lines is the number of unknown parameters, concentration or thickness, minus the number of layers that have more than one unknown concentration. As stated earlier, a layer with only one unknown concentration can be solved by setting its fraction equal to 1 minus the sum of all the others in the layer.

For each unknown concentration or thickness, there should be at least one analytical line whose intensity changes with the value of the concentration or thickness. This holds for the initial guess and also for the final solution. For example, the thickness of a very thick layer cannot be determined because the count rates will not change significantly when the thickness changes. In the other way around, the thickness of a thin layer cannot be efficiently determined if the initial guess of the thickness is infinitely thick.

To determine the thickness of a layer, a line coming from this layer can be used, and also a line coming from a layer below the layer. The count rate of the line coming from the layer increases with the thickness of the layer, but the count rate of the line coming from a layer below the layer of interest will decrease with the thickness of the layer.

### 5.5.5 Applications

*Zinc coating on iron.* The simplest application of a multilayered sample is a thin layer on a substrate. As an example a pure zinc layer on an infinitely thick pure iron substrate is discussed. The thickness of the coating is the only unknown parameter because the concentrations of both zinc and iron are 100%.

The count rate of zinc lines increases exponentially with the thickness of the zinc layer and the count rate of iron lines decreases exponentially with the thickness of the zinc layer. Figure 5.15a shows the intensity of the Zn L$\alpha$ and Fe L$\alpha$ lines as function of the zinc layer thickness and Fig. 5.15b shows the same for the Zn K$\alpha$ and Fe K$\alpha$ lines.

These figures show that all four lines can be used to determine the thickness of the zinc layer, because the intensity of all four lines changes with the layer thickness. Using K lines, thicknesses of up to about 20 $\mu$ can be determined, but using L lines, the limit is about 0.75 $\mu$. For K$\alpha$ lines the intensity changes significantly up to 20 $\mu$, while for L$\alpha$ this is limited to 0.75 $\mu$.

*WSi$_x$ on silicon.* A more difficult application is a thin WSi$_x$ layer on a pure silicon substrate. The objective is to calculate both the thickness and the chemical composition of the WSi$_x$, which means determining the value of $x$. The thickness varies between 20 and 200 nm and the value of $x$ varies between 2.2 and 2.8, which means that the weight fraction Si in the top layer varies between about 0.70 and 0.75. There are two unknown parameters: the thickness of the top layer and the weight fraction of one of the elements in the top layer. The weight fraction of the other element can be determined

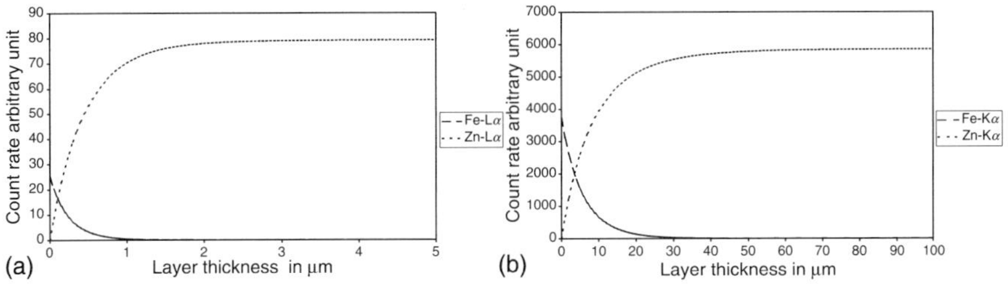


**Fig. 5.15.** (**a**) Intensity of the Zn L$\alpha$ and Fe L$\alpha$ lines versus the zinc layer thickness. (**b**) Intensity of the Zn K$\alpha$ and Fe K$\alpha$ lines versus the zinc layer thickness

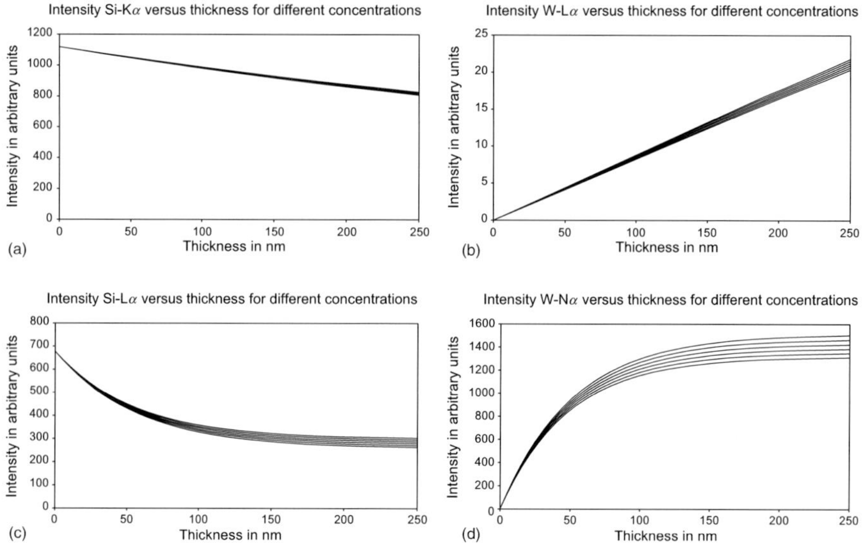


**Fig. 5.16.** $WSi_x$ layer on silicon. Intensities as a function of layer thickness. (**a**) Si K$\alpha$, (**b**) W L$\alpha$, (**c**) Si L$\alpha$, (**d**) W N$\alpha$

by difference. This means that at least two lines are required to solve the problem.

Figure 5.16 shows the intensities of the Si K$\alpha$ and W L$\alpha$ lines as function of the thickness of the top layer, for W concentrations varying between 0.70 and 0.75. The Si K$\alpha$ intensity *decreases* when the thickness *increases* (Fig. 5.16a). This is the case because the main contribution to Si K$\alpha$ comes from the substrate, and the thicker the top layer, the less of the substrate will be seen. The concentration of Si in the top has a very low influence on the Si K$\alpha$ intensity. Figure 5.16b shows that the W L$\alpha$ intensity increases when the thickness of the top-layer increases and also that the influence of the W concentration is very low.

Because of the low influence of the W (and Si) concentration on the Si K$\alpha$ and W L$\alpha$ intensities, the composition of the top layer cannot be determined accurately with these lines. Only the thickness can be determined with these lines.

Figure 5.16c,d shows that the Si L$\alpha$ and W N$\alpha$ intensities are much more sensitive to the Si concentration than Si K$\alpha$ and W L$\alpha$, and also that larger thicknesses of the top layer cannot be determined accurately with them, because their intensities do not change significantly with thickness.

With a combination of, e.g., Si K$\alpha$/Si L$\alpha$ or Si K$\alpha$/W N$\alpha$, it is possible to determine both the thickness and the composition of the top layer. From the Si K$\alpha$ intensity the thickness can be determined. After that, the Si concentration can be computed from the Si L$\alpha$ or W N$\alpha$ intensity. While this is described here as two sequential steps, the discussed back-calculation schemes for FPs allow this to be done simultaneously.

It is also possible to use all four lines simultaneously. This has the advantage that the accuracy of the analysis can be estimated, because there are four degrees of freedom but only two unknown parameters.

## 5.6 Complex Excitation Effects and Light Elements

*N. Kawahara*

### 5.6.1 Indirect Excitation Processes in the Low Energy Region

Led by the advances of the techniques in modern instruments such as synthetic multilayer monochromators or thin detector windows, XRF applications spread out to lower energy regions and lighter element analysis. There are particularly an increasing number of analytical applications in the field of thin and layered samples which require measurements of L or M (and even N-) lines at low energies.

This section discusses some unusual indirect excitation processes in the regime of low energies, which are mostly negligible at higher energies where direct excitation is efficient and the dominating factor. At lower energies, however, excitation by conventional X-ray tubes is inefficient because the low energy limit of the available tube photons is often much higher than the binding energies of the shells of interest, and indirect excitation effects gain (relative) importance. They may even by far outweigh direct excitation.

### 5.6.2 Secondary Excitation by Electrons

When a primary X-ray photon excites an atom, a photo-electron and an Auger-electron are generated as well as a fluorescent photon. For light elements such as C, B, and Be, the wide gap between their absorption edge energies and the low energy limit of the spectral distribution of tube photons results in a very low fluorescent photon emission probability and the emission of photo-electrons with relatively high energy. Considering the fact that a single electron can originate a long sequence of (inner shell) ionizations while photoelectric excitation is basically a single-event process, the excitation by photo-electrons can cause a rather large contribution compared to direct excitation. When light element analytes are contained in a heavy element matrix, photo-electrons may be abundantly emitted by the matrix elements as well and often Auger-electrons with relatively high energies cause further enhancement.

Mathematical expressions for the characteristic X-ray excitation by electrons have been reported mainly in the field of electron probe microanalysis (EPMA). The number of photons, $n_i(E_0)$, emitted from the $i$-shell of analyte atoms by interaction with a single electron of initial kinetic energy $E_0$ is expressed in the following equation, which was originally reported by Green and Cosslett [22].

$$n_i(E_0) = \frac{N_A \rho W_i}{A_i} \int_{E_0}^{E_{\text{Edge},\,i}} Q_i(E) \frac{1}{\mathrm{d}E/\mathrm{d}s} \mathrm{d}E. \tag{5.114}$$

Here, $Q_i(E)$ and $\mathrm{d}E/\mathrm{d}s$ are the inner shell ionization cross-section and the stopping power (energy loss function), respectively, of electrons in a material. Note that the lower limit of the integral is the edge energy of the shell of interest, $E_{\text{Edge,i}}$. The inner shell ionization cross section, $Q_i(E)$, is given by:

$$Q_i(E) = \frac{Z_{\text{subshell},\,i} \pi e^4}{E\, E_{ip}} b \ln \frac{4E}{B}. \tag{5.115}$$

Taking $U = E/E_{\text{Edge},i}$,

$$Q_{ip}(E) = 6.51 \times 10^{-20} \frac{Z_{\text{subshell},\,i}}{E_{\text{Edge},\,i}{}^2} b \frac{\ln(cU)}{U} \ (\text{cm}^2), \tag{5.116}$$

where $Z_{\text{subshell},i}$ is the number of electrons in the $i$ subshell; $b$ and $c$ are parameters for which several values are suggested for each element [43].

Based on Bethe's expression [7], several expressions for the stopping power $\mathrm{d}E/\mathrm{d}s$ have been proposed, for example by Love et al. [35]:

$$\frac{\mathrm{d}E}{\mathrm{d}s} = -\frac{\rho}{J'} \sum_k \frac{W_k Z_k}{A_k} \frac{1}{1.18 \times 10^{-5}\sqrt{E/J'} + 1.47 \times 10^{-6}(E/J')} \tag{5.117}$$

$$J' = \frac{\exp \sum_k \dfrac{w_k Z_k}{A_k} \ln J_k}{\sum_k \dfrac{w_k Z_k}{A_k}},$$

where $J_k$ is the mean ionization potential of element $k$, and $Z_k$ its atomic number. The mean ionization potential has been discussed by several authors. The following equation is by Wilson [61]:

$$J_k = 0.0115\, Z_k \ (\text{keV}).$$

Figure 5.17 shows the resulting number of ionizations, $n_i(E_0)$, caused by a single electron as a function of the initial electron energy while slowing down, for several elements. It particularly illustrates the huge enhancement effect for very light elements such as carbon, which is mainly due to the low binding energy of K-electrons and multiple excitations.

## Secondary Excitation by Photo-Electrons

The photon number of secondary photons excited by interaction with a photo-electron can be expressed by combining the $n_i(E)$ of (5.114) with (5.12) for direct excitation. Thereby it is assumed that the secondary excitations by an electron generated within the specimen occur in the vicinity of the point

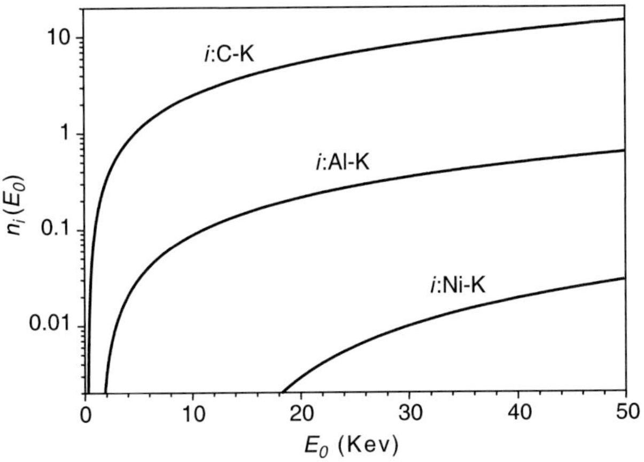


**Fig. 5.17.** Number of secondary ionizations, $n_i(E_0)$, in elements C, Al, and Ni, excited per electron as a function of its initial energy. This is also the ratio of secondary to primary photons, because the transition probabilities, fluorescent yields, etc. are common factors

where the electron is generated, i.e., that all secondary photons originate from practically the same depth as the directly excited photon. The equations are:

$$
\begin{aligned}
\mathrm{d}N_{i,\,\text{photo-electron}} &= \mathrm{d}N_0(E)\,G_iC_i\frac{1}{\mu^*}\sum_j \tau_j n_i(E-E_{\text{Edge},\,j})\\
N_{i,\,\text{photo-electron}} &= G_iC_i\sum_j\int_{E=E_{\text{Edge},\,j}}^{E=E_{\max}} \frac{N_0(E)\tau_j n_i(E-E_{\text{Edge},\,j})\,\mathrm{d}E}{\mu^*}\,.
\end{aligned}
\tag{5.118}
$$

Comparison of (5.118) with (5.12) for primary excitation shows that the contribution of $N_{i,\text{photo-electron}}$ exceeds $N_{i,\text{primary}}$ for light elements when $n_i(E)$ is large. For example, $N_{C\ K\alpha,\text{photo-electron}}$ for graphite is five times larger than $N_{C\ K\alpha,\text{primary}}$ for 20 keV monochromatic source (Fig. 5.17). Furthermore, when the light element is contained in a matrix consisting largely of heavy elements with a consequently large $\tau_j$, $N_{i,\text{photo-electron}}$ may become much larger than $N_{i,\text{primary}}$.

## Secondary Excitation by Auger-Electrons

If a light element analyte constitutes a monoelement sample, it will not be excited by its own Auger-electrons. However, when the light element analyte is embedded in a heavy element matrix, the Auger-electrons from the accompanying elements can excite light element fluorescence. For example, when carbon in steel is analyzed, a Fe KLL Auger-electron with a kinetic energy of 6.3 keV can excite multiple carbon K-shells.

The number characteristic photons generated by an Auger-electron is expressed using the $n_i(E)$ as follows.

$$\mathrm{d}N_{i,\,\mathrm{Auger-electron}} = \mathrm{d}N_0(E)G_iC_i\frac{1}{\mu^*}\sum_j \tau_j(1-\omega_j)\sum_k n_i(E_{\mathrm{Auger-electron},\,jk})$$

$$N_{i,\,\mathrm{Auger-electron}} = G_iC_i \sum_j \int_{E=E_{\mathrm{Edge},\,j}}^{E=E_{\max}} \frac{N_0(E)\tau_j(1-\omega_j)\sum_k n_i(E_{\mathrm{Auger-electron},\,jk})}{\mu^*}\mathrm{d}E. \tag{5.119}$$

### Secondary Excitation by Electrons in Thin Films

On extension of the equation for thin films (including multilayer structures and films on substrates), the assumption that "secondary excitations by an electron occurs in vicinity of the point where the electron is generated" must be re-examined. The maximum range of traveling paths of electrons is obtained by integrating the stopping power (5.117). Figure 5.18 shows the resulting maximum ranges of electrons in several materials and illustrates how far photo-electrons penetrate into other layers in thin film structures. Accounting for such electron traveling paths in calculation is reported by Kataoka et al. [28].

## 5.6.3 Cascade Effect

An indirect excitation process, which is of importance in low energy L-series fluorescent measurements, is the *cascade effect*, or emission of L-series photon following radiative and nonradiative relaxations of the K-shell ionization.

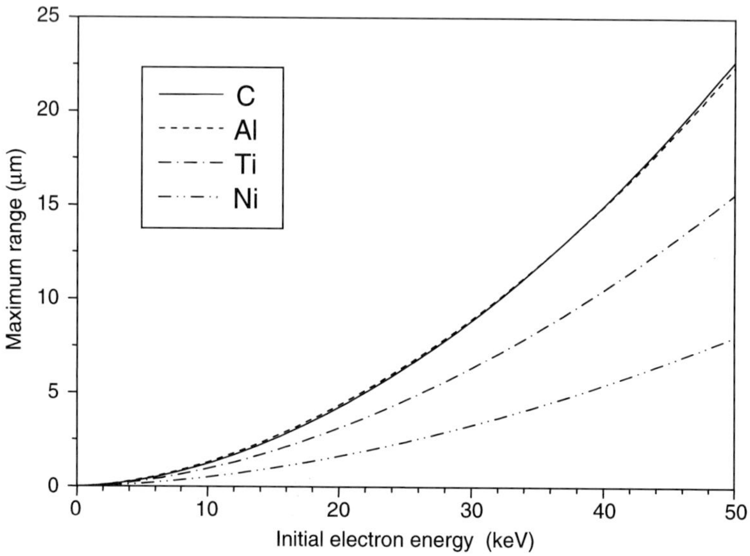


**Fig. 5.18.** Maximum ranges for electrons in C, Al, Ti, and Ni, computed by using (5.117)

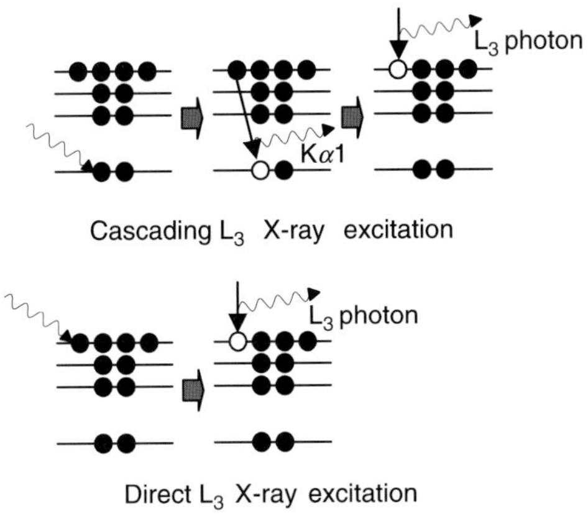


**Fig. 5.19.** Principle of indirect excitation by photons from cascading relaxation

This process occurs when the incident photon energy is higher than the K-shell binding energy. It occurs in M- (and outer) series fluorescent measurements as well.

Figure 5.19 is a schematic view of this effect taking the case of $L\alpha$ emission: when an atom is ionized in the K shell by an incident photon, it can be de-excited by emitting a $K\alpha_1$ photon, whereby an $L_3$ shell hole is generated. This state is virtually same as after direct $L_3$ excitation by an incident photon. Then an X-ray photon belonging to the $L_3$ series, such as $L\alpha$ or $L\beta_2$, can be emitted. As a matter of course, radiative relaxation processes of the K shell other than $K\alpha_1$ emission and nonradiative or Auger-transition processes are possible and probable [1, 2].

This cascade effect occurs when the energy of the incident X-ray photon is higher than the binding energy of the K shell. It has been studied and already implemented in some commercial software. Figure 5.20 shows the experimental

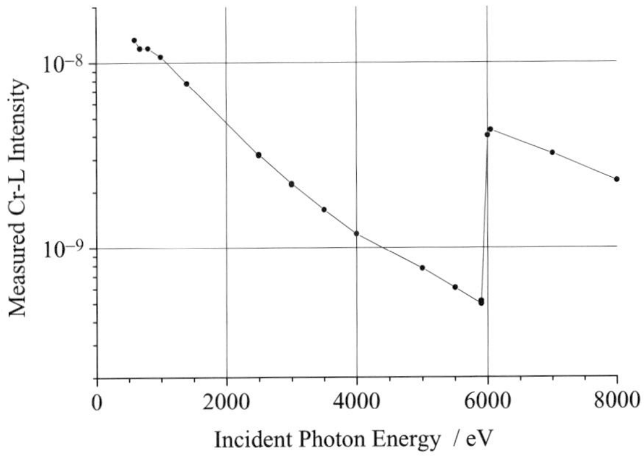


**Fig. 5.20.** Experimental count rates of Cr-L fluorescent photons excited by a monochromatic source with variable energy. Note the jump near 6000 eV (Cr K-edge)

evidence by measured Cr-L fluorescent photons excited by a tunable monochromatic source (using a synchrotron radiation facility). The drastic jump at a primary photon energy near 5.99 keV (Cr-K edge energy) illustrates the effect excellently [29].

## 5.7 Standardless Methods

*K.-E. Mauser*

### 5.7.1 Introduction

For some decades, wavelength-dispersive X-ray fluorescence analysis (WDX) was strictly divided into two quite different approaches: one for *qualitative analysis*, and the other for *quantitative analysis.*

*Qualitative analysis* was possible only with sequential systems: a spectrum was recorded; lines and elements were identified and sometimes qualified as majors, minors, or traces. For many years this had to be done manually, software for automatic or interactive evaluation of spectra was offered very late by the manufacturers of wavelength-dispersive spectrometers. Therefore, energy-dispersive XRF (EDX) was seen the preferred technique to obtain a qualitative overview.

*Quantitative analysis*, on the other hand, was the domain of wavelength-dispersive systems. Simultaneous (multichannel) instruments could measure just the peak intensities at fixed reflection angles; sequential spectrometers followed the same strategy, but added the option to measure background intensities at one or two fixed angles for improved trace analysis. The calibrations were set up for specific materials with narrow concentration ranges and based on standards, which matched the unknown samples as closely as possible. Thus the reputation has been created that quantitative WDX analysis is a very accurate method with relative errors down to (in special cases even below) 0.1%.

There was clearly a gap between these two approaches: how to get more than just qualitative information for samples which do not fit to a calibration, but are too singular as to justify the amount of work required for a dedicated calibration.

### 5.7.2 Semiquantitative Analysis

In the 1980s, several manufacturers of WDX instruments and software began to develop methods of quantitative XRF analysis for all kinds of samples without the need for any specific calibration process or any specific standard samples.

In the beginning the developers of these methods compared the quality of the results to the high accuracy of specialized calibrations and therefore used the term *semiquantitative analysis.* But as the accuracy of these methods improved and approached typical errors of 1% relative for many applications, *semiquantitative* was no longer an appropriate term and was replaced by *standardless methods.*

### 5.7.3 Requirements for a Standardless Method

Any X-ray spectrometer will determine just "intensities," i.e., the number of photons of a given energy which hit the detector in a given time. In other words, XRF is a relative analytical method and will always require an analytical function, which relates the measured intensities to concentrations of elements or compounds. A method without specific standards has to cover a maximum range of elements in as wide a variety of materials as possible, and there is clearly a need for a universal and comprehensive matrix correction method. Empirical methods cannot fulfill these requirements; they are too limited in concentration range and element coverage. The only choice is the use of matrix corrections based on FPs.

Look at a typical article on FP or influence coefficients and the task may seem relatively simple: for each element (in fact: each X-ray emission line to be used) there is just one empirical parameter related to the instrument, typically expressed as the intensity of the pure analyte, everything else is handled by the FP model. However, a closer examination shows immediately that there are some more difficulties to overcome.

Any article on FPs or influence coefficients has a small print somewhere: the FP calculations are based on pure net intensities, i.e., on intensities already corrected for dead time, line overlap, spectral contamination, and background.

Most publications on FPs use metals as examples, prepared with a nearly perfect surface, and all relevant elements determined by XRF. The typical analyst applying a standardless method has every kind of samples, including powders, small or oddly shaped pieces, liquids, etc. Therefore any standardless method has to face a series of challenges. First of all, correct net intensities have to be determined. This task requires:

- A set of measurement conditions covering concentrations from trace to 100% for all elements to be analyzed by XRF;
- A reliable method for background determination and background subtraction;
- A universal correction method for spectral interference (line overlaps);
- Absorption corrections for films used as a support for liquid samples or loose powders;
- Subtraction of contaminations with (nearly) constant intensity, e.g., elements present in the supporting film.

Once these intensities are available, the concentrations can be determined. This will require:

- The calculation of concentrations in the specimen as measured based on a universal calibration;
- Ways to handle nonmeasured elements;
- Correction procedures for size and thickness of the specimen;
- Subtraction of contaminations with (nearly) constant concentrations, e.g., tungsten from a grinding vessel or contaminations in a flux;
- Calculation of the composition of the original sample (without preparation agents, including loss on ignition (LOI)).

A closer look at these demands will help to understand performance and limitations of the commercially available software packages for standardless analysis. The focus will be on wavelength-dispersive systems, but most of the comments are valid for EDX instruments, too.

## Universal Measurement Conditions

Quantitative XRF analysis in general can be applied for concentrations between 0.1 ppm and 100%. However, quantitative methods for dedicated materials typically cover just one or two orders of magnitude in concentrations. Sequential WDX systems are typically optimized element by element: crystals, collimators, detectors, and discriminators are selected for the limited intensity range in that specific material.

Measurement conditions for universal standardless methods have to cover the maximum possible range of concentrations and intensities, including the pure analyte on one side and the lowest possible traces on the other side. One way of extending the dynamic range is to automatically adjust the tube power depending on the intensity of the current sample. A different tube voltage would modify the shape of the excitation spectrum and therefore interfere with the whole calibration and matrix correction procedure. The intensity of all lines is strictly proportional to the tube current; therefore a reduced current can be handled easily by the software.

## Determination of the Background Intensity

The net intensity is just the peak intensity minus the background intensity at the peak position. Unfortunately the background intensity at the peak position cannot be determined directly, because normally there is no "blank" sample available (a true "blank" sample for this purpose has to have exactly the same scattering properties, i.e., the blank sample has to be identical to the unknown sample except for the analyte and traces).

Traditionally sequential WDX systems measure the background intensity at one or more reflection angles adjacent to the line and calculate the background under the line by a simple linear or curved approximation. This method has severe drawbacks for a universal standardless method: It is practically impossible to define fixed background positions, which are never disturbed by a weak line of an exotic element that shows up in high concentration in one of the samples.

There are various solutions available:

- Limit the scope and create specialized standardless methods, e.g., for geological applications or for petrochemistry. Undisturbed background positions adjacent to the peak can then be found, and the net intensity can be determined reliably as long as these methods are applied only to the dedicated materials;
- Measure the whole spectrum (or at least the spectral ranges around the lines of interest) and calculate the background under the line using an angular range of several degrees and not just one or two fixed positions. This method is not sensitive to individual lines of other elements within the background range;
- Measure at fixed peak positions and at a few well-selected fixed background positions and fit these points with a sophisticated calculation of the background based on the scattered tube spectrum and the sample matrix.

The scan method has the advantage of a straightforward background determination independent of any concentration-related calculations and allows graphical checks and interactive evaluation. Small differences in peak position by chemical shift for the light elements or instrumental effects are corrected automatically. The disadvantage of the scan method compared to peak-background methods is the longer measuring time or poorer counting statistics: a peak-background method *concentrates* the measuring time on the peaks and dedicated background positions, whereas the scan method *distributes* the measurement time over the whole spectral range. On the other hand, any peak-background method is a black box without the chance to see more than intensity numbers, and therefore without the possibility to detect and correct errors in the automatic background calculations.

### Spectral Interferences

Typically software for wavelength-dispersive spectrometers corrects line overlaps in a very simple way. Take as an example the line overlap of the Mo Ll line on P K$\alpha$: The measured intensity of Mo K$\alpha$ is multiplied by an empirically determined factor and subtracted from the intensity of P K$\alpha$. The energy of Mo K$\alpha$ is 17.5 keV, the energy of the overlapping Mo Ll line is 2.0 keV. The simple model assumes a constant ratio of these two lines, but any varying absorption edge between the two energies will change the ratio. Therefore this method can be applied only for dedicated calibrations covering narrow concentration ranges. It will not work for a universal standardless method.

A way to handle line overlaps more universally is the use of calculated intensities: the FP model applied to calculate the concentration of Mo based on the intensity of Mo K$\alpha$ can easily calculate the theoretical intensity for the overlapping Mo Ll line as well—individually for the current sample including all absorption edges. The empirical factor specific for the instrument is now related to the calculated intensity of the overlapping line itself, not to the measured intensity of a line totally different in energy, and therefore applicable over the whole concentration range for all elements.

Even when theoretical intensities are used, there are still empirical factors which have to be determined, one for each overlap which may happen between around 80 elements, each in a range of up to 100%. There are two options to handle these empirical factors:

- The line overlap factors are determined individually for each spectrometer. This will require the measurement of a well-selected set of samples including all the elements of interest;
- The line overlap factors are assumed to be the same for all spectrometers of one type. The manufacturer of the standardless method determines these factors once and delivers it with the software. Thus, installation of the standardless method is fast and simple.

Typically different crystals and collimators of the same type are fairly uniform, so that overlaps of first-order lines are nearly constant. Overlaps of higher orders are strongly influenced by the resolution of the detector, which has to be controlled regularly especially for flow proportional counters. As a rule, prefixed overlap coefficients will do the job as long as there is no demand to reach the frontiers in detection limits (DLs) and accuracy.

There is one other important difference in using calculated intensities (or concentrations) for line overlap corrections instead of measured intensities. An overlap correction with measured intensities is a separate step, independent of any concentration value. The use of calculated intensities for line overlaps combines the determination of spectral interference with the calculation of concentrations. Any error in a calibration, in the definition of the matrix or preparation agents may immediately influence the net intensities of overlapped lines.

### Corrections for Supporting Films and Intensity-Related Contaminations

Liquid samples and loose powders are measured in sample cells with a thin film at the bottom. The selection of the appropriate film is a trade-off between stability, resistance to X-rays, and chemical resistance on one side, and absorption of long wavelengths on the other side. Standardless methods will allow using any kind of film, as long as composition, thickness and density are known. The absorption effects can be calculated and the measured intensities corrected accordingly. These films may have contaminations of elements used as catalysts in the production process. Such an element in the film emits an intensity (nearly) independent of the material to be analyzed, i.e., this kind of contamination has to be handled at the intensity level.

### Calculation of Element Concentrations in the Specimen

Once net intensities have been determined, the element concentrations in the specimen can be calculated. For FP methods there is just one empirical coefficient required for each spectral line to be used. This factor can be expressed as *intensity of the pure analyte.* This does not mean that the pure analyte has to be measured; in fact, the factor can be determined by any sample of known composition containing a significant amount of the analyte.

Again there are two options to determine these calibration factors for a specific instrument:

- The calibration factors are determined individually for each spectrometer. This will require the measurement of a well-selected set of samples including all the elements of interest;
- The calibration factors are assumed to be very similar for all spectrometers of one type. The manufacturer of the standardless method determines these factors once and delivers them with the software. A procedure following the scheme of drift corrections is used to adapt these calibration factors to the specific spectrometer. Adjacent lines measured under the same conditions except for the reflection angles are assumed to have the same correction factors between the "master" instrument and the specific spectrometer. Thus the standardless method can be installed and maintained with just a few very stable glass samples.

### Handling of Preparation Agents

There are many materials which cannot be analyzed in their original form. Preparation methods have to be used which may require adding a preparation agent, e.g., a binder for a pressed pellet or a flux for a fused "glass" bead. These preparation agents are part of the specimen as analyzed in the spectrometer, they influence the absorption properties and therefore have to be handled by the FP method in the same way as any element present in the original material. Therefore for the use of any standardless method

- The preparation agents have to be defined correctly (elements, concentrations, amount of sample and of preparation agent);
- The primary result of the FP calculation is the composition of the specimen including the preparation agents;
- The standardless method will calculate the composition of the original material in a second step.

**Table 5.7.** Influence of preparation agents

| | | Pressed pellet | | Fused bead | |
|---|---|---|---|---|---|
| Element/ oxide | Certified concentration | Elements in prepared specimen | Oxides in original material | Elements in prepared specimen | Oxides in original material |
| H/Li | | 0.76* | | 0.07* | |
| B | | 2.70* | | 22.01* | |
| O | | 50.4* | | 63.33* | |
| $Na_2O$ | 0.43 | 0.22 | 0.36 | 0.048 | 0.46 |
| $Al_2O_3$ | 17.34 | 7.80 | 17.4 | 1.25 | 16.9 |
| $SiO_2$ | 63.63 | 24.85 | 62.9 | 4.17 | 63.8 |
| $K_2O$ | 15.35 | 10.78 | 15.3 | 1.77 | 15.2 |
| CaO | 0.68 | 0.41 | 0.68 | 0.067 | 0.67 |

*Values calculated based on binder, flux, dilution, and chemistry

This procedure is illustrated in Table 5.7. The same material is measured as a pressed pellet prepared with a binder of boric acid and as a fused bead melted with lithium tetraborate. The primary concentrations in the two specimens are quite different, but in both cases the final results are close to the certified concentrations.

### The Handling of Nonmeasured Elements

X-ray fluorescence cannot determine all elements of the periodic table. H and Li do not emit radiation of accessible wavelength, the measurement of Be has nearly no practical application. The radiation of B, C, N, O, and F is absorbed by the supporting film required for liquid samples or loose powders. These elements can be measured in solid and dry samples, but the information depth of less than 1 $\mu$m requires meticulous sample preparation. This demand is especially difficult to meet because fusion is typically not possible (it will add B and O, and lose C as $CO_2$) and any binder for pressed pellets contains at least some of these light elements.

On the other hand, standardless methods like all evaluations based on FPs will require including all elements with a significant matrix influence in the calculations. There are various ways to handle nonmeasured elements; each of them requires some knowledge about the sample to be evaluated:

- By definition of compounds, e.g., oxides or carbonates: define $CaCO_3$ and $MgCO_3$, and the standardless method will determine the concentrations of C and O based on the measurement of Ca and Mg;
- By balance: the concentration of an unmeasured element or compound is calculated as 100% minus the sum of all measured concentrations. This calculation by balance has to be part of the iteration process; it cannot be

**Table 5.8.** Influence of various light matrices

| | Certified concentrations | Calculated concentrations for matrix | | |
|---|---|---|---|---|
| Matrix | Oil ($CH_2$) | Oil ($CH_2$) | Ethanol ($C_2H_5OH$) | Water ($H_2O$) |
| Mg (mg $g^{-1}$) | 300 | 250 | 364 | 538 |
| Si (mg $g^{-1}$) | 300 | 279 | 414 | 620 |
| Ni (mg $g^{-1}$) | 300 | 298 | 477 | 748 |
| Pb (mg $g^{-1}$) | 300 | 299 | 457 | 697 |
| Mo (mg $g^{-1}$) | 300 | 297 | 365 | 470 |
| Compton ratio | 1 | 0.99 | 0.80 | 0.59 |

done as a second step after the measured elements have been determined. Typical compounds defined as balance are $CH_2$ in oils and plastics, and $H_2O$ in aqueous solutions;

- By inserting a known value determined with a different method, e.g., LOI inserted as $CO_2$.

A small variation in a light matrix may have a huge effect on the results. This is illustrated in Table 5.8. An oil sample containing 300 ppm of various elements is evaluated with the input of oil, ethanol and water as a matrix. As expected the results are correct for oil ($CH_2$), but already the addition of just a single oxygen atom per molecule in ethanol changes the values significantly.

## Size and Thickness of the Specimen

Typically, an X-ray spectrometer sees only some milligrams of material close to the surface of the specimen. As the major precondition for reliable results, this volume has to be representative of the material to be analyzed.

The layer on the sample surface, which emits X-rays, can be less than 1 $\mu$m for light elements in a heavy matrix. For heavy elements in a light matrix, the penetration of the X-rays can go up to several centimeters, so it may happen that a specimen is not "infinitely thick." Standardless methods are able to correct for sample thickness. The value required is the mass density (e.g., in (g $cm^{-2}$)). This can be easily defined by diameter and mass of the specimen, or by thickness and density.

Specimens which do not cover the whole open area of a sample cup are more difficult to handle. The intensity of the primary radiation coming from the X-ray tube is not homogeneously distributed over the irradiated area. The exact size of the specimen may not be known. Additionally such irregularly shaped samples may not be flat, so the distance to tube and detector is not correct for part of the sample.

A crude empirical approach is to "normalize" the results. The simple idea of normalizing the final concentrations does not work at all. Imagine that the intensities of a small sample are too low by a factor of 100. The relations between intensities and concentrations are not linear, so the concentrations determined with the wrong intensities will not be the correct ones divided by 100, but differ by random factors influenced by wrongly corrected matrix influence. The correct way is to modify the intensities by a common factor, calculate the concentrations and iterate this process until the final result gives a total of 100%.

### Unknown Parameters

Sales literature on commercial standardless software packages may stress the point that one can determine automatically unknown parameters like matrix elements, LOI, sample thickness, sample size or dilution.

There is no hidden secret behind these statements; they are just based on the fact that the sum of concentrations in the specimen and in the original material has to be 100%. If that is not the case for the sum of measured elements in the specimen, the software can modify various parameters to achieve this goal: add nonmeasured elements, change dilution, vary thickness or size of the specimen. Since there is just one condition (the sum equals 100%), but several parameters, which can be modified, there is no unequivocal solution. The software does not know "automatically," which parameter to modify; the user has to provide additional information about the sample.

### Scattered Tube Radiation

The X-ray spectrum offers some additional information, which can be used in standardless methods: the X-rays emitted from the tube are scattered in the specimen; this scattering effect strongly depends on the composition of the specimen, especially the amount of light elements. X-rays are scattered without loss of energy (Rayleigh or elastic scattering) and with a certain loss of energy (Compton or inelastic scattering). The spectrum therefore shows the Rayleigh peak at exactly the wavelength of the characteristic tube radiation, e.g., Rh $K\alpha$, whereas the Compton peak is a broad peak shifted to lower energy.

Standardless methods can use the intensity of the Compton peak, the Rayleigh peak or the ratio of both as additional information. One way to do this is the comparison of theoretical Compton intensities with the measured values.

For known composition of a sample, a theoretical Compton intensity can be calculated. Measured Compton intensities in standard samples can be used to scale the theoretical intensities in a way that the ratio of theoretical and measured Compton intensities (the *Compton factor*) is equal to 1. This factor can be calculated for any sample evaluated with the standardless method. A significant deviation from 1 indicates that something is wrong with the evaluation. Table 5.8 gives an example: the correct input of oil as the matrix results in Compton ratio of 0.99, i.e., very close to 1. Calculated with the wrong inputs of ethanol or water, the Compton ratio clearly differs from 1.

### Standardless Methods Versus Special Calibrations

There are still various good reasons for the use of specialized quantitative methods for certain applications:

- The measuring conditions can be optimized for a certain material with narrow concentration ranges. As a result the measurement time can be reduced or lower limits of detection can be achieved;
- The measurement can be limited to just the element(s) of interest. Either it is assumed that the matrix effect is constant for all samples or the matrix effect is handled in a global way using an internal standard (an added element or the Compton intensity, as discussed in Sect. 5.4). Thus the total measurement time can be significantly lower than for a standardless method, which will always require to measure (or define in some other way) all elements with a significant matrix influence;
- The standard samples can closely match the unknowns to be determined. Effects causing an inhomogeneous distribution of the elements in the specimen (mineralogical effects, particle size effects, surface effects) may not be significant as long as the standards are very similar, but these effects are practically impossible to handle in a general standardless method. Inaccuracies in the fundamental coefficients may just cancel out when standards of similar composition are used for the calibration. Thus a specialized method can be more accurate than a standardless method.

On the other hand, the better accuracy of specialized methods is just a possibility, but not the rule. Especially calibrations with empirical matrix correction coefficients tend to be "overcorrected." The standard deviation of the calibration itself very often looks perfect, but too many coefficients have been determined with too low a number of standards or with standards having strong correlations in their composition. In any case, the accuracy of such a calibration should be checked by the analysis of samples with known composition, which are not used for the calibration.

### Summary

Standardless methods are now delivered with most of the new X-ray spectrometers. These methods extend the applications of X-ray fluorescence analysis and make it much easier to work with the spectrometer, especially for the newcomer in XRF. Nevertheless standardless methods do not replace the traditional dedicated methods in general. Especially in process and quality control special quantitative methods can be developed, which are faster and more accurate than a universal standardless approach.

## 5.8 Monte Carlo Methods

*M. Mantler*

Monte Carlo methods are numerical methods of statistical simulation. In XRF, they are sometimes used when a problem can be only (or more successfully) described by statistical models rather than by closed equations. Examples are inhomogeneous samples, samples with irregular shapes, or excitation of fluorescent radiation by electrons.

The key components of Monte Carlo computations are a random number generator, functions (probability density functions, PDF) that relate the generated random numbers to observable (countable) events, and counters for these events. A simple example is the selection of a random penetration event for a photon in absorbing material, i.e., the position (depth) of interaction. Simplest, a small discrete set of random numbers, $R$, could be obtained by casting dice, where each numbered face shows with equal probability: $R = \{1, 2, 3, 4, 5, 6\}$, $P(R) = 1/6$. Then the thickness of the specimen (assume infinite thickness) is subdivided into a corresponding number, $m$, of regions ($m = 6$ in this case) in such a way that in each of them the same number of photons ($N_0/m$) is absorbed (this is the important point: each randomly generated number, $R$, must correspond to a set of equally probable events, in this case to an equal number of absorbed photons). From the absorption law $N/N_0 = \exp(-\mu D)$ follows for the $i$th region from $D_{i-1}$ to $D_i$:

$$
\begin{aligned}
&\frac{N_{i-1} - N_i}{N_0} = \frac{N_0/m}{N_0} = \frac{1}{m} = \exp(-D_{i-1}\mu) - \exp(-D_i\mu) \\
&\exp(-D_1\mu) = 1 - \frac{1}{m}, \quad \cdots, \quad \exp(-D_i\mu) = 1 - \frac{i}{m}, \cdots \\
&\text{and} \\
&D_i = -\frac{1}{\mu}\log\left(1 - \frac{i}{m}\right) \quad \text{with} \quad i = 0,\ 1,\ 2, \ldots,\ m \quad \text{and} \quad D_0 = 0
\end{aligned}
\tag{5.120}
$$

For $\mu = 1$ one obtains $D_0 = 0$, $D_1 \approx 0.18\,\text{cm}$, $D_2 \approx 0.40\,\text{cm}$, $D_3 \approx 0.69\,\text{cm}$, $D_4 \approx 1.09\,\text{cm}$, $D_5 \approx 1.79\,\text{cm}$, and $D_6 = \infty$; reading four eyes on the dice would then be interpreted as the absorption of a photon somewhere within interval 4 (0.69–1.09 cm). In practice, this is perhaps not sufficiently accurate, and a continuous relationship with infinitely small intervals is preferred. This is easily achieved by replacing the discrete term $i/m$ in the equation for $D_i$ above by a continuous, equally distributed random number, $R$, which is also normalized to 1 and defined within the interval $(0, 1)$. Note that a sequence of such random numbers $\{R_i\}$ is equivalent to the sequence $\{1 - R_i\}$, and therefore

$$
D(R) = -\frac{1}{\mu}\log(R) = -\Lambda\log(R) \tag{5.121}
$$

$\Lambda$ is the mean free path length. Given the starting point of a photon, this defines the distance to travel to the point of the expected interaction.

As soon as the point of interaction is reached, the question arises, which of the possible types of interaction occurs. For XRF, these are photo-absorption, coherent, and incoherent scattering with probabilities $\tau/\mu$, $\sigma_{\text{coh}}/\mu$, and $\sigma_{\text{incoh}}/\mu$. Keeping in mind that $\tau/\mu + \sigma_{\text{coh}}/\mu + \sigma_{\text{incoh}}/\mu = 1$ allows a selection to be made by a single random number, $R$:

Photo-absorption for $0 \leq R \leq \tau/\mu$
Coherent scattering for $\tau/\mu < R \leq \tau/\mu + \sigma_{\text{coh}}/\mu$
Incoherent scattering for $\tau/\mu + \sigma_{\text{coh}}/\mu < R \leq 1$.

A further subdivision of $\tau/\mu = \tau_{\text{K}}/\mu + \tau_{L1}/\mu + \tau_{L2}/\mu + \tau_{L3}/\mu + \ldots$ can be made to select an edge for photo-absorption (K, $\text{L}_1$, $\text{L}_2$, $\text{L}_3$,...), depending on the energy of the photon. Additional random selections are necessary to determine the type of relaxation after photo-absorption, the probability of an Auger-effect, and in case of a photon emission, its direction and next distance to travel. This process is repeated until the photon leaves the specimen and enters a detector. In principle, all indirect excitation processes can be included in such a scheme, even complex combinations such as scattering events and multiple indirect excitations (like tertiary excitation and higher). An important aspect is that the element, with which the photon interacts at a specific geometrical point in the specimen, can be defined by a distribution map or also be subject to random selection, so that a specimen must not necessarily be homogeneous. In a similar way, irregularly shaped specimens can be treated [17]. Excitation of photons by electrons (for example in microprobe analysis or for light elements) can be modeled by similar schemes.

While such possibilities appear fascinating and useful, there are disadvantages as well. The most important is that virtual photons – after their histories are computed and they possibly end up in a virtual detector – are counted, and are thereby subject to counting statistics just as in a real measurement. A large number of histories are required for reasonably accurate results, and powerful computers are required in order to keep computing times at an acceptable level.

## 5.9 Errors and Reliability Issues

*M. Mantler*

The determination of the *true value* of a physical quantity, for example of an element concentration, is in principle impossible. The difference between the true value and a measured value is the error and expresses the experimental uncertainty. Certain errors vary statistically when the experiment is repeated under identical circumstances and are then called statistical error, while others show a bias toward higher of lower values, do not vary, and are called systematic errors. Systematic errors (*determinate errors*) have theoretically a known reason and are known in size and sign, and they can be corrected, at least in principle.

Only few observed events in X-ray analysis are of true statistical nature, but among them is one of highest importance: The emission of photons from an excited atom. The measured variable, the number of counts in a given time span (*count rate*), is therefore affected by a truly statistical error. Most other errors are in fact systematic errors, where a variety of parameters is related to the measured quantity in a much too complex way as to be mathematically addressed in practice (*indeterminate errors*). In most cases such apparent statistical errors can be treated by the same mathematical approach as true statistical errors, but their statistical parameters (such as the standard deviation, $\sigma$) can be only derived from experimental findings, while for the truly statistically varying emission rates they are accurately predictable from physical principles (*counting statistics*).

## Definitions [24]

**Accuracy** is the closeness of agreement between a test result and the accepted reference value [ISO 5725-1] or the closeness of the agreement between the result of a measurement and a true value of the measurand. Accuracy is a qualitative concept [VIM:1993].

**Certified reference material (CRM)** is a reference material, accompanied by a certificate, one or more of whose property values are certified by a procedure which establishes its traceability to an accurate realization of the unit in which the property values are expressed, and for which each certified value is accompanied by an uncertainty at a stated level of confidence [ISO Guide 30:1992].

**Reference material (RM)** is a material or substance one or more of whose property values are sufficiently homogeneous, stable, and well established to be used for the calibration of an apparatus, the assessment of a measurement method, or for assigning values to materials [ISO Guide 30:1992].

**Reference material certificate** is the document accompanying a certified reference material stating one or more property values and their uncertainties, and confirming that the necessary procedures have been carried out to ensure their validity and traceability [ISO Guide 30:1992].

**Uncertainty of a certified value** is the estimate attached to a certified value of a quantity, which characterizes the range of values within which the true value is asserted to lie with a stated level of confidence [ISO Guide 30:1992], or the parameter, associated with the result of a measurement, that characterizes the dispersion of the values that could reasonably be attributed to the measurand [VIM:1993].

**Traceability** is the property of the result of a measurement or the value of a standard whereby it can be related to stated references, usually national or international standards, through an unbroken chain of comparisons all having stated uncertainties. The concept is often expressed by the adjective traceable. The unbroken chain of comparisons is often called a traceability chain [VIM:1993]. This definition applies to both physical and chemical measurements. Traceability only exists when metrological evidence, appropriate for the application, is collected on a continuing basis to document the traceability chain and quantify its associated measurement uncertainties.

### 5.9.1 Mathematical Treatment of Statistical Errors

Statistical errors vary from measurement to measurement and are by their nature unpredictable in size and sign. A frequent assumption is that a Gaussian distribution, $G(x)$ describes frequency (or probability) of the occurrence of a specific value of a measured variable, $x$. This is equivalent to the probability of obtaining a result with an error $\Delta x = (x - x_o)$:

$$N(x_1, x_2) = \int_{x=x_1}^{x=x_2} G(x)\,\mathrm{d}x = \frac{1}{\sqrt{2\pi}\sigma} \int_{x=x_1}^{x=x_2} e^{-\frac{(x-x_0)^2}{2\sigma^2}}\,\mathrm{d}x \tag{5.122}$$

$N$ is the probability to measure a value, $x$, between $x_1$ and $x_2$. Parameters of $G(x)$ are the true value, $x_o$, and the standard deviation, $\sigma$. In real measurements, a random subset (*statistical sample*) of all possible $x$-values (*statistical population*) is collected: $x_1, x_2, \ldots, x_n$.

The mean value, $\bar{x}$, is the best approximation which can be obtained from the statistical sample for the true value, $x_0$:

$$x_0 \approx \bar{x} = \frac{\sum x_i}{n}. \tag{5.123}$$

The standard deviation of the statistical sample, $\sigma$, can be obtained from the measured data and specifies the quality of the measurement. It represents a specific statistical sample and therefore varies when another set of data is collected (i.e., when the measurement is repeated). Larger numbers, $n$, of measurements (observations) improve the result.

*Propagation of statistical errors.* Frequently a result, $r$, is computed from several measured parameters, $p_1, \ldots, p_n$, by a function $f(p_1, \ldots, p_n)$. The standard deviation of the result, $\sigma_r$, can be derived from the standard deviations of the measured parameters, $\sigma_{p1}, \ldots, \sigma_{pn}$:

$$\sigma_r = \sqrt{\sum_i \left( \frac{\partial f}{\partial p_i} \sigma_{p_i} \right)^2} \tag{5.124}$$

### 5.9.2 Counting Statistics

The emission of a photon by a quantum-mechanical relaxation process of an excited atom is a statistical event in the sense that the exact point in time cannot be predicted: the time intervals between the individual emission events vary statistically. $P(x)$, the probability to measure $x$ photons in a given timespan, is

$$P(x) = \frac{\overline{N}^x e^{-\overline{N}}}{x!} \qquad \sigma = \sqrt{\overline{N}} \tag{5.125}$$

$\overline{N}$ is the average number of counted photons in the same span. Note that $x$ is defined as an integral positive number. The time span is not a parameter of the statistical distribution, and $\overline{N}$ and $x$ are therefore dimensionless numbers, not count rates (of dimension $s^{-1}$). Also the standard deviation, $\sigma$, is not a parameter of the distribution function, but a necessary consequence.

For large numbers of $\overline{N}$ the distribution can be approximated by a Gaussian. This can be safely used for most practical applications.

$$P(x) \approx \frac{1}{\sigma\sqrt{2\pi}} e^{-\frac{x-\overline{N}}{2\sigma^2}} \quad \text{where} \quad \sigma = \sqrt{\overline{N}}. \tag{5.126}$$

The error-propagation law is the same as for general statistical errors. For the frequently used sum of counts or difference of counts, the resulting error is

$$\begin{aligned} N_r &= N_1 \pm N_2 \\ \sigma_r^2 &= N_1 + N_2. \end{aligned} \tag{5.127}$$

### 5.9.3 Detection Limits

The detection limit (DL) in the context of chemical element analysis is the minimum amount of an element that can be detected within a given period of time with a given experimental setup and for a specific matrix. While from a practical point of view the problem of analyzing traces of an element in an abundantly available specimen (e.g., mercury in urban drinking water) is distinguished from detecting a small quantity of a pure element material, the statistical treatment is the same for both cases.

The principal limitation of the detectability of an element is the background noise in the measurement. The DL is reached at the point where the signal from the analyte element is just on the verge of significantly protruding above the background.

The measured total number of photons from an element $i, N_{i,\text{total}}$, is the sum of the characteristic photons, $N_i$, and the background photons, $N_{\text{background}}$. In the case of trace analysis, $N_i$ will be small, hence $N_{i,\text{total}} \approx N_{\text{background}}$.

$$\begin{aligned} N_{\text{total}} &= N_i + N_{\text{background}} \\ N_i &= N_{\text{total}} - N_{\text{background}} \\ \sigma_i^2 &= N_{\text{total}} + N_{\text{background}} \approx 2N_{\text{background}} \end{aligned}$$

$N_{\text{background}} + N_i$ differs significantly from $N_{\text{background}}$, when $N_i > X\sigma_i$ where $X$ is a matter of convention: a choice of $X = 1$ would give the correct result in 67% of all cases, $X = 2$ in 95%, and $X = 3$ in 99.7%. Assuming $X = 2$, which is most commonly used, defines the DL as a minimum number of counts of

$$N_i^{\text{DL}} = 2\sigma_{\text{background}} = 2\sqrt{2N_{\text{background}}} \approx 3\sqrt{N_{\text{background}}}. \tag{5.128}$$

In the range of trace analysis the counts are proportional to the mass (or mass fraction, $C_i$) of the analyzed element $i$, as can be seen for example by using a simple $\alpha$-coefficient algorithm for the relationship between $N_i$ and $C_i$ for $C_i \ll 1$:

$$\begin{aligned} \frac{C_i}{R_i} &= \frac{C_i}{N_i} N_{(i)} = 1 + \sum_j \alpha_{ij} C_j \\ N_i &= \frac{C_i}{1+\sum_j \alpha_{ij} C_j} N_{(i)} \approx \frac{C_i}{\text{const}} N_{(i)} = \beta C_i \end{aligned}.$$

The proportionality factor, $\beta$, is the specific count rate for element $i$. It follows for the DL in terms of the concentration of the analyte element:

$$C_i^{\text{DL}} = \frac{3}{\beta}\sqrt{N_{\text{background}}}. \tag{5.129a}$$

So far it has been assumed that the background counts and the total counts have been accumulated within the same unspecified counting time, for example $t_0 = 1\,\text{s}$. Other counting times require substitution of $N_{\text{total}}$, $N_i$, and $N_{\text{background}}$ by $tN_{\text{total}}$, $tN_i$, and $tN_{\text{background}}$:

$$C_i^{\text{DL}} = \frac{3}{\beta}\sqrt{\frac{N_{\text{background}}}{t/t_0}}. \tag{5.129b}$$

In order to improve the DL to its half value it is obviously necessary to extend the counting times by a factor 4. It should be noted that the determination of the DL does not require preparing and measuring a specimen with an actual concentration $C_i^{\text{DL}}$. It is sufficient to provide a specimen with $C_i$ for which $N_i = \beta C_i$ is applicable (linear range) and compute $C_i^{\text{DL}}$ from $\beta$ and $N_{\text{background}}$. Equation (5.129b) might suggest that the DL $C_i^{\text{DL}}$ can be lowered to any desired value by simply increasing the counting time, $t$. There is, however, a limit, which is given by the long-term stability of the analytical instrument.

The DL is a property of a specific method carried out with a specific instrument, specific settings (for example, of the counting times), and sample preparation techniques. For example, preconcentration methods may in certain cases improve the DLs drastically. It is essential to provide such particulars when reporting DLs. Many literature data refer to counting times of 100 s or 1000 s.

# 5.10 Standardized Methods

*K.-E. Mauser*

## 5.10.1 Introduction

"X-ray fluorescence is the best method to get wrong results in a perfectly reproducible way." Whoever made this statement has not been totally wrong. The measurement process itself is extremely stable, but a misfit between standards and unknowns or an overcorrected empirical calibration may nevertheless cause wrong results for certain kinds of samples.

X-ray fluorescence analysis is an important tool for quality and process control in industry. For these applications, very often consistency of results is much more important than the *true value*. The results of various laboratories have to match within a clearly defined experimental error. This demand can be achieved by carefully tested and well-documented procedures, which are followed by all laboratories working in a certain field of applications.

National and international bodies coordinate the development of standardized methods for XRF analysis. National standards are defined, e.g., by American Society for Testing and Materials (ASTM) in the USA, or Deutsches Institut für Normung (DIN) in Germany. Such national standards are increasingly replaced by European (EN) or worldwide international (ISO) standards.

## 5.10.2 General Features of Standardized Methods

There is a demand for the development of standardized methods when XRF is used for the same material at many places and the results have to match within a clearly defined tolerance. In some cases, the creation of a standardized method is driven by governmental regulations specifying a maximum amount of an element in a product, e.g., Pb in gasoline or S in petroleum products.

Standardized procedures always have a clearly defined scope and specify the material(s) to be analyzed, the elements, and the range of concentrations to be determined. Standardized procedures may explicitly exclude certain materials or the presence of certain elements.

Standardized methods have to be applicable in many laboratories with a wide range of instruments and software packages, including very old ones. Therefore, standardized methods are typically based on well-established instrument features and very simple mathematical models, which can be handled by all software packages. Typically, matrix effects are reduced by the preparation method (dilution) or handled by an internal standard, so there is no need for sophisticated matrix correction procedures.

Standardized methods are designed, tested, discussed, and decided by groups of specialists in technical committees. This process usually takes several years and may increase the risk of a significant gap between a standardized method and the current state of the art for instrument and software.

Most of the standardized methods for X-ray fluorescence deal with the analysis of petrochemical products or the analysis of oxidic raw materials or finished products, e.g., in the cement industry.

### What is Standardized?

Standardized procedures sometimes define each detail from instrument type to be used up to the reproducibility of the final results. Typically, the following items are included:

- Method, instrument and measurement process;
- Standard samples and sample preparation;
- Calculation methods;
- Precision of results.

A closer look at typical specifications will help to understand the advantages and drawbacks of standardized methods.

### Method, Instrument, and Measurement Process

A standardized method will specify X-ray fluorescence in general or, more specifically, an energy-dispersive (EDX) or a wavelength-dispersive spectrometer (WDX). Sometimes EDX or WDX is not stated explicitly, but follows from the description of the instrument details.

Such instrument details can include the anode material and the minimum power of the X-ray tube, the analyzer crystal to be used in a wavelength-dispersive system or the specific kind of detector. The more detailed these descriptions, the higher is the risk to exclude new developments. For example, a modern medium-power wavelength-dispersive spectrometer with a sealed proportional detector may easily outperform old high-power instruments, but will possibly not fulfill the specifications of a minimum tube power and a flow proportional counter. For future standardized methods, it may be more appropriate to define parameters, which are directly relevant for the quality of the results, e.g., the minimum counting statistical error or the minimum resolution of a detector instead of technical specifications subject to innovations.

Standardized methods typically specify the X-ray line to be used and special measurement parameters like the energy or wavelength for background measurements.

### Sample Preparation and Standard Samples

Typically, an X-ray spectrometer sees only some milligrams of material close to the surface of the specimen. As an important precondition for reliable results, this volume has to be representative of the material to be analyzed. Therefore, all standardized methods describe sample preparation in detail including, e.g., the purity of reagents to be used.

Many standardized methods deal with the analysis of liquid samples or materials prepared as fused beads. For these applications synthetic standards can be used. The standardized methods prescribe in detail how to mix these standards by using pure reagents. When such synthetic standards are prepared exactly as defined, various laboratories will base their calibrations on exactly the same kind of standards and therefore ensure consistency of results.

Sometimes certified reference materials are an option or have to be used. These materials have to be acquired from a national standards body or an accredited supplier.

## Calculation of Results

The primary result of a measurement is a spectrum or a set of intensities at predefined reflection angles. Various steps are required to calculate the final result:

- Determination and subtraction of the background;
- Correction for instrumental drift;
- Correction for spectral overlaps;
- Subtraction of a blank sample;
- Division by an internal standard;
- Correction for matrix influence;
- Calculation of concentrations.

Standardized methods widely differ in the description of the calculation methods. Some methods just mention to follow the operating manual, while others recommend to check the corrections by analyzing certified reference materials and not to accept the manufacturer's software *at face value.*

Other standardized methods strictly define each calculation step from intensities to final concentrations, and look like a textbook on the mathematics of correction procedures. Again, a standardized method defining every detail may establish an out-of-date procedure and block-up the use of innovative methods.

## Precision of Results

All standard methods set minimum requirements for the precision of the results. There are two different parameters used to specify the precision: repeatability and reproducibility.

The *repeatability* sets the permissible tolerances within a laboratory. It is defined as the difference between successive results, obtained by the same operator with the same apparatus under constant operating conditions on identical material. The repeatability can be easily checked individually in each laboratory.

The *reproducibility* deals with the permissible tolerances between laboratories and is defined as the difference between two single and independent results obtained by different operators working in different laboratories on identical material. It follows from the definition that the reproducibility cannot be determined by a single laboratory. A comprehensive way to determine the reproducibility of a method is round-robin tests often organized by the technical committees defining the standardized methods.

All standardized methods define minimum precision requirements and thus make sure that the results of all laboratories using the same method are consistent within the permissible tolerance. Does this automatically mean that the result hits the "true" value within the permissible tolerance? In many standardized methods, this question is not raised at all. Some standardized methods mention that there are no accepted and certified reference materials available and therefore the *bias* cannot be determined. Only a few standardized methods show the results of inter-laboratory studies comparing the results of the standardized methods with the certified values of reference materials.

### 5.10.3 Standardized Methods Versus Universal Calibrations and Standardless Methods

Universal calibrations and standardless methods based on fundamental parameters can handle a wide range of elements and concentrations and do not require a complicated calibration process or the mixture of the sample with an internal standard. Such standardless methods can be optimized for a certain kind of matrix, e.g., for a light matrix consisting of H, C, and O with all the other elements in the trace range. The limits of detection and the reproducibility of such optimized standardless methods are similar to those of the standardized methods. On the other hand, all universal methods using fundamental parameters will require that all elements with a significant interelement effect are measured.

In general, the criteria for a decision between a standardized method and a standardless method are as follows:

- A standardized method (ISO, ASTM, DIN) has to be used when required by legislation or the rules of the quality management system;
- If only a few elements have to be determined (or just one as in the case of S and Pb in petrochemistry), a standardized method may offer much shorter measurement times. A more universal method based on fundamental parameters will not allow to skip the measurement of elements with a significant matrix influence;
- A universal calibration based on fundamental parameters is the best procedure to be applied when no standardized method is available or when a higher number of elements have to be analyzed.

### 5.10.4 Summary

Standardized methods are an important tool to ensure consistent results in X-ray fluorescence analysis. However, if these methods are too detailed in their technical specifications, they may delay innovations in hardware and software. For many applications, optimized standardless methods can ensure the same quality of results without the need for a special calibration process.

Such methods are an option when a standardized procedure is not demanded by legislation or quality management, and when a longer measurement time does not matter.

## Symbols and Terminology

| | |
|---|---|
| $i,\ j,\ k,\ \ldots,\ n$ | Subscripts for elements (or compounds) in any given sample/specimen. Subscript $i$ usually designates the analyte and $j$ the remaining elements (or compounds) in the specimen (referred to as matrix elements or compounds). The subscripts are also used to indicate the (analyte or otherwise referred to) line of that element, or the absorption edge belonging to that line |
| S, St, s, u | Subscripts for specimen (could be a reference standard, any specimen, or a specimen of unknown composition) |
| $Z$ | Atomic number |
| $\Sigma_n$ | Summation of all elements, $i = 1,\ \ldots,\ n$, in a specimen |
| $\Sigma_M$ | Summation of all the matrix elements (also $\Sigma_{j\neq i}$) |
| $\Sigma_J$ | Summation over all lines of element $j$ that cause indirect enhancement |
| $M_i$ | Term quantifying matrix effects on analyte $i$ (also $M_{i,St}$, etc.) |
| $\lambda$ | Wavelength |
| $\lambda_i$ | Wavelength of a characteristic line of element $i$, usually the analyte line |
| $\lambda_{i,\,\mathrm{K}\alpha}, \lambda_{i,\,\mathrm{KL2}}$ | Wavelength, element $i$ (Siegbahn or TLD) |
| $\lambda_{\mathrm{min}},\ \lambda_0$ | Minimum wavelength in the X-ray tube continuum |
| $\lambda_{\mathrm{abs}},\ \lambda_{\mathrm{Edge},\,i}$ | Wavelength of an absorption edge; element can be indicated by subscript, $i,\ j,\ k$, S, St, s, u, ..., $\lambda_{i,\mathrm{abs}}$. |
| $E$ | Energy (same subscripts as for $\lambda$). $E = hc/\lambda$ |
| $\psi',\ \psi''$ | Incidence angle, emergence angle |
| $g_i$ | Proportionality constant, instrument dependent |
| $G_i$ | Number of $i$-photons (analyte line) emitted after absorption of a primary (tube) photon in element $i$ |
| $p_{\lambda i}$ | Transition probability into a vacant shell (Also with TLD, e.g., $p_{\mathrm{KL}_{II}}$, or with element/line subscript, $p_i$) |
| $\bar{\mu}_{i\lambda}$ | Linear absorption coefficient, for wavelength $\lambda$ in element $i$. Instead of $\lambda$, the energy $E$ or an indicator for a line is used, e.g., $j$. |
| $\mu(\mathrm{or}\ \bar{\mu}_{i\lambda}/\rho)$ | Mass absorption coefficient in units of $\mathrm{cm^2\,g^{-1}}$ |
| $\mu_{i\lambda}$ | Mass absorption coefficient for element $i$ for a given $\lambda$ (or $E$ or line $j$) |
| $\mu_{S\lambda}$ | Mass absorption coefficient of a specimen for a given $\lambda$, $\mu_{S\lambda} = \sum_n \mu_n C_n$ |

| | |
|---|---|
| $\mu'_i$ | Effective mass absorption coefficient for element i for incident $\lambda$, $\mu'_i = \mu_{i\lambda}\csc\psi'$ |
| $\mu'_S$ | Effective mass absorption coefficient for a specimen for incident $\lambda$, $\mu'_S = \mu_{S\lambda}\csc\psi'$ |
| $\mu''_i$ | Effective mass absorption coefficient for element $i$ for emitted $\lambda_i$, $\mu''_i = \mu_{ii}\csc\psi'' = \mu_{i\lambda_i}\csc\psi''$ |
| $\mu''_S$ | Effective mass absorption coefficient for a specimen for emitted $\lambda_i$, $\mu''_S = \mu_{Si}\csc\psi'' = \mu_{S\lambda_i}\csc\psi''$ |
| $\mu^*_i$ | Total effective mass absorption coefficient for element $i$, $\mu^*_i = \mu'_i + \mu''_i$ |
| $\mu^*_\mathrm{S}, \mu^*$ | Total effective mass absorption coefficient for a specimen, $\mu^*_\mathrm{S} = \mu'_\mathrm{S} + \mu''_\mathrm{S}$ (Wavelengths or energies can be specified in all above cases: $\mu'_{i,\lambda}$, $\mu'_{\mathrm{S},\lambda}$, $\mu'_{i,\lambda_i}$, ...) |
| $N_i$ | Count rate from analyte line, $i$, usually expressed in counts per second (counts/s or cps) |
| $N_{i\lambda}$ | Same as $N_i$, but the count rate is generated by monochromatic incident radiation |
| $N_{(i)}$ | Measured net count rate for element $i$ for a pure specimen of element $i$ |
| $N_{(i)\lambda}$ | Same as $N_{(i)}$, but the count rate is generated by monochromatic incident radiation |
| $N(\lambda)\,\mathrm{d}\lambda$ | Incident photons within interval($\lambda$, $\lambda+\mathrm{d}\lambda$), per cm$^2$ and second (flux) |
| $R_i$ | Ratio of count rates of element $i$ in a specimen relative to that of the pure element $i$; $R_i = N_i/N_{(i)}$ ($R_i$ can be calculated from measured intensities (cps) or from theoretical intensities) |
| $C_i$ | Concentration (weight fraction) of analyte $i$ |
| $C_j$ | Concentration (weight fraction) of matrix element $j$ |
| $C_M$ | Concentration (weight fraction) of matrix, $C_M = \Sigma_j C_j$ |
| $\tau$ | Mass photo-absorption coefficient (subscripts as for $\mu$) |
| $\sigma$ | Mass-scattering coefficient, subscripts as for $\mu$ |
| $\sigma$ | Also used for standard deviation |
| $n$ | A (total) number of items, generally of elements in a sample |
| $\wedge$ | Indicates items "per unit area," such as $\hat{m} = m/\Lambda$ (mass/area) or $\hat{N}_\lambda = \hat{N}_{\lambda,t}/t$ (photon flux) |
| $\Lambda$ | Area, used while $A$ denotes atomic weights |
| Flux | Number of particles or photons, per square centimeters per second |
| (Real) intensity | $I_\lambda = \hat{N}_\lambda hc/\lambda$ Physical intensity in units of W m$^{-2}$ or equivalent |
| $T$ | Thickness (of thin layers) |

| | |
|---|---|
| $t$ | (Counting) time (seconds) |
| $\omega$ | Fluorescent yield |
| $S$ | Absorption edge jump ratio, |

$$S = \frac{\text{Photo-absorption coefficient at high energy side of edge}}{\text{Photo-absorption coefficient at low energy side of edge}}$$

## Symbols and Terminology Used in the Influence Coefficients Methods

In the **Rousseau algorithm**:

| | |
|---|---|
| $\alpha_{ij}$ | Theoretical multielement influence coefficient for correcting absorption effect of element $j$ on analyte $i$ |
| $\varepsilon_{ij}$ | Theoretical multielement influence coefficient for correcting enhancement effect of element $j$ on analyte $i$ |
| $W_i(\lambda_k)$ | Weighting factor for each incident $\lambda_k$ for analyte $i$ |
| $\beta_{ij}(\lambda_k)$ | Theoretical binary influence coefficient for correcting absorption effect of element $j$ on analyte $i$ |
| $\delta_{ij}(\lambda_k)$ | Theoretical binary influence coefficient for correcting enhancement effect of element $j$ on analyte $i$ |

In the **Lachance–Traill algorithm**:

*For monochromatic incident radiation:*

| | |
|---|---|
| $a_{ij\lambda}$ | Theoretical binary influence coefficient for correcting absorption effect of matrix element $j$ on analyte $i$; $a_{ij\lambda} \equiv \beta_{ij}(\lambda_k)$ |
| $e_{ij\lambda}$ | Theoretical binary influence coefficient for correcting enhancement effect of matrix element $j$ on analyte $i$; $e_{ij\lambda} = \delta_{ij}(\lambda_k)$ |
| $m_{ij\lambda}$ | Theoretical binary influence coefficient for correcting for net matrix effect (absorption – enhancement) of element $j$ on analyte $i$ |

*For polychromatic incident radiation:*

| | |
|---|---|
| $a_{ij}$ | Theoretical multielement influence coefficient for correcting absorption effect of matrix element $j$ on analyte $i$ |
| $e_{ij}$ | Theoretical multielement coefficient for correcting enhancement effect of matrix element $j$ on analyte $i$ |
| $m_{ij}$ | Theoretical multielement influence coefficient for correcting for net matrix effect of element $j$ on analyte $i$ |

where

| | |
|---|---|
| $\{\ldots\}_{ij} = a_{ij}$ | if no enhancement by element $j$ (absorption is always present) |
| $\{\ldots\}_{ij} = a_{ij} - e_{ij}$ | if enhancement by element $j$ |
| $\{\ldots\}_{ij} = m_{ij}$ | net effect of element $j$, i.e., $m_{ij} = a_{ij} - e_{ij}$ |

*In general, for the Lachance–Traill algorithm:*

$\{\ldots\}_{ij} = m_{ij}$ If the influence coefficient is calculated from fundamental theory

$\{\ldots\}_{ij} = a_{ij}$ If the influence coefficient is calculated from fundamental theory involving minor approximations (N.B.: different from Rousseau's $\alpha_{ij}$)

$\{\ldots\}_{ij} = r_{ij}$ If the influence coefficient is calculated from least squares regression of experimental data, i.e., an empirical coefficient.

## References


1. Bambynek W, Crasemann B, Fink RW, Freund HU, Mark H, Swift CD, Price RE, Rao PV, X-ray fluorescent yields, Auger, and Coster–Kronig transition probabilities. Rev Mod Phys **44**, 716–813 (1972)
2. Bambynek W, Crasemann B, Fink RW, Freund HU, Mark H, Swift CD, Price RE, Rao PV, Erratum: X-ray fluorescent yields, Auger, and Coster–Kronig transition probabilities. Rev Mod Phys **46**, 853 (1974)
3. Barkla CG, Sadler CA, Homogeneous secondary Röntgen radiations. Philos Mag **16**, 550 (1908)
4. Beattie HJ, Brissey RM, Calibration method for X-ray fluorescence spectrometry. Anal Chem **26**, 980–983 (1954)
5. Beatty RT, The direct production of characteristic Röntgen radiations by cathode particles. Proc R Soc Lond A **87**, 511 (1912)
6. Bertin E, Principles and Practice of X-Ray Spectrometric Analysis. Plenum Press, New York (1970)
7. Bethe HA, Zur Theorie des Durchgangs schneller Korpuskularstrahlen durch Materie. Ann Phys **5**, 325–400 (1930)
8. Broll N, Quantitative X-ray fluorescence analysis. Theory and practice of the fundamental coefficient method. X-Ray Spectrom **15**, 271–285 (1986)
9. Broll N, Caussin P, Peter M, Matrix correction in X-ray fluorescence analysis by the effective coefficient method. X-Ray Spectrom **21**, 43–49 (1992)
10. Broll N, Tertian R, Quantitative X-ray fluorescence analysis by use of fundamental influence coefficients. X-Ray Spectrom **12**, 30–37 (1983)
11. Claisse F, Quintin M, Generalization of the Lachance–Traill method for the correction of the matrix effect in X-ray fluorescence analysis. Can Spectrosc **12**, 129–134 (1967)
12. Criss JW, Birks LS, Calculation methods for fluorescent X-ray spectrometry: empirical coefficients vs. fundamental parameters. Anal Chem **40**, 1080–1086 (1968)
13. de Boer DKG, Brouwer PN, Fundamental parameter-based X-ray fluorescence: analysis of thin and multilayer samples. Adv X-Ray Anal **33**, 237–245 (1990)
14. de Jongh WK, X-ray fluorescence analysis applying theoretical matrix corrections. Stainless Steel. X-Ray Spectrom **2**, 151–158 (1973)
15. Ebel H, Svagera R, Ebel MF, Shaltout A, Hubbell J, Numerical description of photoelectric absorption coefficients for fundamental parameter programs. X-Ray Spectrom **32**, 442–451 (2003)

16. Elam WT, Ravel BD, Sieber JR, A new atomic database for X-ray spectroscopic calculations. Radiat Phys Chem, **63**, 121–128 (2002)
17. Gardner RP, Hawthorne AR, Monte Carlo, simulation of the X-ray fluorescence excited by discrete energy photons in homogeneous samples including tertiary inter-element effects. X-Ray Spectrom **4**, 138–148 (1975)
18. Gilfrich JV, X-ray fluorescence analysis at the naval research laboratory. X-Ray Spectrom **20**, 203–211 (2001)
19. Gilfrich JV, Birks LS, Spectral distribution of X-ray tubes for quantitative X-ray fluorescence analysis. Anal Chem **40**, 1077 (1968)
20. Gillam E, Heal HT, J Appl Phys **3**, 352 (1953)
21. Glocker R, Schreiber H, Quantitative Röntgenspektralanalyse mit Kalterregung des Spektrums. Ann Phys **85**, 1089–1102 (1928)
22. Green M, Cosslett VE, The efficiency of production of characteristic X-radiation in thick targets of a pure element. Proc Phys Soc **76**, 1206–1214 (1961)
23. Hower J, Matrix corrections in the X-ray spectrographic trace element analysis of rocks and Soils. Am Miner **44**, 19 (1959)
24. International Organization for Standardization (ISO), Geneva. ISO-Guide 30 (1992) Terms and Definitions Used in Connection with Reference Materials. VIM (1993) International Vocabulary of Base and General Terms in Metrology. ISO-14597 (1995) Determination of nickel and vanadium in liquid fuels – Wavelength-dispersive X-ray fluorescence method
25. Ito M, Sato S, Narita M, Comparison of the Japanese industrial standard and $\alpha$-correction methods for X-ray fluorescence analysis of steels. X-ray Spectrom **10**, 103–108 (1981)
26. JIS G, 1256-1997, Iron and steel – method for X-ray fluorescence spectrometric analysis. Japanese Standards Association, Tokyo (1997)
27. Kalman ZH, Heller L, Theoretical study of X-ray fluorescent determination of traces of heavy elements in a light matrix – application to rocks and soils, Anal Chem 946–951 (1962)
28. Kataoka Y, Kawahara N, Arai T, Uda M, Study of light element analysis of thin films for fundamental parameter method. Adv X-ray Anal **41**, 76–83 (1998)
29. Kawahara N, Shoji T, Yamada T, Kataoka Y, Beckhoff B, Ulm G, Mantler M, Fundamental parameter method for the low energy region including cascade effect and photoelectron excitation. Adv X-ray Anal **45**, 511–516 (2002)
30. Kramers HA, On the theory of X-ray absorption and of the continuous X-ray spectrum. Philos Mag **46**, 836–871 (1923)
31. Lachance GR, Fundamental coefficients for X-ray spectrochemical analysis. Can Spectrosc **15**, 64–72 (1970)
32. Lachance GR, The role of alpha coefficients in X-ray spectrometry, International Conference on Industrial Inorganic Elemental Analysis, June 3, Metz, France (1981)
33. Lachance GR, Traill RJ, A practical solution to the matrix problem in X-ray analysis. Can Spectrosc vol **11**, Part I: pp 43–48; Part II: pp 63–71 (1966)
34. Lachance GR, Claisse F, Quantitative X-ray fluorescence analysis: Theory and application. Wiley, Chichester (1995)
35. Love G, Cox MG, Scott VD, A versatile atomic number correction for electron-probe microanalysis. J Phys D: Appl Phys **11**, 7–21 (1978)

36. Mantler M, LAMA III – a computer program for quantitative XRFA of bulk specimens and thin film layers. Adv X-ray Anal **27**, 433–440 (1983)
37. Mantler M, Recent methods and applications of X-ray fluorescence analysis. Prog Crystal Growth Charact **14**, 213–261 (1987)
38. Mantler M, Quantitative XRFA of light elements by the fundamental parameter method. Adv X-Ray Anal **36**, 27–33 (1993)
39. Martin RL, NBSGSC – Fundamental Parameters – X-Ray Analysis for Personal Computers, Rainier Software, WA, USA (1987)
40. Moseley HGJ, The high frequency spectra of the elements. Philos Mag **26**, 1024–1034 (Part 1) and Philos Mag **27**, 703–714 (Part 2) (1913, 1914)
41. Nesbitt RW, Mastins H, Stolz GW, Bruce DR, Matrix corrections in trace element analysis by X-ray fluorescence: an extension of the Compton scattering technique to long wavelengths. Chem Geol **18**, 203–213 (1976)
42. Pella PA, Feng L, Small JA, An analytical algorithm for calculation of spectral distributions of X-ray tubes for quantitative X-ray fluorescence analysis. X-Ray Spectrom **14**, 125–135 (1985)
43. Powell CJ, Cross sections for ionization of inner-shell electrons by electrons. Rev Mod Phys **48**, 33–46 (1976)
44. Rasberry SD, Heinrich KFJ, Calibration for interelement effects in X-ray fluorescence analysis. Anal Chem **46**, 81–89 (1974)
45. Reynolds R, Estimation of mass absorption coefficients by Compton scattering: improvements and extensions of the method. Am Miner **52**, 1493–1502 (1967)
46. Rousseau RM, Fundamental algorithm between concentrations and intensity in XRF analysis. Part 1 - theory. X-ray Spectrom **13**, 115–120 (1984)
47. Rousseau RM, Fundamental algorithm between concentrations and intensity in XRF analysis. Part 2 - practical application. X-ray Spectrom **13**, 121–125 (1984)
48. Rousseau RM, Claisse F, Theoretical alpha coefficients for the Claisse–Quintin relation for X-ray spectrochemical analysis. X-Ray Spectrom **3**, 31–36 (1974)
49. Rousseau RM, Willis JP, Duncan AR, Practical XRF calibration procedures for major and trace elements. X-ray Spectrom **25**, 179–189 (1996)
50. Rousseau RM, Bouchard M, Fundamental algorithm between concentration and intensity in XRF analysis. Part 3 - experimental verification. X-Ray Spectrom. **15**, 207–215 (1986)
51. Sherman J, The correlation between fluorescent X-ray intensity and chemical composition. ASTM Spec Tech Publ **157**, 27–33 (1954)
52. Sherman J, The theoretical derivation of fluorescent X-ray intensities from mixtures. Spectrochim Acta **7**, 283–306 (1955)
53. Shiraiwa T, Fujino N, Theoretical calculation of fluorescent X-ray intensities in fluorescent X-ray spectrochemical analysis. Jpn J Appl Phys **5**, 886–899 (1966)
54. Siegbahn M, The Spectroscopy of X-rays, Oxford University Press, London (1925)
55. Tao GY, Pella PA, Rousseau RM, NBSGSC – a Fortran program for quantitative X-ray fluorescence analysis. NBS Technical Note 1213, National Institute of Standards and Technology, Gaithersburg, USA (1985)
56. Tertian R, A new approach to the study and control of interelement effects in the X-ray fluorescence analysis of metal alloys and other multi-component systems. X-Ray Spectrom **2**, 95–108 (1973)
57. Tertian R, An accurate coefficient method for X-ray fluorescence analysis. Adv X-Ray Anal **19**, 85–109 (1976)

58. Tertian R, Vie le Sage R, The equivalent wavelength notion: definition, properties, applications. X-Ray Spectrom **4**, 73–83 (1975)
59. von Hevesy G, Chemical Analysis by X-Rays and Its Applications, McGraw-Hill, New York (1932)
60. Walker D, Behavior of X-ray mass absorption coefficients near absorption edges: Reynolds' method revisited. Am Miner **58**, 1069–1072 (1973)
61. Wilson RR, Range and ionization measurements on high speed protons. Phys Rev **60**, 749–753 (1941)

# 6

# Specimen Preparation

*J. Injuk, R. Van Grieken, A. Blank, L. Eksperiandova and V. Buhrke*

## 6.1 Introduction

The objective of XRF analysis is to obtain accurate analytical data about the composition of a specific material, and this cannot be done unless the specimen is properly prepared. It is, of course, axiomatic that in any analytical procedure it is incumbent upon an analyst to minimize, or limit random errors and to exclude systematic ones which can occur during an analysis. Avoiding these errors requires a good deal of time and effort. XRF analysis can be, and often is, a simple procedure, and the time to perform an analysis can be very short; still, it can also be a daunting task to prepare specimens of certain materials which are difficult to convert from the bulk into a specimen suitable for analysis. What is a suitable specimen? A suitable specimen is one which contains a homogeneous distribution of all elements on the surface to be analysed, and is in the proper form, such as a smooth surface, before being exposed to the primary excitation radiation.

There is no question that the source of the largest, single, potential error during an XRF analysis is the result of improper preparation of the specimen. There are no software programs available to avoid or correct for specimen preparation errors. The analyst must either know how to prepare a specimen, or he/she must learn how to do it. The literature should always be consulted for this information before launching on a program to develop a method. The references at the end of this Chapter contain several books which contain much information about specimen preparation, and one of the references [1] contains an extensive bibliography on the subject of specimen preparation.

Specimen preparation involves several operations. It often begins by extracting a sample from a bulk material (sometimes a bulk is not available, and the specimen is prepared directly from a sample). The next step is to prepare a specimen from the sample. The specimen and sample differ in that the term specimen refers to a sample which has undergone the necessary treatment to make it suitable for analysis, namely, enable to produce accurate data about the composition of the material from which the specimen was derived.

The specimen is defined as the object which is placed in the specimen holder of the X-ray spectrometer, and subsequently used to produce the data for analysis. The desired end result of specimen preparation is therefore to make sure that the relatively small volume of specimen analysed by the X-ray beam (the top layer of the specimen from which the secondary radiation from the analyte escapes the specimen and is counted by the detection system) is representative of the bulk.

In XRF, the precision of an analysis is described by the Poisson law with a dispersion $S^2$ equal to the average number of pulses $N$. Therefore at $N = 2 \times 10^6$ counts, and an insignificant background level, the relative standard deviation ($S_r$) is $1/\sqrt{N} = 7 \times 10^{-4}$ that is 0.07% (rel.). In most XRF analyses, the counting error is very small compared to that caused by improper sampling (not taking a representative sample from the bulk) and specimen preparation. The analyst should always remember that XRF sampling and specimen preparation errors can be several orders of magnitude greater than any other errors which can occur during an analysis.

XRF is a versatile method of analysis. Samples can be liquid, solid, or powder. However, they must not be damaging to the spectrometer. In most cases, the specimen is brought to the spectrometer; however, there are portable instruments which allow the analyst to take the instrument to the specimen.

The literature on XRF is quite large. A more extensive description of specimen preparation procedures can be found in some earlier books [1–3]. Several reviews which deal with specimen preparation of liquids [4], water [5], aerosols [6], pulps [7], films [8], ferrous metals [9, 10], rocks [11] and various materials of natural origins [12] have been published. Other methods of specimen preparation are presented in numerous papers [13–17]. We wish to emphasize that this Chapter discusses only the most common procedures.

## 6.2 Liquids

### 6.2.1 Direct Analysis of Liquids and Solutions

In many cases, liquids can be analysed as received. However, a serious problem occurs when bubbles are formed in the liquid. Bubbles can cause problems such as incomplete filling, or expansion of the film covering the holder. In the latter case, the expansion changes the distance between the anode of the X-ray tube (source) and the specimen: resulting in a counting error. Liquid specimens require special holders made of stainless steel, polyethylene or polytetrafluorethylene (PTFE or Teflon). These holders have a thin (3–6 μm) window made of a low-Z material such as Mylar or Kapton. Some windows contain micro-pores to allow gas to escape the holder, thereby eliminating bubble problems. Some spectrometers have the X-ray tube located below the specimen holder. This geometry requires that the liquid holder window be located at the bottom. The window material absorbs the secondary radiation, especially that from low-Z elements, and also produces background radiation. Liquids are usually analysed at atmospheric pressure in the specimen chamber, and this air also absorbs the low-Z radiation. The result is that the analysis of low concentrations of very few parts per million of low-Z elements (such as B) is not possible. Liquids have an advantage, namely, minimum matrix effects, and it is easy to add internal standards to a liquid. However, in contrast to standards prepared by fusion or pelletizing, liquid standards deteriorate with time due to adsorption and precipitation effects. There are recent publications [18, 19] which discuss the use of on-line XRF, in which there is a continuous flow of the specimen through the specimen chamber. This sort of holder is also useful for the analysis of oils, thin greases, petroleum products such as used oils. Pitches and asphalt samples can also be analysed by special preparation techniques (see Sect. 6.2.2).

Although the analysis of solutions is a common occurrence, it is more difficult than the analysis of solid specimens and errors are usually larger for liquids than for solids.

### 6.2.2 Conversion of Liquids into Quasi-Solid Specimens

Most solid specimens are easier to analyse than liquids. Therefore, it is often advisable to convert a liquid into a quasi-solid by introducing an additive to increase the viscosity of the liquid, or by freezing it [20–22]. Alkaline solutions can be transformed into a quasi-solid state by adding a mixture of amylopectine and ethylene glycol. When eicosane is added to fuel oil, it produces a homogeneous mass which is solid at room temperature.

Gelatinizing substances (such as gelatine or agar), when added to solutions, can produce a quasi-solid (jelly-like) specimen of suitable size and shape for use in different types of spectrometers [13]. For example, a 10% (mass) of gelatin or 3% (mass) of agar can be added, in powder form, to the solution and left to stand to expand. The mixture is then heated at 60–80° for a few minutes to obtain a homogeneous solution. The hot solution is transferred to a cuvette made of Teflon, and left until it starts to stiffen. Then the specimen to fit to the specimen chamber of the spectrometer is chapped: the cuvette with the specimen is turned upside down onto a smooth plate covered by a Teflon film. As a result of the contact with the hydrophobic surface, an elastic gelatinous specimen is formed with a polished surface suitable for XRF analysis. Such a specimen can be obtained from aqueous solutions having a pH of 1–10, as well as from organo-aqueous solutions containing a water-miscible solvent (e.g., ethanol, methanol or acetone). Gelatin specimens tend to melt if the X-ray beam is too intense. The analyst must experiment to find the proper working conditions to avoid melting or destruction of the specimen by the X-ray beam. We recommend the addition of a 0.25% (mass) of quinhydrone to the specimens. This will cause oxidation of the gelatine heteroatoms and cause bonding of macromolecules together, thus favouring formation of a polymer with a higher thermal stability. Quinhydrone is a low-Z material and it does not absorb much but does increase the viscosity of the liquid. A microscopic examination of the specimen showed that the formation of bubbles is not a problem at an exposure of 10–100 s, current of 20 mA and voltage of 20–40 kV.

Physical–chemical investigations of the process of obtaining quasi-solid specimens for X-ray fluorescence analysis were carried out [23]. To prepare specimens from organic solutions, in particular from organic extracts containing elements to be determined, it is proposed [24, 25] to make an organogel. In this procedure, 1 mL of a hot (60–80°C) aqueous 10% gelatine and 0.5 mL of surfactant solution (e.g., 5% aqueous solution of sodium oleate) are added to 1 mL of the extract; this is thoroughly mixed and the specimen is left until gelation begins. A flat specimen surface is formed when the above mixture contacts a Teflon film placed on a smooth, flat plate. The organogels produced look like an emulsion of oil in water. External standard, internal standard and background standard methods can be used to create calibration curves. Jelly-like specimens have several advantages: a flat surface; the possibility of analysing unsaturated layers, suspensions, aqueous solutions of different acidity, aqueous-organic systems and organic solutions. Producing quasi-solid specimens is not very difficult.

### 6.2.3 Conversion of Liquids into Organic Glassy Polymer Specimens

When only a small volume of aqueous solution is available, we recommend the use of the saccharose (sucrose) [26] procedure to produce an organic glassy polymer specimen. The procedure is: 1 g of saccharose is added to 2–3 $cm^3$ of solution having a pH of 2–9. The mixture is then heated to 130–150°C for 3–5 min. Caramelization takes place as the water evaporates. Specimens of a suitable form are made by cooling the caramelized mass on the smooth, flat surface of a Teflon plate, or on a film inside a Teflon mold, or on another hydrophobic material. The resulting mass is only about 1/2 or 1/3 the size of the original solution. This procedure produces a specimen with a higher concentration than the original solution. Mechanism and kinetics for saccharose specimen formation as well as their physical–chemical properties were studied [27]. These saccharose specimens were, e.g. used for the determination of selenium in wastewater.

### 6.2.4 Conversion of Liquids into Thin Films

When films correspond to the criterion of a "thin" layer [28], the absorption of primary and secondary radiation in the film is negligible. In this case, the chemical composition (matrix effect) of the film and its thickness do not influence the intensity of the analytical line, and a linear relationship between the intensity of the analytical line and the elemental concentration is observed. For clarity, in the thin-film specimen, the final film thickness is typically a few micrometers down to a few nanometers in thickness. However, not all specimens correspond to the criterion of "thin" layer and often it is necessary to work under the conditions of "unsaturated" layers [28], when the specimen thickness is to be strictly fixed. These films contain a small mass of material, and therefore, only elements at high concentration produce enough signal to produce accurate data.

One of the methods to analyse a thin film is to dissolve carboxymethyl cellulose (CMC) in the solution to be analysed [29]; then to pour the resulting solution into a plastic ring on a polyethylene substrate; and then to dry it. The CMC film can be removed from the substrate together with the ring to serve as a specimen.

A total-reflection XRF method (TXRF), reported more than 20 years ago, demonstrates how to analyse thin film specimens. TXRF allows the analysis of elements down to the ppb level, whereas normal XRF has detection limits in the ppm range. Incident and scattered radiation impinge upon the specimen at an angle low enough to permit total reflection and therefore do not enter the detector. The 1996 reference [30] can be consulted for a review of articles on TXRF. Optically smooth plates of quartz, glass, silicon and some other materials are used as substrates. Two papers [31, 32] discuss the analysis of microliters of solution by TXRF. How to obtain thin films for TRXRF analysis of archeological ceramics by means of suspension preparation and using internal standards is proposed [33].

### 6.2.5 Analysis of Solutions after Preconcentration of Microimpurities

A major drawback during the analysis of liquids is the occurrence of a high background and scattering radiation which makes it difficult to analyse for low concentrations of low-Z elements. In general, the analysis of aqueous samples without preconcentration results in limits of detection in the high ppm range; this is a very limiting factor for the analysis of most natural waters.

A method to improve detection limits is to remove the matrix material. Producing quasi-solid specimens as concentrates, as well as a thin film, is one way of increasing the concentration of the analytes so that they can produce an adequate signal above background. Many preconcentration methods exist for the analysis of water by XRF. As long ago as 1982, Van Grieken [34] published a comprehensive review on the subject, with more than 500 references already.

A fairly simple method is to evaporate solutions on a solid substrate or paper filter. This is possible if samples have a low salinity and low water hardness. To achieve detection limits at the ppb level, evaporation of about 100 mL of water is necessary. A large water sample can also be freeze-dried and the evaporated residue can then be made into a pellet, possibly after mixing with an organic binder to reduce matrix effects. Freeze-drying of a 250 mL quantity of wastewater on 100 mg graphite followed by grinding and pelletizing of the residue can lead to detection limits of a few $\mu g\,L^{-1}$. In the evaporated residue, the non-volatile elements are collected quantitatively and the risk of contamination is minimal [35].

Unfortunately, taking samples for dryness causes some experimental problems, such as fractional crystallization, splashing, etc. Preconcentration by evaporation is not a preferred method, unless it is used in TXRF analysis. In some cases, a simple evaporation of the solution onto a filter paper including a confining ring can produce good results. The advantage of this method is that it is a relatively simple and quick way to produce a suitable specimen from a small mass of sample. On the other hand, the disadvantage of the method is that the elements of interest may not be homogeneously distributed on the filter paper because the distribution is affected by the drying property of the filter paper. Such experimental errors can be minimized by drying at low temperature and by using an internal standard. Recently the micro-droplet preparation method has become more popular. For example, the determination of U and Ce in nuclear fuel can be done if the sample is dissolved in a $HNO_3$ solution and then microdroplets of this solution are placed onto a filter paper [36]. A new thin film sample support was designed [37] to allow a small amount of sample solution (about 50 μL) to accumulate into a small spot. By a special surface treatment, this newly developed film (AP2TM, Process Analytics) creates approximately a 1.5-mm diameter hydrophilic spot in the centre of the hydrophobic field, which minimizes the scattering of X-rays from the matrix of the sample. The authors report that the precision of the method is from 2% to 9% for transition metals at the 50 ppb concentration level, and the analysis only takes about 15 min. In contrast to commonly used sample support films (like, e.g., Prolene, Kapton, Mylar), this film has a high temperature resistance, and it is therefore suitable for analysis of high boiling acids and organics.

Precipitation or coprecipitation from solutions is widely used to concentrate elements before determining them by XRF analysis. Various coprecipitation agents (collectors) have been proposed in the literature. Hydroxides of iron, aluminium, indium (III) or zirconium, diethyl- or pyrrolidine-dithiocarbamates are often used as collectors for heavy metals. Sometimes cocrystallization on 1-(2-pyridilazo)-2-naphtol, or coprecipitation on complicated collectors, such as, e.g., on indium (III) oxiginate in the presence of tannin, are used. Separating phosphate and arsenate ions from aqueous solutions as low-soluble quinidinium phosphoro-(arseno-)molybdate [38] or bromides as AgBr [39] in sea water are examples of using precipitation as a preconcentration method for subsequent XRF analysis.

In addition to paper filters and membranes, nuclear (Nuclepore) and glass fibre filters are used to collect precipitates from solutions. The concentrates obtained are separated from the filters and compacted or placed between two Mylar films in a special holder [40].

Concentration by means of precipitation or coprecipitation can be used to analyse many different materials, but this procedure is labour intensive and difficult to automate. Separation of the precipitate from the filter may also result in an underestimation of the concentrations.

A procedure such as cake compaction (pressing, pelletizing) is a good way to produce a specimen with a smooth surface [13].

A method to analyse for low concentrations in solution is to preconcentrate with ion exchange resins or adsorbents and then elute and evaporate onto filter paper discs. If elements of a low concentration are to be determined in a solid, the solid can be put into solution and the solution then passed through an ion exchange resin column for adsorption and subsequent analysis. Preconcentration factors of up to $4 \times 10^4$ can be achieved by using about 100 mg of a suitable ion-exchange resin [41]. An ion-exchange resin can also be used as a sample support, either by pelletizing the exchange resin or by using ion-exchange filter papers. This method is useful for the determination of metals in very dilute solutions. The procedure involves shaking the solution with a known amount of resin for several minutes, followed by filtering, drying, and pelletizing. By this method, e.g., 0.1 ppm of Au in a solution can be determined within less than 15 min [42]. Another possible method involves concentration of a litre of water on a cation-exchange resin column and the direct analysis of the resin by energy-dispersive XRF. This method can be used to measure trace metal contaminations of a few parts per billion in drinking water. Anion-exchange filters can be used to concentrate or selectively remove a certain component. An example is the determination of Cr(III) and Cr(VI) in an acid medium: an anion-exchange filter paper will collect Cr(VI), while Cr(III) passes through. Subsequently, toxic Cr(VI) can be directly analysed on the filter paper [43].

Among the most often used adsorbents are those containing immobilized complexing reagents: polyacrylonitrile modified by polyethylene polyamine [44], naphthalene modified by 1-(2-pyridilazo)-2-naphtol [45], aminostyrene-divinyl benzene copolymer with grafted hydroxyaminodiacetic acid [46], cellulose with grafted amino [47], mercapto [48], aminocarboxy [49] groups, silica gel modified by dodecylaminoethyl-b-tridecylammonium iodide [50], polyurethane foam impregnated by some reagents, as well as some commercially available chelating adsorbents, for example, POLIORGS. There are many cases where elements to be determined can be complexed by introducing organic agents into the sample solution, and then concentrated using an unselective adsorbent such as activated charcoal. For example, in [51], when determining Cu, Co, Ni, Fe, Zn, Pb, Cd, Mn, Hg, and Bi in water, the complexes of the elements with 1-(2-pyridylazo)-2-naphtol were at first obtained and then adsorbed on acetyl cellulose filter of "VLADIPOR" type. The authors of [52] proposed to use unsaturated tungsten–phosphorus compounds as adsorbents for a series of transition metals; in this case, such compounds act as macroligands. When a finely dispersed adsorbent was used, the process was performed mainly in a static mode, that is, by shaking the solution with a known amount of resin. The same mode may be used for adsorbents shaped as tablets, discs or fibrous textiles. Adsorption on filter membranes in a dynamic mode (using a peristaltic pump) was found to be very effective, since in such cases the adsorbent capacity is used to its maximum extent. When the process is run in an automatic mode, the method takes less time. Preconcentration on a column appears to be less effective in this case [13].

Another possible method is to collect finely dispersed adsorbents on a filter paper. The obtained concentrate is then fixed on the filter by an aerosol varnish [53] or mixed with polystyrene and poured into a holder for powders [54]. The concentrates obtained are sometimes compacted by adding a binder such as boric acid or cellulose. The use of a compacted specimen substantially lowers the detection limits for the elements to be determined, and enhances the reproducibility by an order of magnitude compared to the method using uncompacted concentrates. In [44], concentrates on an organic adsorbent are wetted with alcohol, charred, and then mixed with polystyrene and pressed.

A seldom used method is to elute the elements after adsorption/absorption preconcentration on a column. An original technique proposed in [52] is to produce a specimen after the adsorption preconcentration: the finely dispersed concentrate is mixed with an epoxy resin, and the half-liquid mixture obtained is placed onto a disc made from a polycarbonate, and allowed to spread uniformly over the surface of the disc prior to hardening. Adsorption/absorption preconcentration is often used because the method allows one to obtain a solid concentrate suitable for direct application in XRF analysis. The results seem to be the best when preconcentration is performed in the dynamic mode on a filter (membrane) using a peristaltic pump [13].

An examination of the literature reveals that there are several XRF papers which refer to methods using a preliminary extraction and concentration of the elements to be determined. These papers can be divided into several groups:

- Liquid extraction of the elements to be determined; specimens can be obtained by evaporation of an organic solvent or aqueous phase on a suitable solid support [55], or formation of a thin film after extraction onto polymeric films such as Chlorin [56] or Carbosil-70 followed by solvent evaporation [57];
- Preparation of a gel-like specimen (organogel) from an organic extract [22, 58] by the addition of a surfactant and aqueous gelatin solution.

According to the papers of the latter group [59–61], elements to be determined are extracted from solutions by using a low-melting organic compound which solidifies at room temperature to permit the extracts to be used without additional treatment. The low-melting extractants used include: C17–C20 aliphatic monocarboxylic acids; and long-chain alcohols with admixtures of various reagents (2,2′-dipyridyl, 1,10-phenantroline, trialkyl amine, etc.).

Another preconcentration method which has been proposed to combine with XRF is crystallization preconcentration. It implies the segregation of micro components by directed crystallization of the solution or melt, that is, crystallization of liquid into a given direction which is perpendicular to the plate interphase boundary (the crystallization front) [62, 63]. Low-temperature directed crystallization by the Bridgman–Stockbarger method may be automated [62]. The method is a directed crystallization of a liquid from down to top. This method consists in lowering an ampoule which contains the solution at a pre-set constant rate, from the zone of a circular heater to a refrigerating thermostat chamber filled with a liquid antifreeze agent [62]. Different variants of preconcentration techniques with XRF analysis of the concentrates were applied for impurity determination in wastewater [25]. At that a low-temperature directed crystallization and a liquid–liquid extraction as preconcentration methods were used.

Humic substances presenting in natural water often prevent preconcentration of metal ions. It was proposed [64] to use the ultrasonic pretreatment before analysis of water to eliminate this drawback.

The analysis of solid materials in aqueous samples requires that the solution be filtered through Nuclepore or Millipore membranes with a 0.4 or 0.45 μm pore-size (this is the conventional limit between "dissolved" and "particulate" matter in environmental waters) and dried; the filter membrane is then analysed as a solid. The thin films prepared by chemofiltering the complexes of rare earth elements with Arsenaso III or Thorin through membrane polyamide filter papers were applied for XRF analysis using a conventional tube system or synchrotron radiation source [65].

## 6.3 Solid Specimens

### 6.3.1 Metallic Specimens

XRF is a widely used technique in metal production. This industry stresses the need for a simple and fast method; therefore, the preparation of a specimen has to be fast, simple and reproducible. The most universal specimen preparation procedures are cutting, milling and casting. XRF analysis of metals allows to determine the composition of a relatively thin surface layer of the investigated specimen. The surface must therefore be representative of the bulk sample. The major surface preparation methods are machining (milling, turning), mechanical grinding, as well as polishing and etching. In any surface preparation technique special care must be taken to prevent any surface contamination during specimen preparation. Special care is very important in the case of relatively soft metals such as aluminium, where particles of the grinding agent may penetrate into the sample surface during surface finishing. The selection of a suitable abrasive is difficult, particularly when elements of interest are silicon, aluminium or iron. For instance, SiC and $Al_2O_3$ are very effective abrasives, but both of them contain elements of interest to the metals industry. In such cases electrolytic polishing or etching is recommended. After each grinding or polishing step, it is important to make sure that the surfaces are kept clean. Hence, all traces of lubricant, cutting fluid or finger marks must be removed by cleaning with isopropyl alcohol prior to examination. Occasionally, the surface of several metals is affected by corrosion which progresses with time. Therefore, the best results are always obtained when analysing a specimen without much delay after surface preparation [1].

When the surface of the specimen is rough and common methods such as polishing and milling are not effective, the samples can be prepared by obtaining turnings of the samples and then pressing the turnings into pellets with a hydraulic press at several hundreds of mega Pascal. When this procedure is done properly, the turnings are pressed to produce a surface that is smooth enough for analysis. Specimen preparation by pressing is discussed in more detail in Sect. 6.3.2. Selection of the most suitable surface preparation technique depends on the characteristics of an alloy or group of alloys. For example, hard alloys resistant to cutting and abrasion, such as cast iron, can be ground and pelletized prior to analysis. On the other hand, if a sample is composed of soft, malleable, multiphase alloys, smearing effects of the softer components, like lead and aluminum, cause serious problems. These soft elements will concentrate on the surface. That is why the fluorescence intensities of these soft metals are too high, while those of the harder metals, and metals with lower Z, are too low. In such cases, when the sample surface does not contain a representative concentration of elements in the bulk, electrolytic polishing or etching should be employed. For some archaeological materials, the etching procedure might be too excessive since many inclusions are dissolved during etching. For bronze alloys the recommended etching reagent is a solution of alcoholic ferric chloride (120 ml $C_2H_5OH$, 30 ml HCl and 10 g $FeCl_3$), whereas for brass alloys a solution of aqueous ammonium persulfate (100 ml $H_2O$ and 10 g $(NH_4)_2S_2O_8$) is preferred. Ancient metals etch very fast; only a 5–10 s etching time is required.

The XRF analysis of semiconductors is often carried out by total reflection X-ray fluorescence (TXRF) or synchrotron radiation X-ray fluorescence (SR-XRF). However, there are also some interesting applications with XRF such as determination of nickel in silicon wafers. Nickel implantation in wafers can be measured directly with XRF, by simply placing the wafer in the spectrometer. More difficult is the preparation of calibration standards for this application. They can be obtained in an unconventional way by spiking a gelatin containing a nickel standard on parts of the silicon wafer and drying prior to analysis [66].

Irregularly shaped metallic samples can be prepared for analysis by embedding a piece of the metal in a special wax resin (e.g., acrylic resin and methyl acrylic resin). This block can be cut to the proper shape and then polished using a SiC and diamond paste prior to analysis [67].

Another, relatively new technique is re-melting of the sample. Re-melting is used, as a rule, to prepare specimens of not very high-melting point materials, including primary metals, to homogenize the sample and to obtain ultimately a specimen of desirable shape and with a smooth surface. Such a specimen preparation technique was used to analyse bronzes, blister and electrolytic copper, aluminium and other metals. The original ferroalloy is ground to a grain size of 2 mm with a steel disc mill, followed by a tungsten mill, after previous crushing with a jaw breaker. Subsequently, the grained particles are mixed with a portion of iron (15 g FeMo-alloy + 25 g pure Fe) and this final mixture is melted inside an induction furnace. Afterwards the melted metal is centrifuged and cast in an appropriate mold prior to analysis. Calibration samples can easily be obtained in the same way. The determination of chromium in Mo-ferroalloy and Nb-ferroalloy requires something like re-melting because due to their inhomogeneity and brittleness, direct analysis by XRF would not produce accurate data [68, 69].

Similar to metals, organic substances (e.g., resins) can be re-melted if the process does not cause a decomposition of the material. The so-called "hot casting" procedure completely eliminates the use of mixing and grinding equipment and awkward cleanups. Pitch or asphalt is directly "hot cast" into a plastic sample cup fitted with a thick Mylar film. After cooling, the film is carefully removed from the surface of the sample to expose a perfectly flat surface ready for XRF analysis [1].

A number of metals and alloys that are otherwise very difficult to deal with can be brought into solution by use of aqua-regia (a mixture of one part of concentrated $HNO_3$ with three parts of concentrated HCl) with gentle heating to 60°C. Aqua regia is used to dissolve a number of metals (e.g., Pd, Pt, Sn, Ge), and alloys, including steels, Mo-, Rh-, Ir-, Ru- and high-temperature alloys. The ability of this solvent to dissolve metals is due to the formation of chloro-complexes as well as to the catalytic effect of $Cl_2$ and NOCl. The solution becomes more corrosive if it is allowed to stand for 10–20 min before heating. Fe-ore and V-ore can be dissolved by aqua-regia. The same is true for some sulfides (e.g., pyrites and copper sulfide ores), although some sulfur is lost as $H_2S$. $HNO_3$ and HCl are sometimes mixed in a ratio of 3:1 rather than 1:3. This ratio is sometimes called an inverted aqua-regia and is used to oxidize sulfur and pyrites. Another corrosive reagent consists of $HNO_3/HF$. The complexing effect of the fluoride ion is utilized in this solution, and with this combination a number of metals (e.g., Si, Nb, Ta, Zr, Hf, W, Ti) and alloys (e.g., Nb–Sn, Al–Cr, Cu–Si, Ca–Si) can be dissolved. However, some elements may be lost during the digestion step, especially elements like Se, Hg and Sn which can form volatile components with various kinds of acid mixtures. Samples treated with a hot mixture of $HNO_3/HF/HClO_4$, e.g., lose all Se and Cr, and some Hg, As, Ge, Te, Re, Os and Ru can also be lost [70]. Methods which lead to losses of elements during preparation should, of course, not be used unless the elements are not of interest. Methods for analysing solutions are discussed above in Sect. 6.2.

### 6.3.2 Powder Specimens

#### General

A common technique for geological, industrial and biological materials is the preparation of powders and pellets. A powder specimen is prepared when the original sample is too heterogeneous for direct analysis, too brittle to form a self-supporting disc, or when a suitably smooth, flat surface is not available. XRF analysis of powder samples requires to have a uniform particle size, and to have a grain size small enough to avoid particle-size effect problems. A dependence of elemental analytical line intensity on the density of powders, both in bulk and pelleted (see Sect. 6.3.2) at constant pressure, that was defined by the variation in particle size and surface interaction forces was found [71]. Crushing, pulverizing and milling procedures can be used to achieve these goals.

Additives (see below) are sometimes used to provide a proper grinding medium for the powders. In general, 2–10% (weight) of additive is used, and is adequate, for the grinding, blending and briquetting. A powder specimen has to fit into the specimen chamber of the X-ray spectrometer. A round, flat disk is ideal. Commercial presses and dies are available for this purpose. In some cases, loose (unpressed) powder can be placed in a powder holder which has a thin window, and is then inserted directly into the specimen chamber of the spectrometer.

The powder can also be packed in cells or spread out on a thin film material (so-called "slurry" technique). The slurry technique works for water insoluble materials. A water slurry is prepared out of a few milligrams of powder and a few milliliters of water. A turbid suspension is made and then filtered through, e.g., a Nuclepore filter. The objective is to produce a uniformly thick, homogenous specimen. Processing large bulk samples poses the biggest problem. The first major challenge is to remove a portion of the bulk which is representative of the bulk material. This is not an easy task and requires great care when there is a large bulk of material to work with. The next challenge is to pulverize the sample so as to produce fine particles of uniform size and without contaminating the sample or segregating it. The sample must also have a homogenous distribution of all elements throughout the entire sample.

### Grinding

A major objective during the preparation of a sample from a bulk is the reduction of the powder to a uniform particle size. For a routine trace XRF analysis, a particle size of less than 60 μm is commonly accepted. Reduction of particle size is usually accomplished by grinding. Various methods for grinding samples are available. For routine work, manual grinding, using a mortar and pestle (agate, corundum, mullite), is traditionally done, while for finer (1–10 μm) and controlled particle sizes, a variety of mills are commercially available. SPEX Industries (see e.g. www.spexcsp.com) produces a grinder, called a mixer mill, which does an excellent job of producing particles of a uniform size. It is necessary to note that this grinder can only be used when there are several grams of material to work with. When dealing with extremely small amounts of sample, hand grinding is the best option although it is tedious and time-consuming, and requires care to produce particles with a uniform size distribution. Materials that are difficult to treat, like rocks and ores, are ground better after the introduction of an additive such as sodium stearate. It is found that manual grinding of, e.g., inorganic aluminosilicates by using a mortar and pestle reduces the particle size to about 40 μm which is acceptable even for X-ray diffraction methods (XRD). However, as mentioned above, the particle size distribution must be controlled or else preferential absorption of secondary radiation can occur in the particles of different size. In the case of coal, samples are prepared as a powder (5 g sample ground in a tungsten carbide rotary swing mill together with 1 g of a boric acid binder and 100 mg of sodium stearate). A grinding time of about 6 min will reduce the particle size to 50 μm. In some analyses, it is important to use a uniform grinding time for all samples to assure reproducible results. For materials that may be damaged or undergo alteration during size reduction, grinding is usually carried out in ethyl or isopropyl alcohol using a pestle and mortar. Grinding in liquid nitrogen with a SPEX freezer/mill can also be effective for some materials, in particular, polymers and plastics.

Mechanical grinders can also be used under dry and wet conditions. Grinding samples which contain a mixture of particles with different physical properties (such as different phases) may lead to significantly different grinding of the different phases. Experience has shown that all grinding tools are a potential source of contamination. Therefore, the composition of the grinding media should be selected with care to avoid materials which contain elements of interest in the analysis. Agate, for instance, introduces traces of $SiO_2$, Mg and Ca to the specimen. This might be of less importance for geological materials, but for biological materials these blank values represent a major source of errors.

The effect of surface roughness of particles on XRF analysis of powdered samples is discussed in [72]. The authors show that errors due to surface roughness can be minimized if measurements are performed at a steep take-off angle $\geq 70°$. Unfortunately, most wavelength-dispersive spectrometers now in production do not allow the take-off angle to be varied.

## Pelletizing

Pelletizing a powder sample is necessary to reduce surface effects and to yield a better precision than that obtained by the analysis of loose powders. In general, provided that the powder particles are of a uniform size distribution, and are less than 50 μm in diameter (300 mesh), the sample should be pressed into a disc (pelletized) at 600–800 MPa (6–8 tons per $cm^2$). If the self-bonding properties of the powder are good, low pressures of perhaps up to 100–400 MPa (1–4 tons per $cm^2$) can be employed. High-pressure pelletizing in a die, or directly in a sample cup, often results in a fracture of the pellet following the removal of pressure from the die. The release of pressure should be slow and always at the same rate in order to produce reproducible data.

It is sometimes necessary to add a binder before pelletizing to form stable pellets, and prevent caking of the sample on the surface of the die. Binders are normally composed of low Z-elements or organic materials. As mentioned above, these types of materials can produce scattered X-rays and thereby increase the background counts. This effect can become important in the determination of trace elements in the low-Z region. Therefore, the choice of the binding agent should be made with special care. It must be free of contamination and should have a low absorption coefficient for the analyte lines. The most useful binders are starch, cellulose, lucite, polyvinyl, urea, boric acid, graphite, etc. Liquid binders like ethyl alcohol or diethyl ether can be mixed with the sample manually, whereas powdered binders are recommended to be mixed mechanically to form a homogeneous mixture. A recommended procedure is the addition of 2–10% by weight of binder to the sample. Further stability can be achieved by spraying the pellet with a 1% solution of Formvar in chloroform. However, this should not be done unless the pellet is unstable.

The particle size of the powder to be compacted can influence the emission intensity of the elements under determination, therefore, the sample must be ground to a specified particle size and have a uniform size distribution. It has been shown [73] that, when compacted specimens are used, the emission intensity increases with an increase in compaction pressure. Calibration with internal standard was applied to eliminate particle size effects for powder samples of copper ore, concentrates and copper slag prepared in the form of pellets [74]. Still another source of systematic errors is when a phase composition distinction exists between specimens to be analysed and reference materials. Compaction of powdered samples under constant pressure results in a reduction of errors, but does not eliminate them completely. To eliminate the error mentioned above, it was proposed [13] to compact specimens of constant volume which, provided the compact material mass is constant, results in specimens of a specified density, irrespective of the phase composition of the initial powders. For this purpose, a mould was used where the plunger movement was limited by a ring made of an essentially incompressible material.

### 6.3.3 Fused Specimens

Samples which do not go into solution easily, cannot be re-melted, or tend to remain heterogeneous after grinding and pelletizing, are often treated by the Claisse fusion technique. Many useful recipes for fusion are given in a book by Buhrke et al. [1]. Disadvantages of the fusion techniques are the time and material costs involved, and the dilution effect of the sample which makes it difficult, or even impossible, to determine trace elements in a specimen. Probably the most effective way of preparing a homogeneous powder sample is the borax fusion method [75]. In principle, it involves fusion of the sample with an excess of sodium or lithium tetraborate and casting into a solid bead with a flat, mirror like surface. Chemical reaction in the melt converts the phases present in the sample into glass-like borates, giving a homogeneous bead of dimensions ideal for direct placement in a spectrometer. Manual application of the technique is rather time-consuming, while a number of automated and semi-automated borax bead-making machines are commercially available.

A critical aspect of the method is the mass ratio of sample to fusion mixture because it controls several factors, such as the speed and degree of the chemical reaction, final mass absorption coefficient and the actual dilution factor applied to the element to be determined. By using fusion aids (like iodides and peroxides) and high atomic number absorbers as a part of the fusion mixture (like barium or lanthanum salts), these factors can be somehow controlled. Claisse's original method proposed sodium tetraborate to sample mass ratio of 100:1 [75]. In 1962, Rose et al. [76] suggested the use of lithium tetraborate in a ratio of 4:1. A low dilution fusion technique was developed, where the flux-to-sample mass ratio is 2:1.

Mostly $LiBO_2$ is applied as a flux, as it is more reactive and forms fluxes with higher fluidity. A critical review on this subject was done by Bower and Valentine [77]. It is shown [78] that the using of fusion beads ($Na_2CO_3$, $Li_2B_4O_7$ as flux with addition of NaI) produces more precise results than the using of compressed pellets (boric acid as plastifier) under XRF determination of major and minor elements in ferroalloys.

Borate fusion is widely used; however, the method requires slow cooling of the disk to avoid possible spontaneous cracking due to residual thermoelastic stresses. Fluxing with lithium metaphosphate eliminates this problem by the formation of phosphate complexes of transient metals which are the components of the substances under analysis. It was shown [79] that the use of lithium metaphosphate as a flux for high-precision XRF analysis of oxide materials allows one to produce specimens in the form of homogeneous glass-like disks which do not require prolonged annealing or subsequent machining. It is possible to use this technique to produce specimens of various oxide materials by varying the mass ratio of the materials in the flux. As shown by a special investigation [80], reference materials in the form of glass-like phosphatic specimens can be used after having been stored for a year and a half or even longer, provided they are kept in a desiccator over phosphorus pentoxide. Using this specimen preparation technique, the authors [81] managed to develop a procedure for Ca, Sr, Bi, Pb, and Cu determination in bismuth-containing superconducting ceramics (the flux-to-sample mass ratio being 2:1). Such a procedure is characterized by insignificant systematic errors and very low random errors (for an individual result, $S_r$ is $5\times10^{-3}-7\times10^{-3}$).

The actual fusion reaction can be performed at 800 to 1000° C in a crucible made of Pt, Ni or quartz glass. All of these materials suffer from the disadvantage that the melt tends to wet the sides of the crucible and it is impossible to achieve complete recovery of the fused mixture. The use of graphite crucibles partially avoids this problem, but the best results are obtained by the use of a crucible made of Pt and 3% Au. Sometimes, the specimen is taken directly from the crucible in which it is melted. However, a more common procedure is to transfer the melt into another crucible-mould or into a ring placed on a smooth, flat plate [82]. Good results are attained when a glassy carbon crucible is used as a mould. This is cheaper than using Pt crucibles; does not interact with metal oxides and thus makes it possible to obtain ingots having ideally smooth surfaces. Unfortunately, this material is subject to oxidation by air oxygen at high temperatures. The most practical procedure is to flux in a platinum crucible, and then use a glassy carbon holder to serve as a mould to produce the specimen disk [13]. Special matrix correction coefficients are proposed for taking into account losses of volatile components (e.g., $CO_2$, $NO_2$, $O_2$) during the melting procedure [83].

## 6.4 Biological Samples

Solid biological materials are generally heterogeneous. Therefore, drying, powdering, homogenizing and homogeneity testing might be necessary before preparing the samples for measurement. Various treatments of biological materials include further on lyophilization, ashing and wet digestion. The digestion method should be optimized and the recovery of the elements and precision of the procedure should be thoroughly tested.

Specimen preparation for XRF of plants, leaves and vegetation in general, is simple and fast. The representative samples are collected (approximately 5 g), dried at 85°C, pulverized and pressed into a pellet without any contamination. Usually 10–20 elements can be determined quantitatively in a matter of minutes [84].

Most human and animal tissues are very soft and hence, they must first be stabilized and strengthened before sections can be cut. One possibility is to freeze-dry the tissue and pulverize it at liquid nitrogen temperatures. A small amount of that powder can be fixed to a thin membrane with a solution of 1% polystyrene in benzene and measured. Frozen organic tissue can be cut with a microtome, and if the sections are not too thin, they can be dried and irradiated as self-supporting targets. To quantify the results, it is sufficient to add an internal standard before the analysis and to determine the dry mass of the sections after analysis.

The interest in clinical applications increases from year to year. In vivo analysis, particularly for Pb, is a major issue for XRF. The direct determination of Pb in bone gives an indication of the ongoing accumulation, whereas conventional blood studies only reflect recent exposures. Other heavy metals of interest are As, Cd, Hg and U. Further in vivo investigations dealt with the concentration of various elements in different organs, mainly in the field of kidney, liver and lung research. However, the sensitivity and detection limits of the in vivo applications are still rather poor.

Other applications of XRF analysis include hair, teeth, nails, biological liquids (like blood and serum), drugs and medicines. Blood samples, e.g., can be pipetted in 100-μL portions onto a filter paper. Serum samples can be treated almost in the same way but instead of a filter paper, polypropylene foil should be selected as a carrier. To avoid losses of volatile components, urine and blood samples need to be dried in a refrigerator at temperatures about 2–4°C. Detection limits between $1.2\,\mathrm{ng\,L^{-1}}$ for Ca in urine and $50\,\mathrm{\mu g\,L^{-1}}$ for Sr in blood are achieved with such treatment [85, 86].

Like for the preconcentration of traces from water samples, various kinds of coprecipitation or ion exchange methods have been proposed for biological samples. A number of chemical modifiers, such as magnesium nitrate, nickel nitrate and ammonium pyrrolidine dithio carbamate (APDC) coupled with magnesium nitrate, and air-drying of the sample, make an effective sample preparation method for the determination of a wide range of elements at different concentration levels, from ppb's to ppm's [87] For example, a very low concentration of Cr ($0.3\,\mathrm{ng\,mL^{-1}}$) in plasma could be determined by complexation with APDC and extraction with methyl isobutyl ketone. After evaporation of organic solvent, the residue is dissolved in acid and deposited on a thin polycarbonate foil. Hair samples (0.1 g) can be digested with a 1:5 mixture of $HClO_4$ and $HNO_3$ in a Teflon bomb and the trace elements are concentrated using APDC at pH 3–4. In this way, elements such as Fe, Ni, Cu, Zn and Pb could be determined down to $0.4\,\mathrm{\mu g\,g^{-1}}$. The same limits of detection could be achieved if, e.g., 1 g of hair sample is pelletized with 0.1 g CaO and 20 g cellulose.

The wet digestion methods, by microwave or high pressure, are very effective. With a mixture of $HNO_3$, $H_2SO_4$ and $HClO_4$, most organic material can be attacked. Even more effective is a combination of $HNO_3$ and HF as also silicates in organic materials are dissolved. Care should be taken to work, whenever possible, in closed conditions to avoid contamination and/or losses of elements. The combination of microwave and pressure digestion allows quantitative recovery of elements that may be volatilized in open digests (like Ge, Se, Hg), as well as the recovery of noble metals (Pt, Os, and Ir) to the ppb level in solid samples even in the extremely difficult digestion of organic material. For a basic understanding of the microwave acid digestion theory, including safety guidelines and dissolution methods for geological, metallurgical, biological, food and other samples, we recommend the professional reference book edited by Kingston and Jassie [88].

## 6.5 Aerosol and Dust Specimens

XRF is frequently invoked for trace analysis in air pollution studies. In the air, trace elements are almost exclusively in the particulate phase at typical concentrations of $50\text{–}500\,\mathrm{ng\,m^{-3}}$. By simply drawing a large air volume through a filter, large preconcentration factors are easily achieved. Adequately loaded filters with air particulate matter are presented directly to the XRF unit. Unless the filters are extremely loaded, no correction for the X-ray absorption in the aerosol material is usually necessary.

The part of the filter with a sample actually analysed by XRF has an area of only a few square centimeters and hence it must be representative for the entire sample. Therefore, a critical step is the selection of an appropriate filter material. Teflon and polycarbonate (Nuclepore) filters are ideal because of their high purity and because they are surface collectors. Whatman-41 cellulose filters have been used widely because of their low cost, but they are rather thick (around $9\,\mathrm{mg\,cm^{-2}}$ versus $1.1\,\mathrm{mg\,cm^{-2}}$ for Nuclepore), which leads to more X-ray scatter and higher detection limits. They also partially collect particulate matter within their depth, so that X-ray absorption corrections become more complicated. Glass fibre filters should be avoided in any case due to their high inorganic impurity. High-purity quartz fibre filters have recently been proposed. Before selecting a filter material for a particular application, the blank count of the filter or background level of the material to be analysed must be determined since all filters contain various elements as major, minor and trace constituents.

Regarding aerosol analysis, several factors that affect X-ray intensity must be considered: attenuation of X-rays within the individual aerosol particle is only important for particles larger than a few micrometer but accumulation of particles can have an effect (in the case of heavily loaded filters) and the filter itself can attenuate incident and emerging X-rays (if particles are not collected on the surface). An appropriate correction involves knowledge of the particle size distribution, and selecting an adequate sampling time and an appropriate filter material. For example, if thin filters, like Nuclepore or Millipore, are selected for the collection, the absorption effects in the filter itself are small and often negligible for X-rays above 3 keV, since normally all particles should be retained at the filter surface. However, for light elements such as Si, P, S and Cl, the absorption effect needs to be corrected by using a suitable correction procedure.

Besides conventional XRF, also TXRF is nowadays more frequently applied for aerosol analysis [89]. To take advantage of the low detection limits and to reduce the matrix effects, originating from the collection material, filter samples need to be digested. Normally, aerosol loaded filters are dissolved with supra pure concentrated (70%) $HNO_3$ and HF acids in a high-pressure digestion vessel. A standard is normally added prior to digestion. With this procedure, element concentrations between $0.2\,\mathrm{ng\,m^{-3}}$ for Cu and $1\,\mathrm{ng\,m^{-3}}$ for Mn are easily determined.

## 6.6 Standards

X-ray spectrometry is essentially a comparative method of analysis. Therefore, it is necessary that all standards and unknowns have a high degree of similarity (approximately the same matrix composition) and that they are presented to the spectrometer in a reproducible and identical way. Standards may be a certified reference material or a sample itself but analysed by another analytical technique. Many specimens in use may be prepared also in the laboratory. They are usually composed of one or more elements at different levels of concentrations within different matrices. It is essential for standards to provide high-count rates for the element to be determined. Sometimes use is made of an internal standard. Standards can be a liquid, a gas or a solid. It can be pure, mixed powder, natural or synthetic materials.

Today, there are three thousand reference standards listed in catalogues and hundreds more in preparation. These standards include ores, metals and alloys, rocks, minerals, waste products and dusts collected by electrostatic precipitators. Environmental problems have increased the demand for standards for tailings, industrial wastes and isotopic materials.

Usually the number of reference materials for XRF analysis is from 4–5 to 18–20. This number rises as the number of the determined elements and their mutual effects become larger. The higher the demanded necessary accuracy of the analysis results, the more reference materials are needed. But the method of $\alpha$-correction [90] allows to decrease the number of reference materials to one to two without essentially lessening the accuracy [91].

## List of Abbreviations

APDC Ammonium pyrrolidine-dithiocarbamate
DDTC Sodium diethyldithio-carbamates
EDXRF Energy dispersive X-ray fluorescence
PTFE Polytetrafluorethylene
SR-XR Synchrotron radiation X-ray fluorescence
TXRF Total reflection X-ray fluorescence
XRD X-Ray diffraction
XRF X-Ray fluorescence

## References

1. Buhrke VE, Jenkins R, Smith DK, A Practical Guide for the Preparation of Specimens for X-Ray Fluorescence and X-Ray Diffraction Analysis, Wiley, New York (1998)
2. Injuk J, Van Grieken R, Sample preparation for XRF, In: Van Grieken R, Markowicz A (eds), Handbook of X-Ray Spectrometry, Marcel Dekker, New York, 657–692 (1993)
3. Oliver AJ, Bennett HH, *XRF Analysis of Ceramics, Minerals and Other Materials*, Wiley, New York, (1992)
4. Uhlig S, Peters M, Analysis **20**, 48 (1992)
5. Van Grieken R, Anal Chim Acta **143**, 3 (1982)
6. Smagunova AN, Gunicheva TN, Karpukova OM, Kozlov VA, Zavod Lab **59**(No. 4), 20 (1993)
7. Volodin SA, Revenko AG, Zh Anal Khim **44**, 1541 (1989)

8. Tsvetyansky AL, Yeritenko AN, Zavod Lab **56**(No. 4), 25 (1990)
9. Smagunova AN, Rozova OF, Skribko NN, Zavod Lab **56**(No. 9), 28 (1990)
10. Koch KH, Ohls K, Flock J, Lab Prax **14**, 1022 (1990)
11. Afonin VP, X-ray fluorescence analysis of rocks, Fresenius Z Anal Chem **335**, 54–57 (1989)
12. Revenko AG, Zavod Lab **60**(No. 11), 16 (1994)
13. Blank AB, Eksperiandova LP, X-Ray Spectrom **27**, 147 (1998)
14. Török SB, Van Grieken RE, Anal Chem **64**, 180R (1992)
15. Török SB, Van Grieken RE, Anal Chem **66**, 186R (1994)
16. Török SB, Labar J, Injuk J, Van Grieken R, Anal Chem **68**, 467R (1996)
17. Szalóki I, Török Sz, Ro C-U, Injuk J, Van Grieken R, Anal Chem **72**, 211R (2000)
18. Ceasy LE, Adv X-Ray Anal **37**, 729 (1994)
19. Davidson RM, Fuel **73**, 988 (1994)
20. Lyamina OI, Kuprianova TA, Gimelfarb FA, Zh Anal Khim **50**, 271 (1995)
21. Eksperiandova LP, Spolnik ZM, Blank AB, Adv X-Ray Anal **38**, 735 (1995)
22. Eksperiandova LP, Blank AB, Makarovskaya YN, X-Ray Spectrom **28**, 24–26 (1999)
23. Eksperiandova LP, Makarovskaya YN, Blank AB, Panikarskaya VD, Kolotiy OD, Kharkov University Bulletin. No. 437, Chemical series, Issue 3, 148 (1999)
24. Eksperiandova LP, Makarovskaya YN, Blank AB, Anal Chim Acta **371**, 105 (1998)
25. Eksperiandova LP, Blank AB, Makarovskaya YN, X-Ray Spectrom **31**, 259 (2002)
26. Eksperiandova LP, Blank AB, Fokina II, Fresenius' J Anal Chem **361**, 287 (1998)
27. Eksperiandova LP, Fokina II, Panikarskaya VD, Blank AB, Kharkov University Bulletin. No. 532, Chemical series, Issue 7, 91 (2001)
28. Jenkins R, De Vries JL, *Practical X-Ray Spectrometry,* 2nd edn, Macmillan, London (1983)
29. Volkov VF, Semenova EB, Gerasimov SA, Sinitsin VN, Zavod Lab **54**(No. 12), 46 (1988)
30. Wobrauschek P, Streli C, Kregsamer P, Ladisich W, Rieder R, J Trace and Microprobe Techn **14**(No. 1), 103 (1996)
31. Klockenkämper R, von Bohlen A, J Anal Atom Spectrom 7, 273 (1992)
32. Injuk J, Van Grieken R, Spectrochim Acta B **50**(No. 14), 1787 (1995)
33. Cariati F, Fermo P, Gifardoni S, Galli A, Milazzo M, Spectrochim Acta B **58**, 177 (2003)
34. Van Grieken R, Anal Chim Acta **143**, 3 (1982)
35. Smits J, Nelissen J, Van Grieken R, Anal Chim Acta **111**, 215 (1979)
36. Hanif I, Hanif J, Hasany SM, Iqbal MZ, X-Ray Spectrom **24**, 298 (1995)
37. Wilson AF, Turner DC, Robbins AA, Process Analytics, Orem, UT (1998)
38. Engel G, Ostner R, Bebner H, Chem Ztg **113**, 382 (1989)
39. Taira Y, Maeda Y, Azumi T, Nippon Kaisui Gakkaishi **42**, 109 (1988)
40. Peräniemi S, Vepsäläinen, Mustalahti H, Ahlgrén M, Determination of phosphorus in waste water by EDXRF, Fresenius J Anal Chem **344**, 118–122 (1992)
41. Leyden DE, Patterson TA, Alberts JJ, Anal Chem **47**, 733 (1975)
42. Cojocaru V, Spiridon S, J Radioanal Nucl Chem **170**, 259 (1993)
43. Prange A, Knöchel A, Anal Chim Acta **172**, 79 (1985)

44. Andreeva IY, Polivanova NG, Bogomolova DN, Zh Anal Khim **47**, 383 (1992)
45. Bhagavathy V, Reddy MLP, Prasada Rao T, Damodaran AD, J Radioanal Nucl Chem **149**, 35 (1991)
46. Moriyasu K, Mizuta H, Nishikawa Y, Bunseki Kahaku **40**, 175 (1991)
47. Kulagina NV, Tikhomirova TI, Sorokina HM, Fadeeva VI, Tsisin GI, Zolotov YA, Vestn Mosk Univ Khimia **35**, 167 (1994)
48. Zhao E, Qiu L, Lihua Jianyan Huaxue Fence **30**, 106 (1994)
49. Seregina IF, Tsisin GI, Shilnikov AM, Formanovsky AA, Zolotov YA, Zh Anal Khim **48**, 166 (1993)
50. Belikov KN, Blank AB, Shevtsov NI, Nadzhabova OYu, Tananaiko MM, Anal Chim Acta **383**, 277 (1999)
51. Terletskaya AV, Bogoslavskaya TA, Borzakovsky AE, Stikhar OL, Ageeva IA, Khim Tekhnol ody **15**, 435 (1993)
52. Khuder A, Maslov LP, Kjutvitsky VA, Zh Anal Khim **51**, 121 (1996)
53. Frolova MM, Golentovskaya IP, Smagunova AN, Zagumennova VF, Morozova LB, Trofimov BA, Zavod ab **57**(No. 8), 33 (1991)
54. Khvostova VP, Trunova VA, Baryshev VB, Zolotarev KV, Nucl Instrum Meth Phys Res A **308**, 315 (1991)
55. Gainutdinova DI, Kryukov SV, Garifzyanov AR, Toropova VF, Budnikov GK, Al'-Khatib D, Zavod Lab Diagn Mater **67**(8), 6 (2001)
56. Revenko AG, Volodin SA, Uvarov AI, Zavod Lab **57**(No. 4), 29 (1991)
57. Makarovskaya YN, Eksperiandova LP, Blank AB, Zh Anal Khim **54**, 1167 (1999)
58. Eksperiandova LP, Blank AB, Makarovskaya, YN, X-ray fluorescence analysis of organic extracts having the form of organogels. In: *Book of Abstracts*, 45th Annual Denver X-Ray Conference, Denver, Colorado, 35 (1996)
59. Zebreva AI, Kurilovich YE, Lobanov FI, Andreeva NN, Zh Anal Khim **47**, 629 (1992)
60. Logunova SA, Lobanov FI, Sarzhenko IN, Karpova LA, Akberdina ES, Zh Anal Khim **50**, 994 (1995)
61. Logunova SA, Lobanov FI, Markin VA, Zavod Lab **62**(No. 9), 24 (1996)
62. Blank AB, Pure Substances Analysis Using Crystallization Preconcentration Khimia, Moscow (1986) [in Russian]
63. Blank AB, Eksperiandova LP, Aliseichik VV, Shevtsov NI, Zh Anal Khim **45**, 1847 (1990)
64. Makarovskaya YaN, Eksperiandova LP, Blank AB, J Anal Chem **57**, 110 (2003)
65. De Viti, Perez CA, Maratti MI, Bueno S, Olsina R, Masi A, X-Ray Spectrom **30**, 308 (2001)
66. Bubert H, Burba P, Klockenkämper R, Schönborn A, Wielunski M, Fresenius' J nal Chem **341**, 245 (1991)
67. Wybenga F, X-Ray Spectrom **8**, 182 (1979)
68. Coedo AG, Dorado MT, Rivero CJ, Cobo IG, Analyst **119**, 421 (1994)
69. Cobo IG, Dorado MT, Coedo AG, J Trace Microprobe Technol **13**, 431 (1995)
70. Tölg GZ, Anal Chem **190**, 161 (1962)
71. Zagorodniy VV, X-Ray Spectrom **32**, 40 (2003)
72. Kitov BI, Smagunova AV, Portnov MA, Zh Anal Khim **45**, 1927 (1990)
73. Smagunova AN, Rozova OF, Prekina IM, Zh Anal Khim **49**(No. 7), 711 (1994)
74. Mzyk Z, Baranovska I, Mzyk J, X-Ray Spectrom **31**, 39 (2002).
75. Tertian R, Claisse F, *Principles of Quantitative X-Ray Fluorescence Analysis* Wiley-Heyden, London (1982)

76. Rose H, Adler I, Flanagan FJ, US Geol Survey Prof Pap 450-B, 80 (1960)
77. Bower NW, Valentine G, X-Ray Spectrom **15**, 73 (1986)
78. Coedo AG, Dorado T, Padilla I, Alguacil FJ, X-Ray Spectrom **31**, 424 (2002)
79. Blank AB, Shevtsov NI, Mirenskaya II, Nartova ZM, Zh Anal Khim **45**, 1724 (1990)
80. Spolnik ZM, Mateichenko PV, Shevtsov NI, Blank AB, Zh Anal Khim **48**, 1599 (1993)
81. Shevtsov NI, Nartova ZM, Kvichko LA, Blank AB, Zh Anal Khim **46**, 591 (1991)
82. Finkelshtein AL, Zh Anal Khim **47**, 1709 (1992)
83. Hettipathirana T, Grey N, Naidu R, Spectrochim Acta B **58**, 85 (2003)
84. Kocman V, Peel TE, Tomlinson GH, Commun Soil Sci Plant Anal **22** (19, 20) 2063 (1991)
85. Hong NT, Ha HV, X-Ray Spectrom **25**, 3 (1996)
86. Viksna A, Mwiruki D, Jagner D, Selin E, X-Ray Spectrom **24**, 76 (1995)
87. Savage I, Haswell SJ, J Anal Atom Spectrom **13**, 1119 (1998)
88. Kingston HM, Jassie LB, *Introduction to Microwave Sample Preparation, Theory and Practice* ACS Reference Book Series, American Chemical Society, Washington, D.C. (1988)
89. Injuk J, Van Grieken R, Spectrochim Acta B **50**, 1787 (1995)
90. Lee RF, McConchie DM, X-Ray Spectrom **11**, 55 (1982)
91. Blank AB, Golovko SA, Spolnik ZM, Shevtsov NI, Zh Anal Khim **48**, 320 (1993)

# 7

# Methodological Developments and Applications

## 7.1 Micro X-Ray Fluorescence Spectroscopy

*B. Kanngießer, M. Haschke*

### 7.1.1 Introduction

Micro X-ray fluorescence spectroscopy (micro-XRF) is one of the latest and very dynamic branches of X-ray fluorescence spectroscopy. Here "micro" indicates that localised regions of a sample are investigated with spatial resolution on the micrometer scale. Micro-XRF has developed very rapidly in the last decade, mainly due to the use of synchrotron radiation. The reason for that preference is the by far higher intensity of synchrotron radiation in comparison to X-ray tube radiation. The first arrangements employed collimators that could only use a small fraction of the emitted radiation for the excitation. The next step involved the use of mono-capillary optics, but also in this case the amount of radiation brought onto the sample is limited by the small entrance diameter of the capillary.

Despite the lower intensity delivered by X-ray tubes bench-top instruments with tube excitation have come into use for micro-XRF over the last years. For these instruments it is necessary to closely couple X-ray optics to the tube window in order to concentrate a large portion of the tube radiation into a small sample area [1–3, 5]. Typically, capillary optics–especially polycapillary lenses–are in use for that purpose.

With micro-XRF, not only the elemental composition of a sample is accessible, but also the related spatial distribution. Furthermore, the use of synchrotron radiation extends the possibilities of micro-XRF with respect to the elemental range to light elements and increases both the elemental sensitivity and the spatial resolution. Another advantage of using synchrotron radiation is the possibility to combine micro-XRF with other microanalytical methods. A first overview of micro-XRF techniques with the focus on micro-XRF at synchrotron sources can be found in the monograph by Janssens et al. [4].

This chapter focuses on laboratory units and their various applications. It delivers an overview on existing micro-XRF instruments and discusses the instrument components according to their practical relevance. Various fields of applications are also discussed giving an impression on the wide range of different analytical tasks that can be solved with a micro-XRF bench-top unit. The last part of this chapter describes the newest development in micro-XRF, which is the extension of micro-XRF into a depth-sensitive method. This results in a true three-dimensional analytical method. The new set-up is described for the laboratory as well as for the synchrotron source, and examples are given for both arrangements to demonstrate their application. This last part also serves as a transition to the following chapter by Simionovici and Chevallier in which the actual developments in micro-XRF at synchrotron sources are described.

### 7.1.2 General Description of Micro-XRF Laboratory Units

There is a wide range of analytical methods available with which the chemical composition of a sample can be determined. They differ in the covered range of concentration and of elements, in information depth, in accuracy etc. But only a few of them are able to selectively analyse small sample areas. Table 7.1 provides a summary of the variety of analytical methods currently available for this purpose. Properties, such as the elemental range covered, together with the depth of the analysed sample volume are summarised for every method.

The relation between the detectable concentration range and the spatial resolution is displayed in Fig. 7.1. Only the method with a spatial resolution better than 1 mm can be regarded as microanalytical method. From Fig. 7.1

**Table 7.1.** Analytical methods currently available for chemical composition together with their spatial resolution

| Method | Elements | Spatial resolution | Sample quality |
|---|---|---|---|
| AAS | Li–U | | Solution |
| ICP-OES | Li–U | | Solution, slurry |
| ICP-MS | Li–U | | Solution |
| LA-ICP-MS | Li–U | µm | Solids |
| OES | Li–U | | Solid (electrically conductive) |
| GDS | Li–U | nm–µm | Solids |
| WD-XRF | Li–U | µm–mm | Solids, liquids, powder |
| ED-XRF | Na–U | µm–mm | Solids, liquids, powder |
| Micro-XRF | Na–U | µm–mm | Solids, liquids, powder |
| EPMA | Be–U | µm | Solid (electrically conductive) |
| PIXE | C–U | nm–µm | Solids |
| SIMS | Li–U | nm | Solids |
| AES | B–U | nm | Solids |

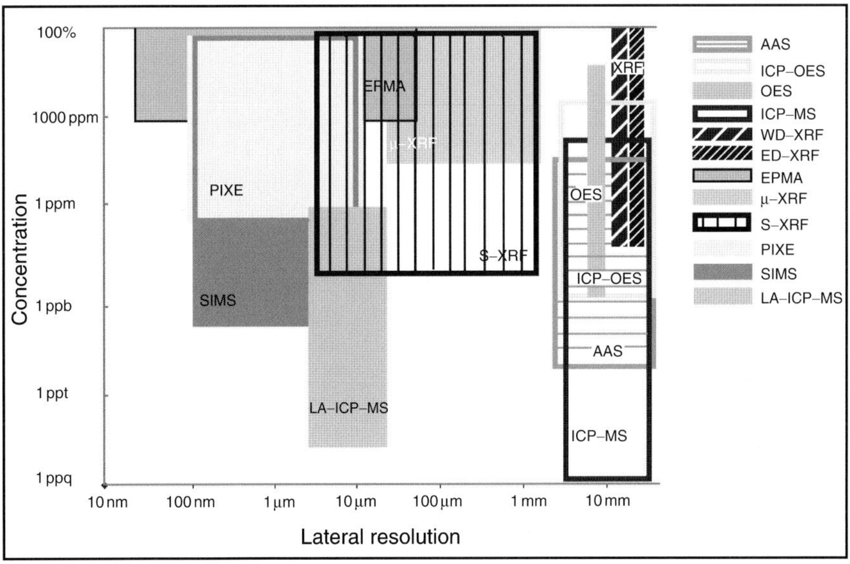


**Fig. 7.1.** Analytical methods and their spatial resolution and concentration range

it can be seen that micro-XRF already covers a relatively large concentration range together with an acceptable range of spatial resolution. Hence, it can be expected that this method will find a wide range of applications in the near future.

With the availability of compact X-ray optics, laboratory instruments were made possible for micro-XRF [6–10]. All these instruments have a very similar configuration–an X-ray tube as source irradiates the sample and an energy dispersive detector system performs the spectrum acquisition. X-ray optical components can be found either between source and sample or between sample and detector. In the first case, a large solid angle of the X-ray tube radiation is captured and concentrated by the X-ray optic onto a small area of the sample. If the optic is placed between sample and detector a large area of the sample is irradiated but the fluorescence radiation is collected by the detector from only a small area of the sample. In both cases, the use of an X-ray optic defines a small sample volume analysed. In the case of using an X-ray optic between "source—sample" and "sample—detector" a confocal set-up is established which enables depth-sensitive investigations. This will be described in the last part of this chapter.

The concentration of the source radiation onto a small sample area can be achieved very effectively by using a low power micro-focus X-ray tube. On the other hand when the X-ray optic is placed between sample and detector a high power X-ray tube is necessary in order to get sufficient fluorescence intensity from a small sample area [11].

In general, for the concentration of the source radiation onto the sample capillary optics (i.e. mono-capillaries or polycapillary lenses) are used. Also for collecting the fluorescence radiation from a selected part of the sample, capillary optics are the best choice. However, in special cases it could be advantageous to use mirrors for collecting the fluorescence radiation. The peak-to-background ratio might be improved which enhances the sensitivity significantly.

It should be noted that all the X-ray optics have an energy dependent transmission function and will, therefore, influence the spectrum finally measured. For BRAGG-optics it is obvious that only the BRAGG-reflected radiation can be used. Placing a BRAGG-optic between source and sample a quasi-monochromatic excitation is produced which might be of interest for special applications. An example is the micro-XRF laboratory unit of XOS/Horiba which is especially designed for the analysis of sulphur in oil. It excites the S fluorescence radiation with a high efficiency but produces a low background yielding very low limits of detection [12].

Total reflection optics deliver a broader energy bandwidth but the transmitted intensity is limited by the critical angle for total reflection. A good compromise is possible by combining BRAGG-reflection and diffuse-scattering properties with a suitable material such as highly oriented pyrolytic graphite (HOPG). Herewith, BRAGG-reflected radiation of higher intensity and the diffuse scattered Bremsstrahlung of lower intensity can be used for excitation [13, 14]. As already mentioned, capillary optics influence the energy distribution of the transmitted spectrum, that is, the transmission is a function of energy. The reason is the energy dependence of the critical angle for total reflection within the capillaries and the number of reflections in the capillaries [15]. These parameters are influenced by a variety of capillary properties like inner diameter of the capillary, material, curvature, surface roughness, etc.

The transmission function varies from capillary optic to capillary optic and, additionally, depends on the alignment of the optic. This influence has to be considered in the case of quantification (especially for fundamental parameter-based determinations) because changes in the energy distribution of the excitation spectrum will influence the excitation efficiency of the different elements in a sample.

There is another important parameter that has to be considered for micro-beam analysis–the spot size and the spectral and spatial distribution of the exciting beam. If a collimator is used, the spot size is typically larger than 50–100 μm (FWHM), otherwise the transmitted radiation from an X-ray tube is too small. The spot size produced by a collimator is given by the diameter of the aperture and the distance from the source to the sample and the spot size is constant for all energies.

For capillary optics, the spot size can be as small as a few μm for shaped mono-capillaries and around 10 μm for polycapillary lenses. In contrast to collimators the spot size depends on the photon energy. The lower the photon energy the larger the spot size. Therefore, the intensity distribution in the spot is energy-dependent, as well. Furthermore, for very high energies it occurs that the tube radiation penetrates the capillary without being concentrated. In this case, a halo effect can be observed that enlarges the spot size again.

Measurements of the spectral and spatial intensity distribution are scarcely available. Here, more extended examinations are necessary.

The energy dispersive detector mostly used in bench-top micro-XRF instruments is a Si(Li) detector, but also LN-free detectors like PIN-diodes or drift chamber detectors (SDD) are in use [see Section 4.2]. Conventional Si(Li)-detectors have the advantage that they absorb high-energy radiation more efficiently due to their larger crystal thickness. However, the demand for continuous cooling of these detectors limits their applicability. The most restricting property of LN-free detectors is their small crystal thickness, which is at present typically 0.5 mm for PIN-diodes and 0.3 mm for SDDs. This means that radiation with energy up to around 12 keV can be detected with satisfying efficiency, while radiation with higher energy is only partly absorbed by the detector. PIN-diodes and SDDs are not completely vacuum-sealed but are enclosed in a reduced $N_2$ atmosphere. Therefore, an additional absorption is introduced which is important for low energy radiation detection. The question of detector efficiency is discussed in more detail in Section 4.2.

The pulse throughput performance also varies for different detectors. The throughput is influenced by the charge mobility in the detector and the counting electronics. For PIN-diodes the pulse rate limit is caused by the rise time of the charge pulse. For Si(Li) detectors the thickness limits the collection time, even though SDDs reach the highest pulse load capacity which is at present possible. Pulse rates can exceed 100 000 counts per seconds with a dead time in the range of 30% and less (SDD). For an SDD a wide range of pulse rates can be covered with only a few shaping times. This in turn simplifies the handling of the pulse processor.

At present most micro-XRF units are used as laboratory equipment in research laboratories [15, 16]. They serve in a wide range of interesting applications and also for methodological development.

There are also several commercial instruments on the market which can be used for micro-XRF spectrometry. Some of them still employ pinhole apertures for the creation of an excitation spot in the micrometer regime. The distance of the X-ray tube anode to the sample has to be very short in order to get sufficient excitation intensity and to produce a small spot size. Most of the instruments, however, are using X-ray optics for the concentration of the tube radiation onto the sample which leads to an improved analytical performance.

An overview of the instruments present in the market is given in Table 7.2. In addition to bench-top laboratory units there are also portable (or at least mobile) instruments available with which an onsite analysis can be performed. This feature becomes important if the samples to be investigated are very valuable, that is, if it is not possible to move the sample for the analysis to the laboratory-especially if these samples are very large like paintings, art objects, etc. In these instruments, the measuring unit is mounted on a special support and can be moved in different positions. Because these instruments are only partially shielded they demand special precautions with regard to radiation safety.

**Table 7.2.** Summary of commercial micro XRF instruments

| Parameter | KEVEX OMICRON | HORIBA XGT 5000 | COX ITRAX 5$\mu$ | EDAX EAGLE $\mu$Probe | Seiko SFT 5100/5120 |
|---|---|---|---|---|---|
| tube | microfocus | microfocus | microfocus | microfocus | microfocus |
| target | Rh, Mo, W | Rh | Ti,Cr,Cu,Mo,Ag | Rh, Mo, Ag, Cr | Mo |
| voltage | 50 kV | 50 kV | 60 kV | up to 50 kV | Fix 15,45,50kV |
| power | 50 W | 50 W | 3 kW | 40 W | |
| X-ray optics | collimator diff.sizes | mono-capillary | mono-capillary | capillary | 2 collimators |
| spot size | Ø30–3000 µm | 10 µm (tapered) 100 µm (cylind) | 10 µm (tapered) camera (cylind) | 300 µm (mono) 40 µm (lens) | Ø25 µm, Ø100 µm Ø100x1800 µm |
| detector type | Si(Li) | HP-Si | Si(Li) | Si(Li) | Si(Li) |
| area [mm$^2$] | 30 | | 10 | 80 mm$^2$, 150 eV | |
| ener. resol. [eV] | | 145 eV | 135 eV | 30 mm$^2$, 135 eV | |
| working-distance | ≈ 10 mm | ≈ 4 mm | ≈ 2 mm | 22 mm | 5/15 mm |
| transmission-Det. | | scint. counter | | | |
| stage movement [mm] | | 100 × 100 × 40 or 350 × 400 × 40 | Up to 300 × 400 × 100 | 100 × 100 × 100 or 300 × 300 × 400 | 75 × 75 × 35 or 220 × 150 × 40 |
| positioning | stepping motor | DC motor | stepping motor | stepping motor | manual |
| step size | | 2 µm | 2 µm | 5 µm,(rotation opt) | |
| vacuum | evacuable | detection path | detection path | Evacuable | evacuable |
| sample view | zoom-optics | 2 video-cameras | video-camera no meas-position | 2 video-cameras | video-camera |
| magnification | 40–200 | 2–25, 100 opt | 50 | 10, 100 or 200 | 20 |
| colour | colour | colour | b/w | colour | colour |
| excitation direction | from bottom | from top | from side | from top | from top |
| focussing | fixed by stage | no autofocus | no autofocus | autofocus | video system |
| quantification models | | FP w/o standard | | FP w/o standard | FP w/o standard |
| bulk sample | w/ standard | semi-empirical | w/ standard | semi-empirical | semi-empirical |
| layer sample | w/ standard | no | no | w/o standard | w/ standard |
| Homepage | | http://www.horibalab.com/Products/xgt5000.html | http://www.coxsys.se/XRF_microscope.htm | http://www.edax.com/products/XRFmicro/index.html | http://www.siint.com/en/products/xrf_analysis.html |

**Table 7.2.** *Continued*

| Parameter | Veeco MXR | Jeol JSX-3600 | Shimadzu µEDX 1200/1400 | SPECTRO Midex and Midex M |
|---|---|---|---|---|
| tube | microfocus | microfocus | microfocus | side window |
| target | Rh, | Mo | Rh | Mo |
| voltage | up to 50 kV | up to 50 kV | up to 50 kV | Up to 45 kV |
| power | 50 W | 75 W | 50 W | 30 W |
| X-ray optics | collimator, diff.sizes | collimator, diff.sizes | poly-capillary lens | collimators of different sizes |
| spot size | poly-cap 50 µm | 50,120,500 µm | 50 µm | 0.1, 0.3, 0.5, 1mm x 1.6 due to divergency |
| detector type | Prop-Counter, | Si(Li) | Si(Li), 30 $mm^2$ | SDD |
| area [$mm^2$] | 1000 $mm^2$ | | 150 eV | |
| ener. resol. [eV] | 1 keV (PIN, 10 $mm^2$, 180 eV Si(Li),30 $mm^2$, 140 eV)* | 165 eV | (SDD, 5,135 eV)* | 5 $mm^2$ 150 eV |
| working-distance transmission-Det. | | | 6.5/1.5 mm | 5 mm |
| stage | | 75 × 75 × 50 | 100 × 100 × 40 | 200 × 150 × 50 250 × 250 × 150 (M) |
| movement [mm] positioning | stepping motor | motor | motor | motor |
| step size | | | | 2.5 µm |
| vacuum | detection path | evacuable | no | no, He flush |
| sample view | video-camera | zoom optics | 2 video-cameras | 2 video-cameras |
| magnification | 30 or 90 | 40–120 | | 2, 100 |
| colour | colour | colour | colour | colour |
| excitation direction | from top | from top | from top | from top |
| focussing | | autofocus | autofocus | no autofocus |
| quantification models | FP w/o standard | FP w/o standard | FP w/o standard | FP w/o standard |
| bulk sample | no | semi-empirical | semi-empirical | |
| layer sample | w/ standard | w/ standard | w/ standard | no |
| Homepage | http://www.thermo.com/com/cda/product/detail/1,1055,10120344,00.html | http://www.jeol.co.jp/english/products/product/jsx-3600/ | http://www.kratos.com/XRF/muEDX.html | http://www.spectro-ai.com/pages/e/p010511.htm |

* These instruments are available with different detectors

There are two instruments commercially available (see Table 7.3) but some other prototypes have been developed, manufactured and used for selected applications in art analysis [17, 18].

All these instruments employ X-ray optics between the tube and the sample in order to collimate or concentrate the tube radiation onto the sample. This type of excitation is also possible in a scanning electron microscope (SEM). An external low-power X-ray tube combined with an X-ray optic excites the sample inside the electron microscope. If the SEM is equipped with an EPMA-detector it can be used for the detection of fluorescence radiation. The arrangement has the advantage of a low spectral background, that is, high sensitivity for trace elements. The upgrade of an SEM is offered by two companies [19, 20]. In both cases, low power X-ray tubes are used in connection with capillary optics. The measuring position has to be defined with the help of the SEM. The area analysed with the X-ray beam is larger than that analysed with the electron microscope.

Despite the fact that micro-XRF is a relative new analytical method it has already found a number of different, interesting applications. A short summary is given in Table 7.4.

**Table 7.3.** Summary of commercial portable micro-XRF instruments

| Parameter | Jordan Valley EX3600/6600LS-OA | Manufacturer Röntec - Bruker AXS Microanalysis GmbH |
|---|---|---|
| tube | microfocus | microfocus |
| target | Rh, Mo, W | Rh, W, Mo, other targets on request |
| voltage | 50 kV | 50 kV |
| power | 50 W | 50 W |
| X-ray optics | collimator | poly-capillary lens |
| Spot size | Ø30 – 3000 µm | 300 µm |
| detector type | Si(Li) LN2 | SDD |
| area [$mm^2$] | 30 $mm^2$ | 5 $mm^2$ |
| energy resolution [eV] | 145 eV | 145 eV |
| working distance | ≈10 mm | ≈4 mm |
| elemental range | Al – U | Al – U |
| sample positioning | stepping motor | coarse: manually fine: stepping motor |
| sample view | zoom-optics | 2 video-cameras |
| magnification | 40–200 | 2–25, 100 (opt) |
| colour | Colour | Colour |
| excitation direction | Horizontal | flexible |
| focussing | crossed laser | crossed laser |
| quantification | FP w/o standard | semi-empirical |
| model (bulk) | semi-empirical | |
| Homepage | www.jordanvalley-apd.com | www.bruker-axs-microanalysis.de |

**Table 7.4.** Applications fields of micro-XRF bench-top units

| Applications | |
|---|---|
| forensic | particles, gun shot residue, pigments, money (paper, ink), drugs, |
| archeometry | pigments, alloys, paintings (lacquer) |
| research | material research, photosensitive films, ceramics, |
| quality control | medical devices (composition of filigree parts), pharmaceutical (contaminations), solar cells (layer structure), microelectronics (thin layers, contaminations), plastics (filler), wear debris |
| failure analysis | contaminations on small sample areas, particles |
| geology | distribution of elements in minerals or sediment cores |
| biomedical, medical | distribution of heavy elements in bones, tissues etc. |

In all instruments the sample is fixed on a stage that moves the sample into the measuring position. The sample stage is motor driven with a step size of a few micrometers (typically 2–10 μm). This is sufficient with respect to the typical spot size diameter of 20–300 μm. The step size should be 6–10 times smaller than the spot size diameter in order to get a precise sample positioning. Depending upon the instrument, it is possible to use the stage for positioning of a single measuring point and also for running a scan with a complex pattern of measuring points. Repeated measurements with the same pattern on different samples become possible. This pattern can also be used for distribution analysis—either on a line or over an area.

In contrast to the capabilities of an electron microscope, with a micro-XRF spectrometer it is not possible to produce an "image" of the sample with the exciting beam. Therefore, additional instrumental components are necessary for sample observation. Typically video-microscopes serve this purpose. The optical magnification of the video system has to be variable in order to see specific details of the sample–from a general low magnification overview in order to quickly locate the desired measuring position up to a high magnification for close-up details. The variation in optical magnification has to cover the range from $\approx 2$ up to $\approx 200$. This is achieved either with zoom-optics or with several video-camera systems of different magnifications.

Another important component of micro-XRF spectrometers is the sample chamber itself. The size and the possibility for evacuation are main characteristics of the chamber. The sample chamber dimensions determine the maximum size of the measurable specimen. In the case of the "open" portable/mobile instrument there is no such restriction. The possibility to evacuate the sample chamber determines the lightest element that can be analysed; for an evacuated sample chamber this is typically Na. Even though it is possible to detect fluorescence radiation from lighter elements, usually it cannot be used for quantitative analysis. For instruments without an evacuable sample chamber the lightest measurable element depends on the "sample–detector" distance. A He-flush may be used to improve the sensitivity for light elements.

Of course, one of the main requirements for a microanalytical method is a quantification procedure. Most of the instruments possess quantification models for both bulk materials and coating systems. Quantification procedures including the additional measurements of reference standards or standard free procedures are available. In particular for microanalysis it is difficult to use semi-empirical calibration models with reference standards because the sample is often inhomogeneous. Hence, the sample composition may change from measuring position to measuring position. It would imply that many different calibrations are being prepared with the same set of reference standards. Therefore, a standardless model has a higher flexibility, albeit it yields results of a slightly lower accuracy.

Spectrometers employing X-ray optics between sample and detector exist up to now only as set-ups at synchrotron sources [11]. The irradiation of the sample is carried out with synchrotron radiation in order to get a high excitation intensity. As collection optics a polycapillary half lens is used having a large solid angle for collecting the fluorescence radiation and producing a parallel beam impinging onto the detector; or a monolithic polycapillary cone (poly CCC)(see chap. 3.2.2) is used for the fluorescence collection which has a wider solid angle of detection and gives a slightly better spatial resolution. In dependence of the collection optics, spatial resolution of a few micrometers is possible.

### 7.1.3 Applications of Micro X-Ray Fluorescence Analysis

#### Heading Comparison of the Analytical Performance of Micro-XRF with other Analytical Methods

The small area investigated in micro-XRF opens the possibility for a wide variety of applications. It is possible to examine a single small sample such as an individual particle or an inclusion in a relatively homogeneous material as well as a set of several measuring points along a line or over an area (i.e. matrix). In that way the distribution of elements along a line or an area can be determined as a line scan or as a mapping, respectively. Even though a small analyte is examined it may also be used to determine analytical parameters such as the composition for "bulk" material or the thickness and/or the composition of a coating layer on a sample.

Some of these applications are also possible using electron-beam microanalysis (EPMA) but there are some important differences between these two analytical methods. These differences apply to

1. sample preparation and
2. analytical performance.

Regarding *sample preparation*, with X-ray excitation the sample does not have to be electrically conductive, that is, non-metals like minerals, glasses, plastics etc. can be measured without the need of an electrically conductive coating. This simplifies and shortens the effort for the analysis.

Another difference is that there is no need of a vacuum for X-ray excitation measurements. If the working distance between sample and detector is small, it is possible also to measure light elements like chlorine, sulphur, phosphor or silicon with sufficient efficiency. Thus the sample itself will not be modified by the vacuum, for example, wet samples like pastes or biological samples can be analysed. If the analysis of light elements is necessary with a high sensitivity, it is possible to evacuate the sample chamber to reduce the absorption of the fluorescent radiation from the sample. In this case only a low vacuum is necessary, that is, not better than 0.1 Torr. Such a vacuum can be achieved within relatively short times with roughing vacuum pumps.

The energy absorbed by the sample is much smaller in the case of X-ray excitation compared to electron excitation and the absorption occurs within a larger volume of the sample (not just in a very thin surface layer as for electrons). Very sensitive samples can be analysed with high excitation intensity risking minimal or no damage at all during the measurement period. The possibility to measure in air together with the negligible heating of the sample is valuable not only for organic samples or thin foils but also for minerals if they contain, for example, bonded water. For such samples, the heating induced by electron excitation may reduce the sample volume by evaporation of the water during the measurement. The same sample can be measured for a much longer time without any damaging effects when X-ray beams are used for excitation. Remember that in micro-XRF systems the X-ray source is provided by a low power X-ray tube.

There are also differences in the *analytical performance* between EPMA and micro-XRF with advantages and disadvantages for both methods. EPMA has advantages regarding spatial resolution because the electron beam can be focussed to a much smaller spot size than the X-ray beam. The smallest spot size for SEMs is in the nm-range whereas for micro-XRF bench top units the limit is in the range of around 10 μm.

In addition, with EPMA the analysis of light elements down to Boron is possible. This is due to the different excitation modes. With an electron beam instrument it is quite straightforward to reduce the excitation voltage to be close to the absorption edge of the element of interest enhancing the excitation probability. Furthermore, the cross sections for electron and X-ray excitation differ by more than two orders of magnitude for light elements rendering the excitation by electrons more efficient at these energies. Additionally, electrons are absorbed in the upper surface of the sample which is at the same time the part of the sample that contributes mainly to the fluorescence radiation of light elements. Fluorescence radiation that is excited in deeper layers will be absorbed in the sample itself. In the case of X-ray excitation the absorption of the primary beam also occurs in deeper layers but the fluorescence radiation of light elements does not reach the sample surface.

**Table 7.5.** Typical limits of detection for micro-XRF and EPMA for a selection of elements in ppm

| Element | Micro-XRF | EPMA |
|---|---|---|
| Ti | 100 | 1000 |
| Cr | 80 | 800 |
| Mn | 50 | 800 |
| Fe | 40 | 800 |
| Ni | 30 | 900 |
| Cu | 20 | 1000 |
| Mo | 200 | 2000 |
| Sn | 300 | 4000 |
| Pb | 200 | 5000 |

On the other hand, the fact that X-rays penetrate the sample deeper may also be advantageous for certain applications. For the analysis of bulk materials a deeper penetration of the exciting beam implies that a larger representative sample volume will contribute to the detected signal. In the case of coated materials, it is possible to measure thicker coatings or multiple coatings. Another important advantage is the significantly higher sensitivity for trace element analysis. This is a result of the better peak-to-background ratio produced by X-ray excitation. With electron excitation both fluorescence and Bremsstrahlung radiation is present in the spectrum. The continuous Bremsstrahlung is superimposed upon the characteristic radiation of the sample's analyte elements as an unwanted background noise and, thus, reducing the sensitivity. Excitation with X-rays does not produce this Bremsstrahlung background and, therefore, the sensitivity for trace elements is higher by a factor of 10–50 (see Table 7.5).

The same or, for special excitation conditions, even a higher sensitivity is achieved with conventional EDXRF but these are designed to analyse large sample areas, only. Here, the sample should be homogenous for reliable quantitative measurements. By comparison, micro-XRF is designed to investigate and analyse the inhomogeneities of materials which typically represent the structure of many real samples.

### Identification of Inclusions in Homogeneous Material

Homogeneous materials may contain inclusions originating, for example, from environmental contaminations or from a production process. Such inhomogeneities (inclusions) can adversely influence the properties of the material and prohibit its intended use. Examples are plastic films covering a photographic paper; technical glasses such as monitor screens and optical lenses; the plastic body of a compact disk. In each case, the presence of an inhomogeneity or an inclusion will influence the intended performance. An inhomogeneity in the photographic film can be seen on the picture and influence the quality of the image; an inhomogeneity in a technical glass, for example, a TV-screen, may disturb the image quality of the TV; an inclusion in the plastic body of a compact disk influences the transmission and reflection of the light in the CD and it may disturb the read-out process.

Identification of such inclusions will identify their source and enable appropriate remedial action to be taken such as using cleaner production environment (clean room quality) or by renewing the production equipment in case of wear debris contamination (harder materials, new coatings in case of abraded coatings).

## Inhomogeneities in a CD

The plastic body of a compact disk has to be very clean because every inhomogeneity influences the penetration and reflection of the laser light on the metallic mirror. Therefore, a complete control of the CD's quality is necessary. Any observed inclusion has to be analysed quickly and its origin has to be identified. Fig. 7.2a and b show different inclusions in a CD. Their different

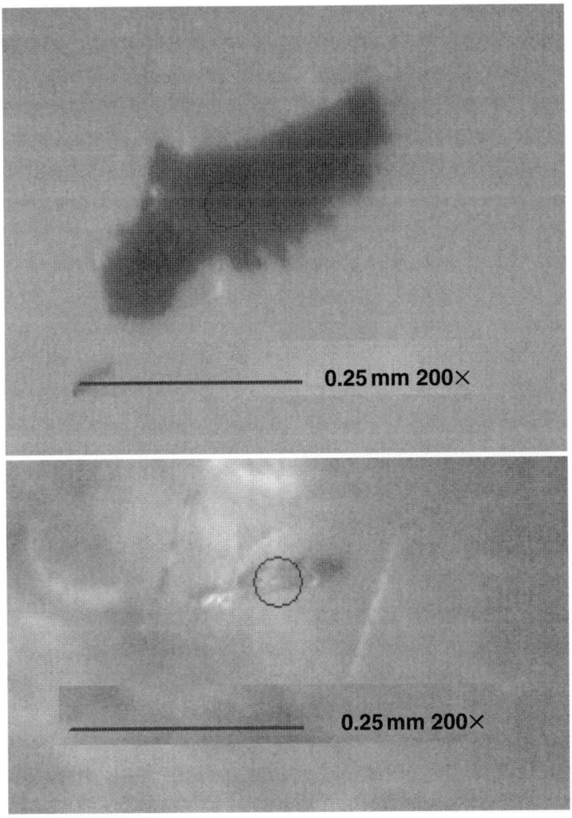


**Fig. 7.2.** Image of an inclusion in a CD of **(a)** environmental origin; **(b)** production line origins

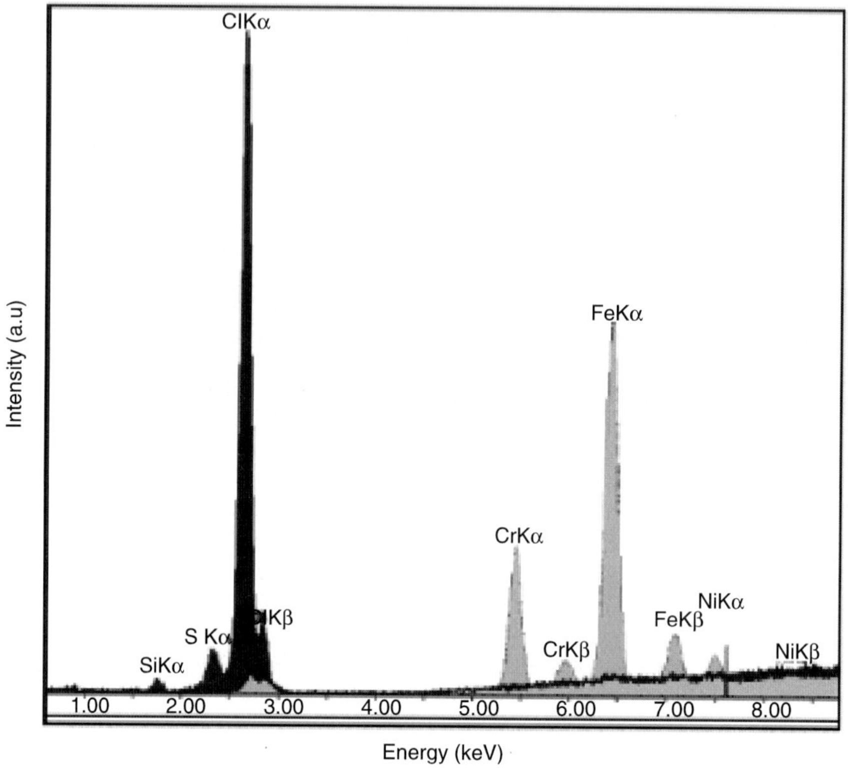


**Fig. 7.3.** Spectra of different particles of inclusion in a CD shown in Fig. 7.2. Dark spectrum – Fig. 7.2a; light spectrum – Fig. 7.2b. Measuring conditions: 40 kV, 150 µA, 60 s, 50 µm spot size

origin can be seen from the spectra measured in the displayed spot of these images of the disk. The spectra are displayed in Fig. 7.3. In the first case (Fig. 7.2a, dark spectrum in Fig. 7.3) Si, S and Cl can be identified. These are elements that are characteristically found in the environment. Hence, it is necessary to improve the overall cleanliness or protection of the production environment.

The other inclusion (Fig. 7.2b, light spectrum in Fig. 7.3) shows very clearly the elements Cr, Fe, and Ni with intensity ratios indicating stainless steel as the material of origin. This particle was an abrasion particle from the production equipment. In this case it had to be checked if the abrasion is too high in some places of the equipment or if it would be necessary to coat them with a hard material like TiN. Such measurements can be performed fast and without extensive sample preparation. The results provide information about the situation of a production line and identify the cause of manufacturing faults. With EPMA this kind of measurements would not have been possible in such an easy way, because the sample is not electrically conductive and the inclusions are within the plastic body. The sample would have to be polished or sectioned to bring the inclusion to the surface in order to be analysed and a conductive coating must be applied.

## Single Particle Analysis in Wear Debris

The identification of single particles of different kinds in a set of particles is another example for particle analysis. This analytical task arises for wear debris. The identification of individual particles provides information about its particular source that can be helpful for specific maintenance requirements of big engines from planes or ships. The result of small particle analysis may be influenced by their shape and size. As these geometric parameters can vary in a wide range, it is necessary to use special identification strategies.

If the particle size is smaller than the depth from which the fluorescence radiation is also originating, the measured intensity depends both on concentration and on sample size (thickness). Then, the shape of the sample has to be taken into account, as well. In a first approximation the shape of the sample can be derived from an optical image which demands an image processing stage in addition to the XRF analysis.

Another alternative is the use of internal standards for the different samples [21] or the use of chemometric methods for particle identification [22]. As an internal standard a main component of the sample may be used for normalisation. Assuming a fixed concentration of that main component for all samples, a normalisation procedure will reduce the influence of the sample geometry. However, such an approach is valid only in cases for which the internal standard and the element of interest have similar matrix absorption parameters. This is applicable only in situations for which the atomic numbers of these elements are not too different.

Another possibility is the use of chemometric methods. A simple procedure is spectra matching for which the intensities of identified elements of an unknown sample are compared to the identified elements of a set of reference materials. The variation coefficient is calculated resulting in a ranking of the probability of agreement of the unknown with the reference sample. This is a fast and easy way to handle method for material identification without a high analytical effort.

Another chemometric method in use for material identification is principal component analysis (PCA). Also for that method it is necessary to prepare a substantial library of particle spectra but for various compositions and also for various sizes and shapes. It requires considerable effort to "train" the instrument but the identification of the particle's composition and size will, then, be possible. In PCA the complete spectrum measured is evaluated. Different components of the spectrum are used for the identification—the peak intensities and the intensity of the scattered primary radiation. This facilitates the identification of different materials. Very often it is not necessary to know the exact composition of a material but it is sufficient to identify its type. The identification of a material can be carried out with a higher confidence level than the analysis of the whole composition.

PCA has been demonstrated for glass particles of various compositions and sizes [23]. Reference glasses were first measured as bulk material. Then the glasses were broken and the different pieces were sieved and classified according to their sizes. Measurements were performed on a wide range of glass particles of different sizes. Glass was used because of its relatively small matrix absorption. This, in turn, enables a significant influence of particle size and shape on the fluorescence intensity. Conventional analysis was not successful in this case. The evaluation of the spectra correcting for particle size resulted in a wide distribution of the composition for particles with the same size classification. This has been probably caused by the sample shape.

However, the PCA plots displayed in Fig. 7.4a show that particles of the same composition but of different size are concentrated in clusters and in Fig. 7.4b that particle of the same composition but different size are arranged in a line according to the particle size with different lines for different compositions. The scattering of the values is relatively large but definite identification was possible.

This example shows that especially for micro-XRF spectroscopy it is useful to consider new and alternative ways for quantification or identification. Using comparative methods, a higher effort for the "quantification" is necessary. A lot of reference materials of known composition and size need to be measured. Also some test analysis are necessary to define the best "principal components" for the different specimen parameters. After that the whole procedure, the identification of the unknown material, is very fast and has sufficient reliability.

### Distribution Analysis

If there is a possibility to analyse a small sample area then it is also possible to measure a sequence of small sample areas. These measuring points can be arranged in different patterns such as a line or a matrix (array). A distribution analysis can be performed as line scan or mapping. This offers a new dimension of information for material analysis but it does require appropriate automatic collection and data presentation procedures.

### Multiple Point Analysis

It should be possible to arrange a pattern of measuring points on one or various samples that are placed in a sample holder. In this way, automatic measurements on large sample arrays are feasible and the sample holder may be used like a sample magazine. An interesting example is the use of micro-XRF for *combinatorial chemistry* [24]. This is a way to develop new materials such as catalysts, high temperature super conductors or magnetic materials,

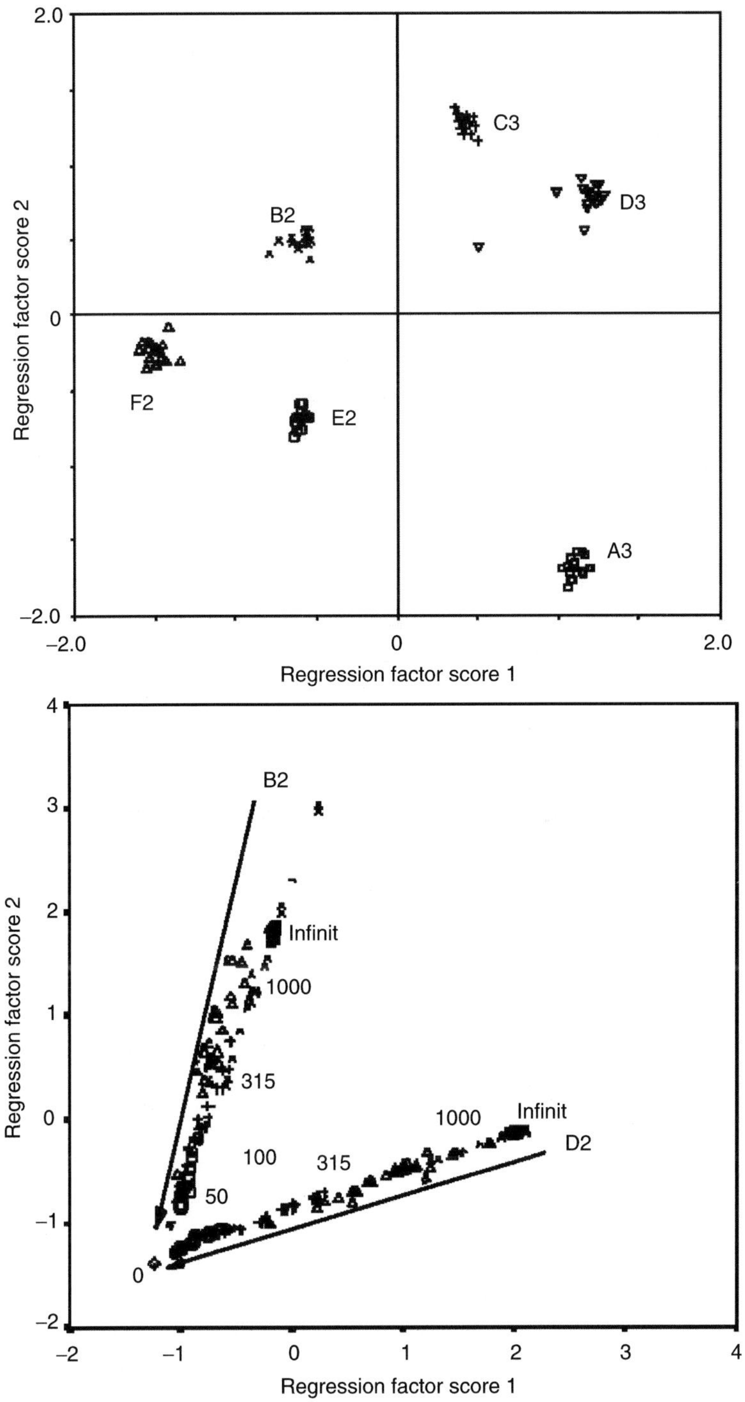


**Fig. 7.4.** Plot of main components **(a)** for six reference glasses; **(b)** for two different particle composition of various sizes

that is, materials where the relation between material composition and sample treatment and the intended property is still not completely understood. Often a "trial and error" method for material development is necessary. Conventionally, this will be done sequentially, that is, a new material will be generated, characterised and tested for a selected property. The procedure is iteratively repeated until the optimum performance is achieved. The procedure can be accelerated drastically if complete material banks with varying compositions are prepared instead of one composition alone. These materials may then be analysed and tested in parallel which minimises the time needed. The material bank is arranged in an array that can also be used as a sample magazine (as displayed in Fig. 7.5). However, the quantity for each specimen has to be reduced, that is, only a small volume of each sample is available but the number of samples is enhanced.

These samples are analysed, either for their composition or homogeneity in order to assess the "correct" sample preparation. For the determination of the inorganic composition, micro-XRF is a very useful tool. The required amount of the sample for an analysis is small and the high sample number conventionally arranged in arrays can be measured automatically with the appropriate stage programming. The results are stored in a database system that rapidly allows access to and further treatment of the data.

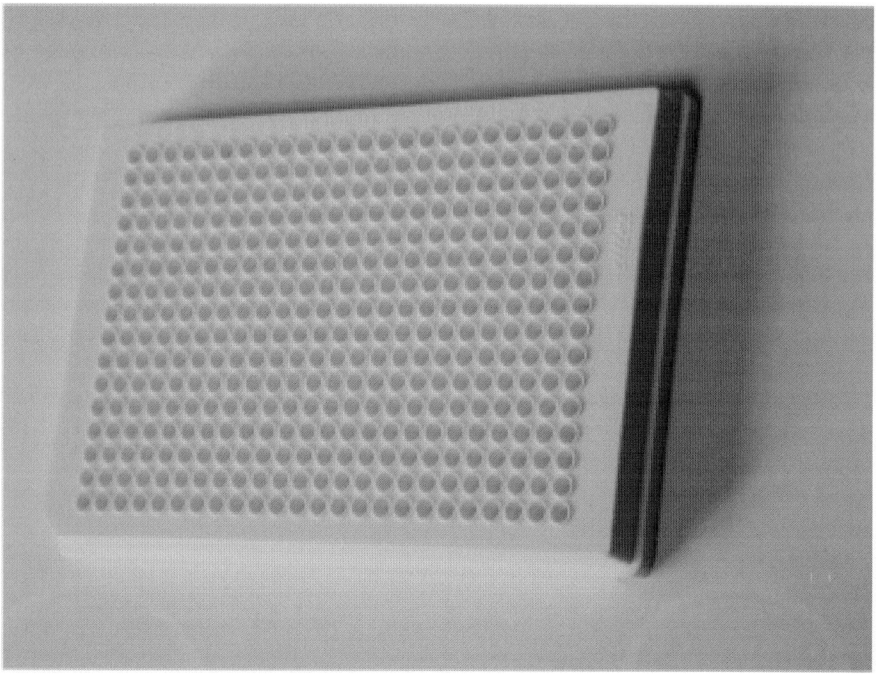

**Fig. 7.5.** Sample magazine for combinatorial chemistry

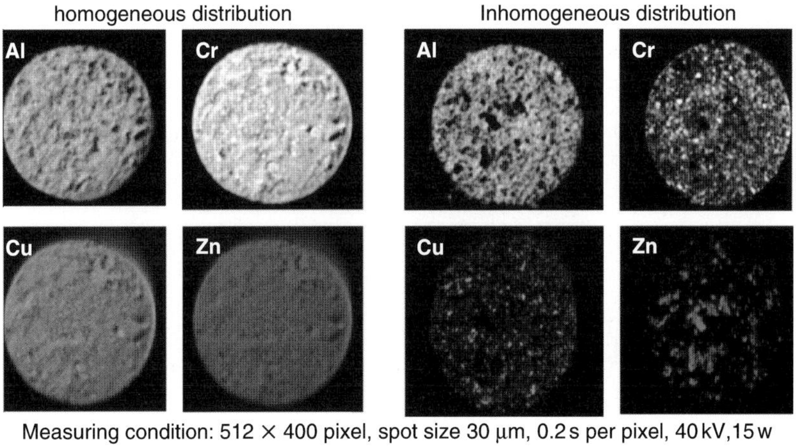


**Fig. 7.6.** Homogeneous and inhomogeneous elemental distributions of elements in a sample

Fig. 7.6 shows the result of a homogeneity test for the elemental distribution of a catalyst. The two distributions are very different; one preparation method yields homogeneous samples whereas the other method yields inhomogeneous ones. It is obvious that the appropriate preparation method needed to produce the desired homogeneous samples for any subsequent test preparation has been found rapidly.

## Line Scan

The arrangement of measuring points along a line provides the basis for line scans. The distribution of different elements is analysed along any line "drawn" on the sample surface. The size of the area analysed and the step size of the measurement interval determine the spatial resolution.

Such types of measurements provide linear elemental distribution information. Two examples will be presented here. Other interesting applications include the determination and analysis of a sequence of lacquer layers in a paint layer, the measurement of special elements over a distance after a heat treatment in order to determine diffusion coefficients and, the distribution of elements in tree rings or in sediment layers.

*Distribution of Phosphorus in a Roll of a Roller Bearing*

In a used roll of a roller bearing it occurs that there are different colours on the surface. The question is where do these colours come from and how that influences the function of the rolls. A line scan on the roll surface shows that the concentration of phosphorus is clearly correlated to the intensity of the colour as displayed in Fig. 7.7. The measurement was performed at 40 kV, 200 μA with a spot size of 300 μm and a step size of 300 μm.

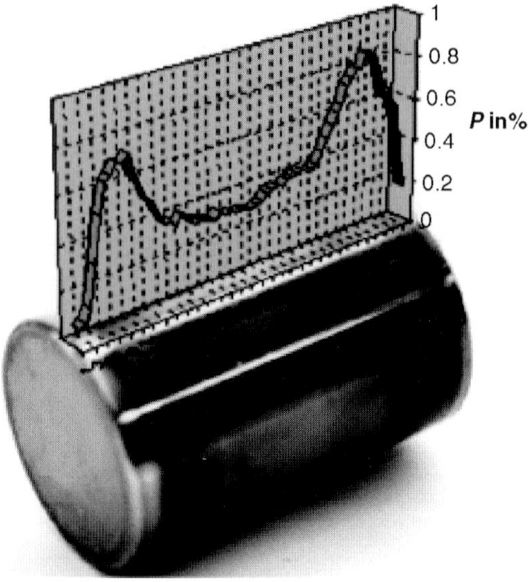


**Fig. 7.7.** Distribution of phosphorus on a roll of a roller bearing

Phosphorus was a part of the grease formulation used in this roller bearing. During operation, the rolls undergo local heating, in particular in contact regions. At these positions a diffusion of phosphorus into the steel is possible. The intensity of phosphorus, therefore, is a projection of the heating of the roll and points to friction during running in the bearing. It can be concluded from this measurement that the shape of the rolls should be changed to be slightly convex to have a contact over the complete surface of the roll and reduce abrasion and friction in this way.

*Elemental Distribution in a Sediment Bore Core*

Sediments in the ocean or in deep lakes give valuable information on the geological or climatological history over a long period of time. Their layers are undisturbed and, thus, provide a picture of geological and/or pollution changes contained within the sedimentation deposited in the past.

For that reason sediment bore cores are collected from different locations of the ocean bed but also from special lakes that are very deep or are in special geologically interesting areas. A lot of these bore cores are collected in the frame of the ocean drilling program (ODP).

The analysis of these bore cores requires a lot of instrumental effort (a large number of collected data points) because a high spatial resolution is mandatory in order to obtain the corresponding high time resolution for the climatological changes. The conventional analytical way is to cut the bore cores into thin layers, dissolve the different cuts and carry out a wet chemical analysis.

These methods are destructive and damage the bore cores which are no longer available for further examination. With micro-XRF a non-destructive analysis can be performed and even a higher spatial/time resolution is obtained.

Results are presented from an investigation that was performed for a bore core of the Cariaco basin of the southern Caribbean Sea drilled to a depth of 893 m [25]. Investigation of this bore core may contribute to an explanation for the collapse of the Maya civilization.

In the Cariaco basin there is a rapid sedimental deposition of approximately 30 cm per 1000 years or 0.3 mm per year. These organic-rich sediments are visibly laminated and have a very low disturbance. With a spatial resolution of 50 μm an equivalent of approximately six data points per year can be measured.

On the bore core the Ti-distribution was used as an indicator for the regional hydrological situation because this element was delivered as sediments by the rivers to the Cariaco basin. The measuring conditions were: spot and step size: 50 μm, 40 kV, 800 μA, 60 s per point. The dating of the layers were performed with the C14-method.

The spatial resolution corresponds to a time resolution of around two months. The annual signal of Ti can clearly be detected and interpreted as the change of rainfall over the year. All other measurements up to this time had a resolution of not less than 1–2 mm, which corresponds to a time resolution of approximately 2 years. The distribution of the Ti-content of the sediment is shown for a long period of time in Fig. 7.8 It shows annual changes that depend on the position of the Intercontinental Convergence Zone (ITCZ). The position of that zone changes between summer and winter time and determines the amount of rain. It can be explained by the fact that most of the Ti-content in the sediment is generated by a Ti-mineral that is brought to the sea by rivers only. The fluctuation of the Ti content is, therefore, related to the amount of the river water and to the amount of rain.

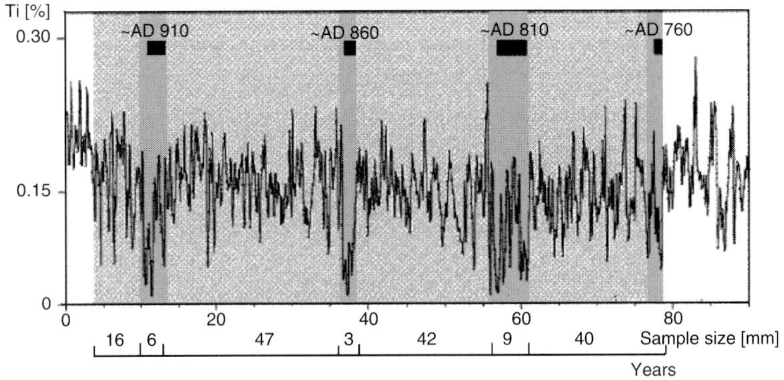


**Fig. 7.8.** Distribution of Ti-intensity in a bore core of the Cariaco basin [from 25]

From the distribution in Fig. 7.8 it may be concluded that in certain periods of time—especially around 910, 860, 810, 760 (highlighted in the darker grey), the Ti-content is significantly lower for a few years than in other periods. During these periods the amount of rain was reduced significantly.

The population of the Mayas on Yucatan was already relatively high by approximately 12 million people on a small area. A high agricultural productivity was needed to produce a sufficient food supply. If the amount of precipitation was drastically reduced for long periods of time, then it could also be expected that the agricultural productivity suffered. During such periods there was not enough food available to feed the population. The multiple step collapse of the Maya civilization can be explained in terms of these food problems.

This explanation is in agreement with other archaeological observations indicating that the Maya civilization collapsed in several steps, that there were no wars or other disastrous event that damaged the people.

*Mapping*

Mapping means that measurements are performed on a large matrix and the elemental distribution is displayed for an area. The combination of various elemental distributions may provide further information on material composition, for example, phase distributions.

*Investigation of Malignant Cells*

Malignant cells are very similar to healthy cells. Destroying them in a selective way is very difficult. One way of killing malignant cells is irradiation with ionising radiation but radiation cannot distinguish between healthy and malignant cells. By marking malignant cells they can be identified. The marking depends on the type of irradiation that is used.

Currently, a new method has been developed—the Neutron Capture Cancer Therapy (NCT) [26]. This method is a non-invasive therapy for malignant cells that is performed in two steps:

1. Injection of a substance that will preferentially adhere to cancerous cells. This substance contains an isotope with a large cross section for thermal neutrons producing, for example, $\alpha$-particles.
2. The patient will be irradiated with thermal neutrons. In subsequent nuclear reactions $\alpha$-particles are generated which destroy the surrounding malignant cells.

**Table 7.6.** Cross sections for thermal neutrons

| Element | $^{1}H$ | $^{12}C$ | $^{14}N$ | $^{16}O$ | $^{10}B$ | $^{157}Gd$ |
|---|---|---|---|---|---|---|
| cross section [barn] | 0.333 | 0.0035 | 1.83 | $2 \times 10^{-4}$ | 3.84 | $2.5 \times 10^{5}$ |

For NCT, different elements can be used. An important selection criterion is a significant higher cross section for thermal neutrons than that for the typical elements of the cells and a nuclear reaction must be possible. The cross sections of some typical elements are summarised in Table 7.6.

It can be seen that for Boron, but more significantly for Gadolinium, the cross sections are higher than for the typical elements of a cell. For Boron the nuclear reaction B(n,$\alpha$)Li generates $\alpha$-radiation. This radiation ($\alpha$-particles) will be absorbed in the immediate environment of the emission and, hence, preferably destroys the malignant cells. Compounds with Boron are already in use for therapy. Unfortunately, the difference in the cross section to ones for the other components of the cells is relatively low, that is, the selectivity in this case is not satisfactory.

Gadolinium has a significantly larger cross section for thermal neutrons. Gd is already used for tracering of cerebral tumours. In this case the large magnetic moment of $Gd^{3+}$ is used for NMR-tomography. But $Gd^{3+}$ is toxic; therefore the Gd is embedded in the compound Gd-Diethylenetriaminepentaacetic acid (Gd-DTPA).

The reaction $^{157}$Gd(n,$\gamma$) $^{158}$Gd generates $\gamma$-radiation with energies up to 7.8 MeV. The energy of the $\gamma$-radiation is too high to influence neighbouring cells. In addition to the $\gamma$-radiation, Auger-electrons with an energy of around 41 keV are generated. These electrons will be absorbed in the neighbourhood of the malignant cells and may destroy them. The problem here is to ensure that Gd-DTPA will be selectively added to malignant cells. It is possible to identify cancerous cells in the histology with the help of an optical microscope but it is not possible to identify the Gd-atoms at the same time. Using micro-XRF it might be possible to identify the distribution of Gd in a tissue but it is not clear if these are the malignant cells. A combination of both methods may provide the answer.

Therefore, tests were performed to identify the distribution of Gd in the tissue of a rat aorta. The tissue was positioned on a thin plastic film. The scattering of the penetrating radiation is small, the spectral background is reduced, and any spectral interference from elements in a glass sample support (slide) are avoided. This enhances the sensitivity for the small concentration of Gd in the thin organic layer.

The measurement results are displayed in Fig. 7.9. In this case, the area analysed was around $3.5 \times 2.8$ mm. This was measured in a matrix size of $128 \times 100$, that is, the step size was around 30 μm. The excitation conditions were 40 kV, 1 mA with a measuring time per pixel of 1 s (resulting in a measuring time of 3.5 h for the complete distribution). The image shows that the Gd is not homogeneously distributed. There is an enhanced concentration in the middle of the aorta wall. A histological examination with an optical microscope may clarify if the malignant cells are concentrated in the middle of the aorta's wall as well.

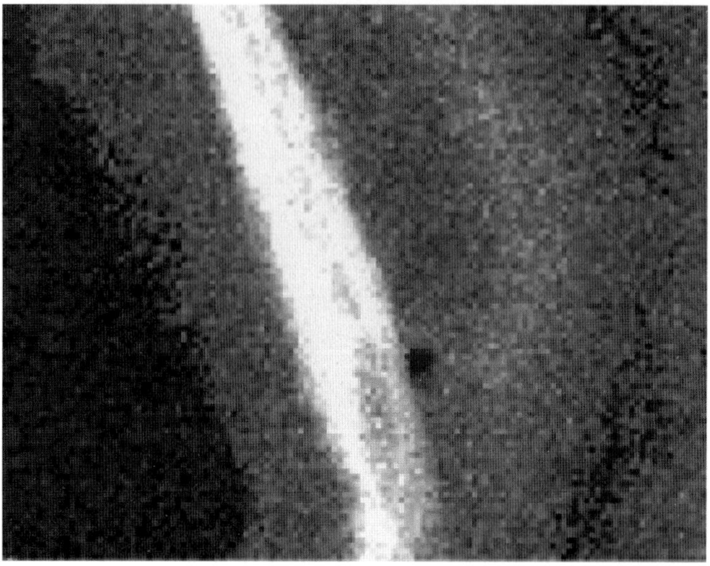

**Fig. 7.9.** Gd-distribution in a section of a rat aorta. The higher the Gd-intensity, the brighter the image

*Restoration of Frescoes*

Nowadays, medieval frescoes show a risk to be damaged by high concentrations of air pollutants. The lime upon which frescoes have been painted chemically changes in its composition. The $SO_3$ in the air combines with the lime ($CaCO_3$) to form Gypsum ($CaSO_4$). Gypsum is softer than lime and more easily damaged by mechanical influences. Therefore, different techniques for restoration are under review. One of the conservation procedures is carried out with a Barium-two-step-technique [27]. The fresco is initially treated with a compress of cellulose pulp saturated with $(NH_4)_2CO_3$ in water. This removes the sulphate and transforms the gypsum ($CaSO_4$) back into lime ($CaCO_3$) according to

$$(NH_4)_2CO_3 + CaSO_4 \Rightarrow CaCO_3 + (NH_4)_2SO_4 .$$

In a second step, a compress of a cellulose pulp with a barium-hydroxide $Ba(OH)_2$ solution is used to stabilise the lime by substituting Ca by Ba. This change is described by the following processes:

$$Ba(OH)_2 + CaCO_3 \Rightarrow BaSO_4 \Downarrow + Ca(OH)_2$$
$$Ba(OH)_2 + (NH_4)_2SO_4 \Rightarrow BaSO_4 \Downarrow + NH_3 + H_2O$$
$$Ba(OH)_2 + CO_2 \Rightarrow BaCO_3 \Downarrow + H_2O .$$

The barium salts are much more stable against the influence of $SO_3$ pollutants than the Ca-compounds. The frescoes will not only be restored but also protected against further damage from similar environmental influences.

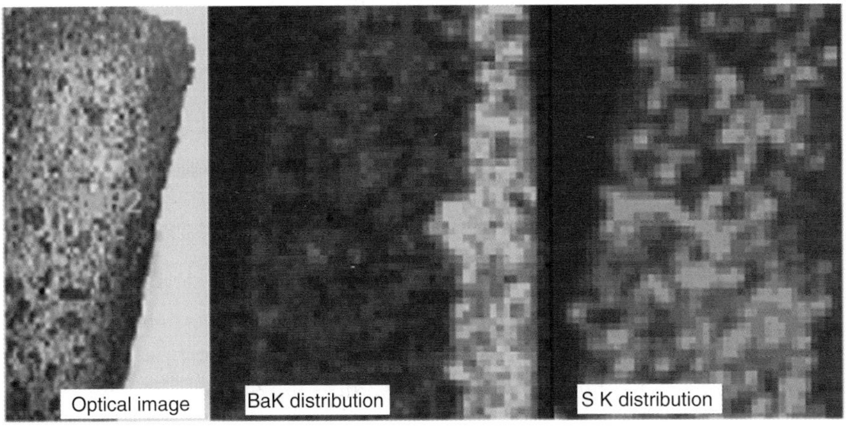


**Fig. 7.10.** Distribution of Ba and S in a fresco cut

The arising analytical problem is to determine the penetration depth of the barium into the fresco and to investigate the de-enrichment of sulphur on the surface. This task can be carried out with micro-XRF. A piece of a fresco was cut perpendicular to the surface and the distribution of Ba and S were measured over this area. The results are displayed in Fig. 7.10. The surface of the fresco is at the right-hand side. The figure shows both the optical image and the elemental distributions of Ba and S. It can be seen that Ba shows a high intensity close to the surface with a penetration depth of around 1.5 mm. On the other hand, the distribution of S clearly shows that S has a very low intensity at the surface. Only beyond a depth of around 1.5 mm the S concentration does become higher being similar to that normally found in lime. The area analysed was around $20 \times 16\,\mathrm{mm}^2$ and the matrix size was $64 \times 50$, that is, the step size was approx. 300 µm with a spot size of 300 µm. The measurement conditions were: 40 kV, 1000 µA, 1 s per pixel.

If the mapping is performed as a spectral mapping, that is, for all pixels analysed, the complete spectrum is saved, the data can be reprocessed and results may be displayed in various forms. The data can be evaluated as a line scan using the saved spectral mapping data. For any given line defined within the mapping area the data stored within the spectra are associated to the pixels closest to this line. They are used to calculate the desired intensity distribution. If more than one line is drawn, an averaging of the distribution in this direction is possible. The intensities of the elements of interest for the defined line scan are calculated in this way and displayed in Fig. 7.11. The intensities are normalised to the maximum for each element. These two distributions clearly show the enrichment of Ba close to the surface (right hand side of the distance axis) to a thickness of around 1.5 mm and a corresponding depletion of S.

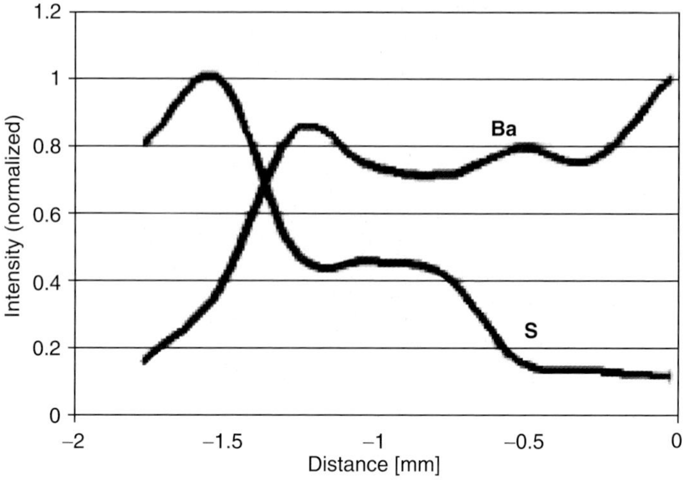


**Fig. 7.11.** Line scan for Ba and S perpendicular to the frescos surface (at 0 mm)

*Coating Thickness Analysis*

X-ray fluorescence is already being used, for a long time now, to determine coating thickness. Both the thickness and, in case of alloyed layers, the composition can be non-destructively analysed with X-rays (see Sect. 7.4). A typical requirement for this analysis is that only small areas are available for the measurement. Hence, the excitation beam has to be collimated into a small spot size with dimensions in the range of 100–500 μm. This is achieved with pinhole apertures. However, with the enhancement of integration especially in microelectronics but also in nanotechnologies, the available areas for the measurement are reduced. Now the spot size needs to be only 30 μm or even less. Also the coatings have become thinner, which requires to enhance the sensitivity of the method. Coating thickness in the lower nanometer (nm) range is common.

These two conditions, smaller spot size and higher sensitivity, require changes to the instrumental concept. The smaller spot size needs an X-ray optic that concentrates enough primary radiation onto the sample. Thin layers in the nm-range correspond to bulk sensitivity in the range of around 10 ppm, which is achievable only by using high-resolution detectors with a high peak-to-background ratio like Si(Li), PIN-diodes or SDDs (see Sects. 7.4.2 and 7.4.3).

An example is presented for a thin multilayer structure. In electronics fabrication, lead frames are used to glue the chips onto a support and connect the pads on the chip to the contacts of the lead frame by bonding wires. The contacts of the lead frame are used for soldering on the printed circuit board. The lead frame has to be coated with precious metals in order to form reliable bond contacts.

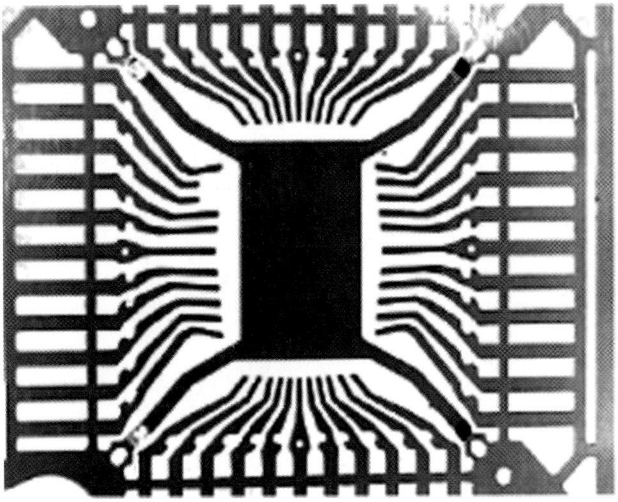

**Fig. 7.12.** Part of an AuPdNi-plated lead frame

In the earlier days, only the contacts of the lead frames were selectively plated with a Ag-layer having a thickness of around 2–4 μm. This selective coating needs an additional step in production—appropriate masking of all but the tips. The masks had to be manufactured and correctly positioned for every lead frame. This expensive step can be avoided now by a full cover of the lead frames with an Au on Pd coating. Of course these materials are much more expensive than Ag but the coating thickness is in the lower nm-range. The costs of the material are less than for selective Ag-plating because of the low material consumption.

Figure 7.12 shows a part of such a modern lead frame. In Fig. 7.13 the corresponding spectrum collected at a point is displayed. The spectrum in Fig. 7.13 shows the response obtained for both the Au M- and Au L-series of characteristic fluorescence lines. From the Pd-layer only the L fluorescence radiation is displayed. It can be seen that the Au M fluorescence radiation exhibits a higher intensity for this thin layer than the Au L fluorescence radiation. This is due to the stronger self-absorption of Au M fluorescence radiation in the Au-layer. This gives a higher sensitivity for coating thickness determination. The limit of detection can be estimated from this measurement in the range of 0.3 nm. However, not only the sensitivity but also repeatability and stability are important parameters for an analytical method, especially for industrial applications because the product quality has to be controlled with high reproducibility.

Repeatability can be determined by repeated measurements of the same sample. A typical result for the Au-coating is displayed in Fig. 7.14, which shows a statistical distribution of the thickness determination in the region of around 16 nm for 1000 measurements of 30 s. The lines in the diagram correspond to the $2\sigma$ standard deviation values of the distribution. Stability

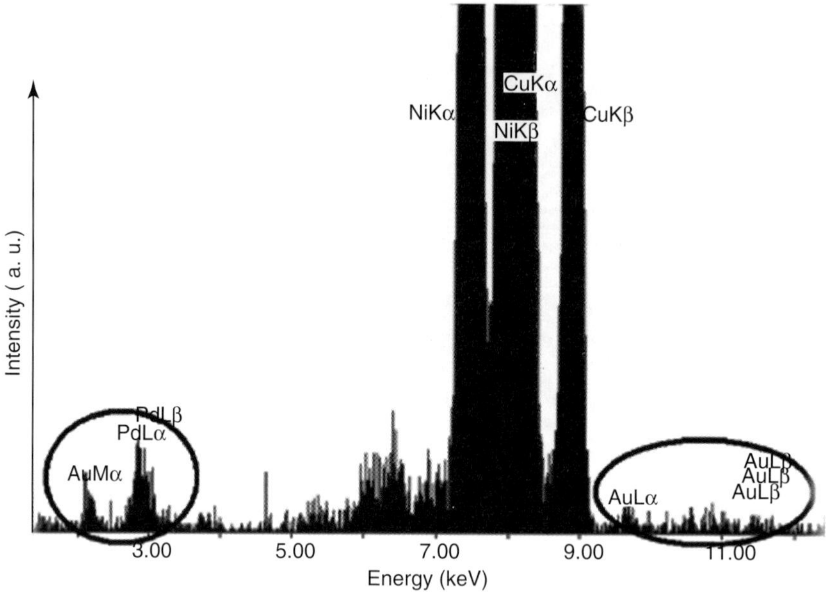


**Fig. 7.13.** Spectrum of a coating of 16 nm Au on 36 nm Pd on 1.4 μm Ni measured with 20 kV, 100 μA for 20 s

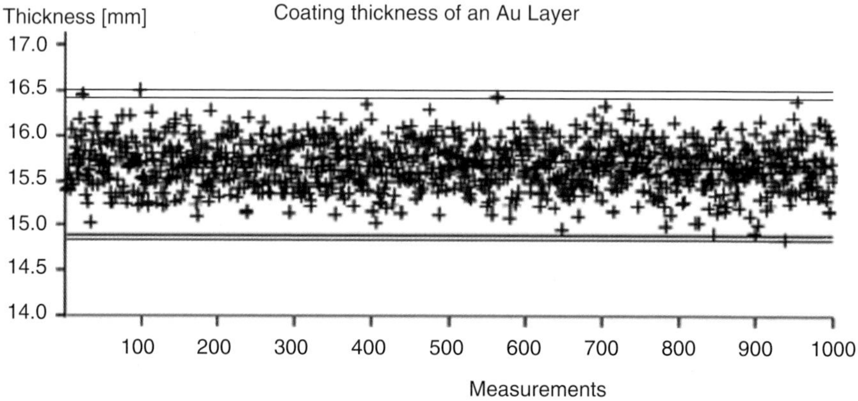


**Fig. 7.14.** Au-thickness determined for 1000 repeated 30 s measurements

means the possibility to get the same result over a longer period of time. If the deviations of the mean values of several such measuring series are in the range of the statistical value of the repeated measurements the stability is very good and fulfils the requirements.

*Ultra Trace Elemental Analysis with Micro-XRF*

Micro-XRF can also be used for ultra trace analysis if a special sample preparation technique is employed. This special sample preparation is called dried spot sampling technique and involves a deposition of small drops of a liquid sample onto a thin film. If the drop is dried out to a solid residue it can be analysed by a collimated beam. The scattering of the excitation radiation on the sample is very low because of the small sample volume and because the sample consist of elements with higher atomic number that have a small scattering amplitude [28]. It results in a high peak-to-background ratio and, therefore, a high sensitivity for trace elements. Limits of detection in the lower ppb range can be achieved. The disadvantage of the method is the long drying time that can be in the range of hours. Also evaporation of volatile elements might be a problem.

The sample preparation technique can be improved by reducing the size of the sample drop. Hence, the drying time is reduced significantly [29]. The drop size reduction is possible with a nanoliter injector similar in operation to that of an ink-jet printer that generates drops down to 2 nl. The drying time is reduced down to a few seconds because of the relatively larger surface area of the drops. Smaller dried spots can be generated by smaller drops, as well. Then the dried sample can be excited completely with the micro-beam of the spectrometer. In addition, the loss of volatile elements is minimised by the short evaporation time. The recovery rates should be better than for the larger droplets. Extended examinations of this nanodroplet technique [29] show another advantage—the sensitivity is enhanced, especially for the determination of absolute concentrations. Some typical limits of detection are displayed in Fig. 7.15.

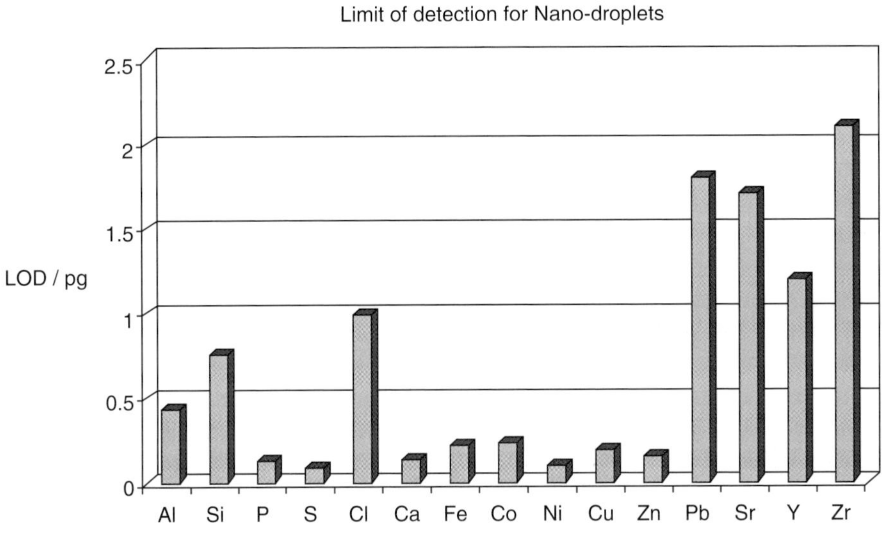


**Fig. 7.15.** Limits of detection obtained with micro-XRF using the "nano-droplet" technique [from 29]

This application shows that by using a special sample preparation technique micro-XRF becomes a very sensitive analytical method that is applicable for ultra trace elemental analysis. The limits of detection are in the low and sub-pg range.

### 7.1.4 3D Micro X-Ray Fluorescence Spectroscopy

The previous sub-chapters described the use of the microscopic excitation spot formed with the help of X-ray optics to perform very local analysis by lateral two-dimensional (2D) elemental mapping or line-scanning along the sample surface. The information retrieved has the disadvantage that it is not explicitly depth sensitive, rather a convoluted product of the incident X-rays, penetration capability and the self-absorption correction depending on both the energy of the exciting radiation and the energy of the fluorescence radiation.

Not long ago, Kanngießer et al. [30] and other groups expanded the capabilities of the micro-XRF spectroscopy to the depth, thus creating a true three-dimensional (3D) micro X-ray fluorescence spectroscopy (3D micro-XRF). The 3D micro-XRF is realised by a confocal arrangement, which consists of X-ray optics in the excitation as well as in the detection channel. A microvolume is defined by the overlap of the foci of both X-ray optics. If the sample is moved through this microvolume, its chemical composition can be non-destructively investigated not only laterally along the surface but also within the sample at the depth of interest, see Fig. 7.16 Furthermore, if synchrotron radiation is used as exciting radiation, the excitation energy is tunable, which facilitates micro-XAFS investigation for chemical speciation at the same spot in the sample. Thus, not only depth information concerning the elemental distribution but also chemical speciation and phase information are obtainable at the same point of interest in the sample.

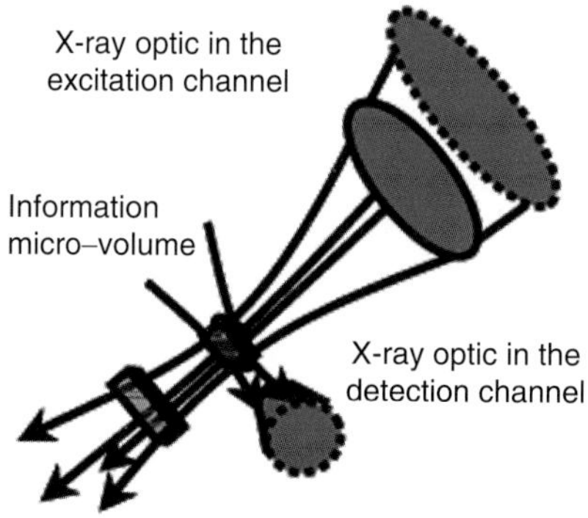


**Fig. 7.16.** Scheme of the confocal set-up for 3D micro-XRF spectroscopy

The information depth of the microvolume depends on the energy of the exciting radiation, the energy of the fluorescence radiation, the incidence angle, the angle of reflection, and the sample composition. The spatial resolution of this 3D micro-XRF set-up depends on the FWHM values of the overlapping focal profiles. In addition, a better peak-to-background ratio can be achieved by restricting the detector field of view.

For reducing the scattered radiation due to the restriction of the detector's field of view several other groups have already suggested the confocal set-up [31–33]. One step further is the introduction of depth profiling with such a set-up. Soon after the first proof of principle measurements in 2002, Proost et al. [34] realised a similar set-up at the beamline L at HASYLAB. The same group built up the confocal set-up at the ID 22 of the ESRF for the investigation of inclusions in geological samples [35].

*Characterisation*

The 3D micro-XRF set-up of Kanngießer et al. was realised at the BAMline located at a 7T wavelength shifter at BESSY [30]. Figure 7.17 shows a photograph of the set-up at the BAMline. To create the microvolume a polycapillary halflens with a focus of about 30 μm at a working distance of 16 mm was used in the excitation channel. In the detection channel a polycapillary conical collimator (poly CCC) was adjusted directly on the snout of the Si(Li)

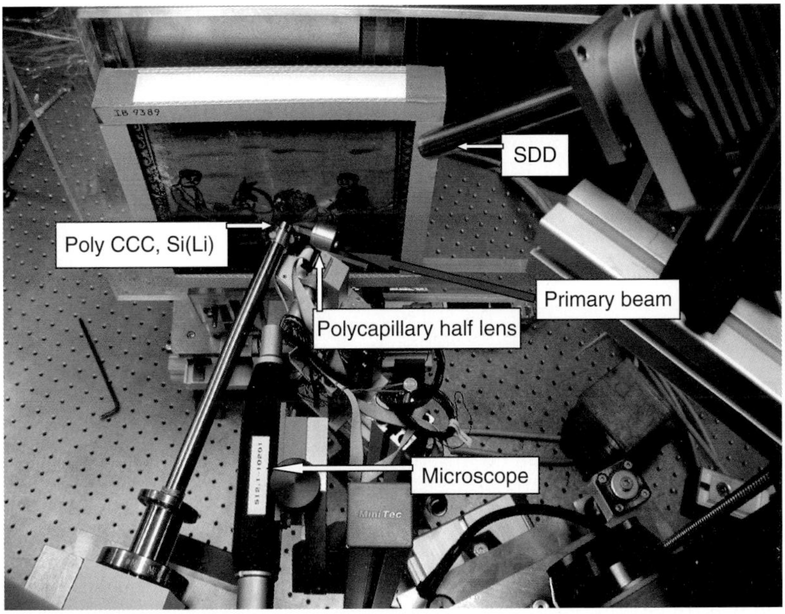


**Fig. 7.17.** Photograph of the set-up for 3D micro-X-ray fluorenscence spectroscopy at the BAMline, BESSY in Berlin

detector. The focus of the poly CCC has an FWHM of 20 μm at a working distance of 1.2 mm. The second detector of the set-up, the drift chamber detector, was used to monitor the global X-ray spectrum without depth resolution.

The microvolume formed was characterised by scanning the volume with a 2 μm thick Cu foil. The Cu foil was moved through the beam in 5 μm wide steps in the horizontal plane with respect to the storage ring and in the vertical direction. The intensity of the Cu K$\alpha$ radiation plotted in dependence of the steps is proportional to a sensitivity curve for Cu of the set-up. Figure 7.19 shows the Cu intensity curve of the confocal set-up in the horizontal plane for an excitation energy of 17 keV. The FWHM of the curve is about 20 μm, which can be regarded as the spatial resolution of the set-up. The intensity curve in the vertical direction showed the same FWHM. The 20 μm FWHM presents the lower limit of resolution achievable with this set-up, because the minimum spatial resolution is mainly determined by the FWHM of the focus of the X-ray optics in the detection channel. In addition to the FWHM of the intensity curve, the steepness of the slope is decisive for the evaluation of the spatial resolution of the method. The steepness of the sensitivity curve in Fig. 7.18 is given as the relative distance for 90% of the slope intensity. The use of a poly CCC in the detection channel has two advantages in comparison to the use of a polycapillary full lens, as proposed by Kumakhov [31]. The most important one is the greater acceptance angle of around 0.2 sr in comparison to an average acceptance angle for a polycapillary full or half lens of about 0.02 sr, leading to a higher transmittance of the radiation. The other advantage is the focus size for radiation with an energy below around 8 keV of the poly CCC, which is smaller than any polycapillary which can actually be produced. The disadvantage of using a poly CCC is its length, which might restrict the energy regime. If the length of the poly CCC is too short, the concentration of radiation with a higher energy is less efficient. With higher energies a substantial part of the incoming radiation might pass directly the walls of the class capillaries without being totally reflected. With the poly CCC used in the above confocal set-up, measurements are restricted to energies under 20 keV.

The linear mass attenuation coefficient of this special capillary glass for Cu K$\alpha$ ($\mu_{\text{lin}} = 392/\text{cm}^{-1}$) is about two times higher than the respective coefficient for Ag K$\alpha$ ($\mu_{\text{lin}} = 187/\text{cm}^{-1}$). Hence, the poly CCC must be two times longer in order to concentrate the Ag K$\alpha$ radiation as efficiently as the Cu K$\alpha$ radiation. The production of longer poly CCCs to circumvent this problem has not been possible up to now. Additionally, longer poly CCCs will have a smaller acceptance angle. Hence, for energies higher than around 20 keV the use of polycapillary half lenses is advisable.

*Measurements*

The first application of the new 3D micro-XRF method was the investigation of ancient Indian Mughal miniatures from the Museum of Indian Art in Berlin [30]. The Mughal paintings were believed to consist of several well-separated

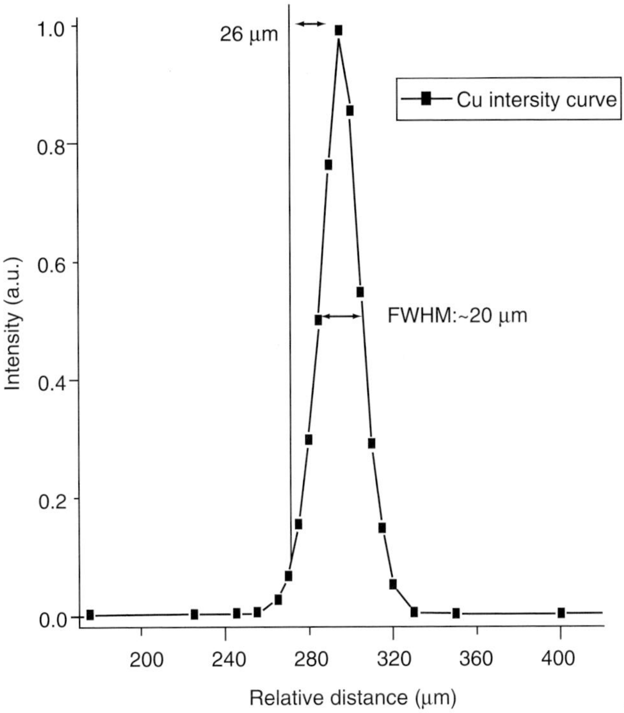


**Fig. 7.18.** Intensity curve of the confocal set-up for a thin Cu foil with an excitation energy of 17 keV

polished pigment layers on paper. Until now, only scarce information was available on this painting technique and the artist's pigment palette. With the help of depth profiling measurements the layered structure of the paintings could be evaluated. Figure 7.19 shows an example of the depth profiling in a miniature. It also demonstrates the depth resolution which could be achieved with the confocal set-up at the BAMline. The element depth profiles shown were obtained by moving the miniature perpendicular to its surface through the microvolume in 5 µm wide steps. At each step a fluorescence spectrum was taken and the net peak areas of the L$\alpha$ line of Hg and Pb were evaluated. To facilitate a better comparison, the two intensity depth profiles are scaled to the same height.

The depth profiles reveal a layered structure of the painting with a 10 µm thick cinnabar (HgS) layer painted on a lead white ($2PbCO_3{\cdot}Pb(OH)_2$) ground layer of the same thickness. Hence, the successive layers could be distinguished with a resolution of about 10 µm. Further investigations showed that it is also possible to distinguish the same element in different layers which can neither be achieved by conventional XRF nor by any other non-destructive analytical method. As an example, Fig. 7.20 shows the elemental depth profiles at two

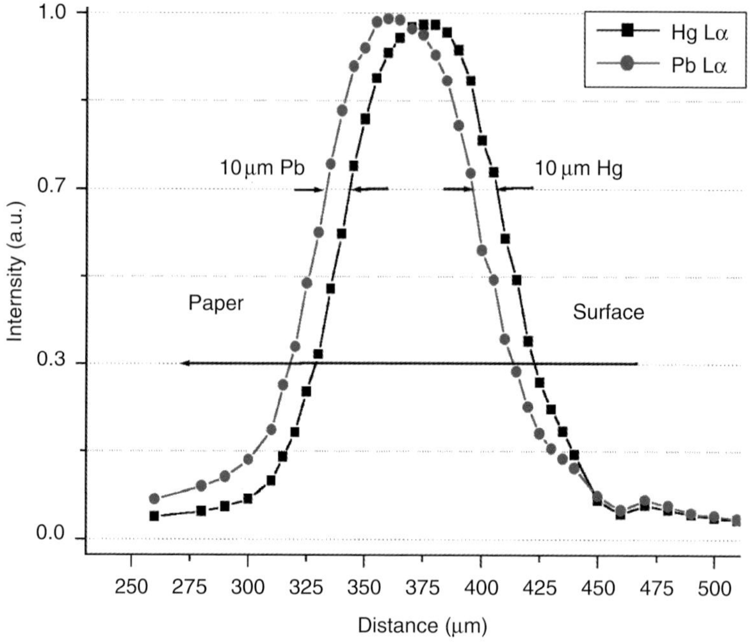


**Fig. 7.19.** Micro-XRF depth profiles of a cinnabar (HgS) and a lead white $((2PbCO_3 \cdot Pb(OH)_2)$ layer system of a Mughal miniature

different locations of a Mughal miniature. The first point of measurement was on the green background of the painting, whereas the second measurement was carried out on the face of the Mughal.

The depth profiles of the Pb $L\beta$ line and of the Ba L multiplet on the face, depicted in the upper graph, exhibit only one layer. The fluorescence depth profiles of the green background clearly exhibit also one single paint layer. In contrast, the depth profile of the scattered radiation, the sum of the coherent (Rayleigh) and incoherent (Compton) scattered radiation, shows prominent intensity fluctuations below the green paint layer. As the intensity of the sum of the scattered radiation is proportional to the average atomic number of the material, the fluctuations indicate a composite paper structure. This is corroborated by art historians who report the gluing of thin paper to produce a cardboard for these paintings. Hence, the evaluation of the depth profiles of the Compton and Rayleigh scattered exciting radiation delivers additional information on the sample investigated. These depth profiles can be transformed into a figure of merit for the average density and/or for the average $Z$ of the local point investigated in the depth of interest. For a full quantification of the elemental depth profiles a deconvolution from the sensitivity profile and the correction for absorption effects have to be the next steps in the development of the method.

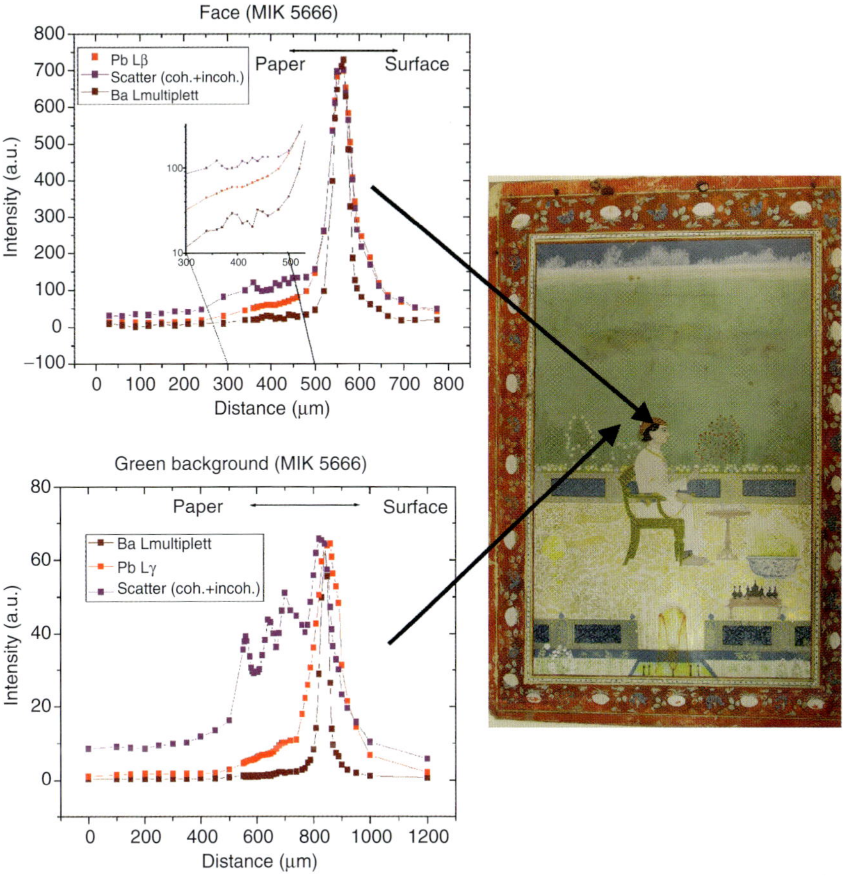


**Fig. 7.20.** Depth profiles of fluorescence and scattered radiation at two locations of the Mughal miniature MIK I 5666 dated from the 18th century

*Comparison with a Tabletop 3D micro-XRF Set-up*

3D micro-XRF is not only possible at synchrotron radiation sources. At the Technical University of Berlin, Malzer and Kanngießer constructed a tabletop 3D micro-XRF set-up, which is shown in the photographs of Fig. 7.21. Here, the source is a Mo micro-focus X-ray tube combined with a polycapillary full lens in the excitation channel. The micro-focus tube has a FWHM spot size of 100 μm; the polycapillary lens has an 80 μm FWHM spot size with a working distance of 22 mm. In the detection channel the same poly CCC and the same Si(Li) detector is used as was used for the set-up at the synchrotron source BESSY. A second Si(Li) detector is positioned downstream on the optical axis of the polycapillary lens to ensure the alignment of the lens and to monitor the alignment of the two foci of the X-ray optics to produce the microvolume.

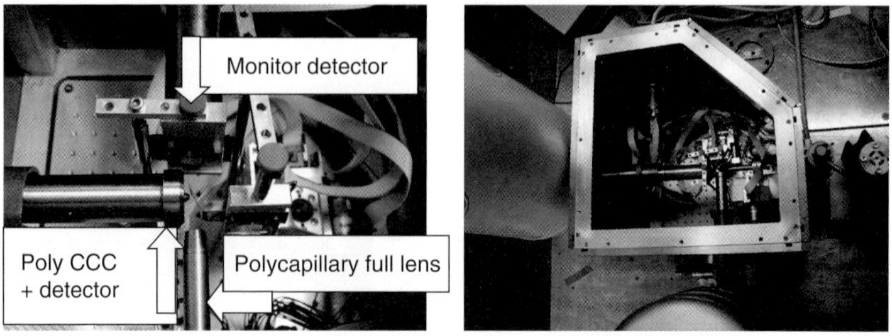


**Fig. 7.21.** Photographs of the 3D micro-XRF set-up, depicting a closer view on the confocal set-up on the left hand side and a top-view of the whole set-up on the right-hand side

The whole set-up is located in a vacuum chamber which facilitates the measurement of light elements. Both X-ray tube and Si(Li) detector are equipped with thin windows for the same reason.

With the tabletop 3D micro-XRF set-up pigment layers have been investigated for their homogeneity [36]. Now it is possible to image the elemental distribution in a certain depth of the sample, which helps to differentiate between various kinds of inhomogeneities, like inclusions and particles. Figure 7.22 shows the intensity depth profiles of a green chrysocolla ($CuSiO_3 \cdot nH_2O$) pigment layer on a lead white ($2PbCO_3{\cdot}Pb(OH)_2$) ground layer painted on paper. The depth profiles are scaled to the same height. The pigment layers were scanned with a step width of 10 μm. The measurement lifetime at each step was 100 s with a tube current of 0.6 mA and a voltage of 35 kV.

The two pigment layers are clearly distinguishable in the intensity depth profiles of their characteristic elements Cu and Pb. Both curves have a FWHM of about 100 μm, which correspond to the layer thickness. Furthermore, the depth profiles for Si and Fe indicate inhomogeneities in and on the green chrysocolla pigment layer. The fact that the main part of the two curves is lying in the Cu intensity curves indicates inhomogeneities rather than a thin top layer. The different FWHMs of 30 μm for Fe and 77 μm for Si, respectively, are another indicator for inhomogeneities in a layer. And indeed the chrysocolla pigment is known to have inclusions containing iron and quartz particles. A full 3D scan, that is, a lateral scan in addition to the depth profiles, corroborates the results.

The 3D scan clearly demonstrated that the fluorescence was coming from an Fe particle of about 30 μm in diameter. The contour plot of the Si fluorescence showed furthermore that the Fe particle is partly embedded in a quartz inclusion of the top pigment layer. At the site of the Fe particle the Si fluorescence is diminished due to the absorption of the particle. The latter finding could only be stated unambiguously with the lateral scan in combination with the Fe depth profiles.

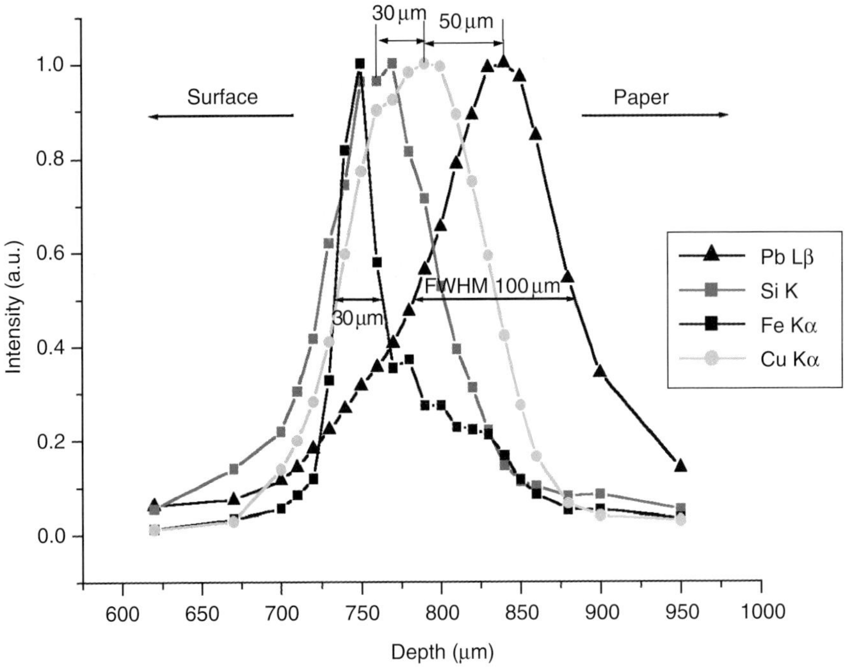


**Fig. 7.22.** Intensity depth profiles of a green chrysocolla ($CuSiO_3 \cdot nH_2O$) pigment layer on a lead white ($2PbCO_3 \cdot Pb(OH)_2$) ground layer. The depth profiles are scaled to the same height for better comparison

In order to compare the capacities of the synchrotron based set-up and of the tabletop set-up a glass standard was investigated with both arrangements with the same main excitation energy of 17 keV. Yet, one has to keep in mind that the excitation of the sample in the tabletop set-up is polychromatic by using an X-ray tube. For the comparison of the two set-ups the lower limit of detection (LLD) can be taken as a figure of merit. The LLD gives a direct impression of the analytical possibilities. Table 7.7 shows the lower limit of detection (LLD) for selected elements of the glass standard.

The comparison shows that the detection sensitivity of the synchrotron-based set-up is at least one to two orders of magnitude higher. The smaller

**Table 7.7.** Comparison of lower limits of detection for synchrotron and tabletop set-up

| Element | LLD /ppm for synchrotron set-up | LLD /ppm for Mo tabletop set-up |
|---|---|---|
| K | 1500 | 6000 |
| Cu | 60 | 2000 |
| Y | 5 | 380 |

difference in the detection limits for the lighter elements can be explained by the additional excitation of the tube's bremsstrahlung spectrum lying closer to the absorption edges of these elements. The better detection limits achieved with synchrotron radiation are due to its considerably higher flux, its linear polarization and monochromacity. At 17 keV the flux at the BAMline is about three orders of magnitude higher than the one of a microfocus Mo X-ray tube. The higher brightness of the synchrotron radiation is not important when the set-up consists of polycapillary lenses in the excitation channel. The linear polarization and monochromacity improve the detection sensitivity indirectly due to the better peak-to-background ratio produced. Hence, it may be stated that with the synchrotron arrangement major and minor elements, with elemental concentrations in the higher ppm range to the percent range, in favourable situations even trace elements, can be detected into the depth. With the tabletop arrangement 3D investigations are restricted to the detection of major elements in the sample; in favourable cases also minor elements are detectable.

For a depth-sensitive method the information depth is of major interest. It can be defined as follows:

$$x_{\mathrm{LLD}} = \frac{2.3}{\mu} \lg \left( \frac{w}{\mathrm{LLD}} \right) ,$$

with $\mu$ being the linear mass absorption coefficient, $w$ being the weight fraction of the element, and LLD being the lower limit of detection for the element at the surface. The equation was derived by assuming a negligible self-absorption at the surface. Hence, the information depth is linear dependent on the elemental composition of the sample and logarithmic dependent on the flux of the excitation and on the concentration of the element investigated. For example, the information depth for $Y$ in the glass standard investigated with the synchrotron based set-up ranges from 250 to 300 µm. The margin for the information depth indicates the margin for the matrix of the glass standard. In general, the synchrotron based 3D micro-XRF set-up gains a 2–20 times higher information depth.

In order to gain more insight into the depth dependence of the fluorescence intensity produced we have calculated absolute countrates for fluorescence line intensities of a layer system. The layer system chosen is one investigated with the 3D confocal set-up (see Fig. 7.19) and is composed of two pigment layers, a cinnabar (HgS) layer on top of a lead white ($2PbCO_3{\cdot}Pb(OH)_2$) layer. In two series of calculations the thickness of both layers is varied from 2 to 100 µm, respectively, whereas the other layer thickness is kept constant at 10 µm. The incidence flux is taken to be $2 \times 10^8$ photons/s which is about the flux delivered by the BAMline. The incident angle $\Phi$ is 25° and the emergence angle $\Psi$ is 65°. Both beams are supposed to be parallel. The solid angle of acceptance of the poly CCC (0.2 sr) was taken into account for the restricted field of view of a detector. The countrates falling onto the detector were calculated for two different excitation energies, namely 17.4 and 27 keV. The fundamental parameters needed for the calculations are taken from the actual compilations of Elam [37], Henke [38], and McMaster [39].

Figure 7.23 shows the thickness variation of the top-layer, the Cinnabar layer. The thickness of the bottom-layer, the lead white layer, is kept constant at 10 µm.

The fluorescence line intensities of the top-layer elements Hg and S increase with increasing layer thickness until a thickness is reached for which the self-absorption becomes so strong that the additional thickness does not contribute to an additional fluorescence signal. For the Hg L$\alpha$ line this critical thickness is reached at about 25 µm for the 17.4 keV excitation energy and at about 40 µm for the 27 keV excitation energy. The difference in the critical thickness for the two excitation energies can be explained by the lower attenuation for the 27 keV excitation energy. For all thickness the absolute countrates for the 17.4 keV excitation energy are higher which is due to the higher photoelectric cross section for this energy in comparison to the one for 27 keV. A similar behaviour shows the S K$\alpha$ fluorescence line for both excitation energies. Here, the critical thickness is reached at about 15 µm for the 17.4 keV excitation energy, whereas the 27 keV excitation energy is associated by a critical thickness of about 30 µm. The critical thickness is thinner for both excitation energies

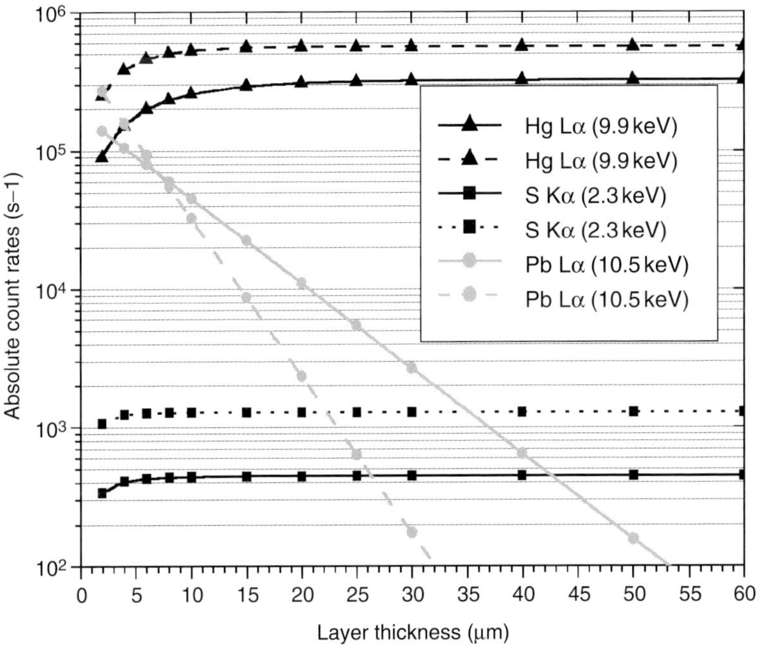


**Fig. 7.23.** Calculated fluorescence line intensities for a thickness variation of the top-layer (HgS) with two excitation energies, at 27 keV (*solid line*) and at 17.4 keV (*dashed line*), respectively

than the ones for the Hg L$\alpha$ line because of the mass attenuation coefficient for the S K$\alpha$ line, which is twice as high in the case of the cinnabar layer. Also the absolute countrates are two orders of magnitude smaller for the S K$\alpha$ line in comparison to the Hg L$\alpha$ line, even though the elements have almost the same density. The main reason lies in the photoelectric cross section for the S K-shell, which is already two orders of magnitude smaller than the one for the Hg LIII-shell for the same excitation energy. Additionally, the fluorescence yield of the S K-shell is one order of magnitude smaller than the one for the Hg LIII-shell. On the other hand, the S K$\alpha$ line is enhanced by more fluorescence lines than the Hg L$\alpha$ line. In the case of the S K$\alpha$ line all L-Lines from the Hg as well as from the Pb of the bottom layer produce secondary enhancement. In the case of the Hg L$\alpha$ line only the Pb fluorescence lines from the LII- and LI-series (except the L$\beta_2$ line) have energies which can excite the Hg L$\alpha$ line. Thus, secondary enhancement is included in the calculations, as can be seen in Fig. 7.23.

Pb is the only detectable fluorescence element in air from the bottom-layer ($2PbCO_3{\cdot}Pb(OH)_2$). Both curves of the respective excitation energy show the characteristic increasing attenuation with increasing thickness of the top-layer until the attenuation is so strong that the Pb fluorescence line is not detectable anymore. The crossing of both curves at a top-layer thickness of about 8 µm shows the trade-off between the higher penetration depth of the 27 keV excitation radiation and the higher photoelectric cross section for the 17.4 keV excitation radiation. At top-layer thickness below 8 µm the higher photoelectric cross section surpasses the higher penetration depth and vice versa for layer thickness over 8 µm.

This example gives the range of depth profiling for a heavy matrix. Information from the bottom layer can be gained only if the top layer thickness is in the order of several ten micrometers. In the case of a light matrix, for example, biological samples, the information depth can reach the mm range.

In conclusion it can be said that with the new method elemental distributions in 3D objects can now be measured in a non-destructive manner with a 3D resolution. At synchrotrons a resolution of a few micrometers is currently achievable by carrying out depth profiling on layered structures. Herewith major and minor elements can be distinguished in different layers, even if the same element is present in successive layers. The latter is the most severe restriction on conventional 2D micro-XRF. The same holds for the tabletop set-up if the sensitivity of the arrangement is taken one to two orders of magnitude lower. Further on, with a first evaluation and quantification approach the determination of layer thickness and local elemental densities becomes feasible. The latter has to be emphasized because it is a distinct feature of 3D micro-XRF in comparison to ordinary XRF or micro-XRF. Whereas in ordinary XRF and micro-XRF arrangements weight fractions of the elements of interest are determined, 3D micro-XRF always delivers elemental densities. Basically, this is due to the fact that the number of photons detected in 3D micro-XRF depends on the number of atoms inside the microvolume defined by the confocal set-up.

Furthermore, with the confocal set-up not only can information on the elemental distribution but also on the local chemical state of the sample be gained. This is possible at synchrotron sources where the excitation energy can be tuned over an absorption edge facilitating X-ray absorption fine structure spectroscopy. Hence, micro-XANES and micro-EXAFS investigation are feasible at the same location where elemental depth profiles are obtained. First depth-sensitive micro-XANES investigations have already been carried out. For a reliable interpretation of the spectra an evaluation code has still to be created which takes the attenuation of the micro-XANES spectra gained in the depth into account.

The 3D micro X-ray spectroscopy can contribute considerably to a variety of fields of application, such as environmental science and quality control in materials science. Another very interesting research field is the analysis of histological samples in life science. In first measurements, we could show that it is possible to investigate the elemental distribution in a complete rat eye, suggesting that in some cases it is possible to investigate histological samples without having to cut the samples into sections. Furthermore, in vivo measurements may be conceivable in the future.

A comparable method which also directly provides 3D information on the elemental distribution is fluorescence tomography. This special form of computed microtomography was already installed at second generation synchrotron facilities in the 1990s [40, 41]. A qualitative jump is obtained at third generation synchrotron facilities, where the counting time for a full tomographic scan has been reduced to a reasonable and feasible time of half a day. The research at the ID22 of the ESRF is a forerunner for methodological development of X-ray fluorescence techniques. Simionovici et al. reduced the data collection time for a full 3D fluorescence tomographic scan to less than eight hours [42]. Nevertheless, the speed of the data collection remains a serious obstacle because, besides the two dimensional lateral scan, the objects under investigation have to be rotated as well. With the 3D micro-XRF method the samples, instead of being rotated, have to be moved perpendicular to their surface to get depth information. This reduces the data collection time in comparison to the 3D microtomography, especially when layered structures are investigated. Furthermore, the rotation of the sample restricts possible sample sizes. On the other hand, the 3D microtomography yields by far better spatial resolution down to 50 nm, whereas the actual limit of the 3D micro-XRS is about several μm. This limit could only be overcome when X-ray optics other than the capillary optics are used. Actual X-ray optics which could deliver a smaller spot size, like Fresnel zone plates, would restrict the 3D micro-XRS to elemental analysis as they also have monochromatizing properties. Hence, the 3D micro-XRS would lose its second advantage in comparison to the 3D microtomography, which is the combination with XAFS.

Both methods have their limitations with respect to the information depth of the fluorescent radiation as well as a reliable correction for self-absorption. Recently, Golosio et al. [43] proposed a methodological approach to overcome reconstruction problems in tomography due to the self-absorption effects in the sample. They integrate the information obtained by fluorescence, Compton and transmission tomography, which seems to be a fruitful approach. A similar approach to solve the self-absorption problem for 3D micro-XRF is conceivable. The additional information on mass density and an average $Z$ provided by Compton- and Rayleigh scattering can also be used.

## 7.2 Micro-XRF with Synchrotron Radiation

*A. Simionovici, P. Chevallier*

### 7.2.1 Introduction

There is a growing need for trace element analysis at levels less than one part per million (ppm) by weight in nearly every scientific field. This first step of trace element analysis in the bulk is rapidly becoming insufficient and techniques capable of mapping the distribution of elements at the highest spatial resolution are desired.

For qualitative and quantitative analysis, X-ray fluorescence (XRF) is often used as a "finger printing" technique for characterizing the bulk elemental composition of a material. Mapping the lateral distribution of the elemental composition has been performed using X-ray emission microprobes. The great majority of these microprobes uses electrons as projectiles; a few others employ high-energy protons (PIXE) as a mean of excitation. Only recently, photon microprobes have become available, installed on various synchrotron radiation (SR) facilities.

As we shall see, the photon is a very interesting projectile, and should undoubtedly lead to the best microprobe if enough photons can be concentrated into few $\mu m^2$ spots. Due to the very small aperture of all hard X-ray optics this cannot be obtained with the best conventional X-ray tubes but only by using SR, the characteristics of which have initiated the development of new X-ray optical devices. Thus, efficient photon microprobes can only be installed on modern (third generation) SR facilities and it seems that they will always remain a limited analytical tool that will never supplant the wide use of other kinds of microprobes.

Lateral resolution at the micron scale is now reached with a wide variety of modern techniques. Secondary ion mass spectroscopy (SIMS) [44, 45] offers this resolution with excellent depth resolution as well, but is destructive and quantification is difficult in many cases. Laser ablation microprobe mass analysis (LAMMA) [46, 47] is a very useful technique but suffers from the nonreproductive character of the laser-induced ionization and difficulties are often encountered in interpreting the mass spectra. Electron probe microanalysis (EPMA) [48] presents extremely high lateral resolution (few nm).

Unfortunately, due to the spread of the electron beam below the surface of the sample, the X-ray fluorescence spectrum arises from a sphere of about 1 μm radius or larger. Moreover, this technique is limited to the determination of major constituents because of the high background generated by the incident and secondary electrons as they slow down through the sample and only probe the surface layers due to the limited penetration depth of electrons. Micro-PIXE offers high sensitivity (down to the ppm range) and can reach the micron scale in lateral resolution [49–51] but with severe damage to sensitive samples. Micro-PIXE, however, can be easily coupled with nuclear reaction analysis (NRA) for the determination of light elements, and with Rutherford back scattering (RBS) for depth profile determination, which is of great interest in the characterization of many samples. Finally, all these techniques require the sample to be in vacuum which may be impossible with some very volatile, too large, or living samples.

Subμm spatial resolution is now achievable in the hard X-ray range thanks to the advent of new X-ray optics in conjunction with SR. Thus microsynchrotron radiation X-ray fluorescence (micro-SRXRF) has appeared in recent years as an interesting quantitative microanalytical technique with high sensitivity that can be used on any kind of samples in air (except for the lightest elements). Moreover, by associating it to other techniques such as extended X-ray absorption fine structure (EXAFS), X-ray absorption near edge structure (XANES), diffraction or computed tomography (CT), very useful structural, chemical and morphological information can be obtained.

### 7.2.2 The General Setup

The general setup of about all photon microprobes installed on synchrotron beamlines is presented in Fig. 7.24 Most of the components have been described in great detail in this book so we will only briefly justify the choices made.

#### The Photon Source

In this chapter, we only consider photon microprobes installed on SR facilities. As a matter of fact the brilliance of these third generation machines is so high that they represent the best sources for a very sensitive photon microprobe. Nevertheless, bending magnets or insertion devices deliver quite different X-ray beams and this has to be taken into account.

Nowadays, bending magnets are less in demand and these sources may be used for a permanent installation of a microprobe although the brilliance is far from what can be reached on undulators. The continuous energy spectrum that extends more or less in the hard X-ray regime can be an advantage. To our knowledge, wigglers which allow the selection of much higher energies are never used because of their lower brilliance although they allow to excite heavy elements in the K shell, leading to a much simpler X-ray spectrum than for the L shell.

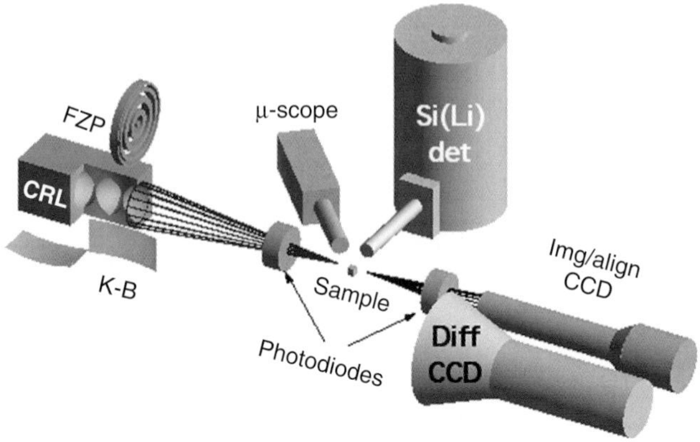


**Fig. 7.24.** Schematic setup of a microprobe

Undulators with their exceptional brilliance and nearly monochromatic energy spectrum are undoubtedly the best sources that can be thought of especially when they are tunable in energy and of low beta type (see Sect. 2.4). The size of the beam, even many tens of meters from the electron source, perfectly fits the usually very small aperture of the focusing X-ray optics. Recently, undulators of a very small gap, without a vacuum chamber, so-called "in vacuum" undulators, have appeared and they allow reaching very high energies (>100 keV) so far reserved for wigglers or wavelength shifters. Thus, a single device can satisfy the needs of many communities providing high flux, high brilliance up to high energies. Unfortunately, these sources are in high demand and in most cases the beam time will have to be shared with other kinds of experiments.

## X-Ray Optics

This field has known tremendous development since the advent of SR and figures that were only dreamt of a few decades ago are now routinely used. Here we will only compare the various kinds of optics in use now on different beam lines all over the world with the only goal of microfluorescence analysis. For more details the reader should refer to earlier chapters of this book.

Tapered monocapillaries [52, 53] seem not very reliable and present a rather disappointingly small gain in photon flux. They are achromatic systems that can be interesting for microLaue experiments or can be used after a high-resolution monochromator for micro-XAS (X-ray Absorption Spectroscopy). Positioning is rather simple but the outgoing beam is quite divergent so that the sample has to be placed right at the end of the capillary which may be awkward if special environment conditions (temperature, pressure) are required.

Elliptically bent mirrors used in the Kirkpatrick–Baez geometry [54, 55] are known for decades to focus X-ray but it is only quite recently [56] that one reached small enough slope errors to produce micrometer size beams with the use of specially designed benders. This is also a totally *achromatic* system, which as mentioned above, is very interesting for many microanalytical techniques. Alignment of the two orthogonal mirrors is quite delicate and has to be assisted by computer and the whole set is quite expensive which may be their only inconvenience. By coating the two mirrors with multilayers, the incident Bragg angle becomes about ten times bigger than the critical angle needed for total reflection on plane mirrors. Thus, the aperture of this kind of optics becomes very interesting and in conjunction with the large bandpass of multilayers, the photon flux on the sample can be extremely high. However, in this case the system is no longer achromatic and this combination should be reserved for microfluorescence analysis of ultratrace elements. Recently [57], K-B systems of very high flux ($> 1E11\,\mathrm{ph\,s^{-1}}$) or very small beamspots ($\approx 100\,\mathrm{nm}$) have been commissioned at ESRF and thanks to their achromatic capability have become the "workhorse" of the installed microprobes.

Fresnel zone plate lenses were another approach for fine X-ray focusing as was demonstrated in the soft X-ray regime in Gottingen by the group of Schmahl [58]. Unfortunately, in the hard energy regime these optics [59] could not be produced thick enough to produce a good contrast between absorbing and transmitting zones. This led Aristoff [60] to propose using these diffractive optics in the reflection mode and benefiting of the lens substrate as a monochromator. This solution (Bragg–Fresnel lenses etched on multilayers) was retained to build the LURE-IMT photon microprobe [61] on the DCI (Orsay, France) synchrotron ring as for this first generation machine the small brilliance had to be compensated by a broadband pass. Now Fresnel zone plates are available for hard X-rays and are much in favor for the simplicity in alignment. Thanks to the significant progress in etching techniques these lenses have now a diameter quite comparable with the incident beam size delivered by low-beta undulators and are thus perfectly adapted to these photon sources. Contrary to BFL, these lenses can be used after a high-resolution monochromator for microabsorption experiments but the focal distance will change with energy and this has to be taken into account when scanning over a large energy range like for EXAFS experiments. This is a disadvantage over K-B optics that might be compensated by the usually large space left around the sample as their focal length is typically between 20 and 100 cm and by the compact lens holder.

Quite recently, compound refractive lenses [62, 63] were developed and are now routinely used at ESRF. Their relatively large aperture are well matched to the undulators sources, and their large focal length and the easy alignment procedure are very interesting features for a photon microprobe. When low-Z materials are used (Be [64], Li, CVD diamond [65], Si [66]) these lenses have very high transmission and can be used efficiently down to a few keV energies. When used after a monochromator for absorption experiments they will also have a change in the focal distance but due to the usually large focal length this may not be too dramatic when scanning in a small energy range like for XANES experiments.

### Sample Environment

This represents the ensemble of sample holders and chambers used for routine control and devoted to special kinds of samples. Devices like ovens, cryostats, magnetic field chambers, diamond anvil cells (DAC), etc. with their respective windows are among the customary equipment. The drawbacks of this type of environment is the appearance of multiple Compton scattering resulting in an increase of the background with a resulting pourer detection limit. When samples are in air, the absorption in the path between the sample and the detector will reduce the sensitivity for light elements (up to about $Z = 18$) and even preclude the observation of elements lighter than aluminum. In particular, the argon peak will interfere severely with the L spectrum of elements rhodium to cadmium and in certain conditions even krypton may be a nuisance and its K line may be confused with the $L_\alpha$ lead spectrum.

As already stated, one of the goals of XRF analysis is to deal with most kinds of samples. Nevertheless, in order to extract quantitative information the sample is supposed to be flat. Quasiuniversally, the sample is set in the vertical plane at 45° from the incident beam, so that the detector can be placed in the orbital plane of the synchrotron ring and at right angles to it. As the synchrotron beam has a very high linear polarization in this plane, this geometry insures obtaining the highest signal/noise ratio as the elastic (Rayleigh) and inelastic (Compton) scattering are at a minimum in the horizontal plane at 90° from the incident beam.

The sample holder has to be miniaturized and remote controlled and must respond satisfy two different constraints: a large range of displacement (few cm) and a very high accuracy in the positioning (better than 100 nm for the best microbeams) with backlash free motion. The large displacement is achieved with rather conventional steppers or continuous current (CC) motors. Minimum steps can be on the order of 0.1 μm. The absolute accuracy may not be better than 2 or 3 μm but the bidirectional repeatability is much better: around 0.2 or 0.3 μm. The rather poor absolute accuracy is not such a problem since these motors are used solely for positioning or low resolution scans. Then, for the final backlash free scanning a second set of highly precise drives such as piezo-actuators are the best choice because of their very fast movement ($\leq 20$ Hz), resolution ($\geq 5$ nm) and accuracy and repeatability well below the micron. Their small range (typically limited to 400 μm) is not really a disadvantage since at this scale it is doubtful that larger maps than $200 \times 200$ μm can be thought of.

#### *Sample Chamber*

The sample chamber has the role of reducing the amount of air in the vicinity of the sample or in the paths of the source-sample and sample-detector thus minimizing the scattering or the nefarious fluorescences of the rare gasses present in the air (Ar, Kr). Alternatively, the sample chamber must be isolated from the oxidizing influence of the $O_2$ from the air. Two alternative approaches exist: the vacuum-compatible chamber, pumped by a primary ($\approx$ mbar) or secondary ($\approx 10^{-6}$ bars) pumping system and the He chamber. The former is quite cumbersome as it requires real o-rings or gaskets and a rigid structure which cannot be miniaturized. Very often, the whole microprobe setup is enclosed in such a chamber of large dimensions, which puts a strict requirement on all the motor drives which must be vacuum-compatible and not heat up. The He chamber, on the other hand can be miniaturized as it does not need a rigid structure. Miniature plastic chambers can be easily adapted to the topmost sample stages and filled with He at atmospheric pressure. The chambers feature windows for the incident and transmitted beams made of Kapton or Mylar films and sometimes the detector is also inserted in the chamber through a window. The whole chamber can be made of a polyethylene bag surrounding the sample, the detector snout and the beam in/out windows.

*Sample Holder*

Sample holders exist in a very large variety, both commercial and homemade. The most frequent ones are systems from Nonius or Huber, which consist of goniometer heads which can accept cryo-loops or magnetic fixtures for fast lock-in. Alternatively, the samples can be glued or sandwiched between thin ultrapure plastic film such as Kapton or Ultralene of a few microns thickness and fixed on a 35 mm film slide. The slide can then be easily manuipulated and clamped into position on a special clamp which allows easy mounting on the sample topmost stages. For micropositioning, sometimes sample holders are transfered from light microscopes which allow comfortable positioning on indexed remote-controlled stages identical to the ones on the beamline. Once a particular zone has been aligned on the microscope, the whole sample holder can be transfered to the beamline drives and the position of the microscope cross-hairs will be at the center of the X-ray microbeam.

*Microscope*

A horizontal conventional microscope of high magnification ($\geq 100$) is routinely installed on the beamline, linked to the outside high resolution one. The beamline microscope must have a large working distance, normally 3.5–10 cm, so as to allow a variety of relatively bulky sample environments (oven, DAC, cryostats, etc.). It must also have a remotely controlled zoom allowing to change magnification on the sample face from outside the experimental hutch and gradually focus on the zone of interest. Commercial systems from Mitutoyo, Olympus or Questar are routinely used.

## Detector

This undoubtedly is the most critical link in the whole chain of items found in a photon microprobe. First of all, to avoid useless saturation of the counting system, the detection geometry takes into account the large horizontal polarization of the synchrotron beam. To minimize scattering from the sample, the detector is placed in the orbital plane and perpendicular to the beam direction.

To record the X-ray fluorescence spectrum, most facilities work in the energy dispersive mode (EDS) and out of them, most have retained the now very conventional Si(Li) solid-state detectors. Their main advantage is the high efficiency up to about 20 keV. The energy resolution of about 140 eV at 6 keV is usually sufficiently small to separate the X-ray lines of neighboring elements but still quite too large to avoid overlapping of the $K_\alpha$ and $K_\beta$ lines of successive elements or various interferences between K lines of light elements and L lines of heavy ones. This will impose one to use quite sophisticated programs to achieve a reliable deconvolution of the whole spectrum. The best resolution is only obtained for long shaping time constants of the main amplifier (typically around 10 μs, which means pulse durations of about 60 μs). This limits the count rate to about 5000–10000 counts per second (cps). Such values are readily reached, even with microbeams, in the case of highly fluorescing samples, like metallic objects. Pulse pileup is another drawback of this long pulse duration. This can be avoided by the use of pileup rejection systems, which are quite efficient. Nevertheless, care has to be exercised because the true sum peak is always recorded and dead time corrections must be made for the count loss in case of quantitative measurements. To support higher count rates, the use of multielement detectors is beginning to spread, however spectra from all detectors cannot be readily summed up as they have quite different characteristics. Instead, they must be treated separately, multiplying the data analysis effort by 10–30 depending on the number of individual elements.

This problem may well become obsolete with the arrival of new detectors like the silicon drift chambers (SDD) which exhibit excellent energy resolution even at very high count rates. Another advantage is that these detectors are cooled by Peltier effect leading to much more compact and light detectors than the traditional Si(Li) with its bulky liquid nitrogen reservoir. The main disadvantage is the small thickness of the detector (no more than 300 μm) which limits the efficiency. However, as is the case with the Vertex SDD from Radiant Detectors, these systems are capable of extremely high countrates of up to 1 million cps, without spectrum distortions. Alternatively, these SDD could be made out of Ge, or 0.5–1 mm thick Si thus considerably increasing their high energy efficiency.

Until now "high purity germanium" detectors (HpGe) were not used for fluorescence analysis although they have better resolution than Si-based detectors (as low as 110 eV at 6 keV). This is mainly due to the very important escape peak associated with the total absorption peak that will complicate spectrum analysis. However, if analysis of K lines from heavy elements is expected, they will become a must.

Contrary to laboratory XRF, wavelength-dispersive devices (WDX) using crystal spectrometers are rarely used in SXRF due to their low transmission (except in the case of direct excitation by a white beam). Their use in a scanning Bragg angle mode is not fully satisfactory for multi-element analysis. The loss of time required to get a complete spectrum may then not be compensated by the gain in energy resolution that will remove most interferences.

A good compromise could be the use of filters. Such filters can be positioned between the sample and the detector. An aluminum filter would absorb low energy photons more strongly than high energy ones but the filter purity is mandatory.

A better solution is to use wavelength-dispersive elements as a filter. This can be done in two different ways. The first one, suggested by Grigorieva [67], uses cylindrical highly oriented mosaic graphite (HOPG) crystals placed between the sample and the detector. A beam stop on the system axis allows only diffracted photons to reach the detector. Depending on the sample to detector distance, a $\Delta\theta$ domain around the mean angle $\theta$ is transmitted which means that the detector will only receive photons in an energy domain $\Delta$E around a mean energy $E$. The energy resolution is that of the Si(Li) detector but saturation can be avoided when covering an energy range without major elements. The major peaks are produced by major element fluorescences or Compton scattering and they render direct EDS analysis impossible, such as for instance measuring Ni in a stainless steel alloy or trace metal elements in a thick low Z matrix.

Another arrangement proposed by Chevallier [68] tries to ally the high resolution of WDS systems with the high efficiency of EDS devices. It uses a flat highly oriented mosaic graphite crystal and a position sensitive proportional counter as detector. Depending on the distance and the mean angle, an energy domain is analyzed. It can be shown that this system is in first order focusing versus the acceptance angle. Consequently, for a sufficient sample-detector separation $d$, the count loss does not follow a $1/\mathrm{d}^2$ law and the energy resolution is that of the crystal which can be of only a few eV.

### 7.2.3 Quantitative Aspect

The integrated peak area $A$ of a characteristic line in the fluorescence spectrum is representative of the concentration of the corresponding element in the sample. If the incident angle of the primary beam and of the outgoing fluorescence photon on the sample are both equal to alpha, the relation between this area and the concentration is derived through the classical fundamental parameter method and is of the form:

$$A_{ijk} = I\left(E_0\right) C_i F_{ijk} \frac{\tau_{ij}\left(E_0\right)}{M} \left[1 - \exp\left(-M \frac{\rho z}{\sin\alpha}\right)\right], \qquad (7.1)$$

where $I(E_0)$ is the incident flux in photons per second, $C_i$ the concentration of element $i$ (in $\mathrm{gg}^{-1}$), $F_{ijk}$ the probability that an ionization in shell $j$ of element $i$ leads to the emission of line k, $\tau_{ij}$ the photoelectric absorption coefficient of element $i$ in shell $j$, $\rho$ the density and $z$ the thickness of the sample. $M = \mu(E_0) + \mu(E_{ijk})$, where the various coefficients are the mass absorption coefficients of the matrix (excluding trace elements) at energy $E_0$ and $E_{ijk}$ respectively.

This general equation shows that if we know the area mass of the sample ($\rho z$ in g cm$^{-2}$) and the major element content (**M**), the concentration of element $i$ can be calculated from the area of the $ij$ line. Nevertheless, we must keep in mind that this result is obtained after integration of the contribution of an infinitesimal volume of the sample over the whole thickness and this implies intrinsically that the sample composition is homogenous. We normally perform elemental mapping of the sample with a microbeam because of its inhomogeneity but then we can no longer rely on this equation.

Two extreme cases of this finite thickness target equation can be thought off: the thin target case where we can use a first order approximation of the exponential term, and at the opposite, the infinitely thick target approximation where the exponential term is null:

$$A_{ijk}^{\text{thin}} = I(E_0)\, C_i F_{ijk} \tau_{ij}(E_0) \frac{\rho z}{\sin\alpha} \quad \text{and} \quad A_{ijk}^{\infty} = I(E_0)\, C_i F_{ijk} \frac{\tau_{ij}(E_0)}{M}. \qquad (7.2)$$

If we introduce the notion of fluorescence half depth ($D_{1/2}$), that is the depth from which half of the fluorescence signal of element $i$ comes from, we can represent these equations on a universal curve whatever the sample (see Fig. 7.25). If we allow a 10% deviation from the exact formula we see that the thin target approximation is valid up to 0.3 $D_{1/2}$ and the infinitely thick target can be retained for $D \geq 0.3\ D_{1/2}$. The problem is that the same sample can

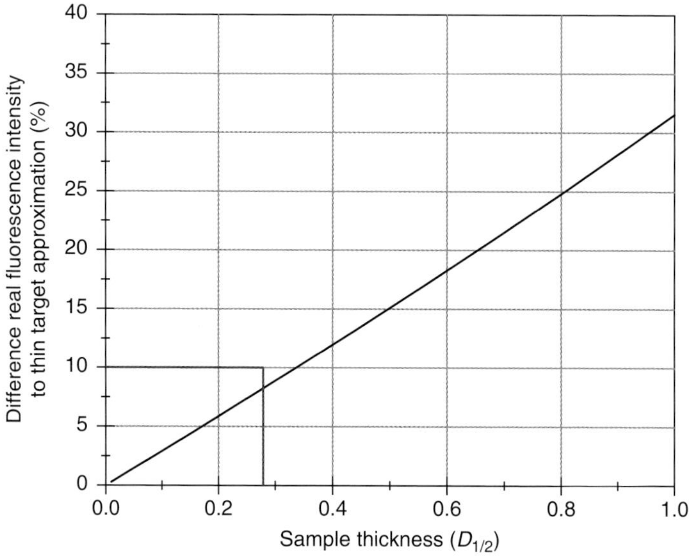


**Fig. 7.25.** Real fluorescence to thin target approx. difference as a function of the half-depth

be considered as thin, thick or infinitely thick depending on the element under consideration. This is shown in Fig. 7.26 where we have reported on a bar chart the limits of these three domains for element Si to Zn in a representative soil sample (we used the Clark [69] evaluation of natural abundance of elements on Earth). We see that a 10-μm thick sample is infinitely thick for elements up to silicon, can be considered as thick from potassium to copper and becomes thin for zinc.

In the infinitely thick approach the knowledge of **M** is mandatory. This can sometimes be calculated for each pixel of the map if all elements are detected which is absolutely not true in many cases. For example, C, O, B and Li that can be in quite high concentrations are very unlikely to be detected with conventional Si(Li) detectors even in vacuum whereas Na, Mg, Al and Si are not usually detected when operating in air. Of course, in this case the hypothesis of homogeneity is much more difficult to comply with. Furthermore, the lateral resolution which can be as low as one μm or less, is flared out by the penetration depth of X-rays in case of heterogeneous samples.

In effect, the thickness of the sample should fit the lateral resolution fixed by the microbeam size. This means that if 1-μm resolution is required we will quite often be in the thin target approximation. Referring to Fig. 7.26 we see that the thin target approximation ($z < 0.3\ D_{1/2}$) is valid for practically all elements in biological samples where precisely one can make such thin slices. It would be the case for most elements in geological samples if 1-μm thick slices could be obtained which is quite uncertain for soil samples for example. For metallurgical samples (Fe matrix) we shall nearly always need samples of less than 1-μm thick to be in this approximation.

Quantitative analysis seems easier in the thin target approximation. Nevertheless the area mass of the sample is needed which may seem trivial but which is not so in many cases. The uncertainty on thin target thicknesses is

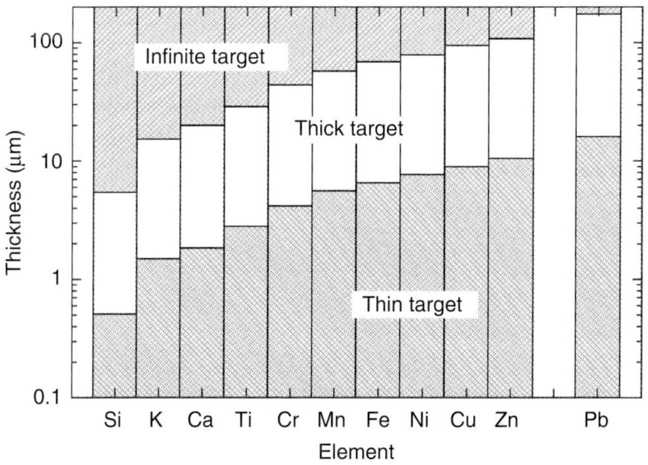


**Fig. 7.26.** Thin, thick and infinite target thicknesses as a function of elements

often quite large and significant variations from point to point can exist. In other samples the heterogeneity is so large that even a good knowledge of the thickness is insufficient to have a rough idea of the area mass. This is the case of soil samples where important void fractions makes the problem even more difficult. In laboratory XRF, the intensity of the Compton and Rayleigh scattered radiation from the sample is often used for quantitative purposes. Unfortunately, it is not as simple as that for polarized synchrotron radiation and this technique has not been validated yet. As the beam polarization can change from one experimental run to another or even during the same run if the orbital plane changes only very slightly, it cannot yet be considered as a reliable parameter. Anyhow, for such thin samples we will have to take into account the contribution of the scattered radiation from the sample's supporting substrate, which is often a thick silicon slice.

In conclusion, contrary to what is often praised, absolute quantitative microscopic fluorescence analysis is still very difficult to achieve in most cases.

### 7.2.4 Elemental Mapping

#### Acquisition Procedure

The software used during the acquisition should allow to automatically make point analysis where the coordinates of each points of interest are stored in a file, line scans (not restricted to horizontal or vertical lines) simply by setting the coordinates of the end points and of course polygonal maps of various areas of the sample.

The program places the sample in the starting position, already referenced on the indexed external microscope, then starts the acquisition for a preset time, stores the data, and repeats this cycle of operations until all points have been analyzed.

For some positions in the sample we may find locally very high concentrations of highly fluorescent elements. This will introduce dead time corrections and we must record for each analyzed point the exact real and live (active) times. Normalization to the incident flux measured right in front of the sample by a PiN diode (photodiode) is of course, mandatory.

#### Acquisition Time and Data Storage

In practice, with $10^{10}$ photons per second in the beamspot (a reasonable value after monochromatization on third generation machines) we can observe transition elements in a biological matrix at the ppm level in less than 10 s/ point. For a $64 \times 64$ map with a 1-µm resolution this takes about 10 h of beam time. However, the dwell time per pixel depends also on the sample's matrix. For example, if we wish to observe zinc at the ppm level in an iron matrix that means that we shall record about one million iron $K$ lines for a single one of zinc. Then if we can tolerate a 10 kcps count rate in the detector (a maximum for most Si(Li)) we shall only record one zinc $K$ line every 100 s. Such an analysis is unfeasible with Si(Li) detectors and here only WDS devices can be used. So far we only considered counting times, but when mapping out a sample, the motor finite moving time must also be added as well as overhead times required by the control protocol communicating with the ADC and storing the files on the associated hard drive. This can be up to 0.5 s/point. Recently, at ESRF a new raster scan, whereby the two scanning directions are sampled while being continuously scanned, was implemented. This allows reducing the overhead time to a few ms when using piezoelectric drives, and also acquiring full maps in very short times. For samples undergoing radiation damage such as biological samples or very thin substrates, it is necessary to perform several "raster" scans finally summed up to obtain the necessary statistics, while allowing the sample to "recover" and cool down between two successive passes.

For each pixel we must store the two coordinates, the real and live times, the content of the region of interest (ROI) corresponding to the various elements searched as well as the complete fluorescence spectrum. In the energy range of typical X-ray fluorescence spectra (0–20 keV), 2048 channels for the ADC are sufficient so that each spectrum, stored as a binary file takes about 4–5 Kbytes of memory. For a $64 \times 64$ map this represents about 16–20 MB, and about two-three times more if we store them in the plain ASCII format.

### Data Reduction

The ideal way to proceed would be to analyze the spectrum of pixel $i$ while recording the fluorescence spectra of pixel $i + 1$. Thanks to the actual speed of individual computers this is not a problem. In fact the problem is usually not linked to computation time but rather to the ability of the program to recognize the exact (complete) list of elements, escape peaks, and sum peaks to be considered in each pixel. The experimentalist has to establish the list of lines to be fitted as well as the region of the full spectrum where the fit is applied before calculation can start. Furthermore, the background shape has to be chosen from a list of customary shapes. All these specifications are stored in an input file. However, if an element not reported in this file appears in a given pixel, this element will never be considered and, worst, in order to minimize the $\chi^2$ value the program will wrongly estimate the area of the other peaks. This problem can be minimized if at the same time that we save the energy spectrum of each pixel we record a special file, which contains the sum of all individual spectra. It is this sum spectrum that will be used for the first analysis because in this case the list of element is much more likely to be complete and the input file thus established will match all the observed lines. Current software codes publicly available include Qxas[1] from the IAEA, Seibersdorf and PyMCA[2] from the ESRF, Grenoble.

[1]QXAS - Manual for Quantitative X-ray Analysis System. (1996)

[2]http://www.esrf.fr/computing/bliss/downloads/index.html

## Data Analysis

Usually when the acquisition is finished the scientist is left with a file containing a table where for each pixel are reported the acquisition and total elapsed time and the content of each selected ROI. After correction from beam intensity decay, this file can directly be used by the appropriate software (Igor, Surfer) to produce maps in various representations for each element.

For cases when the full spectrum is acquired, a special fitting program allows their deconvolution and produces a table very similar to the preceding. Here the true area of each peak replaces the ROI content and a special column collects the goodness of fit $\chi^2$ value of each computation. An abnormally high $\chi^2$ value will reveal a problem during deconvolution (most of the time the omission of a pertinent element for this particular pixel).

It is the comparison of the maps of each element that is important. This often reveals a more or less strong correlation (or anticorrelation) between some elements. This is an interesting step toward the characterization of the species encountered, especially if we keep in mind that we can work at the ppm level. The elemental map images can be further treated using high level image processing tools such as principal component analysis (PCA) or factor analysis in order to isolate and classify by order of importance their most relevant features. As said above, maps are most easily and rapidly drawn from the table of the various ROI contents. These maps are very useful for an immediate estimate of the result of the experiment and will help deciding the opportunity to perform other kind of analyses, during the same run and on the same sample.

For this reason the photon microprobe is a quite unique tool due to the variety of interaction processes with matter. In particular the possible correlation brought to evidence, will show the exact position on the sample where absorption (EXAFS or XANES) or diffraction measurements would bring invaluable information on the chemical state or structural environment of an element. We thus recommend that micro-XRF be associated, on the same beamline, without removing the sample, to simultaneous or subsequent absorption or diffraction measurements.

As an example we present in Fig. 7.27 the map of the distribution of iron and manganese in a polluted soil sample. The dark area represents high el-

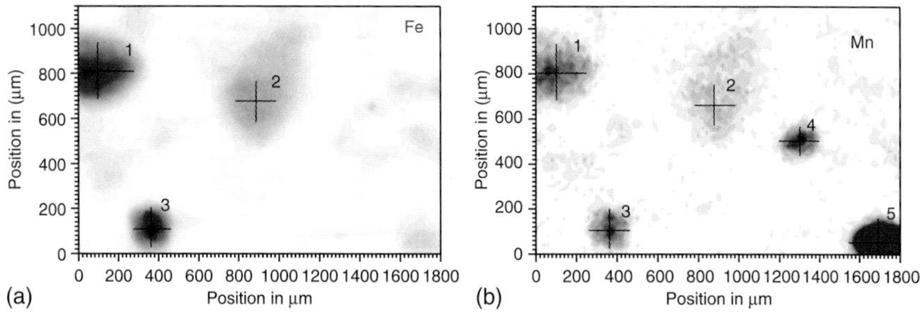


**Fig. 7.27.** Fe **(a)** and Mn **(b)** maps in a soil sample using ROIs

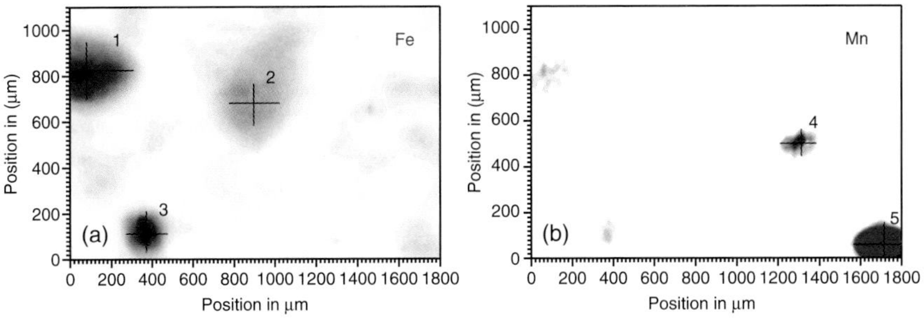


**Fig. 7.28.** Fe **(a)** and Mn **(b)** maps in a soil sample using fitted lines

emental concentrations. In map (a), Fe seems concentrated in three distinct regions marked 1, 2 and 3. In map (b), Mn appears to be concentrated in the same regions (1,2,3) plus two other marked 4 and 5 (very intense). This suggests that in this sample at least two different forms of Mn cohabit: one associated to iron (1,2,3) and another uncorrelated with this element (4 and 5).

Now if we look at Fig. 7.28, which represents the same map but generated from the true areas of the fitted $K_\alpha$ lines of Fe and Mn on each spectrum the situation is completely different. Iron still appears at positions 1, 2 and 3 but manganese is now only found in regions 4 and 5. What has happened is obviously explained in Fig. 7.29, which is the fluorescence spectrum of a pure iron sample. We see that the fluorescence peak recorded with a Si(Li) detector shows a low energy tail that extends quite far from the main peak and in particular in the ROI devoted to Mn.

Consequently, there is always a given fraction of the Fe intensity in the Mn ROI. This is why Fe and Mn seem so strongly correlated in regions 1, 2

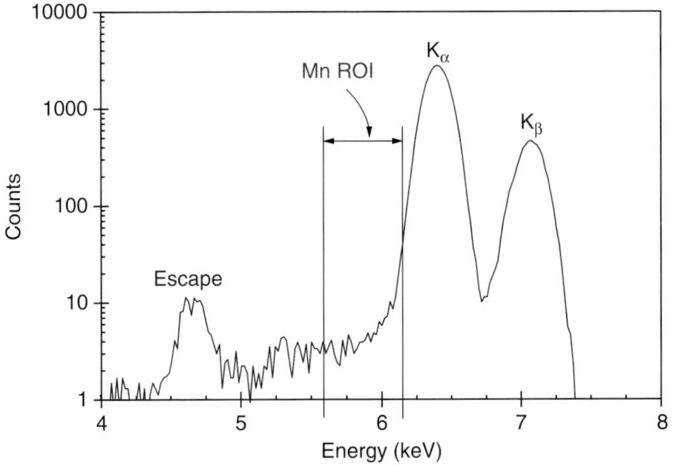


**Fig. 7.29.** Fe spectrum in a soil sample

and 3 of Fig. 7.27, whereas Mn has completely disappeared from these regions in Fig. 7.28. Another interesting feature is seen in the lower right corner of the iron map of Fig. 7.27 where we can observe a slight iron concentration corresponding to region 5 of the Mn map. In regions 1, 2 and 3 Fe is dominant and Mn in low concentration while it is the opposite in region 5. In Fig. 7.28, this has disappeared because there is effectively no iron in region 5. If there is a high concentration of Mn in this area its $K_\beta$ line, which falls in the Fe $K_\alpha$ ROI will misleadingly suggest the presence of iron in this position.

### 7.2.5 Examples of Applications

#### Single Cell Spectroscopy Using PINK Beams

deeper insights into biological processes and cellular analysis require nondestructive quantitative techniques with high elemental and chemical sensitivity and high spatial resolution, such as synchrotron induced X-ray fluorescence (SXRF). In this work [70], we performed SXRF microanalysis of single cells by imaging the intracellular distribution of trace elements and pharmacological doses of the anticancer drug, 4′-iodo-4′-deoxydoxorubicin (IDX). Spatial distribution and concentration of trace elements in tissues are important, as they are involved in some pathological conditions and in many biological functions of living organisms like metabolism and nutrition. CRL lenses offer several advantages compared to other focusing systems, mainly high flux tolerance, efficiency for focusing high energy X-rays up to 60 keV, and almost no spherical aberrations. These lenses used with a "pink" (polychromatic) excitation provide a fast acquisition rate and sub-ppm limits of detection. Additionally, the CRL lenses are currently the only focusing devices capable of sustaining the high flux, high heatload of a PINK beam without noticeable damage.

A human ovarian adenocarcinoma (IGROV1) cell line was used for this study. The cells, grown directly on 0.2-μm Formvar film were incubated with complete culture medium and exponentially growing cells were exposed to 5 μM IDX. Cell monolayers were rinsed, then cryofixed and freeze-dried at $-30°$ C. Analyses were performed at 14 keV using either pink or monochromatic excitation. Pink excitation is produced by direct, high intensity, medium bandwidth beams from the undulator impinging onto a flat mirror. The multistrip mirror spans several full undulator harmonics and decreases the beam heatload by a horizontal deflection at a grazing angle of 2.6 mrad which gives an energy cutoff of 24 keV for the Pd strip. In order to remove the contribution of lower energy harmonics, a 2-mm Al filter was used. In the pink beam, CRLs produce a flux of about $5 \times 10^{10}$ photon $s^{-1}$ $\mu m^{-2}$ and a beam spot of 10 μm horizontally by 1 μm vertically for 50 Al-lenses, focal distance 713 mm. For the same spot size, CRLs with monochromatic excitation give a flux about 10 times less. The minimal detection limits (MDL) evaluated using a NIST standard reference material (SRM 1833) yielded about 30 ppb for elements such as Zn. Data analysis was performed using the WinAxil software [71] in order to correct for X-ray photon background and fit elemental X-ray lines detected in the sample.

In the case of a cell monolayer, the thin sample approximation can be applied for samples previously measured by RBS analyses to have a mean surface mass around 260 μg cm$^{-2}$. With polychromatic pink beam a close look at the undulator spectrum reveals a contribution of photons of higher energy harmonics at 16.9, 19.7, and 22.4 keV, which are greatly reduced by the lens and 10-μm pinhole assembly installed before the sample. The $\boldsymbol{k}$ ratio of the number of photons relative to the 14 keV harmonic is obtained from the undulator spectrum calculated using the ESRF Synchrotron Radiation Workshop (SRW) code [72]. Only the 14 keV harmonic is effectively focused inside the 10-μm pinhole; the other three higher energy harmonics before the mirror cutoff at 24 keV spread out over a large area and give a reduced contribution through the pinhole. A PIN diode placed before the sample and used for normalization, generates a total current $I_0$ for a given flux of $N_0$ photons s$^{-1}$ of energy $E_0$ as:

$$I_0 \propto N_0(1 - \mathrm{e}^{-\mu_0 d}) \tag{7.3}$$

with $\mu_0$ the energy deposition coefficient for Si, and $d$ the Si PIN-diode thickness. For a given PIN diode, the current is in fact:

$$I = I_0(1 + k_1 f_1 + k_2 f_2 + k_3 f_3) \tag{7.4}$$

with $k$ ratios previously calculated and $f_i$ obtained according to PIN-diode calculations using the energy deposition coefficient for each contributing harmonics of energy $Ei$ . Since $I$ measured in pA is known, the number of 14 keV photons $N_0$ is estimated, then those of higher and lower energy harmonics. From these values and those of fluorescence cross-sections $\sigma_F(Ei,\ Z)$, corrected quantitative elemental maps were generated. The following equation was used for quantization, based on the thin sample approximation:

$$N_Z = \frac{N_0\ t\ C_Z\ \sigma_\mathrm{F}\ \exp(-\mu_\mathrm{air}\ (E_\mathrm{F})\rho_\mathrm{air} d)\varepsilon_\mathrm{det}\ \Omega}{4\Pi\ \sin\alpha} \times \int_0^T \exp\left[-x\rho_\mathrm{S}\left(\frac{\mu_\mathrm{s}(E_0)}{\sin\alpha} + \frac{\mu_\mathrm{s}(E_\mathrm{F})}{\sin\beta}\right)\right]\mathrm{d}x, \tag{7.5}$$

where $N_Z$ is the number of counts in the characteristic line of element $Z$, $N_0$ is the incident photons s$^{-1}$, $t$ is the integration time, $C_Z$ is the concentration of element $Z$, $\sigma_F$ the fluorescence cross-section in cm$^2$ g$^{-1}$, $\mu$ the mass absorption coefficients of air or sample for the respective energies, $\rho$ the density (g cm$^{-3}$) of air or sample, $\varepsilon_{det}$ the detector efficiency, $\Omega$ the detection solid angle, $\alpha$ and $\beta$ the beam incident and take-off angles, $x$ the sample depth coordinate, integrated from 0 to $T$ (sample thickness). Spectra taken from single-cells treated with 5 μM of IDX are shown in Fig. 7.28. The spectrum taken using a monochromatic beam with 120 s acquisition time (Fig. 7.30(a)) is still of poor

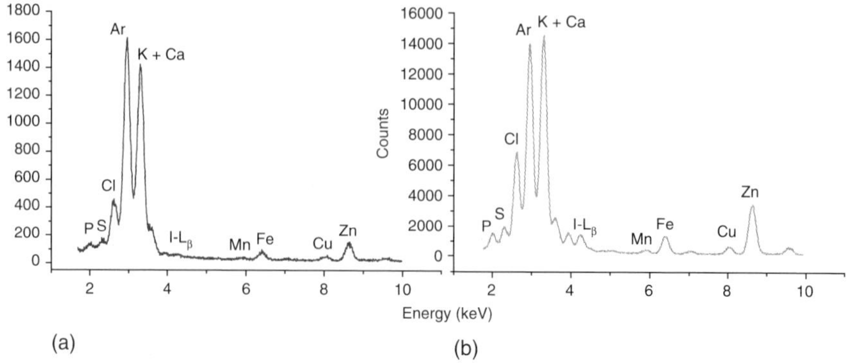


**Fig. 7.30.** Spectra obtained from a IGROV1 cell treated with 5 μM of iododeoxy-doxorubicin and freeze-dried. **(a)** Spectrum obtained using a 14 keV monochromatic focused beam, acquisition time is 120 s. **(b)** Spectrum obtained using a focused 14 keV polychromatic pink excitation, acquisition time is 100 s

quality while the one from pink beam (Fig. 7.30(b)) 100 s counting time shows well-defined X-ray peaks of intracellular elements P, S, Cl, Ca, K, Mn, Fe, Cu, Zn, and I from drug treatment.

Potassium is the major element in cells and gives elemental maps with the highest counting statistics. Compared to light microscopy cell visualization, potassium maps depict the cell boundaries roughly, particularly in the nuclear region. Iodine imaging of cells treated with 5 μM of IDX was performed in this study and yielded intracellular distributions of trace elements comparable to previous results obtained by micro-PIXE (proton induced X-ray emission) for quite higher doses of IDX of about 20 μM. Particularly, the colocalization of iron and iodine within the cell nucleus is still observed. The results obtained on the iodine distribution, in comparison with potassium and iron, are presented in Fig. 7.31 This was also found using monochromatic excitation (data not shown), but more than 12 h of mapping were necessary. From quantitative analysis, the surface concentrations displayed a maximum of 0.02 $\mu g\,cm^{-2}$ for iron and 0.15 $\mu g\,cm^{-2}$ for iodine in the nuclear region. Using a mean value of freeze-dried cells surface mass of 260 $\mu g\,cm^{-2}$ obtained by RBS, maximum concentrations for cells treated with 5 μM IDX were 10340 ppm for potassium, 274 ppm for zinc, 76 ppm for iron, and 580 ppm for iodine.

These results are in agreement with previously published data on IGROV1 cells trace element content [73]. When estimating error propagation the main contributing source of error on the calculated concentration C will be given by the standard deviation on the cell thickness $T$. The radiation dose deposited within a pixel is roughly about $10^6$ Grays with pink excitation and can be down to $10^4$ Grays using monochromatic excitation which is several orders of magnitude less than scanning ion microprobe, and still acceptable for freeze-dried samples. No radiation damages (shrinkage or change in color) were observed at the cellular level but it can be noted that under the pink

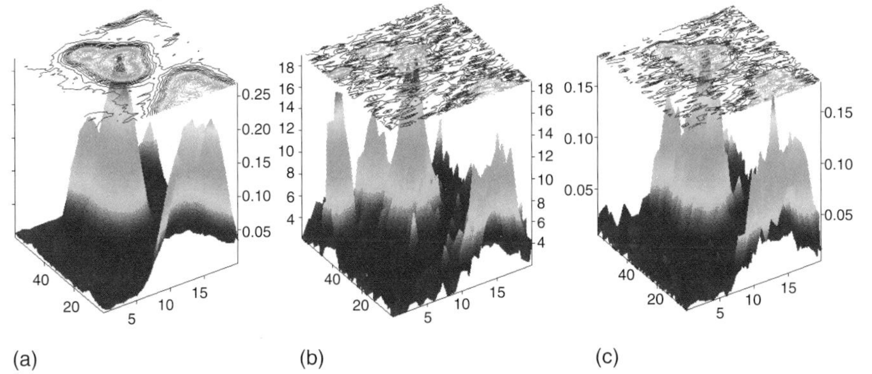


**Fig. 7.31.** Two-dimensional elemental distribution of a freeze-dried cancer cell treated with 5 μM of iodo-deoxydoxorubicin. The cell was mapped with a 14 keV polychromatic pink excitation, stepsize $1 \times 3\,\mu m^2$ (V × H) and 2.5 s acquisition time/step, scan size: $60 \times 60\,\mu m^2$, around 2 h total acquisition time. **(a)** Potassium distribution ($K_\alpha$ X-ray line), **(b)** Iron ($K_\alpha$ X-ray line), **(c)** Iodine ($L_\beta$ X-ray line)

beam, the thin Formvar films can be damaged mainly due to the high number of ionizations rather than from the rise in temperature. A compromise is required between the desired resolution, mapping time, and the dose that the sample can sustain. Improvements in limits of detection can be reached working above the $K$ edge of iodine $B_K \approx 33.2\,\text{keV}$ with micron-sized beams of high energy and this will be achieved by the future development of CRL lenses. Alternatively, the samples can be raster-scanned using a piezo YZ assembly, spreading out the heatload over several successive scans. This experiment opens a new way toward possibilities of mapping intracellular distribution of drugs used at pharmacological doses. SXRF is complementary to confocal microscopy or nuclear microprobe analysis, and brings missing information not accessible by any other technique. Finally, simultaneous chemical speciation by X-ray absorption spectroscopy and microanalysis of living cells are exciting perspectives under active investigation.

## Element Specific Tomography

X-ray fluorescence computed microtomography (XFCMT) is a nondestructive, noninvasive imaging method which was introduced over 9 years ago [74] and started to play an increasing role in microanalysis [75–77]. XFCMT is an excellent complementary technique to phase contrast imaging in that it offers the much-needed elemental sensitivity down to trace element concentrations with the same micron-sized spatial resolution. In order to retrieve the quantitative 2D/3D elemental information at the end of the tomographic scan, reconstruction techniques are used as opposed to the direct imaging methods associated with 2D mapping. As it requires a pencil beam as its probe, synchrotron radiation fluorescence tomography expanded and became a precise and relatively straightforward method of microanalysis only with the advent of third-generation synchrotron sources such as the ESRF, APS and SPRING 8 that provide high energy/high flux beams.

In the following, we describe high precision experiments performed at the ID22 beamline of the ESRF on real samples featuring inhomogeneous elemental distributions, from the fields of Earth and Space Sciences [78].

Depending on the desired resolution, either of the vertical or horizontal scanning geometries are used, with the associated detectors positioned vertically, respectively horizontally at 90° to the beam. The horizontal focusing/scanning geometry exhibits a significant decrease in flux necessary to demagnify the rather large horizontal source size by closing down the beamline horizontal slits but has a better spectral purity, as it features Rayleigh and Compton scattering a few tens of times less than the vertical geometry, thanks to the 90° angle between incident and outgoing photons in the orbit plane and the high degree of linear polarization in the horizontal plane.

The two movements used for the tomographic scans are $Z/Y$ (vertical/horizonatal, precision 0.1 μm) and $R_Y/R_Z$ (rotation around a horizonatal/vertical axis, precision 0.001°). The other movements ($X$, $R_X$, $y$) are used to align the sample rotation axis in the beam. The sample is mounted on a Huber goniometer head which is prealigned on a visible microscope setup in order to bring its rotation axis perpendicular to the beam and to reduce precession of the tomographic rotation axis at the beam position.

Combining XFCMT with the transmission-tomography which provides the spatial distribution of the total absorption coefficient inside the sample is done routinely. However, coupling to these two probes the Compton tomography allows accessing the spatial distribution of the electron density, which is closely related to the mass-density distribution. The transmitted, fluorescent and Compton signals are acquired simultaneously. While all these three techniques are well known, up to now the problem of integrating the information that they provide has never been discussed in-depth. Furthermore, the reconstruction problem for fluorescence tomography is much more difficult than it is for transmission tomography, due to the absorption of the photons on the excitation and detection paths, and a general solution to this problem has not yet been found.This work introduces a completely new perspective on the problem of integrating the information from the three different types of signals. A new reconstruction method is used for the reconstruction of a grain of the martian meteorite NWA817, which is an excellent test case, as it is composed to a great extent, as all such meteorites, of elements inaccessible in fluorescence such as O, Mg, and Si.

The modification in the setup from the one in Fig. 7.24 is the fluorescence detector, located in the vertical plane at 90° from the incident beam. The sample is placed on a rotation-translation stepper system for translation along a vertical direction perpendicular to the beam and rotation around an axis perpendicular both to the beam and to the translation.

*Transmission Tomography*

The signal used in the transmission tomography is that collected by the photodiodes set before and after the sample. For a given value of the angle $\theta$ and of the translation $s$, the intensity of the beam that is transmitted across the sample is:

$$I(s,\theta) = I_0 \exp\left(-\int \mu(s,u,E_0)\,\mathrm{d}u\right), \tag{7.6}$$

where $I$ is the intensity of the incident beam and $\mu$ is the linear absorption coefficient at the point of coordinates $(s,u)$ and at the energy of the primary beam $E_0$. (see Fig. 7.32). The projection function $P$ is defined as:

$$P(s,\theta) = \log\left(\frac{I_0}{I(s,\theta)}\right) = \int \mu(s,u,E_0)\,\mathrm{d}u = \Re\mu, \tag{7.7}$$

where we defined $\Re$ as the Radon transform operator [79]. The reconstruction problem of transmission tomography consists in the inversion of the Radon transform operator, i.e., given a set of measured projections $P(s,\theta)$ the reconstruction aims at finding a distribution of the absorption coefficient $\mu$ whose Radon transform $\Re\mu$ is equal to the measured projections. Using various reconstruction algorithms it is possible to reconstruct the spatial distribution of the linear absorption coefficient at the energy of the primary beam.

*Compton/Rayleigh Tomography*

The Compton/Rayleigh tomography is based on the signal produced on the energy dispersive detector at 90° by the photons that come from elastic

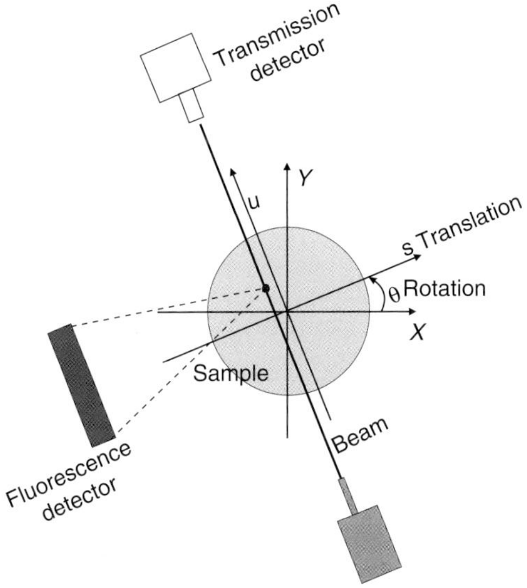


**Fig. 7.32.** Experimental setup for tomography experiments

(Rayleigh) and inelastic (Compton) single-scattering events. As these two effects produce quite different lines, resolved by the energy dispersive detector only for high energies, we will sum these contributions as "scat." Referring to Fig. 7.32, the contribution of a small path d$u$ along the beam to the detected signal is given by

$$\mathrm{d}N = I_0 \varepsilon_{\mathrm{D}} f(s,u) p_{\mathrm{scat}}(s,u) g(s,u)\,\mathrm{d}u \tag{7.8}$$

with $I_0$ the intensity of the incident beam, $\varepsilon_D$ the detector efficiency and $f(s,u)$ the probability that a photon entering the sample reaches the point $(s,u)$:

$$f(s,u) = \exp\left(-\int^{u} \mu(s,u',E_0)\,\mathrm{d}u'\right) \tag{7.9}$$

and $p_{scat}$ is the scattering probability per unit path-length per unit solid angle

$$p_{\mathrm{scat}}(s,u) = \sum_i N_A \frac{\rho_i}{A_i} \frac{d\sigma_{\mathrm{scat}}}{d\Omega}(Z_i, E_0, \Theta), \tag{7.10}$$

where $N_A$ is the Avogadro number, $\rho_i$, $A_i$ and $Z_i$ are the density, the atomic weight and the atomic number of the $i$th element, $\mathrm{d}\sigma/\mathrm{d}\Omega$ is the differential scattering cross-section (Compton + Rayleigh) and $\Theta$ is the scattering polar angle of the detector with respect to the incident beam direction. $g(s,u)$ is the probability that a photon scattered from $(s,u)$ reaches the detector

$$g(s,u) = \int_{\Omega_{\mathrm{D}}} \mathrm{d}\Omega \exp\left(-\int_{(s,u)}^{\mathrm{Det}} \mu(l,E')\mathrm{d}l\right), \tag{7.11}$$

where $\Omega_D$ is the solid angle from the interaction point to the detector surface and $\mu(l,E')$ is the absorption coefficient at the energy of the scattered photon. The radiation reaching the detector, normalized by the factor $\varepsilon_D$ $I_0$, can be obtained by an integration along the path d$u$:

$$S_{\mathrm{scat}}(s,u) = \int f(s,u)\, g(s,u)\, p_{\mathrm{scat}}(s,u)\,\mathrm{d}u. \tag{7.12}$$

This equation is often referred to as the generalized Radon transform of $p_{scat}$ and lacks a general analytical solution, although several algorithms to numerically solve it have been proposed [80, 81]. At high energies and not-too-low angles the scattering cross-section is dominated by the Compton contribution, through the momentum transfer $q = E_0 \sin\left(\frac{\Theta}{2}\right)$. For high values of $q$ the free-electrons approximation holds, and the Compton scattering probability per unit path-length per unit solid angle can be written simply as the product of the electron density $n$ and of the Klein–Nishina cross-section. Expanding $n$ we obtain:

$$p_{\mathrm{Compt}}(E_0,\Theta) = \sum_i N_{\mathrm{A}} \rho_i \frac{Z_i}{A_i} \frac{\mathrm{d}\sigma_{\mathrm{KN}}}{\mathrm{d}\Omega} \approx \frac{1}{2} N_{\mathrm{A}} \rho \frac{\mathrm{d}\sigma_{\mathrm{KN}}}{\mathrm{d}\Omega} \tag{7.13}$$

This approximate relationship can be used for estimating the spatial distribution of the mass density from the Compton tomography reconstruction.

*Fluorescence Tomography*

Fluorescence tomography is based on the signal produced on an energy sensitive detector, generally placed at 90° to the incident beam, by the photons coming from fluorescent emission. For the element $m$, the expression of the fluorescent signal is formally identical to that of the Compton/Rayleigh signal with the difference that the cross-section of scattering is replaced by the fluorescence one and the absorption coefficient at the scattered energy (close to the incident energy $E_0$) is replaced by the one at the fluorescence emission line $E'$. Note that while $\mu(E_0)$ can be reconstructed from the transmission tomography, $\mu(E')$ is a difficult quantity to evaluate. As a practical solution, the absorption at the fluorescent line energies is sometimes scaled as $E^{-3}$, independent of the composition, under the assumption that the main contribution to the absorption is the photoelectric effect. This approximation is justified only if the scaling interval is not in the vicinity of the absorption edges of the sample elements or is much larger than them. Another approach is to try to solve a more complex numerical problem, where the dependency of $\mu$ on all the element concentrations is taken into account . A method that provides a numerical solution to the reconstruction problem by combining the information that comes from transmission, Compton/Rayleigh and fluorescence tomography has been developed [78]. This method uses so-called "optimal-estimate functions" to extrapolate the absorption coefficient $\mu(E')$ at all fluorescence lines energies from $\mu(E_0)$ and $p_{\text{scat}}(E_0, \Theta)$. For the fluorescence reconstruction it is necessary to estimate $\mu(E')$ at the energies of elements not accessible in fluorescence ($Z \leq 15$). When the mass-density distribution of some elements is known, one can separate the contribution of the known elements to $\mu(E_0)$ and $p_{\text{scat}}(E_0, \Theta)$ from the contribution of the unknown treated as a whole.

The reconstruction algorithm starts from an initial guess of the distributions of all the elements with $Z \geq 15$ which can be obtained by a reconstruction with no self-absorption corrections. Further details of the reconstruction algorithm, combining the three tomographic signals can be found in the paper by Golosio et al. [78]. This method was applied to the study of a grain extracted from the martian meteorite NWA817 and stored in a silica container. Special care was given to the determination of the Fe and Mn contents since they are key values to discuss the martian origin.

The grain was placed inside a pure silica capillary having a 170 µm-internal diameter and 45-µm-thick walls. A monochromatic beam of 25 keV was used, with a transverse beamsize of $3 \times 5 \mu m^2$ (Vert. × Hor.). The beamspot was obtained using compound refractive lenses for focusing coupled to a 5-µm pinhole to reduce the horizontal size. The measured flux on the sample was about $10^{10}$ ph s$^{-1}$ in the focused beam. The capillary was placed on a system

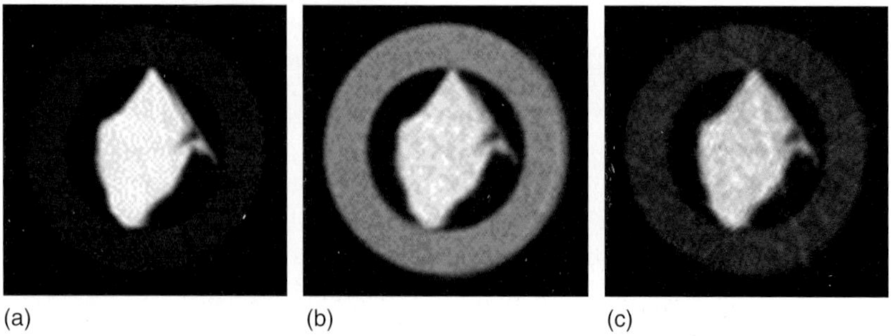


**Fig. 7.33.** NWA 817 grain reconstructed in **(a)** transmission, **(b)** Compton and **(c)** Rayleigh

for vertical translation and rotation around a horizontal axis (vertical scanning geometry). A Si-drift detector placed at 90° in the vertical plane was used for measuring the intensity of the fluorescent/scattered signals. The sample was scanned with 120 angular steps of 3° and 100 translation steps for each angle. Fitting and background subtraction for the X-ray characteristic lines as well as for the Compton and Rayleigh signal were performed using the Axil program. All distributions were reconstructed on 100 × 100 square grids of 3 μm pixel size.

The distribution of $\mu(E_0)$ was reconstructed from transmission tomography using the filtered back-projection algorithm and is presented in Fig. 7.33a. The distributions of $p_{Compt}$ and $p_{Ray}$ have been reconstructed from the Compton and Rayleigh signals using the SART algorithm, modified for absorption correction (Fig. 7.33b and c).

As expected from the Compton, Rayleigh and photo-absorption cross-sections, respectively proportional to $Z$, $Z^2$ and $Z^4$, the relative contrast between the pure silica capillary and the ferrous silicate grain increases from the Compton to the transmission reconstructions. Iron and manganese contributed significantly to the fluorescent signal and their reconstructions with/without self-absorption corrections are presented in Fig. 7.34a and 7.34b (Fe) and 12a and b(Mn).

Self-absorption effects are clearly visible in the image without self-absorption corrections, as an underestimation of the gray-level inside the sample. The average densities of iron and manganese, calculated from the distributions shown in Fig. 7.34b and Fig. 7.35b are, respectively $\rho_{\mathrm{Fe}} = 1.45 \pm 0.07\,\mathrm{g\,cm^{-3}}$ and $0.058 \pm 0.017\,\mathrm{g\,cm^{-3}}$. The average total density, evaluated from the distributions of Fe, Mn and unmeasured fluorescences is $3.89 \pm 0.3\,\mathrm{g\,cm^{-3}}$ while the corresponding weight fractions of iron $\rho_{\mathrm{Fe}}/\rho_{\mathrm{tot}}$ respectively manganese $\rho_{\mathrm{Mn}}/\rho_{\mathrm{tot}}$ are approximately $37.3 \pm 3$ wt% and $1.5 \pm 0.5$ wt%. All these results were compared to an EPMA study [82] of thin sections of the same meteorite and discrepancies between the Fe estimations and the average

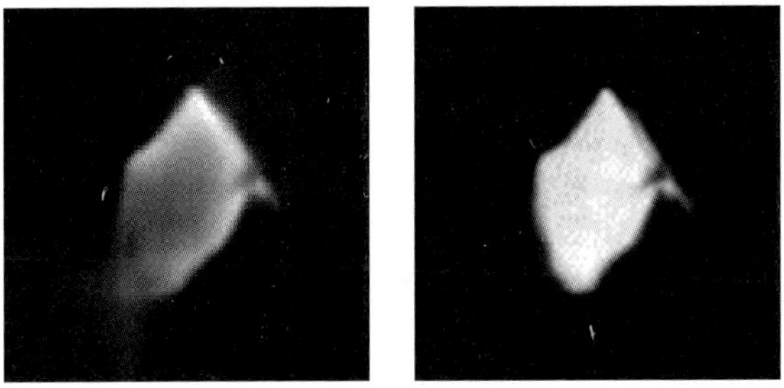

(a) (b)

**Fig. 7.34.** Fe reconstructions (a) without self-absorption, (b) with self-absorption corrections

bulk chemical composition of the NWA817 olivine grains appear on the order of 1 wt%. For Mn, the discrepancies are of about 0.5 wt% and they appear largely due to the Mn/Fe overlap, already discussed in Fig. 7.29.

Improvements on the collection statistics, achievable by using a multielement solid state detector, and by reducing the acquisition electronics overhead time and alignment time through automatic alignment procedures are direct ways of significantly improving our accuracy.

To conclude, a new technique has been developed that combines the information from transmission, fluorescence and scattering tomography. The studied reconstruction methods applied to submillimeter silicate grains are efficient to determine in the volume of the sample, the average densities and the weight fractions of the major and detectable elements with precisions better than 10 wt%. Analysis of the minor elements however, remains semi-quantitative.

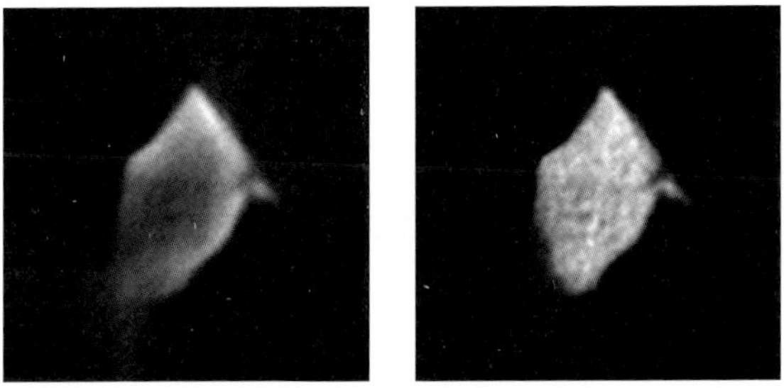

(a) (b)

**Fig. 7.35.** Mn reconstructions **(a)** without self-absorption, **(b)** with self-absorption corrections

## 7.3 Total-Reflection X-Ray Fluorescence (TXRF) Wafer Analysis

*C. Streli, P. Wobrauschek, L. Fabry, S. Pahlke, F. Comin, R. Barrett, P. Pianetta, K. Lüning*[3], *B. Beckhoff*

### 7.3.1 Introduction

The semiconductor industry must earn profits by manufacturing quickly changing, high-performance, and high quality mass products for rapidly decreasing prices. Contamination causes quantifiable losses in yields; therefore, expenditures on defect diagnostics and monitoring of process cleanliness are necessary but not sufficient preconditions for profitable operation [83]. By understanding how to avoid metallic and nonmetallic contamination, material defects, and process flaws, the objectives of economic growth and income became attainable in the semiconductor industry. Though smart engineering measures lessen the pressure on ultra pure process conditions [84], the reproducible cleanliness of starting materials well below $10^{11}$ atoms/cm$^2$ or ions/cm$^2$, and for gate oxide layers in the range of $10^9$ atoms/cm$^2$, is an absolute requirement. The purposeful use of sophisticated diagnostic tools resulted in quantifiable defect reduction, advanced knowledge of materials and processes, and in steep learning curves. Achieving a yield of 98% took 7 years for 256 Kb technology but only 2 years for the crucially more complex 16 Mb generations. Analytical tools must be integrated in the fab operation system, must provide relevant and reliable technology information on the production floor as an at-line or in-line metrology tool.

Providing high uptime and reasonable throughput, total reflection X-ray fluorescence (TXRF) became a standard at-line and in-line inspection tool of monitoring wafer cleanliness up to 300 mm diameter in the last decade. Even beyond the aspect of laboratory operation, TXRF systems are ideal tools for the analysis of surface contamination because the various technical elements that generate a TXRF spectrum are much easier to control than those of the other analytical tools of contamination analysis on the surface. Another merit of TXRF is its capability in the area of layered structures to inspect depth profiles, layer compositions, and/or thicknesses because the intensity of the fluorescence radiation correlates with the thickness of the layer as a function of the glancing angle. The function of fluorescence intensity vs. glancing angle depends upon the optical form of the predominant contamination: particulate, surface film, or near surface bulk contamination exhibits characteristic curvature due to the Fresnel formulas of reflection. Therefore, the optical surface conditions on practical samples must match the optical conditions of the reference materials [85].

[3] Née Baur

Spot-wise TXRF data cannot be generalized for the whole surface without performing statistically valid measurements that permit generalization. However, whole-surface results can be provided after a vapor phase decomposition (VPD) preparation of the native silicon oxide that has become a standard preparation procedure of wafer surface analysis. Combining TXRF with VPD leads to improved detection limits (LOD) of dissolved surface contamination and provides reproducible data on overall surface cleanliness, particularly, when the VPD preparation is automated. In order to increase the reproducibility of the VPD method, an automated surface preparation system has been recently developed by our group[4] and accepted by the international semiconductor community. Pursuing goals of higher sensitivity, synchrotron radiation (SR) TXRF and other novel X-ray sources have already penetrated the frontiers of classical TXRF [86].

Comparative methods of surface analysis such as inductively coupled plasma mass spectrometry after VPD preparation (VPD-ICP-MS), secondary ion mass spectrometry (SIMS), scanning Auger microprobe (SAM), scanning electron microscopy with energy dispersive X-ray analysis (SEM-EDX), high resolution electron energy loss spectroscopy (HR-EELS), Rutherford back scattering (RBS), etc. require equipment of the same or higher complexity and are troubled by more tedious calibration, data management, and quality control procedures.

Data management of VPD-ICP-MS requires, e.g., a memory capacity 60% larger than required for VPD-TXRF. However, under optimized plasma and nebulizer conditions, ICP-MS, an ideal tool for ultra pure water, can be readily applied to monitoring a limited number of light metal surface contaminants after VPD preparation [87]. The main difference between hyphenated VPD and SIMS, SAM, EDX, EELS, or RBS is that the VPD methods are integrating and thus average the surface contamination whereas the other methods are probing a spot of the surface. As a result, VPD-ICP-MS is incapable of providing mapping or spot-wise information [88].

Here we provide a review of TXRF as a paramount part of an industrial ultra trace analytical inspection system for process media and novel microanalytical methods for monitoring the cleanliness of silicon wafer. The methodical design had to balance affordable costs and severe requirements of robust, ultra trace analytical routine procedures.

The validation of mono-isotopic elements such as Na, K, and Al has been an issue to be solved because the complementary graphite furnace atomic absorption spectrometry (GFAAS) is not sensitive enough to be used as an independent method and statistics sets the limits of interlaboratory correlation at the lower range of ng/ml (ppb) [89]. Consequently, interlaboratory evaluation of ultra trace analytical methods cannot be applied to validation in the range of lower ppts (ng/l) either. In the sub-ppb range, we had to go back to intralaboratory, independent, and complementary methods of validation,

[4]Siltronic AG, Central Research & Development, Burghausen, Germany

such as instrumental neutron activation analysis (INAA) [90], and laser diode atomic absorption spectroscopy (LAAS) [91].

By applying modulation techniques and second harmonic generation (SHG) connected with the use of $LiIO_3$ laser diode radiation source, the relative standard deviation of the background can be reduced due to the missing flicker noise of the hollow cathode lamp. In close cooperation with our laboratory,[5] LAAS has been developed by LaserSpec Analytik GmbH. 80807 München, eventually in fusion with ATOMIKA Instruments and became a very sensitive analytical tool for Al at 396.15 nm (SHG) and K at 776.7 nm wavelength down to 5 ng/l and, respectively, 0.1 ng/l. LAAS is also an ideal tool of in situ analysis of various cleaning bathes on the production floor. In the referenced application, LAAS has surpassed the classical GFAAS [92] by one order of magnitude. Until an efficient SHG laser diode for Na is commercially available, INAA will remain the most suitable validation method for Na down to an absolute amount of about 1 ng.

### 7.3.2 Analysis of Metallic Surface Contamination by Means of TXRF

In the daily analytical routine, the merit of TXRF is the simplicity of the TXRF spectra and the broad range of linear response over four orders of magnitude. The spectral simplicity allows straightforward quality control of the analytical results and easy data management. The broad range of linear response provides facile ways of calibration control combined with automated sample preparation, measurement, and data evaluation. In this regard, competing microanalytical methods cannot defeat TXRF. Providing high uptime and reasonable throughput, TXRF proved to be an ideal at- and in-line microanalytical tool of monitoring front-end cleanliness and process hygiene of wafer processes. The disadvantages of commercially available TXRF are the high investment costs, unsatisfactory detection limits (LOD) for light elements, and the inability to analyze the target X-ray source metal. The latter issue was recently circumvented by applying bright synchrotron radiation (SR) sources [93]. Further development requests advanced sensitivity, particularly in spot-wise surface analytical application, and a broader choice of analytes including the light elements, as suggested by the presently valid International Technology Roadmap for Semiconductors [94].

### 7.3.3 Historic Background

Compton reported in 1923 that reflectivity on a flat target was increasing below a critical angle of 0.1° under conditions of total X-ray reflection. The high reflectivity of the sample support reduced the spectral background of the support and improved the LOD down to picogram levels in the early 1970s

[5]Siltronic AG, Central Research & Development, Burghausen, Germany

when Yoneda and Horiuchi applied the principle of TXRF, mainly, to ultra trace elemental microanalyses of biological samples in 1971 [95]. For more details of the history of TXRF please refer to [96–98]. Since then, TXRF has become the standard tool of surface and subsurface microanalysis [99–101]. In 1983, Becker reported on the angular dependence of X-ray fluorescence intensities in the range of total reflection [102]. By deploying SR X-ray sources, the LODs have been recently improved [103] in combination with VPD preparation techniques [104, 105], even to ultra traces (pg) of light elements such as Al as described in Sect. 7.3.10, speciation of different chemical states [106], novel optical arrangements [107], innovative X-ray sources [108, 109] and, particularly, sensitive novel detectors (cf. Chap. 4) have set the pace of further development in the field of TXRF.

Beside the characteristic fluorescence, simultaneously, a competing process of Auger emission also takes place, with higher yield for light elements $Z < 20$ than for heavy elements. If the primary X-rays are not monochromatic and exhibit a continuous spectrum, without band filters, the X-rays will be strongly scattered. The scattered X-ray photons inconveniently increase the background, when the X-ray source is utilized at photon energies of 5 keV to 30 keV.

According to (7.14), the *critical* angle $\alpha_{\mathrm{crit}}$ depends upon the photon energy $E$ and the atomic mass of medium $A$, its atomic number $Z$ and density $\rho$. Below the *critical* angle of incidence, the penetration depth approaches a minimum, particularly in the case of reflective surfaces such as chemi-mechanically polished monocrystalline Si. Total reflection principally disappears on rough surfaces. An estimate for the critical angle of total reflection is

$$\alpha_{\mathrm{crit}} \approx \frac{1.65}{E}\sqrt{\frac{Z}{A}\rho}, \qquad (7.14)$$

where $\rho$ is expressed in $\mathrm{g\,cm^{-3}}$, $E$ in keV, $\alpha$ in degrees.

For practical applications, multiple-beam interferences of the incident and reflected beams are relevant both on stratified multilayers and above optically flat samples. Periodic, plane-parallel, isotopic, and homogeneous layers are applied as optical filters, i.e., monochromators for short wavelengths in TXRF instruments. At certain X-ray wavelengths, steep incidence angles, and azimuthal orientations, the optical conditions satisfy the reflections of Bragg reflections. Stratified multilayers and/or monocrystalline silicon specimen reflect with varying background intensity when turning around to measure at different $\Theta$ azimuthals. At local Bragg maxima, spurious peaks of increased background signal emerge [110–112]. Thus, monocrystalline silicon surfaces can only be analyzed at $\Theta$ orientations of local Bragg minima. Our laboratory also tested the effect of off-axis crystal orientations on the Bragg reflection because realistic wafers are off-axis oriented by 0.6°, 0.3°, and <0.1° and proved that background variations were minimal at positions $30° < \Theta < 42°$ (Fig. 7.36).

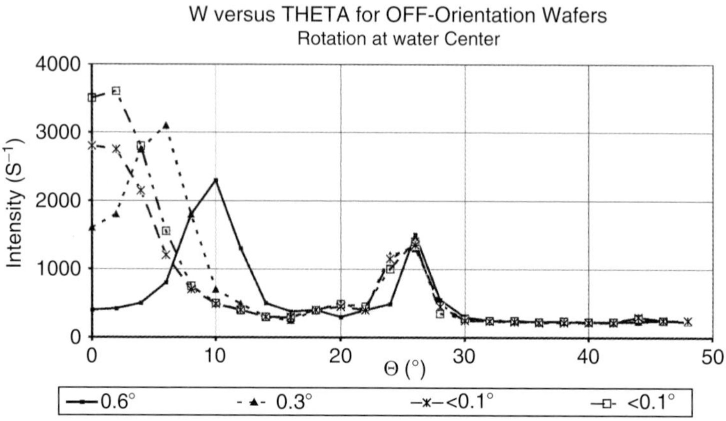


**Fig. 7.36.** Azimuthal $\Theta$ (THETA ANGLE) intensity function of Bragg reflection maxima counts per second (cps) depending on varying off-axis orientations of a (1 0 0) silicon crystal. X-ray source: W-$L_{\beta}$,9.6 keV, Bragg reflection at the (2 2 4) and (4 4 4) planes. Test measurements using a Rigaku 3726 TXRF system

### 7.3.4 Instrumentation of Total Reflection X-Ray Fluorescence Analysis

Though the construction of a TXRF system, including the X-ray source, energy-dispersive detector, and pulse-processing electronics is similar to that of the classical XRF, the geometric arrangement must provide total reflection of the monochromatic primary beam. The detector is positioned above the specimen surface. At the same time, the totally reflected beam interferes with the incident primary beam. This interference leads to the formation of standing waves above the surface of a homogeneous or in a multiple-layered sample and, simultaneously, a part of the primary beam fades away in an evanescent wave field in the subsurface layers [113]. In the investigations, reported in this chapter, ATOMIKA Instruments [114], Model TXRF 8010/TXRF 8030 W have been used.

For preferable economic operation, TXRF instruments consist of a fine-focused, sealed anode X-ray source, a low pass filter, collimator, monochromator, sample holder, detector, and an electronic registration unit. The system runs with an uptime of >90%. Various target materials, e.g., Mo, W for heavy metals Cr, Cu, Au for light elements, or alloys, e.g., Mo–W are available. The X-ray tubes are used in fine-focused, sealed-anode tubes that operate with a generator of 3.5 kW and need water-cooling (5 L min $^{-1}$) applying a power of about 2, max. 3 kW and they operate for 2000–6000 h. Rotating anode sources increase operation costs. They are built for 2000 h of operation and powered by up to 30 kW under strong water-cooling (15 L min$^{-1}$).

Currently, a curved double reflector is used for the low-pass filter. Tunable monochromators in combination with an alloy target are also commercially available. Monochromators are made of LiF, highly oriented pyrolytic graphite

(HOPG), or multilayers of W/Si, W/C, Mo/$B_4$C to provide intense and monochromatic excitation, by, e.g., Mo-$K_\alpha$ or W-$L_\beta$. For further advancement of TXRF and rotatable multielement anode please refer to [115].

The sample-positioning device must operate at high geometrical reproducibility in all three dimensions with automated sample load and unload stations. The sample must be in an evacuated or helium-flushed chamber to avoid absorption in the air that Ar-$K_\alpha$ interferes with Ag-$L_\alpha$, Cd-$L_\alpha$. Investigations on DL improvements have long been focusing on new, bright, and monochromatic X-ray sources such as SR-TXRF (Chap. 2.4.5) and femtosecond laser-induced X-ray sources [108, 109].

The energy-dispersive (EDX) solid state detector (SSD) is made of a lithium-drifted Si crystal (Si(Li)). Between a thin p-type and an n-type layer lies a large Si crystal of high resistivity. The front and the end plane of the crystal are coated with Au and serve as electrodes. By inverse dc voltage the crystal is polarized and forms a p-i-n diode in reverse bias. It must be cooled at 77 K by liquid nitrogen. Incident X-ray photons >2 keV from the fluorescing analyte readily impinge the crystal through a Be window and ionize the crystal by generating photoelectrons, Auger electrons, and Compton electrons. These electrons excite outer electrons from the valence band into the conduction band of Si. A cascade of electron—hole pairs emerge until the energy of incident photons get completely converted. Due to the reverse bias, electrons rapidly drift to the positive n-type electrode, and holes to the negative p-type electrode. The number of the electron-hole pairs is proportional to the energy of impinging photons, thus, the charge pulse is a measure of the characteristic photon energy. Optimal resolution, i.e., a sharp full width at half-maximum (FWHM ≈0.1 keV) is a trade-off between high count rate, i.e., low dead-time and sharp spectral resolution. The analyst must always be aware of potential spectral interferences that, occasionally, cannot be resolved with standard EDX detectors [94]. The central component of electronic data collection and management systems is a multi-channel analyzer (MCA). Figure 7.37 shows the flow chart of an EDXRS spectrometer.

Detector systems are the most dynamic field of current TXRF development. Silicon drift detector arrays and new electronic concepts involving preamplifier multiplexing and digital signal processing will result in further improvements of sensitivity and simpler operation without liquid nitrogen cooling [115, 116]. For monitoring cleanliness, the robust ultra trace analytical systems must become an integrated part of the operation and data management system of the factory (Fig. 7.38) [88].

### 7.3.5 Quantification of TXRF Analysis

Beside the effects of Bragg reflection on the quantification, we have systematically studied less obvious pitfalls of quantitative TXRF analysis [85]. The bulk-type fluorescence yield curve depends also on the surface microroughness [117]. The background noise is a function of the optical quality of the

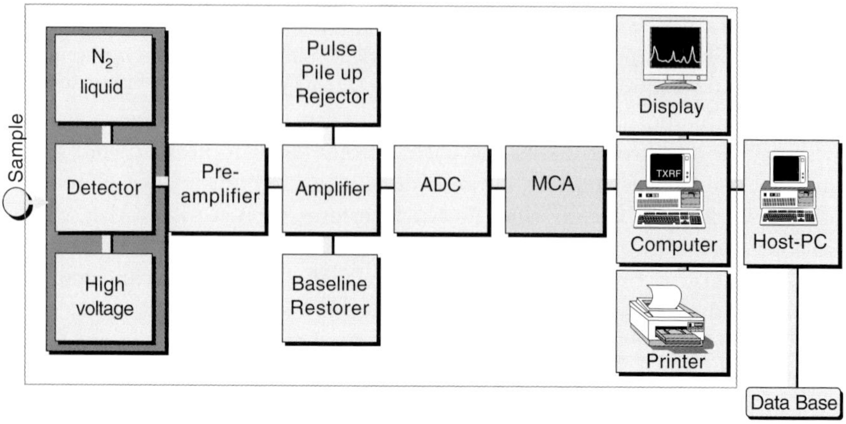


**Fig. 7.37.** Flow chart of a TXRF energy-dispersive spectrometer (ADC: analog/digital converter; MCA: multichannel analyzer). The quality control of the results on a host PC and the integration in the factory data management, including the calibration records, are necessary features of the monitoring methods. Communication between data management between host PC and database is a crucial feature of the operational data management in the fabrication line

surface, the off-axis orientation of the crystal and, at certain excitation energies, of the azimuthal crystal orientation. Therefore, reference materials must be carefully investigated by angle-scan prior to use. Angle-scan characteristics of the sample may not deviate from the angle-scan characteristics of the reference.

Real silicon surfaces can be covered by both particulates and impurity films. Some of the common metal impurities, e.g., Zn, Fe, etc. occur in both

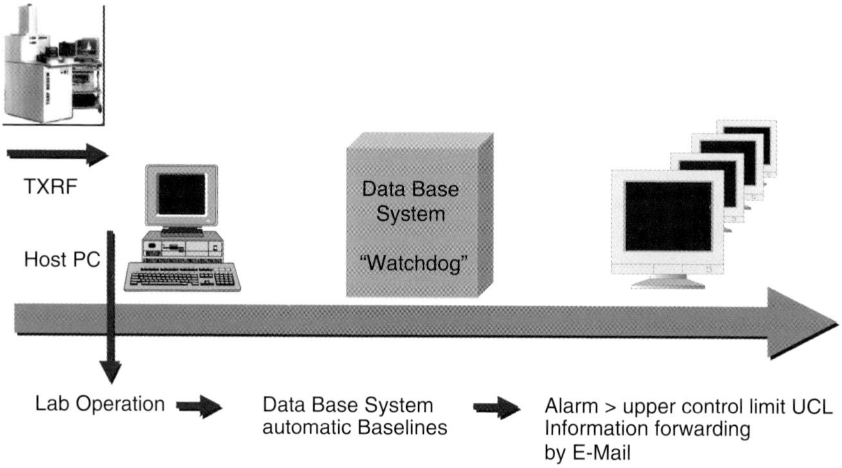


**Fig. 7.38.** Watchdog, intelligent data management includes an automated alarm function whenever data surpass the upper control limit

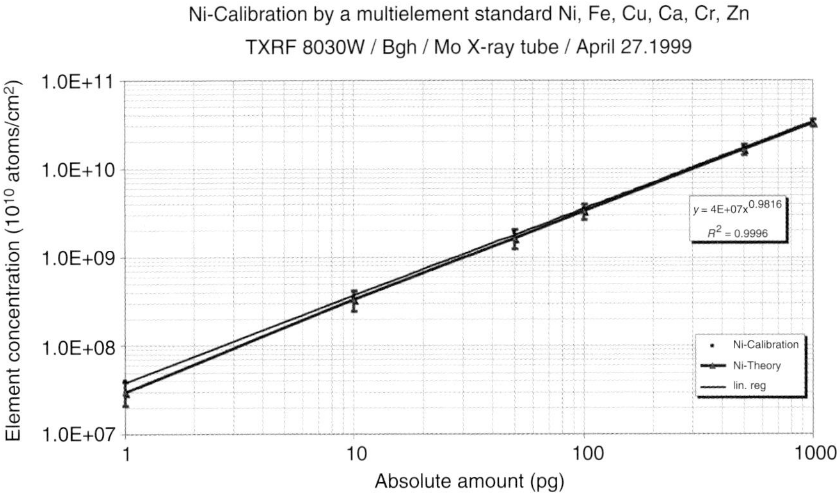


**Fig. 7.39.** Regular control of calibration linearity using residual microdroplets on a pure as-polished silicon wafer

particulate and film forms, but the angle-dependence of fluorescence yield of optically thick particulates and thin films are different: counts on plated films increase with rising incident angles, whereas counts on particulate residues are independent of the incident angle up to higher degrees of relevant penetration. Some other transition metals, such as Cu and Ni easily migrate in Si even at room temperature and may remain undetectable by surface analysis. Thus, calibration must take into account the reproducibility of both optical and chemical conditions of the reference analyte on the surface. Therefore, measurement on particulate cannot be calibrated using film-type reference materials. The error of the reference samples must also be accounted for in all results.

Although the linearity range of TXRF is broad over an areal density range of $10^{10} - 10^{13}$ atoms/cm$^2$ and it allows a controlled one-point calibration using (1 pg to 1 ng residual spot of Ni in $Ni(NO_3)_2$ on a hydrophobic wafer surface, Fig. 7.39), practical analyses must rely on a calibration over the whole range of interest. Schwenke et al. elaborated on the absolute calibration procedure using a polished nickel plate [118]. Calibration and measurement are standardized for the referenced VPD-TXRF applications [119]. The linearity of the one-point calibration must be controlled by a series of residual droplets of multielement stock solutions containing 1 pg to 1 ng of Fe-, Cu-, Ni-, Cr-, Zn-, Ca-nitrate on a hydrophobic wafer (Fig. 7.40). The corresponding relative standard deviations are shown in Fig. 7.41 in dependence on the absolute amount of the analyte.

Some spurious peaks can emerge due to contamination along the beam path, others can be assigned to escape peaks in the energy-dispersive spectra.

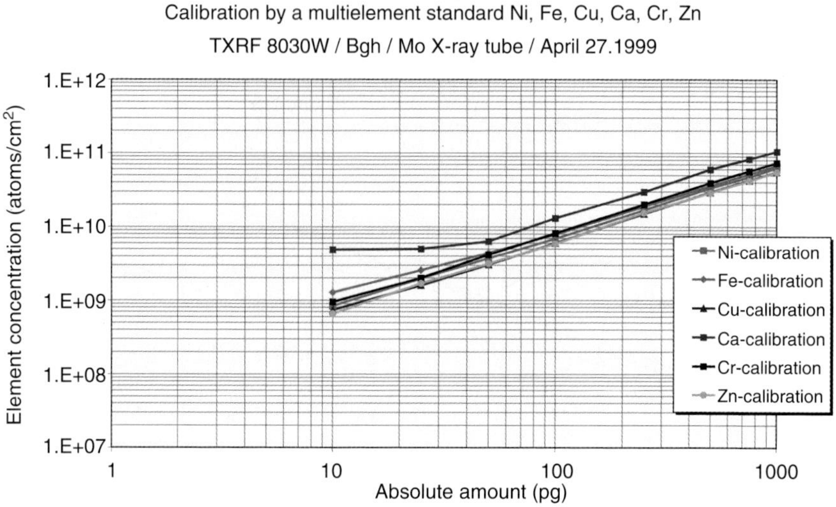


**Fig. 7.40.** TRXF calibration control curve of multielement standard (Ni, Fe, Cu, Ca, Cr, Zn) using droplet residues on silicon wafers

An escape peak emerges when an intense fluorescence *mother* peak of sufficiently high energy is recorded by the SSD and photoelectrons can leave the inner shell of Si atoms of the SSD crystal lattice. The excited Si emits an X-ray photon that is usually reabsorbed by the crystal and the generated photoelectrons trigger a chain-hole cascade, i.e., a charge pulse in the same way like the fluorescence from the sample. However, photons near the crystal

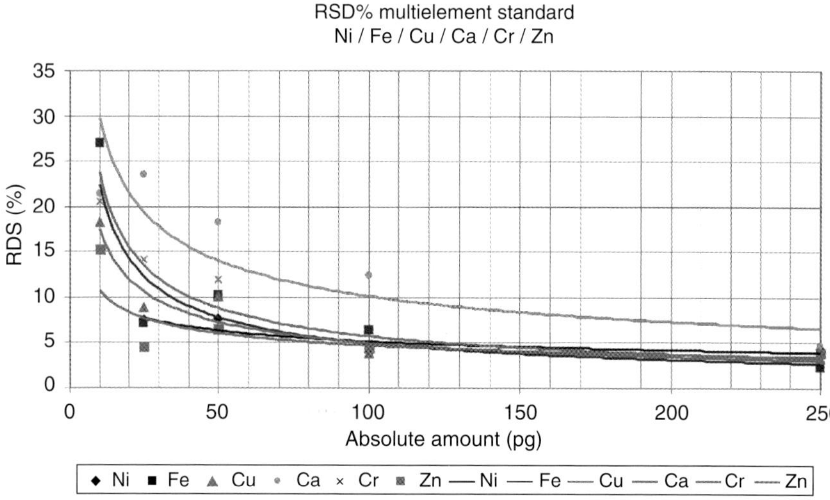


**Fig. 7.41.** Relative standard deviation (RSD) depending on the absolute amount of analyte

**Table 7.8.** Analytes whose $K_\alpha$ peaks interfere with escape peaks of various other elements

| Analyte | Interfering escape peaks |
|---|---|
| V | Co-$K_\alpha$, Mn-$K_\beta$, Dy-$L_\alpha$, Ho-$L_\alpha$, Er-$L_\alpha$, Gd-$L_\beta$ |
| Cr | Co-$K_\alpha$, Fe-$K_\beta$, Er-$L_\alpha$, Tm-$L_\alpha$, Tb-$L_\beta$, Dy-$L_\beta$ |
| Mn | Ni-$K_\alpha$, Co-$K_\beta$, Yb-$L_\alpha$, Lu-$L_\alpha$, Hf-$L_\alpha$, Ho-$L_\beta$ |
| Fe | Cu-$K_\alpha$, W-$L_\alpha$, Ta-$L_\alpha$, Hf-$L_\alpha$, Ni-$K_\beta$, Tm-$L_\beta$, Er-$L_\beta$ |
| Co | Zn-$K_\alpha$,W-$L_\alpha$, Cu-$K_\beta$, By-$L_\beta$, Lu-$L_\beta$, Re-$L_\alpha$, Os-$L_\alpha$ |
| Ni | Ga-$K_\alpha$, Pt-$L_\alpha$, Ir-$L_\alpha$, Ta-$L_\beta$, Hf-$L_\beta$ |
| Cu | Zn-$K_\beta$, Au-$L_\alpha$,W-$L_\beta$, Hg-$L_\alpha$, Ge-$K_\alpha$ |

surface might also escape from the crystal without producing the cascade of charge pulse and carry off the energy difference between the energy of the mother peak and the energy of the generated inner photoelectron. This energy difference appears to be a separate *daughter* peak in the spectrum. Using SSD, certain escape peaks of various elements interfere with the element of interest as listed in Table 7.8 [120].

Spurious peaks can also appear by energy-doubling, when the charge collection induced by the first fluorescence photon has not been completed while a second photon from the same fluorescence source is impinging and the detector registers the double number of charges and then a sum-up peak emerges. Sum-up peaks can be reduced by pulse Pile-Up-Rejector (PUR, Fig. 7.37). Contamination, along the beam path, e.g., Fe in the detector window (Be) may also result in spurious peaks and limit the detection capability of certain elements involved [121].

Reliable TXRF quantification is based on peak-search software that combines calculations on relative intensities, peak localization, identification, an accurate measurement of relative fluorescence intensities, and the element deduction. In this regard, element deduction means that, for unambiguous detection, at least two of the principal peaks can be detected for each analyte of interest. In ultra trace analysis, only the strongest $\alpha$ peaks can be detected. Therefore, special attention must be paid to interfering satellite and spurious peaks.

Using the optically flat, chemically pure Ni plate, TXRF systems can be also calibrated in the sense of absolute calibration that is traceable to accepted reference materials [122]:

$$N_x = B_x C_x . \tag{7.15}$$

$N_x$ is the background-corrected net intensity of the principal peak of an analyte $x$, $B_x$ a proportionality factor of absolute sensitivity of the standard reference, e.g., a Ni plate, $C_x$ the concentration of the analyte $x$. Multielement analyses are based on known relative sensitivities (Fig. 7.40):

$$S_j = \frac{N_j/c_j}{N_{\mathrm{rf}}/c_{\mathrm{rf}}} S_{\mathrm{rf}} . \tag{7.16}$$

$S$ is the relative sensitivity, $N$ the net intensity, $c$ the concentration relative to the "rf" reference values. Figure 7.42 shows the LODs in the 5 pg to 10 pg range

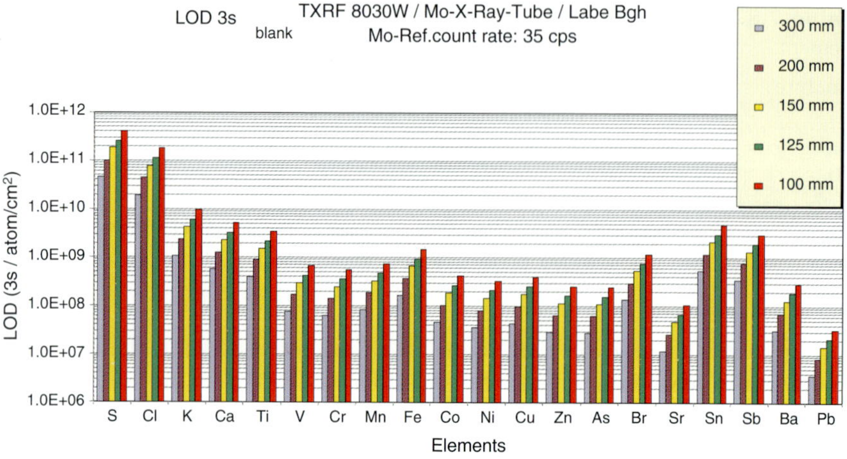


**Fig. 7.42.** Detection limits per 3s/atoms/cm$^2$ of VPD-TXRF measurements for wafers (100 mm to 300 mm) using Wafer surface preparation system (WSPS) for sample preparation as measured

of multielement droplets, the threefold values of standard deviation, when 10 residual droplet of blank solution measured $\leq$atoms/cm$^2$, non-detected values were not included. An internal standard can also be used to reduce matrix effects. Values of relative intensities can be adjusted to theoretical calculations including, e.g., the impact of the mass absorption effects [123]. The propagation of errors in $c_{\mathrm{rf}}$ values must be taken into consideration because declared standard reference materials can vary by 3% to 20% [124] and the accuracy of the reference concentrations strongly affect the accuracy of the results [125]. Using monochromatic Au-L$_\beta$ (11.44 keV) excitation, the migration effects of Cu in monocrystalline Si reference materials were studied by TXRF [126] and it was found that not only the optical compatibility, but also the thermal and chemical stability of the reference must be approved when using reference materials of unknown physicochemical behavior.

For VPD-TXRF application, we calibrate and/or control our TXRF by measuring a series of diluted, multicomponent standard reference solutions as droplet residues in the same manner like atomic spectrometers are calibrated with standard stock addition (Fig. 7.40). The reliable preparation of controlled, spin-on spiking has been recently reported by our laboratory [127]. For general issues of external calibration of TXRF systems and validation with complementary methods [128, 129] please refer also to Diebold [130] and Hockett [131].

### 7.3.6 Surface Analysis

Semiconductor applications of TXRF go back to the late 1980s and were introduced by Eichinger et al. [132]. Due to its high sensitivity, wide linear range, the facile deconvolution of TXRF spectra, and the simplicity of its calibration control, TXRF soon became the affordable workhorse of monitoring cleaning efficiency and front-end wafer cleanliness [133, 134]. TXRF provides the mandatory quality control with reliable data to be easily integrated in the factory data management as shown in Fig. 7.38 [88]. Applying automated operation with repeated one-point calibration every 8 h, automated data management and *up-load*, high uptime, *at-* and *in-line* applications of TXRF are common in the semiconductor industry [85].

Besides that TXRF offers some more features that can be useful for the investigation of semiconductors.

### Particulate and Film-Type Surface Contamination on Wafer Surfaces

The penetration depth of impinging X-rays is limited to the upper nanometers under the condition of total reflection when using light substrates such as Si, quartz, or poly(methylmethacrylate). The fluorescence signal arises from the uppermost layer, therefore, the scattered intensity, i.e., background is about six-times lower than in a conventional XRF arrangement. Positioning thin (<100 nm), even precisely dried residues of a solution, suspension or dispersion, metallic smear or organic tissues within the standing wave front above the substrate in the angular range of total reflection, the signal intensity strongly oscillates. However, the oscillation virtually diminishes with increasing grain size or film thickness of ca. 1000 nm [135].

As shown in Fig. 7.43, the particulate (b) and thin-layer type (c) of angle-characteristics do not differ at larger angles [136, 137]. Upon the angle-characteristics of thin films above or of embedded analytes below the

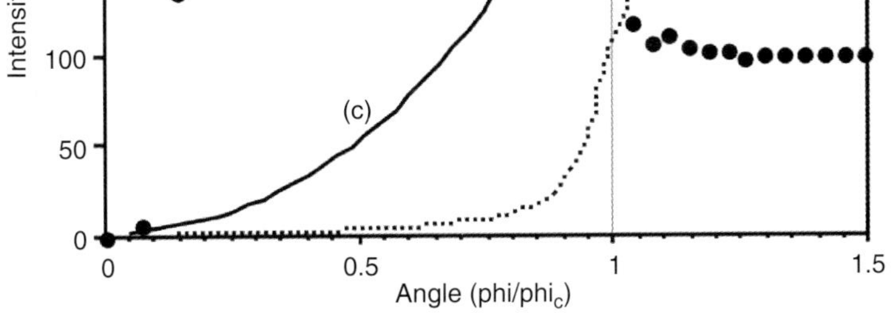


**Fig. 7.43.** Different shapes of the angle dependence of the fluorescence signal for bulk, surface layer and residue on reflecting surface

subsurface (c), thick films can be clearly distinguished from particulate analytes by recording the fluorescence intensity at more than one glancing angles below the angle of critical incidence. However, the surface roughness crucially modifies the peak height, as if it was a stratified structure [138].

A reliable quantification for granular residues can be reached by setting the glancing angle to approximately 70% of the critical angle of total reflection because the fluorescence yield of particulates is nearly equal to the fluorescence yield of film-type contamination, cf. (b) and (c) in Fig. 7.43, in this angular range.

### Depth Profiling by TXRF

Buried layers and shallow dopant implants are important passive and/or active structures of integrated microcircuits. On the other hand, TXRF is a sensitive microanalytical tool of inspection of buried layers. In a thick substrate, the angle characteristics are a function of the layer thickness [139, 140]. Using angle-scan TXRF, stratified microstructures can be accurately analyzed for element composition, layer thickness, and density [135].

Shallow doping profiles, particularly those of As, require reliable nanoscale information on the dopant distribution. Particularly at high concentration, As forms clusters in monocrystalline Si and, therefore, not all the As atoms are electrically active donors. We have evaluated various analytical techniques and found that optimized secondary ion mass spectrometry (SIMS) methods can be reliably applied to layers at 5 nm and below, but for the upper subsurface, the technique of choice is angle-dependent TXRF, as shown in Fig. 7.44 [141]. The angle-scan TXRF is a nondestructive method, whereas the combination of TXRF with layer-by-layer chemical etching provides reproducible results by a destructive type of stratigraphy [104, 142–144].

### Characterization of Layers on Si Wafers and Implants in Si Wafers by TXRF

TXRF allows the nondestructive element analysis of particulate spot size samples but also of near surface layers, layers on top of a reflecting surface, as well as so-called buried layers and depth profiles below a reflecting surface. This is a consequence of the variation of the primary intensity above and below the surface with the angle of incidence. The variation results from an interference caused by the superposition of incoming and reflected beam as can be seen in Fig. 7.45 for Si and Mo-K$\alpha$ radiation.

Above the surface a standing wave is formed in the intersection of incident and reflected beam. For the critical angle 1.8 mrad above the surface for Mo-K$\alpha$ on Si nodes and antinodes follow with a distance of about $d = 18$ nm and the first antinode coincides with the surface. Assuming a reflectivity of 90%, the antinodes have a 3.6-fold intensity of the primary beam. Below the surface the intensity is damped exponentially within a depth of some 10 nm

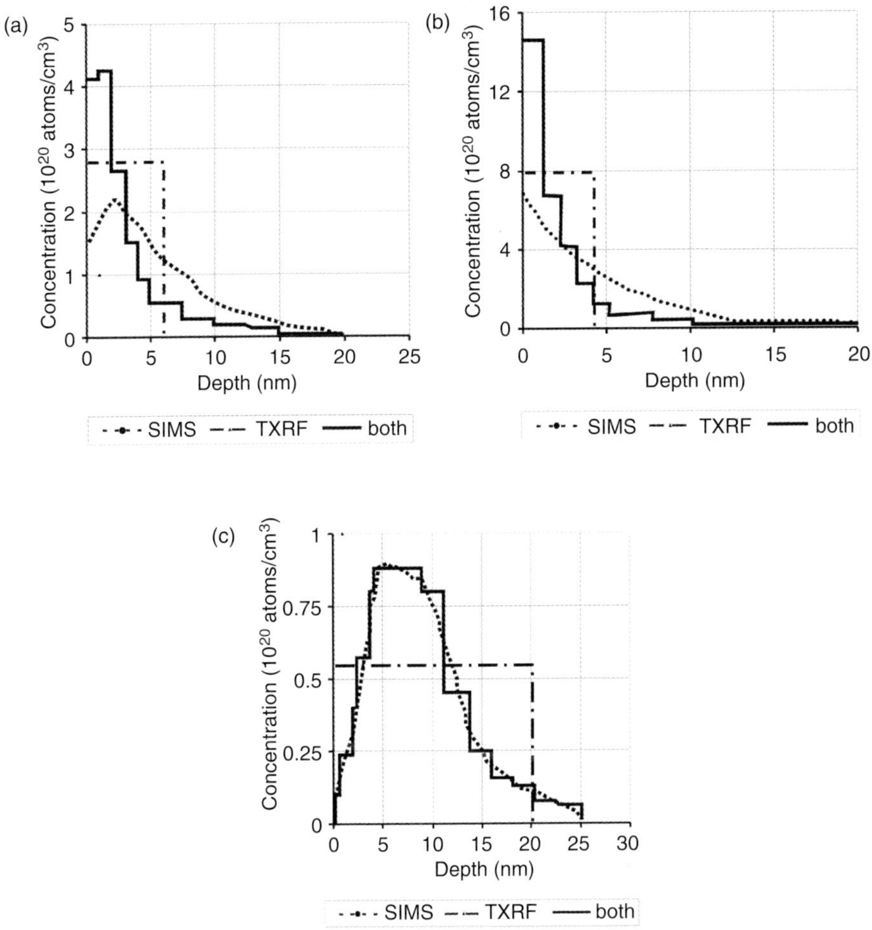


**Fig. 7.44.** Implantation profiles of shallow As doping with implantation energy (**a**) 500 eV, (**b**) 1 keV and (**c**) 5 keV using SIMS, TXRF, and SIMS in combination with TXRF

(penetration depth). For angles smaller than the critical angle, the distance $d$ is stretched and the first antinode moves away from the surface. Inside the substrate, the intensity is damped within a few nanometers. For angles larger than the critical angle $d$ is compressed, the oscillations vanish, and intensity approaches unity; also inside the substrate to some micrometers depth, the penetration depth increases. The big advantage of excitation close to the critical angle for the investigations of depth profiles and buried layers close to the surface is the fact that the intensity of the exciting radiation due to the standing wave phenomenon is up to a factor of 4 higher than excited at an angle higher than the critical angle.

As the primary field varies with the angle of incidence also the intensity of the fluorescence signal shows this variation. From the obtained shape it is possible to distinguish between film type samples, residue samples, thin layers,

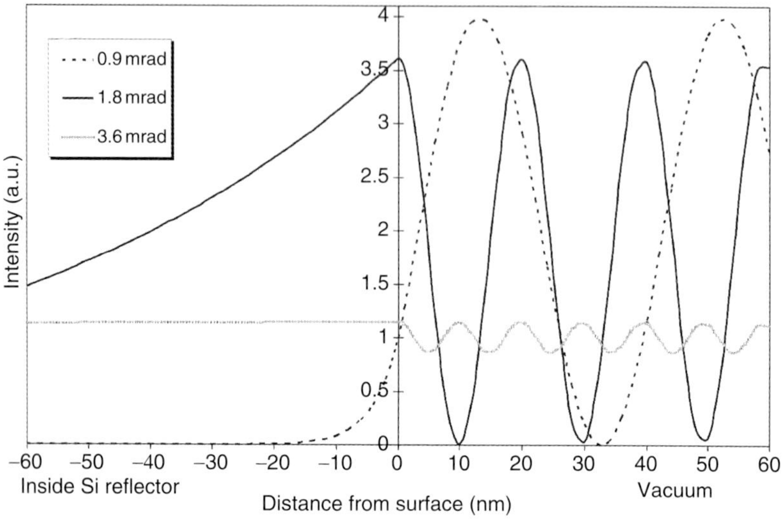


**Fig. 7.45.** Intensity above and below the Si surface for Mo-K$\alpha$ radiation (17.5 keV) for various angles of incidence, 1.8 mrad is the critical angle

and buried layers. The theory behind this is discussed in detail in [145–155] as well as in [113]. The differences have been shown in Fig. 7.43.

The step-like function (Sc) is obtained, if the contamination forms particles on the surface of the wafer (equivalent chemical analysis TXRF). The peaking curve (Ni) is obtained, when the atoms are evenly distributed within a layer of a few nanometers thickness placed on the wafer surface. Most of the real samples do not show one of these extreme cases. If one does not want to measure the complete angle-dependent behavior, only a single measurement performed at the operating angle, where the two curves (Sc, Ni) are crossing, allows already accurate quantification. For completeness, the typical curve for the bulk material (silicon of the wafer) or bulk contamination is given. Usually the bulk Si signal is used for control of the angular adjustment. Quantification differs for the types of contaminants.

For the case of *granular residues* on a substrate (particulate type)—which is equivalent to chemical analysis by using TXRF—the intensity above the critical angle is constant, because the thin, small "sample" is completely excited. The intensity doubles at the critical angle in a step-like fashion and remains at the twofold value down to very small angles due to total reflection:

$$I_i^{\text{particle}}(\varphi_0) = k^{\text{particle}} \cdot I_0 \cdot c_i \cdot (1 + R(\varphi_0)), \tag{4a}$$

following the angular behavior of the reflection coefficient $R(\varphi_0)$. This intensity is proportional to the primary intensity $I_0$ and the interesting area-related concentration $c_i$. For incidence angles $\varphi_\text{w}$ below $\varphi_\text{crit}$, (4a) can be simplified to

$$I_i^{\text{particle}} \cong k^{\text{particle}} \cdot I_0 \cdot c_i \cdot 2. \tag{4b}$$

For *buried layers* (film like) in a substrate the intensity far above the critical angle becomes constant. The asymptotic behavior of the intensities for the particulate and film like contaminations can even be equal, if both concentration values are the same and the appropriate scaling factors $k^{\text{particle}}$ and $k^{\text{film}}$ are chosen. But the intensity for the buried layers steadily increases with decreasing angle and can reach (theoretically) the fourfold value at the critical angle. For the smaller incidence angles the intensity approaches zero, according to

$$I_i^{\text{film}}(\varphi_0) = k^{\text{film}} \cdot I_0 \cdot c_i \cdot (1 - R(\varphi_0)) \cdot \frac{\varphi}{z_P} = k^{\text{film}} \cdot I_0 \cdot c_i \cdot T(\varphi_0) \cdot \frac{\varphi_0}{\varphi_T(\varphi_0)} \cdot \rho \cdot \tau_m, \quad (7.5)$$

where $Z_P$ is the penetration depth

$$z_p = \frac{1}{\rho \tau_m} \varphi_T .$$

The angular behavior is caused by the counter effect of the transmission coefficient $T(\varphi_0)$ and the refraction angle $\varphi_T$.

The intensities of the characteristic X-rays for homogeneously distributed contaminations and major constituent(s) in an infinitely thick and flat substrate can be described by:

$$I_i^{\text{bulk}}(\varphi_0) = k^{\text{bulk}} \cdot I_0 \cdot c_i \cdot (1 - R(\varphi_0)) \cdot \varphi_0 = k^{\text{bulk}} \cdot I_0 \cdot c_i \cdot T(\varphi_0) \cdot \varphi_0 . \quad (7.6)$$

Due to the geometry factor, represented by the direct proportionality to $\varphi_0$ , there is a monotonous increase in intensity for angles above the critical angle.

For practical purposes the divergence of the primary beam has to be considered, as it influences the signal intensity and thus the shape of the $I/\phi$ curve.

The scaling (calibrating) factors $k$ must be determined during calibration of an instrument. Layer type external standards produced by immersion or spin-coating of a wafer with a spiked solution are recommended [156, 157] but also particulate type [158] and bulk type standards have been used [159, 160]. All types of standards are commercially available. Calibration by internal standard is not permissible.

In the case of residue on a surface (droplet pipetted on reflector, liquid evaporated), the signal is proportional to the factor $1+R$, where $R$ as usually means the reflection coefficient. This is only valid, if the grain size of the residue is in the range of 100 nm and larger because the oscillations vanish. Therefore, the addition of an internal standard homogeneously distributed is a successful solution for calibration. If the reflectivity is assumed to be nearly unity, the fluorescence signal is doubled for angles lower than the critical angle. The angle dependence of the bulk signal—atoms exceeding some 1000 nm depth—the signal is low for small angles and increases rapidly above the critical angle due to increase of the penetration depth. For the near surface impurities the angle dependence leads to a strong peaking at the critical angle.

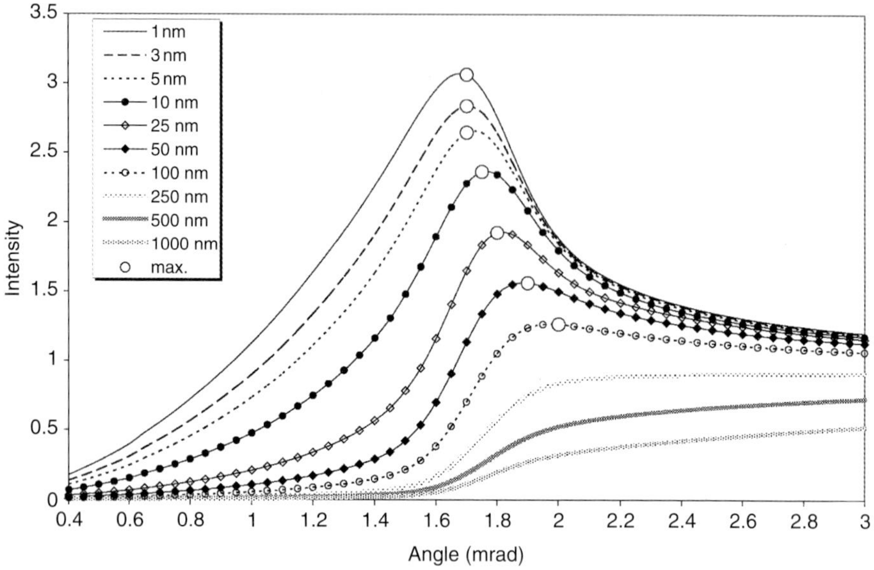


**Fig. 7.46.** Variation of the shape of the fluorescence intensity vs. angle of incidence with variation of the depth of the implanted Ni atoms. Rectangular depth profile assumed

Looking at an analyte below the surface, so-called "buried layers" or depth profiles, the fluorescence signal shows a behavior like the one seen in Fig. 7.46 where a rectangular profile of Ni atoms in Si is assumed for depth varying from 1 nm to 1000 nm.

The fluorescence intensity $I_i(\varphi_0)$ of an element $i$, implanted in a flat smooth substrate, with a depth profile $c_i(z)$, showing a variation of its concentration over depth $z$ (normal to the surface), is a function of the incident angle $\varphi_0$:

$$I_i(\varphi_0) \quad \propto \quad T(\varphi_0) \cdot \frac{\varphi_0}{\varphi_T(\varphi_0)} \cdot \int_{z=0}^{\infty} c_i(z) \cdot \exp\left[-\left(\frac{\mu_1}{\varphi_T(\varphi_0)} + \mu_2\right) \cdot \rho \cdot z\right] \cdot \rho \cdot \mathrm{d}z, \tag{7.7}$$

which is the generalized form of (7.6). The mass attenuation coefficients $\mu_1$ and $\mu_2$ describe the absorption of the exciting and the fluorescence radiation of the element $i$, respectively, $\rho$ is the density of the substrate. The transmission coefficient $T(\varphi_0)$ takes into account the refracted part of the incoming radiation and $\varphi_T$ is the refraction angle inside the substrate.

The data evaluation is performed by comparing the measured data points with theoretical calculations and the adjustment of the parameters like assumptions about the shape of profile, depth, and roughness to achieve the best fit. Figure 7.47 shows the differences in the response for different depths and the fit to the measured data points. The resolution is less than 5 nm.

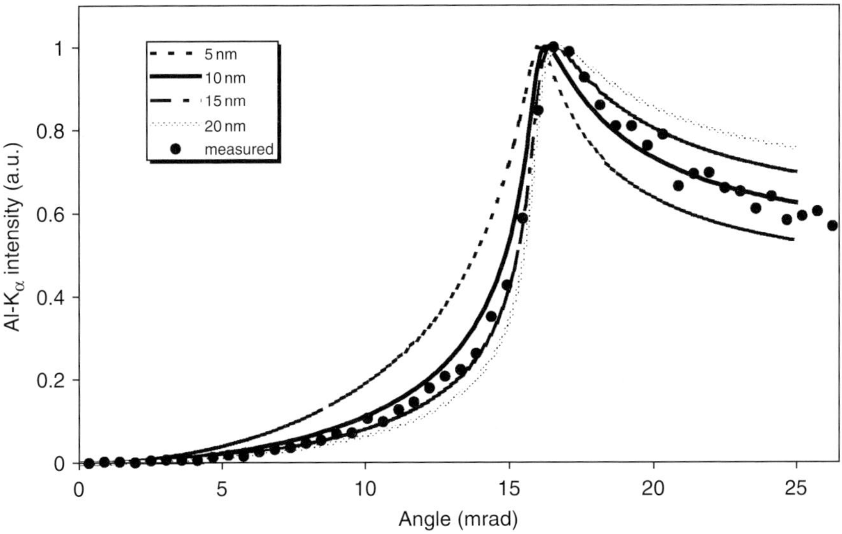


**Fig. 7.47.** Angle dependence of the Al fluorescence signal from a rectangular profile with a depth of 10 nm

Reflecting layers on a substrate induce some oscillations due to the standing wave. Figure 7.48 shows various Ni layers on Si excited with Mo-K$_\alpha$ radiation. It is recognizable that the critical angle changes from that of Si to that of Co with increasing Co thickness.

The composition, density, and thickness of single as well as periodic and non-periodic multiple films can be analyzed [161, 162]. The fluorescence intensity $I_i(\varphi_0)$ of an element $i$ (with concentration $c_i$) embedded in a single thin film with thickness $d$ on top of a flat smooth substrate, is a function of the incident angle $\varphi_0$ and is described by [163–166].

$$I_i(\varphi_0) \propto c_i \cdot \left\{ \begin{array}{l} |E_T|^2 \cdot \dfrac{1-\exp\left[-(\mu_1/\varphi_T+\mu_2)\cdot\rho\cdot d\right]}{\mu_1/\varphi_T+\mu_2} \\ +|E_R|^2 \cdot \dfrac{1-\exp\left[-(-\mu_1/\varphi_T+\mu_2)\cdot\rho\cdot d\right]}{-\mu_1/\varphi_T+\mu_2} \\ +2\cdot\text{Real part}\left(E_T\cdot E_R\cdot\dfrac{1-\exp\left[-\left(-\mathrm{i}\cdot 4\pi\cdot\frac{\varphi_T}{\lambda\cdot\rho}+\mu_2\right)\cdot\rho\cdot d\right]}{-\mathrm{i}\cdot 4\pi\cdot\frac{\varphi_T}{\lambda\cdot\rho}+\mu_2}\right) \end{array} \right\}. \tag{7.8}$$

The electric fields $E_T$ (transmitted by the film surface) and $E_R$ (reflected from the substrate) are derived from Fresnel's equations and are complex quantities. The wavelength of the exciting radiation is $\lambda$. The formation of standing wave above a reflecting surface is already taken into account by this formalism. An example is given for a Ni thin film (Fig. 7.48).

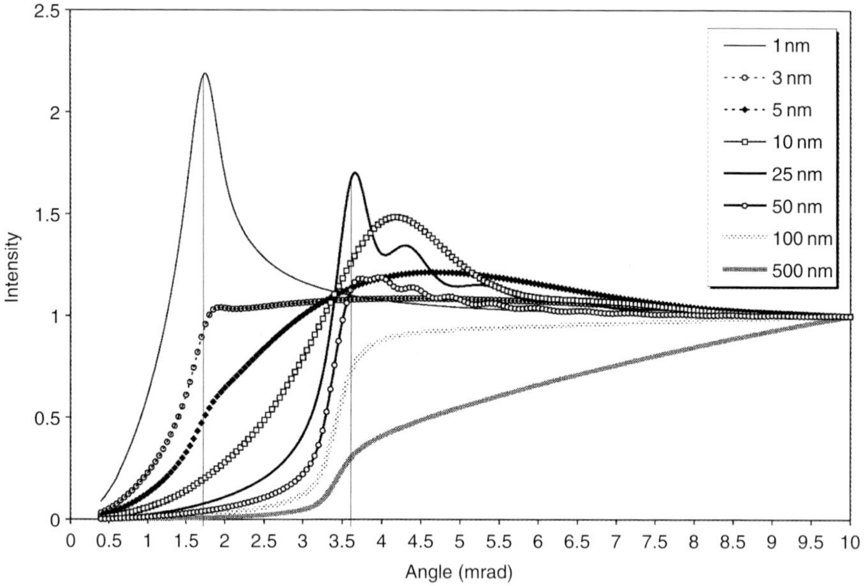


**Fig. 7.48.** Angle dependence of the Ni-K$_\alpha$ fluorescence signal for various layer thicknesses, excited with Mo-K$_\alpha$ radiation

The layer(s) and the substrate must be homogeneous, flat, and smooth, with perfectly sharp interfaces. With some effort, even a certain roughness of the interfaces can be taken into account in the calculations [167]. A review on grazing incidence X-ray spectrometry and reflectometry has been provided by Stoev and Sakurai [168].

### 7.3.7 Statistical Process Control (SPC)

Concerning the pitfalls of practical TXRF analysis, the reliable quantification of TXRF is crucially dependent on accurate, reproducible, and stable standards of the same optical reflection as the specimen to be analyzed. Since the TXRF response is linear over a broad range of contamination levels and the relative fluorescence intensities of the elements are known, it is possible to use a Ni reference at relatively high areal concentration such as $10^{14}$ atoms/cm$^2$ and quantify four orders of magnitude lower levels with the same one-point calibration [137]. As depicted in Fig. 7.49, this way a routine control of calibration status can be regularly carried out as requested by international quality standards such as DIN EN ISO 9001 Sects. 4.11 and 4.12.

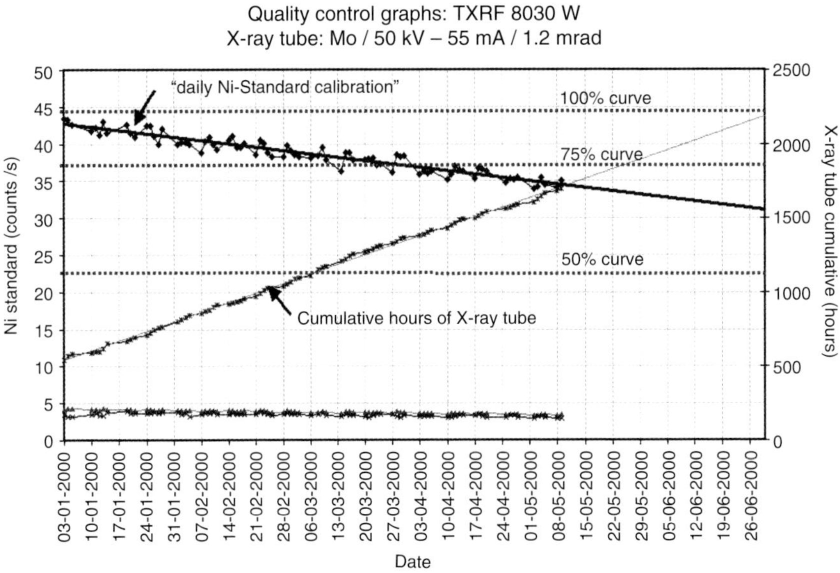


**Fig. 7.49.** SPC Chart for TXRF measurement systems. Upon daily calibrations with a Ni standard reference sample the sensitivity of the system can be controlled. When the X-ray intensity sinks, e.g., below 75% or 50% of the original intensity of 5-day average, the tube must be replaced

### 7.3.8 Automated Vapor Phase Decomposition (VPD) Preparation

VPD-TXRF is a standard, invasive but nondestructive, multi-stage method of wafer surface analysis [119, 169]. The native oxide on wafer surfaces readily reacts with HF vapor and forms small droplets on the hydrophobic wafer surface at room temperature [170]. These droplets can be collected with a scanning droplet that contains all dissolved contamination [171–175]. The droplet must be reproducibly dried forming a concentrated spot without losing analytes and measured against the droplet residue of the blank scanning solution using an automated preparation system as depicted in [105]. Depending on the scanned surface area, the enrichment factor VPD preparation lies between 78 and 720 for 100 mm and 300 mm diameter, respectively.

Applying statistical design-of-experiment principles, we carefully optimized the VPD preparation for robust reproducibility and high recovery rate [176]. The preparation method has been also correlated with accepted and traceable reference materials [127–129]. Later, in order to increase the reproducibility of the method, an automated, 300 mm capable, multi-stage system[6] was developed by Wacker Siltronic central analytical laboratory in

[6]Other VPD systems are: Liquid Drop Wafer Inspection manufactured by Mesa-Tek Inc., Morganville NJ 07751; Wafer Surface Scanner of Sankyo Engineering Co., Ltd,

**Fig. 7.50.** Wafer surface preparation system modules (WSPS): a novel system for vapor phase decomposition preparation for wafer diameters 100 up to 300 mm

cooperation with GeMeTec [177], Germany. Since then, this reliable modulated system has dominated the market of automated VPD preparation systems.

The operator loads the wafers in a carrier on the load/unload module (for follow up on the procedure please refer to Fig. 7.50) beginning at module no. 4, counting the modules from the left. A robot arm takes the wafers one by one into the HF vapor phase treatment module no. 2, then onto the droplet collection scanner module no. 1. Both the scanning pattern and the scanned area can be adjusted to individual requirements. The splitting of the droplet into two parts and placing the aliquot into a vial is optional and affords economical, simultaneous processing of expensive samples and evaluation by complementary methods. Then the robot transfers the wafer with the dried scanning droplet residue and a blank droplet to the drying station module no. 3. After controlled drying, the wafer arrives at the carrier unload module no. 4. The droplet residue remains concentrated in an area of ∅ 200 μm without losing analyte, as shown in Fig. 7.51 [178]. By means of the automated VPD equipment both the LOD, at present in the upper range of areal density of $10^7$ atoms/cm$^2$, the procedural reproducibility (standard deviation <30%) and the operational reliability (>90 %) of the VPD-TXRF procedure are improved [105].

The accuracy of VPD-TXRF is a sensitive function of the reliability of the standard reference material and of the reproducibility of the droplet drying

Tachikawa-shi, 190 Japan; Purex Preparation Method for Wafer Analysis of Purex Engineering Co., Ltd., Tokyo, 103 Japan.

(a) (b)

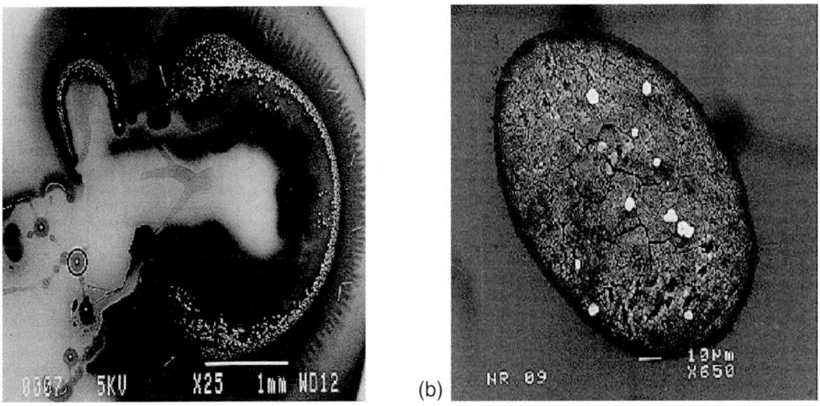


**Fig. 7.51.** Dried scanning droplets, worst case (**a**): exploded droplet under enforced drying using an infrared lamp, droplet size of a few mm; optimized case (**b**): scanning droplet dried under controlled conditions using vacuum and carrier gas in the drying module of WSPS

procedure [179]. Mass absorption sets the upper limit of reliable analyses at $10^{14}$ atoms/cm$^2$ [123]. Concerning the recovery of less soluble fluorides such as $CaF_2$, silicides, e.g., $NiSi_2$ and electroplating noble metals, solubility in aqueous HF solutions is the only matrix limitation of VPD-TXRF in any VPD preparation and hyphenated VPD-methods. Though the VPD preparation has been optimized for maximum recovery >90% even for the electropositive Cu [105, 176], noble metals can only be dissolved when the oxidation potential of the scanning solution is higher than the reduction potential of the bare Si surface. Consequently, we suggest a vigilant approach in the diagnostic use of VPD, applying correlation with other, independent, and complementary methods. For this purpose, optional droplet-splitting facilitates simultaneous complementary analyses [180]. Current research focuses on issues of sample holders [181] and light element capabilities [130].

### 7.3.9 Low $Z$ Determination – Problems – Solutions and Results

The requirements of the SEMATEC road map also includes the determination of Na and Al at ultra trace levels ($1.5 \times 10^9$ atoms/cm$^2$ for Al and $2 \times 10^8$ atoms/cm$^2$ for Na in 2003). The measurement of low $Z$ elements at the required ultra-trace levels is possible with other methods like ICP-MS. However, ICP-MS not only requires sample preparation but also requires more tedious data evaluation and data management than TXRF. Measurement of low $Z$ elements with TXRF would also allow in situ mapping that is not viable by VPD-ICP-MS. But these light elements are more difficult to excite as well as to detect; furthermore fluorescence yield is very low for low $Z$ elements. Special detectors offering high efficiency in the low energy region have to be used in combination with special sources like special X-ray tubes

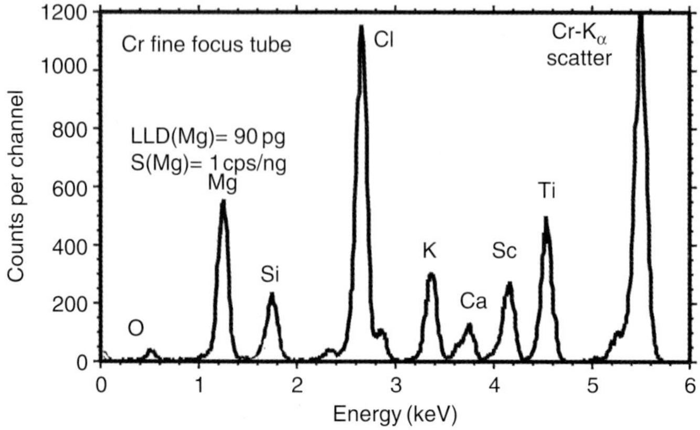


**Fig. 7.52.** Spectrum of a droplet sample on a Si wafer, excited with Cr fine focus X-ray tube, equipped with an ML monochromator, 40 kV, 33 mA, 100 s. Detection limits obtained for Mg are 90 pg with a sensitivity of 1 cps $ng^{-1}$

or synchrotron radiation emitting a sufficient flux of low energy photons for efficient excitation.

The anode material with the lowest $Z$ element as target material for fine focus X-ray tubes commercially available is Cr (Cr-$K_\alpha = 5.41$ keV). Figure 7.52 shows the spectrum obtained with monochromatic Cr-$K_\alpha$ excitation measured in the wafer chamber of the Atominstitut (ATI) which was equipped with a 30 $mm^2$ GeHP detector with an ultrathin Diamond window. The monochromator used was a W/C multilayer monochromator. The detection limits obtained are 90 pg for Mg. This corresponds to about $2 \times 10^{12}$ atoms/$cm^2$ assuming an inspected area of 1 $cm^2$, which is realized by the solid angle of the setup.

To improve the detection limits further, the X-ray group at ATI developed a windowless Si anode X-ray tube [103], which leads to a reduction in detection limits down to the low pg level for Mg. The tube can only be operated up to 500 W. Si as anode material offers the advantage of not exciting Si with the monochromatized Si-$K_\alpha$ line ($E = 1.83$ keV), so the detector will not be saturated by the Si signal from the wafer. As it was a setup for research there was no window between the sample chamber and tube so that the tube was flushed during every sample change as can be seen from Fig. 7.53. A very important step was to insert a magnetic electron trap to capture the electrons scattered on the anode before they can reach the detector area. Another step was the insertion of a thin C foil as optical filter because the ultra thin window of the detector was transmitting light of the glowing cathode which was reflected by the polished surface of the Si anode onto the detector crystal.

Preliminary measurements were performed in the first prototype of the ATI low $Z$ spectrometer which is equipped with a simple cut-off reflector

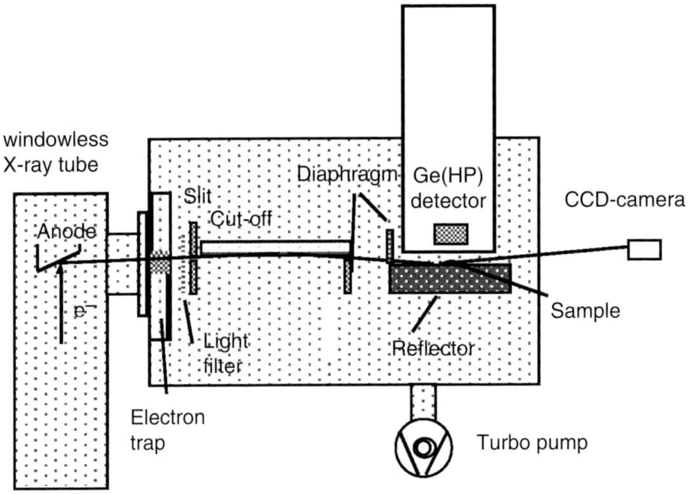


**Fig. 7.53.** Scheme of the ATI TXRF spectrometer with the windowless X-ray tube with Si anode

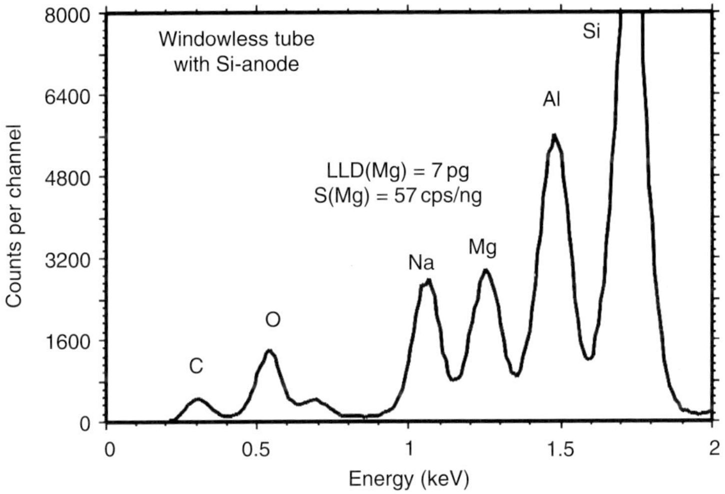


**Fig. 7.54.** Spectrum of a droplet sample on an Si wafer, excited with the windowless Si anode tube

instead of a multilayer monochromator to suppress the high energy part of the exciting spectrum [182]. The 30 $mm^2$ GeHP detector with a 400-μm Diamond window was used. The detection limit obtained with this windowless Si anode X-ray tube was 7 pg for Mg. Figure 7.54 shows a spectrum.

A similar idea for the efficient excitation of low $Z$ elements was realized by T. Arai from RIGAKU [183], using a W rotating anode tube for excitation. The exit window is thin enough to transmit the W-$M_\beta$ characteristic line with

1.83 keV, the monochromator can be selected among three possibilities, i.e., W-$M_\beta$ or W-$L_\beta$ characteristic radiation or higher photon energies from the W bremsstrahlung continuous spectrum. This configuration is commercially available, the RIGAKU 3750 for 300 mm wafer [184]. The detection limits for Al in straight TXRF are reported to be $5 \times 10^{10}$ atoms/cm$^2$. This instrument allows mapping of the wafer. After VPD procedure detection limits of $3 \times 10^{10}$ for Na and $2 \times 10^9$ for Al for a 6 -in. wafer were reported [185]. Another recently commercially available instrument is from GEMETEC, Munich [186]. The detection limits are $8 \times 10^{10}$ atoms/cm$^2$ for Na and $1.5 \times 10^{10}$ for Al for a 200 -mm wafer after VPD sample preparation. The excitation is done with a standard 2 kW Cr-fine focus tube. In a presentation at the TXRF conference 2000 in Vienna, this company presented detection limits of $2 \times 10^{10}$ atoms/cm$^2$ for Na and $2 \times 10^9$ atoms/cm$^2$ for Al [187].

### 7.3.10 Synchrotron Radiation Induced TXRF

#### General Remarks

If the detection limits should be further improved it is helpful to look at the definition of the lower limits of detection (LLD):

$$\mathrm{LLD} = \frac{3 \cdot \sqrt{I_B/t}}{I_N} \cdot m_{\mathrm{sample}}, \tag{7.9}$$

where $I_B$ is the background intensity, $I_N$ is the net intensity and $m_{\mathrm{sample}}$ is the sample mass and $t$ is the measuring time.

One can easily see from (7.9) that there are different ways of improving the detection limits, namely increasing the sensitivity $S$, reducing the background and increasing the measuring time, which, however, is limited for practical reasons.

Besides using total reflection geometry for reducing the spectral background and doubling the fluorescence signal from the sample, a further possibility to reduce scatter contributions from the sample itself is the use of polarized primary radiation [188, 189]. The scheme can be seen in Fig. 7.55.

Due to the anisotropic emission characteristics of the scattered radiation based on the classical dipole radiation it is advantageous to place a detector in such a position that only the isotropic emission of the fluorescence signal is detected. So the combination of TXRF with polarized radiation leads to a lower background. Moreover, the use of monochromatic primary radiation improves the background conditions because only photons of one energy can be scattered. An increase in sensitivity can be attained by using a tunable intense excitation source, enabling the exciting energy to be adjusted to just above the absorption edge of the element of interest.

Synchrotron radiation with its outstanding properties offers new possibilities for improving the performance of TXRF. The intense beam with a

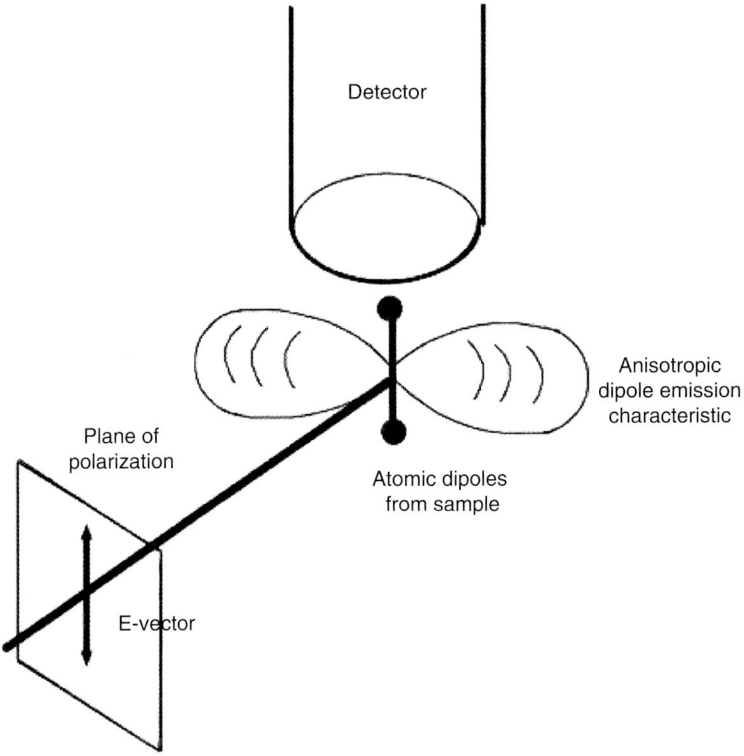


**Fig. 7.55.** Schematic view of the effect of using polarized radiation

continuous spectral distribution from photon energies in the infrared to high energy X-ray photons as well as the linear polarization in the orbit plane and its natural collimation are features best suited for excitation in total reflection geometry. Figure 7.56 shows the spectral brightness of Beamline L at HASYLAB allowing excitation by photons between a few eV and 100 keV.

For optimal excitation conditions the spectral distribution can be modified by elements like cut-off mirrors, monochromators, and filters. Details can be found in [190]. Multilayer monochromators are best suited for the combination with synchrotron radiation excited XRF [191, 192]. In comparison to crystal monochromators they offer a larger bandwidth ($\Delta E/E \approx 0.01$), which leads to a much larger photon flux on the sample. Another advantage is the possibility of selecting the excitation energy just below a matrix element with high concentration and just above the absorption edge of the element of interest ("selective excitation"), with the drawback, however, of an increased background due to Raman scattering.

The combination of TXRF with synchrotron radiation allows various geometrical arrangements for reflector and detector. Figure 7.57 shows three possibilities. For geometry A the polarization effect is fully utilized by positioning

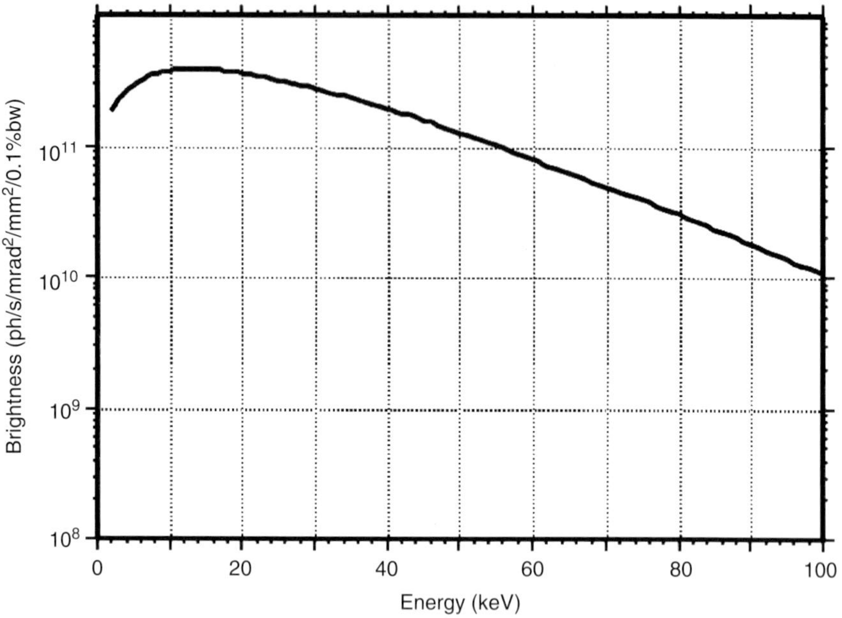


**Fig. 7.56.** Spectral distribution of Beamline L at HASYLAB

the detector axis in the plane of the orbit. Scattered radiation is not emitted in that direction. The sample is excited efficiently, and full homogenous illumination of the sample by the width of the beam in the horizontal plane is given. There are hardly any losses due to the collimators because the beam is naturally collimated in the vertical plane with 0.1 mrad to 0.2 mrad divergence depending on the energy. The detection of the fluorescence signal is not optimal because the detector must be sidelooking to take advantage of the polarization effect. The fluorescent radiation has a long path in the sample to reach the detector.

The excitation conditions for the arrangement as displayed in Fig. 7.57 (b) are poor. Most of the photons in the horizontal plane are absorbed in the collimation system. The intensity distribution in the vertical plane drops drastically and therefore the fluorescence intensity also drops with the deviation of sample regions from the plane of reference. Thus a restriction to 2 mm to 4 mm sample diameters is advisable due to the intensity and polarization distribution in the vertical plane. However, the detection efficiency is perfect because of the large solid angle. Excellent excitation and detection will be achieved with arrangement (c). This combination of sample–detector position results in a complete loss of the use of the polarization effect. If the sample is small, which is the case in ultra trace analysis, the scattering contribution from the sample itself is negligible. Scattering from the substrate is reduced by total reflection.

If the experiments are performed in air, scattering of the exciting radiation contributes to the background. So all measurements have to be done in a vacuum chamber and thus led to excellent results.

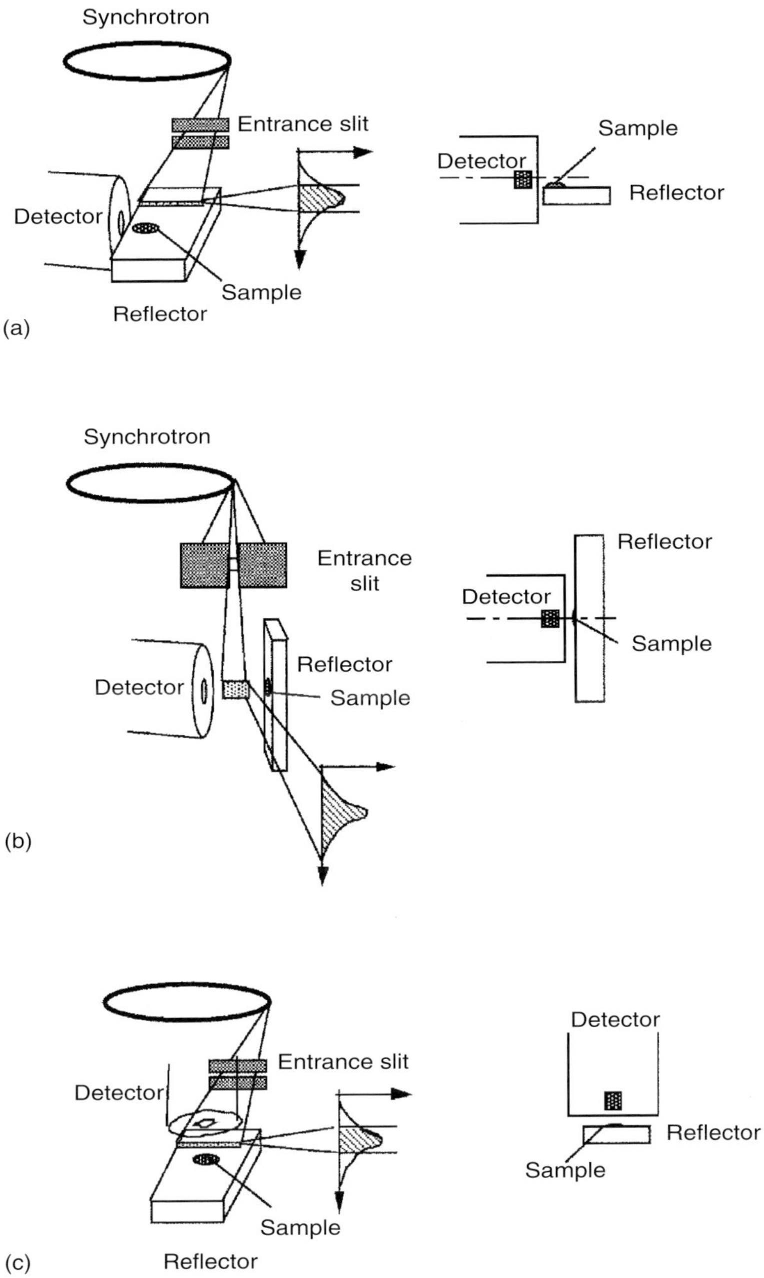


**Fig. 7.57.** Three possibilities of arranging wafer and detector for SR-TXRF

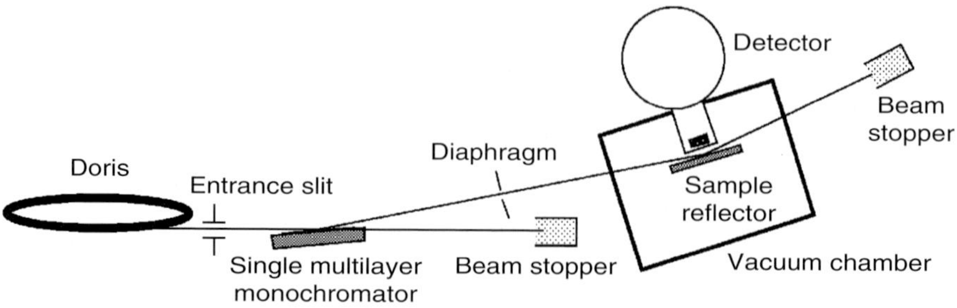


**Fig. 7.58.** SR-TXRF setup at HASYLAB, beamline L from the ATI group

## Results From the ATI Group

First experiments of the group from Atominstitut, Vienna, have been performed in 1994 at HASYLAB, DESY Hamburg, Beamline L, which is a bending magnet beamline. Details are described in [86]. The arrangement can be seen in Fig. 7.58 following the geometry (b) of Fig. 7.57 with vertical reflector and sidelooking detector. The beam is collimated by a primary slit system and then impinges on the multilayer monochromator; the primary beam after the monochromator is absorbed in a beamstopper. The shielding of the vacuum chamber is extremely important because there are always multiple scattered high energy photons which would penetrate into the measuring chamber through the Al wall. Detailed information concerning the experiments are given in [193, 194]. Detection limits for Ni with excitation energy of 10 keV were found to be 13 fg or $1.3 \times 10^8$ atoms/cm$^2$ for an inspected area of 1 cm$^2$. An Al filter was used to prevent the detector to be

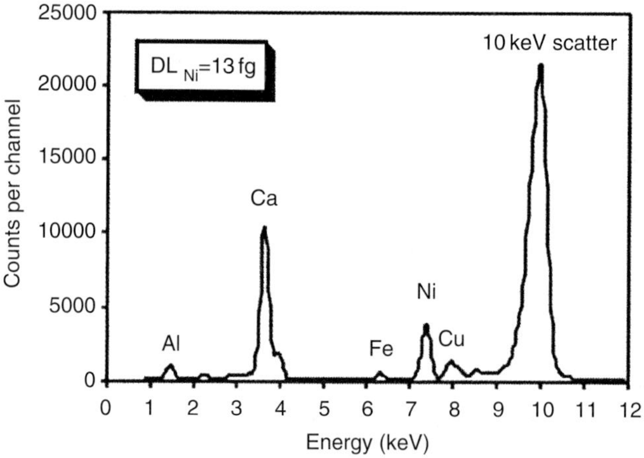


**Fig. 7.59.** Spectrum of 10 pg Ni sample on a Si wafer, measured at HASYLAB, Beamline L in 1994

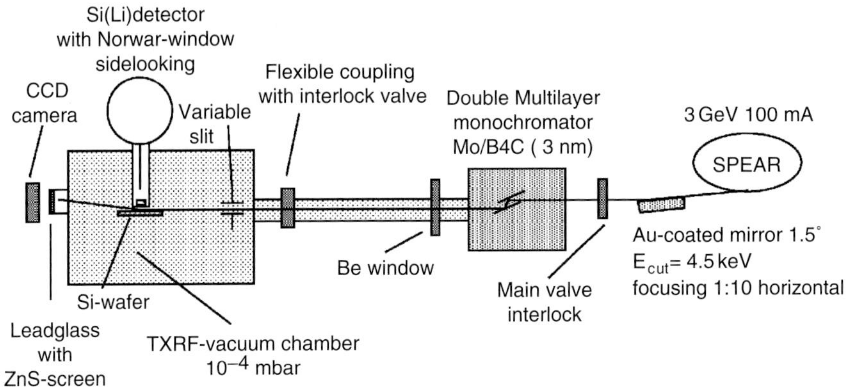


**Fig. 7.60.** SR-TXRF setup at SSRL, Beamline 3-3

saturated by the Si signal. Figure 7.59 shows the respective spectrum. From the group at Atominstitut a comparison of the various geometries have been performed [195].

The ideal source for efficient excitation of the low $Z$ elements is definitely synchrotron radiation due to its high intensity also in the low energy region. Experiments have been performed at SSRL, BESSY II, and ESRF, starting with the work of Madden et al. in 1993 [196], at SSRL, Beamline III-4 with filtered white radiation. The group of the Atominstitut started their experiments also in 1993 at Beamline III-4 and investigated in the following years the spectral distribution and geometries [103, 197, 198]. The best suitable beamline at SSRL turned out to be bending magnet beamline BL 3-3 in combination with a multilayer monochromator. Figure 7.60 shows the setup of the Beamline 3-3 at SSRL. Figure 7.61 shows the spectrum obtained from a sample of 25 pg Na, Mg, and Al. Detection limits of 127 fg for Na have been obtained for droplet samples simulating the VPD droplets. Details can be found in [199]. In 1999, the ATI group started experiments at the plane grating monochromator beamline for undulator radiation of the PTB at BESSY II. Figure 7.62 shows the setup. The undulator of the beamline could also be operated in wiggler mode, so an efficient excitation of Na, Mg, and Al was possible, the detection limits obtained were in the low pg range. As this beamline is best suited for providing a low energy (<1 keV) beam with extremely high spectral purity, detection limits for C and N could be determined. Figure 7.63 shows the respective spectrum. The detection limits are below 1 pg for carbon and oxygen. Also a comparison between the possible geometries have been performed, the vertical arrangement of the wafer and sidelooking detector turned out to provide better results. Details can be found in [200]. Recent improvements were achieved using the undulator U49 (cf. the subsection facility of the PTB at BESSY).

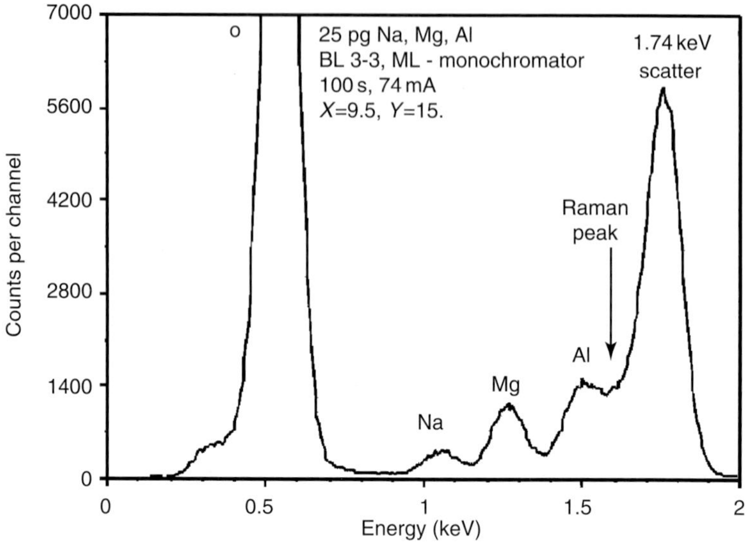


**Fig. 7.61.** Spectrum of a sample containing 25 pg Na, Mg and Al, measured at BL 3-3 in vertical geometry

## Facility at SSRL

As mentioned in the previous sections, TXRF in conjunction with synchrotron radiation had demonstrated enhanced sensitivities which, for transition metals, are approximately 50 times better than what is possible with conventional X-ray sources. The efforts at SSRL began with industrial interest to both extend the TXRF technique to low-$Z$ elements such as aluminum [196] as well as improving the sensitivity for transition metal detection [201–203]. The ensuing development effort resulted in a multi-company collaboration under the

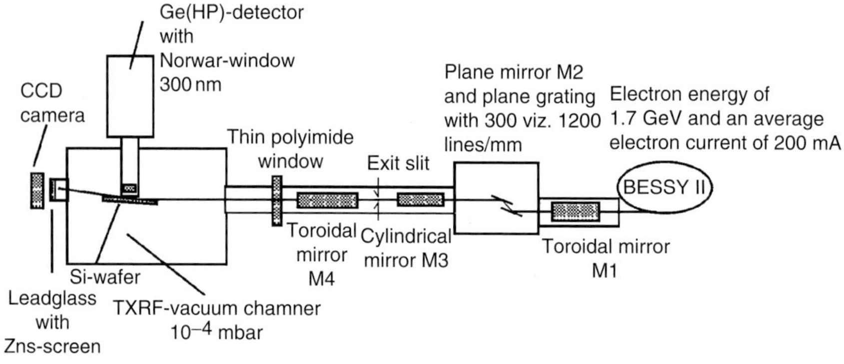


**Fig. 7.62.** TXRF setup at the PGM beamline of PTB for undulator radiation at BESSY

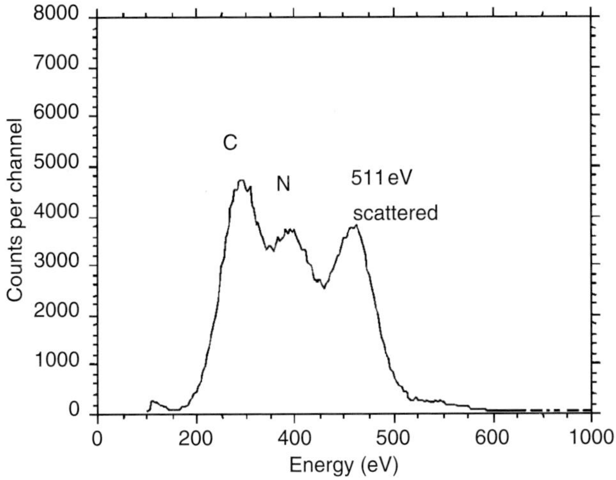


**Fig. 7.63.** Spectrum of the same wafer on the droplet of 1 ng C,1.2 ng N, 1.3 ng O, and 1.9 ng Na with radiation of 511 eV photon energy at an electron beam current of 66 mA during a counting time of 100 s

auspices of Sematech that resulted in the establishment of a facility at SSRL capable of routinely performing well-calibrated, industrially relevant measurements as well as carrying out research to understand the basic deposition mechanisms of metals on semiconductor surfaces [204].

Important considerations for the development of an industrially useful synchrotron TXRF facility include: a high brightness X-ray source, a suitable detector, a sample chamber in a cleanrom mini-environment, as well as automated measurement/analysis software that allows for calibration of concentrations and wafer mapping. All of these elements are equally important in providing high sensitivity measurements useful in problem solving and process development.

One of the primary features that distinguish SR-TXRF from what can be done using conventional sources is the availability of a linearly polarized beam with high flux and low divergence. This leads to increased fluorescence signals while reducing the background due to elastic and inelastic scattering of the X-rays. In addition, by exploiting the tunability of synchrotron radiation, the photoabsorption cross-section can be optimized for detection of transition metals or low $Z$ elements. Synchrotron radiation provides three different types of sources: bending magnets, wigglers, and undulators. This section will focus on the use of wiggler sources with comparisons made to bending magnet sources. Details of an undulator-based beamline will be given in the next section, which describes the TXRF facility at the ESRF.

Synchrotron radiation is emitted tangentially to the electron orbit in the bending magnets, and its properties are fully defined by the energy of the electrons in the storage ring and the radius of the bending magnet. A wiggler, however, consists of a period array of magnets in a storage ring straight section that provides a source that can be described as a sum of $2n$ individual bending magnet sources where $n$ is the number of magnetic periods in the wiggler. This significantly influences the intensity, collimation, and spectral distribution of the emitted radiation. One example of this is the characteristic energy of the wiggler radiation, or its critical energy, which is shifted toward higher energies as compared to the one of a bending magnet source due to the higher magnetic field and thus smaller electron bending radii in the wiggler [205]. Figure 7.64 shows the horizontal brightness obtained from the 27 period wiggler at beamline 6.2 at the Stanford Synchrotron Radiation Laboratory (SSRL). For a typical excitation energy of 11 keV for transition metal detection, the wiggler shows a higher horizontal brightness of about 2 orders of magnitude relative to a typical SSRL bending magnet source. It should be noted that the storage ring, SPEAR2, being used at SSRL is currently a second-generation synchrotron source in which the electron beam emittance is fairly large. As a result, a wiggler source is required to achieve the high fluxes needed for state-of-the-art detection limits. However, SPEAR2 is being upgraded to a third generation storage ring known as SPEAR3. Once that upgrade is completed in January 2004, the reduced radius of curvature and reduced electron beam size of an SPEAR3 bending magnet will make such a source competitive with the present SPEAR2 wiggler for excitation energies above 10 keV. In addition, the SPEAR3 wiggler will become even more favorable with a factor of 45 advantage in horizontal brightness over the SPEAR2 wiggler. These comparisons are also included in Fig. 7.64.

Optimized undulator sources are in many ways more favorable than either wiggler or bending magnet sources but require a third generation, high energy (>6 GeV) storage ring. The undulator provides a beam with higher natural collimation and linear polarization affording the possibility of better MDLs. However, with counting rate limitations of the presently available detector technologies, all advanced synchrotron implementations to date (on wiggler and undulator sources) have achieved approximately the same detection limits.

Since the wiggler spectrum shown in Fig. 7.64 is continuous, it is very important that the X-rays be monochromatized to minimize scattered radiation under the peaks of interest which would severely degrade the detection limits. Multilayers are ideal broadband and high throughput monochromator "crystals" for TXRF provided that they are designed to have a high peak reflectivity (reflectivities ~70% are commercially available) and low backgrounds. The calculated reflectivity as a function of energy for a single and double reflection for the multilayers used in this work is shown in Fig. 7.65. Note that the reflectivity in the region below the Bragg peak for a single reflection is approximately $1 \times 10^{-3}$. Even at this low level, there are enough transmitted photons in the region below the Bragg peak that diffraction from the wafer into the detector results in small but significant peaks that can interfere with the signals of interest. As a result, two multilayers, which are normally

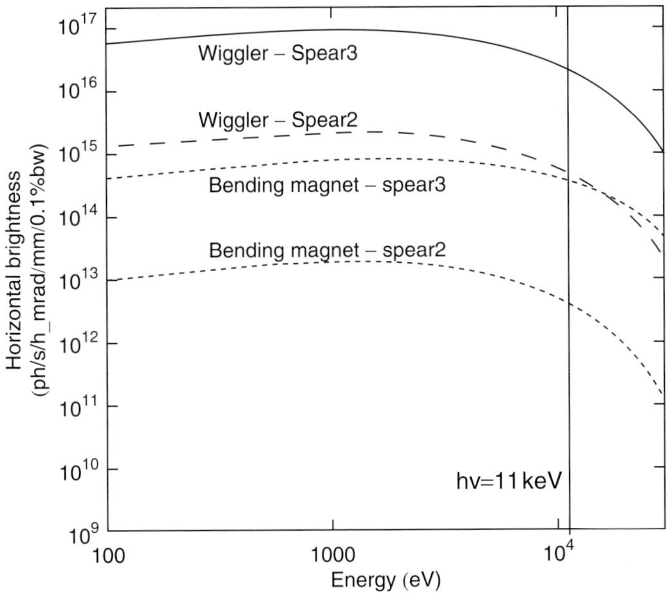


**Fig. 7.64.** Comparison of the horizontal brightness (brightness integrated over the vertical source size and divergence) for both SPEAR2 and SPEAR3 bending magnet and wiggler sources

placed in a parallel geometry, are used to eliminate this problem [201]. To further increase the flux density on the sample, the X-ray beam is focused onto the Si wafer using a 1:1 geometry. Figure 7.66 shows the resulting beamline configuration including the detector and sample chamber.

As shown in Fig. 7.66, the semiconductor detector and Si wafer are mounted in configuration B as shown in Fig. 7.57 in which the detector is oriented horizontally, along the polarization vector of the incident radiation, and the wafer is mounted vertically. As mentioned above, this configuration allows the detector to collect a large solid angle of fluorescent radiation as well as maintaining the advantage of reduced scattering by collecting the signal along the linear polarization vector of the synchrotron radiation.

It is also very important that the detector does not add any parasitic fluorescence that could contribute to the measured signal. This is a problem arising primarily for the detection of transition elements in which the high intensity scatter signal enters the detector striking components made of elements with fluorescence peaks in the regions of interest. An example is shown in Fig. 7.67 which shows the spectrum obtained from a clean wafer for three different internal configurations of the SSRL detector.

The top spectrum shows definite signals from Cu, Ni, and Fe. The Cu and Ni signals were identified by the detector manufacturer (Kevex) as being due to the presence of these elements in the internal components of the original

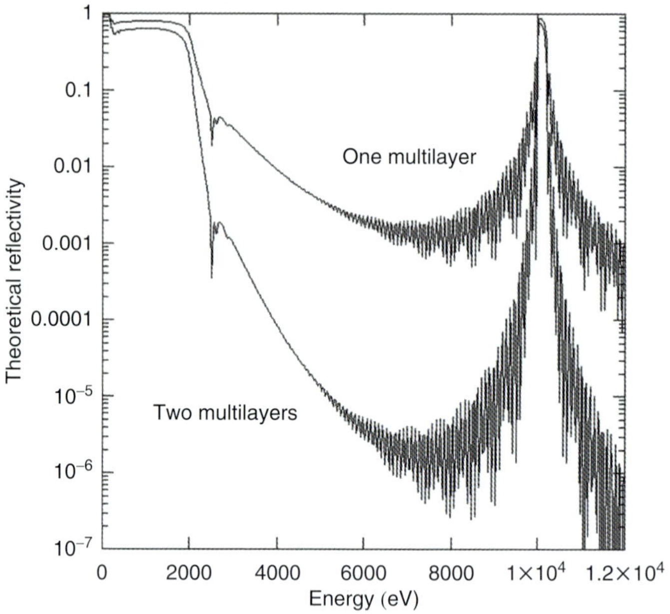


**Fig. 7.65.** Theoretical reflectivity of a single 29 Å period $Mo/B_4C$ multilayer as well as the double multilayer configuration used in the SSRL TXRF monochromator

detector. Once these components were replaced with materials whose fluorescent lines were outside of the region of interest, the second spectrum was obtained. This one still shows a small but significant Cu signal at a level equivalent to $4 \times 10^9$ atoms/cm$^2$ contaminant atoms on a sample wafer. This level of parasitic fluorescence must clearly be eliminated if reliable sensitivities of less than $1 \times 10^8$ atoms/cm$^2$ are to be obtained. In this case, it was determined

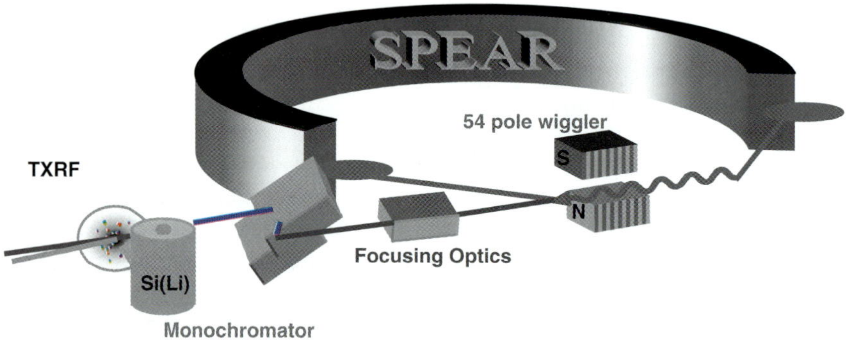


**Fig. 7.66.** Schematic of the SSRL TXRF facility showing the SPEAR2 storage ring, BL6-2 54 pole wiggler, toroidal focusing mirror, double multilayer monochromator, Si(Li) detector oriented along the polarization vector of the incident radiation, and the vertically mounted wafer

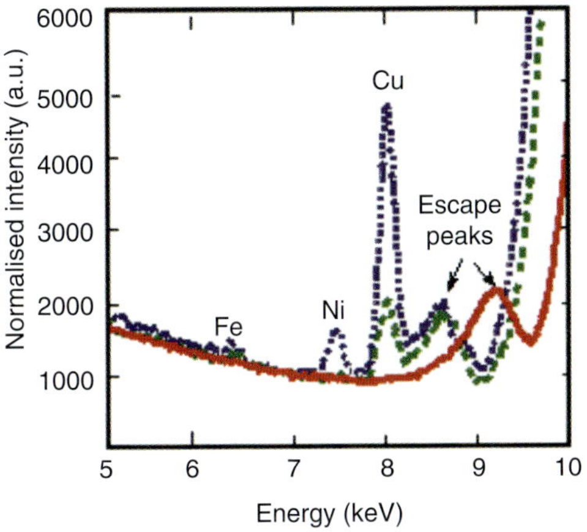


**Fig. 7.67.** Spectrum obtained from a clean wafer for three different internal detector configurations. The spectra in dashed and dotted lines were from the first two configurations used and showed significant parasitic signals. The spectrum in full line was taken after all offending elements were removed from the detector. Note that the red spectrum was taken at higher excitation energy than the first two

that the Cu signal was due to Cu in the Al alloy being used to fabricate the detector housing. Once this housing was replaced with 99.999% pure aluminum, the third spectrum with no parasitic signals was obtained and the detector could be used under the most rigorous conditions. Although the occurrence of parasitic signals became evident due to the very high intensities made available with synchrotron radiation, the solutions developed as a result of the synchrotron experiments were transferred to the conventional TXRF instrumentation and now such "clean" detectors are available commercially [206].

Figure 7.68 gives a clear appreciation for the power of a synchrotron source for TXRF applications. Both spectra are from a standard made by dip preparation and are measured for 1000 s counting time. The top spectrum is from a rotating anode TXRF system and the bottom taken at the SSRL facility. The most apparent difference is the peak intensity: 65 counts for the conventional spectrum and 16000 counts for the synchrotron spectrum. This alone accounts for a factor of 16 improved MDL. Furthermore, the background in the synchrotron spectrum is proportionally lower than in the conventional spectrum resulting in an additional factor of 3 and thus yielding an overall improvement in MDL of approximately a factor of 50. The predominant background in the 3 keV to 7 keV region in the synchrotron spectrum is due to photoelectron Bremsstrahlung and is one of the major factors limiting the sensitivity in the synchrotron case whereas in the case of the conventional spectrum, the background in the same region is primarily statistical noise. That means that in the conventional spectrum the intensity of the Bremsstrahlung is below the one of the statistical noise. The background under the higher energy

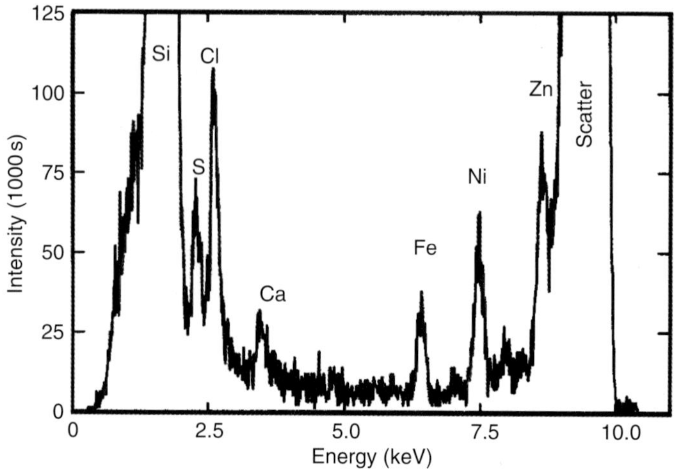


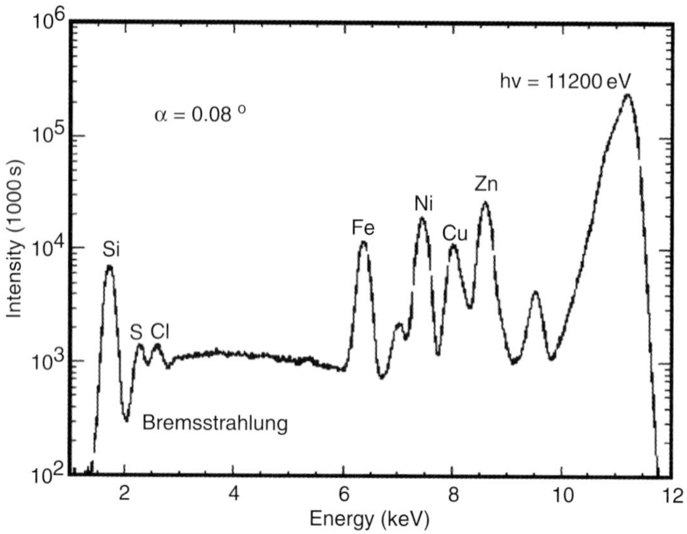


**Fig. 7.68.** Conventional (top) and synchrotron (bottom) TXRF spectra from $1 \times 10^{11}$ atom/cm$^2$ Fe, Ni, and Zn standards made by dip contamination [202]

fluorescence signals such as Ni and Zn is formed by the tail of the inelastic scattering peak which is reduced in a relative sense for the synchrotron case due to the linear polarization of the radiation. In addition, since the synchrotron excitation source can be tuned in energy, the scatter peak and its associated escape peak can be easily shifted to higher energy away from the Zn peak greatly improving the Zn MDL. Finally, it should also be noted that a 25 μm Teflon filter is placed in front of the detector to reduce the intensity of the Si substrate signal and prevent saturation of the detector.

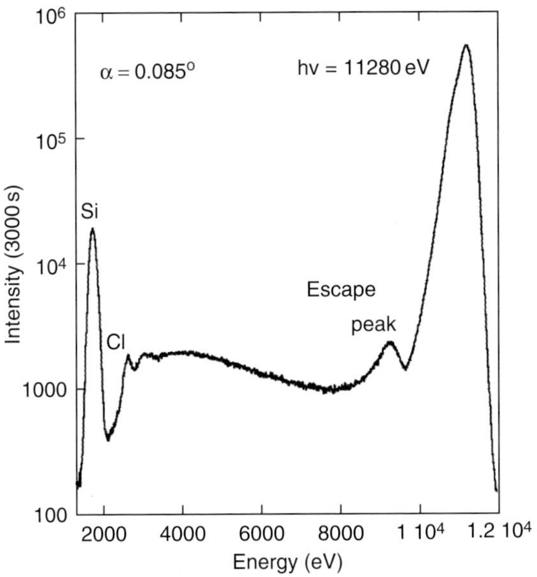


**Fig. 7.69.** TXRF spectrum of an ultraclean wafer taken for 3000 s at an excitation energy of 11.28 keV. The MDL for this spectrum is $4.4 \times 10^7$ atoms/cm$^2$

This results in a decrease in intensity between 2 keV and 2.5 keV, but does not significantly affect the fluorescence intensity of the transition elements of interest [202, 203, 206].

Although the MDL can be easily obtained from the standard spectrum of Fig. 7.68, the background which arises from the higher signals associated with the standard gives a higher apparent MDL than one measured using a sample with lower concentrations. Therefore the MDLs quoted here are typically obtained using samples with low overall levels whose concentration has been calibrated using a standard.

The most advanced processes for wafer cleaning available in the industry are truly able to produce wafers whose levels of surface contamination are below what can be observed even with SR-TXRF. The spectrum of such an ultraclean wafer taken with 11.28 keV excitation energy is shown in Fig. 7.69.

The only features observed are the Si, Cl, Ag, and the escape and scatter peaks. The Cl comes from the cleaning solution and is viewed as a benign component that does not pose problems for gate oxidation and the Ag is an artifact from the detector collimator. Ag was intentionally chosen as the collimator material because the Ag L fluorescence peaks were small and out of the region of interest. This spectrum was taken for a count time of 3000 s and corresponds to an MDL of $4.4 \times 10^7$ atoms/cm$^2$. Note that this concentration corresponds to a detected mass of approximately 0.1 fg when the $2 \times 7\,\mathrm{mm}^2$ sampling area is taken into account. The SSRL TXRF facility is sufficiently

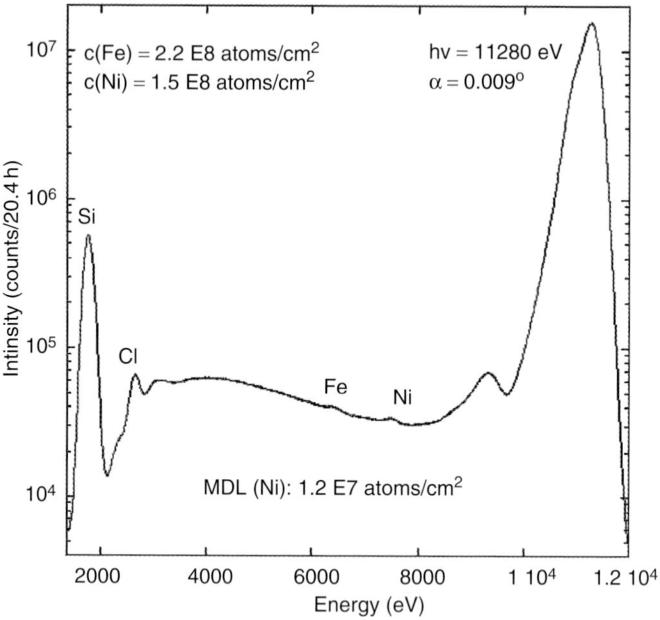


**Fig. 7.70.** TXRF spectrum of an unintentionally contaminated wafer taken for 20.4 h of counting time at an excitation energy of 11.28 keV. The MDL for this spectrum is $1.2 \times 10^7$ atoms/cm$^2$

well behaved that the use of extended counting times to improve detection limits for ultra clean wafers can be routinely performed. In an effort to determine the extent to which this is true, a very lightly contaminated wafer was measured for 1000 s, then repeatedly measured using a series of 5000 s spectra with a total counting time of approximately 48 h. Figure 7.70 shows the sum of the individual spectra after 20.4 of the 48 h.

If one were to compare the first spectrum of this series (1000 s counting time) with that of Fig. 7.70, the same features at the same concentrations would be observed with the only difference being the respective MDLs. The 1000 s spectrum yields an MDL of $8.9 \times 10^7$ atoms/cm$^2$ for Ni while the sum spectrum of 20.4 h yields a proportionately better MDL of $1.2 \times 10^7$ atoms/cm$^2$. The relative MDLs are consistent with the improvement expected from the longer counting time, i.e., approximately equal to the ratio of the square root of the counting times. Beyond the 20 h counting time, the appearance of a Zn peak as well as variations in the intensity of the Fe and Ni peaks were observed. Such changes could be the result of contaminants from the vacuum chamber being deposited onto the surface as well as diffusion of the contaminants within the wafer. This example demonstrates that the practical limits on the MDL is simply related to the number of signal counts which is what would be expected in the case of a well-behaved system. Higher counting rates rather than longer counting times would be the ideal way to obtain more counts in the signal peaks; however, detector saturation stands in the way of achieving these results. For example, in the case of the SPEAR3 storage ring, the expected increase in flux density of a factor 45 should improve the MDL by a factor of 6.7 resulting in a minimum detection limit of about $1.2 \times 10^7$ atoms/cm$^2$ for Ni after a standard 1000 s counting time. However, this would require a detector whose count rate capability were as high as 500 kHz while maintaining a resolution of better than 200 eV.

With such low MDLs being made possible with synchrotron TXRF, it is natural to think about pushing these detection limits even further by incorporating a VPD preconcentration as is currently being done with conventional TXRF. However, the contribution of impurities in the ultrapure water or the chemicals used in the VPD process as well as the handling of the wafers cannot be ignored when trying to perform such extrapolations. A recent study has shown that a baseline is introduced by the VPD process itself and that the MDLs for VPD-SR-TXRF with a 1000 s counting time will range from a low of $1.2 \cdot 10^5$ atoms/cm$^2$ for Co to a high of $2.7 \times 10^7$ atoms/cm$^2$ for Cr [207]. The MDLs for the other transition metals are within this range: Fe ($2.2 \cdot 10^7$ atoms/cm$^2$); Ni ($6.7 \times 10^5$ atoms/cm$^2$, Cu ($9.8 \times 10^6$ atoms/cm$^2$), and Zn ($3.1 \times 10^5$ atoms/cm$^2$). This baseline is well within the sensitivity range of the SR-TXRF measurement as can be shown for the case of Zn which has one of the lower VPD-SR-TXRF MDLs observed. In this case, the Zn MDL of $2.2 \times 10^7$ atoms/cm$^2$ corresponds to a concentration in the VPD droplet of $7 \times 10^8$ atoms/cm$^2$ which is an order of magnitude higher than the MDL of SR-TXRF.

The mechanisms associated with the introduction of impurities onto the silicon wafer surface during the VPD process are identical to what happens, for example, during chemical cleaning of wafers prior to gate oxidation. In these situations, metal impurities present in the etching solutions or in the UPW used in wafer processing will become readily ionized. As a result the solutions behave like an electrochemical bath in which the impurities can be deposited on the wafer surface. The deposition mechanism for a particular impurity can be affected by a variety of parameters including the amount of dissolved oxygen in or the pH value of the solution as well as the doping in the semiconductor. Important insights into the chemical reactions leading to metal deposition can be obtained by understanding the oxidation state or speciation of the deposited metal atoms. Fortunately, the oxidation state of the trace impurities can be obtained from X-ray Absorption Near Edge Spectroscopy (XANES) [208] in conjunction with a SR-TXRF measurement where the photon energy is now scanned across the absorption edge of the element of interest. This requires the use of a silicon crystal monochromator vs. multilayers so that some sensitivity is sacrificed but the resulting MDL of $1 \times 10^9$ atoms/cm$^2$ is still adequate for most of these studies. By using XANES spectra of reference materials as fingerprints, specific details about the chemical environment can be derived. For example, Fig. 7.71 shows the Cu 1s XANES spectrum of an Si wafer dipped into Cu-contaminated (1000 ppb) 2% HF (solid dots) with a

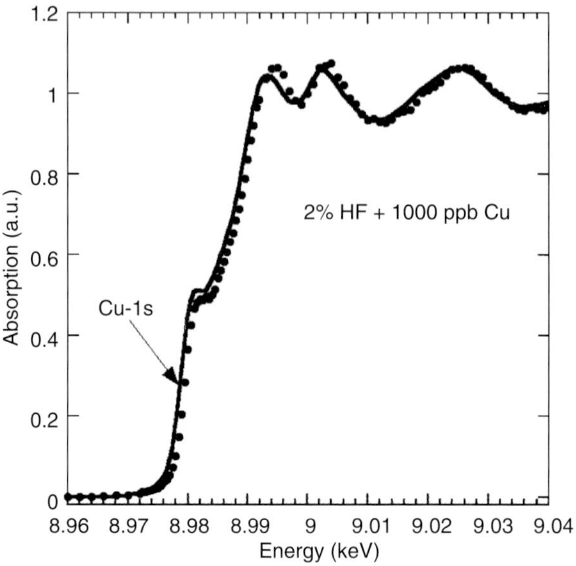


**Fig. 7.71.** Cu 1s XANES spectrum of a Si wafer dipped into Cu-contaminated (1000 ppb), 2% HF (dots) with a resulting surface concentration of $6 \times 10^{13}$ atoms/cm$^2$. The solid curve is the XANES spectrum from a Cu metal foil standard

resulting Cu surface concentration of $6 \times 10^{13}$ atoms/cm$^2$. This is identical to the spectrum of bulk Cu metal (solid line) also shown in Fig. 7.71 except for a small but real 0.35 eV shift. The use of this simple fingerprinting method verifies that the deposited Cu exists as clusters, which are mainly metallic in character. This is also in line with electrochemical models predicting that the reaction pathway for metal deposition in a low pH solution such as HF will be reductive [209]. Such studies have also been extended to more complex solutions involving oxidative depositions in high pH solutions [204].

As mentioned in the previous section, the detection of light elements such as Al presents a challenging situation where the tunability of synchrotron radiation can be used to suppress the excitation of the substrate Si $K_\alpha$ fluorescence while enhancing the cross section for the Al K$\alpha$. Results from a bending magnet at SSRL (BL 3-3) from a Si wafer intentionally contaminated with $3 \times 10^{11}$ atoms/cm$^2$ of Al are shown in Fig. 7.72. The spectrum was taken with an excitation energy of 1720 eV at an angle of incidence of 0.1° and measured for 10000 s. The low energy peak is the Al $K_\alpha$ fluorescence signal at 1487 eV and the high energy signal results from the elastically scattered synchrotron radiation. This spectrum also shows the Raman contribution to the background as an asymmetric peak with a tail extending under the entire Al peak. Baur et al. have shown the Raman signal to be one of the primary factors that limit the MDL for Al [210]. This work

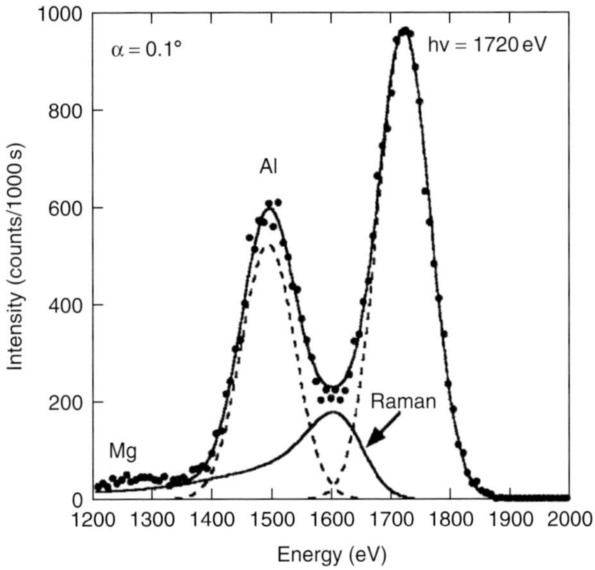


**Fig. 7.72.** TXRF spectrum of an Si wafer intentionally contaminated with $3 \times 10^{11}$ atoms/cm$^2$ Al taken for 10000 s at an excitation energy of 1.72 keV and an incidence angle of 0.1°. The corresponding MDL for a 1000 s counting time is $5.7 \times 10^9$ atoms/cm$^2$

has also provided a detailed analysis of the Raman contribution to the overall spectrum that must be taken into account when determining the MDL. With this analysis method, the MDL for the spectrum of Fig. 7.72 is found to be $5.7 \times 10^9$ atoms/cm$^2$ for a counting time of 1000 s. As described above, laboratory instruments are also able to provide acceptable sensitivities for Al detection by using the W-$M_\alpha$ fluorescence line (1780 eV) for primary excitation [183]. Figure 7.73 shows the fluorescence spectrum obtained from a wafer with an intentional Al contamination of $8 \times 10^{12}$ atoms/cm$^2$ using a TXRF 300 system from Rigaku [211]. This spectrum has been integrated for 3600 s at an angle of incidence of 0.45°. The fluorescence signature of the Al $K_\alpha$ line is observed in the low energy region of the spectrum with the dominant contribution at 1780 eV being due to elastic scattering of the W-$M_\alpha$ excitation line. By comparing the elastic scattering contributions in Figs. 7.72 and 7.73, the advantage of using a linearly polarized synchrotron source becomes apparent. The Raman contribution is also seen to provide a significant contribution in this spectrum and results in an MDL of $4.7 \times 10^{10}$ atoms/cm$^2$ for a standard 1000 s counting time [211]. It is worthy to note that this type of data analysis, which resulted from the synchrotron experiments, has currently been implemented in the analysis software for the Rigaku TXRF 300 instrument. Finally, as in the case of the transition

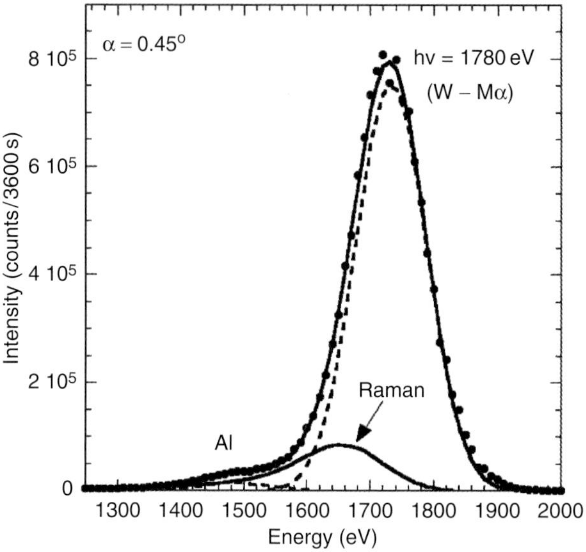


**Fig. 7.73.** Conventional TXRF spectrum from a Rigaku TXRF 300 system of a Si wafer intentionally contaminated with $8 \times 10^{12}$ atoms/cm$^2$ Al taken for 3600 s at an excitation energy of 1780 eV (W-$M_\alpha$ excitation) and an incidence angle of 0.45°. The corresponding MDL for a 1000 s counting time is $4.7 \times 10^{10}$ atoms/cm$^2$

elements, the MDL for Al and other light elements can be improved with the increased horizontal brightness as available from third generation synchrotron sources.

In addition to the examples discussed above, the list of applications carried out by industry using the SSRL facility has been quite varied and include evaluation of advanced cleaning methodologies, qualification of wet stations used in actual wafer manufacturing, verification of repairs to manufacturing lines, qualification of cleaning chemicals, and quality control on VPD processes. In all these cases, the high sensitivities and flexibility afforded by synchrotron TXRF were critical in addressing the problems of interest.

#### Facility at ESRF

The needs of the semiconductor industry for a more strict control on wafer contamination and the availability of synchrotron radiation sources for uses other than pure research has stimulated the development of test facilities at synchrotrons and primarily at SSRL in Stanford [203].

In Europe, semiconductor companies encouraged similar developments at the existing synchrotron installations. At the ESRF, the European Synchrotron Radiation Facility in Grenoble, a feasibility test in 1996 [212] showed that third generation machines, as the fully dedicated and application-oriented

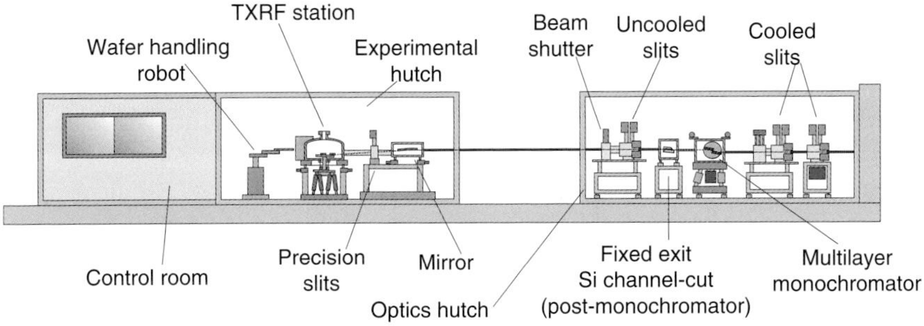


**Fig. 7.74.** Layout of the ESRF beamline dedicated to TXRF mapping of impurities at the wafer surface

installations like ESRF are called nowadays, offer new possibilities for mapping the distribution of metallic contaminants over the wafer surface with LLD in the $10^8$ atoms/cm$^2$ range. These possibilities are the combined result of the increased flux of X-ray photon that can be shined on the wafer surface and of the linear polarization of the radiation. The gain in flux alone, in fact would not be directly exploitable if the elastic scattering contribution that would saturate the solid-state detector could not be strongly suppressed. The early experiments at ESRF showed that by exploiting the linear polarization of the ESRF beams it was possible to reduce the elastic scattering peak to values comparable to the fluorescence signals from $10^{12}$ atoms/cm$^2$ by sensibly increasing the distance between wafer and detector. Following this test, an agreement was signed in 1997 by European semiconductor related organizations[7] to sponsor a four-year program aimed at the construction at ESRF of a facility completely dedicated to TXRF analysis of 200 mm and 300 mm diameter wafers.

*Experimental Layout*

The facility was opened to users in 2000. The layout of Fig. 7.74 shows the essential elements that compose the 50 m long installation: the X-ray beams produced on the right by insertion devices are conditioned in an Optics Hutch by slits and photon absorber. The hundreds of watt of power emitted by the undulators would not be directly exploitable without any further conditioning. For this, out of the energy spectrum of the emitted radiation, a cooled multilayer monochromator selects a bandpass continuously tunable between 800 eV and 20000 eV. Typical flux at the output of this monochromator is of the order of $10^{14}$ photon per second.

The X-ray beam passes then in the Experimental Hutch: a lead shielded enclosure kept in class 100 by filters and laminar flow hoods and that hosts

[7]Personnel funding has been provided through the MEDEA platform, while capital funding has been provided by the LETI and TELECOM companies

the TXRF measuring chamber and the ancillary wafer-handling devices for automatic loading and unloading of the wafers.

Because of the need to handle photon energies as low as 800 eV, the entire beamline, TXRF chamber, and detection system are in vacuum with no window to intercept the beam.

The optical layout of the installation is completed by a silicon post-monochromator that can be inserted in the beam path after the multilayer monochromator; this monochromator selects a bandpass of few eV for performing, when required, XAS measurements around selected absorption edges for analyzing chemical and geometrical atomic structure around the contaminant species.

Finally, a mirror for rejecting the higher harmonics transmitted by the monochromator is installed upstream the wafer analysis station. This mirror is of the bimorph type and can be piezoelectrically bent longitudinally to focus the radiation in the vertical direction and increase then the photon density on a particular area of the wafer.

*The TXRF Station*

The TXRF end-station is at the core of the beamline. The station encompasses an atmospheric loading robot that transfers the wafers from standard cassettes to an indexer to azimuthally orient and center it before passing introduction into the load-lock vessel. The load-lock can host up to five 200 mm and five 300 mm wafers. After pump-down of the load-lock, one wafer at a time can be transferred to the main TXRF chamber for analysis. The TXRF system is sketched in Fig. 7.74: a large, rugged hexapode actuator installed in air just below the vacuum chamber is coupled through a bellow and a rotary feedthrough to the electrostatic chuck that flattens and holds the wafer in vacuum. All the alignment procedures of the wafer on the X-ray beam are performed via the air operated hexapode and transmitted through the bellow and the rotary feedthrough to the vacuum chuck.

*The Detection Scheme*

The operation scheme of the station was selected on the basis of a few geometrical constraints; on the one hand it was desirable to have the wafer loading, unloading, and alignment procedures to be performed in an horizontal plane for ease of operation; on the other, the fluorescence detectors should look along the horizontal polarization plane of the radiation to minimize the collection of elastically scattered photons. A third important constraint is that the radiation beam cannot be focused down into a small region of the wafer because any single element detector would be easily saturated. As a result the unfocused beam of radiation is left to impinge on the wafer surface at the desired angle of incidence along an entire diameter. The wafer stays then in an essentially horizontal geometry: the side tilt of about 5° toward the two arrays of six Si(Li) solid state detectors permits a close approach of the silicon elements to the impinging line. Each element of the arrays is provided of a

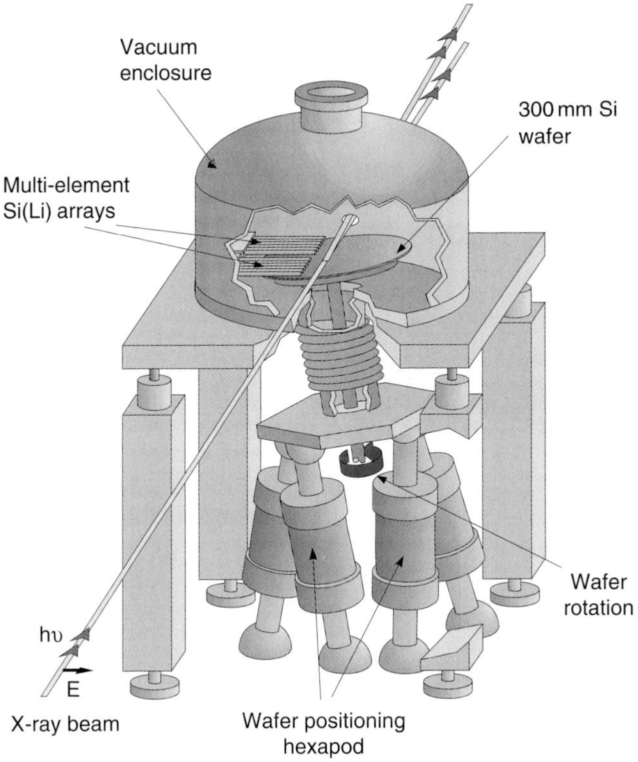


**Fig. 7.75.** Structure of the TXRF end station. The 200 mm or 300 mm wafer sits at the center of the vacuum enclosure; two arrays of solid-state detectors approach sideways the line illuminated by the beam; an external goniometer rotates the wafer for mapping the complete surface. All the alignments are performed from the outside by an external heaxapode through the bellow and the rotary feedthrough

collimator that limits the detection footprint to a segment of about 12.5 mm of the illuminated diameter. The geometry is shown in Fig. 7.76. A mapping of the contaminant distribution along this line is then performed in parallel by the detector array and a complete rotation of the wafer around its axis will permit a complete mapping of the impurity distribution. Special care has been put on the initial alignment of the chuck with respect to the external axis of rotation in order to keep the precession angle below $50''$; this avoids the continuous readjustment of the angle of incidence during the rotation scan. Of course it is always possible to trade lateral resolution with detection limits by adding spectra from contiguous pixels.

*Performance and Limitations*

With processing time per wafers of the order of one hour the attainable LLDs for transition metals are in the range of $10^8$ atoms/cm$^2$. Much longer integration times have been chosen to analyze an ultra-clean wafer borrowed from

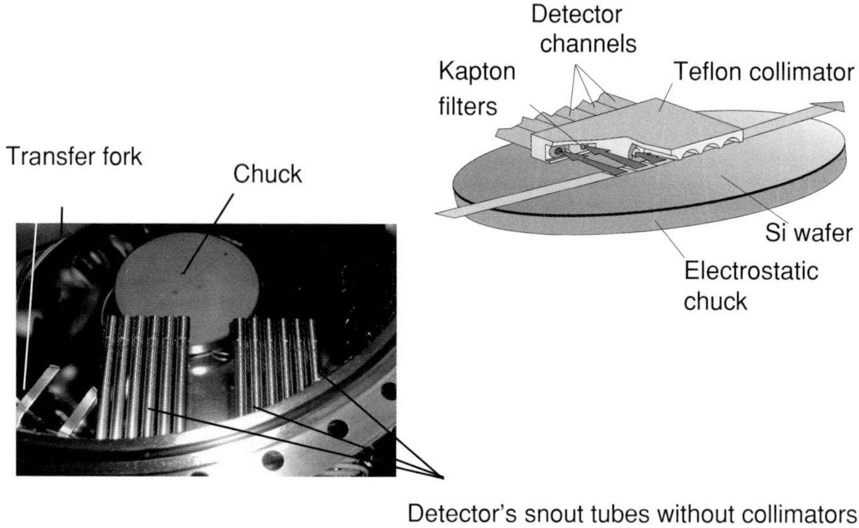


**Fig. 7.76.** Parallel detection scheme. The drawing at the top shows the overall geometrical wafer-detector arrangement. The distance of the beam impinging points from the 20 mm$^2$ active detector crystal is 85 mm. On the left a top view of the chamber with the lid removed: the two sets of snout tubes of the detectors are clearly visible above the chuck

the SSRL facility (see preceding section), in order to asses the stability of the instrument and the absence of any stray signal from the system environment. The results obtained of LLD in the lower $10^7$ range are in line with the SSRL findings (see Fig. 7.77).

The detection of low $Z$ atoms presents instead additional problems that have been addressed with the development of Synchrotron Radiation TXRF [210] and that establish new, intrinsic limits to the lowest detectable concentration of impurities. When detecting elements lighter than silicon, in fact, it is necessary to utilize excitation radiation below the Si K-edge threshold in order to cut down the intense silicon fluorescence that with its low energy tail would bury the emission from aluminum. However, when exciting below the silicon threshold, the emission spectra is dominated by a resonant Raman background. This spectral distribution of inelastically scattered photons is produced by incoming photons that lose the energies to excite 2s electrons in the continuum. This process is responsible for a triangular-shaped background peaked at the exciting energy decrease of 100 eV and far extending in the low energy tail. This background, at variance with the elastic scattering, does not present a pronounced variation with the detection geometry and limits the LLD for Al to a value of $7 \cdot 10^9$ atoms/cm$^2$. Figure 7.78 shows the spectra from two reference wafers with $\sim 10^{11}$ and $\sim 10^{12}$ Al atoms/cm$^2$ as compared to the spectrum from a clean wafer. The implication of the Raman effect in limiting the LLD for Al is quite evident.

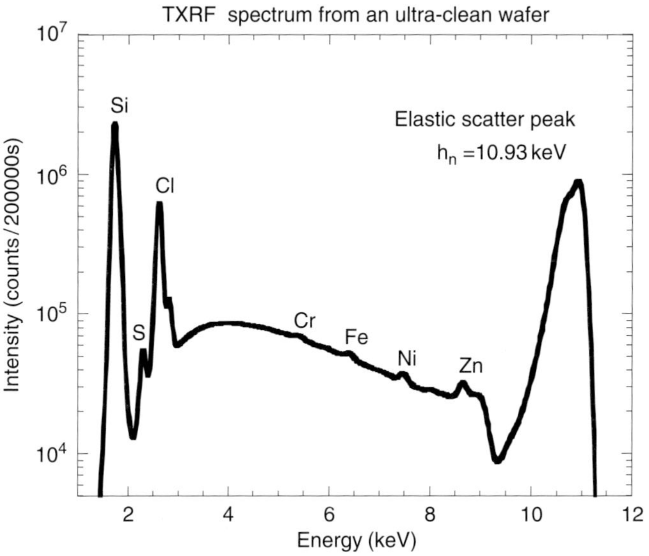


**Fig. 7.77.** SR-TXRF spectrum from an ultra-clean wafer from Motorola used as cross calibration test between SSRL and ESRF TXRF facilities. The Cr, Fe, and Ni contaminants are in the order of few $10^8$ atoms/cm$^2$ and the lower limit of detection in $10^7$ atoms/cm$^2$ scale

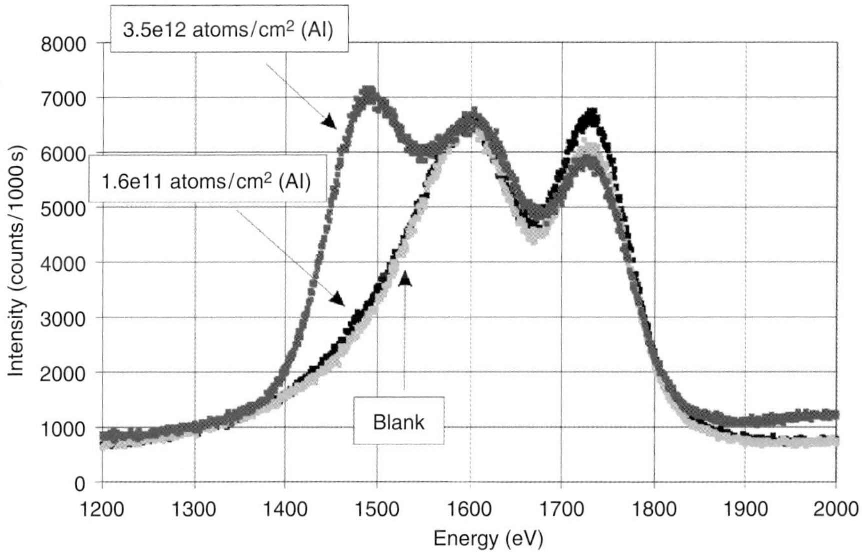


**Fig. 7.78.** Spectra of Al contaminated ($1.6 \times \cdot 10^{11}$ at/cm$^2$ and $3.5 \times \cdot 10^{12}$ at/cm$^2$) and blank wafers (energy 1730 eV, incidence angle 8 mrad). The rightmost peak at 1740 eV is residual Si fluorescence excited by the high energy tail of the multilayer pair transmission. The central peak at 1600 eV is the Raman background that extends at lower energies interfering considerably with the Al peak at 1490 eV

In the case that the required mapping detection limits for impurities of low $Z$ elements would be lower than $7 \times 10^9$ atoms/cm$^2$ it would be necessary to move from energy dispersive detection toward wavelength dispersive solutions, that because of the higher energy resolution of optical elements such as multilayers would increase the relevance of the low energy fluorescence peaks with respect to the Raman background.

*Automation*

The facility has been designed with a high degree of automation in mind to cope with the high throughput that can be envisaged for the years to come.

*Acknowledgments*

The realization of the facility has been possible by the dedication of Monique Navizet, the work of Paolo Mangiagalli and Giorgio Apostolo, and the skills of Guy Dumoulin. The important effort on the software aspects of automation has been supported by Emmanuel Papillon.

## Facility of the PTB at BESSY II

Using undulator radiation at electron storage rings, very high photon fluxes of linear polarized radiation for the TXRF excitation can be achieved. The photon energy of the undulator radiation depends on the energy of the stored electrons as well as on both the strength and periodicity of the magnetic field generated by the undulator. At the 1.7 GeV electron storage ring BESSY II in Berlin, Germany, photon energies of undulator radiation range from the VUV well into the soft X-ray regime, up to about 2 keV, thus allowing for the effective inner-shell excitation of light elements. The Physikalisch-Technische Bundesanstalt (PTB), Germany's national institute for metrology, operates a laboratory [213] at BESSY II, including a plane grating monochromator (PGM) beamline [214] for undulator radiation. Complementary to the work at SSRL and ESRF, the PTB lays emphasis on the methodological development of soft X-ray TXRF analysis, providing a means for the quantitation of low $Z$ contaminants on silicon wafer surfaces at its PGM beamline. Initial TXRF investigations were carried out in cooperation with the ATI and the semiconductor company Siltronic in order to determine the absolute LOD values of TXRF for some low Z elements [200, 215] such as C, N, Na, Mg and Al, which ranged between 0.3 pg and 1.3 pg with respect to a measuring time of 1000 s. The calculated VPD detection limits, based on the assumption that the sample droplets were collected from a 200 mm wafer, ranged from $2 \times 10^7$ atoms/cm$^2$ to about $10^8$ atoms/cm$^2$ for Na, Mg, and Al, thus fulfilling current analytical requirements of the semiconductor roadmap. For the explicit purpose of the semiconductor industry, the PTB has extended its TXRF instrumentation [216] and, to date, can handle 200 and 300 mm silicon wafers. In recent investigations employing again small droplets containing the elements Na [216] and Al deposited on 200 mm Si wafers considerably improved LOD values of less than 100 fg were achieved by further optimizing the excitation conditions.

Compared to the analysis of medium $Z$ elements, such as transition metals for which excitation energies in the 5 keV to 15 keV range are needed to induce K fluorescence, some drawbacks exist when exciting K fluorescence radiation of low $Z$ elements below 2 keV. The measured fluorescence count rates are proportional to the fluorescence yields when the self-absorption effects are negligible, which is the case when only minute amounts of a low $Z$ element are present. Since the fluorescence yield rapidly decreases as the atomic number $Z$ decreases, the fluorescence yields of light elements are about one to two orders of magnitude smaller than those for medium $Z$ elements. Additionally, the detection efficiency decreases with decreasing photon energy due to the inevitable absorption of the fluorescence radiation in the entrance windows of an energy-dispersive semiconductor detector, leading to a reduction in the detected fluorescence count rates. Another complication is that the recorded spectra are more difficult to deconvolute because the energy resolution of conventional semiconductor detectors does not, in general, decrease quickly enough with respect to the decreasing energy in order to allow for a sufficient separation of the K fluorescence lines of neighboring elements. The situation becomes even more unwieldy when L fluorescence lines of medium $Z$ elements overlap with K fluorescence lines of low $Z$ elements. Therefore, when considering a specific fluorescence line of an element, the detected peak shapes and the relative intensities of the adjacent fluorescence lines of other elements strongly influence the lower level of detection (LLD or LOD).

The first step in the quantitation of TXRF spectra is to determine the detected fluorescence count rates. Here, experimental detector response functions [217] can most effectively contribute to deconvoluting the TXRF spectra [200]. The detected count rates are then divided by the efficiency of the detector, leading to the fluorescence intensities. To ensure reliable values, the energy dependence of the detection efficiency should be well known, for example by means of an absolute calibration [213].

In the following, the PGM beamline characteristics and the TXRF instrumentation of the PTB will be explained in detail. The PGM beamline provides high photon flux of high spectral purity in the soft X-ray range. The photon flux at the PGM ranges from $6.0 \times 10^{11}\,\mathrm{s}^{-1}$ at about 400 eV to $6.0 \times 10^{9}\,\mathrm{s}^{-1}$ at 1700 eV using the U180 undulator with respect to a typical stored electron current of 200 mA. The U49 undulator produces an even higher flux, for example about $8 \times 10^{11}\,\mathrm{s}^{-1}$ at the photon energy of 400 eV. The PGM resolving power $E/\Delta E$, ranging from about 1000 to 9000, depends on both the exit slit size and the trigonometric ratio of the incident to the exit angle at the 1200 l/mm grating. The horizontal beam profile in the focal plane of the PGM beamline is about 140 µm (FWHM) wide, whereas the vertical extension of the beam profile is generally equal to the size of the exit slit, which can be varied between 20 µm and about 600 µm.

**Fig. 7.79.** Picture of the UHV-compatible TXRF irradiation chamber of the PTB for wafer samples with diameters ranging from 25 mm to 75 mm located at the PGM experimental station. The mounting of equipment such as the Si(Li) detector as well as samples changes can be performed below a class 100 cleanroom

The first irradiation chamber [200] of the PTB for TXRF and XRF analysis is shown in Fig. 7.79. This UHV chamber can simultaneously accommodate three wafer samples with areas ranging from $25 \times 25\,\mathrm{mm}^2$ to $30 \times 30\,\mathrm{mm}^2$ or two wafer samples, each with an area of up to $50 \times 50\,\mathrm{mm}^2$, or one wafer sample with a diameter ranging between 75 and 100 mm. The individual wafer samples can be scanned completely in the horizontal direction. Scanning along their vertical axes can lead to some restrictions that depend on the specific sample holder employed. With respect to the incident radiation, the possible angles of incidence range from $0.001°$ to $5°$. At an angle of incidence of $2.5°$ and a typical exit slit size of 40 µm, for example, the PGM beam profile results in an illuminated area of $3.2\,\mathrm{mm} \times 40\,\mu\mathrm{m}$. The solid angle of detection is defined by the distance between the Si(Li) detector, which has a 4.4 mm large diameter of its silicon entrance diaphragm, and the illuminated area on the silicon wafer. This distance can be varied from a few millimeters up to several centimeters, depending on the total fluorescence count rate to be detected. The Si(Li) detector is calibrated absolutely, i.e., both its counting efficiency and the energy dependence of its response to monochromatic radiation was determined prior to its use for detecting fluorescence photons.

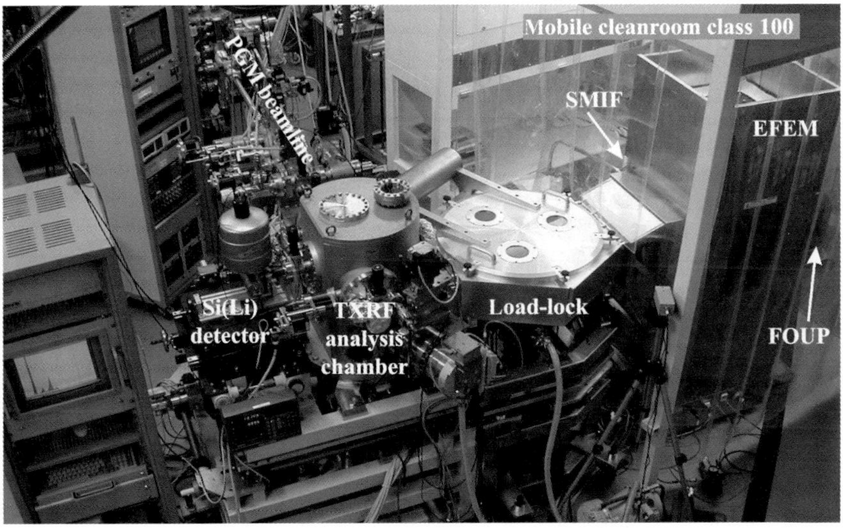


**Fig. 7.80.** Picture of the entire experimental (T)XRF arrangement for 200 mm and 300 mm wafers, including an equipment front-end module (EFEM) placed below a class 100 cleanroom, a high-vacuum load-lock, the ultra-high vacuum TXRF analysis chamber, and a conventional Si(Li) detector

A novel instrumentation [216], recently designed and constructed by the PTB for the explicit purpose of the semiconductor industry, fully utilizes the excitation conditions at the PGM beamline: 300 mm Si wafers, as well as 200 mm wafers, are transported directly from their shipping cassettes (FOUP or SMIF) via a prealigner into a high vacuum load-lock by an adapted commercial equipment front-end module (EFEM). After the pump down, a vacuum robot located inside the load-lock takes the wafer and places it inside the UHV irradiation chamber on an electrostatic chuck (ESC) mounted on an 8-axis manipulator. Figure 7.80 shows a picture of the entire arrangement. To take advantage of the linear polarization of the exciting radiation, the ESC is moved into a vertical orientation during the measurements. The whole surface of a 200 mm or a 300 mm wafer can be scanned. To extend the capability of the system from the TXRF into the XRF regime, the angle of incidence can be increased from grazing incidence up to 45°, allowing for the analysis of thin multi-elemental, multi-layered structures. Both conventional Si(Li) detectors or superconducting tunnel junction (STJ) detectors [218, 219] can be used. The EFEM, including a class 1 mini-environment, is surrounded by a mobile class 100 environment to further reduce the risk of unintentional cross-contaminations of the samples.

The commissioning of this arrangement was initiated in late 2002 with the investigation of a 200 mm silicon wafer, which was intentionally contaminated with 100 μl droplets containing metal and light element contamination in the

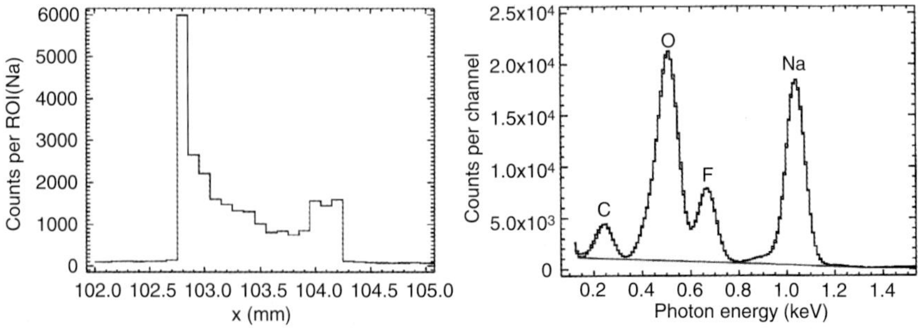


**Fig. 7.81.** TXRF investigation of a 100 μl droplet containing 500 pg Na dried on a 200 mm silicon wafer. The figure to the left shows the lateral distribution of Na, deducted from the number of events in the corresponding region of interest (ROI) in 10 s spectra, using a 150 μm wide excitation beam. The TXRF spectrum to the right was recorded at the respective maximum Na position during 100 s real time

pg range. The droplet depositions were prepared by the Central Analytical Laboratories of the company Siltronic. The first element of interest was Na, which was excited with monochromatic radiation having a photon energy of 1206 eV at an angle of incidence of one degree. The left part of Fig. 7.81 shows the lateral scan across the 500 pg Na droplet [216]. The TXRF spectrum recorded at the maximum position of the lateral scans was deconvoluted by means of experimentally determined detector response functions. The remaining background, which was used to calculate the LOD value, is depicted by the gray line to the right in Fig. 7.81. For the accumulation time of 100 s, the conservative estimate of the LOD value for Na is 1.0 pg. Taking into account the ratio of the respective Na deposition present in the lateral maximum position to the total Na deposition in the droplet, the LOD value for Na is only 170 fg resulting in an LOD well below 80 fg for a measuring time of 1000 s. Employing various exchangeable thin absorption filters in front of the Si(Li) detector it becomes possible to modify the spectral distribution detected in favor of selected fluorescence lines of interest, thus improving the related detection limits, e.g., of transition metals such as Ni, Cu and Zn, by factors ranging from about 3 to 6.

A study [220] concerning the resonant Raman scattering effect [210] which is a limiter for the determination of Al at ultra trace levels, was performed with respect to both the energetic and the angular variation of the incident radiation employing a droplet deposited on a 100 mm wafer in the TXRF chamber of the ATI. In a recent investigation employing a droplet deposition on a 200 mm wafer in the new PTB arrangement a LOD value of about 90 fg could be achieved with respect to a measuring time of 1000 s by means of further varying the excitation conditions.

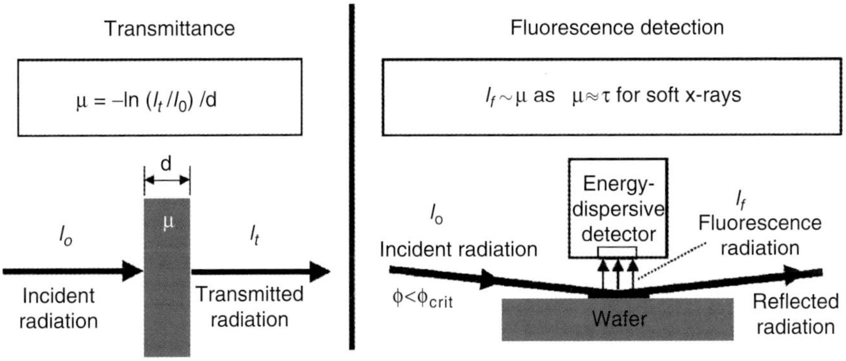


**Fig. 7.82.** Principle of transmission-based NEXAFS vs. fluorescence-based TXRF-NEXAFS

In additional studies employing the ATI arrangement for 100 mm wafers at the PGM beamline, a LOD [220] of 7 ng for B was found. Furthermore, by varying the angle of incidence, a TXRF investigation [220] of both a 1.6 nm C – 2.1 nm Ni – 1.6 nm C multi-layered structure and a 5 nm thick C layer was performed, confirming the respective layer thicknesses.

Organic contaminants are starting to play an important role in the production and quality control of silicon wafers. A near-edge X-ray absorption fine structure (NEXAFS) investigation in conjunction with TXRF is able to contribute to the speciation of low $Z$ and organic compounds. The PGM beamline is an appropriate source for TXRF-NEXAFS experiments, which require the tunability of the excitation radiation and a relatively high resolving power while ensuring both a sufficient photon flux to allow for trace analysis of low $Z$ elements and a high spectral purity. Figure 7.82 shows the principle of NEXAFS measurements obtained by transmittance measurements of thin foils and the TXRF-NEXAFS based on the detection of element specific fluorescence radiation. The contaminants were diluted and deposited as droplets on wafer surfaces. The K-edges of C, N, O have been examined in initial experiments using the two TXRF chambers [216, 221] of the PTB. Apart from the K fluorescence count rates of the element of interest also the incident radiant power was recorded by means of a reference measurement employing a calibrated photo diode [213] allowing for an absolute normalization procedure. During the TXRF spectrum acquisition, the incident radiant power can be monitored by both the measurement of the stored electron beam current as well as the measurement of the radiant power of the beam reflected at the wafer sample. For a typical energetic scan at the N-K edge the incident photon energy was varied from 395 eV to 415 eV in steps of 125 meV or 250 meV. At each point of the scan a TXRF spectrum was recorded during 15 s to 40 s and the detected K fluorescence count rate deduced. TXRF-NEXAFS spectra recorded from droplets containing 12 ng N deposited on a 200 mm silicon wafer are shown

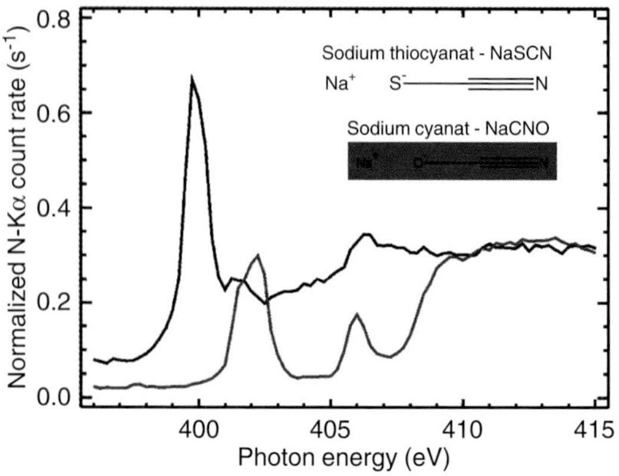


**Fig. 7.83.** TXRF-NEXAFS spectra of about 500 pg of two different compounds of light elements deposited as small droplets on a 200 mm silicon wafer surface. In these measurements, the energy of monochromatized undulator radiation is varied in the range around the K absorption edge of N in steps of 0.25 eV while recording the N-$K_\alpha$ fluorescence count rate normalized to the incident radiant power (nW)

in Fig. 7.83. Here about 500 pg of each compound was excited by the incident beam. The reproducibility of these measurements at the N edge was good, whereas unintentional contaminations of the wafer at the C-K edge and the native oxide layer at the O-K edge can interfere, thus requiring TXRF-NEXAFS scans at both the on-droplet and a reliable off-droplet position. Self-absorption effects and dependencies of the absorption on the adsorbate orientation will require further investigation. vs. TXRF-NEXAFS [222] has also been employed to investigate the chemical state of nitrogen compounds in aerosols. The aerosol samples of different size fractions were deposited on $30 \times 30\,\mathrm{mm}^2$ large silicon wafer surfaces in relatively narrow (100 µm to 300 µm) lines by means of a May impactor. The TXRF detection limits for nitrogen being in the upper fg range enable analysis of aerosol samples taken during 10-minute intervals with acceptable accuracy. The applicability of the technique to real aerosol samples has been used to compare nitrogen oxidation state in suburban and rural aerosols. For example, A reference spectrum of ammonium sulfate at the N K-edge contains a typical multiplet of $\pi^*$ resonances between 400.0 eV and 401.5 eV. The N K-edge NEXAFS spectra of the nano-sized aerosol fractions showed multiplet $\pi^*$ resonances similar to that of ammonium sulfate.

Both the first TXRF irradiation chamber and the novel (T)XRF instrumentation for analyzing contamination and depositions on 200 mm and 300 mm silicon wafer surfaces offer off-line reference measurements to assess wafer cleaning procedures and multi-layered systems on wafers, the latter also in a completely reference-free XRF mode [144]. The various experimental investigations demonstrate the promising prospects of this new instrument in soft X-ray TXRF analysis for ultra-trace contamination of light elements and their speciation by combing NEXAFS with TXRF. Further investigations will also focus on various kinds of contamination such as surface clusters and particles, intentionally deposited aerosols as well as on a reliable analysis of nano-layered materials on wafers exploring the transition regime between TXRF and XRF in the soft X-ray range.

*Acknowledgments*

The construction and operation of the (T)XRF instrumentation for 200 mm and 300 mm wafers has been possible by the dedication of Rolf Fliegauf as well as the work of Jan Weser and Michael Bock. Recent experiments and quantitation improvements were performed by Michael Kolbe and by Matthias Müller. Special thanks are due to Gerhard Ulm for his support in all of these activities.

### 7.3.11 Conclusion and Outlook

In a well-balanced system of highly motivated, well-trained personnel, and automated equipment, pure reagents and bulk media, cleanrooms and integrated data management, TXRF can and must contribute to quality assurance and process stability, support, and canalize creative engineering by continuous learning about materials and processes. TXRF has the advantage of controlled one-point calibration, a linear dynamic range of three orders of magnitude, high grade of automation in operation and data management, high up-time, and a simple control of data plausibility.

For that hotly pursued goal, we have been developing and combining novel instrumental procedures of ultra trace contamination analysis for the silicon industry for a decade. Ten years is a long time when experiencing the revolutionary changes in the silicon industry and in integrated device manufacturing. It is yet a short period for international standardization of new ultra trace analytical procedures in the daily routine of a global, corporate quality assurance system. Our ultra trace analytical equipment and methods are integrated in the factory operation system. We established an internationally accepted, automated surface preparation and standard test method that applies our vapor phase decomposition/total reflection X-ray fluorescence for semiconductor applications.

We anticipate further TXRF development in the field of in situ instrumentation, particularly in the field of the ultra trace analysis of nanostructures, including mapping and speciation. Sensitive detector systems will remain the most dynamic field of TXRF research. The resulting new atomic-level knowledge will be necessary to be able to integrate emerging new materials and process technologies in silicon wafer and integrated device manufacturing.

## 7.4 Analysis of Layers

*V. Rößiger and B. Nensel*

### 7.4.1 Introduction to the Analysis of Layers

Viewed historically, we can recognize two sources for layer analysis with X-ray fluorescence practiced today. First, we notice the classical materials analysis [224–226], which typically assumes thick samples but easily misses this prerequisite with light matrixes. In this case, the non-saturation thickness layer was viewed as an unwanted disturbance that led to an even greater complexity of the already rather complicated evaluation formalisms. However, methods have been developed that take into account the finite sample thickness [225] and therefore can be counted among the layer analysis methods.

On the other hand, the rapid development of the surface technology in the second half of the last century created numerous analytical problems. To name but one example, galvanically deposited gold and solder layers on printed circuit boards and on contacts required large-scale measurements. For a number of reasons, this was asking too much of the beta backscatter method, the dominating method up into the 1970s – and the energy-dispersive X-ray fluorescence method experienced a rebound. Complementing the already well- established energy-dispersive XRFA (X-ray fluorescence analysis) with Si(Li)-detectors [226, 227], comparatively simple instruments with proportional counter tubes were developed and employed [228–230]. Initially, they were specialized only for the measurement of the layer thickness of coatings made of one or at most two elements. From the measured spectra, the count rates in certain windows ("regions of interest" = ROI) were converted into layer thicknesses, using simple model equations. They could be calibrated using standards. Setting up physically sound models for industrial applications was not yet feasible with these instruments, mainly due to the limited computation capacity of the 8-bit computers used at that time. Still, they were instruments that were suitable for process control. It was necessary to solve very specific problems that did not exist in the general field of materials analysis. One of the problems was the small size of the samples. It was necessary to use a measurement spot size significantly below 1 mm. The samples needed to be positioned properly and monitored. Finally yet importantly, the main application of this measurement method in industrial manufacturing rather than in research labs required a high degree of robustness and user-friendliness.

There are also several practical reasons why especially the XRFA method has found such broad acceptance alongside other radiometric methods. First of which is the typical information depth in a magnitude of barely one micrometer to several tens of micrometers. This corresponds to the thicknesses produced by most chemical and physical coating methods (electroplating,

CVD, PVD). Electron beam microanalysis [231] and ion beam technologies such as PIXE [232] or RBS [233], on the other hand, cannot analyze to such depths—and high vacuum is required. XRFA measures in air, significantly simplifying instrument operation and keeping costs comparatively low. The fact that the method is non-destructive and non-contact does not need to be emphasized here.

This chapter is not an exhaustive presentation of the field with all its historical developments, its different theoretical approaches, and its countless applications. To minimize the risk of information overflow, an abstract formalism utilizing the development of the so-called fundamental parameter (FP) calculation is described, which will then allow the theoretical and application-relevant questions to become transparent.

The FP method refers to the fact that, for a well-defined situation (sample, instrument), the measurement signal can be computed on the foundation of well-known physical processes and constants. Section 7.4.2 will provide an overview of this formalism while Sect. 7.4.3 will cover the computation of the desired sample parameters from the measurement data. Mathematically, this can be described as parameter estimation. From today's perspective, a general solution is not possible, such that approximation methods known from numerical mathematics must be used.

Section 7.4.4 will introduce an implementation of the FP method for the practical analysis of layers and materials (WinFTM [234]). In the same context, the traceability of the measurement results will be discussed as well.

The presentations in Sects. 7.4.2 and 7.4.3 are generalized; subsequent sections are essentially limited to the energy-dispersive method. The description of the experimental part (Sect. 7.4.5) is limited to today's common instrument types, excluding some interesting modifications or separate methods, such as TXRF, for example. This becomes obvious with the selection of application examples (Sect. 7.4.6) being influenced by the practical experience of the authors.

### 7.4.2 Theory of the Quantitative Layer Analysis: Yield Calculation

#### The Layer Model

In general, the layer to be analyzed is defined by its plane-parallel boundary planes. Their distance is the thickness $d$ of the layer. It is one of the most sought-after measurement quantities, yet strictly speaking cannot be measured using X-ray fluorescence because the measurement requires knowledge of the density of the layer material. If a layer consists of several elements, we always need to assume a homogenous distribution of these elements. Within the scope of this model, a layer system or a sequence of layers is characterized fully by a mass per unit area matrix $X_{ij}$, where the first index $i$ designates the layer number and the second index $j$ the type of element. The entire mass per unit area $F_i$ of a layer $i$ is then

$$F_i = \sum_j X_{ij}. \tag{7.10}$$

The thickness of the $i$-th layer is

$$d_i = F_i/\rho_i, \tag{7.11}$$

where $\rho_i$ is the density of the layer $i$. The concentration $C_{ij}$ of the element $j$ in the layer $i$ is simply

$$C_{ij} = X_{ij}/F_i. \tag{7.12}$$

The value of the density is a function of the composition of the layer for which, however, no general term can be stated. As an approximation, we use

$$1/\rho_i \approx \sum_j C_{ij}/\rho_{ij}, \tag{7.13}$$

where $\rho_{ij}$ is the density of the $j$-th element of the $i$-th layer. It is assumed that each element in the composition retains its own density, as if this were a mixture of microscopically finely distributed pure elements. However, this in principle indefiniteness of the density is significant only if indeed geometric thicknesses are to be measured. The WinFTM® software [234] utilizes (7.13) as a pre-setting with tabulated $\rho_{ij}$, where the densities can be overwritten, cf. Fig. 7.92. From the perspective of the X-ray fluorescence, characterizing the sample using the mass per unit area $X_{ij}$ is sufficient. Aside from specific optical appearances at the total reflection, XRFA is blind for quantities such as thickness or density of a layer.

### The Equipment

From the perspective of the formal layer analysis, the instrument is characterized by the primary radiation ("spectral density function") and by the geometry parameters, which define the sample position with regard to radiation source and detector. To simplify matters, a few applicable assumptions are made:

1. The total thickness of the sample is small compared to the distance from the detector.
2. The lateral expansion of the interactive zone (measurement spot size) is small compared to the distance from the detector and the primary radiation is parallel.
3. The detector window is small compared to the distance between sample and detector.

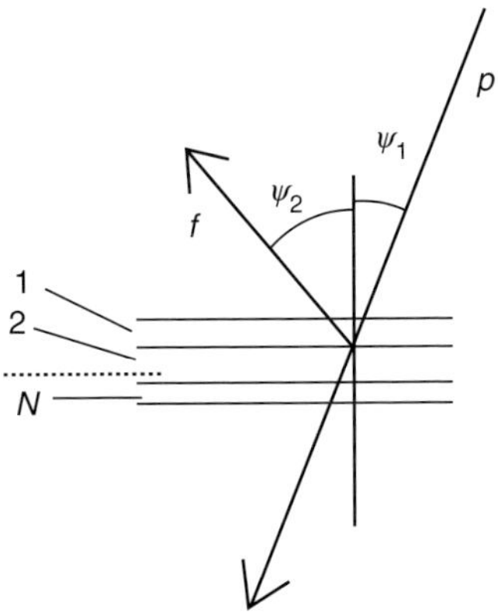


**Fig. 7.84.** The model

With these assumptions, we can count on a fixed incidence and take-off angle, $\psi_1$ an $\psi_2$ for each partial layer $i$; cf. Fig. 7.84.

The properties of the radiation source required for the quantitative computation are fully described through the spectral composition of the primary radiation $g(E)\,\mathrm{d}E$, where $g(E)$ stands for the number of photons in the energy interval $E$ ... $E+\mathrm{d}E$ that strike the sample surface per second. Potential modifications by means of air path absorption or by special filters are included. With an X-ray tube as radiation source, the spectral distribution reflects the influence of the operating voltage, the anode material and anode angle, the window material, and possibly additional absorbers or monochromators.

The properties of the detector with regard to solid angle, detector efficiency, and absorption in air should be de-coupled from yield calculation itself. This is possible when we normalize the intensity of any radiation component $k$ to that of the solid pure element. We then speak of normalized count rates $R_k$. Generally, $k$ applies across all measurable fluorescence components of the layer system.

With $g(E)$, $\psi_1$ and $\psi_2$, the measuring equipment is sufficiently characterized for computation of the intensity of the fluorescence radiation.

## Formalism of the Yield Calculations

The calculation of the relevant fluorescence component $R_k$ is the prerequisite for the determination of the unknown $X_{ij}$. The final objective is to find a formalism that allows for a normalized countrate

$$R_k = f(X_{ij}, g(E), \psi_1, \psi_2) \tag{7.14}$$

to be calculated as a function of the partial masses per unit area $X_{ij}$ and of the instrument properties ($g(E)$, $\psi_1$, $\psi_2$).

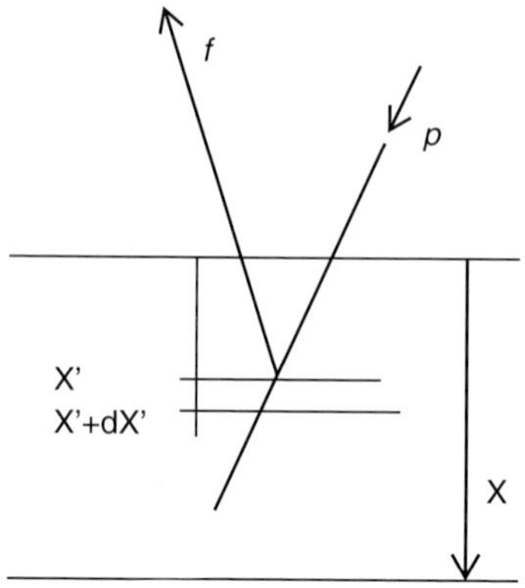


**Fig. 7.85.** Primary fluorescence excitation of layer with mass per unit area $X$

*Single Layer—Primary Excitation*

The probability of generating a fluorescence quantum with the quality $k$ (e.g., Cu-Kα) for a layer with an infinitesimal thickness[1] ($\mathrm{d}X'$) in the depth $X'$ is proportional to the interaction length $\mathrm{d}X'/\cos\psi_1$ and proportional to the concentration $C_j$ of the respective element $j$. Different from (7.12), here we can omit the index $i$ for the single layer example. For the mono-energetic excitation of the component $k$ with the energy $E$, the effective cross-section [224, 226] is designated with $\tau_k^o$. The index $k$ designates both the element and the type of radiation, i.e., $K, L, \ldots$ and $\alpha, \beta, \ldots$, etc. If we also take into account the absorption[2] of the primary radiation $p$ on its path to the location of interaction and that of the fluorescence radiation $f$ on its path to the detector (Fig. 7.85), we obtain as infinitesimal yield

$$\mathrm{d}y_k = \exp(-\mu_0 X'/\cos\psi_1) * (C_j \tau_k^0 \,\mathrm{d}X'/\cos\psi_1) * \exp(-\mu_k X'/\cos\psi_2), \quad (7.15)$$

which we only need to integrate across the total mass per unit area $X$ in order to arrive at the total yield:

$$y_k = (C_j \tau_k^0/\mu' \cos\psi_1) * (1 - \exp(-\mu' X)), \quad (7.16)$$

with

$$\mu' = \mu_0/\cos\psi_1 + \mu_k/\cos\psi_2. \quad (7.17)$$

For simplicity, in this case the geometry factor—the probability that a fluorescence quantum will be detected—can be ignored, since we will, as mentioned

[1]To eliminate density from the equations, mass per unit area dimensions are used in the calculation , when units of length are used in graphical presentation, they always should be interpreted as length*density.

[2]The absorption cross-sections $\mu_0$ and $\mu_k$ are, dependant on the elemental compositions of the layer [226, 227].

before, normalize to the yield of the pure elements. From (7.16), we receive for $C_j = 1$ and $X \to \infty$ the yield of the pure saturation thick element $j$.

$$y_{k0} = \tau_k^o/\mu' \cos\psi_1 \tag{7.18}$$

*Single Layer — Secondary Excitation (Enhancement)*

We call interactions that occur in addition to excitation by the radiation source (i.e., the primary excitation) secondary excitation. Any radiation component that has experienced primary excitation and that is energetically capable of secondary excitation must be taken into account. We differentiate between secondary excitation within a layer (this section) and that between layers (next section). For illustration purposes, Fig. 7.86 shows a representative example for the process for the case of a NiZn alloy layer. Ni–K radiation with an ionization energy of 8.331 keV is excited very effectively by Zn-K radiation, which has energies of 8.63 keV (Kα) and 9.57 (Kβ1). In this example and for thicker layers, the secondary excitation is even more effective than the primary excitation. When ignoring this effect, the Ni concentration would be calculated wrong by more than 100%.

Within a layer, the radiation component $k$ of the element $j$ (concentration $C_j$) experiences secondary fluorescence excitation by a radiation component

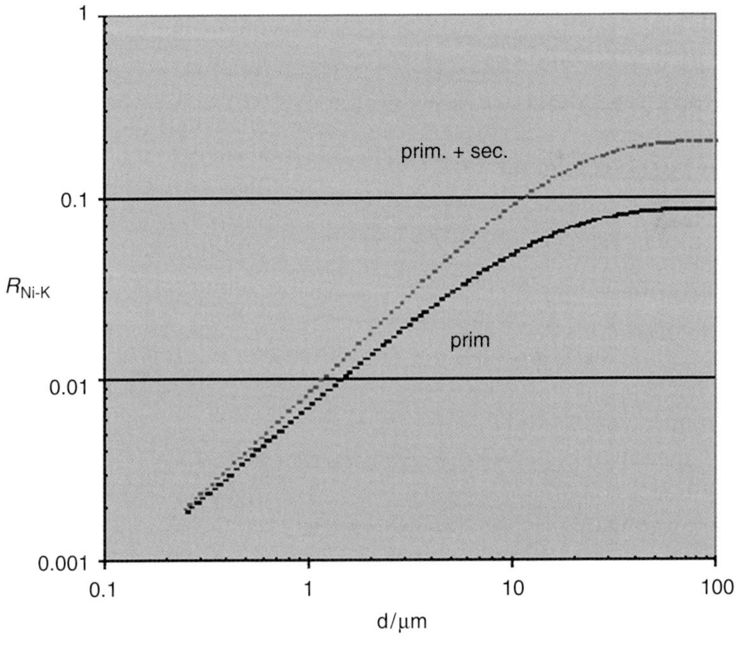


**Fig. 7.86.** Relative Ni–K yield $R_{Ni-K}$ of an Ni10Zn90 alloy layer as a function of coating thickness $d$. Excitation using an X-ray tube (W-anode, 50 kV, $\psi_1 = 0°$, $\psi_2 = 37°$)

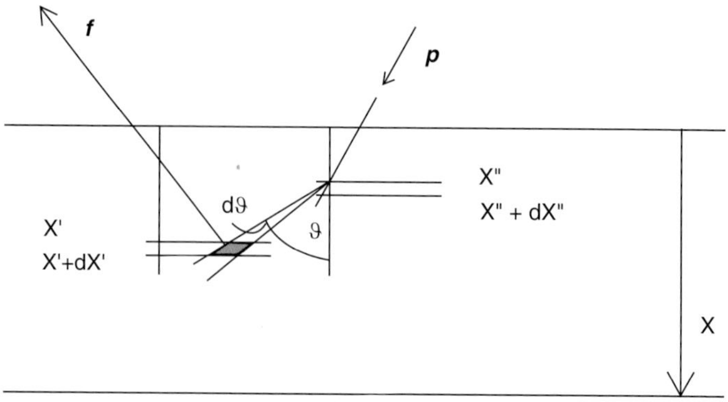


**Fig. 7.87.** Secondary excitation within a layer. $p$ designates the primary radiation and $f$ the fluorescence radiation

$kk$ of another element $jj$ (concentration $C_{jj}$ (Fig. 7.87)). Again, we observe an infinitesimal layer between $X'$ and $X' + \mathrm{d}X'$ for the excitation of the component $k$ by component $kk$. We designate the respective interaction cross-section with $\tau_k^{kk}$ (the upper index characterizes the exciting radiation). The source of the $kk$-component emits isotropically, such that the probability $p(\vartheta)\mathrm{d}\vartheta$, to be emitted in the angle range $\vartheta, \ldots, \vartheta + \mathrm{d}\vartheta$, is given by

$$p(\vartheta)\mathrm{d}\vartheta = (\sin(\vartheta)/2)\mathrm{d}\vartheta \tag{7.19}$$

(the integral of $0, \ldots, \pi$ results in $= 1$). Thus, the differential yield $\mathrm{d}y_k^{kk}$ can be described as a product of probabilities:

$$\begin{aligned} \mathrm{d}y_k^{kk} &= \exp(-\mu_0 X''/\cos\psi_1) * (C_{jj}\mathrm{d}X''/\cos\psi_1) * p(\vartheta)\mathrm{d}\vartheta \\ &\quad * \exp(-\mu_{kk}(X' - X'')/\cos\vartheta) * (C_j\tau_k^{kk}\mathrm{d}X'/\cos\vartheta) \\ &\quad * \exp(-\mu_k X'/\cos\psi_2). \end{aligned} \tag{7.20}$$

Three integrations over $X''$, $X'$ and $\vartheta$ must be carried out. Strictly speaking, (7.20) applies only to $X' > X''$ and $\vartheta < \pi/2$ (Fig. 7.87). An analogous relation applies to the reverse case ($X'' > X'$ and $\pi/2 < \vartheta < \pi$), such that we can write

$$\begin{aligned} y_k^{kk} &= (C_{jj}\tau_{kk}^{o} C_j \tau_k^{kk}/2\cos\psi_1) * \int_0^X \mathrm{d}X' * \exp(-\mu_k X'/\cos\psi_2) \\ &\quad * \left( \int_0^{X'} \mathrm{d}X'' * \exp(-\mu_0 X''/\cos\psi_1) \int_0^{\pi/2} \mathrm{d}\vartheta \tan\vartheta \exp(-\mu_{kk}(X' - X'')/\cos\vartheta) \right. \\ &\quad \left. + \int_{X'}^{X} \mathrm{d}X'' * \exp(-\mu_0 X''/\cos\psi_1) \int_{\pi/2}^{\pi} \mathrm{d}\vartheta \tan\vartheta \exp(-\mu_{kk}(X'' - X')/\cos\vartheta) \right). \end{aligned} \tag{7.21}$$

This integral cannot be solved fundamentally. With the substitution $t = -\alpha/\cos\vartheta$, the $\vartheta$-integral can at least be converted to an integral exponential function

$$\int_{0}^{\pi/2} \mathrm{d}\vartheta \tan\vartheta \exp(-\alpha/\cos\vartheta) = \int_{-\alpha}^{-\infty} \mathrm{d}t \exp(t)/t = -Ei(-\alpha) \tag{7.22}$$

and then being determined numerically [235]. Up to now, we acted as if the examined layer were the first layer (top layer). If this is not the case, then in addition the absorption by the top and any possible interim layers must be taken into account for both primary radiation as well as secondary radiation.

*Secondary Excitation — Enhancement by Another Layer*

Secondary excitation by a radiation component of another layer is determined similarly to the one described above (Fig. 7.88). If, for the sake of simplicity, we again do not concern ourselves with the absorption of the top and interim layers, we obtain for the situation where the exciting layer $ii$ is located above the excited layer $i$ the differential yield

$$\begin{aligned} \mathrm{d}y_k^{kk} &= \exp(-\mu_0 X''_{ii}/\cos\psi_1) * (C_{jj}\tau_{kk}^0 \mathrm{d}X''_{ii}/\cos\psi_1) * p(\vartheta)\mathrm{d}\vartheta \\ &\quad * \exp(-\mu_{kk}^{ii}(X_{ii} - X''_{ii})/\cos\vartheta - \mu_{kk}^{i} X'_i/\cos\vartheta) * (C_j \tau_k^{kk} \mathrm{d}X'/\cos\vartheta) \\ &\quad * \exp(-\mu_k^i X'_i/\cos\psi_2), \end{aligned} \tag{7.23}$$

which analogous to (7.21) is to be integrated over $\vartheta(0,\ldots,\pi/2)$, $X'_{ii}$, $(0,\ldots,X_{ii})$ and $X'_i(0,\ldots,X_i)$. Here as well we need to rely on numeric approximation methods.

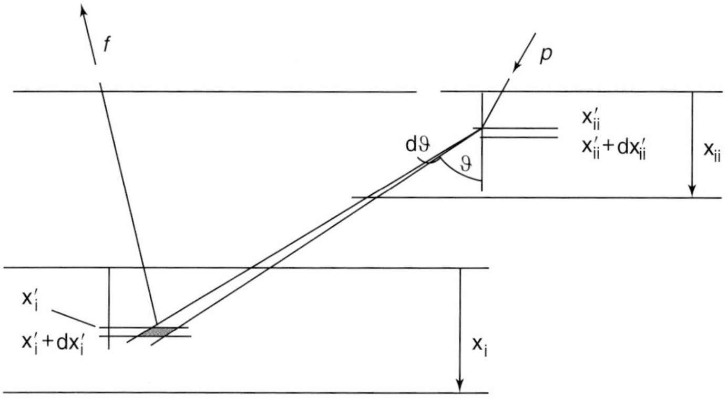


**Fig. 7.88.** Model for secondary excitation between layers. Exciting layer $ii$ is located above excited layer I; $p$ designates primary radiation, $f$ fluorescence radiation

**Table 7.9.** Secondary excitation. Plus (+) and Minus (−) indicate which radiation component of the layer system Cr/Ni/Cu/Fe yields secondary excitation for which element. The enhancement effect occurs when the radiation energy is greater than the respective ionization energy. Only K-radiation components that are easily available for the measurement are shown

| Exciting radiation component and radiation energy in keV | | Excited atomic shell (ionization energy in keV) Cr–K (5.988) | Fe–K (7.057) | Ni–K (8.331) | Cu–K (8.980) |
|---|---|---|---|---|---|
| Cr-Kα | 5.41 | − | − | − | − |
| Cr-Kβ | 5.95 | − | − | − | − |
| Fe-Kα | 6.40 | + | − | − | − |
| Fe-Kβ | 7.06 | + | − | − | − |
| Ni-Kα | 7.47 | + | + | − | − |
| Ni-Kβ | 8.28 | + | + | − | − |
| Cu-Kα | 8.04 | + | + | − | − |
| Cu-Kβ | 8.93 | + | + | + | − |

For illustration purposes, Table 7.9 shows the excitation channels (which fluorescence component excites which element) for the example Cr/Ni/Cu/Fe[3] (cf. Sect. 7.4.6 Decorative Multi-coatings). Similar to Fig. 7.86, we can again observe situations where the secondary excitation dominates in comparison to the primary excitation. This primarily concerns Cr. Eleven different kinds of secondary excitation need to be included because we need to treat the K$\alpha$ and the K$\beta$ radiation components separately.

*Summary – Polychromatic Excitation*

It is the objective to calculate all relevant fluorescence components $R_k$ of (7.14). A complete description of all equations involved has limited usefulness; here rather the procedure shall be described in a summary taking into account primary and secondary excitation but not tertiary or those of an even higher order.

Analogous to the example in Table 7.9, at first, all elements have to be determined that are capable of producing secondary excitation with any fluorescence component, either in the same or in a different partial layer $i$. With an increasing number of elements, or layers, a considerable number is quickly generated. For each individual secondary excitation process, the absorption of both the primary radiation and that of the secondary excitation radiation must be taken into account. Finally the absorption of the analyzed component $k$ on its path to the detector must be considered

[3]We designate the individual layers with the elements that are contained in them, beginning from the top.

For polychromatic excitation, which is defined by the so-called spectral density function $g(E)$, we need to mathematically convolute the mono-energetic yields $y(E)$, in which the integral can be approximated well by a sum expression

$$\int_{0}^{\mathrm{Emax}} y(E) * g(E)\mathrm{d}E \approx \sum_{n} y(E_n)g(E_n), \tag{7.24}$$

which corresponds to a discretization of the primary energy distribution.

### 7.4.3 Calculation of the Unknown Measurement Quantities $X_{ij}$

The objective of this section is to calculate from the measured spectrum the unknown sample parameters *Xij*, i.e., the individual element masses per unit area in each layer. The interface to the FP theory is established by the normalized countrates $R_k$, which can be computed from the measured spectrum with some effort. Figure 7.89 provides an example.

#### Spectra Evaluation

The basis for the analysis is an essentially complete and error-free evaluation of the experimental spectrum. Meaning, to determine the normalized count rates $R_k$. As a good approximation, we can view a (energy-dispersive) acquired pulse height spectrum $s_1$ (l= channel number) as a superposition of detector-typical response functions $B_{kl}(E_k)$,

$$s_l = \sum_{k} R_k^* B_{kl}(E_k). \tag{7.25}$$

Summation is over all detectable components $k$ with energy $E_k$.

For the pure element that belongs to $k$ (all $R_k = 1$), the measured spectrum is exactly the sum of the related response functions. Unfortunately, we cannot measure the components $k$ individually, and therefore determine the response functions directly – this, however, is not the only difficulty when evaluating the spectra. Equation (7.25) neglects the spectral noise caused by scatter and such artifacts as pulse pile-up (cf. [227], for example). Since this is a general problem of energy-dispersive XRFA, and not specific to the analysis of layers, a reference is made of the state-of-the-art description of this problem [236], otherwise the solution of the linear equation system (7.25) is treated as known with regard to the unknown $R_k$.

#### The Inversion Problem

We can view the conversion of the measured spectrum into the $R_k$-countrates as information compression. The other conversion, into the actual measurement quantities of interest, the masses per unit area $X_{ij}$, is a comparatively

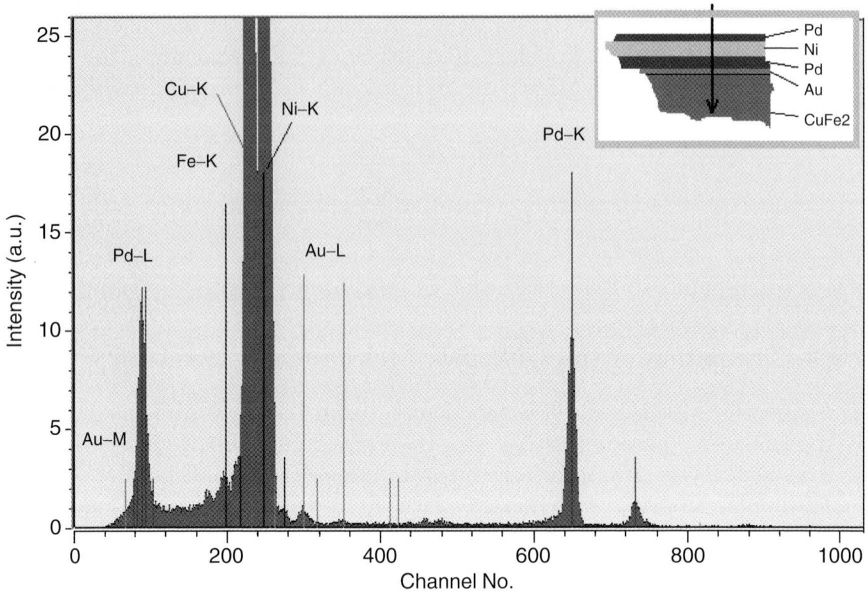


**Fig. 7.89.** Spectrum of a sample with the layer sequence Pd/Ni/Pd/Au/CuFe2 (top right), recorded using a Fischerscope® X-Ray XDAL (50 kV, W-anode, primary filter 8 μm Ni, $\psi_1 = 0°$, $\psi_2 = 45°$, detector = PIN-diode, cf. Sect. 7.4.5). The fluorescence components (peaks) are indicated by lines. For the fine structure of the fluorescence radiation spectrum, cf. [226]. The objective of the spectra evaluation is to determine the area of each individual peak. Potential overlap with adjacent peaks and the continuous radiation background due to scattering of the primary radiation to the detector must be taken into account. In addition, one must pay attention to detector-specific artifacts such as pulse pile-up and escape-peaks. Here, as in the subsequent spectra, the x-axis denotes the channel number and the y-axis the number of the counted events per second

more complicated problem. If we identify in (7.14) the left sides with the experiential, normalized countrates from the spectra evaluation $R_k^{\text{exp}}$, we receive for each measurable component $k$ an equation for $X_{ij}$.

$$R_k^{\text{exp}} \cong (R_k^{\text{theor}} =) f(X_{ij}, g(E), \psi_1, \psi_2). \tag{7.26}$$

In this manner, (7.26) defines a non-linear equation system, where the solution cannot be provided directly, or by analytical means. As a solution, we view those values *Xij*, for which the values $R_k^{\text{theor}}$ calculated according to (7.14) correspond best with the experimental countrates. Generally, the minimum of a Least Squares calculation

$$F(X_{ij}) = \sum_k ((R_k^{\text{theor}} - R_k^{\text{exp}})/\Delta R_k)^2 \rightarrow \text{Min} \tag{7.27}$$

is considered to be the best fit. At the same time the differences are normalized with regard to the uncertainty of the experiential countrates $\Delta R_k$. This is absolutely necessary in order for the radiation components with smaller intensity, and thus with greater uncertainty, to not falsify the result. The squared sum F is a function of the unknown $X_{ij}$. This transforms the problem of solving the equation system into a minimum search, for which a number of mathematic/numeric algorithms is available, cf. [237], or Sect. 5.5.3 (Back-Calculation Schemes)

### Characteristics

The general solution of the problem presented above is more of theoretical than practical importance. Most often, the tasks are of a rather simple nature, such that simple equations are sufficient to convert the measured signal quantities into the desired parameters. If one does not want to or is not able to proceed on the difficult path of a complete solution to the problem, two shortcuts are available that often lead to the goal as well. The first shortcut (Parameterized Model Equations) concerns only the inversion problem, the more radical second shortcut (Empiric Characteristics) also circumvents to a large degree the difficulties of the theoretical yield calculations.

### Parameterized Model Equations

For a specified $X_{ij}$ , the program calculates the $R_k$ values, assuming that the intensity calculation according to (7.14) works correctly, the values are also accessible to the measurement. Thus, one is able to calculate one or more measurable countrates for any desired $X_{ij}$-combinations (parameter table). On this basis, model equations F are created for each $X_{ij}$ of interest

$$X_{ij} = F(\mathbf{P}; R_1, R_2, \dots R_K). \tag{7.28}$$

These equations are parameterized and the respective parameters $\mathbf{P}$ are in essence determined beforehand—specifically through the adaptation to the computed parameter table. An example is the often-used "linear mode," which offers good results for small layer thicknesses. In the case of a single coating

$$X_1 = p_1 + p_2 \cdot R_1 \qquad (if R_1 << 1). \tag{7.29}$$

Equation (7.29) simply describes the linear correlation between $R$ and $X$, which applies to a small range only. The parameters can be predicted through FP calculations, such that no calibration standards are required. Of course, this procedure may also be expanded to more complex situations. Selection of the function F (7.28) is typically not disclosed in commercial evaluation programs, with the exception of trivial correlations. Such parameterized characteristics are useful even without pre-calculated points for constructing empiric characteristics.

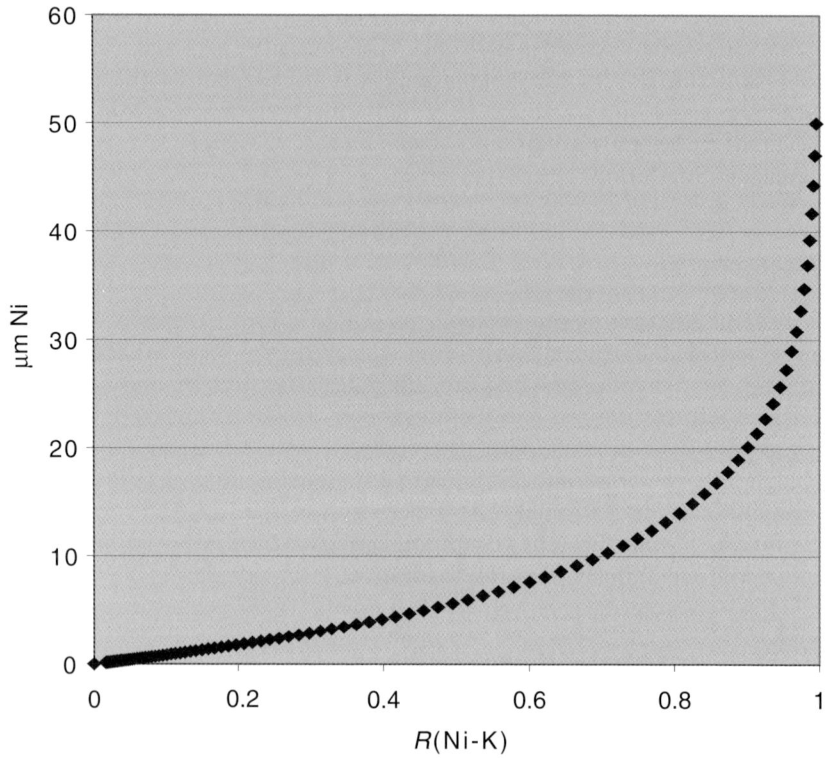


**Fig. 7.90.** Plot for Ni thickness as a function of Ni–K-intensity, computed using the FP method can be approximated well using a logarithmic function (7.30). Per definition the R values for absent Ni and for saturation thick Ni are known, the plot is governed by a single parameter. One standard would be sufficient to determine such a characteristic

*Empiric Characteristics*

Qualitatively the functional plot profile is known from theory. If this plot can be described quantitatively with only a few free parameters, the adaptation using a calibration (i.e., with the measurement of samples with known $X_{ij}$) is a suitable means to circumvent the difficult inversion problem. For example, one can solve (7.16) from Sect. 7.4.2, which strictly speaking applies only to primary excitation with mono-energetic radiation, for $X$ and obtain a quite suitable (empiric) characteristic with two free parameters

$$X_1 = \text{Const1} * \log(1 - \text{Const2} * R). \tag{7.30}$$

If we are dealing with only one layer, we can approximate the actual plot profile with such an equation. Only through verification with many precisely measured standards will one notice that there are some differences. The reasons may be not consider polychromatic excitation. This is often the beginning point of trial-based re-adjustments. The situation is less clear as layer structures become more complex. In any case, it requires many standards to measure "correctly."

From a historic perspective, such empirically determined characteristics were the foundation for the first XRF layer thickness instruments. Often, so-called window countrates are used as independent variables and not $R_k$ values, which must be determined per spectra evaluation. These window countrates are channel contents of the spectrum that are added up across a particular range (ROI).

Today, the use of FP-supported algorithms is more common. The next section will offer an example of such an FP-supported program.

### 7.4.4 The WinFTM® Program

This software program utilizes for the evaluation of the measured spectrum the fundamentals described above. The current version (WinFTM®-V.6) has been developed on the foundation of the previous software version WinFTM®-V.3 described in [238], which was limited to a maximum of three unknown measurement quantities. WinFTM®-V.6 is capable of computing any number of measurement quantities, practically the number is only limited on a formal basis to 24, which, considering physical conditions, at least today, does not constitute a serious limitation. The software program is designed such that it cooperates with any type of instrument (defined by the quantities described in 7.4.2 The Equipment p. 556), regardless of the detector or radiation source that is being used. An additional generalization in comparison to the previous version is related to the so-called measurement mode. It determines whether the individual measurement quantities are coating thicknesses or concentrations. In WinFTM®-V.6, any desired combination is allowed, for example, a single layer thickness, a combination of layer thicknesses, as well as a layer on a solid material, which composition can be determined simultaneously. While WinFTM®-V.3 operated with a fixed number of specified measurement modes, the measurement mode in V.6 is only a derived variable, which only needs to meet the formal condition that the total number of measurement quantities and the total number of elements is $\leq$24. In addition, the assumptions of 7.4.2 The Layer Model (p. 555) and 7.4.2 The Equipment (p. 556) apply, namely that the sample exhibits a plane-parallel layer structure in the area of the interaction and that the elements are evenly distributed within each layer. Of course, in addition to the measurement of layered structures, regular material analysis is included as a special case (a single layer with thickness = infinite).

#### DefMA

This acronym can be read as "Definition of Measurement and Application conditions". It designates the basic conditions of what is to be measured and

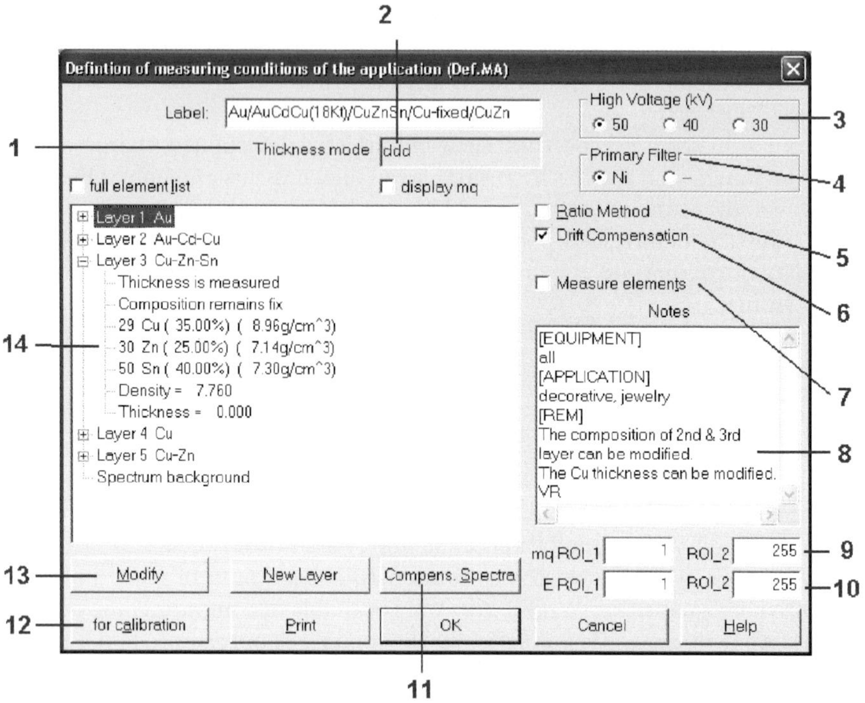


**Fig. 7.91.** Example of a DefMA. The marked points are explained in the text

how. This includes both the structure of the sample, that is, which elements are contained in which layer, and the parameters of excitation (high voltage of the X-ray tube, primary filter). Figure 7.91 shows the respective screen template. The explanations for the fields shown in this figure are used to point to basic properties and features of the software program.

*Feature 1*

"*Thickness mode*". Alternatively, "Mass per unit area mode relative," "Mass per unit area mode absolute," "Solution analysis," and "Component mode" are available.

"Thickness mode" is the most common setting. For layers, the result is presented in thickness units, e.g., in "µm". As described in Sect. 7.4.2 The Layer Model (p. 555), a density is determined that is then used to convert the initially determined partial mass per unit area $X_{ij}$ into the thickness units desired in practical applications.

This is skipped in the "Mass per unit area modes." A layer is characterized by its mass per unit area, e.g., in "mg/cm$^2$". The difference between "relative" and "absolute" refers simply to the readout of the concentration for alloy layers: either in % or each portion is displayed as a mass per unit area value. Bulk concentrations (sub-layers with thicknesses = infinite) are depicted in %, in general.

The mode "Solution analysis" is a special version of the "Mass per unit area mode absolute." Here, we deal with the concentration readout of a liquid layer with a fixed thickness which is given through the measurement cell [239] that is being used. The display is in units of mass per volume, e.g., in "$g/l^{-1}$".

With the "Component mode," the analysis is not carried out according to elements but according to pre-defined components. These are, for example, chemical compounds such as oxides.

*Feature 2*

Here, the measurement mode is displayed. It is derived from the actual definition of the sample structure (Feature 14). A series of letters, $d$ for thickness, $g$ for mass per unit area and $C$ or $c$ for the concentration (mass portion) informs the user about the determined parameter and the units of measurement.

*Features 3 and 4*

Setting of the operating voltage of the X-ray tube and primary filter. They are dependent on the actual instrument type.

*Feature 5*

"*Ratio method*". This toggle activates an internal spectra normalization. During evaluation, only the relative plot profile of the spectrum is taken into account. Cf. also [238].

*Feature 6*

"*Drift compensation.*" This feature is used to correct a potential drift of the peak position in the spectrum. This prevents a misrepresentation of the result, especially when critical peak overlap is present.

*Feature 7*

"*Measure elements.*" Because internally all spectral components are based on the intensity of the respective pure element, cf. Sect. 7.4.2 The Equipment seq., the knowledge of the pure element spectrum is of fundamental importance. If the respective toggle is activated, all elements that occur in the sample must be measured as solid samples. Otherwise, one can resort to an element library that is calibrated during installation. It contains the spectra of 12 or 14 pre-defined elements, depending on the type of detector. Based on these spectra, the program computes all other required pure element spectra. This function of the program is of utmost significance, because not all elements are available as measurement samples. Excluded are light elements, where the fluorescence radiation cannot be measured. Which ones these are depends on the instrument (detector).

*Feature 8*

"*Notepad.*" All DefMAs released by the supplier contain notes regarding the specifications of the respective application. Each user equipped with respective privileges can make entries that can then be viewed in a ready set-up product (cf. Sect. 7.4.4 Measurement).

*Features 9 and 10*

"*mq-ROI*" *and* "*A-ROI.*" These entries specify regions ("regions of interest") for the spectra evaluation and the computation of the mq quality factor. Regarding the specific meaning, cf. Sect. 7.4.4 mq-A Measure for Measurement Quality. Typically, the full spectrum is used. Reasons for limiting these factors are various, their use for interesting applications is described in [240, 241].

*Feature 11*

"*Compensation spectra.*" This feature is used to specify provisions for correcting the radiation background.

*Feature 12*

"*For calibration.*" As described previously in [238], WinFTM® allows for differences in the sample structure (see Features 13 and 14 below) between the specimen and the calibration standard. Therefore, two parallel definitions are available for either of the two specimen and calibration standards. Consequently, during measurement, specimens and standards, which may differ from one another, are administered separately. The extent of the admissible difference must be decided responsibly in each individual case by the user itself. The software will only perform formal checks .

*Features 13 and 14*

*Input of the sample structure.* This is the actual core of the DefMA. The sample consists of layers. A solid material is a layer, where the known thickness is = infinite (saturation). Window 14 lists all entered partial layers, and the key "Modify" (13) opens the entry template of Fig. 7.92, where detailed information can be found for each respective layer (here it is the second, counted

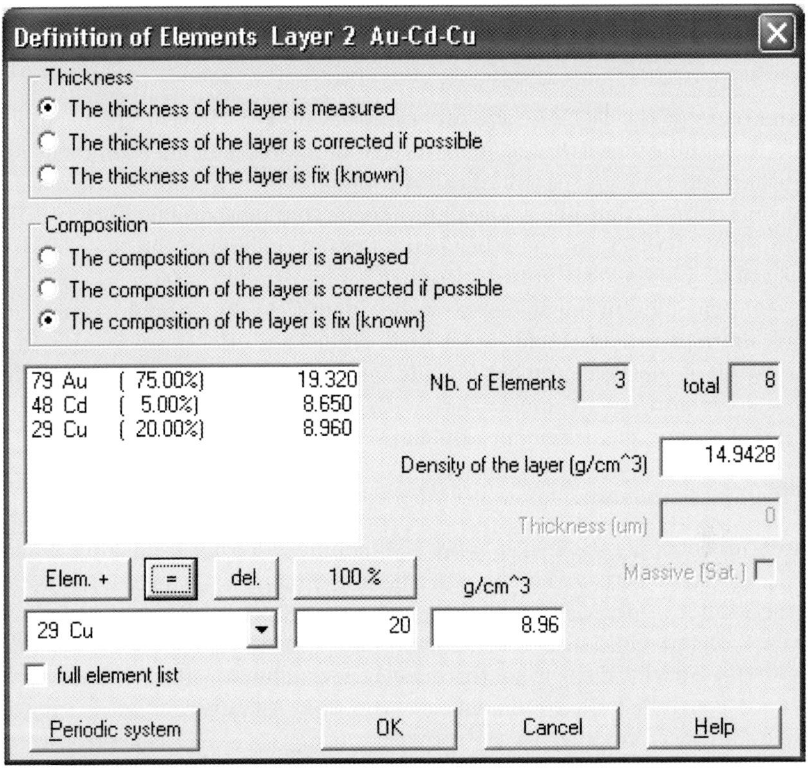


**Fig. 7.92.** Entering sample structure and measurement plan. Here, an 18 kt layer with a fixed composition, consisting of 75% Au, 5% Cd and 20% Cu, is entered. The program calculates the density using internal pure element densities according to (7.13), Sect. 7.4.2 The Layer Model, and takes this as the preset value, which can be manually overwritten in this field

from the top, cf. also Fig. 7.84): the elements or components that comprise the layer and what shall be measured. Initially, if more than one element or more than one component is present, the thickness or the total mass per unit area as well as the composition can be measured. However, it is also possible to keep the composition as fixed and to measure only the thickness. As another option, both can be set to specific fixed quantities. The shown example Au/AuCdCu(18 Kt)/CuZnSn/Cu/CuZn is a simple triple layer measurement, where the 18 kt layer is composed of Au, Cd, and Cu and the white bronze layer of Cu, Sn, and Zn with a fixed composition. The substrate material is brass (CuZn), which is copper-plated. Due to the multiple occurrence of Cu, the pre-copper-plating cannot be measured and is therefore entered as a fixed quantity.

## Measurement

*General*

On the basis of the DefMA, the program sets up the "measurement application." Here, all other settings, such as size of aperture or measurement spot, anode current, type and extent of calibration, and presentation of the measurement uncertainty, that physically define the actual measurement are specified. Administrative tasks, on the other hand, are not defined in the "measurement application" but instead in the "product." The product relates to everything that concerns measurement display type, identification through certain comments, units of measurement, statistical evaluation, and possibly aids in the automation of the measurement (saving measurement positions). The product accesses a measurement application, such that even several of such products can use the same measurement application (including the calibration).

*Making Measurements*

All instruments operating with WinFTM include video optics making the sample visible on the screen. The measurement spot shown in the crosshairs takes into account the divergence of the primary radiation (cf. Sect. 7.4.5 X-Ray Source – Beam Guidance and Measurement Spot). By focusing the video image, the correct measuring distance is set. The so-called DCM method (*d*istance *c*ontrolled *m*easurement) ensures that the current measuring distance is always provided to the program using an ADC converter that is coupled with the focusing mechanism, even for variable measuring distances. This is important for example if one needs to measure into a recess with an instrument that measures from below.

In instruments with fixed measuring distances, the measuring head is moved to keep it constant. An auto-focus system that maximizes the contrast in the selected video range, ensures a sufficiently constant measuring distance. In addition to the simple measurement that is triggered by a mouse click or a push of a button, numerous automatic sequences are possible that are programmed by the user.

*mq - A Measure for Measurement Quality*

The FP calculation of the fluorescence components (cf. Sect. 7.4.2) allows the software to carry out a test calculation, in order to check whether the presented measurement result is indeed consistent with the measured spectrum. Supposedly a sample, that is not consistent with the definition of the sample structure of the DefMA, had been placed on the instrument for measurement. In such a case, the program must be able to advise the operator of the mismatch, instead of just calculating and displaying a meaningless numeric. The mentioned test calculation provides a number with a value that is an effective measure for the correspondence of the measured spectrum $s_l^{\text{exp}}$ with the theoretically calculated $s_l^{\text{theor}}$ ($l$ = channel number, runs across all $N$ according to Feature 9 set channels). The mq-value is a measure of the mean square deviation of the measured versus theoretical spectrum, whereby the individual channel contents are normalized according to their Poisson uncertainty $u_l$ (uncertainty of the difference due to the counting statistics, cf. [226], for example):

$$mq^2 = (1/N) \sum_l (s_l^{\text{exp}} - s_l^{\text{theor}})^2 / u_l^2. \qquad (7.31)$$

Thus, a value mq $\approx$ 1 is relatively good, while a value above 2 already indicates that there are certain small discrepancies present. The program accepts a maximum value of mq = 5 (default setting). This limit may be changed in the application.

*Evaluation*

The use of XRF instruments in process control or quality assurance requires a customized presentation of the measurement results. WinFTM offers numerous variants, even including an analysis of variance (ANOVA). Not all options can be presented here. Mention shall be made of only one option, which is set up factory side for each instrument: the stability test. It is used by the instrument to monitor itself with regard to intensity of the primary radiation, resolution, and amplification factor of the spectrometer (detector and electronics), and in doing so it utilizes some of the existing evaluation options. Figure 7.93 gives an example. Additional options for displaying the evaluated measurement result are included in Sect. 7.4.6 (Application examples).

**Calibration—Measurement Uncertainty—Traceability**

With a consistent error calculation, WinFTM secures the traceability of the individual measurement result. Distinguishing it from other evaluation programs, which are, for example, based on empiric characteristics (cf. Sect. 7.4.3 Characteristics), WinFTM already provides standard-free measurement results, which, as a rule, are already relatively accurate. This measurement result is the solution of the non-linear equation system described in Sect. 7.4.3 Characteristics, with the boundary conditions given by the instrument and geometry parameters. To secure traceability, the theoretical (standard-free) measurement result is compared to the "correct" result. To this end, samples (standards) with known "correct" values are required. Based on this comparison (Fig. 7.94), the program carries out a correction of the theoretical values. Due to this so-called calibration, the measurement results are then traceable to the used standards. As shown below WinFTM can indicate how "exact" or uncertain these corrected measurement results are.

Peak position: the mean of each block (5 readings) is displayed

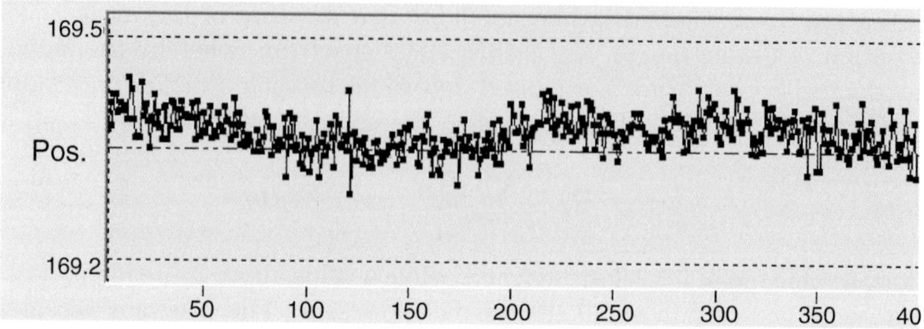


Statistical evaluation of 400 blocks

| Block Size n = 5 | | Int. | Pos. | fwhm |
|---|---|---|---|---|
| | | 0 | 0 | 0 |
| Mean value | x.. | 5203.2 | 169.4 | 13.65 |
| Mean Std.Dev. | s. | 17.84 | 0.044 | 0.166 |
| Mean Range | R. | 44.08 | 0.109 | 0.411 |
| Number of Blocks | nbl | 400 | 400 | 400 |
| Min.Block Mean Value | min | 5179 | 169.3 | 13.4 |
| Max.Block Mean Value | max | 5203 | 169.4 | 13.9 |
| Std.Dev | sa | | 0.016 | |

**Fig. 7.93.** Stability test. An instrument with a proportional counter tube, the Ag-K peak as measured for about 4 h using a defined cycle of measurements in fixed blocks and breaks after several blocks. The statistical evaluation provides mean values and scatter values of total intensity, peak position, and full width half maximum (fwhm). The analysis of variance detects drifts and characterizes them with a value sa that characterizes the variance between the blocks. In the example case, a small drift by 0.02 channels has been established for the Ag-K$\alpha$ peak (at channel 170.6). The upper portion shows the corresponding block mean values in an SPC chart. Intensity and full width half maximum are stable. The 100 evaluated blocks have been taken from a longer test series

### *The Term "Measurement Uncertainty"*

"Measurement uncertainty" characterizes the difference of the displayed measurement value to the true value. Samples that represent a "correct" value are called "standards." Their specified values are called nominal values or engraved values. However, only if their uncertainty is stated are they valid

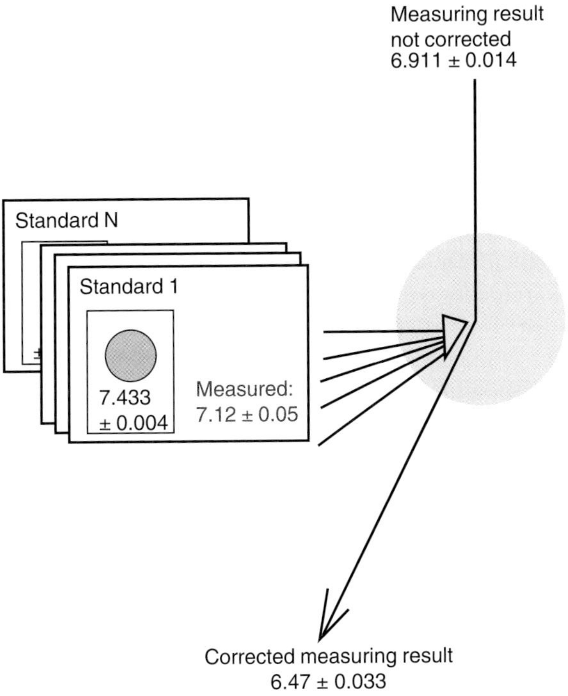


**Fig. 7.94.** Representation of the correction using a calibration

and can be used for calibrations. The uncertainty is a measure for the difference to the unknown "true" value, and it is, naturally, a probability statement. Its numeric value $\delta X$ characterizes a range $X \pm \delta X$. It is dependent on the probability of having the "true" value included in this range. The term "level of confidence," is typically used in statistics literature to characterize this probability [242–244, 246]. Based on a probability density with normal distribution for the differences to the true value, one identifies the distribution parameter $\sigma$ with the "standard measurement uncertainty" $u$, and can specify well-defined measurement uncertainties using $k$ multiples thereof. The often-recommended expansion factor $k$ should be 2 (then referred to as an "expanded measurement uncertainty," such that a range of $X \pm 2 * u$ includes the true value with a probability of 95%).

Using $k = 1$ makes sense for several reasons. For a large number of repeatability measurements, the standard deviation will here approach the random (statistical) portion (see below) of the measurement uncertainty. The level of confidence for $k = 1$ is 68%.

To permit only a small probability of error, $k = 3$ can be selected, where the confidence interval covers 99.7%. In the end, the unit in which the measurement error is stated does not matter as long as it is known. To master all requirements, WinFTM allows for the input of $k$. The default value is set to $k = 2$.

Regarding the terminology: "measurement uncertainty" is the term used in national and international standards. In everyday common language, terms such as "error" and "accuracy" are common. In cases where these terms are occasionally used in this paper, one always should refer to them as "measurement uncertainty."

*Random Errors (Precision) [244, 245]*

If one were to repeat a measurement $N$ times under repeatability conditions, the single measurements will scatter around a mean value. The cause for this is the random distribution of the individual channel contents of the spectrum, the law it abides is known, the so-called Poisson distribution. This is an important special case in measurement technology that allows us, or respectively the program, to make a prediction about the repeatability precision based on only one measurement, or one single spectrum. In reality, there are, however, other causes for random fluctuations that cannot be grasped theoretically, such as the influence of the gas amplification of a proportional counter in connection with the software drift compensation, for example (cf. Feature 6 in Sect. 7.4.4 DefMA). In individual cases, this can lead the measured deviation (standard deviation) to exceed the theoretical prediction. In any case, the random error component (precision) can be determined using statistical means. WinFTM-V.6 computes the uncertainty of the mean value from the repeatability distribution, whereby the numeric value is dependent on the selected expansion factor $k$, and therefore on the confidence interval. Random influences except the ones from repeatability are covered by the term "reproducibility." The program cannot estimate the reproducibility. In the literature (cf. [245, 246] for example) are described certain test procedures to evaluate both repeatability and reproduciblity, so called R&R studies. If the individual measurements, for example, are always connected with re-positioning of the sample, possibly even with different operators, then the distribution of the measurements will reflect the influences of repeatability and reproduciblity. The confidence interval of the mean value will then simply also characterize the reproducibility based on the analyzed influencing factors. In any case, these measurement errors are greater than the random (repeatability) uncertainties attributed to the X-ray fluorescence method due to the Poisson distribution of the spectral intensities.

*Systematic Error and Calibration*

At least in a thought experiment, the random error portion can be reduced to Zero, even if it were through a very large number of consecutive measurements and mean value generation. We call the remaining difference to the true value the systematic uncertainty $u_{\text{sys}}$. This systematic measurement uncertainty must be corrected, for the instrument to measure correctly. The basis of this procedure, known as calibration, is initially the determination of the deviation from the correct value. The evaluation software then makes the appropriate corrections (Fig. 7.94). Standards are the basis for the calibration. They are

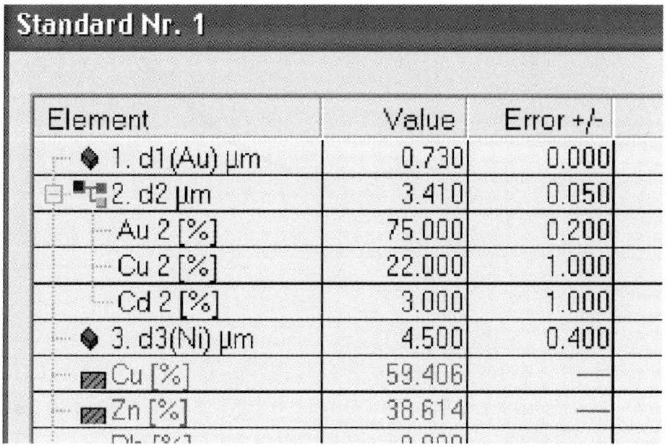
Standard Nr. 1

| Element | Value | Error +/- |
|---|---|---|
| 1. d1(Au) µm | 0.730 | 0.000 |
| 2. d2 µm | 3.410 | 0.050 |
| Au 2 [%] | 75.000 | 0.200 |
| Cu 2 [%] | 22.000 | 1.000 |
| Cd 2 [%] | 3.000 | 1.000 |
| 3. d3(Ni) µm | 4.500 | 0.400 |
| Cu [%] | 59.406 | — |
| Zn [%] | 38.614 | — |

**Fig. 7.95.** Calibration standard set. Input of nominal values of the standard. Uncertainties (errors) must be entered as well. If default value Zero is not changed, the software will automatically use 1% from the nominal value for layer thicknesses and masses per unit area, 1% of mass as uncertainty for concentrations, referenced to $k = 1$ (68% confidence interval). Example applies to measurement application Au/AuCuCd/Ni/CuZnPb; Fig. 7.92 shows the DefMA for this application

samples that correspond in structure to the measurement application and have values that are known within a specified uncertainty. The systematic error portion can be determined by measuring these standards. However, one has to take into account that this measurement has a random measurement uncertainty associated with it and the nominal value of the standard only has a finite accuracy.

While the random errors described above can be characterized using statistical means, for the systematic error portion this is only possible by comparing the measurement result with standard samples. It would be possible to describe and even correct the systematic error if the nominal values of the used standard samples were exact and it were possible to measure them without a random failure portion. Neither one of these is possible. Both the uncertainty of the nominal value and the (random) measurement uncertainty of the calibration measurement determine the error of the calibration – and therefore the uncertainty of the correction that is based on the calibration (Fig. 7.96). Particularly, it must be taken into account that from experience, the uncertainty of the calibration increases for measurement results that are distant from the values of the standards. The uncertainty of the correction is of course dependent on the size of the correction itself. Due to the quality of the FP-supported evaluation program, the required corrections are rather small, such that this fact dominates the measurement error only in rare cases. It is important that this error can now be stated, therefore making the "risk" known. Details for determining the systematic measurement uncertainty are described in [247].

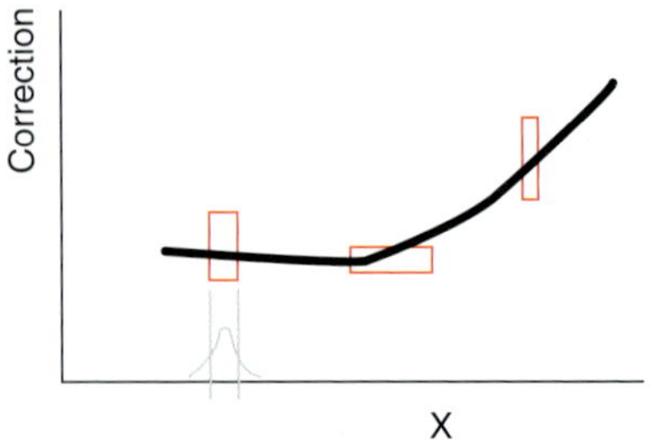


**Fig. 7.96.** Principle of calibration. Three standards define a correction function. X is the theoretical uncorrected value. Its uncertainty is reflected by the width of the red rectangle. Its height equals the uncertainty of the standard value according to the inputs of Fig. 7.95, and the uncertainty of the calibration measurement

A prerequisite for calculating the systematic measurement uncertainty is the knowledge of the uncertainties of the standards. This takes place in the definition the calibration standard set, where the corresponding information is provided to the program, cf. Fig. 7.95. If a calibration is not performed, the systematic portion cannot be determined and the program will default it to Zero.

WinFTM V.6 can display the uncertainties for single readings and for the mean values (either on the screen or through the so-called online export function) [248], namely the random and the systematic portion, or both together. Table 7.10 provides an overview.

**Table 7.10.** Measurement uncertainty with WinFTM-V.6

| Toggle "Display of the measurement uncertainty" | Error information for the single reading | Error information for the mean value |
|---|---|---|
| Random portion | Theoretical error estimate from the measured spectrum | Confidence interval of the mean value according to $t$-distribution (Student-Distribution.), calculated from the standard distribution of the block |
| Systematic portion | Computed from the random error of the calibration measurements and the uncertainties of the nominal values of the standards, as entered into the "Calibration Standard Set" | Uses the error value of the last single reading of the block |
| Systematic + random portion | Root of the square sum of the random and the systematic portion | Root of the square sum of the random and of the systematic portion |

### 7.4.5 Instruments

The foremost requirement for any instrument is that a high reproducibility of the results is ensured. To this end, a good repeatability is required, which for XRFA instruments is essentially determined by counting statistics and therefore by the X-ray source, the beam guidance and the detector.

For many applications in the field of layer analysis, good knowledge of the measurement location is also necessary for a good reproducibility. Thus, next to X-ray source and detector, the beam guidance and the observation of the measurement spot are important components within the instruments (cf. Fig. 7.97). The instruments are designed such that the specimen can be viewed using a video camera. The observer has a vertical view (or at least one under a steep angle) of the sample surface, and the measurement spot is indicated by the crosshairs on the video image (Fig. 7.98).

Accurate positioning using the video image is in many instances an important prerequisite for correct measurements.

#### Detectors

Selection of a suitable detector type is governed by the requirements of the application. The objective is essentially good repeatability. Because the structures to be analyzed are typically very small, the active measurement areas must be very small as well, leading to very low intensities of the radiation to be detected. To balance this, the detectors should operate very efficiently, and in particular cover a very large solid angle. Size of the detector window and distance from the sample play an important role in this respect. Table 7.11 offers some values as a guideline.

In many cases, the energy resolution of the detector is of secondary importance. However, with strong overlap of element lines in the spectrum, or when measuring very thin layers or low element contents, a better energy resolution will result in a better repeatability. Of course, in addition, cost and handling (cooling, durability, stability) are of great significance in industrial manufacturing or quality control.

Considering what has been said thus far, it is understandable that proportional counter tubes are commonly used. Although the energy resolution is relatively poor, with low costs they generally exhibit large active areas. Semiconductor detectors (in layer analysis almost exclusively Si-PIN diodes and Si-drift chamber detectors are used) have a significantly better energy resolution, however, costs are higher and the active areas are significantly smaller. Their use is therefore limited to applications where a high energy resolution, due to strong overlap of peaks in the spectrum or due to low detection limits, is required. The increasing complexity of the measurement applications and ever thinner layers will most likely lead to greater use of semiconductor detectors in the future.

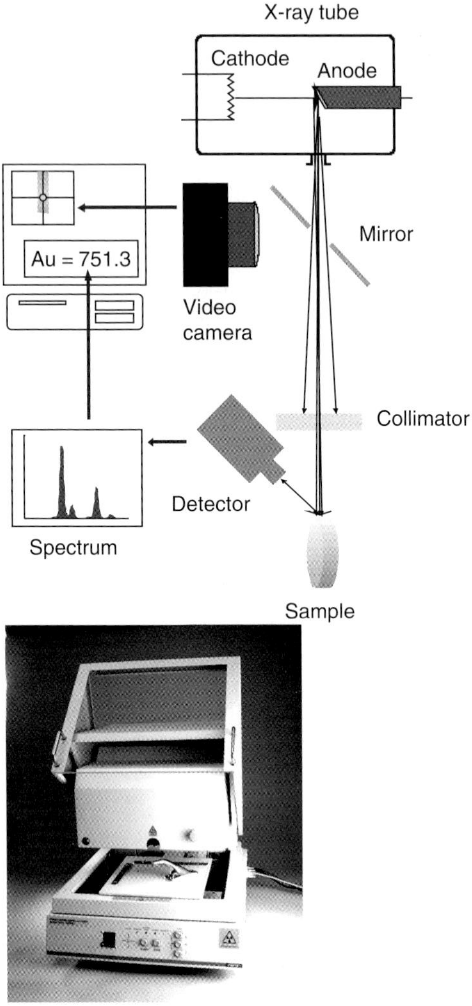


**Fig. 7.97.** XRFA instrument for layer analysis. Primary radiation generated by the X-ray tube is laterally limited by a collimator or an X-ray optics, such that a defined area of the sample is excited. Using a video camera, the sample can be observed during measurement. Instrument equipped with a programmable, motorized measuring stage, measurement can be easily automated to cover several samples or several spots of one sample. Detector registers fluorescence radiation that is converted into a spectrum by suitable electronics. Spectrum is used by evaluation program as the basis for the computation of the result values. Representing a typical instrument, the Fischerscope X-ray XDAL is shown above. Equipped with a PIN diode with Peltier cooling as the detector. X-ray source is a microfocus tube with a W-anode that operates at a maximum of 50 kV and 50 W. Measurement example of Fig. 7.89 as well as several applications of Sect. 7.4.4 refer to this instrument

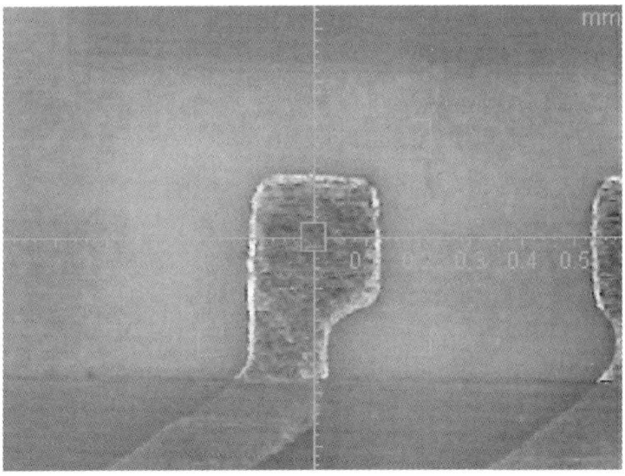


**Fig. 7.98.** Video image of a bond pad. Instrument: FISCHERSCOPE XDVM-μ. The size of the shown measurement spot is 0.04 mm × 0.05 mm

**Table 7.11.** Comparison of proportional counter tubes and semiconductor detectors. The given numbers are typical values

| Property | Proportional counter | Si PIN | Si drift chamber | Si(Li) |
|---|---|---|---|---|
| Energy resolution (FWHM) for Mn-K$\alpha$ | 900 eV | 180 eV | 150 eV | 125 eV |
| Effective entrance window | 10 cm$^2$ | 5–25 mm$^2$ | 5–10 mm$^2$ | 10–100 mm$^2$ |
| Solid angle $\Omega$ | 0.05 sr | 0.005 sr | 0.002 sr | 0.01 sr |
| Cooling | None | Peltier | Peltier | Liquid Nitrogen |

## X-Ray Source – Beam Guidance and Measurement Spot

The system consisting of X-ray source and beam guidance system determines the intensity of the fluorescence radiation emitted by the sample, and therefore the repeatability of the measurement. At the same time, they define size and position of the measurement spot and therefore influence the reproducibility. The following section will provide more detailed information about the individual components.

### *X-Ray Tubes*

The spectral distribution of the radiation of the X-ray tube controls the excitation of the fluorescence radiation. To achieve optimum excitation of fluorescence line, as much radiation as possible should be in the energy range just above the absorption edge that belongs to the particular fluorescence line. However, most often, several fluorescence lines need to be excited because either several elements are present in a sample or various samples need to be analyzed. The energy range of the lines for the elements that are most common in layer analysis stretch from approx. 3 to 30 keV. In most cases, therefore, a tube with a W anode is employed as a good compromise. Certain adjustment of the primary spectrum can be achieved by the applied high voltage or by additional primary filters (cf. Sect. 7.4.4 DefMA, Features 3 and 4). For special applications, anode materials with a high fraction of characteristic radiation are employed, for example Cr for elements with an atomic number of $\leq 22$ or Mo for the excitation of Au-L and Pt-L lines.

*Apertures*

The size of the measurement spot $M$ is dependent on the size of the focal spot $L$ on the anode. Micro-focus tubes with focal spot sizes of less than 0.2 mm diameter are used to realize very small measurement spots. Viewing the linear expansion of the measurement spot $M$ (Fig. 7.99), one can see that it depends on the dimension of the X-ray source $L$, the size of the collimator $d$ and the distances source to collimator $a$ and collimator to measurement plane $b$:

$$M = d \cdot (a + b)/a + L \cdot b/a. \tag{7.32}$$

$M$ specifies the maximum extension of the illuminated area at the measurement plane. No radiation is present outside of this area. The intensity distribution within this area, in particular the transition at the border depends on the lateral intensity distribution of the X-ray source. A clearly delimited center area $P$ with a nearly even distribution results only with a point source. To obtain small, well-defined measurement spots, the distance from the collimator to the measurement plane and the size of the X-ray source should be as small as possible. If these conditions are fulfilled, the size of the measurement spot is then essentially determined by the collimator size. In practical applications,

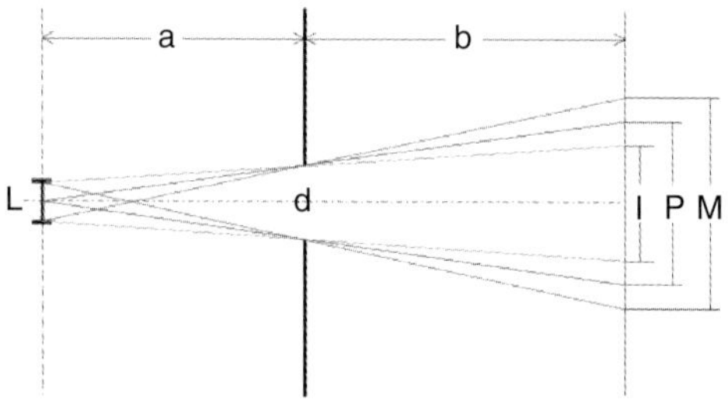


**Fig. 7.99.** Size of measurement spot M(I, P) depending on distances a ,b, aperture opening d (collimator) and size L of the focal spot of the tube

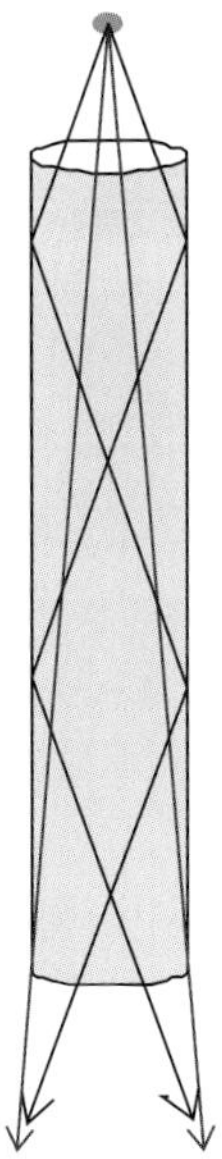

**Fig. 7.100.** Monocapillary

collimators with round or rectangular shapes are used and dimensions down to less than 50 µm are realized. A lower limit is given only by the resulting intensity of the primary radiation.

*Focusing Elements (X-Ray optics)*

Small measuring spots and high primary intensities can be achieved using focusing elements. Because the refractive index of X-rays for almost all materials is very close to 1, special X-ray optics must be used for focusing purposes, cf. also Chapter 3 "X-ray Optics" (pp. 85–198). In instruments used for layer analysis, they are all based on the principle of a light conductor by means of total internal reflection.

Figure 7.100 shows the principle of a monocapillary. X-rays are reflected at the interior walls of a glass capillary at the transition from air as optically denser medium to glass as optically thinner medium, if the angle of incidence is smaller than the critical angle for total reflection. Through repeated reflection, transport of the radiation through the capillary occurs almost without loss. The size of the measurement spot is determined by the inside diameter of the capillary, the divergence according to the critical angle of the total reflection and the distance to the measurement plane. The amplification of the intensity results from the ratio of the covered solid angle of a collimator at the exit location of the capillary and of the solid angle covered by the capillary. However, since the refractive index depends also on the energy of the radiation, in addition to the material of the capillary, the size of the measurement

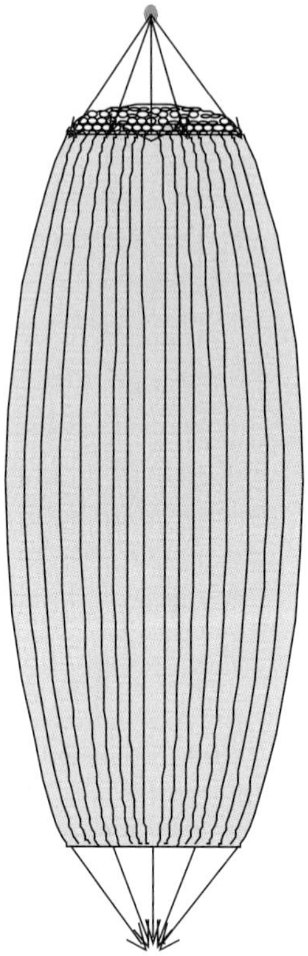

**Fig. 7.101.** Polycapillary lens

spot and the intensity amplification depend on the energy of the radiation. Polycapillary lenses can be created by bundling many individual capillaries (Fig. 7.101). They offer the advantage that a significantly greater solid angle is covered compared to a monocapillary, which increases the intensity amplification significantly. Using a suitable shape, focusing of the radiation can be achieved, which results in an increase in the radiation density. The size of the measurement spot is determined by the critical angle of total reflection for a given energy, the inside diameter of the individual capillaries, and the output focal length of the optic.

Figure 7.102 shows a system consisting of two X-ray mirrors that serve as X-ray guides in one spatial direction. The generated measurement spot is determined in one direction by the size of the exit slot of the mirror, the critical angle of total reflection, and again the distance optic output to measurement plane. The other direction is defined by a classic aperture. In contrast to the capillary optics, one obtains a measurement spot of rectangular shape that is sharply defined in the direction in which the aperture is effective, regardless of the X-ray energy. The intensity amplification that can be achieved is comparable to that achieved with a monocapillary.

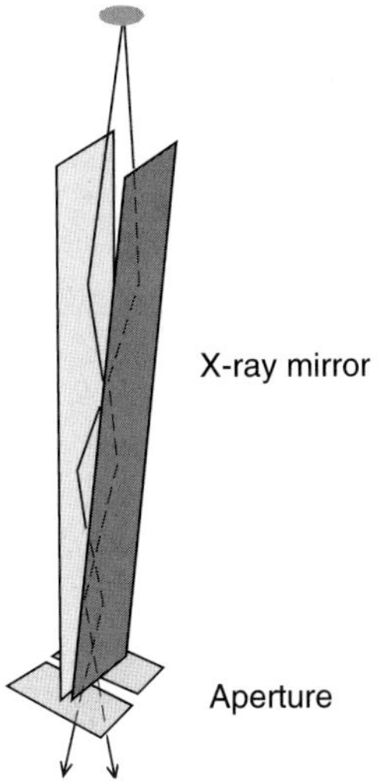


**Fig. 7.102.** X-ray mirror. Figure 7.104 shows an application example

**Table 7.12.** Comparison of various X-ray optics

| | Monocapillary | Polycapillary | X-ray mirror |
|---|---|---|---|
| Amplification | ~10 | ~500 | ~10 |
| Measurement spot (FWHM) | 10–300 µm | 20–50 µm | x≥20 µm<br>y 40–200 µm |

A comparison of the various X-ray optics shows that the polycapillary provides by far the highest intensity gain for small measurement spots (Table 7.12).

## Measurements From the Top or From the Bottom (Fig. 7.103)

Placing the measuring head, consisting of X-ray source, detector, and beam collimation, above the sample, such that the X-ray beam strikes the sample from above, allows for the analysis of samples with very different forms and geometries. The measuring head can travel in the z-direction, enabling measurements of samples of different size and in particular in recesses and indentations. In addition, the sample can be placed on a stage that can move in X-Y-direction. If the X-Y-Z-axes are motorized and programmable, measurements on several locations of a sample or on several samples can be automated.

If the measuring head is located below the sample, then the distance between sample and detector becomes constant simply by placing the sample onto the specimen support. This enables easy handling of flat samples.

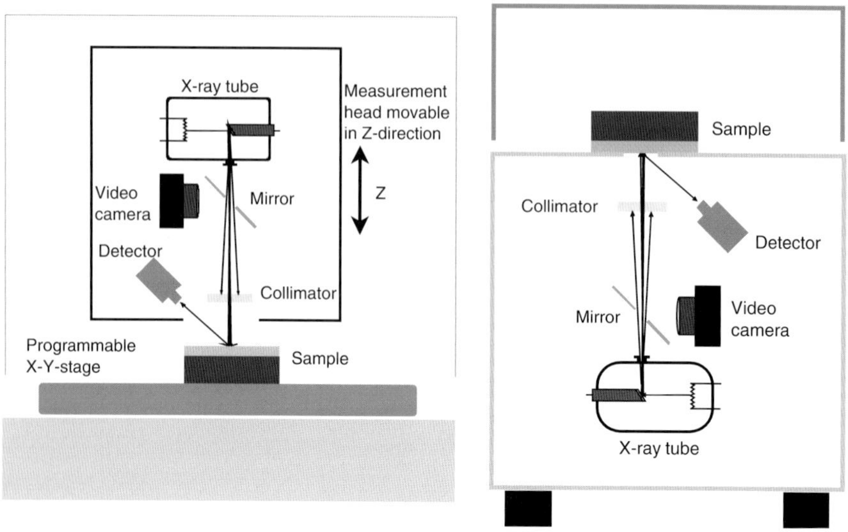


**Fig. 7.103.** Basic design of measuring instruments for measurement from top or from bottom

However, automatic sample positioning is hardly possible with instrument geometry from the bottom.

### 7.4.6 Application Examples

Covered are a few typical applications that are possible using the shelf production models.

#### Single Layer, Zinc and Zinc Alloys, Zinc Flake Coatings

Zinc layers provide effective corrosion protection for steel surfaces. They are, therefore, used in many different ways, for example, especially in the automotive industry. A sufficient layer thickness is required for a quality coating. The zinc layer can be measured by the magnetic induction method on large-area, flat components using hand-held probes [249]. XRFA further enables layer thickness measurement even on components with complicated shapes, such that the method has been widely accepted in the electroplating industry. The simple measurement application zinc on iron (Zn/Fe) rarely causes problems for the operator; at most, the shape of the specimen may require careful positioning. Measurement on threads or in indentations may not be so simple due to potential obstruction of the path between the location were the X-ray fluorescence is emanating from the sample and the detector.

Coatings of a Zn-alloy significantly improve the properties of the coating. On the other hand, the coating process and the characterization of the layer

**Table 7.13.** Evaluation of the measurement of a chromate passivation layer on CoZn on Fe

| Measurements | Cr | CoZn2 | Co2(%) |
|---|---|---|---|
| n=1 | 25.85 | 9.586 | 0.932 |
| n=2 | 23.70 | 9.602 | 0.786 |
| n=3 | 24.39 | 9.549 | 0.825 |
| n=5 | 23.77 | 9.630 | 0.840 |
| | Cr($\mu$g/cm$^2$) | CoZn($\mu$m) | %Co |
| Mean value | 25.08 | 9.617 | 0.871 |
| Standard deviation | 1.336 | 0.053 | 0.082 |
| Stat. meas. uncertainty (68%) | 0.357 | 0.014 | 0.022 |
| Number of measurements | 15 | 15 | 15 |

A Fischerscope XDAL with a PIN diode as detector has been used (cf. Sect. 7.4.5 Detectors). The aperture has a diameter of 0.6 mm. Additional instrument details: X-ray tube with W-anode, 50 kV, primary filter 10 µm Ni, measuring time 30 s

become more difficult. Alloy elements may be Ni (6–15%), Fe (about 1%) or Co (about 1%). In the case of Ni, the composition can be measured well using instruments with proportional counter tubes. For Fe and Co they cannot be used because of the severe peak overlap caused by the poor energy resolution of the detector. Of course, it is still possible to measure the thickness for those alloy coatings. Instruments with semiconductor detectors are well capable of measuring the other alloy elements, except for Fe due to the substrate material of steel. Table 7.13 shows an example.

This application measures simultaneously the Cr-containing passivation layer. These typically very thin layers additionally improve the corrosion resistance, such that they are very common in the automotive and other industries. Many types of passivations are available [250]. For XRFA, it is important that only the Cr mass per unit area can be measured and not the oxidation state of the Cr. For effective passivation a minimum layer of Cr is necessary, although the Cr mass per unit area by itself is not sufficient [251]. The instrument must be sensitive enough due to the small amount of Cr; therefore only instruments with semiconductor detectors can be used. On the other hand, the chromatization does not influence the "normal" measurement of the zinc layer [252].

In addition to the galvanic Zn layers, Zinc flake coatings have found acceptance in certain applications [253]. They are applied as an aqueous suspension and cured when heat-treated. For XRFA, they present a mixture of Zn and Al, often dye and other organic additives are added as well. In principle, X-ray fluorescence can only determine the zinc amounts. However, since the signal of the Fe substrate material is also available as information, such a layer may be measured as a two-component "alloy layer" AlZn. If $TiO_2$ is added as pigment, the model AlTiZn/Fe will provide reproducible results. The density must be determined with a different layer thickness measuring method, e.g., by cross-section or by the magnetic induction method. Typically, these corrosion protection layers are combined with a functional top layer. These top layers also consist of light elements, and to a degree again of Zn and Ti. This poses a difficult if not hopeless situation: both layers consist of the same elements. However, they have a different composition. The problem could indeed be solved if the concentrations in the individual layers were fixed or at were least within very narrow limits. After all, the top layer is low in Zn and the layer below rich in Zn. If samples of the individual layers, deposited on Fe, are available, their compositions can be measured using the alloy layer measurement application AlTiZn/Fe, the density of the individual coatings can be measured using one of the above mentioned methods. A combination of such top and interim layers can then be measured reliably. Unfortunately, the requirement of a somewhat constant composition of the individual layers is not fulfilled in practical applications. When analyzing a large number of actual parts, a Zn content was found in a range from 25% to 75%, while lab samples only had a range from 65% to 70%.

### Single Layer Approximation

Under these circumstances, the dual coating measurement is not possible, because the systematic deviations become unacceptably large due to the different compositions of the partial layers. The solution is to consider the entire layer system as a 3-component single layer. In this manner, composition and thickness can be approximated. One problem here is the correct determination of the alloy density that must be made due to the many possible variations of top layers. Considering all the problems that occur with Zinc flake coatings the resulting thickness error with less than 10% is sufficiently small. The Zn mass per unit area itself, which is responsible for the corrosion protection, can be measured much more accurately. It is not influenced by the uncertainty of the density.

### Local Distribution

One advantage of XRFA is that the size of the primary beam determines the analyzed area and that the local resolution of the measurement can be very sensitive when using suitable collimators or X-ray optics (cf. Sect. 7.4.5). One example is presented in Fig. 7.104, where the local resolution is very high. A square with sides of only 0.5 mm had been scanned. The measurement application was set such that all layer components (Al, Ti and Zn) were analyzed with regard to their mass per unit area, of course only in single-layer mode. The sample is a bolt that is coated using a procedure employed around the world. The scanned section was selected at random and is in the center of the coated surface of the cylinder, not on the edge.

If one considers that the corrosion protection is accomplished by sacrificing Zn, which cannot be "replenished" then the locations with the smallest Zn plating are, of course, the first that will corrode. If one only focuses on the

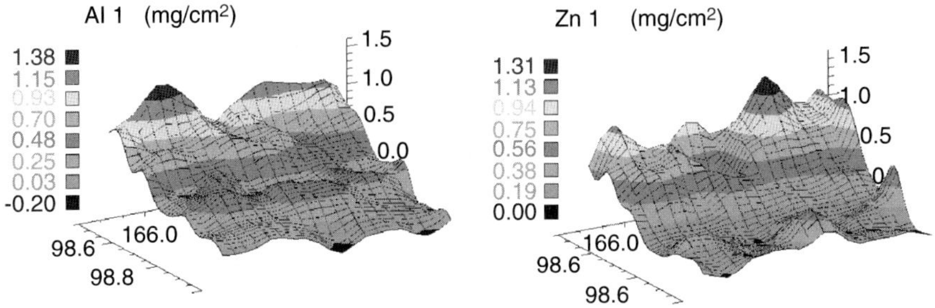


**Fig. 7.104.** Distribution of Al and Zn mass per unit area within an area of 0.5 mm × 0.5 mm, Fischerscope® XDVM-$\mu$, equipped with an X-ray mirror as described in Sect. 7.4.5 Focusing Elements (X-ray optics) as focusing element. Local resolution approximately 20 µm× 50 µm

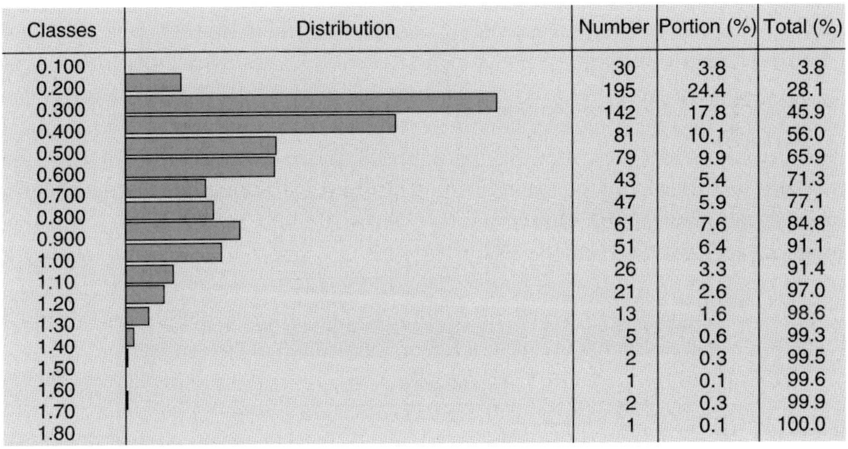

| Classes | Number | Portion (%) | Total (%) |
|---|---|---|---|
| 0.100 | 30 | 3.8 | 3.8 |
| 0.200 | 195 | 24.4 | 28.1 |
| 0.300 | 142 | 17.8 | 45.9 |
| 0.400 | 81 | 10.1 | 56.0 |
| 0.500 | 79 | 9.9 | 65.9 |
| 0.600 | 43 | 5.4 | 71.3 |
| 0.700 | 47 | 5.9 | 77.1 |
| 0.800 | 61 | 7.6 | 84.8 |
| 0.900 | 51 | 6.4 | 91.1 |
| 1.00 | 26 | 3.3 | 91.4 |
| 1.10 | 21 | 2.6 | 97.0 |
| 1.20 | 13 | 1.6 | 98.6 |
| 1.30 | 5 | 0.6 | 99.3 |
| 1.40 | 2 | 0.3 | 99.5 |
| 1.50 | 1 | 0.1 | 99.6 |
| 1.60 | 2 | 0.3 | 99.9 |
| 1.70 | 1 | 0.1 | 100.0 |
| 1.80 | | | |



**Fig. 7.105.** Histogram of Zn mass per unit area distribution (in mg cm$^2$) for the measurement according to Fig. 7.104. Maximum of the distribution is at 0.25 mg cm$^{-2}$. Mean value 0.54 mg cm$^{-2}$

feature "layer thickness," which actually means the mean layer thickness, one is "blind" for the critical "thin spots." While Fig. 7.104 is quite illustrative, the histogram of the Zn distribution (Fig. 7.105) is more informative for a quantitative evaluation of the coating, especially when the measurement locations are truly representative (which is not being claimed for this example). Here, we have the rare case that mean value and most frequent value are apart by more than 100% (!).

## Decorative Multi-Coatings

Lustrous, semi-lustrous and dark lustrous surfaces play a big role with consumer goods, though of course certain functional properties such as hardness, abrasion resistance, and corrosion protection are important as well. Classic nickel-plating in conjunction with a lustrous chrome layer is still a popular application. The measurement application is Cr/Ni/Cu/Fe. The Cu layer is indispensable as a bonding agent; Ni provides the corrosion protection, and Cr the luster. For XRFA, this layer system is not trivial; because we find here a rather complicated interaction of secondary excitation, cf. Sect. 7.4.2 Secondary Excitation–Enhancement by another layer Cr-K is excited by all other involved elements, and the magnitude of this enhancement effect is significantly influenced by the initially unknown interim layers Ni and Cu. This clearly defined measurement application is a good test example of whether a yield calculation according to Sect. 7.4.2 functions properly. The finite thickness measurement range creates another difficulty. There are applications with very thick Ni and Cu layers that often illustrate the limits of our method. While the primary interaction reaches sufficient depths due to the high operating voltage of the X-ray tube, the self-absorption of the layers has a limiting effect on the measurement range. The maximum can only be obtained through a "steep" geometry (i.e., an essentially vertical emission angle $\psi_2$, cf. Fig. 7.84). It is increasingly difficult to measure the buried layers as the thickness of the top layer increases. This trivial rule can be quantitatively challenged using WinFTM V.6. The program calculates the random measurement uncertainty (repeatability) for each measurement application, such that predictions can be made without trial about what can and what cannot be measured. Figure 7.106 shows the calculated relative measurement uncertainty for the Cu thickness for two different Cr/Ni variations.

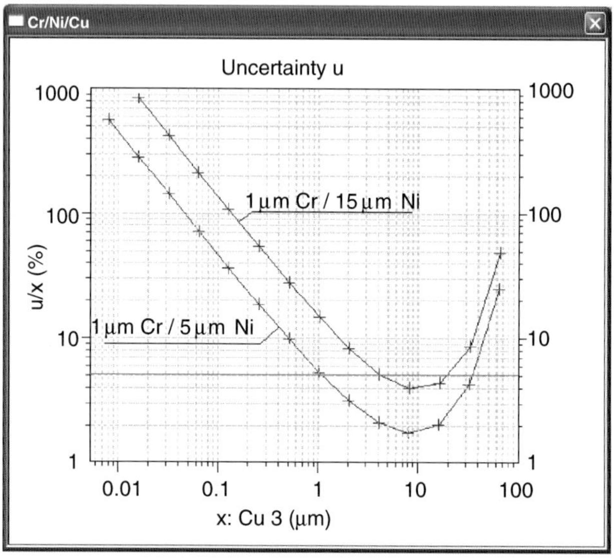


**Fig. 7.106.** Calculation of the relative measurement uncertainty for Cu due to counting statistics only. Cr/Ni/Cu/Fe application, Fischerscope® XULM WinFTM®V.6. Parameters: 50 kV, W-anode, 1 mm Al-filter, $\psi_1 = 0°$, $\psi_2 = 40°$. The two curves refer to top layers Cr + Ni with different thicknesses. The green line indicates a relative measurement uncertainty of 5% (confidence interval = 68%)

## Chemical Tin and Chemical Silver – The Influence of the Substrate Material and the Density Problem

Electroless deposited Sn or Ag layers are used for printed circuit boards. Ag or Sn is applied directly to the Cu conductor as a bonding agent. The measurement with X-ray fluorescence should actually present no difficulty when one observes the usual details for measuring pc-boards (possibly Cu with a finite thickness, Br-K-fluorescence from the epoxy resin carrier). Still, we often encounter significant differences when comparing measurements made with different instruments, even if a calibration had been performed prior to the measurement.

Recommended for the calibration are foils on various pc-board blanks (Cu/epoxy resin), because these samples differ only slightly from actual specimens. Solid plated standards with a solid Cu substrate material show a different scatter background as Fig. 7.107 shows very effectively. Also diffusion of Sn into Cu takes place and intermetallic phases are formed.

The relatively thin Ag layer absorbs Cu-K only slightly. The Ag-K peak is overlaid by a scatter radiation background, which can be even stronger with thinner Cu materials. Therefore its correction with a compensation spectrum (cf. Sect. 7.4.4 DefMA) as specified in the DefMA is necessary. WinFTM automatically takes the overlap of the Ag-L peak with the Cu-K escape peak at channel 24 into account.

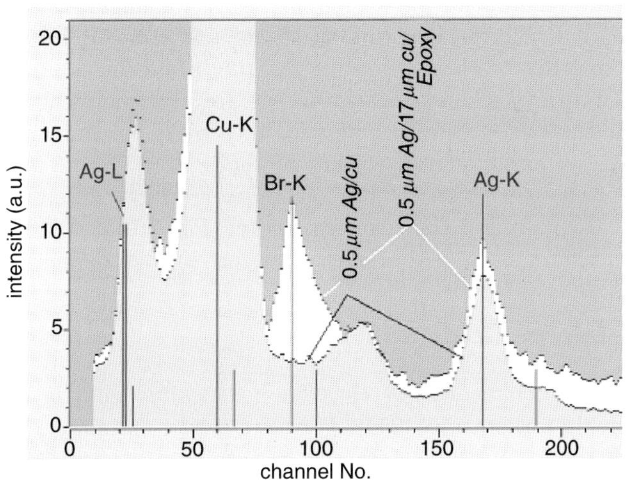


**Fig. 7.107.** Spectrum of Ag layer with thickness of 0.5 μm, once on massive Cu, the other time on a typical pc-board material with about 17 μm Cu. Its Br-containing epoxy resin is glass-fiber reinforced, which leads to an elevated scatter background and to emission of Br-K-radiation. Although the latter does not influence the Ag or the Sn measurement, a correction of the continuous radiation background is very important. Disregard can lead to errors in the range of several tenth of a micrometer. Instrument: Fischerscope X-Ray XDVM-W. Detector = proportional counter tube

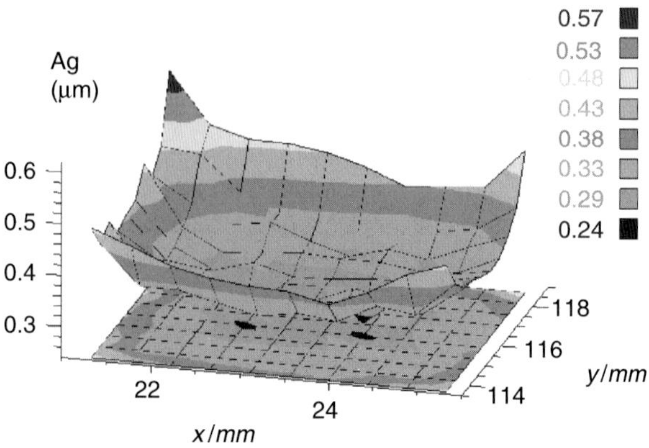


**Fig. 7.108.** Ag thickness distribution over Cu, scanned area of about 4.5 mm × 5.5 mm, using Fischerscope X-Ray XDVM-W

One reason that can influence the mentioned difficulty when comparing measurement results between different instruments is that the background correction is handled differently by various instruments or evaluation programs. Additional reasons are the layer thickness distribution or the use of different densities for the chemically (electroless) deposited metals Ag or Sn. The density of the layer is usually less than that of the pure metal. Bath manufacturers sometimes provide a density range, e.g., 3–4 $g\,cm^{-3}$ for chemical Sn. Due to the impossibility of measuring the density using X-ray fluorescence, we recommend to use the densities of the metals (for Ag 10.5, and for Sn 7.3 $g\,cm^{-3}$), in order to obtain a secure starting point for comparison. If one considers that true thickness measurements are extremely difficult or inaccurate for thicknesses $< 1\,\mu m$, a reference to a mass per unit area secures a traceable and comparable measurement method.

In spite of the electroless deposition, there are large differences between the center of the area and the edge regions (cf. Fig. 7.108). When comparing results the measurement position must therefore be well defined.

## Functional Layers. Gold and PdNi

Electrolytically (galvanically) or electroless deposited gold layers are common in the field of electronics [254]. They are used to minimize the electrical resistance for plug-type connectors and as tarnish protection for solder and bonding surfaces. The base material is Cu (printed circuit boards) or a Cu alloy (plug-type connectors). Due to the material costs, coating is carried out using narrow tolerances and secondly only the functional surfaces are coated. Furthermore, due to increased degree of integration, the packing density on pc-board and circuits (lead frames) decreases, such that for the coated surfaces

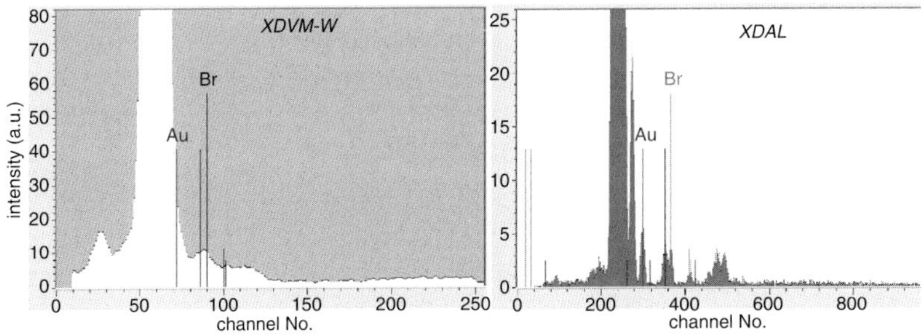


**Fig. 7.109.** Spectrum of a solder area that is coated with Au and NiP on a pc-board, using Fischerscope XDAL (detector = PIN diode) and XDVM-W (Detector = proportional counter tube). Table 7.14 provides the evaluation of the same sample using both instruments

become ever smaller. It is exactly this application Au/Ni/Cu that provided the decisive breakthrough of the XRF layer analysis measurement method in the 1980s. Replacing Au with Pd or PdNi [255] was originally the inexpensive alternative to save Au. In addition one found excellent tribologic layer properties, such that PdNi is still in use today, independent of the Pd and Au price. Both materials are combined such that the analytically not easy layer sequence Au/PdNi/Ni/Cu alloy is a frequent application for the XRF layer analysis. Here the PdNi composition must be known, of course.

*Coated Conductor Paths. pc-boards*

Similar to the chemical Ag or chemical Sn case, the interaction of the primary beam with the carrier material made of epoxy resin or synthetics causes certain problems for this application. Because bromine salts are often dissolved in the epoxy resin as a fire retardant, Br-K radiation is excited, which energetically is close to the Au-L lines (cf. Fig. 7.109.). Demands on peak evaluation routines are particularly challenging when using a proportional counter for radiation detection. The influence of the scatter radiation and of the Br-K radiation becomes larger as Cu layers become thinner. Separating the Ni and Cu peaks is also more difficult with the proportional counter than with the semiconductor detector. On the other hand, if the Au layer thickness is not too small (>50 nm), the proportional counter tube is a reliable tool for measuring Au and Ni on pc-boards, as the comparison of Table 7.14 shows. Although the Cu thickness is not of interest, it has to be measured to correctly calculate the spectral overlap of the Cu and Ni peaks. WinFTM can suppress the display of the Cu result if desired. Flexible pc-boards do not contain Br additives, but the design of the sample support requires particular care [256].

**Table 7.14.** Measurement evaluation of two samples using instruments that differ in detector $D$ and size of measurement spot $M$

| Sample | Measuring time | Measured quantity (layer) | Standard deviation from a minimum of 20 single readings | | |
|---|---|---|---|---|---|
| | | | XDVM-WD: Proportional counter tube $M: \emptyset\ 0.27\,\text{mm}$ | XDALD: PIN diode $M: \emptyset\ 0.7\,\text{mm}$ | XDVM-$\mu$ SSD-MoD: PIN diode M : 0.1 mm × 0.1 mm |
| Pc-board | 20 s | 86 nm Au | 5 nm | 4 nm | 3.3 nm |
| Au/NiP/Cu/* | | 4.4 µm NiP | 0.16 µm | 0.05 µm | 0.05 µm |
| Lead frame | 100 s | 5 nm Au | 1.2 nm | 0.8 nm | 0.5 nm |
| Au/Pd/Ni/* | | 30 nm Pd | 2 nm | 3.8 nm | 3 nm |
| | | 0.6 µm Ni | 0.02 µm | 0.005 µm | 0.01 µm |

*Small Measurement Spots for Lead-frames and pc-boards*

Miniaturization of Si chips and their large number of lead connections requires a large number of small bonding surfaces that are coated with chemical nickel and thin Au (Fig. 7.110). The XDVM-µ is well suited to perform these measurements; it uses an X-ray mirror as described in Fig. 7.102 for beam focusing. Due to the proportional counter tube with its large entry window, the fluorescence radiation can be detected very efficiently such that one can expect to achieve about the same precision as with the XDVM-W, in spite of the very small measurement spot (cf. Table 7.14). Using the small spot, which is very well defined in the x-direction, one can still analyze the lateral thickness distribution even on such small areas as shown in Fig. 7.111.

*Plug-Type Connector Contacts*

The layer sequence is an Au/Ni/Cu-alloy or an Au/PdNi/Ni/Cu-alloy. This traditional application has changed in the course of recent times. Not only have the effective areas become smaller, the geometry may have become more complex as well [257]. For this reason, the requirement of a plane-parallel layer in the area of interaction (cf. Sect. 7.4.2 The Equipment) is essentially no longer fulfilled. With a radius of curvature $< 1\,\text{mm}$, problems arise which however can be resolved taking certain provisions [258]. These are primarily proper positioning and an adapted measurement spot size.

Another difficulty with analyzing contacts is the frequent change in substrate materials. Depending on the desired material properties, various alloys such as CuFe, CuZn, CuNiZn, or CuSn with varying compositions are used. This means a new "substrate material correction" (normalization for a new

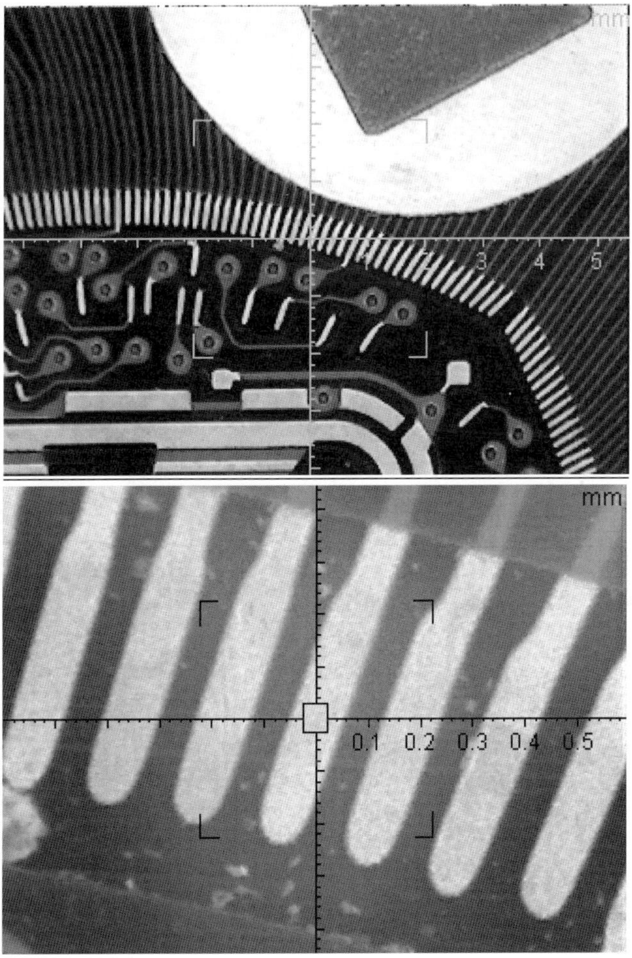


**Fig. 7.110.** Bonding surfaces of a printed circuit, overview above and detail below. The shown measurement spot of the XDVM-$\mu$ is here 40 µm×60 µm

substrate) has to be performed with every change of the substrate material. WinFTM V.6 offers an elegant simplification. According to the basic philosophy presented in Sect. 7.4.4, the substrate material can be analyzed in addition to the measurement of the actual layers of interest above. In this way, measurement applications can be combined and the effort for calibrations and inspections can be minimized significantly. Of course, there are limits, which are, however, surprisingly liberal. Examples are described in [259]. This capability offers advantages especially for incoming inspection, aside of just the use for contacts. An initial measurement of the uncoated substrate material is not required.

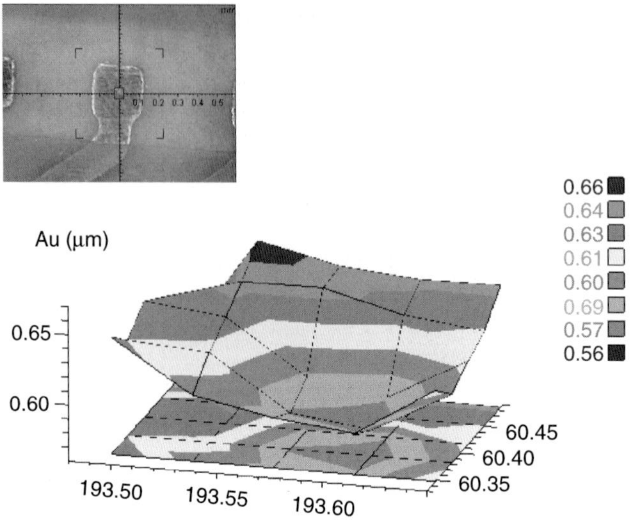


**Fig. 7.111.** Lateral distribution of the Au thickness of the bonding area shown above, measured with the XDVM.μ

## Hard Material Layers

Machining tools are often coated with hard materials. They consist of specialty steels (HSS = high speed steel) or of a hard material, which often is tungsten carbide (WC) cemented with Co. The coatings are Ti composites (TiC, TiN, among others) or even diamond like coatings. In addition, special layer sequences are employed as well with structures that are not entirely known. During chemical vapor deposition (CVD), certain parameters are set in succession, such that combinations of harder and softer regions are achieved as desired.

X-ray fluorescence can only measure the Ti coating. $Al_2O_3$ layers may be characterized under certain conditions [260]. For hard material coatings, typically process-specific standards are used, for which the layer thickness has been determined by cross-section. Of course, it is not possible to measure diamond coatings using XRF. Either the beta backscatter or the magnetic induction methods can be used.

A particularity of the named WC hard metal consists of a near-surface Co-leaching that has various causes. Thus, the stabilizing cement between the individual hard metal grains is missing. This Co-leached layer must be ground off prior to applying of the hard material coating, otherwise the coating will bond very poorly, leading to dramatic reductions in the service life of the tools. Here, XRFA can contribute in detecting this Co-leaching. The

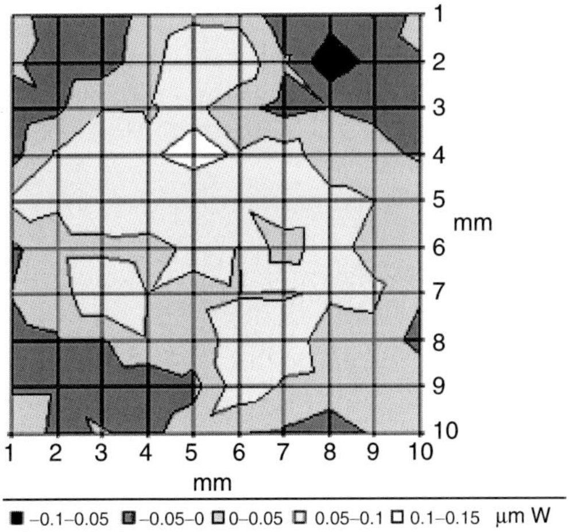


**Fig. 7.112.** Lateral distribution Co-leaching depth. The sample is a square cutting tool that has been cleaned at the corners using a "micro beam." The untreated central zone exhibits significantly greater Co-leaching. The difference is about 0.2 μm. Sample from Balzers, Liechtenstein

layer thickness measurement application is WC/Co-WC, with normalization on the unharmed substrate material without Co-leaching, being of utmost importance (Fig. 7.112).

## Precious Metal Layers. Jewelry

Covered here are Au alloy layers (mostly CuCdAu) with a high Au content (>12 kt). Substrate materials are brass (CuZn) but also CrNi-steel (stainless steel). For Cu alloys, a bonding agent is necessary between the Au alloy layer and the substrate material. This is generally Ni, often with a copper backing. Due to common Ni allergies in humans, for jewelry with insufficient CuCdAu-layer thicknesses, Ni is often replaced by so-called white bronze (CuZnSn). This does not make it easier to analyze the Au alloy coating, especially since to top it off a thin hard gold coating (< 1 μm) is often applied as scratch protection. Thus, the layer system is (from top to bottom, thickness and concentration values are approximate values)

1. Au (1 μm, with 1% Co or Ni),
2. CuCdAu (2–5 μm, 20% Cu, 5% Cd, rest Au),
3. Ni (2 μm) or CuZnSn (2 μm, 40% Cu, 20% Zn, rest Sn),
4. Cu (3 μm),
5. Cu70Zn30 (saturation thickness).

Aside from the fact that we have here a quadruple coating system, where two partial layers consist of several elements, some elements occur multiple times: Au in the 1st and 2nd layer, Zn in the 3rd layer and in the substrate material, and Cu even in the 2nd, 3rd, and 4th layer as well as in the substrate material.

In view of the limited information from the XRF spectrum, it seems that simplifying assumptions are unavoidable when analyzing this system. The composition of the partial alloy layers must be entered with fixed values (cf. Sect. 7.4.4, DefMA Features 13 and 14); otherwise, the evaluation will become unstable or ambiguous. The equation system (7.26), cf. Sect. 7.4.3, will then have no explicit solution. This means that various values of the AuCdCu thickness with different compositions have an equally good spectra adaptation. In other words, the method fails here because the task is too complex. It is definitely possible that measurements with different geometries and/or different excitation conditions provide additional information in order to enable an unambiguous solution. Now, with WinFTM-V.6 and such complex cases we can only analyze how the simplification of the DefMA influences the results.

As an example, we will consider the somewhat simpler layer system AuCdCu/Ni/Cu/CuZn. The Cu-content of the substrate material shall be known. Thus, the Cu of the 3rd layer is here the problem. The interim Ni layer is not thick enough to suppress the Cu signal of the 3rd layer, which interferes with the alloy measurement. The Cu, which is rather uninteresting, provides a fluorescence signal, which the evaluation cannot differentiate from the Cu of the Au-alloy layer. Expressed more precisely: the differentiation is not sensitive enough to support a dependable measurement. What remains is an essentially realistic input of the Cu thickness fixed by WinFTM when evaluating the other measured quantities. The influence of this Cu value, on the composition and thickness of the Au alloy layer is shown in Fig. 7.113.

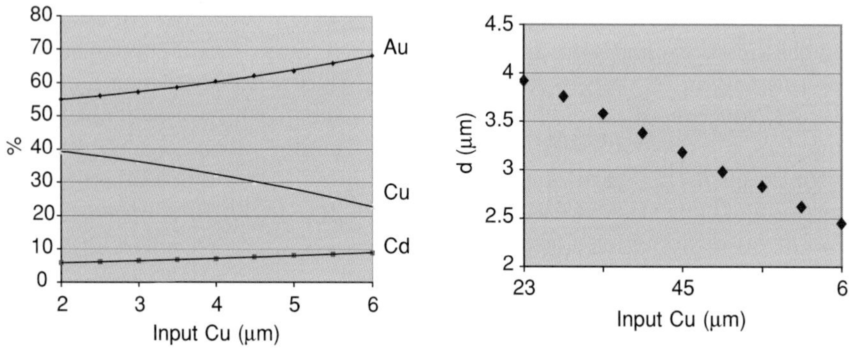


**Fig. 7.113.** Dependence of the calculated measurement quantities (concentration, thickness d) on the input of Cu layer thickness, defined as fixed by the DefMA. Same spectrum evaluated using different DefMA inputs.Using Fischercope® XDAL

What has been expected qualitatively is supported numerically: 1 µm more Cu in the interim layer reduces the Cu content of the alloy layer by 4%. The Au content is overestimated, the influence on Cd is less. The layer thickness is also affected (diagram to the right). This dependency on the Cu thickness is significant, in particular if one considers the typical process control of the Cu deposit. Certain mitigation results from the opposite tendency of the %Au and the thickness dependency for the Au mass per unit area: while the displayed Au content increases by 6% for a Cu change from 3 to 4 µm, and the displayed thickness is reduced by 10%, the displayed Au mass per unit area[4] decreases by only 4%.

### Layer Analysis as a Means

Here examples are described, where secondary samples from liquids and solids samples are analyzed as layers. In this way, the methods developed for the analysis of layers can be transferred to other analytical tasks without effort.

*Analysis of Liquids*

It is possible to simply measure the liquid in a beaker; this however has disadvantages. From greater depths, only scatter radiation is being emitted, which reduces the detection sensitivity. In addition, the assumption that the thickness of the sample shall be small versus the distance to the detector is not fulfilled, which makes the quantification more difficult. A simple measurement cell that sets a fixed liquid level has proven well for the analysis of main components in galvanic baths (metal concentrations in a range of a few g/l)). The bottom of the cell is made of a material which emits a fluorescence radiation that provides an additional source of information for characterizing the solution. In this manner, the analysis of a liquid becomes a layer analysis. The actually measured mass per unit area X of the individual components (e.g., in $g\,cm^{-2}$) can be converted easily into a volume concentration (e.g., $g\,cm^{-3}$) using the specified thickness $d$ of the liquid layer. In contrast to the analysis of an alloy layer, the spectral background is here significantly more severe due to the scattering of the primary radiation. Correct spectra evaluation is again of highest importance.

Through evaporation of the solution, e.g., on a thin foil, the solvent can be removed to a large extent, which reduces the distorting scatter radiation background. One disadvantage that results is the inhomogeneous lateral distribution of the liquid residue. Averaging is accomplished with a sufficiently large measurement spot (large collimator).

[4]With the *mass per unit area mode, absolute* (cf. Sect.7.4.4), the partial Au mass per unit area is displayed as the result. With known densities, it can also be determined from the thickness and the composition.

One interesting option for increasing the detection sensitivity is to bind the metal ions to a surface using chemisorption, and to subsequently measure them "in the dry" [261]. However, this type of enrichment is very matrix-dependent, such that the bound components do not clearly reflect the conditions in the solution.

*Analysis of Abrasion*

Using an abrasive paper or similar, one removes a small amount of the component to be analyzed. This is then either placed directly into the instrument or onto a thin, adhesive foil. In either case, one obtains a thin sample with a composition that is representative for the actual sample. This rather simple "trick" has often been successful when the object to be analyzed cannot be easily placed into the measurement chamber easily.

### 7.4.7 Summary and Outlook

We have discussed many application examples and even the theoretical part in principle did not indicate a limitation concerning the number of the layers to be analyzed and the number of elements in each layer. Aside from a limitation through absorption and matters related to the instrument, diverse applications are possible.

However, a few fundamental prerequisites must be fulfilled, which are described in Sect. 7.4.2. First, there is the "one-dimensionality," that is, the infinite expansion of the sample, at least in the area of interaction. With very small objects, e.g., at some coated conductor paths this is not the case. Lateral expansion and thickness are not of the same dimension. We, therefore, have to deal with 3-dimensional samples, where our term layer no longer applies.

Furthermore, we assumed the layer to be plane-parallel and homogenous. Concentration profiles such as the ones that come into existence by diffusion, are, therefore, excluded. Most technically generated alloy layers are, only roughly homogeneous. Thus, we only measure the composition for the model, with a mean value that is certainly not too far off; however, it does not fully correspond to the "real" value. Still, at least for large-area samples, it is indeed possible to obtain additional information by measuring under various angles and arrive at least at a coarse depth profile. Measurements with different excitation conditions may be of help as well. In the future, even for production model instruments, a measurement may consist of obtaining not only one single spectrum.

Concerning the effort to obtain as much information as possible from the experiment, one should note that we have evaluated "only" the fluorescence radiation. So far, the scattered primary radiation had been considered as radiation background; it made the evaluation of the spectrum with regard to the fluorescence component more difficult. Yet, it provides information and as in the bulk analysis, it could be a welcome source of information in the layer analysis. On the other hand, the theoretical treatment is rather difficult and the question remains, whether we will be successful in calculating the scatter spectrum for any layer system, as is the case with the fluorescence spectrum. If the development of the computer technology continues, in particular with regard to the increase in floating number operations per unit of time, it will be possible to integrate more sophisticated algorithms in the "normal" software of production mode instruments – such that today's state of the art could not be achieved with PCs of ten years ago.

## 7.5 Environmental Studies

*S. Kurunczi, J. Osán, S. Török, M. Betti*

### 7.5.1 Introduction

With the political turmoil of the late 1960s the concern of human kind was shifted from passive observer and sufferer of environmental problems to taking an active role in policy making. During the 1980s and 1990s most democratic societies formulated their fourth generation constitutions declaring the right of the population to a healthy environment. Besides, the national-level international movements (like the different environmental action plans of the EC) had started. This social environment triggered very much the new applications in environmental sciences. While in the 1970s the applications demonstrated some amateurism, by the end of the century environmental applications were the most published X-ray analysis results in the scientific literature.

This boom of XRF application was also related to the commercial availability of semiconductor detectors (lower price) that enabled simultaneous and fast multielemental trace analysis as well as the analytical electron microscopes (EPMA) and the loss of interest of nuclear science in the MeV energy charged particles (PIXE). An important further step was the availability of spectrum deconvolution and matrix correction codes (AXIL, Gupix, Sampoo) that was mandatory for quantitative analysis [262, 263].

Our definition of the environment has changed since the scouting experiments of XRF and PIXE [264]. In the 1970s we were only concerned with a confined environment directly related to human health and life quality.

Nowadays environmental science in a broad sense means a science of complex interactions that occur among terrestrial, atmospheric, and aquatic ecosystems (http://www.britannica.com). This shift in apprehension was generated by the transformation processes recognized in compartments (e.g., reduction–oxidation, biomethylation, bioaccumulation). Although these are indirect effects they have a strong influence finally on human health. The physical environments (air, water, and earth) are tied closely with living systems, among them human beings. These four compartments have a mutual interaction with technology. With the desire for a higher material standard of life and consumption a massive worldwide pollution was developed. Most environmental analysis tasks have the objective of studying the effects of such pollution by monitoring the chemical composition or structure of samples. More specifically the objectives of monitoring can be the assessment of the effect of pollution on humans, the identification of the cause and effect relationship, the evaluation of pollutant interaction, source apportionment, or the need for legislative control on emission of pollutants. It is quite unfortunate that XRF has never been rewarded to be a standard method having legal force for environmental application in Europe. However, in the United States most aerosol monitoring analysis is carried our by XRF. A remarkable observation was made in a recent study that investigated the distribution of general applications in the field of X-ray spectrometry [265]. Analytical abstract database in 1998 showed that environmental and geological applications of XRS make up more than 80% of the total which means that at least in the academic sectors most applications are carried out in this field. As can be seen in Fig. 7.114, water, aerosol, and soil analysis are the major applications in this field. For this reason, this chapter will concentrate on the analyses techniques and recent results in these popular fields. Interested readers might find overviews and references on other environmental applications in the annular *X-ray spectrometry reviews of Analytical Chemistry* [266–268] and *Journal of Analytical Atomic Spectrometry* [269, 270].

From the above definition (of the complex environment) we have to conclude that environmental applications involve large sample numbers and the samples themselves have a large variability in concentration. For such tasks X-ray fluorescence and in some applications (like aerosol analysis) PIXE was an ideal method due to the low cost/sample/element on the one hand and because of the high throughput capability on the other.

Characteristic X-ray emission spectroscopies from an analytical point of view are atomic number sensitive methods. Most routine analyses in the last decades have been carried out for geological explorations [272]. Environmental applications merit basically from the non-destructiveness and the wide dynamic range of the technique. However, conventional XRF is often criticized of having relatively high detection limits.

Since the beginning of the 1990s many other techniques have emerged, some of them with much lower minimum detection limits (MDL) even for

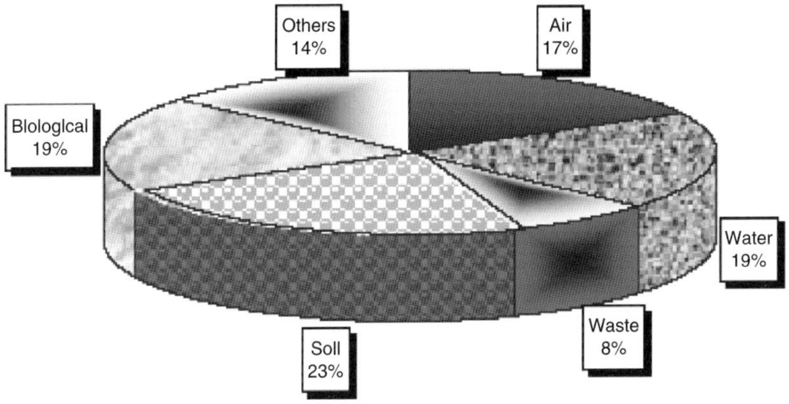


**Fig. 7.114.** Relative contributions of published manuscripts in the different application fields of X-ray spectrometry in 1998. Reproduced from Injuk and Van Grieken [265] by permission of John Wiley & Sons

bulk analysis. Figure 7.115 displays the detection limits of XRF techniques at this time.

The high performance in terms of smallest absolute mass was already obvious at that time. With the new developments of the past 10 years in X-ray optics, micro beam techniques have become very competitive in performance

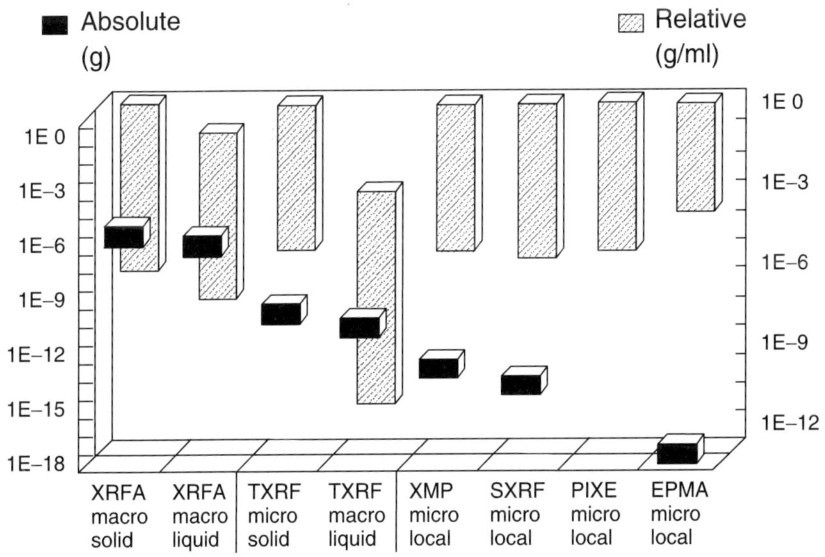


**Fig. 7.115.** MDL in $\mathrm{g\,ml^{-1}}$ and in absolute mass. Reproduced from Tölg and Klockenkämper [271] by permission of Elsevier Science BV

**Table 7.15.** Some analysis characteristics of micro beam X-ray techniques

| Exciting beam | Beam size | Lateral resolution (μm) | Penetration (μm) | MDL (ppm) | Relocation | Quantitation |
|---|---|---|---|---|---|---|
| Electron (15–25 keV) | 50 nm | 1–2 | 1–5 | 100–10000 | difficult | established methods |
| Proton (2–3 MeV) | 0.3–5 μm | 1–2 | 5–20 | 1–10 | difficult | easy for most samples |
| X-ray | 0.5–20 μm | 0.5–10 | 10–1000 | 1–100 | possible[a] | advanced |

[a] In intense synchrotron beams valence state or structure might change

for small volume samples (see Table 7.15). However, microscopic volumes are not really representative of environmental analysis that anticipates monitoring of large masses or volumes with temporal resolution. From this reasoning it follows that samples with high level of heterogeneity are most suitable for microscopic volume analysis and usually are kind of complementary to bulk results. As a complementary technique to other instrumental methods (AAS, ICP-AES, ICP-MS), XRF has the advantage that it does not destroy the sample and very little or no sample preparation is needed. In many applications there is always an attempt to analyze the sample for an alternative method and either use the same sample preparation or relocate the sample for the next analysis. Often the simple fact that a technique is non-destructive does not imply that the sample can straightforwardly be analyzed with various techniques. When performance of the different analytical methods is compared this property plays an important role. In view of these the performance of micro beam X-ray emission techniques is compared in Table 7.15.

The following paragraphs will describe some novel and competitive applications and analytical problems solved by X-ray fluorescence in the field of environmental analysis.

### 7.5.2 Water

#### General

The goal of aquatic chemistry is to provide a basic understanding of the behavior of elements in water. It is important to have data sets for concentrations and for environmental parameters to further understand the bio- and geochemistry of elements. The utility of water research is found in water management and environmental protection, since the extensive exploitation of natural waters leads to several detrimental effects. First, we would like to refer to basic handbooks as starting points for those readers who wish to intensify their knowledge in water chemistry, see e.g. [273–275]. In the following only selected topics will be retrieved and discussed, applying a practical approach to inorganic analysis of natural waters where XRF and TXRF methods have been successfully used.

Natural waters are complex systems, because dissolved metals, dissolved gases, and organic compounds can interact with each other, and with the non-soluble fine particulates and colloids. It is of great importance to maintain stability of species and to avoid contamination during water sampling and storage. The inorganic analysis of water requires the filtration of the particulate phase by means of simple membrane filters (see: Chap. 6: Specimen preparation). The usual practice is to use a 0.5 μm (or 0.45 μm) pore-size filter, which removes most of the microorganisms, particulates, and colloids. In this way the dissolved phase of the water sample will still contain salt, sugar, amino acid, protein, and polysaccharide. It is worth noting that the 0.5 μm pore-size is not a sharp cutting (especially for fiber-filters or porous membrane filters). Also, the 0.5 μm pore size is arbitrarily selected and one can find smaller values down to 0.1 μm. The filtration spectrum can be extended with other techniques, such as ultrafiltration, nanofiltration or reverse osmosis. Sometimes it is important to have preliminary knowledge about the sample to be analyzed and to refine the objective of the analysis in order to select the appropriate filtration method [276]. The filtrate is acidified and stored in refrigerator (at 4°C) to preserve it until measurement.

Seawater can be characterized by a higher salt content and a usually lower transition metal concentration than freshwaters. It can be considered as a constant ionic medium. The treatment of seawater as a two-component system, water and salt, has several advantages. Although the idea of "constant composition" is recognized as being not strictly accurate, it is a useful approach in simplifying the study of the various physicochemical properties of seawater. The major components of seawater are $Na^+$–10000 ppm; $Mg^{2+}$–1200 ppm; $Ca^{2+}$ – 400 ppm; $K^+$ – 400 ppm [273]. Many of the remaining ions, despite their low concentration, may influence significantly global cycling as being micronutrients in the enzymes and electron transport system present in the living organisms. Trace element measurements at ppt ($pg\,ml^{-1}$) concentrations represent a significant challenge for instrumental analysis. A periodic table of the elements displaying the distribution of elements in the North Pacific has been compiled by Nozaki [277].

Following Livingstone [278], the mean concentrations of major ions for world average river waters are $Na^+$ – 6.3 ppm; $Mg^{2+}$ – 4.1 ppm; $Ca^{2+}$ – 15 ppm and $K^+$ – 2.3 ppm. Lakes have similar salt composition, but the actual geochemical properties of the catchment area may significantly alter the concentrations. Acid rain or acid mine drainage effluents from nearby sources further contribute to the observed concentrations of salt and trace metals. Rivers and lakes can be affected by seawater; when there is a hydrological connection between them an increased salt concentration will be found in freshwaters. As a consequence of these factors, the salt concentration can reach even values

10–20 times higher than the world average. Trace metal concentrations of riverine water are normally higher than that of seawater representing a ppb ($ng\,ml^{-1}$) level. The concentrations appear in a wide range for the same element because of the similar effects mentioned above for salt. Furthermore, different elements exhibit about 6 or 7 orders of concentration range in a given freshwater sample. For example, Fe concentration is often high relative to other trace metals reaching even few ppm, while Co concentration is usually some tens of ppt. Rare earth elements are at the low ppt level. The toxicity of many heavy metals to fish is inversely related to water hardness, with $Ca^{2+}$ competing with free metal ions for binding sites in biological system.

Rain water collects its pollutants largely by dissolving particulate matter in the atmosphere (upper troposphere) when droplets of water nucleate on atmospheric particulates, and secondarily by dissolving gasses from the atmosphere. In open oceans and coastal areas a salt content essentially like that of seawater (same ionic proportions but much more dilute) will be characteristic. Terrestrial rain compositions vary significantly from place to place because the regional geology can greatly affect the types of particulates that get added to the atmosphere. The main composition in northern Europe is: $Na^+ - < 0.4\,ppm$; $Mg^{2+} - < 0.1\,ppm$; $Ca^{2+} - < 0.15\,ppm$; $K^+ - < 0.17\,ppm$ [279]. These values are significantly lower than that of lakes or rivers. Taking into account the way of rain formation, the low salt concentration is not surprising and similar behavior may be expected for trace metals as well. Reimann et al. [279] found, e.g., Fe usually below 10 ppb, Cd below 0.1 ppb, Ni with a highest concentration of 57 ppb but normally $<1$ ppb and Pb at $\sim 1$ ppb or less.

Drinking water quality is of course of the highest concern for human health. This does not mean that the lowest concentrations should be desirable, because some trace minerals have been shown to be needed in supplementing diets. These common trace elements are Fe, Cu, Zn, Mn, Co, I, and Se. They are needed in very small amounts, or traces, in the diet, and hence their name "trace minerals". Furthermore, such beneficial effect is not only for these trace minerals, but also for salt. A certain minimum content of Mg and Ca in drinking water is crucial from both the technical (water corrosivity) and public health points of view [280]. There are recommendations for a minimum hardness, especially when water is softened or desalinated for public use. An environmental standard is, however, the maximum concentration of a pollutant in the water. It is set with regard to scientific and medical evidence in relation to impacts upon public health or upon natural ecosystem at a level of minimum or zero risk. Regarding the sample preparation, unfiltered samples may be more appropriate for health-related studies, as most of the particulate matter will ultimately yield bioavailable concentrations of heavy metals in the hydrochloric acid digestion system, that is, the human stomach. Again, trace metal concentrations in drinking water are shown in a broad range, depending on the source of water (lakes, rivers, etc.) or the way of transport. For example, concentrations of Pb in drinking water vary greatly, related particularly to the presence of lead in the plumbing system. This feature is taken into account by the US EPA regulation,[1] setting an action level at 15 ppb and similarly with EC regulation at 10 ppb. The objective of the European Commission is to simplify the many individual directives covering water into a Water Resources Framework Directive. The EC Drinking water Directive 80/778/EEC has been updated in 98/83/EC and can be found on the WWW.[2] Also the World Health Organization (WHO) guidelines are public with many other relevant data.[3]

From the above summary one can conclude some basic requirements for the analytical methods being used for water analysis. The most important are that the method should be a multielemental one if possible, exhibit a wide dynamic measurement range and be free from matrix effect along with a detection limit (DL) at least of ppb. The analytical methods most commonly used for such tasks are atomic absorption spectrometry (AAS) or inductively coupled plasma mass spectrometry (ICP-MS). As it will be discussed later, total reflection X-ray fluorescence (TXRF) method has performance characteristics very similar to that of inductively coupled plasma atomic emission spectrometry (ICP-AES). The next paragraphs will demonstrate that only the TXRF method meets the above requirements, while the conventional XRF has serious limitations and can be applied only in combination with preconcentration.

Preconcentration and matrix removal may be required where the concentrations of elements in the water samples are particularly low and the interferences caused by the matrix dominate any direct analysis. This is especially the case for conventional XRF due to its moderate sensitivity and high detection limits. From the point of view of trace metal analysis, salt content represents the matrix and it is detrimental owing to the strong absorption effect caused in both exciting and fluorescence radiation. Actually, salt removal is performed together with preconcentration, making it a difficult problem when extremely efficient salt separation is needed with a quantitative trace metal recovery. There is a large selection of materials for separation-preconcentration: activated carbon, liquid extraction, polyurethane foam loaded with chelating agents, ion-exchange to mention only those that are most popular (see also Chap. 6). There is a trend in analytical methods for water analysis with a greater use of solid-phase extraction (SPE) [281]. Naturally, new SPE materials are being developed until now and it seems to be a never-ending story; however, present efforts are aimed mostly at finding an optimum material for a particular application rather than a universal sorbent suitable for every purpose. The analytical practice may be divided into two fields regarding specificity. A basic question of screening or specific target analysis demands for different approach and this point should be carefully defined at the outset of the work.

---

[1] http://www.epa.gov/safewater.

[2] http://www.europa.eu.int/comm/environment/water/water-drink/index_en.

[3] http://www.who.int/water_sanitation_health/dwq/guidelines/en.

It has become apparent that any mistake occurring in collecting and processing a water sample or contamination during analysis could lead to a substantial error in the final result regardless of the performance of any state-of-the-art analytical technique applied subsequently. Therefore, it is important to emphasize the need for a clean environment during sample preparation and analysis [282]. In the case of TXRF a specimen is prepared on a flat carrier drying a liquid drop on it. The absolute amount of this film-like deposit is usually below 1 μg and it is measured in the laboratory air. In the case of XRF, mainly thicker solid samples are prepared from the water sample, e.g., after preconcentration on ion-exchange resin. More strict rules of cleanliness are therefore required for the TXRF laboratory, applying Class-100 level or better air quality.[4]

## Conventional-XRF Applications

Soon after the introduction of the energy dispersive (ED) spectrometers [283], several methods were explored for a suitable sample preparation. Various preconcentration techniques were applied to wavelength dispersive spectrometers earlier, thus the idea of upgrading the sample preparation with the much efficient ED detection system seemed promising. Luke's paper [284] is based on measurements with a wavelength dispersive spectrometer and described a comprehensive view of precipitation methods mainly with carbamate, cupferrate, ammonia, and hydrogen sulfide. The paper discussed different coprecipitating elements that were necessary to obtain quantitative recovery. The precipitates were filtered on a paper filter from aqueous solution and the calibration was established in the low μg range. A selective preconcentration was developed using ammonium pyrrolidinedithiocarbamate and the subsequent liquid extraction of the organic complexes [285]. This method is based on the simple fact that organic reagents containing cyanide, heavy halides, and sulfur donors form much stable complexes with transition metals than with alkali or alkaline earths [286]. Seawater matrix (alkali metals, alkaline earth metals, and halogens) do not form extractable complexes with dithiocarbamates. The complexed transition metals are extracted into an organic solvent (e.g., methyl-isobutyl-ketone or chloroform) while the matrix is retained in the aqueous solution. This liquid–liquid extraction procedure is normally performed in funnels, hence the handling of large volume of water samples ultimately limits the accessible preconcentration factor (typically between 10 and 50). The use of ion exchange resin-loaded paper as Au-preconcentrator was discussed in another early paper [287]. Electrodeposition was also explored for water preconcentration using a pyrolytic graphite electrode [288]. The constant current mode of electrolysis was applied at 0.5 mA and then a 0.2-mm thick disk was cleaved and covered by lacquer. The authors emphasized the problem associated with the stability in electrolysis operation compared to other factors.

[4]Class 100: less than 3,500 particles/$m^3$ with diameters >= 0.5 μm; no particles >= 5 μm (U.S.Stand. 209D).

One solution is to completely deposit all the metal in the sample, but this would be time consuming. Typical working condition was 90-min electrodeposition ($U = 15\,\mathrm{V}$) from 15 ml solution, providing accurate analysis at the low ppm level. From the foregoing, a DL of some ppm can be generalized for typical WD systems.

Although liquids can be analyzed directly in the EDXRF spectrometer using a special cell, the water mass will cause a high background in the spectrum and will increase the detection limits to several ppm. In 1982, Van Grieken [289] reviewed the preconcentration methods applied for water analysis and the subsequent XRF analysis, taking into account mainly ED spectrometers. Two important groups of the preconcentration methods discussed are precipitation (typically chelate forming agents) and collection on ion-exchange resins.

Traditional chelating reagents such as sodium diethyldithiocarbamate (NaDDTC), dibenzyl dithiocarbamate (DBDTC), ammonium pyrrolidinedithiocarbamate (APDC) form strong, insoluble chelates in aqueous solution with many transition metals [290, 291]. The resulting precipitate is then collected on a filter, dried and subjected to XRF. A typical method was described in detail for rain water, a challenging topic because of the very low concentrations involved [292]. The authors investigated the influence of pH, stability and concentration of APDC, concentration of carrier ion, and the concentration of an oxidizing agent on the quantitative preconcentration. Radioisotope ($^{109}$Cd) excitation source and a common Si(Li) detector were used in the spectrometer; DL were obtained at low ppb (Fe–1.7, Co–0.8, Ni–0.6, Cu–0.9, Zn–0.8, Pb–1.3).

For chelating ion exchange, Chelex-100 is still the most commonly used resin as cationic exchanger. A chelating sorbent essentially consists of two components: the chelate forming functional group and the polymeric matrix or the support; the properties of both components determine the features and the applications of the respective material. The active group in the resin is iminodiacetate fixed on a poly(styrene-divinylbenzene) support. Chelating provides more specificity toward transition metals compared to simple ion exchangers. A typical preconcentration method involves column or disk arrangement of the resin and then the water sample is passed through it. Following this preconcentration step, the resin is either measured directly or after pelletizing [293]. An improved detection limit can be obtained this way: DL below 1 ppb was presented for Zn, Cd, Hg, and Pb taking 1 l water sample and 1000 s measurement time.

A variant of chelating preconcentration is the polyurethane foam disk loaded with an appropriate agent (e.g., diethylammonium diethyldithiocarbamate) [294]. In optimizing the performance of the analysis, sample (disk) thickness proved to be essential. Increasing the mass (thickness) probed by the X-ray will result in higher fluorescence intensity and lower DL up to a certain extent. For much thicker samples, however, the background will increase drastically, leading to worse DL. This implies that a minimum DL can be found as a function of sample thickness [295]. Optimizing the polyurethane foam thickness the authors obtained DL for Co and Zn at 10 and 3 ppb, respectively after preconcentration from 25 ml and taking 600 s counting time. A simplified method of chelating disk method has been published recently preconcentrating 1 l drinking water [296]. The detection limits for Cd, Ni, Cu, and Pb were 3.8, 0.6, 0.4, and 0.3 ppb, respectively. The preconcentration can be conducted in the field, eliminating the transport of large volume water samples.

As seen from above, absorption often occurs in conventional XRF, thus a correction is needed to obtain a reliable result. There are different approaches (external standards, internal standards, dilution method, emission-transmission method and mathematical corrections) to the absorption correction and these approaches request a given time depending on the difficulty of the analytical problem [297]. Using the afore-mentioned methods, quantification is usually straightforward. Thin film technique represents a special case where no absorption takes place, making the quantification much easier. The simplest method for thin film preparation is to place a liquid drop on a membrane filter or backing foil and dry it. The fluorescence intensities from extremely thin residues can be very low because of the low amount of mass probed, resulting in a high DL. However, with an optimized method it is possible to approach the results obtained by TXRF [298]. In another study, Meltzer and King [299] concluded that direct excitation is more advantageous than secondary target mode because of the gain in intensity. Detection limits were obtained at the low ppb level, enabling direct analysis of reference waters with 5–10% accuracy. At this concentration level the purity of backing material is also an important factor; Mylar and Formvar films have been proved to be acceptable, clean, and economic materials. One can instantly conclude the method of choice that would be the evaporation of large volumes for thin sample preparation. Because of the high salt content possibly present in fresh waters and especially in sea water, evaporation can be done only for rain and some low salinity waters. An alternative way of evaporation can be designed when the dried residue is taken up into a small volume acidic solution. The possible contamination during a long evaporation process will impede this procedure. Freeze-drying with faster evaporation avoids the loss of volatile elements, and careful correction for matrix effects can be achieved by internal standards or by means of Compton scattered lines [300]. However, freeze-drying as a simple preconcentration technique is recommended for waters with low matrix content.

The development of micro-XRF methods, based on the efficient beam focusing, opened the door to a novel approach [301]. Microfocus X-ray tubes or conventional diffraction tubes with polycapillary optics can be used as excitation sources for the analysis of small-area sample; see: Fig. 7.116. Sample preparation from water should meet the geometry requirement for the excitation; a special hydrophobic foil is used here to obtain a small spot of dried residue, which is then excited with the focused intensive beam. According to the standard procedure 50 μl of liquid is dropped on the film (2-μm-thick

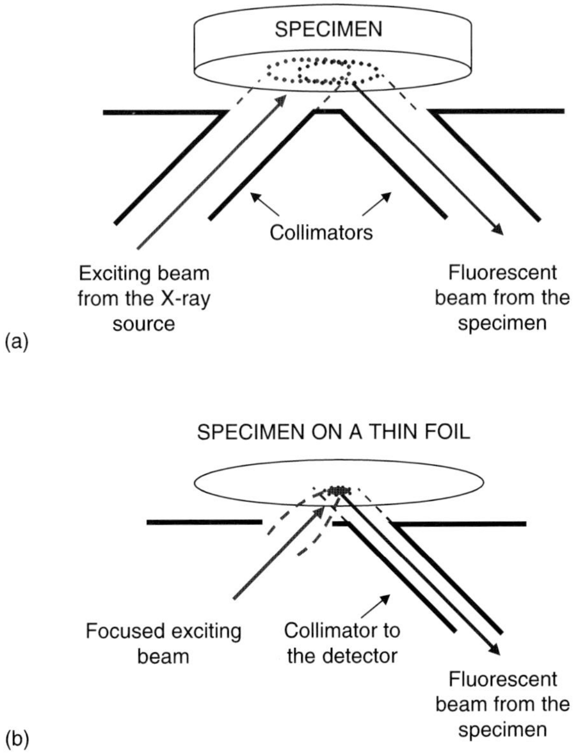


**Fig. 7.116.** Excitation-detection geometry of XRF. (**a**) The conventional ED-XRF spectrometers are equipped with collimators of typical diameter of few millimeters. (**b**) A new approach of excitation delivers focused beam on the specimen surface. In this case a polycapillary or a microfocus X-ray tube produces a small-spot beam, typically with a diameter less than 100 µm. There is a demand for specimen presentation in the case of focused-beam excitation in order to exploit the X-ray flux available in a restricted area compared with the conventional case

AP1$^{TM}$ film), allowed to evaporate resulting in a small solid residue of less than 2 mm in diameter [302]. Instrumental DLs were obtained at 2–3 ppb in environmental water sample. By using a thinner sample support (0.15-µm film) it was possible to reach DL below ppb concentration [303].

When membrane filters are used instead of thin foils, the preparation method may be much simpler. Chromium, arsenic, and lead were determined by EDXRF after adsorbing the species on hydrous ferric oxide [304]. In practice Whatman filters were first impregnated with the hydrous ferric oxide and then were placed in 100 ml water and shaken for 24 h. The filter collects the analytes from the original water into the impregnated middle part. Although not extremely thin, this method can also be considered to be a variant of "thin-layer" technique, since there is no need for absorption correction. The DLs are attainable at the low-ppb level.

Recently, there were relatively few papers in which new approaches to XRF were explored. Analysts prefer techniques that are simple, consume fewer reagents and a simple operation to minimize contamination or loss. On site preconcentration or sample preparation may offer the advantage to prevent any change in composition during transportation of the sample to the laboratory. Ion exchange membranes were used for river water sampling [305]. Portable XRF spectrometers are novel techniques enabling in site measurement which further simplifies the analysis. A filter for liquids was designed for selective extraction of heavy metals [306] combined with a portable spectrometer. Using the filtration process for Pb in tap water a DL of 8 ppb Pb for a 500-ml sample was obtained. An extraction efficiency of 85–95% for a 4 ml/min flow rate was observed.

XRF was applied to a slightly different problem in water purification. Contaminated waters are usually treated by activated carbon to remove harmful substances. Gharaibeh et al. [307] examined the utility of the solid residue of olive mill product in removing heavy metals from waters searching for a cheap alternative. They recommended the use of the olive by-product on the basis of the experimentally determined absorption curves. Transition metals such as Cr, Ni, Pb, Cd, and Zn were absorbed on olive residue from solutions in batch experiments. The concentrations of these metals in the olive absorbent were determined at the m/m% level.

## TXRF Applications

Total-reflection X-ray fluorescence analysis may be considered as a variant of XRF. The basic components are the same for TXRF and XRF: an X-ray source, an energy dispersive detector and signal processing electronics. In the case of TXRF, however, the exciting beam hits the sample at a very small incident angle so that the beam is totally reflected. The sample is presented for analysis on optically flat surfaces (e.g., quartz, Si-wafer, plexiglass). It can be shown that X-rays totally reflected at a plane surface penetrate only the upper atomic layers of the medium, the penetration depth is about a few nanometers below the critical angle. Only these non-totally reflected photons can interact and be scattered. If the sample is a thin material layer on the flat carrier surface, then the background originated from the scattering of the exciting beam will be extremely small compared to the conventional XRF. The second main point that contributes to the power of TXRF is the excitation by not only the incident beam, but the totally reflected one as well. The third important effect that characterizes TXRF against XRF is the formation of standing waves above the flat carrier surface, resulting in stronger field intensity and excitation. As a summary of the effects outlined above, the background is decreased while the fluorescent intensity is increased in TXRF mode as compared to conventional XRF. It is worthwhile to mention a technical advantage of the TXRF measurement geometry: the detector can be placed near to the specimen resulting in a large solid angle and thereby increasing the geometrical efficiency of the fluorescence detection.

The TXRF method is a micro-technique in contrast to the conventional-XRF, since only a very small amount of sample is needed for analysis (typically 1 ng–1 μg of water residue). Itas a lot in common with ICP-AES rather than with the conventional-XRF. The range of elements from $Z = 11$ to $Z = 92$ can be analyzed and it can be further extended toward lower $Z$ with some instrumental modifications [308]. The TXRF method is a multielement technique like ICP-AES, but the fluorescent signals are collected simultaneously in the case of TXRF. Sample preparation is the most time-consuming step in the analysis, spectrum collection, and evaluation is straightforward and rapid. Fortunately, well-established sample preparation methods can be directly adapted from AAS or ICP-AES. Another similarity with the ICP-AES is the accessible DL.

The instrumentation for the TXRF spectrometer is described in detail by Klockenkämper [309] [see Fig. 7.117]. Table top laboratory instruments are available on the market, having the basic design very similar for all systems. The X-ray source is usually a fine-focus diffraction tube, with molybdenum anode for a wide energy range of excitation. The line-focus window is used for TXRF because a strip-like beam is needed for excitation. For the Mo-K$\alpha$ line (17.4 keV) total reflection condition can be set at a small angle of incidence ($\alpha_{\text{crit}}$=0.1° for a quartz reflector as specimen holder). Almost all TXRF spectrometers are equipped with energy dispersive detectors, dominantly with Si(Li) demanding a regular liquid nitrogen supply, but the number of Peltier-cooled detectors is increasing. The whole TXRF instrument consists only of a few parts, this simplicity is an advantage in operation and maintenance.

X-rays from the tube are passed through a thin Mo-foil before reaching the first reflector working as a low-pass filter. The first reflector allows the operation of the X-ray tube in its most effective mode, setting at ~50 kV high voltage. The high energy bremsstrahlung will then be cut by the first reflector.

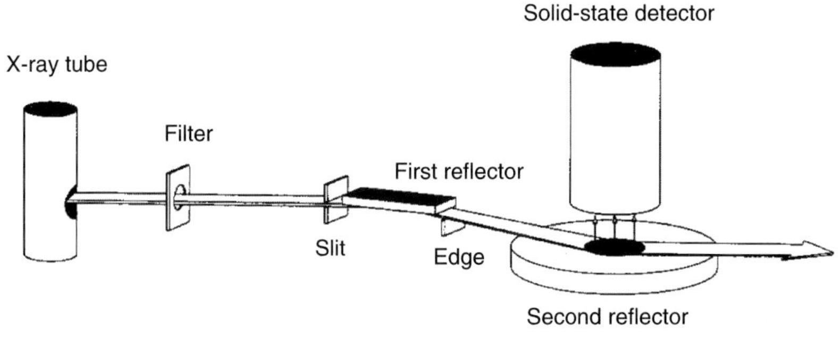


**Fig. 7.117.** The geometrical arrangement of TXRF instrumentation illustrates the simplicity of this method. Reproduced from: Klockenkämper [309] by permission of John Wiley & Sons

Finally a small slit or edge is placed in front of the sample reflector to define the beam impinging on the specimen by blocking scattered rays. Although the beam-adapting unit is simple, the alignment requires some experience. For trace analysis of water, only one angle position is used slightly below the critical angle (for Mo-K$\alpha$ 0.07° is preferred).

An important analytical characteristic is the instrumental detection limit (DL) which is $Z$-dependent and is around the low ppb level (or some tens of picograms in absolute amount). The corresponding absolute DL is 1–20 pg. The DL can be improved by decreasing the background or increasing the fluorescent intensity. Whereas variation in instrumental parameters has relatively little influence in general, the main effort has been extended to chemical preconcentration.

There are, however, certain cases where the instrumental settings may have a beneficial effect. By modifying the excitation spectrum with monochromators and collimators, the spectrometer can be tuned to a specific problem. Considering an intermediate-$Z$ specimen, the lowest detection limit can be obtained by means of a simple cut-off reflector. In this case, the background is not intensive because of the low scatter, and it can be further decreased by choosing an appropriate distance between the specimen and the cut-off reflector. The fluorescent signal is stronger than in the case of monochromatic excitation from single crystal monochromators, double reflectors, or multilayers. Therefore, the simple cut-off reflector is preferred in normal laboratory experiments. The reflector is usually aligned by manually setting the proper position of the baseplate for the total reflection. When high-scattering low-$Z$ matrix specimen is aimed, multilayers may be advantageous by decreasing the background and hence improving the DL. In this respect salt separation offers an alternative in water analysis.

Quantitative analysis of TXRF data is accomplished by standard addition, internal standardization, or by means of theoretical calculation using physical and geometrical data (this is the so-called fundamental parameter calculation). When the fundamental parameter method is used, an exact knowledge of the specimen mass, the excitation spectrum is required as well as the excitation and the detection solid angle. A comparison of the fundamental parameter calculation with the internal standard (empirical) calibration showed an excellent agreement [310].

Standard addition is well known from other analytical methods, it is often used in AAS. A known amount of analyte is added into the unknown sample so that the added concentration should be approximately the same concentration as the unknown [311]. The complete analysis includes a series of measurements of double and triple-added amounts and then regression fit on the intensity–concentration graph. Internal standard is normally added into the water or other solution to be analyzed in the first step. Recovery of the applied procedure can be taken into account in this way. The choice of a suitable standard is first determined by the composition of the unknown sample: element which is not expected in the unknown may only be used (e.g.,

Ga, Y, Co). The liquid sample (5–50 μl) is then pipetted onto the pre-cleaned quartz reflector and dried. The drying process is performed in a clean environment, taking only a few minutes to avoid contamination. Applying a hot plate or an infrared lamp can reduce the drying time. The residue formed has a size of 2–6 mm depending on the droplet volume and content. Furthermore, the horizontal dimensions can be reduced with surface preconditioning of the sample reflector with silicon oil to make it hydrophobic. The vertical extension of the residue is normally a few hundred nanometers, again depending largely on the actual composition, volume, and drying procedure. Uniform distribution of the internal standard along with the analyte, as well as the minimum residue height (~50 nm) is the two main prerequisites of the accurate analysis [312]. For calibration, a multi-element standard solution is prepared taking into account a limited number of elements in order to avoid line overlapping. It is recommended to use two standard solutions each containing five to seven elements, but one common element should be present (normally the internal standard). Spectra are acquired from the calibration samples and calculation is based on the relative sensitivities with respect to the internal standard and is determined from the calibration spectrum by the formula:

$$S_x = \frac{N_x/c_x}{N_{\text{std}}/c_{\text{std}}}, \tag{7.33}$$

where $S$ is the relative sensitivity of the elements ($x$) with respect to the internal standard (std). $N$ and $c$ stand for the net peak area and the concentration of the given element. This elegant method is widely used in TXRF quantification because it is reliable for all types of samples and matrices, and a single measurement is sufficient for the multielement calibration [313]. The calibration curve is a plot from the individual sensitivity values versus atomic number. A polynomial fit of the experimental sensitivities makes possible the determination of those sensitivity values of elements, which were not present in the standard sample. Elements to be analyzed can be divided into two groups; one for those that can be detected from the K$\alpha$ line and the others detected from the L$\alpha$ line. Sensitivity values of L$\alpha$ lines are about one order smaller than those values of K$\alpha$ lines, therefore two separate calibration curves are established. Once performed, the calibration remains valid for several weeks and only a routine check is recommended before measurement. The sample to be analyzed is spiked with the internal standard as mentioned before, and measured with the same instrumental settings as the calibration. The unknown concentrations are calculated by:

$$c_x = \frac{N_x/S_x}{N_{\text{std}}} c_{\text{std}}, \tag{7.34}$$

where $N$ is the net peak area of the analyte ($x$) and the internal standard (std); $S$ the relative sensitivity; $c$ the concentration.

Water samples are ideal for TXRF because the dried residue can be easily prepared from droplets using micropipettes. In a pioneering work [314], it has been shown that the instrumental DL is ~1 ppb, rendering TXRF suitable for direct water analysis in many cases. The authors have already discussed the effect from inorganic and organic matrix and recommended 0.5% as an upper limit for salt. If the matrix content is higher than this value, the water should be separated/preconcentrated in order to keep the DL appropriate for the given sample. It is important to emphasize that TXRF is suitable for direct determination of transition elements such as Cr, Fe, Ni, Cu, As in dilute solutions, like drinking water [315]. A study of drinking water has shown how the elemental content of water depends on geomorphology and local contamination [316]. The authors presented DL without preconcentration, maximum allowed values and recommended values according to the Portuguese law. The direct DL values were in general much lower than the limiting values legally established.

Generally, all the separation methods can be applied for TXRF sample preparation which has been discussed in the XRF section. Finally, the sample may be either in liquid or solid form, though liquid is preferred for the sake of droplet preparation on the sample reflector. Liquid–liquid extraction or solid phase extraction methods are often considered, but there are other unique methods, such as electrodeposition.

Regarding the liquid extraction methods, NaDDTC or APDC complexing agents are used for trace metal coordination as described in the previous section, see also Fig. 7.118. The trace metals complexed will then be extracted from the original water using a few ml of MBIK (organic solvent). The solvent of trace metals is either used for direct droplet preparation or back-extracted into Milli-Q water. By dropping 30 μl organic extractum on the reflector in 2-μl steps, Holynska et al. [317] obtained DLs of 0.04, 0.06, and 0.06 $\mu g\,l^{-1}$ for Cu, Hg, and Pb respectively. Furthermore, it is possible to filter the trace metals in complexed form through a Nuclepore filter and subsequently analyze a portion of the deposit [318]. With this method the DLs were claimed to be 0.6, 0.5, 0.4, 1, and 0.8 $ng\,ml^{-1}$ for Fe, Ni, Cu, Zn, and Pb. A powerful preconcentration based on reverse-phase separation of the dithiocarbamate–metal complexes is described by Prange et al. [319]. The original water was spiked with internal standard (Se) and buffered to pH 4.5–5. Then 1 ml of a methanolic 4% (w/v) solution of NaDDTC was added and shaken to mix. After the reverse-phase column had been pretreated with a few milliliters of methanol and ultrapure water, the reaction mixture was sucked through the column within 10–15 min, the carbamate complexes being adsorbed on the column (packing: Chromosorb AW-DMCS). After the column was sucked dry, the complexes were eluted with 3–4 ml of purified chloroform/methanol (1:1 v/v). For 200 ml water sample and 1000 s measurement, DLs of 5–20 ppt were achieved for the first row transition metals.

The reverse-phase method was applied for the analysis of Mo, U, and V in Baltic Sea waters [320]. DLs were reported as 14.3 ppt for U, 17.3 ppt for V, and 9.6 ppt for Mo. Since, for example, Mo average ocean concentration is 10.5 ppb, the method was sufficiently powerful for this task. The

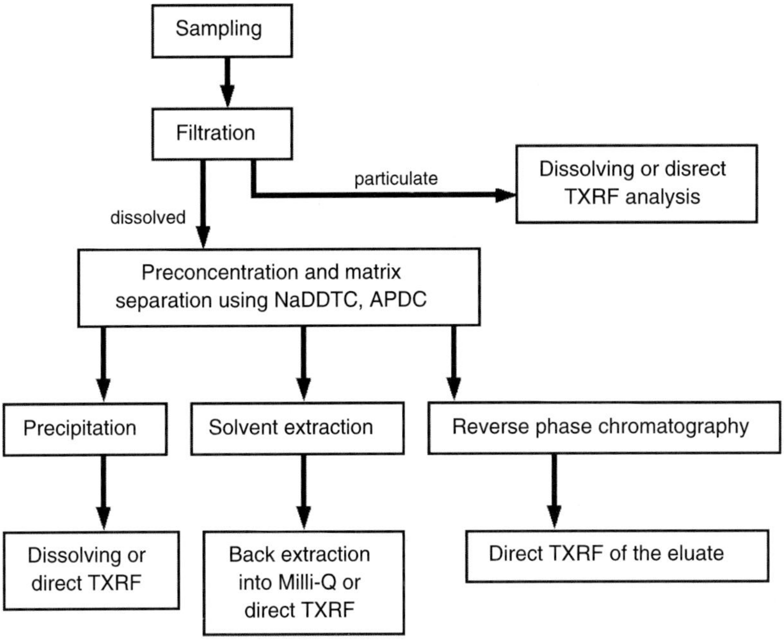


**Fig. 7.118.** Overview of the preparation methods using NaDDTC or APDC

concentrations of Mo and U were found to be controlled through conservative mixing with North Atlantic waters. Scavenging seemed to be the main determining factor for V distribution; the Baltic waters were depleted in dissolved V by more than 60% relative to the Atlantic waters. The total average concentration was obtained at 1.4 ppb, and there were no systematic changes in the concentration profile of V with depth. Schmidt and his colleagues [321] improved the reverse-phase separation method. They designed a more compact system containing the sample reservoir vessel, the chromatographic column, and the eluant tube. Trace metals were enriched in the eluate by a concentration factor of 45. Seawater was measured after the preconcentration, and depth concentration profiles were obtained for the eastern part of the Atlantic Ocean.

Introduction of commercially made on-line SPE applications eliminated the need for laborious preparation of laboratory-made columns, and speeded up the tedious handling procedures in environmental laboratories. Most analysts still prefer laboratory made columns and packings, possibly as a result of greater selection of materials, hence tailoring the method to special needs. The analyte flow in TXRF is inherently discontinuous, therefore the on-line method cannot be exploited to its full extent. Preliminary investigations showed that the advantageous points are the low sample volume required and a high sample throughput [322]. An SPE method always consists of three to four successive steps, which is to be considered in optimization. First, the solid sorbent should be conditioned using an appropriate solvent, with the aim of wetting the sorbent, solvation of the functional groups, and removing the impurities. The second step is the percolation of the sample through the solid sorbent (loading). The flow rate should be set to a sufficiently low value so that the metal ions are efficiently adsorbed. The third step (which is optional) may be the washing of the sorbent to eliminate matrix components. The last step is the elution of trace metals, usually with a strong solvent (acid, base or organic). The key parameters are the solvent flow rate and total volume to obtain a quantitative recovery and high preconcentration factor. The reverse-phase extraction method discussed above was improved continuously. Recently, an automated SPE extraction version has been reported [323]. The authors reached good precision and accuracy with the replicate analysis of certified reference waters, such as CASS-2 and NASS-4. It was, however, impossible to decrease the total analysis time compared with manual operation.

A promising preconcentration method for seawater has been described [324] basing on electrodeposition of the metal ions from saline solution. Glassy carbon was chosen as sample carrier, which fulfils the requirements for both TXRF and electrochemistry. Duration of electrodeposition was set to 30 min at a potential of 1800 mV and the pH of the solution was adjusted to 4.3. Applying Mo-K$\alpha$ excitation (or W continuum) and 500 s counting, typical DL of 5–20 ng/l were obtained. From the experimental results on some artificial saline waters, the authors found little salinity dependence of the deposition yield. This study clearly shows the utility of the electrodeposition in seawater analysis.

The low volume of sample needed for TXRF analysis may be important when the availability is limited. In the case of precious or biological samples this situation is quite common. When considering natural water analysis, the liquid volume is normally not restricted. The small water volume, however, makes the filtration and preparation fast and could prove efficient in reducing contamination too. Lofthouse et al. [325] examined the analysis accuracy for microvolumes of reference water samples. They compared TXRF with ICP-MS equipped with a microconcentric nebulizer. ICP-MS offered better detection limits for a wider range of elements but at least 90 µl of sample was required and digestion of the sample was essential for organic solutions. TXRF provided much simpler and faster preparation for these samples, but the instrumental analysis took usually longer than with the ICP-MS.

Trace metals can exist in a number of physicochemical forms or species in natural waters. These include free aquatic ionic forms, dissolved inorganic, or organic complexes, complexes with colloidal particulate matter (inorganic or organic), and complexes associated with the biota. The speciation can have a major influence on the transport, behavior, bioavailability, and toxicity of a trace metal. The free ionic form of a number of trace elements (e.g. Cd, Cu, Pb, Zn) has been found to be considerably more toxic to aquatic biota than complexed, colloidal, or particulate forms. However, for elements such as Hg and Sn, the methylated forms are more toxic than the ionic forms.

TXRF is not able to identify different molecular species or moieties. Hence, speciation analysis can be performed after an appropriate chemical preparation and separation of such moieties. Because of this necessary additional step(s), the application of TXRF is restricted to the field of speciation. A successful work has been presented [326] determining metals and their species in aquatic humic substances. Humic substances (HS) were preconcentrated from water samples, both by the XAD 8 procedure and by preparative ultrafiltration. The HS sample was subsequently filtered through a multistage ultrafiltration system, off-line, and the fractions were measured by TXRF. Several studies have related bioavailability or toxicity of trace metals to fractions retained on different ion exchange resins. For example Figura and McDuffie [327] determined lability of metals using voltammetric and ion-exchange (Chelex-100) techniques and operationally defined metal speciation into very labile, moderately labile, slowly labile, and inert species, see also [328]. As we discussed earlier, the feasibility of on-line preconcentration in TXRF is limited by the inherent discontinous nature of droplet preparation on reflectors. Therefore, TXRF speciation would be possible to simply divide and collect different chromatographic fractions and then being presented for elemental analysis.

Synchrotron sources are unique facilities, offering high brilliance X-rays for highlighted research. Further advantages are the tunability of the excitation energy and linear polarization. With the increasing number of synchrotron storage rings in operation, more beamtime would be expected for TXRF analysis. The routine profile of water research is demanding and requires an easily accessible beamline and clean environment appropriate to the ultratrace analysis. Generally, it is difficult to find a beamline which would fulfill requirements for routine water analysis. Wobrauschek and co-workers [329–331] systematically studied the excitation conditions and sample geometries for synchrotron radiation TXRF analysis. Streli [332] has reviewed the field of low energy TXRF, which has a direct relevance to synchrotron sources as a consequence of the strong flux in the low energy region being available from common storage rings of low critical energy. This can "compensate" for the loss of weak fluorescence yield for low $Z$ elements, providing DL in the low pg level. Nevertheless, synchrotron radiation TXRF found most of its application in semiconductor quality control (see Chap. 7.3, TXRF wafer analysis by Streli et al.).

Recently, a new instrumental improvement in wavelength-dispersive detection appeared and made possible WD-TXRF measurement. Earlier it was thought that the poor detection efficiency of the WD systems would not allow a DL acceptable for trace element analysis. With the advent of extremely brilliant beams from third generation synchrotron sources, and with the improvement in detection efficiency of the WD systems, compatible DLs were obtained. The technical ways to downsize a Johansson type WD spectrometer were discussed [333], and the authors presented the first results for dilute standard solutions and serum sample [334]. Since the spectrometer uses a focusing optics for X-ray fluorescence detection, it is important to prepare a very small specimen. Using this Johansson spectrometer, a sample preparation method was developed [335] enabling a small spot of dried residue obtained on a Si-wafer. There is another novel detection device, the so-called superconducting tunnel junction (STJ) detector, which offers high counting rate capability and excellent energy resolution of 10–20 eV [336, 337]. These detectors may contribute significantly in up-grading TXRF analysis at synchrotron sources.

#### Conclusion

For many applications XRF analysis is considered to be inaccurate, and to have poor detection limits. This opinion has changed fundamentally with the successful instrumental improvements and methodological developments in sample preparation and quantification. Thin film techniques provide a simple method for conventional XRF specimen preparation, giving a straightforward and reliable quantification of the spectra as well. Especially, micro-droplet preparation proved to be successful in combination with a focused beam of excitation. After a boom in TXRF research and application in the early 1990s, the speed rapidly dropped mainly as a result of the strong competitive techniques (ICP-AES, ICP-MS). The overall trends in water research shifted toward the understanding of organic contamination. Research on trace metals focused on speciation with the development of improved analytical methods. TXRF seems to be losing the competition, but it is absolutely not inevitable. Many of the natural water analysis tasks can be performed by direct TXRF virtually without any preconcentration. In the case of higher dissolved content in water samples, the matrix does not change fundamentally the spectra, only raises the DL. The thrust of instrumental analytical methods has been shifting toward speciation and other unique problems such as the low-level measurement of lanthanides and actinides. TXRF can meet these demands with the combination of appropriate sample preparation developments and high-resolution, new detectors. An important advantage of TXRF to be emphasized is the very small amount of liquid needed for specimen preparation, consequently a potential in biology. Low cost of TXRF should be recalled as well as the simple instrumentation and practice.

A growing number of synchrotron applications can be anticipated in the future. These brilliant excitation sources contributed to push down DL and offered complementary absorption measurements (XANES, EXAFS) for speciation. Instrumental development of detectors (efficiency, resolution and counting rate) will surely continue, serving as a rich source of new applications in water chemistry.

### 7.5.3 Atmospheric Aerosol

Particulate pollutants in the atmosphere consist of a mixture of suspended solid or liquid particles originating from different natural and anthropogenic sources. Knowledge of the chemical composition of atmospheric aerosol is required for air quality characterization at a given location. The ultimate goal of aerosol studies is to determine the particle sources and to study the possible physical and chemical processes occurring in the particles during their atmospheric transportation. The most common method to derive information about the sources is quantitative bulk analysis of particles collected sequentially onto filters, combined with factor analysis on the time or location variance of the elemental concentrations [338]. The sources of air pollution have specific chemical fingerprints, source apportionment studies are also based on accurate knowledge on the chemical composition of the aerosol and its time variation. The chemical characterization of atmospheric particulate matter also provides an insight to the processes in atmospheric chemistry.

Earlier air quality standards only regulated the mass of total suspended particulates (TSP), and concentration of some hazardous components without size fractionation. The atmospheric transportation behavior of the particles, however, strongly depends on their morphology and size. Also, the deposition of inhaled dust in the human respiratory tract is strongly dependent on the size and compositional distribution of the particles, since one particle having elevated concentrations of toxic elements can cause DNA change in one pulmonary cell, as a precursor of lung cancer [339]. The size of the particles can range form a few nanometers up to several 100 μm [340], but for source apportionment studies the accumulation (respirable aerosol, 0.1–2.5 μm) and sedimentation type particles (2.5–100 μm) are the subject of study. The long range transport of heavy metal pollutants is mostly determined by particles smaller than 2.5 μm ($PM_{2.5}$).

In the past decade, however, limiting values were also reported for $PM_{10}$, $PM_{2.5}$ (particulates of maximum 10 or 2.5 μm aerodynamic diameter). For Europe council directive 1999/30/EC set limiting values for $PM_{10}$ and lead in air [341]. For this reason, bulk analysis is mostly carried out on size-fractionated samples collected by stacked filter units or cascade impactors [342].

Several analytical methods are in use for chemical characterization of aerosols. As far as the elemental composition is concerned, XRF can provide partial analysis with a medium sensitivity. Its advantages are that it does not require difficult sample preparation, it is low cost and it is non-destructive. XRF spectroscopy is a useful analytical tool for determining the sample composition down to trace concentration levels. The method is appreciated for its sufficient sensitivity, high dynamic range, non-destructiveness, and because of its relatively simple relation to the fundamental physics of atom – radiation interaction [343]. In order to minimize the matrix effects in XRF, atmospheric particulate samples are collected on membrane filters or thin foils in the impactors. Depending on the sampling time sampled particles form a thin layer on the surface of the substrate materials that enables simple quantitation since the response function will be in the linear range. The substrate should be formed of low-$Z$ elements and should be free from contaminating elements. Teflon membrane filters are ideal for XRF because they are capable of higher particle load than polycarbonate membrane filters and the presence of F is not interfering with the analysis. Since the filter samples can be regarded as thin samples, therefore a simple procedure for quantitative analysis can be performed in order to calculate the elemental concentrations.

The sampling strategy should be optimized to the requirements of the study. For air quality monitoring, the limiting or target values are in the range of $\mu g\,m^{-3}$ or for some very toxic components in $ng\,m^{-3}$. The minimum detection limits of XRF, depending on the measurement setup and the analyzed element, are between 2 and $100\,ng\,cm^{-2}$. Nowadays, contemporary table top spectrometers have sufficiently low MDL to detect heavy metals in aerosol samples down to $ng\,cm^{-2}$ [344], which means that with $10\,m^3$ volume sampled sub $ng\,m^{-3}$ can be reached, that is far below the limiting value for most elements and close to the ambient concentrations. The area of filter covered by aerosol particles, the used flow rate and the duration of sampling should be optimized in order to obtain sufficient material on the sample for analysis. The area covered by particles should not be much larger than the sample surface illuminated by the exciting beam. A homogeneous distribution of particles on the filter surface is critical for accurate elemental analysis. This condition is mostly fulfilled, however, there might be a radial concentration gradient decreasing from the center. In most of the cases the best compromise for the sampling strategy lead to a reasonable time resolution of one day. The source apportionment studies based on XRF analysis are mostly based on daily averages of elemental concentration on the sites of interest. A standard filtered tube excitation or secondary target EDXRF is capable of providing partial analysis of $23 < Z < 90$, so habitually complementary analyses (mostly by ion chromatography) for sulfate, nitrate, ammonium, and organic pollutants (OC, BC) has to be carried out. Some of these analyses can be performed on the same filters after XRF analysis due to the non-destructiveness of the analysis.

## Energy-Dispersive X-Ray Fluorescence Analysis of Aerosols

Samples collected on membrane filters can be regarded as thin samples, because X-ray attenuation by the filter material can be neglected, since the particles are deposited on the surface of the filter. However, the X-ray attenuation within the single particles causes the so-called particle size effect, that is most pronounced for light elements ($Z < 20$) present in the coarse size fraction. The magnitude of the particle size effect depends on the size distribution of the particles deposited on the filter and the energies of the exciting and especially the emitting characteristic X-rays. Correction factors for aerosol particles of various sizes and compositions have been published [345]. For conventional EDXRF of $PM_{10}$, it can be estimated that the correction remains limited to at most 25% for Cl and maximum 15% for K and Ca, therefore reasonable quantitative results can be obtained when the particle size effect is neglected [346]. The sensitivity curve for the analyzed elements can be obtained by measurement of a series of single-element thin film standards (e.g., Micromatter) in identical conditions as the unknown aerosol samples. Multielemental standard reference materials (SRMs) prepared as aerosol particulate matter deposited on polycarbonate filters can also be utilized. If the composition and the size distribution of the particles in standards are similar to those of the unknown aerosol samples, the particle size effect can be cancelled out. Different institutes distribute such SRMs, but the homogeneity of the deposit is very critical for utilizing the certified concentration values for different elements. Due to the extreme homogeneity required, the loading of the filters by airborne particulate matter is simulated using filtering of a suspension of homogenized material through the filters [347].

Using a standard diffraction tube with secondary targets, or direct excitation with a low-power X-ray tube, reasonable detection limits allowing a time resolution of 24 h can be reached for a reasonable counting time (5 s to 10000 s). As an example, TSP samples collected at Siófok (Lake Balaton, Hungary) in four campaigns in the time period of 1999–2000 were analyzed by means of EDXRF using a lower power Rh-anode X-ray tube (17.5 W) with direct excitation of the filter samples under vacuum conditions [348]. The average and maximum concentrations of trace elements are shown in Table 7.16, in comparison with the limiting values according to the Hungarian Standard. The obtained concentrations are similar to those found in other

**Table 7.16.** Elemental concentrations of TSP samples compared to the limiting values in Hungary

| | | Concentration ($\mu g\,m^{-3}$) | |
|---|---|---|---|
| Element | Average | Maximum | Limiting value |
| Al | 0.624 | 1.75 | 30 |
| Si | 1.65 | 4.96 | 23.3 |
| P | 0.0738 | 0.149 | 21.8 |
| S | 2.15 | 5.17 | 20 |
| Cl | 0.0736 | 0.27 | 30 |
| K | 0.419 | 1.12 | 22.6 |
| Ca | 1.2 | 3.38 | 21.4 |
| Ti | 0.042 | 0.124 | n.a. |
| V | 0.00179 | 0.00724 | 2 |
| Cr | 0.00442 | 0.0128 | 1.5 |
| Mn | 0.0113 | 0.0455 | 1 |
| Fe | 0.597 | 2.14 | 200 |
| Ni | 0.0023 | 0.00497 | 1 |
| Cu | 0.01 | 0.0436 | 2 |
| Zn | 0.0384 | 0.0886 | 50 |
| Br | 0.012 | 0.0193 | 10 |
| Rb | 0.00271 | 0.00589 | n.a. |
| Sr | 0.0058 | 0.0128 | n.a. |
| Pb | 0.0371 | 0.106 | 0.3 |

areas of Hungary [349], and are in accordance with results obtained by wet chemical methods for the time period of 1995—1998 [350]. The average concentrations are far below the limiting values, and lead is the only element where the maximum concentration is close to the limiting value.

There are no large industrial point sources of air pollution near the lake. The nearest power plant, which burns coal, is about 30 km to the northwest. Local sources include motor vehicle emissions and dust suspended by vehicles on paved and unpaved roads or as the result of construction or by wind stress. The time variation of the concentrations did not show characteristic seasonal variations but it was in correlation with the incoming air mass.

A temporal resolution of as low as 1 h can be reached using a combination of streaker sampler and EDXRF using capillary optics [346]. The single-stage streaker device draws the air through an $8 \times 1\,\text{mm}^2$ rectangular orifice through a polycarbonate membrane filter with an average flow rate of $0.16\,\text{L}\,\text{min}^{-1}$. The filter is mounted to a circular frame rotating at a speed of 45°/day, resulting in an 8-mm wide circular deposit of aerosol particles. The pre-impactor with an aerodynamic cut-off diameter of 10 μm restricted the aerosol collection to $PM_{10}$. The radiation of an X-ray tube was concentrated to a beam diameter of around 1 mm using a glass capillary. A Mo anode X-ray tube was used for measurement of high atomic number elements ($19 < Z$), while light elements were determined using a Cr anode tube, in a He purged sample chamber. The analysis was performed using 1-mm steps corresponding to a 1 h of aerosol exposure. Using a 100 s measurement time at each spot, a scan of a complete sample collected during 1 week took about 6 h. The measurement geometry and the filter exposed in the streaker sampler are shown in Fig. 7.119.

The detection limits expressed in $\mu\text{g}\,\text{m}^{-3}$ compared to that of particle-induced X-ray emission (PIXE) using 2.4 MeV proton excitation, calculated

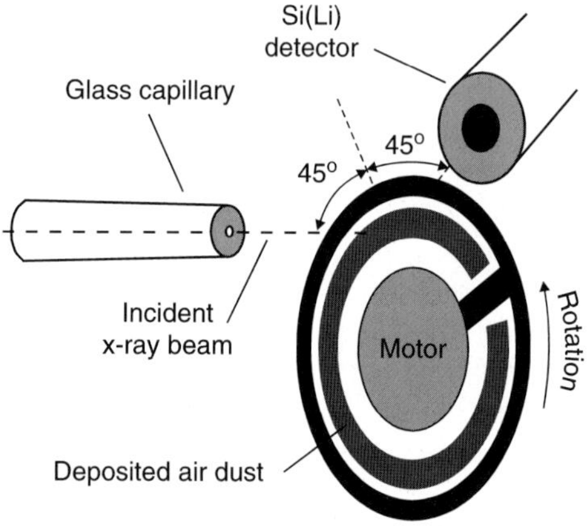


**Fig. 7.119.** Geometrical arrangement for the X-ray fluorescence analysis of a filter sample taken with the streaker collection device. Reproduced from Holynska et al. [346] by permission of Elsevier Science B.V.

**Table 7.17.** Interference-free detection limits (DLs) in capillary optics EDXRF and PIXE analysis of aerosol streaker samples collected on Nuclepore polycarbonate filters

| | DL (µg m$^{-3}$) | | Range of concentrations |
|---|---|---|---|
| Element | EDXRF | PIXE | in urban areas (µg m$^{-3}$) |
| S | 0.8[a] | 0.23 | 2–15 |
| Cl | 0.5[a] | 0.2 | 0.1–10 |
| K | 0.2[a] | 0.14 | 0.2–4 |
| Ca | 0.1[a] | 0.11 | 1–20 |
| Fe | 0.08[b] | 0.014 | 0.5–20 |
| Cu | 0.03[b] | 0.008 | 0.02–5 |
| Zn | 0.03[b] | 0.008 | 0.02–10 |
| Br | 0.04[b] | 0.014 | 0.005–1 |
| Pb | 0.07[b] | 0.028 | 0.1–10 |

Comparison with typical ranges of particulate element concentrations in urban areas [345].
[a]Measurement with Cr X-ray tube.
[b]Measurement with Mo X-ray tube.

for a 1-mm$^2$ beam size and 100 s measuring time, are shown in Table 7.17. The detection limits are in the same range of that of PIXE, and are in the lower end of the expected urban aerosol concentration ranges for most of the elements studied.

### Total Reflection X-Ray Fluorescence Analysis of Aerosols

The sensitivity of TXRF is close to that of ICP-AES, therefore it is commonly applied to the analysis of liquid samples. However, the TXRF technique has been recently successfully applied for elemental analysis of atmospheric aerosol. One of the most common possibilities of aerosol analysis using TXRF is utilizing digested samples and the analytical procedure for liquid samples. If particles are collected on non-reflecting backing like filters, only this method is applicable [351–353]. Schmeling [354] reported on a study concerning the seasonal variation of aerosol composition and its correlation with the wind directions and air mass backward trajectories for Chicago. Sampling volumes of 2 m$^3$ provided sufficient material for analysis that enabled a 2-h time resolution using a moderate, 16.7 Lpm flow rate. Detection limits from 1 to 8 ng g$^{-1}$ could be achieved for elements Pb–K. Seasonal variations in elemental concentrations could be obtained by TXRF, that were found to be most pronounced in summer, where concentrations of almost all elements detected were lower. Influences of anthropogenic pollution could be found in all samples. Sulfur, however, was enriched in summer due to photochemical production and removal of large particles containing mostly soil dust elements by deposition into Lake Michigan (east wind). However, the pre-treatment may change the elemental composition of the sample, increase the content of some elements by contamination and decrease the content of some elements by losses. Another option for using TXRF in aerosol analysis is when particles are directly deposited on reflector surfaces. In this case the only possibility for air sampling is the impaction method, where reflector surfaces suitable for TXRF analysis are directly placed as impaction plates into the device resulting in samples that can be analyzed directly after sampling. Schneider [355] first used sample carriers for TXRF analysis as impaction plates in a cascade impactor. Such direct measurements need very critical settings of the device to ensure that after sample changing the same total reflection conditions are fulfilled and the physical position of all the samples is the same as that of the standards. Usually the addition of an internal standard is needed for quantification (to calibrate for the illuminated area). This procedure has to be performed carefully in order to cover the impaction spot with the dried deposit of the internal standard solution. This step requires a skilled analyst. For this reason impactors having circular orifices are recommended for sampling, providing an aerosol deposit of larger than 3 mm$^2$. Using direct analysis, the risks of contamination and the loss of material to be analyzed are reduced. Different reflector materials were investigated as impaction plates. Injuk and Van Grieken [356] investigated the detection limits for quartz glass, siliconized quartz glass, and plexiglass reflectors, where the siliconized quartz glass reflectors were prepared by dispersing a drop of silicon solution on the quartz glass surfaces. Their results indicated that the use of quartz glass carriers provided the best detection limits. Among different reflector materials, the use of polished silicon substrates have been evaluated by Esaka et al. [357]. They pointed out that an important factor determining the detection limits achievable by the reflectors is their surface roughness. The detection limits of elements for the polished Si carriers in TXRF analysis were superior to those for glassy carbon (two orders of magnitude) and even for quartz glass carriers (three times, see Fig. 7.120). Molybdenum anode X-ray tube operating at 40 kV and 40 mA was used for irradiation of the samples.

Si carriers were used for the elemental analysis of aerosol particles collected at Ibaraki, Japan from January to November 2002. From the variations in the elemental concentrations with time, the origin of the particles was deduced. Four groups of elements could be distinguished, which could be related to crustal, yellow dust, and anthropogenic sources, because of the availability of size-segregated samples.

The Si-wafer substrates were also used as impaction surfaces by Török et al. [358]. For particles smaller than 0.5 µm, 0.02 m$^3$ of air was sufficient for TXRF analysis of low-$Z$ elements at the BESSY synchrotron source. Also it was feasible to investigate the chemical environment of nitrogen using TXRF-NEXAFS from nanoscopic amount of material. This capability can improve the time resolution to below 10 min that is commonly expected from air quality monitoring devices of "classical pollutants" such as $SO_2$, $NO_X$, $PM_{10}$. There is of course the inherent limitation of SR measurement that the sample number that can be analyzed is very restricted.

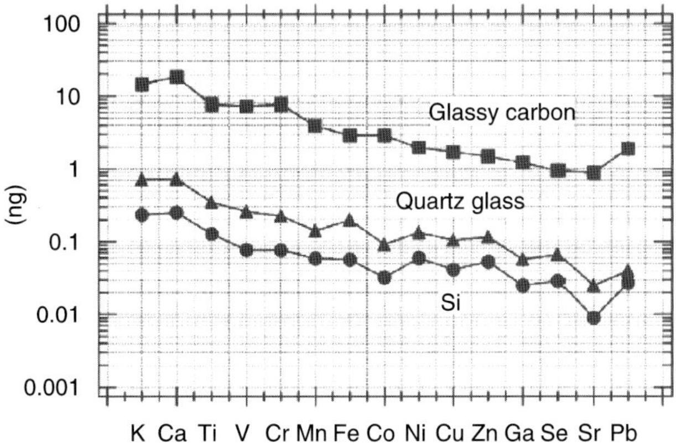


**Fig. 7.120.** The detection limits of elements for the glassy carbon, the quartz glass, and the Si carriers. Reproduced from Esaka et al. [357] by permission of Elsevier Science B.V.

### 7.5.4 Monte Carlo Based Quantitative Methods for Single Particles

Many environmental applications demand for the quantitative characterization of particulate matter. There are two major approaches in the analysis of particulate matter: bulk and single-particle analysis. By using bulk methods, only the average composition of the particulate sample can be obtained. However, these samples are heterogeneous mixtures of different types of particles, so the average composition and average diameter do not describe well the dispersion of the particles in the sample. In order to derive environmentally relevant information on the sources of the environmental particles like aerosols, soils, or sediments – that are the most investigated by XRF – accurate chemical information is necessary including major, minor, and trace elements. The major composition of individual microparticles including low atomic number (low-$Z$) elements can be quantitatively determined using electron-probe X-ray microanalysis (EPMA) [359, 360]. However, on the basis of the major element composition natural and anthropogenic particles cannot be identified unambiguously (consider e.g., soil-derived and fly-ash particles in atmospheric aerosol). Therefore, the accurate knowledge of the minor and trace composition of individual particles is of high importance for source profiling studies, as they offer more distinctive features for the identification of different particle sources. As X-rays originating from conventional diffraction X-ray tubes or synchrotron sources can be focused to form micrometer-sized beams [361], the examination of microparticles became possible using microbeam XRF [362]. Minor elements can also be detected using micro-PIXE in individual particles, but for achieving fully quantified results, it has to be combined with other methods [363].

Using photons or charged particles for excitation, the measured X-ray intensities of elements depend non-linearly on the concentrations, and they are not independent from each other. For this reason, no explicit mathematical expression can be given for calculating concentrations from measured intensities. The problem has to be solved in an iterative way. The conventional quantification methods [364] are not suitable for single-particle analysis, because they are not flexible enough for various experimental and sample conditions, such as the analysis of microparticles having irregular shapes and heterogeneous compositions.

The most widely used analysis technique for single particle analysis is EPMA and the quantitation methods for such application were developed earlier and will be briefly summarized here. The quantification methods in EPMA, such as classical ZAF and $\phi(\rho z)$-based procedures, aim to correct for matrix and geometric effects, observed in particulate matter, that are even the most pronounced for light-element X-rays. The most reliable and widely used quantification method for microparticles is the so-called particle-ZAF algorithm developed by Armstrong and Buseck [365]. However, the particle-ZAF method based on the use of bulk standards introduces large errors for light element analysis (that are major components of environmental samples). This is mostly because of the large absorption correction needed for low-$Z$ element quantification and the difference between the bulk standard and single particle in their behavior under electron bombardment. In addition, when the average atomic number of the substrate significantly differs from that of the particle, the side-scattering correction of the $\phi(\rho z)$ function is reasonable only if the electron excitation volume is smaller than the particle itself. It has been shown that with the use of micrometer-sized X-ray beams, the topological variation can be distinguished from chemical inhomogeneities by looking at the correlations between the intensities from different elements [366]. Thus, by using microbeam XRF in a scanning mode it should be possible to separate the impact of chemical inhomogeneities and sample topology. However, for the quantitative trace element analysis the different matrix and attenuation effects caused by the differences in the size and shape of the particle have to be taken into account accurately. Also, the intensity variation over the X-ray microbeam has to be described well.

It is well known that Monte Carlo simulation is a reliable tool for modeling the matrix and geometry effects in X-ray generation using excitation either using photons or electrons for excitation of the microscopic volume of interest. Quantification can be achieved by iterative adaptation of the simulated elemental concentrations until the deviations between the simulated and experimental peak areas fall within the statistical uncertainties of the recorded fluorescent lines [360, 367].

A significant advantage of the MC simulation-based quantification scheme compared to other methods, such as fundamental parameter (FP) algorithms, is that the simulated spectrum can be compared directly to the experimental data in its entirety, taking into account not only the fluorescence line intensities, but also the scattered background of the XRF spectra. This is coupled with the fact that MC simulations are not limited to first or second order approximations and to ideal geometries.

### Simulation Models in XRF and EPMA

For a typical XRF experiment, the simulation code must consider the three most important interaction types in the X-ray energy range of 1–100 keV, i.e., (i) photoelectric absorption followed by X-ray fluorescence or Auger electron emission, (ii) Rayleigh, and (iii) Compton scattering. The simulation of these interactions within the sample and its environment (e.g., surrounding air) and calculating the probability of subsequent fluorescence or scattered X-ray emission within the solid angle of the modelled detector allow building up the complete spectral response. The exciting radiation can have any given (numerically defined) energy distribution in the above energy range and can either be unpolarized or linearly polarized. The modeled multi-element sample can contain maximum 92 elements from H to U. The detailed description of the code can be found elsewhere [368–370].

Several Monte Carlo programs exist that allow the simulation of electron trajectories in bulk samples and the production and absorption of X-rays. The ANSI standard C code CASINO program, developed by Hovington et al. [371] is a single-scattering Monte Carlo calculation, especially designed for low-energy beam interactions in bulk, thin foil, and the spherical inclusions with homogeneous composition. Tabulated Mott elastic cross sections and experimentally determined stopping powers are employed for description of the electron–matter interactions. The program can simulate several types of signals, such as X-rays and secondary and backscattered electrons, as a point analysis, as a line scan, or as an image format, for accelerating voltages between 0.1 and 30 kV. For modeling the fluorescence intensity emitted by microparticles sitting on a flat substrate, the CASINO Monte Carlo program was specifically modified and extended [360, 372]. The simulation of entire X-ray spectra and Bremsstrahlung background are implemented using the parametrization of Kirkpatrick and Weidmann [373] for the background calculation. This extension makes the comparison of experimental and simulated X-ray spectra possible, including characteristic X-ray lines and Bremsstrahlung background. The reliability of MC simulations is demonstrated through comparison of measured and simulated X-ray spectra of SRMs for micro-XRF (Fig. 7.121) and EPMA (Fig. 7.122). The agreement between the measured and simulated spectra is reasonably good for both cases.

### The Quantification Algorithm

As MC simulation can reliably model characteristic X-ray intensities as well as continuum background for microparticles of a given composition, size, shape and density, it can be used for quantitative analysis. The input data of the quantification procedure are the measured characteristic X-ray intensities of the elements involved in the particle, the sizes, the shape and the average

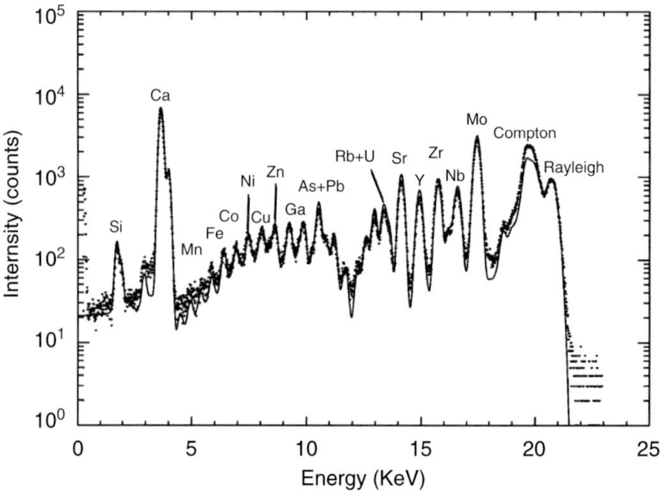


**Fig. 7.121.** Experimental and simulated XRF spectra from an NIST SRM 612 multi-element glass calibration standard having a thickness of 100 μm. The experimental spectra correspond to monochromatic ($E_0 = 20.8\,\text{keV}$) excitation at HASYLAB Beam Line L. *Dotted curves*: experimental data; *solid lines*: simulation. Reproduced from Török et al. [374] by permission of the Royal Society of Chemistry

density ($\rho$) of the particle. The $I_{i,\text{meas}}$ measured characteristic intensity values can be determined by fitting the experimental spectra using non-linear least squares fitting using the AXIL code [375]. Although the zero-approximation and the a priori consideration are different for the two techniques, exactly the same iterative procedure can be used to refine the concentration of each element. The $(k+1)$th approximation for the concentration of the $i$th element $C_i^{(k+1)}$ is described as follows:

$$C_i^{(k+1)} = C_i^{(k)} \cdot \frac{I_{i,\text{meas}}}{I_{i,\text{calc}} \cdot \sum\limits_{j=1}^{n} C_j^{(k)} \cdot \frac{I_{j,\text{meas}}}{I_{j,\text{calc}}}}, \tag{7.35}$$

where $I_{i,\text{meas}}$ and $I_{i,\text{calc}}$ are the measured and the simulated X-ray intensities and $n$ is the number of elements in the particle, respectively. Including the normalization in the numerical procedure assumes that practically all elements are observed, i.e. $\sum_{i=1}^{n} C_i = 1$.

The procedure is terminated, if either of the following conditions is met:

$$\left| C_i^{(k+1)} - C_i^{(k)} \right| < \varepsilon, \tag{7.36}$$

$$\sum_{i=1}^{n} \frac{(I_{i,\text{meas}} - I_{i,\text{calc}})^2}{\sigma_{i,\,\text{meas}}^2} < \delta, \tag{7.37}$$

where $\varepsilon$ and $\delta$ are given as input values for limitation of the iteration steps, and $\sigma_j$ is the standard deviation of the measured X-ray intensities of the $j$th

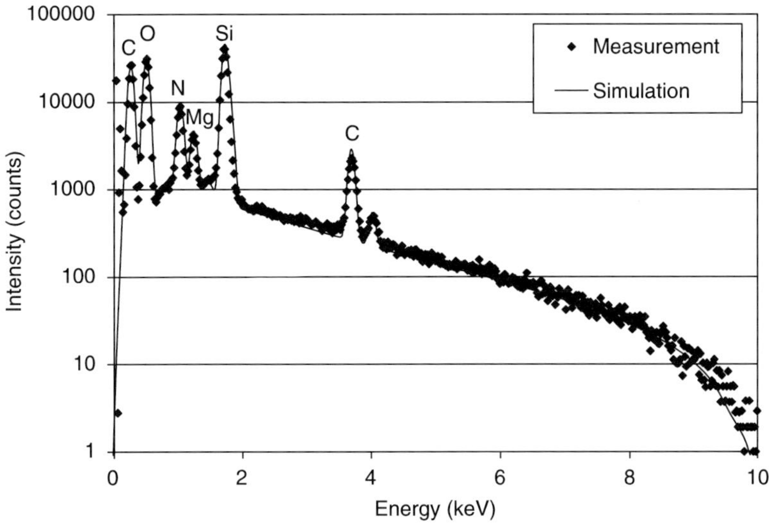


**Fig. 7.122.** Measured and simulated EPMA spectra of a 20-μm diameter soda-lime glass standard sphere (SPI #2716) with a 100-nm carbon layer

element. According to our experience, the convergence speed of the numerical approximation depends strongly on the number of simulated electrons and the number of elements considered in the sample. The average number of steps is between 5 and 10 for accuracy values $0.005 < \varepsilon < 0.01$ and $1.5 < \delta < 2.0$.

The zero-approximation of the elemental concentrations $C_i^0$ for EPMA can be assumed as the normalized measured intensity of element $i$, because practically all elements are observed (except H, He, Li, and Be):

$$C_{i,0} = \frac{I_{i,\text{meas}}}{\sum_{j=1}^{n} I_{j,\text{meas}}}. \tag{7.38}$$

In micro-XRF, however, the irradiated mass, the shape, and the matrix of the particles are not known and it is very difficult to determine these parameters experimentally. Therefore it is necessary to make some a priori assumptions on the basis of independent bulk measurements. The shape of the particles can be assumed to be spherical. The diameter or mass of the particle-sphere can be estimated by using optical microscopy or from the intensity of the scatter peaks, respectively. It can be assumed that each element with a detectable X-ray line is present in its most common oxide form in the particle, such as $SiO_2$, $SO_3$, $K_2O$, CaO, $TiO_2$, and $Fe_2O_3$. All other elements can be assumed to be present in atomic form. The sum of the elemental and oxide concentrations can be assumed to be 100% in the case of coal fly ash and sediment particles.

The initial concentration of analyte $i$, $C_{i,0}$ with detectable X-ray line was calculated as:

$$C_{i,0} = \frac{I_{i,\mathrm{meas}}/S_i}{\sum\limits_{j=m+1}^{n} I_{j,\mathrm{meas}}/S_j} \times \left(1 - \sum_{j=1}^{m} C_j\right), \tag{7.39}$$

where $S_i$ is the sensitivity for element $i$, $n$ is the total number of elements in the sample, and the first $m$ elements constitute the dark matrix (which can be derived from stoichiometry or known from other, bulk measurements). If EPMA results are available for exactly the same particle or the same particle type (including light elements such as C, N, and O), the dark matrix composition considered for micro-XRF can be calculated from the major elemental composition obtained from EPMA.

Using the Monte Carlo based quantification method, the minor element content of individual microparticles can be determined using laboratory scale micro-XRF setups based on a standard diffraction X-ray tube and capillary optics. If the trace element content of the particles is demanded, micro-SRXRF measurements are necessary, using the matrix composition determined for each individual particle separately using the laboratory micro-XRF system or EPMA measurements.

The Monte Carlo based standardless quantification method for both EPMA and micro-XRF was tested using particulate standards with known composition. Relative deviations in the range of 5–20% have been achieved by the Monte Carlo quantification scheme, depending on the analyzed element and sample type [359, 360, 367, 376].

Errors in the quantitative results are mostly due to the uncertainties in the physical constants (cross sections, fluorescence yields, transition probabilities, etc.) applied in the simulations and due to uncertainties concerning various instrumental parameters. The latter includes uncertainties in the employed excitation spectrum in case of a given polychromatic X-ray source, and often insufficient knowledge on the used detector response-function characterizing the energy dispersive detector in the experiment.

### Application Examples of Single Particle Analysis

The applicability of the MC code for the quantitative trace-element analysis was demonstrated on low-density particles [367]. Examples for the combined application of micro-XRF and EPMA are shown for individual soil [377] and river sediment [378] particles.

As the simulation code can predict reliably the measured XRF intensities (and sensitivities) for particulate standards [367], the model could be applied for the quantitative analysis of unknown soil particles originating from Kosovo, based on the iterative adaptation scheme discussed above. The micro-XRF measurements were carried out at the D09B-XRF beamline of LNLS

**Table 7.18.** Elemental composition of three soil particles calculated using the Monte Carlo quantification scheme for micro-XRF

| | Concentration (wt%) | | |
|---|---|---|---|
| Elements | Particle 1 | Particle 2 | Particle 3 |
| O[a] | 47.2 | 49.2 | 37.6 |
| Mg[a] | 4.7 | 4.8 | 3.9 |
| Al[a] | 15.3 | 15.9 | 11.4 |
| Si[a] | 20.5 | 21.3 | 17.11 |
| P | 0.4 | <DL | <DL |
| S | 0.3 | <DL | <DL |
| K | 1.6 | 2.1 | 7.8 |
| Ca | 7.6 | 2.9 | 1.7 |
| Ti | 0.3 | 0.5 | 0.7 |
| V | 0.007 | 0.006 | 0.02 |
| Cr | 0.006 | 0.01 | 0.01 |
| Mn | 0.03 | 0.13 | 0.005 |
| Fe | 1.91 | 2.98 | 0.09 |
| Co | 0.007 | <DL | <DL |
| Ni | 0.002 | 0.004 | <DL |
| Cu | 0.002 | 0.003 | 0.003 |
| Zn | 0.004 | 0.008 | 0.002 |
| U | 0.13 | 0.20 | 18.2 |

The matrix composition (O, Mg, Al, Si) was assumed as the average composition of soil obtained from electron probe microanalysis

Relative standard deviations are in the range of 2–15%.

[a]Calculated from EPMA.

(Campinas, Brazil), using white beam excitation. As an example, the calculated elemental compositions of three typical individual particles are shown in Table 7.18, corresponding to particles having elevated U concentrations. In these calculations, the matrix composition was estimated by EPMA, based on a Monte Carlo quantification procedure for electron interactions [359, 360]. In Fig. 7.123, the experimental and the corresponding simulated XRF spectra of a soil particle (Particle 1 in Table 7.18) measured by the LNLS micro-XRF setup is shown. The agreement between the measured and simulated spectra is satisfactory for this particle; both with respect to the fluorescence line and scatter background intensities.

As illustrated in Table 7.18, the U concentration within these particles range from 0.12% to 18.4%, which represent concentration values averaged over the entire intersection volume between the beam and the particle agglomerate in question. The EPMA results showed that most of the concentrated DU (depleted uranium)particles were in the 0.8–1.5 μm size range, which were often agglomerated with larger soil particles [379]. The particles in the 20–50 μm diameter range were assumed to be heterogeneous, different number of small DU particles were attached to them, therefore the average uranium concentration over the particles could vary in a wide range [377].

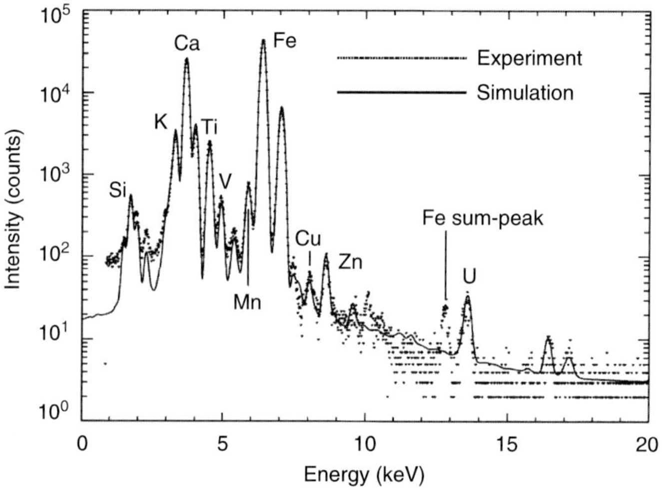


**Fig. 7.123.** Comparison of experimental and simulated micro-XRF spectra for an individual soil particle collected from Kosovo. The spectra were collected at LNLS, using white beam excitation. Reproduced from Török et al. [377] by permission of Elsevier Science B.V.

The trace composition of anthropogenic particles after a transboundary water pollution event of the Tisza river (Hungary) was studied using white beam micro-XRF at HASYLAB Beamline L (Hamburg, Germany). The used experimental setup allowed the quantification of trace lanthanide elements using their K lines. Although the white beam micro-XRF setup allowed only the investigation of particles in the sedimentation size range, it provided useful information on the trace content of the particles, which is an average value for the whole area of a particle. Table 7.19 summarizes the result of 33 selected particles in the diameter range of 40–80 μm, the maximum and average concentrations of trace elements, and the average trace composition of five clusters obtained using hierarchical cluster analysis (HCA). The composition of the dark matrix was estimated using the EPMA results of similar particle classes. The average and maximum concentrations of copper and zinc are in the same range as it was found for the polluted bulk samples, which is in accordance with the small size of the grains enriched with these elements. For lead, however, an order of magnitude higher maximum concentration was observed, indicating the presence of larger (5–10 μm) lead-containing grains than it was presumed from the EPMA results [380]. The maximum observed concentration of cadmium (>700 $\mu g\,g^{-1}$) indicates that this highly toxic heavy metal is also inhomogeneously distributed among the anthropogenic particles. The determination of this metal was not possible using EPMA and tube-excited

**Table 7.19.** Summary of the micro-XRF point analysis of 33 large (60–100 μm) individual particles

| | Cluster 1 | Cluster 2 | Cluster 3 | Cluster 4 | Cluster 5 | Average | Maximum |
|---|---|---|---|---|---|---|---|
| Number of particles | 18 | 9 | 3 | 2 | 1 | 33 | 33 |
| Element | Concentration ($\mu g\, g^{-1}$) | | | | | | |
| Mn | 476 | 107 | 3520 | 559 | - | 643 | 8375 |
| Fe | 20950 | 5940 | 22940 | 29000 | 4810 | 17060 | 72000 |
| Cu | 322 | 42 | 414 | 25 | 89 | 229 | 1259 |
| Zn | 1119 | 117 | 1079 | 225 | 2703 | 836 | 4332 |
| Rb | 92 | 155 | 96 | 8 | 18 | 102 | 648 |
| Sr | 77 | 116 | 44 | 5 | 5 | 78 | 668 |
| Y | 16 | 10 | 134 | 1 | 8 | 24 | 357 |
| Zr | 44 | 166 | 48 | 3 | 514 | 89 | 1382 |
| Nb | 11 | 5 | 14 | 1 | 1 | 9 | 50 |
| Pd | - | - | 49 | - | - | 4 | 148 |
| Ag | 4 | - | 19 | 1 | 5 | 4 | 50 |
| Cd | 11 | - | 263 | 2 | 46 | 32 | 723 |
| Sn | 197 | 21 | 9 | 6 | 11 | 115 | 1591 |
| Sb | 7 | 33 | 61 | 2 | 5 | 19 | 291 |
| Cs | 7 | 21 | 17 | 2 | 3 | 11 | 99 |
| Ba | 486 | 3284 | 173 | 47 | 92 | 1182 | 12160 |
| La | 13 | 3 | 21 | 4 | 5 | 10 | 59 |
| Ce | 38 | 5 | 56 | 4 | 10 | 27 | 187 |
| Pb | 605 | 78 | 6652 | 36 | 189 | 964 | 16570 |

The concentrations were calculated using the MC based quantification method. Five clusters were obtained using HCA

bulk XRF. As it is shown in Table 7.19, Clusters 2 and 4 can be regarded as natural sediment particles, because they were formed from particles with very low copper, zinc, and lead content. A comparison of simulated and measured X-ray spectra of a typical sediment particle classified to Cluster 2 is presented in Fig. 7.124, showing good agreement. The particles however, which could be associated with the mine pollution accident (Clusters 1, 3, and 5), contained La, Ce, and Nd also in an order of magnitude higher than natural sediment particles [378].

### 7.5.5 Radionuclides and Radioactive Materials

Radionuclides may be introduced into the environment through a variety of systems and processes. Human activities involving nuclear weapons and nuclear fuel cycle (including mining, milling, fuel enrichment, fabrication, reactor operation, spent fuel stores, reprocessing facilities, medical applications and waste storage) are important, leading to a significant creation and release of radioactivity. Human technology also releases pre-existing natural

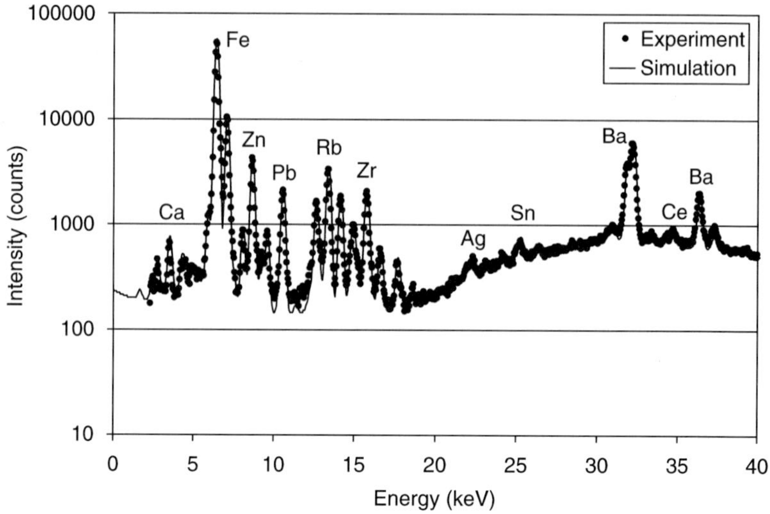


**Fig. 7.124.** Measured and simulated white-beam micro-XRF spectra of a typical surface sediment particle collected from the Tisza main riverbed. The diameter of the particle is 60 μm. Reproduced from Osán et al. [378] by permission of Elsevier Science B.V.

radionuclides, which would otherwise remain trapped in the earth's crust. For instance, burning of fossil fuel (oil and coal) dominates direct atmospheric release at pre-existing natural radioactivity

The fallout distribution pattern depends on the weather conditions (i.e., wet or dry). Variation in the degree of interception of fallout is found depending on the nature of the surface and the physical–chemical form of radionuclide which may vary, depending on release and transport conditions in addition to elements' properties. A general distinction can be made between gases, aerosol, and particulate material. Particles with higher activity concentration, known as "hot particles", may result from atmospheric nuclear weapon tests or nuclear reactor accidents. This activity is diluted as material is transferred to soil and water directly or via vegetation and movement through other biota.

The ability to develop adequate models for predicting the fate of inorganic contaminants, including radionuclides, in both surface and subsurface environments, is highly dependent on accurate knowledge of the partitioning of these constituents between the solid and solutions phases and ultimately on the capability to provide molecular-level information on chemical species distributions in both theses phases. Furthermore, the development of environmentally sound, yet cost effective remediation strategies requires an understanding of the chemical speciation of the contaminants within the sediment and soil matrices in which they are contained.

The analytical techniques used for the monitoring of the isotopes dispersed in the environment as manmade isotopes or natural isotopes are radiometric methods as well as mass spectrometric methods, which are mainly used for the longer-lived isotopes. The first are based on the nuclear properties of the radionuclides. The proper selection of a particular radiation detector or method of radioactivity analysis requires a good understanding of the properties of nuclear radiation, the mechanisms of interaction of radiation with matter, half-life, decay schemes, decay abundances, and energy of decay. Sometimes radiation is classified into ionizing radiation, which produces ionization or ion pairs as it travels through matter, and non-ionizing radiation. These may be further grouped into particulate radiation that may have a definite mass and non-particulate radiation or radioelectromagnetic radiation. Radioactivity, depending on the type, may emanate directly from the nucleus of an unstable atom or from its electron. The forms of ionizing radiation of most concern at detectable decay products of radionuclides are particulate radiation, including $\alpha$-particles and $\beta$-particles (negatrons and positrons), which are emitted from the nuclei of atoms, at internal conversion and Auger electrons, and non-particulate electromagnetic radiation, including Bremsstrahlung, X-radiation and $\gamma$-radiation.

X-ray fluorescence can be exploited for the determination of the bulk concentrations of actinides in environmental samples like soil, sediments, and nuclear waste disposals. The technique is limited since it is only responding to the atomic number of the elements and therefore isotopic information that is essential for determination of the origin of the material is lost. However, XRF has been exploited to study chemical distribution of actinides as released from spent nuclear fuel disposal and storage both in bulk and in particulate matter.

Subsurface transport of groundwater contaminants is greatly influenced by chemical speciation, precipitation, and sorption processes. The transport of Pu potentially released from spent nuclear fuel disposal and storage sites is dependent on its interaction with mineral surfaces and speciation in the subsurface. Plutonium is a key component of spent nuclear fuel that requires careful strategies for environmentally safe management and long-term geological storage. Studies investigated sorbed Pu on a natural heterogeneous material (tuff) using in situ spatially resolved synchrotron micro-X-ray fluorescence (micro-SXRF) [381]. Micro-SXRF spectroscopy can be used to produce images of elemental distributions on a microscopic level and therefore results as an excellent technique for determining the microdistributions of sorbed Pu among various minerals in tuff material. Figure 7.125 shows an SXRF spectrum obtained for a region on the tuff that contained a highly elevated sorbed Pu. Once identified by SXRF imaging, isolated regions of Pu enrichment may be probed with micro-X-ray absorption near-edge structure (micro-XANES) and extended-X-ray absorption fine structure (EXAFS) spectroscopies. Hence, measurements of the bonding environment of particular species on individual mineral microphases can be made on these materials. In situ micro-XANES and – EXAFS techniques have been utilized to study the local molecular bonding environments of metal contaminants in heterogeneous environmental systems on a micrometer scale [382, 383].

X-ray fluorescence spectrometry has been used combined with vibrational microscopy to assess highly heterogeneous actinides contaminated

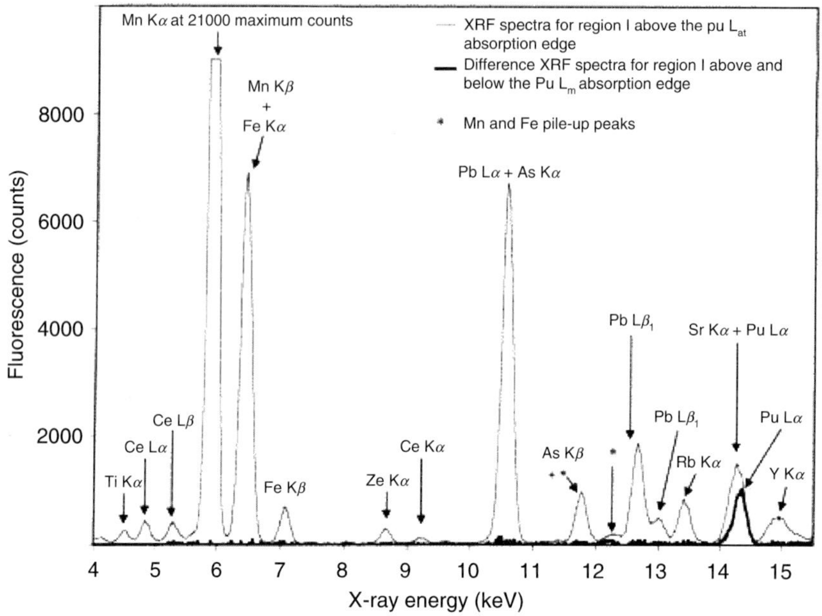


**Fig. 7.125.** SXRF spectrum from a region of tuff that contained highly elevated sorbed Pu. Reproduced from Duff et al. [381] by permission of American Chemical Society

materials [384]. In the context of disposition and treatment of various waste forms contaminated with actinides, the study of chemical interactions and structural characteristics of uranium, neptunium, plutonium, and americium species in different substrates provides vital information in exploring stabilization, processing, and storage scenarios. Interactions of actinides with various waste forms and media, as well as the fate and the transport of actinides, are areas of particular interest. When actual transuranic waste has been in contact for a period of time with a brine solution, a precipitate is formed. The authors [384] have examined this precipitate for gross elemental composition and semiquantitation of Mg, Al, Ca, Fe, Nd, Th, U, Np, and Pu by WDXRF. Additionally, energy-dispersive micro-XRF spectrometry has been used to map and correlate elemental composition within these precipitates. In order to gain further information on the molecular species present, Raman and infrared measurements were used in combination and molecular and elemental information for highly heterogeneous materials could be obtained.

Micro-X-ray fluorescence has turned out to be an experimental approach having considerable potential to provide unique problem-solving capabilities [385–393]. The distinguishing feature of micro-XRF compared to other XRF approaches as bulk XRF and high-resolving ED or WD micro-XRF using a scanning electron beam is that it fills the analytical gap between these two methods. The principle advantages of micro-XRF are the non-destructive analysis and the need for little or no sample preparation. In addition, most problems can be solved with qualitative rather than quantitative information. The capacity of micro-XRF for quantifying elemental concentration distributions across a large sample surface area has been demonstrated investigating depleted uranium(DU) containing 6% niobium [394]. Nb is commonly alloyed with uranium to prevent surface oxidation, and determining how the niobium concentration is distributed throughout a sample is useful in explaining observed material properties. A process was developed to convert qualitative micro-XRF niobium distribution images from depleted uranium samples into images displaying concentration values. Thus, micro-XRF was utilized to determine elemental concentrations across a surface in a manner similar to that of the established method of EPMA. However, even laboratory scale micro-XRF can provide information from relatively large sample areas (over 100 $cm^2$ in size) using a 500-µm beam.

Characterization of environmental hot particles in terms of chemical and isotopic composition is essential in order to understand transport mechanisms in the environment as well as their bioavailability. Microanalytical techniques such as EPMA and secondary ion mass spectrometry (SIMS) combined with scanning electron microscopy (SEM) EDX or WDX system have been largely exploited for these applications. Recently, micro-SXRF has been applied for the determination of trace element constituents in hot particles. In particular, the technique has been very effective for the determination of the chemical composition of microparticles containing depleted uranium [377]. Measurements performed by SIMS on the same particles to determine their isotopic composition confirmed that the particles contained reprocessed depleted uranium. Other authors [379] have studied microparticles bearing depleted uranium. They have exploited micro-XRF imaging and SEM-EDX. Their objective was to test the suitability of these techniques to identify the presence of small DU particles and measure their size distribution and the $^{235}U/^{238}U$ isotopic ratio by SIMS. By micro-XRF different points of the sample were irradiated using a narrow primary X-ray beam. At each point an X-ray fluorescence spectrum was collected, the concentration of a particular element versus the position was obtained, and a concentration map was drawn. The technique did not provide information on the isotopic composition and only gives information on the size of the particles if they were smaller than the beam diameter. The authors, in order to confirm the presence of uranium, analyzed two samples with a large beam (0.7 mm in diameter). The spectra at each different pixel were summed up to produce a cumulative spectrum. The uranium $L_\alpha$ characteristic X-ray confirmed the presence of uranium in both samples. The elemental maps of uranium were then constructed by analyzing the two samples with a finer beam (35 µm) by summing up at each pixel the counts in the energy range 13.4–13.6 keV. Although SEM-EDX is unable to provide information on isotopic composition of the sample, it gives excellent information on the particles' dimensions and size distribution.

In a sediment sample stemming from a reprocessing plant, micro-SXRF could reveal the association of high concentration of Zr (1100 pm) and Hf (70 ppm) in a particle containing 15 ppm of U [374]. From this finding it was supposed that uranium contained in the particle was anthropogenic. Since no isotopic ratio measurements were performed no final conclusions on the origin of the particle could be drawn.

## 7.6 Geology, Mining, Metallurgy

*D. Rammlmair, M. Wilke, K. Rickers, R.A. Schwarzer, A. Möller, A. Wittenberg*

### 7.6.1 Introduction

Geology, mineralogy, exploration, and mining require an interdisciplinary approach involving geological mapping, tectonics, petrography, geochemistry, sedimentology, remote sensing, geophysics, microbiology, engineering, mineral treatment, and melioration. Within this frame geochemistry is one of the basic tools used in characterizing rocks of magmatic, metamorphic, and sedimentological origin, as well as soils and weathering products of these. The obtained geochemical information allows us to classify the rock either in terms of an internationally approved rock name classification [395] (Le Maitre et al. and references), genetic origin, such as the classification of magmatic rocks to geotectonic environments [396] (Pearce and references) or parental magma series. Moreover and more importantly, genetic processes like fractionation and/or mixing and/or assimilation during transport in the earth's interior become elucidated through the geochemical characterization of the source rock, the interim, and the final product. Even chemical changes within apparent homogeneous successions due to cryptic layering, metamorphism, metasomatism, alteration, weathering and/or mineralization can be visualized.

The characterization of soil samples is fundamental for understanding soil-forming processes and transport processes within soils and their contribution to the groundwater quality. Soil characterization on a regional scale is embedded and used in soil information systems. The data obtained are used for monitoring the amount and type of fertilizer to be adopted in the high-technology GPS (global positioning system)-controlled agricultural industry. For these applications, high accuracy of the obtained geochemical data is crucial. Regarding exploration and mining, the rapid characterization of prospective zones due to relative enrichment or even depletion of diagnostic or pathfinder elements is fundamental for finding and developing a mine site.

In the field of metallurgy, online quality control of the products is relevant at every point of the production process: starting at the mining site itself, blocks of rock are selected and mixed properly to obtain a homogeneous ore concentration used in the treatment plant. Within the treatment plant, quality monitoring of the separation and concentration process, and the final products is crucial. For these applications, speed of analysis is of major importance. In situ online data acquisition is the optimal solution for decision-making processes. X-ray fluorescence (XRF) was [397, 398] and is the method applied to solve these most basic problems of geologists, mining engineers, and metallurgists.

Several scales of chemical information are useful for the geoscientist.

- The macroscale involving direct field measurements, laboratory investigation of homogeneous bulk samples, and product online control.
- The mesoscale refers to the hand specimen size, down to a spatial resolution of 30–100 μm. Element 1D scans and 2D distribution maps provide zones of interest for further microanalytical research.
- The microscale focuses on single minerals, mineral zoning, single particles, fluid and melt inclusions, and compositions within experimental apparatus, e.g., diamond anvil cells (DAC).

The principles and advantages of wavelength-dispersive-(WDXRF), energy-dispersive-(EDXRF), nondispersive-(NDXRF), total reflection-(TXRF), and grazing emission X-ray fluorescence(GEXRF) have been described in detail by Bertin [399] and Lachance [400] and are the subject in some of the chapters (Chaps. 2–5) of this book. Hahn-Weinheimer et al. [401] and Weber-Diefenbach (and references) [402] refer in detail to the applicability to geosciences. Brill [403] compared WDXRF and EDXRF systems. Kramar [404] discussed the progress in EDXRF technology. Several manufacturers of XRF instruments provide basic information on instrumentation and applications via the internet. Depending on the field of application, the individual systems are designed to fulfill certain conditions. Spectral resolution, excitation conditions, and spatial resolution are thereby the most relevant aspects. A number of XRF instruments on the market meet the requirements of application in the field or outcrop, in the laboratory, and for online monitoring of technical processes.

Robust and highly accurate laboratory systems facilitate bulk analyses, fast online systems help to control production quality, handheld systems enable analysis in the field and help to reduce the number of samples by selective sampling, and high spatially resolving systems support meso- and microanalytical approaches. However, for each scientific application an adequate method should be chosen based on the need of qualitative, semiquantitative, or quantitative information and the required spatial resolution.

This chapter does not refer much to the methodological and instrumental background, tubes, filters, detectors, analyzer crystals, optics, evaluation and correction methods, since these are treated in detail in Chap. 2–5 of this volume. However it aims to give an insight on the application of the XRF methods within the fields of geology, mineralogy, mining, and metallurgy. Among the large variety of possible applications, special emphasis is laid on examples using EDXRF, and on analytical methods with spatial resolution that cover analyses on the hand specimen scale or even on the microscopic scale. Investigations on the microscopic scale have been strongly boosted by the use of synchrotron radiation. Hence, special treatments of such applications are given in Sect. 7.6.4 (Microanalysis using synchrotron radiation XRF). Since geologists study processes in the history of the earth's evolution, dating methods are crucial for the understanding of the earth's history, and an example of chemical dating using U–Th–Pb contents in monazite is presented in Sect. 7.6.4 (U–Th–Pb chemical dating). As a valuable extension to XRF methods, geological applications of X-ray absorption fine structure spectroscopy (XAFS) are shown in Sect. 7.6.4 (XAFS in earth sciences). Finally, the application of imaging XRF, combined with texture and lattice strain measurements in deformed metals, gives an insight on applications in metallurgical problems (Sect. 7.6.3 (Imaging Micro XRF, Texture and Lattice Strain Mapping, p. 658)).

### 7.6.2 Macroscale

#### In Situ Analysis in the Field

The field geologist has to decide right in the field on the most adequate sample for the geological, stratigraphic, exploration, or tectonic work that has to be performed.

Portable EDXRF systems based on radioisotope sources ($^{55}$Fe, $^{109}$Cd, $^{244}$Cm, $^{241}$Am) have been in use since decades. Recently handheld minute size laser induced X-ray sources equipped with energy-dispersive pin-diode detectors have been used to provide basic field information for a wide range of elements, simultaneously allowing to identify rock types for mapping purposes [405], differentiate within stratigraphic sequences by element finger-print patterns, and obtain geochemical information from areas where sampling is not possible. Heavy metals are especially suitable for this method as they have low detection limits. The development of minute-size low-energy consuming systems is motivated by the interest to obtain information from planets such as Mars. The prototype has landed on Mars' surface and provided information on the Martian geochemistry [406, 407].

Investigations of mine waste [408] and contaminated soils in the vicinity of industrial sites are also suitable applications of this type of rapid screening [409–412].

The method has been applied in a qualitative evaluation of heavy metals in soils by McCain [413] using an isotope-source energy-dispersive XRF analyzer for rapid onsite screening.

Onsite data acquisition has been used in addition to outline zones of interest in exploration areas previous to further detailed representative sampling for accurate XRF analyses [414, 415]. Konstantinov et al. [416] developed a new indicator method as an effective exploration method for blind ore. The method takes advantage of the partial redeposition of ore and vein matter in the overlying rocks as a result of the magmatic, hydrothermal, and tectonic processes in the form of shears, veinlets, and/or dikes. A number of indicator elements are analyzed to localize anomalous zones. The indicator method uses special sampling of the post-ore uplift structures. The proposed technique was successfully used to prospect blind ore mineralization and new ore bodies in various geological environments and is applicable over a vertical range of more than 200 m. The resolution identifies individual blind ore bodies as well as groups of ore bodies.

The rapid element screening can further be applied to characterize ore types, drill cores, and hand specimens. The fast turnover of information is a very valuable contribution to improve decisions and save resources.

Within the metallurgical section the onsite characterization of stainless steel samples [417], metal alloys, and metal scrap allows gaining higher purity of the end products.

Sample preparation for the screening method in general is minimized. The best available surfaces are used. Analyses aiming for better reproducibility might use homogenization methods in the field. The data arrived at are qualitative to semiquantitative since comparable sample geometry and homogeneity cannot be provided in the field. Quantitative values are hard to be obtained by this fast method and are not really necessary since element fingerprint pattern and element ratios generally provide the necessary information needed for making decisions in the field.

Of late emphasis has been on the development of downhole-EDXRF systems [418] applied in drillholes to monitor in situ chemical patterns of the drilled soils and rock units.

## Quantitative Analysis in the Laboratory and in the Field

After the selection of the most adequate sample, or a random or a grid-oriented sample series in the field, high-quality representative analyses are performed by standard XRF techniques in the laboratory or benchtop-equipped mobile field laboratories. Depending on the amount and type of sample, WDXRF EDXRF, and for special occasions even TXRF are used to get the appropriate geochemical data for interpretation of the field data.

Various sample preparation techniques are used on geological samples to obtain the optimum results. Detailed descriptions of the individual techniques, and advantages and problems are available in literature [399, 400, 402], (van Grieken et al. Chap. 6) and internet (www.spectroscopynow.com; www.learnxrf.com). The most important requirements for excellent reproducibility are homogeneity, grain size and minimum thickness of the sample, and limitation of matrix effects.

## Applications

### Classification of Rocks and Plate Tectonic Implications

Identification and nomenclature of rock samples is usually done by the field geologist on the basis of macroscopic and microscopic observations. However, the magmatic equivalent of altered and metamorphosed rocks is generally identified by geochemical investigations. The classical schemes of Streckeisen [419], Le Bas et al. [420], and Le Maitre et al. (and references) [395] have a mineralogical background and are based on the relative proportions of minerals such as alkali feldspar, plagioclase, quartz and feldspatoids in volcanic rocks, and plutonic equivalents. There are various methods (e.g., the Niggli values [421] and the CIPW norm [422, 423]) to convert the chemical raw data into classifications that relate the chemical composition to the mineralogy of the rocks. In these norms collective components or contents of normative minerals are calculated from the chemical composition that enables an improved classification of the rocks where mineral content is not directly determinable (e.g., fine grained or altered rocks). In parallel, some classifications directly use element concentrations to classify rocks as in the TAS diagram, where the total alkali content versus the silica content is plotted [420].

Classification schemes such as SEDNORM [424] and MINLITH [425] are applied for characterization of sediments.

Besides the classification of samples into rock types, a number of classification schemes on the basis of major and trace elements are available, which may be used to establish genetic features of a given rock unit, such as the geotectonic environment [396] (Pearce and references). These schemes have been applied in numerous cases to classical ophiolite sequences, komatiitic successions, and granitoid systems.

Field observations often reach their limits with regard to stratigraphy within sediments, salt successions, or layered intrusions due to the macroscopic conformity of the rock units. Detailed geochemical investigations – mainly XRF based - provide a fast tool for correlation of units from different localities (e.g., drill cores from marine environments, exploration targets) and even provide up and down criteria for overturned sequences in highly tectonized areas. The documentation of saw-teeth patterns in cryptically layered magmatic successions provides insight into magma-mixing episodes and processes. For example, investigations on the Bushveld and Stillwater complexes or the Skaergaard and Dufek intrusions have shown the importance of detailed geochemical approaches for the interpretation of their complex history [426, 427].

### Mineral Chemistry

In combination with methods that provide insights into the crystallographic structure, such as X-ray diffraction analysis (XRD). XRF is used for characterizing the composition and purity of minerals and mineral concentrates [397].

Analyses of mineral concentrates are performed in the same way as for the whole rocks. Large single grains of micas, hornblendes, or zircons, to name a few, have been investigated by WDXRF for elements like K, Rb, Sr, Nd, Sm, U, and Th for classification or as a base for age-dating purposes. The beauty of this method is its nondestructiveness. Hence, the same samples can be used for further analyses. This application competes with electron microprobe analysis (EMPA) of minerals, which is also capable of simultaneous major and trace element analysis of multielement samples and has the additional advantage of achieving a high spatial resolution in the micrometer range. However, EMPA requires additional time-consuming preparation steps, as polished thin sections of mineral and rock samples are generally used for analysis.

More recently, multielement EDXRF designed for microprobes named XRF microprobe or EMMA (energy-dispersion miniprobe multielement analyzer) have been developed [428] for special applications (see U–Th–Pb chemical dating in Sect. 7.6.4).

The XRF, a fast, precise and low-priced method when compared to other multielement analytical techniques, is now a routine technique in industrial applications [429]. As an in- or rather on-time method for routine bulk analyses during industrial production, WDXRF is used [430]. In particular, it is the industrial manufacturing of high-quality products, which requires very pure minerals either of natural or synthetic origin [431]. For this purpose pressed pellets of either pure minerals or mineral concentrates are analyzed. While using pressed pellets most of the interesting trace elements can be analyzed due to the lack of dilution. The disadvantages of pressed pellets are related to grain effects and chemically zoned single grains. Since the last few years, problems related to grain effects seem to be fully understood and hence can be taken into account by corrections [432].

While EDXRF and WDXRF are widely utilized it is especially the TXRF technique which is of great interest for the purest material chemical analysis, since the detection limits can be reduced to those of the very expensive and sophisticated inductively coupled plasma mass spectrometer (ICP-MS) technique. For further information, in particular about the TXRF used for wafer monitoring, see Sect. 7.3.6.

### Geochemical Mapping, Prospection, and Exploration

Large-scale low-density geochemical surveys provide basic knowledge on geochemical provinces that can be used to outline prospective areas. A reliable and comprehensive database of the natural variability of the geochemical background of diagnostic elements is the basis for administrative and legal issues concerning sustainable long-term management of environmental and mineral resources [433].

Data acquisition by XRF is a valuable basis for geochemical mapping. Numerous element distribution maps have been produced in Europe and worldwide at regional and local scale for characterization of catchment areas,

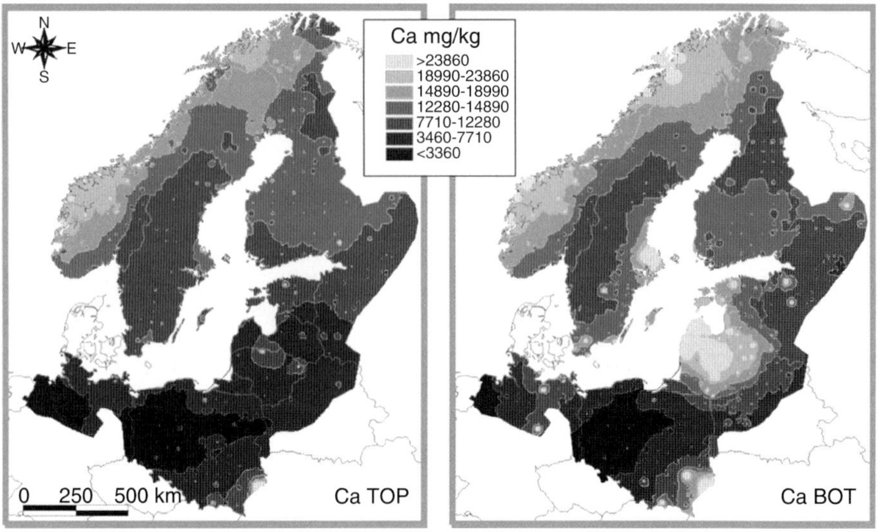


**Fig. 7.126.** Ca in agricultural soil, Ap (TOP) and B/C-horizon (BOT), total concentrations, WD-XRF (Courtesy of Reimann et al. [435])

soils, rock units, and mineralization for the support of mapping, exploration, and environmental investigations. Reimann et al. [434, 435] have shown in the Baltic Soil Survey that top and bottom soil (TOP and BOT) provide basic information on regional distribution of major and trace elements. The Ca maps (Fig. 7.126) for example show that both TOP- and BOT maps contain the highest concentrations above the Caledonides, intermediate values for most of Sweden and Finland, which may be assigned to granitic rocks, and high Ca occurs above sediments in the SE of Sweden.

Chemical reconnaissance mapping and exploration used the same method as described earlier. In general, depending on the requirements a 2-mm and a 60-$\mu$m fraction is obtained from the soil or stream sediments. Alternatively, heavy mineral concentrates, chip or channel samples, as well as hard rock samples along base lines, grids, and streams may be used to identify areas for detailed investigation. In special cases, plants (e.g., for Au, Cu, Ni) and termite hills (e.g., for kimberlites) are used for prospection to outline zones of interest under thick sedimentary cover [436, 437]. X-ray fluorescence analyses provide over 30 elements for purposes at reasonable costs and fast turnover, crucial for the exploration geologist. Anomalous zones are subsequently mapped in detail, including drilling at sites of major importance to increase the knowledge on the anomalous zone of interest.

## Mining, Whole Rock, and Mineral Treatment

In the mining sector recognition of wallrock, subeconomic, economic, and high-grade ore portions within a mining site is done on a visual basis as long as the lithology provides this information. However, significant differences often cannot be recognized this way onsite. Fast sampling has to be performed on an in situ basis via handheld EDXRF systems or by quick turnover via the laboratory. Often drill arrays are performed before the mining reaches the zone of interest. Detailed (1 m down to 10 cm) sampling of the drill core enables the mining engineer to outline blocks of similar ore content. In modern open pit mining the operators of trucks and Caterpillars are remote controlled onsite via GPS, with the adequate information of block mining in the proper ratio providing a continuous flow of the standard product for optimum recovery of the ore or other raw material of interest.

From mining to final product all processes have to be monitored by mineralogical and geochemical methods to enhance the recovery and the quality of the mineral products or concentrates. All steps are accompanied by sampling and eventually, like in the cement, glass, or ceramic industry, even online continuous measurement in the production chain of the components can be performed.

### Coal, Oil, Bitumen

Coals, oil shales, hydrocarbons, bitumens, and ashes are analyzed by XRF for their carbon, sulfur, and trace element contents [438–440]. Genetic implications can be drawn from the chemical composition of coals as influenced by mineralogical changes in the primary environment of deposition, by epigenetic and supergene alterations. Only a few elements (e.g., P, REE (rare earth elements)) vary with the rank of coalification [441]. The trace elements are believed to be associated with either organic compounds (Mo, Hg, and U) or minerals such as aluminum-iron-silicates (Sc, Ge, and Bi), aluminum-silicates (Cs, Be, Th, Pb, Ga, and REE), iron-phosphates (Zn, Rb, and Zr), iron-sulfides (As, Cd, Cr, Cu, Hg, Ni, Tl, and V), and carbonates (Sr, Mn, and W).

Some coals are strongly enriched in the trace elements U, Pb, and Hg due to volcanic activity. Epigenetic sulfide and carbonate mineralization are common in some coalfields. Elements such as Zn, As, Sb, Hg can be enriched in micropores and microfractures [442]. Hence, the hazardous potential of a coal due to its combustion has to be estimated by analyzing the ashes in the laboratory and even onsite [443].

### Potential Hosts for Nuclear Waste Depositories

The nuclear industry, basically the electric power industry, is in need of finding a lasting and durable depository for high radioactive nuclear waste. Until now no location has been equipped. Rock salt sequences, clay formations, and granitoids have proved to fulfill the requirements. In part they do. Again detailed geochemical investigations in the underground and via drill cores are necessary to understand the stratigraphic background (Plate II 2.1-2), recognize weak zones, learn about the behavior of the material in contact with the nuclear waste bearing containers and about its neutralization potential for radionuclides that may have escaped from the container.

To check the possibility of final save storage in former rock salt mines samples of rock salt have been analyzed by WDXRF for Rb, Sr, Br to characterize the contents of Br in view of primary distribution and redistribution due to recrystallization and tectonic overprint [444].

### 7.6.3 Mesoscale

#### Analyses of Inhomogeneous Samples

Hand specimen rock samples are usually heterogeneous and cannot be characterized in an appropriate way macroscopically. Therefore the samples are either homogenized for bulk analyses, or small parts are selected to be analyzed in detail by microanalytical methods such as microscopy, EMPA, scanning electron microscopy (SEM), cathodoluminescence (CL), infrared (IR) or Raman spectroscopy. With the adaptation of fast simultaneous EDXRF systems with high spatial resolution, hand specimen and drill cores of sample sizes of 10 cm by 10 cm and more can be analyzed in a relatively short time for major and trace elements at a spatial resolution of 100 μm to 10 μm. The obtained elemental scans provide a facility to select appropriate areas for further analyses by EMPA, SEM, and other microanalytical systems.

Time is the crucial part for sample characterization. Due to the application of powerful microfocus X-ray tubes and advanced X-ray optics such as mono- or polycapillaries, highly oriented pyrolytic graphite (HOPG) [445], and multilayers for EDXRF, the photon flux has been increased in such a way that even 8 W X-ray tubes provide enough flux when using sensitive detectors to allow reduction on the acquisition time to parts per second for qualitative and semiquantitative measurements. Some instruments with fast scanning devices are able to provide an overview of major element distributions for large samples within minutes. Statistical improvements will be obtained by accumulation of repeated measurements. The application of pin diodes or Peltier-cooled silicon drift detectors enables continuous measurements at reasonable costs. Commercial scanning facilities generally work in air. Some benchtop systems can switch to vacuum conditions in order to lower the detection limits and to enable the analysis of light elements such as Na, Mg, Al. Zhdanov [446] provides details on manufacturers of micro-XRF systems in the internet.

#### Sample Preparation Techniques

In general, all types of samples are suitable for qualitative analyses, but to obtain semiquantitative or quantitative information, samples have to fulfill certain conditions. A relatively smooth polished surface is required. However, grain, pore and crystal size distribution, water and organic content, crystal orientations in the sample are properties that strongly influence the signal and may hinder the access of quantitative information [399].

**Fig. 7.127.** Geoscanner equipped with columns

## Instrumentation and Methods

The applications shown later have been performed on two individual instruments, both designed by COX Analytical Systems AB, Sweden. The Itrax geoscanner (XRGS), a special design by the Bundesanstalt für Geowissenschaften und Rohstoffe (BGR), Germany (Figs. 7.127 and 7.128), operates with flat beam optics suitable for 1D profiles, while the standard ITRAX X-ray microscope using a 100-μm monocapillary is applicable for point measurements, 1D profiles, and 2D scans. Both systems are based on a Rich. Seifert & Co. ISO-DEBEYEFLEX 3003 HV-generator and the standard operation voltage for the data acquisition is 45 kV at 30 mA with long fine-focused Philips XRD Mo tube type PW 2275/20 (3 kW, 60 kV).

Within the XRGS the beam is focused by a glass slit capillary at 50 μm height and 2000 μm length. The detector is placed at an angle of 45° to the sample surface. It is an energy-dispersive, Peltier-cooled RÖNTEC

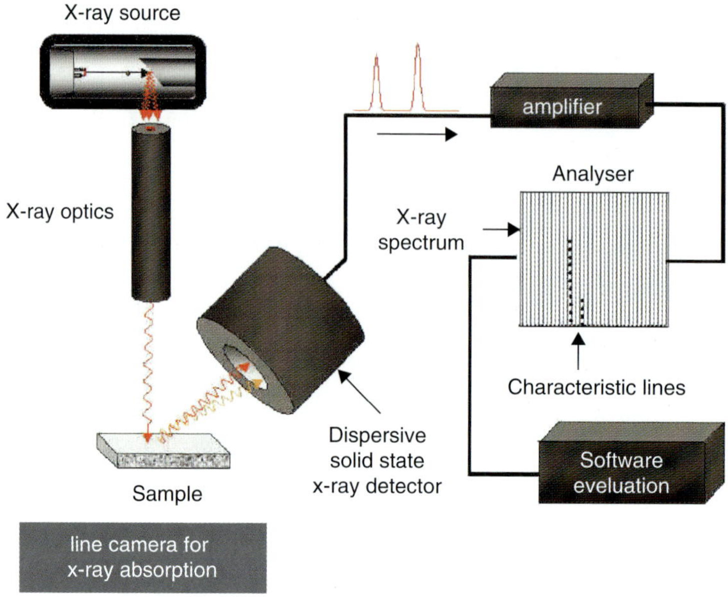


**Fig. 7.128.** Schematic view of the EDXRF geoscanner (courtesy of COX Analytical Systems AB, Sweden)

drift-chamber detector, with an energy resolution of 149 eV provided by a vacuum nozzle. The XRGS is further equipped with a rotating stage for six samples. The samples are scanned vertically by a line camera to find the area of interest, and after rotation to the measurement position, the sample is focused and scanned by moving the X-ray tube and detector unit. Step sizes range from 50 to 200 μm and time of exposure from 1 to 30 s according to the individual requirements.

A number of elements can be chosen to be displayed on the monitor showing the variation of elements during the XRF scan. For each step the complete energy-dispersive spectrum is stored in 1024 channels for later evaluation. The obtained spectra consist of the characteristic element lines of the sample, besides the tube characteristics, e.g., the Mo lines, the incoherent Compton radiation, and the Bremsstrahlung. Due to the glass slit capillary the intensity of the X-rays with higher energy is increased and the Bremsstrahlung at lower energies is minimized (up to the Fe K$\alpha$). Relative concentrations are calculated by the fundamental parameter method and are made available on a spreadsheet [447].

The ITRAX X-ray microscope is based on the same principle, but is equipped with a glass monocapillary of 100 μm diameter and an energy-dispersive RÖNTEC Si(Li) detector mounted at an angle of 45° to the sample surface with an energy resolution of 139 eV and a nozzle which may be evacuated. An optical camera and a sample stage complete the system.

The obtained spectra from either point, line or area scan measurements, are saved and elemental concentrations are determined using the fundamental parameter approach. A number of individual points, profiles, or areas can be selected for analysis using the optical camera. The elements of interest can be marked, and the time for each analysis (minimum 0.5 s) and the individual step size (100 μm standard) can be selected. The data from the selected element channels are stored as text files. The spatial distribution of each selected element can be viewed and printed. The data can be further processed in a spreadsheet or by image analysis [447].

Imaging is one of the key aspects of geochemical mapping. Visualization of chemical patterns at any scale provides a quick overview of huge amounts of data. The EDXRF images can provide a data cube of elements with x columns and y rows. An automated segmentation of EDXRF image sets previsualizes heterogeneities of samples in an adequate way [448]. Element mapping generally requires stepwise sequential data acquisition. With EDXRF the accumulation time can be reduced due to the simultaneous spectra acquisition. A new method for simultaneous X-ray imaging using a microchannel plate X-ray optics [449] has opened a new dimension of element mapping.

One aspect in EDXRF mapping is the data acquisition, which is dependent on a number of factors mentioned earlier, but major problems arise from the time-consuming data presentation, and manipulation and adaptation of the coordinate systems to other microanalytical systems to verify anomalies in the sample. Since the chemical data acquisition is free of distortion, the combination with optical scans achieved by optical desktop scanner or by slide scanner is a basic approach to interpret chemical and optical data.

## Applications

### Stratigraphic Control

Within uniform rock sequences the identification of characteristic marker horizons is generally a difficult task. Dating by absolute and relative methods might solve the problem, but it is time consuming and would require repetition for each locality. Chemical bulk analyses of samples taken in equidistant steps do not provide a continuous image but deliver bulk information on an integrated sample only. The adaptation of laser ablation (LA)-ICP-MS techniques along 1D profiles and micro-EDXRF 2D-mapping gives very detailed information for relatively small sample sets. Using the EDXRF scanner described earlier, meters of drill cores can be scanned almost continuously at a 50-μm spatial resolution with elevated statistical significance as compared to a point-like source. Within fine-grained laminated marine and lake sediments this method provides excellent spatial information on seasonal compositional changes of varves [450–452], turbidites, and flood-derived sediments [453].

Quantitative analyses are difficult to obtain, since geometry, grain size, and water content of the samples are in general unknown. On the other hand it is not really needed, since the semiquantitative information obtained reflects the variability of certain element concentrations for individual lamina. A comparison of sets of individual summer or winter layers shows that the differences in element contents are basically lithologically controlled. Tilting with respect to the horizontal X-ray beam of thin lamina appears to be a problem locally. Nevertheless, the information obtained from a 1-cm sample provides details of 10 to 30 years in finely varved sequences, composed of winter and summer layers [452]. The spatial resolution of 50 μm is high enough to outline even activities of algae and diatomea adding silica to the sediment in spring and summer, whilst Ca precipitates are related to summer, and Mn and Fe precipitates to autumn and winter (Plate I 3). Sediment cores of the Lake Constance are used to reconstruct the environmental history and processes within the lake and its catchment area, such as heavy metal contamination, eutrophication, and climate history. The sediment input is governed by the annual runoff due to the melting of snow in the catchment area of the alpine Rhine River, which drains some 12000 $km^2$ of the Alps. This generates a maximum sediment input during the summer. Additionally, single events of heavy rainfall generate strong flood events that lead to annually laminated sediments. The sedimentary record is composed of flood events of the Rhine River system (grayish layers) and the molasse-catchment basin (yellowish–brownish layers), mixtures of both, as well as individual carbonate rich lamina (whitish layers), which can be observed within the background sedimentation (Plate I 4).

## Alteration, Mineralization, and Tectonic Overprint

The understanding of alteration and mineralization patterns in brecciated, hydrothermally altered rocks is a complex problem. Alteration and mineralization textures are often difficult to interpret by eye. Microscopy of thin and polished sections provide insight on the chemical composition of small portions only. The XRD patterns of the bulk or small selected areas give information on primary and secondary products but not much detail on the spatial distribution.

Micro-XRF enables the visualization of chemical composition distribution patterns on the hand specimen scale providing valuable data for understanding rock–water interaction processes. For example in andesitic rock units the fragmented rock might undergo all stages of degradation, namely propylitization, sericitization, alunitization, ferric and silicic alteration, and finally weathering. All these stages are documented by the redistribution of elements (i.e., Fe, Ca, K, Si, Cu) within the rock. Enrichment and depletion can be recognized (Plate II 1.1-6) and related to respective alteration processes. Some elements such as Ti (Plate II 1.3) remain immobile and outline the distribution of the andesite fragments. Alteration might go so far that only small relics (see Plate II 1.2: elevated Ca) or even traces of fragments enable the identification of the primary rock type. At the rim of the fragments potassic and ferric alteration can be noticed by high K and Fe concentrations (Plate II 1.2-3). Within the interfragmental area precipitation of sulfates, e.g., alunite ( Plate II 1b, d-g: K, Al, S) and sulfur (Plate II 1.3, 1.5-6: S), takes place. Silica accumulates in small interstitial areas (Plate II 1.4, 1.6).

A major field of interest is the distinct identification of mineralized zones in rock sequences, especially in drill core samples. This sample type provides considerable advantages for the 1D-EDXRF scanning devices: a more or less continuous succession, a relatively smooth surface, and in the case of a half-core even a flat surface. Depending on the desired detection limits one hour or a few days are needed for the analysis of a sample of 1 m length. For most exploration approaches 20 to 30 s per 50 μm step (20000 points/m) are sufficient. Scanning hard rock samples, for example chromite, chalcopyrite, or pentlandite distribution, focuses on the distribution of Cr, Cu, and Ni, respectively. Searching for anomalous zones of PGE or Au content is a quick and successful approach (Plate I 2) to detect areas of element enrichment. While the localization of Cr, Cu, and Ni due to a good peak/background ratio can be quite easily attributed visually to zones of mineralization, the distribution of Au and PGE is frequently obscured by diffraction peaks due to low concentrations, grain size effects, and background noise variation because of porosity, etc., and requires more detailed information by microscopical follow up or 2D-EDXRF mapping to differentiate between point-like PGE anomalies or silicate-shaped diffusive patterns on the same energy level. Besides these difficulties the localization of PGE [454], Ge-bearing phases [455], or gold in banded iron formation (Plate II 3.1-2) is possible within a short period of time.

A further aspect which deserves attention is the possibility of determining mineral ratios of/in the sample by identifying and attributing mineral specific element patterns based on element ratios (Plate II 5.1-2). This approach has been applied for EDX-SEM systems for many years in the mining industry to monitor all steps within the treatment plants to obtain the best possible concentrate. The determination of the degree of liberation of a certain mineral, the inter-grown phases, the characterization of the size and shape and finally of the consequences for the mineral processing such as milling, screening, magnetic or gravitative separation and flotation, mainly forwarded by the QEM $\times$ SEM (quantitative evaluation of minerals by scanning electron microscopy), have been developed by Gottlieb and scientists of CSIRO, Australia [456, 457]. With some modifications such a system could be applied to the mesoscopic scale to identify individual minerals [458]. Michibayashi et al. [460] characterized silicate minerals from ductile shear zones by EDXRF microscopy.

Tectonization, fluid transport, alteration, and mineralization often interact in such a way that a proper attribution of the succession of the events appears to be difficult on a macroscopic scale. Often thin section work goes so much into detail that peculiarities are missed. By EDXRF mesoscale scanning this problem can be overcome. Zones of enrichment or depletion can be outlined,

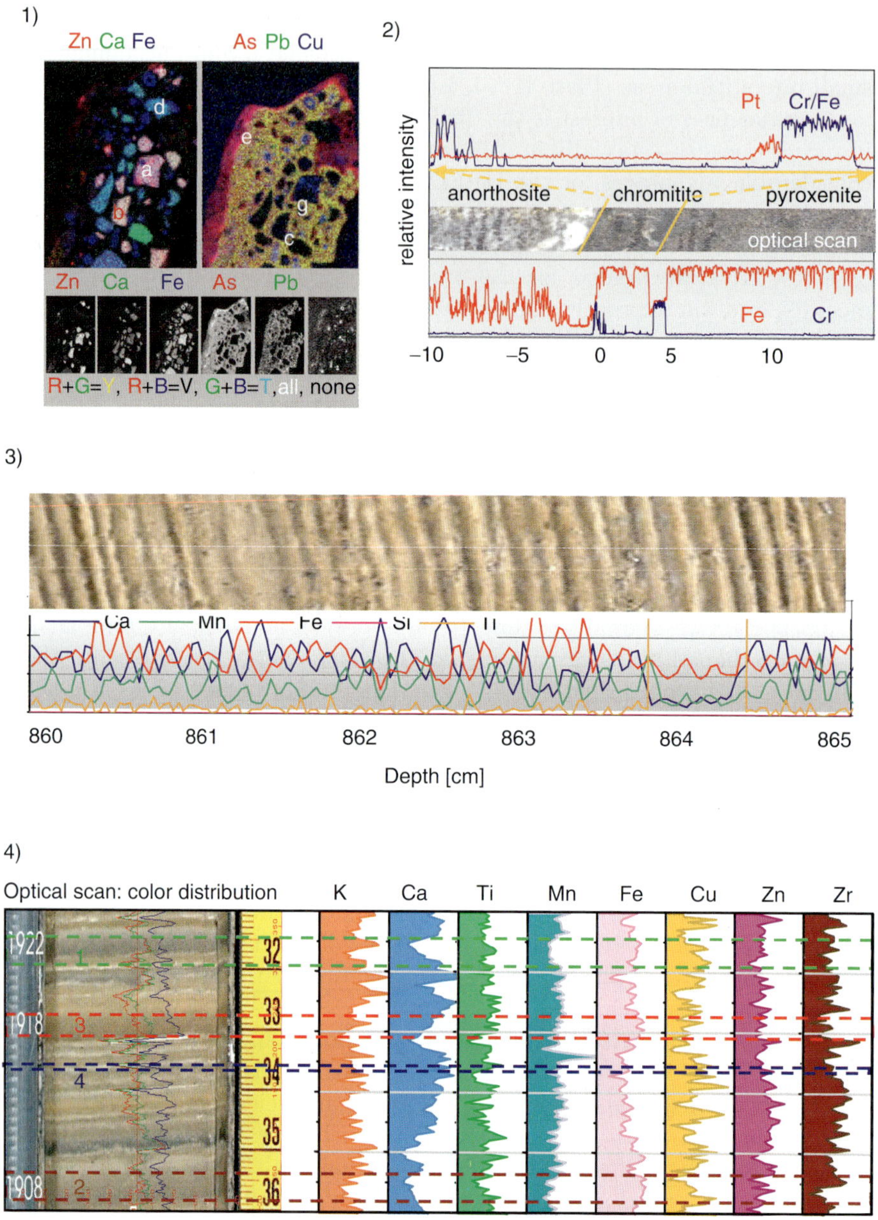


**Plate I. 1–4.**

1. Consists of silicate slag with sulfide spheres (g), fly ash (d), and roasted arsenopyrite have partly been altered and agglutinated by an As–Pb-rich gel phase (f) showing a rim effect (e). Zn occurs partly within the glass being highly dissoluble (a), and also in stable Zn spinel (b) (*Figure caption continued next page*)

multiple tectonic or recrystallization events can probably be identified and used to identify a typical chemical signature (Plate II 4.1-4).

## In Situ XRF Analyses During Experiments

In many cases geological investigations require experiments under controlled laboratory conditions to understand natural processes. The XRGS setup was designed to run column tests accompanied by repeated EDXRF scans. The aim of the project is to obtain kinetic information on the behavior of certain elements within capillary force-driven transport in columns of sedimentary material and, of the result of this transport, to develop a newly formed hardpan. The experiment was performed to elucidate the transport conditions in slag heaps from metallurgical plants and fines from tailings dams, but it can also be applied to monitor transport in soils or weathering processes in sediments. Figure 7.129 shows the result of an experiment that was run for five months. The image exhibits mushroom-like features in the center of the sand-filled column, documented by both optical and chemical (Fe) images, respectively. These features are believed to be derived from pulse-like availability of the transported matter in solution from the water-saturated level filled with mining residues. At the capillary fringe at the top of the column, alternating hardpan layers of Ca- and Fe-rich precipitates (gels) can be observed.

The column material is filled in a polyethylene tube ($\phi = 5$ cm; 80 μm thick) and is water saturated over a large part of the column. Due to these circumstances the fluorescence yield for Ca is reduced by 40–50% and for Fe by about 10%. However, enough information is provided to visualize the

---

**Plate I. 1–4.** *Continued.*
2. XRGS profile across the anorthosite–chromitite–pyroxenite boundary of the Merensky Reef, Bushveld Complex, SA. Fe and Cr highlight the chromite and pyroxenite content. Low Fe values outline the plagioclase. The detailed view of the chromitite zone shows elevated pT values adjacent to the upper chromitite layer
3. Laminated lake sediments (Belauersee, northern Germany) show alternate dark and bright layers, which are interpreted to document summer (Ca) and winter (Fe, organic) precipitation, respectively. Mn is slightly shifted to the autumn. Ti and Si indicate influence from other sources such as wind transported sand or dust
4. Flood records: Example from Lake Constance showing various flood events with a distinct chemical signature. The 1922 flood of the Rhine river system (1) shows a typical increase of Fe and decrease of Ca in the beginning. The irregularities toward the top might be due to the contribution of other tributaries recruiting from the molasse–catchment (brown and yellowish layers). The 1908 flood (2) can be regarded as a mix of the Rhine River system and probably Bregenzer Ache River. The 1918 flood of the molasse–catchment (3) shows a strong increase of Ca, K, Ti, and Mn and a decrease of Fe. Between the flood events some whitish layers appear, which show elevated Ca due to precipitated calcite (4). Zr, Zn, Ti, Mn peaks will need more detailed investigation for proper explanation

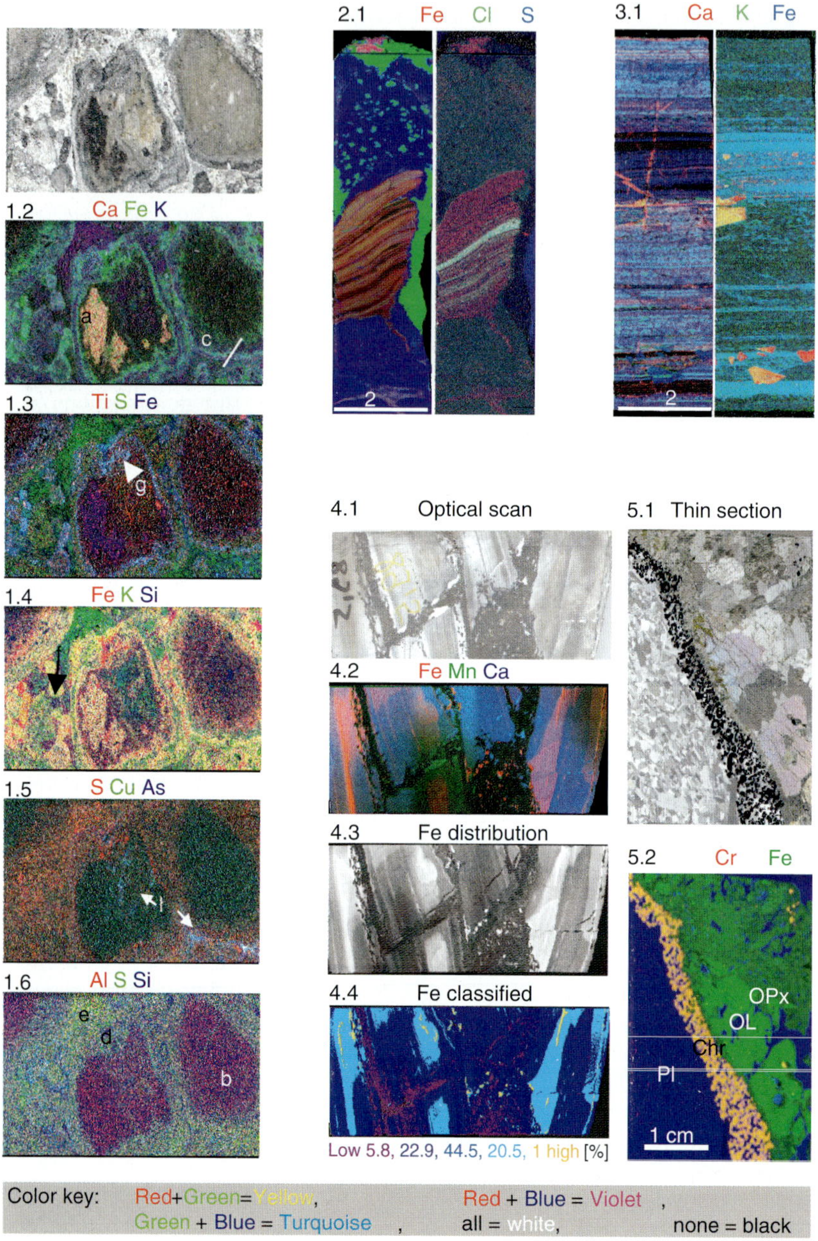


**Plate II. 1–5.**
Visualization of chemical alteration pattern of andesitic breccia:
**1.1-6** (**a**) relatively fresh Ca rich relic with ilmenite, (**b**) altered with Ti, (**c**) alteration rim with Si, Ca depletion, and Fe and K enrichment rims. The interstitial filling consists of (**d**) alunite with (**e**) some native sulfur and (**f**) silicified zones. (**g**) fragments contain Fe, S filled fractures, whilst (**h**) Cu–As-mineralization occurs along cross-cutting features. Some elements are enhanced relative to others
**2.1-2.** Ore genesis: Mn-bearing banded iron formation (**a**) with calcite (**b**) and quartz veinlets (**c**). Pyrite blasts (**d**) are concentrated in two zones with variable grain sizes. Some pyrite grains show elevated Au values (**e**) (*Figure caption continued next page*)

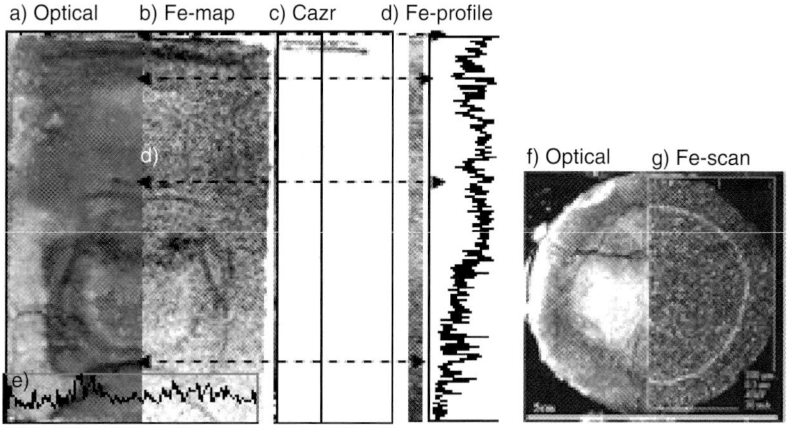


**Fig. 7.129.** (**a**–**g**) Visualization of transport conditions in a column experiment. Image of multiple mushroom-like oxidation fronts in a longitudinal section of the upper 9 cm quartz filling of a pyrite column showing favored transport pathways. The Fe distribution across the mushroom texture mirrors the photograph (**a**) The Ca distribution shows a crust with distinct layers (**c**). The upper crust bends around the edge and incrustrates the outer wall for about 4 cm down the column. The black dots are isolated zircons (**c**) in the sand matrix. A central vertical profile (**d**) shows the intensities of the Fe distribution gaining toward the crust. The cross-section at 9 cm depth (**e**) outlines the dark Fe-rich zone. A horizontal section through the central part of the column (**f**) shows an outer rim in gray, a highly oxidized zone in black and a white core for the photo. The EDXRF Fe map (**g**) shows an enriched zone (white) about 0.5 cm from the rim (see black zone in a). White dots in the center and rim correspond to Fe-rich minerals in the quartz filling. The gray scattered distribution is attributed to Fe-gel coating of quartz grains

---

**Plate II 1–5.** *Continued.*
**3.1-2.** Stratigraphic control: Within salt diapirs a stratigraphic correlation of individual lamina is often complicated. Detailed EDXRF mapping provides a characterization of sequences. Folded anhydrite with rock salt cubes with laminated clay and salt injections. The clay sequence shows characteristic Zr-enriched layers. Tectonic injections of clay and rock salt into anhydrite and salt soaked clay layers can be observed.
**4.1-4.** Optical (4.1) and EDXRF scans of a mineralized breccia from Namibia (Melcher [455]). A composite elemental map (4.2) shows the PbZnCu ore in black, and disseminated and patchy Fe sulfide in red. Mn forms a halo in the dolomite and precipitates as manganiferous–calcite in the ore zone. Earlier calcite veinlets had displaced pulse-like Fe-infiltration zones. The Fe distribution (4.3) shows when classified (4.4) several zones. Dolomite background (blue), infiltration turquoise (yellow highest concentrations), zones of leaching and Cu ore (dark blue), and very low values in Mn-carbonate veinlets and halo zones (purple) are seen.
**5.1-2.** EDXRF combined elemental distribution map mirroring the mineralogical distribution. For comparison an optical scan outlining plagioclase (Pl), chromitite (Chr), olivine (Ol), and orthopyroxene (OPx) is shown.

transport process. The main transport takes place in the center, but chemical signature at the rim follows the central trend except at a lower concentration.

## Imaging Micro-XRF, Texture, and Lattice Strain Mapping

XRF analysis determines the relative amounts of elements within the probed sample volume. Material properties of solids, however, not only depend on the concentration of elements, but also on the phases as well as on the arrangement of crystallites and their orientation distribution, the *crystal texture*. Polymorphism for instance, i.e., crystal forms of the same chemical composition but of different crystal lattices, cannot be detected by any analytical method. Therefore, analytical investigations must often be complemented by using additional techniques to reveal the morphology, the grain orientations, and the distribution of internal stress. In the 1990s, SEM in combination with elemental analysis by energy-dispersive spectroscopy (EDS) and backscatter Kikuchi diffraction (BKD, also known as "Automated EBSD (electron backscatter diffraction)") has evolved into the standard technique to provide conveniently these complementary aspects of microstructure [460]. Typical spatial resolutions are 5 nm in SEM images with secondary electrons, 0.5 μm in EDS microanalysis, and 50 nm in BKD. Better resolutions of about one order of magnitude are obtained from thin specimens in the transmission electron microscope (TEM) using EDS analysis and transmission Kikuchi diffraction [461].

Electron microscopy is not a universal technique to meet all requirements of microstructure characterization. Many applications in mineralogy only require, at a moderate resolution from macroscopic areas in the specimen surface or from cross-sections, the knowledge of the element distribution and the spatial distribution of certain crystal lattice directions (i.e., pole densities rather than the full crystal lattice orientation of each grain). In these cases, XRF and XRD are the methods of choice. These X-ray techniques are principally superior by about two orders of magnitude in accuracy over their electron microscopical counterparts, namely, EDS analysis and Kikuchi diffraction. Continuous background is almost negligible in XRF. The dispersion, $\Delta\vartheta/\vartheta$s in XRD is excellent as compared to electron diffraction, since spectral linewidth, $\Delta\lambda/\lambda$, is extremely small and the characteristic X-ray wavelengths, $\lambda$, are in the range of the lattice spacings, thus producing large Bragg angles, $\vartheta$.

Almost all X-ray analyses in the laboratory are nonimaging. Conventional lenses are not available for X-rays. To obtain microstructural images from a localized specimen region of the morphology, the spatial distributions of elements, the type of crystal lattice (i.e., phases), grain orientations, and residual stress rather than an average over the whole illuminated area, a mapping technique is required whereby a small spot-size X-ray beam is generated, and the specimen is mechanically scanned under the stationary beam spot. A synchrotron source is ideal for this purpose [462], but availability still demands using conventional X-ray generators as a more feasible alternative. X-ray microbeam scanning instruments for imaging XRF have demonstrated their high potential for many years [463–466], and commercial systems have been developed [447, 467]. Modern X-ray texture goniometers are often equipped with a computer-controlled x-y-z specimen stage on the Eulerian cradle. They enable, in principle, texture mapping, if appropriate software for controlling the mechanical x-y-z sample scan, the acquisition, interpretation and representation of the diffraction data is available. Spatially resolved texture analysis has been demonstrated [468] on a commercial X-ray texture goniometer by calculating offline orientation densities, texture symmetry, texture index, and the Taylor factor in several raster points.

### The X-Ray Scanning Instrument Based on a Conventional X-Ray Source and EDS

Inhomogeneities of morphology, element and phase distribution, texture and residual stress in the submillimeter range are quite common in geology and mineralogy. They have to be accessed in quantitative microstructural analysis since they bear information about the history of rock forming and sedimentation, and are often decisive in the reliability and performance of a product in metallurgy. Spatial resolutions of conventional XRF and XRD are insufficient for studying local features and gradients. To avoid some limitations of electron microscopy as well as conventional X-ray techniques, an X-ray scanning instrument (XSI), based on energy-dispersive X-ray spectroscopy (EDXRS) and diffraction (EDXRD), has been developed [469, 470]. The system is unique in that it combines four techniques: spatially resolved pole-figure measurement [470, 471] as well as, on a grain-specific level, imaging micro-XRF analysis, texture mapping, and lattice strain mapping. Although the initial applications have focused on metals and alloys, and their response to plastic deformation, a wide field regarding deformation of rocks and minerals, in general can be tackled.

In our design, shown schematically in Fig. 7.130, the XSI consists of a conventional X-ray source (PHILIPS generator PW 1830/25) equipped with a fine-focus tube, a two-circle goniometer base (PHILIPS PW 1835), and an open Euler cradle (PHILIPS X'Pert MRD goniometer). Most applications have been carried out with a tungsten anode in order to increase the continuous background. The *white* primary radiation is not filtered, thus enabling a wide range of elements to be excited for fluorescence emission and hence setting almost no practical restriction on the range of lattice spacings for diffraction. A circular pinhole collimator is used to produce a small primary beam spot. The diffracted secondary spectrum is detected with either a conventional proportional counter and postspecimen filter for pole-figure measurement or an energy-dispersive solid-state spectrometer system (SPECTRACE 6100 and TRACOR NORAN Si(Li) detector) for acquiring XRF spectra and texture maps.

The instrument is fully controlled by a personal computer. The specimen is translated step by step across a user-defined grid pattern in the x-y-z stage

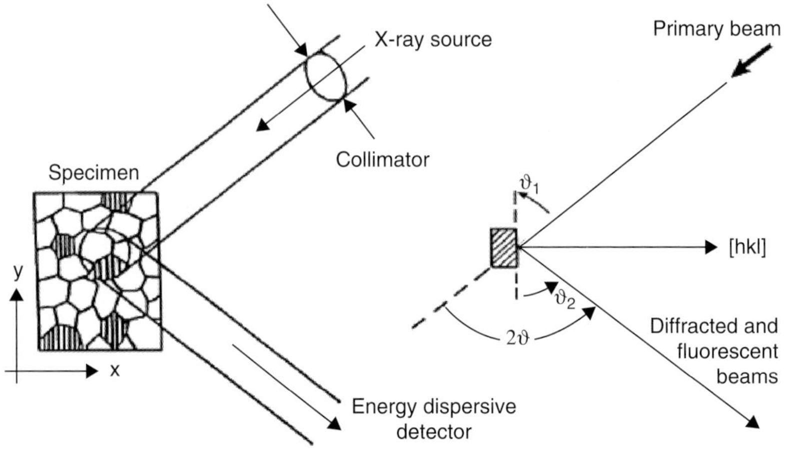


**Fig. 7.130.** Schematics of microbeam X-ray mapping

(Fig. 7.131). Grid patterns of any shape are possible, in order to match the scanned field to the area to be sampled, e.g., scans along a line, across rectangles or polygons in 1-µm wide steps over an area of 100 mm by 100 mm wide. The current sample position under the beam spot is monitored with a video camera, and so the corner points of the sampled region are easily defined.

### Energy-Dispersive Spectroscopy and Data Acquisition

Identification of spectral components in the secondary beam must take into account the following features:

- a low background of continuous radiation

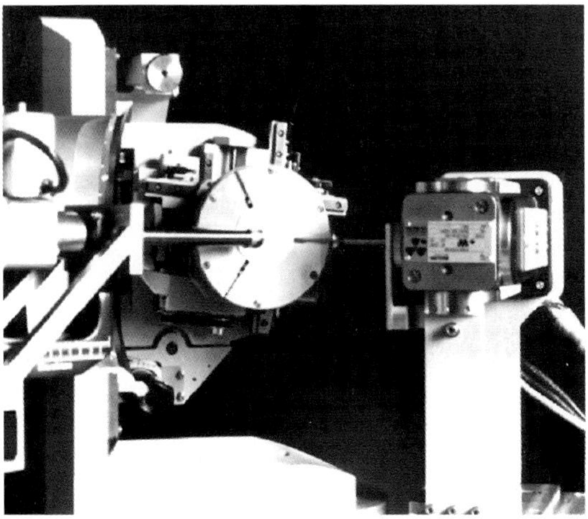

**Fig. 7.131.** The experimental setup of the X-ray scanning instrument

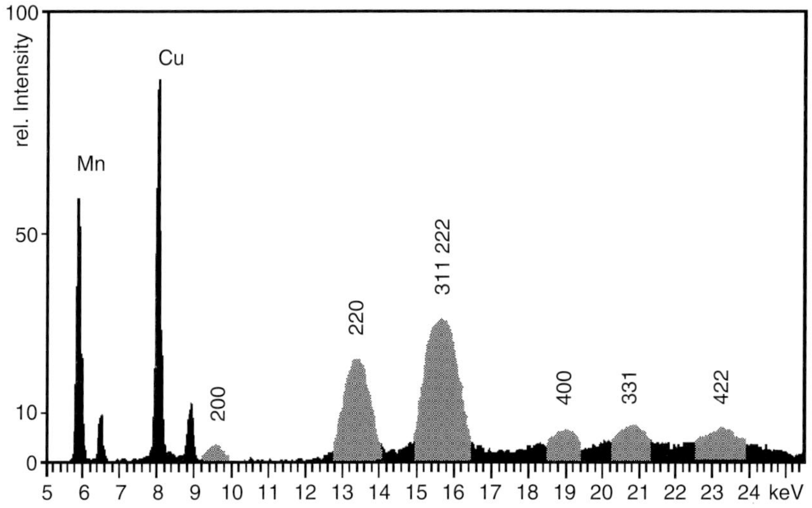


**Fig. 7.132.** Energy-dispersive X-ray spectrum from an AlMn 1% metal sheet

- sharp characteristic lines due to scattering from the primary beam
- sharp characteristic fluorescence lines from the specimen
- broad diffraction peaks from the specimen (Fig. 7.132)

Absorption in air has to be considered for quantum energies $E_x < 3\,\text{keV}$. Hence, a helium atmosphere or an evacuated specimen chamber will be indispensable if light elements are to be detected by XRF or if materials with large lattice constants have to be studied by XRD. Very high quantum energies, on the other hand, are not adequate, since Bragg's angles decrease unduly up to grazing incidence with hard X-rays. Then the spot size is extremely foreshortened and spatial resolution is impaired. A high excitation of fluorescence lines is obtained with an overvoltage of the primary beam of about twice the energy of the characteristic lines of the elements to be detected. A good compromise is obtained for most samples with an angle of incidence of $\vartheta_1 \sim 45°$ and quantum energies of the primary beam between 8 keV and 15 keV.

A comprehensive PC software package has been developed. One part contains the hardware control of the goniometer, the Eulerian cradle, the x-y-z sample stage and the proportional counter, as well as the EDS system. The other part has been developed for the graphical representation of measured pole figures and scanning data of selected sample areas. The experimental pole-figure data can be used for ODF analysis, irrespective of the chosen angular sampling grid and specimen symmetry, and for the calculation of average tensorial material properties by using a dedicated ODF program [472]. An analysis tool identifies the fluorescence and the diffraction peaks in the spectrum. Several XRF elemental as well as texture distribution maps of selected crystallographic directions can be obtained at the same time.

Regions of interest (ROI) are set on selected peaks for evaluating the fractions of counts in each window separately by integrating the ROI intensities. The spatial distribution of residual lattice strain is obtained by interpreting the widths and positions of diffraction peaks in the spectrum of every image point. For mapping the microstructure pixel by pixel, the count rates falling in a selected spectral window are represented on the digital image by attributing specific pseudocolors.

### Basics of Energy-Dispersive X-Ray Diffraction

Energy-dispersive X-ray diffraction (EDXRD) is based on Bragg diffraction of a collimated *white* primary X-ray beam in dependence on the orientation distribution of the grains in the surface. If a grain fulfills the Bragg condition (7.40):

$$E_{hkl} = n \cdot h \cdot c / (2 \cdot d_{hkl} \cdot \sin \vartheta_{hkl}) \quad (7.40)$$

at an angle of incidence, $\vartheta_1$, due to an appropriate orientation of lattice planes at spacings $d_{hkl}$ and for a photon energy $E_{hkl}$ of the primary beam, a partial ray of this energy is reflected. It is collected with a solid-state Si(Li) detector at a take-off angle, $\vartheta_2.\vartheta_{hkl} = (\vartheta_1 + \vartheta_2)/2$ is known as Bragg angle (Fig. 7.130). In general, several sets of lattice planes will meet the Bragg condition at the same time for a broad *white* spectrum of primaries. They produce several diffraction peaks on the ED spectrum. If the illuminated specimen volume is composed of grains of different orientations, each crystallite may reflect its own set of partial beams of different energies into the take-off angle, $\vartheta_2$. All partial beams emitted under the same angle $\vartheta_2$ merge in the secondary spectrum. So several diffraction peaks are produced whose intensities are a measure of the volumes of all grains that fulfill the Bragg condition for the corresponding hkl lattice directions, irrespective of their spatial arrangement in the irradiated sample volume. They resemble traditional hkl diffraction peaks. The intensity of an hkl diffraction peak corresponds to the pole density on the hkl reference sphere of texture at the particular angular setting, $(\alpha, \beta)$, of the specimen with respect to a reference system.

The measured pole densities are proportional not only to the volume fractions of the related grains, but also to the intensities falling in the corresponding spectral windows of primary X-ray radiation. They are further affected by the spectral absorption of the X-rays on their way from the source to the detector (this is in particular the case within the sample), and by possible nonlinearities of the detector system. In quantitative texture analysis, corrections are applied to the measured pole densities.

If the Bragg angle is moved from $\vartheta_{hkl}$ to $\vartheta_{hkl}{}^*$, the centers of the diffraction peaks on the ED spectrum shift from energy channels $E_{hkl}$ to $E_{hkl}{}^*$, according to (7.40), whereas the characteristic X-ray lines that are scattered from the primary radiation, and the fluorescence lines from the specimen, stay on fixed positions on the energy scale. Diffraction peaks and characteristic lines can be easily discriminated. A weak diffraction peak, due to a low pole density, can be enhanced considerably by adjusting the Bragg angle such that the energy of the diffraction peak coincides with a characteristic line in the primary beam spectrum.

Lattice strain is analyzed and mapped with the same setup, in addition to texture and elemental mapping, by evaluating the shift and width of diffraction peaks in the spectrum of each image point [473]. The shift $\Delta\vartheta$ of a diffraction peak, $\vartheta_{\mathrm{hkl}}$, as compared to the peak position $\vartheta_0$ with a relaxed specimen with lattice spacing $\mathrm{d}_0$, follows from Bragg's equation:

$$\Delta\vartheta = \vartheta_{\mathrm{hkl}} - \vartheta_0 = -\tan\vartheta_0(d_{\mathrm{hkl}} - d_0)/d_0, \tag{7.41}$$

or transformed to the scale of photon energy, E, which is more adequate for energy-dispersive diffraction:

$$\Delta\mathrm{E} = \mathrm{E}_{\mathrm{hkl}} - \mathrm{E}_0 = -\mathrm{E}_0(\mathrm{d}_{\mathrm{hkl}} - \mathrm{d}_0)/\mathrm{d}_0. \tag{7.42}$$

Line-profile analysis is performed by Gauss fitting after background correction. The width of a diffraction line reflects the variations of lattice spacing, $\Delta\mathrm{d}$, residual strain from grain to grain in the diffracting volume, strain gradients across grains, or microstructural imperfections in the grains (residual stress of the second and third kind). For a (semi-) quantitative estimate of residual strain, it is common practice to take, under the same experimental conditions, the difference of the full widths at half maximum (FWHM) of the line profiles with the strained and with an annealed reference specimen, respectively. More sophisticated quantitative approaches are based on line-profile analysis by Fourier deconvolution or on profile-matching simulations. Peak broadening due to the finite primary beam and detector apertures, defocusing effects, small grain size, and limited spectral resolution of the detector have to be considered. A well-collimated parallel primary beam and a narrow (Soller slit) collimator on the diffracted beam side in front of the detector can drastically improve $\Delta\mathrm{a}/\mathrm{a}$ resolution, at the expense of intensity and necessary dwell time per pixel. Though being very intense, narrow characteristic lines are, unfortunately, inadequate to excite and enhance weak diffraction peaks for lattice strain analysis, since the primary beam has to supply a sufficiently broad and flat spectrum of primary radiation for ED diffraction. Otherwise the shape of the diffraction peak simply reflects the profile of the exciting characteristic primary beam line.

In the XSI, the sensitivity and the $\Delta\mathrm{a}/\mathrm{a}$ resolution of residual strain measurement are finally limited by the spectral resolution of the solid-state spectrometer system. A standard Si(Li) detector has been used here with a nominal spectral resolution of 140 eV at 5.89 keV (Mn K$\alpha$ line). A tradeoff has to be made between intensity and peak width by using rather wide primary beam and detector apertures. At $2\vartheta = 40°$ (corresponding to $\mathrm{E}_{\mathrm{hkl}} \approx 10$ to 13 keV for low index planes of metals) apertures of 0.35° produce an FWHM between 400 eV and 500 eV, which is distributed over about 100 spectral channels. The apertures are adjusted by using appropriate pinhole diaphragms in the primary beam and a sufficient distance of the Si(Li) detector from the specimen. Peak position is not affected by the aperture size in a symmetric goniometer setup ($\vartheta 1 = \vartheta 2$). By a Gauss fit, the peak center can be located to $\Delta E \leq 0.5\,eV$. Hence residual strains of the first and second kind are well resolved down to $\Delta a/a \approx \cdot 10^{-4}$.

A noticeable improvement in lattice strain analysis is expected from a reduction of the energy interval per spectral channel and a higher energy resolution of the detector system. However, a practical limitation still rests on the need of excessively large apertures for gaining sufficient peak intensity. A position sensitive detector (PSD) with a standard Bragg-Brentano setup and monochromatic radiation is still superior by one order of magnitude or more as far as high dispersion is concerned for residual strain analysis. This would enable a reliable peak-width analysis and also the mapping of residual strain distributions even of the third kind. The replacement of the ED system, however, is at the expense of simultaneous acquisition of several peaks/maps at the same time, and stringent restrictions would be imposed in addition to satisfy the Rowland-focusing condition for the Bragg-Brentano geometry.

Spatial resolution is limited by the size of the X-ray spot on the tilted specimen surface. Since no conventional lenses are available for white X-rays, a beam collimator has been used here, which consists of two small circular diaphragms of diameter d at distance a and b, respectively, from the sample. The effective spot size is then:

$$d_s = (1 + 2a/b) \text{ x } d. \quad (7.43)$$

The main drawback of a diaphragm collimator is the loss of intensity. An increase in spot brightness is expected from glass capillary collimators [474, 475], which would enable a higher spatial resolution and a shorter acquisition time. The apertures, $\Delta\vartheta$, of the primary and secondary beams are less crucial, unless lattice strain mapping is of concern. Since texture in most materials is hardly sharper than a few degrees, a width of diffraction peaks in this range, according to $\Delta\vartheta$ in (7.41), is quite acceptable. The width of fluorescence peaks is not affected by the aperture of the beam spot.

The oblique incidence of the primary beam results in an elliptical foreshortening of the spot on the sample surface. Spatial resolution decreases with angular tilt, $\gamma$, of the specimen with respect to the incoming beam. A circular primary X-ray spot is elongated by projection on the tilted specimen surface to an ellipse of aspect ratio $1/\cos\gamma$. The practical limit of spatial resolution is presently at 0.1 mm for pole-figure measurement and lattice strain mapping, and 0.05 mm for element and texture mapping, due to the low intensity of the collimated primary beam.

Pole-figure measurement has to precede texture mapping on the same specimen area in order to recognize significant texture components. Thinned angular grids, equal area grids, or relevant sections on the pole sphere, as well as areas of any shape on the specimen, can be selected to reduce acquisition time by the combination of pole-figure measurement with a stepwise-controlled specimen tilt through the pole-figure grid and a controlled spatial x-y translation of the specimen. Large tilts, however, may result in inconsistent pole-figure data if the texture is not uniform in depth beneath the surface, since XRD becomes more surface sensitive with increasing incidence angle $\gamma$. Information depth decreases linearly with $\cos\gamma = \sin\vartheta\ \cos\alpha$.

Pole density can be measured alternatively in the conventional way with filtered – *monochromatic* – primary beam radiation and a proportional counter set to the appropriate $\vartheta$–$2\vartheta$ positions for each measured hkl pole figure, one after another. The specimen is tilted to the specified pole-figure points for intensity acquisition step by step. Energy-dispersive microdiffraction, however, using *white* (i.e., not filtered) primary beam radiation and a lightweight Peltier-cooled solid-state detector [476] enables a significantly faster and more accurate acquisition of several pole figures at the same time.

Any specimen reference direction ($\alpha$, $\beta$) can be used for mapping provided that pole density (i.e., diffracted intensity) is high enough – which of course depends on local texture. It is worth noting that texture, as well as residual strain, is probed exclusively only for those grains which satisfy the Bragg condition for the chosen specimen-fixed reference direction, e.g., the specimen normal direction in the case of the symmetric goniometer setup with the specimen not tilted. There is no information acquired on grains at other orientations.

## Examples of Application

As a test specimen for spatial resolution, regular gold patches ($420 \times 420\,\mu$m wide) have been deposited on a glass substrate by sputtercoating through a TEM specimen support grid. Using 30-μm wide apertures and the Au L$\alpha$ fluorescence lines for imaging, the free bars ($<60\,\mu$m wide gaps) between the patches are clearly resolved in two perpendicular directions in the XRF map (Fig. 7.133) [460]. A spatial resolution of approximately $30\,\mu$m is thus demonstrated.

A deformation pattern has been formed by impressing the jaws of a mechanical testing machine into the coarse grain surface of a recrystallized aluminum sheet metal. Finally, the sample surface was ground and mechanically polished such that the pattern was no longer visible. The stamp process, however, has changed the local texture. In the texture maps (Fig. 7.134), high pole densities can be recognized in regions where the corrugating teeth of the jaws caused a high deformation, while the grooves in the jaws induced a deformation low enough to preserve the original matrix texture. The texture maps of Fig. 7.134 were acquired simultaneously from the same sample area at symmetrical diffraction settings ($\vartheta 1 = \vartheta 2$). Hence, the pole densities in the sample normal direction are shown. The matrix texture did not contribute to these selected diffraction peaks. If a texture component of the matrix is of interest, the specimen has to be tilted off the symmetrical geometry to an

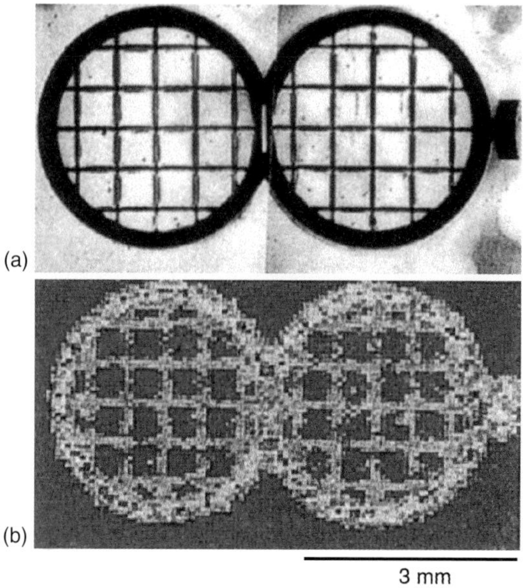


**Fig. 7.133.** Gold pattern sputtered through a TEM support grid on a glass substrate. (**a**) Image in the light microscope.
(**b**) X-ray micro-fluorescence map with the Au L$\alpha$ line

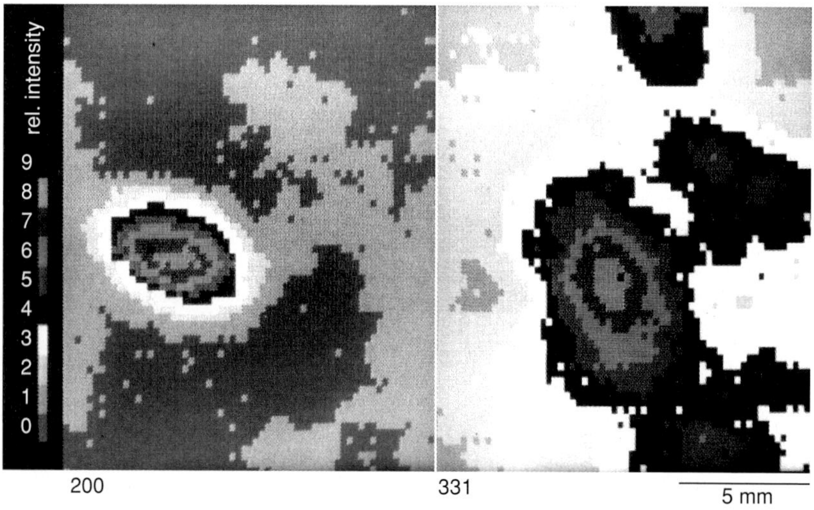


**Fig. 7.134.** Texture maps of a deformation pattern in aluminum revealing local plastic deformation. Left: {200} reflections; right: {331} reflections

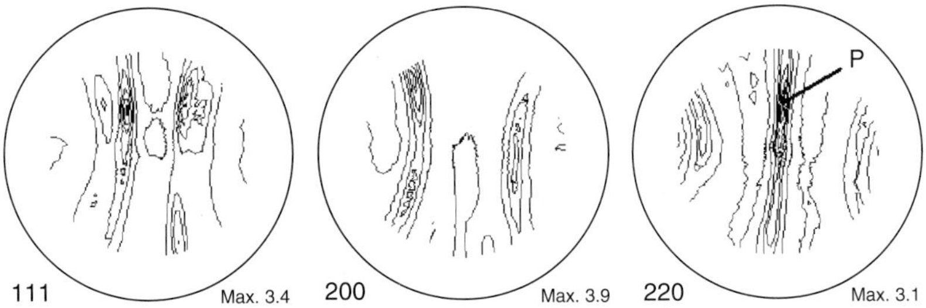


**Fig. 7.135.** Pole figures acquired across the rivet bolt of Fig. 7.136. The polar specimen direction P ($\alpha = 35°$, $\beta = 83°$) has been used for texture and residual strain mapping

appropriate pole-figure point. The corresponding angles of tilt can be read from X-ray pole figures that have to be acquired in advance.

The X-ray spectrum (Fig. 7.132) contains two sets of characteristic lines: a set of Cu K$\alpha$ and Cu K$\beta$ of scattered primary beam radiation (Cu anode, 40 keV), as well as a strong set of Mn K$\alpha$ and Mn K$\beta$ fluorescence radiation, which has been emitted from the rather small content of manganese in the AlMn 1% alloy. Al–K fluorescence radiation, as a possible characteristic line from an Al sample, is not present. The Al–K fluorescent line is excited rather weakly by Cu K radiation and, furthermore, is strongly absorbed in air on the way from the sample to the EDS detector. In contrast to the sharp characteristic lines, the diffraction peaks are broad, due to the wide beam aperture (see (7.41)). In Fig. 7.132, the windows of the diffraction peaks (200), (220), (311)/(222), (400), (331), and (422) are marked by hatching. The (311) and (222) reflections are coincident, but the contribution of (222) is negligible in this particular point on the sample because the (111) peak is missing.

A cross-section of an aluminum rivet was carefully polished flat by gentle mechanical grinding and finishing. A final electrochemical polishing was applied to remove any mechanical damage introduced during preparation. Before mapping, several pole figures had been acquired in steps along lines across the head and the bolt of the rivet to recognize relevant texture components. A strong fiber texture has been found in the bolt region (Fig. 7.135) [473]. The (220) diffraction peak P ($\vartheta 1 = \vartheta 2 = 20.1°$, $E_{220} = 12.62$ keV; $\alpha = 35°$ and $\beta = 83°$) has been used for evaluating at the same time the spatial distributions of pole density (texture map) and strain (residual strain map). Dwell time extended over 10 s/pixel to yield a sufficient count statistics in view of the low intensity. At a step width of 120 µm 2,232 image points had been acquired.

A clear inhomogeneity is revealed in the texture (Fig. 7.136a) as well as in the strain maps (Fig. 7.136b). Two almost parallel strips of high pole density run along the axis of the bolt with a slight shift to the right-hand side. They are supposed to form a tube of about one-third of the bolt diameter in the

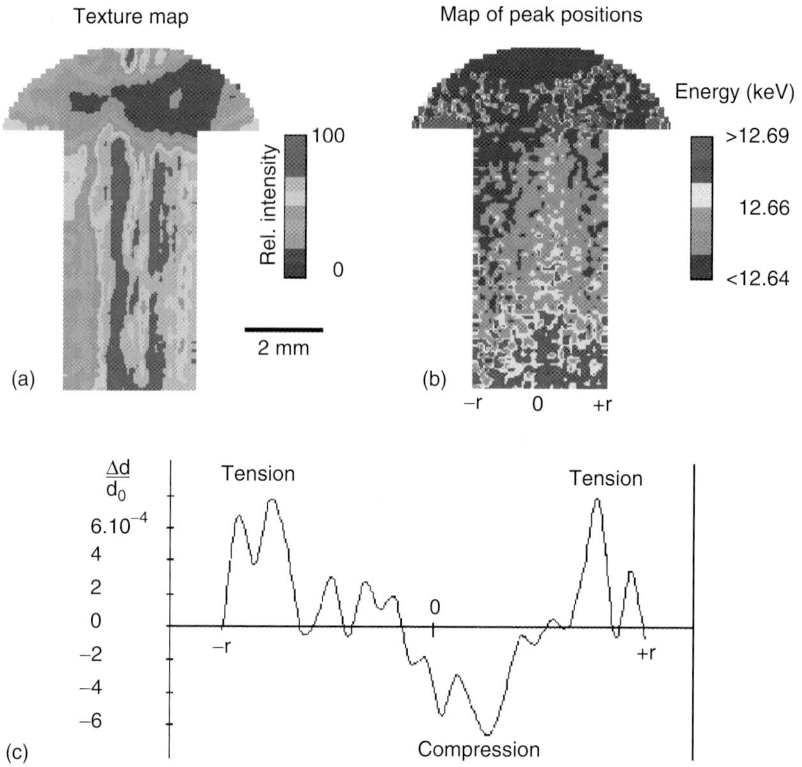


**Fig. 7.136.** Cross-section of an aluminum rivet. Distribution maps (**a**) of the intensity (pole density, texture) and (**b**) of the shift (residual strain) of the 220 peak at pole P ($\alpha = 35°$, $\beta = 83°$). (**c**) The lattice strain $\Delta d/d$ across the bolt

3-dimensional rivet. The head of the rivet shows a much lower density of this pole. It is worth noting that it forms a butterfly-shaped minimum, which might result from punching the head during production.

Along the bolt is a gradient in compressive residual strain to the head. A dish-shaped maximum of compressive strain (dark) is found in the head, which is related to the butterfly-shaped minimum in the (220) texture map. There is an increased compressive residual strain in the interior of the texture tube (dark) and a tensile strain (bright) to the bolt surface. Statistics has been improved by averaging the peak shifts along a section of the shaft. So the radial distribution of residual strain across the bolt is clearly seen in the graph line (Fig. 7.136c).

### 7.6.4 Microscale

Chemical analyses with high spatial resolution are of major interest in geological applications as minerals are often chemically heterogeneous on the microscopic scale and these compositional variations are of major importance for understanding the formation processes. Most of this microanalytical work is performed by EMPA due to the spatial resolution in the micrometer range, well-established quantification procedures, and the widespread availability of instruments. While this method is well suited for the determination of major elements it is not sensitive enough to detect trace elements at levels below approximately 100 ppm. To fill this gap, micro-XRF instruments using X-rays for excitation for multielement EDXRF analyses may be used and an instrument has been developed especially for trace element analyses on single mineral grains [428]. The detection limit of such an instrument is about 5–10 times better than with EMPA but the spatial resolution is about 10–100 times lower. A promising application is the U–Th–Pb dating of grains of monazite as described later. For trace element analyses with higher spatial resolution and/or element contents in the lower ppm range synchrotron-based micro-XRF methods have to be applied.

### Sample Preparation

For trace element analysis with XRF at high spatial resolution sample preparation is of major importance. For positioning the beam exactly on the sample and precisely defining the analyzed sample volume, its surface should be smooth or even polished. The X-ray beam penetrates deep into most geologic materials. For an exact quantification of the XRF spectra it is crucial to know the thickness of the analyzed sample volume. For those two reasons, double-sided polished samples of consistent thicknesses are most favored for micro-XRF analysis. Another consequence of the penetrative nature of the beam concerning the sample preparation is that the spatial resolution is controlled by the sample thickness. Any backing of the sample should be minimized or avoided since the backing material will contribute to the scattered radiation and will increase the background of the spectra leading to higher detection limits and potential additional peaks. Samples should be prepared as thin as possible while still mechanically stable. Typical thicknesses of geological samples for microanalysis vary between 30 and 100 μm.

### U–Th–Pb Chemical Dating

Radiometric dating provides absolute constraints on the timing and duration of (long-lived) geological processes and is therefore of fundamental importance within the geosciences. U–Th–Pb dating is one of most used tools for radiometric dating. It is based on the decay of radioactive U and Th isotope to stable isotopes of Pb in three independent decay schemes (see later). Chemical dating, the calculation of the age from elemental abundances of U, Th, and Pb, is one of the earliest absolute geochronological techniques. It was even used on minerals containing abundant U or Th (e.g., uraninite – $UO_2$; [477]) before the development of mass spectrometry. Geochronology by isotope analysis is a modern and more accurate technique allowing the comparison of the three differently calculated ages and corrections for nonradiogenic Pb in the mineral, but it is expensive and labor intensive.

Renewed interest in chemical dating techniques arose from the application of relatively cheap and readily available EMPA [478, 479], which has the additional advantage of a spatial resolution in the range of 2–3 µm. This spatial resolution allows the study of complexly zoned minerals. The application of nondestructive analyses on Th- and U-rich accessory minerals, especially monazite, became the focus of these studies because in situ microanalysis in geological samples allows the correlation of age data with the textural development of the main rock constituents.

Monazite ($CePO_4$) combines several characteristics that makes it very favorable for chemical U–Th–Pb dating. First, it is a common accessory mineral in many crustal rocks, granitoids, gneisses, as well as metasediments. Second, it typically incorporates little Pb upon crystallization, but contains very high amounts of $ThO_2$ (up to 30 wt%) and $UO_2$ (up to 3 wt%), which leads to a fast accumulation of significant amounts of radiogenic Pb. Quite importantly, radiation damage that could cause loss of Pb by diffusion is usually negligible because the crystal lattice anneals quickly. Zoning within monazite can be controlled with backscatter electron images due to the large compositional variability (heavy elements like Th, U, Y, and REE) in the monazite solid-solution series toward huttonite ($ThSiO_4$) or brabantite ($CaTh(PO_4)_2$). Monitoring trace elements in monazite (e.g., Y, REE) has been used to link monazite ages with specific metamorphic episodes [480] and pressure–temperature paths of its host rock [481]. This has made in situ monazite studies a versatile tool in the interpretation of complex metamorphic rocks. Chemical U–Th–Pb dating may also be applied to other minerals, such as xenotime, zircon, or allanite, but most of them do not fulfill all requirements for a suitable geochronometer or just contain much less radiogenic Pb than monazite.

## Comparison of Analytical Approaches

In chemical dating, an age can be calculated iteratively, by assuming that the total amount of Pb can be divided into thorogenic and uranogenic portions of the same age (e.g., [478]) or by calculating independent Th–Pb and U–Pb ages based on the algorithm of Rhede et al. [482]. The most commonly used analytical method for chemical U–Th–Pb dating is wavelength dispersive X-ray detection with an electron microprobe (e.g., [478, 479, 482–485]). At 20 kV acceleration voltage and 50–100 nA intensity the electron beam can be focused to give an excitation volume of approximately 2–3 µm width and 3–5 µm depth. However, the long half-lives of U ($^{238}U = 4.47 \times 10^9$ a, $^{235}U = 0.704 \times 10^9$ a) and Th ($^{232}Th = 14.01 \times 10^9$ a) mean that accumulation of significant amounts of radiogenic Pb takes a long time and the typical detection limits of approximately 100 ppm for Pb prohibit precise determination of ages younger than approximately 200 Ma on the EMP of the main rock constituents (see Figure 7.137).

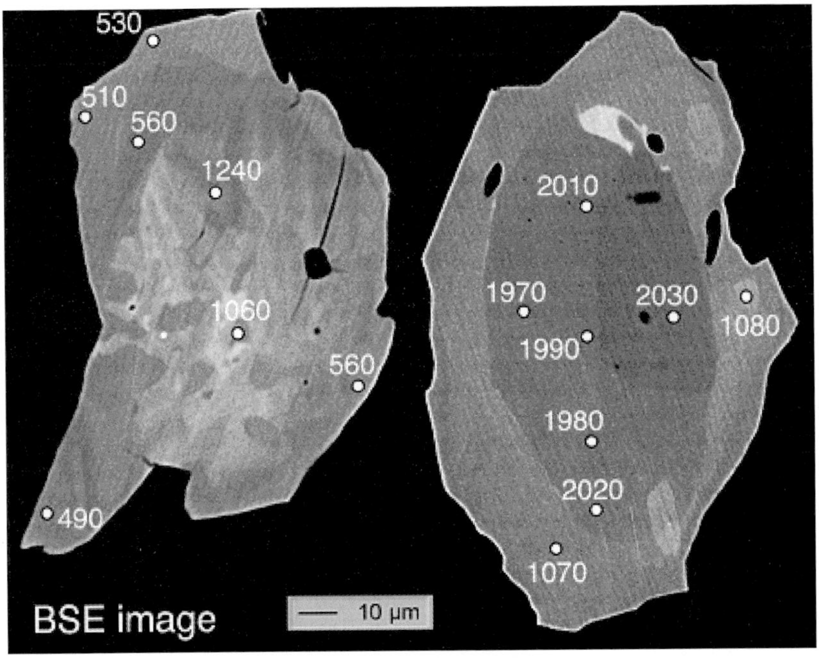


**Fig. 7.137.** Monazite from the Zambezi Belt, Zambia [490]. Brightness of the backscatter-electron (BSE) images is a measure of the average atomic number of the material. The images are used to detect compositional zoning of target minerals prior to analysis. Areas with higher Th and U enclose internally complex grain cores with highly variable Th in the left image and lower U contents in dark zones in the right image. Lobate boundaries of the rims impinging on the inner parts of grains are typical for fluid-assisted recrystallization. Areas analyzed by electron microprobe are indicated by circles of approximately 3 µm diameter, with the approximate calculated ages given in million years

Bruhn et al. [486] used a proton microprobe (PIXE) to date monazite, with lateral spot size of 20–30 µm but high penetration depth (30 µm), and they were able to reproduce isotopic ages of monazite standard material. A 1.5-nA sample current and acquisition times of 5–10 min yielded detection limits of approximately 10–20 ppm for Pb, Th, and U and $<$ 1000 ppm for the REE. Frei et al. [487] have shown the feasibility of micro-PIXE chemical dating on a 1 Ga old high uranium (265 ppm) titanite. Structural damage to the minerals due to the large mass of the protons in PIXE analyses has not been investigated but cannot be ruled out.

Analytical techniques using X-ray induced excitation of fluorescence (XRF) also have good potential in providing lower detection limits due to the better peak-to-background ratio of this technique. A micro-XRF method was developed by Cheburkin et al. [428] with a set-up using monochromatic X-ray radiation. Scherrer et al. [488] modified this technique, using it to date monazite with a lateral resolution between 38 and 90 µm in doubly polished sections. These methods use EDX detection, enabling the acquisition of the complete XRF spectrum at once in a relatively short time. However, at the energies needed (approximately 17–20 keV), the beam transects the complete thickness of a 30-µm-thick thin-section (i.e., thickness of the sample determines the analyzed volume) and therefore spatial control is limited. Detection limits of approximately 10 ppm were achieved with a beam of 50 µm on whole grains at 40-min data collection times, sufficient to date monazite grains as young as 15 Ma [489].

Additional improvements may be achieved by the use of synchrotron radiation XRF (SRXRF). The higher photon flux and brilliance of such X-ray sources provide the possibility of focusing the beam down to 2–3 µm lateral diameter or less. The high photon flux of the beam still ensures very low detection limits down to ppm range or better, thus pushing the theoretical limits for the youngest datable monazite below 1 Ma even for monazite with low contents of $ThO_2$ (4 wt%). In other words monazite with an average $ThO_2$ of 10 wt%, yielding 20 ppm of radiogenic Pb within approximately five million years could be dated relatively precisely. However, small amounts of nonradiogenic Pb may cause some concern at this level of detection.

The lower detection limits in PIXE, micro-XRF, and SRXRF analyses make the investigation of other phases or minerals with lower concentrations of U and Th that accumulate less radiogenic Pb viable (e.g., zircon, titanite), of course only if nonradiogenic Pb is negligible.

### Microanalysis Using Synchrotron Radiation XRF

The use of SR as a photon source in contrast to a lab source has enabled studies on trace element contents and compositional changes of tracements in minerals and glasses on the micrometer scale. The high flux of synchrotron radiation X-rays additionally enables analyses of buried materials like fluid and melt inclusions and in situ studies in sample volumes inside an experimental apparatus like a diamond anvil cell. The wide continuous energy spectrum provides multielement information.

Elements of interest are dependent on the given geological question and may cover the whole periodic table. While major element compositions in glasses and minerals are easily accessible by EMPA, trace element studies with high spatial resolution of as many elements as possible are a major application of µ-SRXRF. As the synchrotron radiation spectrum is continuous the excitation spectrum may be efficiently adjusted to the sample and the given analytical problem by the use of filters or monochromators. Additionally, elements with Z up to 82 may be detected via K-shell excitations (Fig. 7.138), which is a significant improvement compared to conventional X-ray tubes (maximum excitation conditions generally below 20 keV), e.g., solving the problem of strong overlaps of REE-L lines with K-lines of lower Z elements. In order

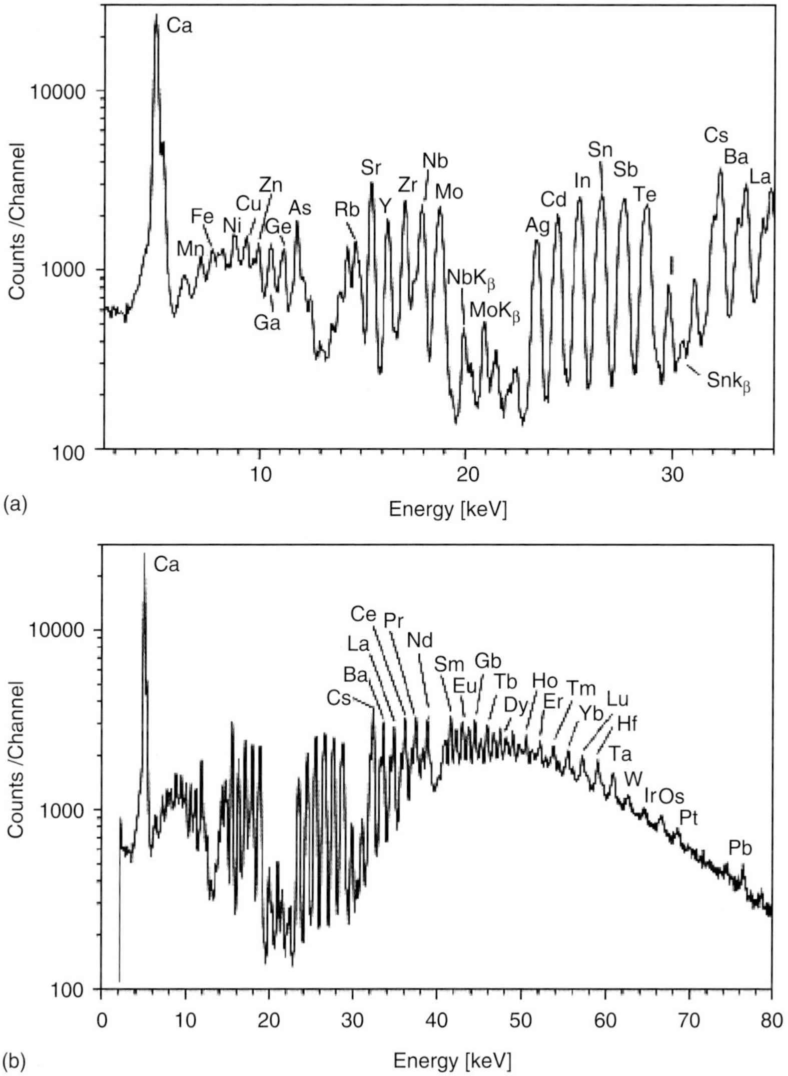


**Fig. 7.138.** (**a–b**) Fluorescence spectrum of multielement 100 μm thick doubly polished standard reference material NIST612. The glass matrix is doped with trace elements with Z between 22 and 92 in the concentration range between 30 and 80 ppm. The most important K-lines are labeled. Note that Pb may be detected via K-shell excitation. A 20-mm thick Al absorber was inserted for spectrum acquisition to adjust excitation conditions for elements with Z around 45–60. The live time was 300 s and the beam size 50 x 50 $\mu m^2$. Figure 7.138a shows an enlarged view of the low energy part of the spectrum displayed in Fig. 7.138b

to achieve efficient simultaneous excitation toward high-energy K-shells, the polychromatic beam is generally used to achieve satisfactory flux. However, elastically scattered photons contribute to the background at all energy levels resulting in lower sensitivity and may lead in some cases to detector overloading. Peak-to-background ratios of the low-energy part of the spectra may be improved by inserting absorbers for low-energy cut-off. Another, more effective, method to reduce the background contribution is the use of monochromatic or small bandwidth excitation resulting in lower background and thus higher sensitivities. An additional advantage is the possibility of selecting the optimum excitation conditions for the element of interest, i.e., an excitation just above its absorption edge. Because of the lower photon flux when using just a small bandwidth of the spectrum, the detection of heavy elements via K-shell excitation is limited to third-generation synchrotron radiation sources and beamlines optimized for hard X-rays. Nakai et al. [491] have shown an example of using 116-keV X-rays as an excitation source of X-ray fluorescence analyses. The detection limits were minimal for W with 0.1 ppm in the SRM NIST612. Those measurements, however, were not achieved with a microbeam. This is a future option, which can be achieved by use of compound refractive lenses (CRL) that enable focusing even of hard X-rays down to the 100 nm scale (at 24 keV, [492]) or to the micrometer scale or better (up to 120 keV, Schroer, pers. com.) at third-generation synchrotron sources.

## Applications

*Mineral, Particle, and Glass Compositions.* In cases when especially low limits of detection at high spatial resolution are needed, XRF analysis using synchrotron radiation is applied. Quantitative information is generally obtained by using fundamental parameter approaches or Monte Carlo simulations [493]. However, the quantification via comparison to standard reference materials is also a common approach. Recently, a new set of standard reference glasses from representative geological rock types has been produced and found to be suitable for use in μ-SRXRF [494]. The achieved accuracy of quantitative trace element analysis by using iterative Monte Carlo simulations may reach 8–10% even at concentration levels as low as 50 ppm [493].

Synchrotron radiation microanalysis was applied to run products of experimental geochemistry to determine trace element diffusivities in andesite melts [495, 496] and trace element partitioning behavior between garnet and melt [496]. Trace elemental maps on manganese nodules were determined with μ-SRXRF to investigate long-term environmental changes as evidenced by trace element compositions [497]. Feldspar trace element compositions determined with μ-SRXRF in combination with textural observations were studied, e.g., on ignimbrite "A," Gran Canaria, Canary Islands, and used to reconstruct dynamic processes such as magma mixing and contamination in evolving rhyolitic magma chambers [498]. Analyses of chevkinite and coexisting glass using μ-SRXRF revealed the importance of chevkinite as an accessory phase for the REE budget of evolving felsic peralkaline magmas as it concentrates Zr, Nb, and the LREE [499]. Trace element analysis of minerals and interstitial glasses in xenoliths from the East Eifel, Germany, revealed Pleistocene underplating and metasomatism processes of the lower continental crust [500]. By μ-SRXRF determinations on microtektites of the Cretaceous-Tertiary boundary from various localities, the tektites could be linked to their probable source within the Chicxulub impact crater by trace element fingerprints [501]. Besides analyses of glass and minerals another growing field of application for SRXRF is major and trace element analysis of particles from aerosols or dust in the light of their environmental health potential. The quantitative analysis of particles, however, is difficult due to their complexity in terms of chemical heterogeneity, crystallinity, and surface topology [502]. Vincze et al. [503] have shown that the accuracy of quantitative analysis of coal fly ash particles using μ-SRXRF and Monte Carlo simulations ranges between 5 and 30%. Typical airborne particles like fly ash particles [503] and fuel particles [504] were analyzed for potentially toxic elements. Another field of application which also has the aim of tracing environmental problems is the analysis of sediment-derived particles. Osán et al. [505] were able to learn about the origin of the toxicity of river sediment pollution particles by combined μ-SRXRF and μ-XANES (X-ray absorption near-edge structure). Salbu et al. [506] determined the oxidation states of uranium in depleted uranium particles (average size: 2 μm or less) from soil samples by μ-XANES and μ-SRXRF. The results are fundamental for understanding the potential remobilization of U within the contaminated area. The combined study of chemical composition and speciation is of major importance in particle research as the speciation has a major influence on the transport and dissolution behavior of the metal ions from the particle, biological uptake, and the toxicity of the particle. To identify the areas of interest for μ-XANES, μ-SRXRF is commonly applied.

*Fluid and Melt Inclusion Analysis.* Fluid and melt inclusions in mineral hosts are formed during crystal growth or crack annealing and – if their composition is conserved after the time of trapping – uniquely record directly the fluid and melt phase that was present during the rock genesis. Typically, many generations of fluids are included within the same mineral host (e.g., Roedder [507]). Thus, the analysis of single inclusions allows studying the evolution of crustal rocks, element transport in the crust, and enrichment processes of economically important elements. Analysis of single inclusions is ambitious because of their small size in the micrometer range and the fact that they are shielded by their host mineral. Trace element analysis of single inclusions is possible with PIXE, LA-ICP-MS, and SRXRF. Laser ablation ICP-MS allows the multielement analysis of major and trace elements with very high accuracy and low limits of detection [508, 509]. The sensitivity varies from very high to low. Typical spot sizes have a diameter from a few to tens of micrometers. Audétat et al. [510] have used LA-ICP-MS of single fluid inclusions to trace the formation of a magmatic-hydrothermal ore deposit. Heinrich et al. [511] were able to retrieve information on the metal fractionation between magmatic brine and vapor using the same method. However, LA-ICP-MS is destructive and does not permit reproducibility measurements which are possible with PIXE and SRXRF. For inclusion studies, both methods have been proven to be complementary. The proton microprobe was shown to be best for near-surface inclusions and light elements ($Z < 22$; e.g., Ménez et al. [512, 513]). With a spatial resolution of about 2 μm, it is well suited for element distribution maps (e.g., Kamenetsky et al. [514], Ryan et al. [515, 516]). Detection limits are in the 10 ppm range which is significantly higher than those for SRXRF. Synchrotron radiation XRF facilitates multielement analysis of major and trace elements for heavy elements with Z between 15 and 92 at high spatial resolution. The determination of light elements in most geological samples is mainly semiquantitative because the fluorescence is strongly absorbed in the dense geologic materials. Possible beam sizes are extremely variable with cross-sections between 1 and 2 mm and diameters in the micrometer range and can easily be adjusted to the size of the inclusion of interest. Minimum sensitivities are in the sub-ppm range. Element distribution maps of multiphase single inclusions enable to study the element distribution among vapor, liquid, and crystals. Bühn et al. [517] used a beam diameter of only 2 μm and were able to distinguish between different optically identical carbonates in three different types of fluid inclusions. With this information they could reveal the complex heterogeneous entrapment and necking-down process in an evolving carbonatitic fluid system. The capability of 2D SRXRF elemental maps of fluid inclusion studies is also documented in Fig. 7.139.

Area scans display an enrichment of Sn and Sb within the inclusion, which may be correlated to a tiny opaque daughter phase, possibly $SnO_2$. Philippot et al. [518] as well as Ménez et al. [513] pointed out the possible identification of optically invisible minerals within individual fluid inclusions by performing 2D SRXRF scans of single inclusions using a beam in the micrometer range (size of beamspot on the sample: $2 \times 7 - 10\,\mu m$). This demonstrates the importance of the analysis of complete inclusions either by 2D scans or by using a beam size covering the complete inclusion for exact quantitative analysis. Using 2D elemental maps Ménez et al. [513] could demonstrate the preferential partitioning of Br into the vapor phase (beamspot $2 \times 7\,\mu m$) of a liquid and vapor bearing inclusion. This finding supports the results of Ryan et al. [519], Damman et al. [520], and Heinrich et al. [511] using PIXE analysis and – as Br has the capability of complexing metals – is of major importance in the study of segregation and enrichment processes in ore-forming systems. Recent developments of a confocal set-up of SRXRF [521] have enabled 3D imaging of single inclusions that significantly improves the possibilities to distinct elemental distributions between different phases within inclusions.

Besides the potential for qualitative analysis of trace elements in fluid and melt inclusions and element distributions, the method allows quantitative analyses of the inclusion fluorescence spectra. One peculiarity of inclusion analyses is the heterogeneous geometry of the samples. This means that the

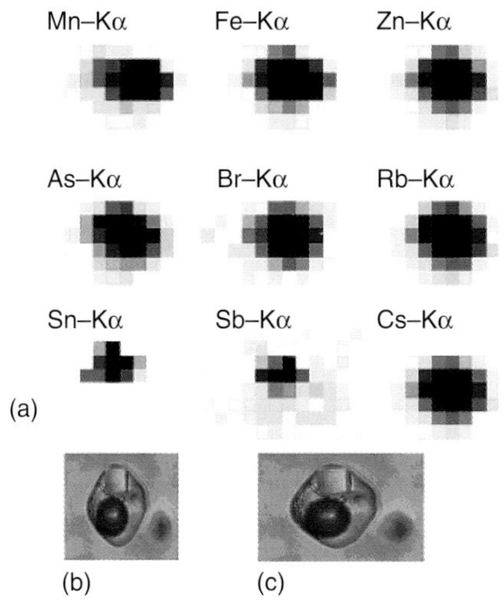


**Fig. 7.139.** (**a**) Example of 2D measurement of single fluid inclusion. Step sizes are 7 μm in both directions, the diameter of the beam is 10 μm. The spatial resolution, however, is lower than step sizes due to the thickness of the sample and the alignment of sample to detector in 45° geometry. For each element, the relative Kα line intensities are shown; maximum and minimum values refer to black and white pixels, respectively. Manganese, Fe, Zn, As, Br, Rb, and Cs display a rather uniform distribution throughout the inclusion. Antimony and Sn display clusters that may be related to a black daughter crystal within the inclusion. (**b**) Photomicrograph of the inclusion. (**c**) Projection of the inclusion onto a plane parallel to the detector surface

primary beam as well as the fluorescence radiation is absorbed in the sample and air surrounding the inclusion. This absorption is energy- and consequently element-specific (Fig. 7.140). Therefore, for inclusion analysis, the thickness and composition of the top layer has to be exactly known for accurate depth correction and quantification. A second peculiarity of inclusion analyses is that internal standardization is generally not applicable as there is no possibility to determine the concentration of one element within the inclusion, which is at the same time sensitive for SRXRF by an alternate method. One exception is Cl in high-salinity flat-lying inclusions, which may be quantified by cryometric measurements of inclusions. However, Vanko et al. [522] demonstrated that an error in the thickness estimation of only 1 μm of the overlaying matrix yields an error of several tens of percentages for elements with $Z < 20$. This means that results from an internal calibration to Cl, which has a Z of 17, are highly dependent on precise thickness estimation. In cases where internal calibration is impossible, the spectra may be calibrated externally by the use of thin-film NIST reference materials [522, 523] or synthetic fluid inclusions [513]. Quantitative results may also be obtained in a standardless mode [524] by comparison of peak areas of measured spectra and full XRF spectra simulated using a Monte Carlo code [525–528] that is applicable

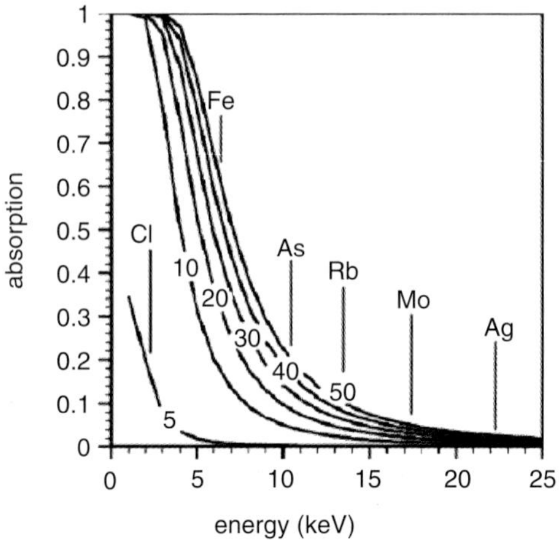


**Fig. 7.140.** Energy-dependent X-ray absorption in quartz of different thicknesses. The numbers on the lines refer to thicknesses in micrometer. The position of some interesting K$\alpha$ lines is marked

to a multilayer geometry. The achieved accuracy for this approach tested on synthetic fluid inclusions was better than approaches determining absolute concentrations by using a calibration on reference materials 17% error for concentrated inclusions and 30% close to the lower limit of detection [524].

Single inclusion analyses are extremely powerful in ore deposit studies and allow the tracing of enrichment processes. One recent example is the study of the chemical evolution of an ore-forming pegmatite-hydrothermal system at the Ehrenfriedersdorf Complex, Germany, which is genetically related to important tin–tungsten deposits [524]. Arsenic and Cu were found to preferentially fractionate into the aqueous phase during phase separation of silicate melts and coexisting aqueous fluids. The enrichment processes relevant for the tin deposit take place in the hydrothermal stage at temperatures around 460°C. Early hydrothermal fluids are enriched in Cu, As, and Sn. The Rb/Cs ratios are low as are the contents of Zn, Fe, and Mn. Later fluids (inclusion temperature around 430°C) are brines and display extreme enrichments of Zn, Fe, Mn, Rb, and Cs (Fig. 7.141).

*In Situ Determination of Mineral Solubilities.* One recent application of XRF with synchrotron radiation has been the determination of mineral solubilities at high temperatures and pressures using a hydrothermal diamond anvil cell (HDAC) [529]. Several experimental methods have been used to study fluid-mineral equilibria that control the mobility of elements during geological processes. All those techniques, however, rely on indirect observation. The HDAC method is the first to allow real in situ measurements of the solubility of solid phases by observation of the XRF signal of dissolved elements in a fluid inside the sample volume with temperature and pressure. The fluorescence

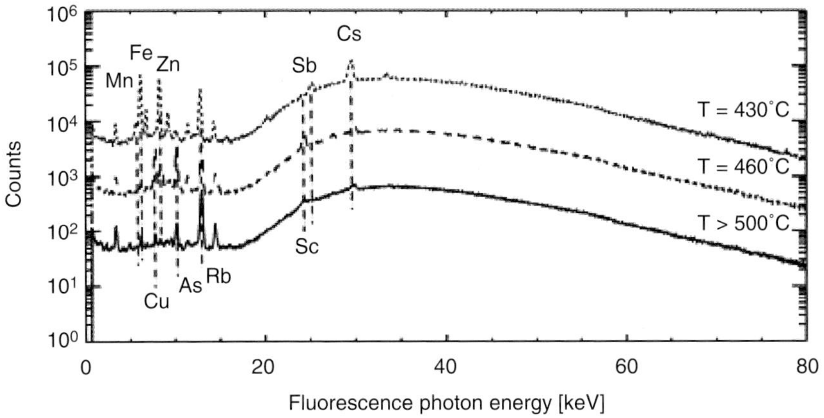


**Fig. 7.141.** Fluorescence spectra of three fluid inclusions representing the chemical stages relevant for ore enrichment processes of the pegmatite-hydrothermal system of the Ehrenfriedersdorf complex. The important elements are marked. The indicated temperatures refer to the approximate inclusion temperatures of the fluids from which the succession of the fluids may be deduced. For better representation, the spectra have been shifted along the y axis

signal is calibrated against standards, and concentrations of the element of interest may be determined at discrete P–T conditions. Up to now, two different cell set-ups have been proposed: both suggestions rely on externally heated DAC. One experimental setup was suggested using a Bassett-type HDAC [530], which was adapted to allow recording of the fluorescence of the sample volume in optimum geometry which is at an angle of 90° to the incident beam. Detection limits are below 80 ppm for elements with $Z = 37 - 65$ even for experiments performed in a second-generation synchrotron source [531]. The method is thus applicable to relatively low solubilities and at pressures up to 1.1 GPa and temperatures to 800°C. A first example of application was the determination of the solubility of AgCl in water at 300 to 450°C and pressures to 0.76 GPa [529].

A second experimental set-up has been proposed without modifying the original design of the Bassett-type cell achieving higher pressures of up to 3.6 GPa [532]. However, the conservative set-up of the HDAC demands for collection of the fluorescence spectra in forward scattering geometry at an angle of only 15° to the incident beam, which dramatically increases the detection limit to approximately 6 ppm for Sr [532] for experiments in a third-generation synchrotron source. Using this set-up, Sanchez-Valle et al. [532] investigated the P–T dependence of strontianite solubility in $H_2O$ and found an increase of at least a factor of 2 between 0.4 and 3.6 GPa.

These data help to constrain better models on quantitative mass transfer processes in subduction zones. A second, extremely useful, application of DAC and SRXRF is the time-dependent measurements of the solubility of minerals at high P and T. Initial experiments were performed with this technique on strontianite [532] and monazite [533]. These studies help in understanding the dissolution kinetics of minerals at conditions in the earth's crust.

## X-Ray Absorption Fine Structure Spectroscopy (XAFS) in Earth Sciences

While XRF analyses provide information on the concentration of elements in a given sample, XAFS is able to add information on the chemical state or structural environment of these elements in the sample. The following is intended to give a brief overview on the technique of XAFS together with a few examples of applications relevant to geological problems. For a more thorough introduction to the field of XAFS and applications in earth sciences, the reader is referred to more comprehensive articles and books (e.g., Brown et al. [533, 534], Schulze et al. [535], Fenter et al. [536], Henderson and Baker [537]).

XAFS is performed by scanning the energy of the incident X-ray beam across a certain energy threshold needed for excitation of a core-level electron, typically at K- or L-edge of a given element. If the energy bandwidth of the incoming beam is small enough, small variation or a fine structure present in the otherwise smoothly varying X-ray absorption coefficient are observed, which are dependent on the chemical state and structural environment of the absorbing element. The observed fine structure arises from transitions of the photo-electron to higher bound-state energy levels (before or at the edge) or from backscattering of the photoelectron from the atoms surrounding the absorbing element (at the edge and after the edge). The method is basically grouped into two methods (see also Fig. 7.142):

- XANES: using the fine structure occurring just before and up to approximately 100–150 eV after the edge. This region in the spectrum is sensitive to the valence state of the absorbing element and to the geometry and the type of surrounding atoms.
- EXAFS (extended X-ray absorption fine structure): the fine structure occurring far beyond the edge (starting from approximately 100 up to 1000 eV or further) provides quantitative information on the distance and number of atoms surrounding the absorbing atom.

Because XAFS works as a local probe, the aforementioned information is also accessible in systems lacking long-range order as silicate glasses and melts or aqueous solutions and hydrothermal or even supercritical fluids. As XAFS is element specific it also allows investigation of elements that are only near or at the surface of a mineral or phase, thus providing a helpful tool for the investigation of processes at the mineral–water interface.

XAFS spectrometers usually rely on the continuous spectrum at high photon flux available at synchrotron radiation facilities. In order to detect the fine structure at the absorption edge a monochromatic beam with a sufficiently small bandwidth is needed. Typically used in the energy regime suitable for

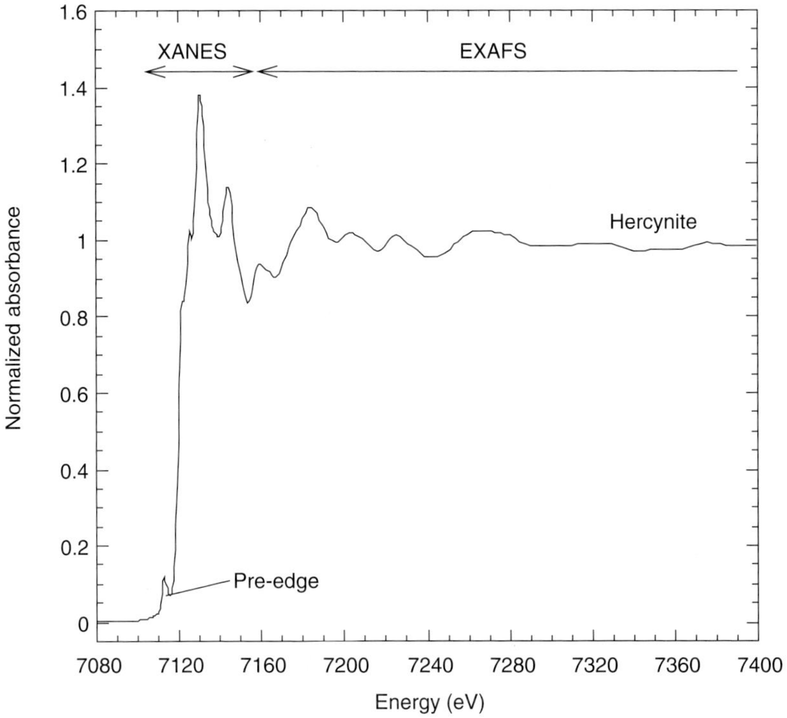


**Fig. 7.142.** XAFS-spectrum at the Fe K-edge of hercynite (Fe–Al-spinel) showing the respective regions within the spectrum as explained in the text. Note that the EXAFS region extends much further than shown here

investigations of the 3d-transition elements are Si (111) based double-crystal monochromators providing a resolving power E/ΔE of approx. 4000 at 7 keV. While such a resolution is adequate enough for investigations of the EXAFS region it may be insufficient to resolve all spectral features present in the XANES region so that monochromator crystals with higher intrinsic resolution have to be used, e.g., Si (220) or Si (311). The actual resolution needed is determined by the energy width of the transitions that are probed. The width is a function of the core-hole lifetime and increases with increasing atomic number. Use of a small bandwidth monochromator, however, may be limited due the accompanied loss in photon flux that potentially makes high-resolution measurements unfeasible due to low counting rates at the detector.

### XAFS and Fluorescence Yield

Originally X-ray absorption spectroscopy is performed by measuring the intensity of the X-ray beam before and after the sample ($I_0$ and $I_{tr}$, respectively), so that the absorption coefficient $\mu = \ln I_0/I_{tr}$. A further possibility is the indirect measurement of $\mu$ through detection of the fluorescence yield as $\mu \approx I_f/I_0$. This relation holds as long as the concentration of the absorbing element is low or the sample is very thin. For thick and concentrated samples the oscillations and resonances above the edge are significantly damped, and peaks in the pre-edge region artificially are enlarged due to self-absorption of the incoming beam. These effects are dependent on the detection angle and are minimized by detection in grazing exit geometry (see Tröger et al. [538]; Manceau et al. [539]).

Detection of the fluorescence can be performed in several ways. The simplest way is by using a gas-filled ionization chamber [540]. Such a detector enables the detection of a large solid angle although the dynamic range is limited. The sensitivity for the energy used can be adjusted by the use of the appropriate gas filling. Use of this detector is only possible if only the element of interest is excited due to the low-energy resolution of this type of detector. In a similar way silicon pin diodes can be used to provide a higher sensitivity and much larger dynamic range [541] although the solid angle of detection is smaller. Use of solid-state detectors such as Si(Li), high-purity Ge, or Si-drift detectors provide additional advantages due to the high-energy resolution that can be achieved with these devices. Processing of the detector signal by a multichannel analyzer provides the possibility of separating the fluorescence signal from the elastically scattered background or from fluorescence radiation produced by other elements contained in the sample. Sophisticated treatment of each XRF spectrum that is saved at each energy position of a XANES scan enables the collection of XANES even at very low concentration levels [542]. In such a data treatment the fluorescence yield is separated from the background by fitting of the XRF spectrum. This procedure provides a much better separation of background and signal and reduces the noise contribution from elastic and Compton scattering.

### Micro-XAFS

Use of a microbeam enables the investigation of samples that are inhomogeneous even on a microscopic scale. Analyses can be performed at specific sites, grains, etc. of a given sample or even mappings of the chemical state can be performed. Combined with concentration-distribution maps (from micro-XRF or EMPA) this will provide information on the distribution of the chemical state or structural environment of the mapped element in the sample. The synchrotron beam is usually strongly polarized. Therefore, spectra acquired by micro-XAFS on single crystals of minerals as occurring in rock thin-sections may be affected by the orientation of the crystal toward the beam, especially in strongly anisotropic phases [543, 544]. This additional uncertainty has to be taken into account for a robust interpretation of micro-XAFS data.

If elements show gradual lateral change in oxidation state within a sample the distribution of the oxidation state can be mapped by superimposition of XRF maps excited by X-ray energies at prominent maxima of the XANES of each oxidation state. The position of these maxima varies with oxidation state by a few electron volts and therefore setting the energy at either position will lead to preferential excitation of either oxidation state (e.g., compare spectra in Fig. 7.143a). Normalization of the intensity at these energy positions to the height of the edge jump and subtraction of the two maps collected at the two energies will yield the distribution of oxidation state within the sample. Thus, for each position on the sample, the fluorescence yield at three energies has to be measured: one before the edge, one after, and one at the redox-sensitive maximum (see also Sutton et al. [545]).

## Applications

As the position of the absorption edge is dependent on the oxidation state XANES spectra can be used for the determination of the valence state. This can be done by using the edge position, i.e., determination of position in comparison to the variation observed for the end member valence states. However, the observed position of the edge may be affected by resonances occurring in the edge region that are due to multiple scattering or transitions of the photoelectrons to localized energy levels (these features are very sensitive to the local geometry around the absorber and therefore structure dependent). Another possibility is the use of the whole XANES spectra: if the end member spectra of each oxidation state are clearly known from available model compounds the oxidation state of the unknown sample can be determined by linear combinations of spectra of the model compounds (e.g., Garvie and Buseck [546]).

As the structure of the XANES is also a "fingerprint" of the absorbing element in a given phase (e.g., examples in Fig. 7.143a), linear combinations of end member spectra can be used to determine the host phase of a given element and its relative abundance in a sample (e.g., Welter et al. [547]; Calmano et al. [548]; Vincze et al. [549], Salbu et al. [506]).

The pre-edge feature that probes 1s->3d transitions (especially sensitive to oxidation state) is only affected by the first co-ordination shell around the absorber (usually oxygen). Thus, it will show much less dependence on changes in the medium-range environment or bulk structure than the structure of the spectra occurring at the main edge. The pre-edge position varies significantly with oxidation state and shows variation in intensity due to the geometry of oxygen-coordination polyhedron (octahedral, tetrahedral, etc. see Fig. 7.143b; Wilke et al. [550]). In the case of Fe, the difference between the pre-edge position (measured as centroid) of ferric and ferrous Fe does not vary with the coordination geometry. Therefore, a linear variation of the position with redox ratio can be expected if the coordination does not change with oxidation state. Exploiting this effect, it was shown to give moderately precise estimates for the oxidation state (approx. $\geq 10-15\%$ error) by the use of general calibration curves (Fig. 7.143c) (Bajt et al. [551] 1995; Wilke et al. [550]). These calibrations were applied to several geological problems where knowledge of

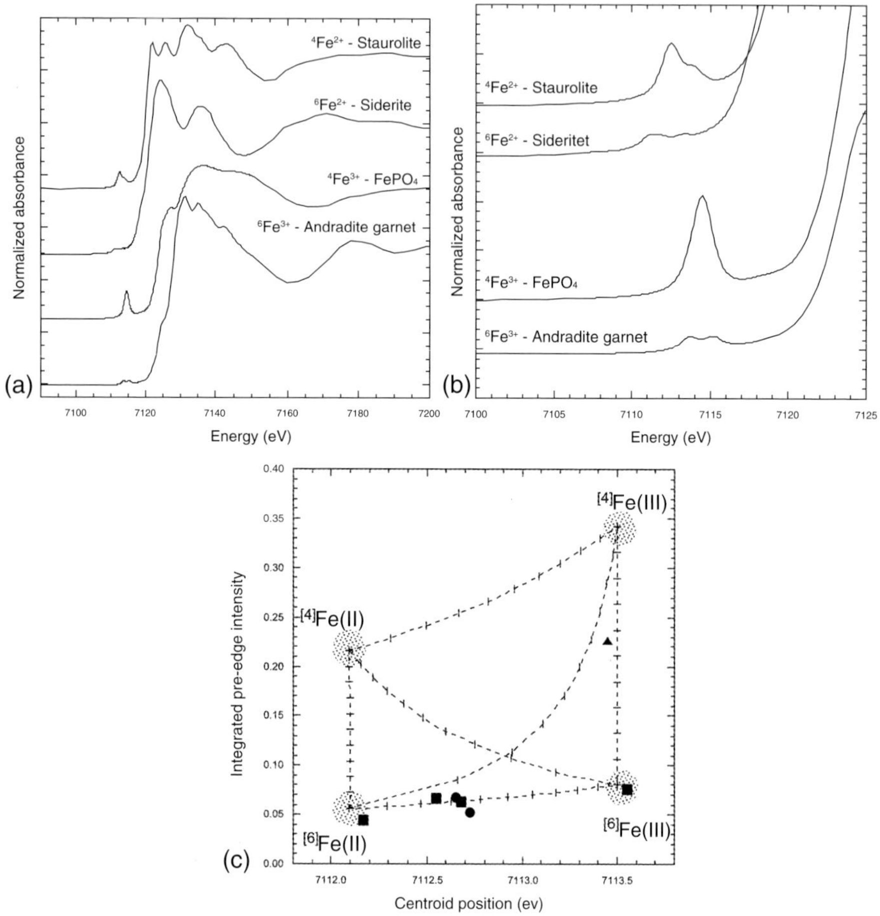


**Fig. 7.143.** (**a**) XANES spectra collected at the Fe K-edge of several model compounds (staurolite: Fe–Mg–Al nesosilicate; siderite: Fe-carbonate; andradite garnet: Ca–Fe nesosilicate) containing ferrous and ferric iron in various coordination geometries as indicated. The position of the main edge is dependent on the oxidation state. The fine structure, observed at the main crest of the edge, provides a fingerprint of the local and medium-range structural environment of Fe in these phases. (**b**) A zoom of the pre-edge region: The position of the pre-edge depends on the oxidation state while the structure and intensity of the pre-edge is mainly dependent on the geometry of the first coordination shell of Fe (oxygen in these examples). (**c**) Variation diagram after Wilke et al. [550] showing the systematics of the pre-edge variation with oxidation state and coordination geometry, taking into account tetrahedral and octahedral site geometries. Integrated intensity and centroid position are determined on background corrected pre-edges. Plotted are examples of pre-edges of several other phases: Triangle: Maghemite ($\gamma$-$Fe_2O_3$) with 2/3 $^4Fe^{3+}$ and 1/3 $^6Fe^{3+}$; Circles: Amphiboles with $^6Fe$, $Fe^{3+}/\sum Fe = 0.4$; Squares: Clinopyroxenes with $^6Fe$, $Fe^{3+}/\sum Fe = 0$, 1 and 0.3 (Data taken from Wilke et al. [550] and Schmid et al. [544])

$Fe^{3+}$ was essential (e.g., Dyar et al. [552]; Schmid et al. [544]). In these studies the $Fe^{3+}$ content of minerals was determined by micro-XAFS on 30-μm-thick thin sections of rock samples, providing information on the oxidation state of Fe that was not available from the chemical microanalysis with the electron microprobe.

Apart from investigations at ambient conditions XAFS also provides the possibility of characterizing samples at high temperatures and/or high pressures. Such studies are important if the samples are not stable at ambient conditions. For example, it was shown by high-temperature XANES and EXAFS that the speciation of elements in silicate melts at high temperature may differ from the one found in glass (Jackson et al. [553]; Farges et al. [554, 555]; and Fig. 7.144). Such differences have strong implications in the relationship between melt structure and macroscopic physical properties such as viscosity. Investigation of aqueous solutions at high temperature and high pressure using X-ray-transparent pressure cells (e.g., Bassett et al. [556]; Anderson et al. [557]; Seward et al. [558]) provides a direct view on the speciation of elements in such systems. Such data are important for the understanding of the processes involved during mass transport by fluids and fluid–rock interactions in the earth's crust, or in any physico-chemical process occurring at high temperatures and pressures.

To improve the results as compared to the examples by Dyar et al. [552] and Schmid et al. [544] a high-resolution monochromator should be used in order to better separate the pre-edge feature from the edge background (in case of Fe: Si(220) or (311) monochromator). Using high-resolution spectra and phase/mineral-specific calibrations errors below 10% can probably be achieved.

XAFS may also be used to perform time-dependent investigations to study the dynamic behavior of chemical and physical processes, such as redox reactions or phase transitions. For rather slow reactions this can be achieved by acquiring a time series of normal XAFS scans (10–20 min/spectrum). Such a study not only provides data on the kinetics of a reaction but may also reveal information on transient intermediate states that might be formed during reaction (e.g., Bargar et al. [559]). For fast reactions in the minute or even second regime, so-called QEXAFS methods have to be used, where spectra are collected continuously within seconds or even fraction of a second (e.g., Frahm [560], Lützenkirchen-Hecht et al. [561])

### 7.6.5 Conclusions

XRF analysis is one of the most important analytical methods for geoscientists. The method can be applied manually, in automated mode, and even by remote control in space, at any scale, on other planets, in the field, in boreholes, at industrial sites, in the laboratory and even under submicroscopic conditions in all fields of geoscience geological mapping, tectonics, petrography, geochemistry, sedimentology, remote-sensing, geophysics, microbiology,

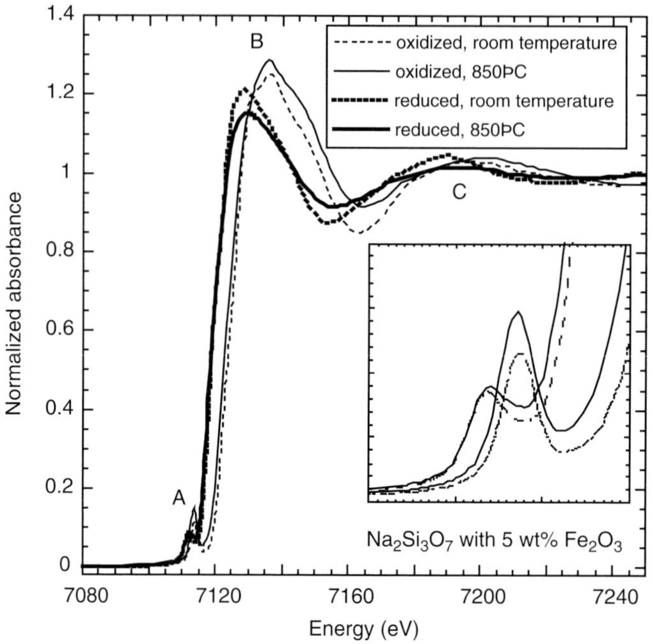


**Fig. 7.144.** XANES at Fe–K–edge acquired on glass at room temperature and on melt at 850°C at oxidizing and reducing conditions in fluorescence mode (see Brown et al. [534] for experimental setup). Note the changes occurring at pre-edge (A), main crest (B) and first EXAFS-oscillation (C). Pre-edge intensity of the high-temperature spectrum of the oxidized sample is significantly higher than the one of the room temperature spectrum, indicating change in coordination toward tetrahedral geometry at high temperature. This interpretation is consistent with an observed shift of the first EXAFS-oscillation toward higher energies that indicate shortening of the Fe–O distance (as expected for change from higher to lower oxygen coordination number). The change in the pre-edge of the reduced sample between high and room temperature is much smaller and almost negligible after background subtraction. This indicates only small changes in the local structure at reducing conditions consistent with only a small shift of the first EXAFS oscillation. The loss in the EXAFS amplitude of high temperature is related to the thermally induced vibrational disorder at high temperature

engineering, exploration, mining, mineral treatment, melioration, and metallurgy. Standard quantitative analyses by WDXRF, EDXRF, and TXRF are well developed for homogenized samples, but have to be individually modified for materials of anomalous chemical composition such as ores, salt, minerals, crude raw oils, fluids, and plants.

The method, if modified properly, is suitable for the analysis of all elements from carbon to uranium depending on the analytical conditions. Besides the analysis of homogeneous samples, geoscientists require spatially resolved information of inhomogeneous samples. Knowledge of the type, shape, and frequency of an anomaly that may appear as a certain mineral (Au, PGM, etc.), a lamina (winter/summer), or a hot spot of a contaminant (Pb, Cu, Zn) is crucial for the interpretation of mineralization, climate impact/influence, contamination, etc. During industrial processes like mining and metallurgy the online control of the product at all stages of the production line is a key parameter since costs can be reduced, and the performance and quality of the end product is improved. Besides the manual specimen acquisition, in situ online systems are installed to monitor the flow at sensitive steps of the treatment process.

The miniaturization of systems that may be used within drill cores bore holes and as handheld instruments in the field and the advancement in element mapping and imaging systems provide very detailed information for the geoscientist onsite for fast adequate decision and precision sampling.

On the other hand, the improvement of X-ray optical systems at reasonable costs and highly sensitive detectors will push forward micro-XRF systems in the laboratory to solve analytical problems at any fields.

The scanning X-ray instrument based on energy-dispersive spectroscopy enables a comprehensive material characterization. It combines simultaneous imaging of element distributions by X-ray microfluorescence analysis, texture mapping, and residual lattice strain mapping on a grain specific scale, as well as quantitative texture analysis.

## 7.7 Applications in Arts and Archaeology

*O. Hahn, I. Reiche, H. Stege**

### 7.7.1 General Remarks

Archaeological and art-historical research is usually concerned with the question of origin, dating, attribution, and reception of cultural objects. Stylistic and art-historical examinations in combination with the investigation of technological treatises and recipes can answer many questions but often the analysis of the physical properties and the chemical composition of the artefacts is also necessary. According to Mantler and Schreiner "Styles were sometimes copied at locations and periods completely different from those of their origin", and therefore physico-chemical investigations "are helpful and increasingly applied to allocate an object to a particular historic or prehistoric

*Née Bronk

context, to determine the correctness of the claimed provenance or to explore the stechnology used for manufacturing" [564]. Furthermore, the development of reversible restoration or conservation concepts requires the knowledge of material compositions and ageing phenomena of the artefacts. Finally, objects that were submitted to a conservation treatment may be investigated to distinguish original from later additions or to assess the impact of restoration.

A particular requirement for the investigation of works of art is the use of techniques that are non-destructive or only need minimal sampling [562]. After analysis, the unchanged sample should preferably be available for further studies. In accordance to Lahanier et al. [563], the ideal procedure for analysing art, historical, or archaeological objects should be non-destructive (respecting the physical integrity of the object), but also fast (to analyse large numbers of similar objects or to investigate a single object at various locations), universal (to analyse many materials and objects of various shapes and dimensions), versatile (allowing to obtain average compositional information but also permitting local analysis of small areas), sensitive (so that object grouping and other types of provenance analysis can be done not only by means of major elements but also by means of trace element fingerprints), and multielemental (to obtain simultaneously information on many elements in a single measurement).

As X-ray techniques meet most of these requirements, analyses of objects of artistic or archaeological value with XRF [564], TXRF [565, 566], SRXRF [567], and micro-XRF [562] are now quite frequent. The importance of X-ray techniques for the study of art and archaeological objects was for example emphasized by a special millennium edition of the journal X-Ray Spectrometry on Cultural Heritage (2000 Vol. 29 No. 1). Indeed, XRF represents one of the most suitable methods to obtain qualitative and semiquantitative information on a great diversity of materials.

In conventional XRF equipment a vacuum chamber is usually part of the spectrometer to determine elements between $Z = 11$ and about 19 (Na to K). Using vacuum limits the possible object size and may cause damage to fragile pieces. As Schreiner [568] first demonstrated, a purging of the detection path with helium effectively reduces the absorption of low-energy lines even in an open arrangement. Unfortunately, many spectrometers have the drawback of being stationary, which may limit their application to art objects. The transport of fragile objects into a laboratory may involve unacceptable risks, or in certain cases (e.g., wall paintings) is even impossible. First, portable XRF spectrometers of the last decades, which were primarily developed for geochemical and metallurgical purposes, are usually equipped with a radionuclide source and, because of low excitation intensities, are restricted to large spot sizes of the incident beam. In the last years, a new generation of compact and portable XRF set-ups have been designed consisting of air-cooled low-power X-ray tubes, detectors without the need of liquid nitrogen cooling, such as silicon drift chamber detectors and miniaturized electronics [569, 570].

A high lateral resolution is often a crucial factor in the analysis of art and archaeological objects. The actual requirement of the spot size strongly depends on the type of object and material: as a rule, about 1 cm for metals, stones and ceramics, about 0.5 mm for paintings and 0.1–0.05 mm for filigree objects like, e.g., miniature paintings, jewellery, or painted enamels are appropriate. Some special applications require a particular lateral resolution of 1–5 μm to analyse crystallite inclusions in gemstones or biogenic materials like hair, bone and dentine and to carry out element mappings or line scans for the determination of concentration profiles. Recent developments of total-reflection glass-capillary optics in combination with air-cooled X-ray tubes [569] and silicon drift detectors without need for liquid nitrogen cooling provide a high lateral resolution associated with a maximum of mobility [571, 572].

A very promising new micro-XRF development consists in the design of a confocal XRF set-up by using two focusing devices as for instance capillaries, one in the excitation and another one in the detection channel. The set-up permits to perform non-invasive depth-sensitive analyses with a resolution of about 10 μm. This new 3D micro-XRF method opens up the way for many non-destructive archaeometric investigations of complex layered or inhomogeneous materials [573].

The nature of art and archaeological artefacts raises particular problems which may affect the interpretation of the analyses. The objects are usually not "ideal" but of complex shape, irregular morphology, heterogeneous composition and may show surface alteration or several layers. However, it should be kept in mind that because of the uniqueness of art and archaeological artefacts most analyses have to be regarded as single case studies where routine procedures are not easily applicable and require experienced scientists for measurement and data interpretation.

The following section aims at summarizing relevant applications (1998-2003) of XRF in the field of archaeometry. The selection was made to demonstrate the diversity of material groups and problems involved and does not claim to completely cover all published work. The order, in which the material groups are discussed, roughly corresponds to the frequency of XRF applications.

### 7.7.2 Materials Groups

#### Pigments and Inks

A substantial formal principle of art objects is their wealth of colours. Coloured surfaces are found in wall or panel paintings, in illuminated manuscripts, but also in wall papers and painted sculptures. Usually paints are made by dispersing pigments in a liquid binding media (e.g., drying oil, gum Arabic, or egg white). After applying a layer of paint, it is generally dried and eventually other layers are added. In several cases, a final coating of varnish protects paintings. The thickness for the whole system ranges from a few micrometres up to 1 mm and more. A list of some commonly used pigments and their chemical composition is given in Table 7.20 [574, 575].

**Table 7.20.** Selected historical inorganic pigments and their chemical formula or main component. The left columns contain pigments that were used since antiquity and Medival ages. Synthetic pigments since 1700 are listed in the right columns, indicating their earliest use in parentheses

| | | | |
|---|---|---|---|
| **Red pigments** | | | |
| Minium | $Pb_3O_4$ | Cadmium red (1910) | $CdS + CdSe$ |
| Red ochre | e.g., $Fe_2O_3$ (haematite) | Chrome red (1809) | $PbCrO_4 \times Pb(OH)_2$ |
| Realgar | $As_2S_2$(AsS) | | |
| Cinnabar | HgS | | |
| **White pigments** | | | |
| Chalk | $CaCO_3$ | Anatase (1920) | $TiO_2$ |
| Gypsum | $CaSO_4 \times 2H_2O$ | Antimony white (1920) | $Sb_2O_3$ |
| Lead white | $2PbCO_3 \times Pb(OH)_2$ | Lithopone (1853) | $ZnS + BaSO_4$ |
| | | Permanent white (1830) | $BaSO_4$ |
| | | Rutile (1938) | $TiO_2$ |
| | | Zinc white (1835) | ZnO |
| **Blue pigments** | | | |
| Azurite | $2CuCO_3 \times Cu(OH)_2$ | Cobalt blue (1775) | $CoO \times Al_2O_3$ |
| Egyptian blue | $CaCuSi_4O_{10}$ | Manganese blue (1907) | $BaSO_4 \times Ba(MnO_4)_2$ |
| Lapis Lazuli | $Na_{8-10}Al_6Si_6O_{24}S_{2-4}$ | Ultramarine | $Na_{8-10}Al_6Si_6O_{24}S_{2-4}$ |
| Smalt | Coglass | Prussian blue (1708) | $Fe_4[Fe(CN)_6]_3$ |
| **Black Pigments** | | | |
| Antimony black | $Sb_2S_3$ | Black iron oxide (1920) | $Fe_3O_4$ |
| Ivory black | $C + Ca_3(PO_4)_2$ | Manganese Oxide (1920) | $MnO_2 + Mn_2O_3$ |
| **Yellow pigments** | | | |
| Orpiment | $As_2S_3$ | Cadmium yellow (1825) | CdS |
| Lead–tin yellow | $Pb_2SnO_4$ (type I) $PbSnO_3$ (type II) | Chrome yellow (1809) | $2PbSO_4 \times PbCrO_4$ |
| Massicot | PbO | Cobalt yellow (1848) | $K_3[Co(NO_2)_6] \times H_2O$ |
| Naples yellow | $Pb(SbO_3)_2/Pb_3(SbO_4)_2$ | | |
| Yellow ochre | e.g., FeOOH (goethite) | | |
| **Green pigments** | | | |
| Malachite | $CuCO_3 \times Cu(OH)_2$ | Chrome green (1809) | $Cr_2O_3$ |
| Posnjakite | $Cu_4[(OH)_6(SO_4)] \times H_2O$ | Cobalt green (1780) | $CoO \times 5ZnO$ |
| Verdigris | $Cu(CH_3COO)_2 \times nCu(OH)_2 \times mH_2O$ | Emerald green (1814) | $Cu(CH_3COO)_2 \times 3Cu(AsO_2)_2$ |

Natural minerals with a high colouring power and a wide stability against changes of temperature, weather, and light have been known since prehistoric times (e.g., ochre, iron oxide) and have been used as pigments. Artificial pigments (e.g., lead white, Egyptian blue) have also been known since antiquity. Since the chronological use of most pigments is known today, it is possible to determine an approximate dating criterion for the genesis of painted historical objects. Some pigments appeared on artists' palettes after a certain time (*post quem*) others disappeared before that time (*ante quem*). Figure 7.145 shows two copper plates from Albrecht Dürer "*Petrus und Johannes heilen den Lahmen*" (Kupferstichkabinett (KK), Berlin, SMB), both additionally coloured. It could be shown that the first one was coloured in the sixteenth century, whereas the coloration of the second was added in the nineteenth century [576].

The palette of inorganic pigments includes several hundreds of different types with well-known chemical composition. Each pigment is characterized by its colour and by a particular set of different elements which, if detected in correct proportions, allow the identification of the pigment. Some pigments (e.g., ultramarine) are found as either natural or artificial which can be distinguished by their impurities. Furthermore, the investigation of minor phases or inclusions and trace elements permits conclusions about the origin or the

**Fig. 7.145.** (**a**) Kupferstichkabinett Berlin (Inv.-Nr. A 137). The coloration was likely to have been created in the sixteenth century, containing azurite, malachite, lead white, cinnabar, ochre, minium, gold ink, and calcite. (**b**) Kupferstichkabinett Berlin (Inv.-Nr. Am 551). The coloration was added in the nineteenth century, containing zinc white and chrome green

manufacturing of particular pigments. The analysis of special compounds (e.g., gypsum, calcite, or particular salts) which were used as substrate or mordant refers to organic dyes and pigments. Varnishes and binding media consist in general of organic compounds and thereby predominantly of light elements (C, H, O, N, S, P), which give only modest fluorescent signals. Although XRF investigations on polychrome objects especially of larger dimension are usually done non-destructively with open spectrometer set-ups, TXRF is also applicable: Microsamples can be prepared by rubbing a dry cotton–wood bud, a so-called Q-tip, over the painted surface [577]. However, this method does not apply for paintings covered with a varnish.

Some difficulties may affect the identification of pigments with XRF. Occasionally components which differ only by the mass fraction of some constituent elements can form the same pigment (e.g., lead–tin yellow (type I/II) see Table 7.20). Others may also have different crystallographic phases, e.g., titanium oxide, which can appear as either anatase (produced since 1920) or rutile (since 1938). Further difficulty occurs for a couple of pigments, which share colour and characteristic elements but only differ in the occurrence of light elements (C, H, O) whose detection or attribution is not realizable in air (e.g., green pigments such as malachite and verdigris). In multiple layer systems it is generally possible to distinguish information coming from different layers by means of characteristic ratios of the $K$ to $L$ and $L_\alpha$ to $L_\beta$ lines, respectively. If one pigment occurs several times in different paint layers, it is not possible to assign clearly the experimental results, except when using the newly established method based on a confocal 3D micro-XRF set-up. This method was first applied to the non-destructive study of paint layers of Indian Mughal miniature paintings [573].

However, the identification of pigments in paintings [578] and inks in manuscripts [579, 580] is successful in many cases and the method also allows the determination of the mixing ratios of pigments. By means of adequate quantification procedures taking into account inhomogeneous ink paper layer systems, it is possible to discriminate between different historical iron gall inks in manuscripts and compositions [581].

### Metals and Archaeometallurgical Samples

XRF analyses of ancient metals seek to increase the knowledge of the exact nature of the raw material used, of technological features, of its fabrication as a function of the period and region of origin, and of its history of use as well as of corrosion processes. Information on the chemical composition would also allow the definition of provenance criteria for the extracted raw material or of indirect dating of the manufacturing period. Since classical XRF is a surface analytical method with a relatively small analysis depth (some tens of micrometres in metals), various studies also focus on metal surface alteration phenomena such as patina formation. Conventional XRF coupled with energy (EDXRF) or wavelength dispersive (WDXRF) X-ray analysing systems seems to be best suited for metal surface analyses as the X-ray yield of metallic elements is relatively important. SRXRF is only more appropriate when information on deeper analysis depth, on bulk chemical composition of the metal, is searched or when only very small amounts in the microgram range are available for analysis. Microscopic XRF is useful when specific requirements exist with respect to the morphology of the objects. Problems encountered with XRF analyses of metals are strong matrix effects (interferences with matrix lines and self-absorption). For instance, Zn is very difficult to determine by classical XRF in Cu alloys [582]. However, comparison with other analytical methods such as atomic absorption spectroscopy (AAS), optical emission spectroscopy (OES), inductively coupled plasma mass spectrometry (ICP-MS) and Neutron activation analysis (NAA) showed the reliability of XRF for metal analysis when sufficient attention is paid to its limitations.

Important archaeological and historical metal alloys are especially based on Cu (bronze and brass), Ag, Au, Sn, Pb–Sn, Fe, and last but not least Hg. Metal objects in archaeology and history comprise artefacts of practical use such as coins, plates, cups, fibulae, but also arms, shields, as well as art and representative objects like statues or crowns.

A recent application of XRF on archaeometallurgical questions is the study of patinas formed on archaeological bronze alloys by Wadsak et al. [583]. The samples originate from the archaeological site of Bliesbruck-Reinheim (Germany). Micro-EDXRF was used to perform line scans on prepared cross sections. These analyses were coupled with TOF-SIMS and SEM-EDX to show differential loss of Cu content and increase of Sn and Pb due to surface alteration.

Klockenkämper et al. [584] examined Roman imperial Ag coins (from Augustus, 30 BC to Diocletian, AD 300) by SEM-EDX and WDXRF allowing analysis of near-surface layers of only 3 μm and 30 μm thickness, respectively. These analyses revealed inhomogeneities of the obverse and the reverse of a coin and near-surface Ag enrichment. Different Ag contents could be attributed to particular Roman periods, even if the Ag content can be scattered for coins of the same emperor period due to a variable abrasion of Ag-enriched surface layers. Linke et al. [585] compared EDXRF, PIXE, and SEM-EDX for the analysis of medieval Ag coins in order to establish a provenance criterion.

Spatially resolved SRXRF was adapted for the analyses of ancient silverpoint drawings, a graphical technique very much appreciated in the Renaissance, by Reiche et al. [586]. The amount of the deposited silver on the paper does not exceed some hundreds of microgram per square centimetre. Thus, a very sensitive analytical and non-destructive method was required for the analyses of these drawings. First investigations of several sheets of Albrecht Dürer's famous sketchbook that he drew during his journey in the Netherlands in 1520/21 confirm that all drawings but one were made with the same silverpoint containing about 11 wt% of Cu. Beside Ag and Cu, Hg could be detected in the silver marks. Its presence is probably due to contamination from the atmosphere. Further analyses on silverpoint drawings of other artists as Jan van Eyck or Hans Holbein the Elder are in progress in order to get further insights in this graphical technique.

Milazzo and Cicardi [587] analysed the Au alloy of the famous Crown of Monza, Holy Crown, or Corona Ferrea with a portable device using a radioactive $^{241}$Am X-ray source. It was expected to facilitate chronological and historical interpretations of the objects from the Monza Cathedral. These investigations also illustrate general quantification problems of the chemical composition of irregular objects with a non-flat surface with portable instruments. Using a standard related quantification, the metal composition could be determined with a standard deviation of 2%. Pernicka et al. [588] analysed the Au inlays of the Nebra sky disk, the oldest representation of the sky dating from early bronze age (end of seventeenth – first half of sixteenth century BC) found in Saxony-Anhalt (Germany) in 1999, by SRXRF. These investigations showed differences in the Sn content of the Au inlays and allowed the reconstruction of different fabrication steps of the ornamentation of the disk.

Adriens et al. [589] analysed nine powder samples originating from the archaeometallurgical cassiterite ($SnO_2$) site in Göltepe, 100 km north of the Mediterranean coastal city, Tarsus, dating from the Early Bronze Age. The aim of this study was to determine if the hypothesis of mining and smelting of Sn at this site was realistic at such an early time and if it was perhaps associated with gold winning. Bulk analyses by XRF and photoelectron spectroscopy allowed determination of the sample Sn content and Sn oxidation state, respectively. The distinction between unprocessed ground ore material, residues from an ore concentration process and waste material such as slag was possible. This investigation showed that only high Sn-containing materials were selectively transported to Göltepe for ore dressing and smelting.

The study of entrapped slag inclusions in archaeological iron from India, the 1600-year old Delhi iron pillar with an excellent corrosion resistance, was realized using a combination of X-ray diffraction (XRD) and micro-XRF by Dillman and Balasubramaniam [590]. Information on iron fabrication and properties could be obtained from these analyses. This study as well as the one mentioned above represents a good example for the use of complementary techniques for the understanding of the evolution of metallurgical processes and for archaeometric applications in general.

### Ceramics and Porcelain

Quantitative XRF (both EDX and WDX techniques) probably found its most widespread archaeometrical use in this materials group. Fifty to several hundred milligram of powdered sample may easily be taken by drilling.

The main focus of quantitative investigations on pottery and bricks is the question of provenance distinction and distinguishing different workshops. This is nowadays mainly achieved by applying methods of multivariate statistics, such as principal component analysis. Moreover, XRF is a routine method for the quantitative determination of U, Th, and K contents in samples which undergo dating by thermoluminescence or optical stimulated luminescence.

Since the 1970s, there have been numerous major provenance studies dealing with pottery from very different periods and origins. Some recent works are by Crocetti et al. [591] on pottery ware from the Marecchia valley (Italy), Morandi et al. [592] on Etruscan ceramics from Marzabotto, Pillay et al. [593] on Iron Age pottery from South Africa, and Bakraji et al. [594] on archaeological ceramics from Mar-Takla, Syria. Usually petrographic investigations, XRD, thermoanalytical and other methods complete the XRF results.

A major investigation with XRF beside macroscopic and microscopic studies was devoted to stone axes in Ireland and revealed the geochemical sources and petrography of the rock type porcellanite as the dominant source [595]. Several studies have been devoted to historical building materials such as the characterization of mortars in buildings of Pisa (Italy) by Franzini et al. [596], nineteenth-century bricks made in Brandenburg (Germany) by Röhrs [597] or stone decay phenomena at the cathedral of Bari (Italy) by Torfs et al. [598].

In the last years, porcelain ware and porcelain painting were in the focus of several Chinese workgroups: Leung et al. [599] undertook EDXRF analyses of porcelain body and glazes of the Song–Yuan dynasties (tenth to fourteenth century.) classifying different styles by principal component analysis and distinguishing imitation of Ding porcelain. The work of Yu and Miao [600] focused on the characterization of cobalt-blue and white porcelains mainly of different periods (fifteenth to eighteenth century AD) using the Mn/Fe ratios.

XRF has been somehow overshadowed by INAA (instrumental neutron activation analysis) as routine technique for archaeological provenance problems in the last two decades thus raising questions of data comparability.

Hein et al. [601] tested the interlaboratory results on pottery for different techniques (XRF, INAA, ICP-MS, ICP-OES). They attempted to establish calibration factors between pairs of analytical set-ups to smooth systematic differences among the results. Adan-Bayewitz et al. [602] suggested a high-precision XRF technique using Compton and Rayleigh scatter peak intensities as internal standards and achieve similar precision and accuracy ranges as with best INAA techniques. A case study of pottery from Roman Galilee even showed provenance results superior to INAA at the same samples.

TXRF, SRXRF, and micro-XRF are rarely or not applied because of the sufficient sensitivity of conventional XRF and, on the other hand, because of the heterogeneity of ceramic material. A comparison of semiquantitative TXRF and INAA assessing the potential for provenance discrimination of pottery was made by Garcia-Heras et al. [603] at a case study of Late Iron Age Celtiberian ceramics.

### Glass, Glazes, Enamels

Glass and its low-melting relatives (enamels, glass painters' fluxes, glazes on faience, and ceramics) in general have a complex and variable composition. As they are made of at least two but often four or five raw materials (quartz, fluxes, colourers, opacifiers, etc.) the multielemental analysis of major as well as trace elements is highly desirable.

However, the minimum need of 50 mg sample for a fully quantitative analysis has hampered the application of conventional XRF for historical glasses in the past as this requirement can never be fulfilled when analysing intact museum objects and is even problematic for archaeological samples. Therefore, for over 20 years mainly SEM–EDX has been (and still is) the routine technique in this field.

The only exception with respect to sample availability is the analysis of obsidian, a natural glass from volcanic sources: Seelenfreund et al. [604] recently demonstrated the possibility of provenance discrimination of obsidian from different Chilean sites based on trace element comparison.

Usually, the determination of light elements requires vacuum conditions or, alternatively, helium purging. However, even non-quantitative measurements in air can lead to valuable insights into, e.g., the technological background and provenance discrimination as demonstrated by studies of Wobrauschek et al. [605] on Celtic glass artefacts and by Kunicki-Goldfinger et al. [606] on Baroque crystal glass. Heck and Hoffmann [607] used XRF, micro-XRF, and SEM–EDX for the determination of the glass matrix as well as colouring oxides and traces in opaque glass beads from Merovingian times. The non-destructiveness of XRF may be valuable for investigations of multiple-layered glass objects or thin colour applications, such as in stained glass. While qualitative determination of the colour type and elements is feasible, in general no quantitative determination can be performed due to insufficient layer thicknesses. As an example, Jembrih-Simbürger et al. [608, 609] analysed nineteenth-century *Art Nouveau* lustre glass to distinguish the products of different glass-houses and the technique of silver stained glasses. In these studies EDXRF was combined with ion beam techniques (PIXE, PIGE, and RBS) as well as SEM/EDX and TEM. Bonizzoni et al. [610] characterized fragments of stained glass of the Certosa di Pavia, Italy for developing strategies of further restoration treatment.

The use of synchrotron sources for trace and ultra-trace determinations is a promising way for providing refined provenance classification of glasses, although the panoply of raw materials is much more complex than for ceramics. Up to now, Janssens et al. [562] have reported two studies on Roman glass finds from Qumrân, Israel and sixteenth and seventeenth-century Venetian and *Façon-de-Venise* glass from the Netherlands. In addition to SEM–EDX analysis of main and minor element oxides, medium and heavy elements (Fe–Ba) were determined by micro-SRXRF down to concentration levels of 1–10 ppm.

TXRF has so far only been tested in one study on post-medieval German glass by Wegstein et al. [611]. The need for removing a small sample and the difficulty to get full quantitative data make a more extensive use for glass analysis rather doubtful.

The actual "revival" of XRF in terms of capillary optics, non-cryogenic detectors and miniaturized devices has significantly improved the possibilities of non-destructive glass analysis by laboratory micro-XRF set-ups and let us expect a by far more extended application in the future (see General Remarks for references). Röhrs and Stege [612] used a mobile micro-XRF spectrometer for in-museum measurements of sixteenth and seventeenth-century painted enamels from Limoges (France). The study aimed to yield more precise information on chronological and workshop specificities on about 150 objects from different German collections. Micro-XRF proved to be a convenient method to identify replicas and forgeries made in the nineteenth century. Beside newly discovered elements for glassmaking (Cr, U, Ir) significantly higher lead oxide contents in transparent colours, lead arsenate as a new opacifier and a reduced content of minor and trace elements (Ti, Ba, Rb, Sr, P, Cl, etc.) were found to be the main characteristics of enamels that were not genuine.

Surface deterioration in its various appearances from slight *craquelé* to thick corrosion layers has to be carefully considered. Indeed, a true non-destructive analysis of the original bulk glass by XRF (or other methods) is per se not possible. Even visually unchanged surfaces of historical glasses generally have undergone considerable compositional changes, first of all alkali leaching and enrichment of silicon. Because the information depth of the light-element K-lines is only a few micrometres, non-destructive XRF may lead to erroneous results, e.g., for sodium and silicon. A slight abrasion as minor surface preparation by local polishing with (water-free) diamond pastes is sufficient to remove a few micrometres of the corrosion layer. However, polishing is not recommendable in case of heavy deteriorated glasses because more original material would have to be removed than required by sampling for SEM.

When the investigation of valuable art objects made of glass, enamel, glazed ceramics, or porcelain is planned, the possible risk of irradiation damages should be carefully considered and tested in advance with the set-up in use. Depending on the total dose (and therefore on the intensity of the primary beam and the acquisition time), brown discoloration may occur which is especially visible on opaque white silicate materials (such as enamel, porcelain) but also transparent glasses. Although this effect is reversible due to the formation of metastable energy levels of photoelectrons in non-conductors (so called "colour centres") it may disturb the appearance of the object.

### Minerals, Gemstones, Rocks

Beside pigments, minerals, gemstones, or rocks and other composite materials are very important in art and archaeology. Since prehistory, these materials have been transformed into tools, art objects, and objects of common use or applied as decoration on other materials. The chemical analysis of these materials can provide insights into the provenance of the objects, information on manufacturing and on alteration phenomena. The non-destructiveness of XRF is particularly appreciated when analysing precious gemstones for indications on provenance, on differentiation of natural and synthetic gems and on genesis of these minerals.

Douglas and Chase [613] studied ten jade beads and ornaments of a pectoral, a rare archaeological find in China dating from the Eastern Zhou dynasty, in order to determine if the different pieces are of the same origin. The jade compositions are similar but do not appear to belong to the current configuration of the pectoral. The analysis suggested that the necklace is in fact a reconfiguration using original chain, additional wires and jades, probably original, to construct a pastiche.

Joseph et al. [614] characterized rubies and sapphires from different sources by EDXRF in order to distinguish naturally occurring stones from synthetic ones. Since major light elements present in gemstones are not appropriate for "fingerprint analysis", one has to rely on heavy trace elements [615].

Archaeological jade objects are currently covered with a white powdery layer. Adams et al. [616] studied the jade corrosion phenomenon consisting of a surface patina on buried nephrite objects by means of micro-SRXRF. The alteration can be correlated with the surface state of the objects. It occurs more often on unpolished or less hard parts of the jade. Organic deposits, calcification or leaching of minerals in the medium of decomposing bodies are supposed to be the cause of surface corrosion of buried jade objects.

Potts et al. [617] analysed the effects of surface irregularity on the quantitative analysis of irregular-shaped archaeological rock samples by portable XRF with a radionuclide source. Indeed, the provenance study of Roman granite columns and British stone axes needs quantitative data for comparison with available geochemical data on known geological sources. They showed non-negligible problems with the quantitative analysis of irregular-shaped archaeological rock samples by XRF, even for small air gaps between 1 mm and 2 mm, unless an appropriate correction procedure is applied to peak intensities. A correction procedure based on the Compton and Rayleigh scatter peak intensities derived from characteristic X-rays from radioisotope excitation permits the determination of a normalization factor to compensate for surface irregularity effects.

## Organic Artefacts and Biomaterials

Organic and biomaterials belong to the materials group that usually cannot be properly analysed by means of conventional XRF because they are mainly composed of light elements. However, organic compounds as varnishes are often complex mixtures that also contain additives as drying agents or pigments. Van Bohlen and Meyer [618] studied historic varnishes (1550–1950) used for the coating of music instruments, especially violins, by means of TXRF. TXRF only needs microsampling (sample mass below 20 μg) and allows the analysis of a large number of elements with high sensitivity. The study revealed different elemental distributions for certain workshops and manufacturing dates. Some elements could also be related to contamination of the materials during the preparation of varnishes and also the varnishing process itself by metal tools.

The detection of inorganic salts may be of interest for historical silks which in the past underwent a process called "silk weightening". The remaining crystals are supposed to decrease the stability of the silk and therefore cause conservation problems. XRF conveniently allows the detection of the remains of weightening, first of all Sn, Si, and P on the textiles.

Bone, teeth, or ivory are composite biomaterials that register in their chemical composition on a trace element level a wealth of information on palaeodiet, palaeoclimate and they can be used for dating. As these materials are subject to complex alteration processes during their burial time, it is important to understand the diagenetic modifications in order to evaluate the informative potential. Especially, the distribution of REE by measuring concentration profiles (mapping, line scans) can provide information on palaeoenvironment during fossilization processes. Janssens et al. [619] determined the trace concentrations of REE in fossilized bones with associated sediments from Olduvai Gorge (Tanzania) from contrasting terrestrial depositional environments by micro-SRXRF with a limit of detection of 10 ppm. The micro-SRXRF experiments were conducted at HASYLAB beamline L providing photons in the very hard X-ray region (70–100 keV) which permits the analysis of heavy element at trace concentrations by means of K-fluorescent X-rays. The determination of REE concentration profiles enabled to address fossilization process involved in anoxic lacustrine depositional and in oxic fluvial palaeoenvironments.

Petrified wood is an important geological witness of earlier biological activity. Information about seasonal changes, i.e., temperature, could be obtained if spatially resolved analysis allowed the penetration of intraring structure. Petrified wood is mostly found in the silicified form, however specimens transformed into calcite, pyrite, charcoal are also known. Complementary analysis using electron microprobe and tabletop capillary and SR-X-ray microprobes as well as micro-SR-XRD allowed the detection of elements and mineral phases on cross sections [620] and the distinction of two types of fossilizations: (a) a replacement process leading to dark coloured wood and (b) an impregnation process with the preservation of rings and cells resulting in beige coloured wood.

An ongoing study on the decontamination of wood objects that were treated with organochlorine wood preservatives makes also use of XRF as a non-destructive and fast control method. Remains of formerly widely common wood preservatives containing DDT (dichlorodiphenyltrichloroethane), Lindan ($\gamma$-1,2,3,4,5,6-hexachlorocyclohexane), and PCP (pentachlorophenol) may present a major health problem and affect the appearance of wooden objects, e.g., by massive white recrystallization products. Such organochlorine agents were used on sculptures, easel paintings, frames, furniture, altars, and organs to control wood-destroying organisms. Different approaches are currently tested to remove the preservatives without damaging the object itself. A mobile micro-XRF system is used in addition to more detailed, but destructive analytical methods such as GC-MS to determine a sum factor for the preservative decrease during extraction by measuring the change of the chlorine signal [621].

### 7.7.3 Conclusions and Perspectives

The study of the chemical composition of art and archaeological objects is a real interdisciplinary research. Therefore, it needs the cooperation of researchers of different fields (chemists, physicists, conservators, archaeologists, and art-historians) in order to interpret the analytical results correctly. The solution of a specific problem must be discussed on the basis of complex multidisciplinary parameters available on an object such as information on the chemical composition, on possible surface alterations depending on particular conservation conditions, on archaeological site specificities and former restoration treatments. In addition, analysis with complementary techniques such as neutron, X-ray or electron diffraction, X-ray absorption fine structure, Raman or IR spectroscopy giving additional structural and molecular information at different structural levels have to be performed in order to get reliable and complementary information on the different materials.

The technological developments of the last ten years raised in a revolutionary manner the number of possible XRF applications in art and archaeology and the potential of information that can be deduced from XRF studies. The most promising perspectives for the determination of new criteria of provenance, of manufacturing and (indirectly) of dating are the increased use of microfocus beamlines at synchrotron facilities for archaeometric studies, analyses with "routine" mobile microspectrometers permitting a better knowledge of the material used for many museum objects and the application of confocal 3D micro-XRF.

In contrast to the rapid improvement of the XRF devices, at the moment, there is still a lack of quantification procedures that are accurate, flexible, and easy to use. Especially for laboratory micro-XRF set-ups where usually the polychrome spectrum leaving a polycapillary is used for excitation, reliable standard-bound and standardless quantification routines have still to be established. Hopefully, this drawback will be overcome in the not too distant future.

## 7.8 XRF-Application in Numismatics

*J. Engelhardt*

### 7.8.1 Introduction

The object coin appears in history at approximately 700 BC and is still to be found in large quantities. Very early, numismatists were interested in the composition of their coins. The composition of coins can provide information on the alloy employed, melting technology, mints, dating, relative metallic values and the melting down of previously struck coins. The analyses also allow correlating variations in alloy composition with debasement and distinguishing imitations from authentic specimens. Section 7.8.7 provides some examples on how XRF was used to solve the above-mentioned tasks.

In the last centuries only wet chemical analysis (WCA) [622] was available so there emerged the need for a nondestructive, fast and simple method. The nondestructive nature of XRF investigations offers this simple and fast method. Since 1950, every type of XRF device (WDXRF [623, 631] EDXRF [624], EDAX [659], PIXE [625], SEM/EDX [626]) has been used to investigate coin compositions with variant success.

Though XRF investigations compete with other methods such as neutron activation analysis (NAA) [627], proton activation analysis (PAA) [628] and WCA, the XRF methods are validated by use for investigation into coins (see Carter et al. [629] who compared six methods and see Sect. 7.8.6). Precautions due to the surface structure of the coin have to be considered for coin investigations with XRF (see Sect. 7.8.3).

The Royal Numismatic Society developed a Guide to analyse coins through a uniform and comparable selection of information [630]. This code of practice combines both the numismatist and the investigator aspects.

### 7.8.2 History of XRF Investigations of Coins

Wavelength-disperse X-ray fluorescence analysers have been used for the investigation of museum objects since their introduction. In 1960, Hall [631] developed a method to investigate ancient coins and found good agreement between his results and the results of chemical analysis. During the 1970's Hall [631] and later on Metcalf and Merrick [623] used the "Milliprobe" (an XRF device with a spot of $1\,mm^2$) to investigate thousands of ancient and Middle Age coins.

In the early 1980's the PIXE was tested by Ferreira and Gil [632] and Mommsen and Schmittinger [633] for analysing gold and silver coins. The PIXE method proved to be especially successful by eliminating standard reference materials and by using high energetic protons (68 MeV), which led to a better insight into the whole coin.

A new generation of transportable μ-XRF systems is available with the development of smaller detectors (PIN diode) and new kinds of optics (glass capillary optics). Also the use of synchrotron beamlines as excitation sources has entered the field of XRF investigations of coins [634].

### 7.8.3 General Remarks

Coins are wonderful objects for direct investigations with XRF because the shape and size of common coins allow simple transportation and mounting into each kind of EDXRF device. In addition, coins are normally available in large amounts, so that good statistical evaluations can be executed.

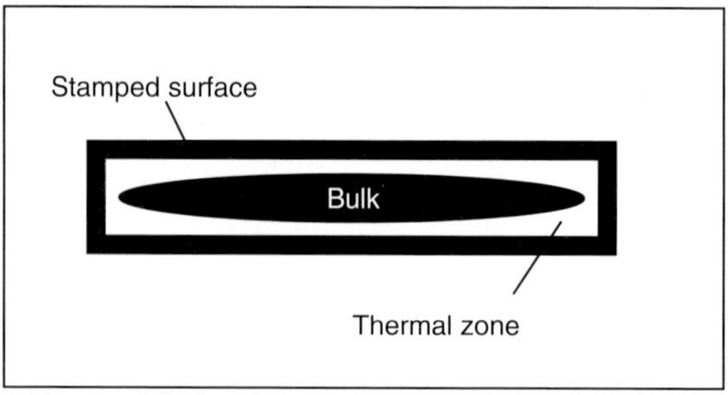


**Fig. 7.146.** Schema of a stamped coin [637]

An additional advantage in using coins is that often the used alloys are homogenous and the only inhomogeneities are related with different phases (e.g. segregates, grain boundaries) and corrosion phenomena. The excavation and storage history of older coins complicate the situation for the analyses, e.g. surface enrichment effects can be found [635]. Mantler and Schreiner [636] mentioned that the irradiation of a brilliant copper surface with X-rays can lead to darkening of the surface due to the energy deposition, but normally coins did not have such brilliant surfaces.

The object coin can be simplified by the model in Fig. 7.146 in accordance with Condermain and Picon [637]. It is differentiated between the stamped surface, the influence zone of the stamping (thermal zone with diffusion to yield in separation effects) and finally the bulk body. The corrosion effect influences not only the stamped surface but also the composition of the subsurface layers. A remaining undisturbed zone might be found only in the middle of the coin. The power which is needed to strike a coin was calculated by Delamare et al. [638] for Byzantine gold solidi. He attributes the stress of a coin to coin composition.

To reduce the stress in coins they had to be annealed. The stress corrosion results in corrosion of the bulk material and leads to a lack of undisturbed material.

Furthermore, due to the roughness of the stamped coin surface, the emitted X-ray fluorescence intensities depend on the irradiated position at the surface, which leads to a decrease in the precision of the concentration determination.

The corrosion phenomenon depends on the composition of the used alloy. This leads to surface enrichment effects, e.g. the noble metal silver for silver coins. A dezincification process is well known for coins consisting of bronze or orichalcum (brass). Calliari et al. [639] showed this effect by using EDXRF and SEM/EDX for Roman assets. The dezincification process led to a more than 70 μm thick layer and, consequently, the composition of the surface did not reflect the composition of the bulk (determined by Atomic Absorption Spectroscopy [AAS]).

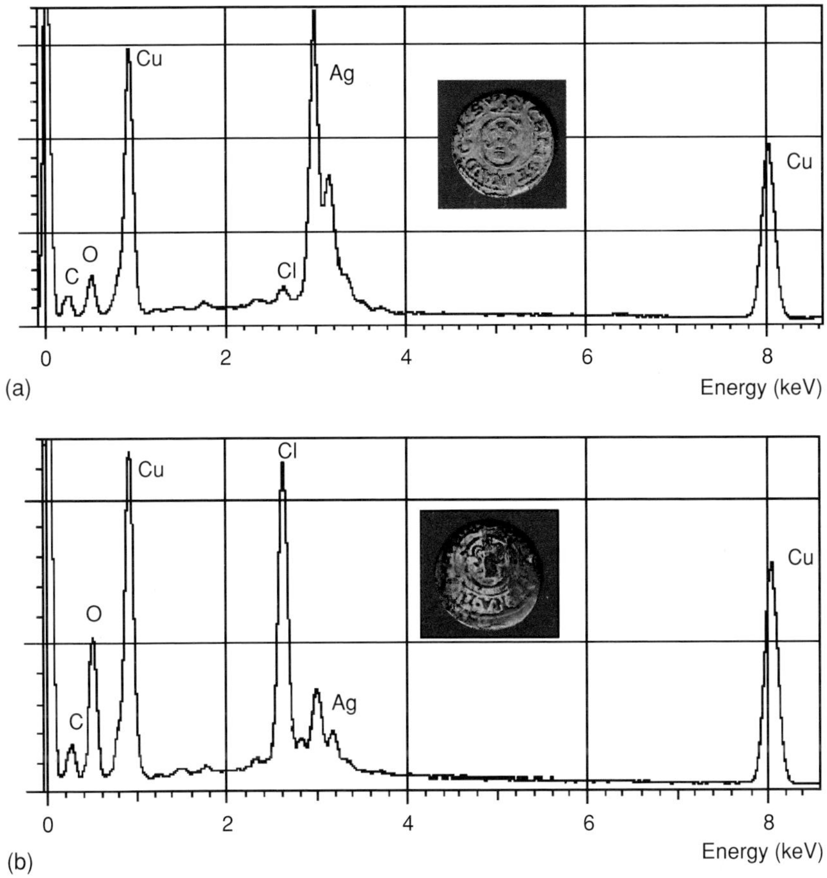


**Fig. 7.147.** Pictures and EDAX spectra of a noncorroded (**a**) and a corroded coin (**b**) solidus minted in Livonia in 1652 under the reign of Christina

Two EDX spectra of two different solidus coins from Livonia struck in 1652 under the reign of Christina of Sweden are shown in Fig. 7.147. The spectrum A represents a coin that is not corroded, the spectrum B a strongly corroded coin. Both spectra do not show the original composition of the coins, which are made of a copper/silver alloy with 10% silver [640]. In spectrum A, the silver content is too high due to the silver enrichment effect resulting from blanching. The typically green corrosion products resulting from $CuCO_3$, $Ag_2S$ and AgCl, which form always the patina of silver coins, can be found in spectrum B. The silver content in spectrum B is more realistic.

The lack of XRF investigations into coins is related to the very low information depth, which ranges for a silver-based coin for copper from 3 μm (Cu $K\alpha$) to nearly 30 μm (Ag $K\alpha$) for silver. The XRF methods are in concurrence with other nondestructive methods such as NAA or PAA, which offer a complete elemental analysis of the entire coin.

Precipitates or segregates of lead appear often in bronze alloys, which can be large. Beauchesne et al. [641] obtained doubtful results investigating lead-bronze coins with PIXE due to the depth profile of the surface layers.

Special production procedures enrich the noble metal in the surface. Such techniques are well known and have been used in all centuries up to now.

Examples for determining the silver content in modern coins using XRF are given in Fig. 7.148 (struck between 1900 and 1970). A good correlation between the measured and theoretical content can be found for most coins but not for silver plated coins as indicated by triangles in Fig. 7.148.

The XRF techniques are suitable for the determination of whether a coin is silver-plated or not because the possibility of using the different X-ray fluorescence series leads to information of different depth zones.

The interpretation of the result of the analysis should be performed with caution due to the appearance of enriched or debased zones resulting from corrosion or the production step. It is always necessary to relate the analyses results derived from the surface to results from bulk material investigations.

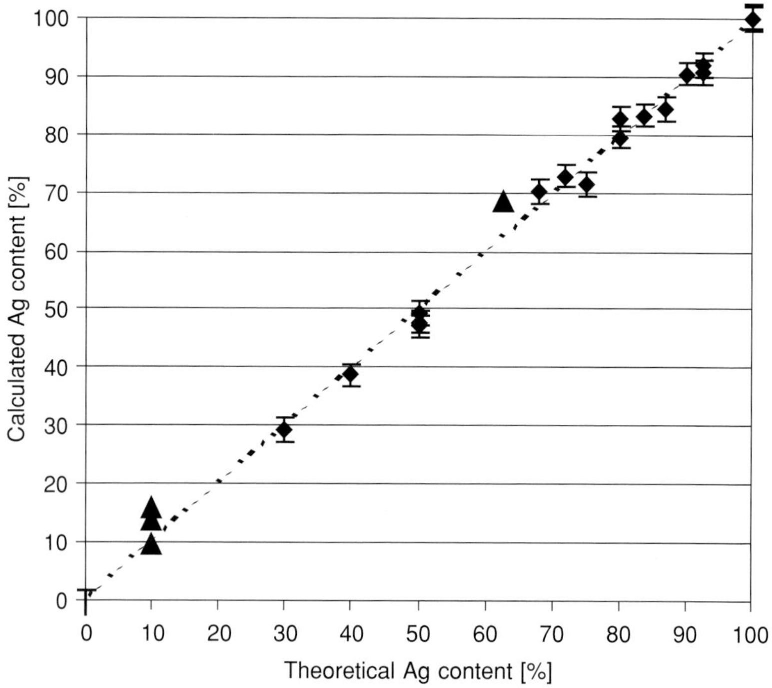


**Fig. 7.148.** Correlation between measured silver content (XRF) and the theoretical silver content for modern silver coins; ♦ good fit; ▲ worse fit due to silver plating

Linke et al. [642] tried to relate the surface enrichment of silver to the silver content of the bulk material and a time depending etching with sulphuric acid. They determined the intensity relation of the Ag-L and the Ag-K series resulting from combined investigations with EDXRF and SEM/EDX. However, no relation between etching time, bulk and surface concentration was found to derive the silver content of the bulk material by knowing the intensity relation.

### 7.8.4 Preparation of Coins for Surface and Bulk Analysis

In the preparation of coins for XRF analysis, the investigator is confronted with the numismatist's view [643]. The aim is to get information about the bulk content, which requires a cleaning step due to the surface enrichment or debasement effects.

Most authors [645, 647] favoured a slight destroying step, which can be divided into two groups (Table 7.21). Carter [647] compared the success of different cleaning methods with respect to the surface enrichment effect of copper in recently manufactured silver–copper alloys. In all of these procedures care must be taken during the cleaning process because some elements (e.g. lead) can be smeared on the surface, which results in excessive content. In grinding procedures the undisturbed blank is obtained when the concentrations remain constant.

Some cleaning procedures for coin surfaces are well known from numismatics literature [648] but the aim of this cleaning method is to get a blank and brilliant surface. The procedures mostly include a chemical etching, which can lead to a depletion of the less noble metals within the surface. For example, the copper content of silver coins will be reduced when using a silver cleaning bath. Other cleaning operations use an ultrasonic bath with ethanol or soft cleaning with soap but these procedures do not remove the corrosion products completely.

The work of Al-Kofahi and Al-Tarawneh [649] is a good example of cleaning that is not thorough. They cleaned the surfaces of silver Dirhams with soap, water and a steel brush only. Based on the pictures shown of the heavily corroded Dirhams in their report the results of the XRF investigation are doubtful (e.g. the silver content varies between 8% and 52%).

**Table 7.21.** Cleaning procedures used for coins

| Common procedures | Alloy depending procedures |
|---|---|
| 1. Total abrasion of the surface up to a blank disk [644] | 1. Electrochemical polishing [646] |
| 2. Grinding with emery paper [624] | 2. Etching [645] |
| 3. Blasting with pure aluminium oxide/air mixture [646] | |
| 4. Microgrinding and polishing of a very small part [634] | |

### 7.8.5 Metals and Standards

Mandal et al. [650] and Al-Kofahi et al. [651] made attempts to use fundamental parameters to analyse coins but this implies that all elements are detectable. Good results can be obtained only for main elements. Subsequently, Al-Kofahi et al. [649] used a well-investigated coin to improve their results in addition to the fundamental parameters analyses.

The use of external reference materials is helpful for most XRF investigations of coins. If no standard materials are available, one has to produce one's own standards based on metallurgical phase diagrams. The use of well-investigated coins makes it possible to reduce the effect of surface roughness. Stahl et al. [652] used cheap modern silver coins as calibration standards. The main and minor elements, which appear in silver, gold, copper, bronze and brass coins, are listed in Table 7.22. More than 30 elements are detectable in addition to the trace elements. The trace elements are of special interest concerning the identification of different production techniques.

### 7.8.6 Accuracy and Precision

The advantage of nondestructive investigations into coins is often coupled with loss of precision. Rough stamped surfaces show a non-ideal behavior since the surface profile depth is often bigger than the information depth of XRF methods. Lutz and Pernicka [653] compared the results of EDXRF and AAS/NAA investigations and showed that the surface roughness influences the results only by 10% for brass coins. Similar results were obtained by Cowell [654] for gold coins and Stankiewicz [655] for copper alloys.

Additionally, the surface composition varies from the obverse to the reverse side of a coin due to the corrosion phenomenon, so that the elemental concentration could only be used as an average value. Two possibilities to reduce the surface roughness effects are used: the excitation spot (μ-XRF) can be reduced, so that a scan over the surface is possible, or the excitation spot can be enlarged (some square centimetres), so that the whole surface is excited.

The precision for the main elements (1–100%) is 1–5% and for the minor elements (0.1–1%) 10–15% if the surface is free of disturbed layers. Carter et al. [661] had shown that the precision of main elements in ancient Roman

**Table 7.22.** Compositions of coins from [630]

| Type of coin | Main elements | Minor elements |
|---|---|---|
| Gold | Au, Ag, Cu | Pb, Sn, Bi, Fe, Hg |
| Silver | Ag, Cu, Pb, Hg | Bi, Fe, Sn, As, Sb |
| Brass, bronze or copper | Cu, Zn, Sn, Pb, | As, Sb, Co, Fe, Ni, Cr, S, Ag, Au, Br |

coins can be reduced to 1%, so that the precision is better than the variation, which appears during the production process. The precision and accuracy in coin analyses using EDXRF is compared by Lutz and Pernicka [653] with NAA and AAS. The accuracy is within 5% for the main elements when analysing a large and clean surface. Carter et al. [629] compared different XRF techniques with NAA, PAA, WCA and AAS on Roman orichalcum coins and obtained good accuracies within the standard deviation. Further uncertainties may arise from the treatment of coins in museums for restoration.

### 7.8.7 Some Examples of Typical Questions of the Numismatist

#### False or Genuine

The question whether a coin is false or genuine routinely appears for expensive coins. Answering this question can be quite difficult due to the expertise involved in forgeries. Forgeries, which are made to mislead the numismatic specialist are often of such good quality that the question of whether a coin is false or genuine cannot be that easily answered by optical investigations of the coin by specialists.

Klockenkämper et al. [656] developed a method, which combines EDXRF and EDAX, to decide between original and forged gold coins. They took the total relation of the main and trace elements of surface and subsurface into account and obtained clear results applying complex criteria schema. Modern alloys are too pure in comparison to former times and, e.g. no Hg or Sb can be found in the forgeries. A similar approach to identify forgeries was made by Auer et al. [657] with platinum coins issued by the Russian tsars between 1824 and 1828.

Poorly produced forgeries are easy to identify. Numismatists are more interested in the study of contemporary forgeries. Reiff et al. [658] investigated the techniques to produce gilded gold forgeries. They could decide between three different techniques by identifying the composition and thickness of a gold cover by applying EDXRF, electron probe micro analysis (EPMA) and SEM/EDX.

#### Influence of the Economy or Political Situations on Coin Constitution

In ancient times, the noble metal content of a coin reflected the real value, so that the denomination of coins reflects directly the economic and political situation. A debasement of coins during longer time periods can be found in all epochs, which reflects the normal inflation. The results of EDXRF investigations into the silver content of grossus pragensis (1300–1547) using a $^{241}$Am source are shown in Fig. 7.149. The results are in between the obverse and reverse values. The decrease of the silver coins seems to be linear but, in fact, it happened in steps.

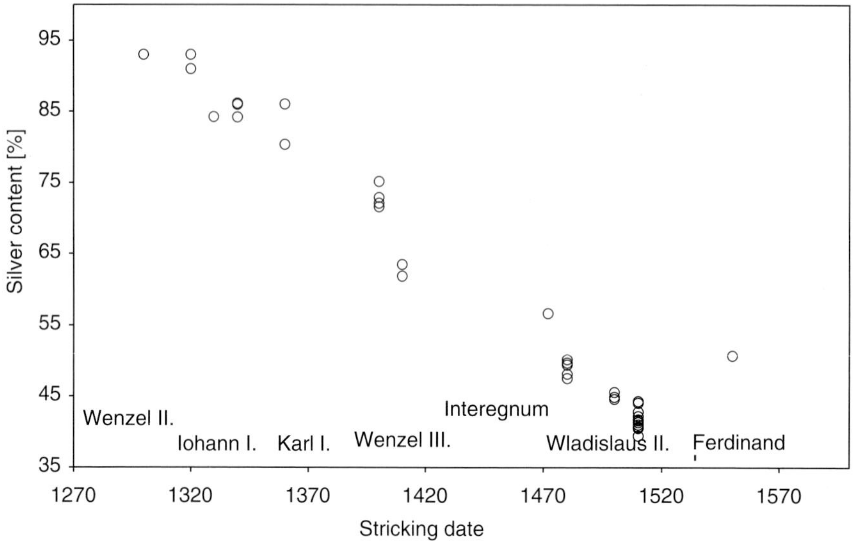


**Fig. 7.149.** Silver content of grossus pragensis, Ag content was determined using a $^{241}$Am source and a Si(Li) detector

Political instable epochs, e.g. civil wars, often led to poorly minted coins. The necessary ores or metals from mines did not reach the mint during these times and substitute elements had to be used. As a result unusual metal concentrations could appear in coins. For instance the determination of "white bronze" in drachmae was related, according to Constantinescu et al. [666] to the Roman Civil wars (first century BC). They found a thin (submicronic) layer of tin in EDXRF and PIXE investigations, which can easily be mistaken for silver.

Imitations were made during all epochs. The aim of imitations might be in the simple forgery or the economic need of a monetary system. The barbarian imitations fall in the last category. Most of these coins had no readable inscription, so that the composition can be helpful for identifying the period. Barbarian imitations have often comparable noble metal content but different concentrations of the minor or trace elements. Uzonyi et al. [625] investigated Greek Dyrrhachium silver coins and imitations with micro-PIXE. They found that the imitations had a significantly higher tin content but the same content of silver.

## Production Techniques

The composition of coins varies widely with the different time periods due to the production technique of coins. Two remarkable points remain: the industrialization around 1900 and the refinement of pure metals at the end of the seventeenth century. Silver coins consist only of 99% silver prior to 1700. The coins were produced using Cu and the noble metals Ag and Au which mostly contain Pb, Sn, Bi, Hg as additional elements before 1900.

Great interest is evinced in determining the production techniques. Special techniques are blanching, plating and dipping of the coin. All methods aim to enrich the silver content in the surface, so that finally the coin seems to be more valuable than the actual content. Klockenkämper et al. [659] developed a method using two different XRF techniques to obtain a relationship between the silver in the surface layer (EDAX, 5 μm, Ag-L line series) and the subsurfaces (WDXRF, 50 μm, Ag-K line series), with which they investigated Roman coins up to the Diocletian reform (50 BC to 300 AD). A simpler EDXRF study was done by Flokowski et al. [644] who used two different radionuclide sources for excitation of the Ag-K and Ag-L line series. They were able to decide which Polish coins of the fifteenth and sixteenth century are surface enriched.

The blanching was performed by boiling coins in a water solution of wine lees ($KHC_4H_4O_6$). The surface enrichment (e.g. by plating the surface with purer silver) leads to a time-depending wear, which is demonstrated in Fig. 7.150 for Mexican silver pesos of 1967. Fig. 7.151 shows the obverse of the pesos used for the EDXRF investigation. From left to right the increasing wear can be deduced. The copper content increases more drastically than the silver content decreases with increasing wear off (not normalised to 100%) due to the low information depth for copper. A similar effect also appears for ancient coins.

Especially in ancient times the production of coin alloys depended on the available ores, so the content can vary. For instance the results of the silver

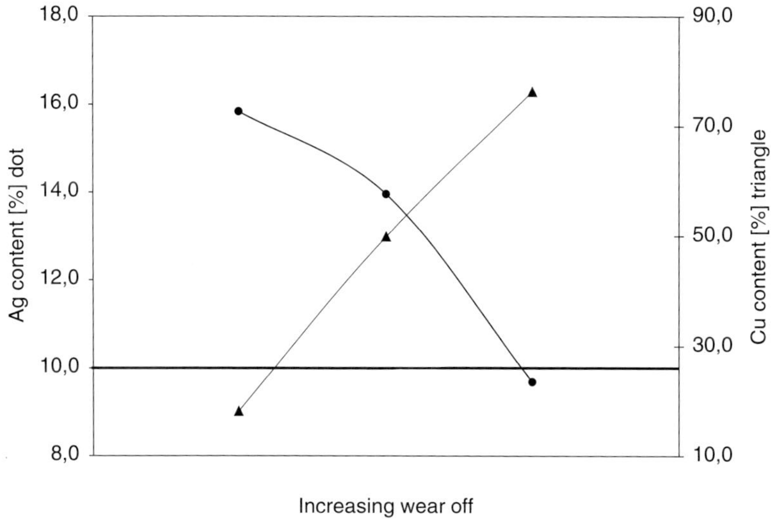


**Fig. 7.150.** Silver plating of Mexican Pesos of 1967; the big line at 10% silver represents the bulk Ag concentration (•), and the respective copper content (▲)

**Fig. 7.151.** The obverse of the silver plated Mexican Pesos; from left to right increasing wear off of the surface layer as determined by the XRF investigation depicted in Fig. 7.150

content of Roman denarii of the year 196 AD in Fig. 7.152 are given, in which the silver content varies between 47% and 62%. Later, the same denarii were analysed again by Butcher [660] using AAS. He observed a better agreement (around 48%) in his results and attributed the less accurate results of Carter to a surface treatment step.

The silver is used as amalgam for plating copper-silver coins with purer silver, so that Hg can be found only in the surface layer. A similar technique was used for gilding coins with gold (see Reiff et al. [658]). More information about production techniques and the final compositions can be found in Gale et al. [662] who investigated early Greek coinage.

## Relation to the Ores Used

Many attempts were made to identify the ore used in coins, so that a decision can be made whether the ore was mined near the mint or got by trade to the mint. Metcalf and Schweizer [663] tried to identify the Au content in English silver coins using the "Milliprobe" (an XRF device with a spot of 1 $mm^2$). The English ores have an Au content of 2–3% in contrast to the European

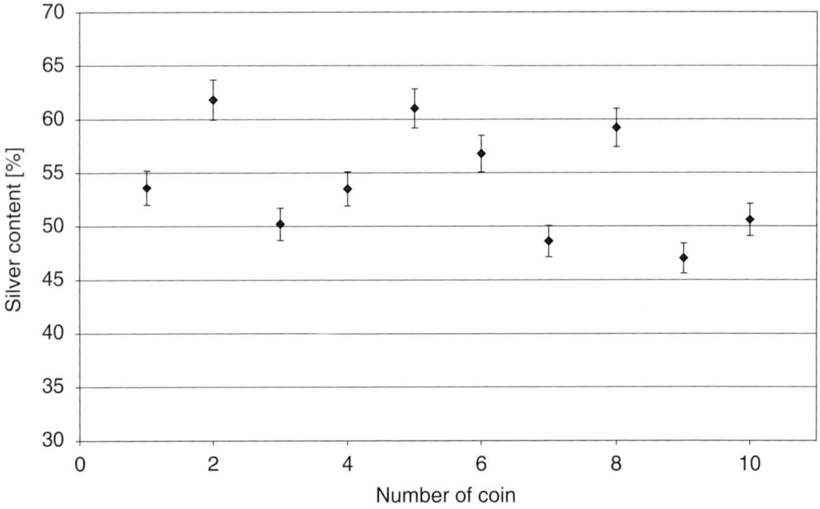


**Fig. 7.152.** Variation of the silver content of Septimus severus denarii in the year 196 AD from Carter and Carter [661]

ores with Au content of only 0.3%. But they were not able to find any relation to English coins of the twelfth century. They explained this with the denomination of the current coin every 23 years. The coins were remelted and new ones produced.

Hrabanek [664] tried to find the relationship between the Bi content of grossus pragensis and the different ores of the Erzgebirge using EDXRF and EMPA. Two groups could be identified with high or low Bi content, but a clear relation could not be achieved due to the low number (25) of coins investigated. For unalloyed silver coins, the Sb and Au content is mostly unaffected by the cupellation process (melting the rough metal with bone ash in cups). Cowell and Lowick [665] were able to determine in AAS and EDXRF investigations that the silver ores used for minting were from different panjher mines. As criteria for unalloyed coins they used the remaining copper content, which must be below 0.5%.

Au ores can be identified by the fingerprint of the platinum elements, which Constantinescu et al. [666] showed on the gold coins from the Dacian, using PIXE and EDXRF. For copper ores, Bowman et al. [667] could identify different types with the assumption that the traces can be found in the artefacts too.

#### Identification of Mints

The identification of a new unknown type of coin and relating it to a mint is a task of the numismatists. In the ancient and Middle Ages often the coin had no inscription. Here, the determination of the production technique in combination with the composition of the coin is helpful in solving the problem. A total analysis is necessary including the bulk and surface investigations.

Most ores had to be refined and alloyed before they could be used as coin blanks. Both are time consuming processes and need high technical standards. A sign of a bad mint master is the appearance of elements as the residues of ores in the coins. The determination of Br in silver tetradrachme is a typical example (using PIXE, EDXRF and PAA [666, 668]) and could be attributed to a Barbarian workshop.

### 7.8.8 Conclusion

The investigation of coins are often performed using XRF methods due to the demand for a nondestructive method. But the information about the metal content is reduced to the near surface layers, which can differ from the bulk material. Here, the results of XRF investigation are critical with respect to determining the excavation history and production technique of the coins. It is helpful to evaluate the XRF results with other methods such as the NAA, PAA or WCA. The determination of the specific gravity is a frequently used method [669] for the main elements Au and Ag in gold coins, which can be easily added to the determination by XRF.

If precaution is taken during the preparation in consideration, the XRF is a favourable method to analyse coins. The near surface nature of EDXRF investigations is excellent for the detection of blanching and plating techniques.

### 7.8.9 Recommended Reading

Most literature deals with the numismatic interpretation of the analysis results. Very helpful journals are "Archaeometry" and "The American Journal of Numismatics". In addition, the publications of the Royal Numismatic Society are of interest because four special publications have the titles: "Metallurgy in Numismatics". A French series of monographs is the Cahiers Ernest-Babelon, which are sister-publications to Metallurgy in Numismatics.

## 7.9 Analysis for Forensic Investigations

*J. Zięba-Palus*

### 7.9.1 The Specificity of Forensic Research

One can mention a number of characteristic features of forensic chemical analysis – strictly related to its forensic aspect – which differentiate it from other kinds of analyses. The aim of forensic chemical investigations is, above all, the identification of various materials, constituting evidence in court cases. Thus, it can be thought of as a kind of material science, differing in the fact that usually only very small quantities of material are examined and that obtained results must meet criminalistic and legal requirements.

#### The Material

Great advances in the natural sciences mean that identification of various materials is currently being carried out with increasing accuracy and precision. In general, determining the chemical components of examined materials and categorizing them is not usually difficult. Additionally, in the case of some materials such as car paint or fibres, the possession of a broad database allows us to determine their producer and the time of their production. There are, however, limitations, which an expert should be aware of when performing chemical examinations. Information about materials obtained in the course of analysis allows group identification, but not individual identification. In other words, analytical chemical methods will help to answer whether examined materials are the same, or different. However, these methods do not provide an answer as to whether these materials were part of one and the same material system, i.e. from the same object.

The analysed samples are usually very small, often having a mass of the order of micrograms or less. The examined material is not only quantitatively limited, but also unique, and must be sufficient for carrying out all of the required examinations. Samples delivered for examinations are usually impure and often difficult to separate from the base in which they occur. The smaller the dimensions of the examined materials and the more subtle their structure, the greater is the influence of the base on the results of the examinations. Thus, careful separation of the examined material from the base is necessary, and if this is not possible, its influence on the results of the analysis should be taken into account.

Materials sent for chemical research seldom make up a homogeneous system. Usually they are complex, and all components are mixed together. In every case these materials are multiple-component mixtures.

Materials that are the subject of criminalistic investigations can be divided into so-called evidence materials, those found and secured during inspection of the scene of the event (criminalistic traces) and comparative materials, secured during various actions under the aegis of legal proceedings, that are similar in use and properties to the evidence material.

## Methods

In order to identify the examined materials, it is necessary to determine their chemical composition, mostly qualitative, and establish some of their physical and chemical properties. For this purpose, several different methods of chemical analysis can be applied. The fundamental rule of cross-checking methods is obligatory when carrying out investigations. The results obtained with one method should be confirmed with other techniques.

Until quite recently, many traces, even some of those that were visible to the naked eye at the scene of the event, were not secured, because sufficiently precise methods for their analysis did not exist. Only the elaboration by science of micro-research methods and their introduction into practice by forensic scientists has made possible the analysis of these traces and their utilization as a valuable source of information about the event, i.e. its course and persons taking part in it.

Usually, in crime detection, one applies such methods that do not destroy samples, but on the contrary, allow repeated examination by means of the same or another method. Nowadays, classical chemistry methods are being replaced by instrumental methods, making it possible to obtain results quickly with an ever-smaller quantity of sample required for analysis, and with significantly lower detection limits.

The basic method of trace analysis beside microscopy is microspectrometry in the full range of electromagnetic radiation (UV/VIS, IR, X). Its advantages, such as the possibility of examining very small quantities of the sample, eliminating the arduous process of sample preparation, high sensitivity and the possibility of multiple repetition of measurements without destruction of the sample, have resulted in these methods being used today in the majority of criminalistic laboratories in the examination of traces revealed at the scene of an event.

### Interpretation of Results

The most difficult stage of the entire process of identification of materials constituting evidence in court cases is the interpretation of the obtained results of the examinations. Most importantly, the raw results – burdened with measurement errors – have to be processed. The dispersion of results, measurement error and confidence range are evaluated. Correlation analysis is also often performed. Furthermore, more advanced statistical and chemometric methods are also sometimes used.

The criminalistic interpretation of the obtained results of investigations is a separate issue. A match between examined materials in terms of chemical composition and properties is not sufficient for establishing that they are identical. It is necessary to have knowledge about the different types of the given material that exist, about its variability within each type, about its use and distribution in the surrounding world. Sometimes, being acquainted with the circumstances surrounding the course of the event itself is also useful. That is why databases are being created concerning defined types of materials secured at the scene of the event.

Generally, if, as a result of the performed comparative analysis, a match in terms of properties and chemical composition between the material forming the trace and the reference material is established, it can be concluded on this basis that these materials may have a common origin. In practice, the probability that they originate from the same source is very great. If, however, a difference in properties or chemical composition is revealed, it can be accepted that the examined materials are significantly different.

## 7.9.2 The XRF Method in Forensic Research

Among analytical techniques applied in determination of the chemical composition of material forming a criminalistic trace, methods of elemental analysis have a special place. X-ray microspectrometry has replaced methods such as atomic emission spectrometry (AES) or atomic absorption spectrometry (AAS) commonly used earlier. The relatively easy preparation of the sample for analysis, the small quantity necessary to perform measurements, the nondestructive and multielemental character of the analysis, the speed of the analysis process, the high sensitivity, the automation of measurement are the main advantages of this method of establishing elemental composition, leading to its application in forensic analysis. Moreover, the method can provide both an average compositional information on the whole sample and in its particular small areas. Three kinds of measuring techniques that utilize X-rays are most often used in the analysis of traces: XRF, μ-XRF and SEM/EDX.

So-called contact traces like paint chips, glass fragments, oil stains, fibres, metal attrition, soils, building materials, loose powdered materials, appear in the traditional areas of forensic science. Their identification and comparison can help to establish if the suspected person was involved in a crime. In the following sections the application of X-ray analysis in the examination of selected contact traces is shown.

### Paint Traces

One of the kinds of criminalistic traces collected for physico-chemical examinations is particles of paint coats, being revealed most often in connection with events such as car accidents, robberies or burglaries. They occur in the form of microfragments of paint coat, frequently with an area of several square millimeter or less, or visible smears of paint in the form of coloured streaks on the clothing of persons being involved in these events or on other substrates. The aim of paint examination is to establish the degree of similarity between the sample forming the paint trace and the sample originating from the suspect (from his vehicle, tools used in the act, etc.). Identification analysis is also carried out, which leads to the determination of the type of paint product, its use, the producer and the year of production. Routine examination of the paint encompasses establishing the colour and shade of the sample, the structure of the paint fragment and also analysis of the chemical composition.

Most often fragments of paint have a multilayer structure. Each layer (about 10–50 μm thick) is made up of painting material and is a mixture of many chemical compounds. Paint smears, on the other hand, are, as a rule, made up of one or two layers of painting material mixed together and sunk into the base (e.g. among the fibres of the fabric). Every paint (except for binder, which is composed of synthetic resins and additions) contains a combination of organic and inorganic colouring pigments and also extenders and decorative (effect) pigments. Pigments provide the paint coat with its colour, whereas extenders are responsible for the decorative effects of paint coats (e.g. covering and polish) and its resistance to the activity of atmospheric factors. Samples of paint coats can have the same binder and differ in the composition of pigments. This is especially true of paints that are applied in the motor industry, where one type of car paint is offered in various colours and shades. Paints of the same colour can also contain a different set of pigments and extenders, which depend on the use of the paint and the producer of the article. Thus, when comparing paint samples it is necessary to identify pigments and extenders.

Pigments and extenders are usually identified on the basis of the elemental composition of the paint sample, making use of IR spectra and data on possible pigment sets used in the paint industry. The method of scanning electron microscopy coupled with energy dispersive X-ray spectrometry (SEM/EDX) has been applied in the analysis of pigments and extenders since the mid-1970s [674, 679]. SEM enables imaging of the sample and serves as the source of its excitation. The interaction of electrons with the sample produces characteristic X-rays, forming a "fingerprint" of the element, which is detected and identified by an EDX spectrometer connected to an SEM. The results of elemental analysis can be registered in the form of X-ray spectra or presented in the form of a so-called elemental map of the sample, which shows the kinds of detected elements and their distribution in the sample in a graphic and clear way (Fig. 7.153). However, elemental maps are generally not quantitative and may lack the sensitivity to demonstrate minor sample differences.

Because of the small size of fragments of paint coat and their layer structure, a number of sample preparation techniques for examination by this method are applied, making possible interaction of the beam with every layer, and thus ensuring registration of the X-ray signal for each of them. Most often the fragment of paint coat becomes embedded in the resin, creating a block, which is then cut into slices perpendicular to the surface of the coat fragment, using a microtome. The preparation obtained in this manner is a cross-section of the examined sample. In order to obtain a smooth surface on the sample, it is polished and then placed on a stub inside the sample chamber of the electron microscope. Another method is to separate each layer from the fragment of paint coat and then place each one on an SEM stub to be individually analysed. One can also prepare paint chip by exposing portions of each layer in a stair-step fashion using a scalpel. It is also common to excise a thin peel from each layer and mount the thin peels on a stub.

By the techniques of embedding and polishing, one obtains samples with a smooth and flat surface, ideal for examination by the SEM/EDX and μ-XRF methods. However, this technique has a number of disadvantages. The most fundamental one is the fact that after embedding the sample in resin it is difficult to recover it for examinations by other methods. Furthermore, when the paint layer (visible in the cross-section of the sample) is very thin ($<10\,\mu$m), one must be certain that just this one layer is being analysed, and not the neighbouring ones as well. When polishing a sample there is always the possibility of contaminating a paint layer by other layers. All this means that embedding of the sample is often rejected in favour of cutting it with a scalpel in order to expose all paint layers. However, determination of the elemental composition of paint forming smears requires the isolation of a paint particle from the substrate.

Identification of an element in the sample is carried out by comparison of the obtained spectrum with the spectrum of a standard sample of this element. The lack of homogeneity within a particular layer forces one to perform measurements in many places on the sample and to take the mean value of the obtained results. Data on different possible kinds of compounds used in the composition of various paint coats are very helpful in the identification of a pigment or an extender. Sometimes, interpretation of the obtained results is made more difficult by the fact that detected elements may also originate from the organic components of the paint (the binder and organic pigments). The qualitative elemental composition of paint is usually determined by SEM/EDX and SEM/WDX methods [671, 672, 674, 681]. WDX is

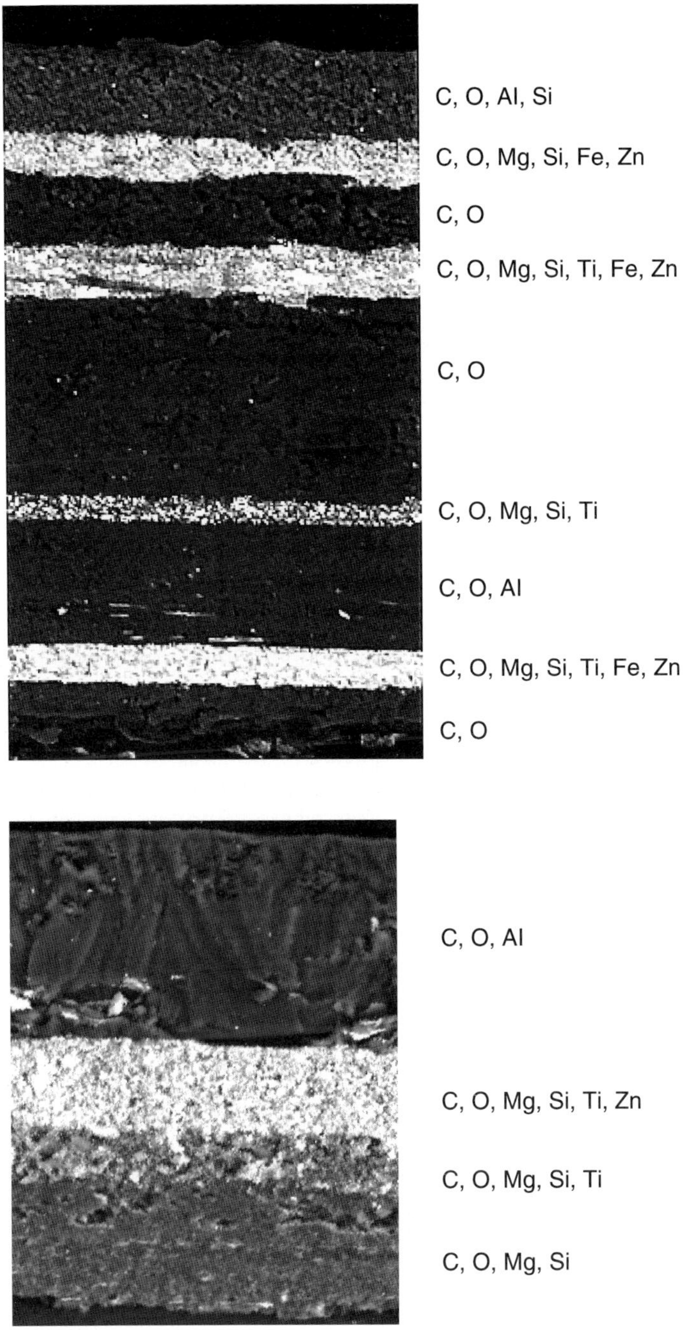


**Fig. 7.153.** Cross-sections of paint fragments (picture in the back scattered electron mode of SEM) and elemental content of each layer

a technique that can allow better resolution of the X-ray lines and so, more specific interpretation of the qualitative elemental data.

X-ray fluorescence spectrometry (XRF) is a technique similar to SEM/EDX. It is also capable of nondestructive analysis of a small paint sample. The difference lies in the application of an X-ray beam as the source of excitation instead of an electron beam. It is a complementary method to SEM/EDX, XRF being more sensitive for elements of high atomic weight and less sensitive for those of lower atomic weight. Utilization of an X-ray beam provides a source of higher energy and so, exposure to higher excitation potentials results in the detection of higher-energy K and L lines, which makes identification of heavy elements easier, or even in some cases makes possible the identification of otherwise unidentifiable elements. Thus it allows us to identify, e.g. pigments and extenders containing Co and Mn, and also to determine elements included in organic pigments, e.g. metal complexes-Cu in phthalocyanine [678, 679]. Moreover, replacing the electron beam by the X-ray beam eliminates the necessity of covering samples with a conductive layer. The detection limits of XRF for most elements are ten times better.

The main disadvantage of the XRF method is that it is not suitable for analysis of very small samples, e.g. individual layers of paint. The diameter of the X-ray beam in older models of spectrometers was many times greater than the thickness of a single layer of paint. The newest spectrometers have the X-ray beam collimated to about 30 μm, but beam collimation lowers the beam intensity and lengthens the time of analysis [674]. In the case of layers less than 20-μm thick, the detected signal comes not only from the examined layer but also from neighbouring ones. Furthermore, the beam penetrates deep into the sample – deeper than the electron beam – and so, the obtained signal also originates from the deeper layers of the examined sample. Thus, in order to obtain reliable results, a good definition of the examined region of the sample is necessary. The differences in the thickness of the examined sample, due to the nonhomogeneity of the material under consideration, can cause differences in the obtained spectra, i.e. in the element concentrations. It is also possible to obtain the same bulk analysis for two paint chips having different layer structure. Fisher [673] has presented a general discussion of the forensic applications of X-ray fluorescence.

Massonnet [676] described in detail the advantages and disadvantages of XRF and XRD techniques in examinations of samples of car paint coats. X-ray diffraction (XRD) is another method occasionally used by forensic paint examiners, which analyses the crystalline structure of the material rather than its elemental content and has the ability to provide definitive identification of inorganic components.

Zeichner et al. [680] applied X-ray analysis in characterization of spray paints. They found large variations in the characteristics of coatings depending on whether the cans were shaken before the spray was applied.

Zięba [682] examined sixty paint fragments taken from new and repainted cars by the use of both infrared spectroscopy and X-ray microanalysis. It was found that in most cases it was possible to differentiate between the fragments of the same colour of the external layer, taking into consideration the number, colour and the chemical composition of each layer in the paint chip for comparison.

Bayard and Stoney [670] have designed and constructed prototypes of an electron microprobe instrument that is intended to improve efficiency, reduce costs and extend the capability of the forensic examination of trace evidence. They have coupled a polarized light microscope with an electron microprobe, allowing the determination of elemental composition along with direct observation of the sample's optical properties and visible cathodoluminescence. Data observed on paint chips and mineral grains from soil are shown.

Currently, the most suitable method for trace evidence analysis in most forensic laboratories is SEM equipped with an energy dispersive X-ray spectrometer. Many forensic scientists have integrated new X-ray spectrometry equipment into their laboratories in recent years, e.g. capillary optics for microfluorescence analysis, and total reflection X-ray fluorescence (TXRF), but no publications have been presented up till now [677]. However, some interesting papers on the application of X-ray spectrometry to cultural and heritage samples have been published. Mantler et al. [675] presented examples of analysis by X-ray fluorescence spectrometry in art and archaeology, including pigments in paint layers. Theoretical aspects of information depths and shielding effects in layered materials were discussed. Elemental maps were experimentally obtained by a specially designed X-ray spectrometer and electron excited XRF.

## Oil Stains

A material that is commonly subjected to forensic investigation in road accident cases is motor oil. Oil traces are revealed on the clothing of victims as greasy stains, arising as a result of contact between the victim and the chassis of a vehicle. In the course of a criminalistic examination, an oil sample extracted from the fabric is compared with oil taken from the suspected car in order to establish their similarity. Chromatography, IR spectrometry and different methods of elemental analysis are routinely applied in the examination of oils.

Elemental analysis of oil helps both in its identification and in assessment of its degree of use. Refining additives suspended in the hydrocarbon base are organic compounds most often from the following groups: alkylosulphonates, alkylonaphtenates, alkylophenolates and alkylodithiophosphates mainly of zinc, calcium, magnesium and barium. They are added to the oil in order to improve its usage properties (they influence the stickiness, lubrication, density and freezing point, retard the process of oxidation and decrease corrosive activity). Particular types of oils differ among themselves both in the hydrocarbon base, and in the kind and concentration of additives. In some oils, additive concentration reaches 20%. When the engine is working during use of the vehicle, the composition of the oil changes. Dissolved or suspended products of decomposition and oxidation of its components appear, and also small bits of metal originating from wearing in e.g. cylinders, valves and other parts of the oiled engine. Their quantity in the oil depends mostly on usage conditions and on the technical state of the vehicle.

The concentration of these metals increases while the engine is working. Thus, analysis of used oil in order to determine the concentration of metals has been successfully applied for many years in the monitoring of degree of use of the engine, often making possible the detection of damage or faulty working of the engine [686].

In criminalistic practice, a sample for identification and comparative examination is obtained by extraction of a lubricant from blots visible on clothing, obtaining, in effect, one to several droplets of the sample. The high content of organic compounds and the considerable viscosity of the oil have always caused significant difficulties in the preparation of a sample for elemental analysis [683, 690, 691]. The main difficulty in obtaining repeatable results of analysis stems from the fact that the sample is a suspension of metal particles that in time undergo sedimentation. Therefore, analytical results are burdened with some errors. Better results can be achieved by preparing oil samples for examination by means of the incineration technique and then dissolving the obtained residue in a mineral acid.

Formerly, atomic absorption spectrometry (AAS) or atomic emission spectrometry (AES) were commonly used. Nowadays, a convenient method of elemental analysis of oil is X-ray fluorescence (XRF). Less sensitive in comparison to AAS, it does, however, have one big advantage. This method makes it possible to analyse an oil sample directly in the unchanged state or possibly after dilution with an organic solvent [688]. However, the XRF method for the quantitative determination of elements requires complicated calibration with standard samples. So, comparison of the composition of oil samples is often performed by a nonquantitative method such as peak rationing.

A model experiment was performed in order to establish the possibility of discriminating between used oil samples on the basis of their elemental composition [691]. Oil samples were withdrawn from the sumps of two cars after various time of exploitation. Elements originating both from the additives and from the wearing away parts of the engine were determined in an oil sample. The semiquantitative method was applied, comparing characteristic signals for particular elements with the signal of a chosen element in the sample. It was observed that as the oil in the car was used up, the concentrations of, among other things, iron, cadmium, lead and copper increased. The concentration of these metals in oil is related to its degree of use, whereas the concentration of calcium, barium, magnesium, zinc and phosphorus is stable. Providing information on the kind and quantities of refining additives can be helpful in differentiating between types of oil. A schema of criminalistic inference on the basis of comparison of metal contents was proposed as well.

Espinosa [684] stated that the presence of zinc dithiophosphate is the basis for differentiating between automotive and locomotive oils. In environmental protection studies he applied XRF techniques to the determination of Zn originating from this additive in various kinds of oils.

Comparing fresh oils the presence of Zn, Ca, P and S was detected by means of XFR method [692]. Although the spectra obtained manifested to be generally similar, the peak intensities differed slightly in fact. The calculated sets of peak integral ratios: S/Zn, Ca/Zn were in general systematically different for the studied oil samples. It was found that the elemental analysis providing information on the additives is complementary to IR spectroscopy in differentiating among oil samples.

Freitag [685] reported on the application of TXRF to the determination of trace elements in lubricating oils. It is useful for analysis of wear particles that occur only in amounts of the order of micrograms – such quantities being sufficient for complete analysis.

When comparing analytical results and inferring about the similarity of examined oil samples, it is necessary to use some statistical methods. Observed differences are small and only application of statistical analysis in the evaluation of obtained results allows us to decide whether they are caused by real differences in the chemical composition of these samples or by random errors. However, this requires many repetitions of measurements for each sample, which in practice can be a problem due to the large overall number of measurements necessary.

Lloyd [687] and Siegel [689] applied with success synchrotron X-ray fluorescence to differentiation between petroleum products.

The XRF method is also used in the analysis of other heavy products of the petroleum industry, e.g. oils, greases, and bituminous masses, detected in the form of blots visible to the naked eye on the clothing of car accident victims or persons suspected of starting fires.

## Glass Microtraces

Glass fragments are known to transfer to the clothing of a person breaking a window. These fragments may be used as evidence, associating the breaker with the crime. Traces in the form of glass fragments are also revealed in cases of traffic accidents, fights or robberies. Fragments of the glass have various sizes. Those found at the scene of the event are larger fragments, whereas those revealed on the clothing, hair or body of persons are small – with linear dimension less than 1 mm. Routine examination of glass fragments encompasses establishing their elemental composition and determining some physical properties, such as the refractive index and density. These data can be used by a forensic scientist for comparison of samples of glass, and also for ascertaining the kind of object they could have been broken off from, hence, establishing their origin.

It is worth noting that the chemical composition and properties of glass are very similar, irrespective of the type and application of the glass. Technological advances in glass manufacture have led to less variability in physical and optical properties between products manufactured by different companies, and also to less variability between the different types made by the same manufacturer. Consequently, the discrimination potential (the ability to distinguish between glass fragments) has been diminished and sole reliance on measurements such as density and refractive index can lead to overstating the value of positive matches [693, 694]. Elemental analysis, however, is promising.

The major raw materials employed for the manufacture of soda-lime-silica glasses are soda ash ($Na_2CO_3$), limestone (CaO) and sand ($SiO_2$). The other components of glass are different for different types of glass. So, the main elements such as Na, Ca, Si and Al are present in all glass categories at nearly the same level. The differences concern other elements originating from various additives, which are added to improve the usage properties of the glass or in connection with its later application or else originate from impurities in raw materials used in the production process. Their concentration is significantly lower (at trace level).

The chemical composition of glass can be determined by many methods [697, 698]. Forensic sciences prefer nondestructive methods, allowing the sample to be examined using two or more analytical methods. Another desirable feature is the possibility of simultaneous determination of several elements (analytes), using the smallest possible amount of studied material. In the case of analysis of glass microfragments, these requirements are fulfilled by SEM/EDX and XRF methods. They permit identification of elements occurring in glass in large quantities (in raw materials: oxides and modifiers). Such methods as inductively coupled plasma-atomic emission spectrometry (ICP-AES), atomic absorption spectrometry (AAS) with flameless atomization and inductively coupled plasma/mass spectrometry (ICP-MS) are also applied to the analysis of glass. The main disadvantage of these methods is that the examined sample is destroyed in the process, i.e. dissolved or incinerated. The great advantage of these methods, however, is the possibility of detection of elements present in trace quantities in the examined sample.

Two new instrumental techniques, known as total reflection fluorescence (TXRF) and microbeam XRF are becoming increasingly popular, while radiochemical methods such as neutron activation analysis (NAA) have, for practical reasons, been replaced by ICP-MS or laser ablation-inductively coupled plasma/mass spectrometry (LA-ICP-MS) [700, 701].

Terry et al. [706] pioneered use of the SEM/EDX method on its own for the determination of elemental composition of different groups of glass used in Australia. Currently, SEM/EDX and SEM/WDX are applied in the routine examination and evaluation of glass evidence in the majority of forensic laboratories. For qualitative determination, a glass fragment is placed on a stub and, after sputtering with carbon, assuring its conductivity, the elemental composition is determined. In the case of quantitative analysis, sample preparation is required, involving embedding the glass fragment in a plastic resin and then polishing the surface until it is flat, using grinding methods.

The surface is usually coated with a carbon layer and the fragment is sampled at different locations. Glass samples for examination by the XRF method are prepared in a similar manner.

The technique of X-ray fluorescence is widely used for the qualitative and quantitative analysis of elements having atomic numbers greater than oxygen. Instrumentation to detect the emission can be divided into two types, wavelength dispersive and energy dispersive. Quantitative elemental analysis is possible thanks to the linear correlation between the intensity of the characteristic X-radiation of the element generated in the sample by electrons and the concentration of this element. This is a relative method involving comparison of the measured intensity of the characteristic radiation of a given element in the sample with the intensity of radiation measured for a reference (standard) sample.

In practice, quantitative analysis of forensic glass samples is best achieved by an evaluation of the ratios of elements rather than by measurement of absolute concentrations. Very small and irregularly shaped samples, which are commonly found in forensic casework, are not suitable for this type of analysis. Samples with flat surfaces and known working angles are necessary for quantitative determinations. Quantitative analysis with or without standards can be used for the analysis of elements such as Na, Mg, Al, Si, K and Ca, assuming the rest of the sample matrix is oxygen.

The measurement of major, minor and trace elemental composition of glass is very important (valuable) for its discrimination and classification into glass types. It is usually helpful to be able to classify the questioned glass into one of a number of possible categories, such as sheet, container, vehicle window, vehicle headlamp or tableware. It is necessary to apply statistical methods in the characterization of glass evidence by its elemental composition.

Traditional treatment of the data involves determining the mean concentration and the standard deviation for each element and then comparing the means using a "3 sigma rule" or testing the match criteria with a strict range overlap for each of the elements. Advanced methods of comparison use statistical tests, cluster analysis or Bayesian approach [694–696, 707].

A number of publications have been devoted to the evaluation of different instrumental methods for the classification and discrimination of forensic glass fragments using chemical composition data. Reeve first reported the use of SEM/EDX as a means to further discrimination of glass samples that were not distinguishable by refractive index and density only. Studying glass samples, he compared the ratio of the concentration of each element to the concentration of calcium. He achieved very good discrimination.

Ryland [705] described a classification scheme for sheet vs. container glass samples using SEM microprobe determination of Ca/Mg intensity ratios and XRF Ca/Fe concentration ratios.

Total reflection X-ray fluorescence (TXRF) is a relatively new technique designed for surface analysis, which has found some application in forensic science. This technique uses a primary beam with a very low glancing angle, so that X-rays mainly reflect from the surface, penetrating only the top layer (several tens of Ångstroms). This reduces scattering and improves the signal-to-noise ratios. A detailed description of the TXRF technique can be found in a paper by Klockenkämper et al. [700]. Kubic et al. [702] reported the application of TXRF and a related technique to the analysis of small glass fragments. The detection level was less than or equal to 10 pg for 50 elements, from phosphorus to uranium. Little sample preparation is needed for semiquantitative screening, while dissolved samples give results comparable to other quantitative analysis techniques.

A promising X-ray technique for the analysis of small irregularly shaped glass fragments is the μ-beam XRF technique developed in Sweden and reported by Rindby et al. [699, 704]. This technique incorporates a very narrow beam of X-rays focused by conical capillaries. Application of capillary optics enabled the development of microbeam instruments of sensitivity comparable to instruments using conventional optics.

Using this technique they detected, e.g. calcium in samples in quantities of 40 fg. They ascertained that the precision of determinations is considerably higher compared to SEM. Analysing the results of the classification of glass samples on the basis of determination of the elemental composition by both methods, they concluded that measurements with the XRF method have greater discrimination power and divide the examined group of glass samples into a greater number of subgroups.

Kuisma-Kursula [703] has reported a study on accuracy, precision and detection limits of SEM/WDX, SEM/EDX, and proton induced X-ray emission (PIXE) spectroscopy in the multielemental analysis of medieval glass. The SEM/WDX and SEM/EDX methods were suitable for analysing major and minor components of glass samples. Trace element ($<1$ wt%) analysis was possible using the WDX or PIXE method. The PIXE method proved to be the most sensitive. Hicks et al. [722] assess the potential of μ-XRF in classification of glasses according to their respective origin using different statistical tools and in discrimination among glasses with similar refractive index (RI). The results obtained show that it is possible to analyse non-destructively and reproducibly small glass fragments ($>0.3$ mm) with the μ-XRF spectrometer. Glass types of the same RI can be differentiated by this technique as well. Neural networks and linear discriminant analysis using qualitative and semiquantitative data allow the classification of glass specimens with high degree of reliability.

Brożek-Mucha et al. [696] applied statistical and chemometric methods in the interpretation of analytical data obtained during examinations of glass fragments by elemental analysis. They proposed a scheme of glass classification based on results of quantitative analysis of 153 glass objects. Car window glass, car headlamps, external glass of car light bulbs, internal glass of car light bulbs, internal glass of ordinary light bulbs, and sheet glass were investigated by SEM/EDX. Concentrations of aluminium, barium, calcium, iron, lead, magnesium, potassium, sodium and zinc were determined. A nonstatistical method and the use of cluster analysis led to correct classification of most of the studied glass samples.

Becker [695] presented a proficiency test that was carried out in order to evaluate differences between the results of elemental analysis of float glass. Three pairs of glass samples with similar refractive indices had to be examined using refractive index measurement and elemental analysis. Different methods of elemental analysis, i.e. SEM/EDX, μ-XRF and ICP-MS were applied. Based on refractive index measurements, complete differentiation of the glasses was not possible. After the use of elemental techniques, full differentiation of the six glass types was possible. The strategies used by different laboratories in order to discriminate between samples and the merits of the various techniques applied were discussed.

## Gunshot Residue

The identification of metal traces secured for criminalistic examination might not appear to pose particular problems. Indeed, when the trace is in the form of a fragment of a metal object, dust or shavings, every method of determining the chemical composition can be utilized. However, when the metal trace is revealed on the body or clothing in the form of little particles as a result of contact with a metal object (e.g. a knife or a bullet along the edge of a wound) or in the form of metallic particles originating from the primer of handgun ammunition, then its identification requires use of specialized sensitive microanalytical techniques. Gunshot residue (GSR) examination plays an important role in establishing some circumstances of a crime with the use of a firearm. This kind of examination is complementary to ballistic examinations of weapons and ammunition.

Powder gases leaving the barrel of a firearm contain products of explosive reactions of the primer, and then the propellant, and also of interactions of these materials with other parts of the cartridge and weapon. The chemical composition and properties of gunshot residue depend directly on the kinds of materials used in the production of the ammunition. The most characteristic gunshot residues are metallic particles arising from components of the primer, demonstrating characteristic morphology (size of the order of micrometers, approximately spherical shape) and also specific chemical composition (lead, antimony and barium in the case of lead ammunition) [716]. Found around the gunshot hole and on the clothing and body of the shooter, they provide information on which to base, among other things, inferences about the shooting distance and the kind of ammunition used (and thus the weapon) and, most importantly, they serve to link the suspect to the shooting.

For many years there was no sufficiently specific method for the identification of characteristic gunshot residues. One could not see metallic particles due to their size (5–50 $\mu$m) and their presence was ascertained indirectly by means of colouring chemical reactions or such instrumental methods as AAS, NAA or XRF. These methods, however, are not specific and record all impurities, independent of their source of origin. The most successful technique to date for the analysis of GSR particles is, without doubt, scanning electron microscopy with an energy dispersive X-ray analyser (SEM/EDX). With most of the other techniques that have been used for GSR analysis, the sample is destroyed during examination. With SEM/EDX, however, the sample is virtually unaffected by the analysis and can be re-examined, if necessary, many times. Robin Keeley wrote a general introduction on the application of electron microscopy to GSR examination [712]. It laid down the basic techniques for the collection, examination and identification of GSR tapings taken from the hands of suspects. Other papers followed in profusion [711, 719, 720], all of them providing valuable contribution to the science. The basic techniques for obtaining samples and examining them on the SEM have, however, remained the same.

Mostly, half-inch diameter aluminium stubs with an adhesive layer of double-sided tape are used for sampling. When the sample has been secured on the stub, one then searches through it for spherical metallic particles of defined diameter and chemical composition. The method has many advantages, but its basic drawback is that it is time-consuming if the investigation is carried out manually. Suitable software for automatically searching through the secured material on the stub (in order to detect particles with specific features) shortens the time of investigation many times [717].

Since the introduction of scanning electron microscopy/energy dispersive X-ray examination of gunshot residue in the 1970s, it has been considered that the firing of a gun produces materials that are unique to that process. Wolten [718], in the most comprehensive study of GSR analysis by SEM, stated that micron-sized particles, having a morphology consistent with rapid cooling, formed a liquid state. They contained the elemental combustion of either lead/barium/antimony or barium/antimony and were unique to the detonation of the primer of a round of ammunition.

Metallic particle classification was carried out in the following way: the most characteristic (unique) particles were singled out, followed by indicative, one and two component particles, e.g. antimony, lead, lead/antimony, always accompanying the former and occurring in considerably greater quantities. [721]. Finding GSR on material received from a suspect confirms his or her participation in the event.

Only a few experienced specialists undertake the difficult task of identifying used ammunition on the basis of qualitative assessment of differences in the chemical composition of GSR [712]. Recently, some attempts were made at a systematic study of primer discharge residue originating from various brands of ammunition. Evaluation of the obtained analytical results was carried out by, among other things, chemometric methods [708, 709]. These studies showed significant differences between gunshot residues originating from ammunition types, and thus it was possible to achieve at least group identification of ammunition (and hence the type of firearm used) from information on GSR alone. In order to elaborate a reliable scheme of individual classification of ammunition on the basis of GSR alone, it cannot be excluded that it will be necessary to analyse the internal structure and chemical composition of individual particles [714].

The XRF method, however, is irreplaceable in investigations of gunshot holes occurring on various bases. A fragment of fabric cut from clothing or a fragment of skin or bones together with the gunshot wound can be placed directly in the measuring-dish of the X-ray spectrometer and analysed for the presence of metals originating from the primer, the bullet and the cartridge. Their disclosure confirms the use of a firearm and contributes to inferences about estimations of the shooting distance [716].

Recently, the first attempt to use μ-XRF in examinations of gunshot residues was carried out [710]. Charpentier et al. applied this method to the analysis of gunshot residues from primers of lead-free ammunition and ammunition in the area surrounding bullet holes, and found it to be a very important tool for determining shooting distance. They determined the elemental composition of a 500-μm zone near the bullet hole. In addition, as the method has a mapping system, it provided data on elemental distribution in a given area. The method allows the detection and quantification of strontium residues on the target up to a distance of 45 cm from the target.

Many improvised explosive devices contain a pyrotechnic or home-made explosive mixture as the explosive charge. Pyrotechnic residues contain a complex mixture of combustion products and unconsumed material. Residues are often analysed by scanning electron microscopy with energy X-ray spectrometry. Residues from unconfined burning studies are comprised of spheroid particles with an elemental composition characteristic of the original material. Confined burning of flash powders yields residues whose spheroid particles are mainly composed of metal fuel with iron from the steel confinement pipe. The detection of characteristic spheroid particles in residues, particularly in the absence of any unconsumed material and in combination with other types of analysis, may help forensic scientists to determine the original filling.

SEM/EDX only identifies the elements present in a material; it does not provide information on the specific compounds into which these elements are combined. Another analysis technique needs to be used to comprehensively study the composition of the particles [715].

A survey of the pyrotechnics literature indicates that various compounds of lead, barium, and antimony are used in the manufacture of pyrotechnic devices. Mosher [713] investigated the possibility of finding particles in firework compositions or combustion residues that might resemble those found during routine GSR analysis. The presence of magnesium in the majority of PbBaSb particles found would alert a GSR analyst to the possibility of an untypical ammunition or pyrotechnic device being the source of the particles.

**Miscellaneous**

XRF spectrometry is applied in screening analysis of complex materials, such as sediments from waters, the soil, body fluids, hair or stomach contents, with the aim of quickly establishing the qualitative elemental composition (especially heavy metals being components of inorganic poisons) of the material examined. Although, quantitative analysis is usually performed by means of AAS or ICP-MS methods, XRF being able to analyse the material in situ allows us to obtain information on all elements in the sample, and then to select those that are interesting for further examination.

### 7.9.3 Conclusions

The process of forensic identification of a material forming a trace at a crime scene requires the application of reliable and sensitive analytical methods. Forensic science frequently adapts new analytical techniques to its own needs, if they allow analysis of the chemical composition of small samples in a simple way without destruction of the analysed material. X-ray spectrometry, either alone or in conjunction with scanning electron microscopy, has a well-established position in the forensic analysis of elemental composition. Great expectations are currently linked to the μ-XRF technique. The small diameter of the X-ray beam focused on the sample allows analysis of single microfragments of glass objects or individual paint layers. It does not involve complicated sample preparation prior to the examinations. It is possible that further development of this technique will enable forensic experts to examine evidential traces directly on the substrate on which they occur.

## 7.10 X-Ray Fluorescence Analysis in the Life Sciences

*G. Weseloh, S. Staub and J. Feuerborn*

### 7.10.1 Introduction

Physiologists became aware of the fundamental constitution of biological matter and the significance of relationships between life and environment in the first half of the nineteenth century and this marks the beginning of modern reception of the mechanistic nature of life processes. Since then qualitative and quantitative aspects constitute the fundamentals of physiological sciences. Meanwhile, promoted by an extreme advancement of measurement techniques, the common interest increasingly has shifted from the major to the minor constituents of living organisms. Today, it is commonly accepted that 11 major and at least 10 minor or "trace" elements, most of them being metals or metalloids, are essential for one or another form of life. Above all, with the application of sufficient sensitive methods for elemental analysis, nearly all of the 90 stable elements of the periodic system can be detected in every organism.

Trace elements display important biological functions and have an impact on all life processes. A depletion of essential elements such as manganese, iron, copper, nickel, zinc, selenium, or iodine will lead to various human deficiency diseases. An accumulation of elements can lead to toxic symptoms or even poisoning and is frequently caused by heavy metals like cadmium, mercury, or lead as environmental contaminants. Therefore, there is a high demand for trace analytical information in the medical and clinical field.

Most of the essential trace elements present in the mammalian organisms are bound to proteins or peptides where they play a central role as structural or catalytic components. The specificity of their biological function is reflected by the inhomogeneous distribution of the trace elements among the tissues. Therefore, total concentrations, which are obtained by bulk analytical methods, have to be complemented by information about local element distributions as they are delivered by microprobe investigations. In the majority of cases, the nature of the metal–protein interaction and their function is not yet known, and their identification and the elucidation of their function is of great interest in many fields of life sciences.

Trace analyses are often carried out on different materials used as monitors for organs or the whole organism. Suitable monitors are body fluids such as blood, serum, plasma, and urine or tissue samples from biopsies of organs, as well as bones, hair, or nails. Hair, bone, and brain are considered to be the preferred target for the deposition of heavy metals, e.g. lead and mercury, from environmental sources. They are indicators for an estimation of long-term human exposure to these elements.

Today X-ray fluorescence analysis is playing an important role in the determination of the elemental content of biological matter. While "classical" energy- or wavelength dispersive X-ray fluorescence analysis with conventional X-ray tubes or radioisotopes as excitation sources has attained some interest in routine- and process analytics, more recent set-ups like micro focus- XRF and total reflection X-ray fluorescence analysis (TXRF) show unique features and special advantages over concurrent methods in the analysis of biological specimens. Although restricted to large facilities, X-ray fluorescence analysis with synchrotron radiation (SRXRF) is admitting all methods of X-ray fluorescence analysis with unmatched sensitivity, speed and, in the case of micro focus applications, spatial resolution.

### 7.10.2 X-Ray Fluorescence Analysis by Means of X-Ray Tubes and Radioisotopes

Although methods like AAS, ICP-OES or ICP-MS dominate the analysis of biological materials due to their low detection limits and their capability for elemental speciation when linked to separation techniques, the use of XRF spectrometry is widespread in life sciences.

XRF analysis is a fast, quantitative, nondestructive multi-element method, and these features together render it the preferred method for many applications. Furthermore, it is an inexpensive technique which is easy to operate. A disadvantage of "common" XRF analysis is the relatively high limit of detection. For investigations of elements in sub-$\mu g\,g^{-1}$ concentrations, accumulation procedures are necessary.

In many cases, it is possible to analyse the samples directly without any pretreatment, but then, for quantification, as internal standardization by addition of elements in a known concentration is not possible, often external calibration procedures in connection with *Monte Carlo* simulations, matrix corrections and other calibration procedures are necessary. For most applications only little sample pretreatment like drying (e.g. by lyophilization) and grinding is necessary, thus the risk of contamination is low. Due to its high water content, the mass of biological samples is reduced by a factor of nearly 5 by these steps, which leads to reduced background and improved sensitivity. Additionally, during preparation an internal standard can be added easily.

Today, silicon-based energy dispersive detectors are of central importance for the analysis in the field of life sciences, because of their multi-element capability they are able to reveal present elements as well as absent ones. Therefore, in standard procedures, samples are dried and homogenized before measurements, and EDXRS-detectors are used in most of the applications. Thus, these commonly used sample pretreatment and measurement procedures are not explicitly mentioned in the following section.

In the last years, instrumentations have been developed, like new radiation sources, focussing optics, and set-ups for nearly monochromatic radiation and high polarization that led to lower detection limits, small focuses and the employment of portable systems. The mentioned features make X-ray fluorescence analysis suitable for the investigation of organ tissues, biological materials, medical research and biotechnological processes.

## Medical Investigations in Tissues

X-ray fluorescence technique is widely applied for the analysis of inorganic constituents in light matrices, such as tissues, body fluids and hair. Carvalho et al. [723] examined human tissues (bone, hair, liver, and kidney) from 61 individuals in order to investigate the correlation between elemental composition and the dependence of these concentrations on age and gender. The XRF spectrometer consisted of a three-axial geometry with a tungsten X-ray tube and a molybdenum secondary target. The studied elements were Mn, Fe, Cu, Zn, Se, As, Sr, Rb, and Pb. For hair, a significant difference between male and female was found for zinc, whereas in bone there was a positive correlation with age for Zn, Sr, and Pb.

Geraki and Farquharson investigated the elemental composition of healthy and pathological breast tissues to aid the diagnosis of breast cancer [724], and found a significant elevation of zinc in the pathological tissue. The elemental changes in normal and cancerous tissue of mice were reported by Feldstein et al. [725]. The detection limit for solid tissue was 2 $\mu g\,g^{-1}$ for Zn and Fe. They obtained an increase in Rb level in the tumour by a factor between 4 and 10 relative to normal tissue. In another study [726], Carvalho and Marques showed the use of XRF spectrometry for the quantitative analysis of 14 elements in healthy liver, brain and kidney tissues, and the corresponding tissues from subjects suffering from liver cirrhosis. Higher concentrations of Co, Fe, Pb and lower values of Zn and Se were found in cirrhotic liver in comparison with healthy liver. Brain samples from subjects with cirrhosis showed enhanced levels of Pb (3.4 $mg\,g^{-1}$ dry weight). They determined the limits of detection from 0.6 $\mu g\,g^{-1}$ (Ni, As, Se, Sr) to 10 $\mu g\,g^{-1}$ (K) or 20 $\mu g\,g^{-1}$ (Ca). Carvalho et al. [727] also investigated the concentration of heavy metals in muscle, liver, fat tissue and skin in dolphins from the Atlantic Ocean. The samples were lyophilized and grinded in a liquid nitrogen cooled freezer mill before being pressed to pellets. The highest amounts of Mn, Cu, and Hg were always found in the liver, whereas Zn and Se concentrations were higher in skin. The obtained limits of detection ranged from 3.1 $\mu g\,g^{-1}$ (Fe) to 0.5 $\mu g\,g^{-1}$ (As).

The uptake of arsenic in the organs and tissue of mice was reported by Sanchez et al. [728]. An external standard for calibration and an internal standard (addition) were used to obtain concentrations, which showed uncertainties of 2% (urine), 7% (tissue) and 20% (blood). They reported concentrations of arsenic in the blood of 2.9 and in urine of 42 $\mu g\,g^{-1}$. Boman et al. [729] measured the elemental concentrations of S, Cl, K, Ca, Fe, Cu, Zn, As, Se, Br, Sr, and Pb in a number of Vietnamese animals in order to establish a background set of trace element concentration data with an XRF set-up in three-axial geometry. The detection limits ranged from 0.9 (Br, Se) to 5 (Mn) and 20 (Ca) $\mu g\,g^{-1}$. The effect of lithium augmentation on the trace elemental profile in diabetic rats was examined by Dhawan et al. [730]. The lithium treatment showed a significant depression in K and Rb levels and an elevation in the Br level. In other applications, the calcium and phosphor contents in osteoporosis model rats [731], and the distribution of trace elements of emboli in the cardiovascular system [732] were examined. To clear up the beneficial effects of cave therapy, Alfoldy et al. [733] characterized aerosol particles and their deposition in the respiratory tract. They suggested using the relevant chemical information for lung deposition modelling.

The influence of zinc on the toxic effects of organophosphorous insecticides (chlorpyrifos) in rats was examined by Goel et al. [734]. It as shown that the co-administration of zinc restored serum and liver marker enzymes to normal limits and could be used as a mediator of toxic effects. Liu et al. [735] studied the distribution patterns of trace elements in an experimental model of iodine deficiency using XRF and other techniques. They found that I and I plus Se supplementation improved thyroid hormone metabolism, but also affected the contents of Cu, Mn, Rb, and Zn in erythrocytes and of Br in the brain. The removal of iodine contrast media (iopromide or iomeprol) used in radiographic procedures on renal patients was monitored by Schindler et al. [736]. They concluded that high-flux haemodialysis and online haemodiafiltration remove the examined contrast media most effectively.

The analysis of hair can show relations to diseases or poisons, environmental exposure and nutrition status. Dede et al. [737] determined trace element levels in human scalp hair and correlated them to environmental exposure effects. The samples were dried, ashed and pelleted. For excitation, a $^{109}$Cd radioisotope source was used and detection limits for Fe, Cu, Zn, and Pb $< 1\,\mu g\,g^{-1}$ (dry hair) and 2–5 $\mu g\,g^{-1}$ (ashed hair) were obtained. In a study of MacPherson and Bacso [738], the relation between calcium concentration in hair and the mortality of coronary heart disease (CHD) in the UK was examined. The analysis was carried out with an annular $^{55}$Fe radioisotope source in co-axial source-sample-detector geometry. The results showed a correlation between hair calcium content, water hardness, sunshine hours and standardized mortality ratios with CHD. A method for quantitative determination of various elements (S, Cl, K, Ca, Fe, Cu, Zn, Br, Sr, and Pb) in hair samples of 102 deceased persons, mainly of cardiovascular diseases, was reported by Meitin et al. [739]. In a preponderant forensic work, Toribara analysed directly a single hair by linear scanning and determined the circumstances that led to a fatal exposure of dimethylmercury [740]. He could associate two areas of high Hg amount with the intake date and the time when chelator therapy was started 5 months later.

The death of a nineteenth-century Portuguese king could be explained by ingestion of compounds of As and Pb by Carvalho et al. [741]. They analysed some viscera remains that were kept in a porcelain container since then, using a tungsten X-ray tube and a secondary target (normally Mo) to obtain a monochromatic source, a three-axial geometry and an Ag collimator to improve the detection limits. The As concentration in the soft tissues was found to be increased by a factor of nearly 500 compared to healthy organs and could be explained, in agreement with the anatomical-pathological description, by acute toxicosis.

## In Vivo Determination of Toxic Elements and Diagnostic Facilities

Metal concentrations in humans are usually analysed in blood or urine, and can be used to assess the metal amount in an internal organ or the body, but in general, these relationships are not that simple. As there is a strong need to control the levels of toxic elements, especially for people who are occupationally exposed, and biopsies are tried to be avoided, the in vivo analysis of some elements has become of more interest.

In several papers, Todd and colleagues described the development of Pb analysis in bone and made some efforts in theoretical considerations, calculating correction factors, accuracy and precision. It was shown that a secondary target gave greater Pb X-ray peak signal-to-background ratios than partially plane polarized XRF. Substantial matrix effects from calcium-rich areas and attenuation by skin and adipose tissue were reported [742]. Theoretical considerations regarding in vivo measurements of lead in bone by L-shell XRF (method of correcting for attenuation, contributions to the measurement uncertainty, depth of bone sampled, signal strength and interferences) are reviewed in [743]. Mathematical techniques for uncertainty calculations are described by Kondrashov and Rothenburg [744]. In [745], the accuracy and precision of in vivo XRF measurements of Pb in bone were examined, comparing L-shell XRF analysis of cadaver bones with an ETAAS method based on acid digestion. The Pb content was estimated for intact legs and for dissected tibia following removal of overlying tissue. The agreement between L-shell XRF and AAS was reasonably good for bare bone measurements, but poor for intact legs. Nevertheless, there have been several applications of lead K-shell XRF measurements. $^{109}$Cd excited K XRF was used by Brito et al. [746] to analyse tibia lead and calcaneus lead concentrations in 101 workers of a lead smelter to study the long-term relationship between bone lead levels and the cumulative blood lead index. Cheng et al. [747] studied the relationship between blood and bone lead concentrations and the development of hypertension. In an extensive study of more than 800 subjects, a positive association between baseline bone lead level and the incidence of hypertension was found, but no correlation to blood lead level. The availability of lead in bone to chelation was examined by finger bone measurements in [748]. Lead K-shell XRF in tibia was investigated with regard to the relationship between low level Pb exposure and electrocardiographic (ECG) conduction disturbances [749]. Bone Pb levels were positively associated with ECG disturbances, particularly in younger men.

In [750], Börjesson et al. described in vivo studies to determine Cd, Hg and Pb concentrations in liver, kidney and bone of occupationally exposed workers and smokers. They reported minimum detectable concentrations of Hg in the kidneys of 12–45 $\mu g\,g^{-1}$ (depending on the position in the kidney) and 10 $\mu g\,g^{-1}$ in bone. In another study, a noninvasive method for the measurement of Pt concentration in kidneys of patients receiving chemotherapy with cisplatin platinum and analogues was developed [751]. It was shown that clinical measurements can be made with a bilayer of copper and silicon as polarizer, and a 0.25 mm tin filter between the X-ray beam and the polarizer to achieve a detection limit of 16 $\mu g\,g^{-1}$ at a measurement time of 2000 s.

McLean, Robertson and Jay studied the release of iodinated bovine serum albumin from several adjuvants protecting antigens from rapid degradation or disposal after injection in rats [752]. They found that $^{241}$Am excited XRF is a reliable noninvasive method to monitor the retention of antigens in these adjuvant solutions. Farquharson et al. demonstrated the determination of Fe, Cu, and Zn in skin phantoms [753], and studied the feasibility of a low-dose in vivo evaluation of Fe in skin as a monitor treatment in cases of iron overload [754, 755]. Using a tungsten dual focus tube with a 1.5 mm aluminium filter and germanium as second target, they reported minimum detectable concentrations of $< 10\,\mu g\,g^{-1}$ for iron and $5\,\mu g\,g^{-1}$ for copper and zinc, which are comparable to the levels found in healthy human skin.

Since teeth are also part of the skeleton, but more accessible, there have been studies on the development of a method to estimate essential and toxic trace elements by in vivo XRF measurements [756, 757]. They describe a $^{109}$Cd XRF facility for determination of Ca, Pb, Sr, and Zn in tooth enamel, and suggest the use of the Ca content of enamel as internal standard.

### Other Biological Samples and Biomonitoring

The trace element content of plants and animals depend strongly on the environment (air, ground, water) as well as their feed and nutrition. Therefore, it is possible to use them for biomonitoring studies and to differentiate between healthy and sick organisms.

Ranaweera et al. [758] investigated the uptake of As, Cd, Cu, Fe, Pb, Mn, and Ni in some native plant species in Sri Lanka to assess their potential to remove heavy metals from waste water. They found *Eichhornia crassipes* to be the most efficient of the tested species. The application of XRF in the analysis of microalgae, marine algea and corals was reported by Mendoza et al. [759]. They examined external standardization, a fundamental parameter method with Fe as internal standard and an absolute method based on elemental sensitivity using the fluorescence excitation in the backside of the sample. The range of elements in 19 different species of sponges from eastern Atlantic coastal waters were studied by Araujo et al. [760]. Some species showed enhanced accumulation, especially zinc and nickel, unrelated to the sampling location. High concentrations of bromine were found, particularly in sponges with low silicon content, which was, as expected, the dominant element.

The muscles and livers of farm cultivated trouts and river rainbow trouts were analysed by Akyuz [761]. They used radioisotope XRF to determine a range of elements, including Sc, Au, La, and Ce, and detected a higher Fe content in farmed fish. In a study of Senhou et al., the sensitivity of six biomonitors (lichens, moss, and barks) and local variations of several element concentrations were compared [762]. Lichens were found to be most sensitive for the fixation and accumulation of polluting elements. In a further study, the performances and limitations of 14 MeV NAA, thermal neutron activation analysis and radioisotopic XRF analysis were discussed for the determination of 43 elements in biomonitors [763]. XRF was found to be more flexible and less expensive, but showed higher detection limits ($> 8\,\mu g\,g^{-1}$). However, S, Ni, Cu, Y, Sn, and Pb could only be determined by XRF. For other elements, a good comparability of XRF and NAA was shown and the complementarity of these techniques was concluded.

Larsson and Helmisaari examined the accumulation of eight elements in the annual rings of Scots pine trees in the vicinity of a coppe–rnickel smelter with a spatial resolution of $0.3 \times 1$ mm [764]. They found changes in short and long time scales and showed the possibility to investigate the affection of nutrition status by reduction of air pollutants by position-sensitive XRF.

The nutritional status of pine trees was examined by Viksna et al. [765]. They analysed the needles in 1-mm steps along the length with scanning XRF in a three-axial set-up. Large differences in the distribution patterns were found for some elements, e.g. Zn and Mn, between needles of different age and variations correlated with the soil pH, whereas it was almost constant for Cu and Fe. The results were validated by ETAAS. Anjos et al. [766] used XRF analysis with a Ti-filter to investigate the influence of organic compost from urban garbage on the concentrations of 12 elements in radish plants at $\mu g\,g^{-1}$ levels. They stated a significant increase of some elements (K, Ca, Mn, Zn, Rb and Pb), and suggested the use of radish plants as a bioindicator of contamination in agricultural soils. The influence of a phytopathology called "sooty mould" on the Fe content in olive trees was studied by Aragao et al. [767]. A significant increase in infected leaves was found ($80\,mg\,g^{-1}$ average), whereas in noninfected samples no Fe was detected.

### Biotechnological and Other Applications

In many processes, it is of great importance to survey the amounts of trace elements in the involved substances. The level of toxic elements arising from contamination of the raw material or during processing has to be kept below the boundary limit, i.e. food and pharmaceutical products, and the content of intended elements, like Ca in powdered milk or the active sites in metalloproteins, has to be assured.

Salvador et al. [768] used XRF for the quality control of commercial tea by evaluating the inorganic constitution profile. Fe, Co, Ni, Cu, and Zn were detected in every sample, and Ti, Cr, and Mn in some of them. The limits of detection ranged from 5 (Zn) to 14 (Fe) $\mu g\,g^{-1}$. The authors suggest to use the combination of elements and their concentrations as fingerprints in the production process. Essential and trace element contents of 14 Nigerian medicinal plants were analysed by Obiajunwa et al. [769]. They reported the absence of toxic elements like Cd, As, Pb, and Hg in all samples, whereas 14 essential and trace elements in a broad variety of concentrations were found. Cesium and selenium uptake from enriched culture medium of cultivated mushrooms was monitored by Racz et al. [770] with radioisotope XRF from the point of view of health and environmental protection.

Cevik et al. analysed various elements in tobacco and its ash [771], and quantified them with the help of standard addition. The concentrations of K, Ca, Mn, Fe, Cu, Zn, Rb, and Sr in Indian spices (pepper, clove, cardamom, cinnamon and cumin) were determined by Joseph et al. [772]. The samples were dried, powdered, and mixed with cellulose binder to obtain self-supporting pellets, and then analysed by XRF. Almeida et al. [773] examined ten rum samples from Brazil by XRF, using ammonium pyrrolidinedithiocarbamate preconcentration, to determine Fe, Cu, and Zn concentrations. The standard deviations in triplicate analysis were below 10% at the $\mu g\,g^{-1}$ level and the detection limits ranged from 2 (Cu) to 7 (Fe) $ng\,mL^{-1}$. In two samples, the measured Cu concentrations exceeded the maximum allowable concentration. Twelve elements, including P and Si, were analysed in some varieties of rice by Rewatkar et al. [774] without any chemical treatment. Miah et al. described a rapid XRF determination of Ca, Cl, and K in barley leaves [775] by selective excitation of areas smaller 100 μm. Samples were collected from plants that had been fertilized chemically or treated with different mixtures of sewage sludge. They suggested this method for the diagnosis of the nutrition status of plants. Lopez de Ruiz et al. studied several methods of matrix removal of solid and extracted samples of medicinal herbs for the quantification of trace chromium by WDXRF after preconcentration [776] and reported a limit of detection of $0.027\,\mu g\,mL^{-1}$. The optimized methods were verified by ICP-AES.

The role of portable XRF spectrometers in Russian veterinary toxicology, as for their capability to determine simultaneously the majority of toxic elements at low costs, was reported by Pukhovskij and Kovalev [777]. Another application of portable XRF systems was described by Wheeler and Warren [778]: To improve sampling methods for monitoring whole-body exposure to toxic substances, they studied a standardized approach for the measurement of contamination over the entire coveralls of employees exposed to copper-based biocides. This approach was compared to conventional patch sampling and whole-suit digestion methods. The authors reported deviations of up to 20% to the whole-suit digestion method as reference benchmark (60% for the patch method), but much better spatial resolution, and therefore more detailed information for chemical risk assessment.

A focused beam of 16-μm diameter collimated by a pyrex glass capillary (10-μm diameter at the beam exit end) was used by Fukumoto et al. for the elemental mapping of biological samples [779]. They examined the distribution of K, Ca, and Mn in a living leaf after exposure to high dose of X-rays and a drop of artificial acid rain. No radiation damages or changes in the distribution patterns were found after 50 h irradiation, whereas acid solution decreased the amounts of the three elements significantly after 1 h. Spectra sampling time was 1 or 4 s per pixel.

Another application of μ-XRF was described by Brazeau and Wong [780]: They analysed gunshot residues on human tissues and clothings with a beam of 300-μm diameter from a rhodium anode tube. The elements of interest were Pb, Sb, and Ba (to confirm a gunshot wound) as well as Cu, Zn, and Fe (for information of the shooting distance) to help the interpretation of the event.

### 7.10.3 Total Reflection X-Ray Fluorescence Analysis (TXRF)

Total reflection X-ray fluorescence analysis (TXRF) is an energy dispersive XRF technique with sample excitation in grazing incidence geometry. In contrast to XRF, where the primary beam strikes the sample at an angle of around

45°, the incident beam is totally reflected by using a glancing angle of less than 0.1° [781]. General capabilities and limitations of TXRF are described by Klockenkämper [782] in the first book about TXRF.

TXRF has been developed into a powerful tool for chemical micro, trace, and ultra trace elemental determination [783, 784]. Analyses of elements with $Z \geq 11$ (sodium) are possible. In contrast to conventional X-ray fluorescence analysis, $ng\,g^{-1}$ relative detection limits and absolute detection limits in the picogram range in light sample matrices from biological and medical origins can be reached with TXRF. It is the high reflectivity of the sample carrier that nearly eliminates the spectral background. With preconcentration techniques relative detection limits in the order of picogram per gram are attained [785].

In addition, depth profiling by grazing emission X-ray fluorescence spectrometry (GEXRF) is described. The emitted X-ray fluorescence photons are detected thereby below the critical angle. Wiener et al. [786] used the ion beam sputtering technique in which samples were etched to a bevel shape in combination with GEXRF, giving a depth resolution of 2.5 nm. In case of different support materials, a variable incident angle allows optimization of the signal-to-background ratios and, hence, improved sensitivity. Variable incident angles in TXRF open new possibilities and more reliable quantification in the field of analytical quality assurance of TXRF measurements as well as the possibility of optimizing the measurement angle with respect to the sample carrier [787]. Furthermore GEXRF is of much interest in material sciences.

Biomaterials in particular can be treated by various methods of sample preparation. Some of them can be analysed directly. Therefore, droplets of liquids are placed on the centre of circular sample carriers, dried by evaporation and analysed directly. The sample carriers used have to be coated by a hydrophobic film in order to avoid bleeding of droplets. Quartz glass sample carriers were utilized and alternatives such as germanium [788] or boron nitride [789], thin foils of Mylar® (polyethylene terephthalate) [790], Kapton® (polyimide) [791], polyester [792], and Perspex® (polymethylmethacrylate) [793] have also been studied. The advantage of Perspex® is the usage without precleaning and higher stability in contrast to thin foils. About 100 µL liquid, some 200 µg of an inorganic solid sample, or about 50 µg of an organic biotic sample can be used at maximum. Besides liquids, fine powders or pulverized materials can be applied as a few grains or prepared as suspension [782]. Furthermore, tissue or plant material can be applied as thin frozen sections [794].

Mostly, a suitable sample preparation step is carried out previously, which decomposes the sample matrix and transforms it into a clear colourless solution. Techniques of sample preparation with shorter decomposition times, e.g. microwave digestion with nitric acid [795] or freeze-cutting by a microtome [796] have been utilized. Klockenkämper [782] has summarized various methods of homogenization and preconcentration.

Using several certified reference materials and intercomparison tests the reliability of determinations by TXRF was proved. They showed high performance for the certification of reference materials [797–800] and competed well with established analytical techniques such as neutron activation analysis

(NAA) [801], atomic absorption spectrometry (AAS) [801–804], inductively coupled plasma optical emission spectrometry (ICP-OES), or inductively coupled plasma mass spectrometry (ICP-MS), as shown in several papers [805, 806, 848].

Altogether, TXRF is a quick, specific, and inexpensive technique, which requires a very small amount of sample. In addition, TXRF is a suitable technique for investigating widely spread problems concerning a wide range of biochemical systems because of its multi-elemental capability and power of detection. In recent studies, sample materials of many varieties have been investigated by TXRF. The method was shown to be especially suitable for trace analysis of whole blood, blood serum, blood plasma, urine, and amniotic fluid, as well as for micro-analysis of organ tissues such as lung, brain, hair, tooth, and dental plaque, and analyses of various biological matrices such as enzymes, polysaccharides, proteins, and biofilms.

## Organ Tissues

TXRF is suitable for the analysis of different kinds of tissues such as lung and hair. Organ tissue can be analysed either by direct excitation or after certain sample preparation steps such as ashing [843], dilution [809–811], pressure digestion [812], or microwave digestion [795]. However, a fast analysis down to the lower microgram per gram (ppm) range after freeze-cutting of micrometer thick sections tissue placed directly on the TXRF carrier may be recommended [813]. Apart from simplicity, the direct method prevents contamination and losses, which may be caused by a chemical preparation. Tissue samples were prepared by a freezing microtome. At a temperature of about $-20°C$, thin sections were cut with a thickness of about $10\,\mu m$, $10\,mm$ in diameter, and about $200\,\mu g$ [796]. After a short time, the thin slices can be prepared on a sample carrier, which must be clean but not hydrophobic (as required for the preparation of solutions). Tissue samples are often available in paraffin wax. These embedded thin tissue slices in paraffin wax have to be stretched on a water bath just under the melting point of paraffin [814]. A droplet of $5\,\mu L$ internal standard element solution should be placed on the section and dried under an infrared lamp.

Benninghoff and von Czarnowski et al. [812, 815] used TXRF to determine the trace elemental distribution in cancerous and normal human tissue. Trace elemental load in cancerous and normal human lung tissue was also investigated by Kubala-Kukuś et al. [816] using TXRF and PIXE. Therefore a small piece of 0.08–0.6 g was mixed with 1.5 mL of high purity nitric acid and with $100\,\mu L$ of $Y(NO_3)_2$ in water solution, which was added as internal standard. After a further mineralization during 3 min in a microwave oven $5\,\mu L$ of the mixture were dried on a quartz sample carrier in infrared for TXRF analysis. Significant enhancements of titanium (48%), chromium (20%), and manganese (36%) were found in cancerous lung tissue from female patients compared with tissue from male patients. Conversely, levels of lead and strontium were found to be up to 30% lower in female than male cancerous lung

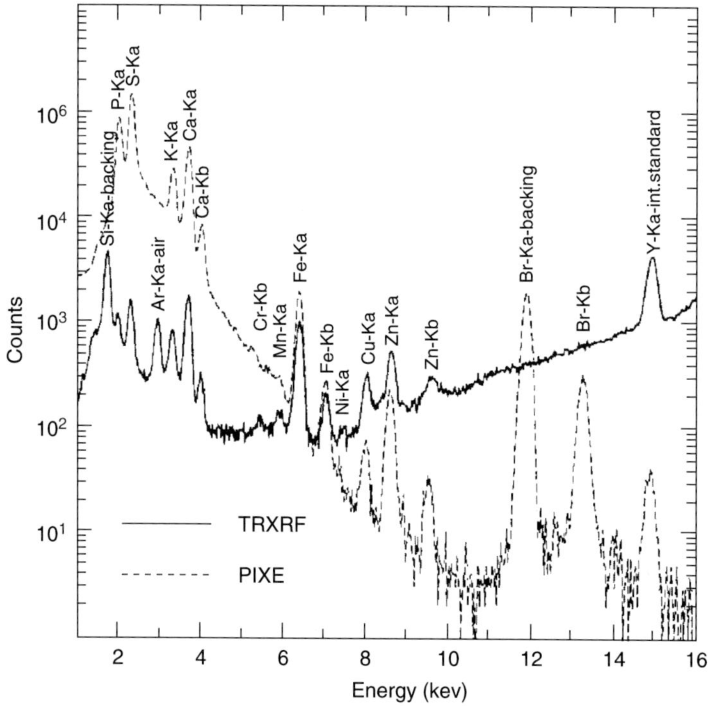


**Fig. 7.154.** Typical TXRF spectrum of a cancerous lung tissue with Mo K excitation in comparison with PIXE spectrum recorded for 2.0 MeV protons (Reprinted from Nucl. Instr. Meth. Phys. Res. B (150), A. Kubala-Kukuś, J. Braziewicz, D. Banaś, U. Majewska, S. Góźdź, A. Urbaniak: Trace element load in cancer and normal lung tissue, 193–199 (1999), [816], with permission from Elsevier)

tissue. A spectrum of a cancerous *lung tissue* sample with yttrium as internal standard is shown in Fig. 7.154.

The statistical distribution of trace elements in *full-term placenta* samples taken from 120 women living in two different places in Poland regarded as "polluted" and "unpolluted" were studied by Majewska et al. [817] and compared with PIXE. The samples were just taken after normal delivery and frozen. The frozen samples were analysed as described by Kubala-Kukuś et al. [816].

Suzuki et al. [818] determined trace amounts of cisplatin platinum (*cis*-diamminedichloro-platinum (II), *cis*-DDP) in biopsy samples as well as of blood serum samples with detection limits of $0.1\,\mu g\,g^{-1}$ in biological samples. *Cis*-DDP is one of the most widely used drugs in the treatment of cancer, because it shows remarkable activity alone or in combination with drugs in the treatment of several tumours. Ten milligram samples were mixed with $500\,\mu L$ of nitric acid and heated with 1 mL water for 8 min in a microwave oven. After cooling, the sample solution was evaporated on the hot plate to about 10 μL. After addition of 200 μL nitric acid, the sample was evaporated again to about 10 μL. After another cooling, the sample solution standardized with strontium (1 μL, 10 $\mu g\, g^{-1}$) was analysed.

Pérez et al. [819] investigated *trans*-[$PtCl_2$($N,N$-dimethylamine)(isopropylamine)], a novel *trans*-platinum compound that shows cytotoxic activity in several cisplatin platinum-resistant cell lines as well as *cis*- and *trans*-DDP by means of molecular cell biology techniques and TXRF. The results show that this drug is more cytotoxic and induces a higher amount of apoptotic cells than cisplatin platinum in A2780*cis*R cells. However, the intracellular accumulation and extent of binding to DNA of *trans*-[$PtCl_2$($N,N$-dimethylamine)(isopropylamine)] is lower than that of *cis*-DDP. TXRF determinations were carried out by introducing 100 μL sample of either cell supernatants or cellular DNAs standardized with 100 $ng\, mL^{-1}$ of vanadium standard solution into a high-purity nitrogen flow concentrator at a temperature of 70°C until the volume was reduced five times.

Using Compton peak standardization Marco et al. [810] determined copper and zinc in *brain* tissue as well as in human serum. The Compton peak was used as an internal standard because of its relation to the mass absorption parameters of the sample material. Results were in good agreement with results obtained by AAS.

Tapia et al. [820] determined the cellular content of copper, iron, and zinc in different mammalian *cell lines* by TXRF as described by Günther and von Bohlen [821] and González et al. [822]. All cell lines displayed the same pattern of relative intracellular abundance of these trace metals: Cu $<$ Fe $<$ Zn. Increasing the extra cellular copper concentration this pattern is shifted to Fe $<$ Zn $<$ Cu [820]. For TXRF analysis one fifths of the digested sample volumes were standardized with 1 $\mu g\, mL^{-1}$ of selenium standard solution.

*Hair* is an example of tissue that can be analysed directly. Prange et al. [823] measured human hair directly and after oxygen plasma treatment, which yields better results due to the background reduction.

In contrast, Trunova et al. [824] analysed trace elements in human hair by TXRF and SRXRF. For the TXRF measurements eight to ten hair strands of about 1 mg were pulled out with the roots and dissolved in 100 μL concentrated nitric acid with an internal vanadium standard added. Results for the elements nickel, copper, zinc, selenium, and bromine are identical for both methods within the limits of error. In contrast, for other elements as potassium, iron, and strontium more deviations are reported.

*Dental plaque* taken near amalgam fillings, gold crowns, and intact teeth were analysed by von Bohlen et al. [825]. The relative mass fractions of phosphor, sulphur, chlorine, potassium, calcium, manganese, iron, bromine, and rubidium are similar in plaque of the three sets of samples. A significant correlation between mercury accumulation in plaque and the amalgam fillings was established. The results for other elements, e.g. gold, are less significant.

Bernasconi et al. [826] determined titanium levels in *rabbit* organs after endosseous implants insertion after microwave digestion and compared them with PIXE measurements in order to evaluate a possible diffusion and accumulation of this metal in the tissues.

Pollen, bee tissue, and honey samples collected from selected areas of Kenya were analysed by Wamwangi et al. [827] using EDXRF and TXRF. Iron levels in the bee tissues were higher than in pollen and honey samples, which agrees with the results of Kump et al. [828], who used TXRF and radioisotope XRF to determine trace elements in bee honey, pollen, and tissue.

### Whole Blood

Direct excitation of blood samples is not so efficient in comparison to their analysis after chemical treatment, because of the high organic content and resulting intensive Compton scattering [799]. However, direct excitation is simple and rapid and most of all minimizes the risk of sample contamination. In contrast, Prange et al. [795] used microwave digestion of 1 mL samples with 5 mL high purity concentrated nitric acid (first step: 550 W, 15 min, second step: 400 W, 20 min). Cobalt and gallium were used as internal standards reaching a final concentration of 20 mg $L^{-1}$ in the digested sample. Peaks of iron, potassium, and calcium dominate the spectra of whole blood samples. Their great concentration impedes the detection of, e.g. manganese, nickel, and lead. Detection limits of these elements can be improved by separation of major elements. Iron can be removed from strong acid solutions of hydrochloric acid by extraction with MIBK (methylisobutylketone). For this purpose the digestion solution is evaporated to dryness, the residue is taken up with 1 mL high purity 6 N hydrochloric acid and spiked with cobalt as internal standard to a 2 μg $g^{-1}$ end-concentration and heated for 15 min. Then the solution is extracted with 1 mL MIBK, which is pre-extracted with 6 N hydrochloric acid; 10 μL from the aqueous phase is measured by TXRF.

Lead is mainly stored in bone tissue (78–90%), but blood transports the lead through the body. Although it contains only 2% of the total load present in the body [829], blood lead is a monitor for lead in a body [830].

Ayala et al. [831] developed a fast sample preparation technique for the analysis of lead in whole human blood from donors occupationally exposed to lead contamination in a car battery factory and from unexposed donors. Two microlitres of fresh blood on a quartz glass carrier was dried in vacuum and ashed in a low-temperature oxygen plasma asher (18 min at 50 mW). The residue was analysed by TXRF with a cut-off reflector giving a significant background reduction. Detection limits of about 0.03 μg $mL^{-1}$ for lead were achieved.

Lead in blood was also investigated by Martinez et al. [829] in a randomly sampled population living in the metropolitan zone of Mexico Valley. Samples were treated following the technique of Prange [795]. The levels of blood lead correlate with the decrease of lead concentration in air and soils.

Whole blood TXRF analysis and comparison with conventional XRF measurements were investigated by Viksna and Selin [832]. Detection limits are about twice as low for TXRF as compared with conventional XRF. Therefore 60 µL of 10% tetramethylammonium hydroxide (TMAH) was added to 180 µL of whole blood in a 1.5 mL polypropylene container. The solution was well mixed and after 5 min 120 µL of internal standard was added. Afterwards the sample solution was carefully mixed again and 3 µL were analysed.

Zarkadas et al. [799] inserted a 100 mg cm$^{-2}$ molybdenum filter behind the cut-off reflector, which significantly improved the peak-to-background ratio for direct measurements of 5 or 10 µL quantities whole blood, especially for the elements copper, zinc, selenium, and bromine. The influence of self-absorption phenomena in the quantification procedure was investigated and validated by using gallium as internal standard. Slowly drying under an infrared lamp compared to drying under vacuum showed better results, because the foaming of serum in vacuum prevented the formation of a thin film. Precision of about 4% resulted by analysing the biological reference sample A-13 (IAEA, animal blood reference material).

### Blood Serum

Knoth and Schwenke [833] analysed human blood serum directly by drying 10 µL serum to a thin film with detection limits of about 1.5 mmol L$^{-1}$ and 3–5% precision in the 20 mmol L$^{-1}$ range of the metals. Concentrations of iron, zinc, and bromine in human blood serum were determined directly by Yap et al. [834], too. They analysed a random sampling of 62 individuals in an essentially Chinese population.

In contrast, Knöchel and Bethel [835, 836] developed a rapid sample preparation technique for the analysis of iron, copper, zinc, nickel, selenium, rubidium, and tin with detection limits between 40 ng mL$^{-1}$ for calcium and 220 ng mL$^{-1}$ for iron and good reproducibility. After separation of the blood serum from the plasma by centrifugation and subsequently freeze-drying, the protein matrix of 100 mg was separated by cold ashing in oxygen plasma. Separation of the electrolyte elements by chelating with ion exchanger cellulose HYPHAN reduces detection limits by a factor of 10. One millilitre serum was digested, diluted to 10 mL and pH was adjusted to 3.8 with ammonia. After addition of 100 µL 1 mol L$^{-1}$ acetic acetate buffer, it was percolated through 100 mg cellulose HYPHAN and eluted with 2 mL 6 mmol L$^{-1}$ hydrochloric acid.

Furthermore, microwave digestion after Prange et al. [795] also is well suited for multi-elemental determination in whole blood and blood serum.

Dogan et al. [837] determined the distribution of trace elements iron, copper, zinc, selenium, and rubidium in sera from patients with the complete type of Behçet disease in the active stage as well as from healthy controls. Therefore, digestion by nitric acid under high pressure and normal pressure led to appropriate sample preparations for TXRF.

Greaves et al. [809] developed a TXRF procedure to monitor drug-related platinum levels in serum and urine samples from cancer patients after undergoing chemotherapy with platinum-containing drugs. Samples were standardized by adding known trace amounts of elements absent in these matrices, diluting 1:1 with water and then irradiating with cut-off filtered or monochromatic X-rays. The method was shown to be appropriate to routine monitoring of platinum in patients.

For routine clinical analysis Marcó et al. [810] proposed a quantitative approach by using the Compton peak standardization. Copper, iron, platinum, selenium, and zinc in human serum were determined with a precision ranging from 6% to 16% relative standard deviation. The method was applied in the routine analysis of blood samples of cancer patients and normal blood donors. From their results from cancer patients, which were significantly different from normal, they suggested that the determination of the copper concentration and the zinc to copper ratio in serum samples could be a useful screening tool for cancer diagnosis [802].

Suzuki et al. [818] analysed blood serum by addition of 2 mL nitric acid to 200 µL blood serum and digestion for 24 h at 110°C on a hot plate. After cooling, the resultant solution was diluted with water until the total volume of the solution was 5 mL. Hundred microlitres were evaporated on the hot plate to about 10 µL. After cooling, the sample solution standardized with strontium (1 µL, 10 $\mu g\,g^{-1}$) was analysed.

Serum selenium concentrations in a population of 78 sucking 2-month-old calves were determined by Buoso et al. [838] directly by TXRF analysis of 10 µL vacuum dried sample in order to corroborate a clinical diagnosis of severe deficiency status.

## Blood Plasma

The high protein content of blood plasma can lead to the generation of an unevenly distributed film on the TXRF sample carrier, which produces a characteristically high detector dead time.

Concerning the analysis of clinical samples, Savage and Haswell [839] developed a method for direct determination of trace elements in blood plasma samples by using chemical modifiers to compensate analyte loss or sample modifications during drying. In addition to the usage of chemical modifiers as magnesium nitrate, nickel nitrate, and ammonium pyrrolidinedithiocarbamate (APDC), the suitability of using a microwave ashing system to reduce the organic component in sample deposits on the sample carriers were studied. The most suitable method is air-drying of blood plasma samples with the addition of APDC, as reported by Knoth and Schwenke [840]. Savage and Haswell [839] used a mixture of magnesium nitrate and APDC. A typical precision of less than 8% relative standard deviation and recoveries of 90–100% were reported.

Bellisola et al. [841] investigated selenium and other trace elements, as copper, zinc, bromine, and rubidium, in human blood plasma with the aim of studying both the intestinal absorption and the rate of selenoprotein synthesis in the liver after a single oral dose ingestion of sodium selenite. The results obtained confirm those already observed in experiments with rats, where a peak of selenium becomes evident in the plasma after 3–4 h and then decreases [808]. The anticoagulant EDTAK3 was present during the blood sample collection. After centrifugation (1500 $\times g$ for 10 min at room temperature) 100 µL of separated plasma sample was mixed with gallium as internal standard and thereof 20 µL were analysed on a silica sample holder by TXRF.

Mansoor et al. [842] found a correlation between blood plasma total homocysteine and copper in patients with peripheral vascular disease and concluded that the atherogenicity of homocysteine may be related to copper-dependent interactions. More studies are required to investigate the combined pathophysical mechanisms of homocysteine and copper pathogenicity.

## Urine

Biological samples were analysed by Messerschmidt et al. [843] after separation and enrichment of palladium and gold by reductive coprecipitation with mercury, followed by evaporation of the mercury. Detection limits of about 2 ng $L^{-1}$ in *urine* were determined corresponding to absolute amounts of about 50 pg. Other sample matrices, e.g. plant material and other biological tissues can be treated by the same procedure which consists of decomposition of the sample, the separation of palladium and gold, and subsequent determination of the elements by TXRF.

For the decomposition 20 mL of urine and 1 mL of concentrated nitric acid were mixed and slowly heated in order to concentrate the volume to about 2 mL. After addition of further 4 mL of concentrated nitric acid and 0.5 mL of 12 mol $L^{-1}$ hydrochloric acid, the sample was digested in a high-pressure asher (50 min at 320°C and 130 hPa). In order to remove the excess of acids from the resulting solution the volume is finally reduced to about 0.5 mL.

For the separation of palladium and gold 2 mL de-ionized water is added to the clear digestion solution in a 10-mL tube. After addition of 0.2 mL of a mercury solution (corresponding to 30 mg of mercury) and 0.5 mL of formic acid, the tubes are heated up to 80–85°C in an aluminium block. About 10 min later a second addition of 0.5 mL of formic acid becomes necessary to complete the reduction step. After further heating (30 min at 80–85°, subsequently increasing heating to 115–120°C, 10 min at 115–120°C), the reduction step is finished. The mercury droplets are washed with 3 mL of 0.4 mol $L^{-1}$ hydrochloric acid and centrifuged. The resulting one mercury droplet is washed with water and propanol. After transfer onto a siliconized quartz-glass target suitable as sample carrier, the mercury is evaporated by heating at 300°C and analysed by TXRF.

Zarkadas et al. [844] developed a method for the quantitative determination of uranium in human urine. Chemical treatment and preconcentration is necessary, but the method is only suitable for measuring uranium concentrations above normal. Therefore, 200 mL urine obtained from a 24-h pool urine sample was spiked with cobalt as internal standard to a final concentration of 100 μg $L^{-1}$. After addition of 30 mL concentrated nitric acid, the sample was heated at 95°C for approximately 3 h with continuous stirring until the initial volume was reduced to approximately 10–20 mL. With additional 30 mL nitric acid the solution was slowly heated to the boiling point within 1 h, and maintained at this temperature for approximately another 10–15 min. After removing the digest from the heater, a few drops of hydrogen peroxide were added and the solution was brought again to the boiling point. The last step has to be repeated until no fumes of nitrogen dioxide are observed. For separation of alkaline and alkaline-earth elements a similar preconcentration procedure as described by Knöchel et al. [845] was used. Thereby the solution was evaporated to dryness and diluted with bidistilled water to a final volume of 200 mL. After cooling and ultrasonic homogenization, the pH was adjusted to approximately 5 by addition of ammonia. Thereby, high calcium concentrations in urine lead to the formation and precipitation of insoluble calcium hydroxide. Later the solution was filtered through a polyester membrane and readjusted to pH = 5 if necessary. Afterwards 1 mL freshly prepared 4% methanolic sodium dibenzyldithiocarbamate (NaDBDTC) solution was added and the sample was well mixed by stirring and ultrasonic treatment. The sample was left to stand for 5 min, filtered, and the insoluble carbamates were washed off the polyester membrane by elution with 4 mL of chloroform.

TXRF analysis of drug-related platinum levels in urine and serum samples from cancer patients after undergoing chemotherapy with platinum-containing drugs were made by Greaves et al. [809] (see *Blood Serum*).

The metabolism and excretion of selenium and other trace elements in human urine was investigated without sample preparation by Bellisola et al. [841]. In contrast to human blood plasma, urinary selenium excretion progressively increases after a single oral dose ingestion of sodium selenite.

## Amniotic Fluid

Zinc concentrations in amniotic fluids from 38 pregnant Venezuelan women were investigated by Greaves et al. using TXRF. In order to test the accuracy of this TXRF method Liendo et al. [846] compared PIXE to TXRF using a common group of 10 amniotic fluid samples. The samples were analysed directly without sample preparation. Different degrees of consistency between PIXE and TXRF for the elemental analysis of a complex matrix such as amniotic fluids were observed.

Trace elemental content in human amniotic fluid was investigated also by Carvalho et al. [800]. Therefore, organic matrix was removed by treatment of a 5 μL sample with 20 μL nitric acid followed by oxygen plasma ashing for 30 min as described by Prange and Schwenke [823]. Thereby yttrium was used as internal standard for TXRF analysis. The study showed calcium and iron increase in amniotic fluid and placenta samples with maternal age measured by EDXRF. Zinc and copper also increase in amniotic fluid in contrast to placenta.

## Other Biological Samples

Determination of the elemental composition of cerebrospinal fluid (CSF) in the neurodegenerative disease amyotrophic lateral sclerosis was applied by Ostachowicz [847]. CSF is a medium for transportation of chemicals to and from the brain. It is a clear liquid, which is very similar to the liquid portion of blood (plasma) and contains various trace elements and other substances such as enzymes and carbohydrates. First 2 µL of CSF were analysed directly with addition of gallium standard solution ($1\,\mu g\,g^{-1}$) as an internal standard. Then the sample was treated with nitric acid 1:1 and measured again. Removal of chlorine from the sample with dilute nitric acid leads to an improvement in the detection limits owing to reduction of the background, especially for the light elements.

Trace elements in drugs were investigated by Wagner et al. [807]. These element concentrations provide element fingerprints, which offer the possibility to discriminate between different batches of *lecithin, insulin, procaine*, and *tryptophan* originating from different production or purification processes. Substances of a good solubility such as insulin were dissolved in $0.1\,mol\,L^{-1}$ nitric acid. The solution was dried after addition of rubidium as internal standard. Sparingly soluble drugs were digested in closed vessels under high pressure at 160°C for 12 h. Twenty to thirty-five milligram lecithin, 33–39 mg procaine, or 10–50 mg tryptophan were weighed and mineralized by addition of 1 mL concentrated nitric acid. The resulting 50µL solution was mixed with 5 µL of internal standard (rubidium, $1\,mg\,L^{-1}$).

Kelkó-Lévai et al. [803] determined trace elements in pharmaceutical substances as sugars, sorbitol, mannitol, paracetamol, amidopyrine, and chloral hydrate after a tenfold preconcentration in 1–5% $m/v$ solution. For substances not soluble in water or in water–methanol (ethanol) mixtures a direct TXRF method can be used. As example, the trace metal level in indomethacin was determined using a 1% $m/v$ solution of the drug in a mixture of 98% acetone and 2% $1\,mol\,L^{-1}$ hydrochloric acid.

As cofactors of proteins and enzymes metal-ions such as iron, nickel, copper, manganese, or molybdenum are of biochemical relevance. Wittershagen et al. [848] determined metals in enzyme complexes, in particular the determination of molar fractions of the metals in enzymes. For this 5 µL of the enzymatic solution (enzyme concentration $10–50\,\mu mol\,L^{-1}$, corresponding to about $1–7\,mg\,L^{-1}$ metal concentration and $20–60\,mg\,L^{-1}$ sulphur concentration) are required without extensive pretreatment.

A further aspect is the determination of light elements. In particular, sulphur is of interest as it is found in the two sulphur-containing amino acids cysteine and methionine, and occurs in some enzymes in the form of ironsulphur clusters. Steinmeyer and Kolbesen [849] described sulphur determination with a sulphur concentration ranging from $0.5\,mg\,L^{-1}$ to $20\,mg\,L^{-1}$ in biological samples like *methionine* and *insulin* by TXRF. By using smoothing detergents such as 1% hydrofluoric acid, 1% malic acid, and 2% hydrazine hydrate the recovery rate of the sulphur determination range about ±10%. This offers possibilities for screening of proteins or other sulphur-containing macromolecules for confirmation of mutagenesis steps and provision of structural characterizations.

Mertens et al. [850] described sulphur determination for easy and simultaneous calculation of the *protein* concentrations in enzymes. The purified enzymes were transferred into a buffer solution containing Tris (hydroxymethyl)-aminomethane (TRIS) and (TRIS)acetate as described by Wittershagen et al. [848]. Four microlitres of the enzyme samples was dried in air at room temperature onto quartz glass carriers for TXRF analysis. A detection limit of 0.5 ng sulphur in diisopropylfluorophosphatase (DFPase) was achieved. For the larger enzyme NADH:Q oxidoreductase, an enzyme of the respiratory chain of bovine heart mitochondria, a detection limit of 5 ng sulphur was reached.

Quantification of platinum bound to DNA was studied by Ruiz et al. [811]. Platinum concentrations from 3 ng $mL^{-1}$ to 30 ng $mL^{-1}$ were determined with relative standard deviation between 2% and 8%. Hundred microlitres of DNA in water suspension with vanadium standardization was reduced by a factor of five in a high-purity nitrogen flow concentrator at a temperature of 70°C. Afterwards an aliquot of 5 µL was analysed after hot drying.

Due to its multi-elemental capability and high sensitivity TXRF is a valuable tool for biomonitoring studies of metal pollution in plant material of any kind as well as fish or animal tissue, and a variety of foodstuffs.

Michaelis et al. [851] analysed, e.g. *fine roots* which were prepared with masses ranging from 40 µg to 160 µg. Fine roots of trees are responsible for the main uptake of nutrients and water uptake from the soil solution. Roots smaller than 2 mm in diameter are called fine roots. At high heavy metal concentrations in the soil, the growth and survival of the fine roots are disturbed and base cations are replaced by heavy metals. Samples were prepared by digesting the carefully washed fine roots with 100 µL concentrated nitric acid at a temperature of about 70°C before spiking the acid with a cobalt standard solution. Distinctive differences in frequency distribution of the element concentrations have been observed in fine roots from sound and diseases trees.

Viksna and Olsson [804, 852] analysed *Scots pine needles* from four different sites in Latvia as well as fine roots of Scots pine collected at different distances from a coppernickel smelter in southwest Finland. For monitoring the nutritional status of forest trees and studies of deposition and impact of air pollutants, the chemical composition of pine needles is used. First, the pine needles were washed twice with chloroform for the removal of airborne deposits from the surface of the needles. After that they were washed with deionized water twice, dried for 20 h at 60°C and for 2 h at 80°C, and then digested with concentrated nitric acid for TXRF analysis.

Xylem sap of nickel contaminated *cucumber plants* growing in hydroponics containing urea as the sole nitrogen source was characterized by Mihucz et al. [853]. For the direct determination of metals in the xylem sap and for the investigations of chromatographic fractions eluted from a size-exclusion column, 100 μL of xylem sample was spiked with gallium standard solution. Thereof 25 μL were analysed by TXRF after drying on a hot plate. Nickel addition seemed to decrease the urea toxicity to these plants.

Metals in macrozoobenthos samples as *Caddisfly* and *mayfly larvae* from a river in western Austria were analysed by Miesbauer et al. [854]. For metal analysis the dried macrozoobenthos samples were digested with nitric acid and hydrogen peroxide in a microwave oven. The TXRF results show significant higher metal concentrations in aquatic insect larvae from industrially contaminated site than in reference samples.

Mages et al. [855] developed a dry and a wet preparation technique for direct element determination of single freshwater crustacean specimens. In case of the dry method specimens were washed with 0.2 μm filtered lake water, frozen in liquid nitrogen and placed on a TXRF sample carrier. In case of the wet method, specimens were washed with 0.2 μm filtered lake water, placed on a TXRF sample carrier and dried in air. Using both preparation methods in combination with TXRF for analysis of *Daphnia pulex* (body size: 0.5–2.5 mm) from a deep, oligotrophic freshwater lake located in southern Chile, no significant differences concerning the elements calcium, potassium, iron, zinc, bromine, phosphorus, and copper were observed. However, the determination of manganese, sulphur, and strontium revealed significant differences between the two methods. It seemed that the dry method yields more precise results and is a good alternative to the classical digestion method.

Heavy metal contents of aquatic *biofilms* isolated from stones collected from the River Elbe and ceramic plates exposed in the River Elbe were determined by Friese et al. [856]. Bioconcentration factors between about $10^2$ and $6 \times 10^4$ depending on the element were obtained in the biofilm matrix compared to the variations observed in element concentrations of the bulk water phase of the River Elbe.

Another example is the multi-elemental speciation in *lambs lettuce* and *cauliflowers* with gel permeation chromatography and TXRF analysis by Günther and von Bohlen [821, 857]. Samples of lambs lettuce were freeze-dried and pulverized in a mortar. The resulting homogenous samples of 1.5–2 g were mixed with purified quartz and extracted with 20 mM TRIS® hydrochloric acid buffer (pH = 8). The extractions were cooled and by centrifugation cytosol was obtained. Exemplary cytosol of lambs lettuce and cauliflowers were separated by GPC and the obtained fractions were analysed by TXRF.

### 7.10.4 Synchrotron Radiation Induced TXRF

Synchrotron radiation as a highly intensive and brilliant, naturally collimated X-ray beam with a continuous spectral distribution from electron volt to 500 keV being linearly polarized in the orbital plane offers advantages for use as an excitation source in XRF [785, 858]. By combination of TXRF with synchrotron radiation detection limits are drastically reduced, compared with X-ray tubes, and approached the femtogram level.

Synchrotron radiation induced TXRF can be achieved with different geometrical arrangements of reflector and detector. Three possible geometries, their advantages and disadvantages are described [858, 859].

A main field of application for SRTXRF is wafer surface analysis and many papers considered this topic, but analyses of biological materials are also reported.

### Organ Tissues

Kondo et al. determined copper and zinc content in carcinoma tissue using 11 keV monochromatic synchrotron radiation excitation. The copper level in carcinoma tissue was found to be significantly smaller than in noncarcinoma tissue, whereas no clear tendency was observed for the level of zinc [860]. Unfortunately, the original article is available only in the Japanese language.

At the Beijing synchrotron radiation source further investigations of trace elements in *cells of lung and cervical cancer* before and after preliminarily apoptosis were performed by Yuying et al. [861]. Approximately 0.2 to 2.0 μL of diluted standard specimen or cell solution was dropped on a polished sample carrier and dried in a super clean air environment. Before analysis, the number of the cells within the thin dried liquid drop residue was counted using a microscope.

### Other Biological Samples

The iron, copper, and zinc content in the *colostrums milk* was studied by da Costa et al. [862] with SRTXRF at the National Laboratory of Synchrotron Light (LNLS) in Campinas, Brazil. The composition of human milk has attracted a lot of interest, since it represents the most suitable pattern of nutrients for the younger infant. Therefore, 0.5 mL samples of milk from 50 Brazilian mothers were treated with 0.5 mL of nitric acid and 0.5 mL hydrogen peroxide at 60°C for 5 days. Afterwards the solution was left to dry and the volume recovered to 0.5 mL 0.1 mol $L^{-1}$ nitric acid as diluting agent. Five microlitres of the solution with gallium as internal standard were analysed on a Perspex sample carrier by SRTXRF.

Ultra trace elements in small amounts of various *drugs* of abuse were detected by Muratsu et al. [863] using SRTXRF analysis. The experiments were performed at the BL24XU undulator beam line of the Synchrotron Radiation Facility SPring-8 in Hyogo, Japan. One microlitre solutions containing 10 μg of various drugs as methamphetamine, amphetamine, 3,4-methylenedioxymethamphetamine, cocaine, and heroin were directly analysed. For example, in a seized methamphetamine sample, iodine was found, which

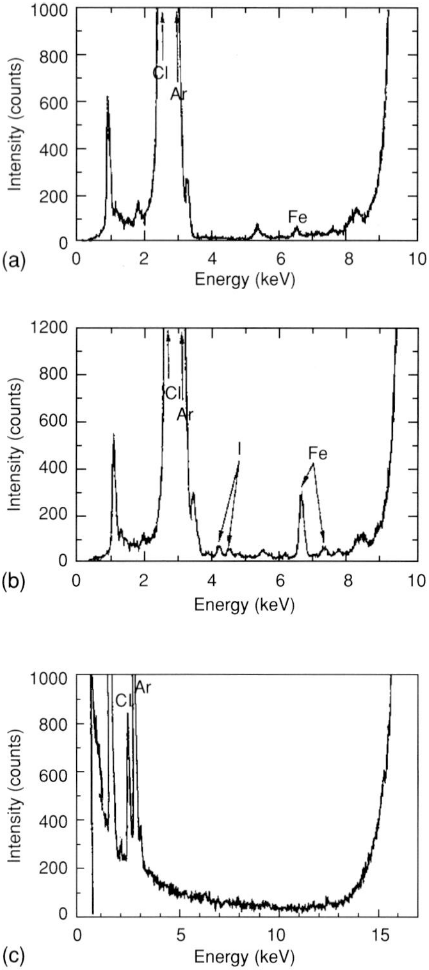


**Fig. 7.155.** Comparison of two SRTXRF spectra (**a**), (**b**), and a conventional TXRF spectrum (**c**) of methaphetamine: (**a**) a control methaphetamine HCl; (**b**), (**c**) a seized methaphetamine HCl (Reprinted from J. Forensic Sciences 47(5), S. Muratsu, T. Ninomiya, Y. Kagoshima, J. Matsui, 944–949 (2002) [863]), Copyright ASTM INTERNATIONAL, Reprinted with permission)

could be indicative of synthetic route (Fig. 7.155). Thereby, different drugs can be identified by means of their trace elemental fingerprints.

Using SRTXRF picogram amounts of each trace element in 10 μg drug samples were easily detectable. Detection limits are lower by a factor between 100 and 1000 compared to standard TXRF.

### 7.10.5 X-Ray Fluorescence Analysis Using Synchrotron Radiation

Synchrotron radiation today is the most powerful source of X-rays. The prominent features of synchrotron radiation are as follows:

- High intensity and high brilliance.
- High degree of linear polarization in the orbit plane.
- Spectral range and bandwidth of radiation from bending magnets and insertion devices can be adjusted to the desired task by monochromators.
- The energy of the exciting beam can range from some hundred electronvolt to well above 50 keV (depending on synchrotron radiation source).

These features render synchrotron radiation a efficient tool for synchrotron radiation X-ray fluorescence analysis (SRXRF) and make it very well suited for the investigation of biological samples. The high brilliance of third generation sources of synchrotron radiation make them especially suited for microfocus applications. Consequently, microfocus applications constitute the majority of activities in this area.

While only recently spatial resolutions of some micrometers or even submicrometers have become available on a routine basis at third generation sources, the term "X-ray microprobe" traditionally refers to spatial resolutions of 10 or even several hundreds of micrometers.

#### General Aspects

*Spatial Resolution*

Biological matter generally is highly organized and is functionally, structurally, and chemically heterogeneous at least in a 1:$10^6$ ratio on a scale from $10^{-8}$ m to $10^{-2}$ m. From this it is obvious that the spatial resolution of a microprobe experiment has to be adapted to the problem. Currently, best X-ray microprobes are able to produce microbeams with submicrometer dimensions employing zone plates [864] or compound refractive lenses (CRLs) [865]. Thus, by these microprobes subcellular structures in eukaryotic cells as well as individual bacteria can be resolved. Today in the majority of X-ray microprobes, "microbeam" dimensions in the range from 5 µm to 100 µm are available on a routine basis. So micro beam experiments with truly subcellular resolution, currently, are more the exception than the rule.

*Limit of Detection*

Detection limits depend on sample matrix, excitation and detection conditions, and sampling time. With few exceptions (teeth and bones), biological samples mainly consists of light matrix elements with low $Z$. Therefore, Compton scattering dominates over Rayleigh scattering, and the relatively high Compton- background has a pronounced influence on the detection limit. Generally, monochromatic excitation with reasonable bandwidth ($\Delta E/E = 0.01$

($\sim$ 100 eV at 10 keV)) by multilayer monochromators, if at all [866], is decreasing the limit of detection only by factor 2 to 20 [867], independent from atomic number ($Z$) and sample properties. For SRXRF-microprobes in the hard X-ray region (2–80 keV), relative detection limits at about 0.01 to 10 $\mu g\, g^{-1}$ matrix for elements from $Z = 17$ (Cl) to $Z = 92$(U) are cited for various beam lines [868–874]. Recalculated for a sample density of 1 $g\, cm^{-3}$, the area related detection limit in these terms is about 0.01–10 $ng\, cm^{-2}$. The absolute detection limit is strongly dependent on the irradiated sample area. In microfocus experiments, absolute detection limits as low as $10^{-18}$ g can be calculated [875].

In experiments with synchrotron radiation, time is often a limiting factor. While sampling times of some hundred seconds are tolerable in single point analysis, successful area scans with thousands of pixels require sampling times below 60 s or less. Fortunately, the limit of detection is proportional to the square root of the sampling time, so that a reduction of the sampling time by factor 10 only leads to a worsening of the detection limit by factor 3.3. Nevertheless, counting statistics play an important role for peak identification and quantification especially in cases of low count numbers, so that there is a lower limit for the sampling time, which is highly dependent on the element concentration of the sample, the photon flux of the exciting beam and the efficiency of the detection system. Currently, for microprobe experiments with spatial resolutions of a few micrometers, sampling times down to 1s/pixel had been achieved at third generation sources [876, 877, 899, 909].

The element content of biological matter varies widely in respect to distribution and concentration. The so-called trace elements show average elemental concentrations below 50 $mg\, kg^{-1}$ body weight [878]. In many tissues and even for some essential trace elements, the natural concentration falls below 1 $\mu g\, g^{-1}$ matrix, which is below the detection limit of most X-ray microprobes under short-time sampling conditions. Thus, in higher organisms under unexposed environmental conditions usually only 12 elements can be detected (K, Ca, Mn, Fe, Cu, Zn, P, S, Cl, Se, Br, Rb) on a typical hard X-ray setup, seven of them are metals. Only manganese, iron, copper, zinc, selenium, bromine and rubidium can be regarded as "trace elements", the others have to be referred as "main elements" or macrominerals [878]. Not unexpectedly, toxicological and pharmacological topics are dominating the X-ray microprobe experiments, because the elemental content under "exposed conditions" normally is much higher and easily detectable.

### *Sample Preparation*

Sampling and sample preparation play an important role for successful microprobe analysis. Due to the penetrating nature of hard X-rays, the sample thickness has to be adapted to the desired spatial resolution and the expected distribution of the analytes. Generally, sample thickness is chosen in the order of the diameter of the incident beam. One of the benefits of X-ray fluorescence analysis is that often there is no need for intensive pretreatment of the sample. For microprobe experiments, it is absolutely necessary that the native distribution and the concentration of the analytes is not disturbed. Therefore, freeze-cut techniques of native specimens with cryomicrotoms are the standard procedure for sample preparation of soft tissues. Sometimes, paraffin [879, 880] or resin-embedded specimens [877] had been used. In the case of single cells, there is no need of sectioning and direct preparation on thin support films is also possible [864].

The sample support has to be selected carefully. It should be as thin as possible and must be free from contaminating elements. Due to the high penetration depth of hard X-rays, thick sample supports produce excessive Compton scatter, which impairs the detection limits particularly in the case of thin specimens. Very few thin film supports are able to fulfil the demands of SRXRF for biological samples in respect of purity, as the concentration of analytes in biological specimens often is very low. Conventional sample foils for X-ray fluorescence analysis, especially Mylar® or Kapton® foils, generally are heavily contaminated by zinc, iron and calcium.

Frequently, the concentration of analytes in biological samples lies in the region of 10 to 100 $\mu g\, g^{-1}$ or even lower. In the case of thin samples (ca. 10 μm and below), the coherent and incoherent scatter by far is the dominating signal. If the experiment, as usual in most experimental set-ups, is conducted in air, the argon $K_\alpha$- line is interfering with the fluorescence lines of light elements. Figure 7.156 shows an SRXRF spectrum of a 30 μm hamster brain thin cut.

Both, scatter and the Ar fluorescence contribute to the incoming count rate of the detector and elevate the dead time to a great extent. Minimizing the contribution of scatter (both sample and air derived) therefore is very significant for sensitive analysis. The position of the detector exactly in the orbit-plane of the synchrotron and right-angled in respect to the incident beam, to take advantage of the linear polarisation of the synchrotron beam, is an important prerequisite to minimize the influence of Rayleigh and Compton scatter. In polychromatic excitation modes, the selection of high-pass filters for tailor-made optimal excitation conditions is of high importance. Currently, most experimental set-ups operate under ambient air. Nevertheless, the measurement in helium atmosphere or under vacuum conditions eliminates the air-derived parts of scatter and fluorescence (Ar, Xe) signals.

*Radiation Damage*

In X-ray fluorescence analysis, at a given sensitivity, radiation damage of a sample is several orders of magnitude lower than with scanning ion microprobes [881]. As the mass absorption coefficient depends on the energy of the radiation, lower energies cause more sample damage. Under monochromatic excitation conditions, the total photon flux is much lower than under polychromatic conditions, so that sample damage in terms of "destruction"

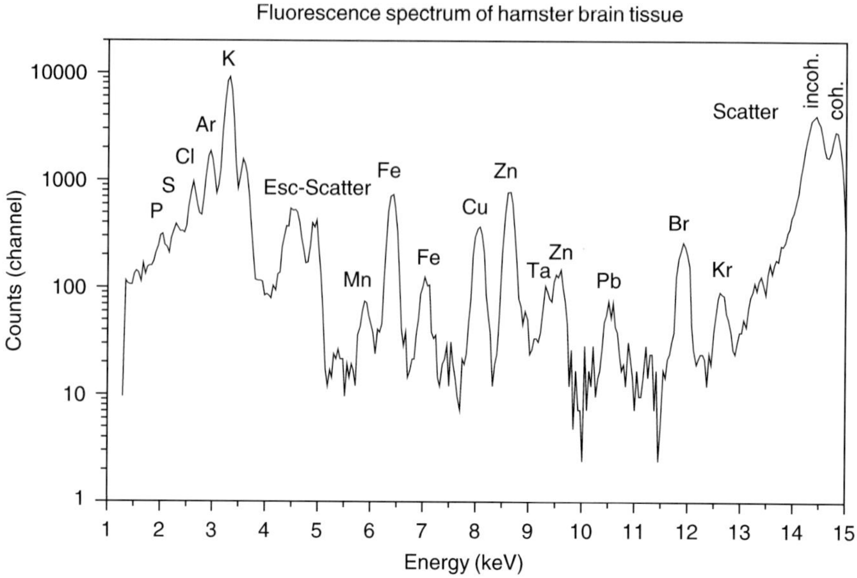


**Fig. 7.156.** SXRF spectrum of a hamster brain tissue cryo cut (30 μm) on polycarbonate film, 2 μm: excitation: Si(111) monochromator 14.8 keV with polycapillary optics. Detector: HPGe 30 mm$^2$, acquisition time 300 s (real). Pb and Ta fluorescence signals are derived from the detector collimator. (M. Kühbacher and G. Weseloh (Hahn-Meitner-Institute Berlin), beam line L, HASYLAB, unpublished)

is negligible. In a microbeam experiment at an ESRF undulator, Bohic et al. calculated a radiation dose of about $10^6$ Gray in "pink" excitation mode and $10^4$ Gray in monochromatic excitation mode [882], which imposed no observable radiation damage on the sample. However, at that point, it has to be stressed that a chemical modification, rather than a destruction of the sample by radiation is not easily detected and may cause false results in X-ray absorption fine structure experiments (XAFS).

## Examples

Due to the limited availability of beam time and the commitment to large and expensive installations synchrotron radiation X-ray fluorescence analysis by far has not become a routine tool of research in the life sciences. The annual review of Potts and Ellis et al. in the *Journal of Analytical Atomic Spectrometry* [883] counts only 39 original papers in this area between 1998 and 2003. Moreover, only about a dozen groups worldwide regularly show activities in this area. In section ***homogenous samples*** we give a short survey over actual research activities on various types of samples, which had been investigated by SRXRF. In most of the investigations, only qualitative results in respect to mass distributions or concentrations of the analytes were obtained, despite the potential of the method to obtain quantitative results.

Generally, there are two main groups of applications of synchrotron radiation X-ray fluorescence in life sciences: "microfocus" applications, in which a small collimated or focussed beam is applied with the objective of attaining spatial resolved data of the elemental composition of the sample, and applications, in which spatial resolution is unnecessary or only of secondary importance. In the subsequent section, we emphasize the former type of applications, because they are typical for the field of synchrotron radiation X-ray fluorescence. The applications can be further divided in respect to the sample properties, which can be of hard matter (teeth, bones, wood) and soft matter (tissues, cells, hair). A particular case is the investigation of single cells, because sample preparation and measurement conditions (especially the diameter of the exciting beam) are different from other applications.

*Homogenous Samples*

For biomonitoring purposes, Koutzenogii et al. investigated a multiplicity of samples from humans, animals and plants from the Altai region (Siberia) at the VEPP-3 storage ring (Novosibirsk) [884] with a monochromatized beam with photon energies between 22 keV and 25 keV and a beam diameter of 3–4 mm. For analysis, blood samples were placed on a filter, other samples were prepared as tablet. By this method, 27 elements from K to U were quantified either assuming the "thin sample approximation" for blood samples or, for tablets, after application of several correction factors, after peak fitting by AXIL. For blood samples, detection limits between $8\,\mu g\,g^{-1}$ for K, $0.08\,\mu g\,g^{-1}$ for Zn and $0.14\,\mu g\,g^{-1}$ for Pb are cited. The authors conclude that SRXRFA has a high importance and efficiency for multi-element analysis of biological objects for medical and ecological purposes.

Schulmann-Choron and coworkers determined the selenium content of digested samples of rat brain at the bending magnet beam line D15 at LURE (Orsay, France) [885]. The beam was monochromomatized and focused by highly orientated pyrolytic graphite (HOPG) crystals to a beam spot of about 1000 µm diameter at an energy around 14.5 keV. Sample drops were placed on 1.5-µm Mylar® film and dried at room temperature. Quantification was performed after addition of vanadium as internal standard element after correction of air absorption and detector efficiency. The minimum detectable limit (MDL) for selenium was found to be $20\,ng\,g^{-1}$ sample wet weight at an acquisition time of 500 s.

The zinc content of purified recombinant zinc binding proteins (glutathione S-transferase zinc-finger fusion proteins) was studied by Pfister et al. at the bending magnet beam line X26 at the NSLS (Brookhaven, NY) [886]. Droplets of 5 µL were spotted onto a hydrophobic AP1 membrane (thickness 0.15 µm) and analysed in wet state at an energy of 9.6 keV with a beamsize of $1000 \times 150\,\mu m^2$ (horizontal × vertical) with an acquisition time of 10 min. Besides Zn, also small amounts of Ni and Fe were found to be present in the preparations. The minimum detectable limit (MDL) for zinc was calculated from the independently determined Zn content of a protein sample to

49 ng $g^{-1}$ solution. The results were discussed in respect to the application to microarray techniques with sample volumes of only 0.5 nl.

*Speciation Analysis on Proteins*

The metal content of proteins after separation by isoelectric focusing (IEF) [887] and sodium dodecyl sulphate polyacrylamide gel electrophoresis (SDS-PAGE) [888] was investigated by Gao et al. at the bending magnet beam line 4W1A of the Beijing Electron and Positron Collider National Laboratory. After protein separation, the dried, cellophane covered electrophoresis gels were scanned with a filtered, collimated white beam spot of $500 \times 100\,\mu m^2$ (horizontal × vertical) along the separation path with a step size between 0.4 mm and 1 mm and an acquisition time of 300 s per step. On the dried gels, Fe, Cu, and Zn could be determined qualitatively. The authors conclude that this method would be suitable for lighter trace elements like Fe, Cu, and Zn. At the present stage, the sensitivity of the method was found to be insufficient for the detection of other trace elements with lower abundances in tissues. The quantification of the metal content of the separated proteins was found to be impeded by the beam size, which was much smaller than the dimensions of the protein bands in the electrophoresis gels.

In a similar approach, Weseloh et al. studied the metal content of a metalloprotein standard (Zn-loaded apoazurin) and the selenium content of a homogenate from rat testes after separation by sodium dodecylsulphate polyacrylamide gel electrophoresis (SDS-PAGE) [889]. The investigations were carried out at the bending magnet beam line L of DORIS III at HASYLAB (Hamburg, Germany) with either white beam or monochromatic excitation at 15.5 keV (multilayer monochromator). The collimated beam had a dimension of either $40 \times 40\,\mu m^2$ (white beam) or $500 \times 500\,\mu m^2$ (monochromatic beam). In contrast to the method of Gao et al. [887, 888] the separated proteins were electroblotted onto a 200 μm thick polyvinylidendifluoride (PVDF) membrane. Under both excitation conditions, the acquisition time was 300 s (realtime). The blotting membrane was scanned with a step size of 2 mm for the standard protein and 1 mm for the rat testis homogenate. By this method, the main selenoprotein of rat testis (phospholipid hydroperoxide glutathione peroxidase (PHGPx)) could be successfully separated and detected. The authors conclude that improvements of this method have still to be made with regard to detection limits and contamination control.

*Hard Materials (Teeth, Bones, and Scales)*

Teeth and Bones are hard samples and normally are cross-sectioned and polished prior to measurement. Carvalho et al. investigated the content of Hg, Bi, Cu and Zn in amalgam treated and control teeth [890]. Prior to measurement, the amalgam was completely removed from the restored teeth by a stainless steel tool, the teeth were cross-sectioned along the vertical median plane by a diamond saw and sliced into specimens of about 0.5-mm thickness. SRXRFA was carried out at the bending magnet beam line $D_1$ of the DCI storage ring of LURE. The monochromatic beam with energy of 18 keV was generated by Bragg–Fresnel multilayer lenses and focussed to 100 μm. Again concentrations were calculated using a program package from LURE with Ca as internal standard. The detection limit was estimated to be 0.5 $\mu g\,g^{-1}$. Samples were scanned with a lateral resolution of 100 μm. In amalgam restored teeth, the content of Hg, Zn, Cu, and in some cases also Bi was found to be elevated compared to control teeth. The highest values for Hg, Cu, and Zn concentrations were found in the regions surrounding the amalgam fillings and reached 600 $\mu g\,g^{-1}$ for Hg, 100 $\mu g\,g^{-1}$ for Cu, and 1600 $\mu g\,g^{-1}$ for Zn. Mean values of these elements were elevated throughout the tooth in comparison to control teeth. Bismuth was found only in teeth, which were restored more than 40 years ago and then was used for base fillings in the inner part of the tooth, surrounding the nerve canal. No drift of Hg and Bi could be observed for the whole tooth, while elevations of the Cu and Zn concentrations were also found in dentine and root regions. No differences were found for the concentration of Mn, Fe, and Pb for restored and control teeth. Mean values of about 50 $\mu g\,g^{-1}$ for lead are cited. Interestingly, for both restored and control samples the lead concentration was elevated in the inner part of the teeth, probably reflecting elevated blood lead concentrations.

The migration of eel was studied by Nakai et al. by SRXRF analysis of the Sr/Ca ratio of otoliths [891]. Otoliths were embedded in polyester resin and polished from both sides to the midplane with polishing papers (sample thickness not given). Two-dimensional SRXRF measurements were carried out at beam line 4A of Photon factory, Japan. Samples were excited by a Si(111) monochromatized beam of 17 keV, which was collimated to 100 μm × 100 μm by a cross-slit system. The intensities of Sr, Ca, and P $K_\alpha$ lines were measured with a counting time of 3 s/pixel and a step size of 100 μm. The Sr/Ca ratio reflects the Sr/Ca ratio of the water environment, which is much higher for salt water than for fresh water. As the otolith consists of daily growth layers, the life cycles of the eels, which migrate between salt- and freshwater, is preserved in the layer structure of the otolith in different Sr/Ca ratios.

The distribution of Pb and Ca in the growth region of regenerating and ontogenic scales of lead-exposed carp was investigated by Yoshitomi et al. at beam line 4A at the Photon Factory, Institute of Materials Structure Science, Tsukuba (Japan) [892]. The excised scales were dried, attached to a Mylar® foil and SRXRFA was carried out under vacuum conditions with a monochromatized beam of 13.5 keV, which was generated by a Si(111) double crystal monochromator. Line scans were performed with a beam size of $300 \times 300\,\mu m^2$ and a step size of 100 μm/step at an acquisition time of 100 s/pixel. The intensities of Pb $L_\alpha$ and Ca $K_\alpha + K_\beta$ lines were normalized to the intensity of the elastic and inelastic scatter peaks. The main elements observed were Ca, P, Zn, and Pb. From the similar distribution of Ca and Pb in regenerating scale it was concluded that the region of Pb accumulation corresponds well with the calcifying region of the scales, and Pb, due to its chemical similarity, is processed in the same way as Ca. As the distribution of Pb in the ontogenic and regenerating scales is highly related to the history of exposure to the element, the authors suggest that SRXRFA can be used to estimate the time course of Pb contamination of fish in natural environments.

Ektessabi et al. studied bone and tissue material from failed hip replacement prosthesis and compared SRXRF with PIXE [893]. The bone material from ground 50-μm thin sections was investigated by a monochromatized beam (photon energy 10 keV, beam dimensions not given) at a small synchrotron (ring radius 1.0 m). SXRFA demonstrated the presence of different metals, such as Al (!), Ti, Fe, and Cr in the tissue. A comparison between the synchrotron radiation XRF-spectrum and the PIXE spectra shows, that under the given experimental conditions, XRF using synchrotron radiation has a better sensitivity for detecting the low mass elements, especially aluminium. Later they expanded the investigations at the hard X-ray facility SPring-8 with a spatial resolution up to 2 μm [894]. In this study also XANES experiments for chemical speciation analysis were conducted, and the findings of light microscopic images were correlated with the elemental distributions.

Thin sections of bone tumours were investigated by Huang et al. at the Beijing synchrotron radiation facility [895] with a 60 μm × 60 μm collimated polychromatic beam. They reported detection limits between $0.30\,\mu g\,g^{-1}$ for Y and $70\,\mu g\,g^{-1}$ for P. Values for S, Ca, Fe, Cu, and Zn and the ratio Cu/Zn were increasing between normal and tumour tissue. The authors conclude that SRXRF can be used to analyse the concentration of trace elements within a few cells and to estimate the levels of trace elements in them.

### *Plant Material*

In order to test the suitability as bioindicators for environmetal pollution, the element content of leaves of *Nerium oleander* from different urban and one rural location was investigated by de Jesus et al. at the National Synchrotron Light Laboratory (LNLS), Campinas (Brazil). Washed and unwashed leaves were subjected to acid digestion by $HNO_3/HClO_4$ mixtures, mixed with ammonium pyrrolidine dithiocarbamate and absorbed onto a cellulose membrane. SRXRFA was carried out with a polychromatic beam by means of an HPGe detector. The element concentrations were determined using the "thin sample approximation" after calibration with commercial element standard foils. At an acquisition time of 500 s detection limits between $0.2\,ng\,g^{-1}$ and $0.6\,ng\,g^{-1}$ could be obtained for a multi-element solution. While the contents of Ti, V, and Fe were higher in the urban sites with heavy traffic, the content of Pb, Zn, and Cu was not clearly dependent on the traffic flow. The content of Mn, Ni, and Co did not vary with the traffic level. In almost all samples, the content of washed samples was lower than the unwashed samples.

The application of the SRXRF microprobe on dendroanalysis was studied by Martin et al. The authors analysed the distribution of Ca, Mn, Fe, Ni,

Zn, Cr, and Cu in a cylindrical wood core of 4-mm diameter of a 15-year-old red pine at a location that had been exposed to airborne metals and acid rain from local smelters [896]. The metal distribution was determined both by secondary ion mass spectrometry (SIMS) and by SRXRFA. SRXRFA was carried out with a white beam in an area of $2.5 \times 12\,\mathrm{mm}^2$ with a step size of $50\,\mu\mathrm{m} \times 500\,\mu\mathrm{m}$ at beam line X26A at the NSLS (Brookhaven). The beam was focused by a Kirkpatrick-Baez optic to a spot size of $10 \times 15\,\mu\mathrm{m}^2$. The sampling time was not given. No effort was made to quantify the results in terms of element concentrations. While Ca and Mn as well as Zn an Ni were found to match closely the location of the annual rings, suggesting a variable uptake throughout any growing season, the distribution of Cr, Fe, and Cu showed a much more irregular distribution across the tree rings. The authors conclude that the distribution patterns of elements in the submillimetre range do not reflect bulk metal concentrations, and that small area analysis is better suited to study the physiological activity of metals, when their distribution patterns can be correlated to identifiable wood components. In respect to SRXRFA, SIMS had the advantage of a higher surface sensitivity for more elements and a higher spatial resolution. The same group, in a preliminary and qualitative approach, investigated the distribution of K, Ca, Zn, Mn, Fe, and Cu across a section of a black spruce under monochromatic conditions with a Kirkpatrick-Baez optics microprobe generating a spot-size of $15\,\mu\mathrm{m}$ [897]. The sample was thin-sectioned and polished (thickness not given). The authors found that each of the elements under study showed a different distribution across the diameter of the stem wood and conclude that micro-SRXRF provides an effective means for surveying changes in metal content of wood sections in a reasonable time throughout the entire diameter of tree stem wood.

*Hair*

The benefit of X-ray microanalysis for the investigation of uptake kinetics of heavy metals in the organism was demonstrated by Shimojo et al. by determining the Hg content in hair of rats exposed to methylmercury chloride [898]. The mercury compound was administered orally for five consecutive days in a nonlethal dose of $5\,\mathrm{mg\,Hg\,kg^{-1}}$ body weight; and hair samples were collected, either by shaving on several successive days up to 30 days for bulk analysis by flameless atomic absorption spectrometry (FAAS), or by extraction of some hairs, up to 14 days after the first administration for SRXRF analysis. The results of the endogenous exposure of the hair were compared to exogenous exposure by immersion of hair of a nontreated rat in a methylmercury chloride solution ($1\,\mu\mathrm{g\,g^{-1}}$) for 90 min. Hair samples for SRXRF analysis were embedded in acrylic resin and cut into 25-μm-thick cross-sections by a microtome with a tungsten knife. SRXRF measurements of Hg, S, and Zn were made at the Photon factory, National Laboratory for High Energy Physics (Tsukuba, Japan) utilizing a monochromatic X-ray microbeam of 14.38 keV and a beam size of $5 \times 6\,\mu\mathrm{m}^2$ obtained by a multilayer monochromator and a KirkpatrickBaez type focussing optics. Scanning step size was 5 µm horizontally and vertically and counting time was 10 s/pixel. For the comparison of Hg distribution between endogenous and exogenous exposure the scanning step size was set to 2 µm. No attempts were made for quantitative determination of the Hg content. By comparison of the Hg X-ray fluorescence intensity of the single hair sections and the results of the bulk analysis with FAAS it could be shown that SRXRF is suitable to determine a degree of Hg exposure using only a single hair. The two-dimensional distribution of Hg on hair cross-sections showed a pronounced difference between endogenous and exogenous Hg exposure. The endogenous exposure resulted in a preferential accumulation of Hg in the hair cortex than in the medulla or cuticle as well, while exogenous exposure resulted in a Hg accumulation on the surface of the hair cuticle, but not in the cortex. In the endogenous exposure model, the distribution of Hg corresponded well to the distribution of sulphur, while in case of exogenous exposure, Hg accumulated in a more outer area than S, indicating that the Hg distribution originated from external adsorption. The authors conclude that SRXRFA is one of the most promising techniques for differentiation between external and internal body exposure of heavy metals in biological monitoring studies of mammalians including humans.

Calcium distribution in cross-sections of human hair was analysed by Briki et al. on a submicrometer scale at beam line ID21 of the ESRF (Grenoble, France) [899]. Hair samples were cut perpendicular to the hair axis and analysed by a monochromatized beam above the Ca K-edge, where the Fresnel zone plate generated beam spot size was $0.3 \times 0.3\,\mu m^2$. Calcium and sulphur mappings of hair with low and high calcium content were generated with a pixel size of $1 \times 1\,\mu m^2$ and a collection time of 1 s/pixel. The hairs were analysed without and after leaching by hydrochloric acid. Calcium concentration was calculated from the fluorescence intensity ratios of Ca and S, assuming a constant sulphur concentration from hair to hair. Unlike sulphur, the calcium content was found to be not uniformly distributed in hair cross-section with higher concentrations within the central medulla, small (µm) calcium-rich zones within the cortex and narrow calcium rich zones in the outer cuticle. X-ray diffraction, carried out at station D43 of LURE (Université Paris-Sud) revealed the presence of crystallized lipids in Ca-rich hair. From these experiments and additional information of infrared spectromicroscopy the authors conclude that two distinct types of calcium are present in human hair: easy to remove lipid-bound calcium (calcium soaps) and putatively protein-bound calcium, which resists removal by hydrochloric acid.

### *Tissues*

The elemental distribution in bovine growth plate cartilage was investigated by Vittur et al. at the NSLS with an $8\,\mu m \times 8\,\mu m$ collimated beam [880]. The tissue slice was prepared by conventional methods for light microscopes. Five micrometer thick sections were embedded in paraffin and prepared "carrier free" on a hole between two sheets of cellulose acetate. The scan area was up to 400 µm × 400 µm with an acquisition time of 30 s/pixel. The distributions of S, Cl, K, Ca, Sr, Cu, Zn, and Ni were recorded. Problems were reported due to the inhomogeneous nature of the sample (different types of tissue with different compositions). The authors discuss the significance of their findings in respect to differences in the distribution of Zn, Cu, and Ni in different zones of bone in the biochemical context of bone metabolism.

Kawai et al. investigated the cadmium content of thin sections (3 µm) of rat kidney with monochromatic excitation and a beam size of about 200 µm [879] at SPring-8. The tissue, which contained 7 $\mu g\, g^{-1}$ Cd, was mounted on a 15-µm polypropylene foil. For proper spectra an acquisition time of 100 s was found to be necessary.

The uptake of chromium into the testis of mice was investigated by Ortega et al. [900]. Male mice were treated with 1 mmol $CrCl_3$ per kilogram of body weight by intraperitoneal injection. After 24 h, the organs were excised and frozen in liquid nitrogen. Freeze-dried tissue sections, 20-µm thick, were prepared using a cryomicrotome with a tungsten carbide knife to avoid chromium contamination. Cryosections were mounted onto 0.2-µm formvar foils and freeze-dried. Adjacent tissue sections were prepared for cytology and histology for comparison with nuclear microprobe and microfocus-SRXRF-images. SRXRFA was performed at the ESRF Beam line ID21 (Grenoble, France) using a monochromatized beam at 6.13 keV which was generated by an Si[111] crystal monochromator. The photon beam was focused down to $1 \times 3\,\mu m^2$ by a condenser zone plate. Single point analysis was performed at distinguished isolated sample points. In respect to SRXRFA, only qualitative results were reported (chromium was detected or not). Chromium distributions show a strong accumulation within the testes limiting membrane (tunica albuginea), and within the interstitial cells inside the organ.

In a combined approach, Paluszkiewicz and Kwiatek analysed human cancer prostate tissues by means of Fourier-transform infrared (FTIR) microspectrometry and synchrotron X-ray fluorescence analysis [901]. The excised tissues were frozen in liquid nitrogen, and sliced into 12 µm thick sections by a cryomicrotom and mounted on 3 µm thick Mylar® foils for analysis by SRXRFA and FTIR microspectrometry. For comparison, adjacent slices were mounted on conventional specimen slides and KBr-discs for FTIR spectrometry. The SRXRFA-experiments were conducted at the X26A beam line at NSLS, Brookhaven with a monochromatic beam of 17 keV and $16 \times 14\,\mu m^2$ beam size. Single point analysis was done on cancerous and healthy tissues with an acquisition time of 300 s (live). Quantification for Ca, Fe and Zn was carried out by comparison with self-prepared standard samples under the same conditions. The authors found profound differences in the element concentration between cancerous and healthy tissues. In the IR-region between $2800\,cm^{-1}$ and $3000\,cm^{-1}$ cancerous and healthy tissues are distinguished by different band intensities for $CH_2$ and $CH_3$ groups. This pioneering work shows the possibility of FTIR spectroscopy application to biological tissues placed on Mylar® foils, which offers the possible study of the same biological sample by means of two different nondestructive techniques giving complementary information on sample composition in a micrometer scale. At the same time the authors claim that the samples placed on Mylar® foils can be stained after the measurements by conventional histochemical methods for direct correlation with histopathological findings. Therefore, the preparation of adjacent thin cuts of tissue by a microtome for gaining additional information on the microscopy structure of the sample is not always necessary and can be substituted by application of histochemical staining methods after the X-ray analysis directly on the sample. It has to be mentioned that infrared microspectrometry also can be realized at infrared beamlines at some synchrotrons, which offers higher spatial resolution and higher sensitivity.

In another investigation, Kwiatek et al. again at NSLS investigated cancerous and noncancerous prostate tissues by a combination of µ-SRXRFA and X-ray absorption near edge structure analysis (XANES) for Fe [902]. The element concentration of Ca, Ti, V, Cr, Mn, Fe, Zn, and Se was determined for cancerous and noncancerous tissue by single point analysis with a monochromatic beam of 17 and 7.2 keV with a collection time of 300 s (live). 2D-mapping was performed for the main elements K, Cl, and Fe with an acquisition time of only 2 s/pixel with 15 µm step size in scan areas at $300 \times 300$ and $500 \times 500\,\mu m^2$. Beam size was $16 \times 14\,\mu m^2$. The concentration of the elements was estimated by comparison with self-prepared thin multi-element standards embedded in polyacrylamide. In the cancerous tissue, the concentration of Ca, Cr, Fe and Se was elevated significantly, while the concentration of Ti, V and Zn was reduced in relation to healthy tissue. From XANES investigations of selected points in cancerous and noncancerous tissue it was deduced that iron occurs in the +3 oxidation state in the cancerous part of the prostate tissue, while in the noncancerous part it is present at a lower oxidation state.

Pinheiro et al. compared µ-SRXRFA and nuclear microprobe (PIXE) by investigating thin sections (10–15 µm) of lung-associated lymph nodes [903]. Cryosections were mounted onto a polycarbonate film of 1.5 µm thickness and freeze-dried. Microprobe analyses were first carried out using SRXFA, because of its nondestructive nature, and then Rutherford backscattering (RBS) and PIXE were carried out in the same regions in order to determine area density and to produce elemental maps and concentrations. The SRXRF analyses were carried out on a microprobe set-up at the ESRF ID22 beam line using a monochromatic 14.3 keV beam with beam spot of $10 \times 1\,\mu m^2$ produced by compound refractive lenses (CRL). Spectra were collected for 6 s/pixel. Spectra analysis and element quantitation was performed with the WinAxil software assuming thin target criteria. The area density was determined with the help of RBS. A good agreement was achieved for some of the elements detected by both SRXRF and PIXE using micro-RBS areal density results to normalize elemental data. The combination of the methods permitted to quantify elements (Ca, Ti, Cr, Fe, Ni, Cu, Zn, Pb) in thin sections of biological tissues.

Brain and central nervous system (CNS) tissue has been one major target of biological synchrotron-radiation fluorescence analysis. The main pathological conditions in the focus of interest are Alzheimer's disease (AD), Parkinson's disease (PD) and Amyotrophic lateral sclerosis (ALS).

The role of iron in Parkinson's disease and parkinsonism–dementia complex (PDC) in patients from Japan and Guam was studied by Yoshida et al. with a SRXRF-microprobe and XANES investigations [904]. Paraffin embedded sections of 8 μm thickness from altogether four patients were mouned on Mylar® film and investigated by a collimated beam of 9.6 μm diameter with monochromatic undulator radiation at 7.16 keV (above the Fe K-edge) at beam line 39XU of SPRING 8, Hyogo (Japan). Acquisition time for iron mapping was 8 s/pixel. In the presented examples, the scanned area comprised a region of 500 × 240 μm. XANES spectra were measured at several discrete points in the sample. The quantification procedure was not stated in this work. X-ray fluorescence and XANES analysis demonstrated that the localizations and chemical states of iron (Fe(II) and Fe(III)) in the neuromelanin granules and in the neurons of the pars compacta of substantia nigra as well as in the glial cells were different between PD, PDC and controls. The differences were discussed in detail in respect to "oxidative stress" by generation of free radicals in these diseases. Additionally, variations in the distribution of titanium were found between Japanese and Guamanian patients.

In a preliminary approach, the localization of several elements in cryosections of scrapie-infected hamster brain was investigated by Kühbacher et al. at the bending magnet beam line L of DORIS III at HASYLAB (Hamburg, Germany) [905]. Ten micrometer cryo-sections of hamster brain tissue were thaw-mounted on thin polypropylene foil, air dried and sealed with a second polypropylene foil in order to avoid open handling of the possibly human pathogen tissue sample at the synchrotron. Adjacent sections were used for histological staining and immunohistochemical analysis. The SRXRF analyses were performed using a multilayer-monochromatized beam at an incident X-ray energy of 11.3 keV with a beam spot diameter of 15 μm, which was generated by a polycapillary half lens (Kumakov lens). Area scans were performed with 26 × 26 pixels at 15-μm step size with an acquisition time of 10 s real time per pixel. Spectra analyses were carried out with the AXIL-QXAS package. Relative detection limits are cited as 0.1 $\mu g\,g^{-1}$ sample for Fe, Cu and Zn at 10 s sample time and were determined by means of a standard reference sample (NIST SRM 1577, bovine liver). The authors found a significant difference in the distribution pattern of copper between the samples from the infected brains and the controls. They state that these findings demonstrated the suitability of SRXRFA in the investigation of disease-associated changes in the metal content of tissue sections. Furthermore, microfocus-SRXRFA has been shown to be a useful instrument for the investigation of metals and metalloids in infectious biological material, which has to be handled in sealed compartments.

Szczerbowska-Boruchowska et al. analysed human central nervous system tissue at the undulator beam line ID22 of the ESRF, Grenoble (France) [906]. Autopsy samples from the substantia nigra (brain) and from ventral roots of thoracic spinal cord of patients with PD and ALS and one control were cut into sections of 20-μm thickness in a cryomicrotome, mounted on AP1 foil and freeze-dried. Corresponding slices were prepared for histopathological investigations. Micro-SRXRF was carried out with a monochromatic beam at 17 keV and with polychromatic "pink" undulator radiation. The beam was focused with parabolic compound refractive lenses (CRL) for a beamspot dimension of $10 \times 0.2\,\mu m^2$ (horizontal × vertical) in monochromatic and $5 \times 2\,\mu m^2$ in pink mode. Typical scanning areas of $500 \times 500\,\mu m^2$ were mapped by steps of 10 μm (horizontal) and 5 μm (vertical). The sampling time was 3 s/pixel. Additionally, investigations of single neurons were performed at areas of $100 \times 100\,\mu m^2$ in steps of $5 \times 2\,\mu m^2$. Quantitative analysis was done by comparison with standard reference samples (NIST SRM 1832 and 1833) after fitting the spectrum with AXIL. The detected elements were P, S, Cl, K, Ca, Fe, Cu, Zn, Se, Br, Rb, and Sr at an area related limit of detection down to 1 ng $cm^{-2}$. The differences in the distribution of the elements between the PD and ALS cases and the control sample may be related to the functions of the elements in the central nervous system tissue and correspond to existing hypotheses of neurodegeneration. The authors conclude that μ-SRXRFA is a sensitive technique for elemental mapping in thin tissue sections at the single cell level.

*Cells*

A very stable beam at micron or submicron dimensions with sufficient intensity is indispensable for successful analysis of single cells with subcellular resolution. Much care has to be taken in respect to sample preparation and fixation, as the natural distribution and content of elements in the living cell has to be preserved. This is especially true if the trace element distribution within the cell is to be analysed. Thus, the investigation of single cells is one of the most demanding tasks of micro X-ray fluorescence analysis. Only recently, with the advent of efficient hard X-ray zone plates and compound refractive lenses (CRL) at third generation sources of synchrotron radiation in the hard X-ray region, progress in μ-SRXRF analysis has led to beam dimensions in the submicrometer range with sufficient intensity. As for cells, penetration lengths for X-rays with photon energies in the 4–10 keV region are in the range of hundreds of micrometers to millimeters, and the absorption length for fluorescence photons of biologically important elements, such as P, S, Cl and atoms with higher atomic weight, are in the range of 10 μm and above. Therefore, there is no need for thinning or sectioning of the sample. Self-absorption plays a negligible role in quantifying measured concentrations. For geometric considerations, though, the thickness of the object should be in the range of the lateral beam dimension. In single cell analysis in a standard micro-SRXRF set-up, like in applications with lower resolution, discrimination in vertical direction (that means, in beam direction) is not possible. Therefore, lateral changes in element distribution can result from changes inside the cell and those which are located at the outer cell membrane. To avoid these shortcomings, thin-slicing in most cases is thought to be an important prerequisite for high-resolution imaging of single cells by SRXRFA.

The accumulation of $Mn^{2+}$, $Nd^{3+}$, $Ce^{3+}$, and other elements (Ca, Fe, Zn) in the fresh water flagellate *Euglena gracilis* was investigated by Shen et al. at the wiggler beam line 3W1A of the Beijing Synchrotron Radiation Facility (BSRF) [907]. The cells were incubated in solutions of $6\,\mu g\,mL^{-1}$ $MnCl_2$, $3\,\mu g\,mL^{-1}$ $CeCl_3$ or $1.6\,\mu g\,mL^{-1}$ $NdCl_3$ in deionized water, washed by repeated resuspension, fixed by 0.25% glutaraldehyde and dropped onto Mylar® film. X-ray fluorescence analysis was carried out with a collimated white beam and a focal size of $15 \times 20\,\mu m^2$, which corresponds to the size of a single cell of *E. gracilis*. Spectra were analysed by AXIL, and quantification of the content of Ca, Mn, Fe, and Zn was achieved by comparison to single element standards. Additionally, XANES experiments were conducted at beam line 4W1B of the BSRF. The analytical results were compared to PIXE investigations of the ashed samples. As expected, the content of Mn and Nd in the cells increased after incubation with the corresponding metals, while the cellular content of the essential metal cations Ca, Fe, and Zn decreased in the treated cells. The metal content of the cells was given in a range between 1.4 pg/cell for Ca and 0.03 pg/cell for Mn.

Ishihara and coworkers studied the uptake and distribution of chromium, some main elements (P, S, Cl, K) and zinc in chromium(III)-chloride-treated mouse macrophages by SRXRFA and transmission electron microscope energy dispersive X-ray analysis (TEM-EDX) [908]. SRXRFA was carried out at the bending magnet beam line 4A at the Photo Factory (Tsukuba, Japan). The beam was monochromatized by a multilayer monochromator to 14.2 keV and focused by a Kirkpatrick-Baez mirror to 5 μm in diameter. Distribution maps of Cr, Ca and Zn across one single macrophage were generated by scanning an area of $42 \times 42\,\mu m^2$ with a step size of 1 μm. The macrophages were incubated with $CrCl_3$ in various concentrations for various times, harvested by centrifugation, dehydrated with ethanol and dried on PET films at room temperature. Because of the relative large beam size, a clearly detailed distribution pattern of the elements within the cell was not possible. Only relative concentrations of Cr, P, Ca, and Zn dependent on the incubation period could be evaluated.

Zone plates for hard X-ray microimaging of single cells were used by Kemner et al. at the APS (Argonne, IL) [864]. The authors determined the distribution of 3d-elements, particularly Cu, in a single hydrated *Pseudomonas fluorescens* bacterium, which was adhered to a Kapton® film. The zone plate produced a beam of $0.15\,\mu m^2$ cross-section at 10 keV. The samples were mounted on a piezo stage at only 5° to the incident beam. An image map of $6 \times 8\,\mu m^2$ was produced with a pixel size of $0.15\,\mu m^2$ and an acquisition time of 5 s/pixel. The total sampling time was approximately 8 h. The authors conclude that a cryostat would be required to freeze the samples in order to reduce the effects of radiation damage for X-ray absorption fine structure studies (XAFS).

The uptake of platinum anticancer drugs in individual ovarian cancer cells was investigated by Cholewa et al. with a submicron beam (0.5 μm) generated by a hard X-ray Fresnel zone-plate optics at the Advanced Photon Source (APS), Argonne, IL. [909]. The authors report an "imaging sensitivity" of 80 attograms/$\mu m^2$ at a count rate of 10 counts/s. The method was compared with a 3 MeV nuclear microprobe. The cells were exposed to various concentrations of the platinum compounds in culture vessels, harvested by centrifugation, resuspended in a volatile buffer (ammonium acetate), frozen in liquid Freon on a thin nylon foil and freeze-dried. Element concentrations were calculated by comparison of the characteristic X-ray fluorescent intensities to those from a standard specimen. Cells were scanned with a step-width of 1.0 μm and a collection time of 1–2 s/pixel. The authors conclude that micro-SXRFA at APS can be done with higher resolution and minium limits of detection and several orders of magnitude (factor 1000) better sensitivity than with nuclear microprobes. In the same publication, they report on an investigation on the uptake of chromium compounds in human V79 cells with a focused beam with a focal spot of 40 μm diameter, which was generated by a paraboloidally tapered glass monocapillary (intensity gain 120 at 8 keV) at the NSLS at Brookhaven. In these experiments, the conversion of the chromium compounds from initially chromium(VI) or chromium(V) to chromium(III) inside the cell could be observed on a single cell level.

For the fluorescence imaging of single cells Bohic et al. employed CRLs at beam line ID22 at the ESRF (Grenoble) [882]. Human ovarian adenocarcinoma cells (IGROV1) were grown directly on thin (0.2 μm) Formvar film supports, which were treated with an attachment factor. The cell monolayers were incubated with 4′-iodo-4′deoxy doxorubicin (IDX), rinsed, cryofixed into liquid nitrogen chilled isopentane and freeze-dryed at −30°C. Analysis was performed at monochromatic (14 keV) or "pink" undulator radiation. The CRLs produced a beam spot of $10 \times 1\,\mu m^2$ (horizontally × vertically). In "pink" beam mode, 2.5 s counting time proved to be sufficient for elemental mapping, whereas an acquisition time of 120 s (monochromatic) or 100 s ("pink") produced well-defined peaks of P, S, Cl, Ca, K, Mn, Fe, Cu, Zn, and I from drug treatment. Area-related concentrations were calculated after data fitting by WinAxil software on the basis of the number of incident photons obtained by a pin-diode, the known energy distribution of the exciting beam and the "thin sample". Mass related concentrations were derived from the determination of the cells surface mass obtained by separate Rutherford backscattering spectrometry (RBS) analyses. The imaging maps showed a colocalization of iron and iodine within the cell nucleus.

The oxidation state of chromium in cryofixed and freeze-dried human ovarian carcinoma cells (IGROV1) and Chinese Hamster Ovary cells (CHO-AA8) was studied also by Ortega et al. at the ESRF beam line ID21 (Grenoble) with a monochromatized focal spot of 1.5 μm in diameter generated by zone plates [910]. Oxidation state mapping was performed by scanning the sample first at an energy above the chromium(VI) pre-edge (5993 eV) and then above the absorption edge of chromium(III) and chromium(VI) (6020 eV). Prior to preparation, the cells were incubated for 1 h with a soluble Cr(VI) compound ($Na_2CrO_4$) or for 24 h with a Cr(VI) compound of low solubility ($PbCrO_4$). After the incubation period, no chromium (VI) could be detected inside the cells, suggesting a rapid intracellular reduction of chromium (VI). However, a relatively high concentration of chromium was located at the nucleus.

The distribution and oxidation state of vanadium in blood cells of ascidians was studied by Takemoto et al. at beam line ID21 at the ESRF (Grenoble) [911]. The beam spot was focused by zone plates down to 0.5 μm. Various blood cells from two ascidians (*Phallusia mammillata* and *Ascidia sydneiensis samea*) were prepared by sophisticated procedures. The cells were studied under three conditions: either under wet conditions in a special wet specimen holder, frozen in a cryo-fixation holder or freeze-dried on a EM micro grid. Frozen or freeze-dried samples were prepared by immersion in liquid isopentane. To maintain cryo-fixed samples at low temperature during the measurements, the sample holder was cooled down to −120°C by a nitrogen flow cryo-stream. By this procedure radiation damage of the frozen sample could be minimized. Sample adjustment was achieved by X-ray transmission imaging. The vanadium content was detected after excitation above the vanadium K-edge by a monochromatic beam of 5500 eV. The oxidation state was determined on the cryo-fixed samples by XANES spectrometry. It could be shown, that vanadium is distributed uniformly in the vacuole of "signet ring cells" of the ascidians. The valence state of the vanadium inside the cells was found to be +3 and +4, as could be judged by comparison with vanadium compounds of various oxidation states.

Dillon et al. investigated the biotransformation of chromium compounds of various oxidation states (Cr(III), Cr(V), and $Na_2Cr_2O_7$) by V79 Chinese hamster lung cells at beam line 2ID-D of the Advanced Photon Source (APS), Argonne, IL [912]. Cells were incubated with the chromium complexes, harvested by centrifugation, and either washed with ammonium acetate buffer and freeze-dried onto nylon films or embedded in Spurr's resin after chemical fixation/dehydration and cut into thin slices by means of a microtome. All measurements were conducted under helium atmosphere using a monochromatic 10 keV incident beam. The beam was focused by a phase-zone-plate based optics to either 1 μm diameter for the analysis of whole freeze-dried cells or 0.3 μm diameter for the analysis of thin sectioned resin-embedded cells. Elements analysed were P, Cl, K, Ca, Cr, Fe, Cu, and Zn. Individual cells were located by collecting an X-ray transmission image of the sample with a CCD camera placed behind a scintillator. Whole cells were scanned in 1-μm steps with a total area between $20 \times 20\,\mu m^2$ and $30 \times 30\,\mu m^2$ with an acquisition time of 3 s/pixel. Microprobe measurements of thin-sectioned cells were performed using a 0.3 µm-step size with a scanning area of $12 \times 12\,\mu m^2$ with an acquisition time of 6 s/pixel. Data analysis of the elemental maps was performed by normalization to the number of total counts of a given fluorescence signal and subtraction of background (which was defined by the signal outside the cells). The chromium concentration of one single cell was estimated by comparison with a thin-film NIST standard SRM 1832 and the integrated number of total counts of chromium per cell. No increases in intracellular Cr were observed for cells that had been treated with the Cr(III) and Cr(V) complexes, while chromate was readily taken up by the cells. Careful analysis of the mappings of the thin sectioned cells revealed the existence of three zones of different element composition within the cell. Large amounts of phosphate were attributed to the presence of DNA and, hence, the nucleus of the cell. After incubation with Cr(VI), chromium was found to be distributed in various concentrations within the cell boundary, with a high amount of chromium being localized in the phosphate-rich area. The authors conclude that by SRXRF microprobe technique unprecedented information on the distribution of both endogenous and exogenous elements within a cell can be obtained, which can provide new insights into disease processes at the subcellular level.

The elemental distribution of P, S, Cl, K, Ca, Mn, Fe, Ni, Cu, and Zn in nonvirulent and virulent, phagocytosed and external mycobacteria (*Mycobacterium smegmatis* and *M. avium*) under various conditions was determined by Maser et al. at beam line 2-ID-D at the APS (Argonne, IL) [875]. The focus of 0.15 µm diameter at 10 keV (monochromatized) again was generated by a zone plate optics. Macrophages (as hosts for mycobacteria) were cultivated on formvar-coated gold grits, infected with *M. smegmatis* or *M. avium* resp., washed with buffer solution and fixed in paraformaldehyde. Finally, the grids were again washed with buffer solution, rinsed with water and air-dried. Areas of interest (individual, phagocytosed mycobacteria) were identified with an optical microscope on a calibrated $x/y$ stage. SRXRF- scans were performed under helium environment with an area of typically $3 \times 3\,\mu m$ and a step size of 0.2 µm at each bacterium. Acquisition times between 10 s and 20 s were required to achieve the photon statistics needed to separate the signal of the phagocytosed mycobacterium from the background signal resulting from the macrophage. For quantification, the microprobe was calibrated using NIST thin-film standards (SRM 1832 and SRM 1833). Quantitative data were obtained as area density. In respect to the different strains of mycobacteria, a significant increase of the Fe concentration 24 h after infection was observed in virulent *M. avium*, whereas a significant decrease of the Fe concentrations occurred in nonvirulent *M. smegmatis*. The authors summarize that the use of a hard X-ray microprobe in combination with optical microscopy and standard biochemical methods is a powerful approach for the study of the role of metals in the infection by intracellular pathogens.

# 7.11 Non-Invasive Identification of Chemical Compounds by EDXRS

*P. Hoffmann*

## 7.11.1 Introduction

X-ray spectrometry (XRS) or X-ray fluorescence (XRF) analysis is one of the most widely used analytical methods. It is applied mainly for the characterization of solid samples (e.g., powdered, compact). For quantification on the one hand calibration samples of similar composition are needed and on the other hand calculation programmes are used for fundamental parameter evaluation [913]. Both, often simultaneously used procedures are complex but important to obtain a reliable result.

However, identification of the elements in a sample (qualitative analysis) by XRS using Moseley law is very simple. The difference in quantitative and in qualitative analysis is the reason for the different aspects with regard to the application of this method.

In an XRS spectrum in addition to the characteristic X-ray lines a lot of information can be found, which characterize the excitation source (e.g., position of the Rayleigh and Compton scattering lines), the detector (e.g., escape-peaks, sum peaks), and the sample (e.g., background, intensity and width of the scattering peaks). All chemical elements in the sample contribute to the resulting spectrum, also those which cannot be identified by a fluorescence signal.

The main mechanisms of the contributions are the Rayleigh (elastic, coherent) and Compton (inelastic, incoherent) scattering effects of the excitation radiation in the sample. Rayleigh scattering results from the interaction of radiation with strongly bound inner orbital electrons. High-atomic-number elements have a larger proportion of such electrons and produce relatively intensive Rayleigh signals. The scattered X-rays are monoenergetic. Compton scatter results from the interaction of radiation with the more loosely bound outer electrons of an atom. The energy of the scattered photons is spread over a large range and is a function of the scattering angle $\Theta$. As in XRS instruments $\Theta$ is fixed by the design of the instrumental geometry, a relatively narrow line is produced with a low-energy tail. The intensity of the Compton scattering spectrum is strongly related to the mean atomic number of the sample. Samples containing light elements are much more effective than samples of heavier elements in scattering the excitation radiation [914]. This phenomenon is the basis of matrix correction procedures, described in the text books. The intensity of the Compton scatter peak is inversely proportional to the mass attenuation coefficient [915, 916]. The procedure to correct matrix effects by the intensities of Compton peaks is reviewed in [917]. A special application of the Compton scattering correction method is described for powdered samples with not identical bulk density [917].

This knowledge gave rise to the concept to apply XRS as a non-invasive, non-destructive, simple, and rapid method of identification. Examples of application are: identification of unused chemicals returned to the production industry, certification of chemical compounds in a packing plant, control for hidden toxic, hazardous, and explosive material.

### 7.11.2 Experimental Part

The main advantages of the application of XRS in identity control are the non-destruction of the sample, the short analysis time, the relatively great depth of information, the accuracy of analysis, the simple operability, the economic efficiency, and the possible high sample throughput [918, 919] .

For this application the energy-dispersive mode of XRS (EDXRS) is preferred, because the total spectrum is recorded simultaneously.

#### Experimental Set-ups for EDXRS Analysis

Three set-ups of different type were used.

*Method A: Self-made Set-up [920]*

This EDXRS set-up, designed for industrial identification of a great number of packed samples, was based on commercially available equipment parts. An AEG Ag-anode with maximum power operating conditions of 60 kV and 60 mA was chosen as excitation source. A Canberra Si(Li)-detector and suitable electronics for treating the signals with an overall resolution of 165 eV at the Mn Kα-line was used to measure the spectra. An excitation geometry of 60° was fixed between X-ray tube, sample and detector for two reasons: (1) sufficiently good resolution of Compton and Rayleigh scattering lines, and (2) relative high intensity of the Compton scattering line [921]. The cross-section of the spectrometer is shown in Fig. 7.157. A triple-layered shielding of the sample chamber with Al/Cu/Pb was used to reduce the scattering intensity at the construction materials.

The chemical compounds were measured through the PE-bottle side wall 20 mm above the bottom. The operating conditions were adjusted to a high voltage of the tube of 45 kV, a tube current of 2 mA, a primary X-ray beam collimator with a diameter of 2 mm made of aluminium, a Ag primary X-ray beam filter with a thickness of 0.12 mm and a detector Al-aperture with an opening of 3 mm (Fig. 7.157). These parameters were chosen for an effective excitation and detection of the entire energy range of interest from 3 to 40 keV.

*Method B: XLab2000 (SPECTRO A.I., Kleve, Germany)*

The commercially available laboratory device (XLab2000) was equipped at first with a Rh-anode X-ray tube working at 44 kV and at an automatic adjustable current resulting in a deadtime of 50% (the tube was replaced after usual decreasing of the excitation intensity by a Pd-anode tube). For a

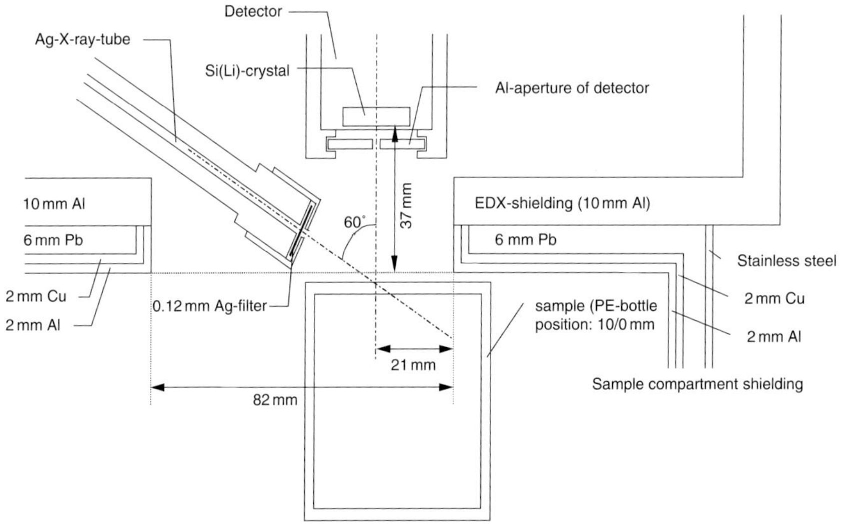


**Fig. 7.157.** Set-up of the XRF system for non-invasive identification of chemical compounds (method A)

comparison with the self-constructed device a Ag secondary target was installed. A Si(Li)-detector with a resolution of 145 eV (MnKα at an integral counting rate of 10000 counts per second). The excitation geometry was fixed at 90° as usual for minimizing the inelastic scattering probability.

*Method C: XEPOS (SPECTRO A.I., Kleve, Germany)*

Additionally, the desktop version XEPOS was used. This apparatus was equipped with a Pd-anode X-ray tube working at a voltage of 45 kV and a current of 0.5–1.0 mA. For excitation a Mo secondary target was installed and for recording the spectra a Si drift chamber with a resolution of 165 eV (MnKα at an integral counting rate of 10000 counts per second) was used. The excitation geometry was fixed at 90°.

Both commercially available devices were designed for minimizing the scattering effects. In order to achieve minimum scattering radiation both systems are arranged in a so-called Cartesian set-up producing a polarized excitation radiation not being scattered into the detector direction [922]. That is very effective for optimizing the lower limits of detection. On the other hand, the decrease of the scattering intensity and a shift of the Compton and Rayleigh peaks are disadvantages with regard to the purposes this paper is focussed on.

### 7.11.3 Results

#### Test Measurements

Test measurements were necessary to show and to optimize the functions. At first a test sample was measured to exhibit the spectral region for the

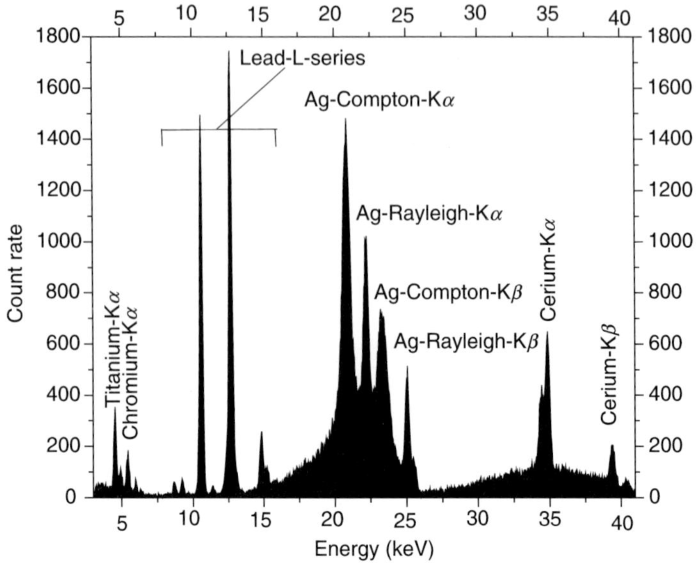


**Fig. 7.158.** EDXRF-spectrum of a synthetic sample ($K_2Cr_2O_7$, $Ce(SO_4)_2{\cdot}4H_2O$, Pb) for the investigation of detectable fluorescence lines through a PE-bottle wall (method A, counting time: 30 s)

evaluation of the signals. The measurement of a sample containing the chemical compounds $K_2Cr_2O_7$, $Ce(SO_4)_2{\cdot}4H_2O$, and Pb with method A is presented (Fig. 7.158). The fluorescence lines CrKα/β, CeKα/β and the PbL-series are clearly observed (additionally TiKα/β can be observed, generated by the pigment of the PE reagent bottle). Furthermore, the Compton and Rayleigh scattering lines of the Ag-anode tube radiation (15–26 keV) are seen, which give very valuable information about the compound, as will be explained in the following. It can be seen that the detection sensitivity rapidly decreases for X-ray energies at and below 5 keV. Hence, elements with $Z < 22$ (Ti) cannot be detected through the PE-bottle wall and at atmospheric pressure. The other end of the detectable X-ray energy range is given by the tube voltage of 45 kV, so that X-ray-lines with energies up to 40 keV are detectable.

The behaviour of the Compton and Rayleigh scattering lines is shown in Fig. 7.159. In this figure are shown the recorded spectra of three Ca-containing compounds. With decreasing intensity of the Ca fluorescence signals (proportional to the amount of Ca, not to the concentration of Ca) the intensity of the Compton scattering lines increases (Rayleigh scattering lines exhibit another behaviour). From Fig. 7.159 it can be concluded that the intensity of the Compton scattering peak is highly influenced by all the elements in the sample. That is proved by the reciprocal correlation between the mean atomic number and the Compton peak intensity. From this observation it is

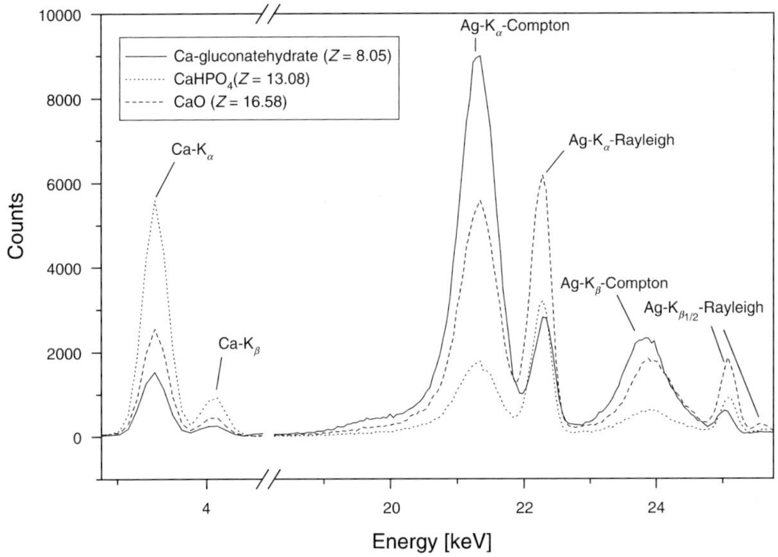


**Fig. 7.159.** Three calcium compounds and their ranges of fluorescence and scattering (method B, Ag secondary target, 30 s)

obvious to test, if the intensities of the fluorescence lines, and the Compton and Rayleigh scattering peaks can be used for identification of a chemical compound.

## First Approach to Identification Measurements

Several authors found a relation between the intensity ratio Rayleigh-line to Compton-line of the Kα X-ray target line and the mean atomic number [923–925]. It was, therefore, tried to use this dependence for the classification of a group of chemical compounds in PE-bottles without any detectable fluorescence signal by "method A". This relationship is shown for 14 compounds in Fig. 7.160. Linearity of this dependence is observable with a correlation coefficient $R^2 = 0.814$, but a compound classification is not feasible due to the pronounced scatter of the data and the resulting overlap for the various compounds.

## Multivariate Data Analysis of the EDXRS-Spectra

It was mentioned in Sect. 7.11.2 that there is some useful information for compound identification using the peak areas of characteristic fluorescence lines, and Compton- and Rayleigh-scattering lines. All these lines represent only a small part of an EDXRF-spectrum. Applying the methods of multivariate data analysis it is easy to make use either of much wider spectrum parts

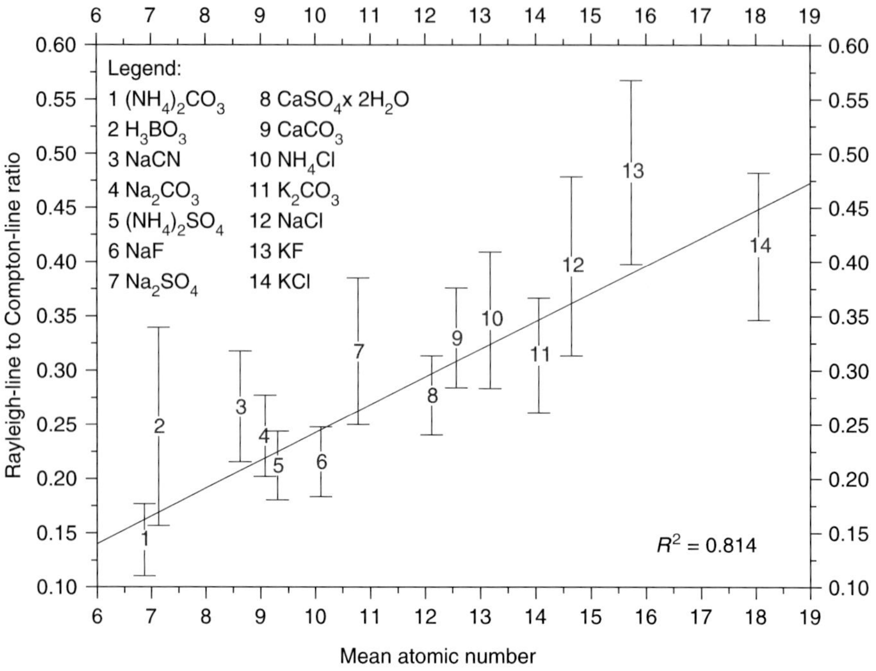


**Fig. 7.160.** Rayleigh-line to Compton-line intensity ratio vs. mean atomic number for the EDXRF-spectra of compounds not exhibiting fluorescence lines (error bars = 1 s for $n = 10$)

or even of the whole content of information available in a spectrum. It was, therefore, tried to apply principle component analysis (PCA) and regularized discriminant analysis (RDA) for the classification of compounds under the given experimental conditions.

*Principal Component Analysis (PCA)*

Multivariate data analysis and PCA are widely used in EDXRS to investigate influences of various physical parameters on the results of quantitative X-ray analysis [926–930]. PCA is known as a method for transforming a data set from the original space of variables (e.g., ranges of single energy channels of an EDXRF-spectrum) into a new space which is orthogonally related to the original space of variables. The principle of transformation is the extraction of maximum variance for each successive new variable. This procedure leads to a separation of valuable information from noise and to a selection of a few influential and statistically significant variables. The new coordinates of the data in the main component space are called scores. The relation between the original variables to the new space of variables is expressed by the loadings. The principal components (PCs) reflect the influence of physical parameters on the data sets.

The energy ranges of the detected characteristic fluorescence lines, of the Compton-scattering lines and of the Rayleigh-scattering lines in the recorded EDXRF-spectra for various compound groups were subjected to PCA after centroiding and autoscaling the respective data sets.

Peak areas were calculated using the software MCDWIN. The start and end of the respective integration windows were determined from the first derivative of the spectra. Spectrum smoothing was tested, but it was found that spectrum evaluation without application of any smoothing procedure yielded the best results. All of the processing steps were performed in the program SCAM (Scan for Windows 1995) [931]. The results of PCA for three groups of compounds of the elements lead, copper and chromium are shown in the Figs. 7.161a–c. These score plots show the position of the compounds in relation to the first two principal components.

The score plots of Figs. 7.161a–c exhibit a better resolution of the investigated spectral ranges along the first two PCs with decreasing atomic number $Z$. This is mainly due to the increasing penetration depth for the exciting X-ray radiation with decreasing mean atomic number of the matrix (substance) and could mean that an increasing amount of useful information

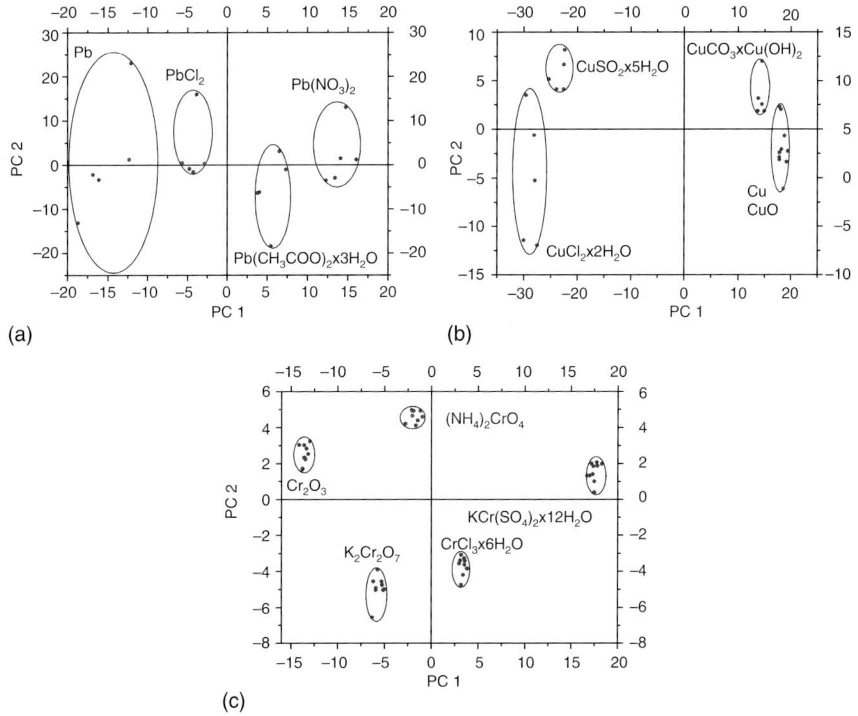


**Fig. 7.161. (a–c).** Score plots of the fluorescence lines, the Compton-lines, and the Rayleigh-lines ranges of the EDXRF-spectra of 4 lead, 5 copper, and 5 chromium compounds (method A)

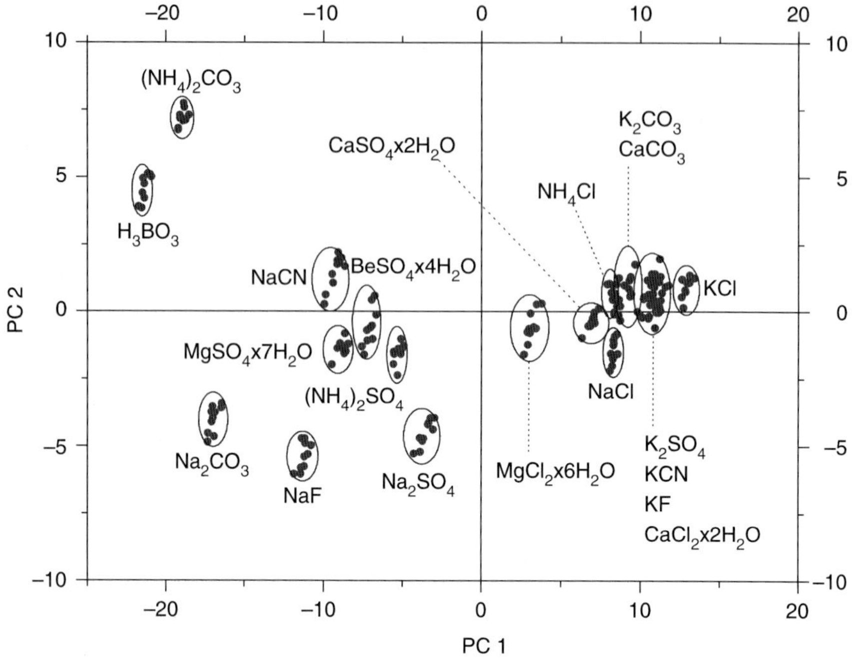


**Fig. 7.162.** Score plot of 20 compounds which exhibit no detectable element lines through a PE bottle wall (method A)

might be available also for the group of compounds which do not exhibit fluorescence lines. Therefore, 20 such compounds (those mentioned in Fig. 7.160 and some others) were subjected to EDXRF-spectrum registration and subsequent processing of the data by PCA of the Compton- and Rayleigh-scattering regions of the spectra. The results are shown in Fig. 7.162. It can be deduced that essential differences in the relevant spectra are present and can be separated from noise by PCA, thereby making possible a good differentiation of most inspected compounds in the score plot of the two first principal components. The insufficient separation of some of the samples is due to the very similar mean atomic number of the corresponding compounds.

*Regularized Discriminant Analysis (RDA)*

The method described in Sect. 7.11.3 cannot be used for the classification of unknown compounds because of the change of the PCA model by adding new compounds to the collective. For this purpose, classification methods have been designed [932–934]. Such methods are generating models based on data of samples of known composition, which are grouped into classes. Discriminant methods of analysis use centroids of these classes as well as the covariance matrix of the whole data set or of each class as model parameters. Unknown samples can then be assigned to these classes by calculating the distance of the spectrum of the unknown sample to the class centroid. This distance is corrected for the covariance between the variables (Mahalanobis distance). The RDA is a classification method designed for applications in which less observations than variables are available [932].

This is the case in EDXRF-spectra evaluation. It was shown that the predictive capabilities of RDA are superior to those of other procedures [933, 934]. The RDA-calculation uses two regularization factors which optimize the discriminant function for each model. The numerical values for these factors $\rho$ and $\gamma$ are between 0 and 1. According to the literature [935] different values for the two factors represent different kinds of discrimination models such as the quadratic discriminant analysis (QDA), the linear discriminant analysis (LDA), or the nearest mean classification (NM). The EDXRF-spectra of chromium and iron containing compounds, respectively, as well as the 20 compounds not exhibiting fluorescence signals (see Fig. 7.162) were classified by RDA [920]. The results comprise the best regularization factors, the optimal type of classification method, the misclassification risk after a cross validation and the interference of single compounds in a group.

#### Interferences in the Chemical Compound Identification

As outlined in the introduction, the intensity and the width of the scattering lines are not only a function of the mean atomic number of a sample, but other physical parameters influence the spectral pattern, too: e.g., the grain size distribution of the sample, the packing wall thickness, and the position of the sample. Therefore, the influence of these parameters was examined separately [936].

*Influence of the Size Fraction*

To illustrate the influence of the grain size, the spectra of seven size fractions of $SiO_2$ were measured fivefold each (100 s measuring time) and a PCA was calculated (Fig. 7.163). The separation shown in Fig. 7.163 plays an important role in the identification of compounds. That is exhibited in Fig. 7.164, where the results for the seven grain size fractions of $SiO_2$ are displayed in a score plot of Ca containing compounds (these are defined in Table 7.23). For this presentation, first the PCA was calculated for the grain size spectra. Using the resulting loadings, the PCA was calculated with the Ca compound spectra. The result shows a linear dependence of the position in the PCA score plot of calcium compounds measured by "method C". It should be noted that the PC scores of $SiO_2$ (mean $Z = 10.8$) are situated around the score of calcium glycerophosphate hydrate with nearly the same mean atomic number = 10.74.

*Influence of the Packing Wall Thickness*

In XRF measurements the sample is usually applied in a container consisting of a plastic ring, a thin foil and a lid (so called "spectro cup"). Chemical compounds are generally available in closed bottles with varying wall thickness.

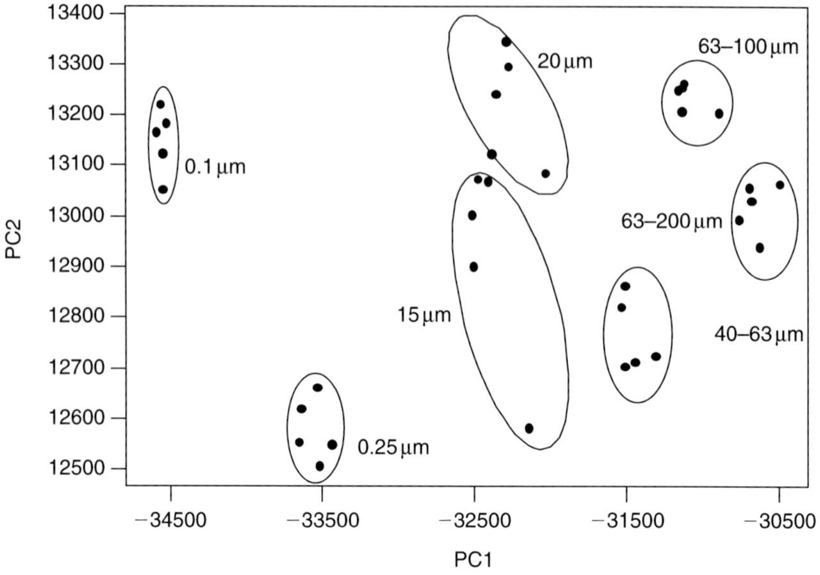


**Fig. 7.163.** Score plot illustrating seven grain size fractions of $SiO_2$ (method C, 100 s)

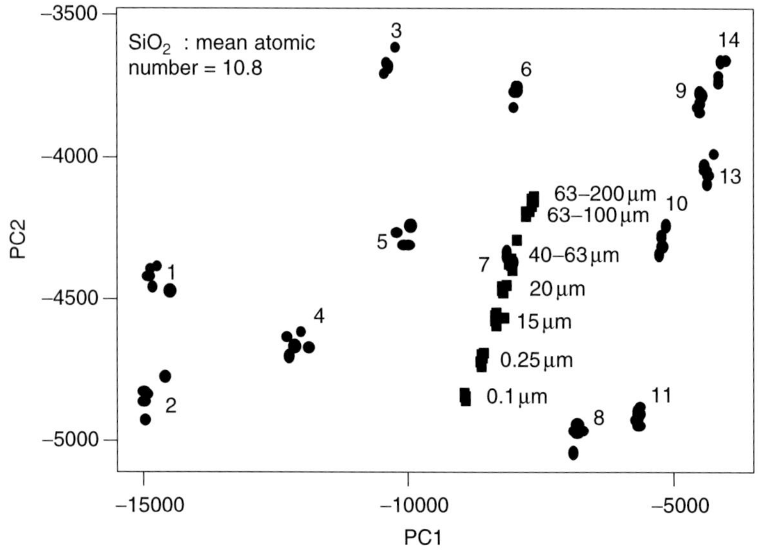


**Fig. 7.164.** Score plot of seven grain size fractions of $SiO_2$ displayed in a score plot of Ca compounds (method C)

**Table 7.23.** Compounds of calcium with the number in the PCA plot, the compound name, the Art. number (Merck), the mean atomic number and remarks (cg, compact-grained; cc, coarse-crystalline; c, clotted; p, packing)

| No. | Compound name | Art. no. | Mean atomic number | Remarks |
|---|---|---|---|---|
| 1 | calcium gluconate hydrate | 102094 | 8.05 | cg |
| 2 | calcium L-ascorbate dehydrate | 500579 | 8.15 | cg |
| 3 | calcium lactate pentahydrate | 102102 | 8.63 | cg |
| 4 | calcium D-saccharate | 102148 | 8.70 | cg |
| 5 | calcium diacetate acetic acid dehydrate | 102077 | 8.93 | c |
| 6 | Tri-calcium dicitrate tetrahydrate | 102092 | 9.80 | cg |
| 7 | calcium glycerophosphate hydrate | 104112 | 10.74 | cg |
| 8 | calcium hydrogenphosphate dehydrate | 102146 | 11.85 | cg |
| 9 | calcium carbonate (marble) | 105986 | 12.56 | cc |
| 10 | calcium carbonate | 102066 | 12.56 | cg |
| 11 | calcium hydrogenphosphate | 102144 | 13.08 | cg |
| 12 | tri-calcium phosphate | 102143 | 14.05 | cg |
| 13 | calcium hydroxide | 102110 | 14.30 | cg |
| 14 | calcium oxide | 102109 | 16.58 | c |
| 15 | calcium-Titriplex® dehydrate | 108439 | 8.58 | p |
| 16 | calcium nitrate tetrahydrate | 102121 | 9.68 | p |
| 17 | calcium sulfate dehydrate | 102160 | 12.11 | p |
| 18 | calcium sulfate dehydrate | 102161 | 12.11 | p |
| 19 | calcium chloride hexahydrate | 102072 | 12.72 | p |
| 20 | calcium fluoride | 102842 | 14.64 | p |
| 21 | calcium chloride dehydrate | 102383 | 15.43 | p |
| 22 | calcium chloride | 102389 | 15.43 | p |
| 23 | calcium bromide hydrate | 102055 | 18.09 | p |

All these container materials influence the XRF-spectra. In order to investigate the effect of the wall thickness on the reliable identification of chemical compounds, a solid titanium block was measured in eight PE containers of varying wall thickness (2.5 µm–2.5 mm) with "method A".

After calculating the PCA of the eight spectra the score plot was displayed in a PCA plot for 16 calcium compounds defined in Table 7.23 (Fig. 7.165). First the PCA of the samples with varying wall thickness was calculated. Using the resulting loadings a new PCA was calculated with the spectra of the Ca compounds. The scattering signals region was used as input variable for multivariate analysis. It is evident from Fig. 7.165 that the impact of the packing wall thickness in the interval 2.5 µm–2.5 mm is large enough to mix up different chemicals by measurements through the packing. However, a slight variation of the packing wall thickness (e.g., $1 \pm 0.1$ mm) should be acceptable to prevent a misclassification. Furthermore, one can derive that uncertainties

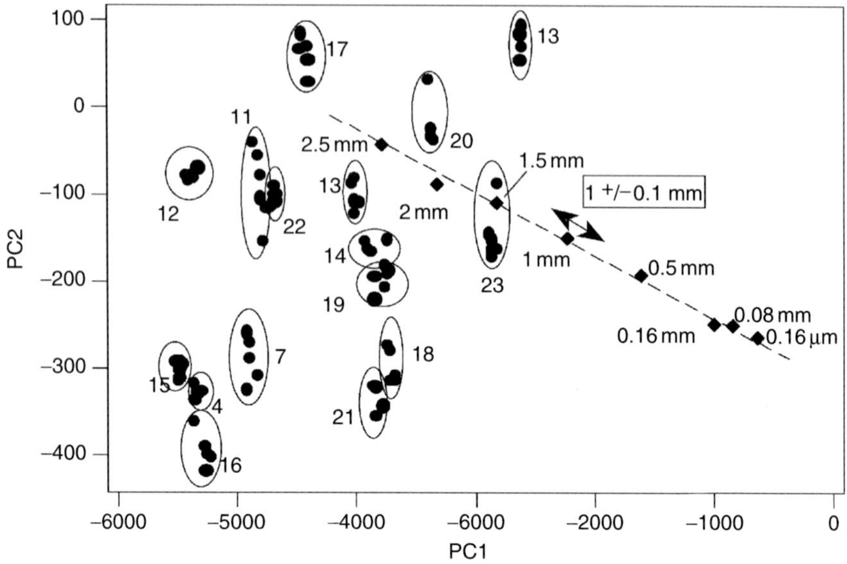


**Fig. 7.165.** Score plot for a titanium block with varying package wall thickness displayed in a score plot of Ca compounds (method A)

in the production of spectro cups have no effect on the identification or classification.

*Influence of Variable Positions*

In a commercial XRF apparatus the sample position is usually fixed strongly. That is important, as a precisely constant angle "excitation source-sample-detector" is inevitable for reliable comparisons of spectra. This strong constancy of the angle could not be established in "method A", as containers of various size and various form are to be examined. In this device the already mentioned sample consisting of a titanium block in a PE container (square reagent packing with rounded edges) was measured at different rotating positions. This influence is shown in Fig. 7.166. The score plot for measurements at different angles is displayed in a plot for calcium compounds (defined in Table 7.23). First the PCA was calculated for the Ti samples at different positions. Subsequently, using the resulting loadings, the PCA was calculated with the Ca compounds spectra. Only a slight rotation of 2.5° counter clockwise leads to a totally different position in the PCA score plot.

### 7.11.4 Discussion

The described investigations show that a fast non-invasive identification and classification of inorganic solid compounds is feasible by EDXRF-spectrum evaluation. As it was pointed out, the limit for the resolution of compounds

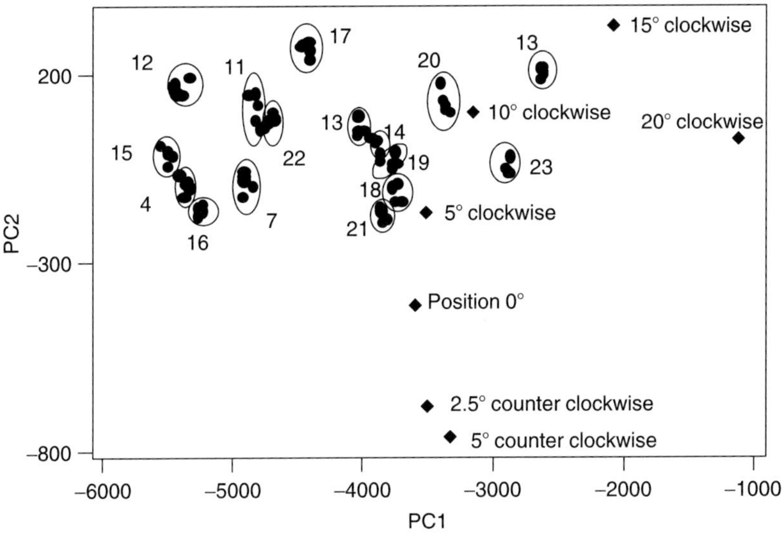


**Fig. 7.166.** Score plot of samples in varying positions displayed in a score plot of Ca compounds (method A)

in a PC score plot is at first a function of the differences in the mean atomic number. Beside interferences by the grain size, by changing water content of the compound, by inaccurate positioning of the sample, and by variations in the packing wall thickness, the resolution becomes better by increasing measuring time and, simultaneously, decreasing statistical uncertainty. In order to test the limits of "method B" 21 organic compounds, given in Table 7.24 (with mean atomic numbers in the range of 5.48 and 6.88) were measured fivefold. The resulting PCA plot is given in Fig. 7.167. The diagram shows that, principally, it is possible to separate regions for compounds with even very similar mean atomic numbers.

An example of the practical application of the fast identification procedure using XRF and subsequent statistical evaluation is the identity control of powdered chemicals in packing plants [937]. In this case a more simple method was used by comparing the measured spectra with those of a library.

Such non-destructive identification procedures can be applied, furthermore, in the characterization of e.g., alloys [938], organic halides, narcotics in packing materials, explosives in mail envelopes, and for pharmaceuticals.

The characterization of chemicals without unpacking from customers' return was successful by defining first the group. An identification of all the possible samples in one run was hard to realize, as many variables had to be taken into account, e.g., mean atomic number; grain size; humidity; position of the sample; size, thickness and form of the packing; packing material (e.g., glass, quartz, plastic (PE, PP), metal (Al, tin-plate), paper or cardboard). For this reason enormous expense was necessary to build up a reliable library.

**Table 7.24.** Organic compounds with the number in the PCA plot, the compound name, the article number (Merck Company), the mean atomic number, and remarks (cg, compact-grained; cc, coarse-crystalline)

| No. | Compound name | Art. no | Mean atomic number | Remarks |
|---|---|---|---|---|
| 1 | starch (triticitale) | 111046 | default | cg |
| 2 | cholesterin | 103672 | 5.48 | cg |
| 3 | sorbic acid | 100662 | 6.21 | cg |
| 4 | salicylic acid | 100631 | 6.47 | cg |
| 5 | salicylic acid | 818731 | 6.47 | cc |
| 6 | glutamin | 100289 | 6.50 | cg |
| 7 | ammonium acetate | 101116 | 6.55 | cc |
| 8 | L-glutaminic acid | 100290 | 6.65 | cg |
| 9 | D-glutaminic acid | 100291 | 6.65 | cg |
| 10 | urea | 108486 | 6.66 | cc |
| 11 | mannite | 108486 | 6.66 | cc |
| 12 | sorbite | 105988 | 6.66 | cg |
| 13 | sorbite | 103140 | 6.66 | cg |
| 14 | glycin | 500190 | 6.70 | cc |
| 15 | saccarose | 107654 | 6.70 | cg |
| 16 | glucose | 108337 | 6.73 | cg |
| 17 | glucose-monohydrate | 108346 | 6.77 | cg |
| 18 | ascorbic acid | 500069 | 6.86 | cc |
| 19 | ascorbic acid | 500074 | 6.86 | cg |
| 20 | ammonium carbonate | 101131 | 6.88 | cc |
| 21 | ammonium carbonate | 101136 | 6.88 | cg |

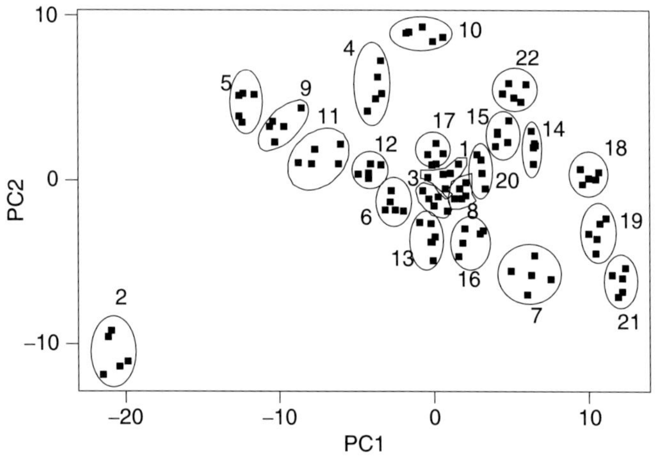


**Fig. 7.167.** Score plot of 21 organic compounds (method B, Ag secondary target, 30 s)

If the experimental relationship between the physical parameters of chemical compounds and the location of their spectra in the PCA data space is known, it should be possible to anticipate their position in the space of the library data set, thus predicting the spectra without further measurements.

## References


1. Carpenter DA, Improved laboratory X-ray source for microfluorescence analysis. X-Ray Spectrom **18**, 253–257 (1989)
2. Yamamoto Y, Hosokawa Y, Development of an innovative 5 μm focused X-ray beam energy-dispersive spectrometer and its applications. Japan J of Appl Phys **27**, L2203–L2206 (1988)
3. Wittry DB, Focussing X-rays for microprobe X-ray fluorescence analysis. In: Proceedings of the 50th Annual Meeting of the Electron Microscopy Society of America (EMSA/MAS), Boston (1992)
4. Janssens KHA, Adam FCV, Rindby A (eds), Microscopic X-Ray Fluorescence Analysis, Wiley, Chichester (2000)
5. Rindby A, Janssens KHA, Microbeam XRF In: Van Grieken RE, Markowicz AA (eds) Handbook of X-ray spectrometry, Second edition revised and expanded, Marcel Dekker Inc, New York, 631–718 (2002)
6. Yinming Y, Ding X, An investigation of X-ray fluorescence analysis with an X-ray focusing system (X-ray lens). Nucl Instrum Methods Phys Res **82B**, 121–124 (1993)
7. Carpenter DA, Taylor MA, Holcombe CE, Applications of a laboratory X-ray microprobe to materials analysis. Adv in X-Ray Anal **32**, 115–120 (1989)
8. Wherry CD, Cross BJ, Briggs TH, An automated X-ray microfluorescence materials analysis system. Adv X-Ray Anal **31**, 93–98 (1988)
9. Rindby A, Engström P, Janssens K, Osan J, Micro-distribution of heavy elements in highly inhomogeneous particles generated from μ-beam XRF/XRD analysis. Nucl Instrum Methods Phys Res **124B**, 591–604 (1997)
10. Haschke M, Scholz W, Theis U, Nicolosi J, Scruggs B, Herczeg L, Description of a new Micro-Xray-Spectrometer. J de Physique IV **12**, 83–96 (2002)
11. Bzhaumikhov AA, Langhoff N, Schmalz J, Wedell R, Beloglazov VI, Lebedev NF, Polycapillary collimator for micro-XRF. Proc SPIE **3444**, 430–435 (1998)
12. XOS, company information, http://www.xrayoptics.com/analyzer/Sindie%20Bench-top%2010-15.pdf.
13. Heckel H, Ryon D, Polarized beam X-ray fluorescence analysis. In: Van Grieken RE, Markowicz AA (eds) Handbook of X-ray spectrometry, Second edition revised and expanded, Marcel Dekker Inc, New York, 603–630 (2002)
14. Spectro, company instrument information XEPOS, http://www.spectro-ai.com/pages/e/p010503.htm.
15. Haschke, M, Haller M, Examination of polycapillary lenses for their use in micro-XRF spectrometers. X-Ray Spectrom **32**, 239–247 (2003)
16. Holynska B, Ostachowicz J, Ostrowski A, Ptasinski D, Wegrzynek D, J Trace Microprobe Techniq **13**, 163 (1995)

17. Bichlmeier S, Janssens K, Heckel J, Gibson D, Hoffmann P, Ortner HM Component selection for a compact micro-XRF spectrometer. X-Ray Spectrom **30**, 8–14 (2001)
18. Longoni A, Fiorini C, Leutenegger P, Sciuti, Fronterotta G, Strüder L, Lechner P, A portable XRF spectrometer fon non-destructive analyses in archeometry. Nucl Instrum Methods Phys Res **409A**, 407–409 (1998)
19. IfG, company instrument information iMOXS, http://www.ifg-adlershof.de/imoxs.pdf.
20. IXRF, company instrument information fx option, http://www.ixrfsystems.com/whatsnew.html.
21. Rindby A, Voglis P, Attaelmanan A, Analysis of inhomogeneous and irregularly shaped samples by the use of XRF micro-beam correlation analysis. X-Ray Spectrom **25**, 39–49 (1996)
22. Rödel C, Diplomarbeit TU Berlin (2000)
23. Rödel C, Bronk H, Haschke M, Investigation of the influence of particle size on the quantitative analyses of glasses by energy-dispersive micro x-ray fluorescence spectrometry. X-Ray Spectrom **31**, 16–26 (2002)
24. Haschke M, Klein J, Stichert W, Vietze U, High Throughput Experimentation (HTE) mit der µ-Röntgen-Fluoreszenz GIT Labor-Fachzeitschrift **7**, 780–783 (2002)
25. Haug GH, Günther D, Peterson LC, Sigman DM, Hughen KA, Aeschlimann B, Climate and the collapse of Maya civilization. Science **299**: 1731–1735 (2003)
26. de Stasio G, Casalbore P, Pallini R, Gilbert B, Sanita F, Ciotti MT, Rosi G, Festinesi A, Larocca LM, Rinelli A, Perret D, Mogk DW, Perfetti P, Mehta MP, Mercanti D, Gadolinium in Human Glioblastoma Cells for Gadolinium Neutron Capture Therapy. Cancer Res **61**, 4273 (2001)
27. Ioele, M, Instituto Centrale per il Restaurio, personal communication.
28. Meltzer C, King BS, Trace element analysis of solutions at the pbb level. Advances in X-Ray Anal **34**, 41–55 (1991)
29. Miller TC, Havrilla GJ, Nanodroplets: A new method for dried spot preparation and analysis. X-Ray Spectrom **33**, 101–106 (2004)
30. Kanngießer B, Malzer W, Reiche I, A new 3D micro X-ray fluorescence analysis set-up- first archaeometric applications. Nucl Instrum Methods Phys Res **211/2B**, 259–264 (2003)
31. Kumakhov MA, Capillary optics and their use in x-ray analysis. X-Ray Spectrom **29**, 343–348 (2000)
32. Ding X, Gao N, Havrilla GJ, Advances in laboratory-based X-ray sources and optics. Proc SPIE **4144**, 174–182 (2000)
33. Fiorini C, Longoni A, Bjeoumikhov A, A new detection system with polycapillary conic collimator for high-localized analysis of x-ray fluorescence emission. IEEE Trans Nucl Sci **48**, 268–271 (2001)
34. Proost K, Janssens K, Vincze L, Falkenberg G, Gao N, Bly P, Feasibility and characteristics of confocal micro-XRF at Hasylab BL L. HASYLAB Jahresbericht 2002, available online (2003)
35. Vincze L, Vekemans B, Brenker FE, Falkenberg G, Rickers K, Somogyi A, Kersten M, Adams F, Three-dimensional trace element analysis by confocal X-ray microfluorescence imaging. Anal Chem **76**, 6786–6791 (2004)
36. Kanngießer B, Malzer W, Rodriguez AF, Reiche I, 3D micro-XRF Investigations of Paint Layers with a tabletop set-up. Spectrochim Acta B **60**, 41–47 (2005)

37. Elam WT, Ravel BD, Sieber JR, A new atomic database for x-ray spectroscopy calculations. Rad Phys and Chem **63**, 121–128 (2002)
38. Henke BL, Gullikson EM, Davis JC, X-ray interactions: photoabsorption, scattering, transmission, and reflection at E = 50–30000 eV, Z = 1–92. At and Nucl Data Tables **54**, 181-342 (1993)
39. McMaster WH, Kerr del Grande N, Mallet JH, Hubbel JH, Compilation of x-ray cross sections, Lawrence Radiation Laboratory, University of California (1969)
40. Boisseau P, Ph.D. thesis, Massachusetts Institute of Technology (1986)
41. Janssens K, Vincze L, Vekemans B, Adams F, Haller M, Knöchel A, The use of ellipsoidal lead-glass capillaries for micro-focussing of highly energetic (0-60 keV) synchrotron radiation. J Anal Atom Spectrom **13**, 339–350 (1998)
42. Simionovici AS, Chukalina M, Schroer CG, Drakopoulos M, Snigirev A, Snigireva I, Lengeler B, Janssens K, Adams F, High-resolution X-ray fluorescence microtomography of homogeneous samples. IEEE Trans Nucl Sci **47**, 2736–2740 (2000)
43. Golosio B, Simionovici A, Somogyi A, Lamelle L, Chukalina M, Brunetti A, Internal elemental microanalysis combining x-ray fluorescence, Compton and transmission tomography. J Appl Phys **94**, 145–156 (2003)
44. Castaing R, Slodzian GJ, Microscopie **1**, 395 (1962)
45. Michiels F, Van Hoolst W, Van Espen P, Adams F, J Am Soc Mass Spectrom **1**, 37 (1990)
46. Denoyer E, Van Grieken R, Adams F, Natush D, Anal Chem **26A**, 54 (1982)
47. Adams F, Mauney T, Adv Mass Spectrom **10**, 507 (1986)
48. Goldstein SI, Newbury DE, Echlin P, Joy DC, Fiori C, Lifshin E Scanning electron microscopy and X-ray microanalysis, Plenum Press, New York (1984)
49. Cookson JA, Principles & Applications of High-energy Ion Microbeams, (eds), Watt F and Grime GW, Adam Hilger, Bristol (1987)
50. Grime GW, Dawson M, Marsh M, McArthur IC, Watt F, The Oxford submicron nuclear microscopy facility. Nucl Instrum Methods in Phys Res Part B **54**, 52–63 (1991)
51. Jamieson DN, New generation nuclear microprobe systems. Nucl Instrum Methods in Phys Res Part B **181**, 1–11 (2001)
52. Bilderback D, Hoffman SA, Thiel D, Science **263**, 201 (1994)
53. Hoffman SA, Thiel DJ, Bilderback DH, Nucl Instrum Methods A **347**, 384 (1994)
54. Kirkpatrick P, Baez A, J Opt Soc Am **38**, 766 (1948)
55. Suzuki Y, Uchida F, Rev Sci Instrum **63**, 578 (1992)
56. Hignette O, Rostaing G, Cloetens P, Rommeveaux A, Ludwig W, Freund A, In McNulty I, (ed.), X-ray Micro- and Nano-Focusing: Applications and Techniques II, 4499 of Proceedings of the SPIE, 105 (2001)
57. Hignette O, Cloetens P, Lee WK, Ludwig W, Rostaing G, J Phys IV **104**, 231 (2003)
58. Niemann B, Rudolph G, Schmahl G, Opt Commun **12**, 160 (1974)
59. Lai B, Yun W, Legnini D, Xiao Y, Chrzas J, Viccaro P, White V, Bajikar S, Denton D, Cerrina F, Fabrizio E, Gentili M, Grella L, Baciocchi M, Appl Phys Lett **61**, 1877 (1992)
60. Aristov VV, Basov YA, Kulipanov GN, Pindyurin VF, Snigirev AA, Sokolov AS, Focusing properties of a Bragg–Fresnel lens in the white spectrum of synchrotron radiation. Nucl Instrum Methods in Phys Res Part A **274**, 390–393 (1989)

61. Chevallier P, Dhez P, Legrand F, Erko A, Agafonov Yu, Panchenko LA, Yakshin A, J Trace Microprobe Tech **14**(3), 517–539 (1996)
62. Snigirev A, Kohn V, Snigireva I, Lengeler B, Nature (Lond) **384**, 49 (1996)
63. Lengeler B, Schroer CG, Richwin M, Tümmler J, Drakopoulos M, Snigirev A, Snigireva I, Appl Phys Lett **74**, 3924 (1999)
64. Lengeler B, Schroer CG, Kuhlmann M, Benner B, Günzler TF, Kurapova O, Rau C, Weitkamp T, Simionovici AS, Snigirev A, Snigireva I, Beryllium parabolic refractive X-ray lenses for full field imaging and scanning microscopy with hard X-rays, J Phys IV **104**, 221–224 (2003)
65. Nöhammer B, David C, Rothuizen H, Hoszowska J, Simionovici A, Deep reactive ion etching of silicon and diamond for the fabrication of planar refractive hard X-ray lenses, Microelectr Eng **67–68**, 453–460 (2003)
66. Schroer CG, Kuhlmann M, Hunger UT, Gunzler TF, Kurapova O, Feste S, Frehse F, Lengeler B, Drakopoulos M, Somogyi A, Simionovici AS, Snigirev A, Snigireva I, Schug C, Schröder WH, Nanofocusing parabolic refractive X-ray lenses, Appl Phys Lett **82**(9), 1485–1487 (2003)
67. Grigorieva IG, Antonov AA, X-Ray Spectrom **32**, 64–68 (2003)
68. Chevallier P, Populus P, private communication, LURE (2000)
69. Mason B, Principles of Geochemistry, Wiley, New York (1952)
70. Bohic S, Simionovici A, Ortega R, Snigirev A, Synchrotron hard X-Ray microprobe for fluorescence imaging of single cells. Appl Phys Lett **78**, 3544 (2001)
71. Vekemans B, Janssens K, Vincze L, Adams F, Van Espen P, X-ray Spectrom **23**, 278 (1994)
72. See Section 2.4 Synchrotron Radiation Sources, (66–80). Package available at http://www.esrf.fr/machine/support/ids/Public/index.html
73. Ortega R, Moretto P, Fajac A, Benard J, Llabador Y, Simonoff M, Cell Mol Biol **42**, 77 (1996)
74. Boisseau P, Ph.D. Dissertation, MIT, Cambridge (1986)
75. Rust G-F, Weigelt J, IEEE Trans Nucl Sci **45**, 75 (1998)
76. Simionovici A, Chukalina M, Schroer Ch, Drakopoulos M, Snigirev A, Snigireva I, Lengeler B, Janssens K, Adams F, High resolution X-ray fluorescence tomography on homogenous samples, IEEE Trans Nucl Sci **47**(6), 2736 (2000)
77. Chukalina M, Simionovici A, Snigirev A, Jeffries T, Quantitative characterization of microsamples by X-ray fluorescence tomography, X-Ray Spectrom **31**(6), 448–450 (2002)
78. Golosio B, Simionovici A, Somogyi A, Lemelle L, Chukalina M, Brunetti A, J Appl Phys **94**, 145–157 (2003)
79. Kak C, Slaney M, Principles of Computerized Tomographic Imaging, IEEE Press, New York (1988)
80. Hogan JP, Gonsalves RA, Krieger AS, IEEE Trans Nucl Sci **38**, 1721 (1991)
81. Brunetti A, Golosio B, Comp Phys Comm **141**, 412 (2001)
82. Sautter V, Barrat JA, Jambon A, Lorand JP, Gillet Ph, Javoy M, Joron JL, Lesourd M, A new martian meteorite from Morocco: the nakhlite North West Africa 817. Earth Planet Sci Lett **195**, 223–238 (2002)
83. Fabry L, Matsushita Y, in Semiconductor Silicon/1998 (8th International Symposium), Huff HR, Tsuya H, and Gösele U, (eds), San Diego, California, Spring 1998, Proc Electrochem Soc Vol. **98-1**, 1459 (1998)
84. Fabry L, Trace analytical performance, epistemology and engineering creativity. Fresenius J Anal Chem **357**, 148–150 (1997)

85. Fabry L, Köster L, Pahlke S, Kotz L, Hage J, IEEE Trans Semicond Manuf **9**, 428 (1996)
86. Wobrauschek P, Kregsamer P, Ladisich W, Streli C, Pahlke S, Fabry L, Garbe SA, Haller M, Knöchel A, Radtke M, Nucl Instrum Methods A **363**, 619 (1995)
87. Fucsko J, Tan SH, Balazs MK, J Electrochem Soc **114**, 1105 (1993)
88. Fabry L, Accreditation and Quality Assurance, Springer, Berlin Heidelberg, New York, **1**, 99 (1996)
89. Thompson M, A natural history of analytical methods, Analyst **124**, 991 (1999)
90. Huber A, Böhm G, Pahlke S, J Radioanal Nucl Chem **169**, 93 (1993)
91. Schnürer-Patschan Ch, Hrosch W, Pahlke S, Kotz L, Fabry L, in: Crystalline Defects and Contamination Control: Their Impact and Control in Device Manufacturing II, Kolbesen BO, P. Stallhofer P, Claeys CL, and Tardiff F, (eds), Paris, France - September 1997 Proc The Electrochem Soc Vol. **97-22**, 496 (1997)
92. Wenz B, Sperling M, Atomabsorptionsspektrometrie, Wiley-VCH, (1997)
93. Wobrauschek P, Görgl R, Kregsamer P, Streli Ch, Pahlke S, Fabry L, Haller M, Knöchel A, Radtke, Spectrochim Acta, Part B **52**, 901 (1997)
94. International Technology Roadmap for Semiconductors, Semiconductor Industry Association, San Jose CA, 95110 (1999)
95. Yoneda Y, Horiuchi T, Rev Sci Instrum **42**, 1069 (1971)
96. Streli C, X-Ray Spectrom **29**, 203 (2000)
97. Kregsamer P, Spectrochim Acta B **46**, 1333 (1991)
98. Aiginger H, Spectrochim Acta B **46**, 1313 (1991)
99. Aiginger H, Wobrauschek P, Nucl Instrum Methods **114**, 157 (1974)
100. Knoth J, Schwenke H, Fresenius' Z. Anal Chem **291**, 200 (1978)
101. Schwenke H, Knoth J, Nucl Instrum Methods **193**, 239 (1982)
102. Becker RS, Golovchenko JA, Patel JR, Phys Rev Lett **50**, 153 (1983)
103. Streli C, Wobrauschek P, Bauer V, Kregsamer P, Görgl R, Pianetta P, Ryon R, Pahlke S, Fabry L, Spectrochim Acta Part B **52**, 861 (1997)
104. Klockenkämper R, von Bohlen A, A new method for depth profiling of shallow layers in silicon wafers by repeated chemical etching and total reflection X-ray fluorescence analysis. Spectrochim Acta B **54**, 1385–1392 (1999)
105. Fabry L, Pahlke S, Kotz L, Wobrauschek P, Streli Ch, Eichinger P, Fresenius' J Anal Chem **363**, 98 (1999)
106. Xiao X, Hayakawa S, Goshi Y, Oshima M, Anal Sci (Jpn) **14**, 1139 (1998)
107. deBokx PK, Kidd SJ, Wiener G, Urbach HP, De Gendt S, Mertens PW, Heyns MM, I: Silicon Mater Sci Technol, Huff H, Tsuya H, Gösele U, (eds), The Electrochem Soc, PV **98-1**, p. 1511 (1998)
108. Schwenke H, Knoth J, Eur Patent Appl, 199.24 204.6 05/27/2000
109. Wobrauschek P, Streli C, Advanced Light Sources, ADLIS, TU Wien, 284 (1999)
110. Yakushiji K, Ohkawa S, Yoshinaga A, Harada J, Jpn J Appl Phys **32**, 1191 (1993)
111. Yakushiji K, Ohkawa S, Yoshinaga A, Harada J, Jpn J Appl Phys **33**, 1130 (1994)
112. Liou BW, Lee CL, Chin J Phys **37**, 623–30 (1999)
113. Klockenkämper R: Total-Reflection X-Ray Fluorescence Analysis, Vol. **140**, Chem Anal, Winefordner JD, (ed.), Wiley (1997)

114. ATOMIKA, Sales Department, Bruckmannring 40, D-85764 Oberschleissheim, Germany
115. Szaloki I, Török SzB, Chul-Un Ro, Injuk J, Van Grieken RE, Anal Chem **72**, 211R (2000)
116. Strüder L, Meidinger N, Stotter D, Kemmer J, Lechner P, Leutenegger P, Soltau H, Eggert F, Rohde M, Schülein T, Microscopy Microanal **4**, 622–31 (1998)
117. Tsuji K, Yamada T, Utaka T, Hirokawa K, Appl J Phys **78**, 969 (1995)
118. Knoth J, Beaven P, Schwenke H, Dobler M, Reus U, Spectrochimi Acta B **56**, 2275–2281 (2001)
119. Standard Test Method M 33: Test Method for the Determination of Residual Surface Contamination on Silicon Wafers, Semiconductor Equipment and Materials International, San Jose, CA 95134, attached to Chapter 8
120. Streli Ch, Kregsamer P, Wobrauschek P, Gatterbauer H, Pianetta P, Pahlke S, Fabry L, Palmetshofer L, Schmeling M, Spectrochim Acta Part B **54**, 1433 (1999)
121. Yakushiji K, Ohkawa S, Yoshinaga A, Harada J, Anal Sci (Jpn) **11**, 505 (1995)
122. Knoth J, Schwenke H, Eichinger P, In: Proc. 2nd International Symp. Ultra Clean Processing of Silicon Surfaces (UCPSS '94 IMEC, 19–21 September, 1994, Bruges), Heyns M, Heuris M, Mertens P, (eds), ACCO Leuven, 107 (1994)
123. Klockenkämper R, von Bohlen A, Spectrochim Acta B **44**, 461 (1989)
124. Mori Y, Shimanoe K, Sakon T, Anal Sci (Jpn) **11**, 499 (1995)
125. Rink I, Rostam-Khani P, Knoth J, Schwenke H, de Gendt S, Wortelboer R, Calibration of straight total reflection x-ray fluorescence spectrometry - results of a European Round Robin test, Spectrochim. Acta B **56**, 2283–2292 (2001)
126. Mori Y, Shimanoe K, Anal Sci (Jpn) **12**, 277 (1996)
127. Hölzl R, Range K-J, Fabry L, Huber D, J Electrochem Soc **146**, 2245 (1999)
128. Hackl B, Range K-J, Stallhofer P, Fabry L, J Electrochem Soc **139**(5), 1495 (1992)
129. Fabry L, Pahlke S, Kotz L, Schmemmel E, Berneike W, Crystalline Defects and Contamination: Their Impact and Control in Device Manufacturing, Kolbesen BO, Stallhofer P, Cor Claeys, Tardif F. eds., Proc Vol. **93-15**, 232 (1993)
130. Diebold AC, J Vac Sci Technol A **14**, 1919 (1996)
131. Hockett RS, Anal Sci (Jpn) **11**, 511 (1995)
132. Eichinger P, Rath H-J, Schwenke H, In: Semicond. Fabrication: Technology and Metrology, ASTM STP 990, Gupta DC, (ed.), ASTM 305 (1989)
133. Penka V, Hub W, Fresen Z Anal Chem **333**, 586 (1989)
134. Sees JA, Hall LH, J Electrochem Soc **142**, 1238 (1995)
135. Schwenke H, Knoth J, Weisbrod U, X-Ray Spectrom **20**, 277 (1991)
136. Weisbrod U, Gutschke R, Knoth J, Schwenke H, Fresenius' J Anal Chem **341**, 83 (1991)
137. Berneike W, Spectrochim Acta B **48**, 269 (1993)
138. Schwenke H, Gutschke R, Knoth J, Kock M, Appl Phys A **54**, 460 (1992)
139. Yamagami M, Nonoguchi M, Yamada T, Shoji T, Utaka T, Nomura S, Taniguchi K, Wakita H, Ikeda S, VDP/TXRF analysis of trace elements on a silicon wafer, X-Ray Spectrom **28**, 451–455 (1999)
140. Schwenke H, Knoth J, Total Reflection XRF in: Van Grieken RE, Markowicz AA (eds) Handbook on X-Ray Spectrometry, Marcel Dekker, New York, 453–490 (1993)

141. Schwenke H, Frey L, Pahlke S, Fabry L, J Electrochem Soc **144**, 3979 (1997)
142. Klockenkämper R, von Bohlen A, Becker HW, Palmetshofer L, Surf Interface Anal **27**, 1003 (1999)
143. Klockenkämper R, von Bohlen A, J Anal At Spectrom **14**, 571 (1999)
144. Klockenkämper R, von Bohlen A, Anal Commun **36**, 27 (1999)
145. Schwenke H, Knoth J, Total Reflection XRF in: Van Grieken RE, Markiwicz AA (eds) Handbook on X-Ray Spectrometry, Marcel Dekker, New York, 453–490 (1993)
146. Weisbrod U, Gutschke R, Knoth J, Schwenke H, Fresen Z Anal Chem **341**, 83 (1991)
147. Weisbrod U, Gutschke R, Knoth J, Schwenke H, Appl Phys **A53**, 449 (1991)
148. Knoth J, Schwenke H, Weisbrod U, Spectrochim Acta **44B**(5), 477 (1989)
149. Schwenke H, Gutschke R, Knoth R, Adv X-Ray Anal **35**, 941 (1992)
150. Knoth J, Bormann R, Gutschke R, Michaelsen C, Schwenke H, Spectrochim Acta **48B**, 285 (1993)
151. Schwenke H, Bormann R, Knoth J, Prange A, Spectrochim Acta **48B**, 293 (1993)
152. De Boer DKG, Van Den Hoogenhof WW, Adv X-Ray Anal **34**, 35 (1991)
153. De Boer DKG, Van Den Hoogenhof WW, Spectrochim Acta **46B**, 1323 (1991)
154. de Boer DKG, Leenaers AJG, Van den Hoogenhof WW, The profile of layered materials reflected by glancing incidence X-ray analysis. Appl Phys A **58**, 169–172 (1994)
155. de Boer DKG, Leenaers AJG, Van den Hoogenhof WW, Glancing incidence X-ray analysis of thin-layered materials: A review. X-Ray Spectrom **24**, 91–102 (1995)
156. Torcheux L, Degraeve B, Mayeux A, Delamar M, Calibration procedure for quantitative surface analysis by total-reflection X-ray fluorescence. SIA, Surf Interface Anal **21**, 192–198 (1994)
157. Mori Y, Shimanoe K, Sakon T, Adv X-ray Chem Anal Jpn **26s**, 69 (1995)
158. Fabry L, Pahlke S, Kotz L, Adachi Y, Furukawa S, Adv X-ray Chem Anal Jpn **26s**, 19 (1995)
159. Gutschke R, Diploma Thesis, University of Hamburg (1991)
160. Schwenke H, Knoth J, Adv X-ray Chem Anal Jpn **26s**, 137 (1995)
161. Weisbrod U, Gutschke R, Knoth J, Schwenke H, Appl Phys A **53**, 449 (1991)
162. Schwenke H, Gutschke R, Knoth J, Adv X-Ray Anal **35**, 941 (1992)
163. de Boer DKG, Spectrochim Acta B **46**, 1433 (1991)
de Boer DKG, van den Hoogenhof WW, Adv X-ray Anal **34**, 35 (1991)
164. de Boer DKG, van den Hoogenhof WW, Spectrochim Acta B **46**, 1323 (1991)
165. de Boer DKG, Phys Rev B **44**, 498 (1991)
166. de Boer DKG, van den Hoogenhof WW, Adv X-ray Anal **34**, 35 (1991)
167. van den Hoogenhof WW, de Boer DKG, Spectrochim Acta B **48**, 277 (1993)
168. Stoev KN, Sakurai K, Spectrochim Acta B **54**, 41 (1999)
169. Standard Test Method E 45: Test Method for the Determination of Inorganic Contamination form Minienvironments, Semiconductor, Semiconductor Equipment and Materials International, San Jose CA 95134
170. Shiraiwa T, Fujino N, Sumita S, Tanizoe Y, in: Semiconductor Fabrication: Technology and Metrology ASTM STP 990, Gupta DC (ed.), ASTM, 314 (1989)

171. Huber A, Rath H-J, Eichinger P, Bauer T, Kotz L, Staudigl R, in: Diagnostic Techniques for Semiconductor Materials and Devices, Schaffner TJ, Schroder DK (eds), Proc Vol **88–20**, 109 (1988)
172. Neumann C, Eichinger P, Spectrochim Acta B **467**, 1369 (1991)
173. Buhrer G, Application of vapor phase decomposition/total reflection X-ray fluorescence in the silicon semiconductor manufacturing environment, Spectrochim Acta B **54**, 1399–1407 (1999)
174. Diebold AC, Maillot P, Gordon M, Baylis J, Chacon J, Witowski R, Arlinghaus HF, Knapp JA, Doyle BL, J Vac Sci Technol A **10**, 2942 (1992)
175. Metz S, Kilian G, Mainka G, Angelkort C, Rittmeyer C, Stelter H, Fester A, Kolbesen BO, in: Crystalline Defects and Contamination Control: Their Impact and Control in Device Manufacturing II, Kolbesen BO, P. Stallhofer P, Claeys CL, and Tardiff F (eds), Paris, France - September 1997 Proc Electrochem Soc, Vol **97–22**, 458 (1997)
176. Fabry L, Pahlke S, Kotz L, Rüfer H, Ehmann T, Bächmann K, Crystalline Defects and Contamination Control: Their Impact and Control in Device Manufacturing II, Kolbesen BO, P. Stallhofer P, Claeys CL, Tardif F (eds), Paris, France - September 1997 Proc Electrochem Soc, Vol **97–22**, 468 (1977)
177. GeMeTec GmbH, D-81379 München
178. Fabry L, Pahlke S, Kotz L, Fresenius' J Anal Chem **354**, 266 (1996)
179. Kondo H, Ryuta J, Morita E, Yoshimi T, Shimanuki Y, Jpn J Appl Phys, Part 2, 1A/B **31**, L11 (1992)
180. Tölg G, Analyst **112**, 365 (1987)
181. Yamaguchi H, Itoh S, Igarashi S, Naitoh K, Hasegawa R, Anal Sci (Jpn) **14**, 909 (1998)
182. Streli C, Bauer V, Wobrauschek P, Adv X-ray Anal **39**, 771–779 (1997)
183. Fukuda T, Shoji T, Funabashi M, Utaka T, Arai T, Miyazaki K, Shimazaki A, Wilson R, Adv X-ray Anal **39**, 781 (1997)
184. http://www.rigaku.com/Semi-XRF/txrf3750
185. Funabashi M, Matsuo M, Kawada N, Yamagami M, Wilson R, Spectrochim Acta **B543**, 1409 (1999)
186. http://www. Gemetec.de
187. Proceedings of the 8th Conference on TXRF and related Techniques, Vienna September (2000)
188. Ryon RW, Zahrt JD, Polarized Beam X-Ray Fluorescence in: Van Grieken RE, Markowicz AA (eds) Handbook of X-Ray Spectrometry, Marcel Dekker, New York, 491–516 (1993)
189. Aiginger H, Wobrauschek P, Brauner C, Nucl Instrum Methods **120**, 541 (1974)
190. Wobrauschek P, Streli C, X-ray and Inner-Shell Processes, (eds), Johnson RL, Schmidt-Böcking H, Sonntag B, AIP Conf Proc **389**, 233 (1997)
191. Bilderback DH, Lairson BM, Barbee Jr TW, Ice GE, Sparks Jr CJ, Nucl Instrum Methods **208**, 251 (1983)
192. Pianetta P, Barbee Jr TW, Applications of multilayers to synchrotron radiation, Nucl Instrum Methods in Phys Res A **266**, 441–446 (1988)
193. Rieder R, Wobrauschek P, Ladisich W, Streli C, Aiginger H, Garbe S, Gaul G, Knöchel A, Lechtenberg F, Nucl Instrum Methods **355**(2,3), 648 (1995)
194. Wobrauschek P, Kregsamer P, Ladisich W, Rieder R, Streli C, Adv X-ray Anal **39**, 755 (1997)
195. Görgl R, Wobrauschek P, Kregsamer P, Streli C, Haller M, Knöchel A, Radtke M, X-ray Spectrom **26**(4), 189 (1997)

196. Madden M, Wherry D, Pianetta P, Brennan S, Mat Res Soc Symp Proc **307**, 125 (1993)
197. Streli C, Wobrauschek P, Ladisich W, Rieder R, Aiginger H, Ryon R, Pianetta P, Nucl Instr Meth **A345**, 399 (1994)
198. Streli C, Kregsamer P, Wobrauschek P, Gatterbauer, Pianetta P, Pahlke S, Fabry L, Palmetsdorfer L, Schmeling M, Low Z total reflection analysis - challenges and answers. Spectrochim Acta **54B**(10), 1433–1441 (1999)
199. Streli C, Wobrauschek P, Kregsamer P, Pepponi G, Pianetta P, Pahlke S, Fabry L, Spectrochim Acta **56B**, 2085–2094 (2001)
200. Beckhoff B, Fliegauf R, Ulm G, Pepponi G, Streli C, Wobrauschek P, Fabry L, Pahlke S, Spectrochim Acta B **56**, 2073–2083 (2001)
201. Brennan S, Tompkins W, Takaura N, Pianetta P, Laderman SS, Fischer-Colbrie A, Kortright JB, Madden MC, Wherry DC, Nucl Instrum Methods **347**, 417 (1994)
202. Laderman SS, Fischer-Colbrie A, Shimazaki A, Miyazaki K, Brennan S, Takaura N, Pianetta P, Kortright JB, Anal Sci **11**, 515 (1995)
203. Pianetta P, Takaura N, Brennan S, Tompkins W, Laderman SS, Fischer-Colbrie A, Shimazaki A, Miyazaki K, Madden M, Wherry DC, Kortright JB, Rev Sci Instrum **66**, 1293 (1995)
204. Singh A, Baur K, Brennan S, Homma T, Kubo N, Pianetta P, Mater Res Soc Symp Proc **716**, 23 (2002)
205. Winick H, Doniach S, Synchrotron Radiation Research, Plenum, New York (1979)
206. Takaura N, Brennan S, Pianetta P, Laderman SS, Fischer-Colbrie A, Kortright JB, Wherry DC, Miyazaki K, Shimazaki A, Adv X-ray Chem Anal **26s**, 113 (1995)
207. Wang J, Balazs MK, Pianetta P, Baur K, Brennan S, Bonne T, Rosamila JM, The Limit of Transition Metal Detection on Silicon Wafers. Semiconductor Fabtech **16**, 195–198 (2002)
208. Bianconi A, in: X-ray Absorption: Principles, Applications, Techniques of EXAFS, SEXAFS, and XANES, edited by Koningsberger DC, Prins R Wiley, New York, 618–624 (1988)
209. Homma T, Chidsey W, J Phys Chem B **102** 7919 (1998)
210. Baur K, Kerner J, Brennan S, Singh A, Pianetta P, J Appl Phys **88**, 4642 (2000)
211. Baur K, Brennan, Burrow B, Werho D, Pianetta P, Spectrochim Acta B **56**, 2049 (2001)
212. Ortega L, Comin F, Formoso V, Stierle A, J Synchr Radat **5**, 1064–1066 (1998)
213. Beckhoff B, Klein R, Krumrey M, Scholze F, Thornagel R, Ulm G, Nucl Instrum Methods A **444**, 480–483 (2000)
214. Senf F, Eggenstein F, Flechsig U, Gudat W, Klein R, Rabus H, Ulm G, J Synchr Rad **5**, 780–782 (1998).
215. Streli C, Wobrauschek P, Beckhoff B, Ulm G, Fabry L, Pahlke S, X-Ray Spectrom **30**, 24–31 (2001)
216. Beckhoff B, Fliegauf R, Ulm G, Weser J, Pepponi G, Streli C, Wobrauschek P, Ehmann Th, Fabry L, Mantler C, Pahlke S, Kanngießer B, Malzer W, Ultratrace Analysis of Light Elements and Speciation of Minute Organic Contaminants on Silicon Wafer Surfaces by means of TXRF in Combination with

NEXAFS in: Analytical and Diagnostic Techniques for Semiconductor Materials, Devices and Processes, Kolbeson BO, P. Stallhofer P, Claeys CL, and Tardiff F (eds), Proc Electochem Soc Vol **2003–03**, 120–128 (2003)
217. Scholze F, Procop M, X-Ray Spectrom **30**, 24–31 (2004)
218. Beckhoff B, Fliegauf R, Ulm G, Spectrochim Acta Part B **58**, 615–626 (2003)
219. Bechstein S, Beckhoff B, Fliegauf R, Weser J, Ulm G, Spectrochim Acta Part B **59**, 215–221 (2004)
220. Streli C, Pepponi G, Wobrauschek P, Beckhoff B, Ulm G, Pahlke S, Fabry L, Ehmann Th, Kanngießer B, Malzer W, Jark W, Spectrochim Acta B **58**, 2113–2121 (2003)
221. Pepponi G, Beckhoff B, Ehmann T, Ulm G, Streli C, Fabry L, Pahlke S, Wobrauschek P, Spectrochim Acta B **58**, 2245–2253 (2003)
222. Török S, Osán J, Beckhoff B, Ulm G, J Diffraction **19**, 81–86 (2004)
223. Kolbe M, Beckhoff B, Krumrey M, Ulm G, Spectrochim Acta Part B **60**, 505–510 (2005)
224. Broll N, Tertian R, X-Ray Spectrom **12**, 30 (1983)
225. Van Dyck PM, Török S, Van Grieken R, Anal Chem **58**, 1761 (1986)
226. Bertin EP, Principles and Practice of X-ray Spectrometric Analysis, Plenum, New York (1970)
227. Jenkins R, Gould RW, Gedcke D, Quantitative X-Ray Spectrometry, Marcel Dekker, New York (1981)
228. Behncke HH, Metal Finishing **83**, 33 (1984)
229. Ferrandino F, Metal Finishing **85**, 29 (1986)
230. Shan R, Sandberg A, Advances in X-ray Analysis Plenum, New York, vol. **26**, (1983)
231. Colliex C, Proc 8th European Congress on Electron Microscopy 57 (1984)
232. Pierce TB, J Trace Microprobe Tech **2**, 291 (1984/1985)
233. Palmetshöfer L, in: Surface and Thin Film Analysis, Wiley-VCH, Weinheim, (2002)
234. Product of Helmut-Fischer GmbH + Co. KG, Sindelfingen, Germany.
235. de Boer DKG, X-Ray Spectrom **19**, 145 (1990)
236. Van Espen PJM, Janssens KHA, Spectrum Evaluation in: Van Grieken RE, Markowicz AA (eds) Handbook of X-Ray Spectrometry, Marcel Dekker, New York 181–294 (1993)
237. Press WH, Teukolski SA, Vetterling WT, Flannery BP, Numerical Recipes in C, Cambridge University Press (1992)
238. Rößiger V, Kaiser K-H, in: Jahrbuch Oberflächentechnik 1998, Metall Verlag Heidelberg, 313 (1998)
239. Fischer Application Report vr9605
240. Fischer Application Report vr0227
241. Rößiger V, Galvanotechnik 92, **8**, 2070 (2001)
242. Taylor BN, Kuyatt CE, NIST Technical Note 1297, 1994 Edition (Supersedes 1993 Edition)
243. DIN 1319, Part 3, Grundbegriffe der Messtechnik.
244. DIN 55350–13, Begriffe zur Genauigkeit von Ermittlungsverfahren und Ermittlungsergebnissen.
245. Barrentine LB, Concepts for R&R Studies, ASQ Quality Press (2003)
246. Weise K, Wöger W, Messunsicherheit und Messdatenauswertung, Wiley-VCH, Weinheim (1999)

247. Fischer Application Report vr0103
248. Fischer Application Report vr0213
249. Neumaier P, Schichtdickenmessung an kleinen Teilen. QZ-Qualität and Zuverlässigkeit **48(05)**, 493–495 (2003)
250. Jansen R, Preischkat P, Metalloberfläche München **56**, 4, 26 (2002)
251. Hülser P, Metalloberfläche München **57**, 7–8, 18 (2003)
252. Rößiger V, JOT, Issue **7**, 62 (2002)
253. Stöckert D, Stahl, Issue **6**, 70 (2002)
254. Simon F, Manz U, Galvanotechnik **94**, 2169 (2003)
255. Kaiser H, (Hrsg.) Edelmetallschichten, Leuze Verlag, Bad Saulgau, 47 (2002)
256. Fischer Application Report vr0223
257. Wingenfeld P, Galvanotechnik **94**, 11 2664 (2003)
258. Fischer Application Report vr0304
259. Rößiger V, Galvanotechnik **95**, 888 (2004)
260. Rößiger V, Metalloberfläche München **53**, 4 (1999)
261. Kallithrakas-Kontos N, TXRF heavy metal analysis after immobilization as PAN Complexes. X-Ray Spectrom **33**, 12–15 (2004)
262. Van Espen P, Nullens H, Adams F, Method for accurate description of full energy peaks in nonlinear least-square analysis of X-ray spectra. Nucl Instrum Methods **145**, 579–582 (1977)
263. Van Espen P, Janssens K, Nobels J, AXIL–PC, software for the analysis of complex X-ray spectra Chemom Intell Lab Syst **1**, 109–114 (1986)
264. Johansson TB, Akselsson KR, Johansson SAE, X-ray analysis: elemental trace analysis at the $10^{-12}$ g level. Nucl Instrum Methods **84**, 141–143 (1970)
265. Injuk J, Van Grieken R, Literature trends in X-ray emission spectrometry in the period 1990–2000 – a review. X-Ray Spectrom **32**, 35–39 (2003)
266. Török SB, Lábár J, Schmeling M, Van Grieken RE, X-Ray Spectrometry. (review) Anal Chem **70**, 495R–517R (1998)
267. Szalóki I, Török SB, Ro CU, Injuk J, Van Grieken RE, X-ray Spectrometry. (review) Anal Chem **72**, 211R–233R (2000)
268. Szalóki I, Török SB, Injuk J, Van Grieken RE, X-ray spectrometry. (review) Anal Chem **74**, 2895–2918 (2002)
269. Potts PJ, Andrew TE, Kregsamer P, Marshall J, Streli C, West M, Wobrauschek P, Atomic spectrometry update. X-ray spectrometry. J Anal At Spectrom **17**, 1439–1455 (2002)
270. Potts PJ, Andrew TE, Kregsamer P, Marshall J, Streli C, West M, Wobrauschek P, Atomic spectrometry update. X-ray spectrometry. J Anal At Spectrom **18**, 1297–1316 (2003)
271. Tölg G, Klockenkämper R, The role of total-reflection X-ray fluorescence in atomic spectroscopy. Spectrochim Acta **48B**, 111–127 (1993)
272. Potts PJ, Webb PC, X-ray-fluorescence spectrometry. (review) J Geo-chem Explor **44**, 251–296 (1992)
273. Stumm W, Morgan JJ, Aquatic Chemistry, Wiley, New York (1996)
274. Salbu B, Steinnes E, Trace elements in Natural Waters, CRC Press, Boca Raton (1995)
275. Chapelle FH, Ground-Water Microbiology and Geochemistry. Wiley, New York (2001)
276. Horowitz AJ Lum KR, Garbarino JR, Hall GEM, Lemieux C, Demas CR, Problems associated with using filtration to define dissolved trace element concentrations in natural water samples. Environ Sci Technol **30**, 954–963 (1996)

277. Nozaki Y, A Fresh Look at Element Distribution in the North Pacific, Eos, Transactions, American Geophysical Union, Electronic Supplement, http://www.agu.org/eos_elecas97025e.html [Posted May 27 1997] (1997)
278. Livingstone DA, Chemical composition of rivers and lakes, in: Data of Geochemistry: US Geol Survey Prof Fleischer M, (ed.), paper 440, G1–G64 (1963)
279. Reimann C, Caritat CD, Halleraker JH, Volden T, Äyräs M, Niskavaara H, Chekushin VA, Pavlov VA, Rainwater composition in eight arctic catchments in Northern Europe [Finland, Norway and Russia]. Atmos Environ **31**, 159–170 (1997)
280. Lacey RF, Shaper AG, Changes in water hardness and cardiovascular death rates. Int J Epidemiol **13**, 18–24 (1984)
281. Camel V, Solid phase extraction of trace elements. Spectrochim Acta **58B**, 1177–1233 (2003)
282. Benoit G, Clean technique measurement of Pb, Ag, and Cd in freshwater: A redefinition of metal pollution. Environ Sci Technol **28**, 1987–1991 (1994)
283. Bowman HR, Hyde EK, Thompson SG, Jared RC, Application of high-resolution semiconductor detectors in X-ray emission spectrography. Science **151**, 562–568 (1966)
284. Luke CL, Determination of trace elements in inorganic and organic materials by X-ray fluorescence spectroscopy. Anal Chim Acta **41**, 237–250 (1968)
285. Kinrade JD, Van Loom JC, Solvent extraction for use with flame atomic absorption spectrometry. Anal Chem **46**, 1894–1898 (1974)
286. Schwarzenbach G, Relationships between metal complex stability and structure of complexing agents. Anal Chem **32**, 6–9 (1960)
287. Green TE, Law SL, Campbell WJ, Use of selective ion exchange paper in X-ray spectrography and neutron activation. Application to the determination of gold. Anal Chem **42**, 1749–1753 (1970)
288. Vassos BH, Hirsch RF, Letterman H, X-ray microdetermination of chromium, cobalt, copper, mercury, nickel and zinc in water using electrochemical preconcentration. Anal Chem **45**, 792–794 (1973)
289. Van Grieken R, Preconcentration methods for the analysis of water by x-ray spectrometric techniques. Anal Chim Acta **143**, 3–34 (1982)
290. Tisue T, Seils C, Keel RT, Preconcentration of subnanogram amounts of metals from natural-waters for X-ray-energy spectrometric determination using pyrrolidinecarbodithioic acid. Anal Chem **57**, 82–87 (1985)
291. Ellis AT, Leyden DE, Wegscheider W, Jablonski BB, Bodnar WB, Preconcentration methods for the determination of trace elements in water by X-ray-fluorescence spectrometry.1. Response characteristics. Anal Chim Acta **142**, 73–87 (1982)
292. Alvarez MA, Alvarez JR, Alvarez RP, Heavy metal analysis of rainwaters by nuclear related techniques: Application of APDC precipitation and energy dispersive X-ray fluorescence. J Radioanal Nucl Chem **245**, 485–489 (2000)
293. Clanet F, Deloncle R, Popoff G, Chelating resin catcher for capture, preconcentration and determination of toxic metal traces (Zn, Cd, Hg, Pb) in waters. Water Res **15**, 591–598 (1981)
294. Braun T, Abbas MN, Török Sz, Szokefalvi-Nagy Z, Radioisotope-induced X-ray fluorescence determination of phenylmercury, methylmercury and inorganic mercury in water after preconcentration on diethylammonium diethyldithiocarbamate-loaded polyurethane foam discs. Anal Chim Acta **160**, 277–282 (1984)

295. Török Sz, Braun T, Dyck PV, Grieken RV, Heterogeneity effects in direct XRF analysis of traces of heavy metals preconcentrated on polyurethane foam sorbents. X-Ray Spectrom **15**, 7–11 (1986)
296. Hou XD Peters HL, Yang Z, Wagner KA, Batchelor JD, Daniel MM, Jones BT, Determination of trace metals in drinking water using solid-phase extraction disks and X-ray fluorescence spectrometry. Appl Spectro **53**, 338–342 (2003)
297. Tertian R, Claisse F, Principles of Quantitative X-Ray Spectrometry. Heyden, London (1982)
298. Rastegar B, Jundt F, Gallmann A, Rastegar F, Leroy MJF, Sample homogeneity in energy-dispersive XRF trace-metal analysis. X-Ray Spectrom **15**, 83–86 (1986)
299. Meltzer C, King BS, Trace element analysis of solutions at the ppb level. Adv X-ray Anal **34**, 41–55 (1991)
300. Fenkart K, Eng E, Frey U, Determination of elements in waste-water by X-ray fluorescence analysis – sample preparation and matrix correction method. Fresenius Z Anal Chem **293**, 364–369 (1978)
301. Gao N, Ponomarev IY, Xiao QF, Gibson WM, Carpenter DA, Enhancement of microbeam X-ray fluorescence analysis using monolithic polycapillary focusing optics. Appl Phys Lett **71**, 3441–3443 (1997)
302. Nielson AJ, Turner DC, Wilson AW, Wherry DC, Wong R, An order of magnitude improvement in detection limits achieved by using a new sample support in small spot XRF analysis. Adv X-Ray Anal **39**, 799–804 (1995)
303. Wilson AW, Turner DC, Robbins AA, Improvements in sample support films for microsample X-ray analysis. Adv X-ray Anal **41**, 301–307 (1997)
304. Hettipathirana TD, Speciation of sub-parts per billion levels of Cr(III) in waters by solid-phase adsorption followed by thin-layer X-ray fluorescence spectrometry. X-Ray Spectrom **30**, 330–337 (2001)
305. Edwards T, Ferrier B, Harriman R, Preliminary investigation on the use of ion-exchange resins for monitoring river water composition. Sci Total Environ **135**, 27–36 (1993)
306. Allen B, Sackett D, Parsons C, Shefsky S, On-site analysis of metals in liquids using portable X-ray fluorescence. Field Anal Methods Hazard. Wastes Toxic Chem, Proc Spec Conf, Air and Waste Management Assoc. Pittsburgh, PA, USA, 838–841 (1997)
307. Gharaibeh SH, Abu-El-Sha'r WY, Al-Kofahi MM, Removal of selected heavy metals from aqueous solutions using processed solid residue of olive mill products. Water Res **32**, 498–502 (1998)
308. Streli C, Aiginger H, Wobrauschek P, Light-element analysis with a new spectrometer for total-reflection X-ray fluorescence. Spectrochim Acta **48B**, 163–170 (1993)
309. Klockenkämper R, Total-Reflection X-ray Fluorescence Analysis. Wiley, New York (1997)
310. Wegrzynek D, Holynska B, Ostachowicz B, A comparison of the performance of a fundamental parameter method for analysis of total reflection X-ray fluorescence spectra and determination of trace elements, versus an empirical quantification procedure. Spectrochim Acta **53B**, 43–48 (1998)
311. Yap CT, X-ray total reflection fluorescence analysis of iron, copper, zinc, and bromine in human-serum. Appl Spectrosc **42**, 1250–1253 (1988)

312. de Boer DKG, X-ray standing waves and the critical sample thickness for total-reflection X-ray fluorescence analysis. Spectrochim Acta **46B**, 1433–1436 (1991)
313. Klockenkämper R, von Bohlen A, Determination of the critical thickness and the sensitivity for thin-film analysis by total reflection X-ray fluorescence spectrometry. Spectrochim Acta **44B**, 461–469 (1989)
314. Knoth J, Schwenke H, X-ray-fluorescence spectrometer with totally reflecting sample support for trace analysis at ppb level. Fresenius Z Anal Chem **291**, 200–204 (1978)
315. Holyńska B, Olko M, Ostachowicz B, Ostachowicz J, Wegrzynek D, Claes M, Van Grieken R, de Bokx P, Kump P, Necemer M, Performance of total reflection and grazing emission X-ray fluorescence spectrometry for the determination of trace metals in drinking water in relation to other analytical techniques. Fresenius J Anal Chem **362**, 294–298 (1998)
316. Barreiros MA, Carvalho ML, Costa MM, Marques MI, Ramos MT, Application of total reflection XRF to elemental studies of drinking water. X-Ray Spectrom **26**, 165–168 (1997)
317. Holyńska B, Ostachowicz B, Wegrzynek D, Simple method of determination of copper, mercury and lead in potable water with preliminary pre-concentration by total reflection X-ray fluores spectrom. Spectrochim Acta **51B**, 769–773 (1996)
318. Kump P, Necemer M, Veber M, Determination of trace elements in mineral water using total reflection X-ray fluorescence spectrometry after preconcentration with ammonium pyrrolidinedithiocarbamate. X-Ray Spectrom **26**, 232–236 (1997)
319. Prange A, Knöchel A, Michaelis W, Multi-element determination of dissolved heavy metal traces in sea water by total-reflection x-ray fluorescence spectrometry. Anal Chim Acta **172**, 79–100 (1985)
320. Prange A, Kremling K, Distribution of dissolved molybdenum, uranium and vanadium in Baltic Sea waters. Mar Chem **16**, 259–274 (1985)
321. Schmidt D, Gerwinski W, Radke I, Trace metal determinations by total-reflection X-ray fluorescence analysis in the open Atlantic Ocean. Spectrochim Acta **48B**, 171–181 (1993)
322. Menegario AA, Pellegrinotti DC, Gine MF, Filho VFN, On-line preconcentration flow system for multi-elemental analysis by total-reflection X-ray fluorescence spectrometry. Spectrochim Acta **58B**, 543–549 (2003)
323. Gerwinski W, Schmidt D, Automated solid-phase extraction for trace metal analysis of sea water: sample preparation for total-reflection X-ray fluorescence measurements. Spectrochim Acta **53B**, 1355–1364 (2001)
324. Griesel S, Reus U, Prange A, Electro-deposition as a sample preparation technique for total-reflection X-ray fluorescence analysis. Spectrochim Acta **56B**, 2107–2115 (2001)
325. Lofthouse SD, Greenway GM, Stephen SC, Comparison of inductively coupled plasma mass spectrometry with a microconcentric nebuliser and total reflection X-ray spectrometry for the analysis of small liquid volume samples. J Anal At Spectrom **13**, 1333–1335 (1998)
326. Aster B, von Bohlen A, Burba P, Determination of metals and their species in aquatic humic substances by using total-reflection X-ray fluorescence spectrometry. Spectrochim Acta **52B**, 1009–1018 (1997)

327. Figura P, McDuffie G, Determination of labilities of soluble trace-metal species in aqueous environmental samples by anodic-stripping voltammetry and chelex column and batch methods. Anal Chem **52**, 1433–1439 (1980)
328. Alvarez MB, Malla ME, Batistoni DA, Performance evaluation of two chelating ion-exchange sorbents for the fractionation of labile and inert metal species from aquatic media. Anal Bioanal Chem **378**, 438–446 (2004)
329. Görgl R, Wobrauschek P, Kregsamer P, Streli C, Haller M, Knöchel A, Radtke M, Total reflection X-ray fluorescence analysis excited by synchrotron radiation (SR-TXRF): Variation of excitation conditions and sample geometries. X-Ray Spectrom **26**, 189–194 (1997)
330. Streli C, Wobrauschek P, Kregsamer P, Pepponi G, Pianetta P, Pahlke S, Fabry L, Synchrotron radiation induced total reflection X-ray fluorescence of low Z elements on Si wafer surfaces at SSRL - comparison of excitation geometries and conditions. Spectrochim Acta **56B**, 2085–2094 (2001)
331. Pepponi G, Streli C, Wobrauschek P, Zamini S, Zoger N, Falkenberg G, Comparison of synchrotron radiation total reflection X-ray fluorescence excitation-detection geometries for samples with differing matrices. Spectrochim Acta B **58**, 2139–2144 (2003)
332. Streli C, Total reflection X-ray fluorescence analysis of light elements. Spectrochim Acta **52B**, 281–293 (1997)
333. Sakurai K, Eba H, Inoue K, Yagi N, Downsizing of Johansson spectrometer for X-ray fluorescence trace analysis with brilliant undulator source. Nucl Instrum Methods Phys Res **467–468A**, 1549–1552 (2001)
334. Sakurai K, Eba H, Inoue K, Yagi N, Wavelength-dispersive total-reflection X-ray fluorescence with an efficient Johansson spectrometer and an undulator X-ray source: detection of 10-16g-level trace metals. Anal Chem **74**, 4532–4535 (2002)
335. Kurunczi S, Sakurai K, Natural Water Specimen Preparation for TXRF analysis, using a Johansson wavelength-dispersive spectrometer. X-Ray Spectrom **34**, 56–58 (2005)
336. Angloher G, Beckhoff B, Bühler M, Feilitzsch Fv, Hertrich T, Hettl P, Höhne J, Huber M, Jochum J, Mößbauer RL, Schnagl J, Scholze F, Ulm G, Development of superconducting tunnel junction detectors for high resolution X-ray spectroscopy. Nucl Instrum Methods Phys Res **444A**, 214–219 (2000)
337. Veldkamp M, Beckhoff B, Fliegauf R, Ulm G, Frank M, Friedrich S, Labov SE, Characterisation of superconducting tunnel junction X-ray detectors by means of monochromatized undulator radiation. Nucl Instrum Methods Phys Res **487A**, 450–456 (2002)
338. Hopke PK, Target transformation factor analysis as an aerosol mass apportionment method – a review and sensitivity study. Atmos Environ **22**, 1777–1792 (1988)
339. Balásházy I, Heistracher T, Hofmann W, Air flow and particle deposition patterns in bronchial airway bifurcations: The effect of different CFD models and bifurcation geometries. J Aerosol Med **9**, 287–301 (1996)
340. Götz G, Mészáros E, Vali G, Atmospheric Particles and Nuclei, Akadémiai Kiadó, Budapest (1991)
341. Councel Directive, 1999/30/EC of 22 April 1999 relating to limit values for sulphur dioxide, nitrogen dioxide and oxides of nitrogen, particulate matter and lead in ambient air. Official Journal of the European Communities L163(29/06/1999), **42**, 41–60 (1999)

342. Howell S, Pszenny AAP, Quinn P, Huebert B, A field intercomparison of three cascade impactors. Aerosol Sci Technol **29**, 475–492 (1998)
343. Jenkins R, Gould RW, Gedcke D, Quantitative X-ray Spectrometry. Marcel Dekker, New York (1981)
344. Samek L, Injuk J, Van Espen P, Van Grieken R, Performance of a new compact EDXRF spectrometer for aerosol analysis. X-Ray Spectrom **31**, 84–86 (2002)
345. Lodge JP, ed, Methods of air sampling and analysis, 3rd edn, Lewis Publishers, Chelsea, MI, USA (1989)
346. Holynska B, Ptasinski J, Maenhaut W, Annegarn HJ, Energy-dispersive x-ray fluorescence spectrometer with capillary optics for the chemical analysis of atmospheric aerosols with high time resolution. J Aerosol Sci **28**, 1455–1463 (1997)
347. Kucera J, Smodiš B, Burns K, De Regge P, Campbell M, Havránek V, Makarewicz M, Toervenyi A, Zeiller E, Preparation and characterization of a set of IAEA reference air filters for quality control in air-pollution studies. Fresenius J Anal Chem **370**, 229–233 (2001)
348. Osán J, Alföldy B, Kurunczi S, Török S, Bozó L, Injuk J, Worobiec A, Van Grieken R, Characterization of atmospheric aerosol particles over Lake Balaton, Hungary, using X-ray emission methods. Idojárás **105**, 145–156 (2001)
349. Borbély-Kiss I, Koltay E, Szabó Gy, Bozó L, Tar K, Composition and sources of urban and rural atmospheric aerosol in Eastern Hungary. J Aerosol Sci **30**, 369–391 (1999)
350. Hlavay J, Polyák K, Weisz M, Monitoring of the natural environment by chemical speciation of elements in aerosol and sediment samples. J Environ Monit **3**, 74–80 (2001)
351. Klockenkämper R, Bayer H, von Bohlen A, Schmeling M, Klockow D, Collection of airborne particulate matter for a subsequent analysis by total reflection X-ray fluorescence. Anal Sci **11**, 495–498 (1995)
352. Salvà A, von Bohlen A, Klockenkämper R, Klockow D, Multielement analysis of airborne particulate matter by total reflection X-ray fluorescence. Quim Anal **12**, 57–62 (1993)
353. Schmeling M, Total-reflection X-ray fluorescence – a tool to obtain information about different air masses and air pollution. Spectrochim Acta **56B**, 2127–2136 (2001)
354. Schmeling M, Seasonal variations in diurnal concentrations of trace elements in atmospheric aerosols in Chicago. Anal Chim Acta **496**, 315–323 (2003)
355. Schneider B, The determination of atmospheric trace metal concentrations by collection of aerosol particles on sample holders for total-reflection X-ray fluorescence. Spectrochim Acta **44B**, 519–523 (1989)
356. Injuk J, Van Grieken R, Optimisation of total-reflection X-ray fluorescence for aerosol analysis. Spectrochim Acta **50B**, 1787–1803 (1995)
357. Esaka F, Watanabe K, Onodera T, Taguchi T, Magara M, Usuda S, The use of Si carriers for aerosol particle collection and subsequent elemental analysis by total-reflection X-ray fluorescence spectrometry. Spectrochim Acta **58B**, 2145–2155 (2003)
358. Török S, Osán J, Beckhoff B, Ulm G, Ultra-trace speciation of nitrogen compounds in aerosols collected on silicon wafer surfaces by means of TXRF-NEXAFS. Powder Diffr **19**, 81–86 (2004)
359. Osán J, Szalóki I, Ro CU, Van Grieken R, Light element analysis of individual microparticles using thin-window EPMA. Mikrochim Acta **132**, 349–355 (2000)

360. Ro CU, Osán J, Szalóki I, de Hoog J, Worobiec A, Van Grieken R, A Monte Carlo program for quantitative electron-induced X-ray analysis of individual particles. Anal Chem **75**, 851–859 (2003)
361. Jones KW, Gordon BM, Trace-element determinations with synchrotron-induced X-ray emission. Anal Chem **61**, 341A–358 (1989)
362. Török S, Sándor S, Xhoffer C, Van Grieken R, Jones KW, Sutton SR, Rivers ML, X-ray microprobe studies of Hungarian background and urban aerosols, Adv X-ray Anal **35**, 1183–1188 (1992)
363. Van Malderen H, Hoornaert S, Injuk J, Przybylowicz WJ, Pineda CA, Prozesky VM, Van Grieken R, Individual particle characterization of Siberian aerosols by micro-PIXE and backscattering spectrometry. X-Ray Spectrom **30**, 320–329 (2001)
364. Markowicz AA, Van Grieken RE, Quantification in XRF analysis of intermediate-thickness samples, in: Van Gieken RE, Markowicz AA (eds) Handbook of X-ray spectrometry. Marcel Dekker, New York, pp. 407–432 (2002)
365. Armstrong JT, Buseck PR, A general characteristic fluorescence correction for the quantitative electron microbeam analysis of thick specimens, thin films and particles. X-Ray Spectrom **14**, 172–182 (1985)
366. Rindby A, Voglis P, Attaelmanan A, Analysis of inhomogeneous and irregularly shaped samples by the use of XRF micro-beam correlation analysis. X-Ray Spectrom **25**, 39–50 (1996)
367. Vincze L, Somogyi A, Osán J, Vekemans B, Török S, Janssens K, Adams F, Quantitative trace-element analysis of individual fly-ash particles by means of micro-XRF. Anal Chem **74**, 1128–1135 (2002)
368. Vincze L, Janssens K, Adams F, A general Monte Carlo simulation of energy-dispersive X-ray fluorescence spectrometers 1. Unpolarized radiation, homogeneous samples. Spectrochim Acta **48B**, 553–573 (1993)
369. Vincze L, Janssens K, Rivers ML, Jones KW, Adams F, A general Monte Carlo simulation of ED-XRF spectrometers. II: Polarized monochromatic radiation, homogeneous samples. Spectrochim Acta **50B**, 127–147 (1995)
370. Vincze L, Janssens K, Vekemans B, Adams F, Modeling of photon scattering at high X-ray energies: experiment versus simulation. J Anal At Spectrom **14**, 529–533 (1999b)
371. Hovington P, Drouin D, Gauvin R, CASINO: A new Monte Carlo code in C language for electron beam interaction – Part I: Description of the program. Scanning **19**, 1–14 (1997)
372. Ro CU, Osán J, Van Grieken R, Determination of low-Z elements in individual environmental particles using windowless EPMA. Anal Chem **71**, 1521–1528 (1999)
373. Kirkpatrick P, Weidmann L, Theoretical continuous X-ray energy and polarization. Phys Rev **67**, 321–339 (1945)
374. Török S, Osán J, Vincze L, Alföldy B, Kerkápoly A, Vajda N, Pérez CA, Falkenberg G, Comparison of nuclear and X-ray techniques for actinide analysis of environmental hot particles. J Anal At Spectrom **18**, 1202–1209 (2003)
375. Vekemans B, Janssens K, Vincze L, Adams F, Van Espen P, Analysis of X-ray spectra by iterative least squares (AXIL): New developments. X-Ray Spectrom **23**, 278–285 (1994)

376. Vincze L, Janssens K, Vekemans B, Adams F, Monte Carlo simulation of X-ray fluorescence spectra: Part 4. Photon scattering at high X-ray energies. Spectrochim Acta **54B**, 1711–1722 (1999)
377. Török S, Osán J, Vincze L, Kurunczi S, Tamborini G, Betti M, Characterization and speciation of depleted uranium in individual soil particles using microanalytical methods. Spectrochim Acta Part B **59**, 689–699 (2004)
378. Osán J, Török S, Alföldy B, Falkenberg G, Characterization of anthropogenic sediment particles after a transboundary water pollution of River Tisza using synchrotron radiation. Spectrochim Acta **59B**, 701–708 (2004)
379. Danesi PR, Markowicz A, Chinea-Cano E, Burkart W, Salbu B, Donohue D, Ruedenauer F, Hedberg M, Vogt S, Zahradnik P, Ciurapinski A, Depleted uranium particles in selected Kosovo samples. J Environ Radioact **64**, 143–154 (2003)
380. Osán J, Kurunczi S, Török S, Van Grieken R, X-ray analysis of riverbank sediment of the Tisza (Hungary): identification of particles from a mine pollution event. Spectrochim Acta **57B**, 413–422 (2002)
381. Duff MC, Hunter DB, Tray IR, Bertsch PM, Reed DT, Sutton SR, Shea-McCarthy G, Kit-ten J, Eng P, Chipera SJ, Vaniman DT, Mineral associations and average oxidation states of sorbed Pu on Tuff. Environ Sci Technol **33**, 2163–2169 (1999)
382. Bertsch PM, Hunter DB, Sutton SR, Bajt S, Rivers ML, In situ chemical speciation of uranium in soils and sediments by micro X-ray absorption spectroscopy. Environ Sci Technol **28**, 980–984 (1994)
383. Tokunaga TK, Brown GE, Pickering IJ, Sutton SR, Bajt S, Selenium redox reactions and transport between ponded waters and sediments. Environ Sci Technol **31**, 1419–1425 (1997)
384. Schoonover JR, Havrilla GJ, Combining X-ray fluorescence spectrometry and vibrational microscopy to assess highly heterogeneous actinide-contaminated materials. Appl Spectrosc **53**, 257–265 (1999)
385. Ice GE, Microbeam-forming methods for synchrotron radiation. X-Ray Spectrom **26**, 315–326 (1997)
386. Pella PA, Lankosz M, Highlights of X-ray spectrometry for microanalysis X-ray Spectrom **26**, 327–332 (1997)
387. Vekemans B, Janssens K, Vincze L, Aerts A. Adams F, Herogen J, Automated segmentation of mu-XRF image sets. X-Ray Spectrom **26**, 333–346 (1997)
388. Lankosz M, Pella PA, A new approach to quantitative X-ray microfluorescence analysis of individual particles. X-Ray Spectrom **26**, 347–349 (1997)
389. Jones KW, Berry WJ, Borsay DJ, Cline HT, Connor WC, Fullmer CS, Applications of synchrotron radiation-induced X-ray emission (SRIXE). X-Ray Spectrom **26**, 350–358 (1997)
390. Iida A, X-ray spectrometric applications of a synchrotron X-ray microbeam. X-Ray Spectrom **26**, 359–363 (1997)
391. Havrilla GJ, Applications of X-ray microfluorescence to materials analysis. X-Ray Spectrom **26**, 364–373 (1997)
392. Ding X, He Y, Yan Y, X-ray source for x-ray microfluorescence using a monolithic X-ray focusing lens combined with aperture optics. X-Ray Spectrom **26**, 374–379 (1997)
393. Hosokawa Y, Ozawa S, Nakazawa H, Nakayama Y, An X-ray guide tube and a desk-top scanning X-ray analytical microscope. X-Ray Spectrom **26**, 380–387 (1997)

394. Worley CG, Havrilla GJ, Dunn PS, Quantification of large scale micro-X-ray fluorescence elemental images. Appl Spectrosc **55**, 1448–1454 (2001)
395. Le Maitre RW (ed.), Streckeisen A, Zanettin B, Le Bas MJ, Bonin B, Bateman P, Bellieni G, Dudek A, Efremova S, Keller J, Lameyre J, Sabine PA, Schmid R, Sorensen H, Woolley AR, Igneous rocks; a classification and glossary of terms; recommendations of the International Union of Geological Sciences Subcommission on the Systematics of Igneous Rocks. 2nd ed Cambridge University Press, Cambridge, United Kingdom, 236 (2002)
396. Pearce JA, An expert system for the tectonic characterization of ancient volcanic rocks. Journ Volc Geotherm Res **32**, 51–65 (1987)
397. Wells N, Smidt RE, Methods for mineral and element analysis. New Zealand Soil Bureau, Sci Rep **10D**, 12 (1978)
398. Hutton JT, Elliott SM, An accurate XRF method for the analysis of geochemical exploration samples for major and trace elements using one glass disc. Chem Geol **29**, 1–11 (1980)
399. Bertin EP, Principles and Practice of X-Ray Spectrometric Analysis. 2nd edn Plenum Press, New York, London, 1079 (1975)
400. Lachance GR, X-Ray Fluorescence Spectrometry, in: Hyman M, Rowe MW (eds), Adv Anal Geochem **1**, 155–192 (1993).
401. Hahn-Weinheimer P, Hirner A, Weber-Diefenbach K, Röntgenfluoreszenzanalytische Methoden. Grundlagen und praktische Anwendung in den Geo-, Material- und Umweltwissenschaften. Analytische Chemie VIII, Vieweg, Braunschweig, 283 (1995)
402. Weber-Diefenbach K, Physikalisch-chemische Untersuchungsmethoden in den Geowissenschaften. In: Pavievi MK Amthauer G (eds) Schweizerbart'sche Verlagsbuchhandlung, Nägele u. Obermiller, Stuttgart, 115–131 (2000)
403. Brill M, Energy-dispersive vs. wavelength-dispersive x-ray fluorescence spectrometry. Metall, Berlin **50**, 662–668 (1996)
404. Kramar U, Advances in energy-dispersive X-ray fluorescence. J Geochem Explor **58** 73–80 (1997)
405. Markham M, Floyd PA, Geochemical fingerprinting of west Cornish greenstones as an aid to provenancing neolithic axes. Proc Ussher Society **9**, 218–223 (1998)
406. Vaniman DT, Bish DL, Blake DF, Elliott ST, Sarrazin P, Collins SA, Chipera S, Landed XRD/XRF analysis of prime targets in the search for past or present Martian life. Journ Geophy Res E **103**, 31, 477–31, 489 (1998)
407. Vaniman DT, Bish DL, Blake DF, Chipera SJ, CCD-based XRD/XRF for determining environmental mineralogy on Mars. In: Hubbard S. (ed), Concepts and approaches for Mars exploration. LPI Contribution, 300–301 (2000)
408. May MT, McCurdy MW, Clement EB, XRF surveys for three mining waste sites. Proc of the 4th International Conference on Tailings and Mine Waste. Balkema AA. Rotterdam, Netherlands, **4**, 461–469 (1997)
409. Martino L, Cho E, Wrobel J, In situ analysis of soil at an open burning/ open detonation disposal facility. Argonne National Laboratory, Argonne, IL, United States, 30 (1994)
410. Mirecki J, Dumars A, Laws B, Tucker J, Comparison of Pb concentrations in contaminated residential soils using portable XRF (field) and ICP-AES (lab) methods. Geol Soc Am **31**, 31 (1999)

411. Jordan MD, Crawford A, Mirecki JE, Analysis of lead-contaminated soil near a residence in Charleston, South Carolina by portable XRF and ICP-AES. Geol Soc Am **32**, 29 (2000)
412. Bawden-Smith J, Reducing contaminated soil rehabilitation costs; review of portable XRF performance on Australian soils. The Aus IMM Bulletin (1990) **6**, 17–19 (2001)
413. McCain RG, Qualitative evaluation of heavy metals in soils using portable XRF instruments. Westinghouse Hanford, Richland, United States, 10 (1993)
414. Zhou S, Sun C, He Z, Wu D, The XRF mark of gold deposit in Beishan, Xinjiang. Chengdu Liguong Xueyuan Xuebao. J Chengdu Inst Technol **24**, 24–28 (1997)
415. Li G, Jia W, Zhou R, Tang H, Fang F, Ma Y, Research on field, highly sensitive X-ray fluorescence system. Chengdu Liguong Xueyuan Xuebao. J Chengdu Inst Technol **25**, 13–20 (1998)
416. Konstantinov MM, Strujkov SF, Application of indicator halos (signs of ore remobilization) in exploration for blind gold and silver deposits. J Geochem Explor **54**, 1–17 (1995)
417. Tiwari MK, Singh AK, Sawhney KJS, Analysis of stainless steel samples by energy dispersive X-ray fluorescence (EDXRF) Spectrometry. Bull Mater Sci **24**, 633–638 (2001)
418. Evans RD, The Atomic Nucleus, Krieger Publ Comp reprint, 972 (1982)
419. Streckeisen A, Classification and Nomenclature of Volcanic Rocks, Lamprophyres, Carbonatites and Melilitic Rocks. IUGS Subcommission on the Systematics of Igneous Rocks. Geol Rundsch **69**, 194–207 (1980)
420. Le Bas MJ, Le Maitre RW, Streckeisen A, Zanettin B, A Chemical Classification of Volcanic Rocks Based on the Total Alkali - Silica Diagram. J Petrol **27**, 745–7750 (1986)
421. Burri C, Petrochemische Berechnungsmethoden auf äquivalenter Grundlage. Birkhäuser, Basel, 343 (1959)
422. Müller P, Von der CIPW-Norm ausgehende Berechnung von Mineralbeständen magmatischer Gesteine in Analogie zu der Modalzusammensetzung plutonischer und vulkanischer Gesteine. Geol Jahrbuch D **55**, (1982) 41 pages
423. Best MG, Igneous and Metamorphic Petrology. Freeman, Appendix E: 616–619 (1982)
424. Cohen D, Ward CR, SEDNORM; a program to calculate a normative mineralogy for sedimentary rocks based on chemical analyses. Comput Geosci **179**, 1235–1253 (1991)
425. Rosen OM, Abbyasov AA, Migdisov AA, Yaroshevskii AA, MINLITH—a Program to Calculate the Normative Mineralogy of Sedimentary Rocks: the Reliability of Results Obtained for Deposits of Old Platforms. Geochem Intern **38**, 388–400 (2000)
426. Wager LR, Rhythmic and cryptic layering in mafic and ultramafic plutons. In: Hess HH Poldervaart A (eds) Basalts - The Poldervaart treatise on rocks of basaltic composition, Interscience, New York, **2**, 573–622 (1968)
427. Wilson JR, Magma chamber processes revealed by studies of layered intrusions. In: Wilson JR Michelsen O (eds) Geoscience in Aarhus. Aarhus University, Aarhus, Denmark, 13–16 (1998)

428. Cheburkin AK, Frei R, Shotyk W, An energy-dispersive miniprobe multielement analyzer (EMMA) for direct analysis of trace elements and chemical age dating of single mineral grains. Chem Geol **135**, 75–87 (1997)
429. Harvey PK, Atkin BP, Automated XRF spectrometric analysis; a rapid, cost-effective analytical technique for the mineral industry. J Geochem Explor **29**, 413–415 (1987)
430. Bach H, Krause D (eds), Analysis of the Composition and Structure of Glass and Glass Ceramics. Schott Series on Glass and Class Ceramics, Springer, Berlin Heidelberg, 528 (1999)
431. Potts PJ, Ellis AT, Kregsamer P, Marshall J, Streli C, West M, Wobrauschek P, Atomic spectrometry update. X-ray fluorescence spectrometry. J Anal At Spectrom **18**, 1297–1316 (2003)
432. Mzyk Z, Baranowska I, Mzyk J, Research on grain size effect in XRF analysis of pelletized samples. X-Ray Spectrom **31**, 39–46 (2000)
433. Darnley A, Björklund A, Bølviken B, Gustavsson N, Koval PV, Plant JA, Steenfelt A, Tauchid M, Xuejing X, A global geochemical database for environmental and resource management. Recommendations for international geochemical mapping. Final report of IGCP-project 259, UNESCO Publishing, Paris, France, 122 (1995)
434. Reimann C, Siewers U, Tarvainen T, Bityukova L, Eriksson J, Gilucis A, Gregorauskiene V, Lukashev V, Matinian NN, Pasieczna A, Baltic soil survey: total concentrations of major and selected trace elements in arable soils from 10 countries around the Baltic Sea. Sci Tot Envir **257**, 155–170 (2000)
435. Reimann C, Siewers U, Tarvainen T, Bityukova L, Eriksson J, Gilucis A, Gregorauskiene V, Lukashev V, Matinian NN, Pasieczna A, Agricultural soils in Northern Europe: A Geochemical Atlas. Geol Jahrbuch Sonderheft **5**, (2003) 279 pages
436. Alexander PO, Biogene indicators of kimberlites and lamproites. In: Science and technology of diamond. Bhatnagar GS (ed) Cambridge Intern Sci Publ Cambridge, UK, 15–37 (1999)
437. Renault S, Sailerova E, Fedikow MAF, Phytoremediation of mine tailings and bio-ore production; results from a study on plant survival at the central Manitoba (Au) minesite (NTS 52L/13), Report of Activities - Manitoba Energy and Mines, Minerals Division, 138–149 (2001)
438. Solodukhin VP, Nuclear-physical methods of analysis in oil geology and industry of Kazakhstan. In: Proc 4th Intern Conf Methods and Applications of Radioanalytical Chemistry; Part 2. J Radioanal Nucl Chem **234**, 101–105 (1998)
439. Parus J, Kierzek J, Malozewska-Bucko B, Determination of the carbon content in coal and ash by XRF. X-Ray Spectrom **29**, 192–195 (2000)
440. Shao L, Jones T, Gayer R, Dai S, Li S, Jiang Y, Zhang P, Petrology and geochemistry of the high-sulphur coals from the Upper Permian carbonate coal measures in the Heshan Coalfield, southern China. Int J Coal Geol **55**, 1–26 (2003)
441. Dill HG, Poehlmann H, Chemical composition and mineral matter of paralic and limnic coal types of lignite through anthracite rank (Germany). In: Hills LV, Henderson CM, Bamber EW (eds) Carboniferous and Permian of the world; XIV ICCP Proc Memoir - Can Soc Petrol Geol **19**, 851–867 (2002)

442. Karayigit AI, Spears DA, Booth CA, Antimony and arsenic anomalies in the coal seams from the Gokler Coalfield, Gediz, Turkey. Int J Coal Geol **44**, 1–17 (2000)
443. Brown FV, Jones SA, On-site determination of ash in coal utilizing a portable XRF analyzer. In: Rhodes J, Barrett CS, Leyden DE, Newkirk JB, Predecki PK, Ruud CO (eds) Twenty-eighth annual conference on applications of X-ray analysis. Adv X-Ray Anal **23**, 57–63 (1980)
444. Schramm M, Bornemann O, Wilke F, Siemann M, Dijk HL, Bromine Analysis - A Powerful Tool to solve Stratigraphical Problems in Exploration Boreholes for Salt Caverns.- SMRI, Technical Paper, Fall 2002 Meeting, Bad Ischl, Austria, 12 (2002)
445. Beckhoff B, Kanngießer B, Improvement of the Detection sensitivity of EDXRF Trace Element Analysis by means of efficient x-ray focusing based on strongly curved HOPG crystals. Adv X-Ray Anal **39**, 109–117 (1997)
446. Zhdanov AY, Catalogue of the manufacturers and suppliers of instrumentation equipment and laboratory equipment. http://www.spectrometer.ru/manufacturers-suppliers/micro-xrf-spectrometers.htm (2003)
447. Rammlmair D, Tacke KD, Jung H, Application of new XRF-scanning techniques to monitor crust formation in column experiments. In: Securing the Future, Proc Intern Conf Mining Envir, Skellefteå, **2**, 683–692 (2001)
448. Vekemans B, Janssens K, Vincze L, Aerts A, Adams F, Hertogen J, Automated Segmentation of $\mu$-XRF Image Sets. X-Ray Spectrom **26**, 333–346 (1997)
449. Price GJ, Fraser GW, Pearson JF, Hutchinson IB, Holland AD, Nussey J, Vernon D, Pullan D, Turner K, Imaging x-ray fluorescence spectroscopy using microchannel plate x-ray optics. Spectros Europe **15**, 8–13 (2003)
450. Rammlmair D, Reinhardt L, Kleinmann A, High resolution EDXRF-scanning deciphering seasonal variation in laminated marine and varved lacustrine sediments. Abstract EGS - AGU - EUG Joint Ass, Nice, France, EAE03-A-12865 (2002)
451. von Rad U, Ali Khan A, Berger WH, Rammlmair D, Treppke U, Varves, turbidites and cycles in upper Holocene sediments (Makran Slope, Northern Arabian Sea). In: Clift PD, Kroon D, Gaedicke C, Craig J (eds) The Tectonic and Climatic Evolution of the Arabian Sea Region. Geol Soc Spec Publ **195**, 387–406 (2002)
452. Kleinmann A, Rammlmair D, Hochauflösende EDXRF-Scans von mittelholozänen gewarvten Sedimenten aus dem Schleinsee und Belauer See. - Terra Nostra **6**, 245–249 (2003)
453. Rammlmair D, Wessels M, Chemical characterization of Lake Constance sediments record by high resolution EDXRF, EGS - AGU - EUG Joint Ass, Nice, France, Abstract EAE03-A-11705 (2003)
454. Oberthür T, Rammlmair D, Kojonen K, Comparison of automated particle search techniques for platinum group elements (PGE) and minerals (PGM): $\mu$-XRF and electron-microprobe. In: Rammlmair D, Mederer J, Oberthür T, Heimann RB, Pentinghaus H (eds) Applied Mineralogy in Research, Economy, Technology, Ecology and Culture, Balkema AA, Rotterdam, Brookfield, **1**, 375–378 (2000)
455. Melcher F, Oberthür T, Rammlmair D, The carbonate-hosted, Ge- and Ag-rich Khusib Springs Cu-Zn-Pb sulphide deposit, Namibia: ore geochemistry and mineralogical variations, IAGOD/Geocongress 34–35 (2002)

456. Gottlieb P, Butcher AR, Ho-Tun E, Sutherland DN, Applications of automated process mineralogy. In: Rammlmair D, Mederer J, Oberthür T, Heimann RB, Pentinghaus H (eds) Applied Mineralogy in Research, Economy, Technology, Ecology and Culture, Balkema AA, Rotterdam, Brookfield, **1**, 321–323 (2000)
457. Butcher AR, Helms TA, Gottlieb P, Bateman R, Ellis S, Johnson NW, Advances in the quantification of gold deportment by QemSCAN. In: 7th Mill Operators' Conference, Kalgoorlie, AusIMM, 267–271 (2000)
458. Togami S, Takano M, Kumazawa M, An Algorithm for the transformation of XRF images into mineral-distribution maps. Can Miner **38**, 1283–1294 (2000)
459. Michibayashi K, Togami S, Takano M, Kumazawa M, Kageyama T, Application of scanning X-ray analytical microscope to the petrographic characterization of a ductile shear zone: an alternative method to image microstructure Tectonophysics, **310**, 55–67 (1999)
460. Schwarzer RA, Automated crystal lattice orientation mapping using a computer-controlled SEM. Micron **28**, 249–265 (1997)
461. Schwarzer RA, Crystallography and microstructure of thin films studied by X-ray and electron diffraction. Mater Sci Forum **287–288**, 23–60 (1998)
462. Wang PC, Cargill III GS, Noyan IC, Liniger EG, Hu CK, Lee KY, Thermal and electromigration strain distributions in 10 $\mu$m-wide aluminum conductor lines measured by X-ray diffraction. Mat Res Soc Symp Proc **473**, 273–278 (1997)
463. Horowith P, Howell JA, A scanning X-ray microscope using synchrotron radiation. Science **178**, 608–611 (1972)
464. Jones KW, Gordon BM, Trace element determinations with synchrotron-induced X-ray emission. Anal Chem **61**, 341A–356A (1989)
465. Rindby A, Engström P, Larsson S, Stocklassa B, Microbeam technique for energy-dispersive X-ray fluorescence. X-Ray Spectrom **18**, 109–112 (1989)
466. Carpenter DA, Improved laboratory X-ray source for microfluorescence analysis. X-Ray Spectrom **18**, 253–257 (1989)
467. Werry D, Cross B, Applications of small-area X-ray fluorescence analysis using a microbeam X-ray source. Spec Int **1**, 58–59 (1989)
468. Klein H, Bunge HJ, Location resolved texture analysis. Z Metallkunde **90**, 103–110 (1999)
469. Schwarzer RA, Scanning X-ray microscopy for texture mapping by energy dispersive diffraction. Proc 12th ICXOM, Cracow **1**, 205–208 (1989)
470. Schwarzer RA, Texture distributions imaged by energy dispersive X-ray diffraction. Steel Res **64**, 570–574 (1993)
471. Fischer AH, Schwarzer RA, X-ray pole figure measurement and texture mapping of selected areas using an X-ray scanning apparatus. Mater Sci Forum **273–275**, 255–262 (1998)
472. Schäfer B, ODF computer program for high-resolution texture analysis of low-symmetry materials. Mater Sci Forum **273–275**, 113–118 (1998)
473. Fischer AH, Schwarzer RA, Mapping of local residual strain with an X-ray scanning apparatus. Mater Sci Forum **273–275**, 673–677 (1998)
474. Maslen EN, An X-ray collimator for single-crystal goniometers. Sci Instrum **35**, 110–111 (1958)
475. Larsson S, Engström P, X-ray microbeam spectroscopy with the use of capillary optics. Adv X-Ray Anal **35**, 1019–1025 (1992)

476. Fischer AH, Weirauch D, Schwarzer RA, Peltier-cooled solid state drift-chamber detector for energy-dispersive X-ray pole figure measurement and texture mapping. Mater Sci Forum **273–275**, 263–270 (1998)
477. Boltwood BB, On the ultimate disintegration products of the radio-active elements. Part II. The disintegration products of uranium. Am J Sci **23**, 77–88 (1907)
478. Suzuki K, Adachi M, The chemical Th-U-total Pb isochron ages of zircon and monazite from the Gray granite of the Hida terrane, Japan. J Earth Sci **38**, 11–37 (1991)
479. Montel JM, Veschambre M, Nicollet C, Datation de la monazite à la microsonde électronique. Comptes Rendus de'l Academie des Sciences Paris, Serie II, Sci Terre Planet. **318**, 1489–1495 (1994)
480. Foster G, Kinny P, Vance D, Prince C, Harris N, The significance of monazite U-Th-Pb age data in metamorphic assemblages; a combined study of monazite and garnet chronometry. Earth Planet Sci Lett **181**, 327–340 (2000)
481. Pyle JM, Spear FS, Wark DA, Electron microprobe analysis of REE in Apatite, Monazite, and Xenotime: Protocols and Pitfalls. In: Phosphates, Kohn MJ, Rakovan J, Hughes JM (eds) Rev Mineral **48**, 337–362 (2002)
482. Rhede D, Wendt I, Förster HJ, A three-dimensional method for calculating independent chemical U/Pb- and Th/Pb-ages of accessory minerals. Chem Geol **130**, 247–253 (1996)
483. Montel JM, Foret S, Veschambre M, Nicollet C, Provost A, Electron microprobe dating of monazite. Chem Geol **131**, 37–53 (1996)
484. Montel JM, Kornprobst J, Vielzeuf D, Preservation of old U-Th-Pb ages in shielded monazite: Example from the Beni Bousera Hercynian kinzigites (Marocco). J Metam Geol **18**, 335–342 (2000)
485. Cocherie A, Albarede F, An improved U-Th-Pb age calculation for electron microprobe dating of monazite. Geochim Cosmochim Acta **65**, 4509–4522 (2001)
486. Bruhn F, Möller A, Sie SH, Hensen BJ, U-Th-Pb Chemical dating of monazites using the proton microprobe. Nucl Instrum Methods Phys Res, Sect B **158**, 616–620 (1999)
487. Frei R, Prozesky VM, Przybzlowicz W, Studies of progressive leaching in single mineral Pb/Pb dating. Nucl Instrum Methods Phys Res, Sect B **130**, 676–681 (1997)
488. Scherrer NC, Engi M, Berger A, Parrish RR, Cheburkin A, Nondestructive chemical dating of young monazite using XRF 2. Context sensitive microanalysis and comparison with Th–Pb laser-ablation mass spectrometric data. Chem Geol **191**, 243–255 (2002)
489. Engi M, Cheburkin AK, Köppel V, Nondestructive chemical dating of young monazite using XRF 1. Design of a mini-probe, age data for samples from the Central Alps, and comparison to U–Pb (TIMS) data. Chem Geol **191**, 225–241 (2002)
490. John T, Subduction and continental collision in the Lufilian Arc - Zambezi Belt orogen: A petrological, geochemical, and geochronological study of eclogites and whiteschists (Zambia), Unpubl. doctorate thesis, Universität Kiel, (2001) 75 pages
491. Nakai I, Terada Y, Itou M, Sakurai Y, Use of highly energetic (116 keV) synchrotron radiation for X-ray fluorescence analysis of trace rare-earth and heavy elements. J Synchrotron Rad **8**, 1078–1081 (2001)

492. Schroer CG, Kuhlmann B, Hunger UT, Günzler TF, Kurapova O, Fest S, Frehse F, Lengeler B, Drakopoulos M, Somogyi A, Simionovici AS, Snigirev A, Snigireva I, Schug C, Schröder WH, Nanofocusing parabolic refractive x-ray lenses. Appl Phys Lett **82**, 1485–1487 (2003)
493. Vekemans B, Vincze L, Somogyi A, Drakopoulos M, Kempenaers L, Simionovici A, Adams F, Quantitative X-ray fluorescence analysis at the ESRF ID18F microprobe. Nucl Instrum Methods Phys Res, Sect B **199**, 396–401 (2003)
494. Jochum KP, Dingwell DB, Rocholl A, Stoll B, Hofmann AW, Becker S, Besmehn A, Bessette D, Dietze HJ, Dulski P, Erzinger J, Hellebrandt E, Hoppe P, Horn I, Janssens K, Jenner GA, Klein M, McDonough WF, Maetz M, Mezger K, Münker C, Nikogosian IK, Pickhard C, Raczek I, Rhede D, Seufert HM, Simakin SG, Sobolev AV, Spettel B, Straub S, Vincze L, Wallianos A, Weckwerth G, Weyer S, Wolf D, Zimmer M, The preparation and preliminary characterization of eight geological MPI-DING reference glasses for in-situ microanalysis. Geostand Newslett **24**, 87–133 (2000)
495. Koepke J, Behrens H, Trace element diffusion in andesitic melts: an application of synchrotron X-ray fluorescence analysis. Geochim Cosmochim Acta **65**, 1481–1498 (2001)
496. Koepke J, Falkenberg G, Rickers K, Diedrich O, Trace element diffusion and element partitioning between garnet and andesite melt using synchrotron X-ray fluorescence microanalysis ($\mu$-SRXRF). Eur J Mineral **15**, 883–892 (2003)
497. Hlawatsch S, Garbe-Schönberg CD, Lechtenberg F, Manceau A, Tamura N, Kulik DA, Kersten M, Trace metal fluxes to ferromanganese nodules from the western Baltic Sea as a record for long-term environmental changes, Chem Geol **182**, 697–709 (2002)
498. Troll VR, Schmincke HU, Magma mixing and crustal recycling recorded in ternary feldspar from compositionally zoned peralkalineingnimbrite A, Gran Canaria, Canary Islands, J Petrol **43**, 243–270 (2002)
499. Troll VR, Sachs PM, Schmincke HU, Sumita M, The REE-Ti mineral chevkinite in comenditic magmas from Gran Canaria, Spain: a SYXRF-probe study, Contrib Miner Petrol **145**, 730–741 (2003)
500. Sachs PM, Hansteen T, Pleistocene underplating and metasomatism of the lower continental crust: a xenolith study. J Petrol **41**, 331–356 (2000)
501. Harting M, Rickers K, Kramar U, Simon R, Staub S, Schulte P, Multielement geochemical investigations by SRXRF microprobe studies on tectite material: evidence from the NE-Mexican Cretaceous/Tertiary record. EOS Trans AGU **83**(47), Abstract P22B-0403 (2002)
502. Rindby A, Engström P, Janssens K, Osán J, Micro-distribution of heavy elements in highly inhomogeneous particles generated from $\mu$-beam XRF/XRD analysis. Nucl Instrum Methods Phys Res, Sect B **124**, 591–604 (1997)
503. Vincze L, Somogyi A, Osán J, Vekemans B, Török S, Janssens K, Adams F, Quantitative trace element analysis of individual fly ash particles by means of X-ray microfluorescence, Anal Chem **74**, 1128–1135 (2002)
504. Salbu B, Kerkling T, Lind OC, Oughton DH, Drakopoulos M, Simionovici A, Snigireva I, Snigirev A, Weitkamp T, Adams F, Janssens K, Kashparov VA, High energy X-ray microscopy for characterisation of fuel particles. Nucl Instrum Methods Phys Res A **467–468**, 1249–1252 (2001)

505. Osán J, Török S, Alföldy B, Falkenberg G, Characterization of anthropogenic sediment particles after a transboundary water pollution of River Tisza using synchrotron radiation. Spectrochim Acta B **59**, 701–708 (2004)
506. Salbu B, Janssens K, Lind OC, Proost K, Danesi PR, Oxidation states of uranium in DU particles from Kosovo. J Environ Radioact **64**, 167–173 (2003)
507. Roedder E, Fluid inclusions. Rev Miner **12**, 644 (1984)
508. Günther D, Audétat A, Frischknecht R, Heinrich CA, Quantitative analysis of major, minor and trace elements in fluid inclusions using laser ablation-inductively coupled plasma mass spectrometry. J Anal At Spectrom **13**, 263–270 (1998)
509. Günther D, Hattendorf B, Audétat A, Multi-element analysis of melt and fluid inclusions with improved detection capabilities for Ca and Fe using laser ablation with a dynamic reaction cell ICP-MS. J Anal At Spectrom **16**, 1085–1090 (2001)
510. Audétat A, Günther D, Heinrich CA, Formation of a magmatic-hydrothermal ore deposit: Insights with LA-ICP-MS analysis of fluid inclusions. Science **279**, 2091–2094 (1998)
511. Heinrich CA, Günther D, Audétat A, Ulrich T, Frischknecht R, Metal fractionation between magmatic brine and vapor, determined by microanalysis of fluid inclusions. Geology **27**, 755–758 (1999)
512. Ménez B, Les microsondes photon et proton appliquées à l'analyse ponctuelle d'inclusions fluides: Un outil pour reconstituer l'évolution des paléosystèmes hydrothermeaux [Individual fluid inclusion microanalysis combining photon and proton microprobes: a tool for reconstructing paleohydrothermal processes]. Ph.D. thesis, Paris VII University (1999)
513. Ménez B, Philippot P, Bonnin-Mosbah M, Simionovici A, Gibert F, Analysis of individual fluid inclusions using synchrotron X-ray fluorescence microprobe: progress toward calibration for trace elements. Geochim Cosmochim Acta **66**, 561–576 (2002)
514. Kamenetsky VS, van Achterberg E, Ryan CG, Naumov VB, Mernagh TP, Davidson P, Extreme chemical heterogeneity of granite-derived hydrothermal fluids: an example from inclusions in a single crystal of miarolitic quartz. Geology **30**, 459–462 (2002)
515. Ryan CG, McInnes BIA, Williams PJ, Dong G, Win TT, Yeats CJ, Imaging fluid inclusion content using the new CSIRO-GEMOC nuclear microprobe. Nucl Instrum Methods Phys Res, Sect B **181**, 570–577 (2001)
516. Ryan CG, Van Achterberg E, Yeats CJ, Drieberg SL, Mark G, McInnes BM, Win TT, Cripps G, Suter GF, Quantitative, high sensitivity, high resolution, nuclear microprobe imaging of fluids, melts and minerals. Nucl Instrum Methods Phys Res, Sect B **188**, 18–27 (2002)
517. Bühn B, Rankin AH, Radtke M, Haller M, Knöchel A, Burbankite, a (Sr, REE, Na, Ca)-carbonate in fluid inclusions from carbonatite-derived fluids; identification and characterization using laser Raman spectroscopy, SEM-EDX, and synchrotron micro-XRF analysis. Am Miner **84**, 1117–1125 (1999)
518. Philippot P, Ménez B, Simionovici A, Chabiron A, Cuney M, Snigirev A, Snigireva I, X-ray imaging of uranium in individual fluid inclusions. Terra Nova **12**, 77–83 (2000)
519. Ryan CG, Heinrich CA, Mernagh TP, PIXE microanalysis of fluid inclusions and its application to study ore metal segregation between magmatic brine and vapor. Nucl Instrum Methods Phys Res, Sect B **77**, 463–471 (1993)

520. Damman AH, Kars SM, Touret JLR, Rieffe EC, Kramer JALM, Vis RD, Pintea I, PIXE and SEM analysis of fluid inclusions in quartz crystals from the K-alteration zone of the Rosia Poieni porphyry-Cu deposit, Apuseni Mountains, Rumania. Eur J Miner **8**, 1081–1096 (1996)
521. Kanngießer B, Malzer W, Reiche I, A new 3D micro X-ray fluorescence analysis set up – First archeometric applications. Nucl Instrum Methods Phys Res, Sect B **211**, 259–264 (2003)
522. Vanko DA, Bonnin-Mosbah M, Philippot P, Roedder E, Sutton S, Fluid inclusions in quartz from oceanic hydrothermal specimens and the Bingham, Utah porphyry-Cu deposit: a study with PIXE and SXRF. Chem Geol **173**, 227–238 (2001)
523. Vanko DA, Mavrogenes JA, Synchrotron-source X-ray fluorescence microprobe: analysis of fluid inclusions. In: McKibben MA, Shanks WC, Ridley WI (eds), Applications of Microanalytical Techniques to Understanding Mineralizing Processes. Rev Econ Geol **7**, 251–263 (1998)
524. Rickers K, Thomas R, Heinrich W, Trace-element analysis of individual synthetic and natural fluid inclusions with synchrotron radiation XRF using Monte Carlo simulations for quantification. Eur J Miner **16**, 23–35 (2004)
525. Vincze L, Janssens K, Adams F, A general Monte Carlo simulation of energy-dispersive X-Ray fluorescence spectrometers – I Unpolarized radiation, homogeneous samples. Spectrochim Acta B **48**, 553–573 (1993)
526. Vincze L, Janssens K, Adams F, Rindby A, A detailed ray-tracing code for capillary X-ray optics. X-ray Spectrom **24**, 27–37 (1995)
527. Vincze L, Janssens K, Adams F, Rivers, ML, Jones KW, A general Monte Carlo simulation of energy-dispersive X-Ray fluorescence spectrometers – II: Polarized monochromatic radiation, homogeneous samples. Spectrochim Acta B **50**, 127–148 (1995)
528. Vincze L, Janssens K, Adams F, Jones KW, A general Monte-Carlo simulation of energy dispersive X-ray fluorescence spectrometers: Part 3. Polarized polychromatic radiation, homogeneous samples. Spectrochim Acta B **50**, 1481–1500 (1995)
529. Schmidt C, Rickers K, In-situ determination of mineral solubilities in fluids using a hydrothermal diamond-anvil cell and SR-XRF: Solubility of AgCl in water as an example. Am Miner **88**, 288–292 (2003)
530. Bassett WA, Shen AH, Bucknum M, Chou IM, A new diamond anvil cell for hydrothermal studies to 2.5 GPa and from −190 to 1200°C. Rev Sci Instr **64**, 2340–2345 (1993)
531. Rickers K, Schmidt C, In situ determination of monazite solubility and dissolution kinetics using a diamond-anvil cell and SR-XRF, Hasylab Annual Report, 937–938 (2002)
532. Sanchez-Valle C, Martinez I, Daniel I, Philippot P, Bohic S, Simionovici A, Dissolution of strontianite at high P-T conditions: an in-situ synchrotron X-ray fluorescence study. Am Miner **88**, 978–985 (2003)
533. Brown GE Jr, Calas G, Waychunas GA, Petiau J, X-ray absorption spectroscopy: Applications in mineralogy and geochemistry. Rev Miner **18**, 431–512 (1988)
534. Brown GE Jr, Farges F, Calas G, X-ray scattering and X-ray spectroscopy studies of silicate melts. Rev Miner **32**, 317–410 (1995)
535. Schulze DG, Stucki JW, Bertsch PM (eds), Synchrotron X-ray methods in clay science. CMS Workshop Lectures **9**, 244 (1999)

536. Fenter PA, Rivers ML, Sturchio NC, Sutton SR (eds), Applications of synchrotron radiation in low-temperature geochemistry and environmental science. Rev Miner **49**, 579 (2002)
537. Henderson GS, Baker DR (eds), Synchrotron Radiation: Earth, Environmental, and material sciences applications. Min Assoc Can, Short Course, SC **30**, 178 (2002)
538. Tröger L, Arvanitis D, Baberschke K, Michaelis, H, Grimm U, Zschech E, Full correction of the self-absorption in soft-fluorescence extended X-ray-absorption fine structure. Phys Rev B **46**, 3283–3289 (1992)
539. Manceau A, Marcus MA, Nobumichi T, Quantitative speciation of heavy metals in soils and sediments by synchrotron X-ray techniques. Rev Miner **49**, 341–428 (2002)
540. Lytle FW, Greegor RB, Sandstrom DR, Marques DR, Wong J, Spiro CL, Huffman GP, Huggins FE, Measurement of soft X-ray absorption spectra with a fluorescence ion chamber detector. Nucl Instrum Methods Phys Res A **226**, 542–548 (1984)
541. Gauthier C, Solé VA, Signorato R, Goulon J, Moguiline E, The ESRF beamline ID26: X-ray absorption on ultra dilute samples. J Synchrotron Rad **6**, 164–166 (1999)
542. Brüggmann U, Haack N, Hermann M, Johnas SKJ, Kappen P, Klementiev K, Welter E, Improvements for absorption spectroscopy at beamlines A1, E4, X1, Hasylab Annual Report 2002, 83–87 (2002)
543. Dyar MD, Lowe EW, Guidotti CV, Delaney JS, $Fe^{3+}$ and $Fe^{2+}$ partitioning in metapelites: A synchrotron micro-XANES study. Am Min **87**, 514–522 (2002a)
544. Schmid R, Wilke M, Oberhänsli R, Janssens K, Falkenberg G, Franz L, Gaab A, Micro-XANES Determination of Ferric Iron and its Application in Thermobarometry. Lithos **70**, 381–392 (2003)
545. Sutton SR, Bertsch PM, Newville M, Rivers ML, Lanzirotti A, Eng P, Microfluorescence and Microtomography. Analyses of heterogeneous earth and environmental materials. Rev Miner **49**, 429–484 (2002)
546. Garvie LAJ, Buseck PR, Ratios of ferrous to ferric iron from nanometre-sized areas in minerals. Nature **396**, 667–670 (1998)
547. Welter E, Calmano W, Mangold S, Tröger L, Chemical speciation of heavy metals in soils by use of XAFS spectroscopy and electron microscopical techniques. Fresenius' J Anal Chem **364**, 238–244 (1999)
548. Calmano W, Mangold S, Welter E, An XAFS investigation of the artefacts caused by sequential extraction analyses of Pb-contaminated soils. Fresenius' J Anal Chem **371**, 823–830 (2001)
549. Vincze L, Wei F, Proost K, Vekemans B, Janssens K, He Y, Yan Y, Falkenberg G, Suitability of polycapillary optics for focusing of monochromatic synchrotron radiation as used in trace level micro-XANES measurements. J Anal At Spectrom **17**, 177–182 (2002b)
550. Wilke M, Farges F, Petit PE, Brown GE Jr, Martin F, Oxidation state and coordination of Fe in minerals: an Fe K XANES spectroscopic study. Am Miner **86**, 714–730 (2001)
551. Bajt S, Sutton SR, Delaney JS, X-ray microprobe analysis of iron oxidation-states in silicates and oxides using x-ray absorption near-edge structure (XANES). Geochim Cosmochim Acta **58**, 5209–5214 (1995)

552. Dyar MD, Gunter ME, Delaney JS, Lanzirotti, Sutton SR, Use of the spindle stage for orientation of single crystals for microXAS: Isotropy and anisotropy in Fe-XANES spectra. Am Miner **87**, 1500–1504 (2002b)
553. Jackson WE, Mustre de Leon J, Brown GE Jr, Waychunas,GA, Conradson SD, Combes J-M, High-temperature XAS study of $Fe_2SiO_4$ liquid: Reduced coordination of ferrous iron. Science **262**, 229–233 (1993)
554. Farges F, Waychunas GA, Brown GE Jr, Calas G, Galoisy L, Structural transformation in Ni-bearing $Na_2Si_2O_5$ glass and melt. Geophy Res Lett **21**, 1931–1934 (1994)
555. Farges F, Hao Gan, Rehr JR, Brown GE Jr, Navrotsky A, A Coordination chemistry of Ti(IV) in silicate glasses and melts: III. Glasses and melts from ambient to high temperatures Geochim Cosmochim Acta **60**, 3055–3065 (1996)
556. Bassett WA, Anderson AJ, Mayanovic RA, Chou IM, Hydrothermal diamond anvil cell XAFS studies of first row transition elements in aqueous solutions up to supercritical conditions. Chem Geol **167**, 3–10 (2000)
557. Anderson AJ, Jayanetti S, Mayanovic RA, Bassett WA, Chou IM, X-ray spectroscopic investigations of fluids in the hydrothermal diamond anvil cell: the hydration structure of aqueous $La^{3+}$ up to 300°C and 1600°C. Am Miner **87**, 262–268 (2002)
558. Seward TM, Henderson CMB, Charnock JM, Driesner T, An EXAFS study of solvation and ion pairing in aqueous strontium solutions to 300°C. Geochim Cosmochim Acta **63**, 2409–2418 (1999)
559. Bargar JR, Tebo BM, Villinski JE, In situ characterization of Mn(II) oxidation by spores of the marine Bacillus sp. Strain SG-1. Geochim Cosmochim Acta **64**, 2775–2778 (2000)
560. Frahm R, New method for time-dependent X-ray absorption studies. J Sci Instr **60**, 2515–2518 (1989)
561. Lützenkirchen-Hecht D, Grundmann S, Frahm R, Piezo-QEXAFS with fluorescence detection: fast time-resolved investigations of dilute specimens. J Synchrotron Rad **8**, 6–9 (2001)
562. Janssens K, Vittiglio G, Dereadt I, Aerts A, Vekemans B, Vincze L, Wei F, Deryck, Schalm O, Adams F, Rindby A, Knöchel A, Simionovici A, Snigirev A, Use of microscopic XRF for non-destructive analysis in art and archeometry. X-Ray Spectrom **29**(1), 73–91 (2000)
563. Lahanier Ch, Preusser FD, Van Zelst, Study and conservation of museum objects: Use of classical analytical techniques. Nuclear Instrum Methods Phys B **14**, 1–9 (1986)
564. Mantler M, Schreiner M, X-Ray fluorescence spectrometry in art and archaeology. X-Ray Spectrom **29**(1), 3–17 (2000)
565. Klockenkämper R, Total-Reflection X-Ray Fluorescence Analysis. Wiley, New York, 208–214 (1997)
566. Vandenabeele P, van Bohlen A, Moens L, Klockenkämper R, Joukes F, Dewispelaere G, Spectroscopic examination of two Egyptian masks—a combined method approach. Anal Lett **33**(15), 3315–3332 (2000)
567. Mommsen H, Beier Th, Dittmann H, Heimermann D, Hein A, Rosenberg A, Boghardt M, X-Ray fluorescence analysis with synchroton radiation on the inks and papers of incunabula. Archaeometry **38**, 347–357 (1996)
568. Schreiner M, Mantler M, Proceedings of the 4th International Conference on Non-destructive Testing of Museum Objects, Berlin, 221–230 (1994)

569. Fiorini C, Longoni A, Application of a new noncryogenic X-ray detector in portable instruments for archaeometric analyses. Rev Sci Instrum **69**(3), 1523–1528 (1998)
570. Vittiglio G, Janssens K, Vekemans B, Adams F, Oost A, A compact small-beam XRF instrument for in-situ analysis of objects of historical and/or artistic value. Spectrochim Acta Part B **54**, 1697–1710 (1999)
571. Bichlmeier S, Janssens K, Heckel J, Gibson D, Hoffmann P, Ortner HM, Component selection for a compact micro-XRF spectrometer. X-Ray Spectrom **30**, 8–14 (2001)
572. Bronk H, Röhrs S, Bjeoumikhov A, Langhoff N, Schmalz, Wedell R, Gorny HE, Herold A, Waldschläger U, ArtTAX®: A new mobile spectrometer for energy dispersive micro X-Ray fluorescence spectrometry on art and archaeological objects. Fresenius' J Anal Chem **371**, 307–316 (2001)
573. Kanngießer B, Malzer W, Reiche I, A new 3D micro X-ray fluorescence analysis set-up – First archaeometric applications. Nucl Instrum Methods Phys B **211**(2), 259–264 (2003)
574. Schramm HP, Hering B, Historische Malmaterialien und ihre Identifizierung. Enke Verlag Stuttgart, 19–78 (1995)
575. Harley R, Artists' Pigments: C. 1600-1835. 2nd Edition, Butterworth, Scientific London, (1982)
576. Hahn O, Oltrogge D, Bevers H, Coloured prints of the 16th century – non-destructive analyses on coloured engravings from Albrecht Dürer and contemporary artists. Archaeometry **46**(1), 273–282 (2004)
577. Klockenkämper R, von Bohlen A, Moens L, Analysis of pigments and inks on oil paintings and historical manuscripts using total reflection X-Ray fluorescence spectrometry. X-Ray Spectrom **29**(1), 119–129 (2000)
578. Neelmeijer C, Brissaud I, Calligaro T, Demortier G, Hautojärvi A, Mäder M, Martinot L, Schreiner M, Tuurnala T, Weber G, Paintings – a challenge for XRF and PIXE analysis. X-Ray Spectrom **29**(1), 101–110 (2000)
579. Wehling B, Vandenabeele P, Moens L, Klockenkämper R, von Bohlen A, Van Hooydonk G, de Reu M, Investigation of pigments in medieval manuscripts by micro Raman spectroscopy and total reflection X-Ray fluorescence spectrometry. Microchim Acta **130**, 253–260 (1999)
580. Hahn O, Gorny HE, Zerstörungsfreie Charakterisierung historischer Eisengallustinten mittels Röntgenfluoreszenzanalyse. Zeitschrift für Kunsttechnologie und Konservierung **14**(2), 384–390 (2000)
581. Malzer W, Hahn O, Kanngießer B, A fingerprint model for inhomogeneous ink paper layer systems measured with micro X-ray fluorescence analysis. X-Ray Spectrom **33**, 229–233 (2004)
582. Lutz J, Pernicka E, Energy dispersive X-ray fluorescence analysis of ancient copper alloys: Empirical values for precision and accuracy. Archaeometry **38**(2), 313–323 (1996)
583. Wadsak M, Constantinides I, Vittiglio G, Adriaens A, Janssens K, Schreiner M, Adams FC, Brunella P, Wuttmann M, Multianalytical study of patina formed on archaeological metal objects from Bliesbruck-Reinheim. Microchim Acta **133**, 159–164 (2000)
584. Klockenkämper R, Bubert H, Hasler K, Detection of near-surface silver enrichment on Roman imperial silver coins by X-ray spectral analysis. Archaeometry **41**(2), 311–320 (1999)

585. Linke R, Schreiner M, Demortier G, Alram M, Determination of the provenance of medieval silver coins: potential and limitations of X-ray analysis using photons, electrons or protons. X-Ray Spectrom **32**, 373–380 (2003)
586. Reiche I, Berger A, Görner W, Merchel S, Radtke M, Riederer J, Riesemeier H, Roth M, Following the traces of Albrecht Dürer: analysis of silverpoint drawings by spatially resolved synchrotron-induced X-ray fluorescence analysis. Nucl Instrum Methods Phys B **226**, 83–91 (2004)
587. Milazzo M, Cicardi C, X-ray fluorescence characterization of the Corona Ferrea. Archaeometry **40**(2), 351–360 (1998)
588. Pernicka E, Radtke M, Riesemeier H, Wunderlich CH, European network of competence at 1600 BC, Highlights of the BESSY Annual Report 2003, 8–9 (2004)
589. Adriens A, Veny P, Adams F, Sporken R, Louette P, Earl B, Özbal H, Yener KA, Analytical investigation of archaeological powders from Göltepe (Turkey), Archaeometry **41**(1), 81–89 (1999)
590. Dillman P, Balasubramaniam R, Characterization of ancient Indian iron and entrapped slag inclusions using electron, photon and nuclear microprobes. Bull Mater Sci **24**(3), 317–322 (2001)
591. Crocetti G, Mattias P, Barrese E, Veggiani A, Mineralogic-petrographic observations and determinations of archaeological and ceramic fragments of Perticara Mountain, Marecchia Valley, Marche [Italy]. Mineral Petrogr Acta **39**, 351–373 (1997)
592. Morandi N, Nannetti MC, Minguzzi V, Monti S, Marchesi M, Mattioli C, Desalvo F, Ceramics from the Etruscan city of Marzabotto: geochemical-mineralogical approach and connections with raw materials. Mineral Petrogr Acta **39**, 341–350 (1997)
593. Pillay AE, Punyadeera C, Jacobson L, Eriksen J, Analysis of ancient pottery and ceramic objects using X-Ray fluorescence spectrometry. X-Ray Spectrom **29**, 53–62 (2000)
594. Bakraji EH, Othman I, Karajou J, Provenance studies of archaeological ceramics from Mar-Takla (Ain-Minin, Syria) using radioisotope X-ray fluorescence method. Nucl Sci Tech **12**(2), 149–153 (2001)
595. Mandal S, Striking the balance: the roles of petrography and geochemistry in stone axe studies in Ireland. Archaeometry **39**(2), 289–308 (1997)
596. Franzini M, Leoni L, Lezzerini M, Sartori F, On the binder of some ancient mortars. Miner Petrol **67**(1–2), 59–69 (1999)
597. Röhrs S, Materialanalyse und Provenienzuntersuchung an Ziegelsteinen des 19. Jahrhunderts aus dem Berliner Umland. Berliner Beiträge zur Archäometrie **16**, 43–89 (1999)
598. Torfs K, Van Grieken R, Zezza F, Garcia N, Macri F, The cathedral of Bari, Italy: evaluation of environmental effects on stone decay phenomena. Stud Conserv **42**(4), 193–206 (1997)
599. Leung PL, Stokes MJ, Chen Tiemei, Quin Dashu, A study of ancient Chinese porcelain wares of the Song-Yuan dynasties from Cizhou and Ding kilns with energy dispersive X-ray fluorescence. Archaeometry **42**(1), 129–140 (2000)
600. Yu K, Miao JM, Characterization of blue and white porcelains using the Mn/Fe ratio from EDXRF, with particular reference to porcelains of the Xuande period (1426 to 1435 A.D.). Appl Radiat Isot **51**(3), 279–283 (1999)

601. Hein A, Tsolakidou A, Iliopoulos I, Mommsen H, Buxeda i Garrigos J, Montana G, Kilikoglou V, Standardisation of elemental analytical techniques applied to provenance studies of archaeological ceramics: an inter laboratory study. Analyst **127**(4), 542–553 (2002)
602. Adan-Bayewitz D, Asaro F, Giauque RD, Determing pottery provenance: Application of a new high-precision X-ray fluorescence method and comparison with instrumental neutron activation analysis. Archaeometry **41**(1), 1–24 (2001)
603. Garcia-Heras M, Blackman MJ, Fernandez-Ruiz R, Bishop RL, Assessing ceramic compositional data: a comparison of total reflection X-ray fluorescence and instrumental neutron activation analysis on late Iron Age Spanish Celtiberian ceramics. Archaeometry **43**(3), 325–347 (2001)
604. Seelenfreund A, Miranda J, Dinator MI, Morales JR, The provenance of archaeological obsidian artifacts from Northern Chile determined by source-induced X-ray fluorescence. J Radioanal Nucl Chem **251**(1), 15–19 (2002)
605. Wobrauschek P, Halmetschlager G, Zamini S, Jokubonis C, Trnka G, Karwoski M, Energy-dispersive X-Ray fluorescence analysis of celtic glasses. X-Ray Spectrom **29**(1), 25–33 (2000)
606. Kunicki-Goldfinger J, Kierzek J, Kasprzak A, Malozewska-Buko B, A study of eighteenth century glass vessels from Central Europe by X-ray fluorescence analysis. X-Ray Spectrom **29**, 310–316 (2000)
607. Heck M, Hoffmann P, Coloured opaque glass beads of the Merovingians. Archaeometry **42**(2), 341–357 (2000)
608. Jembrih D, Schreiner M, Peev M, Krejsa P, Clausen C, Identification and classification of iridescent glass artefacts with XRF and SEM/EDX. Microchim Acta **133**(1–4), 151–157 (2000)
609. Jembrih-Simbürger D, Neelmeijer C, Schalm O, Frederickx P, Schreiner M, De Vis K, Mäder M, Schryvers D, Caen J, The colour of silver stained glass—analytical investigations carried out with XRF, SEM/EDX, TEM and IBA. J Anal Atom Spectrom **4**, 321–328 (2002)
610. Bonizzoni L, Galli A, Milazzo M. Analyses of the stained glasses of Certosa di Pavia. X-Ray Spectrom **31**(1), 35–38 (2002)
611. Wegstein M, Urban H, Rostam-Khani P, Wittershagen A, Kolbesen BO, Total-reflection X-ray fluorescence spectrometry, a powerful tool for semi-quantitative analysis of archaeological glass samples. Spectrochim. Acta Part B **52**, 1057–1061 (1997)
612. Röhrs S, Stege H, Analysis of Limoges painted enamels from the 16th to 19th centuries by using a portable micro X-ray fluorescence spectrometer. X-Ray Spectrom **33**, 396–401 (2004)
613. Douglas JG, Chase WT, Examination of two Eastern Zhou Jade and Gold Assemblages from Jincun, Near Luoyang, Henan Province, China, and implicatons for the original configuration of the Freer Pectoral. Stud Conserv **46**, 35–48 (2001)
614. Joseph D, Lal M, Shinde PS, Padalia BD, Characterization of gem stones (rubies and sapphires) by energy-dispersive X-ray fluorescence spectrometry. X-Ray spectrom **29**, 147–150 (2000)
615. Elen S, Fritsch E, The separation of natural from synthetic colourless sapphire. Gems Gemology **35**(1), 30–41 (1999)

616. Adams F, Adriaens A, Aerts A, De Raedt I, Janssens K, Schalm O, Micro and surface analysis in art and archaeometry. J Anal Atom Spect **12**, 257–265 (1997)
617. Potts PJ, Webb PC, Williams-Thorpe O, Investigation of a correction procedure for surface irregularity effects based on scatter peak intensities in the field analysis of geological and archaeological rock samples by portable X-ray fluorescence spectrometry. J Anal Atom Spect **12**, 769–776 (1997)
618. Van Bohlen A, Meyer F, Microanalysis of old violin varnishes by total-reflection X-ray fluorescence. Spectrochim Acta Part B **52**, 1053–1056 (1997)
619. Janssens K, Vincze L, Vekemans B, Williams CT, Radtke M, Haller M, Knöchel A, The non-destructive determination of REE in fossilized bone using synchrotron radiation induced K-line X-ray microfluorescence analysis. Fresenius' J Anal Chem **363**, 413–420 (1999)
620. Kuszumow A, Vekemans B, Schalm O, Dorriné W, Chevallier P, Dillmann P, Chul-Un Ro, Janssens K, van Grieken R, Analyses of petrified wood by electron, X-ray and optical microprobes. J Anal Atom Spect **13**, 319–333 (1998)
621. Bartoll J, Unger A, Püschner K, Stege H, Micro-XRF investigations of chlorine-containing wood preservatives in art objects. Stud Conserv **48**(3), 195–202 (2003)
622. Schweizer F, Friedman AM, Archaeometry **14**, 103–107 (1972)
623. (a) Merrick JM, Metcalf DM, Archaeometry **11**, 61–65 (1969); (b) Howkes SC, Merrick JM, Metcalf DM, Archaeometry **9**, 98–138 (1966)
624. Klockenkämper R, Hasler K, Fresenius Z Anal Chem **289**, 346–352 (1978)
625. Uzonyi I, Bugoi R, Sasanu A, Kiss AZ, Constantinescu B, Torbagyi M, Nucl Instr Meth in Phys Res **B 161–163**, 748–752 (2000)
626. Archibald MM, Cowell MR, The fineness of Northumbria sceattas. In: Oddy WA (ed),. Metallurgy in Numismatics, **2**, The Royal Numismatic Society Special Publication No. 19, London, 55–64 (1988)
627. (a) Emeleus VM, Archaeometry **1**, 6–15 (1958); (b) Ravetz A, Archaeometry **6**, 46–55 (1963); (c) Blykssen H, Smith PB, Archaeometry **5**, 113–118 (1962); (d) Thiele RW, Aung Khin U, Kyaw U, Archaeometry **14**, 199–219 (1972); (e) Barrandon JN, Callu JP, Brenot C, Archaeometry **19**, 173–186 (1977)
628. (a) Reimers P, Lutz GJ, Segebade C, Archaeometry **19**, 167–172 (1977); (b) Weise HP, Segebade C, J Radioanal Chem **49**, 95–102 (1979); (c) Morrisson C, Barrondon JN, Bendall S. In Oddy WA (ed), Metallurgy in Numismatics, **2**, The Royal Numismatic Society Special Publication No. 19 London, 23–40 (1988)
629. Carter GF, Caley ER, Carlson JH, Carriveau GW, Hughes MJ, Rengan K, Segebade C, Archaeometry **25**, 201–213 (1983)
630. Code of practice for the technical examination of numismatic material. In: Oddy WA (ed), Metallurgy in Numismatics, **2**, The Royal Numismatic Society Special Publication No. 19, London, 129–132 (1988)
631. Hall ET, Archaeometry **3**, 29–35 (1960)
632. Ferreira GP, Gil FB, Archaeometry **23**, 189–197 (1981)
633. Mommsen H, Schmittinger T, Archaeometry **23**, 71–76 (1981)
634. Janssens K, Vittiglio G, Deraedt I, Aerts A, Vekemans B, Vincze L, Wie F, Deryck I, Schalm O, Adams F, Rindby A, Knöchel A, Simionovici A, Snigirev A, X-Ray Spectrom **29**, 73–91 (2000)
635. Hall ET, Archaeometry **4**, 62–66 (1961)

636. Mantler M, Schreiner M, X-Ray Spectrom **29**, 3–17 (2000)
637. (a) Condamin J, Picon M, Archaeometry **7**, 98–105 (1964). (b) Condamin J, Picon M, Changes suffered by coins in the course of time and the influence of these on the results of different methods of analysis. In: Methods of chemical and Metallurgical Investigation of Ancient Coins, Royal Numismatic Society, London (1972)
638. Delamare F, Montmitonnet P, Morrissson C. In: Oddy WA (ed), Metallurgy in Numismatics, 2. The Royal Numismatic Society Special Publication No. 19, 41–53 (1988)
639. Calliari I, Magrini M, Zambon A, Guerriero P, Martini, R, Microstructural and compositional characterization of Roman coins. X-Ray Spectrom **28**, 86–90 (1999)
640. Platzbarzdis A, Die königlich schwedische Münze in Livland, Almquist & Wiksell, Stockholm, 122 (1968)
641. Beauchesne F, Barrandon JN, Alves L, Gil FB, Guerra MF, Archaeometry **30**, 187–197 (1988)
642. Linke R, Schreiner M, Microchim Acta **133**, 165–170 (2000)
643. Cronyn J, Foley K, Numismatics and Conservation: The Conservator's view in The Museums Journal **79**, (1979), Nr. 1 to 3; Pirie E, Numismatics and Conservation: A Numismatics View, in The Museums Journal **79**, (1979), [Nr. 1]
644. Florkowski T, Stós Z, Archaeometry **17**, 165–175 (1975)
645. Zwicky-Sobczyk CN, Stern WB, Archaeometry **39**, 393–405 (1997)
646. Carter GF, Archaeometry **7**, 106–113 (1965)
647. Carter F, Anal Chem **36**, 1264–1268 (1964)
648. Winkowsky H, Münzen pflegen, Kleine numismatische Bibliothek Vol. 3 Battenberg Verlag, München, (1974)
649. Al-Kofahi MM, Al-Tarawneh KF, X-Ray Spectrom **29**, 39–47 (2000)
650. Mandal AC, Sarkar M, Bhattacharya D, Eur Phys J AP **17**, 81–84 (2002)
651. Al-Kofahi MM, Al-Tarawneh KF, Shobaki JM, X-Ray Spectrom **26**, 10–14 (1997)
652. Stahl AM, Gaspar PP, Striegel MF. In: Oddy WA (ed), Metallurgy in Numismatics, **2**, The Royal Numismatic Society Special Publication No. 19, 87–93 (1988)
653. Lutz J, Pernicka E, Archaeometry **38**, 313–323 (1996)
654. Cowell M, In Hackens T, et al. (eds), X-ray Microfluorescence Analysis Applied to Archaeology, PACT 1 Strasbourg, 76–85 (1977)
655. Stankiewicz W, Fudal A, Wojtowicz M, X-ray Spectrom **12**, 92–96 (1983)
656. Klockenkämper R, Becker M, Otto H, Spectrochim Acta **45B**, 1043–1051 (1990)
657. Auer E, Rehren T, von Bohlen A, Kirchner D, Klockenkämper R, Uber die Herstellung der ersten Platinmünzen in Russland. Metalla (Bochum) **5**, 71–90 (1998)
658. Reiff F, Bartels M, Ortner HM, Fresenius J Anal Chem **371**, 1146–1153 (2001)
659. Klockenkämper R, Hasler K, Fresenius Z Anal Chem **290**, 212–216 (1978) Klockenkämper R, Bubert H, Hasler K, Archaeometry **41**, 311–320 (1999)
660. Butcher K Am J Numismat **9**, 17–35 (1997)
661. Carter GF, Carter WH Archaeometry **16**, 201–209 (1974)

662. Gale NH, Gentner W, Wagner GA Mineralogical and geographical Silver sources of archaic Greek coinage. In Metcalf DM and Oddy WA (eds) Metallurgy in Numismatics, **1**. The Royal Numismatic Society Special Publication No. 13, 3–49 (1980)
663. Metcalf DM, Schweizer F Archaeometry **13**, 177–190 (1971)
664. Hrabánek J, Mögliche Silberquellen für die Prägung von Prager Groschen im Mittelalter, Jahrbuch des Nassauischen Vereins für Naturkunde **123**, 105–124 (2002)
665. Cowell MR, Lowick NM, in Oddy WA (ed), Metallurgy in Numismatics **2**, The Royal Numismatic Society Special Publication No. 19, 65–74 (1988)
666. Constantinescu B, Cojocaru V, Bugoi R, Sasianu A, Proc EPAC 2000, Vienna, Austria
667. Bowman R, Friedmann AM, Learner J, Milsted J, Archaeometry **17**, 157–163 (1975)
668. Brissaud I, Wang JX, Chevalier P, Nucl Inst Meth **B49**, 305 (1990)
669. Das HA, Zonderhuis J, Archaeometry **7**, 90–94 (1964)
Oddy WA, Archaeometry **14**, 109–117 (1972)
670. Bayard AM, Stoney AD, The PLM/Microscope: An electron microscope for use on a light microscope stage designed for trace evidence analysis, Microscope **47**, 209–228 (1999)
671. Christensen LH, Drabaek I, Energy-dispersive X-ray fluorescence spectrometry of industrial paint samples, Anal Chimica Acta **188**, 15–24 (1986)
672. Cousins D, Rendle D, Nolan P, The application of microspecrophotometry, X-ray diffraction and electron probe microanalysis to the analysis pigments in paints, J Forens Sci Soc **24**, 348 (1984)
673. Fischer R, Hellmiss G, Principles and forensic applications of X-ray difraction and X-ray flourescence, Advances in Forensic Science, Medical Publishers, Inc., Chicago, IL, Vol. **2**, 129–158 (1989)
674. Henson ML, Jergovich TA, Scanning electron microscopy and energy dispersive X-ray spectrometry (SEM/EDX) for the forensic examination of paints and coatings, in Forensic Examination of Glass and Paint Caddy B (ed.), Taylor & Francis, London, 243–270 (2001)
675. Mantler M, Schreiner M, X-ray fluorescence spectrometry in art and archeology, X-Ray Spectrom **29**, 3–17 (2000)
676. Massonnet G., Comparison of X-ray fluorescence and X-ray diffraction techniques for the forensic analysis of automotive paint, Crime Lab Dig **22**, 321–326 (1995)
677. Stoecklein W, Becker S., Paint and Glass—A review, Report to 13th INTERPOL Forensic Science Symposium, Lyon, October 16 (2001)
678. Suzuki EM, Infrared spectra of U.S. automobile original topcoats (1974–1989): III in situ identification of some organic pigments used in yellow, orange and brown non-metallic and brown metallic finishes-benzimidazolones, J Forens Sci **42**, 619–648 (1997)
679. Thronton JI, Forensic paint examinations, in Forensic Science Handbook (ed. R. Saferstein) Prentice Hall, New Jersey, 529–571 (1982)
680. Zeichner A, Levin N, Landau E, A study of paint coat characteristics produced by spray paints from shaken and nonshaken spray cans, J Forens Sci, Vol. **37**(2), 542–555 (1992)

681. Zięba-Palus J. Selected cases of forensic paint analysis, Science and Justice **39**(2), 123–127 (1999)
682. Zięba-Palus J. Application of micro-Fourier transform infrared spectroscopy to the examination of paint samples, J Mol Struct **511–512**, 327–335 (1999)
683. Barbooti MM, Zaki NS, Baha-Udolin SS, Hassan EB, Use of silica gel in the preparation of used lubricating oil samples for the determination of wear metals by flame AAS, Analyst **115**, 1059–1061 (1990)
684. Espinoza EO, Mann MJ, Leonardo L, Copeland A, Differentiation of automotive and locomotive lubricant oil, J Forens Sci **39**(3), 839–83 (1994)
685. Freitag K, Reus U, Fleischhauer J, The application of TXRF spectrometry for determination of trace metals in lubricating oils, Fresenius Z Anal Chem **334**, 675 (1989)
686. Liu YW, Harding AR, Leyden DE, Determination of wear metals in oil using energy dispersive X-ray spectrometry, Anal Chim Acta **180**, 349–355 (1986)
687. Lloyd JBF, Evett IW, Dubery JM, Examination of petroleum products of high relative molecular mass for forensic science purposes by synchronous fluorescence spectroscopy, J Forens Sci **25**(3), 589–602 (1980)
688. Schäfer HP, Meduna U, RFA-Praxis: Elementanalyse in Frisch- und Altölen. Fresenius Z Anal Chem **334**, 359–360 (1989)
689. Siegel JA, Fisher J, Gilna C, J Forens Sci **30**, s. 741–759 (1985)
690. Vigler MS, Gaylor VF, Trace metals analysis in petroleum products by ASS, Appl Spectroscopy **28**(4), 342–344 (1974)
691. Zięba-Palus J, Examination of used motor oils by flame AAS for criminalistic purposes. A diagnostic study, Forens Sci **91**, 171–179 (1998)
692. Zięba-Palus J, Kościelniak P, Differentiation of motor oils by infrared spectroscopy and elemental analysis for criminalistic purposes, J Mol Struct, Vol. **482–483**, 531–536 (1999)
693. Almirall JR, Elemental analysis of glass fragments in Forensic Examination of Glass and Paint Caddy B (ed.), Taylor & Francis, (2001)
694. Almirall JR, Cole MD, Gettinby G, Furton KG, Discrimination of glass sources using elemental composition and refractive index: development of predictive models, Science Justice **38**(2), 93–100 (1998)
695. Becker S, Gunaratnam L, Hicks T, Stoecklein W, Warman G, The differentiation in the float glass using refractive index and elemental analysis: Comparisons of techniques, Prob Forens Sci **XLVI**, 80–92 (2001)
696. Brożek-Mucha Z., Zadora G., Differentiating between various types of glass using SEM-EDX elemental analysis. A preliminary study, Prob Forens Sci, Vol. **XXXVII**, 68–89 (1998)
697. Buscalia JA, Elemental analysis of small glass fragments in forensic sciences, Anal Chimica Acta **228**, 17–24 (1994)
698. Curran JM, Triggs CM, Almirall JR, Buckleton JS, Walsh KAJ, The interpretation of elemental composition measurements from forensic glass evidence, Sci Justice **37**(4), 241–249 (1997)
699. Engström P, Larsson S, Rindby A, Stocklassa B, A 200 µm X-ray micobeam spectrometer. Nucl. Instrum and Meth in Phys Res, B **36**, 222–226 (1989)
700. Klockenkämper R, Knoth J, Prange A, Schwenke H, Total reflection X-ray fluorescence spectroscopy, Anal Chem **64**, 1115A–1121A (1992)

701. Koons R, Rebbert P, Comparison of refractive index, energy dispersive X-ray fluorescence and inductively coupled plasma atomic emission spectrometry for forensic characterisation of sheet glass fragments, J Anal Atomic Spectrom **6**, 451–456 (1991)
702. Kubic T, Buscalia J, Total reflection X-ray fluorescence spectrometry - further application to elemental profiles on trace evidence. Proceedings of 48$^{th}$ Annual Meeting of the American Academy of Forensic Sciences, Nashville, 19–24, (1996)
703. Kuisma-Kursula P, Accuracy, precision and detection limits of SEM/WDS, SEM/EDS and PIXE in the multielemental analysis of medieval glass, X-ray spectrom **29**, 111–118 (2000)
704. Rindby A, Engström P, Larsson S, Stocklassa B, X-ray spectrom **18**, 109 (1989)
705. Ryland S, Sheet or container? Forensic glass comparisons with an emphasis on source classification, J Forens Sci **31**, 1314–1329 (1986)
706. Terry CM, van Reiscen A, Lynch BF, Quantitative analysis of glasses used within Australia, Forens Sci Int **27**, 247–260 (1985)
707. Zadora G, Brożek-Mucha Z, The use of chosen methods of statistical and chemometric analysis in forensic examinations of glasss, Problems Forens Sci Vol. **XL**, 33–71 (1999)
708. Brożek-Mucha Z, SEM-EDX study of inorganic gunshot residues from Makarov 9 mm ammunition, Problems Forens Sci Vol. **XLI**, 62–89 (2000)
709. Brożek-Mucha Z, Zadora G, Groupping of ammunition types by means of frequency of occurrence of GSR, Foren Sci Int **135**(2), 97–104 (2003)
710. Charpentier B, Desrochers C, Analysis of primer residue from lead free ammunition by X-ray microfluorescence, J Forens Sci **45**(2), 447–452 (2000)
711. DeGaetano D, Siegel JA, Klomparens KL, A comparison of three techniques developed for the sampling and analysis of gunshot residue by SEM/EDX, J Forens Sci **37**, 281–300 (1992)
712. Keeley RH, Nolan PJ, Automatic Particle analysis, Metropolitan Police Lab Rep and Rep To Interpol Meet, Paris, London Metropolitan Police, London, UK (1986)
713. Mosher PV, McVicar MJ, Randall ED, Sild EH, Gunshot residue-similar particles produced by fireworks, Can Soc Forens Sci J **31**(2), 157–168 (1998)
714. Niewoehner L, Wenz W, Application of focused ion beam systems in gunshot residue investigations, J Forens Sci **44**, 105–109 (1999)
715. Phillips SA, Pyrotechnic residues analysis-detection and analysis of characteristic particles by scanning electron microscopy/energy dispersive spectroscopy, Science Justice **41**, 73–80 (2001)
716. Romolo FS, Margot P, Identification of gunshot residue: a critical review, Forens Sci **119**, 195–211 (2001)
717. White RS, Owens AD, Automation of gunshot residue detection and analysis by scanning electron microscopy/energy dispersive X-ray analysis, J Forens Sci **32**, 321–330 (1987)
718. Wolten GM, Nesbitt RS, Calloway AR, Loper GL, Jones PE, Particle analysis for the detection of gun shot residue: Scanning electron microscopy/energy dispersive X-ray characterisation on hand deposits, J Forens Sci **24**(2), 409–422 (1979)

719. Zeichner A, Foner HA, Dvorachek M, Bergman P, Levin N, Concentration techniques for the detection of GSR by SEM/EDX, J Forens Sci **34**(2), 312–320 (1989)
720. Zeichner A, Levin N, Dvorachek M, Gunshot residues formed by using ammunitions that have mercury fulminate based primers, J Forens Sci **37**, 1567–1573 (1992)
721. Zeichner A, Levin N, More on the uniqueness of gunshot residues (GSR) particles, J Forens Sci **41**, 1027–1028 (1997)
722. Hicks T, Sermier FM, Goldmann T, Brunelle A, Champod C, Margot P, The classification and discrimination of glass fragments using non destructive X-ray μ-fluorescence, Forens Sci Int **137**, 107–118 (2003)
723. Carvalho ML, Brito J, Barreiros MA, Study of trace element concentration in human tissues by EDXRF spectrometry. X-Ray Spectrom **27**(3), 198–204 (1998)
724. Geraki K, Farquharson MJ, An x-ray fluorescence system for measuring trace element concentrations in breast tissue. Radiat Phys Chem **61**(1–6), 603–605 (2001)
725. Feldstein H, Cohen Y, Shenberg C, Klein A, Kojller M, Maenhaut W, J. Cafmeyer, Cornelis R, Comparison between levels of trace elements in normal and inoculated mice by XRF and PIXE. Biol Trace Elem Res **61**(2), 169–180 (1998)
726. Carvalho ML, Marques AF, X-ray fluorescence spectrometry: applications in trace elements studies in human tissues from patients with cirrhosis. X-Ray Spectrom **30**(6), 397–402 (2001)
727. Carvalho ML, Pereira RA, Brito J, Heavy metals in soft tissues of Tursiops truncatus and Delphinus delphis from west Atlantic Ocean by X-ray spectrometry. Sci Tot Environ **292**(3), 247–254 (2002)
728. Sanchez HJ, Perez C, Perez RD, XRF analysis of arsenic uptaking in mice organs and tissues. Nucl Instrum Methods Phys Res Sect B **124**(1), 140–142 (1997)
729. Boman J, Wagner A, Brauer H, Binh DV, Trace elements in tissues from Vietnamese animals. X-Ray Spectrom **30**(6), 388–392 (2001)
730. Dhawan D, Singh A, Singh B, Bandhu HK, Chand B, Singh N, Effect of lithium augmentation on the trace elemental profile in diabetic rats. Biometals **12**(4), 375–382 (1999)
731. Senda J, Hashiguchi N, Tanaka Y, Komiyama S, Shimahara M, Kono K, Watanabe T, Dote T, Usuda K, Determination of bone calcium and phosphorus in osteoporosis model rats by X-ray fluorescence analysis. Trace Elem Electrolytes **18**(2), 55–58 (2001)
732. Ates A, Ertugrul M, Determination of trace elements of emboli (clots) in the cardiovascular systems of patients by dispersive x-ray fluorescence analysis. Instrum Sci Technol **30**(4), 449–454 (2002)
733. Alfoldy B, Torok S, Balashazy I, Hofmann W, Winkler-Heil R, EPMA and XRF characterization of therapeutic cave aerosol particles and their deposition in the respiratory system. X-Ray Spectrom **31**(5), 363–367 (2002)
734. Goel A, Chauhan DP, Dhawan DK, Protective effects of zinc in chlorpyrifos induced hepatotoxic biochemical and trace elemental study. Biol Trace Elem Res **74**(2), 171–183 (2000)

735. Liu NQ, Xu Q, Hou XL, Liu PS, Chai ZF, Zhu L, Zhao ZY, Wang ZH, Li YF, The distribution patterns of trace elements in the brain and erythrocytes in a rat experimental model of iodine deficiency. Brain Res Bull **55**(2), 309–312 (2001)
736. Schindler R, Stahl C, Venz S, Ludat K, Krause W, Frei U, Removal of contrast media by different extracorporeal treatments. Nephrol Dial Transplant **16**(7), 1471–1474 (2001)
737. Dede Y, Erten HN, Zararsiz A, Efe N, Determination of trace element levels in human scalp hair in occupationally exposed subjects by XRF. J Radioanal Nucl Chem **247**(2), 393–397 (2001)
738. MacPherson A, Bacso J, Relationship of hair calcium concentration to incidence of coronary heart disease. Sci Tot Environ **255**(1–3), 11–19 (2000)
739. Meitin JJ, De La Fuente F, Mendoza A, Some achievements on EDXRF analysis in Cuba. Applications in industry, medicine and environment. [Nat Cent. Scient. Res. (CNIC), Havana, Cuba]. Proceedings of the European Conference on Energy Dispersive X-Ray Spectrom (EDXRS-98), Bologna, Italy, June 7–12, 351–355 (1999)
740. Toribara TY, Analysis of single hair by XRF discloses mercury intake. Hum Exp Toxicol **20**(4), 185–188 (2001)
741. Carvalho ML, Rodrigues Ferreira FE, Neves MCM, Casaca C, Cunha AS, Marques JP, Amorim P, Marques AF, Marques MI, Arsenic detection in nineteenth century Portuguese King post mortem tissues by energy-dispersive x-ray fluorescence spectrometry. X-Ray Spectrom **31**(4), 305–309 (2002)
742. Todd AC, L-shell x-ray fluorescence measurements of lead in bone: system development. Phys Med Biol **47**(3), 507–522 (2002)
743. Todd AC, L-shell x-ray fluorescence measurements of lead in bone: theoretical considerations. Phys Med Biol **47**(3), 491–505 (2002)
744. Kondrashov VS, Rothenberg SJ, How to calculate lead concentration and concentration uncertainty in XRF in vivo bone lead analysis. Appl Radiat Isot **55**(6), 799–803 (2001)
745. Todd AC, Carroll S, Geraghty C, Khan FA, Moshier EL, Tang S, Parsons PJ, L-shell x-ray fluorescence measurements of lead in bone: accuracy and precision. Phys Med Biol **47**(8), 1399–1419 (2002)
746. Brito JAA, McNeill FE, Chettle DR, Webber CE, Vaillancourt C, Study of the relationships between bone lead levels and its variation with time and the cumulative blood lead index, in a repeated bone lead survey. J Environ Monit **2**(3), 271 (2000)
747. Cheng YW, Schwartz J, Sparrow D, Aro A, Weiss ST, Hu H, Bone lead and blood lead levels in relation to baseline blood pressure and the prospective development of hypertension. (The Normative Aging Study). Am J Epidemiol **153**(2), 164–171 (2001)
748. Gerhardsson L, Borjesson J, Mattsson S, Schutz A, Skerfving S, Chelated lead in relation to lead in bone and ALAD genotype1. Environ Res Sect A **80**(4), 389–398 (1999)
749. Cheng Y, Schwartz J, Vokonas PS, Weiss ST, Aro A, Hu H, Electrocardiographic conduction disturbances in association with low-level lead exposure (The Normative Aging Study). Am J Cardiol **82**(5), 594–599 (1998)

750. Borjesson J, Mattsson S, Alpstein M, Trace element concentrations studied in vivo using X-ray fluorescence analysis. Appl Radiat Isot **49**(5–6), 437–445 (1998)
751. Kadhim R, Al-Hussany A, Ali PA, Hancock DA, El-Sharkawi AM, In vivo measurement of platinum in the kidneys using X-ray fluorescence. In Vivo Body Compos Stud **904**, 263–266 (2000)
752. MacLean DS, Robertson JD, Jay M, Monitoring the retention of a protein antigen in complete Freund's adjuvant, alum, and pluronic F-127 gel formulations by X-ray fluorescence. Pharm Dev Technol **6**(2), 41–246 (2001)
753. Bagshaw AP, Farquharson MF, Simultaneous determination of iron, copper and zinc concentrations in skin phantoms using XRF spectrometry. X-Ray Spectrom **31**, 47–52 (2002)
754. Farquharson MJ, Bradley DA, The feasibility of a sensitive low-dose method for the in vivo evaluation of Fe in skin using K-shell x-ray fluorescence (XRF). Phys Med Biol **44**(4), 955–965 (1999)
755. Farquharson MJ, Bagshaw AP, Porter JB, Abeysinghe RD, The use of skin Fe levels as a surrogate marker for organ Fe levels, to monitor treatment in cases of iron overload. Phys Med Biol **45**(2), 1387–1396 (2000)
756. Zaichick V, Ovchjarenko N, Zaichick S, In vivo energy dispersive x-ray fluorescence for measuring the content of essential and toxic trace elements in teeth. Appl Radiat Isot **50**(2), 283–293 (1999)
757. Bloch P, Shapiro IM, Soule L, Close A, Revich B, Assessment of lead exposure of children from K-XRF measurements of shed teeth. Appl Radiat Isot **49**(5–6), 703–705 (1998)
758. Ranaweera SS, Sudarshanie PMDW, Hewamanna R, X-ray fluorescence technique for the assessment of pollution by heavy metals and their uptake by some inhabitant Flora of Lunawa Lagoon, Sri Lanka. Harmonization Health Relat Environ Meas Using Nucl Isot Tech, Proc Int Symp 1996, International Atomic Energy Agency, Vienna, Austria, 451–456 (1997)
759. Mendoza A, Cesareo R, Valdes M, Meitin JJ, Perez R, Lorente Y, Energy dispersive x-ray fluorescence analysis of marine pollution indicators. J Radioanal Nucl Chem **240**(2), 459–465 (1999)
760. Araujo MF, Cruz A, Humanes M, Lopes MT, Da Silva JAL, Da Silva JJRF, Elemental composition of demospongiae from the eastern Atlantic coastal waters. Chem Speciat Bioavail **11**(1), 25–36 (1999)
761. Akyuz T, Bassari A, Bolcal C, Sener E, Yildiz M, Kucer R, Kaplan Z, Dogan G, Akyuz S, Radioisotope X-ray fluorescence and neutron activation analyses of the trace element concentrations of the rainbow trout. Czech J Phys **49**(Suppl. 1, Pt. 1, 13th Radiochem Conf 1998) 389–393 (1999)
762. Senhou A, Chouak A, Cherkaoui R, Moutia Z, Lferde M, Elyahyaoui A, Khoukhi El T, Bounakhla M, Embarche K, Gaudry A, Ayrault S, Moskura M, Sensitivity of biomonitors and local variations of element concentrations in air pollution biomonitoring. J Radioanal Nucl Chem **254**(2), 343–349 (2002)
763. Senhou A, Chouak A, Cherkaoui R, Lferde M, Elyahyaoui A, El Khoukhi T, Bounakhla M, Embarche K, Bertho X, Gaudry A, Ayrault S, Piccot D, Comparison of 14 MeV-NAA, $K_0$-NAA and ED-XRF for air pollution bio-monitoring. J Radioanal Nucl Chem **253**(2), 247–252 (2002)
764. Larsson C, Helmisaari HS, Accumulation of elements in the annual rings of Scots pine trees in the vicinity of a copper-nickel smelter measured by scanning EDXRF. X-Ray Spectrom **27**(2), 133–139 (1998)

765. Viksna A, Selin Lindgren E, Standzenieks P, Analysis of pine needles by XRF scanning techniques. X-Ray Spectrom **30**(4), 260–266 (2001)
766. Anjos MJ, Lopes RT, Jesus EFO, Simabuco SM, Cesareo R, Quantitative determination of metals in radish using x-ray fluorescence spectrometry. X-Ray Spectrom **31**(2), 120–123 (2002)
767. Aragao PH, Cesareo R, Fernandes EAD, Balogun F, Prota U, Fiori M, Iron in olive tree leaves in the Mediterranean area. J Radioanal Nucl Chem **249**(2), 509–512 (2001)
768. Salvador MJ, Lopes GN, Nascimento VF, Zucchi O, Quality control of commercial tea by x-ray fluorescence. X-Ray Spectrom **31**(2), 141–144 (2002)
769. Obiajunwa EI, Adebajo AC, Omobuwajo OR, Essential and trace element contents of some Nigerian medicinal plants. J Radioanal Nucl Chem **252**(3), 473–476 (2002)
770. Racz L, Bumbalova A, Harangozo M, Tolgyessy J, Tomecek O, Determination of cesium and selenium in cultivated mushrooms using radionuclide x-ray fluorescence technique. J Radioanal Nucl Chem **245**(3), 11–614 (2000)
771. Cevik U, Ergen E, Budak G, Karabulut A, Tirasoglu E, Apaydin G, Kopya AI, Elemental analysis of Akcaabat tobacco and its ash by EDXRF spectrometry. J Quant Spectrosc Radiat Transf **78**(3–4), 409–415 (2003)
772. Joseph D, Lal M, Bajpai HN, Mathur PK, Levels of trace elements of few Indian spices by energy dispersive x-ray fluorescence (EDXRF) method. J Food Sci Technol **36**(3), 264–265 (1999)
773. Almeida E, Nascimento VF, Valencia EPE, Silva R, Concentrations of Fe, Cu and Zn in rum by EDXRF using APDC preconcentration. J Radioanal Nucl Chem **252**(3), 541–544 (2002)
774. Rewatkar V, Saoji AA, Rewatkar KG, Studies of trace elements in some varieties of rice (Oryza sativa) by Energy Dispersive X-ray Fluorescence technique. Asian J Chem **13**(1), 206–210 (2001)
775. Miah MY, Wang MK, Chino M, Energy dispersive x-ray fluorescence for rapid potassium, calcium, and chloride diagnosis in barley. J Plant Nutr **22**(2), 229–235 (1999)
776. Lopez de Ruiz RE, Olsina RA, Masi AN, Different analytical methodologies for the preconcentration and determination of trace chromium by XRF in medicinal herbs with effects on metabolism. X-Ray Spectrom **31**(2), 150–153 (2002)
777. Pukhovskij AV, Kovalev NV, [Central Res Inst Agrochem Service Agric (CINAO), Moscow, Russia], Presented at International Congress on Analytical Chemistry, Moscow, Russia, June 15–21 (1997)
778. Wheeler JP, Warren ND, A dirichlet tessellation-based sampling scheme for measuring whole-body exposure. Ann Occup Hyg **46**(2), 209–217 (2002)
779. Fukumoto N, Kobayashi Y, Kurahashi M, Kojima I, X-ray fluorescent spectroscopy with a focused X-ray beam collimated by a glass capillary guide tube and element mapping of biological samples. Spectrochim Acta, Part B **54**(1), 91–98 (1999)
780. Brazeau J, Wong RK, Analysis of gunshot residues on human tissues and clothing by x-ray microfluorescence. J Forens Sci **42**(3), 424–428 (1997)
781. Wobrauschek P, Use of total reflection X-ray fluorescence analysis in the life sciences. Biol Trace Element Res **43–45** 65–71 (1994)
782. Klockenkämper R, Total reflection x-ray fluorescence analysis. Wiley, New York (1997)

783. Adams F, Janssens K, Snigirev A, Microscopic x-ray fluorescence analysis and related methods with laboratory and synchrotron radiation sources, J Anal At Spectrom **13**, 319–331 (1998)
784. Klockenkämper R, von Bohlen A, Total reflection x-ray fluorescence moving towards nanoanalysis: a survey. Spectrochim Acta B **56**, 2005–2018 (2001)
785. Wobrauschek P, Total reflection x-ray fluorescence spectrometric determination of trace elements in the femtogram region: a survey. J Anal At Spectrom **13**, 333–337 (1998)
786. Wiener G, Günther R, Michaelsen C, Knoth J, Schwenke H, Borman R, Ion beam sputtering techniques for high-resolution concentration depth profiling with glancing-incidence x-ray fluorescence spectrometry. Spectrochim Acta B **52**, 813–821 (1997)
787. Prange A, Reus U, Schwenke H, Knoth J, Optimization of TXRF measurements by variable incident angles. Spectrochim Acta B **54**, 1505–1511 (1999)
788. Wobrauschek P, Aiginger H, X-ray fluorescence analysis in the ng region using total reflection of the primary beam. Spectrochim Acta B **35**, 607–614 (1980)
789. Prange A, Kramer K, Reus U, Boron nitride sample-carriers for total reflection X-ray fluorescence. Spectrochim Acta B **48**(2), 153–161 (1993)
790. Hoffmann P, Lieser KH, Hein M, Flakowski M, Analysis of thin layers by total reflection x-ray fluorescence spectrometry. Spectrochim Acta B **44**(5), 471–476 (1989)
791. Wobrauschek P, Aiginger H, Analytical application of total reflection and polarized x-rays. Fresenius Z Anal Chem **324**, 865–874 (1986)
792. Yamaguchi H, Itoh S, Igarashi S, Naitoh K, Hasegawa R, TXRF analysis of solution samples using polyester film as a disposable sample-carrier cover. Anal Sci **14**, 909–912 (1998)
793. Schmitt M, Hoffmann P, Lieser KH, Perspex as sample-carrier in TXRF. Fresenius Z Anal Chem **328**, 594–595 (1987)
794. Klockenkämper R, von Bohlen A, Elemental analysis of environmental samples by total reflection X-ray fluorescence: a review. X-Ray Spectrom **25**, 156–162 (1996)
795. Prange A, Böddeker H, Michaelis W, Multi-element determination of trace elements in whole blood and blood serum by TXRF. Fresenius Z Anal Chem **335**, 914–918 (1989)
796. Klockenkämper R, von Bohlen A, Wiecken B, Quantification in total reflection X-ray fluorescence analysis of microtome sections. Spectrochim Acta B **44**(5), 511–517 (1989)
797. Greaves ED, Bernasconi G, Wobrauschek P, Streli C, Direct total reflection X-ray fluorescence trace element analysis of organic matrix materials with a semiempirical standard. Spectrochim Acta B **52**, 923–933 (1997)
798. Schmeling M, Alt F, Klockenkämper R, Klockow D, Multi-element analysis by total reflection X-ray fluorescence spectrometry for the certification of lichen research material. Fresenius J Anal Chem **357**, 1042–1044 (1997)
799. Zarkadas C, Karydas AG, Paradellis T, Applicability of direct total reflection X-ray fluorescence analysis in the case of human blood serum samples. Spectrochim Acta B **56**, 2219–2228 (2001)
800. Carvalho ML, Custódio PJ, Reus U, Prange A, Elemental analysis of human amniotic fluid and placenta by total reflection X-ray fluorescence and energy-dispersive X-ray fluorescence: child weight and maternal age dependence. Spectrochim Acta B **56**, 2175–2180 (2001)

801. Prange A, Total reflection x-ray spectrometry: method and applications. Spectrochim Acta B **44**(5), 437–452 (1989)
802. Marcó PLM, Jiménez E, Hernández CEA, Rojas A, Greaves ED, Determination of Zn/Cu ratio and oligoelements in serum samples by total reflection X-ray fluorescence spectrometry for cancer diagnosis. Spectrochim Acta B **56**, 2195–2201 (2001)
803. Kelkó-Lévai Á, Varga I, Zih-Perényi K, Lásztity A, Determination of trace elements in pharmaceutical substances by graphite furnace atomic absorption spectrometry and total reflection X-ray fluorescence after flow injection ion-exchange preconcentration. Spectrochim Acta B **54**, 827–833 (1999)
804. Olsson M, Viksna A, Helmisaari H-S, Multi-element analysis of fine roots of Scots pine by total reflection X-ray fluorescence spectrometry. X-Ray Spectrom **28**, 335–338 (1999)
805. Tölg G, Klockenkämper R, The role of total reflection x-ray fluorescence in atomic spectroscopy. Spectrochim Acta B **48**(2), 111–127 (1993)
806. Dargie M, Markowicz A, Tajani A, Valkovic V, Optimized sample preparation procedures for the analysis of solid materials by total reflection XRF. Fresenius J Anal Chem **357**, 589–593 (1997)
807. Wagner M, Rostam-Khani P, Wittershagen A, Rittmeyer C, Kolbesen BO, Hoffmann H, Trace element determination in drugs by total reflection x-ray fluorescence spectrometry. Spectrochim Acta B **52**, 961–965 (1997)
808. Burk RF, Hill KE, Read R, Bellew T, Response of rat selenoprotein P to selenium administration and fate of its selenium. Am J Physiol **261**, E26–E40 (1991)
809. Greaves ED, Marcó Parra LM, Rojas A, Sajo-Bohus L, Determination of platinum levels in serum and urine samples from pediatric cancer patients by TXRF. X-Ray Spectrom **29**, 349–353 (2000)
810. Marcó LM, Greaves ED, Alvarado J, Analysis of human blood serum and human brain samples by total reflection X-ray fluorescence spectrometry applying Compton peak standardization. Spectrochim Acta B **54**, 1469–1480 (1999)
811. Ruiz RF, Tornero JD, González VM, Alonso C, Quantification of Pt bound to DNA using total reflection X-ray fluorescence (TXRF). Analyst **124**, 583–585 (1999)
812. Benninghoff L, von Czarnowski D, Denkhaus E, Lemke K, Analysis of human tissues by total reflection X-ray fluorescence. Application of chemometrics for diagnostic cancer recognition. Spectrochim Acta B **52**, 1039–1046 (1997)
813. Wobrauschek P, Streli C, Kregsamer P, Ladisich W, Riederer R, Recent developments in TXRF with various excitation source. J Trace Microprobe Technol **14**, 103–117 (1996)
814. Knöchel A, Lechtenberg F, Paulsen M, Ranck A, Staub S, Weseloh G, Ortsabhängige Multielementanalysen in schadstoffbelasteten Lungengeweben von Arbeitern aus dem Uranbergbau, ISBN: 3-924330-41-7, Hamburg, 1–126 (2001)
815. von Czarnowski D, Denkaus E, Lemke H, Determination of trace element distribution in cancerous and normal human tissues by total reflection X-ray fluorescence analysis. Spectrochim Acta B **52**, 1047–1052 (1997)
816. Kubala-Kukuś A, Braziewicz J, Bana D, Majewska U, Gód S, Urbaniak A, Trace element load in cancer and normal lung tissue. Nucl Instr Meth Phys Res B **150**, 193–199 (1999)

817. Majewska U, Braziewicz J, Bana D, Kubala-Kukuś A, Gód S, Pajek M, Zadrozna M, Jaskóla M, Czyzewski T, Some aspects of statistical distribution of trace element concentrations in biomedical samples. Nucl Instr Meth Phys Res B **150**, 254–259 (1999)
818. Suzuki T, Sasaki A, Nakai I, Total reflection x-ray fluorescence analysis of cisplatin platinum in human cancerous tissues obtained by biopsy. J Trace Microprobe Tech **19**(4), 547–561 (2001)
819. Pérez JM, Montero EI, Quiroga AG, Fuertes MA, Alonso C, Navarro-Ranninger C, Cellular uptake, DNA binding and apoptosis induction of cytotoxic trans-[$PtCl_2$(N,N-dimethylamine) (isopropylamine)] in A2780cisR ovarian tumor cells. Metal Based Drugs **8**(1), 29–37 (2001)
820. Tapia L, Suazo M, Hödar C, Cambiazo V, González M, Copper exposure modifies the content and distribution of trace metals in mammalian cultured cells. BioMetals **16**, 169–174 (2003)
821. Günther K, von Bohlen A, Simultanous multi-element determination in vegetable foodstuffs and their respective cell fractions by total reflection X-ray fluorescence (TXRF). Z Lebensm Unters Forsch **190**, 331–335 (1990)
822. González M, Tapia L, Alvarado M, Tornero JD, Fernández R, Intracellular determination of elements in mammalian cultured cells by total reflection X-ray fluorescence spectrometry. J Anal At Spectrom **14**, 885–888 (1999)
823. Prange A, Schwenke H, Trace element analysis using total reflection X-ray fluorescence spectrometry. Adv X-Ray Anal **35**, 899–923 (1992)
824. Trunova V, Parshina N, Kondratyev V, Determination of the distribution of trace elements in human hair as a function of the position on the head by SRXRF and TXRF. J Synchrotron Rad **10**, 371–375 (2003)
825. von Bohlen A, Rechmann P, Tourmann JL, Klockenkämper R, Ultramicro-analysis of dental plaque films by total reflection x-ray fluorescence. J Trace Elem Electrolytes Health Dis **8**, 37–42 (1994)
826. Bernasconi G, Buoso MC, Ceccato D, Cinque G, Dargie M, Galassini S, Moschini G, Passi P, Valkovic V, Varotto S, Zadro A, PIXE and TXRF measurements of titanium levels in rabbit organs after endosseous implants insertion. Nucl Instr Meth Phys Res B **109/110**, 372–374 (1996)
827. Wamwangi DM, Rathore I VS, Katia SK, Mangala MJ, Trace element analyses of pollen, bee tissue, and honey by total reflection and radioisotope excited x-ray fluorescence spectrometry. J Trace Microprobe Tech **18**(2), 303–310 (2000)
828. Kump P, Nečemer M, Šnajder J, Determination of trace elements in bee honey, pollen, and tissue by total reflection and radioisotope x-ray fluorescence spectrometry. Spectrochim Acta B **51**, 499–507 (1996)
829. Martinez T, Lartigue J, Avila-Perez P, Zarazua G, Cabrera L, Tejeda S, Ramirez A, Determination of lead in blood by TXRF and its correlation to environmental lead. Nucl Instr Meth Phys Res B **213**, 584–589 (2004)
830. Carrington CD, Sheehan DM, Bolger PM, Hazard assessment of lead. Food Additives Contam **10**(3), 325–335 (1993)
831. Ayala RE, Alvarez EM, Wobrauschek P, Direct determination of lead in whole human blood by total reflection x-ray fluorescence spectrometry. Spectrochim Acta B **46**(10), 1429–1432 (1991)
832. Viksna A, Selin E, Intercomparison between energy dispersive and total reflection X-ray fluorescence in whole blood analysis. J Trace Microprobe Tech **14**(4), 763–774 (1996)

833. Knoth J, Schwenke H, Marten R, Glauer J, Determination of copper and iron in human blood serum by energy dispersive x-ray analysis. J Clin Chem Clin Biochem **15**, 537–560 (1977)
834. Yap CT, X-ray total reflection fluorescence analysis of iron, copper, zinc, and bromine in human serum. Appl Spectros **42**(7), 1250–1253 (1988)
835. Knöchel A, Bethel U, Hamm V, Multi-element analysis of trace elements in blood serum by TXRF. Fresenius Z Anal Chem **334**, 673 (1989)
836. Bethel U, Hamm V, Knöchel A, Untersuchungen zur Bestimmung von Spurenelementen in Blutserum mit Hilfe der Totalreflexions-Röntgenfluoreszenzanalyse. Fresenius Z Anal Chem **335**, 855–859 (1989)
837. Dogan P, Dogan M, Klockenkämper R, Determination of trace elements in blood serum of patients with Behçet Disease by total reflection x-ray fluorescence analysis. Clin Chem **39**(6), 1037–1041 (1993)
838. Buoso MC, Ceccato D, Moschini G, Bernardini D, Testoni S, Torboli A, Valdes M, Assessment of serum selenium levels in 2-month-old sucking calves using total reflection X-ray fluorescence technique. Spectrochim Acta B **56**, 2181–2186 (2001)
839. Savage I, Haswell SJ, The development of analytical methodology for simultaneous trace elemental analysis of blood plasma samples using total reflection x-ray fluorescence spectrometry. J Anal At Spectrom **13**, 1119–1122 (1998)
840. Knoth J, Schwenke H, An x-ray fluorescence spectrometer with totally reflecting sample support for trace analysis at the ppb level. Fresenius Z Anal Chem **291**, 200–204 (1978)
841. Bellisola G, Pasti F, Valdes M, Torboli A, The use of total reflection X-ray fluorescence to track the metabolism and excretion of selenium in humans. Spectrochim Acta B **54**, 1481–1485 (1999)
842. Mansoor MA, Bergmark C, Haswell SJ, Savage IF, Evans PH, Berge RK, Svardal AM, Kristensen O, Correlation between plasma total homocysteine and copper in patients with Peripheral Vascular Disease. Clin Chem **46**(3), 385–391 (2000)
843. Messerschmidt J, von Bohlen A, Alt F, Kockenkämper R, Separation and enrichment of palladium and gold in biological and environmental samples, adapted to the determination by total reflction x-ray fluorescence. Analyst **125**, 397–399 (2000)
844. Zarkadas C, Karydas AG, Paradellis T, Determination of uranium in human urine by total reflection X-ray fluorescence. Spectrochim Acta B **56**, 2505–2511 (2001)
845. Knöchel A, Prange A, Analytik von Elementspuren in Meerwasser, Fresenius Z Anal Chem **306**, 252–258 (1981)
846. Liendo JA, González AC, Castelli C, Gómez J, Jiménez J, Marcó L, Sajo-Bohus L, Greaves ED, Fletcher NR, Baumann S, Comparison between proton-induced x-ray emission (PIXE) and total reflection x-ray fluorescence (TXRF) spectrometry for the elemental analysis of human amniotic fluid. X-ray Spectrom **28**, 3–8 (1999)
847. Ostachowicz B, Boruchowska M, Lankosz M, Tomik B, Szczudlik A, Adamek D, Application of the TXRF method to the elemental analysis of cerebrospinal fluid in amyotrophic lateral sclerosis. X-ray Spectrom **33**, 46–49 (2004)
848. Wittershagen A, Rostam-Khani P, Zickermann V, Zickermann I, Gemeinhardt S, Ludwig B, Kolbesen BO, Determination of metal-cofactors in

respiratory chain complexes by total reflection x-ray fluorescence spectrometry (TXRF). Fresenius J Anal Chem **361**, 326–328 (1998)
849. Steinmeyer S, Kolbesen BO, Capability and limitations of the determination of sulfur in inorganic and biological matrices by total reflection x-ray fluorescence spectrometry. Spectrochim Acta B **56**, 2165–2173 (2001)
850. Mertens M, Rittmeyer C, Kolbesen BO, Evaluation of the protein concentration in enzymes via determination of sulfur by total reflection x-ray fluorescence spectrometry – limitations of the method. Spectrochim Acta B **56**, 2157–2164 (2001)
851. Michaelis W, Prange A, Trace analysis of geological and environmental samples by total reflection x-ray fluorescence spectrometry. Nucl Geophys **2**(4), 231–245 (1988)
852. Viksna A, Znotina V, Boman J, Concentrations of some elements in and on Scots pine needles. X-Ray Spectrom **28**, 275–281 (1999)
853. Mihucz VG, Tatár E, Varga A, Záray G, Cseh E, Application of total reflection X-ray fluorescence spectrometry and high-performance liquid chromatography for the chemical characterization of xylem saps of nickel contaminated cucumber plants. Spectrochim Acta B **56**, 2235–2246 (2001)
854. Miesbauer H, Köck G, Füreder L, Determination of trace elements in macrozoobenthos samples by total reflection x-ray fluorescence analysis. Spectrochim Acta B **56**, 2203–2207 (2001)
855. Mages M, Woelfl S, Tümpling WV, jun., A method for trace element determination of single Daphnia specimens using total reflection x-ray fluorescence spectrometry. Spectrochim Acta B **56**, 2209–2217 (2001)
856. Friese K, Mages M, Wendt-Potthoff K, Neu TR, Determination of heavy metals in biofilms from the River Elbe by total reflection x-ray fluorescence spectrometry. Spectrochim Acta B **52**, 1019–1025 (1997)
857. Günther K, von Bohlen A, Multi-element speciation in vegetable foodstuffs by gel permeation chromatography (GPC) and total reflection x-ray fluorescence (TXRF). Spectrochim Acta B **46**(10), 1413–1419 (1991)
858. Görgl R, Wobrauschek P, Kregsamer P, Streli C, Haller M, Knöchel A, Radtke M, Total reflection x-ray fluorescence analysis excited by synchrotron radiation (SRTXRF): variation of excitation conditions and sample geometries. X-Ray Spectrom **26**, 189–194 (1997)
859. Streli C, Development of total reflection X-ray fluorescence analysis at the Atominstitute of the Austrian Universities. X-Ray Spectrom **29**, 203–211 (2000)
860. Ellis AT, Holmes M, Kregsamer P, Potts PJ, Streli C, West M, Wobrauschek P, Atomic spectrometry update x-ray fluorescence spectrometry. J Anal At Spectrom **13**, 209R–232R (1998)
861. Yuying H, Yingrong W, Limin Z, Guangcheng L, Wei H, Lizhen Y, Jiapei C, Jiangfu L, Tongcun Z, Enhua C, Beijing synchrotron radiation total reflection x-ray fluorescence analysis facility and its applications on trace element study of cells. Spectrochim Acta B **56**, 2057–2062 (2001)
862. da Costa RSS, do Carmo M das GT, Saunders C, Lopes RT, de Jesus EFO, Simabuco SM, Trace element content of Colostrum milk in Brazil. J Food Compos Anal **15**, 27–33 (2002)
863. Muratsu S, Ninomiya T, Kagoshima Y, Matsui J, Trace elemental analysis of drugs of Abuse using synchrotron radiation total reflection X-ray fluorescence analysis (SRTXRF). J Foren Sci **47**(5), 944–949 (2002)

864. Kemner KM, Yun W, Cai S, Lai B, Lee H-R, Maser J, Legnini DG, Rodrigues W, Jastrow JD, Miller RM, Pratt ST, Schneegurt MA, Kulpa Jr CF, Using zone plates for x-ray microimaging and microspectroscopy in environmental science. J Synchrotron Rad **6**, 639–641 (1999)
865. Lengeler B, Schroer C, Tümmler J, Benner B, Richwin M, Snigirev A, Snigireva I, Drakopoulos M, Imaging by parabolic refractive lenses in the hard x-ray range. J Synchrotron Rad **6**, 1153–1167 (1999)
866. Haller M, Knöchel A, X-ray fluorescence analysis using synchrotron radiation. J Trace Microprobe Technol **14**(3), 461–488 (1996)
867. Adams F, Janssens K, Snigirev A, Microscopic x-ray fluorescence analysis and related methods with laboratory and synchrotron radiation sources. J Anal At Spectrom **13**, 319–331 (1998)
868. Janssens K, Vincze L, Vekemans B, Adams F, Haller M, Knöchel A, Use of leadglass capillaries for micro-focusing of highly-energetic (060 keV) synchrotron radiation. J Anal At Spectrom **13**, 339–350 (1998)
869. Hayakawa S, Tohno S, Takagawa K, Hamamoto A, Nishida Y, Suzuki M, Sato Y, Hirokawa T, Ultra trace characterization using an X-Ray microprobe at SPring-8 BL39XU. Anal Sci **17** (Suppl), i115–i117 (2001)
870. Simon R, Buth G, Hagelstein M, The X-ray-fluorescence facility at ANKA, Karlsruhe: Minimum detection limits and micro probe capabilities. Nucl Instr Methods B **199**, 554–558 (2003)
871. Proost K, Vincze L, Janssens K, Gao N, Bulska E, Schreiner M, Falkenberg G, Characterization of a polycapillary lens for use in micro-XANES experiments. X-Ray Spectrom **32**, 215–222 (2003)
872. Somogyi A, Drakopoulos M, Vincze L, Vekemans B, Camerani C, Janssens K, Snigirev A, Adams F, ID18F: a new micro-x-ray fluorescence end-station at the European Synchrotron Radiation Facility (ESRF): preliminary results. X-Ray Spectrom **30**, 242–252 (2001)
873. Pérez CA, Radtke M, Sánchez H, Tolentino H, Neuenshwander RT, Barg W, Rubio M, Silveira Bueno MI, Raimundo IM, Rohwedder JJR, Synchrotron radiation X-Ray fluorescence at the LNLS: Beamline instrumentation and experiments. X-Ray Spectrom **28**, 320–326 (1999)
874. Jones KW, Berry WJ, Borsay DJ, Cline HT, Conner WC Jr, Fullmer CS, Applications of synchrotron radiation-induced x-ray emission (SRIXE). X-Ray Spectrom **26**, 350–358 (1997)
875. Maser J, Wagner D, Lai B, Legnini D, Moric I, Bermudez L, Trace metals and their relation to bacterial infections studied by x-ray microscopy. J Phys IV France **104**, 283–288 (2003)
876. Szczerbowska-Boruchowska M, Lankosz M, Ostachowicz J, Adamek D, Krygowska-Waijs A, Tomik B, Szczudlik A, Simionovici A, Bohic S, Application of synchrotron radiation for elemental microanalysis of human central nervous system tissue. J Phys IV France **104**, 325–328 (2003)
877. Dillon CT, Lay PA, Kennedy BJ, Stampfl APJ, Cai Z, Ilinski P, Rodrigues W, Legnini DG, Lai B, Maser J, Hard x-ray microprobe studies of chromium(VI)-treated V79 Chinese hamster lung cells : intracellular mapping of the biotransformation products of a chromium carcinogen. J Biol Inorg Chem **7**, 640–645 (2002)
878. Kirchgeßner M, Ernährungsphysiologische Grundlagen der Spurenelementeversorgung in: Spurenelemente und Ernährung, Wolfram G, Kirchgeßner M, (Eds.), Wiss. Verlagsgesellschaft Stuttgart, 135–149 (1990)

879. Kawai J, Takagawa K, Fujisawa S, Ektessabi A, Hayakawa S, Microbeam XANES and x-ray fluorescence analysis of cadmium in kidney. J Trace Microprobe Tech **19**(4), 541–546 (2001)
880. Vittur F, Tuniz C, Paletti S, Rizzo R, Jones KW, Elemental analysis of growth plate cartilage by synchrotron-radiation-induced x-ray emission (SRIXE). Biochem Biophys Res Commun **188**(3), 1010–1017 (1992)
881. Sparks Jr CJ, Synchrotron radiation research, Winick H, Doniach S (Eds.), Plenum Press, New York, 459–512 (1980)
882. Bohic S, Simionovici A, Snigirev A, Ortega R, Devès G, Heymann D, Schroer CG, Synchrotron hard x-ray microprobe: Fluorescence imaging of single cells. Appl Phys Lett **78**(22), 3544–3546 (2001)
883. Potts PJ, Ellis AT, Kregsamer P, Marshall J, Strehli C, West M, Wobrauschek P, Atomic spectrometry update x-ray fluorescence spectrometry. J Anal At Spectrom **18**, 1297–1316 (2003)
884. Koutzenogii KP, Savchenko TI, Chankina OV, Kovalskaya GA, Osipova LP, Bgatov AV, Synchrotron Radiation X-Ray Fluorescence Analysis (SRXRF) for measuring the multielement composition of samples of biogenic nature. J Trace Microprobe Tech **21**(2), 311–325 (2003)
885. Schulmann-Choron N, Chevallier P, de Ceaurriz J, Souleau C, Determination of selenium in rat brain by synchrotron radiation x-ray fluorescence. Analusis **28**, 316–323 (2000)
886. Pfister T, Feng H, Wimmer E, Jones KW, Synchrotron radiation-induced x-ray emission to identify metal ions in preparations of purified protein. BioTechniques **32**, 134–141 (2002)
887. Gao Y, Chen C, Chai Z, Zhao J, Liu J, Zhang P, He W, Huang Y, Detection of metalloproteins in human liver cytosol by synchrotron radiation x-ray fluorescence combined with gel filtration chromatography and isoelectric focusing separation. Analyst **127**, 1700–1704 (2002)
888. Gao Y, Chen C, Zhang P, Chai Z, He W, Huang Y, Detection of metallproteins in human liver cytosol by synchrotron radiation x-ray fluorescence after sodium dodecyl sulphate polyacrylamide gel electrophoresis. Anal Chim Acta **485**, 131–137 (2003)
889. Weseloh G, Kühbacher M, Bertelsmann H, Özaslan M, Kyriakopoulos A, Knöchel A, Behne D, Analysis of metal-containing proteins by gel electrophoresis and synchrotron radiation x-ray fluorescence. J Radioanal Nucl Chem **259**(3), 473–477 (2004)
890. Carvalho ML, Marques JP, Brito J, Casaca C, Cunha AS, Hg, Bi, Cu and Zn distribution in human teeth treated by dental amalgam measured by synchrotron microprobe. Nucl Instr Meth B **196**, 148–154 (2002)
891. Nakai I, Iwata R, Tsukamoto K, Ecological study of the migration of eel by synchrotron radiation induced x-ray fluorescence imaging of otoliths. Spectrochim Acta B **54**, 167–170 (1999)
892. Yoshitomi T, Nakayasu C, Hasegawa S, Iida A, Okamoto N, Site-specific lead distribution in scales of lead-administered carp (cyprinus carpio) by non-destructive SR-XRF analysis. Chemosphere **36**(10), 2305–2310 (1998)
893. Ektessabi AM, Rokkum M, Johansson C, Albrektsson T, Sennerby L, Saisho H, Honda S, Application of synchrotron radiation in investigation of metal-ion release from a hip replacement prosthesis. J Synchrotron Rad **5**, 1136–1138 (1998)

894. Ektessabi A, Shikine S, Kitamura N, Rokkum M, Johannsson C, Distribution and chemical states of iron and chromium released from orthopedic implants into human tissues. X-ray Spectrom **30**, 44–48 (2001)
895. Huang YY, Lu JX, He RG, Zhao LM, Wang ZG, He W, Zhang YX, Study of human bone tumor slice by SRXRF microprobe. Nucl Instr Methods A **467–468**, 1301–1304 (2001)
896. Martin RR, Sham TK, Wong Won G, Jones KW, Feng H, Synchrotron x-ray fluorescence and secondary ion mass spectrometry in tree ring microanalysis: applications to dendroanalysis. X-Ray Spectrom **30**, 338–341 (2001)
897. Martin RR, Naftel SJ, Skinner W, Jones KW, Feng H, Micro-synchrotron x-ray fluorescence of the metal distribution in a black spruce tree stem: evidence for radial mobility. X-Ray Spectrom **32**, 402–407 (2003)
898. Shimojo N, Homma-Takeda S, Ohuchi K, Shinyashiki M, Sun GF, Kumagai Y, Mercury dynamics in hair of rats exposed to methylmercury by synchrotron radiation X-ray fluorescence imaging. Life Sci. **60**(23), 2129–2137 (1997)
899. Briki F, Mérigoux C, Sarrot-Reynauld F, Salomé M, Fayard B, Susini J, Doucet J, Evidence for calcium soaps in human hair shaft revealed by sub-micrometer x-ray fluorescence. J Phys IV France **104**, 337–340 (2003)
900. Ortega R, Devès G, Bonnin-Mosbah M, Salomé M, Susini J, Anderson LM, Kasprzak KS, Chromium mapping in male mice reproductive glands exposed to $CrCl_3$ using proton and x-ray synchrotron radiation microbeams. Nucl Instr Methods B **181**, 485–488 (2001)
901. Paluszkiewicz C, Kwiatek WM, Analysis of human cancer prostate tissues using FTIR microspectroscopy and SRIXE techniques. J Mol Struct **565–566**, 329–334 (2001)
902. Kwiatek WM, Hanson AL, Paluszkiewicz C, Gaka M, Gajda M, Cichocki T, Application of SRIXE and XANES to the determination of the oxidation state of iron in prostate tissue sections. J Alloys Comp **362**, 83–87 (2004)
903. Pinheiro T, Alves LC, Barreiros A, Araujo F, Bohic S, Simionovici A, Imaging and quantification of trace metals in thin biological specimens using microprobe techniques: synchrotron induced x-ray fluorescence microprobe and nuclear microprobe. J. Phys. IV France **104**, 321–324 (2003)
904. Yoshida S, Ide-Ektessabi A, Fujisawa S, Application of synchrotron radiation in neuromicrobiology: role of iron in Parkinson's disease. Struct Chem **14**(1), 85–95 (2003)
905. Kühbacher M, Weseloh G, Thomzig A, Bertelsmann H, Falkenberg G, Radtke M, Riesemeier H, Kyriakopoulos A, Beekes M, Behne D, Analysis and localization of metal- and metalloid-containing proteins by Synchrotron-Radiation X-Ray Fluorescence. X-Ray Spectrom **34**(2), 112–117 (2005)
906. Szczerbowska-Boruchowska M, Lankosz M, Ostachowicz J, Adamek D, Krygowska-Wajs A, Tomik B, Szczudlik A, Simionovici A, Bohic S, Topographic and quantitative microanalysis of human central nervous system tissue using synchrotron radiation. X-Ray Spectrom **33**(1), 3–11 (2003)
907. Shen H, Ren QG, Mi Y, Shi XF, Yao HY, Jin CZ, Huang YY, He W, Zhang J, Liu B, Investigation of metal ion accumulation in Euglena gracilis by fluorescence methods. Nucl Instrum Methods B **189**, 506–510 (2002)
908. Ishihara R, Ide-Ektessabi A, Kitamura N, Fujita Y, Mizuno Y, Ohta T, Investigation of interactions of nano-particles within cells using micro-beam imaging techniques. X-Ray Spectrom **32**, 418–422 (2003)

909. Balaic DX, Barnea Z, Cai Z, Cholewa M, Deacon GB, Dillon CT, Ilinski P, Lay PA, Legnini D, Phillips D, Rainone S, Shea-McCarthy G, Stampfl APJ, Talarico T, Webster LK, Yun W, Lai B, High resolution nuclear and x-ray microprobes and their applications in single cell analysis. Nucl Instr Methods B **181**, 715–722 (2001)
910. Ortega R, Fayard B, Salomé M, Devès G, Susini J, Chromium oxidation state mapping in human cells. J Phys IV France **104**, 289–292 (2003)
911. Takemoto K, Ueki T, Fayard B, Yamamoto A, Salomé M, Scippa S, Susini J, Uyama T, Michibata H, Kihara H, Local distribution of vanadium in living blood cells of ascidians by fluorescence scanning x-ray microscopy ID21 at ESRF, J Phys IV France **104**, 333–336 (2003)
912. Dillon CT, Lay PA, Kennedy BJ, Stampfl APJ, Cai Z, Ilinski P, Rodriguez W, Legnini DC, Lai B, Maser J, Hard x-ray microprobe studies of chromium(VI)-treated V79 Chinese hamster lung cells: intracellular mapping of the biotransformation products of a chromium carcinogen. J Biol Inorg Chem **7**, 640–645 (2002)
913. Van Grieken RE, Markowicz AA, Handbook of X-Ray Spectrometry. 2nd Edn, Marcel Dekker, New York (2002)
914. Potts PJ, A Handbook of Silicate Rock Analysis. Blackie, Glasgow and London, 240 (1992)
915. Reynolds RC, Estimation of mass absorption coefficients by Compton scattering: improvements and extensions of the method. Am Miner **52**, 1493–1502 (1967)
916. Tertian R, Claisse F, Principles of Quantitative X-ray Fluores Anal Heiden, London (1982)
917. Spatz R, Lieser KH, Kritischer Vergleich des Meßbereichs und der Nachweisgrenzen für die energiedispersive Röntgenfluoreszenzanalyse mit Röhrenanregung (Sekundärtarget) und mit Radionuklidanregung ($^{109}$Cd und $^{241}$Am) am Beispiel von Pulverschüttproben auf Kieselgelbasis. Fresenius Z Anal Chem **319**, 732–734 (1977)
918. Strecker H, Harding G, Bomsdorf H, Kanzenbach J, Linde R, Martens G, Detection of Explosives in Airport Baggage Using Coherent X-Ray Scatter Proc SPIE **2092**, 377 (1993)
919. Luggar RD, Farqueharson MJ, Horrocks JA, Multivariate analysis of statistically poor EDXRD spectra for the detection of concealed explosives. X-Ray Spectrom **27**, 87 (1998)
920. Henrich A, Hoffmann P, Ortner HM, Greve T, Itzel H, Non-invasive identification of chemical compounds by energy-dispersive X-ray fluorescence spectrometry, combined with chemometric methods of data evaluation. Fresenius J Anal Chem **368**, 130–138 (2000)
921. Hofmann T, Hoffmann P, Lieser KH, In-line determination of Uranium, Plutonium and Neptunium in process streams by energy-dispersive X-ray fluorescence analysis (EDXRF). J Radioanal Nucl Chem **82**, 201–204 (1984)
922. Heckel J, Ryon RW, Polarized Beam X-ray Fluorescence Analysis, 2002. in Van Grieken RE, Markowicz AA, (Eds.), Handbook of X-Ray Spectrometry, 2nd Edn, Marcel Dekker, New York, 603 (2002)
923. Szegedi S, Tun KM, Ibrahim SM, Determination of ash content in coals by gamma reflection method. J Radioanal Nucl Chem **213**, 403–409 (1996)

924. Confalonieri L, Crippa R, Milazzo M, Experimental approach to analysis of binary alloys by Rayleigh-Compton back-scatter ratio method. Appl Radiat Isot **38**, 139–142 (1987)
925. Kunzendorf H, Quick determination of the average atomic number$\bar{z}$ by X-ray scattering. Nucl Instr Meth **99**, 611–612 (1972)
926. Miranda J, Andrade E, Lopez-Suarez A, Ledesma R, Cahill TA, Wakabayashi PH, A receptor model for atmospheric aerosols from a southwestern site in Mexico City. Atmos Environ **3020**, 3471–3479 (1996)
927. Remola JA, Larrechi MS, Rius FX, Chemometric characterization of 5th century A.D. amphora-producing centres in the Mediterranean. Talanta **40**, 1749–1757 (1993)
928. Benninghoff L, von Czarnowski D, Denkhaus E, Lemke K, Analysis of human tissues by total reflection X-ray fluorescence. Application of chemometrics for diagnostic cancer recognition. Spectrochim Acta Part B **52**, 1039–1046 (1997)
929. Majcen N, Rius FX, Zupan J, Linear and non-linear multivariate analysis in the quality control of industrial titanium dioxide white pigment. Anal Chim Acta **348**, 87–100 (1997)
930. Molt K, Schramm R, Application of factor analysis in EDXRF. Fresenius J Anal Chem **359**, 61–66 (1997)
931. Scan for Windows Users Manual, Version 1.1. Minitab, State College, PA (1995)
932. Friedman JH, Regularized discriminant analysis. J Am Stat Assoc **84**, 165–175 (1989)
933. Frank IE, Friedman JH Classification: oldtimers and newcomers. J Chemom **33**, 463–475 (1989)
934. Wu W, Mallet Y, Walczak B, Penninckx W, Massart DL, Heuerding S, Erni F, Comparison of regularized discriminant analysis, linear discriminant analysis and quadratic discriminant analysis, applied to NIR data. Anal Chim Acta **329**, 257–265 (1996)
935. Baldovin A, Wen W, Massart DL, Turello A, Regularized discriminant analysis RDA—Modelling for the binary discrimination between pollution types. Chemom Intell Lab Syst **381**, 25 (1997)
936. Kessler T, Hoffmann P, Greve T, Ortner HM Optimization of the identification of chemical compounds by energy-dispersive X-ray fluorescence spectrometry. X-Ray Spectrom **31**, 383–390 (2002)
937. Ernst T, Bartels M, Ohm M, Beckenkamp K, Automatische Identitätsprüfung von Feststoffen mit der energiedispersiven Röntgenfluoreszenzanalyse. GIT-Labor-Fachz **47**, 392–396 (2003)
938. Gigante GE, Pedraza LJ, Sciuti S, Analysis of metal alloys by Rayleigh to Compton ratios and X-ray fluorescence peaks in the 50 to 122 keV energy range. Nucl Instr Meth **B12**, 229–234 (1985)

# 8

# Appendix

## 8.1 X-Ray Safety and Protection

*P. Ambrosi*

### 8.1.1 Introduction

X-rays are photons of sufficient energy to ionise atoms or molecules. Such radiation is called ionising radiation. Ionisation is a process to (partially) separate charges, e.g., by separating an electron. The atoms or molecules can be gaseous (e.g., air), liquid (e.g., water), solid (e.g., semiconductors) or a mixture of these (e.g., tissue in the human body). For air, the required energy to produce one ionisation, the ionisation energy, is about 34 eV. Therefore, visible light is not an ionising radiation, because its energy is about 2.5 eV per photon. In liquids and solids the ionisation energy is lower. Photons are uncharged particles which have long pathways through matter without energy deposition and in each energy deposition process they lose large amounts of energy. The main part of this energy is converted into kinetic energy of secondary particles, mainly electrons. These electrons will lose their energy by secondary ionisation processes very close to their location of generation. These processes will occur with defined probability, resulting in an exponential attenuation law. Such radiation is called indirect ionising radiation. The opposite is direct ionising radiation, e.g., electrons, which have short pathways from one energy deposition to the other and in each of the frequent energy deposition processes they lose small but constant amounts of energy. These "quasi" continuous energy deposition processes result in a maximum range of radiation.

In the last few decades, two main developments have taken place in the dosimetry of ionising radiation: the redefinition of the measurands and the lowering of the dose limits.

The measurands are now equally defined for all kinds of radiation and in any case with respect to the human body [1–3]. Therefore, an area dose value is always a good estimate for the individual dose value of a person at the same place, and dose values due to neutron and photon radiation can be added without any difficulty. Another important consequence is that individual dosemeters and area dosemeters must be of different design, whereas for the former quantities such dual use has been possible. The "new" measurands are now used in most countries worldwide.

In all countries the dose limits are prescribed by law and all values are based on recommendations of the ICRP [4]. For all EU Member States, Directive 96/29 [5] forms the basis for any national regulations. In most countries the dose limits have been lowered in the last few years.

This paper will deal only with the dosimetry of external radiation. Other topics such as dosimetry at flight altitudes and internal dosimetry will not be considered.

### 8.1.2 Radiation Protection Quantities

#### General

Radiation protection quantities serve two different purposes, and for each of them particular quantities are available. "Body dose quantities" shall quantify the hazard to the human being as well as possible and in accordance with the state of the art in medicine, e.g. they shall quantify the risk of cancer. It is a disadvantage of the body dose quantities that in most cases they cannot be measured. This is why "operational quantities" are fixed, which are defined for a specific point and thus are measurable. They should safely estimate the protection quantities under as many sets of conditions as possible, in other words, they should be conservative. In addition, they should overestimate the body dose quantities only to a small extent.

The body dose quantities are limited by law because the law-makers intend to limit the health hazards. In radiation protection monitoring, only operational quantities can be measured from which the numerical values of the body dose quantities are to be determined. It is easiest to set the numerical values of the body dose quantities equal to those of the measured operational quantities. This is justified if the measured dose value is much smaller than the dose limits. If the measured dose value is of the same order as the dose limit, the workplace conditions must be taken into account, e.g. whether the whole body or only parts of it are irradiated, what has been the direction of radiation incidence, etc.

Both quantities will be described in some detail in the next section. A more detailed description is given in the recommendations of the ICRP [1].

#### Absorbed Dose

The basic physical quantity is the "absorbed dose $D$" which is a measure of the energy absorbed per unit mass in any medium by any type of ionising radiation. The unit of the absorbed dose is the gray (Gy), $1\,\mathrm{Gy} = 1\,\mathrm{J/kg}$.

### Body Dose Quantities

*Organ Equivalent Dose*

The organ equivalent dose $H_T$ is the product of the organ-absorbed dose $D_T$ and a *radiation weighting factor* $w_R$

$$H_T = w_R \, D_T. \tag{8.1}$$

The organ absorbed dose $D_T$ is the mean absorbed dose in organ T and the radiation weighting factor $w_R$ allows for the biological effectiveness of the kind of radiation involved. For photon and beta radiation it is unity and for neutron radiation it is between 5 and 20 depending on neutron energy.

*Effective Dose*

The effective dose $E$ is the sum of weighted organ equivalent dose values for all relevant organs of the body

$$E = \sum_T w_T \, H_T. \tag{8.2}$$

The *tissue weighting factor* $w_T$ allows for the different radiation sensitivities of the organs in the body. It is between 0.01 for the whole skin and 0.20 for the gonads, see Table 8.1. The unit of the effective dose has a specific name, the sievert (Sv), 1 Sv = 1 J/kg to emphasise the purpose of radiation protection.

The tissue weighting factors $w_T$ are normalised so that

$$\sum_T w_T = 1. \tag{8.3}$$

**Table 8.1.** Tissue weighting factors $w_T$ as given by ICRP [4]

| Organ or tissue | Tissue weighting factor $w_T$ |
|---|---|
| gonads | 0.20 |
| bone marrow (red) | 0.12 |
| colon | 0.12 |
| lung | 0.12 |
| stomach | 0.12 |
| bladder | 0.05 |
| breast | 0.05 |
| liver | 0.05 |
| oesophagus | 0.05 |
| thyroid | 0.05 |
| skin | 0.01 |
| bone surface | 0.01 |
| other tissues and organs | 0.05 |

The effective dose $E$ is an estimate of the *stochastic risk* of hazards occurring. Depending on the dose value, the hazard will occur, with a certain probability, to equally exposed persons, e.g. radiation-induced cancer will afflict 5% of the persons exposed whereas the others will not be affected. In most countries the annual limit for the effective dose for radiation workers is 20 mSv.

*Local Skin Dose*

The local skin dose is the maximum value of the equivalent dose of the basal layer of the skin (depth 0.07 mm), averaged over 1 cm$^2$. The local skin dose is an estimate of the *deterministic risk* of the occurrence of hazards. Depending on the dose, the severity of the hazard will vary, but all equally exposed persons exhibit the same symptoms. In many countries the annual limit of the local skin dose for radiation workers is 500 mSv.

### Operational Quantities

*Dose Equivalent*

To ensure measurability, all operational quantities are defined as point quantities; they are assigned the symbol $H$ and the unit sievert (Sv). In addition, they should take account of the radiation weighting factor $w_R$, but as $w_R$ is not measurable, it is replaced by the quality factor $Q$, which (at least in principle) is measurable. Therefore, the dose equivalent $H$ is the product of $Q$ and the absorbed dose $D$ at that point,

$$H = QD. \tag{8.4}$$

The quality factor is unity for photon and beta radiation and for other types of radiation increases up to 30. All operational quantities are defined in this way, so they are dose equivalents but differ in the point at which they are defined and in the distribution of the surrounding material, e.g. whether the point of definition is surrounded by a person or a phantom (as person replacement).

*Personal Dose Equivalent $H_p(10)$*

The personal dose equivalent $H_\mathrm{p}(10)$ should be a conservative estimate of the effective dose $E$ in the case of whole body exposure. It is defined as the dose equivalent in tissue at 10 mm depth in the person at the location where the dosemeter is worn. As 10 mm is the minimal depth for most organs, $H_\mathrm{p}(10)$ in almost all cases is a conservative estimate of the effective dose $E$. The person himself is part of the definition and will influence the measured dose value. The dosemeter to measure $H_\mathrm{p}(10)$ is called a whole body dosemeter and should (normally) be worn on the chest.

*Personal Dose Equivalent $H_p(0.07)$*

The personal dose equivalent $H_{\mathrm{p}}(0.07)$ should be a conservative estimate of the local skin dose. It is defined as the dose equivalent in tissue at 0.07 mm depth at that location in the person where the dosemeter is worn. This is in line with the definition of the local skin dose. The dosemeter to measure $H_{\mathrm{p}}(0.07)$ is called a partial body dosemeter and should be worn on that part of the body, e.g. a hand, where the largest dose to the skin is expected.

*Ambient Dose Equivalent $H^*(10)$*

The ambient dose equivalent $H^*(10)$ is an operational quantity for area monitoring. It is defined at 10 mm depth, just as $H_{\mathrm{p}}(10)$, but in an artificial phantom made of tissue equivalent material. The definition is rather complicated and should not be given here; for details, see [1]. It is important to note that both, $H^*(10)$ and $H_{\mathrm{p}}(10)$, are defined as conservative estimates of the effective dose $E$.

*Directional Dose Equivalent $H'(0.07)$*

The directional dose equivalent $H'(0.07)$ too is an operational quantity of area monitoring. It is defined at 0.07 mm depth just as $H_{\mathrm{p}}(0.07)$, but again in an artificial phantom made of tissue equivalent material. The rather complicated definition should not be given here; for details, see [1]. It is important to note that both, $H'(0.07)$ and $H_{\mathrm{p}}(0.07)$, are defined as conservative estimates of the local skin dose.

### Summary of Radiation Protection Quantities

Table 8.2 gives an overview of the different types of body doses, the appropriate operational quantities and the respective types of dosemeters and Table 8.3 shows as an example the dose limits for exposed workers given in the EU Directive 96/29 [5].

### 8.1.3 Health Hazards

### Whole Body Exposure

Biological effects of ionising radiation in the body include both damage to the body directly and hereditary (genetic) effects in future generations due

**Table 8.2.** Classification of body dose, operational quantity and type of dosemeter

| body dose (to state limits) | operational quantity | type of dosemeter |
|---|---|---|
| effective dose | $H_{\mathrm{p}}(10)$ | whole body dosemeter |
| | $H^*(10)$ | area dosemeter |
| local skin dose | $H_{\mathrm{p}}(0,07)$ | extremity dosimeter |
| | $H'(0,07)$ | area dosemeter |

**Table 8.3.** Dose limits for exposed workers as stated in the EU Directive 96/29 [5]

| body dose | organ | annual limit for exposed workers |
|---|---|---|
| effective dose | whole body | 20 mSv[a] |
| organ dose | lens of the eye | 150 mSv |
| | hands, forearms, feet and ankles | 500 mSv |
| local skin dose | $1\,cm^2$ of the skin | 500 mSv |

[a]The EU Directive says that the limit shall be 100 mSv over five consecutive years, subject to a maximum effective dose of 50 mSv in any single year. The Member States may fix an annual amount.

to damage to the germ cells in the reproductive organs. Radiation causes ionisation of the cell constituents (mainly water) causing chemical reactions. By these, the complex molecules which form chromosomes can become broken. These effects can cause early death of the cell, prevent or delay cell division, or pass on a permanent modification to the daughter cells. An acute exposure is one in which a large dose is received over a few hours or less. Radiation sickness, which occurs a few hours after an acute exposure, is due to damage to the cells lining the intestine. An acute whole body dose above 1 Sv will produce this effect. At these dose levels the effective dose is no longer an appropriate measurand, therefore, the absorbed dose is used instead. A dose of about 3–4 Gy is lethal for 50% of an exposed population within a few months. For whole body doses from 4 up to 10 Gy, death is due to secondary infections due to depletion of the white blood cells which normally protect against infection. Above 10 Gy gross damage occurs in the lining of the intestine and is followed by severe bacterial invasion.

Whole body doses at these levels are extremely unlikely in case of X-ray fluorescence analytical instrumentation, but low doses may still cause harmful effects in the long term.

Cancer may result from damage to the control system of a single cell, causing it to divide rapidly, the defect being transmitted to the daughter cells. There is a long and variable latent period of 5 to 30 years or more between exposure and appearance of the cancer. The fact that radiation-induced cancers are not normally distinguishable from others leads to much uncertainty as to the risk at low levels of exposure. It is possible that there is a threshold below which cancer induced by radiation does not occur, but it is normally assumed that there is a linear relationship between dose and risk and that any dose, no matter how small, carries some risk. According to ICRP [3], the combined risk of cancer induction and hereditary (genetic) effects is between 5% and 7% per Sv; which means that of 100 persons exposed to 1 Sv between 5 and 7 persons get the mentioned hazard.

### Local Exposure

In case of an accident, using X-ray fluorescence analytical instrumentation, local exposures are much more likely than whole body exposures. Fortunately, the effects of local exposure are much less than those of whole body exposures. A dose of 4 Gy to fingers or the hand may lead to changes in skin pigmentation, blistering and ulceration, but would certainly not be fatal. The hands, forearms, feet and ankles are about 100 times less sensitive to radiation, therefore, the limits given in Table 8.3 are much larger for these extremities.

Other possible late effects of radiation include cataract formation in the lens of the eye. There appears to be a threshold limit of about 15 Gy below which cataracts are not induced. The dose limit for the eye is set so that this value will not be exceeded over a whole working lifetime.

## 8.1.4 Measuring Instruments

### Detectors

Several detection principles are being used, the most important one being the direct detection of ionisation which is the mostly used principle for direct reading instruments:

1. The *ion chamber* is a fixed volume surrounded by walls and filled with gas (mainly air). The charge induced in the gas is measured without any amplification within the detector.
2. The *proportional counter* also consists of a fixed gas-filled volume but due to the higher high voltage applied, the charge pulses generated in the detector are amplified by a fixed factor. Therefore, the charge pulses are proportional to the energy stemming from the radiation particle in the volume.
3. The *Geiger-Müller counter* (GM-counter) also consists of a fixed gas-filled volume but due to the kind of gas and the high voltage applied, the charge pulses generated in the detector are amplified until saturation. Therefore, the charge pulses cannot be used to get information about the energy stemming from the radiation particle in the volume.
4. The *semiconductor detector* consists of a piece of semiconducting material. Similar to an ion chamber, the radiation induces charges which are a measure of the dose. There is no amplification in the detector.

Another detection principle is the indirect detection of ionisation, i.e., the detection of the physical effects occurring during the discharge after ionisation.

1. The *scintillation detector* generates light when exposed to ionising radiation.
2. The *thermoluminescence detector* (TL detector) exhibits light generation when the detector is heated after exposure to ionising radiation.

3. The *radio photoluminescence detector* (RPL detector) exhibits visual light generation when the detector is illuminated with UV-light after exposure to ionising radiation.
4. The *film detector* shows a blackening of the exposed film after development. This is similar to a photographic film.
5. The *radiochromic dye film detector* exhibits colouring (blackening) of the exposed film directly after exposure without development.

A special detection principle is the direct measurement of the energy absorbed in matter. Such a detector is called a *calorimeter* because the increase in temperature (of only a few 1/1,000 K) is measured.

**Dosemeter**

A dosemeter consists of one or more detectors, a reading device, an algorithm to calculate the dose value and sometimes also additional equipment. It must indicate the dose value, the measurand and the unit (e.g., $H_\mathrm{p}(10) = 0.35\,\mathrm{mSv}$). In many cases, the detectors were surrounded by specifically selected materials (plastic or metal) to obtain required properties such as, e.g., a slight dependence of the measured value on the radiation energy and the angle of radiation incidence. Dosemeters based on the same detection principle can have quite different technical specifications.

**Requirements**

There are several international and national requirements which should ensure an appropriate quality of the dosemeters. International requirements are laid down, e.g. by the "International Electrotechnical Commission" (IEC), the "International Organization for Standardization" (ISO) and the "International Atomic Energy Agency" (IAEA) and national requirements e.g., by the "American National Standards Institute" (ANSI) in the US and "Deutsches Institut für Normung e.V." (DIN) in Germany. Most requirements are drawn up to ensure a small dependence of the response on influence quantities (quantities not intended to be measured). Examples of such influence quantities are:

1. Radiation energy and angle of radiation incidence
2. Dose rate for a dose measuring instrument (linearity)
3. Climate (temperature and humidity)
4. Electromagnetic interference (cellular phones)
5. Mechanical shock
6. Light

For practical use it is important to ensure that the dosemeter selected is appropriate for the workplace conditions, e.g. the radiation energy at the workplace is included in the energy range given for the selected dosemeter. The same applies to the temperature.

### 8.1.5 System of Radiation Protection

#### General

In all countries the system of radiation protection is governed by laws, orders and guidelines which differ a lot from one country to another. To give valid information it is required to restrict this chapter to countries with similar regulations. First, the EU Member States are chosen, because for all of them the Council Directive 96/29/Euratom [5] has laid down basic safety standards for radiation protection which must be enforced by every Member State. Secondly, information about the United States of America is given.

#### EU Member States

*Prior authorization* by the competent authorities is required for practices which could result in a significant increase in the exposure of workers or members of the public which cannot be disregarded from the radiation protection point of view. In general, one such practice is the use of X-ray facilities. In the case of prior authorization, a system of radiation protection is required which includes, among other things:

1. The identification of the radiological risk for every workplace
2. The implementation of control measures relating to different working conditions including, where necessary, individual monitoring
3. Medical surveillance
4. Information and training of workers
5. Qualified experts on location

In most countries dosimetry services are available for individual monitoring, which are responsible for the dose monitoring of radiation workers. They issue dosemeters at regular intervals, e.g. every month, and process the dosemeters after the monitoring period to determine the individual dose value. These dose values are stored for several years and summed to get the annual dose value which must not exceed the dose limits stated in Table 8.3.

No prior authorization and *reporting* to the competent authorities are required in some cases listed in article 3 of the Directive [5]. In the case of X-ray facilities, it must be ensured that:

1. They are of a type approved by the competent authorities of the Member State and
2. Under normal operating conditions they do not lead to a dose rate exceeding $1\,\mu\text{Sv h}^{-1}$ at a distance of 0.1 m from any accessible surface of the apparatus

There are X-ray devices for fluorescence analysis available which fulfil these requirements and can be used without reporting to the competent authorities.

### United States of America

The dose limits in the United States of America (US) [6] are similar to those given in Table 8.3, except for the effective dose of the whole body, where 50 mSv per year are permitted instead of 20 mSv per year.

Similar to the *Prior authorization* by the competent authorities, as required by the Directive for the EU Member States, the US Nuclear Regulatory Commission (NRC) [6] states requirements for activities conducted under licenses issued by the Nuclear Regulatory Commission. The main requirements are:

(a) Each licensee shall develop, document, and implement a radiation protection program commensurate with the scope and extent of licensed activities and sufficient to ensure compliance with the provisions of this part. (See Sect. 20.2102 for record keeping requirements relating to these programs.)
(b) The licensee shall use, to the extent practical, procedures and engineering controls based upon sound radiation protection principles to achieve occupational doses and doses to members of the public that are as low as is reasonably achievable (ALARA).
(c) The licensee shall periodically (at least annually) review the radiation protection program content and implementation.

Again, similar to the Directive for the EU Member States, no such licence is required for "cabinet X-ray systems" [7, 8]; but in some states the equipment must be registered, e.g. with Environmental Health and Radiation Safety (EHRS). A cabinet X-ray system is an X-ray system installed in an enclosure. The enclosure is intended to protect people from the X-rays generated and to exclude people from the enclosure's interior. Examples of such cabinet X-ray systems are X-ray fluorescence analytical systems. However, the use of cabinet and other X-ray systems, other than by the US government, is regulated by the states. In the following some representative requirements are listed:

1. A key-activated control must be provided to ensure that X-rays will not be generated when the key is removed.
2. Each door of a cabinet X-ray system must have a minimum of two safety interlocks. Each access panel must have at least one safety interlock.
3. A control, other than the safety interlock, must be provided to resume X-ray generation following X-ray interruption by a safety interlock.
4. Two independent means must be provided to indicate when X-rays are being generated. One may be a milliamp meter labelled to indicate X-ray tube current; the other indicator must consist of an easily seen warning light labelled "X-RAY ON"
5. A clearly legible and visible label bearing the statement: "CAUTION: X-RAYS PRODUCED WHEN ENERGIZED" must be posted near the controls that energize the X-ray tube.

6. Radiation emitted from a cabinet X-ray system must not exceed an exposure rate of 5 μSv/h at any point 5 cm from the external surface at the maximum rated current and voltage.
7. A cabinet X-ray system must contain sufficient shielding and be located so exposure in unrestricted areas does not exceed 20 μSv in one hour and 1 mSv in one year.

The US limit of 5 μSv/h at any point 5 cm from the external surface is larger than the EU limit of 1 μSv/h at any point 10 cm from the external surface, because the radiation source is not the surface but mostly a point several tens of cm inside the cabinet X-ray system. In general, the other requirements are similar between US and EU.

## References


1. ICRU, International Commission on Radiation Units and Measurements, Quantities and Units in Radiation Protection Dosimetry, ICRU Report 51, ICRU Publications, Bethesda, MD (1993)
2. ICRU, International Commission on Radiation Units and Measurements, Conversion Coefficients for Use in Radiological Protection Against External Radiation, ICRU Report 57, ICRU Publications, Bethesda, MD (1998)
3. ICRP, International Commission on Radiological Protection, Recommendations of the International Commission on Radiological Protection, ICRP Publication 60, Annals of ICRP 21, No.1–3 (1991)
4. ICRP, International Commission on Radiological Protection, General Principles for the Radiation Protection of Workers, ICRP Publication 75, Annals of ICRP 27, No.1 (1997)
5. European Commission, Council Directive 96/29/Euratom laying down basic safety standards for the protection of the health of workers and the general public against the dangers arising from ionizing radiation, OJ L 159, 29.6.1996; Bull. 5–1996 (1996)
6. U.S. Nuclear Regulatory Commission, NRC Regulations (10 CFR), Part 20 – Standards for protection against radiation (1994)
7. U.S. Food and Drug Administration, Title 21–Food and drugs, Chapter I Part 1020 – Performance standards for ionizing radiation emitting products, Sec. 1020.40 Cabinet x-ray systems. [CITE: 21CFR1020.40] (2003)
8. U.S. Department of Energy, IMPLEMENTATION GUIDE For Use With Title 10, Code of Federal Regulations, Part 835, Occupational Radiation Protection, Radiation-generating devices. [CITE: G-10 CFR 835/C3 - Rev. 1] (1994)


## 8.2 Useful Data Sources and Links

*R. Wedell, W. Malzer*

Very **useful and currently updated links** can be found on the Website of the European X-Ray Spectrometry Association www.exsa.hu . Especially, for the following items links are given on this site:

1. X-ray emission line energies
   http://xray.uu.se/hypertext/XREmission.html
2. X-ray properties of the elements
   http://www.csrri.iit.edu/periodic-table.html
3. X-ray science related links http://dir.yahoo.com,/Science/Physics/X_Ray/
4. Reference Materials http://www.erm-crm.org/ermcrm
5. Physical reference data
   http://physics.nist.gov/PhysRefData/contents.html
6. X-Ray Optics http://www-cxro.lbl.gov/
7. Photon radiometry http://www.berlin.ptb.de/en/org/7/71/_index.htm
8. Synchrotron radiation facilities http://www.spring8.or.jp/e/other_sr/
9. Physics network http://www.physnet.de/PhysNet/physnet.html
10. Database for Atoms, Molecules, Gases and Plasmas
    http://gaphyor.lpgp.u-psud.fr/gaphyor/index.html
11. Atomic Physics on the Internet http;//plasma-gate.weizmann.ac.il/API.html
12. X-ray data booklet http://xdb.lbl.gov/
13. X-ray matter interaction cross sections for X-ray fluorescence applications http://ftp.esrf.rf/pub/scisoft/xraylib/
14. X-ray Anomalous Scattering http://www.bmsc.washington.edu/scatter/

Furthermore, information about upcoming and earlier conferences on X-ray analytics in and outside Europe, journals and books is also available on that website.

The following Atomic Spectra Database contains spectral reference data which have been critically evaluated and compiled by NIST. The database will be permanently updated. At present data for observed transitions of 99 elements and energy levels of 56 elements are included.

Atomic Spectra Database v. 2.0 (ASD) by D.E. Kelleher, W. L. Wiese, W.C. Martin et al. National Institute of Standards and Technology (NIST) USA: http://physics.nist.gov/asd

*General X-ray data websites and interesting links can be found at*

http://dir.yahoo.com/Science/Physics/X_Ray

*Important XRF conference series*

1. European Conference on X-Ray Spectrometry, EXRS 2006, France, June 2006: www.nucleide.org/exrs2006
2. European Conference on X-Ray Spectrometry, EXRS 2004: www.exrs2004.uniss.it
3. European Conference on Energy Dispersive X-Ray Spectrometry, EDXRS 2002: http://iapf.physik.tu-berlin.de/EDXRS
4. Denver X-Ray Conference : www.dxcicdd.com

*Relevant scientific journals*

1. X-Ray Spectrometry: www3.interscience.wiley.com (see Chemistry/Journals)
2. Spectrochimica Acta Part B: Atomic Spectroscopy: www.sciencedirect.com/science/journal/05848547
3. Radiation Physics and Chemistry: www.sciencedirect.com/science/journal/0969806X
4. Journal of Analytical Atomic Spectrometry: www.rsc.org (see Journals: JAAS)
5. Nuclear Instruments and Methods in Physics Research A: Accelerators, Spectrometers, Detectors and Associated Equipment: : www.sciencedirect.com/science/journal/01689002
6. Nuclear Instruments and Methods in Physics Research B: Interaction with Materials and Atoms: www.elsevier.com/wps/find/journaldescription.cws_home/505674/description#description
7. Atomic Data and Nuclear Data Tables: www.sciencedirect.com/science/journal/0092640X
8. Journal of Physical and Chemical Reference Data: http://jpcrd.aip.org

*Online books on X-Ray spectrometry*

1. Eds: K. Tsuji, J. Injuk, R. Van Grieken, X-Ray Spectrometry: Recent Technological Advances John Wiley & Sons 2004 Print ISBN: 047148640X, Online ISBN: 0470020431: www3.interscience.wiley.com/cgi-bin/bookhome/109075062
2. An article about X-Ray Fluorescence Spectrometry by Ron Jenkins published in Ullmann's Encyclopedia of Industrial Chemistry 2002 can be found at www.mrw.interscience.wiley.com/ueic/articles/b05_675/abstracts-fs.htm
3. The website of the Micro and Trace Analysis Centre at the University of Antwerp contains many interesting publications of this very active research group, which are mostly available free of charge: http://webhost.ua.ac.be/mitac4/
4. For a comparison between micro-XRF and electron probe microanalysis and information about layer analysis the link www.mikroanalytik.de can be recommended.

*Monographs on X-ray spectrometry and related problems*

1. R. E. Van Grieken, A. A. Markowicz (Eds.), Handbook of X-Ray Spectrometry, Second Edition, Revised and Expanded, Marcel Dekker Inc. New York 2002, pp. 36 – 91 (ISBN: 0-8247-0600-5).

2. E. P. Bertin, Principles and Practice of X-Ray Spectrometric Analysis, Plenum Press New York 1975 (ISBN: 0-306-30809-6). (A comprehensive description of the basics of X-ray spectrometry)
3. Röntgenfluoreszenzanalyse Anwendungen in Betriebslaboratorien, von einem Autorenkollektiv unter Federführung von Dr. rer. nat. Helmut Eberhardt, VEB Deutscher Verlag für Grundstoffindustrie Leipzig 1988 (in German) (ISBN: 3-342-00219-0) (The book on X-ray fluorescence analysis gives an introduction to this field especially for students and describes also industrial application of this analytical method).
4. V. Valkovic, X-Ray Spectroscopy in Environmental Sciences, CRC Press 1989 (ISBN: 0-843-34749-1)
5. J. W. Robinson, Practical Handbook of Spectroscopy, CRC Press 1991 (ISBN: 0-849-33708-9)
6. J. Goldstein, D. E. Newbury, D. Williams (Eds.), X-Ray Spectrometry in Electron Beam Instruments, Springer 1995 (ISBN: 0-306-44858-0)
7. B. Markert, Instrumental Element and Multi-Element Analysis of Plant Samples: Methods and Applications, John Wiley & Sons Ltd. 1996 (ISBN: 0-471-95865-4)
8. R. Klockenkämper, Total-Reflection X-Ray Fluorescence Analysis , John Wiley & Sons Ltd. 1996 (ISBN: 0-471-30524-3)
9. V. E. Buhrke, R. Jenkins, D. K. Smith (Eds.), A Practical Guide for the Preparation of Specimens for X-Ray Fluorescence and X-Ray Diffraction Analysis, John Wiley & Sons Ltd. 1998 (ISBN: 0-471-19458-1)
10. E. Lifshin (Ed.), X-Ray Characterization of Materials, John Wiley & Sons Ltd. 1999 (ISBN: 3-527-29657-3)
11. K. H. A. Janssens, F.C.V. Adams, A. Rindby (Eds.), Microscopic X-Ray Fluorescence Analysis, John Wiley & Sons Ltd. 2000 (ISBN: 0-471-97426-9)
12. O. Auciello, A. R. Krauss (Eds), In Situ Real-Time Characterization of Thin Films, John Wiley & Sons Ltd. 2000 (ISBN: 0-471-24141-5)

*In the following reference useful tables of X-ray data are given:*

G. Zschornack, Atomdaten für die Röntgenspektralanalyse (Atomic Data for X-Ray Spectral Analysis in German), Deutscher Verlag für die Grundstoffindustrie, Leipzig 1989 (ISBN:3-342-00308-1)

G. Zschornack, Handbook of X-Ray Data, Springer 2006, (ISBN: 3-540-28618-7) Due August 2006

*The International Union of Pure and Applied Chemistry prepares regularly reports about nomenclature, symbols units and their usage in spectrochemical analysis.*

The following publication contains the IUPAC notation for X-ray emision lines and absorption edges which replaces the Siegbahn notation:

R. Jenkins, R. Manne, R. Robin, C. Senemaud, Nomenclature System for X-ray Spectroscopy, Pure & Applied Chemistry Vol. 63, No. 5, pp. 735-746, 1991 or www.iupac.org/reports/V/spectro/partVIII.pdf

# Index

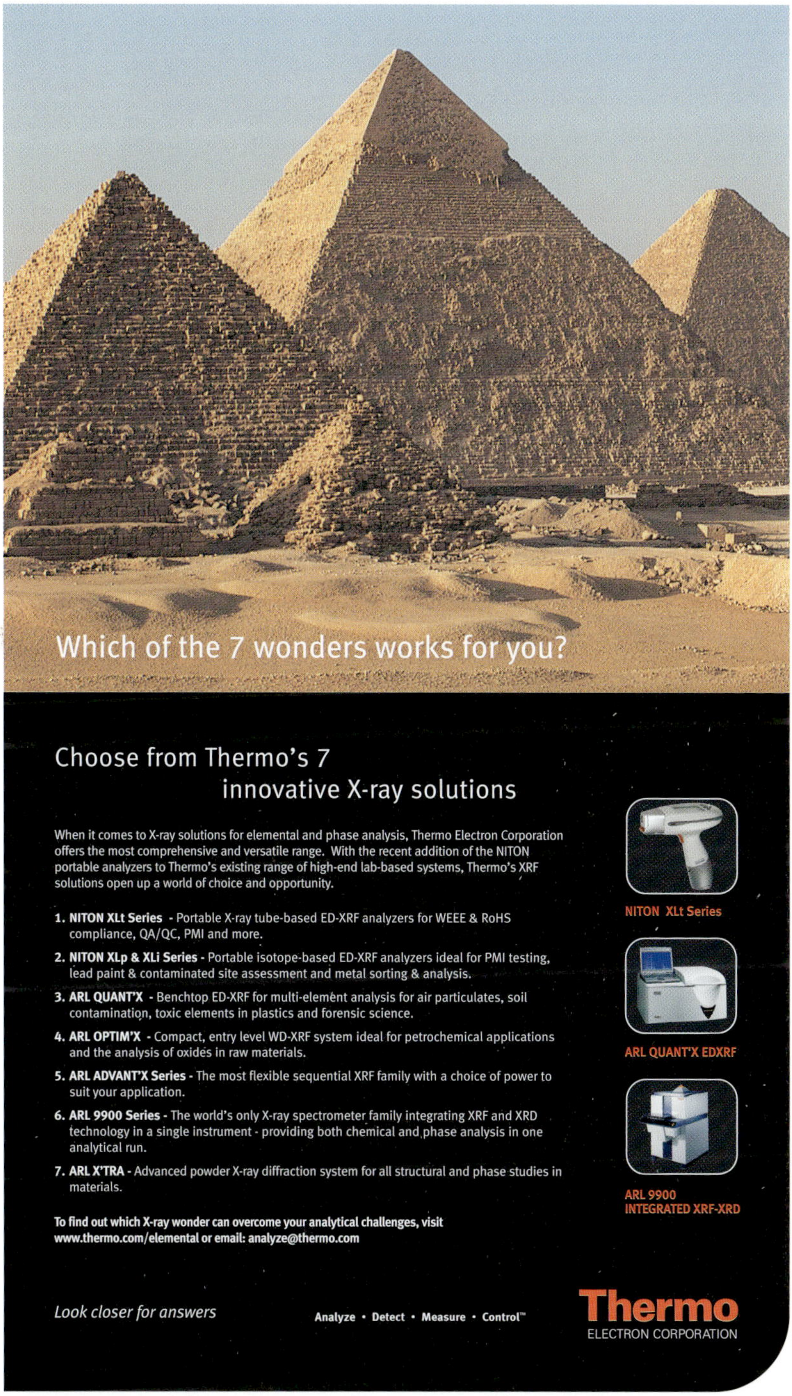
Which of the 7 wonders works for you?
Choose from Thermo's 7 innovative X-ray solutions
When it comes to X-ray solutions for elemental and phase analysis, Thermo Electron Corporation offers the most comprehensive and versatile range. With the recent addition of the NITON portable analyzers to Thermo's existing range of high-end lab-based systems, Thermo's XRF solutions open up a world of choice and opportunity.
1. NITON XLt Series - Portable X-ray tube-based ED-XRF analyzers for WEEE & RoHS compliance, QA/QC, PMI and more.
2. NITON XLp & XLi Series - Portable isotope-based ED-XRF analyzers ideal for PMI testing, lead paint & contaminated site assessment and metal sorting & analysis.
3. ARL QUANT'X - Benchtop ED-XRF for multi-element analysis for air particulates, soil contamination, toxic elements in plastics and forensic science.
4. ARL OPTIM'X - Compact, entry level WD-XRF system ideal for petrochemical applications and the analysis of oxides in raw materials.
5. ARL ADVANT'X Series - The most flexible sequential XRF family with a choice of power to suit your application.
6. ARL 9900 Series - The world's only X-ray spectrometer family integrating XRF and XRD technology in a single instrument - providing both chemical and phase analysis in one analytical run.
7. ARL X'TRA - Advanced powder X-ray diffraction system for all structural and phase studies in materials.
To find out which X-ray wonder can overcome your analytical challenges, visit www.thermo.com/elemental or email: analyze@thermo.com
NITON XLt Series
ARL QUANT'X EDXRF
ARL 9900 INTEGRATED XRF-XRD
Look closer for answers
Analyze • Detect • Measure • Control™
Thermo
ELECTRON CORPORATION

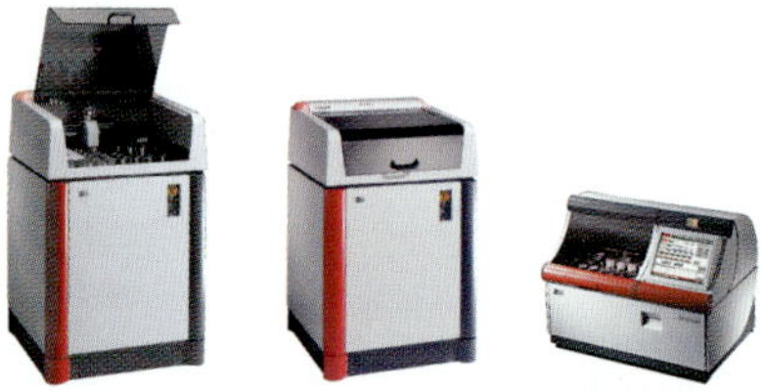

IFG Institute for Scientific Instruments

# Capillary and HOPG X-Ray Optics

## for Research, Industry and Medicine

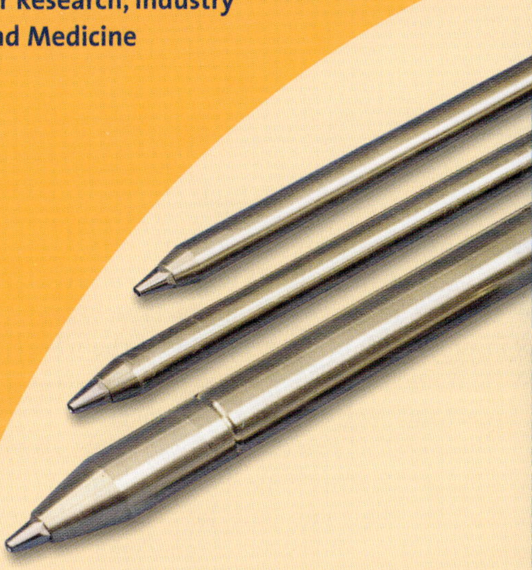

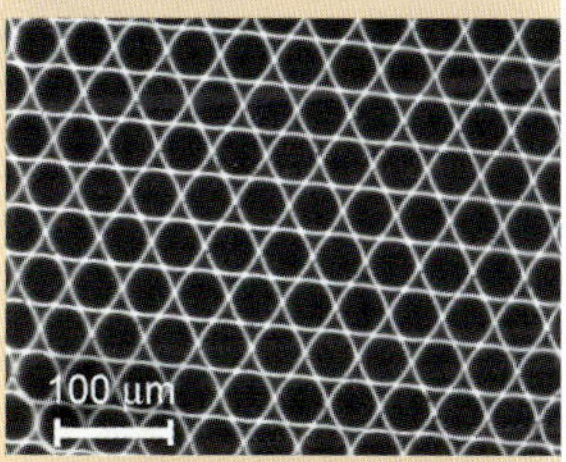


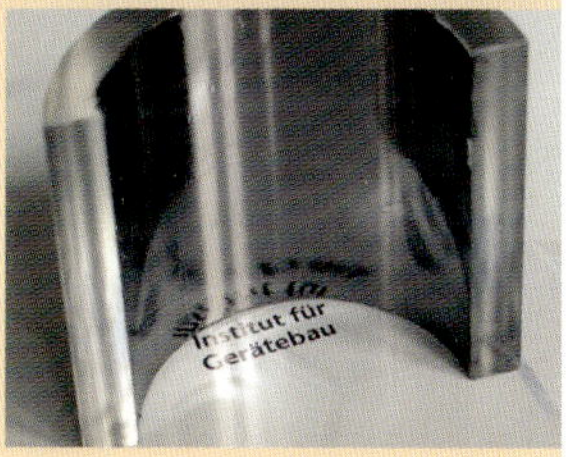


**Cylindrical, elliptical or parabolic mono-capillaries**
for beam shaping providing a medium intensity gain with either small spot size or low divergence

**Polycapillary X-Ray lenses**
for focussing, collimating or parallelising X-Ray beams with a high intensity gain

**HOPG-optics**
for the combination of monochromatisation and beam shaping due to BRAGG-reflection on surfaces with complex geometries

**IfG Institute for Scientific Instruments GmbH**
Rudower Chaussee 29/31
12489 Berlin
Phone: +49-30-6392-6500
Fax: +49-30-6392-6501
info@ifg-adlershof.de
www.ifg-adlershof.de

# Sample Preparation for X-ray Fluorescence Analysis

Dr. Rainer Schramm[1] and Hubert Dzierzawa[2]

[1] FLUXANA
Bonhoefferweg 1, 47551 Bedburg-Hau, Germany
Tel. +49 2821 973875, Fax +49 2821 973876, www.fluxana.com
[2] HD Elektronik und Elektrotechnik GmbH
Sommerdeich 22, 47533 Kleve, Germany
Tel. +49 2821 14810, Fax +49 2821 14809, www.hdelektronik.de

## 1 Introduction

To analyse a sample by X-ray fluorescence analysis (XRF) we have to prepare our sample according to the goal of analysis. XRF is a surface technique, which means that for the element of interest, only a layer with a defined thickness is analysed. This layer must be representative of the whole sample. Therefore the sample must be *homogeneous*.

For solid materials an important fact is that the intensity of the fluorescence radiation depends on the particle size especially for light elements (F to Fe). To achieve precise results, unknowns must show the *same particle size distribution* as calibration samples. This can only be achieved by a defined sample preparation procedure. Table 1 shows the dependence of the sample preparation method on the goal of analysis.

**Table 1.** Sample preparation according to goal of analysis

| goal of analysis | sample preparation |
|---|---|
| Qualitative, test if element is presentr | No preparation, sample is placed directly in a sample cup* |
| Semi quantitative and quantitative | Powder with defined particle size, e.g. 60 µm |
| Quantitative | Pellet w/o binder or fused bead |

*postulates that enough sample material is available to cover the X-ray beam

## 2 Sample Preparation

### Loose Powder / Liquids

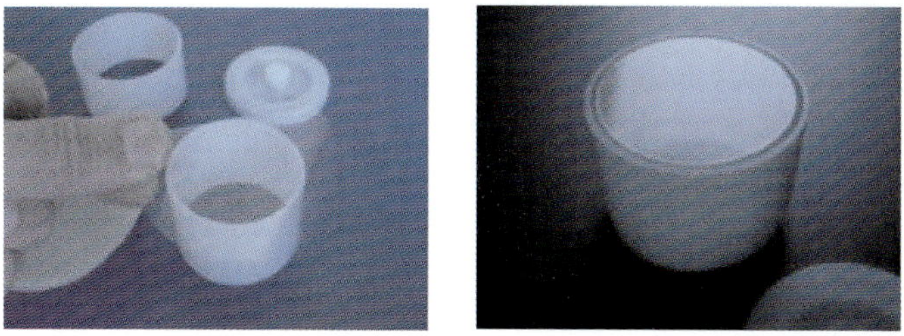

**Fig. 1.** Preparation of a liquid cell or sample cup for the direct analysis of powders or liquids

For the direct measurement of liquids or loose powders, a plastic cup is used (see Fig. 1). This cup consists of an inner ring, an outer ring and a cap (e.g. FLUXANA professional cups). Between the rings is placed an X-ray transparent film which also serves as the bottom of the sample cup. After preparing the sample cup, the sample can be poured into it and then measured.

The choice of the film is defined by the elements of interest (see Fig. 2). It must be noted that some of the films contain different impurities. (For more info: `www.fluxana.com`)

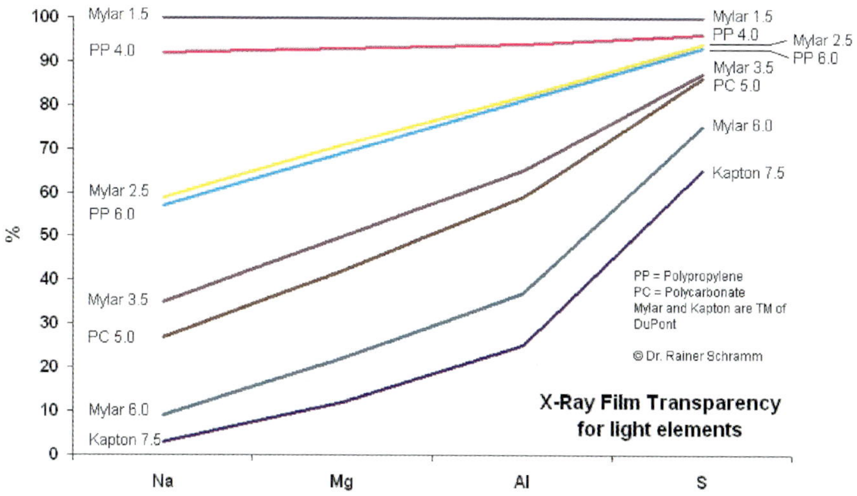


**Fig. 2.** Transparency of different films for X-ray fluorescence radiation of light elements

**Fig. 3.** Preparation of routine cement sample: 45 g of the sample is poured into a 50 ml WC jar and $2 \times 0.25$ g tablets of FLUXANA Multimix are added. A 40-mm pellet is produced by pressing at 25 tons

**Fig. 4.** Preparation of calibration standard with wax: 6 g of the powder is mixed with 2 g of FLUXANA wax. A 40-mm pellet is produced by pressing at 25 tons

## Pressed Pellets

The high error of loose powder analysis can be reduced by milling the powder and reducing the particle size to $< 60\,\mu\text{m}$. This can be achieved with a disc mill, ball mill or "swing" mill (ring and puck mill) using a milling container which is hard enough to grind the material. At this point contaminants in the container material have to be taken into account. Usual materials are agate, aluminium oxide, zirconium oxide or tungsten carbide.
(For more info: `www.fluxana.com`)

To present a good surface and a consistent density to the X-rays, a pressed pellet has to be prepared although. Not all materials will form a stable pellet under pressure. Therefore the use of a binder or grinding additive is recommended.

## Properties of Grinding Additives/Binders for X-ray Fluorescence:

- good grinding performance,
- binding property under pressure with a high variety of materials,
- good stability under X-ray radiation,
- long term stability necessary for standard preparation,
- non-hygroscopic,
- easy to handle.

Depending on the original grain size the sample is milled at $1400\,\text{s}^{-1}$ for 15–60 s. Then a pressed pellet will be produced. There are two ways of adding the binder: (1) a grinding additive like Multimix is added before milling (see Fig. 3). (2) The milling is done with the pure sample and then in a second step a binder like wax is mixed with the sample homogeneously (see Fig. 4).

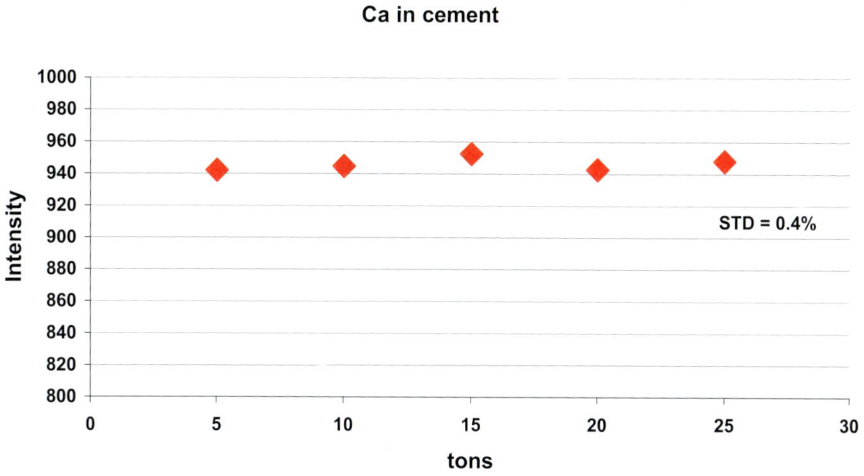


**Fig. 5.** Preparation of 32 mm pressed pellets from cement using different pressure

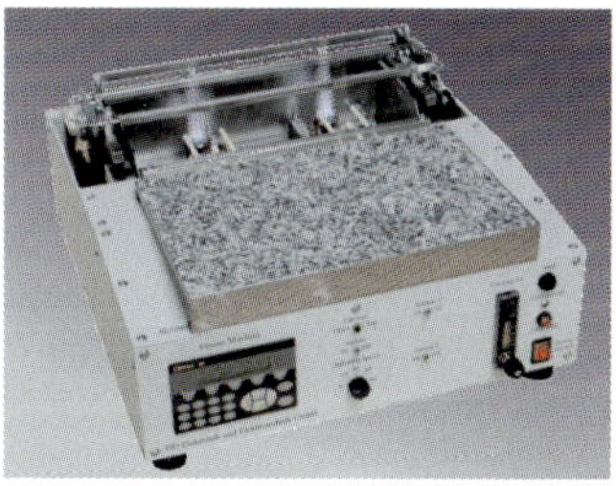

**Fig. 6.** Fully automated fusion machine VULCAN

Figure 5 shows that the influence of the pressure becomes negligible if the particle size is below 60 μm.

## Fused Beads

The minimum error achieved with pressed powder technique is not below 0.5%. For more precise analysis the fusion technique is required. In fusion the sample will be dissolved totally in borate glass. Particle size effects disappear and based on the dilution, matrix effects will be reduced. Also the surface of the bead represents an ideal sample.

## Principle

For a successful fusion the sample must be presented in an oxidic form. This is mostly the case for materials like cement, quartz, clays, ores and limestone. These materials can be fused directly after drying. A typical recipe would be 8 g of lithium tetraborate (FLUXANA FX-X100) and 1 g of sample.

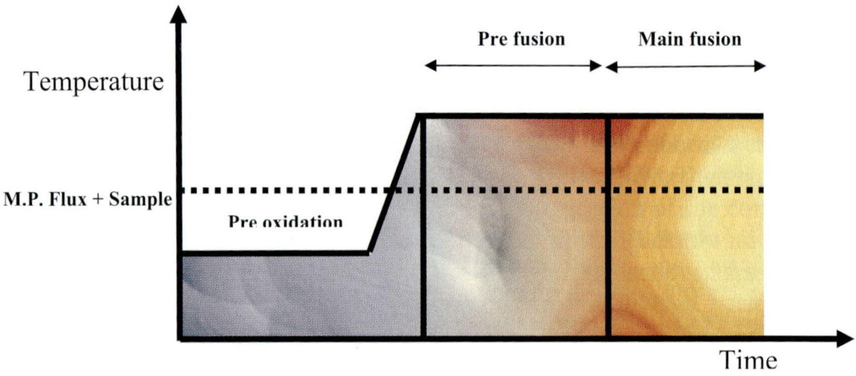


**Fig. 7.** Principle of preparation of a fused bead

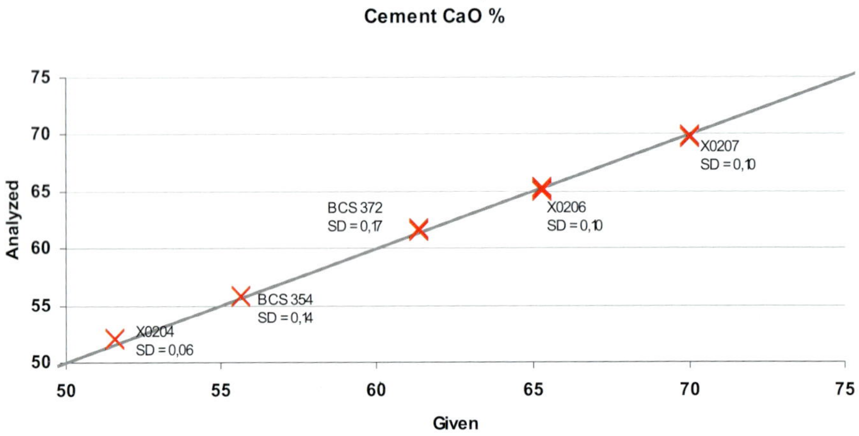


**Fig. 8.** Certified reference materials from BAS, GB (BCS) and Dillinger, D (X) 5 times prepared as fused bead with HD Elektronik VULCAN, SD = Standard Deviation

The mixture is heated in a fully automated fusion machine (HD Elektronik VULCAN, see Fig. 6) to 1050°C for 6 min to become liquid. Another 6 min the sample will be stirred to homogenize the melt which then will be poured in a preheated mould (e.g. FLUXANA OA-439). Figure 7 shows the principle of doing a fusion.

Non oxidic materials like ferro alloys or sulfides can easily be prepared by pre-oxidizing them before fusion.

Figure 8 shows a calibration with certified cements. Each was prepared 5+times. The standard deviation achieved is between 0.05% and 0.02%.

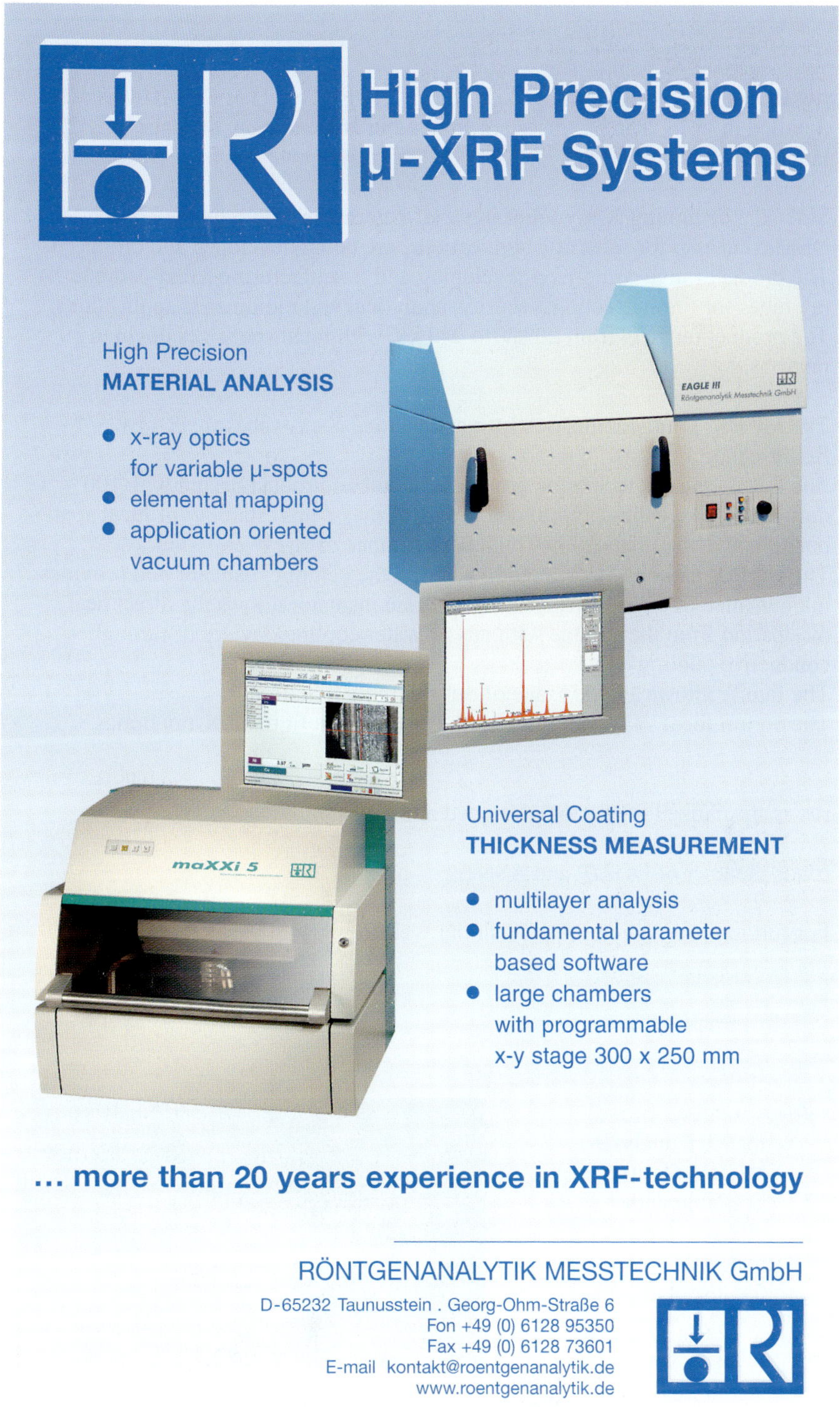
High Precision
µ-XRF Systems
High Precision
MATERIAL ANALYSIS
x-ray optics for variable µ-spots
elemental mapping
application oriented vacuum chambers
EAGLE III
Röntgenanalytik Messtechnik GmbH
maXXi 5
Universal Coating
THICKNESS MEASUREMENT
multilayer analysis
fundamental parameter based software
large chambers with programmable x-y stage 300 x 250 mm
... more than 20 years experience in XRF-technology
RÖNTGENANALYTIK MESSTECHNIK GmbH
D-65232 Taunusstein . Georg-Ohm-Straße 6
Fon +49 (0) 6128 95350
Fax +49 (0) 6128 73601
E-mail kontakt@roentgenanalytik.de
www.roentgenanalytik.de